Log on.

Explore.

Succeed.

P9-DHE-013

# Your steps to success.

## STEP 1: Register

All you need to get started is a valid email address and the access code below. To register, simply:

1. Go to www.mybiology.com
2. Click the appropriate book cover.
   *Cover must match the textbook edition being used for your class.*
3. Click **"Register"** under **"First-Time User?"**
4. Leave **"No, I Am a New User"** selected.
5. Using a coin, scratch off the silver coating below to reveal your access code.
   *Do not use a knife or other sharp object, which can damage the code.*
6. Follow the on-screen instructions to complete registration. You may want to click Log In Now on your registration confirmation screen and bookmark that page.

*During registration, you will establish a personal login name and password to use for logging into the website. You will also be sent a registration confirmation email that contains your login name and password.*

### Your Access Code is:

*Note:* If there is no silver foil covering the access code, it may already have been redeemed, and therefore may no longer be valid. In that case, you can purchase access online using a major credit card. To do so, go to www.mybiology.com and click "Buy Access" and follow the on-screen instructions.

## STEP 2: Log in

1. Go to www.mybiology.com and **"Log In"**.
2. Click **"Log In."**

## STEP 3: (Optional) Join a class

Instructors have the option of creating an online class for you to use with this website. If your instructor decides to do this, you'll need to complete the following steps using the Class ID your instructor provides you. By "joining a class," you enable your instructor to view the scored results of your work on the website in his or her online gradebook.

To join a class:

1. Log into the website. For instructions, see "STEP 2: Log in."
2. Click **"Join a Class"** near the top right.
3. Enter your instructor's **"Class ID"** and then click **"Next."**
4. At the Confirm Class page you will see your instructor's name and class information. If this information is correct, click Next.
5. Click **"Enter Class Now"** from the Class Confirmation page.

- *To confirm your enrollment in the class, check for your instructor and class name at the top right of the page. You will be sent a class enrollment confirmation email.*
- *As you complete quizzes on the website from now through the class end date, your results will post to your instructor's gradebook, in addition to appearing in your personal view of the Results Reporter.*

To log into the class later, follow the instructions under "STEP 2: Log in."

## Got technical questions?

Customer Technical Support: To obtain support, please visit us online anytime at http://247.aw.com where you can search our knowledgebase for common solutions, view product alerts, and review all options for additional assistance.

### SITE REQUIREMENTS

For the latest updates on Site Requirements, go to www.mybiology.com, choose your text cover, and click Site Reqs.

WINDOWS
OS: Windows 2000, XP
Resolution: 1024 x 768
Plugins: Latest Version of Flash/QuickTime/Shockwave (as needed)
Browsers: Internet Explorer 6.0; Firefox 1.0; Netscape 7.2
Internet connection: 56K minimum

MACINTOSH
OS: 10.3.x
Resolution: 1024 x 768
Plugins: Latest Version of Flash/QuickTime/Shockwave (as needed)
Browsers: Netscape 7.2; Firefox 1.0; Safari 1.3
Internet connection: 56K minimum

Log in

Join a class

## Features of the Website for *A Guide to the Natural World*, Fourth Edition

### Chapter Guide
The Chapter Guide organizes media under central concepts that correlate directly to the text.

### Activities
Explore approximately 200 activities, including animations, interactive review exercises, and videos. Test your knowledge with Activities Quizzes that can be e-mailed to instructors.

### Quizzes
Assess your understanding with over 3,600 multiple-choice questions. Each chapter includes a Pre-Test to diagnose current knowledge, an Activities Quiz with graphics that tests understanding of the media Activities in the chapter, and a comprehensive Post Test (50 questions per chapter on average). Each quiz has hints, immediate feedback, grading, and e-mailable results.

### Cumulative Quizzes
Create customized tests on multiple chapters at once, by choosing the chapters and number of questions. Each question provides a hint and reference to the relevant module(s) in the text.

### Word Study Tools
Use electronic Flash Cards to test your knowledge of the key terms and definitions for each chapter or multiple chapters. Includes audio pronunciations for selected terms. Key Terms allow you to study the list of key terms for each chapter, looking up definitions and hearing selected audio pronunciations. Word roots are also provided to improve vocabulary skills. The Glossary provides definitions of boldface terms and selected audio pronunciations.

### Videos
View a collection of 85 videos.

### Web Links and References
Organized by chapter, Web Links and References allow you to extend your knowledge through links to relevant websites, access news links on recent developments in biology, and consult an archive of relevant biology news articles and lists of further readings for each chapter.

# BIOLOGY

## A GUIDE TO THE NATURAL WORLD

### FOURTH EDITION

## DAVID KROGH

PEARSON

Benjamin
Cummings

San Francisco   Boston   New York
Cape Town   Hong Kong   London   Madrid   Mexico City
Montreal   Munich   Paris   Singapore   Sydney   Tokyo   Toronto

Editor-in-Chief: *Beth Wilbur*
Acquisitions Editor: *Star MacKenzie*
Associate Editor: *Benjamin Lau*
Executive Director of Development: *Deborah Gale*
Development Editor: *Anne Scanlan-Rohrer*
Art Development Editor: *Kim Quillin*
Art Editor: *Elisheva Marcus*
Editorial Assistant: *Erin Mann*
Managing Editor: *Michael Early*
Production Supervisor: *Lori Newman*
Production Management: *Patrick Franzen, Pre-Press PMG*
Copy Editor: *Kathleen A. Brown*
Compositor: *Pre-Press PMG*

Art Coordinator: *Crystal Parenteau, Pre-Press PMG and Natalie Davis*
Interior Designer and Design Manager: *Marilyn Perry*
Cover Designer: *Riezebos Holzbaur Design Group*
Illustrators: *Heidi Bertignoll, Imagineering Media Services, Inc.*
Senior Photo and Art Manager: *Travis Amos*
Photo Researchers: *Clare Maxwell, Elaine Soares*
Director, Image Resource Center: *Melinda Patelli*
Image Rights and Permissions Manager: *Zina Arabia*
Image Permissions Coordinator: *Elaine Soares*
Manufacturing Buyer: *Michael Penne*
Senior Marketing Manager: *Jay Jenkins*
Text printer: *RR Donnelley and Sons, Willard*
Cover printer: *Phoenix Color Corp.*

Cover Photo Credit: eagle © Johnny Johnson/The Image Bank; frogs © Gail Shumway/Taxi; fungus © Theo Allofs /Photonica; jellyfish © Gail Shumway/Taxi; ladybug © Martin Ruegner/Stone; plant © Keiji Iwai/Photographer's Choice RR; shell © Andrew Mounter/Taxi

Library of Congress Cataloging-in-Publication Data
Krogh, David, 1949-
  Biology : a guide to the natural world / David Krogh. -- 4th ed.
    p. cm.
  Includes bibliographical references and index.
  ISBN 0-13-225437-9 (alk. paper)
  1. Biology. I. Title.
  QH308.2.K76 2009
  570--dc22
                        2007050887

    ISBN 0132254379 / 9780132254373 (Student edition)
    ISBN 0132254859 / 9780132254854 (Professional copy)
    ISBN 0321556976 / 9780321556974 (Books a la Carte)

2 3 4 5 6 7 8 9 10—DOW—12 11 10 09 08
www.pearsonhighered.com

# BIOLOGY

## A GUIDE TO THE NATURAL WORLD

*For Francis Crick*
*Who never ran out of energy*
*And never ran out of ideas*

# ABOUT THE AUTHOR

### Author

**DAVID KROGH** has been writing about science for 24 years in newspapers, magazines, books, and for educational institutions. He is the author of *Smoking: The Artificial Passion,* an account of the pharmacological and cultural motivations behind the use of tobacco, which was nominated for the *Los Angeles Times* Book Prize in Science and Technology. In 1994, he began work on what would become *Biology: A Guide to the Natural World,* and in 1999 he completed its first edition. Since then, he has produced three more editions of *A Guide to the Natural World* along with a second textbook, *A Brief Guide to Biology.* He holds bachelor's degrees in journalism and history from the University of Missouri.

### Art Development Editor

**Kim Quillin** received her B.A. in Biology at Oberlin College and her Ph.D. in Integrative Biology from the University of California, Berkeley, and has taught undergraduate biology at both institutions. Students and instructors alike have praised the illustration programs she has produced for three textbooks: David Krogh's *Biology: A Guide to the Natural World,* Scott Freeman's *Biological Science,* and Colleen Belk and Virginia Borden's *Biology: Science for Life.*

### Academic Advisors

Tania Baker
*Massachusetts Institute of Technology*

Leon W. Browder
*University of Calgary*

Anu Singh-Cundy
*Western Washington University*

Anthony Ives
*University of Wisconsin*

Greg Podgorski
*Utah State University*

Erica Suchman
*Colorado State University*

Sara Via
*University of Maryland*

Nicholas Wade
*The New York Times*

John Whitmarsh
*National Institute of General Medical Sciences*

# PREFACE

## From the Author

Book titles may be the first thing any reader sees in a book, but they're often the last thing an author ponders. Not so with *Biology: A Guide to the Natural World*. The title arrived fairly early on, courtesy of the muse, and then stuck because it so aptly expresses what I think is special about this book.

Flip through these pages, and you'll see all the elements that students and teachers look for in any modern introductory textbook—rich, full-color art, an extensive study apparatus, and a full complement of digital learning tools. When you leaf slowly through the book and start to read a little of it, however, I think that something a little more subtle starts coming through. This second quality has to do with a sense of connection with students. The sensibility that I hope is apparent in *A Guide to the Natural World* is that there's a wonderful living world to be explored; that we who produced this book would like nothing better than to show this world to students; and that we want to take them on an instructive walk through this world, rather than a difficult march.

All the members of the teams who have produced the four editions of *A Guide to the Natural World* have worked with this idea in mind. We felt that we were taking students on a journey through the living world and that, rather like tour guides, we needed to be mindful of where students were at any given point. Would they remember this term from earlier in the chapter? Had we created enough of a bridge between one subject and the next? The idea was never to leave students with the feeling that they were wandering alone through terrain that lacked signposts. Rather, we aimed to give them the sense that they had a companion—this book—that would guide them through the subject of biology. *A Guide to the Natural World*, then, really is intended as a kind of guide, with its audience being students who are taking biology but not majoring in it.

Biology is complex, however, and if students are to understand it at anything beyond the most superficial level, details are necessary. It won't do to make what one faculty member called "magical leaps" over the difficult parts of complex subjects. Our goal was to make the difficult comprehensible, not to make it disappear altogether. Thus, the reader will find in this book fairly detailed accounts of such subjects as cellular respiration, photosynthesis, immune-system function, and plant reproduction. It was in covering such topics that our concern for student comprehension was put to its greatest test. We like the way we handled these subjects and other key topics, however, and we hope readers will feel the same way.

## What's New in the Fourth Edition?

We've revised the fourth edition of the *Guide* in numerous ways. To mention some of the major changes, readers of this edition will find:

- **A new unit on nutrition.** A growing number of faculty have decided that an important part of biological literacy is having some knowledge of the relationship between diet and health. Accordingly, if you look on page 629, you'll find a new section of the *Guide* that walks students through all the basics of nutrition.

- **A stand-alone human evolution chapter.** In previous editions, the *Guide's* human evolution coverage took the form of a unit within a chapter. With this edition of the *Guide*, human evolution gets its own chapter, beginning on page 376.

- **A new slant on the process of science.** Every edition of the *Guide* has covered the process of science in a series of essays collectively titled *How Did We Learn?* These essays have focused on famous discoveries of the past, but with the fourth edition we've added something new: three essays whose subject matter is the process of science in contemporary research. Thus, Chapter 15 has an account of how a young researcher got stem cells to work in animal subjects (page 282); Chapter 19 has an essay on expeditionary paleontology in Canada's far north (page 370); and Chapter 35 has an account of the steps by which a researcher in Costa Rica became the first scientist to tie global warming to the extinction of individual species (page 746). All the book's essays on scientific discovery now go under the generic title *The Process of Science.*

- **Expanded coverage of diabetes and obesity.** The importance of these two issues has increased greatly since the *Guide's* third edition came out. Accordingly, the book's coverage of both subjects has been strengthened for this edition. If you look on page 540, you'll find the essay "Why Fat Matters and Where It Matters Most" on the metabolic changes that come with being overweight. Then on page 580, there is an account of how blood levels of glucose are controlled in the body. Nearby, you'll find an essay on diabetes, "When Blood Sugar Stays in the Blood."

- **New coverage of other applied topics.** Other applied topics new to the *Guide* with this edition are Avian flu (page 396), methicillin-resistant staph infections (402), blood doping (610), the nature of spinal cord injuries (565), the decibel levels of everyday sounds (571), the means by which blood pressure is measured (614), and the way DNA "fingerprints" are visualized in today's criminal cases (278).

- **New coverage of basic science topics.** Large-scale changes in basic science coverage include a completely redone section on genetic regulation (page 257), a substantially redone section on global warming (page 742), a new section on biogeography as it relates to evolution (page 302), an updated account of Precambrian evolution (page 360), a significantly revised unit on stem cells (page 277), and revised units on three of life's largest taxonomic groups, the archaea, the protists, and the fungi (pages 403, 405, and 417). In our coverage of stem cells, we just managed to get in a bit of information on some important, but late-breaking news in 2007: The announcement by two teams of scientists that, through gene insertion, they had managed to transform ordinary skin cells into cells that seem to have all the developmental potential of embryonic stem cells. The *Guide's* section on this breakthrough can be found on page 282.

Beyond these larger-scale changes, numerous smaller-scale changes have been made throughout the book. To name just a few, if you look on page 112 of the *Guide's* Chapter 6, you'll see that muscle contraction is now used as the example of how ATP powers chemical reactions in the body. Meanwhile, in the Guide's coverage of pedigrees beginning on page 221, you'll find a conventional subject—hemophilia in Europe's royal families—treated in an unconventional way. And if you look at the *Guide's* Chapter 30, beginning on page 643 you'll see that the book's coverage of kidney function has been streamlined in accordance with recommendations from faculty.

- **A new structure for questions.** If you flip through the *Guide*, you'll see that, in each chapter, the text is interrupted in three or four places by small sets of review questions that are headlined *So Far*. These are not questions aimed at testing student abilities to synthesize or think critically; they're questions designed to let students monitor their understanding of each chapter as they move through it. (If they can't answer the *So Far* questions, that's a sure sign they need to do some re-reading before moving on.) Joining these questions are the end-of-chapter Brief Review and Applying Your Knowledge questions that have been a feature of every *Guide* edition. Longtime *Guide* users may then wonder what happened to the *multiple-choice* questions that once came at the end of every chapter. The answer is these questions still exist, but now they're up on the web: Students can access them at **www.aw.com/krogh4**. In sum, the *Guide* now has more questions than ever, but it has them in several different locations.

## Electronic Media and the Fourth Edition: *BioFlix* Animations

With this edition, I have the pleasure of introducing a major new addition to the *Guide's* collection of electronic teaching tools. *Guide* readers will now have access to a series of movie-quality animations that go under the collective title *BioFlix*. These short films use the marvel of computer animation to present such processes as mitosis and cellular respiration in vivid, kinetic detail. Anyone who has ever struggled to explain, say, chromosome movement in cell division need struggle no longer; just have students watch the *BioFlix* film on mitosis. Ten *BioFlix* have been produced so far and more are on the way. Topics covered to date include animal cell structure, plant cell structure, cellular respiration, photosynthesis, mitosis, meiosis, protein synthesis, water transport in plants, nerve signal propagation, and muscle contraction. If you're like me, you may have seen animations on some of these subjects before. Once you see these, however, I think you'll agree that the bar has now been raised on the use of animation

in biology education. *BioFlix* films are identified within *Guide* chapters with a logo that looks like this:

*BioFlix*

A Tour of the Animal Cell

Seeing one of these in a given chapter, students need only boot their web browser and go to **www.aw.com/krogh4** to see the film they're interested in.

Joining the *BioFlix* are a series of 62 smaller-scale Web Animations, retained from the *Guide's* third edition, that cover many of the biological processes reviewed in the book. Each animation comes with its own set of computer-graded exercises and review questions. Web Animations are identified within chapters with logos that look like this:

Web Animation 3.4
Carbohydrates

## The Guide on the Web

Web Animations and *BioFlix* are just two of the learning tools available at the *Guide* website, which has been greatly enhanced for this edition of the book. Students who access the site (at **www.aw.com/krogh4**) now will find not only the graded quizzes and sets of weblinks that have accompanied every edition of the *Guide*, but a clickable list of each chapter's key terms and their definitions; a set of electronic flash cards that can be used to study these key terms; an online e-book version of the entire text; a series of videos from the Discovery Channel on biology topics; and all the art in the book, chapter by chapter in clickable form. Instructors will find all this plus text and image libraries in Power Point, along with tools for using Media Manager.

## *Scientific American Current Issues in Biology*

Another feature of the *Guide* new with this edition is that every new copy of the book will now come bundled with an instructional companion: a bound collection of articles that first appeared in the pages of *Scientific American*. Benjamin Cummings and *Scientific American* have teamed up to issue a series of these bound collections, each one bearing the title *Current Issues in Biology*. New volumes in the series are being issued on a regular basis, which means that every volume of *Current Issues* that comes bundled with a *Guide* copy will contain cutting-edge scientific information. Moreover, the articles in each volume have been carefully selected to meet two criteria: they must appeal to college students and they must be relevant to topics typically covered in college biology courses. Thus, Volume 5 of *Current Issues* has articles on the genetics of alcoholism, on "Lucy" in human evolution, and on "Cancer Cues from Pet Dogs." As you might imagine, articles like these have an ability to spark classroom discussion, as they tend to give an interesting, applied slant to subjects covered in class. And every article is accompanied by a set of faculty-written questions, which should help ensure student comprehension of the articles prior to any

discussion of them. Finally, there's no extra charge for bundling *Current Issues in Biology*; every student who buys a new copy of the *Guide* can have the latest volume of *Current Issues* as part of the bargain.

## Instructor Resource CDs

All the digital resources available for every chapter in the *Guide* can be coordinated from the *Guide's* new Instructor Resource CDs. This means the *BioFlix* offerings, the Discovery Channel videos, all the art and photos in the book, a set of Instructor Animations for each chapter—all these items and more can be accessed through the Instructor CDs. A feature new to the fourth-edition CDs is a Classroom Response System, which has sets of chapter-specific questions that allow instructors to instantly assess how well students are comprehending lectures.

## The *Guide to the Natural World* Team

Textbooks are such enormous collective efforts that it's difficult for an author to thank all the people who've helped out, for the simple reason that the author hasn't *met* a good number of the people who've helped out. The best I can do, therefore, is name a few of the people whose assistance has been valuable to me personally as I've worked on the fourth edition of *A Guide to the Natural World*.

The *Guide's* Developmental Editor, Anne Scanlan-Rohrer, ably guided the book through its early and middle phases, when its third-edition content was being examined and its fourth-edition content was being created. The terrific artist and biologist Kim Quillin continued to produce nearly all the new art for the book, working as usual with her winning combination of good cheer and devotion to clear scientific illustration. My new friends at Benjamin Cummings went above and beyond the call of duty to make sure the book came out right. Among the people at BC I'm indebted to are Production Supervisor Lori Newman, Senior Photo and Art Manager Travis Amos, and Design Manager Marilyn Perry. (The *Guide's* new look is Marilyn's handiwork.) The good folks at Pre-Press PMG took thousands of pieces of *Guide* art and text and made a book out of them. The corporate migration of the *Guide*, from Prentice Hall to Benjamin Cummings, meant that the book had no fewer than four leaders providing overall direction: first Teresa Chung and Jeff Howard at Prentice Hall; and then Chalon Bridges and Star MacKenzie at Benjamin Cummings. All of these editors were committed to a good book and I think all of them will be pleased when they see the final product. The *Guide* had great support at BC right from the start, and this support came from the top—from BC Editor-in-Chief Beth Wilbur. Finally, I owe a great debt of thanks to Associate Editor Ben Lau and Art Editor Ellie Marcus; their good work is on display in nearly every page of the book.

Apart from these publishing team members, more than 330 faculty have now carefully critiqued every word and image you see in the *Guide*. The names of reviewing faculty can be found beginning on page ix. Of these faculty, I need to make special note of the team of academic advisors, listed on page iv, who have provided advice not only on the details of the book, but on its overall structure and coverage. I'd also like to give a special thanks to Professor James Wise of Hampton University in Virginia, whose metaphor of an assembly line can be found enlivening Chapter 7's account of the vital role that oxygen plays in cellular respiration.

Finally, the test of any textbook is its effect in the classroom. As a consequence, for any textbook author, there is no such thing as too much feedback from those who stand at the front of the class. So, if you teach using the *Guide*, or if you've merely looked at it in some detail, here's an invitation: Write me and tell me what you think. Whether your reactions are positive or negative doesn't matter; the book stands to be improved by both kinds of responses. The more faculty I hear from, the better the book's fifth edition stands to be. My e-mail address is: rdkrogh@pacbell.net.

DAVID KROGH
Kensington, California

# Student Supplements

## Study Guide

0-13-225478-6 / 978-0-13-225478-6

This tool is an effective and interactive study guide which helps students concentrate their study efforts. The Study Guide contains quizzes, key term reviews, and helpful summaries.

## Study Card

0-13-225487-5 / 978-0-13-225487-8

This laminated quick reference Study Card is developed specifically for *Biology: A Guide to the Natural World* and can be value packaged with the book at no additional charge. The Study Card provides a brief, chapter-by-chapter review, including key figures, to help students review the most important topics in biology.

## Premium Website

0-13-225480-8 / 978-0-13-225480-9

This comprehensive study resource has been revised to include new BioFlix™ animations on the toughest topics in biology, customized study plans for individual students, and a gradebook that allows students to view their quiz and activity results online. Also included are Web Links and references, Vocabulary Study Tools, Cumulative Tests, Web Animations with built-in quizzes, Discovery Channel video clips, Tutoring Services, Lab Bench, Graph It! Activities and MP3 lectures.

URL: http://www.aw.com/krogh4

## Scientific American *Current Issues in Biology*

*Volume 1:* 0-8053-7507-4 / 978-0-8053-7507-7

*Volume 2:* 0-8053-7108-7 / 978-0-8053-7108-6

*Volume 3:* 0-8053-2146-2 / 978-0-8053-2146-3

*Volume 4:* 0-8053-3566-8 / 978-0-8053-3566-8

*Volume 5:* 0-321-54187-1 / 978-0-321-54187-1

Give your students the best of both worlds—accessible, dynamic, relevant articles from *Scientific American* magazine that present key issues in biology, paired with the authority, reliability, and clarity of Benjamin Cummings' non-majors biology texts. Articles include questions to help students check their comprehension and make connections to science and society. This resource is available at no additional charge when packaged with a new text.

# Instructor Supplements

## Instructor Resource CD-ROM

0-13-225474-3 / 978-0-13-225474-8

The Instructor Resource CD-ROM includes PowerPoint® lectures that integrate figures and new BioFlix™ animations, BLAST animations, Discovery Channel® Videos, all figures, art, and photos in JPEG format, label-edit Powerpoints, and over 200 Instructor Animations accurately depict complex topics and dynamic processes described in the book. New, innovative Classroom Response System questions are offered for each chapter as a way to directly engage students in lectures.

## Instructor Guide

0-13-225476-X / 978-0-13-225476-2

This comprehensive guide provides chapter-by-chapter objectives, lists of common student misconceptions, teaching tips, media tips, additional critical thinking questions, key terms, and word roots. A separate chapter offers suggestions for effective uses of technology in teaching introductory biology. The Instructor Guide is available in print and as a Microsoft® Word document.

## Transparency Acetates

0-13-225489-1 / 978-0-13-225489-2

Full-color acetates include illustrations and tables from the text.

## Printed Test Bank

0-13-225477-8 / 978-0-13-225477-9

## Computerized Test Bank

0-13-225486-7 / 978-0-13-225486-1

All of the exam questions in the Test Bank have been peer reviewed, providing questions that set the standard for quality |and accuracy. Test questions are ranked according to Bloom's taxonomy and improved TestGen® software makes assembling tests easier. The Test Bank is also available in Course Management Systems and in Word® format on the Instructor Resource CD-ROM.

# Course Management

## CourseCompass™

This nationally-hosted, dynamic, interactive online course management system is powered by Blackboard. This easy-to-use and customizable program enables professors to tailor content to meet individual course needs! Every CourseCompass™ course includes preloaded content such as testing and assessment question pools. CourseCompass is also available with an E-Book.

URL: http://www.aw-bc.com/coursecompass

## WebCT

This open-access course management system contains pre-loaded content such as testing and assessment question pools. More content is also available in a premium version.

URL: http://www.aw-bc.com/webct

## Blackboard

This open-access course management system contains preloaded content such as testing and assessment question pools. More content is also available in a premium version.

URL: http://www.aw-bc.com/blackboard

# REVIEWERS

## Fourth Edition Reviewers

Julie Adams, *Ohio Northern University*
Kim Anthony Aaronson, *Truman College*
Bert Atsma, *Union County College*
James W. Bailey, *University of Tennessee*
Nancy Boury, *Iowa State University*
Robert Boyd, *Auburn University*
Peggy Brickman, *University of Georgia*
Carla Bundrick Benejam, *California State University, Monterey Bay*
Don Canfield, *University of Southern Denmark*
Kelly Sue Cartwright, *College of Lake County*
Jean DeSaix, *University of North Carolina, Chapel Hill*
Carolyn Doege, *Cy-Fair College*
JodyLee Duek, *Pima College*
Albia Dugger, *Miami-Dade College*
Teresa Fischer, *Indian River Community College*
Carol Friesen, *Ball State University*
Mita Ghosh, *FCCJ*
David M. Gordon, *Pittsburg State University*
Becky Graham, *The University of West Alabama*
Catherine J. Hurlbut, *Florida Community College*
Karry Kazial, *SUNY Fredonia*
Kirkwood Land, *University of the Pacific*
Brenda Leicht, *University of Iowa*
Lynne Lisenby, *Florida Community College at Jacksonville*
Richard L. Myers, *Missouri State University*
Paul Nolan, *Ithaca College*
Onesimus Otieno, *Oakwood College*
Charles Owens, *King College*
Kathleen Pelkki, *Saginaw Valley State University*
Marceau Ratard, *Delgado Community College*
Robert J. Reinsvold, *University of Northern Colorado*
Amanda Rosenzweig, *Delgado Community College*
Steven Scott, *Merritt College*
Cara Shillington, *Eastern Michigan University*

Bryan Spohn, *Florida Community College at Jacksonville–Kent Campus*
Christine Tachibana, *University of Washington*
Rani Vajravelu, *University of Central Florida*
Jennifer M. Warner, *University of North Carolina, Charlotte*
James Wise, *Hampton University*

## Media Reviewers

Robert S. Boyd, *Auburn University*
Carolyn Glaubensklee, *University of Southern Colorado*
Sherri Gross, *Ithaca College*
David A. Rintoul, *Kansas State University*
Brian Sailer, *Sam Houston State University*
Rebekah J. Thomas, *Saint Leo University*
Jennifer M. Warner, *University of North Carolina, Charlotte*
Jamie Welling, *South Suburban College*

## Past Edition Reviewers

Dawn Adams, *Baylor University*
John Alcock, *Arizona State University*
David L. Alles, *Western Washington University*
Sylvester Allred, *Northern Arizona University*
Gary Anderson, *University of California, Davis*
Marjay A. Anderson, *Howard University*
Michael F. Antolin, *Colorado State University*
Kerri Armstrong, *Community College of Philadelphia*
Mary Ashley, *University of Illinois, Chicago*
Jessica Baack, *Montgomery College*
Kemuel Badger, *Ball State University*
Tania Baker, *Massachusetts Institute of Technology*
Peter Bednekoff, *Eastern Michigan University*
Michael C. Bell, *Richland College*
William J. Bell, *University of Kansas*
David Berrigan, *University of Washington*
Lois A. Bichler, *Stephens College*
Ian M. Bird, *University of Wisconsin, Madison*
A.W. Blackler, *Cornell University*

Andrew Blaustein, *Oregon State University*
Robert S. Boyd, *Auburn University*
Bonnie L. Brenner, *Wilbur Wright College (City College of Chicago)*
Mimi Bres, *Prince George's Community College*
Leon Browder, *University of Calgary*
Arthur L. Buikema, *Virginia Polytechnic Institute and State University (Allegheny College)*
Steven K. Burian, *Southern Connecticut State University*
Janis K. Bush, *University of Texas, San Antonio*
W. Barkley Butler, *Indiana University of Pennsylvania*
Linda Butler, *University of Texas, Austin*
David Byres, *Florida Central Community College, South Campus*
James Cahill, *University of Alberta*
Chantae Calhoun, *Lawson State Community College*
John Capeheart, *University of Houston-Downtown*
Marnie Chapman, *University of Alaska, Southeast Sitka*
Van D. Christman, *Ricks College*
Deborah C. Clark, *Middle Tennessee State University*
William S. Cohen, *University of Kentucky*
Karen A. Conzelman, *Glendale Community College*
Tricia Cooley, *Laredo Community College*
Richard Copping, *Lane College*
Lee Couch, *University of New Mexico*
Patricia Cox, *University of Tennessee, Knoxville*
John Crane, *Washington State University*
Robert Curry, *Villanova University*
Garry Davies, *University of Alaska, Anchorage*
Paula Dedmon, *Gaston College*
Miriam del Campo, *Miami-Dade Community College*
Brent DeMars, *Lakeland Community College*
Llewellyn Densmore, *Texas Technical University*
Jean Dickey, *Clemson University*

Christopher Dobson, *Front Range Community College*
Deborah Dodson, *Vincennes University*
Cathy Donald-Whitney, *Collin County Community College*
Matthew M. Douglas, *Grand Rapids Community College (University of Kansas)*
Lee C. Drickamer, *Southern Illinois University*
Charles Duggins, Jr., *University of South Carolina*
Susan A. Dunford, *University of Cincinnati*
Douglas Eder, *Southern Illinois University*
Ron Edwards, *University of Florida*
Douglas J. Eernisse, *California State University, Fullerton*
Jamin Eisenbach, *Eastern Michigan University*
George Ellmore, *Tufts University*
Patrick E. Elvander, *University of California, Santa Cruz*
Michael Emsley, *George Mason University*
Elyce Ervin, *University of Toledo*
David W. Essar, *Winona State University*
Richard H. Falk, *University of California, Davis*
Michael Farabee, *Estrella Mountain Community College*
Rita Farrar, *Louisiana State University*
John Philip Fawley, *Westminster College*
Eugene J. Fenster, *Longview Community College*
Rebecca Ferrell, *Metropolitan State College at Denver*
Leanne H. Field, *University of Texas, Austin*
Christine M. Foreman, *University of Toledo*
Elizabeth Forston Wells, *George Washington University*
Carl S. Frankel, *Pennsylvania State University*
Ralph Fregosi, *University of Arizona*
Lawrence Friedman, *University of Missouri, St. Louis*
John L. Frola, *University of Akron*
Larry Fulton, *American River College*
Gail E. Gasparich, *Towson University*
Matt Geisler, *University of California, Riverside*
George W. Gilchrist, *College of William and Mary*
Tejendra S. Gill, *University of Houston*
Claudette Giscombe, *University of Southern Indiana*
Carolyn Glaubensklee, *University of Southern Colorado*
Jack M. Goldberg, *University of California, Davis*

Elliott Goldstein, *Arizona State University*
Judith Goodenough, *University of Massachusetts, Amherst*
Glenn A. Gorelick, *Citrus College*
Melvin H. Green, *University of California, San Diego*
G.A. Griffith, *South Suburban College*
Sherri Gross, *Ithaca College*
Edward Hale, *Ball State University*
Gail Hall, *Trinity College*
Madeline Hall, *Cleveland State University*
Linnea S. Hall, *California State University, Sacramento*
Kelly Hamilton, *Shoreline Community College*
James Hampton, *Salt Lake Community College*
Christopher Harendza, *Montgomery County Community College*
Steven C. Harris, *Clarion University*
Barbara Harvey, *Kirkwood Community College*
Allan Hayes Vogel, *Chemeketa Community College*
Steve Heard, *University of Iowa*
Walter Hewitson, *Bridgewater State College*
Jane Aloi Horlings, *Saddleback College*
Eva Horne, *Kansas State University*
Michael Hudecki, *State University of New York, Buffalo*
Michael Hudspeth, *Northern Illinois University*
Terry L. Hufford, *The George Washington University*
Carol A. Hurney, *James Madison University*
Andrea Huvard, *California Lutheran University*
Martin Ikkanda, *Los Angeles Pierce College*
Rose M. Isgrigg, *Ohio University*
Anthony Ives, *University of Wisconsin, Madison*
Rebecca Jann, *Queens University of Charlotte*
Thomas W. Jurik, *Iowa State University*
Arnold J. Karpoff, *University of Louisville*
Anne Keddy-Hector, *Austin Community College*
Kathleen Keeler, *University of Nebraska*
Nancy Keene, *Pellissippi State Technical Community College*
Kevin M. Kelley, *California State University, Long Beach*
Jeanette J. Kiem, *Guilford Technical Community College*
Tom Knoedler, *Ohio State University, Lima Campus*
Don E. Krane, *Wright State University*

Jocelyn Krebs, *University of Alaska, Anchorage*
Kate Lajtha, *Oregon State University*
Erika Ann Lawson, *Columbia College*
Mike Lawson, *Missouri Southern State College*
Harvey Liftin, *Broward Community College*
Ann S. Lumsden, *Florida State University*
Paul Lurquin, *Washington State University*
James Manser, *Harvey Mudd College*
Michael M. Martin, *University of Michigan, Ann Arbor*
Paul Mason, *Butte Community College*
Michel Masson, *Santa Barbara City College*
Andrea Mastro, *Pennsylvania State University*
Mary Victoria McDonald, *University of Central Arkansas*
Mary Colleen McNamara, *Albuquerque T-VI A Community College*
Hugh A. Miller III, *East Tennessee University*
Leslie R. Miller, *Iowa State University*
Lee H. Mitchell, *Mount Hood Community College*
Robert Mitchell, *Community College of Philadelphia*
Jeremy Montague, *Barry University*
Scott M. Moody, *Ohio University*
Janice Moore, *Colorado State University*
Joseph Moore, *California State University*
Jorge A. Moreno, *University of Colorado*
Michael D. Morgan, *University of Wisconsin, Green Bay*
David Mork, *Saint Cloud State University*
Deborah A. Morris, *Portland State University*
Allison Morrison-Shetlar, *Georgia Southern University*
Richard Mortensen, *Albion College*
Christopher B. Mowry, *Berry College*
Michelle Murphy, *University of Notre Dame*
Courtney Murren, *University of Tennessee, Knoxville*
Royden Nakamura, *California Polytechnic State University*
Leann Naughton, *University of Wyoming*
Judy M. Nesmith, *University of Michigan, Dearborn*
Harry Nickla, *Creighton University*
James A. Nienow, *Valdosta State University*
Jane Noble-Harvey, *University of Delaware*
Ray Ochs, *St. John's University*
Hunter O'Reilly, *University of Wisconsin, Milwaukee*
Marcy P. Osgood, *University of Michigan*
Andrea Ostrofsky, *University of Maine*
Maya Patel, *Ithaca College*
Eckle L. Peabody, *Tulsa Community College*

Patricia A. Peroni, *Davidson College*
Rhoda E. Perozzi, *Virginia Commonwealth University*
Carolyn Peters, *Spoon River College*
John S. Peters, *College of Charleston*
Kim M. Peterson, *University of Alaska, Anchorage*
Raleigh K. Pettegrew, *Denison University*
Gary W. Pettibone, *State University of New York, College at Buffalo*
Holly C. Pinkart, *Central Washington University*
Barbara Pleasants, *Iowa State University*
John M. Pleasants, *Iowa State University*
Gregory J. Podgorski, *Utah State University*
Lynn Polasek, *Los Angeles Valley College*
F. Harvey Pough, *Arizona State University, West*
Don Pribor, *University of Toledo*
Michelle Priest, *University of Southern California*
Louis Primavera, *Hawaii Pacific University*
Paul Ramp, *Pellissippi State and Technical Community College*
Ameed Raoof, *University of Michigan Medical School*
Regina Rector, *William Rainey Harper College*
Rick Relyea, *University of Pittsburgh*
Michael H. Renfroe, *James Madison University*
Dennis Richardson, *Quinnipiac University*
Todd Rimkus, *Marymount University*
Sonia J. Ringstrom, *Loyola University*
David A. Rintoul, *Kansas State University*
Darryl Ritter, *Okaloosa-Walton Community College*
Carolyn Roberson, *Roane State Community College*
Laurel Roberts, *University of Pittsburgh*
Jay Robinson, *San Antonio College*
Rodney A. Rogers, *Drake University*
William H. Rohrer, *Union County College*
Leslie Ann Roldan, *Massachusetts Institute of Technology*
Heidi Rottschafer, *University of Notre Dame*
John Rueter, *Portland State University*
Ron Ruppert, *Cuesta College*

Nancy Sanders, *Northeast Missouri Sate University*
Gary Sarinsky, *City University of New York, Kingsborough Community College*
Julie Schroer, *Bismarck State College*
Edna Seaman, *University of Massachusetts, Boston*
Ralph W. Seelke, *University of Wisconsin, Superior*
Prem P. Sehgal, *East Carolina University*
C. Thomas Settlemire, *Bowdoin College*
Robert Shetlar, *Georgia Southern University*
Mark A. Shotwell, *University of Slippery Rock*
Linda Simpson, *University of North Carolina, Charlotte*
Anu Singh-Cundy, *Western Washington University*
Peter Slater, *University of St. Andrews, UK*
Ellen Smith, *Arizona State University, West*
Linda Smith-Staton, *Pellissippi State Technical Community College*
Philip J. Snider, *University of Houston*
Nancy G. Solomon, *Miami University*
Salvatore A. Sparace, *Clemson University*
Frederick W. Spiegel, *University of Arkansas*
Kathleen M. Steinert, *Bellevue Community College*
Allan R. Stevens, *Snow College*
Donald P. Streubel, *Idaho State University*
Erica Lynn Suchman, *Colorado State University*
Gerald Summers, *University of Missouri, Columbia*
Steven Tanner, *University of Missouri*
Todd Templeton, *Metropolitan Community College*
Rebekah J. Thomas, *Saint Leo University*
Joanne Tornow, *University of Southern Mississippi*
Todd T. Tracy, *Colorado State University*
Robin W. Tyser, *University of Wisconsin, LaCrosse*
Joseph W. Vanable, Jr., *Purdue University*
William A. Velhagen, Jr., *Longwood College*
Edward L. Vezey, *Oklahoma State University, Oklahoma City*
Sara Via, *University of Maryland*
Tanya Vickers, *University of Utah*

Janet Vigna, *Grand Valley State University*
Dennis Vrba, *North Iowa Area Community College*
Nicholas Wade, *The New York Times*
Stephen Wagener, *Western Connecticut State University*
Jyoti R. Wagle, *Houston Community College*
John H. Wahlert, *Baruch College, The City University of New York*
Carol Wake, *South Dakota State University*
Timothy S. Wakefield, *John Brown University*
Charles Walcott, *Cornell University*
Gene Walton, *Tallahassee Community College*
Yunqiu Wang, *University of Miami*
Sarah Ward, *Colorado State University*
Jennifer M. Warner, *University of North Carolina*
Cheryl Watson, *Central Connecticut State University*
Richard Weinstein, *University of Tennessee, Knoxville*
Richard Barry Welch, *San Antonio College*
Jamie Welling, *South Suburban College*
Mary Pat Wenderoth, *University of Washington*
John Whitmarsh, *University of Illinois at Urbana-Champaign*
Susan Whittemore, *Keene State University*
Allison Wiedemeier, *University of Missouri-Columbia*
Sandra Winicur, *Indiana University, South Bend*
William Wischusen, *Louisiana State University*
Deborah Wisti-Peterson, *University of Washington*
Rachel Witcher, *University of Central Florida*
Mark A. Woelfe, *Vanderbilt University*
Lorne Wolfe, *Georgia Southern University*
Wade B. Worthen, *Furman University*
Robert Yost, *Indiana University Purdue University Indianapolis*
Calvin Young, *Fullerton College*
Martha C. Zúniga, *University of California, Santa Cruz*
Victoria Zusman, *Miami-Dade Community College*

# A STUDENT'S GUIDE
## through the natural world of biology

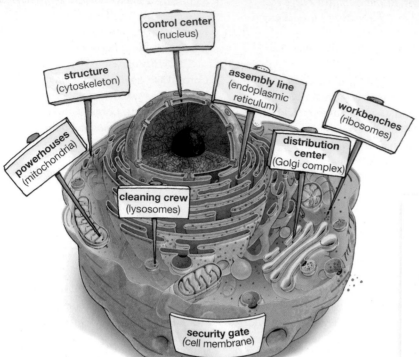

**Figure 4.14**
**The Cell as a Factory**

This comparison may help you to remember some roles of the different parts of the cell.

## Robust Illustration Program

Strong illustrations guide students through figures with clear three-dimensional detail; key information from the text is reinforced in the illustrations.

### In Summary: Structures in the Animal Cell

In your tour of the cell so far, you've pictured a cell as a factory, one that synthesizes proteins in a "production line" that starts with DNA in the nucleus and then goes to the ribosomes (via mRNA), to the rough ER, to the Golgi complex, and finally to the protein's destination (the plasma membrane, export, and so on). You've also seen that cells have other structures, such as lysosomes for digestion and recycling, mitochondria for energy transformation, the smooth endoplasmic reticulum for detoxification and lipid synthesis, and the cytoskeleton for structure and movement. If you look at **Figure 4.14**, you can see, in metaphorical form, a "map" of these component parts within the cell. **Table 4.1** (on page 82 in the dis-

## SO FAR...

1. Plant cells have two organelles and one structure that animal cells do not have. The organelles are _____, which are the sites of photosynthesis, and a large _____, which performs multiple functions, including the storage of nutrients and the degradation of waste products. The structure is the _____, which provides protection and limits the plant cell's uptake of water.

2. Animal cells have only a single organelle that plants cells do not have. This is the _____, which breaks down and recycles materials in the cell.

3. In plants, the channels that facilitate cell-to-cell communication are called _____, while in animals, the tube-like protein structures that facilitate such communication are called _____.

## Accessible Writing Style

Introducing the book your students will actually read! Krogh's accessible writing style walks students step-by-step through complex biological processes and captures their interest with analogies from everyday life, history, art, and literature.

## "So Far..." questions

These questions are interspersed throughout the body of a chapter to give students a moment to reflect on what they have been reading. "So Far..." answers, located on the same spread, serve as a checkpoint for students to assess what they have learned.

### THE PROCESS of SCIENCE
## Plants Make Their Own Food, But How?

S cience constantly is confronted with what are known as "black box" problems. In these problems, researchers know what goes into a process, and they know what comes out of it, but what happens *in between* is a mystery. At the conclusion of World War II, plant science had a doozy of a black box problem. It was clear that plants were starting with carbon dioxide from the air and ending with high-energy carbohydrates, but no one knew what came in between. What were the *intermediate* substances that carbon dioxide was being made part of on the route to food? Getting inside this black box turned out to require a three-part combination: insightful researchers, new scientific techniques, and old-fashioned hard work.

One of the peacetime benefits of World War II's atomic weapons research was that scientists could produce quantities of substances known as radioactive isotopes. As you saw in Chapter 4, isotopes result from variations in the number of neutrons an element has in its nucleus. Radioactivity, meanwhile, is the release of radiant energy from atoms undergoing rapid decay. The element carbon is everywhere in nature, and as it turns out, one of its isotopes, carbon-14, is radioactive. Carbon-14 occurs naturally, but the ability to *manufacture* this substance opened a host of possibilities for research in the mid-1940s.

**Web Animation 8.3**
The Calvin Experiments

Stepping up to the plate in 1946 to take advantage of these possibilities was a University of California, Berkeley, chemist named Melvin Calvin. Then just 35 years old, Calvin was encouraged by the physicist Ernest Lawrence of Berkeley to apply radiocarbon techniques to organic chemistry research. With carbon-14, Calvin realized, he could *tag* carbon dioxide to find out what it becomes part of during photosynthesis, much as a zoologist might tag a bear with a signaling device to find out where it goes during winter. The radiation that carbon-14 emits might be thought of as a signal with a constant message of, "Here I am."

The steps that Calvin and his colleagues followed were simple in conception but very difficult in execution. They first took some photosynthetic algae and put them in a flask that had light shining on it. Like any photosynthetic organism, these algae would, of course, be taking in carbon dioxide to carry out their work. In this case, however, the carbon dioxide in question contained carbon-14. The trick was to allow the algae to perform a little

involved. Stopping photosynthesis at differing points—two seconds after it started, five seconds, and so on—meant stopping the progress of carbon dioxide as it marched through its changes.

All of this would have been for naught, however, without a way to figure out where the carbon was when the process was stopped. This is where the carbon-14 could do its job, but it needed to be joined to another technique that was also newly developed at that time: two-dimensional paper chromatography. If you look at **Figure 1a**, you can see how this technique works. Put a spot of ink on some filter paper and then stick one end of the paper in a solvent—water, for example. The water moves through the paper, carrying different components of the ink along with it at different rates. You can tell something about what the ink is made of by measuring how far its components move through the paper. You can tell still more by doing this in two dimensions. Let the solvent move through the spot one way, then set the paper down at a right angle in a *second* solvent and let it move through in another direction (**Figure 1b**). This is what Calvin and his colleagues did, using as their "ink" the algae that had been killed following photosynthesis.

The paper chromatography left the algae cells drawn out in two directions by the solvents. But, where was the original carbon in this material? Because it was radioactively tagged carbon, it was sending out its signal, saying, "I'm here now," but only a special "receiver" could pick up this signal: photographic film (**Figure 1c**). Setting X-ray film tightly on top of the chromatography paper for long periods—usually about two weeks—caused the radiation from the carbon-14 to expose the film. Spots would show up saying, in effect, "Carbon-14 is now here, in this concentration."

There is a wide gap, however, between seeing a black blotch on a piece of film and *identifying* that blotch as a specific substance. The difficulty the researchers encountered can be measured in time—10 years in this case. That's how long it took these workers to complete their investigation. During that time, Calvin and his principal colleagues—postdoctoral researcher Andrew Benson and graduate student James Bassham—worked extraordinarily long hours in deciphering what the downstream products of photosynthesis were and the order in which they were produced.

This long operation was a great success, however. Calvin's team dismantled the black box of carbon fixation, and the information they uncovered can be used to aid in such things as improved food production. The set of steps these workers elucidated has variously been referred to as the $C_3$ cycle, Calvin-Benson cycle, or Calvin cycle. In 1961, Melvin Calvin received the ultimate in confirmations for a piece of scientific work when he was awarded the Nobel Prize in Chemistry for shining a light on photosynthesis.

---

### ESSAY — When the Cell Cycle Runs Amok: Cancer

T he cell cycle reviewed in this chapter is, in one sense, a common natural process: Cells grow; they duplicate their chromosomes; these chromosomes separate; one cell divides into two. This goes on like clockwork, millions of times a second in each one of us. Then one day we learn that an aunt, a grandfather, or a friend is experiencing an *unrestrained* division of cells. Things aren't explained to us in this way, of course. We are simply told that someone we know has cancer.

At root, all cancers are failures of the cell cycle. Put another way, all cancers represent a failure of cells to limit their multiplication in the cell cycle. What is liver cancer, for example? It is a damaging multiplication of liver cells. First one, then two, then four, then eight liver cells move repeatedly through the cell cycle, and as their numbers increase, they destroy the liver's working tissues. Given that cancer manifests in this way, it's not surprising that a large portion of modern cancer research is *cell cycle* research. The logic here is simple: To the extent that uncontrolled cell division can be stopped, cancer can be stopped.

A number of ideas are being tested for how cancer gets going in the first place, but a common thread that runs through these ideas is that, for cells to be brought to a cancerous state, two things are required: Their accelerators must get stuck and their brakes must fail. The control mechanisms that *induce* cell division must become hyperactive, and the mechanisms that *suppress* cell division must fail to perform. You may have heard a couple of terms used to describe the genetic components of this process. There are normal genes that induce cell division, but that when mutated can cause cancer; these are the stuck-accelerator genes, called oncogenes.

Then, there are genes that normally suppress cell division but that can cause cancer by acting like failed brakes. These are tumor suppressor genes. Note that *both* kinds of genes must malfunction for cancer to get going; indeed, it usually takes a long succession of genetic failures to induce cancer. This is why cancer is most often a disease of the middle-aged and elderly: It can take decades for the required series of mutations to fall into line in a single cell, such that it becomes cancerous.

How do oncogenes interact with tumor suppressor genes? In the normal case, cells will not begin division until prompted to do so by a signal from outside themselves. A protein (called a growth factor) will bind to a cell, setting off a cascade of chemical reactions inside it that triggers division. One of the links in this chemical cascade is a protein called Ras that could be thought of as an old-time railway switch. When Ras is chemically pointed one way (toward "on"), the cell moves through the cell cycle. When it is pointed the other (toward "off"), the cell stays in $G_1$. The gene that codes for Ras can become mutated, however, and when this happens, the Ras protein changes shape and points in the "on" direction all the time—no matter what signals it is getting from the outside. Thus, *ras* is an example of an oncogene. It normally prompts the cell to divide intermittently; but when mutated, it prompts the cell to divide continuously.

A cell with a mutated *ras* gene is not doomed to become cancerous, however. What can save it is a good set of brakes, in the form of tumor suppressor genes. The most important of these genes—one known as *p53*—is so vital to human health that it is sometimes referred to as the "guardian of the genome." In the presence of certain kinds of mutations, *p53* protein levels rise in the cell, and these

# WEB TOOLS
## engage students and assist instructors

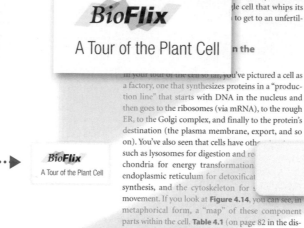

80  CHAPTER 4  **The Cell**

**BioFlix**

A Tour of the Plant Cell

In your tour of the cell so far, you've pictured a cell as a factory, one that synthesizes proteins in a "production line" that starts with DNA in the nucleus and then goes to the ribosomes (via mRNA), to the rough ER, to the Golgi complex, and finally to the protein's destination (the plasma membrane, export, and so on). You've also seen that cells have other structures such as lysosomes for digestion and reproduction for energy transformation, endoplasmic reticulum for detoxification, synthesis, and the cytoskeleton for movement. If you look at **Figure 4.14**, you can see, in metaphorical form, a "map" of these component parts within the cell. **Table 4.1** (on page 82 in the discussion of plant cells) lists cellular elements found in plant and animal cells as well as some elements found only in plant cells.

**SO FAR...**

1. Lipids, including fats, are put together in the organelle called the _____, while worn-out cellular materials are broken down and recycled in the organelles called _____.

2. The organelles called mitochondria transform the _____ from food into a usable chemical form, a molecule called _____.

3. The cell's internal scaffolding, made up of different types of protein fibers, is called the _____.

living thing. If you looked at the cells that make up plants or fungi, for example, you'd see that they have a nucleus, ribosomes, mitochondria, and so forth. But, as you can imagine, the cells of a fungus have to be *somewhat* different from those of an animal since fungi and animals are such different types of organisms. To give you an idea of some of the kinds of differences found in the cells of life's various kingdoms, we'll look briefly now

**Web Animation 4.1**
The Structure of Cells

## Easy-to-locate Web Icons

BioFlix™ and Web Animation icons are placed throughout each chapter and direct students online to view informative animations on the topic they are reading about.

**BioFlix**
A Tour of the Plant Cell

**Web Animation 4.1**
The Structure of Cells

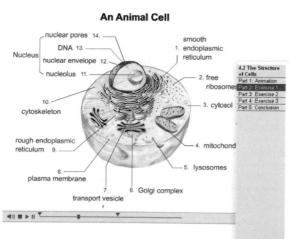

## Web Animations

These web tutorials lead students through a series of biological concepts with animations, interactive exercises, and quizzes for review of difficult concepts.

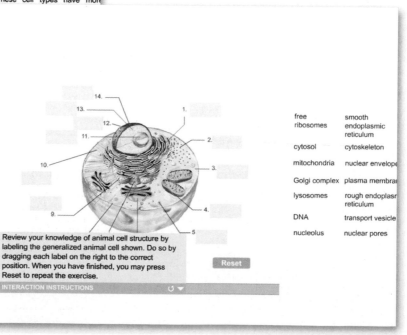

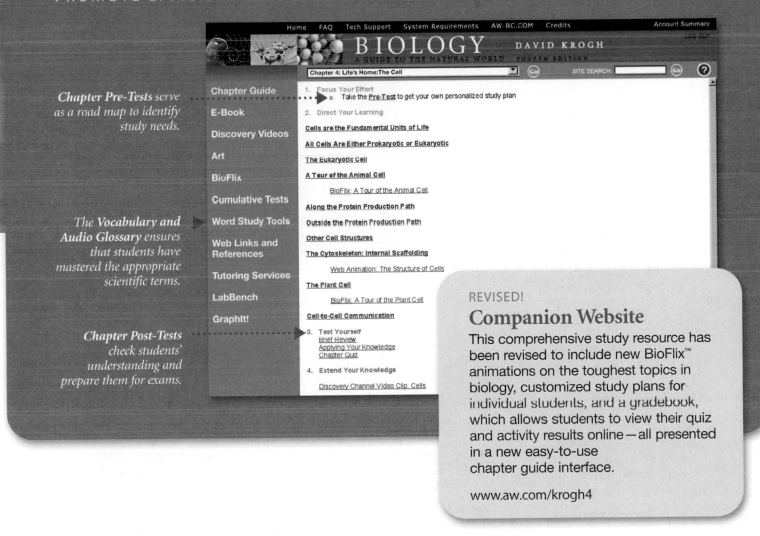

*Chapter Pre-Tests* serve as a road map to identify study needs.

The *Vocabulary and Audio Glossary* ensures that students have mastered the appropriate scientific terms.

*Chapter Post-Tests* check students' understanding and prepare them for exams.

REVISED!

## Companion Website

This comprehensive study resource has been revised to include new BioFlix™ animations on the toughest topics in biology, customized study plans for individual students, and a gradebook, which allows students to view their quiz and activity results online—all presented in a new easy-to-use chapter guide interface.

www.aw.com/krogh4

## SEE SCIENCE IN ACTION WITH DISCOVERY CHANNEL VIDEO CLIPS

These brief 3–5 minute Discovery Channel video clips cover topics from fighting cancer to antibiotic resistance and introduced species. The video clips are located on the companion website.

# GIVE STUDENTS

## interactive tutoring on the toughest topics in biology

**BioFlix**

**NEW! BioFlix™ animations** and student tools invigorate classroom lectures and cover the most difficult biology topics with 3-D movie-quality animations, labeled slide shows, carefully constructed student tutorials, study sheets, and quizzes.

## BioFlix Tutorial Topics

- Tour of an Animal Cell
- Tour of a Plant Cell
- Cellular Respiration
- Photosynthesis
- Mitosis
- Meiosis
- Protein Synthesis
- Water Transport in Plants
- Muscle Contraction
- How Neurons Work

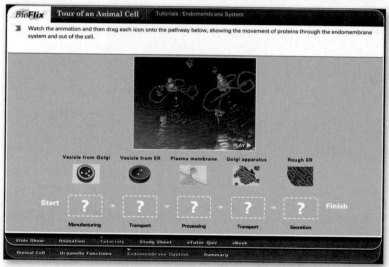

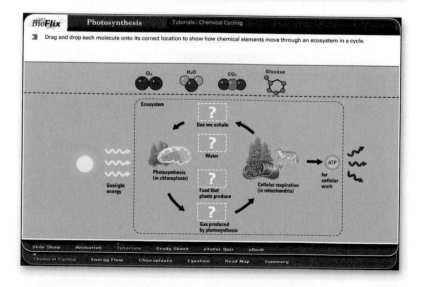

# STUDY TOOLS HELP STUDENTS UNDERSTAND BIOLOGY CONCEPTS

These resources make the toughest topics in biology more accessible.

## Student Quizzes ▶

Keep track of students' progress with these assignable, online quizzes. Scores can be recorded in an online gradebook.

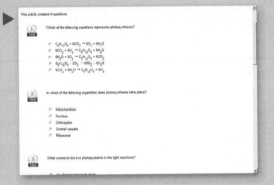

## Tutorials ▶

Carefully constructed student tutorials help students learn about important biological processes.

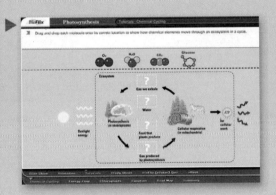

## e-Study Sheets ▲

Students can study for tests by completing printable study sheets.

# PRESENTATION AND COURSE MANAGEMENT TOOLS HELP YOU TEACH

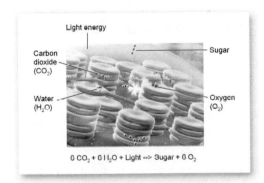

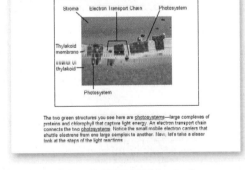

## PowerPoint® Lecture Slides

PowerPoint slides supplement instructors' lectures with chapter outlines and summaries of key concepts.

## Flexible Presentation

Instructors have the flexibility to add their own narration to their BioFlix™ animations by disabling the BioFlix sounds.

## Course Management

Open-access course management systems — BlackBoard, WebCT and CourseCompass™ — contain pre-loaded content such as testing and assessment question pools.

# DYNAMIC RESOURCES
## help you teach your biology course

### Instructor Resource CD-ROM

The **Instructor Resource CD-ROM** includes PowerPoint® lectures that integrate figures and new BioFlix™ animations, BLAST animations, Discovery Channel® Videos, all figures, art, and photos in JPEG format, label-edit PowerPoint slides, and over 200 Instructor Animations that accurately depict complex topics and dynamic processes described in the book. New, innovative Classroom Response System questions are offered for each chapter as a way to directly engage students in lectures.

978-0-13-225474-8 • 0-13-225474-3

*BLAST Animations are 69 scientifically clear and accurate additional animations that instructors can step through in PowerPoint and their narration for lecture.*

*NEW! BioFlix animations invigorate classroom lectures with effective, 3-minute, movie-quality, 3-D graphics.*

*Expanded Test Bank questions are available as downloadable Microsoft® Word® files.*

*Customizable PowerPoint Lectures serve as lecture outlines and are included for each textbook chapter.*

**Chapter 2 Lecture**

**Biology: A Guide to the Natural World**
Fourth Edition

**2.1 Chemistry's Building Block: The Atom**

The fundamental unit of matter is the atom.

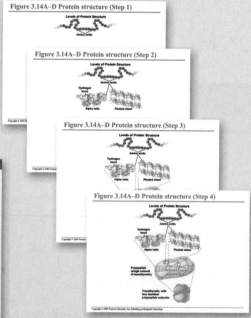

Figure 3.14A–D Protein structure (Step 1)

Figure 3.14A–D Protein structure (Step 2)

Figure 3.14A–D Protein structure (Step 3)

Figure 3.14A–D Protein structure (Step 4)

**Figure 4.16**
**The Plant Cell**
The cell wall, central vacuole, and chloroplasts do not exist in animal cells, but all other components are common to both plant and animal cells.

Plant cells have a cell wall, chloroplasts, and a central vacuole, while animal cells do not.

- cytoskeleton
- cell wall
- chloroplast
- central vacuole

- nuclear envelope
- nuclear pores — nucleus
- DNA
- nucleolus
- rough endoplasmic reticulum
- smooth endoplasmic reticulum
- free ribosomes
- Golgi complex
- cytosol
- plasma membrane
- mitochondrion

**Table 11.1 Pea-Plant Characters Studied by Mendel**

| Character studied | Dominant trait | Recessive trait |
|---|---|---|
| Seed shape | smooth | wrinkled |
| Seed color | yellow | green |
| Pod shape | inflated | wrinkled |
| Pod color | green | yellow |
| Flower color | purple | white |
| | at tip | |
| | | dwarf |

*All of the art, stepped-art, tables, and photos from the book are available with customizable labels and are also available in PowerPoint.*

### Instructor Guide

This comprehensive guide provides chapter-by-chapter objectives, lists of common student misconceptions, teaching tips, media tips, additional critical thinking questions, key terms, and word roots. A separate chapter offers suggestions for effective uses of technology in teaching introductory biology. The **Instructor Guide** is available in print and as a Microsoft® Word® document.

978-0-13-225476-2 • 0-13-225476-X

INSTRUCTOR GUIDE

BIOLOGY

A GUIDE TO THE NATURAL WORLD

DAVID KROGH

### *Scientific American Current Issues in Biology*

Give your students the best of both worlds—accessible, dynamic, relevant articles from *Scientific American* magazine that present key issues in biology, paired with the authority, reliability, and clarity of Benjamin Cummings non-majors biology texts. Articles include questions to help students check their comprehension and make connections to science and society. This resource is available at no additional charge when packaged with a new text.

**Volume 1:** 978-0-8053-7507-7 • 0-8053-7507-4

**Volume 2:** 978-0-8053-7108-6 • 0-8053-7108-7

**Volume 3:** 978-0-8053-7527-5 • 0-8053-7527-9

**Volume 4:** 978-0-8053-3566-8 • 0-8053-3566-8

**Volume 5:** 978-0-321-54187-1 • 0-321-54187-1

# CONTENTS

## CHAPTER 1

### SCIENCE AS A WAY OF LEARNING: A Guide to the Natural World    2

| | | |
|---|---|---|
| **1.1** | **How Does Science Impact the Everyday World?** | **3** |
| | What Do Americans Know about Science? | 6 |
| **1.2** | **What Is Science?** | **6** |
| | Science as a Body of Knowledge | 6 |
| | Science as a Process: Arriving at Scientific Insights | 7 |
| | When Is a Theory Proven? | 9 |
| **1.3** | **The Nature of Biology** | **11** |
| | Life Is Highly Organized in a Hierarchical Manner | 12 |
| | Levels of Organization in Living Things | 12 |
| **1.4** | **Special Qualities of Biology** | **13** |
| | Biology's Chief Unifying Principle | 14 |
| **1.5** | **The Organization Of This Book** | **15** |
| **Chapter Review** | | **16** |
| **Essay** | Lung Cancer, Smoking, and Statistics in Science | **10** |

## Unit 1
## Essential Parts: Atoms, Molecules, and Cells

## CHAPTER 2

### FUNDAMENTAL BUILDING BLOCKS: Chemistry, Water, and pH    18

| | | |
|---|---|---|
| **2.1** | **Chemistry's Building Block: The Atom** | **19** |
| | Protons, Neutrons, and Electrons | 20 |
| | Fundamental Forms of Matter: The Element | 21 |
| **2.2** | **Matter Is Transformed through Chemical Bonding** | **23** |
| | Energy Always Seeks Its Lowest State | 23 |
| | Covalent Bonding | 23 |
| | Reactive and Unreactive Elements | 24 |
| | Polar and Nonpolar Bonding | 26 |
| | Ionic Bonding: When Electrons Are Lost or Gained | 28 |
| | A Third Form of Bonding: Hydrogen Bonding | 29 |
| **2.3** | **Some Qualities of Chemical Compounds** | **29** |
| | Molecules Have a Three-Dimensional Shape | 29 |
| | Molecular Shape Is Very Important in Biology | 30 |

| | | |
|---|---|---|
| **2.4** | **Water and Life** | **30** |
| | Water Is a Major Player in Many of Life's Processes | 31 |
| | Water's Structure Gives It Many Unusual Properties | 32 |
| | Two Important Terms: Hydrophobic and Hydrophilic | 33 |
| **2.5** | **Acids and Bases Are Important to Life** | **34** |
| | Acids Yield Hydrogen Ions in Solution, Bases Accept Them | 34 |
| | Many Common Substances Can Be Ranked According to How Acidic or Basic They Are | 34 |
| | The pH Scale Allows Us to Quantify How Acidic or Basic Compounds Are | 35 |
| | Some Terms Used When Dealing with pH | 35 |
| | **On to Biological Molecules** | **37** |
| **Chapter Review** | | **38** |
| **Essays** | Getting to Know Chemistry's Symbols | **25** |
| | Free Radicals | **27** |

## CHAPTER 3

### LIFE'S COMPONENTS: Biological Molecules    40

| | | |
|---|---|---|
| **3.1** | **Carbon Is Central to the Living World** | **41** |
| **3.2** | **Functional Groups** | **43** |
| **3.3** | **The Molecules of Life: Carbohydrates** | **44** |
| | Kinds of Simple Carbohydrates | 45 |

Complex Carbohydrates Are Made of Chains
of Simple Carbohydrates 45

**3.4 Lipids** 47
One Class of Lipids Is the Glycerides 47
Saturated and Unsaturated Fatty Acids 48
Fatty Acids: From Solid to Liquid 48
Energy Use and Energy Storage via Lipids and
Carbohydrates 49
A Second Class of Lipids Is the Steroids 50
A Third Class of Lipids Is the Phospholipids 51
A Fourth Class of Lipids Is Wax 51

**3.5 Proteins** 53
Proteins Are Made from Chains of Amino Acids 54
A Group of Only 20 Amino Acids Is the Basis for
All Proteins in Living Things 54
Shape Is Critical to the Functioning of All Proteins 55
There are Four Levels of Protein Structure 56
Proteins Can Come Undone 57
Lipoproteins and Glycoproteins 57

**3.6 Nucleic Acids** 57
DNA Provides Information for the Structure
of Proteins 57
The Structural Unit of DNA Is the Nucleotide 59

**On to Cells** 59

**Chapter Review** 60

**Essay** From Trans Fats to Omega-3s: Fats and Health 52


CHAPTER **4**

**LIFE'S HOME: The Cell** 62

**4.1 Cells Are the Fundamental Units of Life** 63

**4.2 All Cells Are Either Prokaryotic or Eukaryotic** 64
Prokaryotic and Eukaryotic Differences 65

**4.3 The Eukaryotic Cell** 65
The Animal Cell 68

**4.4 A Tour of the Animal Cell: Along the Protein
Production Path** 69
Beginning in the Control Center: The Nucleus 69
*Bio*Flix Tour of an Animal Cell 69
Messenger RNA 70
Ribosomes 71
The Rough Endoplasmic Reticulum 72
A Pause for the Nucleolus 72
Elegant Transportation: Transport Vesicles 72
Downstream from the Rough ER: The
Golgi Complex 73
From the Golgi to the Surface 74

**4.5 Outside the Protein Production Path: Other Cell
Structures** 74
The Smooth Endoplasmic Reticulum 74
Tiny Acid Vats: Lysosomes and Cellular Recycling 74
Extracting Energy from Food: Mitochondria 75

**4.6 The Cytoskeleton: Internal Scaffolding** 76
Microfilaments 77
Intermediate Filaments 78
Microtubules 78
Cell Extensions Made of Microtubules:
Cilia and Flagella 79
In Summary: Structures in the Animal Cell 80

**4.7 The Plant Cell** 80
*Bio*Flix Tour of a Plant Cell 80
The Central Vacuole 80
The Cell Wall 82
Chloroplasts 84

**4.8 Cell-to-Cell Communication** 84
Communication among Plant Cells 84
Communication among Animal Cells 85
Leeuwenhoek and the Process of Science 85

**On to the Periphery** 84

**Chapter Review** 88

**Essays** The Size of Cells 66
The Stranger within: Endosymbiosis 78
The Process of Science: First Sightings:
Anton van Leeuwenhoek 86


CHAPTER **5**

**LIFE'S BORDER: The Plasma Membrane** 90

**5.1 The Nature of the Plasma Membrane** 91
First Component: The Phospholipid Bilayer 92
Second Component: Cholesterol 93
Third Component: Proteins 93
Fourth Component: The Glycocalyx 95
The Fluid-Mosaic Membrane Model 95

**5.2 Diffusion, Gradients, and Osmosis** 95
Random Movement and Even Distribution 96
Diffusion through Membranes 96

**5.3 Moving Smaller Substances In and Out** 99
Passive Transport 99
Active Transport 100

**5.4**  **Getting the Big Stuff In and Out**  101
Movement Out: Exocytosis  101
Movement In: Endocytosis  101

**Discovery and the Fluid-Mosaic Model**  103

**On to Energy**  103

**Chapter Review**  106

**Essay**  The Process of Science: The Fluid-Mosaic Model
of the Plasma Membrane  104

## Unit 2
## Energy and Its Transformations

CHAPTER **6**

LIFE'S MAINSPRING:
## An Introduction to Energy  108

**6.1**  **Energy Is Central to Life**  109

**6.2**  **The Nature of Energy**  110
The Forms of Energy  110
The Study of Energy: Thermodynamics  110
The Consequences of Thermodynamics  111

**6.3**  **How Is Energy Used by Living Things?**  112
Up and Down the Great Energy Hill  112
Coupled Reactions  112

**6.4**  **The Energy Dispenser: ATP**  113
The ATP/ADP Cycle  114
Between Food and ATP  114

**6.5**  **Efficient Energy Use in Living Things: Enzymes**  115
Accelerating Reactions  115
Specific Tasks and Metabolic Pathways  116

**6.6**  **Lowering the Activation Barrier through Enzymes**  116
How Do Enzymes Work?  117
An Enzyme in Action: Chymotrypsin  118

**6.7**  **Regulating Enzymatic Activity**  118
Allosteric Regulation of Enzymes  118

**On to Energy Harvesting**  119

**Chapter Review**  120

CHAPTER **7**

VITAL HARVEST:
## Deriving Energy from Food  122

**7.1**  **Energizing ATP: Adding a Phosphate Group to ADP**  124

**7.2**  **Electrons Fall Down the Energy Hill to Drive the
Uphill Production of ATP**  124
Redox Reactions  125
Many Molecules Can Oxidize Other Molecules  125

**7.3**  **The Three Stages of Cellular Respiration:
Glycolysis, the Krebs Cycle, and the Electron
Transport Chain**  127
**BioFlix**  Cellular Respiration  127
Glycolysis: First to Evolve, Less Efficient  127

**7.4**  **First Stage of Respiration: Glycolysis**  127

**7.5**  **Second Stage of Respiration: The Krebs Cycle**  131
Site of Action Moves from the Cytosol to the
Mitochondria  131
Between Glycolysis and the Krebs Cycle, an
Intermediate Step  132
Into the Krebs Cycle: Why Is It a Cycle?  132

**7.6**  **Third Stage of Respiration: The Electron
Transport Chain**  133
Where's the ATP?  134
Finally, Oxygen Is Reduced, Producing Water  135
Bountiful Harvest: ATP Accounting  135

**7.7**  **Other Foods, Other Respiratory Pathways**  135
On to Photosynthesis  137

**Chapter Review**  138

**Essays**  When Energy Harvesting Ends at Glycolysis,
Beer Can Be the Result  130
Energy and Exercise  136

CHAPTER **8**

THE GREEN WORLD'S GIFT:
## Photosynthesis  140

**8.1**  **Photosynthesis and Energy**  141
From Plants, a Great Bounty for Animals  142
Up and Down the Energy Hill Again  142

**8.2**  **The Components of Photosynthesis**  142
Where in the Plant Does Photosynthesis Occur?  143

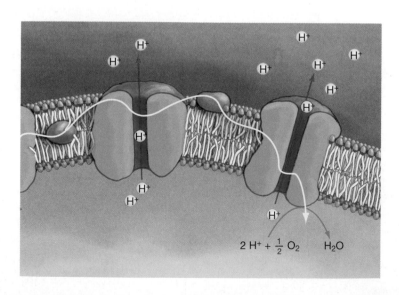

$$2\,H^+ + \tfrac{1}{2}\,O_2 \qquad H_2O$$

There Are Two Essential Stages in Photosynthesis 144
The Working Units of the Light Reactions 144
Energy Transfer in Photosynthesis Works through
   Redox Reactions 145

**8.3 Stage 1: The Steps of the Light Reactions** 146
A Chain of Redox Reactions and Another Boost
   from the Sun 146
The Physical Movement of Electrons in the Light
   Reactions 146

**8.4 What Makes the Light Reactions So Important?** 146
The Splitting of Water: Electrons and Oxygen 146
The Transformation of Solar Energy to
   Chemical Energy 147
Production of ATP 148
*BioFlix* Photosynthesis 148

**8.5 Stage 2: The Calvin Cycle** 148
Energized Sugar Comes from a Cycle of Reactions 148
The Ultimate Product of Photosynthesis 149

**8.6 Photorespiration and the C₄ Pathway** 150
The C₄ Pathway Is Not Always Advantageous 151

**8.7 Another Photosynthetic Variation: CAM Plants** 151
Closing Thoughts on Photosynthesis and Energy 152

**Chapter Review** 156
**Essay** The Process of Science: Plants Make Their Own Food,
   But How? 154

## Unit 3
## How Life Goes On

CHAPTER **9**

## GENETICS AND CELL DIVISION 158

**9.1 An Introduction to Genetics** 159
DNA Contains Instructions for Protein
   Production 159
Genetics as Information Management 161
From One Gene to a Collection 161
The Path of Study in Genetics 162

**9.2 An Introduction to Cell Division** 162
The Replication of DNA 163

**9.3 DNA Is Packaged in Chromosomes** 164
Matched Pairs of Chromosomes 164
Chromosome Duplication as a Part of Cell Division 166

**9.4 Mitosis and Cytokinesis** 168
Mitosis 169
The Phases of Mitosis 168
*BioFlix* Mitosis 169
Cytokinesis 170

**9.5 Variations in Cell Division** 170
Plant Cells 170
Prokaryotes: Bacteria and Archaea 170
On to Meiosis 172

**Chapter Review** 174
**Essay** When the Cell Cycle Runs Amok: Cancer 172

CHAPTER **10**

## PREPARING FOR SEXUAL REPRODUCTION:
## Meiosis 176

**10.1 An Overview of Meiosis** 177
Some Helpful Terms 178

**10.2 The Steps in Meiosis** 178
Meiosis I 178
*BioFlix* Meiosis 181
Meiosis II 182

**10.3 What Is the Significance of Meiosis?** 182
Genetic Diversity through Crossing Over 182
Genetic Diversity through Independent
   Assortment 183
From Genetic Diversity, a Visible Diversity in
   the Living World 183

**10.4 Meiosis and Sex Outcome** 185

**10.5 Gamete Formation in Humans** 186
Sperm Formation 186
Egg Formation: Stem Cells in Females? 187
Egg Formation: From Oogonia to Eggs 187
One Egg, Several Polar Bodies 188

**10.6 Life Cycles: Humans and Other Organisms** 188
Not All Reproduction Is Sexual 188
Variations in Sexual Reproduction 189
On to Patterns of Inheritance 189

**Chapter Review** 190

CHAPTER **11**

## THE FIRST GENETICIST:
## Mendel and His Discoveries 192

**11.1 Mendel and the Black Box** 193

**11.2 The Experimental Subjects: *Pisum sativum*** 194
Phenotype and Genotype 196

**11.3 Starting the Experiments: Yellow and Green Peas** 196
Parental, F₁ and F₂ Generations 196
Interpreting the F₁ and F₂ Results 197

**11.4 Another Generation for Mendel** 198
Mendel's Generations in Pictures 198
The Law of Segregation 200

**11.5**  **Crosses Involving Two Characters**  202
Crosses for Seed Color *and* Seed Shape  202
The Law of Independent Assortment  203

**11.6**  **Reception of Mendel's Ideas**  205

**11.7**  **Incomplete Dominance**  205
Genes Code for Proteins  206

**11.8**  **Lessons from Blood Types: Codominance**  206
Getting Both Types of Surface Proteins  207
Dominant and Recessive Alleles  207

**11.9**  **Multiple Alleles and Polygenic Inheritance**  207

**11.10**  **Genes and Environment**  209

**11.11**  **One Gene, Several Effects: Pleiotropy**  210
On to the Chromosome  210

**Chapter Review**  211

**Essays**  Proportions and Their Causes: The Rules of
Multiplication and Addition  201
Why So Unrecognized?  204

CHAPTER **12**

UNITS OF HEREDITY:
## Chromosomes and Inheritance  214

**12.1**  **X-Linked Inheritance in Humans**  216
The Genetics of Color Vision  216
Alleles and Recessive Disorders  216

**12.2**  **Autosomal Genetic Disorders**  217
Sickle-Cell Anemia  218
Dominant Disorders  218

**12.3**  **Tracking Traits with Pedigrees**  220

**12.4**  **Aberrations in Chromosomal Sets: Polyploidy**  220

**12.5**  **Incorrect Chromosome Number: Aneuploidy**  222
Aneuploidy's Main Cause: Nondisjunction  222

The Consequences of Aneuploidy  223
Abnormal Numbers of Sex Chromosomes  223
Aneuploidy and Cancer  224

**12.6**  **Structural Aberrations in Chromosomes**  225
Deletions  225
Inversions and Translocations  227
Duplications  227
**The Process of Science**  230
**On to DNA**  230

**Chapter Review**  231

**Essays**  PGD: Screening for a Healthy Child  226
The Process of Science: Thomas Hunt Morgan:
Using Fruit Flies to Look More Deeply Into
Genetics  228

CHAPTER **13**

PASSING ON LIFE'S INFORMATION:
## DNA Structure and Replication  234

**13.1**  **What Do Genes Do, and What Are They Made of?**  235
DNA Structure and the Rise of Molecular Biology  236

**13.2**  **Watson and Crick: The Double Helix**  236

**13.3**  **The Components of DNA and Their Arrangement**  237
The Structure of DNA Gives Away the Secret of
Replication  239
DNA Structure and Protein Production  239
The Building Blocks of DNA Replication  240

**13.4**  **Mutations**  240
Cancer and Huntington Mutations  241
Heritable and Nonheritable Mutations  241
What Causes Mutations?  243
Mutations and Evolutionary Adaptation  243
**On to Protein Production**  243

**Chapter Review**  244

**Essay**  The Process of Science: Getting Clear about
What Genes Do: Beadle and Tatum  242

CHAPTER **14**

HOW PROTEINS ARE MADE: **Genetic
Transcription, Translation, and Regulation**  246

**14.1**  **The Structure of Proteins**  247
Synthesizing Many Proteins from 20 Amino Acids  247

**14.2**  **Protein Synthesis in Overview**  248

**14.3**  **A Closer Look at Transcription**  249
Passing on the Message: Base Pairing Again  250
A Triplet Code  250

**14.4**  **A Closer Look at Translation**  252
The Nature of tRNA  252

The Structure of Ribosomes 253

*Bio**Flix*** Protein Synthesis 254

The Steps of Translation 254

**14.5 Genetic Regulation** 257

A Genome Contains Lots More than Genes 257

Promoters, Enhancers, and Proteins that Bind Them 258

More Regulation: Alternative Splicing 259

RNA in Genetic Regulation 260

The Importance of Regulation 261

The Importance of DNA that Doesn't Code
for Proteins 262

What Is Central in Genetics? 262

**14.6 Genetics and Life** 262

Biotechnology Is Next 263

Chapter Review 264

**Essay** Cracking the Genetic Code 256

CHAPTER **15**

THE FUTURE ISN'T WHAT IT USED TO BE:
Biotechnology

**Biotechnology** 266

**15.1 What Is Biotechnology?** 268

**15.2 Transgenic Biotechnology** 268

A Biotech Tool: Restriction Enzymes 269

Another Tool of Biotech: Plasmids 270

Using Biotech's Tools: Getting Human Genes into
Plasmids 270

Getting the Plasmids Back inside Cells,
Turning out Protein 271

A Plasmid Is One Kind of Cloning Vector 271

Real-World Transgenic Biotechnology 272

**15.3 Reproductive Cloning** 272

Reproductive Cloning: How Dolly Was Cloned 273

Cloning and Recombinant DNA 273

Human Cloning 275

**15.4 Forensic Biotechnology** 275

The Use of PCR 276

Finding Individual Patterns 276

**15.5 Stem Cells** 277

Cell Fates: Committed or Not 277

Stem Cells from Embryos 279

Adult Stem Cells 279

The Fight over Embryonic Stem Cells 280

The Potential of Embryonic Stem Cells 280

Stem Cells Meet Cloning: Therapeutic Cloning 281

A Stunning Breakthrough in 2007 282

**15.6 Biotechnology in the Real World** 284

Controversies in Biotechnology 284

Ethical Controversies in Biotechnology 285

On to Evolution 286

Chapter Review 287

**Essays** Reading DNA Profiles 278

The Process of Science: Making Embryonic Stem
Cells Work 282

### Unit 4
# Life's Organizing Principle: Evolution and the Diversity of Life

CHAPTER **16**

AN INTRODUCTION TO EVOLUTION:
Charles Darwin, Evolutionary Thought,
and the Evidence for Evolution 290

**16.1 Evolution and Its Core Principles** 292

Common Descent with Modification 292

Natural Selection 292

The Importance of Evolution as a Concept 292

Evolution Affects Human Perspectives
Regarding Life 292

**16.2 Charles Darwin and the Theory of Evolution** 293

Darwin's Contribution 293

Darwin's Journey of Discovery 293

**16.3 Evolutionary Thinking before Darwin** 294

Charles Lyell and Geology 295

Jean-Baptiste de Lamarck and Evolution 295

Georges Cuvier and Extinction 296

**16.4** Darwin's Insights Following the *Beagle*'s Voyage 296
Perceiving Common Descent with Modification 296
Perceiving Natural Selection 296

**16.5** Alfred Russel Wallace 297

**16.6** Descent with Modification Is Accepted 298

**16.7** Darwin Doubted: The Controversy over Natural Selection 299
Coming to an Understanding of Genetics 299
Vindicating Natural Selection's Role in Evolution 299
Darwin Triumphant: The Modern Synthesis 299

**16.8** Opposition to the Theory of Evolution 300
The False Notion of a Scientific Controversy 300

**16.9** The Evidence for Evolution 301
Radiometric Dating 301
Fossils 301
Comparative Morphology and Embryology 301
Evidence from Biogeography 302
Evidence from Gene Modification 306
Experimental Evidence 306

On to How Evolution Works 307

Chapter Review 308

**Essay** The Evolution of Human Skin Color 304

CHAPTER **17**

THE MEANS OF EVOLUTION:
## Microevolution

310

**17.1** What Is It That Evolves? 311
Populations Are the Essential Units That Evolve 311
Genes Are the Raw Material of Evolution 312

**17.2** Evolution as a Change in the Frequency of Alleles 313

**17.3** Five Agents of Microevolution 313
Mutations: Alterations in the Makeup of DNA 314
Gene Flow: When One Population Joins Another 315
Genetic Drift: The Instability of Small Populations 315
Sexual Selection: When Mating Is Uneven across a Population 318
Natural Selection: Evolution's Adaptive Mechanism 318

**17.4** Natural Selection and Evolutionary Fitness 320
Galapagos Finches: The Studies of Peter and Rosemary Grant 320

**17.5** Three Modes of Natural Selection 322
Stabilizing Selection 323
Directional Selection 323
Disruptive Selection 324

On to the Origin of Species 324

Chapter Review 328

**Essays** Lessons from the Cocker Spaniel: The Price of Inbreeding 318
Detecting Evolution: The Hardy-Weinberg Principle 325

CHAPTER **18**

THE OUTCOMES OF EVOLUTION:
## Macroevolution

330

**18.1** What Is a Species? 331

**18.2** How Do New Species Arise? 333
The Role of Geographic Separation: Allopatric Speciation 333
Reproductive Isolating Mechanisms Are Central to Speciation 334
Six Intrinsic Reproductive Isolating Mechanisms 334
Sympatric Speciation 336

**18.3** Many New Species from One: Adaptive Radiation 337
New Environments and Speciation 338

**18.4** The Pace of Speciation 340

**18.5** The Categorization of Earth's Living Things 341
Taxonomic Classification and the Degree of Relatedness 342
A Taxonomic Example: The Common House Cat 342

**18.6** Classical Taxonomy and Cladistics 345
Another System for Interpreting the Evidence: Cladistics 345
Should Anything but Relatedness Matter in Classification? 346

On to the History of Life 347

Chapter Review 348

**Essay** New Species through Genetic Accidents: Polyploidy 338

CHAPTER **19**

A SLOW UNFOLDING:
## The History of Life on Earth

350

**19.1** The Geological Timescale: Life Marks Earth's Ages 352
Features of the Timescale 353
What Is Notable in Evolution Hinges on Values 354

**19.2** How Did Life Begin? 354
From the Simple to the Complex 355
The Source of Life's Raw Materials 358
Life May Have Begun in Very Hot Water 358
The RNA World 359

**19.3** The Tree of Life 360

**19.4** A Long First Era: The Precambrian 360
Notable Precambrian Events 361

**19.5** The Cambrian Explosion 363

**19.6** The Movement onto the Land: Plants First 365
Adaptations of Plants to the Land 365
Another Plant Innovation: A Vascular System 366

Plants with Seeds: The Gymnosperms and
Angiosperms 366
The Last Plant Revolution So Far: The Angiosperms 366

**19.7 Animals Follow Plants onto the Land** 367
Vertebrates Move onto Land 367
Evolutionary Lines of Land Vertebrates 368
The Primate Mammals 371

**On to Human Evolution** 372

**Chapter Review** 373

**Essays** Physical Forces and Evolution 356
Who Gets a Kingdom? Evolution within
Life's Categories 362
The Process of Science: Going after the Fossils 370

CHAPTER **20**

## ARRIVING LATE, TRAVELING FAR:
## The Evolution of Human Beings

The Evolution of Human Beings 376

**20.1 The Human Family Tree** 377
Connecting the Dots in the Tree 378

**20.2 Human Evolution in Overview** 378

**20.3 Interpreting the Fossil Evidence** 381

**20.4 Snapshots from the Past: Three Hominins** 382

**20.5 The Appearance of Modern Human Beings** 385
Who Lives, Who Doesn't? 385

**20.6 Next-to-Last Standing? The Hobbit People** 387

**On to the Diversity of Life** 387

**Chapter Review** 388

**Essay** Sequencing the Neanderthal Genome 386

CHAPTER **21**

## VIRUSES, BACTERIA, ARCHAEA, AND PROTISTS:
## The Diversity of Life 1

The Diversity of Life 1 390

**21.1 Life's Categories and the Importance of Microbes** 392

**21.2 Viruses: Making a Living by Hijacking Cells** 394
HIV: The AIDS Virus 394
Viral Diversity 396
The Trouble Viruses Cause: Avian Flu 396

**21.3 Bacteria: Masters of Every Environment** 397
Bacterial Numbers and Diversity 399

**21.4 Intimate Strangers: Humans and Bacteria** 400

**21.5 Bacteria and Human Disease** 401
Killing Pathogenic Bacteria: Antibiotics 402
The Threat of Antibiotic Resistance 402

**21.6 Archaea: From Marginal Player to Center Stage** 403
The Separate Status of Domain Archaea 403
Archaea and Their Habitats 404

Extremophiles 404
Prospecting for Extremophiles 405

**21.7 Protists: Pioneers in Diversifying Life** 405

**21.8 Protists and Sexual Reproduction** 406

**21.9 Photosynthesizing Protists** 407

**21.10 Heterotrophic Protists** 408
Heterotrophs with Locomotor Extensions 409
Heterotrophs with Limited Mobility 409

**On to Fungi and Plants** 412

**Chapter Review** 413

**Essays** Unwanted Guest: The Persistence of Herpes 398
Modes of Nutrition: How Organisms Get What
They Need to Survive 400
The Process of Science: The Discovery of Penicillin 410

CHAPTER **22**

## FUNGI AND PLANTS:
## The Diversity of Life 2

The Diversity of Life 2 416

**22.1 The Fungi: Life as a Web of Slender Threads** 417

**22.2 Roles of Fungi in Society and Nature** 419

**22.3 Structure and Reproduction in Fungi** 420
The Life Cycle of a Fungus 421
Reproduction in Other Types of Fungi 422

**22.4 Categories of Fungi** 422
Yeasts: *Saccharomyces cerevisiae* 424

**22.5 Fungal Associations: Lichens and Mycorrhizae** 426
Lichens 426
Mycorrhizae 427

spore

hyphae

**22.6**    **Plants: The Foundation for Much of Life**    427
The Characteristics of Plants    427

**22.7**    **Types of Plants**    429
Bryophytes: Amphibians of the Plant World    430
Seedless Vascular Plants: Ferns and Their
Relatives    431
The First Seed Plants: The Gymnosperms    432
Reproduction through Pollen and Seeds    432

**22.8**    **Angiosperm-Animal Interactions**    434
Seed Endosperm: More Animal Food from
Angiosperms    435
Fruit: An Inducement for Seed Dispersal    435

**On to a Look at Animals**    436

**Chapter Review**    437

**Essay**    A Psychedelic Drug from an Ancient Source    425

CHAPTER **23**

## ANIMALS: **The Diversity of Life 3**    440

**23.1**    **What is an Animal?**    441
**23.2**    **Animal Types: The Family Tree**    442
Additions 1 and 2: Tissues and Symmetry    443
Addition 3: Bilateral Symmetry    444
Addition 4: A Body Cavity    444
A Split in the Animal Kingdom: Protostomes and
Deuterostomes    445
**23.3**    **Phylum Porifera: The Sponges**    446
**23.4**    **Phylum Cnidaria: Jellyfish and Others**    447
The Protostomes    450
**23.5**    **Phylum Platyhelminthes: Flatworms**    450
**23.6**    **Phylum Annelida: Segmented Worms**    451
**23.7**    **Phylum Mollusca: Snails, Oysters, Squid,
and More**    453
**23.8**    **Phylum Nematoda: Roundworms**    455
**23.9**    **Phylum Arthropoda: Insects, Lobsters, Spiders,
and More**    456
Subphylum Uniramia: Insects First    457
Other Uniramians: Millipedes and Centipedes    458
Subphylum Crustacea: Shrimp, Lobsters, Crabs,
Barnacles and More    459
Subphylum Chelicerata: Spiders, Ticks, Mites,
Horseshoe Crabs, and More    459
The Deuterostomes    460
**23.10**    **Phylum Echinodermata: Sea Stars, Sea Urchins,
and More**    460
**23.11**    **Phylum Chordata: Mostly Animals with Backbones**    462
What Is a Vertebrate?    463
Diversity among the Vertebrates    464
**On to Plants**    471
**Chapter Review**    472

petals
sepals
receptacle
pedicel

bud    flower

## Unit 5
## A Bounty That Feeds Us All: Plants

CHAPTER **24**

## THE ANGIOSPERMS: **An Introduction to Flowering Plants**    476

**24.1**    **The Importance of Plants**    477
A Focus on Flowering Plants    478
**24.2**    **The Structure of Flowering Plants**    479
The Basic Division: Roots and Shoots    479
Roots: Absorbing the Vital Water    480
Shoots: Leaves, Stems, and Flowers    481
**24.3**    **Basic Functions in Flowering Plants**    485
Reproduction in Angiosperms    485
Plant Plumbing: The Transport System    486
Communication: Hormones Affect Many Aspects
of Plant Functioning    487
Plant Growth: Indeterminate and at the Tips    489
Defense and Cooperation    490
**24.4**    **Responding to External Signals**    492
Responding to Gravity: Gravitropism    492
Responding to Light: Phototropism    493
Responding to Contact: Thigmotropism    493
Responding to the Passage of the Seasons    493
**On to a More Detailed Picture of Plants**    495
**Chapter Review**    496
**Essays**    What is Plant Food?    484
Keeping Cut Flowers Fresh    488
Ripening Fruit Is a Gas    490

**Essays**  A Tree's History Can Be Seen in Its Wood                512
The Syrup for Your Pancakes Comes from Xylem       516

CHAPTER **25**

## THE ANGIOSPERMS: Form and Function in Flowering Plants

THE ANGIOSPERMS: Form and Function in Flowering Plants                                           498

25.1  Two Ways of Catergorizing Flowering Plants      500
The Life Spans of Angiosperms: Annuals,
Biennials, and Perennials                        500
A Basic Difference among Flowering Plants:
Monocotyledons and Dicotyledons                  500

25.2  There Are Three Fundamental Types of Plant Cells  502
Parenchyma Cells                                 502
Sclerenchyma Cells                               502
Collenchyma Cells                                502
Parenchyma as Starting-State Cells               502

25.3  The Plant Body and Its Tissue Types            503
First: A Distinction between Primary and
Secondary Growth Tissue                          503
Dermal Tissue Is the Plant's Interface with the
Outside World                                    503
Ground Tissue Forms the Bulk of the Primary Plant 504
Vascular Tissue Forms the Plant's Transport System 504
Meristematic Tissue and Primary Plant Growth     506

25.4  How a Plant Grows: Apical Meristems Give Rise to
the Entire Plant                                 506
A Closer Look at Root and Shoot Apical Meristems 507

25.5  Secondary Growth Comes from a Thickening of
Two Types of Tissue                              509
Secondary Growth through the Vascular Cambium:
Secondary Xylem and Phloem                       510
Secondary Xylem Is Responsible for Most of a
Plant's Widening                                 510
Secondary Growth thought the Cork Cambium:
The Plant's Periphery                            511

25.6  How the Plant's Vascular System Functions       513
How the Xylem Conducts Water                      513
*BioFlix*  Water Transport In Plants             514
Food the Plant Makes Is Conducted through
Phloem                                           514

25.7  Sexual Reproduction in Flowering Plants        517
Flowering Plants Reproduce through an
Alternation of Generations                       518
Development of the Male and Female Gametophyte
Generation                                       521
Fertilization of Two Sorts: A New Zygote and Food
for It                                           521

25.8  Embryo, Seed, and Fruit: The Developing Plant   522
The Development of a Seed                         522
The Development of Fruit                          522
Fruits Serve in Protection and in Seed Dispersal 523
Seed Dormancy Can Be Used to a Plant's Advantage 524

On to Animals                                     525

**Chapter Review**                               526

## Unit 6
## What Makes the Organism Tick? Human Anatomy and Physiology

CHAPTER **26**

## INTRODUCTION TO HUMAN ANATOMY AND PHYSIOLOGY: The Integumentary, Skeletal, and Muscular Systems

INTRODUCTION TO HUMAN ANATOMY AND PHYSIOLOGY: The Integumentary, Skeletal, and Muscular Systems                              528

26.1  The Disciplines of Anatomy and Physiology      529

26.2  How Does the Body Regulate Itself?             530
Large-Scale Features of the Body                 531

26.3  Levels of Physical Organization                531

26.4  The Human Body Has Four Basic Tissue Types     532
Epithelial Tissue                                532
Connective Tissue                                532
Muscle Tissue                                    532
Nervous Tissue                                   534

26.5  Organs Are Made of Several Kinds of Tissue     534

26.6  Organs and Tissues Make Up Organ Systems       534
Organ Systems 1: Body Support and Movement—
The Integumentary, Skeletal, and Muscular
Systems                                          534
Organ Systems 2: Coordination, Regulation,
and Defense—The Nervous, Endocrine, and
Immune Systems                                   535
Organ Systems 3: Transport and Exchange with
the Environment—The Cardiovascular,
Respiratory, Digestive, and Urinary Systems      536

26.7  The Integumentary System: Skin and Its Accessories  537
The Structure of Skin                            537
The Outermost Layer of Skin, the Epidermis       537
Beneath the Epidermis: The Dermis and Hypodermis 538
Accessory Structures of the Integumentary System 538

26.8  The Skeletal System                            541
Function and Structure of Bones                  542
Practical Consequences of Bone Dynamics          544
The Human Skeleton                               544
Joints                                           544

26.9  The Muscular System                            546
The Makeup of Muscle                             546
*BioFlix*  Muscle Contraction                    548
How Muscles Work                                 548

On to the Nervous and Endocrine Systems          548

**Chapter Review**                               549

**Essays** Why Fat Matters and Where It Matters Most   **540**
There is No Such Thing as a Fabulous Tan   **542**

CHAPTER **27**

# COMMUNICATION AND CONTROL:
## The Nervous and Endocrine Systems   **552**

| | | |
|---|---|---|
| **27.1** | **Structure of the Nervous System** | **554** |
| **27.2** | **Cells of the Nervous System** | **555** |
| | Anatomy of a Neuron | **557** |
| | The Nature of Glial Cells | **557** |
| | Nerves | **558** |
| **27.3** | **How Nervous System Communication Works** | **558** |
| | Communication within an Axon | **558** |
| | Movement Down the Axon | **560** |
| | *BioFlix* How Neurons Work | **560** |
| | Communication between Cells: The Synapse | **560** |
| | The Importance of Neurotransmitters | **561** |
| **27.4** | **The Spinal Cord** | **561** |
| | The Spinal Cord and the Processing of Information | **562** |
| | Quick, Unconscious Action: Reflexes | **562** |
| **27.5** | **The Autonomic Nervous System** | **563** |
| | Sympathetic and Parasympathetic Divisions | **563** |
| **27.6** | **The Human Brain** | **564** |
| | Six Major Regions of the Brain | **565** |
| **27.7** | **The Nervous System in Action: Our Senses** | **566** |
| **27.8** | **Our Senses of Touch** | **567** |
| **27.9** | **Our Sense of Smell** | **568** |
| **27.10** | **Our Sense of Taste** | **570** |
| **27.11** | **Our Sense of Hearing** | **571** |
| **27.12** | **Our Sense of Vision** | **572** |
| **27.13** | **The Endocrine System** | **576** |
| **27.14** | **Types of Hormones** | **577** |
| **27.15** | **How Is Hormone Secretion Controlled?** | **578** |
| | Hormonal Hierarchy: The Hypothalamus | **578** |
| | The Pituitary Gland | **580** |
| **27.16** | **Hormones in Action: Four Examples** | **580** |
| | Insulin and Glucagon: Keeping a Tight Rein on Glucose | **580** |

| | | |
|---|---|---|
| | Oxytocin: Wide-Ranging Roles for a Single Hormone | **582** |
| | Cortisol: Stress and Illness | **585** |
| | **On to the Immune System** | **585** |
| **Chapter Review** | | **586** |
| **Essays** | Spinal Cord Injuries | **565** |
| | Too Loud: Hair Cell Loss and Hearing | **571** |
| | When Blood Sugar Stays in the Blood: Diabetes | **584** |

CHAPTER **28**

# DEFENDING THE BODY:
## The Immune System   **590**

| | | |
|---|---|---|
| **28.1** | **Two Types of Immune Defense** | **592** |
| **28.2** | **Nonspecific Defenses** | **592** |
| | Nonspecific Cells and Proteins | **593** |
| | Nonspecific Defense and the Inflammatory Response | **593** |
| **28.3** | **Specific Defenses** | **594** |
| | Antibody-Mediated and Cell-Mediated Immunity | **595** |
| **28.4** | **Antibody-Mediated Immunity** | **597** |
| | The Fantastic Diversity of Antibodies | **597** |
| | The Cloning and Differentiation of B Cells | **597** |
| | The Action of the Antibodies | **598** |
| **28.5** | **Cell-Mediated Immunity** | **598** |
| | Cells Bearing Invaders: Antigen-Presenting Cells | **599** |
| | Helper T Cells, Cytotoxic T Cells, and Regulatory T Cells | **599** |
| **28.6** | **AIDS: Attacking the Defenders** | **600** |
| **28.7** | **The Immune System Can Cause Trouble** | **602** |
| | **On to Transport and Exchange** | **603** |
| **Chapter Review** | | **604** |

CHAPTER **29**

# TRANSPORT AND EXCHANGE 1:
## Blood and Breath   **606**

| | | |
|---|---|---|
| **29.1** | **The Cardiovascular System** | **608** |
| **29.2** | **The Composition of Blood** | **608** |
| | Formed Elements | **608** |
| | Blood's Other Major Component: Plasma | **609** |
| **29.3** | **Blood Vessels** | **609** |
| **29.4** | **The Heart and Blood Circulation** | **611** |
| | Following the Path of Circulation | **612** |
| | Valves Control the Flow of Blood | **613** |
| **29.5** | **What is a Heart Attack?** | **613** |
| **29.6** | **Distributing the Goods: The Capillary Beds** | **615** |
| | Forces That Work on Exchange through Capillaries | **616** |
| | Muscles and Valves Work to Return Blood to the Heart | **616** |

**29.7**    **The Respiratory System**                                    616
Structure of the Respiratory System                                        617

**29.8**    **Steps in Respiration**                                       618
First Step: Ventilation                                                    618
Next Steps: Exchange of Gases                                              618

**On to the Digestive and Urinary
Systems**                                                                  619

**Chapter Review**                                                         620

**Essays**  Blood Doping: A Dangerous Way to Cheat                          610
Listening in on Blood Pressure                                             614

CHAPTER **30**

TRANSPORT AND EXCHANGE 2:
**Digestion, Nutrition, and Elimination**                                  622

**30.1**    **The Digestive System**                                       623

**30.2**    **Structure of the Digestive System**                          624
The Digestive Tract in Cross Section                                       624

**30.3**    **Steps in Digestion**                                         625
The Pharynx and Esophagus                                                  625
The Stomach                                                                625
The Small Intestine                                                        626
The Pancreas                                                               627
The Gallbladder                                                            627
The Liver                                                                  627
The Large Intestine                                                        628

**30.4**    **Human Nutrition**                                            629

**30.5**    **Water, Minerals, and Vitamins**                              629
Vitamin and Mineral Sources: Do We Need
Supplements?                                                               632

**30.6**    **Calories and the Energy-Yielding Nutrients**                 633

**30.7**    **Proteins**                                                   634

**30.8**    **Carbohydrates**                                              636
Carbohydrate Choices: Think Fresh and
Whole Grain                                                                637

**30.9**    **Lipids**                                                     639
Lipid Choices: Think Unsaturated                                          640

**30.10**   **Elements of a Healthy Diet**                                 642

**30.11**   **The Urinary System in Overview**                             643

**30.12**   **Structure of the Urinary System**                            643

**30.13**   **How the Kidneys Function**                                   645
Hormonal Control of Water Retention                                        645

**30.14**   **Urine Storage and Excretion**                                646
The Urinary Bladder                                                        646
The Urethra                                                                646
Urination                                                                  646

**On to Development and
Reproduction**                                                             647

**Chapter Review**                                                         648

CHAPTER **31**

AN AMAZINGLY DETAILED SCRIPT:
**Animal Development**                                                     652

**31.1**    **General Processes in Development**                           653
Two Cells Become One: Fertilization                                        653
Three Phases of Early Embryonic Development                                654
Themes in Development: From General to Specific;
Retention of Structures and Processes                                      657

**31.2**    **What Factors Underlie Development?**                         658
The Process of Induction                                                   658
The Interaction of Genes and Proteins                                     659
Three Lessons in One Gene                                                  659

**31.3**    **Unity in Development: Homeobox Genes**                       661

**31.4**    **Developmental Tools: Sculpting the Body**                    662

**31.5**    **Development through Life**                                    663

**On to Human Reproduction**                                               663

**Chapter Review**                                                         664

CHAPTER **32**

HOW THE BABY CAME TO BE:
**Human Reproduction**                                                     666

**32.1**    **Overview of Human Reproduction and
Development**                                                              667
Reproduction in Outline                                                    667

**32.2**    **The Female Reproductive System**                             669
The Female Reproductive Cycle                                              669
How Does an Egg Develop?                                                   670
Changes through the Female Life Span                                       671
The Mystery of Menopause                                                   673

**32.3**    **The Male Reproductive System**                               675
Structure of the Testes                                                    676
Male and Female Gamete Production Compared                                 676
Further Development of Sperm                                               676
Supporting Glands                                                          678

**32.4**    **The Union of Sperm and Egg**                                 678
How Latecomers Are Kept Out                                                679

**32.5**    **Human Development Prior to Birth**                           679
Early Development                                                          680
Development through the Trimesters                                         682

**32.6**   The Birth of the Baby                                    685

On to Ecology                                                        686

**Chapter Review**                                                   687

**Essays**   Hormones and the Female Reproductive Cycle            672
Methods of Contraception                                            677
Sexually Transmitted Disease                                       682

## Unit 7
# The Living World as a Whole: Ecology and Animal Behavior

CHAPTER **33**

## AN INTERACTIVE LIVING WORLD 1:
# Populations in Ecology

Populations in Ecology                                              690

**33.1**   **The Study of Ecology**                                 692
Ecology Is Not Environmentalism                                    692
Path of Study                                                       693

**33.2**   **Populations: Size and Dynamics**                      694
Estimating the Size of a Population                                 694
Growth and Decline of Populations over Time                        694
Calculating Exponential Growth in a Population                     696
Logistical Growth of Populations: Reality Makes
    an Appearance                                                   697

**33.3**   **$r$-Selected and $K$-Selected Species**               698
$K$-Selected, or Equilibrium, Species                              698
$r$-Selected, or Opportunist, Species                              698
Survivorship Curves: At What Point Does Death
    Come in the Life Span?                                          699

**33.4**   **Thinking about Human Populations**                    700
Survivorship Curves Are Constructed from
    Life Tables                                                     700
Population Pyramids: What Proportion of a
    Population Is Young?                                             700
The World's Human Population: Finally Stabilizing  702
Human Population and the Environment                               703

**On to Communities**                                              704

**Chapter Review**                                                 706

CHAPTER **34**

## AN INTERACTIVE LIVING WORLD 2:
# Communities in Ecology

Communities in Ecology                                             708

**34.1**   **Structure in Communities**                            709
Large Numbers of a Few Species: Ecological
    Dominants                                                       709

Importance beyond Numbers: Keystone Species        710
Variety in Communities: What Is Biodiversity?      710

**34.2**   **Types of Interaction among Community Members**   712
Two Important Community Concepts:
    Habitat and Niche                                               712
Competition among Species in a Community           712
Parasites: Making a Living from the Living          716
Beneficial Interactions: Mutualism and
    Commensalism                                                    718
Coevolution: Species Driving Each Other's
    Evolution                                                       718

**34.3**   **Succession in Communities**                           719
An Example of Primary Succession:
    Alaska's Glacier Bay                                            720
Common Elements in Primary Succession             721
Lessons in Succession from Mount St. Helens        722

**On to Ecosystems and Biomes**                                    723

**Chapter Review**                                                 724

**Essays**   Purring Predators: House Cats and Their Prey        715
Why Do Rabid Animals Go Crazy?                     722

CHAPTER **35**

## AN INTERACTIVE LIVING WORLD 3:
# Ecosystems and Biomes

Ecosystems and Biomes                                              726

**35.1**   **The Ecosystem**                                       727

**35.2**   **Abiotic Factors Are a Major Component of
    Any Ecosystem**                                                727
The Cycling of Ecosystem Resources                 728

**35.3**   **How Energy Flows through Ecosystems**                 734
Producers, Consumers, and Trophic Levels           735
Accounting for Energy Flow through the
    Trophic Levels                                                  737
Primary Productivity Varies across the Earth
    by Region                                                       739

**35.4**   **Earth's Physical Environment**                        740
Earth's Atmosphere                                                 740
The Worrisome Issue of Ozone Depletion             740

**35.5    Global Warming**    742
Warming for Certain, Caused by Human Activity    742
What Are the Likely Consequences of Global
    Warming?    743
A Warming We Cannot Stop but Can Lessen    745

**35.6    Earth's Climate**    746
Why Are Some Areas Wet and Some Dry,
    Some Hot and Some Cold?    746
The Circulation of the Atmosphere and Its
    Relation to Rain    748
Mountain Chains Affect Precipitation Patterns    749
The Importance of Climate to Life    749

**35.7    Earth's Biomes**    750
Cold and Lying Low: Tundra    750
Northern Forests: Taiga    751
Hot in Summer, Cold in Winter: Temperate
    Deciduous Forest    752
Dry but Sometimes Very Fertile: Grassland    753
Chaparral: Rainy Winters, Dry Summers    753
The Challenge of Water: Deserts    753
Lush Life, Now Threatened: Tropical Rain Forests    754

**35.8    Life in the Water: Aquatic Ecosystems**    755
Marine Ecosystems    755
Freshwater Systems    759
Life's Largest Scale: The Biosphere    761
**On to Animal Behavior**    762

**Chapter Review**    763

**Essays**  A Cut for the Middleman: Livestock and Food    741
The Process of Science: Global Warming and the
    Harlequin Frogs    746
Good News about the Environment    756
Our Overfished Oceans    758

**36.4    Learning and Behavior**    775
Establishing Relationships: Imprinting    775
The Sensitive Period    776
Other Forms of Learning    776

**36.5    Behavior in Action: How Birds Acquire
    Their Songs**    777

**36.6    Social Behavior**    779
Why Live Alone—or Together?    780
Dominance Hierarchies    780
Territoriality    781
Eusociality: Life in Animal Societies    781

**36.7    Altruism in the Animal Kingdom**    784
Inclusive Fitness at Work    786
Reciprocal Altruism    787
**On to . . . the Rest of Life**    787

**Chapter Review**    790

**Essays**  Biological Rhythms and Sports    774
Are Men "Naturally" Promiscuous and Women
    Reserved?    782
The Process of Science: How Do Sea Turtles
    Find Their Way?    788


Appendix    AP1
Answers to Multiple-Choice and Brief Review
  Questions    A1
Glossary    G1
Credits    C1
Index    I1
Essays    E1
Web Animations    W1

CHAPTER **36**

# ANIMALS AND THEIR ACTIONS:
# Animal Behavior    766

**36.1    The Field of Animal Behavior**    767
Animal Behavior Asks What, Why,
    and How    767
Proximate and Ultimate Causes    768
Ultimate Cause and Natural Selection    769

**36.2    The Web of Behavioral Influences**    770

**36.3    Internal Influences on Behavior**    771
Reflexes    771
Action Patterns    771
Orientation Behavior: Taxis    772
Biological Rhythms: The Internal Clock    773
Longer Internal Cycles: Annual Clocks    773
The Effects of Hormones    774

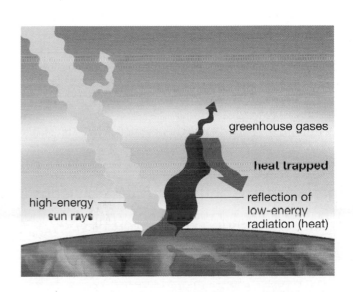

CHAPTER **1**

Measurement plays a part in almost every scientific investigation. Here, an ecologist in Borneo measures a rafflesia plant (Rafflesia keithii), *whose flower is believed to be the largest in the world.*

**1.1**    How Does Science Impact the Everyday World?    **3**

**1.2**    What Is Science?    **6**

**1.3**    The Nature of Biology    **11**

**1.4**    Special Qualities of Biology    **13**

**1.5**    The Organization of This Book    **15**

**Essay**

   Lung Cancer, Smoking, and Statistics in Science    **10**

Science has great impact on our lives now and stands to have greater impact on them in the future. Science is both a body of knowledge and a means of acquiring knowledge. Biology, a branch of science, is the study of life.

# SCIENCE AS A WAY OF LEARNING:
# A Guide to the Natural World

*Biotechnology will be domesticated in the next fifty years as thoroughly as computer technology was in the last fifty years. . . . This means cheap and user-friendly tools and do-it-yourself kits, for gardeners to design their own roses and orchids, and for animal-breeders to design their own lizards and snakes. A new art-form as creative as painting or cinema. It means biotech games for children down to kindergarten age, like computer-games but played with real eggs and seeds instead of with images on a screen. Kids will grow up with an intimate feeling for the organisms that they create.*

— Freeman Dyson

**Y**ou might guess that the words above came from a science fiction novel, but Freeman Dyson was being serious when he made these predictions in 2005. Further, Dyson is no wild-eyed dreamer but a respected physicist who has given a lot of thought to the question of where society is headed. Even if we allow, as baseball player Yogi Berra once said, that "it's hard to make predictions, especially about the future," the fact that someone like Dyson could envision a future in which children design their own animals is a testament to the remarkable times we're living in.

## 1.1 How Does Science Impact the Everyday World?

And what has made our times remarkable? The very engines of change that Dyson was talking about: science, and its sister discipline, technology. In 1900, at the start of the twentieth century, there were democracies and dictatorships; there were corporations and stock markets, and farms, and churches, just as there are now. But there were no powered airplanes. And there was no reliable way to kill infectious organisms inside human beings without killing the humans in the process. And there was no possibility of even *imagining* designer creatures of the sort Dyson sees in our future.

But all this was to change. In 1900, Wilbur and Orville Wright labored in Ohio to understand physical concepts like drag and lift in relation to flight; three years later, Orville took the historic "first flight" in Kitty Hawk, North Carolina. In 1940, Howard Florey and Ernst Chain struggled to find out whether a substance they were working with—a product of a fungus—could kill bacteria that had invaded human beings. The substance worked, and the result was the world's first antibiotic, penicillin. And in 1972, Stanley Cohen of Stanford and Herbert Boyer of the

University of California, San Francisco wondered whether it would be possible to insert a gene into a bacterium, thereby changing the genetic makeup of this tiny creature. The answer was yes, and the result was the world's first designer organism. This initial change in design was admittedly slight, but it led to the designer lineup we have today, which includes rice that contains genes from both bacteria and daffodils (thus allowing the rice to produce its own beta-carotene) and goats that have had human genes inserted into their makeup (thus allowing the goats to produce needed human proteins in their milk). Then, of course, we have *cloned* sheep, kittens, and horses, which are genetic copies of other animals, and are thus designer organisms in the sense that they are like prints made from an original drawing.

If we were to examine a large number of the kind of discoveries we've been talking about, we would find that such breakthroughs are coming our way at a faster pace all the time. To appreciate this change, consider electronic technology. The fundamental discovery that brought about modern electronics, including the computer, was the invention of the

transistor at Bell Labs in 1947. Average Americans then *heard* about computers for 30 years, but not until the mid-1980s were they *using* computers right on their desktops, at home and at work. Once this revolution in circuits got going, however, things came in a torrent. CD players were newfangled devices in the mid-1980s; now, people are downloading MP3 files into their portable global positioning systems. Then again, we need not go back to the 1980s to see how fast technological change is coming. On a typical day, do you e-mail someone, talk on a cell phone, or visit a website? As late as the mid-1990s, the average person would not have done any of these things (**Figure 1.1**).

In a similar vein, the breakthrough that brought about the biotechnology industry was the description of the DNA molecule in 1953 by James Watson and Francis Crick. People heard more about genes in subsequent years, but it took another 25 years for the first genetically engineered medicine (human insulin) to appear. Now, one-quarter of U.S.-grown corn is genetically modified, criminals are regularly convicted (and innocent people freed) through use of DNA "fingerprints," genetically altered salmon are grown to eating weight in half the normal time, and the prediction that, 50 years from now, people will be designing their own organisms is an idea that has to be taken seriously.

It may go without saying that these various innovations and discoveries entail value judgments on the part of society. In 2004, voters in California decided that, over the ensuing 10 years, they would spend $3 billion of state money on stem cell research—mostly on *embryonic* stem cell research, which usually involves removing cells from embryos that would otherwise be discarded by fertility clinics. Meanwhile, on a national level, political leaders came to a very different conclusion about embryonic stem cell research. In 2001, President Bush decided that while the federal government would fund some research with these cells, such funding would be severely restricted. (Funding would only be provided for those "lines" of cells in existence at the time the president made his decision.) Of course, stem cells are not the only science-driven issue that society is grappling with. Think of genetically modified foods, the dizzying possibility of a human clone, and suspects who are listed on arrest warrants not by their names, but by their DNA profiles.

Note that all these issues have been brought to society by science. We might say that science has presented society with *options*, about which society then makes decisions. But, things also work the other way around—society brings issues to science. Why? To get advice. Since scientists are in the business of investigating nature, they function, in effect, as society's eyes

**Figure 1.1**
**Then and Now**

**(a)** A technician enters data into the world's first programmable computer, run initially in 1948. Called "Baby," the computer was more than 2 meters (6 feet) tall and almost 5 meters (15.5 feet) wide but had a total memory of only 128 bytes—less than 1 ten-millionth the capacity of some of today's palm-size flash memory cards.

**(b)** Much smaller, but much more powerful, is the hand-held computer that blogger Steven Fruchter used in 2007 to stream live video from the International Consumer Electronics Show in Las Vegas.

**(a)**

**(b)**

and ears on the natural world. Thus, if we want to know whether the Earth really is warming, or whether we're in danger from an errant comet, then it is not politicians, economists, or business executives who are going to tell us; it is scientists. In line with this, scarcely a week goes by without scientists rendering judgments on vital issues that touch on nature's processes. Do mammograms actually save lives? Does thinning out forests reduce the risk of catastrophic forest fires? What are the risk factors for diabetes? In all these instances, governments and average citizens look to science to provide answers (after which it's up to governments and average citizens to act on the advice they get).

If we take a step back from all this, the message is that science and technology are now woven more tightly into the fabric of society than ever before. Accordingly, to fully participate in the workforce, to make informed choices at the ballot box, or simply to make routine decisions, the average person must now be more scientifically literate than ever before. To get a sense of how this plays out in everyday life, let's look at some biology-related news that came to Americans through one magazine (*Time*) during one short period (August through November 2005). In the process, you will see how learning something about science can provide a foundation for understanding the kind of science-related news that comes our way every day.

**The Evolution Wars**, *Time*'s August 15 cover story, reviewed a controversy then raging over a Pennsylvania school board's decision to present "intelligent design" to district students as an alternative to the theory of evolution (**Figure 1.2**). The board's action (later overturned by a court) was met with outrage from the scientific community, which held that the district wasn't just rejecting a well-grounded scientific theory, but was rejecting science itself. To understand the conflicting ideas in such a controversy, it is important to know what science is—and what it is not. *Time* didn't have the space to go into this topic in detail, but readers of this book can review this subject beginning on page 6 of this chapter. Meanwhile, an account of why Darwin's theory of evolution enjoys near-universal acceptance among biologists can be found in Chapter 16, beginning on page 300.

**How New Heart-Scanning Technology Could Save Your Life** was a September 5 story that focused on a new imaging process that holds promise for providing early warnings of heart attack risk. But what is a heart attack? What factors contribute to this physical calamity, and what events take place during one? For an account, see Chapter 29, "The Heart and Blood Circulation," beginning on page 611.

**Figure 1.2**
**Science in the News**
The importance of science to everyday life is reflected in the large number of news stories that focus on science. Pictured is a cover story on evolution that *Time* magazine ran in August 2005.

**Global Warming: The Culprit?** was an October 3 *Time* story that examined the question of whether there is a connection between global warming and the fact that hurricanes have been increasing in intensity in recent years. Why is there a suspicion that the two things are linked? Hurricanes need warm water to gain strength, and Earth's oceans are warming. The fact that scientists are worried about a connection between global warming and hurricane intensity is one more reason to understand global warming, which is reviewed in Chapter 35, beginning on page 742.

**How Scared Should We Be?** in *Time*'s October 17 issue, asked about the prospects of the world experiencing not just a disease epidemic, but a pandemic—a worldwide outbreak of communicable, lethal illness. In recent years, disease experts have worried about SARS (severe acute respiratory syndrome) but are now more worried about avian flu. As it happens, both illnesses are caused by viruses. But what is a virus? Compared even to bacteria, these tiny entities are so simple that most biologists do not regard them as living things. Yet, look at the power they have. To find out more about them, see "Making a Living by Hijacking Cells," on page 394 of Chapter 21.

**Target: Trans Fats,** appearing on October 24, reported on the 30,000 hours Kraft Foods spent trying to get trans fats out of its Oreo cookies in the period before January 1, 2006, when the Food and Drug Administration began requiring food manufacturers to list trans fats on their Nutrition Facts labels. Of course, Kraft wasn't the only food producer working on trans fats before the 2006 deadline. In the years leading up to it, PepsiCo spent $22 million to remove trans fats from hundreds of production lines in its Frito-Lay snack food division. Trans fats are one kind of "biological molecule" that has an effect on a couple of equally well-known molecules, high-density lipoproteins (HDLs) and low-density lipoproteins (LDLs), better known respectively as good cholesterol and bad cholesterol. To find out more about these substances and their relation to health, look at the "Lipids" section of Chapter 3, beginning on page 47.

**The Down Dilemma**, appearing in *Time*'s November 21 issue, noted that, through 2005, prospective mothers had to wait until the second trimester of their pregnancy to get prenatal testing results for Down syndrome. Late in 2005, however, researchers announced a new test for Down that was not only more accurate than the old test, but that also yielded results in the first trimester. Most cases of Down syndrome result from an embryo that has received three copies of a particular chromosome (chromosome 21) rather than the usual two. But, if neither a mother or father has three copies of chromosome 21, how could an "extra" copy of it get passed along to an embryo? To find out, see "The Consequences of Aneuploidy," in Chapter 12 on page 223.

### What Do Americans Know about Science?

If you knew the scientific background to most of the *Time* stories, that puts you in a pretty select group of Americans. As you can see in the survey results pictured in **Figure 1.3**, the average American adult has what might be called an uneven knowledge of science and the scientific process. Of those who were questioned, 87 percent knew that the oxygen we breathe comes from plants. Yet, one-fourth of this same group said that the sun goes around the Earth, rather than the Earth around the sun. Almost 80 percent of the group understood that "the continents on which we

live have been moving their location for millions of years." Yet more than 50 percent believed that "the earliest humans lived at the same time as the dinosaurs." (In fact, dinosaurs died out more than 64 million years before the first human beings appeared.) What factors go into making people scientifically literate? An important one is the number of science courses they have had in high school or college; as the number of courses goes up, so does the degree to which people feel well informed about scientific issues.

## 1.2   What Is Science?

Having looked a little at the impact that science has, it might be helpful to consider the question of what science is. The point here is to give you some sense of the underpinnings of science—to review something about the how and why of it before you begin looking at the nature of one of its disciplines, biology.

### Science as a Body of Knowledge

Science is in one sense a process—a *way* of learning. In this respect, it is an activity carried out under certain rules, which we'll get to shortly. **Science** is also a body of knowledge about the natural world. It is a collection of unified insights about nature, the evidence for which is an array of facts. The unified insights of science are commonly referred to as *theories*.

It is unfortunate but true that the word "theory" means one thing in everyday speech and something almost completely different in scientific communication. In everyday speech, a theory can be little more than a hunch. It is an unproven idea that may or may not have any supportive evidence. In science, meanwhile, a **theory** is a general set of principles, supported by evidence, that explains some aspect of nature. There is, for example, a Big Bang theory of the universe. It is a general set of principles that explains how our universe came to be and how it developed. Among its principles are that a cataclysmic explosion occurred 13–14 billion years ago; and that, after it, matter first developed in the form of gases,

**Figure 1.3**
**What Do Americans Know about Science?**

Some results regarding Americans' scientific knowledge as published by the National Science Foundation (*Science & Engineering Indicators—2004*).

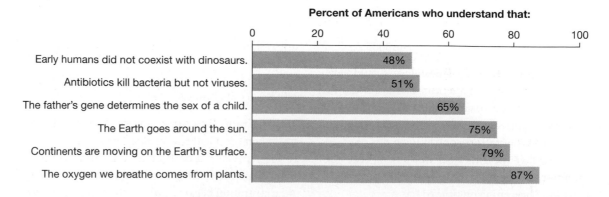

**Percent of Americans who understand that:**

| | |
|---|---|
| Early humans did not coexist with dinosaurs. | 48% |
| Antibiotics kill bacteria but not viruses. | 51% |
| The father's gene determines the sex of a child. | 65% |
| The Earth goes around the sun. | 75% |
| Continents are moving on the Earth's surface. | 79% |
| The oxygen we breathe comes from plants. | 87% |

which then coalesced into the stars we can see all around us. There are numerous facts supporting these principles, such as the current size of the universe and its average temperature.

As you might imagine, with any theory this grand some *pieces* of it are in dispute; some facts don't fit the theory, and scientists disagree about how to interpret this piece of information or that. On the whole, though, these insights have withstood the questioning of critics and stand as a scientific theory.

### The Importance of Theories

Far from being a hunch, a scientific theory actually is a much more valued entity than is a scientific fact because the theory has an *explanatory* power, while a fact generally is an isolated piece of information. That the universe is at least 13 billion years old is a wonderfully interesting fact, but it explains little in comparison with the Big Bang theory. Facts are important; theories could not be supported or refuted without them. But science is first and foremost in the theory-building business, not the fact-finding business.

### Science as a Process: Arriving at Scientific Insights

So how does a body of facts and theories come about? What is the process of scientific investigation in other words? When **science** is viewed as a process, it could be defined as a means of coming to understand the natural world through observation and the testing of hypotheses. This process is referred to as the **scientific method**. The starting state for scientific inquiry is always *observation:* A piece of the natural world is observed to work in a certain way. Then follows the *question*, which broadly speaking is one of three types: a "what" question, a "why" question, or a "how" question. Biologists have asked, for example: What are genes made of? Why does the number of species decrease as we move from the equator to the poles? How does the brain make sense of visual images?

### Formulating Hypotheses, Performing Experiments

Following the formulation of the question, various hypotheses are proposed that might answer it. A **hypothesis** is a tentative, testable explanation for an observed phenomenon. In almost any scientific question, several hypotheses are proposed to account for the same observation. Which one is correct? Most frequently in science, the answer is provided by a series of *experiments*, which are controlled tests of the question at hand (**Figure 1.4**). Scientists don't regard all hypotheses as equally worthy of undergoing experimental test. By the time scientists arrive at the experimental stage, they usually have an idea of which is the most promising hypothesis among the contenders and then proceed to put that hypothesis to the test. Let's see how this worked in an example from history.

### The Test of Experiment: Pasteur and Spontaneous Generation

Does life regularly arise from anything *but* life, or can it be created "spontaneously," through the coming together of basic chemicals? The latter idea had wide acceptance from the time of the ancient Romans, and as late as the nineteenth century it was championed by some of the leading scientists of the day. So, how could the issue be decided? The famous French chemist and medical researcher Louis Pasteur formulated a hypothesis to address this question (**Figure 1.5** on the next page). He believed that many purported examples of life arising spontaneously were simply instances of airborne microscopic organisms landing on a suitable substance and then multiplying in such profusion that they could be seen. Life came from life, in other words, not from spontaneous generation. But how could this be demonstrated? In 1860, Pasteur sterilized a meat broth in glass flasks by heating it, while at the same time, heating the glass *necks* of the flasks, after which he bent the necks into a "swan" or S-shape. The ends of the flasks remained open to the air, but inside the flasks there was not a sign of life. Why? The broth remained sterile because microbe-bearing dust particles got trapped in the bend of the flask's neck. When Pasteur broke the neck off before the bend, however, the flask soon had a riot of bacterial life growing within it. In another test, Pasteur tilted the flask so that the broth *touched* the bend in the neck, a change that likewise got the microbes growing.

### Elements in Pasteur's Experiments

Now, note what was at work here. Pasteur had a preconceived notion of what the truth was, and he designed experiments to test his hypothesis. Critically, he performed the same set of steps several times in the experiments, keeping all elements the same each time—except for one. The nutrient broth was the same in each test; it was heated the same amount of time and in the same kind of flask. What *changed* each time was one critical **variable**: an adjustable condition in an experiment. In this case, the variable was either the shape of the flask neck or the tilt of the flask. Given that all the other elements of the experiments were kept the same, the experiments had rigorous controls: All conditions were held constant over several trials except for a single variable. A control

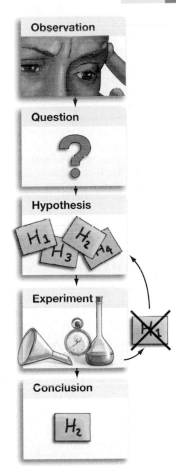

**Figure 1.4**
**Scientific Method**
The scientific method enables us to answer questions by testing hypotheses.

**Web Animation 1.1**
Scientific Method

**Figure 1.5**
**Pasteur's Experiments and the Scientific Method**

Nineteenth-century observation made clear that life would appear in a medium, such as broth, that had been sterilized, but what was the source of this life? One hypothesis was that it arose through *spontaneous generation*, meaning it formed from the simple chemicals in the broth. Conversely, Pasteur hypothesized that it originated from airborne microorganisms. He was able to design an experiment that offered evidence for this hypothesis. The device he used was an S-shaped flask, which enabled air to enter the flask freely while trapping all particles (including invisible microorganisms) in a bend in the neck.

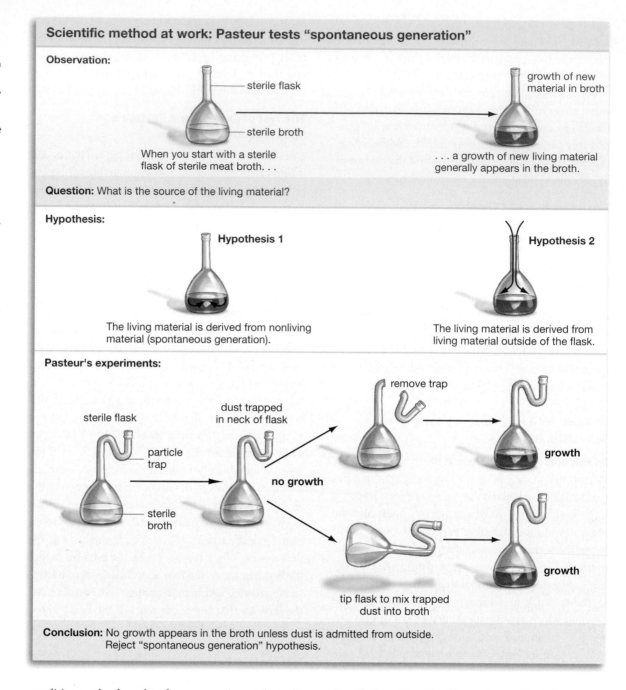

**Scientific method at work: Pasteur tests "spontaneous generation"**

**Observation:**

sterile flask
sterile broth

When you start with a sterile flask of sterile meat broth. . .

growth of new material in broth

. . . a growth of new living material generally appears in the broth.

**Question:** What is the source of the living material?

**Hypothesis:**

**Hypothesis 1**

The living material is derived from nonliving material (spontaneous generation).

**Hypothesis 2**

The living material is derived from living material outside of the flask.

**Pasteur's experiments:**

sterile flask
particle trap
sterile broth

dust trapped in neck of flask

**no growth**

remove trap

**growth**

tip flask to mix trapped dust into broth

**growth**

**Conclusion:** No growth appears in the broth unless dust is admitted from outside. Reject "spontaneous generation" hypothesis.

*condition* can be thought of as an experimental condition that exists prior to the introduction of any variables tested. Pasteur was testing what happened with a broken-necked flask and a tilted flask. The control condition, therefore, was the broth-filled flask left sitting straight up with its particle trap intact. Pasteur's finding that no life grew in this condition is interesting but tells little by itself. We learn something only by comparing this finding to the result Pasteur got when he introduced his variables: that life *did* grow when the neck was broken or the flask was tilted.

Note that the idea of spontaneous generation was not banished with this one set of experiments—nor should it have been. Pasteur's experiments provided one of the *facts* mentioned earlier; in this case the fact

that flasks of liquid will remain sterile under certain conditions. The idea that life arises only from life is, however, is one of the scientific *theories* noted earlier, meaning it is a principle that requires the accumulation of many facts pointing in the same direction.

## Other Kinds of Support for Hypotheses

Some scientific questions are difficult or impossible to test purely through experiment. For example, for years there was a controversy over whether birds are the direct descendants of dinosaurs. What kind of experiment could test this hypothesis? Certain modern-day evidence is available to us—the DNA of living birds, for example—but examining DNA does not amount to an experiment. Instead, it is observation, which is another valid way to test a hypothesis. Evidence from the past can also be observed, which in this case means the observation of dinosaur and bird fossils. Indeed, fossils have been the key evidence in convincing most experts that birds are the descendants of dinosaurs. Experiment and observation are enhanced with the tool of statistics, which frequently is used in science, as you can see in "Lung Cancer, Smoking, and Statistics in Science" on page 10.

---

**SO FAR...**

1. In everyday speech, a theory might be little more than a _____, while in science a theory is a general set of _____, supported by evidence, that explains some aspect of _____.

2. In science, a tentative, testable explanation for an observed phenomenon is referred to as a _____.

3. All properly executed scientific experiments must have rigorous controls, meaning that all aspects of the experiment must be held _____ except for the condition, known as a _____ that is being tested for.

---

## When Is a Theory Proven?

At what point does a theory become proven? An irony of the orderly undertaking called science is that there's nothing orderly about this transition. No scientific supreme court exists to make a decision. Scientists aren't polled for their views on such questions, and even if they were, at what point would we say something had been proven: when more than 50 percent of the experts in the field assent to it? When no dissenters are left?

### Provisional Assent to Findings: Legitimate Evidence and Hypotheses

This lack of finality in science actually fits, however, with one of the central tenets of science, which is that nothing is ever finally proven. Instead, every finding is given only *provisional* assent, meaning it is believed to be true for now pending the addition of new evidence. This principle is so deeply embedded in science that scientists rarely have reason to think about it (just as drivers would seldom contemplate why they are driving on the right side of the road). Yet it is profoundly important because it is the thing that most starkly separates science from belief systems, such as those that operate in politics or religion. Every principle and "fact" in science is subject to modification based solely on the best evidence available. There are no immutable laws and no unquestioned authority

figures. This means there is a paradox in science: its only bedrock is that there is no bedrock; everything scientists "know" is subject to change.

In practice, this is a difficult ideal to live up to. Even when a body of evidence starts to point in a new direction, scientists—like anybody else—may be reluctant to give up old ways of thinking. Recognizing this tendency in human nature, Charles Darwin's friend Thomas Henry Huxley, writing in 1860, gave a beautiful description of the attitude scientists should have when investigating nature:

> Sit down before fact as a little child, be prepared to give up every preconceived notion, follow humbly wherever and whatever abysses nature leads, or you will learn nothing

This principle of science's openness to revision is one of three important scientific principles having to do with the scientific process. Here are all three:

- Every assertion regarding the natural world is subject to challenge and revision based on evidence.

- Any scientific hypothesis or claim must be *falsifiable*, meaning open to negation through scientific inquiry. The assertion that "UFOs are visiting the Earth" does not rise to the level of a scientific claim because there is no way to prove that this is *not* so.

- Scientific inquiry concerns itself only with natural explanations for natural phenomena. Put another way, *supernatural* explanations for the workings of nature lie outside the realm of science and thus cannot be examined through the scientific process.

At first blush, this third principle may seem a little strange. If science is about the testing of hypotheses through open inquiry, then when carrying out such inquiry, why can't scientists test *all* possible explanations—including supernatural explanations? For example, if we find (as we do) that the Hawaiian Islands have a wildly uneven representation of living things—there are 800 species of one kind of fly on the islands but not a single native species of reptile—then why can't scientists investigate a claim such as this: Hawaii has the mix of creatures it does because an intelligent designer arranged things in this way?

The primary answer to this question is that an intelligent designer would, by definition, have abilities that lie outside those in the natural world. Any entity that could arrange species on the Hawaiian Islands as it saw fit would be free to violate *all* the principles that underlie science: the laws of physics and chemistry, to say nothing of biology. And, if this is the case, then sci

# Lung Cancer, Smoking, and Statistics in Science

**V**aluable as they are, experimental and observational tests often are not enough to provide answers to scientific questions. In countless instances, scientists employ an additional tool in coming to comprehend reality—a mathematical tool—as you'll see in the following example.

The evidence that cigarette smoking causes lung cancer (and heart disease and emphysema and on and on) has been around for so long that most people have no idea why smoking was looked into as a health hazard in the first place. You might think that scientists were suspicious of tobacco decades ago and thus began experimenting with it in the laboratory, but this was not the case. Instead, the trail that led to tobacco as a health hazard started with a mystery about disease.

When the lung cancer pioneer Alton Ochsner was in medical school in 1919, his surgery professor brought both the junior and senior classes to see an autopsy of a man who had died of lung cancer. The disease was then so rare that the professor thought the young medical students might never see another case during their professional lifetimes. Prior to the 1920s, lung cancer was among the rarest forms of cancer, because cigarette smoking itself was rare before the twentieth century. It did not become the dominant form of tobacco use in the United States until the 1920s. This made a difference in lung cancer rates because cigarette smoke is inhaled, while pipe and cigar smoke generally are not.

If you look at **Figure 1**, you can see the rise in lung cancer mortality in U.S. males and females from 1930 forward. Note that

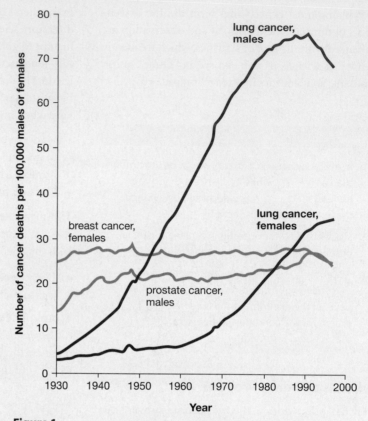

**Figure 1**
**Rise in Lung Cancer Mortality in U.S. Males and Females from 1930 Forward**

entists would have little more ability than, say, artists to investigate intelligent design in Hawaii because scientific principles need not apply to the question at hand.

Apart from this theoretical issue, scientists do not investigate supernatural explanations of natural phenomena for a purely practical reason: there's no percentage in it. To be sure, today's scientists are confronted with a host of unanswered questions about the natural world, but all of the questions that *have* been answered have had natural explanations. To look at this another way, no *supernatural* explanation for a question about the natural world has ever proved persuasive to scientists. So, to put yourself in the place of a scientist, think, by way of analogy, of a car mechanic—one who is confronted with a car that won't start. This is, maybe, the 600th car he has seen in this condition and long experience has taught him to run through a list—check this first, then that, then this. Now, at what point in this process would our mechanic say: "This problem is really difficult; I'm going to start looking for a *ghost*

in this machine." This would never happen, of course, because there's no percentage in it. All the failures the mechanic has encountered before have been understandable as mechanical failures; indeed, all the failures he's ever heard of have been mechanical. So, how far down on his "check this" list would a supernatural explanation be? So far down as to be off the list altogether. For similar reasons, scientists long ago stopped even thinking about the supernatural as an explanation for the natural.

## SO FAR...

1. Every finding in science is subject to _____ based on the accumulation of new _____.

2. Any scientific claim must be falsifiable, meaning open to _____ through means of _____.

3. Science concerns itself only with _____ explanations for natural phenomena, as opposed to _____ explanations for these phenomena.

women show a later rise in lung cancer deaths; this is because women started smoking in large numbers later. (Also note that in the 1990s, lung cancer rates finally began to level off—or drop in the case of men. This was a direct result of a decline in smoking that began in the 1970s.)

Given the lung cancer trends that were apparent in males by the 1930s, the task before scientists was to explain the alarming increase in this disease. What could the cause of this scourge be, the medical detectives wondered? The effects of men being gassed in World War I? Increased road tar? Pollution from power plants? Through the 1940s, cigarette smoking was only one suspect among many.

Laboratory experiments eventually would play a part in fingering tobacco as the lung cancer culprit, but the original indictment of smoking was written in numbers—in statistical tables showing that smokers were contracting lung cancer at much higher rates than nonsmokers.

It has sometimes been said that "science is measurement," and the phrase is a marvel of compact truth. For centuries, people had an idea that smoking might be causing serious harm, but this information fell into the realm of guessing or of *anecdote*, meaning personal stories. The problem with anecdote is that there is no measurement in it; there is no way of judging the validity of one story as opposed to the next. Related to anecdote is the notion of "common sense," which is valuable in many instances but which also had us believing for centuries that the sun moved around the Earth. In the case of smoking, it took the extremely careful measurement provided by a discipline called *epidemiology*—the study of disease distributions—to separate truth from fiction.

### Probability in Science

Note that "measurement" in this instance was a matter of calculating *probability*, which is often the case in science. Epidemiologists found a link between smoking and lung cancer in the sense that those who smoked were more likely to get the disease. Having seen this, scientists then had to ask: Could this result be a matter of pure chance? A person tossing a coin might get heads 5 times in a row, and it might be written off to chance, but would it be the same if the person came up with heads *50* times in a row? No—at that point there would be justification for assuming that some force other than chance was in operation (such as a rigged coin). When the epidemiologists looked at their statistical tables and saw so many more smokers than nonsmokers getting lung cancer, they had to ask whether this result fell into the realm of 5 heads in a row or 50. Even in the earliest studies epidemiologists concluded that more than chance was at work in the results. After many studies, they concluded that smoking was *causing* lung cancer. How did they judge what was probable and what was not in an issue as complicated as this? The researchers relied on techniques developed in the branch of mathematics called *statistics*.

The importance of probability and statistics to science can hardly be overstated. These tools are used frequently in nearly every scientific discipline. Imagine that 10 experimental plots of land are compared, 5 with fertilizer added to them, the other 5 without. The plots with the added fertilizer end up with more growth but fewer kinds of plants. Could the differences between the two kinds of plots be a matter of chance? Here, as in so many other tests, scientists would use the tools of statistics to get at the truth.

## 1.3   The Nature of Biology

Let's shift now from an overview of science to a more narrow focus on **biology**, which can be defined as the study of life. But, what is life? It may surprise you to learn that there is no standard short answer to this question. Indeed, the only agreement among scholars seems to be that there is not, and perhaps cannot be, a short answer to this question. The main impediment to such a definition is that any quality common to all living things is likely to exist in some nonliving things as well. Some living things may "move under their own power," but so does the wind. Living things may grow, but crystals and fire do the same thing. Therefore, biologists generally define life in terms of a group of characteristics possessed by living things. Looked at together, these characteristics are sufficient to separate the living world from the nonliving. We can say that living things:

- Can assimilate (take in) and use energy
- Can respond to their environment
- Can maintain a relatively constant internal environment
- Possess an inherited information base, encoded in DNA, that allows them to function
- Can reproduce through use of the information encoded in DNA
- Are composed of one or more cells
- Evolved from other living things
- Are highly organized compared to inanimate objects

Every one of these qualities exists in all the varieties of Earth's living things. The simplest bacterium needs an energy source no less than any human being. Our energy source is the food that's familiar to us; the bacterium's might be the remains of vegetation in the soil. The bacterium responds to its

••• ANSWERS

1. revision; evidence
2. negation; scientific inquiry
3. natural; supernatural

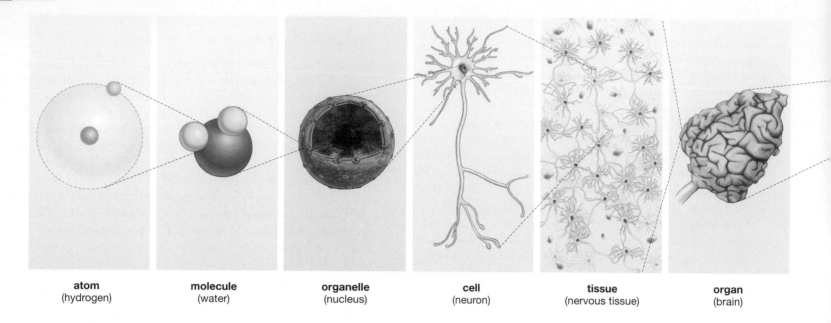

atom
(hydrogen)

molecule
(water)

organelle
(nucleus)

cell
(neuron)

tissue
(nervous tissue)

organ
(brain)

**Figure 1.6**
**Levels of Organization in Living Things**

environment, just as we do. You would take action if you smelled gas in your house; the bacterium would move away if it encountered something it regarded as noxious. Humans maintain **homeostasis** or a relatively stable internal environment by, for example, sweating when they are hot. When the bacterium's external environment becomes too hot, it has certain genes that will switch on to keep it functioning. Both humans and bacteria use the molecule DNA as a repository of the information necessary to allow them to live. Bacteria and human beings both reproduce—bacteria through simple cell division, human beings through the use of two kinds of reproductive cells (egg and sperm). A bacterium is a single-celled life form, while humans are a 10-trillion-celled life form. Bacteria and humans both evolved from complex living things and ultimately share a single common ancestor.

There are some exceptions to these "universals." For example, the overwhelming majority of honeybees and ants are sterile females; they can't reproduce, but no one would doubt that they're alive. In general, however, if something is living, it has all these qualities.

## Life Is Highly Organized in a Hierarchical Manner

One item on the list of qualities requires a little more explanation. Living things are indeed highly organized compared to inanimate objects. But to put a finer point on this, living things are organized in a "hierarchical" manner, meaning one in which lower levels of organization are progressively integrated to make up higher levels. The main levels in this hierarchy could be compared to the organization of a business. In a corporation, there may be several individuals making up an office, several offices making up a

department, several departments making up a division, and so forth. In life, there is one set of organized "building blocks" making up another, as you can see in **Figure 1.6**.

When we begin to think about life in terms of the integration of these many structural levels, we can see that it is not just "highly" organized. Nothing else comes *close* to it in organizational complexity. The sun is a large thing, but it is an uncomplicated thing compared to even the simplest organism. Consider that you have about 10 trillion cells in your body and that, with some exceptions, each of these cells has in it a complement of DNA that is made up of chemical building blocks. How many building blocks? Three billion of them. Now, you probably know that most cells divide regularly, one cell becoming two, the two becoming four, and so on. Each time this happens, each of the 3 billion DNA building blocks must be faithfully *copied* so that both cells resulting from cell division will have their own complete copy of DNA. And this copying of the molecule—before anything is actually done with it—is carried out in almost all the varieties of the 10 trillion cells we have. Complex indeed. Let's see what life's levels of organization are.

## Levels of Organization in Living Things

The building blocks of matter, called *atoms*, lie at the base of life's organizational structure (see Chapter 2 for more about them). Atoms come together to form *molecules*, which are entities consisting of a defined number of atoms that exist in a defined spatial relationship to one another. A molecule of water is one atom of oxygen bonded to two atoms of hydrogen, with these atoms arranged in a precise way. Molecules in turn form what are called *organelles*, meaning

**organ system** (nervous system)  **organism** (sea lion)  **population** (colony)  **community** (giant kelp forest)  **ecosystem** (southern California coast)  **biosphere** (Earth)

"tiny organs" in a cell. Each of your cells has, for example, a structure called a nucleus that contains the cell's primary complement of DNA. Such an organelle is not just a collection of molecules that exist close to one another. It is a highly organized structure, as you can tell just from looking at the rendering of it in Figure 1.6.

At the next step up the organizational chain are entities that are actually living, as opposed to entities that are components of life. *Cells* are units that can do all of the things listed earlier: assimilate energy, reproduce, react to the environment, and so forth. Indeed, most experts would agree that cells are the only place that life exists. You may ask: But isn't there a lot of material in between my cells? The answer is yes; it's mostly water with a good number of other molecules in the mix. But, if all the cells were removed from this watery environment, then there would be nothing resembling life left in it.

The next step up is to a *tissue*, meaning a collection of cells that serve a common function. Your body contains collections of muscle cells, all of which serve the common function of contracting. Each concentration of these cells constitutes muscle tissue. Several kinds of tissues can then come together to form a functioning unit known as an *organ*. Your heart, for example, is a collection of muscle tissue and connective tissue, among other types. Several organs and related tissues then can be integrated into an *organ system*. Contractions of your heart push blood into a system of blood vessels. The heart, blood, and blood vessels form the cardiovascular system. An assemblage of cells, tissues, organs, and organ systems can then form a multicelled *organism*. (However, back down at the cell level, a one-celled bacterium is also an organism; it's just not one with organs, tissues, and so forth.)

From this point out, life's levels of organization all involve many organisms living together. Members of a single type of living thing (a species), living together in a defined area, make up what is known as a *population*. When you look at *all* the kinds of living things in a given area, you are looking at a *community*. When you consider a community and the *non*living elements with which they interact (such as climate and water), the result is an *ecosystem*. Finally, all the communities of the Earth—and the physical environment with which they interact—make up the *biosphere*.

## 1.4 Special Qualities of Biology

Biology traces its origins to the ancient Greeks. In the work of such Greeks as Hippocrates and Galen, we can find the origins of modern medical science. In the work of Aristotle and others, we can find the origins of "natural history," which led to what we think of today as mainstream biology and the larger category of the **life sciences**: a set of disciplines that focus on varying aspects of the living world. Apart from biology, the life sciences include such areas of study as veterinary medicine and forestry.

Despite its ancient origins, biology is, in a sense, a much younger science than, say, physics, which is one of the **physical sciences**, meaning the natural sciences not concerned with life. Western Europe's revolution in the physical sciences probably can be dated from the sixteenth century, when Nicholas Copernicus published his work *On the Revolution of Heavenly Spheres*, which demonstrated that the Earth moves around the sun. Meanwhile, biology did not come into its own as a science until the *nineteenth* century. Prior to the 1800s, biology was almost purely "descriptive," meaning that the naturalists who we would today call biologists largely confined themselves to describing living things—what kinds there were, where they lived, what features they had, and so forth. Beginning in about the 1820s, however, biologists began to formulate biological theories as that term was defined earlier. They began to postulate that all life exists within cells, that life comes only from life, that life is passed on through small packets of information that we now call genes, and so on. To put this another way, biologists in the nineteenth century began describing the *rules* of the living world, whereas before they were largely describing *forms* in the living world.

This change moved biology closer to the same scientific footing as physics, but biology was then, and remains now, a very different kind of science from any of the physical sciences, with physics a clear case in point. One reason for this difference is that the component parts of physics are uniform and far fewer than is the case in biology. Physics deals with only 92 stable elements, such as hydrogen and gold, and to a first approximation, if you've seen one electron, you've seen them all.

Meanwhile, in biology, if you've seen one species, you've seen just that—one species. Each species is at least marginally different from another, and many are greatly dissimilar. Moreover, there are thought to be at least 4 million species on Earth, and each species has all the organizational levels of elements in physics *and more*. (They not only have electrons and atoms, but organelles, cells, tissues, and so on.) Biology is concerned with the rules that govern all species, and

you've seen that there are some biological "universals." However, when cancer researchers are looking for the principles that underlie cell division, they are likely to be looking at only one of two main kinds of cells; when ecologists are looking at what causes dry grassland to turn into desert, their findings are likely to have little relevance to the rain forest. Put simply, the living world is tremendously diverse compared to the nonliving world, and such diversity means that universal rules in biology are likely to be few and far between. Biology is concerned with the particular to a greater degree than is the case in the physical sciences. Note also that "universals" in biology may not apply beyond Earth; we don't know if life even exists anywhere else, much less what its rules are. Meanwhile, the rules of physics truly are universal in that they are equally applicable on Earth or in the farthest reaches of the cosmos.

## Biology's Chief Unifying Principle

Almost all biologists would agree that the most important thread that runs through biology is **evolution**: the gradual modification of populations of living things over time, with this modification sometimes resulting in the development of new species. Evolution is central to biology because every living thing has been shaped by it. (There are no known exceptions to this universal.) Given this, the explanatory power of evolution is immense. Why do peacocks have their finery, or frogs their coloration, or trees their height (**Figure 1.7**)? All these things

**Figure 1.7**
**Evolution Has Shaped the Living World**

**(a)** A peacock displaying his plumage

**(b)** A poison dart frog in Colombia

**(c)** Pine trees in the South Pacific

stand as wonders of nature's diversity, but with knowledge of evolution they are wonders of diversity that *make sense*. For example, why do so many unrelated stinging insects look alike? Evolutionary principles suggest they *evolved* to look alike because of the general protection this provides from predators. Think of yourself for a moment as a bee predator. Having once gotten stung, would you annoy any roundish insect that had a black-and-yellow striped coloration? You probably learned your lesson about this in connection with one species, but many species of insects are now protected from you simply by virtue of the coloration they share (**see Figure 1.8**). Thus, there were reproductive benefits to individual insects that, through genetic chance, happened to get a slightly more striped coloration. They left more offspring because they were bothered less by predators. Over time, entire populations moved in this direction. They evolved, in other words.

The means by which living things can evolve is a topic this book takes up beginning in Chapter 16. For now, just keep in mind that a consideration of evolution is never far from most biological observations. So strong is evolution's explanatory power that, in uncovering something new about, say, a sequence of DNA or the life cycle of a given organism, one of the first things a biologist will ask is: Why would evolution shape things in this way?

**(a)** Golden northern bumblebee

**(b)** Sandhills hornet

**Figure 1.8**
**Similar Enough to Yield a Benefit**
These are two of the many stinging insects that have the black-and-yellow striped coloration that warns away predators.

---

**SO FAR...**

1. Arrange the following levels of organization in living things, going from least inclusive to most inclusive: cell, community, molecule, organ, ecosystem, organism.

2. The most important principle in biology, the theory of evolution, concerns the gradual _____ of populations of living things over time, sometimes resulting in the development of new _____.

3. Biology can be defined as the study of _____.

---

## 1.5 The Organization of This Book

This book has something in common with the levels of organization you looked at earlier in that it too goes from constituent parts to the larger whole. It begins with atoms, moves on to the biological molecules that atoms make up, and then goes to cells. The end of the book covers the highest levels of biological organization: communities, ecosystems, and Earth's biosphere. In between are tours of such facets of life as energy, DNA-encoded information, and reproduction. Even in these sections, however, you'll be moving in a general way from the small to the large because much of the first part of the book is given over to **molecular biology**: the study of individual molecules (such as DNA) as they affect living things. Then, you'll move to information on evolution, which touches on **organismal biology**, meaning the study of whole organisms within biology. Next is the **physiology** or physical functioning of plants and animals, which largely concerns tissues and organs. Finally, there is **ecology**, which is the study of the interactions of organisms with each other and with their physical environment. And so, with this introduction finished, let's begin to look at biology, the study of life.

**... ANSWERS**

1. molecule, cell, organ, organism, community, ecosystem
2. modification; species
3. life

# CHAPTER 1 REVIEW

For study help, activities, and more quiz questions, go to **www.aw.com/krogh4**.

## SUMMARY

### 1.1  How Does Science Impact the Everyday World?

- Science is playing an increasingly important role in the everyday lives of Americans. Scientific advances regularly confront society with choices that have an ethical dimension. Society frequently turns to scientists to answer questions about health, the environment, and other domains of life.

- Americans have an uneven knowledge about science. Almost 80 percent of adult Americans know that the continents are moving about the face of the Earth, for example, but one-quarter think the sun goes around the Earth.

### 1.2  What Is Science?

- In one of its facets, science is a body of knowledge; a collection of unified insights about nature, the evidence for which is an array of facts.

- The unified insights of science are known as theories. A theory is a general set of principles, supported by evidence, that explains some aspect of nature.

- Science can also be defined as a way of learning; a process of coming to understand the natural world through observation and the testing of hypotheses.

- Science works through the scientific method, in which an observation leads to the formulation of a question about the natural world. Then comes a hypothesis—a tentative, testable explanation that has not been proven true. The hypothesis may be tested through observation or through a series of experiments, as aided by statistical procedures. An example of hypothesis testing is Louis Pasteur's experiment regarding the spontaneous generation of life.

- In science, every assertion regarding the natural world is subject to challenge and revision; scientific claims must be falsifiable, meaning open to negation through scientific inquiry; and scientific inquiry is limited to investigating natural (as opposed to supernatural) explanations for natural phenomena.

Web Animation 1.1: Scientific Method

### 1.3  The Nature of Biology

- Biology is the study of life. Life is defined by a group of characteristics possessed by living things. These are that living things can assimilate energy, respond to their environment, maintain a relatively constant internal environment, and reproduce. In addition, living things possess an inherited information base, encoded in DNA, that allows them to function; they are composed of one or more cells; they are evolved from other living things; and they are highly organized compared to inanimate objects.

- Life is organized in a hierarchical manner, running in increasing complexity from atoms to molecules and then to organelles, cells, tissues, organs, organ systems, organisms, populations, communities, ecosystems, and the biosphere.

### 1.4  Special Qualities of Biology

- Until the early nineteenth century, biology was largely a descriptive science, meaning it largely catalogued and described the Earth's living things. Beginning in the 1820s, however, life science researchers began to formulate biological theories, such as that life comes only from life and exists only within cells.

- Biology's subject matter—the living world—is notable for its complexity and diversity compared to other aspects of the natural world (such as stars and atoms). Because of this, biology does not deal in universal rules to the extent that a discipline such as physics does; instead, biological research may focus on particular species, processes, or portions of the living world.

- Biology's chief unifying principle is evolution, which can be defined as the gradual modification of populations of living things over time, with this modification sometimes resulting in the development of new species. Evolution provides the means for making sense of the forms and processes seen in living things on Earth today.

## KEY TERMS

| | | |
|---|---|---|
| biology   11 | organismal biology   15 |
| ecology   15 | physical sciences   13 |
| evolution   14 | physiology   15 |
| homeostasis   12 | science   6, 7 |
| hypothesis   7 | scientific method   7 |
| life sciences   13 | theory   6 |
| molecular biology   15 | variable   7 |

## BRIEF REVIEW

*Answers to Brief Review questions are in the back of the book.*
*For multiple-choice quiz questions, go to* **www.aw.com/krogh4**.

1. What is science? In what ways is science different from a belief system such as religious faith?

2. What is a controlled experiment? Why is it important to keep all variables but one constant in a scientific experiment?

3. How did Louis Pasteur cast doubt on the idea of spontaneous generation?

4. Describe the defining features of life as we know it on Earth.

5. Living things are organized in a hierarchical manner. List all the levels of the biological hierarchy that you can.

## APPLYING YOUR KNOWLEDGE

1. Would you agree that it is valuable for a nation to have a citizenry that is reasonably well versed in science? Give reasons for your answer. Would you say this need has become especially urgent in the last two decades? If so, why?

2. Although science cannot investigate the supposed workings of the supernatural, the scientific method can be used to investigate *claims* that the supernatural has been at work. Can an astrologer know something about you just by knowing the place and time of your birth? A scientific test of this question, carried out in the 1980s, involved providing a group of 30 top-level astrologers with nothing but birth information for a group of people and then seeing whether the astrologers could predict anything about the personalities of these people (as measured by a standard personality "inventory"). The result was that the astrologers did no better than chance in trying to predict personality. Many of the standard tools of science were at work in this test of astrology—for example, controls (the test was the same for each astrologer) and statistical analysis (used to see whether the astrologers did better than chance). Using this test as a case in point, are all claims of supernatural effects open to scientific investigation? Can you think of other claims that could be investigated or any that could not?

3. If you were sent on an interplanetary mission to investigate the presence of life on Mars, what would you look for? Would you explore the land and the atmosphere? Imagine you discover an entity you suspect is a living being. Realizing that life elsewhere in the universe may not be organized by the same rules as on Earth, which of the features of life on Earth, if any, would you insist that the entity display before you would declare it living?

*Life began as a series of chemical reactions in water and water remains indispensable to all living things today. Here, a red-eyed tree frog (Agalychnis callidryas) sits under a plant leaf during a rainstorm.*

2.1    Chemistry's Building Block: The Atom    19

2.2    Matter Is Transformed through Chemical Bonding    23

2.3    Some Qualities of Chemical Compounds    29

2.4    Water and Life    30

2.5    Acids and Bases Are Important to Life    34

**Essays**

Getting to Know Chemistry's Symbols    25

Free Radicals    27

Life is carried on through chains of chemical reactions. The qualities of water have shaped life. The degree to which substances are acidic or basic has strong effects on living things.

# FUNDAMENTAL BUILDING BLOCKS:
# Chemistry, Water, and pH

C ities are made of buildings and buildings are made of bricks; bricks are made of earth, and earth is made of ... ? To answer this question, in this chapter we will look at what the material world is made of.

Biology is our subject, but to fully understand it, you need to learn a little about what underlies biology. You need to learn a little about what *biology* is made of, in a sense. To do this, you need to understand some of the basics in another field: chemistry.

How is chemistry relevant to biology? Well, the average person probably is aware that living things are made up of individual units called cells, but beyond this bit of knowledge, reality fades and a kind of fantasy takes over. In it, the cells that populate people, plants, or birds carry on their activities under the direction of their own low-level consciousness. A cell *decides* to move, it *decides* to divide, and so on. Not so. By the time our story is finished, many chapters from now, it will be clear that the cells that make up complex living things do what they do as the result of a chain of chemical reactions. Repulsion and bonding, latching on and re-forming, depositing and breaking down—what makes people, plants, and birds function at this cellular level is *chemistry*. Given this, the basic principles of chemistry are important to biology—important enough that we'll spend most of this chapter reviewing them. Once this is done, we'll touch on one of life's most important substances, water. Then, we'll finish the chapter with a

look at pH, a chemistry-related concept that has to do with how acidic or "basic" watery solutions are.

## 2.1 Chemistry's Building Block: The Atom

What is chemistry concerned with? Look around you. Do you see a table, light from a lamp, a patch of night or daytime sky? Everything that exists can be viewed as falling into one of two categories: matter or energy. You will learn something about energy in this chapter, but we are most concerned here with *matter and its transformations*, which is the subject of chemistry. Matter can be defined as anything that takes up space and has **mass**. This latter term is a measure of the *quantity* of matter in any given object. How much space does an object occupy—how much volume, to put it another way—and how *dense* is the matter within that space? These are the things that define mass. For our purposes, we may think of mass as equivalent to weight, although physics makes a distinction between these two things.

Beholding matter all around us, it is natural to ask, What is its nature? A child sees a grain of sand, pounds it with a rock, sees the smaller bits that result, and wonders, What is this stuff like at the end of these divisions? Not surprisingly, adults also have wondered about this question—for centuries. About 2,400 years ago, the Greek philosopher Plato accepted the notion that all matter is made up of four primary substances: earth, air, fire, and water. A near-contemporary of his, Democritus, believed that these substances were in turn made up of smaller units that were both invisible and in*divi*sible—they could not be broken down further. He called these units atoms (**Figure 2.1**).

Well, let's give at least one cheer for Democritus because he had it partly right. Centuries of painstaking work between his time and ours has confirmed that matter is indeed composed of tiny pieces of matter, which we still call atoms, but these atoms are not indivisible, as Democritus thought. Rather, they are themselves composed of constituent parts. A superficial account of *all* the parts scientists have discovered to date would go on for pages and still be incomplete. Physicists are continually slamming together parts of atoms with ever-greater force in an effort to determine what else there may be at the heart of matter. This is what the machines called "atom smashers" do. (The physicists who run them could be compared to people who, in trying to find out what parts a watch has, throw it on the ground and record the way its various mechanisms fly out on impact.) This is interesting stuff, but it is purely the business of physics, with little relation to biology. We are not concerned here with what's at the very end of these divisions. We do care a good deal, however, about what's nearly at the end of them.

## Protons, Neutrons, and Electrons

For our purposes, there are three important constituent parts of an atom: **protons, neutrons,** and **electrons**. These three parts exist in a spatial arrangement that is uniform in all matter. Protons and neu-

trons are packed tightly together in a core (the atom's **nucleus**), and electrons move around this core some distance away (**Figure 2.2**). The one variation on this theme is the substance hydrogen, the lightest of all the kinds of matter we will run into. Hydrogen has no neutrons but rather only one proton in its nucleus and one electron in motion around it.

These three "subatomic" particles have mind-bending sizes and proportions. As the chemist P. W. Atkins has pointed out, an atom is so small that 100 million carbon atoms would lie end to end in a line of carbon about this long: ⎯⎯⎯⎯⎯⎯⎯⎯ (3 centimeters). Things are just as disorienting when we consider the size of an atom as a whole relative to the nucleus. The whole atom, with electrons at its edge, is 100,000 times bigger than the nucleus. So, if you were to draw a model of an atom *to scale* and began by sketching a nucleus of, say, half an inch, you'd have to draw some of its electrons more than three-quarters of a mile away.

Although the nucleus accounts for little of the space an atom takes up, it accounts for almost all of the mass an atom has. So negligible are electrons in this regard, in fact, that all of the mass (or weight) of an atom is considered to reside with the protons and neutrons of the nucleus.

The components of atoms have another quality that interests us: electrical charge. Protons are positively charged, and electrons are negatively charged. Neutrons—as their name implies—have no charge; they are electrically neutral. Because all these particles do not exist separately but *combine* to form an atom, as a whole the atom may be electrically neutral as well. The negative charge of the electrons balances out the positive charge of the protons. Why? Because in this state the number of protons an atom has is exactly equal to the number of electrons it has (although we'll see a different, "ionic" state later in this chapter). In contrast, the number of

**Figure 2.1**
**The Building Blocks of Life**
Viewing this sea lion at increasing levels of magnification, we eventually arrive at the building block of all matter, the atom. The atom selected here, from among a multitude that make up the sea lion, is a single hydrogen atom, composed of one proton and one electron.

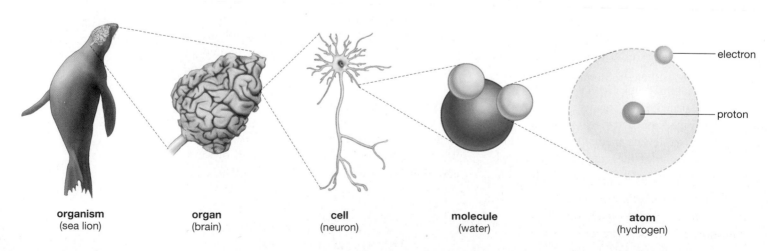

| organism | organ | cell | molecule | atom |
| (sea lion) | (brain) | (neuron) | (water) | (hydrogen) |

electron

proton

neutrons an atom has can vary in relation to the other two particles.

With this picture of atoms in mind, we can begin to answer the question that has been handed down to us through history: What is matter? We certainly have a common sense answer to this question. Matter is any substance that exists in our everyday experience. For example, the iron that goes into cars is matter. But, what is it that differentiates this iron from, say, gold? The answer is that an iron atom has 26 protons in its nucleus, while a gold atom has 79.

## Fundamental Forms of Matter: The Element

Gold is an **element**—a substance that is "pure" in that it cannot be reduced to any simpler set of component substances through chemical processes. And the thing that defines each element is the number of protons it has in its nucleus. A solid-gold bar, then, represents a huge collection of identical atoms, each of which has 79 protons in its nucleus (**Figure 2.3**). In making gold jewelry, an artist may combine gold with another metal such as silver or copper to form an alloy that is stronger than pure gold, but the gold atoms are still present, all retaining their 79-proton nuclei.

Given what you've just read about protons, neutrons, and electrons, you may wonder why gold—or any other element—cannot be reduced to any simpler set of component substances. Aren't protons and neutrons components of atoms? Yes, but they are not component *substances* because they cannot exist by themselves as matter. Rather, protons and neutrons must *combine* with each other to make up atoms.

## Assigning Numbers to the Elements

In the same way that buildings can be defined by a location and thus have a street number assigned to them, elements, which are defined by protons in their nuclei, have an **atomic number** assigned to them. Scientists have constructed the atomic numbering system so that it goes from smallest number of protons to largest. Thus hydrogen, which has only one proton in its nucleus, has the atomic number 1. The next element, helium, has two protons, so it is assigned the atomic number 2. Continuing on this scale all the way through the elements found in nature, we would end with uranium, which has an atomic number of 92.

Given this view of the nature of matter, we are now in a position to answer the question posed at the beginning of the chapter: What is a handful of earth—or anything else—made of? The answer is one or more elements. If you look at **Figure 2.4**, on the next page, you can see the most important elements that go into making up both the Earth's crust and human beings.

## Isotopes

All this seems like a nice, tidy way to identify elements—one element, one atomic number, based on number of protons—except that we're leaving out something. Recall that atoms also have neutrons in their nuclei, that these neutrons add weight to the atom, and that the number of neutrons can vary independently of the number of protons. What this means is that, in thinking about an element in terms of its weight, we have to take neutrons into account. Furthermore, because the number of neutrons in an element's nucleus may vary, we can have various *forms* of elements, called **isotopes**. Most people have heard of one example of an isotope, whether or

**Web Animation 2.1**
Structure of Atoms, Elements, Isotopes

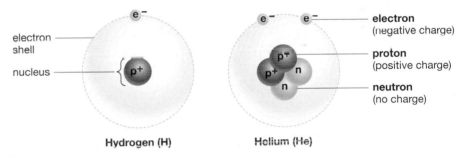

**Figure 2.2**
**Representations of Atoms**

One conceptualization of two separate atoms, hydrogen and helium. The model is not drawn to scale; if it were, the electrons would be perhaps a third of a mile away from the nuclei. The model also is simplified, giving the appearance that electrons exist in track-like orbits around an atom's nucleus. In fact, electrons spend time in volumes of space that have several different shapes.

**Figure 2.3**
**Pure Gold**

Gold is an element because it cannot be reduced to any simpler set of substances through chemical means. Each gold bar is made up of a vast collection of identical atoms—those with 79 protons in their nuclei.

**Figure 2.4**
**Constituent Elements**
The major chemical elements found in Earth's crust (including the oceans and the atmosphere) and in the human body.

Earth's crust

| | |
|---|---|
| other | 8% |
| oxygen (O) | 50% |
| silicon (Si) | 26% |
| aluminum (Al) | 8% |
| calcium (Ca) | 3% |
| iron (Fe) | 5% |

Human body

| | |
|---|---|
| 7% | other |
| 10% | hydrogen (H) |
| 65% | oxygen (O) |
| 18% | carbon (C) |

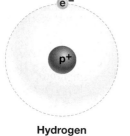

**Hydrogen**

1 proton
0 neutrons

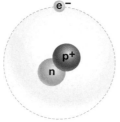

**Deuterium**

1 proton
1 neutron

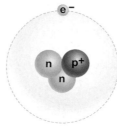

**Tritium**

1 proton
2 neutrons

**Figure 2.5**
**Same Element, Different Forms**
Pictured are three isotopes of hydrogen. Like all isotopes, they differ in their number of neutrons.

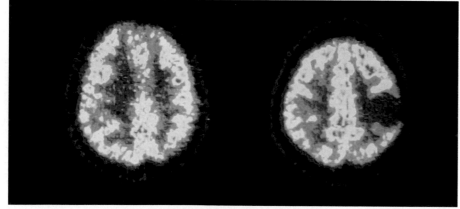

**Figure 2.6**
**Using Isotopes to Spot a Brain Tumor**
Positron emission tomography, or PET scanning, works by means of collisions. A radioactive isotope introduced into the body emits particles called positrons, which collide with electrons that are produced as a result of metabolic activity. The more intense the metabolic activity, the greater the number of collisions that occur. In this PET scan of a human brain (seen from above), differences in metabolic activity register as different colors—red for the greatest amount of activity, then yellow, green, and finally blue for the least amount of activity. In the image at right, the large blue area on the right represents a region that has been rendered inactive by a tumor.

not they recognize it as such. The element carbon has six protons, giving it an atomic number of 6. In its most common form, it also has six neutrons. However, a relatively small amount of carbon exists in a form that has *eight* neutrons. Well, the element is still carbon, and in this form the number of its protons and neutrons equals 14, so the *isotope* is carbon-14, which is used in determining the ages of fossils and geologic samples.

Most elements have several isotopes. Hydrogen, for example, which usually has one proton and one electron, also exists in two other forms: deuterium, which has the proton, electron, and one neutron; and tritium, which has one proton, one electron, and two neutrons (**Figure 2.5**). **Figure 2.6** shows you how isotopes can be used in medicine.

## The Importance of Electrons

In our account so far of the subatomic trio, we have had much to say about protons and neutrons but little to say about electrons. This was necessary because we needed to discuss the nature of matter, but in a sense, you can regard what has been set forth to this point as so much stage-setting because what's most important in biology is the way elements *combine* with other elements. And in this combining, it is the outermost electrons that play a critical role. Just as you come into contact with the world through what lies at your surface—your eyes, your ears, your hands—so atoms link with one another through what lies at their outer edges. An atom's outer electrons exist in a world that can be one of continual forming and breaking of alliances.

## 2.2 Matter Is Transformed through Chemical Bonding

The process of chemical combination and rearrangement is called **chemical bonding,** and for us it represents the heart of the story in chemistry. When the outermost electrons of two atoms come into contact, it becomes possible for these electrons to reshuffle themselves in a way that allows the atoms to become attached to one another. This can take place in two ways: One atom can *give up* one or more electrons to another, or one atom can *share* one or more electrons with another atom. Giving up electrons is called *ionic bonding*; sharing electrons is called *covalent bonding*. A third type of bond, which we'll discuss shortly, is also important for our purposes, the hydrogen bond.

### Energy Always Seeks Its Lowest State

Atoms that undertake bonding with one another do so because they are in a more *stable* state after the bonding than before it. A frequently used phrase is helpful in understanding this kind of stability: Energy always seeks its lowest state. Imagine a boulder perched precariously on a hill. A mere shove might send it rolling toward its lower energy state—at the bottom of the hill. It would not then roll *up* the hill, either spontaneously or with a light shove, because it is now existing in a lower energy state than it did before—one that is clearly more stable than its former precarious perch. With electrons, the energy is not gravitational, but electrical. Atoms bond with one another to the extent that doing so moves them to a lower, more stable energy state. The critical thing for our purposes is that atoms move to this more stable state by filling what is known as their *outer shells*.

### Covalent Bonding

What are the outer shells? Electrons reside in certain well-defined "energy levels" outside the nuclei of atoms. The number of these energy levels varies depending on the element in question. Here, we only need to note the practical effect of these levels on bonding: *Two* electrons are required to fill the first energy level (or shell) of any given atom, but *eight* are required to fill all the levels thereafter in most of the elements that make up the living world. If you look at the electron configurations in **Figure 2.7**, you can see that two elements—hydrogen and helium—have so few electrons in orbit around them that they have nothing *but* a first energy level, while the other elements pictured have two or three energy levels. This means that hydrogen and helium each require only two electrons in orbit around their nuclei to have filled outer shells, but that the other elements

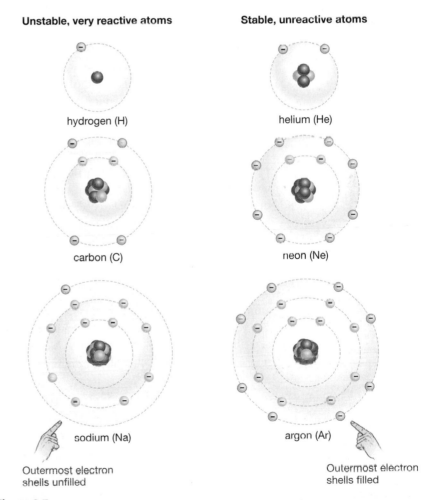

Unstable, very reactive atoms | Stable, unreactive atoms

hydrogen (H)  helium (He)

carbon (C)  neon (Ne)

sodium (Na)  argon (Ar)

Outermost electron shells unfilled | Outermost electron shells filled

**Figure 2.7**
**Electron Configurations in Some Representative Elements**
The concentric rings represent energy levels or "shells" of the elements, and the dots on the rings represent electrons. Hydrogen has but a single shell and a single electron within it, while carbon has two shells with a total of six electrons in them. Helium, neon, and argon have filled outer shells and are thus unreactive. Hydrogen, carbon, and sodium do not have filled outer shells and are thus reactive—they readily combine with other elements.

pictured require eight electrons to complete their outer electron shells.

### How Chemical Bonding Works in One Instance: Water

To see how chemical bonding works in connection with this concept of filled outer shells, take a look at the bonding that occurs with the constituent parts of one of the most simple (and important) substances on Earth, water. In so doing, you'll see one of the kinds of bonding we talked about—covalent bonding.

The familiar chemical symbol for water is $H_2O$. This means that two atoms of hydrogen ($H_2$) have combined with one atom of oxygen (O) to form water. (See "Getting to Know Chemistry's Symbols" on the next page for an explanation of the notation used in chemistry.) Recall that hydrogen has only one electron running around in its single energy level. Also recall, however, that this first level is not completed until it has *two* electrons in it. Next, consider the oxygen atom, which has eight electrons. Looking at **Figure 2.8**, you can see what this means: Two electrons fill oxygen's first energy level, which leaves six for its second. But, remember that the second shells of most atoms are not completed until they hold eight electrons. Thus oxygen, like hydrogen, would welcome a partner. It needs two electrons to fill its outer shell—something that two *atoms* of hydrogen could provide. The outcome is a bonding of two hydrogen atoms with one atom of oxygen. And each of the hydrogen atoms is linked to the oxygen in a **covalent bond**—a chemical bond in which atoms share pairs of electrons. The oxygen atom and first hydrogen atom donate one electron each for the first pair, and these electrons can now be found orbiting the nuclei of both atoms. Then the oxygen and second hydrogen atom each donate one electron for the second pair. The result? Three atoms covalently bonded together, and all of them "satisfied" to be in that condition. (Occasionally in nature, covalent bonding will take place in a way that leaves one atom with an unpaired electron, a potentially harmful phenomenon you can read about in "Free Radicals" on page 27.)

### Matter Is Not Gained or Lost in Chemical Reactions

Note that when this pairing of electrons happens, no matter has been gained or lost. We started with two atoms of hydrogen and one atom of oxygen, and we finish that way. The difference is that these atoms are now bonded. This points up an important principle known as the **law of conservation of mass,** which states that matter is neither created nor destroyed in a chemical reaction.

### What Is a Molecule?

When two or more atoms combine in this kind of covalent reaction, the result is a **molecule:** a compound of a defined number of atoms in a defined spatial relationship. Here, one atom of oxygen has combined with two atoms of hydrogen to create *one water molecule.* (What we commonly think of as water, then, is an enormous, linked collection of these individual water molecules.) A molecule need not be made of two different elements, however. Two hydrogen atoms can covalently bond to form one hydrogen molecule. Conversely, a molecule can contain many different elements bonded together. Consider sucrose, or regular table sugar, which is $C_{12}H_{22}O_{11}$ (12 carbon atoms bonded to 22 hydrogen atoms and 11 oxygen atoms).

## Reactive and Unreactive Elements

The elements considered so far all welcome bonding partners because all of them have incomplete outer shells. This is not true of all elements, however. There is, for example, the helium atom, which has two electrons. It thus *comes equipped*, we might say, with a filled outer shell. As such, it is extremely stable—it is unreactive with other elements. At the opposite end of the spectrum are elements that are extremely reactive.

**Figure 2.8**
**Covalent Bonding**

**(a)** A covalent bond is formed when two atoms share one or more pairs of electrons. The starting state in this reaction is two hydrogen atoms and one oxygen atom. Note that none of these atoms has outer-shell stability—each hydrogen atom would need one more electron to achieve stability, while the oxygen atom would need two more.

**(b)** In bonding, all the atoms achieve this stability; in this case, two pairs of electrons are shared—one pair between one hydrogen atom and the oxygen atom and the other pair between the second hydrogen and the oxygen. The result is the creation of a water molecule.

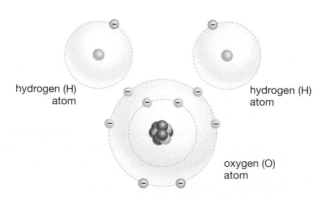

**(a)** Two hydrogen atoms and one oxygen atom

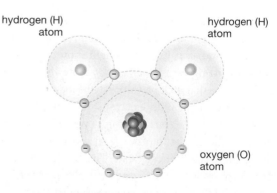

**(b)** One water molecule

# Getting to Know Chemistry's Symbols

One of the "languages of science" is the system of symbols that chemistry uses. Somewhat forbidding at first viewing, this system soon comes to serve its intended purpose of conveying a lot of information quickly.

Our starting place is that each chemical element has its own symbol, so that hydrogen becomes H, carbon C, and platinum Pt.

When we begin to combine these elements into molecules, it is necessary to specify how *many* atoms of each element are part of the molecule. If we have two atoms of oxygen together—which is the way oxygen is usually packaged in our atmosphere—we have the molecule $O_2$. Three molecules of $O_2$ is written as $3O_2$. This kind of notation is known as a **molecular formula**. It is helpful in stipulating the makeup of molecules, from the simple, such as oxygen, to the complex, such as chlorophyll, which is notated $C_{55}H_{72}MgN_4O_5$.

To see a molecular formula is to learn a lot about which atoms are in a molecule but nothing about the way the atoms are *arranged* in relation to one another. (Look at chlorophyll's formula. Is there a line of 55 carbon atoms followed by 72 hydrogen atoms? From the molecular formula, how could you tell?) To convey this ordering information, chemists and biologists use what are known as **structural formulas**—two-dimensional representations of a given molecule. Methane ($CH_4$) is a very simple molecule composed, as the molecular formula shows, of one atom of carbon and four of hydrogen. In a structural formula, these constituent parts are conceptualized like this:

$$H-\overset{\displaystyle H}{\underset{\displaystyle H}{C}}-H$$

methane

Note that there is a single line between each hydrogen atom and the central carbon atom. This signifies that the bond between any of the hydrogen atoms and carbon is a single bond: Each line represents *one* pair of electrons being shared. Thus:

$$H:\overset{..}{\underset{..}{C}}:H$$
H——— electron pairs

There can also be double bonds and triple bonds. When carbon dioxide forms, there are two oxygen atoms. Each shares two pairs of electrons with a lone carbon atom. Here's how we would notate the double bond in a carbon dioxide ($CO_2$) molecule:

$$O=C=O$$
carbon dioxide

Although structural (or "skeletal") formulas can tell us a good deal about the ordering of atoms in a molecule, they tell us little about the *three-dimensional* arrangements of atoms. For this, we rely on two other kinds of representations, the **ball-and-stick model** and the **space-filling model**. An ammonia molecule ($NH_3$) is pictured next in both forms, with the molecular and structural formulas added to show the progression:

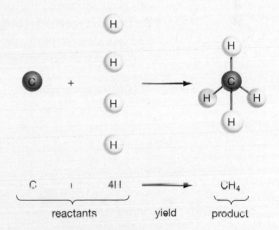

molecular formula    structural formula    ball-and-stick model    space-filling model

Note that the ball-and-stick model gives us a better idea of molecular angles of the atoms, but that the space-filling model gives us a better idea of the relative size of these atoms and how one actually hugs the other. (The dots above the "N" in the structural formula represent an unshared pair of electrons.)

Finally, it is useful to have a way to notate the "before" and "after" stages of a chemical reaction. This is done by employing a simple arrow, as when carbon reacts with hydrogen to form methane: $C + 4H \rightarrow CH_4$. The carbon and hydrogen atoms on the left are **reactants,** the arrow means "yields," and the methane molecule on the right is the **product** of the reaction. Here's a graphic representation of what is happening:

C    +    4H    →    $CH_4$

reactants    yield    product

An important thing to keep in mind about notation goes back to the law of conservation of mass: matter is neither created nor destroyed in a chemical reaction. It follows that reactions such as the one above must be *balanced:* we must have the same number of atoms when the reaction is finished (on the right) as when it started (on the left). Thus, we could not have $C_2 + 4H \rightarrow CH_4$. This is an unbalanced reaction. We started out with two atoms of carbon on the left but somehow ended up with one atom of carbon on the right. Nature doesn't play that way.

Look again at the representation of the sodium atom in Figure 2.7. It has 11 electrons; two in the first shell and eight in the second, which leaves but one electron in the third shell—a very unstable state. Between the extremes of sodium and helium are elements with a range of outer (or *valence*) electrons. Thus, there is a spectrum of stability in the chemical elements, based on the number of outer-shell electrons each element has—from 1 being to 8, with 1 being the most reactive and 8 being the least reactive.

## Polar and Nonpolar Bonding

Not all covalent bonds are created alike. When two hydrogen atoms come together, the result is a hydrogen molecule ($H_2$). Now, in the hydrogen molecule, the electrons are shared *equally.* That is, the two electrons the hydrogen atoms share are equally attracted to each hydrogen atom. This is not the case, however, with the water molecule.

Look at the representation of the water molecule in **Figure 2.9a**. As it turns out, the oxygen atom has greater power to attract electrons to itself than do the hydrogen atoms. The term for measuring this kind of pull is **electronegativity**. Because the oxygen atom has more electronegativity than do the hydrogen atoms, it tends to pull the shared electrons away from the hydrogen and toward itself. When this happens, the molecule takes on a **polarity,** or a difference in electrical charge at one end as opposed to the other. Because electrons are negatively charged and because they can be found closer to the oxygen nucleus, the oxygen end of the molecule becomes slightly negatively charged,

while the hydrogen regions become slightly positively charged. We still have a covalent bond, but it is a specific type: a **polar covalent bond**. Conversely, with the hydrogen molecule—where electrons are shared equally—we have a **nonpolar covalent bond**.

To grasp the importance of this, consider the water molecule with its positive and negative regions. What's going to happen when it comes into contact with *other* polar molecules? The oppositely charged parts of the molecules will attract, and the similarly charged parts will repel. It's like having a bar magnet and trying to bring its positive end into contact with the positive end of another magnet; left on its own, the second magnet just flips around, so that positive is now linked to negative. In the same way, molecules flip around in relation to their polarity.

It is possible for atoms with different electronegativity to link and still have the resulting molecule be nonpolar. Water is polar because the atom with more electronegativity (the oxygen) lies to one *side* of the two hydrogen atoms. Meanwhile, in a molecule such as methane, four hydrogen atoms are arranged in a symmetrical way around a central, and more electronegative, carbon atom (**see Figure 2.9b**). In this arrangement, the differing charges balance each other out, leaving methane with no positive or negative end—meaning it is nonpolar. (Keep this quality of methane in mind as it will be important in the consideration of why oil and water don't mix.) In summary, some molecules are polar while others are nonpolar, and this difference has significant consequences for chemical bonding.

**(a)** Polar water molecule

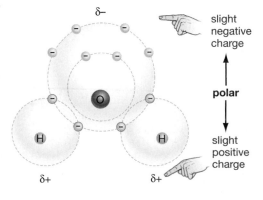

**(b)** Nonpolar methane molecule

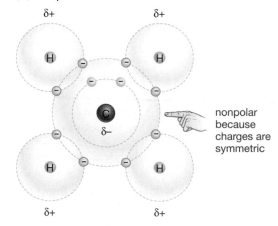

**Figure 2.9**
**Polar and Nonpolar Covalent Bonding**

**(a)** In the water molecule, the oxygen atom exerts greater attraction on the shared electrons than do the hydrogen atoms. Thus, the electrons are shifted toward the oxygen atom, giving the oxygen atom a partial negative charge (because electrons are negatively charged) and the hydrogen atoms a partial positive charge. ("Partial" here is indicated by the Greek delta symbol $\delta$.) The molecule as a whole is polar, meaning it has a difference in charge at one end as opposed to the other.

**(b)** In the methane molecule, the carbon atom is more electronegative than the hydrogen atoms, but the methane molecule as a whole is nonpolar because its hydrogen atoms are arranged symmetrically around the central carbon atom, meaning the partial charges that exist balance each other out. Thus, methane has no difference in charge at one end as opposed to the other.

| ESSAY | Free Radicals |
|---|---|

**W**hy do we get old? What happens to our skin, our hearts, our energy? One factor at work is a set of molecules with a name that makes them sound like a group of 1960s activists now let out of jail. These are the free radicals. As you'll see, these compounds represent a damaging exception to the rules of chemical bonding we've been reviewing.

Every living thing needs a supply of energy to remain alive, and living things such as ourselves get this energy from food. But our cells need to convert the energy contained in food into a form our bodies can use—a substance called ATP. Not surprisingly, then, we have tiny power plants within our cells that do this energy converting. These structures, called mitochondria, take in food and turn out ATP, just as a power plant takes in coal and turns out electricity. And, just as coal cannot be burned in a power plant except in the presence of oxygen, so food cannot be burned in our mitochondria without oxygen.

*Oxygen may sustain us, but it also causes trouble.*

And therein lies the problem. Oxygen may sustain us, but it also causes trouble. If you have ever seen a rusted hinge, then you have seen a piece of metal that has been *oxidized*—oxygen atoms have pulled electrons right off the metal. In the same way, inside our cells oxygen acts like a molecular magnet, pulling electrons toward it and sharing these electrons with other atoms in the process of chemical bonding. But, remember that bonding of this sort means sharing *pairs* of electrons. So, what happens if one atom in a bonded pair ends up with an *un*paired electron—a solo flyer, we might say, that is not paired with any other electron? The resulting molecule is a **free radical**. Oxygen can come together with nitrogen, for example, to form nitric oxide (NO). Recall that oxygen has six outer electrons (and thus needs two for stability), while nitrogen has five outer electrons (and thus needs three). When oxygen and nitrogen hook up, they can share only two electrons with one another before oxygen's outer shell is filled. This, however, leaves nitrogen with an unpaired electron— and a free radical has been created. Although this molecule will be short lived, it begets more of itself: seeking partners, one free radical leads to more, which in turn lead to more.

Where's the harm in this? Well, take a look at the two mice in **Figure 1**. The mouse in the upper photo is just a normal mouse, while the mouse in the lower photo was genetically engineered to lack a protein that repairs damage to the DNA in its mitochondria. You would expect that mice with damaged DNA would suffer some kind of problem, but the fascinating thing is that, when they reach the equivalent of adolescence, these modified mice begin to show signs of *aging*: hair loss, decreased energy, curvature of the spine, and osteoporosis. This is evidence that damage to mitochondrial DNA helps bring about the constellation of afflictions we call

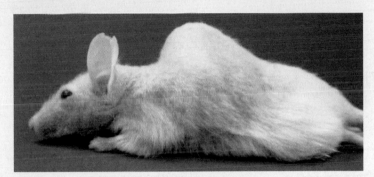

**Figure 1**
**Free Radicals and Aging**

The mice in these photos are both about 40–45 weeks old. Yet, the mouse in the lower photo shows clear signs of aging—its spine is curved, its weight has dropped, and its hair density has been reduced. All this came about because of damage to the DNA in its mitochondria, the tiny structures that are the sites of energy transformation in cells. Such damage is caused largely by free radicals.

aging. And, in a natural state, what can damage mitochondrial DNA? Free radicals. They appear to cause breaks in the DNA and to take it out of its normal shape.

Free radicals are a natural product of metabolism in human beings; they are the price we pay for using oxygen to stay alive. However, they can be created in us in *greater* numbers in accordance with our behavior. Some of the usual suspects seem to be involved here—cigarette smoking, alcohol consumption, and sunlight exposure. Against this production of free radicals, however, nature has also provided its own set of free-radical scavengers, among them beta-carotene and vitamins C and E. (These are so-called antioxidants. Now you can see how they got their name.) Claims are made all the time about the value of buying these substances in pill form. For now, however, experts say that the best bet is to get these free-radical fighters through a diet rich in citrus fruits, whole grains, and vegetables of the green leafy, orange, and yellow variety.

**SO FAR...**

1. Atoms will have a tendency to bond with one another to the extent that they do not have a _____ outer shell, which means having _____ outer shell _____ for most elements, but only _____ outer shell _____ for the simple elements hydrogen and helium.

2. When two atoms share one or more pairs of electrons, they have become linked through a _____ bond.

3. In some of these bonds, one of the linked atoms may exert a greater pull on the shared electrons. When this happens, the result can be a _____ bond, a bond in which differing regions of the resulting molecule take on a difference in _____ in one region, as compared to another.

## Ionic Bonding: When Electrons Are Lost or Gained

So we've gone from nonpolar covalent bonding, in which electrons are shared equally, to polar covalent bonding, in which electrons are pulled to one side of the resulting molecule. What if we carried this just one step further and had instances when the electronegativity differences between two atoms were so extreme that electrons were pulled *off* one atom, only to latch on to the atom that was attracting them? This is what happens in our second type of bonding, **ionic bonding**. The classic illustration of this type of bonding involves the sodium we looked at earlier and the element chlorine. Recall that sodium has 11 electrons, meaning that there is a lone electron flying around in its third electron shell. Chlorine, in contrast, has 17 electrons, meaning it has 7 electrons

**Figure 2.10**
**Ionic Bonding**

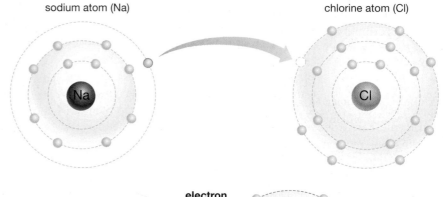

sodium atom (Na)                    chlorine atom (Cl)

**(a)** Initial instability

Sodium has but a single electron in its outer shell, while chlorine has seven, meaning it lacks only a single electron to have a completed outer shell.

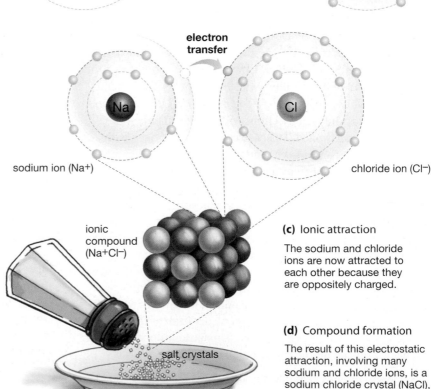

**electron transfer**

**(b)** Electron transfer

When these two atoms come together, sodium loses its third-shell electron to chlorine, in the process becoming a sodium ion with a net positive charge (because it now has more protons than electrons.) Having gained an electron, the chlorine atom becomes a chloride ion, with a net negative charge (because it has more electrons than protons).

sodium ion (Na+)

chloride ion (Cl−)

ionic compound (Na+Cl−)

**(c)** Ionic attraction

The sodium and chloride ions are now attracted to each other because they are oppositely charged.

**(d)** Compound formation

The result of this electrostatic attraction, involving many sodium and chloride ions, is a sodium chloride crystal (NaCl), better known as table salt.

salt crystals

in the third shell. Remember that 8 is the magic number for outer-shell stability. Sodium could get to this number by *losing* one electron, while chlorine could get to it by *gaining* one electron. That's just how this encounter occurs: sodium does in fact lose its one electron, chlorine gains it, and both parties become stable in the process (**Figure 2.10**).

### What Is an Ion?

But this story has a postscript. Having lost an electron (with its negative charge), sodium (Na) then takes on an overall *positive* charge. Having gained an electron, chlorine (Cl) takes on a negative charge. Each is then said to be an **ion**—a charged atom or, to put it another way, an atom with an electron number that differs from its number of protons. We denote the ionized forms of these atoms like this: $Na^+$, $Cl^-$. Were an atom to gain or lose more electrons, a number would be put before the charge sign. For example, to show that the magnesium atom has lost two electrons and thus become a positively charged magnesium ion, we would write $Mg^{2+}$.

Note that we now have two ions, $Na^+$ and $Cl^-$, with differing charges in proximity to one another. They are thus attracted to one another through an *electrostatic attraction* and have thus become bonded through our second form of bonding: an ionic bond. This hardly ever happens with just two atoms, of course. Many billions of atoms are bonded in this way, up, down, and sideways from each other. This whole collection is likewise called an *ion*, or if two or more elements are mixed together this way, an **ionic compound.** The particular ionic compound just described actually is familiar. Sodium and chlorine combine to create sodium chloride, which is better known as table salt. The notation should properly be written $Na^+Cl^-$, but it is usually denoted just as NaCl.

Is it apparent how an ionic compound differs from a molecule? In an ionic compound, there is no fixed number of atoms linked in a defined spatial relationship, as in $H_2O$. Rather, an undefined number of charged atoms are bonded together, as in NaCl.

To take a step back for a second, recall that bonding runs a gamut from the nonpolar covalent bonding (in which electrons are shared equally), to the slightly charged polar covalent bonding (in which electrons are shared somewhat unequally), to the charged ionic bonding (in which electrons are gained or lost altogether). It's important to recognize that there is a spectrum of polarity, and that within it, some bonds are almost completely ionic (as with sodium chloride), while others are completely nonpolar (as with the hydrogen molecule).

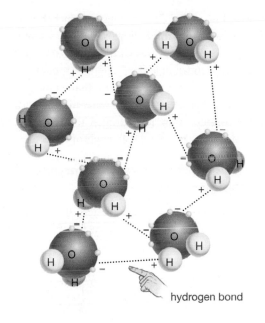

hydrogen bond

**Figure 2.11**
**Hydrogen Bonding**

The hydrogen bond, in this case between water molecules, is indicated by the dotted line. It exists because of the attraction between hydrogen atoms, with their partial positive charge, and the unshared electrons of the oxygen atom, with their partial negative charge.

### A Third Form of Bonding: Hydrogen Bonding

We need to look at one more variant on bonding, called hydrogen bonding. Recall that in any water molecule, the stronger electronegativity of the oxygen atom pulls the electrons shared with the hydrogen atoms toward the oxygen nucleus, giving the oxygen end of the molecule a partial negative charge and the hydrogen end of the molecule a partial positive charge. So, what happens when you place several water molecules together? A positive hydrogen atom of one molecule is weakly attracted to the negative, *unshared* electrons of its oxygen neighbor. Thus is created the **hydrogen bond,** which links an already covalently bonded hydrogen atom with an electronegative atom (in this case, with oxygen; **Figure 2.11**). Hydrogen bonding is a linkage that, for our purposes, nearly always pairs hydrogen with either oxygen or nitrogen. These relatively weak bonds, indicated by a dotted line in illustrations, are important in many of the molecules of life—in deoxyribonucleic acid (DNA), proteins, and elsewhere.

## 2.3 Some Qualities of Chemical Compounds

### Molecules Have a Three-Dimensional Shape

We now need to make more explicit what has been noted only by implication in our diagrams of water molecules: that molecules and ionic compounds have a three-dimensional shape. It is useful to depict them

**Web Animation 2.3**
Geometry, Chemistry, and Biology

... ANSWERS

1. filled; eight; electrons; two; electrons
2. covalent
3. polar covalent; electrical charge

as two-dimensional chains, rings, and such, but in real life a molecule is as three-dimensional as a sculpture. A fair number of shapes are possible; atoms may be lined up in a row, in triangles, or in pyramid shapes. As an example, look at the water and methane molecules in **Figure 2.12**. You can see that in water there is a definite spatial configuration: its hydrogen atoms are splayed out from its oxygen atom at an angle of 104.5°.

## Molecular Shape Is Very Important in Biology

Why does molecular shape matter? It is critical in enabling biological molecules to carry out their activi-

ties. This is because molecular shape determines the capacity of molecules to latch on to or "bind" with one another. When, for example, you smell the aroma of fresh-baked bread, gas molecules wafting off the bread bind with receptor molecules in your nasal passages, thus sending a message to the brain about the presence of bread. It is the precise shape of the gas molecules and nasal receptor molecules that allows them to bind with one another. Look at **Figure 2.13** to see how this works. If you look at **Figure 2.14**, then you can get an idea of how large some biological molecules are relative to the simple molecules considered so far.

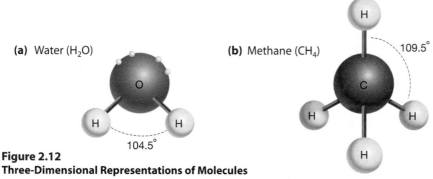

**(a)** Water ($H_2O$)

**(b)** Methane ($CH_4$)

104.5°

109.5°

**Figure 2.12**
**Three-Dimensional Representations of Molecules**
**(a)** In the case of water there is an angle of 104.5° between hydrogen atoms.
**(b)** Methane is a molecule with an angle of 109.5° between hydrogen atoms.

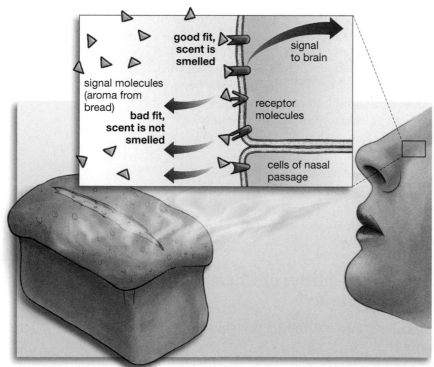

good fit, scent is smelled

signal to brain

signal molecules (aroma from bread)

bad fit, scent is not smelled

receptor molecules

cells of nasal passage

**Figure 2.13**
**The Importance of Shape**
Gas molecules wafting off bread (the triangles) bind with specific receptors on the surface of the cell, thus acting as signaling molecules that set a cellular process in motion. For this binding to take place, the gas and nasal receptor molecules must fit together; this fit is governed by the shape of each molecule.

---

### SO FAR...

1. An ionic bond occurs when one atom _____ one or more _____ to another atom and the resulting ions become attached because of their differing _____.

2. A hydrogen bond links an already covalently bonded _____ with an _____ atom. In biology, such bonds almost always occur between _____ and either _____ or _____.

3. Molecular shape is important in biology because it determines the ability that molecules have to _____ with one another.

---

## 2.4 Water and Life

Having looked at some of the ways in which atoms form the world's substances, we now look in detail at just *one* substance, water. Why such emphasis on water? Because of its great importance to our central subject, which is life.

How are life and water related? Well, consider how the two things have been intertwined over time. Life got started in water and then existed almost nowhere *but* water for eons. More than 3 billion years passed between the time life first got going (in the ancient oceans) and the time living things first came onto land (in the form of ancient plants). Since the movement to land occurred no more than 450 million years ago, this means that life on land came about only in the last 11 percent of life's history. With this in mind, it's not surprising that, when organisms did make the transition to land, they could only manage this feat by carrying a watery environment with them—inside themselves. The plant pioneers had this characteristic 450 million years ago, dinosaurs had it 100 million years ago, and human beings have it today. Human bodies are about 66 percent water by weight, so that if we have, say, a 128-pound person, about 85 pounds of that person will be water. And this proportion is low by some standards: among most vertebrate animals, the water proportion is more like 70 or 80 percent, and for most plant tissues, the figure is 90 percent. Not surprisingly, then, almost all organisms must have a regular supply of water to survive. There are some bacteria that can go into a kind of suspended animation and remain alive for long periods without water, but no living thing can be fully *functional* without water. To reproduce, to move, to obtain nutrients—to carry out most of life's basic processes, in other words—all living things must have water. Given this, if we ask how important water is to life, the answer is that the two things are inseparable.

## Water Is a Major Player in Many of Life's Processes

But, what gives water this status? Recall the notion we touched on earlier of life being, at one level, a series of chemical reactions. It makes sense that these reactions would best take place in some liquid medium—since liquid allows for the easy distribution of materials—and that medium turns out to be water. However, water is not just a passive medium in which reactions take place. It *facilitates* many of these reactions thanks to its chemical structure. To understand this, it's important to know something about three terms that start with an *s*: solution, solute, and solvent.

Pour some salt into a container of water, stir the water, and what happens? The salt quickly disappears. It hasn't actually gone anywhere, of course; it has simply mixed with the water. Now, if it has mixed uniformly, so that there are no lumps of salt here or

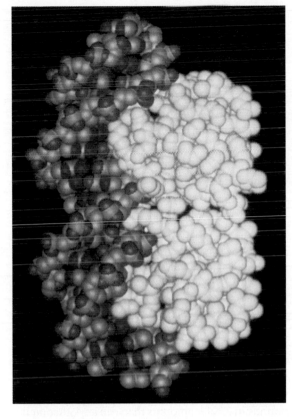

there, you have created a **solution**—a mixture of two or more kinds of molecules, atoms, or ions that is homogeneous throughout. The salt is what's being dissolved, so it is the **solute**. The water is doing the dissolving, so it is the **solvent**. This gets us to the point about water: it's a terrific solvent, which is to say it has a great ability to dissolve other substances.

If you look at **Figure 2.15**, you can get a detailed view of water's solvent power in connection with the

**Figure 2.15**
**Water's Power as a Solvent**

**(a)** Attraction    **(b)** Separation    **(c)** Dispersion

water (solvent)

O

H

Na⁺    Cl⁻

sodium chloride (solute)

Sodium and chloride ions dissolved in water

Sodium chloride's positively charged sodium ions (Na⁺) are attracted to water's negatively charged oxygen atoms, while its negatively charged chloride ions (Cl⁻) are attracted to water's positively charged hydrogen atoms.

Pulled from the crystal, and separated from each other by this attraction, sodium and chloride ions become surrounded by water molecules.

This process of separating sodium and chloride ions repeats until both ions are evenly dispersed, making this an aqueous solution.

... ANSWERS

1. loses; electrons; charges

2. hydrogen atom; electronegative; hydrogen; oxygen; nitrogen

3. bind

salt we just talked about. Remember that water is a polar molecule, which is to say a molecule that has differing electric charges at one end as opposed to the other. Attracted by the polar nature of the water molecule, the sodium and chloride ions that make up a salt crystal separate from the crystal—and from each other. Each ion is then surrounded by several water molecules (Figure 2.15b). These units keep the sodium and chloride ions from getting back together. In other words, they keep the ions evenly dispersed throughout the water, which is what makes this a solution (Figure 2.15c). Water works as a solvent here because the ionic compound sodium chloride carries an electrical charge. What *generally* makes water work as a solvent, however, is its ability to form, with other molecules, the hydrogen bonds we talked about earlier.

## Water's Structure Gives It Many Unusual Properties

When we note water's ability to act as a solvent, we're actually not giving water its due. Water is not just *a* solvent: Over the range of substances, nothing can match it as a solvent. It can dissolve more compounds in greater amounts than can any other liquid.

### Ice Floats Because It Is Less Dense than Water

But solvency power is merely the beginning of water's abilities. It is a multitalented performer. And it achieves this status because it is . . . odd. Compared to other molecules, water is like some zany eccentric whose powers stem precisely from its eccentricity.

Consider the fact that ice floats on water. This is so because the solid form of $H_2O$ is less dense than the liquid form—a strange reversal of nature's normal pattern. Things work this way with $H_2O$ because when water molecules slow their motion in cooling, they are able to form the maximum number of hydrogen bonds with each other. The result is that water molecules are spaced *farther apart* when frozen. Thus, ice is less dense than water. This may seem like some minor, quirky quality, but it actually has the effect of making possible life as we know it. Ice on the surface of water acts to insulate the water beneath it from the freezing surface temperatures and wind above, creating a warmer environment for organisms such as fish (**Figure 2.16**). If ice *sank*, on the other hand, then the entire body of water would freeze solid at colder latitudes, creating an environment in which few living things could survive for long.

### Water Has a Great Capacity to Absorb and Store Heat

Water serves as an insulator not only when it is frozen, but also when it is liquid or gas. This has to do with a quality called **specific heat:** the amount of energy required to raise the temperature of a substance by 1° Celsius. As it turns out, water has a high specific heat. Put a gram container of drinking alcohol (ethyl alcohol) side by side with one of water, heat them both, and it will take almost twice as much energy to raise the temperature of the water 1°C as it will the alcohol. Having absorbed this much heat, however, water then

**Figure 2.16**
**Life Under the Ice**

Water is unusual in that its solid form (ice) is less dense than its liquid form. This means that ice floats, and this in turn means that life can flourish in cold-weather aquatic environments. Pictured are some shrimp-like krill feeding on algae that grow on the underside of ice in the waters off Antarctica.

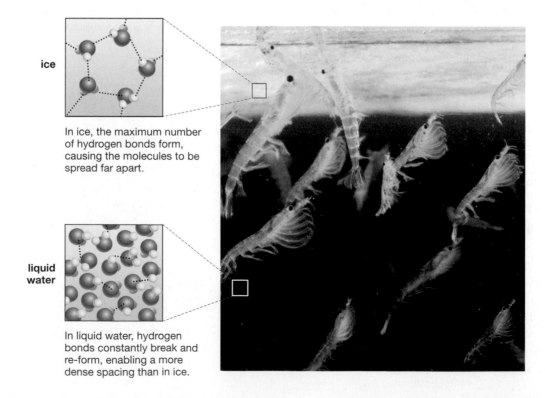

**ice**

In ice, the maximum number of hydrogen bonds form, causing the molecules to be spread far apart.

**liquid water**

In liquid water, hydrogen bonds constantly break and re-form, enabling a more dense spacing than in ice.

has the capacity to *release* it when the environment around it is colder than the water itself. The result? Water acts as a great heat buffer for Earth. The oceans absorb tremendous amounts of radiant energy from the sun only to release this heat when the temperature of the air above the ocean gets colder. Without this buffering, temperature on Earth would be less stable. People who have spent a day and a night in a desert can attest to this effect. The searing heat of the desert day radiates off the desert floor; but at night, with little water vapor in the air to capture this heat, the desert cools dramatically. In the same way, our *internal* temperature can remain much more stable because the water that makes up so much of us is first able to absorb and then to release great amounts of heat. The sweat created in exercise has considerable cooling power because each drop of perspiration carries with it a great deal of heat.

Water derives these powers from its chemical structure, in particular from the hydrogen bonding described earlier. Weak and shifting though individual hydrogen bonds may be, collectively they have great strength. Heat is the motion of molecules. To get molecules moving, though, chemical bonds must first be broken. Because water has a formidable set of hydrogen bonds, it takes a lot of energy to break the bonds and get its molecules moving. This same set of bonds gives water molecules another notable characteristic: *cohesion*, meaning a tendency to stay together. Thanks to this quality, evaporation can draw a chain of water molecules in one continuous column from a plant's roots all the way to its leaves and out into the air as water vapor.

Having looked at the qualities of water, you may be tempted to think how *uncanny* it is that water has all these characteristics that are conducive to life, but remember that life went on solely in water for billions of years before living things ever came onto land. Thus, water has fundamentally conditioned life as we know it. Being surprised about water's life-enhancing qualities is like being surprised that a stage is a good place to put on a play.

### What Water Cannot Do

With its great complexity, life requires molecules that *cannot* be dissolved by water. Such is the case with nonpolar covalent molecules, one example of which is the methane molecule, ($CH_4$). Compounds such as methane are known as **hydrocarbons** because they are made up solely of hydrogen and carbon atoms. Petroleum products are more complex hydrocarbons than methane, but they also are nonpolar, and you can see a vivid demonstration of their

lack of water solubility in oil spills (**Figure 2.17**). Oil doesn't dissolve in water because the oil carries almost no electrical charge that water can bond with; thus, water has no way to separate one oil molecule from another.

### Two Important Terms: Hydrophobic and Hydrophilic

The ability of molecules to form bonds with water has a couple of important names attached to it. Compounds that will interact with water—such as the sodium chloride considered earlier—are known as **hydrophilic** ("water loving"), while compounds that do not interact with water are known as **hydrophobic** ("water fearing"). Both terms are misleading in that no substance has any emotional relationship with water. *Hydrophobic* is particularly off the mark because water does not repel hydrophobic molecules; instead, the strong bonding that water molecules do with each other causes them to form circles around concentrations of hydrophobic molecules, as if they had lassoed them.

The importance of hydrophobic molecules can be illustrated in part by the common milk carton. Why is the milk carton important? Because it can keep milk separate from everything else. We living organisms need some kind of "carton" that can separate the world outside of us from ourselves. Likewise, organisms have great use *within* themselves for compartments that can be sealed off to one degree or another. If water broke down every molecule of life it came in contact with, then it

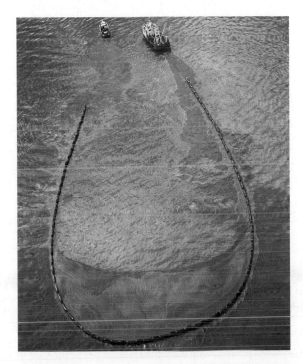

**Figure 2.17**
**Oil and Water Do Not Mix**

When there is an oil spill in the ocean, the oil stays concentrated even as it spreads because oil and water do not form chemical bonds with each other. Here, trawlers are using a boom to clean up after an oil spill in Great Britain.

would break down all these divisions of living systems. Molecules do not have to be completely hydrophilic or hydrophobic. Indeed, as you'll see, a number of important molecules have both hydrophilic and hydrophobic portions.

---

### SO FAR...

1. Pour some salt into a container of water and stir until the salt is no longer visible. In this situation, the salt is the _____, the water is the _____, and the uniform mixture of the two substances is a _____.

2. Thanks to the _____ linking them, water molecules have great _____, meaning a tendency to stay together, and a high _____, with the result that it takes a relatively large amount of _____ to raise their temperature.

3. Oil is an example of a substance that will not readily bond with water and that thus is _____, while sodium chloride is an example of a substance that is _____ because it will readily bond with water.

---

**Web Animation 2.4**
Water and pH

## 2.5 Acids and Bases Are Important to Life

When considering the question of the water-based or "aqueous" solutions so common in nature, an important concept is that of acids, bases, and the pH scale used to measure their levels.

We all have had experience with acids and bases, whether we've called them by these names or not. Acidic substances tend to be a little more familiar: lemon juice, vinegar, tomatoes. Substances that are strongly acidic have a well-deserved reputation for being dangerous: The word *acid* is often used to mean something that can sear human flesh. It might seem to follow that bases are benign, but ammonia is a strong base, as are many oven cleaners. The safe zone for living tissue in general lies with substances that are neither strongly acidic nor strongly basic. Science has developed a way of measuring the degree to which something is acidic or basic—the pH scale. So widespread is pH usage that it pops up from time to time in television advertising ("It's pH-balanced!").

### Acids Yield Hydrogen Ions in Solution, Bases Accept Them

The *H* in pH stands for hydrogen, while the *p* can be thought of as standing for power. Thus, we get "hydrogen power," which describes what lies at the root of

pH. An **acid** is any substance that *yields hydrogen ions* when put in aqueous solution. A **base** is any substance that *accepts* hydrogen ions in aqueous solution.

How might this yielding or accepting come about? Recall first that an ion is a charged atom, and that atoms become charged through the gain or loss of one or more electrons. Because electrons carry a negative charge, the loss of an electron leaves an atom with a net positive charge. Also recall that the hydrogen atom amounts to one central proton and one electron that circles around it. A hydrogen ion, then, is a lone proton that has lost its electron—a positively charged ion whose symbol is $H^+$.

Now, suppose you put an acid—hydrochloric acid (HCl)—into some water. What happens is that HCl *dissociates* or breaks apart into its ionic components, $H^+$ and $Cl^-$. The HCl has therefore yielded a hydrogen ion, ($H^+$). Now, with a greater concentration of hydrogen ions in it, the water is more acidic than it was (**Figure 2.18**).

What about bases? There is a compound called sodium hydroxide (NaOH)—better known as lye—that, when poured into water, dissociates into $Na^+$ (sodium) and $OH^-$ (hydroxide) ions. The place to look here is the $OH^-$ ions. Negatively charged as they are, they would readily bond with positively charged $H^+$ ions. In other words, they would *accept* $H^+$ ions in solution, which is the definition of a base. This accepting of the $H^+$ ions makes the solution more basic—or to look at it another way, less acidic. Thus, acids and bases are something like the two ends of a teeter-totter: when one goes up, the other comes down. As you might have guessed, in the right proportions they can balance each other out perfectly. Look at Figure 2.18 to see how this would play out with the solutions you've looked at so far. Mixing them, the collection of $H^+$ and $OH^-$ ions now comes together in water, and for each pair of ions that interacts, the result is:

$$H^+ + OH^- \longrightarrow H_2O$$

Water, which is neutral on the pH scale. The acid and the base have perfectly balanced one another out. The $OH^-$ ion, generally referred to as the **hydroxide ion,** is important because compounds that yield them in quantity are strongly basic and can be used to move solutions from the acidic toward the basic.

### Many Common Substances Can Be Ranked According to How Acidic or Basic They Are

Look at **Figure 2.19** on page 36 for an idea of how acidic or basic some common substances are. Following from the notion of what pH amounts to, it's clear that battery acid, for example, is strongly acidic because when it dissociates in solution it yields a large number of hydrogen ions. As you move on to lemon juice and then tomatoes, however, you the acids are weaker, which is to say they are substances that yield fewer $H^+$ ions in solution. By the time you get to seawater, you've arrived at substances that *accept* hydrogen ions.

## The pH Scale Allows us to Quantify How Acidic or Basic Compounds Are

The net effect of all this yielding and accepting of hydrogen ions is the concentration of $H^+$ ions in solution. Through this concentration, the notion of pH can be *quantified*—can have numbers attached to it. What is employed here is the **pH scale,** a scale used

in measuring the relative acidity of a substance. Look at Figure 2.19 again, this time in connection with the numbers on the right of the scale. You can see that 0 on the scale is the most acidic, while 14 is the most basic. It's important to note that the pH scale is *logarithmic:* A substance with a pH of 9 is 10 times as basic as a substance with a pH of 8 and 100 times as basic as a substance with a pH of 7.

## Some Terms Used When Dealing with pH

Here are some notes on pH terminology:

- As a solution becomes more basic, its pH *rises.* Thus, the higher the pH, the more basic the solution; the lower the pH, the more acidic the solution. Oven cleaner is said to have a high pH, while lemon juice has a low pH.

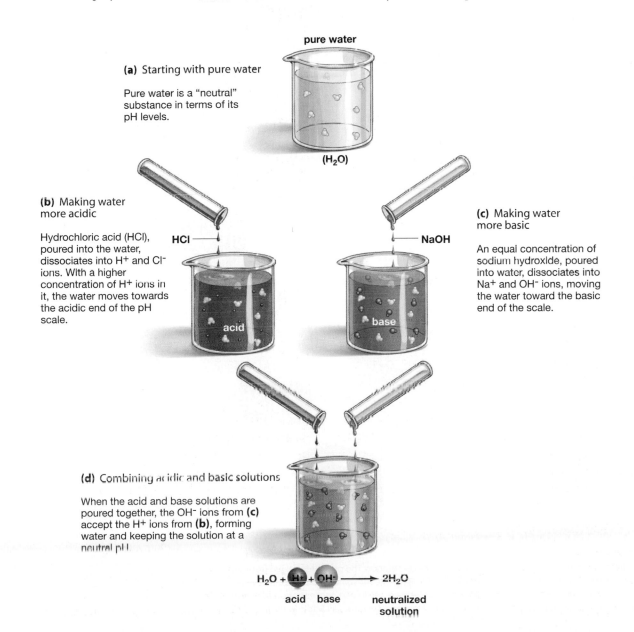

**pure water**

**(a)** Starting with pure water

Pure water is a "neutral" substance in terms of its pH levels.

**(H₂O)**

**(b)** Making water more acidic

Hydrochloric acid (HCl), poured into the water, dissociates into H⁺ and Cl⁻ ions. With a higher concentration of H⁺ ions in it, the water moves towards the acidic end of the pH scale.

HCl

acid

NaOH

**(c)** Making water more basic

An equal concentration of sodium hydroxide, poured into water, dissociates into Na⁺ and OH⁻ ions, moving the water toward the basic end of the scale.

base

**(d)** Combining acidic and basic solutions

When the acid and base solutions are poured together, the OH⁻ ions from **(c)** accept the H⁺ ions from **(b)**, forming water and keeping the solution at a neutral pH.

$$H_2O + H^+ + OH^- \longrightarrow 2H_2O$$

acid      base            neutralized solution

**Figure 2.18
Hydrogen Ions and pH**

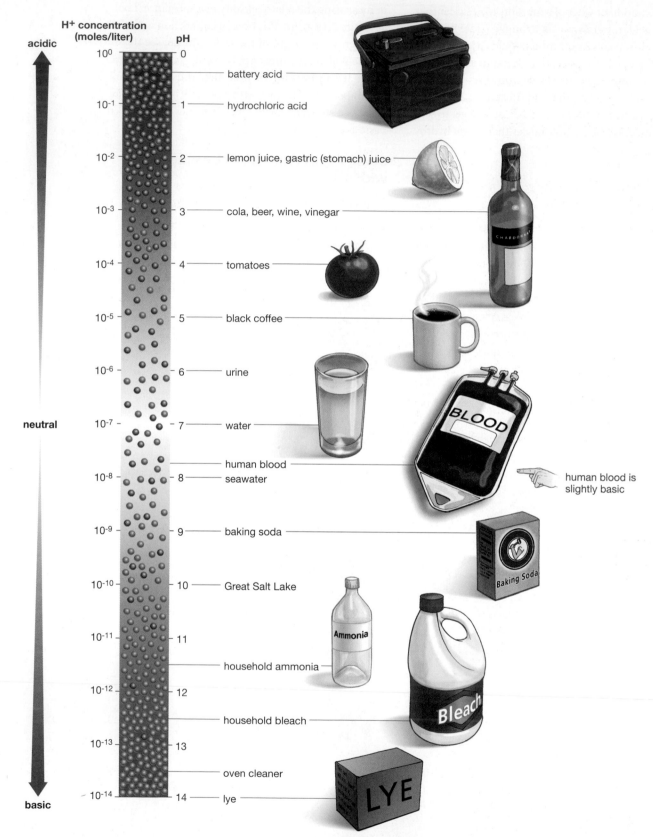

**Figure 2.19**
**Common Substances and the pH Scale**

Chemists use units called moles per liter to measure the concentration of substances in solution. The pH scale, derived from this framework, measures the concentration of hydrogen ions per liter of solution. The most acidic substances on the scale have the greatest concentration of hydrogen ions, while the most basic (or alkaline) substances have the lowest concentration of hydrogen ions. The scale is logarithmic, so that wine, for example, is 10 times as acidic as tomatoes and 100 times as acidic as black coffee.

- Given that a hydrogen ion amounts to a single proton, it is also correct to say that an acid is something that yields *protons* in solution, while a base is something that accepts protons. This is how you will often hear hydrogen ions discussed in biology.

- A solution that is basic is also referred to as an **alkaline** solution.

### Why Does pH Matter?

So, why do we care about pH? The brief answer is because living things are sensitive to its levels in many ways. There is, for example, a class of proteins called enzymes that you'll be reading about in Chapter 3. Enzymes are chemical tools that must retain a specific shape to function. However, if you put an enzyme in a solution with a pH that is too acidic, the enzyme loses it shape. Why? The charged nature of the acidic solution starts breaking down the enzyme's hydrogen bonds. (Remember, lots of positively charged protons are floating around in an acidic solution.) Likewise, cell membranes—which serve the milk carton function noted earlier—can start to break down if pH levels start to go outside normal limits. Membranes and enzymes are so fundamental to life that when they are interfered with, death can result. It is not surprising, then, that many organisms have developed so-called acid-base **buffering systems,** meaning physiological systems that function to keep pH within normal limits. What are these limits? The usual range for living things is about 6–8, with the pH of the human cell being about 7 and that of the blood in our arteries about 7.4. However, some parts of the body have special pH requirements. The interior of your stomach, for example, can have a pH as low as 1—an extremely acidic environment that not only helps break down food, but that also kills most bacteria that ride in on the food. Even a stomach can become too acidic, however (perhaps because of what we've eaten), in which case people may turn to the *phrmaceutical* pH buffers commonly known as antacids. These substances do just what their name implies: they raise pH levels in the stomach, thus helping to alleviate "heartburn."

Beyond its effects on individuals, pH can affect entire communities of living things. You may have heard of acid rain as a phenomenon that affects forests, streams, and lakes. Well, **acid rain** is just what it sounds like: rain whose pH level has been skewed toward the acidic side of the pH scale, with air pollution being the cause of this shift. As you can imagine, when rain itself has an altered pH, there can be widespread trouble for living things. Starting in the 1970s, this began to be the case in many environments worldwide, and this remains true today. (In the United States, the problem has been most severe in an area stretching from Michigan eastward through Pennsylvania and New York State.) Air pollution regulations have lessened the severity of acid rain in recent years, but it remains a long-term problem that requires constant monitoring.

## On to Biological Molecules

In this chapter, you've learned a fair amount about some of nature's most fundamental building blocks: atoms, ions, the simple substance water, and a special ion (the $H^+$ ion) whose levels govern pH. Now, it's time to see how these building blocks fit together to create some of the larger-scale component parts of life. These component parts usually are referred to as *biological molecules*. Everyone has heard of at least some biological molecules. Carbohydrates are one variety of them, for example, and it's easy to name foods that we think of as carbohydrates. But, what *is* a carbohydrate? What is the fundamental nature of this component part of life? To sharpen the point, how does a carbohydrate differ from a protein? In Chapter 3, you'll find out as we explore the materials that make up the living world.

**SO FAR...**

1. An acid is any substance that _____ hydrogen _____ when put into an aqueous solution, while a base is any substance that _____ them.

2. The neutral point on the pH scale is _____, while the most acidic point is _____, and the most basic or alkaline point is _____.

3. Living things tend to have internal pH environments that hover around _____.

••• ANSWERS
1. yields; ions; accepts
2. 7; 0; 14
3. neutral (7)

# CHAPTER 2 REVIEW

For study help, activities, and more quiz questions, go to **www.aw.com/krogh4**.

## SUMMARY

### 2.1 Chemistry's Building Block: The Atom

- The fundamental unit of matter is the atom. The three most important constituent parts of an atom are protons, neutrons, and electrons. Protons and neutrons exist in the atom's nucleus, while electrons move around the nucleus, at some distance from it. Protons are positively charged, and electrons are negatively charged, but neutrons carry no charge.

- An element is any substance that cannot be reduced to any simpler set of constituent substances through chemical means. Each element is defined by the number of protons in its nucleus.

- The number of neutrons in an atom can vary independently of the number of protons. Thus, a single element can exist in various forms, called isotopes, depending on the number of neutrons it possesses.

**Web Animation 2.1:** Structure of Atoms, Elements, Isotopes

### 2.2 Matter Is Transformed through Chemical Bonding

- Atoms can link to one another in the process of chemical bonding. Among the forms this bonding can take are covalent bonding, in which atoms share one or more electrons, and ionic bonding, in which atoms lose and accept electrons from each other.

- Chemical bonding comes about as atoms "seek" their lowest energy state. An atom achieves this state when it has a filled outer electron shell. Hydrogen and helium require two electrons in orbit around their nuclei to have filled outer shells, while most other elements require eight electrons to have filled outer shells.

- A molecule is a compound of a defined number of atoms in a defined spatial relationship. For example, two hydrogen atoms can link with one oxygen atom to form one water molecule.

- Atoms of different elements differ in their power to attract electrons. The term for measuring this power is electronegativity. Through electronegativity, a molecule can take on a polarity, meaning a difference in electrical charge at one end compared to the other. Covalent chemical bonds can be polar or nonpolar. A polar covalent bond exists when shared electrons are not being shared equally among atoms in a molecule, due to electronegativity differences.

- Two atoms will undergo a process of ionization when the electronegativity differences between them are great enough that one atom loses one or more electrons to the other. This process creates ions: atoms whose number of electrons differs from their number of protons. The charge differences that result from ionization can produce an electrostatic attraction between ions. This attraction is an ionic bond. When atoms of two or more elements bond together ionically, the result is an ionic compound.

- Hydrogen bonding links an already covalently bonded hydrogen atom with an electronegative atom. In water, a hydrogen atom of one water molecule will form a hydrogen bond with an unshared oxygen electron of a neighboring water molecule.

**Web Animation 2.2:** Chemical Bonding

### 2.3 Some Qualities of Chemical Compounds

- Three-dimensional molecular shape is important in biology because this shape determines the capacity molecules have to bind with one another.

- A solution is a homogeneous mixture of two or more kinds of molecules, atoms, or ions. The compound dissolved in solution is the solute; the compound doing the dissolving is the solvent.

**Web Animation 2.3:** Geometry, Chemistry, and Biology

### 2.4 Water and Life

- Water has several qualities that have strongly affected life on Earth. It is a powerful solvent, with the ability to dissolve more compounds in greater amounts than any other liquid. Because water's solid form (ice) is less dense than its liquid form, bodies of water in colder climates do not freeze solid in winter, which allows life to flourish under the ice.

- Water has a great capacity to absorb and retain heat. Because of this, the oceans act as heat buffers for the Earth, thus stabilizing Earth's temperature. Water has a high degree of cohesion, which allows water to be drawn up through plants, via evaporation, in one continuous column, from roots through leaves.

- Some compounds do not interact with water. Hydrocarbons such as petroleum are examples of such hydrophobic compounds. Water cannot break down hydrophobic compounds, which is why oil and water don't mix. Compounds that do interact with water are polar or carry an electric charge and are called hydrophilic compounds.

### 2.5 Acids and Bases Are Important to Life

- An acid is any substance that yields hydrogen ions when put in aqueous solution. A base is any substance that accepts hydrogen ions in solution. A base added to an acidic solution makes that solution less acidic, while an acid added to a basic solution makes that solution less basic.

- The concentration of hydrogen ions that a given solution has determines how basic or acidic that solution is, as measured on the pH scale. This scale runs from 0 to 14, with 0 most acidic, 14 most basic, and 7 neutral. The pH scale is logarithmic; a substance with a pH of 9 is 10 times as basic as a substance with a pH of 8. Living things function best in a near-neutral pH, although some systems in living things have different pH requirements.

**Web Animation 2.4:** Water and pH

## KEY TERMS

| | |
|---|---|
| acid   34 | isotope   21 |
| acid rain   37 | law of conservation of |
| alkaline   37 | mass   24 |
| atomic number   21 | mass   19 |
| ball-and-stick model   25 | molecular formula   25 |
| base   34 | molecule   24 |
| buffering system   37 | neutron   20 |
| chemical bonding   23 | nonpolar covalent bond   26 |
| covalent bond   24 | nucleus   20 |
| electron   20 | pH scale   35 |
| electronegativity   26 | polar covalent bond   25 |
| element   21 | polarity   26 |
| free radical   27 | product   25 |
| hydrocarbon   33 | proton   20 |
| hydrogen bond   29 | reactant   25 |
| hydrophilic   33 | specific heat   32 |
| hydrophobic   33 | solute   31 |
| hydroxide ion   34 | solution   31 |
| ion   29 | solvent   31 |
| ionic bonding   28 | space-filling model   25 |
| ionic compound   27 | structural formula   25 |

## BRIEF REVIEW

*Answers to Brief Review questions are in the back of the book.*
*For multiple-choice quiz questions, go to* **www.aw.com/krogh4**.

1. As with most elements, carbon comes in several forms, one of which is carbon-14. What are these forms called, and how does one differ from the other?

2. Compare the size of an atom with the size of its nucleus. Where are the electrons? In light of this, what makes up most of the volume of an atom?

3. Draw a line and label one end "complete + or - charge" and the other end "no charge" to indicate the charges on the molecules or ions after bonding has occurred. Along the line, indicate where polar covalent bonds, nonpolar covalent bonds, and ionic bonds should be placed.

4. Why are atoms unlikely to react when they have their outer shell filled with electrons?

5. Describe two ways that water serves as a heat buffer.

6. Why do living things need to keep such a tight control on their internal pH?

## APPLYING YOUR KNOWLEDGE

1. In the Middle Ages, alchemists labored to turn common materials such as iron into precious metals such as gold. If you could journey back in time, how could you convince an alchemist that iron cannot be changed into gold?

2. Why does a balloon filled with helium float? Hydrogen can make balloons float, but it is not used for this purpose today because it is flammable. Based on chemical principles reviewed in the chapter, can you see why helium is not flammable? (*Hint:* Think what you are adding to a fire when you blow on it.)

3. How is water's importance to life reflected in the normal range of internal pH values that living things are likely to have?

*Living things such as this giant redwood tree (Sequoia sempervirens) can grow to enormous sizes thanks to the strength of materials they themselves produce.*

CHAPTER **3**

**3.1**   Carbon Is Central to the Living World   **41**

**3.2**   Functional Groups   **43**

**3.3**   The Molecules of Life: Carbohydrates   **44**

**3.4**   Lipids   **47**

**3.5**   Proteins   **53**

**3.6**   Nucleic Acids   **57**

**Essay**

From Trans Fats to Omega-3s: Fats and Health   **52**

Four kinds of carbon-based biological molecules form the living world.

# LIFE'S COMPONENTS: Biological Molecules

*We are stardust*

*Billion-year-old carbon*

from "Woodstock," by Joni Mitchell

**S**ongwriters usually specialize in emotional truth, so it's no small feat that, in just a couple of lines, Joni Mitchell could sum up the literal truth of what we are made of. Admittedly, her billion-year-old figure is off—Earth actually is about 4.6 billion years old, and its carbon is older yet. Even so, she was right about the essentials. Are we stardust? To a large extent, yes. All the heavy elements in our bodies—calcium, iron, carbon—were first cooked up in the fiery interior of stars and then dispersed into space when these stars exploded. Over time, these elements became a part of planets such as the Earth.

But surely we're more than carbon, right? Well, by weight we are more oxygen than anything else, mostly because oxygen is a principal element in water, which makes up so much of our bodies. Even so, it's carbon that really is life's element. How so? Well, consider the way in which living things amount to *concentrations* of carbon when compared to the makeup of the non-living world around them. As the environmental scientist Vaclav Smil has noted, Earth's crust is about 0.05 percent carbon, and Earth's atmosphere is about 0.01 percent carbon. Yet, Earth's *organisms* are 18 percent carbon, even when water is taken into account and nearly 50 percent carbon when water is factored out.

Given such figures, it's not surprising that, just as money must flow through an economy for it to operate, so carbon must flow through the living world for it to operate. First, plants and other photosynthesizers take in carbon from the atmosphere—over 100 bil-lion tons of it per year, in the form of atmospheric $CO_2$. Then, the plants, and all the organisms that eat them, use this carbon to make up their tissues and to power their activities. Finally, all these organisms die, and as their remains are broken down, some of the carbon in them is released into the atmosphere. But this is not a one-way trip; sooner or later, this same carbon is likely to move back into the living world, to be made part of a tree or a mushroom or a human being once again.

## 3.1 Carbon is Central to the Living World

So, why does carbon have this special place in life? The answer can be summed up in one word: linkage. Recall from Chapter 2 that carbon has four outer-shell, or "valence," electrons, but that most elements need *eight* outer-shell electrons for maximum stability. This means that carbon achieves maximum stability by linking up with four more electrons, which in turn means that carbon's bonding capacity is great. It can form very large and complex molecules, which is exactly what is needed in connection with the tremendous complexity of living things. Moreover, the bonds that carbon creates are covalent—it is *sharing* electrons with other atoms—which means its

Web Animation 3.1
The Chemistry of Carbon

hydrocarbons, but also when it is part of other molecules as well. If you look at **Table 3.1**, you can see some of the functional groups that are most important in living things.

**Web Animation 3.2**
Monomers and Polymers

### SO FAR...

1. The element _____ is central to the molecules that make up the living world. This element comes to its special status because of the great power it has to form complex and stable _____.

2. A functional group is a group of _____ that confers a special property on a _____ molecule.

### Table 3.1  Functional groups

| Group: | Structural formula: | Found in: |
|---|---|---|
| Carboxyl (—COOH) | —C with =O and —OH | fatty acids, amino acids |
| Hydroxyl (—OH) | —OH | alcohols, carbohydrates |
| Amino (—NH$_2$) | —N with H and H | amino acids |
| Phosphate (—PO$_4$) | —O—P(=O)(—O$^-$)—O$^-$ | DNA, ATP |

### Table 3.2  Monomers, Polymers

| If the monomer is . . . | The polymer is . . . |
|---|---|
| A **monosaccharide** (for example, glucose, fructose) | A **polysaccharide** (for example, starch, glycogen, cellulose) |
| An **amino acid** (for example, arginine, leucine) | A **polypeptide** or **protein** (A- and B-chains of insulin are polypeptides and insulin is a protein) |
| A **nucleotide** (sugar, phosphate, base in combination) | A **nucleic acid** (for example, DNA, RNA) |

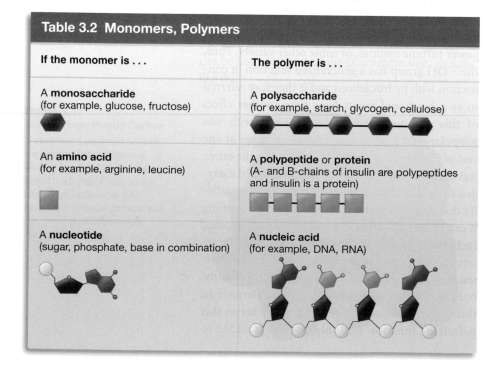

## 3.3  The Molecules of Life: Carbohydrates

With this review of carbon structures under your belt, you're ready to begin looking at some of the classes of carbon-based molecules—the molecules of living things. You'll explore four groupings of these organic compounds: carbohydrates, lipids, proteins, and nucleic acids.

As you get into this section, it will be a great help to keep in mind that *complex* organic molecules often are made from *simpler* molecules. Many of the molecules you'll be reading about have a building-blocks quality to them. Take a simple sugar, or monosaccharide, such as glucose, put it together with another monosaccharide (fructose), and you have a larger *di*saccharide called sucrose (better known as table sugar). Put *many* monosaccharide units together, and you have a polysaccharide, such as starch. The starch is an example of a **polymer**—a large molecule made up of many similar or identical subunits. Meanwhile, the glucose is an example of a **monomer**—a small molecule that can be combined with other similar or identical molecules to make a polymer. Look at **Table 3.2** for examples of both.

### Carbohydrates: From Simple Sugars to Cellulose

Happily, for purposes of memory, the elements in the first molecules we'll look at, carbohydrates, are all hinted at in the name: **Carbohydrates** are organic molecules that always contain carbon, oxygen, and hydrogen and that in many instances contain nothing *but* carbon, oxygen, and hydrogen. Furthermore, they usually contain exactly twice as many hydrogen atoms as oxygen atoms. For example, the carbohydrate glucose ($C_6H_{12}O_6$) contains 12 atoms of hydrogen and 6 of oxygen. Most people think of carbohydrates purely in terms of foods such as breads and pasta (**Figure 3.3**), but as you'll see, carbohydrates have more roles than this in nature.

The building blocks of the carbohydrates are the **monosaccharides** mentioned earlier. These are the monomers of carbohydrates, collectively referred to as **simple sugars**. You've already seen several views of one of these monomers, glucose. Glucose has a use in and of itself: Much of the food we eat is broken down into it, at which point it becomes our most important energy source. Glucose can also bond with other monosaccharides, however, to form more complex carbohydrates. Let's take a look at how this happens. If you look at **Figure 3.4**, you can see an example of two glucose molecules bonding to create the disaccharide called maltose.

**Figure 3.3**
**Carbohydrates in Foods**

Breads, cereals, and pasta make up a significant proportion of our diets. These foods are all rich in carbohydrates, one of the four main types of biological molecules.

Several things are worth noting here. In maltose, as you can see, the link that joins the two glucose monomers is a single oxygen atom linked to carbons of each of the glucose units. To get this, two atoms of hydrogen and one atom of oxygen are split off (on the left side of the equation) from the original glucose molecules. What becomes of these three atoms? Look at the far right-hand side of the reaction and see the $+ H_2O$. The product of this reaction is a molecule of maltose and a molecule of water.

Now note the arrows in the middle of the reaction. You've been used to seeing a single, rightward-pointing arrow, but here the arrows go both ways. This means that this is a reversible reaction—it can proceed in either of the two directions. Maltose can be split apart to yield two *glucose* molecules. Finally, note the existence of the $-OH$ functional group in both the glucose and maltose molecules.

### Kinds of Simple Carbohydrates

You have thus far seen examples of one of the simplest carbohydrates, the monosaccharide glucose, and one slightly more complex carbohydrate, a disaccharide called maltose. There are many kinds

of mono- and disaccharides, however. Among the monosaccharides are, for example, fructose and deoxyribose. Among disaccharides, there are sucrose and lactose. At the risk of pointing out the obvious, note that all these sugars have *–ose* at the end of their name: If it's an *–ose*, it's a sugar (**Figure 3.5**).

### Complex Carbohydrates are Made of Chains of Simple Carbohydrates

If you go another couple of bumps up in complexity from disaccharides, you get to the polymers of carbohydrates, the **polysaccharides**. The *poly* in polysaccharide means "many," while *saccharide* means "sugars," and the term is apt. In the polysaccharide molecule cellulose, for example, there may be 10,000 glucose units linked with one another. The basic unit here is the six-carbon monosaccharide glucose $C_6H_{12}O_6$ from which chains of glucose units are built. The complexity of these molecules gives them their alternate name, *complex carbohydrates*. Four different types of complex carbohydrates interest us: starch, glycogen, cellulose, and chitin (**Figure 3.6** on the next page).

**Starch** is a complex carbohydrate found in plants; it exists in the form of such foods as potatoes, rice, carrots, and corn. In plants, these starches serve as the main form of carbohydrate *storage*, sometimes as seeds (rice and wheat grains) or sometimes as roots (carrots or beets); see **Figure 3.6a**.

**Figure 3.5**
**Sugars Come in Many Forms**

Sucrose or table sugar comes to us from sugarcane or sugar beets, and glucose is found in corn syrup. Fructose comes to us in sweet fruits and in high-fructose corn syrup, which often is used to sweeten soft drinks.

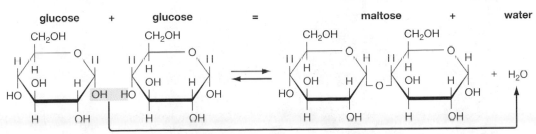

**Figure 3.4**
**Carbohydrates Follow a Building-Blocks Model**

In this example, two units of the monosaccharide (or simple sugar) glucose link to form the disaccharide maltose. In addition to maltose, the reaction yields water. The double arrows indicate that this reaction is reversible; a single maltose molecule can yield two glucose molecules.

Another complex carbohydrate, **glycogen**, serves as the primary form of carbohydrate storage in animals. Thus, glycogen does for animals what starch does for plants. For this reason, glycogen is sometimes called animal starch. The starches or sugars we eat are broken down eventually into glucose, at which point some of this glucose may be used immediately. Some may not be needed right away, however, in which case it's moved into muscle and liver cells to be stored as the more complex carbohydrate glycogen (**Figure 3.6b**).

**Cellulose** is a rigid, complex carbohydrate contained in the cell walls of many organisms. Despite this innocuous-sounding function, cellulose is important because it makes up so much of the natural world. It is easily the most abundant carbohydrate on Earth: Trees, cotton, leaves, and grasses are largely made of it. When cellulose is enmeshed with a hardening compound called *lignin*, the result is a set of cell walls that can hold up giant redwood trees. Because cellulose is so dense and rigid, it is not surprising that human beings and other mammals cannot digest it. This statement may make you wonder about grass-eating *cows*, but in fact cows have cellulose digested for them by special bacteria in their digestive tract. Cellulose is important to humans because it is our major source of insoluble fiber, which helps move foods through the digestive tract. Because cellulose exists in the cell walls of plants, we can get our fiber from such foods as whole grains and fresh fruits (**Figure 3.6c**).

**Chitin** is a complex carbohydrate that forms the external skeleton of the arthropods—all insects, spiders, and crustaceans, for example. In all these animals, chitin plays a structural role similar to that of cellulose in plants: It gives shape and strength to the structure of the organism (**Figure 3.6d**).

## SO FAR...

1. Biological molecules often have a building-block quality to them, with smaller _____ linked together to form larger _____.

2. Carbohydrates are organic molecules that always contain _____, _____, and _____ and that usually contain exactly twice as many _____ atoms as _____ atoms. The building blocks of

• • •

**Figure 3.6**
**Four Examples of Complex Carbohydrates**

All complex carbohydrates are composed of chains of glucose, but they differ in the details of their chemical structure.

**(a)** Starch serves as a form of carbohydrate storage in many plants. Here, starch granules can be seen within the cells of a raw potato slice.

**(b)** Glycogen serves as a form of carbohydrate storage, as shown in this photo of glycogen globules in the liver.

**(c)** Cellulose, running as fibers through cell walls, provides structural support for plants and other organisms. The photo is of sets of cellulose fibers running at right angles to one another in the cell wall of marine algae.

**(d)** Chitin provides structural support for some animals. The outer "skin" or cuticle of arthropods is composed mostly of chitin. The photo shows the exoskeleton of one arthropod, a tick.

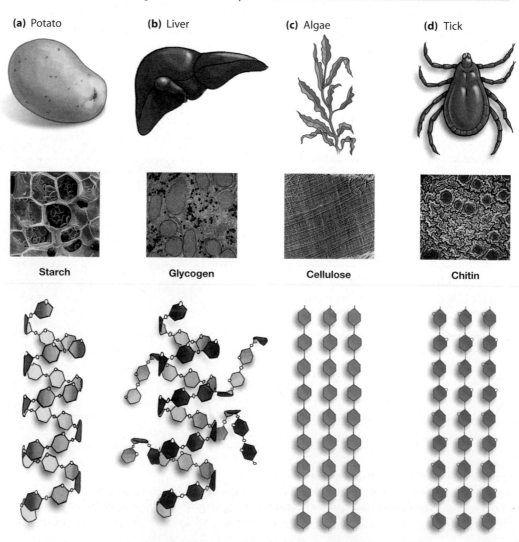

**(a)** Potato   **(b)** Liver   **(c)** Algae   **(d)** Tick

Starch   Glycogen   Cellulose   Chitin

## 3.4 Lipids

We turn now to our second major group of biological molecules, the **lipids**, a class of molecules whose defining characteristic is that they do not readily dissolve in water. It turns out that lipids are made of the same elements as carbohydrates—carbon, hydrogen, and oxygen. But lipids have much more hydrogen, relative to oxygen, than do the carbohydrates. We're all familiar with some lipids; they exist as fats, oils, cholesterol, and as hormones such as testosterone and estrogen. Unlike the other biological molecules you'll be studying, however, a lipid is not a polymer composed of component-part monomers; no single structural unit is common to all lipids. Thus, the one characteristic shared by pure lipids that they do not readily break down in water. Remember a discussion last chapter about the need life has to create internal compartments that are sealed off from one another? Well, thanks to their insolubility, lipids are able to serve this function. In addition, lipids have considerable powers to store energy and to provide insulation (think of the abundant fat on a polar bear).

### One Class of Lipids is the Glycerides

The most common kind of lipid, the glyceride, can be thought of as a molecule in two parts. The first part is a "head" composed of the alcohol glycerol:

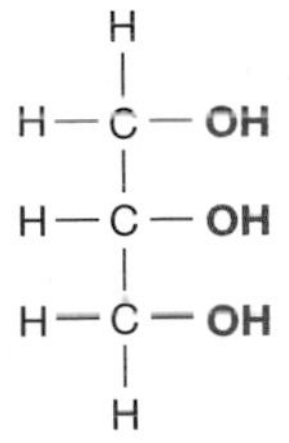

$$
\begin{array}{c}
H \\
| \\
H - C - OH \\
| \\
H - C - OH \\
| \\
H - C - OH \\
| \\
H
\end{array}
$$

The second part is one or more fatty acids, each consisting of a long chain of carbon and hydrogen atoms; an example can be seen in **Figure 3.7**. This is stearic acid, one of the fatty acids in animal fat. As you can see, it amounts to a long chain of hydrogen and carbon atoms—its hydrocarbon portion—that terminates with a COOH or "carboxyl" chemical

group on the left. Now, bringing the carboxyl part of this fatty acid chain together with the −OH group of glycerol is what makes a glyceride. But, you can see that the glycerol has *three* −OH groups on the right. Thus there is, you might say, docking space on glycerol for three fatty acids, and in the synthesis of many glycerides, this linkage takes place: Three fatty acids link with glycerol to form a triglyceride, which is the most important dietary form of lipid. **Figure 3.8** shows how it works schematically.

The Rs on the right end of the COOH group stand for whatever the hydrocarbon chain is of that particular fatty acid. To get a better idea of the shape of a completed triglyceride, look at the space-filling model in **Figure 3.9**. This particular triglyceride, called tristearin, has three stearic fatty acids that stem

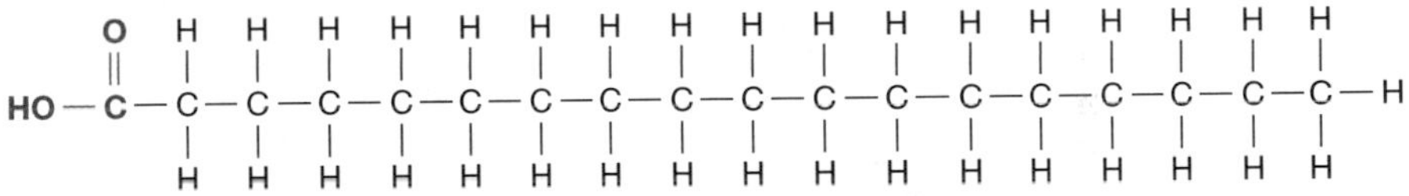

**Figure 3.7**
**Structural Formula for Stearic Acid**

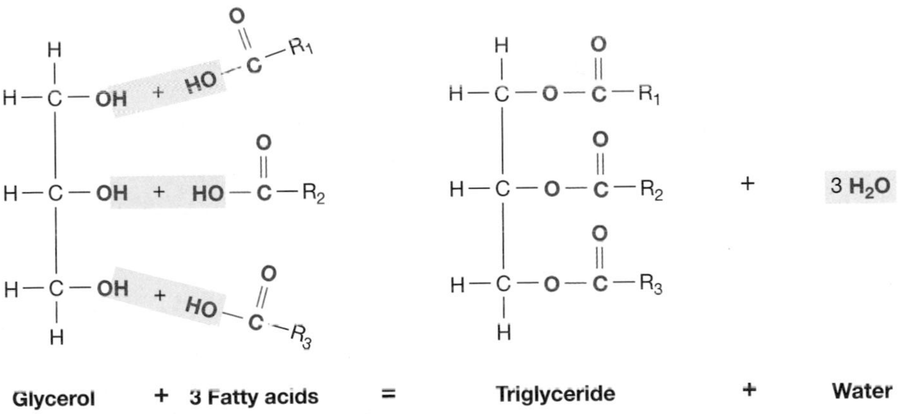

**Figure 3.8**
**Formation of a Triglyceride**

Glycerol + 3 Fatty acids = Triglyceride + Water

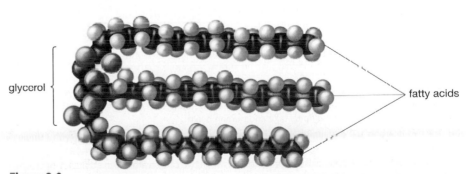

**Figure 3.9**
**The Triglyceride Tristearin**

This lipid molecule is composed of three stearic fatty acids, stemming rightward from the glycerol OH "head." Tristearin is found in both beef fat and the cocoa butter that helps make up chocolate.

like tines on a fork from the glycerol. This is only one possibility among many and is exceptional, rather than usual, in that all three fatty acids are the same. Among the dozens of fatty acids that exist, several *different* kinds of fatty acids generally will hook up in glycerol's three "slots" to form a triglyceride. Actual fat products, such as butter, are composed of different proportions of various fatty acids.

We've seen three fatty acids linking with glycerol to form *tri*glycerides, but monoglycerides (one fatty

acid joined to glycerol) and diglycerides (two fatty acids joined with glycerol) can be formed as well. Triglycerides are the most important of the glycerides, however, because they constitute about 90 percent of the lipid weight in foods.

With all this under your belt, you're ready for a couple of definitions. A **triglyceride** is a lipid molecule formed from three fatty acids bonded to glycerol. A **fatty acid** is a molecule found in many lipids that is composed of a hydrocarbon chain bonded to a carboxyl group.

## Saturated and Unsaturated Fatty Acids

If you now look at **Figure 3.10**, you can see three different fatty acids: palmitic, oleic, and linoleic. There is an obvious difference among them in that the oleic and linoleic hydrocarbon chains are "bent" compared to the straight-line palmitic chain. On closer inspection, you can see that the palmitic acid has an unbroken line of single bonds linking its carbon atoms. Meanwhile, the oleic acid has one double bond between the carbons in its chain, and the linoleic acid has two. Furthermore, note that these double bonds exist precisely where the "kinks" appear in these molecules.

## Fatty Acids: From Solid to Liquid

What these variations describe are the differences between three *kinds* of fatty acids. First, there is a **saturated fatty acid**—a fatty acid with no double bonds between the carbon atoms of its hydrocarbon chain. Then, there is a **monounsaturated fatty acid** (a fatty acid with one double bond between carbon atoms) and a **polyunsaturated fatty acid** (a fatty acid with two or more double bonds between carbon atoms). What a saturated fatty acid is saturated *with* is hydrogen atoms. A given monounsaturated fatty acid could theoretically have its lone carbon-carbon double bond replaced with a *single* bond between the carbons. This change would entail the addition of two more hydrogen atoms at the bond sites. Once this happened, this fatty acid would contain the maximum number of hydrogen atoms possible, meaning it would be a saturated fatty acid.

This may not sound like much of a difference, but it has several important consequences. First, at room temperature, as you move from saturated to unsaturated, you also move from fats in their solid form to fats in their liquid form, which we call **oils**. Why does saturation make this kind of difference? Remember that saturated fatty acids have a straight-line form; as such, they can pack together tightly in a triglyceride, like so many boards in a lumberyard. In contrast, unsaturated fatty acids stick out at varying angles to one

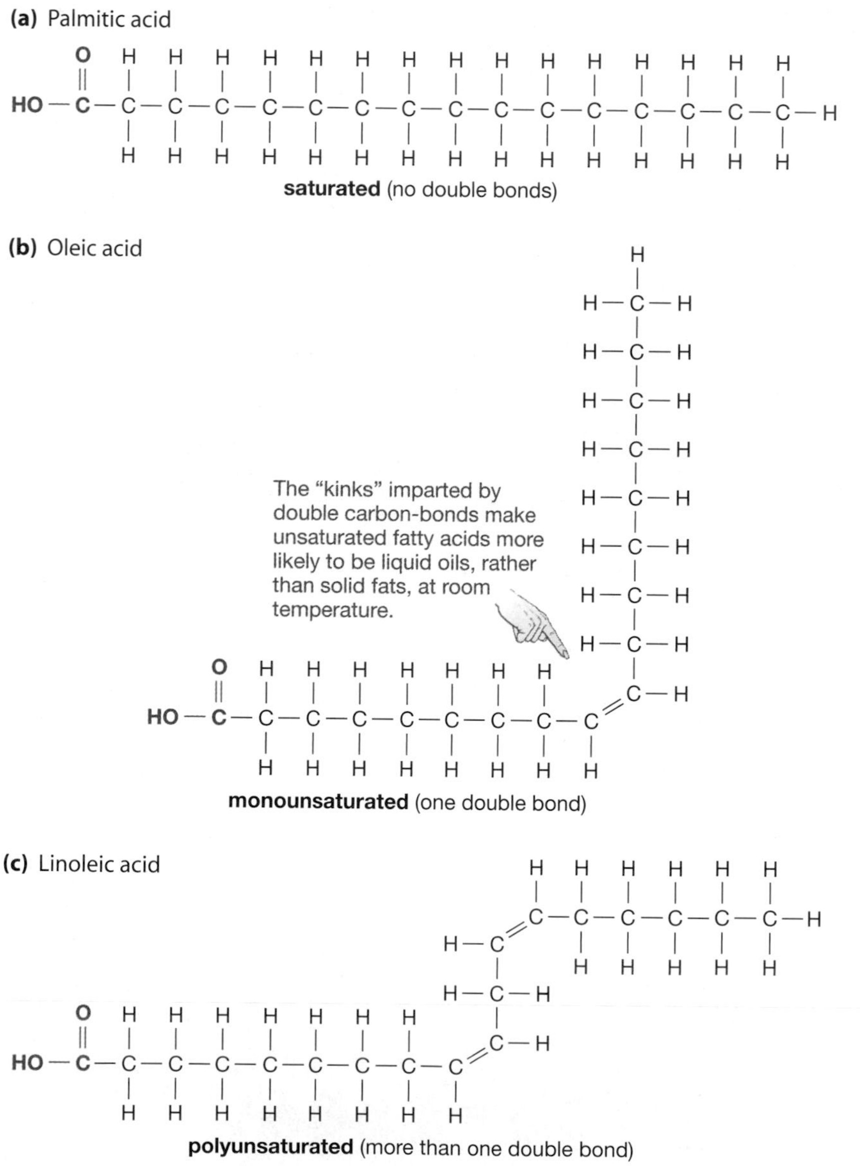

**(a)** Palmitic acid

**saturated** (no double bonds)

**(b)** Oleic acid

The "kinks" imparted by double carbon-bonds make unsaturated fatty acids more likely to be liquid oils, rather than solid fats, at room temperature.

**monounsaturated** (one double bond)

**(c)** Linoleic acid

**polyunsaturated** (more than one double bond)

**Figure 3.10**
**Saturated and Unsaturated Fatty Acids**
The degree to which fatty acid hydrocarbon chains are "saturated" with hydrogen atoms has consequences for both the form these lipids take and human health.

**(a)** The hydrocarbon "tail" in palmitic acid is formed by an unbroken line of carbons, each with a single bond to the next.

**(b)** In oleic acid, a double bond exists at one point between two carbon atoms. An additional hydrogen atom could link to each of these carbon atoms instead, which would make this a saturated fatty acid—saturated with hydrogen atoms. As things stand, this is a monounsaturated fatty acid.

**(c)** The carbons in linoleic acid have double bonds in two locations, making this a polyunsaturated fatty acid.

another; thus, they have a disorder that generally makes them liquid at room temperature. In short, the "kinks" in unsaturated fatty acids have consequences.

The distinction between saturated and unsaturated fatty acids also has another important effect, this one related to human health. Think of it this way: To the degree that *fatty acids* are saturated, the *fats* they make up will be saturated. And consumption of saturated fats has long been linked with heart disease—saturated fats raise total blood cholesterol levels, and high cholesterol levels are a risk factor for heart disease. This seemingly clear picture, constructed from scientific evidence compiled over many years, was called into question in 2006, however, with the publication of the largest study ever conducted on the effects of reducing fats in the diet. You can read more about this in "From Trans Fats to Omega-3s" on page 52.

## Energy Use and Energy Storage via Lipids and Carbohydrates

Lipids have something in common with carbohydrates in connection with energy storage on the one hand and energy use on the other. To be stored, lipids must be in their triglyceride form; to be used to provide energy, however, they must first be broken down into their glycerol and fatty acid components. Carbohydrates, meanwhile, are stored as the complex carbohydrate glycogen, but to be used for energy expenditure, they must first be broken down into their simple carbohydrate, building blocks, usually meaning glucose. This process of alternately building up molecules for energy storage or breaking them down for energy expenditure is a major task of living things. If you look at **Figure 3.11**, you can see how carbohydrates and lipids are alternately stored and used by plants and animals.

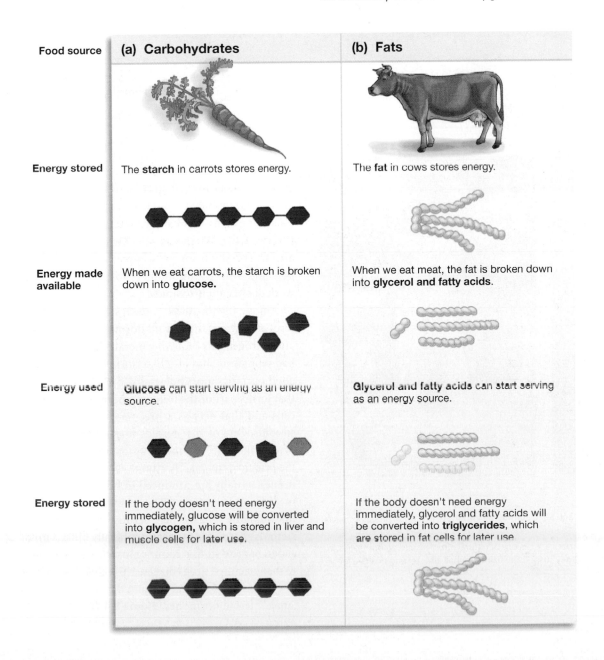

| Food source | **(a) Carbohydrates** | **(b) Fats** |
|---|---|---|
| Energy stored | The **starch** in carrots stores energy. | The **fat** in cows stores energy. |
| Energy made available | When we eat carrots, the starch is broken down into **glucose.** | When we eat meat, the fat is broken down into **glycerol and fatty acids**. |
| Energy used | Glucose can start serving as an energy source. | Glycerol and fatty acids can start serving as an energy source. |
| Energy stored | If the body doesn't need energy immediately, glucose will be converted into **glycogen,** which is stored in liver and muscle cells for later use. | If the body doesn't need energy immediately, glycerol and fatty acids will be converted into **triglycerides,** which are stored in fat cells for later use |

**Figure 3.11
Storage and Use of Carbohydrates and Lipids**
Carbohydrates and lipids generally are stored in one form but used in another.

**(a)** In plants, carbohydrates commonly are stored as starch, a complex carbohydrate that is a major component of such vegetables as carrots. Human beings can consume such a starch, after which it may be broken down into smaller, simple carbohydrate glucose units and used immediately—for exercise or other purposes—or stored in liver and muscle cells as the larger, more complex carbohydrate glycogen.

**(b)** Cows use their fat for energy storage and insulation. The fat portion of the meat that humans consume may be used for energy, but only after it is broken down into its glycerol and fatty acid components. Conversely, humans may convert these components back into triglycerides to be stored in fat cells, providing energy reserves for later.

**Figure 3.12**
**Structure of Steroids**

**(a)** The basic unit of steroids, four interlocked carbon rings.

**(b)** Types of steroids, each differentiated from the other by the side chains that extend from the four-ring skeleton. Testosterone is a principal "male" hormone, while estrogen is a principal "female" hormone. Both of these steroid hormones actually are found in both men and women, though in differing amounts. Cholesterol is also a prevalent and important steroid in both men and women. Although cholesterol has a bad reputation, it has several important functions—for example, breaking down fats.

**(a)** Four-ring steroid structure

**(b)** Side chains make each steroid unique

testosterone

estrogen

cholesterol

**(a)** Natural steroids

**(b)** Pharmaceutical steroids

**Figure 3.13**
**Steroids, in Uses Natural and Unnatural**

**(a)** Some steroid hormones, such as estrogen and testosterone, are important in natural processes, such as reproduction.

**(b)** Other sterioid hormones are laboratory-produced versions of natural muscle-building hormones. In 2005, Rafael Palmeiro, pictured, was given a 10-day suspension from baseball after testing positive for this kind of synthetic steroid.

## A Second Class of Lipids is the Steroids

**Steroids** are a class of lipid molecules that have, as a central element in their structure, four carbon rings. What separates one steroid from another are the various side chains that can be attached to these rings (**Figure 3.12**). When you see how different steroids are structurally from the triglycerides, you can understand why the monomers-to-polymers framework doesn't apply to lipids.

Among the most well-known steroids are cholesterol and two of the steroid hormones, testosterone and estrogen. Like fats in general, cholesterol has a bad reputation, but also like fats in general, it serves good purposes too. **Cholesterol** is a steroid molecule that forms part of the outer membrane of all animal cells and that acts as a precursor for many other steroids. One of the steroids formed from it is the principal "male" hormone testosterone; another is the principal "female" hormone estrogen. (Both hormones actually are produced in both sexes, though in differing amounts.) The term *steroids* by itself undoubtedly rings a bell because the phrase "on steroids" has come to mean artificially bulked up or supercharged. In this common usage, steroids refers to manufactured drugs that are close chemical cousins of one variety of natural steroids, the muscle-building "male" steroid hormones (**Figure 3.13**).

## A Third Class of Lipids is the Phospholipids

Our next class of lipids, phospholipids, has something of the same makeup as triglycerides in that a phospholipid has a glycerol head with fatty acids attached to it. But where triglycerides have three fatty acids stemming from the glycerol, phospholipids have only two (**Figure 3.14**). Linking up with glycerol's third −OH group is a **phosphate group**, which is a phosphorus atom surrounded by four oxygen atoms. Thus, we can define a **phospholipid** as a charged lipid molecule composed of two fatty acids, glycerol, and a phosphate group.

The combination of a phosphate group and fatty acid tails is extremely important because it gives phospholipids a dual nature: Being hydrocarbons, the fatty acid tails are hydrophobic, but the phosphate head is hydro*philic*—it will readily bond with water—because it is *charged*. In Figure 3.14, you can see the effect of this structure in solution. Imagine a phospholipid as a marker buoy in deep water. No matter how you push the hydrocarbon tail around, it's going to end up waving free *out* of the water, while the head is going to be submerged in it. You will learn more about these molecules later, when you study cells; for now, just note that the material on the periphery of cells—the outer membrane of a cell—is largely made of phospholipids. Living things need the kind of partitions described earlier, and these partitions are composed to a significant extent of phospholipids.

## A Fourth Class of Lipids Is Wax

Ever wonder why water rolls off a duck's back? It's because the duck's feathers are coated by another variety of lipid: a wax. Likewise, beehives are made largely of wax (beeswax, of course), and an apple can be polished because its outer surface has a wax coating that takes on a shine when the apple is rubbed. As these examples show, waxes often serve in nature to seal one thing off from another—the duck from water and the apple from the world around it. Waxes are extremely widespread in plants; almost all plant surfaces exposed to air will have a thin covering (called a cuticle) that is composed largely of wax (**Figure 3.15**). These coverings conserve water even as they protect the plant from outside invaders, such as fungi. Waxes are solid yet pliable at lower temperatures, but they will melt at higher temperatures. Like the triglycerides and the phospholipids we looked at, waxes are composed partly of fatty acid chains. However, whereas triglycerides have three

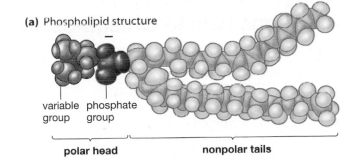

**(a)** Phospholipid structure

variable group   phosphate group

**polar head**      **nonpolar tails**

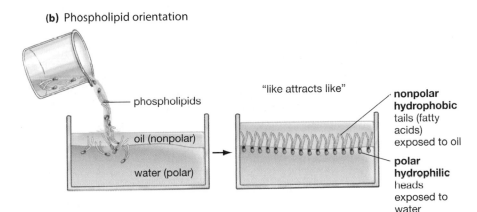

**(b)** Phospholipid orientation

phospholipids

oil (nonpolar)

water (polar)

"like attracts like"

**nonpolar hydrophobic** tails (fatty acids) exposed to oil

**polar hydrophilic** heads exposed to water

**Figure 3.14**
**A Dual-Natured Molecule**

**(a)** Phospholipids are composed of two long fatty acid "tails" attached to a "head" containing a phosphate group (which carries a negative charge) and another variable group (which often carries a charge).

**(b)** Because the head is polarized, it can bond with water and thus will remain submerged in it; the tails have no such bonding capability.

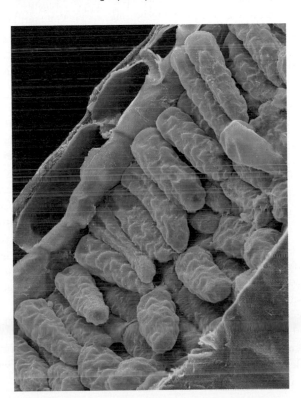

**Figure 3.15**
**A Waxy Outer Covering**

Wax can form a tight seal, which is just what plants need to help them keep water in while keeping microbial invaders out. Pictured is a gardenia leaf which, like almost all leaves, has a layer of outer, epidermal cells (colored blue) that secrete a waxy outer covering called a cuticle (colored yellow). The abundant green cells in the middle are mesophyll cells, which are the sites of photosynthesis in the leaf.

# From Trans Fats to Omega-3s: Fats and Health

**A**fter being told for years to cut down on fats in their diet, Americans got a different message in 2006: that reducing dietary fats seemingly has no effect on whether a person is likely to suffer from heart disease. This was the bottom-line result of the largest study ever conducted on the linkage between low-fat diets and heart disease risk. After eight years of following nearly 49,000 older women—some of them assigned to a low-fat diet, others not—researchers reported that women in the low-fat group suffered heart attacks at the same rate as women in the group who paid no attention to their fat consumption.

These results were immediately challenged, however, by researchers who felt that the study—a part of the national Women's Health Initiative—was answering a question that is now out of date: Is it beneficial to reduce the levels of *all* fats in the diet? The study lumped together all varieties of fats for purposes of measuring a low-fat diet, when in fact there are different kinds of fats that have long been thought to have different effects on health, some of them harmful, but others actually beneficial. Given this, the question that should have been asked, critics said, is whether it is beneficial to lower dietary levels of only two kinds of fats—saturated and trans fats—while holding constant or even increasing consumption of various kinds of unsaturated fats. Pending the outcome of further studies, most experts continue to recommend doing both things as a means of remaining as healthy as possible.

This may seem like a prime example of "easier said than done," given the seeming complexity of this issue; after all, there are trans fats, saturated fats, and several kinds of unsaturated fats to be considered. As it turns out, however, the various categories of dietary fats aren't that many in number, and it's easy to learn which foods fall into which category. Here's a list of the major types of dietary fats, ranked from the most healthful to the most harmful.

- Polyunsaturated fats containing omega-3 fatty acids
- Monounsaturated fats and polyunsaturated fats that do not contain omega-3s
- Saturated fats
- Trans fats

To understand this ranking, it's important to understand the effects these various fats have on two substances produced in the body: low-density lipoproteins or LDLs, and high-density lipoproteins or HDLs. The LDLs (or "bad cholesterol") can be thought of as capsules of protein that carry cholesterol from the liver and small intestines *to* various places in the body, including the heart. Meanwhile, HDLs (or "good cholesterol") can be thought of as protein capsules that carry cholesterol to the liver *from* the heart and other tissues, thus clearing this cholesterol from the system. It is the lodging of LDL-cholesterol molecules in the arteries of the heart that

gets the most prevalent form of heart disease going. Therefore, any food that raises LDL cholesterol levels ought to be avoided, while any food that raises HDL cholesterol levels ought to be sought out. So, where do the various forms of dietary fat rank on this and other measures of health? Here is the rundown on all of them, looking at them again from most healthy to least healthy.

## Polyunsaturated Fats Containing Omega-3s

Standing at the top of the "good fats" list are fats that contain omega-3 fatty acids. Omega-3s are one variety of polyunsaturated fatty acid—fatty acids that have two or more carbon-carbon double bonds somewhere in their hydrocarbon chains. Now, the far end of the hydrocarbon chain (away from the glycerol head) is called the *omega end*. When a carbon double bond is located three carbons up from the omega end, the result is an omega-3 fatty acid. Polyunsaturated fats containing these fatty acids guard against blood clot formation, they reduce fat levels generally in the bloodstream, and they reduce the growth of the fatty deposits that clog heart arteries. Omega-3 fatty acids are found most abundantly in certain kinds of fatty fish, including salmon, albacore tuna, mackerel, lake trout, and sardines; and in lesser amounts in plant-based foods such as walnuts and the soybean-based product tofu.

## Monounsaturated Fats and Other Polyunsaturated Fats

Monounsaturated fats, which are mostly found in oils, leave both LDL and HDL levels unchanged. However, one particular source of monounsaturated fats may actually work to prevent heart disease. Several studies have shown that residents of Mediterranean countries have significantly lower levels of heart disease than people in other parts of the developed world. What these residents have in common, in terms of diet, is consumption of large amounts of the monounsaturated fats in olive oil. In 2005, researchers uncovered a reason that olive oil may be beneficial: It contains a naturally occurring substance (called oleocanthal) that stands to work against heart disease by cutting down on the inflammation that helps it get going. Other sources of monounsaturated fats are canola oil, peanut oil, and avocados.

Polyunsaturated fats that do not contain omega-3s leave LDL levels unchanged, although they do slightly lower the "good" HDL cholesterol levels. Even so, most experts regard them, at worst, as neutral to health. This type of polyunsaturated fat can be found in safflower and corn oil.

## Saturated Fats

No one claims that saturated fats are good for your heart; the question is whether they are bad for it. Most studies have found that saturated fats raise LDL levels, with each 1 percent increase in these fats leading to a 2 percent increase in LDL cholesterol levels. Still, it's unclear what to make of this evidence in light of the Women's Health Initiative study. In it, the women who were on the low-fat diet had hardly any reduction in their LDL levels (and no reduction in their heart disease risk). The usual source for saturated fats is animal fat—primarily fatty meat and dairy products—but they can also be found in a few plant products, such as coconut oil and cocoa butter.

## Trans Fats

Trans fats stand apart from all the other fats we've looked at in that they are produced not by nature, but by an industrial process, called *hydrogenation*, in which hydrogen is bubbled into mono- or polyunsaturated oils. Food manufacturers do this for several reasons: It turns these oils into fats at room temperature, it often gives them a creamy texture that consumers like, and it increases the shelf life of the foods that result. The problem is that hydrogenation changes the chemical structure of fatty acids in ways that make them unhealthful. In most instances, this process has the effect of changing the *orientation* of the hydrogen atoms that already exist at carbon-carbon double bonds. Normally these pairs of hydrogens are on the same side, like this:

cis form (hydrogen atoms on same side of chain)

With so-called partial hydrogenation, they can end up on the opposite or *trans* sides, like this:

trans form (hydrogen atoms on opposite sides of chain)

Thus, we get the term *trans* fatty acids. The trans fats that result from this change are found in a dwindling, but still significant, number of packaged and fast foods: cookies, French fries, cakes, crackers, doughnuts, popcorn, and candy, to name a few. Trans fats earned their reputation as the least healthy fats by having a twin set of bad effects: They raise LDL levels while lowering HDL levels. In addition, they appear to boost fat levels in general in the blood while impairing the ability of blood vessels to open or "dilate."

How can you know how much trans fat you are consuming? For packaged foods, it's easy. Beginning in 2006, the federal Food and Drug Administration (FDA) required food producers to list trans fats separately on the Nutrition Facts labels seen on all packaged products. Check these labels to find out the amount of trans fats you're getting. Chances are, however, that you won't be getting much in the future. The FDA labeling requirement produced a surge of activity among packaged producers to rid their products of trans fats. Hence, what you'll often see on the Nutrition Facts labels these days is "Trans fat 0 g," meaning zero grams of trans fat.

fatty acids linked to a small alcohol (glycerol), a **wax** can be defined as a lipid composed of a single fatty acid linked to a long-chain alcohol.

### SO FAR...

1. The defining characteristic of lipids is that they do not _____ in _____.

2. The fats in foods are composed mostly of triglycerides, which are formed from three _____ bonded to a _____. A saturated fatty acid is one that has no double bonds between the _____ of its _____ chain.

3. Steroids are a class of lipids with a central structure that includes four _____. Examples include _____, which helps make up the outer lining of animal cells, and the hormones _____ and _____. Another class of lipids is important in making up the outer membrane of cells; these are the _____, composed of glycerol, two fatty acids, and a _____ group.

## 3.5 Proteins

Living things must accomplish a great number of tasks just to get through a day, and the diverse biological molecules you've been looking at allow this to happen. You've seen carbohydrates do some things, and you've seen lipids do some more, but in the range of tasks that molecules accomplish, proteins reign supreme. Witness the fact that almost every chemical reaction that takes place in living things is hastened—or, in practical terms, *enabled*—by a particular kind of protein called an enzyme. These molecules function in nature like some vast group of tools, each taking on a specific chemical task. Accordingly, an animal cell might contain up to 4,000 different types of enzymes.

We might marvel at proteins solely because of what enzymes can do, but the amazing thing is that enzymes are only *one class* of proteins. Proteins also form the scaffolding, or structure, of a good deal of tissue; they're active in transporting molecules from one site to another; they allow muscles to contract and cells to

••• ANSWERS

1. readily dissolve; water

2. fatty acids; glycerol; carbon atoms; hydrocarbon

3. carbon rings; cholesterol; estrogen; testosterone; phospholipids; phosphate

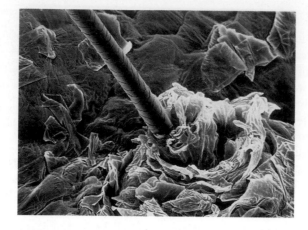

move; some hormones are made from them. If you factor out water, they account for about half the weight of the average cell. In short, it's hard to overestimate the importance of these molecules (**Figure 3.16**). **Table 3.3** lists some of the different kinds of proteins.

## Proteins are Made from Chains of Amino Acids

Proteins are prime examples of the building-block type of molecule described earlier. The monomers in this case are called amino acids. String a minimum number of them together in a chain—some say 10, some say 30—and you have a **polypeptide**, defined as a series of amino acids linked in linear fashion. When the polypeptide chain *folds up* in a specific three-dimensional manner, you have a **protein**, de-

fined as a large, folded chain of amino acids. As a practical matter, proteins are likely to be made of hundreds of amino acids strung together and folded up. As you'll see, it's not unusual for two or more polypeptide chains to be part of a single protein.

**Figure 3.17a** gives the fundamental structural unit for amino acids, followed by a couple of examples. You can see carbon at the center of this unit, an amino group off to its left, and a carboxyl group to its right. What differentiates one amino acid from another is the group of atoms that occupies the R or "side-chain" position. In **Figure 3.17b**, you can see examples of the actual amino acids tyrosine and glutamine with their different occupants of the R position.

## A Group of Only 20 Amino Acids is the Basis for all Proteins in Living Things

Although only two examples are shown here, it is a group of 20 amino acids that is the basis for all proteins in living organisms. The thousands of proteins that exist can be made from a mere 20 amino acids because these amino acids can be strung together in different *orders*. Substitute an alanine here for a glutamine there, and you've got a different protein. In this, amino acids commonly are compared to letters of the alphabet. In English, substituting one letter can take us from *bat* to *hat*. In the natural world, 20 amino acids can be put together in different order to create a multitude of proteins, each with a different function.

### Table 3.3 Types of Proteins

| Type | Role | Examples |
| --- | --- | --- |
| Enzymes | Quicken chemical reactions | Sucrase: Positions sucrose (table sugar) in such a way that it can be broken down into component parts of glucose and fructose |
| Hormones | Chemical messengers | Growth hormone: Stimulates growth of bones |
| Transport | Move other molecules | Hemoglobin: Transports oxygen through blood |
| Contractile | Movement | Myosin and actin: Allow muscles to contract |
| Protective | Healing; defense against invader | Fibrinogen: Stops bleeding<br>Antibodies: Kill bacterial invaders |
| Structural | Mechanical support | Keratin: Hair<br>Collagen: Cartilage |
| Storage | Stores nutrients | Ovalbumin: Egg white, used as nutrient for embryos |
| Toxins | Defense, predation | Bacterial diphtheria toxin |
| Communication | Cell signaling | Glycoprotein: Receptors on cell surface |

The stringing together of amino acids happens in a regular way: The carboxyl group of one amino acid joins to the amino group of another. Look at **Figure 3.18** to see how three amino acids come together.

## Shape is Critical to the Functioning of all Proteins

As a protein's amino acids are strung together in sequence, all the kinds of chemical forces discussed in Chapter 2 begin to work on them. With this, the chain begins to twist, turn, and fold into its unique three-dimensional shape. It turns out that, in the functioning of proteins, this shape, or protein conformation, is utterly crucial. Here's an example of why. If you look at the top illustration in **Figure 3.19**, on the left you can see a computerized model of a protein found on the surface of the influenza virus. If you look on the right, you will see a model of an

immune system protein, called an antibody, that fights foreign invaders such as the influenza virus. Now, how does the antibody interact with the virus?

The linkage of several amino acids...

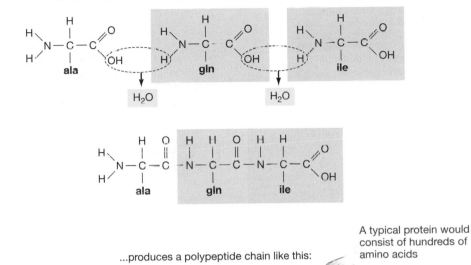

...produces a polypeptide chain like this:

A typical protein would consist of hundreds of amino acids

**Figure 3.18**
**Beginnings of a Protein**
Amino acids join together to form polypeptide chains, which fold up to become proteins. The linking of amino acids yields water as a by-product. In this figure, alanine (ala) first joins with glutamine (gln), which then is linked to isoleucine (ile). (A list of all 20 primary amino acids can be found on page 249.)

**(a)** What all amino acids have in common is an amino group and a carboxyl group attached to a central carbon.

amino group / carboxyl group

side chain

**(b)** What makes the 20 amino acids unique are the side-chains attached to the central carbon.

tyrosine

glutamine

**Figure 3.17**
**Structure of Amino Acids**

**(a)** The structural element common to all amino acids is an amino group and a carboxyl group (on the left and right, respectively) linked by a central carbon with a hydrogen attached to it. What makes one amino acid different from another is the side chain of atoms that occupies the R position.

**(b)** Two examples of actual amino acids; the differing occupants of the R position give us the amino acids tyrosine and glutamine.

(a)

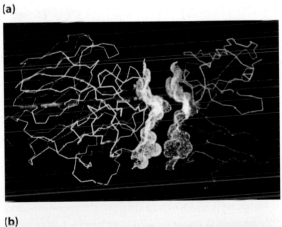

(b)

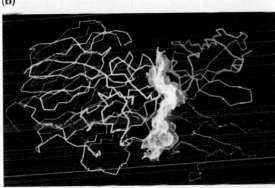

**Figure 3.19**
**Hand-in-Glove Fit**

Molecular shape is a critical element in the functioning of most proteins.

**(a)** The computer-model on the left is of a molecule found on the surface of the influenza virus. On the right is a model of a portion of human protein, called an antibody, that attacks invaders such as viruses.

**(b)** The antibody helps disable the virus by binding with it, as shown. The antibody is able to carry out this binding because it has a shape that is complementary to that of the virus molecule.

## Four Levels of Structure In Proteins

**(a)** Primary structure

The primary structure of any protein is simply its sequence of amino acids. This sequence determines everything else about the protein's final shape.

**(b)** Secondary structure

Structural motifs, such as the corkscrew-like alpha helix, beta pleated sheets, and the less organized "random coils" are parts of many polypeptide chains, forming their secondary structure.

**(c)** Tertiary structure

These motifs may persist through a set of larger-scale turns that make up the tertiary structure of the molecule.

**(d)** Quaternary structure

Several polypeptide chains may be linked together in a given protein, in this case hemoglobin, with their configuration forming its quaternary structure.

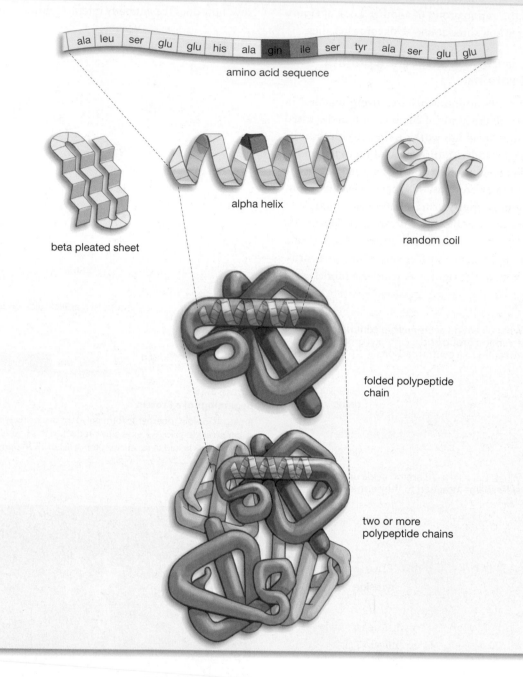

| ala | leu | ser | glu | glu | his | ala | gin | ile | ser | tyr | ala | ser | glu | glu |

amino acid sequence

beta pleated sheet

alpha helix

random coil

folded polypeptide chain

two or more polypeptide chains

**Figure 3.20
Four Levels of Structure in Proteins**

It binds to it as shown in the lower illustration, and this process is made possible by the exact fit of antibody to invader. Look at the shape of the two molecules to see how closely they conform to one another. The binding of an antibody to an invader sounds important enough, but recall that proteins are the chemical enablers called enzymes, and they're transport molecules, and so forth. In all these functions, shape is critical. The architect Frank Lloyd Wright had a famous dictum about his designs: "Form follows function." With proteins, we can turn this around and say that function follows form.

### There are Four Levels of Protein Structure

So, what forms do proteins take? Well, the answer to this depends on what vantage point you adopt in looking at them, as there are four levels of structure in proteins. You can see all four levels in **Figure 3.20**. The first of these, the **primary structure** of a protein, is simply its sequence of amino acids. Everything about the final shape of a protein is dictated by this sequence. Electrochemical bonding and repulsion forces act on this structure, and the result is the folded-up protein.

As it turns out, when these forces begin to operate on the amino acid sequence, a couple of common shapes begin to emerge in the **secondary structure** of proteins, defined as the structure that proteins assume after folding up. The *alpha helix*, a common secondary structure of proteins, has a shape much like a corkscrew. Another common secondary structure in proteins, the *beta pleated sheet*, takes a form like the folds of an accordion. Proteins can be made almost entirely of alpha helices. This is the case with hair, nails, horns, and the like. Likewise, proteins can be made entirely of beta pleated sheets. The most familiar example of this is silk, in which the beta sheets lie pancake-style on one another. Often, however, alpha helices and beta pleated sheets form what we might think of as design motifs within a larger protein structure; they periodically give way to the less-regular segments called random coils. The larger-scale three-dimensional shape that a protein takes is its **tertiary structure**. The way in which *two or more* polypeptide chains come together to form a protein results in that protein's **quaternary structure**.

### Proteins Can Come Undone

As noted, proteins fold up into a precise conformation in order to function. However, proteins can *lose* their shape and thus their functionality. You saw an example of this last chapter in connection with pH and enzymes. In the wrong pH environment, an enzyme can unfold, losing its tertiary structure and thus losing its ability to accelerate a chemical process. Alcohol works as a disinfectant on skin because it *denatures* or alters the shape of the proteins of bacteria.

### Lipoproteins and Glycoproteins

Some molecules in living things are hybrids or combinations of the various types of molecules you've been looking at. **Lipoproteins**, as their name implies, are molecules that are a combination of lipids and proteins. Active in transporting fats throughout the body, lipoproteins are transport molecules that amount to a capsule of protein surrounding a globule of fat.

Two kinds of lipoproteins have managed to enter public consciousness despite the handicap of having long names. These are high-density lipoproteins and low-density lipoproteins, also known as HDLs and LDLs, which were mentioned earlier in connection with dietary fats. What makes them more or less dense is the ratio of protein to lipid in them; lipid is less dense than protein, so a low-density lipoprotein has a relatively large amount of lipid in comparison to protein. The LDLs have acquired a reputation as villains because they carry cholesterol *to* outlying tissues, including the coronary arteries of the heart, where this cholesterol may come to reside, thickening eventually into "plaques" that can block coronary arteries and bring about a heart attack. The HDLs, meanwhile, are regarded as the cavalry; they carry cholesterol *away* from outlying cells to the liver. A high proportion of HDLs in relation to cholesterol is predictive of keeping a healthy heart. (For more on the health effects of these molecules, see "From Omega-3s to Trans Fats" on page 52.)

**Glycoproteins** are combinations of proteins and *carbohydrates*. Where do we find these molecules? One place is the surface of cells, which usually are peppered with a profusion of antenna-like structures, called *receptors,* that allow a cell to receive signals from outside itself. A typical receptor is likely to be composed of both protein and carbohydrate, with the protein forming the "stem" of the receptor and the carbohydrate forming side chains that extend from the stem and serve as the actual binding sites for signaling molecules that may pass by. Some hormones are glycoproteins, along with many other proteins released from cells.

## 3.6  Nucleic Acids

Nucleic acids are the last major type of biological molecule we'll study. The building blocks for them, called nucleotides, can be important molecules in and of themselves, as you'll see in Chapter 6 when we look at an energy-transferring nucleotide called ATP (adenosine triphosphate). Here, however, we'll look primarily at just one nucleic acid, which has the distinction as perhaps the most famous biological molecule of them all.

### DNA Provides Information for the Structure of Proteins

You've learned that proteins perform a large number of biological functions, and that one class of proteins, the enzymes, may be represented with up to 4,000 types in a single animal cell. If you had a factory that turned out 4,000 different kinds of tools, you would obviously need some direction on how each of these tools was to be manufactured: This part of the tool goes here first, and then that goes there, and so on. There is a molecule that in essence provides this kind of information for the construction of proteins. It is **DNA**, or **deoxyribonucleic acid**: the primary information-bearing molecule of life, composed of two linked chains of nucleotides. How many nucleotides? In human beings, about 3 billion of them

# CHAPTER 3 REVIEW

## SUMMARY

### 3.1  Carbon Is Central to the Living World

- Carbon is a central element to life because most biological molecules are built on a carbon framework. Carbon plays this central role because its outer shell has only four of the eight electrons necessary for maximum stability in most elements. Carbon atoms are thus able to form stable, covalent bonds with a wide variety of atoms, including other carbon atoms. The complexity of living things is facilitated by carbon's linkage capacity.

  Web Animation 3.1: The Chemistry of Carbon

### 3.2  Functional Groups

- Groups of atoms known as functional groups can confer special properties on carbon-based molecules. For example, the addition of an –OH group to a hydrocarbon molecule always results in the formation of an alcohol. Functional groups often impart an electrical charge or polarity onto molecules, thus affecting their bonding capacity.

### 3.3  The Molecules of Life: Carbohydrates

- Carbohydrates are formed from the building blocks or monomers of simple sugars, such as glucose. These monomers can be linked to form larger carbohydrate polymers, which are known as polysaccharides or complex carbohydrates. Four polysaccharides are critical in the living world. These are starch, the nutrient storage form of carbohydrates in plants; glycogen, the nutrient storage form of carbohydrates in animals; cellulose, a rigid, structural carbohydrate found in the cells walls of many organisms; and chitin, a tough carbohydrate that forms the external skeleton of arthropods.

  Web Animation 3.2: Monomers and Polymers
  Web Animation 3.3: Carbohydrates

### 3.4  Lipids

- The defining characteristic of all lipids is that they do not readily dissolve in water. Lipids do not possess the monomers-to-polymers structure seen in other biological molecules; no one structural element is common to all lipids.

- Among the most important lipids are the triglycerides, composed of a glyceride and three fatty acids. Most of the fats that human beings consume are triglycerides. Another important variety of lipids is the steroids, all of which have a core of four carbon rings. Examples include cholesterol and such hormones as

testosterone and estrogen. A third class of lipids is the phospholipids, each of which is composed of two fatty acids, glycerol, and a phosphate group. The material forming the outer membrane of cells is largely composed of phospholipids. A fourth class of lipids is the waxes, each of which is composed of a single fatty acid linked to a long-chain alcohol. Waxes have an important "sealing" function in the living world. Almost all plant surfaces exposed to air, for example, have a protective covering made largely of wax.

Web Animation 3.4: Lipids

### 3.5  Proteins

- Proteins are an extremely diverse group of biological molecules composed of the monomers called amino acids. Sequences of amino acids are strung together to produce polypeptide chains, which then fold up into working proteins. Important groups of proteins include enzymes, which hasten chemical reactions, and structural proteins, which make up such structures as hair.

- The primary structure of a protein is its amino acid sequence; this sequence determines a protein's secondary structure—the form a protein assumes after having folded up. The larger-scale three-dimensional shape that a protein assumes is its tertiary structure, and the way two or more polypeptide chains come together to form a protein results in that protein's quaternary structure. The activities of proteins are determined by their final folded shapes.

- Lipoproteins are biological molecules that are combinations of lipids and proteins. High-density and low-density lipoproteins (HDLs and LDLs, respectively), which transport cholesterol in human beings, are important determinants of human heart disease.

- Glycoproteins are combinations of carbohydrates and proteins. The signal-receiving receptors found on cell surfaces often are glycoproteins.

Web Animation 3.5: Proteins

### 3.6  Nucleic Acids

- Nucleic acids are polymers composed of nucleotides. The nucleic acid DNA (deoxyribonucleic acid) is composed of nucleotides that contain a sugar (deoxyribose), a phosphate group, and one of four nitrogen-containing bases. DNA is a repository of genetic information. The sequence of its bases encodes the information for the production of the huge array of proteins produced by living things. A second nucleic acid is RNA (ribonucleic acid), which transports the information encoded in DNA to the sites of protein synthesis—structures called ribosomes—and which helps make up the structure of ribosomes.

Web Animation 3.6: Nucleic Acids

## KEY TERMS

| | |
|---|---|
| carbohydrate 44 | phosphate group 51 |
| cellulose 46 | phospholipid 51 |
| chitin 46 | polymer 44 |
| cholesterol 50 | polypeptide 54 |
| deoxyribonucleic acid (DNA) 57 | polysaccharide 45 |
| fatty acid 48 | polyunsaturated fatty acid 48 |
| functional group 43 | primary structure 56 |
| glycogen 46 | protein 54 |
| glycoprotein 57 | quaternary structure 57 |
| hydrocarbon 42 | ribonucleic acid (RNA) 58 |
| lipid 47 | saturated fatty acid 48 |
| lipoprotein 57 | secondary structure 57 |
| monomer 44 | simple sugar 44 |
| monosaccharide 44 | starch 45 |
| monounsaturated fatty acid 48 | steroid 50 |
| nucleotide 59 | tertiary structure 57 |
| oil 49 | triglyceride 48 |
| organic chemistry 42 | wax 53 |

## BRIEF REVIEW

*Answers to Brief Review questions are in the back of the book. For multiple-choice quiz questions, go to* **www.aw.com/krogh4**.

1. Why is it accurate to look at life as "carbon based," and what quality gives carbon this special status?

2. Both low-density lipoproteins (LDLs) and high-density lipoproteins (HDLs) are involved with carrying fats through the bloodstream. If your LDL count is unusually high, should you be concerned? What if your HDL count is high? Why are they different?

3. What are steroids, as that term is defined in biochemistry? What are steroids in the popular sense of the term?

4. List as many functions of proteins as possible. Why are proteins able to do so many different types of jobs?

5. Why is it accurate to think of DNA as an information-bearing molecule?

## APPLYING YOUR KNOWLEDGE

1. Scientists who look for life outside Earth have a favorite candidate for an element that could take on the role that carbon does here on Earth. The element is silicon, which has 14 electrons, compared to carbon's 6. Based on the rules of chemical bonding you learned about last chapter, why should silicon's electron number make it a candidate for a central biological element?

2. One piece of advice that nutritionists give today is to favor complex carbohydrates, such as those found in whole-grain bread, over simple carbohydrates, such as those found in soft drinks. One reason for this is that simple sugars result in "spikes" in blood sugar levels, while some complex carbohydrates do not. Why should the two kinds of carbohydrates differ in this regard?

3. Many species of beans produce a poisonous group of glycoproteins called lectins, which cause red blood cells to clump together (agglutinate) and cease to function. (Fortunately, cooking destroys most of them.) Using what you learned about the structure and function of glycoproteins, suggest how they might do this. What benefit do you think plants might get just by causing red blood cells to clump together?

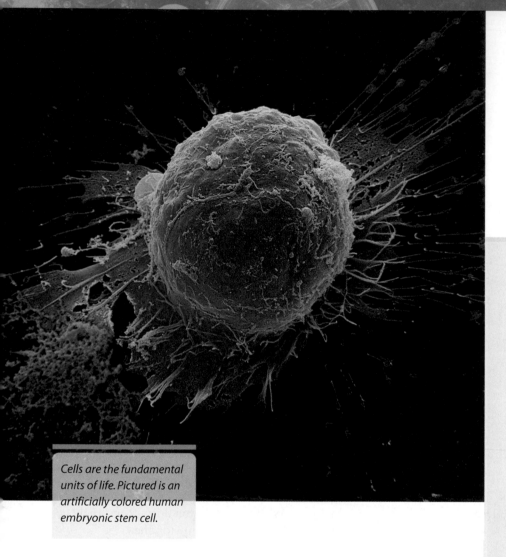

CHAPTER **4**

Cells are the fundamental units of life. Pictured is an artificially colored human embryonic stem cell.

All life exists within cells. These tiny entities can be compared to factories whose products maintain life.

**4.1**   Cells are the Fundamental Units of Life   **63**

**4.2**   All Cells Are Either Prokaryotic or Eukaryotic   **64**

**4.3**   The Eukaryotic Cell   **65**

**4.4**   A Tour of the Animal Cell: Along the Protein Production Path   *Bio***Flix**   **69**

**4.5**   Outside the Protein Production Path: Other Cell Structures   **74**

**4.6**   The Cytoskeleton: Internal Scaffolding   **76**

**4.7**   The Plant Cell   *Bio***Flix**   **80**

**4.8**   Cell-to-Cell Communication   **84**

**Essays**

The Size of Cells   **66**

The Stranger Within: Endosymbiosis   **78**

**THE PROCESS OF SCIENCE:** First Sightings: Anton van Leeuwenhoek   **86**

# LIFE'S HOME:
## The Cell

**T**he United States Senate has 100 members in it, and all of them do occasionally gather to consider given issues. But if you want to know how the Senate actually gets something *done*, you must look else-where. The business of the nation is simply too complex for all Senate members to deal with all issues. Instead, issues are dealt with one at a time within the specialized working units of the Senate known as committees.

Something similar to this goes on in the natural world. If we look at birds, trees, or people and ask how any of these living things actually gets something done, the answer again is through working units—in this case the working units known as cells.

## 4.1 Cells Are the Fundamental Units of Life

Muffins are in the oven, and you are in the living room. Gas molecules from the baking muffins waft into the living room, and some of them happen to make their way to your nose, there to travel a short distance to your upper nasal cavity, and land on a set of ceaselessly waving hair-like projections called cilia. These actually are the extensions of some specialized nerve cells (**Figure 4.1** on the next page). If enough muffin molecules bind with enough of the cilia, an impulse is passed along (through other nerve cells) to trigger not only the *sensation* of smell, but also the *association* of this smell with muffins you have smelled in the past. How do we know muffins are in the oven? Through cells. How do we move our hands or read this page? Through cells. Life's working units are cells, and in our amazingly complex natural world, there is great specialization in them. The nerve cells in human beings are specialists in transmitting nerve

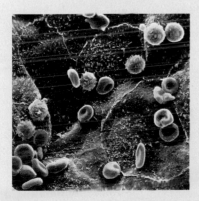
(a)

(b)

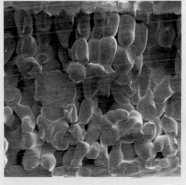
(c)

**Nature's Fundamental Unit**

**(a)** Human red and white blood cells inside a blood vessel. The large, dark ovals that can be seen in the background are flat cells that form the interior lining of the blood vessel.

**(b)** Cells of the fungus brewer's yeast, which is used to make wine, beer, and bread.

**(c)** Cells of a spinach leaf, with the leaf seen in cross section. Two layers of outer or "epidermal" cells can be seen at top and bottom; many of the cells in between perform photosynthesis.

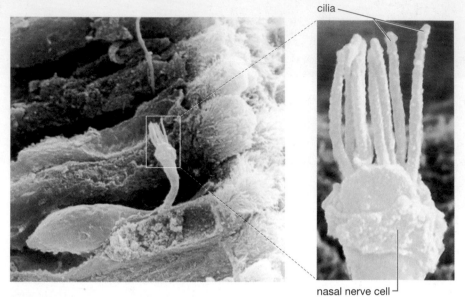

cilia

nasal nerve cell

**Figure 4.1**
**Cells Can Specialize**

In more complex organisms, different cells carry out different functions. In the picture at left, you can see one type of cell—a nerve cell, in this case located in the lining of the nose and surrounded by gray accessory cells. A closer look at one of these nerve cells, in the picture at right, shows a number of hair-like extensions, called cilia, protruding from it. When we smell muffins in the oven, gas molecules that waft off the muffins bind with the cilia, which sets in motion a nerve impulse to the brain about the muffins.

signals, red blood cells are specialists in transporting oxygen, and muscle cells are specialists in contracting.

Running like a thread through this diversity, however, is the unity of cellular life. Every form of life either is a single cell or is composed of cells. The one possible exception to this is viruses, but even they must use the machinery of cells to reproduce. There is unity, too, in the way cells come about: Every cell comes *from* a cell. Human beings are incapable of producing cells from scratch in the laboratory, and so far as we can tell, nature has fashioned cells from simple molecules only once—back when life on Earth got started. The fact that all cells come from cells means that each cell in your body is a link in a cellular chain that stretches back more than 3.5 billion years.

## 4.2  All Cells Are Either Prokaryotic or Eukaryotic

So, what are these tiny working units we call cells? It follows, from the variety of cells that exist, that there is no such thing as a "typical" cell. There are, however,

**Figure 4.2**
**Prokaryotic and Eukaryotic Cells Compared**

A prokaryotic cell is a self-contained organism since the prokaryotes—bacteria and archaea—are all single-celled. Eukaryotic organisms, which may be single- or multicelled, include plants, animals, fungi, and another group called protists.

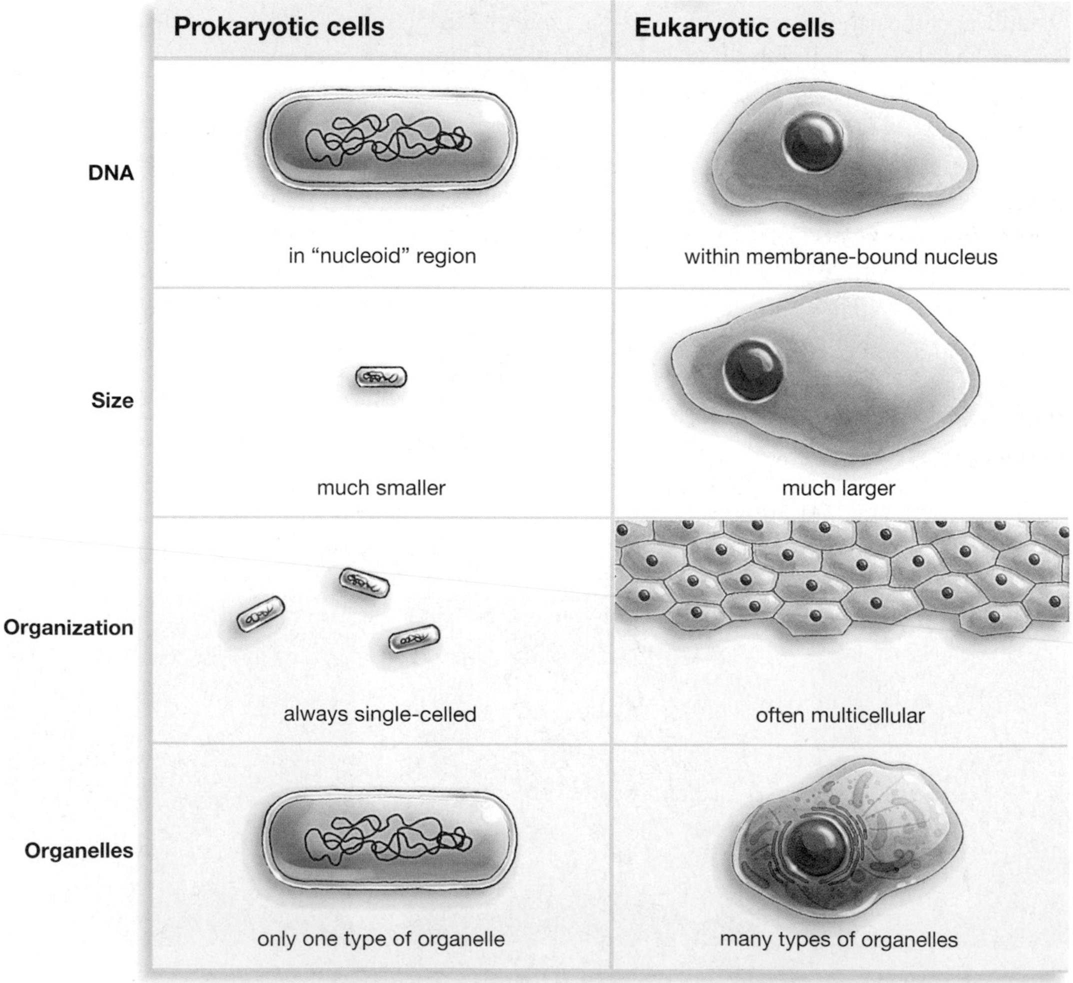

| | Prokaryotic cells | Eukaryotic cells |
|---|---|---|
| **DNA** | in "nucleoid" region | within membrane-bound nucleus |
| **Size** | much smaller | much larger |
| **Organization** | always single-celled | often multicellular |
| **Organelles** | only one type of organelle | many types of organelles |

certain *categories* of cells that are important, and the two most important of these are prokaryotic and eukaryotic cells. Every cell that exists is one or the other, and this simple either-or quality extends to the organisms that fall into these camps. All prokaryotic cells either are bacteria or another microscopic form of life known as archaea. Setting bacteria and archaea aside, *all other cells* are eukaryotic. This means all the cells in plants, in animals, in fungi—all the cells in every living thing except in the bacteria and the archaea.

## Prokaryotic and Eukaryotic Differences

The name *eukaryote* comes from the Greek *eu*, meaning "true," and *karyon*, meaning "nucleus," while *prokaryote* means "before nucleus." These terms describe the most critical distinction between the two cell types. **Eukaryotic cells** are cells whose primary complement of DNA is enclosed within a nucleus. **Prokaryotic cells** are cells whose DNA is not enclosed within a nucleus. To complete this circle, a **nucleus** is a membrane-lined compartment that encloses the primary complement of DNA in eukaryotic cells.

While having a nucleus is the most important difference between eukaryotic and prokaryotic cells, it is not the only difference. It may seem sensible to think of two single-celled creatures—one a prokaryote, the other a eukaryote—as very similar, but the distance between them as life-forms is immense. Human beings and chimpanzees are nearly identical in comparison. Eukaryotic cells tend to be much larger than their prokaryotic counterparts; indeed, thousands of bacteria could easily fit into an average eukaryotic cell (see "The Size of Cells" on page 66). Eukaryotes are often multicelled organisms, while prokaryotes are always single-celled (**Figure 4.2**). Beyond this, eukaryotic cells are *compartmentalized* to a far greater degree than is the case with prokaryotes. The nucleus in eukaryotic cells turns out to be only one variety of **organelle**: a highly organized structure, internal to a cell, that serves some specialized function. Eukaryotic cells contain several different kinds of these "tiny organs." There are organelles called mitochondria, for example, that transform energy from food, and organelles called lysosomes that recycle the raw materials of the cell. In prokaryotic cells, meanwhile, there is only a single type of organelle—a kind of workbench for producing proteins we'll look at later.

If we look at eukaryotes that are familiar to us, such as trees, mushrooms, and horses, it's easy to see that there is a fantastic diversity in the *forms* of eukaryotes compared to the strictly single-celled prokaryotes. It does not follow from this, however, that prokaryotes are uniform or that they are unsuccessful. On the contrary, prokaryotes differ greatly from one another in, say, the way they obtain nutrients, and they are extremely successful, if success is defined as living in a lot of places in huge numbers. As the biologists Lynn Margulis and Karlene Schwartz have observed, more bacteria are living in your mouth right now than the number of people who have ever existed. But for the range of biological structures and processes we want to look at, eukaryotic cells are more diverse and hence are the initial focus of our study.

## 4.3   The Eukaryotic Cell

What are the constituent parts of eukaryotic cells? Here are five larger structures that in turn are composed of component parts. The first two have been mentioned already (**Figure 4.3**):

- The cell's nucleus

- Other organelles (which lie outside the nucleus)

- The **cytosol**, a protein-rich, jelly-like fluid in which the cell's organelles are immersed

- The *cytoskeleton*, a kind of internal scaffolding consisting of three sorts of protein fibers; some of these can be likened to tent poles, others to monorails

- The **plasma membrane**, the outer lining of the cell

**Figure 4.3**
**Eukaryotic Cell**

This cutaway view of a eukaryotic cell displays the elements that nearly all such cells possess: a nucleus, other membrane-bound organelles, a jelly-like cytosol, a cytoskeleton, and an outer plasma membrane.

**Components of eukaryotic cells**

nucleus

other organelles

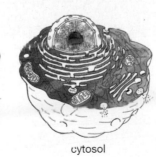
cytosol

cytoskeleton

plasma membrane

# The Size of Cells

**F**or several chapters, you've been going over atoms, ions, molecules, and such—things small enough to be invisible to the unaided eye. For the most part, you've had to imagine what these things look like, simply because most of them are so small that we either have no pictures of them at all (as with electrons) or have few clear pictures (as with atoms). If you flip through the pages of this chapter and those to come, however, you begin to see a fair number of actual photographs. They don't have the same quality as summer vacation snapshots, but they are recognizable as pictures, or more properly as **micrographs**, meaning pictures taken with the aid of a microscope. Micrographs enable us to see surprising things; for example, hundreds of bacteria on the tip of a pin (**Figure 1**). So, with the cells that are introduced in this chapter, there has obviously been a bump up in size into a world that is more easily visible with the help of various kinds of technology.

## How Small Are They?

In taking stock of the microworld, two units of measure are particularly valuable. The smaller of them is the **nanometer**, abbreviated as nm; the larger is the **micrometer**, abbreviated μm (The unfamiliar-looking first letter is the Greek symbol for a small *m*. Scientists are not trying to be purposely obscure in using it; another unit of the metric system, the millimeter, lays claim to the mm

abbreviation.) What these abbreviations stand for is a billionth of a meter (nm) and a millionth of a meter (μm) A meter equals about 39.6 inches, or just over a yard, which gives you some starting sense of physical reality in understanding the rest of these sizes.

Now look at **Figure 2** to see what size various objects are. Atoms are at the bottom of the scale, at about a tenth of a nanometer. Something less than a 10-fold increase gets us to the size of the protein building blocks, called amino acids, that you looked at in Chapter 3. Another 10-fold-plus increase, and you've reached the upper limit on proteins.

You have to go better than 10 times larger than this, however, before you arrive at the size of something that is actually *living* as opposed to something that is a component part of life. Because, as you've seen, life means cells, this means we're talking about the smallest cells in existence, which are bacteria measuring perhaps 200–300 nm. This extreme on the small side of cells has a counterpart on the large side with the single cells we call chicken or ostrich eggs and with certain nerve cells that can stretch out to a meter in length. In general, however, we just cross into the micrometer range with the smaller bacterial cells: about 1–10μm. The cell size for most plants and animals falls in a range that is a little less than 10 times larger than this, about 10–30μm. At 10μm perhaps a billion cells could fit into the tip of your finger. And, how about in your whole body? A standard estimate of the actual number of human cells is 10 trillion.

**(a)** Bacteria on a pin, magnified x 85    **(b)** Magnified x 425    **(c)** Magnified x 2100

 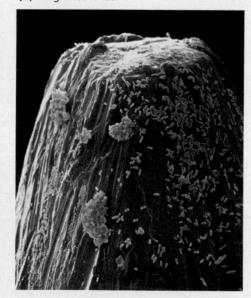 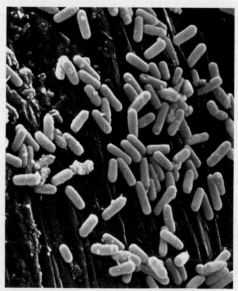

**Figure 1**
**Hidden Life**
Microscope enlargements of the tip of this pin show an abundance of life—in this case bacteria—thriving on an object that we normally think of as devoid of living organisms.

## So, Why So Small?

Having learned how small cells are, you might then well ask: *Why* are they so small? As noted, cells are small chemical factories, and just like any factory, they are constantly shipping things in and out. A primary size-limiting factor for cells is having enough surface area to export and import all that they need.

This constraint comes about because of a fundamental mathematical principle: As the surface area of an object increases, its *volume* increases even more. Say you have a cube 1 inch long on each of its sides. Its surface area is 6 square inches (Length × Width × Number of sides), while its volume is 1 cubic inch (Length × Width × Height). Now, say you increase the side dimension to 8 inches. The surface area goes from 6 to 384 square inches, but the *volume* goes from 1 to 512 cubic inches. At a 1-inch dimension, there were *six* square inches of surface area for every cubic inch of volume; now, there are only *three-quarters* of an inch of surface for every cubic inch of volume. Beyond a certain volume, then, a cell simply would not have enough surface area to import and export all the materials it needs. This effectively sets an upper limit on how big cells can be and helps explain why most of the cells in an elephant are no bigger than those in an ant, although the elephant does have more cells than the ant. **Figure 3** shows you some size comparisons in the micro-world.

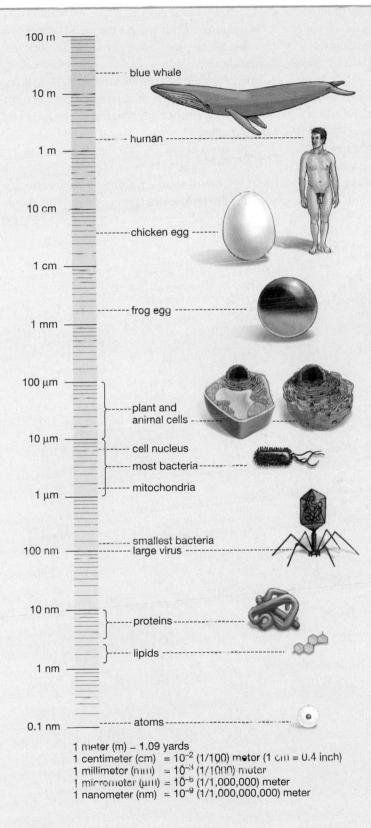

1 meter (m) = 1.09 yards
1 centimeter (cm) = $10^{-2}$ (1/100) meter (1 cm = 0.4 inch)
1 millimeter (mm) = $10^{-3}$ (1/1000) meter
1 micrometer (µm) = $10^{-6}$ (1/1,000,000) meter
1 nanometer (nm) = $10^{-9}$ (1/1,000,000,000) meter

**Figure 2**
**Little and Big**
The sizes of some selected objects in the natural world.

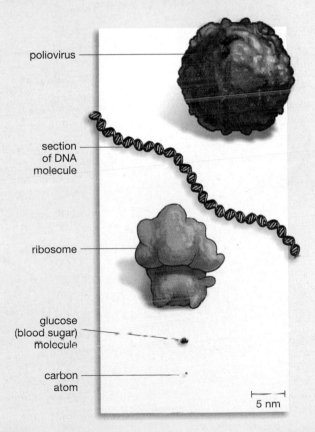

**Figure 3**
**Small Is a Relative Thing**
The sizes and shapes of five natural-world entities, each magnified a million times.

In any discussion of a cell, you are likely to hear the term **cytoplasm**, which simply means the *region* of the cell inside the plasma membrane but outside the nucleus. The cytoplasm is different from the cyto*sol*. If you removed all the structures of the cytoplasm—meaning the organelles and the cytoskeleton—what would be left is the cytosol, which is mostly water. This does not mean that the cytosol is simply a passive medium for the other structures, but it is not an organized structure in the way the organelles are. Almost all the organelles you'll be seeing are encased in their own membranes, just as the whole cell is encased in its plasma membrane. What are membranes? Until you get the formal definition next chapter, think of a membrane as the flexible, chemically active outer lining of a cell or of its compartments. In the balance of this chapter, we'll explore all of the constituent parts listed above except for the plasma membrane, which is so special it gets its own chapter.

## The Animal Cell

To get a sense of what eukaryotic cells are like, it's convenient to look at two types of them: animal cells

**Figure 4.4**
**The Animal Cell**

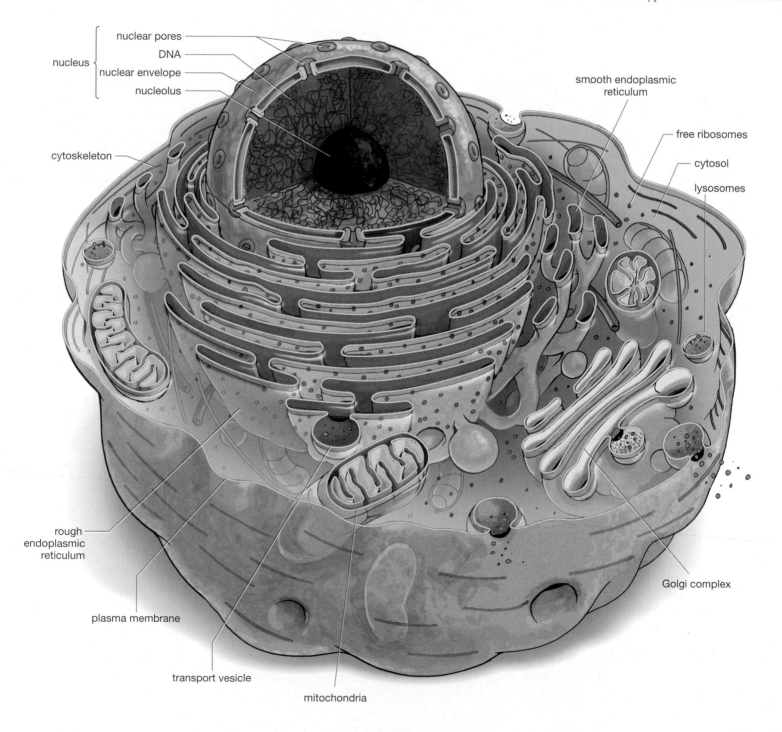

nuclear pores
DNA
nucleus
nuclear envelope
nucleolus

smooth endoplasmic
reticulum

free ribosomes

cytoskeleton

cytosol

lysosomes

rough
endoplasmic
reticulum

Golgi complex

plasma membrane

transport vesicle

mitochondria

and plant cells. These cell types have more similarities than they do differences, but they are different *enough* that it will be helpful to look at them separately. We'll start by examining animal cells and then look at how plant cells differ from them.

Insofar as we can characterize that elusive creature, the "typical" animal cell (**Figure 4.4**), it is roughly spherical, surrounded by, and linked to, cells of similar type, immersed in a watery fluid, and about 25 micrometers (μm) in diameter, meaning that about 30 of them could fit side by side within the period at the end of this sentence.

---

## SO FAR...

1. Every form of life either is a _____ or is composed of _____, each of which can arise only from _____.

2. The two most fundamentally different kinds of cells are _____ cells, each of which has its primary complement of DNA enclosed inside a membrane-lined _____; and _____ cells, with DNA that is not enclosed within this structure.

3. In a eukaryotic cell, the cytoplasm is the region that lies inside the _____ but outside the _____. The jelly-like fluid filling much of this region is called the _____, while the individual highly organized structures within it are called _____.

---

## 4.4 A Tour of the Animal Cell: Along the Protein Production Path

You are now going to take an extended tour of the animal cell, which can be thought of as a living factory. Much of the first part of your trip will be spent tracing the way this factory puts together a product—a protein—for export outside itself. Just as a new employee might tour an assembly line as a means of learning about factory equipment, so you are going to follow the path of protein production as a means of learning about cell equipment.

**Figure 4.5** shows the path you'll be taking, from nucleus to the outer edge of the cell. Don't be bothered by the unfamiliar terms in Figure 4.5 because they'll all be explained in the text. The important thing is that, before we begin our tour, you have some sense of the path that protein production takes.

### Beginning in the Control Center: The Nucleus

As noted in Chapter 3, proteins are critical working molecules in living things, and DNA contains the information for putting proteins together from a starting set of amino acid building blocks. Our entire complement of DNA is like a cookbook that contains individual protein recipes, which we call genes. A given gene's chemical sequence in effect says, "Give me this amino acid, then this one, then this one . . .," the final result being the specifications for a completed

••• ANSWERS
1. cell; cells; another cell
2. eukaryotic; nucleus; prokaryotic
3. plasma membrane; nucleus; cytosol; organelles

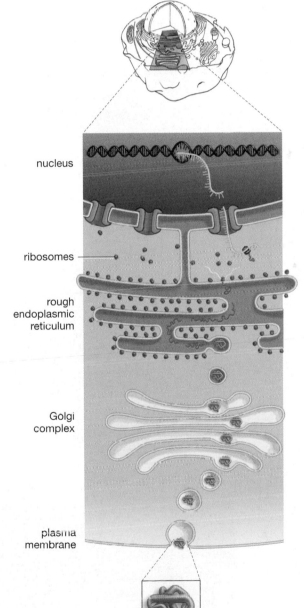

nucleus

ribosomes

rough endoplasmic reticulum

Golgi complex

plasma membrane

1. Instructions from DNA are copied onto mRNA.

2. mRNA moves to ribosome.

3. Ribosome moves to endoplasmic reticulum and "reads" mRNA instructions.

4. Amino acid chain growing from ribosome is dropped inside endoplasmic reticulum membrane. Chain folds into protein.

5. Protein moves to Golgi complex for additional processing and for sorting.

6. Protein moves to plasma membrane for export.

**Figure 4.5**
**Path of Protein Production in Cells**

protein. In the eukaryotic cell, as you've seen, DNA is largely confined within a nucleus bound by a membrane. If you look at **Figure 4.6**, you can see the nature of this membrane. It is the **nuclear envelope**: the double membrane that lines the nucleus in eukaryotic cells.

There comes a point in the life of most cells when they divide, one cell becoming two. Because (with a few exceptions) all cells must possess the set of instructions that are contained in DNA, it follows that when a cell divides, its original complement of DNA must *double*, so that both cells that result from the cell division can have their own DNA. The nucleus, then, is not just the site where DNA exists; it is the site where new DNA is put together, or "synthesized," for this doubling.

## Messenger RNA

At the end of Chapter 3, you saw that, while DNA exists in a cell's nucleus, the proteins that DNA codes for are put together in a cell's cytoplasm. Thus, the information contained in a length of DNA has to be transported from the nucleus to the cytoplasm. How does this happen? DNA's instructions are copied onto a second long-chain molecule, called RNA, and *it* ferries the DNA information out of the nucleus to the site of protein synthesis. RNA actually comes in several forms, but the form that DNA's instructions are copied onto is the well-named *messenger RNA* or mRNA for short. Given that mRNA goes to the cytoplasm, it must, of course, have some way of getting out of the nucleus; its exit points turn out to be thousands of channels that stud the surface of the nuclear envelope—the *nuclear*

**Figure 4.6**
**The Cell's Nucleus**

The DNA of eukaryotic cells is sequestered inside a compartment, the nucleus, which is lined by a double membrane known as the nuclear envelope. Compounds pass into and out of the nucleus through a series of microscopic channels called nuclear pores. The prominent spherical structure within the nucleus is the nucleolus, an area that specializes in the production of ribosomal RNA—the material that helps make up ribosomes. Protein production depends on the information encoded in DNA's sequence of chemical building blocks. This information is copied onto a length of messenger RNA (mRNA), which then exits from the nucleus through a nuclear pore. (Micrograph: ×4,400)

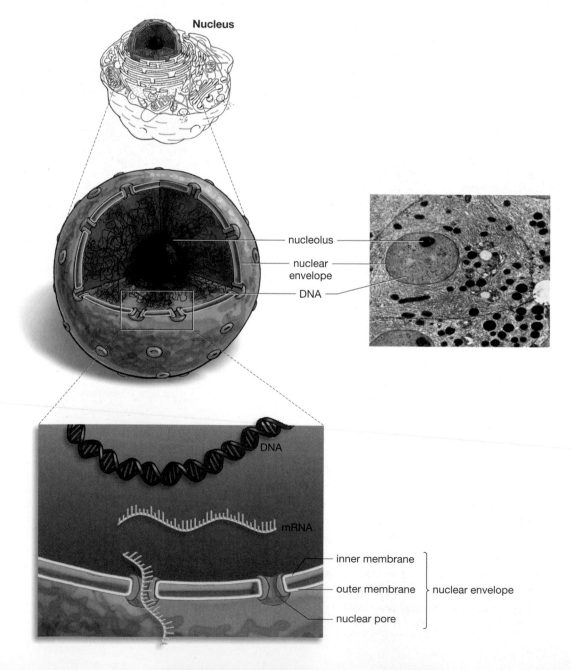

*pores.* Materials can go the other way through the nuclear pores as well; proteins and other materials pass from the cytoplasm into the nucleus by way of them.

## mRNA Moves Out of the Nucleus

Imagine shrinking in size so that the nucleus seems about as big as a house, with you standing outside it. What you would see in protein synthesis is lengths of mRNA rapidly moving out through nuclear pores and dropping off into the cytoplasm. Several varieties of RNA actually come out like this, but for now, let's just follow the trail of the mRNA. You're about ready to leave the nucleus to continue your cell tour, but before you do, note two things. First, DNA contains information for making proteins, and the mRNA chains coming out of the nuclear pores amount to a means of disseminating this information. Thus, it's not hard to conceptualize the nucleus as a control center for the cell. Beyond this, in looking at Figure 4.6, you've probably noticed that there is a rather imposing structure *within* the nucleus, called the nucleolus. For now, just *hold that thought*; the mRNA chains have come out of the nuclear pores and are making a short trip to another part of the cell.

## Ribosomes

Small structures called ribosomes are the destinations for our lengths of mRNA. For now, we can define the **ribosome** as an organelle that serves as the site of protein synthesis in the cell. If an mRNA chain amounts to a set of instructions for putting a protein together, a ribosome amounts to a machine that carries out these instructions. Each ribosome acts as a site through which an mRNA chain is run; as this happens, the ribosome "reads" the chemical sequence in the mRNA chain and starts putting together amino acids in the order specified by the mRNA sequence. The result is a chain of amino acids that grows from the ribosome—a chain that eventually will fold up into the molecule we call a protein (**Figure 4.7**). You may remember that the kind of protein we are tracking will eventually be exported out of the cell altogether. The synthesis of this kind of protein actually stops when only a very short sequence of the amino acid (or "polypeptide") chain has grown from the ribosome. Why? "Export" polypeptide chains need to be processed within other structures in the cell before they can become fully functional proteins. The first step in this process is for

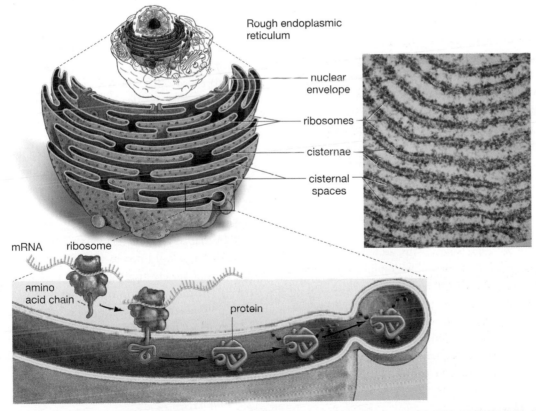

1. mRNA docks on ribosome. Amino acid chain production begins.
2. Ribosome docks on ER. Amino acid chain moves into cisternal space as it is completed.
3. Amino acid chain folds up making a protein.
4. Side chains added to protein.
5. Vesicle formed to house protein while in transport.

**Figure 4.7**
**Proteins Taking Shape: Ribosomes and the Rough Endoplasmic Reticulum**

The steps of the lower figure begin with a length of messenger RNA (mRNA), which has migrated from the nucleus of a cell.

**(1)** The mRNA sequence binds with a ribosome that is freestanding in the cytoplasm. The ribosome then starts to "read" the length of the mRNA, which bears instructions for producing a protein. The result is an amino acid (or polypeptide) chain that begins to grow from one part of the ribosome. The ribosome soon halts this activity, however, and moves toward the rough endoplasmic reticulum (rough ER).

**(2)** The ribosome docks on the outside face of the rough ER and resumes reading the mRNA sequence. When it is completed, the resulting amino acid chain drops into the cisternal space of the rough ER.

**(3)** The amino acid chain folds up, thus becoming a protein.

**(4)** The protein undergoes processing (in this example, by having a side chain added).

**(5)** The protein is then encased in a membrane-lined vesicle, which will bud off from the rough ER and be transported to the Golgi complex. (Micrograph: ×90,500)

the ribosome, and its associated cargo, to migrate a short distance in the cell and then attach to another cell structure.

### The Rough Endoplasmic Reticulum

If you look again at Figure 4.4, you can see that, though the nucleus cuts a roughly spherical figure out of the cell, there is in essence a folded-up continuation of the nuclear envelope on one side. This mass of membrane has a name that is a mouthful: the **rough endoplasmic reticulum**. This structure can be defined as a network of membranes that aids in the processing of proteins in eukaryotic cells. The rough endoplasmic reticulum is rough because it is studded with ribosomes; it is endoplasmic because it lies within (endo) the cyto*plasm*; and it is a reticulum because it is a network, which is what *reticulum* means in Latin. Understandably, it is generally referred to as the rough ER or the RER.

Our ribosome, bound with its attached mRNA and polypeptide chains, will migrate to the rough ER and dock on its outside face, thus joining a multitude of other ribosomes that have done the same thing. As noted, the polypeptide chain that is output from the ribosome is in essence an unfinished protein that needs to go through more processing before it can be exported. The first step in this processing leads only to the other side of the ER wall in which the ribosome is embedded. As the ribosome goes on with its work, the polypeptide chain it is producing drops into chambers inside the rough ER.

If you look at Figure 4.7, you can see that the entire rough ER takes the shape of a set of flattened sacs (called *cisternae*). The membrane that the rough ER is composed of forms the periphery of these sacs. Inside are the *cisternal spaces* of the rough ER. As polypeptide chains enter the cisternal spaces, they first fold up into their protein shapes, as you saw in Chapter 3. Beyond this, most proteins exported from cells have sugar side chains added to them here. Quality control of the production line is in operation in the rough ER as well. Polypeptide chains that have faulty sequences are detected in the rough ER cisternal space and ejected out of the protein production line altogether, after which they are broken down into their component parts. Other proteins will pass the quality control tests, however, and will then move out of the rough ER for more processing.

### Several Locations for Ribosomes

All of the mRNA that comes out of the nuclear pores goes to ribosomes, but only some of these ribosomes end up migrating to the rough ER. A multitude of ribosomes will remain "free ribosomes," which is to say ribosomes that remain free-standing, in the cytosol. What makes the difference? Remember how, in the ribosome we looked at, only a small stretch of the polypeptide chain it was producing emerged before the ribosome first halted its work and then migrated to the rough ER? That small stretch of the chain contained a chemical signal that said, in effect, "rough ER processing needed." The result was the ribosome's move to the rough ER. In general, these rough ER-linked ribosomes produce proteins that will reside in the cell's membranes or that will be exported out of the cell altogether (making them "secretory" proteins). Meanwhile, free ribosomes tend to produce proteins that will be used within the cell's cytoplasm or nucleus.

### A Pause for the Nucleolus

Before you continue on the path of protein processing, think back a bit to the discussion about the nucleus, when you were asked to hold the thought about the large structure *inside* the nucleus called the nucleolus. This is the point where its story can be told because now you know what ribosomes are.

It turns out that ribosomes are mostly made of RNA. (They are made of a mixture of proteins and *ribo*nucleic acid.) So great is the cell's need for ribosomes that a special section of the nucleus is devoted to the synthesis of RNA. This is the nucleolus. The *type* of RNA that's part of the ribosomes is, fittingly enough, called ribosomal RNA or rRNA, and it's one of the multiple varieties of RNA mentioned before. With this in mind, we can define the **nucleolus** as the area within the nucleus of a cell devoted to the production of ribosomal RNA.

Ribosomes are brought only to an unfinished state within the nucleolus, after which they pass through the nuclear pores and into the cytoplasm; when put together in final form there, they begin receiving mRNA chains. They are the one variety of cellular organelle that is not lined by a membrane. (They are also the one variety of organelle that prokaryotic cells have.) The cell's traffic in ribosomes is considerable; with millions of them in existence, lots are going to be wearing out all the time. Perhaps a thousand need to be replaced every minute.

### Elegant Transportation: Transport Vesicles

The proteins that have been processed within the rough ER need to move out of it and to their "downstream" destinations before being exported. But how do proteins move from one location to another within the cell? Recall that the rough ER and the nuclear membrane amount to one long, convoluted membrane. And, as just noted, all the organelles in the cell except ribosomes are "membrane-bound," meaning they have membranes at their periphery. Each of these membranes has its own chemical structure, but collectively they have an amazing ability to work together: A piece of one membrane can *bud off*, as the term goes, carrying inside it some of our proteins-in-process. Moving through the cytosol, this tiny sphere of membrane can then *fuse* with another membrane-bound organelle, releasing its protein cargo in the process. The membrane-lined spheres that move within this network, carrying proteins and other molecules, are called **transport vesicles.** The network itself is known as the **endomembrane system**: an interactive group of membrane-lined organelles and transport vesicles within eukaryotic cells.

This system gives cells a remarkable capability. One minute a piece of membrane may be an integral part of, say, the rough ER; the next, it is separating off as a spheroid vesicle and moving through the cytosol, carrying proteins within it. (These vesicles are not moving under their own power, however. As you'll see later, they are propelled along the "monorails" of the cell's internal skeleton.) Note that many different *kinds* of proteins are being processed at any one time in the rough ER cisternae. It is as if the cellular factory has a lot of different assembly lines working at once. Most of

the proteins under construction, however, are initially bound for the same place—the Golgi complex.

## Downstream from the Rough ER: The Golgi Complex

Once a transport vesicle, bearing proteins, has budded off from the rough ER, it then moves through the cytosol to fuse with the membrane of another organelle, one first noticed by Italian biologist Camillo Golgi at the beginning of the twentieth century. The **Golgi complex** is a network of membranes that processes and distributes proteins that come to it from the rough ER. Some side chains of sugar may be trimmed from proteins here, or phosphate groups may be added, but the Golgi complex does something else as well. Recall that some proteins in this production line are bound for export outside the cell, while others will end up being used within various membranes in the cell. It follows that proteins have to be *sorted and shipped* appropriately, and the Golgi does just this, acting as a kind of distribution center. Chemical "tags" that are part of the proteins often allow for this routing. Remember how, in the rough ER, carbohydrate side chains might be attached to a newly formed protein? Often, these side chains serve as the routing tags; other times, a section of the protein's amino acid sequence will serve this function.

The Golgi is similar to the ER in that it amounts to a series of cisternae or connected membranous sacs with internal spaces (**Figure 4.8**). Proteins arrive

**Figure 4.8**
**Processing and Routing: The Golgi Complex**

Transport vesicles from the rough endoplasmic reticulum (RER) move to the Golgi complex, where they unload their protein contents by fusing with the Golgi membrane. The protein is then passed, within other vesicles, through the layers of disk-shaped Golgi cisternae, where editing of the protein may occur. At the part of the Golgi farthest from the rough ER, the proteins are sorted, packed in vesicles, and shipped to sites mostly in cell membranes or outside the cell altogether. The vesicles in the micrograph are the pink and purple spheres.

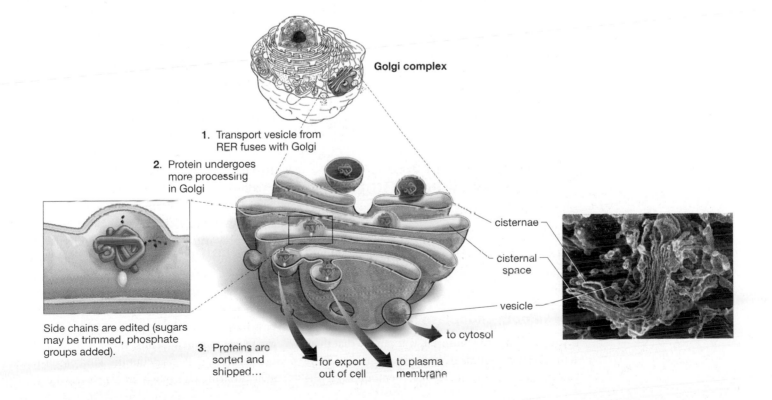

Golgi complex

1. Transport vesicle from RER fuses with Golgi

2. Protein undergoes more processing in Golgi

cisternae

cisternal space

vesicle

to cytosol

Side chains are edited (sugars may be trimmed, phosphate groups added).

3. Proteins are sorted and shipped...

for export out of cell

to plasma membrane

at the Golgi housed in transport vesicles that fuse with the Golgi "face" nearest the rough ER, at which point the vesicles release their protein cargo into the Golgi cisternal sacs for processing. Once processed, proteins of the sort we are following eventually bud off from the outside face of the Golgi, now housed in their final transport vesicles.

### From the Golgi to the Surface

For secretory proteins, the journey that began with the copying of DNA information onto mRNA is almost over. Once a vesicle buds off from the Golgi, all that remains is for it to make its way through the cytosol to the plasma membrane at the outer reaches of the cell. There, the vesicle fuses with the plasma membrane, and the protein is ejected into the extracellular world. This last step, called *exocytosis*, is a process you'll be looking at next chapter. With it, one finished product of the cellular factory has rolled out the door.

---

**SO FAR...**

1. The cell's nucleus holds the cell's primary complement of _____, which contains _____ for the production of proteins.

2. Proteins of the sort followed in the text are put together within tiny organelles called _____ that lie outside the nucleus. During their production, these proteins fold up upon entering the _____, where they undergo editing, after which they move to the _____, where they undergo further editing and are sorted for distribution.

3. Proteins and other materials move through the cell within membrane-lined spheres called _____, which, together with membrane-lined organelles, make up the _____ system.

---

## 4.5 Outside the Protein Production Path: Other Cell Structures

A functioning cell engages in more activities than the protein synthesis and shipment just reviewed for the simple reason that cells do a lot more than produce proteins.

### The Smooth Endoplasmic Reticulum

If you look back to Figure 4.4, you can see that there actually are *two* kinds of endoplasmic reticuli. The part of the ER membrane, farther out from the

nucleus, that has no ribosomes is called the smooth endoplasmic reticulum or smooth ER. It's "smooth" because it is not peppered with ribosomes, and this very quality means it is not a site of protein synthesis. Instead, the **smooth endoplasmic reticulum** is a network of membranes that is the site of the synthesis of various lipids and a site at which potentially harmful substances are detoxified within the cell. The tasks the smooth ER undertakes, however, will vary in accordance with cell type. The lipids we normally think of as "fats" are put together and stored in the smooth ER of liver and fat cells, while the "steroid" lipids reviewed last chapter—testosterone and estrogen—are put together in the smooth ER of the ovaries and testes. The detoxification of potentially harmful substances, such as alcohol, takes place largely in the smooth ER of liver cells.

### Tiny Acid Vats: Lysosomes and Cellular Recycling

Any factory must be able to get rid of some old materials while recycling others. A factory also needs new materials, brought in from the outside, that probably will have to undergo some processing before use. A single organelle in the animal cell aids in doing all these things. It is the **lysosome**: an organelle found in animal cells that digests worn-out cellular materials and foreign materials that enter the cell. Several hundred of these membrane-bound organelles may exist in any given cell. You could think of them as sealed-off acid vats that take in large molecules, break them down, and then return the resulting smaller molecules to the cytosol. What they cannot return, they retain inside themselves or expel outside the cell. They carry out this work not only on molecules entering the cell from the outside (say, invading bacteria) but also on materials that exist inside the cell—on worn-out organelle parts, for example (**Figure 4.9**).

A given lysosome may be filled with as many as 40 different enzymes that can break larger molecules into their component parts—an enzymatic array that allows each lysosome to break down most of what comes its way. A lysosome gets hold of its macromolecule prey through the endomembrane system. A lysosome will fuse with the membrane surrounding a worn-out organelle part; proceeding to engulf it, the lysosome then goes to work breaking the organelle down. The small molecules that result then pass freely out of the lysosome and into the cytosol for reuse elsewhere. Thus, there is recycling at the cellular level. Cells carry out this kind of self-renewal at an amazing rate. Christian de Duve, who with his colleagues discovered lysosomes in the 1950s, has noted the effect of this activity on human brain cells. In an elderly person, he noted,

> such cells have been there for decades. Yet, most of their mitochondria, ribosomes, membranes, and other organelles are less than a month old. Over the years, the cells have destroyed and re-made most of their constituent molecules from hundreds to thousands of times, some even more than 100,000 times.

There are certain things, however, that even lyso-somes cannot digest, and over time these substances can cause lysosomes to leak their acidic contents into the cell, causing great harm. You may have heard of an affliction called black lung disease that preys on miners. One form of this disease, called silicosis, is caused by the inability of lysosomes to digest the silica fibers that miners inhale. Because these fibers have a needle-like shape, they can cause the lyso-somes to leak. The result is an acid spill that can kill cells and produce scar tissue in the lungs, thus reduc-ing the miners' ability to breathe.

## Extracting Energy from Food: Mitochondria

Just as there is no such thing as a free lunch, there is no such thing as free lysosome activity, or ribosomal action, or protein export. There is a price to be paid for all these things, and it is called energy expendi-ture. The fuel for this energy is contained in the food that cells ingest. But the energy in this food has to be

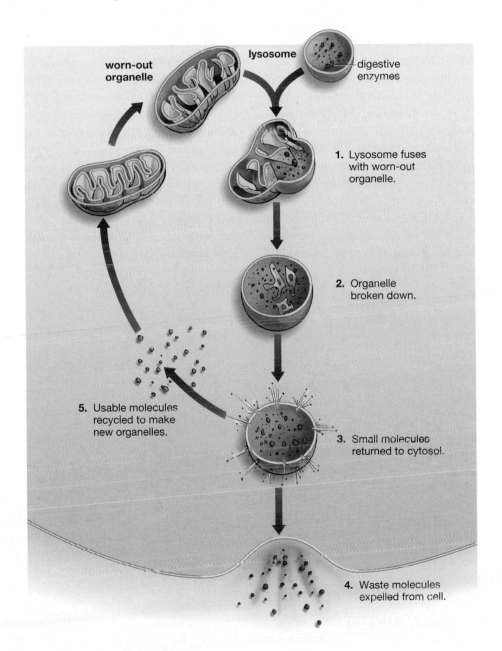

**worn-out organelle**

**lysosome**

digestive enzymes

**1.** Lysosome fuses with worn-out organelle.

**2.** Organelle broken down.

**5.** Usable molecules recycled to make new organelles.

**3.** Small molecules returned to cytosol.

**4.** Waste molecules expelled from cell.

**Figure 4.9**
**Cellular Recycling: Lysosomes**

Lysosomes are membrane-bound organelles that contain potent enzymes capable of di-gesting large molecules, such as worn-out organelles. The useful parts of such organelles will be returned to the cytosol and used elsewhere—a form of cel-lular recycling. If a lysosome cannot digest a given material, it may expel it outside the cell, although in multicelled organ-isms, lysosomes generally will hold on to such materials so structures outside the cell are not harmed. Lysosomes also di-gest small particles, such as food, that come from outside the cell, along with invading or-ganisms, such as bacteria.

••• ANSWERS

**1.** DNA; information

**2.** ribosomes; rough endoplasmic reticulum (RER); Golgi complex

**3.** transport vesicles; endomembrane

extracted from food and converted into a molecular form that cells can easily use—a substance called ATP (adenosine triphosphate). And where does this extraction and conversion take place? Mostly in the tiny "power plant" organelles of the cell, called mitochondria. Just as a city's power plant takes in coal and turns out electricity, so mitochondria take in food and turn out ATP. Thus, **mitochondria** can be defined as organelles that are the primary sites of energy conversion in eukaryotic cells. While cells that don't use much energy might have one or two mitochondria within them, an energy-ravenous liver cell might have a thousand. Most of the heat in our bodies is generated within mitochondria, and almost all the food we eat is ultimately consumed in them.

**Figure 4.10** shows a classical view of a mitochondrion. Note that there is a continuous outer membrane enclosing an inner membrane that has a series of twists in it. The effect of these infoldings is to give mitochondria a larger internal surface area for carrying out their energy transformation activities. This nice, tidy picture may have a problem with it, however, in that scientists are not sure if mitochondria really are the kind of discrete, oval entities shown, or if they actually take the form of a network of membranes—more like the endoplasmic reticulum.

Few details about mitochondria are included here because much of Chapter 7 is devoted to them. Suffice it to say that to carry out their work, mitochondria need not only food, but oxygen. (Ever wonder why you need to breathe?) The *products* of mitochondrial activity, meanwhile, are the ATP we talked about along with water and carbon dioxide. (Ever wonder where the carbon dioxide you exhale comes from?) Another thing to note about mitochondria is that they are the descendants of resident aliens. They started out as bacterial cells that invaded eukaryotic precursor cells more than 1.5 billion years ago only to end up becoming part of these larger cells (see "The Stranger Within," on page 78).

## 4.6 The Cytoskeleton: Internal Scaffolding

When you see a set of cells in action, the word that comes to mind is *hyperactive*. Cells are not passive entities. Jostling, narrowing, expanding, moving about, capturing objects and bringing them in, expelling other objects: This is life as a bubbling cauldron of activity. It was once thought that cells did all these things with pretty much the equipment you've looked at so far, meaning that the Golgi complex, the ribosomes, and so forth were thought to be floating in a featureless soup that was the cytosol. But improved work with microscopes showed that there is, in fact, a complex forest within the cytoplasm. It is the **cytoskeleton:** a network of protein filaments that functions in cell structure, cell movement, and the transport of materials within the cell. This network is found in its full form only in eukaryotic cells, but in the last few years

**Figure 4.10**
**Energy Transformers: Mitochondria**

Just as a power plant converts the energy contained in coal into useful electrical energy, mitochondria convert the energy contained in food into a useful molecular form of energy—the molecule ATP. Cells can contain anywhere from a single mitochondrion to several thousand.

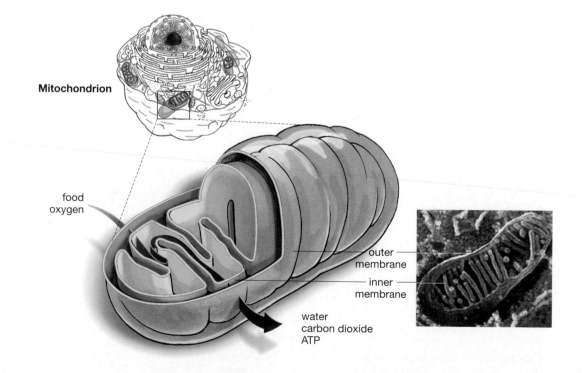

Mitochondrion

food
oxygen

water
carbon dioxide
ATP

outer membrane
inner membrane

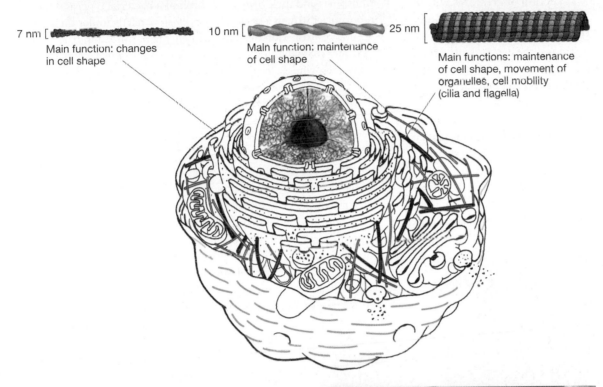

**(a) Microfilaments**

**(b) Intermediate filaments**

**(c) Microtubules**

7 nm

Main function: changes in cell shape

10 nm

Main function: maintenance of cell shape

25 nm

Main functions: maintenance of cell shape, movement of organelles, cell mobility (cilia and flagella)

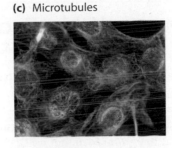

**Figure 4.11**
**Structure and Movement: The Cytoskeleton**

Three types of fibers form the inner scaffolding or cytoskeleton of the eukaryotic cell: microfilaments, intermediate filaments, and microtubules.

scientists have confirmed that bacteria have at least one type of cytoskeletal fiber within them.

Some of the cytoskeleton's fibers are permanent and relatively static, but many are moving, and some are assembled or disassembled very rapidly. The cytoskeleton usually is divided into three component parts. Ordered by size, going from smallest to largest in diameter, these are microfilaments, intermediate filaments, and microtubules (**Figure 4.11**). Let's take a look at some of the characteristics of each.

## Microfilaments

The most slender of the cytoskeletal fibers, **microfilaments** are made of the protein *actin* and serve as a support or "structural" filament in almost all eukaryotic cells. Microfilaments can also help cells move or capture prey, essentially by growing very rapidly at one end—in the direction of the movement or extension—while decomposing rapidly at the other end. You can see a vivid example of microfilament-aided cell extension in **Figure 4.12**.

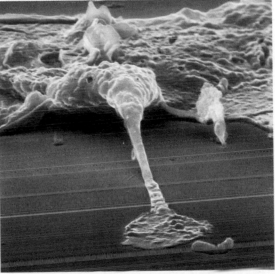

**Figure 4.12**
**Microfilaments in Action**

Certain cells can move or capture prey by sending out extensions of themselves called pseudopodia ("false feet"). It is the rapid construction of actin microfilaments—in the direction of the extension—that makes this possible. Here, a type of blood-borne guard cell called a macrophage is about to use a pseudopodium it is constructing to capture a green bacterium.

ESSAY

## ESSAY | The Stranger Within: Endosymbiosis

**I**f pressed to think about it, most people would probably agree that there's something slightly creepy about being made up of cells. Jostling and dividing as they are, working and dying within us without so much as asking permission, they give us the feeling that the unitary *self* that we so cherish actually amounts to an unruly collection of creatures within.

How sobering it is, then, to realize that we almost certainly are composite beings in more ways than one. Each of our *cells* has within it the vestiges of other living things: the descendants of bacteria that long ago invaded our ancestors' cells, only to take up residence there. In animal cells, the mitochondria that serve as cellular "powerhouses" are almost certainly the descendants of bacteria (**Figure 1**). Plant and algae cells have mitochondria, too, but then go on to have an additional set of bacterial descendants in them: the plastids, which include the chloroplasts that carry out photosynthesis.

The idea that these structures are descended from free-standing bacteria is called the endosymbiotic theory—*endo* for "within" the cell and *symbiotic* for "symbiosis," meaning a situation in which two organisms not of the same species live in close association. What makes us think that endosymbiosis really happened? First, mitochondria and chloroplasts—which, remember, are *organelles* within eukaryotic cells—have many of the characteristics of free-standing *cells*, specifically free-standing bacterial cells. They have their own ribosomes and their own DNA, both of a bacterial type, and they reproduce through a bacteria-like division, partly under their own genetic control. Moreover, the sequencing of mitochondrial DNA indicates that all mitochondria are descended from a single species of bacteria whose closest modern relatives make a living by invading other cells. (As it happens, the single closest relative of the ancient mitochondrial invaders is *Rickettsia prowazekii*, the cause of modern typhus.)

Mitochondria appear to have completed the transition to endosymbiotic living somewhere between 2.2 and 1.5 billion years ago. There are two different schools of thought about who the receiving or "host" cell was for them when they had the form of bacterial invaders. One school says the host was another bacterium, while another school says the host was an early eukaryote. Likewise, there are a couple of different hypotheses about what benefits this merger initially provided to the two organisms.

Today, however, the benefit that mitochondria provide to eukaryotes such as ourselves is clear: They allow us to extract much more energy from food than we could otherwise. Only through mitochondria are we able to use *oxygen* in the energy extraction process and this capacity allows us to wring the maximum amount of energy from every swig of a soft drink or bite of an apple. In line with this, one hypothesis regarding the original merger has to do with oxygen and its use. During a period when oxygen was making up more and more of Earth's atmosphere, this hypothesis holds, the mitochondrial ancestors were able to metabolize oxygen, while the cells they invaded were fairly intolerant of it. The arrangement the two species eventually came to was simple. The host provided food to the bacterial invader, and the invader allowed the host to thrive in an oxygenated world. Then, over time, through a process of internal gene transfer, the invaders made a transition from organisms that were living inside a host to organelles that were simply part of the host.

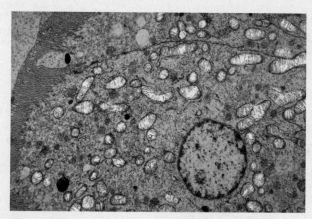

**Figure 1**
**Once Invaders, Now Organelles**
The mitochondria within this cell, colored yellow, are the descendants of bacteria that invaded a set of host cells billions of years ago. Over evolutionary time, these bacteria were transformed into energy-converting organelles that today are found in almost all eukaryotic cells.

Some bacteria have been shown to have an actin-like set of cytoskeleton fibers that helps give them their shape.

### Intermediate Filaments

**Intermediate filaments** are filaments of the cytoskeleton intermediate in diameter between microfilaments and microtubules. These in-between-sized proteins are the most permanent of the cytoskeletal elements, perhaps coming closest to our everyday notion of what a skeleton is like. They stabilize the positions of the nucleus and other organelles within the cell.

### Microtubules

**Microtubules** are the largest of the cytoskeletal filaments, taking the form of tubes composed of the protein tubulin. Microtubules play a structural role in the cell; in fact, theirs seems to be the preeminent structural role in the sense of determining the shape of the cell. But they take on several other tasks in addition. They serve, for example, as the monorails of the cell's internal skeleton discussed earlier. Recall that protein-laden vesicles move from one organelle to another in the cell. These spheres are moving along the "rails" of microtubules, while sitting atop the "engine" of one of the so-called motor proteins (**Figure 4.13**).

## Cell Extensions Made of Microtubules: Cilia and Flagella

Microtubules also form the underlying structure for two kinds of cell extensions, cilia and flagella (see Figure 4.13). **Cilia** are microtubular extensions of cells that take the form of a large number of active, hair-like growths stemming from them. The function of cilia is simple: Move back and forth very rapidly, perhaps 10 to 40 times per second. The effect of this movement can be either to propel a cell or to move material *around* a cell. Cilia are extremely common among single-celled organisms and in some of the cells of simple animals (sponges, jellyfish). You saw an example of cilia in humans earlier in connection with our sense of smell. Our lungs also are lined with cilia whose job it is to sweep the lungs clean of whatever foreign matter has been inhaled. These cilia, like most others, all beat at once in the same direction, acting like rowers in a crew.

Cilia grow from eukaryotic cells in great profusion, but it is a different story with **flagella**—the relatively long, tail-like extensions of some cells that function in cell movement. It is sometimes the case that several flagella will sprout from a given cell, but often there is but a single flagellum. Only one kind of

**Figure 4.13**
**Several Functions for Microtubules**

**(a)** They are the "rails" on which vesicles move through the cell, carried along by "motor proteins."

**(b)** They exist outside the cell in the form of cilia, which are profuse collections of hair-like projections that beat rapidly, forming currents that can propel a cell or move material around it.

**(c)** They are also found outside cells in the form of flagella. The flagellum on this sperm cell is enabling it to seek entry into an egg.

**(a)** Transport monorails

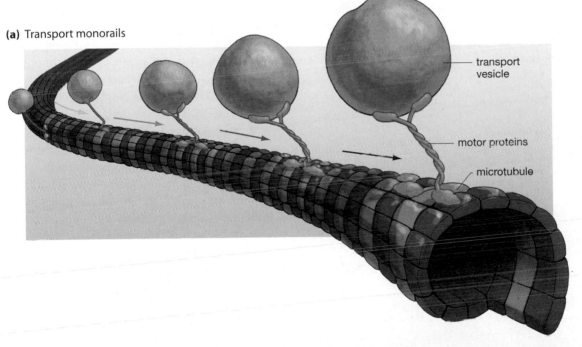

transport vesicle

motor proteins

microtubule

**(b)** Cilia

**(c)** Flagellum

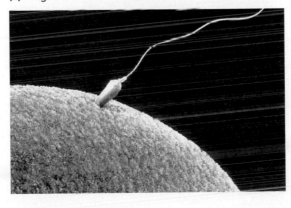

animal cell is flagellated, and it scarcely needs an introduction: A sperm is a single cell that whips its flagellum in a corkscrew motion to get to an unfertilized egg.

### In Summary: Structures in the Animal Cell

In your tour of the cell so far, you've pictured a cell as a factory, one that synthesizes proteins in a "production line" that starts with DNA in the nucleus and then goes to the ribosomes (via mRNA), to the rough ER, to the Golgi complex, and finally to the protein's destination (the plasma membrane, export, and so on). You've also seen that cells have other structures, such as lysosomes for digestion and recycling, mitochondria for energy transformation, the smooth endoplasmic reticulum for detoxification and lipid synthesis, and the cytoskeleton for structure and movement. If you look at **Figure 4.14**, you can see, in metaphorical form, a "map" of these component parts within the cell. **Table 4.1** (on page 82 in the discussion of plant cells) lists cellular elements found in plant and animal cells as well as some elements found only in plant cells.

***BioFlix***
Tour of a Plant Cell

**Web Animation 4.1**
The Structure of Cells

**SO FAR...**

1. Lipids, including fats, are put together in the organelle called the _____, while worn-out cellular materials are broken down and recycled in the organelles called _____.

2. The organelles called mitochondria transform the _____ from food into a usable chemical form, a molecule called _____.

3. The cell's internal scaffolding, made up of different types of protein fibers, is called the _____.

## 4.7 The Plant Cell

The animal cell you just looked at has lots in common with the cells that you'd see in any eukaryotic living thing. If you looked at the cells that make up plants or fungi, for example, you'd see that they have a nucleus, ribosomes, mitochondria, and so forth. But, as you can imagine, the cells of a fungus have to be *somewhat* different from those of an animal since fungi and animals are such different types of organisms. To give you an idea of some of the kinds of differences found in the cells of life's various kingdoms, we'll look briefly now at the plant cell, which differs from an animal cell in several basic ways.

A quick look at **Figure 4.15** will confirm for you the first part of this story—how structurally similar plant and animal cells are. As you see, a plant cell has a nucleus (with a nucleolus), the smooth and rough ERs, a cytoskeleton—most of the things you've just gone over in animal cells. Indeed, there is only one structure present in the animal cells you've looked at that plant cells don't have: the lysosome. What jumps out at you when you look at plant cells is not what they lack compared to animal cells, but what they *have* that animal cells do not. As you can see in **Figure 4.16**, these additions are

- A thick cell wall
- Structures called chloroplasts
- A large structure called a central vacuole

### The Central Vacuole

The central vacuole pictured in Figure 4.16 is so prominent that it appears to be a kind of organelle continent surrounded by a mere moat of cytosol. And, indeed, in a mature plant cell, one or two central vacuoles may comprise 90 percent of cell volume. Although animal cells can have vacuoles, the

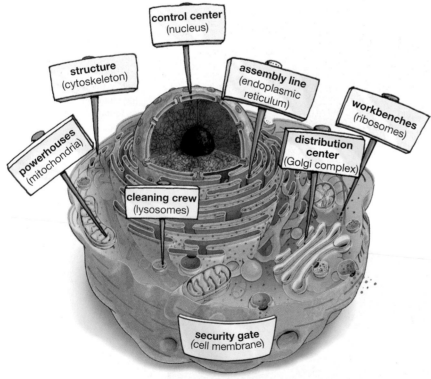

**Figure 4.14**
**The Cell as a Factory**
This comparison may help you to remember some roles of the different parts of the cell.

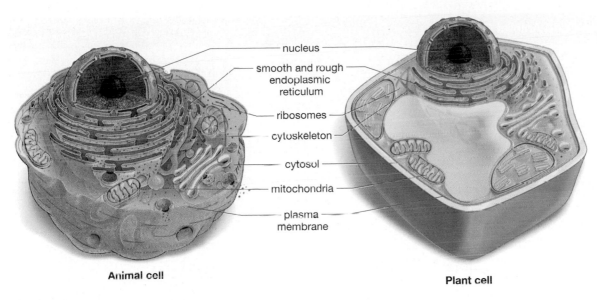

**Animal cell**

**Plant cell**

**Figure 4.15**
**Common Structures in Animal and Plant Cells**

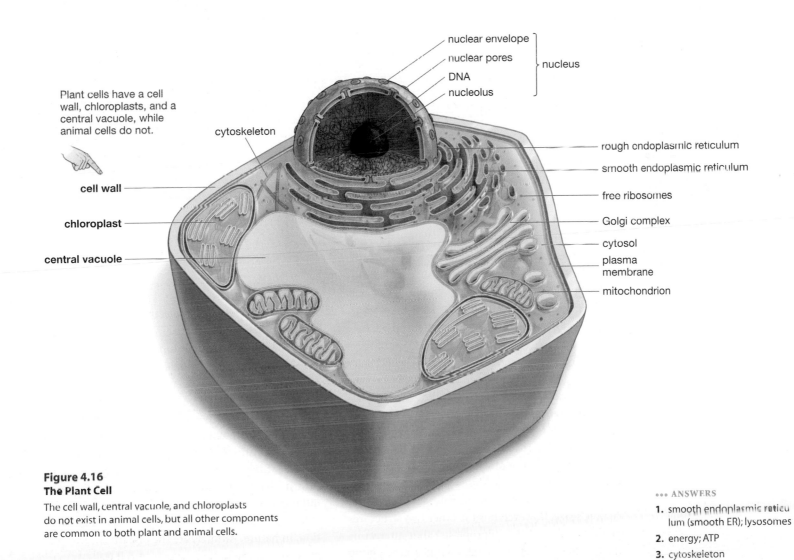

Plant cells have a cell
wall, chloroplasts, and a
central vacuole, while
animal cells do not.

**Figure 4.16**
**The Plant Cell**

The cell wall, central vacuole, and chloroplasts
do not exist in animal cells, but all other components
are common to both plant and animal cells.

••• ANSWERS

1. smooth endoplasmic reticu
lum (smooth ER); lysosomes

2. energy; ATP

3. cytoskeleton

## Table 4.1 Structures in Plant and Animal Cells

| Name | Location | Function |
| --- | --- | --- |
| Cytoskeleton | Cytoplasm | Maintains cell shape, facilitates cell movement and movement of materials within cell |
| Cytosol | Cytoplasm | Protein-rich fluid in which organelles and cytoskeleton are immersed |
| Golgi complex | Cytoplasm | Processing, sorting of proteins |
| Lysosomes (in animal cells only) | Cytoplasm | Digestion of imported materials and cell's own used materials |
| Mitochondria | Cytoplasm | Transform energy from food |
| Nucleolus | Nucleus | Synthesis of ribosomal RNA |
| Nucleus | Inside nuclear envelope | Site of most of the cell's DNA |
| Ribosomes | Rough ER, Free-standing in cytoplasm | Sites of protein synthesis |
| Rough endoplasmic reticulum | Cytoplasm | Protein processing |
| Smooth endoplasmic reticulum | Cytoplasm | Lipid synthesis, storage; detoxification of harmful substances |
| Vesicles | Cytoplasm | Transport of proteins and other cellular materials |
| Cell walls (in plant cells only) | Outside plasma membrane | Limit water uptake; maintain cell membrane shape, protect from outside influences |
| Central vacuole (in plant cells only) | Cytoplasm | Cell metabolism, pH balance, digestion, water maintenance |
| Chloroplasts (in plant cells only) | Cytoplasm | Photosynthesis |

imposing central vacuole in plants is different. For a start, it is composed mostly of water, which demonstrates just how watery plant cells are. A typical animal cell may be 70 percent water, but for plant cells the water proportion is likely to be 90 to 98 percent.

The watery environment of the central vacuole contains hundreds of other substances. Many of these are nutrients; others are waste products. There are also hydrogen ions, pumped in to keep the cell's cytoplasm at a near-neutral pH. Given these diverse materials, the **central vacuole** can be defined as a large, watery plant organelle that has many functions, among them the storage of nutrients and the retention and degradation of waste products. In this last role, the central vacuole uses digestive enzymes, much like the lysosomes in animal cells. Despite the vacuole's size, however, it is the other two structures (the cell wall and the chloroplasts) that interest us most as they tell us the most about differences among

the cells of various types of living things. Let's have a look at these two structures.

### The Cell Wall

Plant cells have an outer protective lining, called a **cell wall**, that makes their plasma membrane—just inside the cell wall—look rather thin and frail by comparison. This is because it *is* thin and frail by comparison; the plasma membrane of a plant cell may be $0.01\mu m$ thick, while the combined units of a cell wall may be 700 times this width—$7\mu m$ or more. Cell walls are nearly always present in plant cells. And they exist in many organisms that are neither plant nor animal—bacteria, fungi, and a group of living things called protists—although the cell walls of these life-forms differ in chemical composition from those of plants. Thus, cell walls are the rule, rather than the exception, in nature. Animals are the one major group of living things whose cells never have cell walls.

What do cell walls do for plant cells? They provide them with structural strength, put a limit on their absorption of water (as you'll see in the next chapter), and generally protect plants from harmful outside influences. So, if they're so useful, why don't *animal* cells have them? Cell walls make for a rather rigid, inflexible organism—like plants, which are stationary. Animals, meanwhile, need to move, both to catch prey and to avoid *being* prey, which means they have to be flexible.

Cell walls in plants can come in several forms, but all such forms will be composed chiefly of a molecule you were introduced to last chapter: cellulose, a complex sugar or "polysaccharide" that is embedded within cell walls in the way reinforcing bars run through concrete. In some cell walls, cellulose is joined by a compound called lignin, which imparts considerable structural strength. You can see a vivid demonstration of this in the material we know as wood, which is largely made of cell walls (**Figure 4.17a**).

Cell walls can serve different functions over the life of an organism. Generally, they are the site of a good deal of metabolic activity and thus should not be considered mere barriers. On the other hand, the outer portion of tree bark consists mostly of *dead* cell walls, which clearly are serving a barrier function (**Figure 4.17b**).

(a)

(b)

secondary xylem (wood)    inner bark    outer bark

**Figure 4.17**
**Great Strength from Small Things**
**(a)** Trees can reach great heights because of the strength of their secondary xylem—better known as wood—which is largely made of cell walls. Pictured is an American basswood tree (*Tilia americana*).
**(b)** Cell walls play important roles in living cells, but they also are valuable as the strong, remaining components of dead cells. The outer bark of the basswood tree, seen in the blow-up at right, is composed mostly of dead cells whose cell walls are infused with a waxy substance that acts to protect the tree from invading insects and microorganisms.

## Chloroplasts

Everyone knows that a major difference between plants and animals is that, while animals must get their food from outside themselves, plants make their own food through the process of photosynthesis. Where does photosynthesis take place in plants? Inside the organelles known as chloroplasts. It's hard to overstate the importance of these tiny, oblong structures, given that they are the food factories for most of the living world. Lions may eat zebras, but zebras eat grass, and grass is essentially produced inside chloroplasts. Thanks to their ability to harness the sun's power, these membrane-laden organelles can start with nothing more than water, a few minerals, and atmospheric carbon dioxide and end up turning out a sugar that builds an entire plant. A byproduct of this operation—a kind of refuse of photosynthesis—is the oxygen that sustains most of Earth's organisms (**Figure 4.18**). **Chloroplasts** are the sites of photosynthesis in algae cells, as well as plant cells, but no other organisms possess these specialized organelles (although certain bacteria are able to perform photosynthesis without them). In plants, chloroplasts are especially abundant in the cells of leaves. A cell that lies toward a leaf's interior might contain hundreds of chloroplasts, each of which is capable of performing photosynthesis on its own.

## 4.8 Cell-to-Cell Communication

Most of what you've seen so far has made the cell seem like an isolated entity, but this is not the case. Single-celled organisms can exist as separate entities, along with certain plant or animal cells (red blood cells, for example), but most plant and animal cells are linked together in organized collections referred to as *tissues*. Not surprisingly, these assemblages of cells—whether plant or animal—have the ability to communicate with one another.

### Communication among Plant Cells

Having noted the thickness of something like the cell walls in plants, you might wonder how one plant cell could interact with another. Communication between plant cells takes place quite readily, however, through a series of tiny channels in the plant cell wall called **plasmodesmata** (singular, plasmodesma). The structure of these channels is such that the cytoplasm of one plant cell is continuous with that of another—so much so that the cytoplasm of an entire plant can be properly looked at as one continuous whole (**Figure 4.19a**). The structure of plasmodesmata is more complex than that of a simple opening, but the basic idea is of a channel-like linkage between two plant cells.

**Figure 4.18**
**Food Source for the World**
Chloroplasts, the tiny organelles that exist in plant and algae cells, are sites of photosynthesis—the process that provides food for most of the living world. Using the starting materials of water, carbon dioxide, and a few minerals, these organisms use energy from sunlight to produce their own food. A double membrane lines the chloroplasts.
(Micrograph: ×13,000)

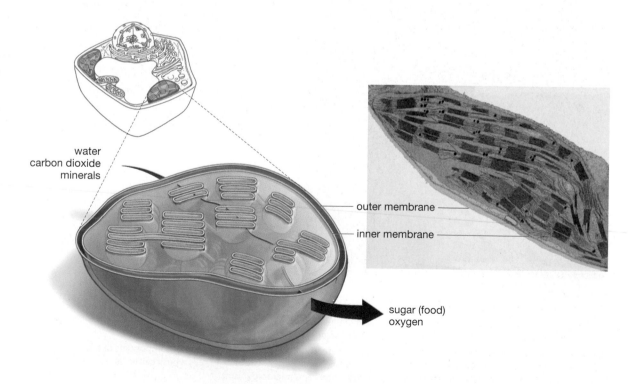

water
carbon dioxide
minerals

outer membrane

inner membrane

sugar (food)
oxygen

## Communication among Animal Cells

There are no plasmodesmata in animal cells, but there are three other kinds of *cell junctions*, or linkages, one of which serves to facilitate cell communication. It is called a gap junction, and it consists of clusters of protein structures that shoot through the plasma membrane of a cell from one side to the other, forming a kind of tube. When these tubes line up in adjacent cells, the result is a channel for passage of small molecules and electrical signals (**Figure 4.19b**). Thus, we can define a **gap junction** as a protein assemblage that forms a communication channel between adjacent animal cells. Note that animal gap junctions and plant plasmodesmata are very different kinds of channels. Plasmodesmata can be thought of as permanent channels between plant cells, whereas gap junctions open only as necessary.

## Leeuwenhoek and the Process of Science

The cellular world you have been reading about was completely unknown to human beings before the seventeenth century. It came as quite a shock back then to learn that the human body had, living within it, a multitude of creatures so small that they could not be perceived with the naked eye. In "First

**Plant tissues**

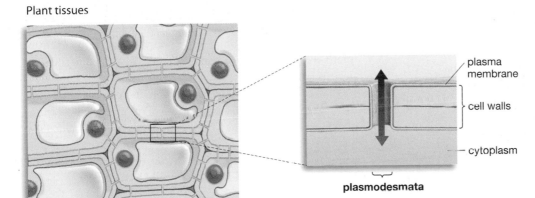

plasmodesmata

plasma membrane

cell walls

cytoplasm

**Figure 4.19
Cell Communication**

### (a) Plasmodesmata

In plants, a series of tiny pores between plant cells, the plasmodesmata, allow for the movement of materials among cells. Thanks to the plasmodesmata channels, the cytoplasm of one cell is continuous with the cytoplasm of the next; the plant as a whole can be thought of as having a single complement of continuous cytoplasm.

**Animal tissues**

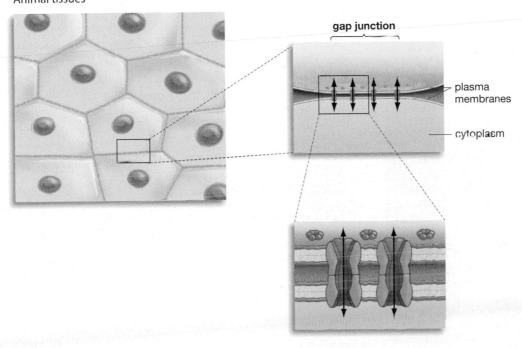

gap junction

plasma membranes

cytoplasm

### (b) Gap junctions

In animals, protein assemblies come into alignment with one another, forming communication channels between cells. A cluster of many such assemblies—perhaps several hundred—is called a gap junction.

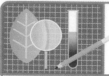

## THE PROCESS OF SCIENCE
# First Sightings: Anton van Leeuwenhoek

**T**o the list of explorers that includes Columbus and Balboa we could add the name Anton van Leeuwenhoek. This unassuming Dutchman was the great early voyager into another world, the micro-world.

It was not until the seventeenth century that human beings realized that things as small as cells existed. Leeuwenhoek began to re-

**(a)** What Leeuwenhoek could see

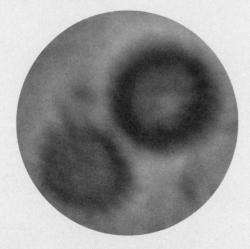

**(b)** Leeuwenhoek with two of his microscopes

port in the 1670s on what he saw with the aid of a device invented at the end of the 1500s—the microscope (**Figure 1**). One of Leeuwenhoek's contemporaries, Englishman Robert Hooke, coined the term *cell* after viewing a slice of cork under a simple microscope, but Hooke's purpose was to reveal the detailed structure of familiar, small objects, such as the flea. Leeuwenhoek, by contrast, revealed the *existence* of creatures unimagined until his time. Moreover, he carried out this work in the most extraordinary fashion: Laboring alone in the small town of Delft with palm-sized magnifiers he himself had created, looking at anything that struck his fancy. (And many things struck his fancy; he once looked at exploding gunpowder under a microscope, nearly blinding himself in the process.) For 50 years, while working as shopkeeper and minor city official, this untrained amateur of boundless curiosity examined the micro-world and reported on it in letters he posted to the Royal Society in London.

Who could believe what he uncovered? How was it possible that there was a buzzing, blooming universe of "animalcules" (little animals) whose existence had been completely unsuspected? Prior to his work, no one thought that any creature smaller than a worm could exist within the human body. Yet, examining scrapings from his own mouth, Leeuwenhoek tells us:

> I saw, with as great a wonderment as ever before, an inconceivably great number of little animalcules, and in so unbelievably small a quantity of the foresaid stuff, that those who didn't see it with their own eyes could scarce credit it.

The animalcules that Leeuwenhoek beheld over his career were single-celled organisms that today are known as bacteria and protists. It would be two hundred years before Leeuwenhoek's findings were fully integrated into a modern theory of cells. Yet, the man from Delft had shown that only a small portion of the world's living things are visible things.

**Figure 1**
**(a) What Leeuwenhoek Could See**
Biologist Brian Ford used an actual 300-year-old Leeuwenhoek microscope to capture this image of spiny spores from a truffle. (×600)
**(b) Leeuwenhoek's Microscopes Were Handheld and Paddle Shape**
Leeuwenhoek revealing the micro-world to Queen Catherine of England, wife of King Charles II (1630–1685).

Sightings: Anton van Leeuwenhoek" on page 86, you can read about the Dutchman who was one of the first visitors to the micro-world.

## On to the Periphery

Having looked at what is inside the cell, you've arrived at the cell's periphery, the plasma membrane, which is where you'll be staying for a while. It may at first seem strange to devote a whole chapter to an outer boundary. How much attention would you pay, after all, to a factory's wall as opposed to its contents? The answer is: A lot, if that wall could facilitate communication with the outer world, continually renew itself, and let some things in while keeping others out. Such is the case with the plasma membrane, a slender lining that manages to make one of the most fundamental distinctions on Earth: Inside, life goes on; outside it does not.

**SO FAR...**

1. Plant cells have two organelles and one structure that animal cells do not have. The organelles are _____, which are the sites of photosynthesis, and a large _____, which performs multiple functions, including the storage of nutrients and the degradation of waste products. The structure is the _____, which provides protection and limits the plant cell's uptake of water.

2. Animal cells have only a single organelle that plants cells do not have. This is the _____, which breaks down and recycles materials in the cell.

3. In plants, the channels that facilitate cell-to-cell communication are called _____, while in animals, the tube-like protein structures that facilitate such communication are called _____.

••• ANSWERS
1. chloroplasts; central vacuole; cell wall
2. lysosome
3. plasmodesmata; gap junctions

# CHAPTER 4 REVIEW

For study help, activities, and more quiz questions, go to **www.aw.com/krogh4**.

## SUMMARY

### 4.1 Cells Are the Fundamental Units of Life

- With the possible exception of viruses, every form of life on Earth either is a cell or is composed of cells. Cells come into existence only through the activity of other cells.

### 4.2 All Cells Are Either Prokaryotic or Eukaryotic

- All cells can be classified as prokaryotic or eukaryotic. Prokaryotic cells either are bacteria or another single-celled life-form called archaea. Setting bacteria and archaea aside, all other cells are eukaryotic.

- Eukaryotic cells have most of their DNA contained in a membrane-lined compartment, called the cell nucleus, whereas prokaryotic cells do not have a nucleus. Eukaryotic cells also tend to be much larger than prokaryotic cells, and they have more of the specialized internal structures called organelles than do prokaryotic cells. Many eukaryotes are multicelled organisms, whereas all prokaryotes are single-celled.

### 4.3 The Eukaryotic Cell

- There are five principal components to the eukaryotic cell: the nucleus, other organelles, the cytosol, the cytoskeleton, and the plasma membrane. Organelles are "tiny organs" within the cell that carry out specialized functions, such as energy transfer and materials recycling. The cytosol is the jelly-like fluid outside the nucleus in which these organelles are immersed. (The cytosol should not be confused with the cytoplasm, which is the region of the cell inside the plasma membrane but outside the nucleus.) The cytoskeleton is a network of protein filaments that function in cell structure, cell movement, and the transport of materials within the cell. The plasma membrane is the outer lining of the cell. A membrane can be defined as the flexible, chemically active outer lining of a cell or of its compartments.

### 4.4 A Tour of the Animal Cell: Along the Protein Production Path

- Information for the construction of proteins is contained in the DNA located in the cell nucleus. This information is copied onto a length of messenger RNA (mRNA) that departs the cell nucleus through its nuclear pores and goes to the sites of protein synthesis, structures called ribosomes, which lie in the cytoplasm. Many ribosomes that receive mRNA chains process only a short stretch of them before migrating to, and then embedding in, one of a series of sacs in a membrane network called the rough endoplasmic reticulum (RER). The polypeptide chains produced by the ribosomal "reading" of the mRNA sequences are dropped from ribosomes into the internal spaces of the RER; there, the polypeptide chains fold up, thus becoming proteins, and undergo editing. Some ribosomes are not embedded in the RER but instead remain free-standing in the cytosol.

- Materials move from one structure to another in the cell via the endomembrane system, in which a piece of membrane, with proteins or other materials inside, can bud off from one organelle, move through the cell, and then fuse with another membrane-lined structure. Membrane-lined structures that carry cellular materials are called transport vesicles.

- Once protein processing is finished in the rough ER, proteins undergoing processing move, via transport vesicles, to the Golgi complex, where they are processed further and marked for shipment to appropriate cellular locations.

  *BioFlix* Tour of an Animal Cell

### 4.5 Outside the Protein Production Path: Other Cell Structures

- The smooth endoplasmic reticulum is a network of membranes that functions to synthesize lipids and to detoxify potentially harmful substances. Lysosomes are organelles that break down worn-out cellular structures or foreign materials that come into the cell. Once this digestion is completed, the lysosomes return the molecular components of these materials to the cytoplasm for further use. Mitochondria are organelles that function to extract energy from food and to transform this energy into a chemical form the cell can use, the molecule ATP.

### 4.6 The Cytoskeleton: Internal Scaffolding

- Cells have within them a web of protein strands, called a cytoskeleton, that provide the cell with structure, facilitate the movement of materials inside the cell, and facilitate cell movement.

- There are three principal types of cytoskeleton elements. Ordered by size, going from smallest to largest in diameter, they are microfilaments, intermediate filaments, and microtubules. Microfilaments are made of the protein actin. They help the cell move and capture prey by forming rapidly in the direction of movement and decomposing rapidly at their other end. Intermediate filaments provide support and structure to the cell. Microtubules play a structural role in cells and facilitate the movement of materials inside the cell by serving as transport "rails."

- Cilia and flagella are extensions of cells composed of micro-tubules. Cilia extend from cells in great numbers, serving to move the cell or to move material around the cell. By contrast, one—or at most a few—flagella extend from cells that have them. The function of flagella is cell movement.

## 4.7 The Plant Cell

- Plant cells have most of the structures found in animal cells—ribosomes, a cell nucleus, a rough ER, and so forth—although plant cells do not have the lysosomes found in animal cells. Plant cells have three structures not found in animal cells: a cell wall, a large central vacuole, and the organelles called chloroplasts. The cell wall gives the plant structural strength and helps regulate the intake and retention of water; the central vacuole stores nutrients and degrades waste products; and chloroplasts are the sites of photosynthesis.

Web Animation 4.1: The Structure of Cells

*BioFlix*   Tour of a Plant Cell

## 4.8 Cell-to-Cell Communication

- Cells are able to communicate with each other through special structures. Plant cells have channels, called plasmodesmata, that are always open and hence have the effect of making the cytoplasm of one plant cell continuous with that of another. Adjacent animal cells have channels, called gap junctions, that are composed of protein assemblages that open only as necessary, allowing the movement of small molecules and electrical signals between cells.

## KEY TERMS

| | |
|---|---|
| cell wall  82 | Golgi complex  73 |
| central vacuole  82 | intermediate filament  78 |
| chloroplast  84 | lysosome  74 |
| cilia  79 | microfilament  77 |
| cytoplasm  68 | micrograph  66 |
| cytoskeleton  76 | micrometer  66 |
| cytosol  65 | microtubule  78 |
| endomembrane system  73 | mitochondria  76 |
| eukaryotic cell  65 | nanometer  66 |
| flagella  79 | nuclear envelope  70 |
| gap junction  85 | nucleolus  72 |

| | |
|---|---|
| nucleus  65 | rough endoplasmic reticulum  72 |
| organelle  65 | smooth endoplasmic reticulum  74 |
| plasma membrane  65 | |
| plasmodesmata  84 | transport vesicle  73 |
| prokaryotic cell  65 | |
| ribosome  71 | |

## BRIEF REVIEW

*Answers to Brief Review questions are in the back of the book.*
*For multiple-choice quiz questions, go to* **www.aw.com/krogh4**.

1. Suppose your entire body were just one gigantic cell with one central nucleus and lots of organelles outside it to perform the various functions. What problems can you envision with this system?

2. What are some of the differences between eukaryotic cells and prokaryotic cells?

3. Insulin, a hormone that reduces blood-sugar levels in the body, is a protein that is exported from special cells in the pancreas. Trace the path from a length of DNA in the nucleus that carries the code for insulin to the release of the hormone in the bloodstream.

4. What are cilia and flagella? What are some of the roles they play in the human body?

5. Consider what plants would be like if they had no plasmodesmata. Explain why this would be a problem.

6. What is a gap junction, and what is its function?

## APPLYING YOUR KNOWLEDGE

1. The text notes that animals could not function as they do if their cells were encased in the thick cell walls that surround plant cells. What about the reverse case? Why would plants be unable to function if their cells lacked cell walls?

2. The earliest organisms in the fossil record are all single-celled. Why do you think life had to "start small" like this?

3. Why do membranes figure so prominently in eukaryotic cells? What essential function do they serve?

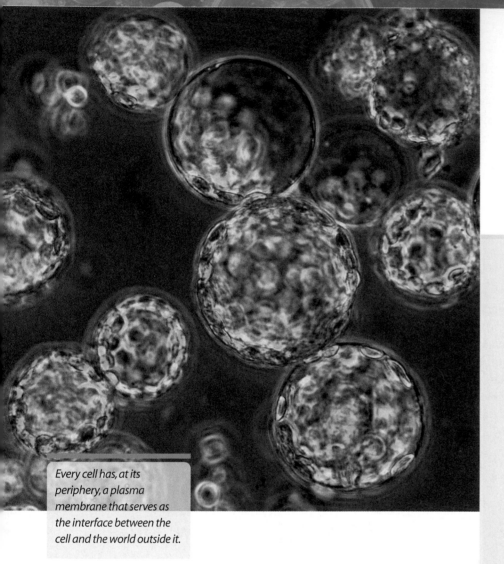

*Every cell has, at its periphery, a plasma membrane that serves as the interface between the cell and the world outside it.*

CHAPTER **5**

5.1   The Nature of the Plasma Membrane          91

5.2   Diffusion, Gradients, and Osmosis          95

5.3   Moving Smaller Substances In and Out       99

5.4   Getting the Big Stuff In and Out          101

**Essay**

**THE PROCESS OF SCIENCE:** The Fluid-Mosaic
Model of the Plasma Membrane              104

The outer lining of cells is in a sense the outer border of life. In its roles as protector, gatekeeper, and message carrier, this lining is indispensable to life.

# LIFE'S BORDER:
# The Plasma Membrane

**T**hings have steadily been getting better for the estimated 30,000 Americans who suffer from the disease cystic fibrosis. At one time, 80 percent of the victims of this illness didn't make it to their first birthday.

But with each passing year, small advances in the treatment of CF have led to small increases in life expectancy. Even so, today's average CF patient lives only to age 35. How does CF bring about this shortening of lives? Although its effects are many, its lethal effects usually have to do with the lungs. More specifically, CF causes a dangerous layer of mucus to accumulate in the passageways of the lungs. The lungs of healthy people are lined by a layer of mucus too. But in their case, the layer is thin and wet—it's pliable enough, we might say, that it can be regularly swept away by the hair-like cilia that function like tiny brooms in the lung passageways. Cystic fibrosis patients, conversely, have a mucus layer that is thicker and drier and that becomes a site for repeated bacterial infections. In time, these infections can result in the destruction of the lung passageways.

So, what accounts for the difference between the lungs of a healthy person and those of a person suffering from CF? Because of a faulty gene, the substance chloride cannot be transported in sufficient quantity from the inside of a CF patient's lung cells to the outside of them (where the mucus builds up). Healthy individuals have a protein that acts as a channel for chloride—a kind of tunnel that spans the cell's outer membrane. In contrast, the cells of CF patients have either defective chloride channels or none at all. The result is a lack of chloride outside their cells, which gets the mucus buildup going.

As noted last chapter, life goes on only inside cells. The fact that the cell's outer, or "plasma" membrane is out on life's edge may prompt the thought that it's *merely* at the edge, as if the real action is taking place deep within the cell. But, consider, as in CF, the effect of having a *defective* plasma membrane. The focus in this chapter is to learn more about this important cell lining.

## 5.1 The Nature of the Plasma Membrane

So, what is the nature of the plasma membrane? First, it's worth noting that it's very much like the membranes described in Chapter 4. Much of what follows about the plasma membrane could also be said of the various membranes that are internal to the cell—those that line lysosomes or that make up the Golgi complex, for example. There are some important differences between the plasma and other membranes, however, because the plasma membrane is constantly interacting with the world outside the cell, whereas internal membranes are not.

When we start to consider the qualities of the plasma membrane, we find that, even in the micro-world of cells, it is an extremely *thin* entity. If you stacked 10,000 of them on top of one another, their combined width would be about that of a sheet of paper. Next, the plasma membrane has a fluid and somewhat fatty makeup. If we looked for a counterpart to it in our everyday world, a soap bubble might come close, meaning this membrane is very flexible. Yet it is stable enough to stay together despite being constantly re-formed, thanks to the endless movement of materials in and out of it. If you imagine half the surface of your skin remaking itself every 30 minutes or so, you begin to get the picture about how dynamic this membrane is. Let's look now at the most important component parts of it. All these components are illustrated in **Figure 5.1**. They are:

- The phospholipid bilayer. Just as bread is made mostly from flour, the plasma membrane is made mostly from two layers of phospholipid molecules, with "tails" that mingle together, but "heads" that point in opposite directions—to the interior of the cell in one direction and to the world outside the cell in the other.

- Cholesterol. Like mortar between bricks, cholesterol acts as a "patching" material for the membrane in animal cells; it also keeps the membrane at an optimal level of fluidity.

- Proteins. These molecules shoot through the membrane and lie on either side of it, serving as support structures, signaling antennas, identification markers, and cellular passageways.

- Glycocalyx. This is the collective term for a set of carbohydrate chains that form a layer on the outside of the membrane.

Now let's look at each of these components in greater detail.

## First Component: The Phospholipid Bilayer

You may recall the discussion in Chapter 3 of phospholipids—molecules that have two long fatty acid chains linked to a phosphate-bearing group (**Figure 5.2**). You may also recall that, because fatty acid chains are lipids, they are hydrophobic or "water fearing," meaning they will not bond with water. Conversely, phosphate groups are charged, hydro*philic* molecules, meaning they are "water loving" and will *readily* bond with water. When such phosphate and lipid components are put together, the resulting phospholipid is a molecule with a dual nature. Its phosphate "head" will seek water, but its fatty acid "tail" will avoid it. Drop a group of

**Figure 5.1**
**The Plasma Membrane**

The cell's outer lining or plasma membrane is composed of four main structural elements, as shown in the figure.

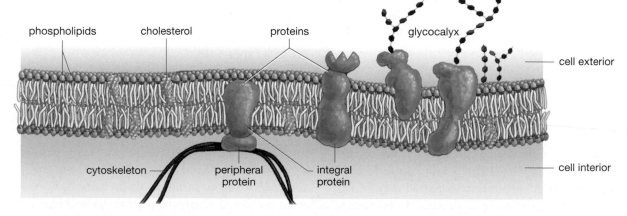

phospholipids   cholesterol   proteins   glycocalyx   cell exterior

cytoskeleton   peripheral protein   integral protein   cell interior

- **Phospholipid bilayer:** a double layer of phospholipid molecules whose hydrophilic "heads" face outward, and whose hydrophobic "tails" point inward, toward each other.

- **Cholesterol** molecules that act as a patching substance and that help the cell maintain an optimal level of fluidity.

- **Proteins**, which are integral, meaning bound to the hydrophobic interior of the membrane, or peripheral, meaning not bound in this way.

- **Glycocalyx:** sugar chains that attach to proteins and phospholipids, serving as protein binding sites and as cell lubrication and adhesion molecules.

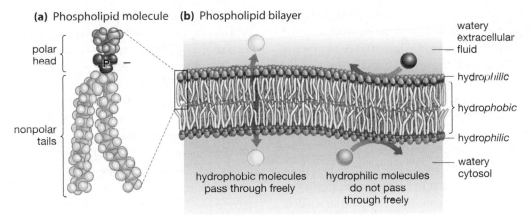

**Figure 5.2**
**Dual-Natured Lining**

**(a)** The essential building block of the cell's plasma membrane is the phospholipid molecule, which has two elements: a hydrophilic "head," containing a charged phosphate group that bonds with water; and hydrophobic "tails," composed of fatty acids that do not bond with water.

**(b)** Two layers of phospholipids sandwich together to form the plasma membrane. The phospholipids' hydrophobic tails form the interior of the membrane, while their hydrophilic heads jut out toward the watery environments that exist on both sides of the membrane. Although the bilayer forms a barrier to all but the smallest hydrophilic molecules, hydrophobic molecules can pass through it fairly easily.

phospholipids in water, and they will arrange themselves into the form you see in Figure 5.2: two *layers* of phospholipids sandwiched together, the tails of each layer pointing inward (thus avoiding water) and the heads pointing outward (thus bonding with the watery environment that lies both inside and outside the cell). With this in mind, the **phospholipid bilayer** can be defined as a chief component of the plasma membrane, composed of two layers of phospholipids, arranged with their fatty acid chains pointing toward each other.

This phospholipid structure has a couple of important consequences for the plasma membrane. First, it helps give the membrane its fluid nature: The bilayer's fatty acid tails have a chemical makeup that is more like that of oil than shortening. Second, because these are two hydrophobic layers of tails, the only substances that can pass through them with ease are other hydrophobic substances or a few very small molecules that aren't hydrophobic. This stems from a rule of thumb in chemistry: Like dissolves like. Fats dissolve in fats, charged molecules dissolve in charged molecules. Thus, various steroid hormones (which, remember, are lipids) gain fairly easy entry into the cell, as do fatty acids, because these substances will dissolve in the lipid bilayer. Conversely, hydrophilic substances—ions, amino acids, all kinds of polar molecules—will not dissolve in the membrane and thus cannot make it past the fatty acid chains without help. What kind of help? Stay tuned.

## Second Component: Cholesterol

Molecules of the lipid material cholesterol nestle between phospholipid molecules throughout the plasma membrane in animal cells, performing two functions. First, they act as a kind of patching substance on the bilayer, keeping some small molecules from getting through. Second, they help keep the membrane at an optimum level of fluidity. Without cholesterol, the plasma membrane would react to low temperatures the same way cooking oil does: It would change from a flexible substance to a hardened one—a dysfunctional condition for a cell. The cholesterol molecules nestled between the phospholipid's fatty acid tails keep this from happening by keeping these chains from packing too closely together. Likewise, they keep the membrane from becoming too fluid when temperatures rise.

## Third Component: Proteins

Embedded within, or lying on, the phospholipid bilayer is a third major group of membrane components, *membrane proteins*, of which there are two principal types, integral and peripheral. **Integral proteins** are plasma membrane proteins that are bound to the membrane's hydrophobic interior. In some cases, these proteins span the entire membrane—popping out on either side of it—while in other cases, they extend only partway in. **Peripheral proteins** are plasma membrane proteins that lie on either side of the membrane but are not bound to its hydrophobic interior. They are usually attached to integral proteins at the surface of the membrane.

It is hard to overstate the importance of proteins to the plasma membrane. By itself, the lipid bilayer has a handful of capabilities. With the addition of proteins, it is turned into a structure with a huge array of capabilities. Here is a selected set of the roles that membrane proteins play (**Figure 5.3**).

### Structural Support

Peripheral proteins that lie on the interior or "cytoplasmic" side of the membrane often are attached to elements of the cell's cytoskeleton. Thus anchored, they help to stabilize various parts of the cell and play a part in giving animal cells their characteristic shape. (Why not plant cells? Because their shape is determined by the cell wall.)

### Recognition

Much like military sentries, some cells need to know the answer to a critical question: Who goes there, friend or foe? The sentries are immune system cells, moving through the body and interacting with proteins that extend from cell surfaces. These proteins have "binding sites" on them with a shape that can convey a message to immune cells: "Self here, pass on by." Conversely, a cell with a *foreign* set of binding sites on its proteins can convey the message: "Invader here, begin an immune system attack." Have you ever wondered what it means for people to have, say, type

A blood? It means that they have type A recognition proteins extending from the plasma membranes of their red blood cells.

### Communication

Cells communicate with one another in various ways. Signals can be sent from one cell to a neighboring cell, and communication over longer distances is possible through the chemical messengers known as hormones. These signals are likely to be channeled through a **receptor protein**—a plasma membrane protein that binds with a signaling molecule.

Receptor proteins—usually just called receptors—have binding sites on them with a shape that is so specific they generally will bind only to a single type of signaling molecule. To take one example of this, the hormone insulin is a signaling molecule. The message it conveys to cells is: "Glucose levels are rising in the bloodstream; construct some channels in your plasma membranes so that you can take some of this glucose in." Now, how does insulin transmit this message? It binds with an insulin receptor protein that spans a given cell's plasma membrane. This binding causes a chemical change on the portion of the protein that lies *inside* the cell. The result is a cascade of cellular reactions that puts in motion the construction of the channels. Note that, through this means, a signaling molecule need not come into the cell to have an effect; it merely needs to bind with a receptor protein that spans the plasma membrane.

**Figure 5.3**
**Roles of Membrane Proteins**
Plasma membrane proteins serve four principal functions, as shown in the figure.

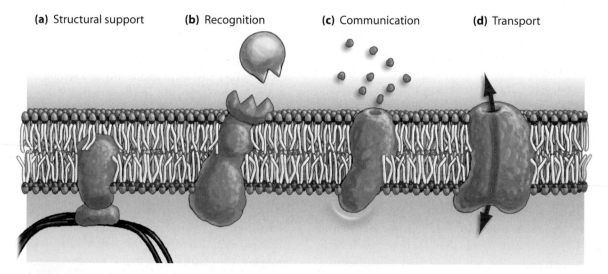

**(a)** Structural support **(b)** Recognition **(c)** Communication **(d)** Transport

Membrane proteins can provide structural support, often when attached to parts of the cell's scaffolding or "cytoskeleton."

Binding sites on some proteins can serve to identify the cell to other cells, such as those of the immune system.

Receptor proteins, protruding out from the plasma membrane, can be the point of contact for signals sent to the cell via traveling molecules, such as hormones.

Proteins can serve as channels through which materials can pass in and out of the cell.

### Transport

You have seen that most materials cannot simply pass through the cell's plasma membrane. Yet, cells need many of these very materials. How do they get in? Different kinds of integral proteins take on a transport task—they act as channels through which materials pass both into cells and out of them. In some cases, these proteins form what might be called passive passageways—microtunnels that simply provide a hydrophilic opening through a hydrophobic environment. In other cases, however, these protein channels are actively involved in moving materials through the membrane. We'll review all the forms of protein-aided transport shortly. For now, just note the existence of **transport proteins**: proteins that facilitate the movement of molecules or ions from one side of the plasma membrane to the other.

### Fourth Component: The Glycocalyx

If you look at Figure 5.1 and focus on the extracellular face of the plasma membrane, you can see, protruding from proteins and phospholipids alike, a number of short, branched extensions. These are simple carbohydrate chains, meaning sugar chains. It turns out that these chains serve as the actual binding sites for many signaling molecules, including insulin. (Remember in Chapter 4 how many of the proteins that were constructed had sugar side chains added to them? Well, now you can see what some of these chains do.) Sugar chains also serve to lubricate cells and allow them to stick to other cells by acting as an adhesion layer. Collectively, all these chains form the **glycocalyx**: an outer layer of the plasma membrane composed of short carbohydrate chains that attach to membrane proteins and phospholipid molecules.

### The Fluid-Mosaic Membrane Model

With this review of component parts under your belt, you're ready for a definition of the **plasma membrane**. It is a membrane, forming the outer boundary of many cells, composed of a phospholipid bilayer that is interspersed with proteins and cholesterol and coated on its exterior face with carbohydrate chains.

Taking a step back, the general message is that the plasma membrane is a loose, lipid structure peppered with proteins and coated with sugars. A better image, however, might be of a *sea* of lipids that is peppered with proteins, some of which are fixed in place, but some of which have a great deal of freedom to move sideways or "laterally" across the membrane, something like icebergs on an ocean. When we overlay this quality of movement on the basic membrane structure just reviewed, the result is the contemporary view of the plasma membrane, the **fluid-mosaic model:** a conceptualization of the plasma membrane as a fluid, phospholipid bilayer that has within it a mosaic of both stationary and mobile proteins. (For information on how scientists developed this conception of the membrane, see "The Process of Science" on page 104.)

---

**SO FAR...**

1. The plasma membrane of an animal cell has four principal components. These are the _____, which is composed of two layers of molecules with fatty acid tails that point toward each other; the lipid molecule _____, which serves as a patching and fluidity control substance; many _____, which serve in such diverse roles as passageways and communication molecules; and the _____, which is a collective term for a set of carbohydrate chains attached to the outside of the membrane.

2. The phospholipid molecules of the plasma membrane are dual natured. The "heads" of these molecules are _____, meaning they will readily bond with _____. Meanwhile, the tails point inward, toward each other, because they are _____, meaning they will not bond with _____.

3. Plasma membrane proteins serve in the role of _____ by binding with signaling molecules that come from outside the cell and in the role of _____ by acting as channels for substances moving into and out of the cell.

---

## 5.2 Diffusion, Gradients, and Osmosis

It may be obvious that one of the main roles of the plasma membrane is to let in substances that are needed within the cell while keeping out substances that aren't needed or that actually are harmful. Having learned something about the membrane's makeup, you will now see how it carries out this task of passage and blockage.

It is necessary to begin this story, however, not with the cell, but with the more general notion of how substances go from places where they are more concentrated to places where they are less concentrated. Anyone who has put some food dye in water has a sense of how this works: A few drops of, say, red dye will tumble into the liquid and start dispersing; eventually, the result is a solution that is uniformly more reddish than plain water (**Figure 5.4**). The question we'll look at is why this takes place.

### Random Movement and Even Distribution

All molecules are constantly in motion, and this motion is random. Because liquid molecules are bonded together so weakly, they can *act* on this motion—they can slide past each other and thus go from one location to another. (Meanwhile, the molecules that make up solids are so tightly bonded together that their motion amounts simply to vibrating in place.) The laws of thermodynamics dictate that, because liquid molecules move randomly, they will naturally move from any initial *ordered* state—in our example, their concentrated state when first dropped in the container—to their most disordered state, meaning evenly distributed throughout a given volume.

What is at work here is **diffusion**: the movement of molecules or ions from a region of their higher concentration to a region of their lower concentration. This notion carries with it the concept of a gradient. A **concentration gradient** is defined as the difference between the highest and lowest concentration of a solute within a given medium. In our example, the solute is the dye, and the medium is water. As with bicycles coasting down a grade, the natural tendency for any solute is to move down its concentration gradient, from higher concentration to lower. Our dye did just that. Bikes and solutes can move up a grade or gradient, but there is a price to be paid, and that price is the expenditure of energy. (On its own, would red dye ever come back to its concentrated state in the container? No. Some work would have to be performed for this to happen, which means that some energy would have to be expended.)

### Diffusion through Membranes

So far, you have looked at molecules diffusing in an undivided container. But the subject here is divisions—those provided by the plasma membrane. Let us consider, therefore, what happens to a solution in a container that is divided by a membrane. If that membrane is *permeable* to both water and the solute—that is, if both water and the solute can freely pass through it—and the solute lies only on one side of the membrane, then the predictable happens: The solute moves down its concentration gradient, diffusing right through the membrane, eventually becoming evenly distributed on both sides of it.

Now let's imagine a *semi*permeable membrane, one that water can freely move through but that solutes cannot. **Figure 5.5** shows you what happens if more solute—in this case, salt—is put on the right

**(a)** Dye is dropped in          **(b)** Diffusion begins          **(c)** Dye is evenly distributed

water molecules

dye molecules

**Figure 5.4**
**From Concentrated to Dispersed**
Diffusion is the movement of molecules or ions from areas of their greater concentration to areas of their lesser concentration. In this sequence of photos and diagrams, a few drops of red dye, added to a beaker of water, are at first heavily concentrated in one area but then begin to diffuse, eventually becoming evenly distributed throughout the solution.

side of the membrane than the left. Water flows through the membrane both ways, but *more* water flows into the right chamber, which has a greater concentration of solutes in it. The result is that the solution on the right side rises to a higher level than the one on the left.

This seems strange on first viewing, as if gravity were taking a vacation on the right side of the container. But what has been demonstrated here is **osmosis**: the net movement of water across a semipermeable membrane from an area of lower solute concentration to an area of higher solute concentration. Why should this occur? In the case of a solute like salt, water molecules will surround and bond with the sodium and chloride ions that salt separates into in solution. Because these solutes are not free to pass through the membrane, the water molecules bound to them will likewise remain confined to the right side. This means that more "free" water will exist on the left side—water that is free to act on the natural tendency of liquid molecules to remain in motion. The result is a net movement of water into the right side (see Figure 5.5b). Looked at one way, this is just another example of diffusion—of a substance moving from an area of its higher concentration to an area of its lower concentration. Once the solute binds with the water molecules on the right side of the container, there is a greater concentration of free water molecules on the left than on the right. By moving to the right, the molecules are simply moving down their concentration gradient.

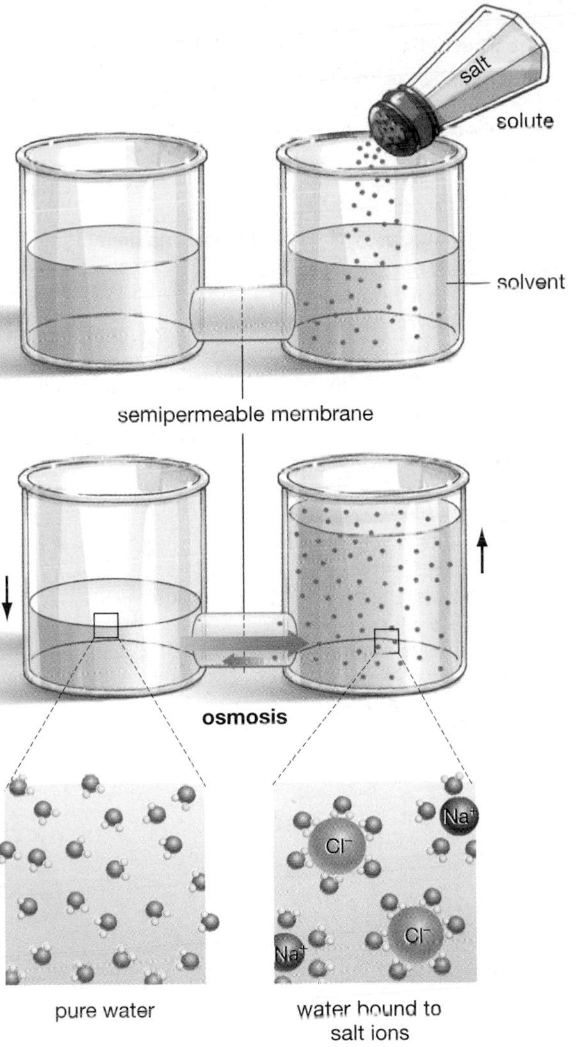

### The Plasma Membrane as a Semipermeable Membrane

So, what does movement of water have to do with the cell's outer lining? The cell's phospholipid bilayer is itself a semipermeable membrane. It is somewhat permeable to water and lipid substances but not permeable to larger charged substances. Thus, osmosis can take place across the plasma membrane—indeed, it does all the time. It is the primary means by which plants *get* water, and it is a player in all sorts of routine metabolic processes in animals. In humans, for example, much of the fluid portion of blood is at one point driven out of the blood vessels called capillaries by the force of blood pressure. How does this fluid get back in? Osmosis. The proteins that remain in the capillaries are like the salt in the container: They bring about an osmosis that pulls the fluid back into the blood vessels.

Or, think about a more extreme example. Why are we always told that someone who is stranded at sea should never drink salt water? Because doing so

**(a)** An aqueous solution divided by a semipermeable membrane has a solute —in this case, salt— poured into its right chamber.

semipermeable membrane

**(b)** As a result, though water continues to flow in both directions through the membrane, there is a net movement of water toward the side with the greater concentration of solutes in it.

**osmosis**

**(c)** Why does this occur? Water molecules that are bonded to the sodium ($Na^+$) and chloride ($Cl^-$) ions that make up salt are not free to pass through the membrane to the left chamber of the container.

pure water          water bound to salt ions

**Figure 5.5
Osmosis in Action**

would provide an enormous concentration of sodium chloride ions in the watery fluid that lies *outside* the cells. This is no different in principle than dumping more solutes into the right side of the container in Figure 5.5. The result would be water flowing out of cells, dehydration of the cells, and in extreme cases, death. (What actually kills people in this situation is a shrinkage of brain cells.)

This is an example of an *osmotic imbalance*, which is a situation in which osmotic pressure is either drying out cells or flooding them. In the case of flooding, plant cells have a great advantage over animal cells—their cell walls. Remember from Chapter 4 that one function of the cell wall is to regulate water uptake. Well, now we can see why this is so valuable. Animal cells, which do not have a cell wall, can expand until they burst when water comes in. Plant cells, conversely, will expand only until their membranes push up against the cell wall with some force, setting up a pressure level that keeps more water from coming in. Such tight quarters actually are an optimal condition

**Web Animation 5.1**

Plasma Membranes and Diffusion

## Figure 5.6
## Osmosis in Cells

**(a)** When solutes (such as salt) exist in greater concentration outside the cell than inside, water moves out of the cell by osmosis, and the cell shrinks. Here, the fluid surrounding the cell is hypertonic to the cell's cytoplasm.

**(b)** When solute concentrations inside and outside the cell are balanced, water movement into and out of the cell is balanced as well. Here, the fluid surrounding the cell is isotonic to the cell's cytoplasm.

**(c)** When solutes exist in greater concentration inside the cell than outside, water moves into the cell by osmosis. This influx may cause animal cells to burst, but plant cells are reinforced with cell walls and thus remain turgid—generally a healthy state for them. Here, the fluid surrounding the cell is hypotonic to the cell's cytoplasm.

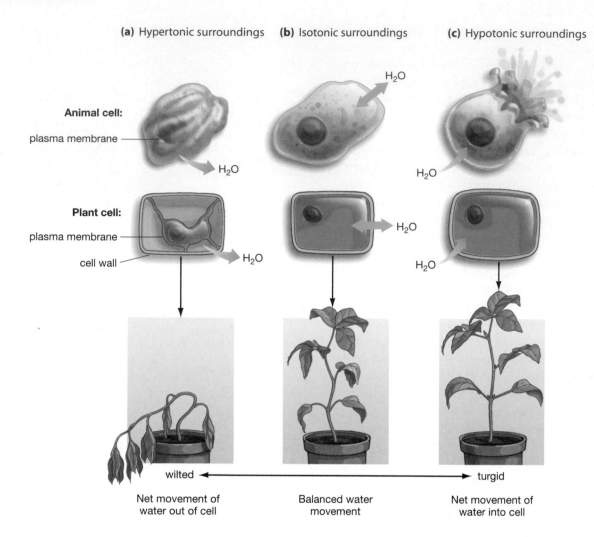

**(a)** Hypertonic surroundings    **(b)** Isotonic surroundings    **(c)** Hypotonic surroundings

Animal cell:
plasma membrane

Plant cell:
plasma membrane
cell wall

wilted ◄—————————————————————————► turgid

Net movement of water out of cell    Balanced water movement    Net movement of water into cell

for plants. A nice, crisp celery stick is one that has achieved this kind of *turgid* state, while a droopy celery stick that has lost this quality has cells that are *flaccid*; flowers or leaves in the latter condition are *wilted* (**Figure 5.6**).

### Osmosis and Cell Environments

Is a given cell likely to lose water to its surroundings, gain water from them, or have a balanced flow back and forth with them? Any of these things is possible, depending on what the solute concentration is outside the cell as opposed to inside. Three terms are helpful in describing the various conditions that can exist. A fluid that has a higher concentration of solutes than another is said to be a **hypertonic solution**. If a cell's surroundings are hypertonic to the cell's cytoplasm, water will flow out of the cell. Two solutions that have equal concentrations of solutes are said to be **isotonic**. If one of these solutions is the cell's cytoplasm and the other is the fluid surrounding the cell, fluid flow will be balanced between cell and surroundings.

Finally, a fluid that has a lower concentration of solutes than another is a **hypotonic solution**. If a cell's surroundings are hypotonic to the cell's cytoplasm, water will flow into the cell. Figure 5.6 gives examples of what can happen to cells in all three types of environments.

### SO FAR...

1.  Diffusion, the movement of molecules or ions from a region of their _____ to a region of their _____ takes place spontaneously in liquids, meaning it takes place without the input of any _____.

2.  Osmosis is the net movement of water across a _____ membrane, from an area of lower _____ to an area of higher _____.

3.  If a cell is surrounded by a fluid that is hypertonic to the cell's cytoplasm, water will flow _____ the cell. If a cell is surrounded by a fluid that is hypotonic to the cell's cytoplasm, water will flow _____ the cell.

# 5.3 Moving Smaller Substances In and Out

This excursion into the land of diffusion and osmosis has prepared us to start looking at the ways materials actually move into and out of the cell, across the plasma membrane. The big picture here is that some molecules are able to cross with no more assistance than is provided by diffusion. Other molecules require diffusion and the protein channels we talked about, and still others require channels and the expenditure of energy to get across. Different terms are applied to these different kinds of transport. **Active transport** is any movement of molecules or ions across a cell membrane that requires the expenditure of energy. **Passive transport** is any movement of molecules or ions across a cell membrane that does not require the expenditure of energy (**Figure 5.7**). Our review of transport begins with two varieties of passive transport.

## Passive Transport

### Simple Diffusion

Gases such as oxygen and carbon dioxide are such small molecules that they need only move down their concentration gradients to pass into or out of the cell. Having been delivered by blood capillaries to an area just outside the cell, oxygen exists there in greater concentration than it does inside the cell. Moving down its concentration gradient, it diffuses through the plasma membrane and emerges on the other side. This is an example of **simple diffusion**, which is diffusion through a cell membrane that does not require a special protein channel. Molecules that are larger than oxygen but that are fat soluble, such as the steroid hormones mentioned before, also move into the cell in this manner because of their ability to dissolve in the bilayer of the membrane. Carbon dioxide, which is formed *in* the cell as a result of cellular respiration, has a net movement *out* of the cell through simple diffusion.

### Facilitated Diffusion: Help from Proteins

Water can traverse the lipid bilayer through simple diffusion, moving through *despite* its hydrophilic nature primarily because it's such a small molecule. Yet, with water itself—and certainly with polar molecules larger than water—protein channels begin to be required for passage in and out of the cell. With this, the method of passage is no longer *simple* diffusion; it is **facilitated diffusion**, which is passage of materials through the plasma membrane that is aided by a transport protein.

With this, we get to the protein specificity referred to earlier. Each transport protein acts as a conduit for only one substance or at most a small family of related substances. Transport through these proteins begins with the kind of binding explained earlier. A circulating molecule of glucose, for example, latches onto the binding site of a glucose transport

**Passive transport**      **Active transport**

simple diffusion      facilitated diffusion

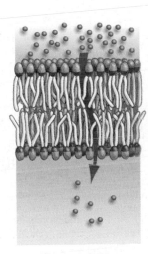

Materials move down their concentration gradient through the phospholipid bilayer.

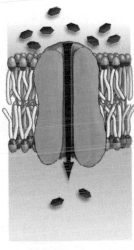

The passage of materials is aided both by a concentration gradient and by a transport protein.

Molecules again move through a transport protein, but now energy must be expended to move them against their concentration gradient.

ATP

**Figure 5.7**
**Transport through the Plasma Membrane**

••• ANSWERS

1. higher concentration; lower concentration; energy

2. semipermeable; solute concentration; solute concentration

3. out of; into

protein, as you can see in **Figure 5.8**. This binding causes the protein to change its shape, thus allowing the glucose molecule to pass through the membrane. In addition to glucose, other hydrophilic molecules, such as amino acids, move through the plasma membrane in this way. Note that this facilitated diffusion does not require the expenditure of energy because it has a concentration gradient working in its favor. Glucose exists in greater concentration outside most cells than inside them. It is moving down its concentration gradient, then, by passing into the cell through a protein channel. So, facilitated diffusion has something in common with simple diffusion. *Both* processes are driven by concentration gradients, meaning that neither requires metabolic energy. Thus, both are specific examples of passive transport.

Although the glucose in our example moved only one way in facilitated diffusion (into the cell), in general transport proteins are a channel for movement either way through the membrane. All that is required is a concentration gradient and binding of the material with the transport protein.

## Active Transport

If passive transport—either in simple or facilitated diffusion—were the only means of membrane passage available, cells would be totally dependent on concentration gradients. If, for example, molecules of a given amino acid come to exist in the same concentration on both sides of the plasma membrane, these amino acid travelers will continue to be transported, but they will be flowing out of the cell at the same rate they are flowing in. For cells, the problem is that some solutes are *needed* in greater concentration, say, inside the cell as opposed to outside. For example, the cells that line your small intestines need to take in a now-familiar substance, glucose, as a means of moving it out of your digestive tract and into your bloodstream. The problem is that, because these cells are in the glucose transport business, the levels of glucose inside them are nearly always higher than the levels of glucose outside them. In this instance, then, there is a concentration gradient working *against* the movement of glucose into the cells.

So, how do cells deal with a situation like this—in this case, how do the intestinal cells get the glucose in? They use pumps: They expend energy and pump the needed materials into themselves. This is active transport at work.

The energy for this transport can come from several sources. In our glucose example, the intestinal cells harness the power of electricity—of oppositely charged ions that attract each other across the cell membrane. Often, however, the energy for a molecular pump comes from a molecule called ATP (adenosine triphosphate), which supplies energy for all kinds of cellular operations. Active transport usually operates through the same kind of shape-changing channel proteins you saw in Figure 5.8 but with the important addition of an energy source.

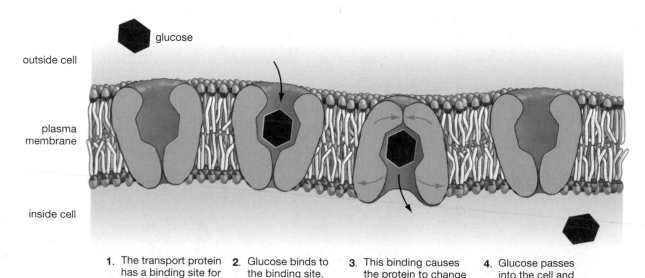

glucose

outside cell

plasma membrane

inside cell

1. The transport protein has a binding site for glucose that is open to the outside of the cell.
2. Glucose binds to the binding site.
3. This binding causes the protein to change shape, exposing glucose to the inside of the cell.
4. Glucose passes into the cell and the protein returns to its original shape.

**Figure 5.8**
**Facilitated Diffusion**
Owing to its size and hydrophilic nature, a molecule such as glucose cannot move into a cell through simple diffusion, even when it has a concentration gradient working in its favor. Its movement must be facilitated by a transport protein, such as the one pictured in the figure. This movement of glucose is passive transport, however, in that no expenditure of cellular energy is required to bring it about.

## 5.4 Getting the Big Stuff In and Out

Pumps and channels, diffusion and osmosis—these mechanisms move substances in and out of the cell, but as it turns out they only move relatively *small* substances. Remember an earlier discussion about the immune system cells that check to see if a given cell is friend or foe? Well, if an immune sentry finds a bacterial foe, it may have to ingest this *whole cell*. As you might guess, a cell cannot do this by employing little channels or pumps. You learned a bit in Chapter 4 about what does happen in these cases; now, you'll be looking at it in somewhat greater detail.

The methods in question here are *endocytosis*, which brings materials into the cell; and *exocytosis*, which sends them out. What these mechanisms have in common is their use of vesicles—the membrane-lined enclosures you learned about last chapter that alternately bud off from membranes or fuse with them.

### Movement Out: Exocytosis

**Exocytosis** is defined as the movement of materials out of the cell through a fusion of a transport vesicle with the plasma membrane. As **Figure 5.9** shows, exocytosis involves a transport vesicle making its way to the plasma membrane and fusing with it, whereupon the vesicle's contents are released into the extracellular fluid. You observed in Chapter 4 that cells use exocytosis when they are exporting proteins. For single-celled creatures, waste products may also be released into the extracellular fluid through this process.

### Movement In: Endocytosis

**Endocytosis** is the movement of relatively large materials into the cell by an infolding of the plasma membrane. It can take two major forms: pinocytosis, which brings in molecules that are larger than those we've looked at so far, and phagocytosis, which brings in much larger molecules yet, as you'll see.

### Pinocytosis

There actually are several varieties of pinocytosis, but we'll only look at one representative form here, called clathrin-mediated endocytosis or CME. This means of bringing materials into cells was once known as receptor-mediated endocytosis, but it turns out that almost all forms of endocytosis employ receptors. The receptors active in CME first bind with molecules of the material that's brought into the cell. Then, while the receptors are holding onto their molecular cargo, they move laterally across the membrane and congregate in a depression referred to as a coated pit. What the pit is coated *with*, on its underside, is the chemically active protein clathrin (hence the term clathrin-mediated endocytosis). Eventually, the pit deepens (or "invaginates") and pinches off, creating the familiar vesicle moving into the cell. CME is very important in getting nutrients and other substances into cells. It is the way cholesterol gets into our cells, for example. Alas, it is

**(a)** Exocytosis

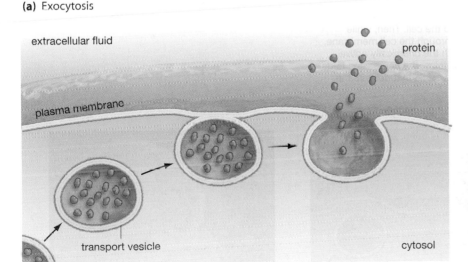

extracellular fluid

protein

plasma membrane

transport vesicle

cytosol

**(b)** Micrograph of exocytosis

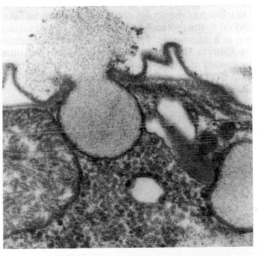

**Figure 5.9**
**Movement Out of the Cell**

**(a)** In exocytosis, a transport vesicle—perhaps loaded with proteins or waste products—moves to the plasma membrane and fuses with it. This section of the membrane then opens, and the contents of the former vesicle are released to the extracellular fluid.

**(b)** Micrograph of material expelled from the cell through exocytosis.

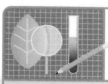

# THE PROCESS *of* SCIENCE
## The Fluid-Mosaic Model of the Plasma Membrane

**T**he richly detailed illustrations found in science textbooks do a great job of presenting information, but they also can be a little misleading. Given their clarity, such illustrations might leave students with the impression that many of the discoveries that scientists make would be apparent to anyone who simply looked in the right place with the right microscope.

**Figure 1a** presents an example of the kind of pictures (or "micrographs") we have of the plasma membrane, while **Figure 1b** presents an example of the kind of artist's rendering you've been seeing of it.

*Our view of the cell membrane was constructed piece by piece in an intellectual chain that stretched out for decades.*

You'll probably agree that no one could have discerned the detail in Figure 1b just by looking at Figure 1a. How, then, do we know that the plasma membrane actually is structured as you see it in Figure 1b? How do we know that it is largely lipid, and that proteins move about on the surface of it like so many icebergs? As it turns out, our view of the cell membrane was constructed piece by piece in an intellectual chain that stretched out for decades.

The idea that the surface of the cell is largely made of lipid material was set forth as early as 1855 by investigator Carl Nägeli. In 1899, Charles E. Overton came to the same conclusion by working with some cellular extensions of plant roots. Overton found that lipid-soluble substances moved quite easily into these "root hairs," while water-soluble substances did not. Remember the saying about like dissolving like? Lipid substances will dissolve within *lipid materials*. Since Overton saw this exact thing happening, his conclusion was that the outside lining of the root cells was composed of lipids.

In 1925, two Dutch scientists, E. G. Gorter and F. Grendel, decided to take red blood cells and measure their lipid content. They knew from the work of another scientist that a given volume of phospholipids would cover a certain *area* when arrayed in a layer. What they found, however, was that the lipids they extracted from the blood cells spread out into an area *twice* as large as would be expected from this benchmark. This doubling, they concluded, came about because the cell membrane was doubled; it existed as the *bi*layer you've been reading about.

Following this advance, however, it was apparent that the plasma membrane had to be composed of more than just two layers of lipids pressed together. For one thing, it was clear that closely related types of hydrophilic molecules passed through the membrane at different rates. This was difficult to reconcile with a uniform phospholipid membrane, which would have let all such substances pass through at nearly the same rate. Given such evidence, James Danielli, writing with Hugh Davson, proposed a view of the plasma membrane that ended up having a long run in science. With modifications over time, the Davson-Danielli model, as it was called, had the honor of being "generally accepted" by scientists from the time it was set forth in 1935 until the early 1970s. It proposed that a *layer of proteins* coated each side of the phospholipid bilayer.

By the 1950s, electron microscopes had become refined enough that they yielded a "railroad track" picture of the plasma membrane: two dark bands separated by a space in the middle (the view you see in **Figure 2**). Seduced, perhaps, by the comforts of what was "known" to be true, scientists took such pictures to be visual proof of the Davson-Danielli model. The two dark lines were protein bands, it was thought, and the space in the middle was the phospholipid bilayer. We now know, however, that the "tracks" actually were electrons that bound both with membrane proteins and with the charged portions of the phospholipid bilayer, thus creating the impression of a continuous track. In reality, there were no protein bands.

Through these years, it was also becoming apparent that there wasn't just one membrane surrounding the cell; instead, there were

**(a)** Micrograph of the plasma membrane

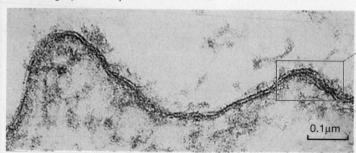

0.1μm

**(b)** Artist's rendering of it

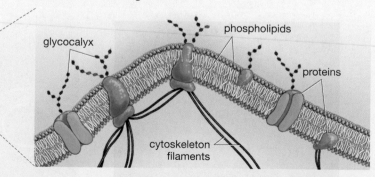

glycocalyx

phospholipids

proteins

cytoskeleton filaments

**Figure 1**
**How Much Can a Picture Tell Us?**

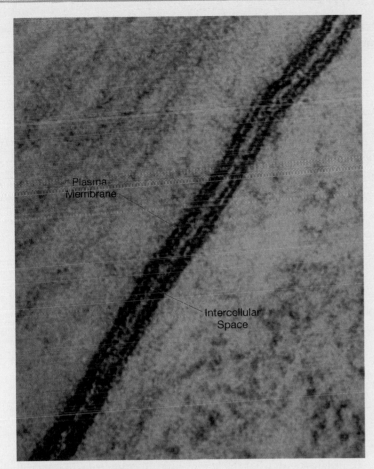

**Figure 2**
**A Close-up of the Plasma Membrane**
Electron micrograph of two adjoining cells, each of which has a plasma membrane with a "railroad track" appearance. The intercellular space is the space between the two cells. (×470,000)

many membranes that were part of a cell—membranes encasing mitochondria, membranes making up the Golgi complex, and so forth. In 1960, J. David Robertson set forth a proposal that saw *all* the cell's membranes as existing within the framework of the Davson-Danielli model. By that time, the venerable model had been modified to include the notion of some proteins *spanning* the bilayer, from cytoplasm to outside face. This, then, was the generally accepted view of cell membranes in the early 1960s: Davson-Danielli-Robertson, which posited a phospholipid bilayer coated on each side with a protein layer and shot through here and there with other proteins.

Into this situation came a chemist, then at Yale, named Jonathan Singer. In 1962, Singer got to thinking about cell membranes and realized that the Davson-Danielli-Robertson model had a real problem: It would require hydro*phobic* parts of the supposed protein coat to be in contact with *water*. This was like assuming you could set one marble on top of another and expect it to stay there. By 1964, Singer was convinced not only that the Davson-Danielli-Robertson model had some shortcomings, but that he had a model that improved on it. It's one thing, however, to believe you're right, based on general chemical principles, and quite another to prove it.

"What happened was that I wasn't going to propose a model without some kind of experiments to go with it," Singer said in an interview. Working with a postdoctoral student of his, John Lenard, Singer published a scientific paper in 1966 that made the basic break with Davson-Danielli-Robertson. The view that emerged from this paper was that the membrane had a series of *individual* proteins that either were embedded in the bilayer or lay on it. By 1971, Singer was able to set forth a case for his model in great detail in a book chapter. The problem was that nobody paid much attention to it. So he set out to write a review of the same subject in *Science* magazine. In the interval between the book chapter and the magazine article, however, one last piece of the membrane puzzle fell into place.

Recall that the modern view of the cell membrane is the fluid-mosaic model. It is *mosaic* because it asserts that there are discrete proteins within the phospholipid bilayer, but it is *fluid* because these proteins (and the phospholipids themselves) can *move* laterally through the membrane. In 1971, Singer had the mosaic part; by 1972, he had the fluid as well.

"I came to Dr. Singer's lab in 1967," said Garth Nicolson, who was a graduate student with Singer in those days and is now president of the Institute for Molecular Medicine in Huntington Beach, California. "I became very interested in the polarity of membranes . . . what I found when doing this work was that the distribution of components [on the surface of the cell] seemed to change dynamically." What Nicolson saw, in other words, was that the constituent parts of the membrane were shifting around on it.

Singer and Nicolson are agreed that a paper by L. D. Frye and Michael Edidin made Singer a believer in the rapid movement of materials across the cell surface. The upshot was an article called "The Fluid Mosaic Model of the Structure of Cell Membranes," which appeared in *Science* magazine in February 1972, with Singer and Nicolson as authors. It turned out to have great impact. Read any biology textbook today, and the account of the cell's plasma membrane is essentially the view set forth in this article. The illustrations contained in textbooks likewise all look like a drawing in the *Science* article. Such was its effect that the fluid-mosaic model is just as likely to be referred to as the "Singer-Nicolson" model, even though, as Garth Nicolson is the first to say, "this is really [Singer's] model . . . he should get the credit for the membrane as far as I'm concerned."

Does this mean that the fluid-mosaic model has at last given us an accurate picture of the membrane? Thus far, the answer seems to be yes. While the original Singer model wasn't perfect, it still wears the mantle of general acceptance by scientists. To be sure, some tweaking of the model has been required. The lateral movement of proteins across the membrane is more restricted than Singer originally thought (though the *means* of this restriction is a matter of great debate at present). And proteins are generally agreed to take up more of the surface of the cell than was believed back in the 1970s. Nevertheless, nothing has come along to replace the Singer-Nicholson model. Given that these scientists proposed this conception of the membrane more than 35 years ago, their model now has had its own long run in science.

# CHAPTER 5 REVIEW

## SUMMARY

### 5.1 The Nature of the Plasma Membrane

- The plasma membrane is a thin, fluid entity that manages to be very flexible and yet is stable enough to stay together despite being continually remade due to the constant movement of materials in and out of it. In animal cells, the plasma membrane has four principal components: (1) a phospholipid bilayer, (2) molecules of cholesterol interspersed within the bilayer, (3) proteins that are embedded in or that lie on the bilayer, and (4) short carbohydrate chains on the cell surface, collectively called the glycocalyx, that function in cell adhesion and as binding sites on proteins.

- Phospholipids are molecules composed of two fatty acid chains linked to a charged phosphate group. The fatty acid chains are hydrophobic, meaning they avoid water, while the phosphate group is hydrophilic, meaning it readily bonds with water. Such phospholipids arrange themselves into bilayers—two layers of phospholipids in which the fatty acid "tails" of each layer point inward (avoiding water), while the phosphate "heads" point outward (bonding with it). Phospholipids take on this configuration in the plasma membrane because a watery environment lies on either side of the membrane.

- The cholesterol molecules that, in animal cells, are interspersed between phospholipid molecules in the plasma membrane perform two functions: They act as a patching material that helps keep some small molecules from moving through the membrane; and they keep the membrane at an optimal level of fluidity.

- Some plasma membrane proteins are integral, meaning they are bound to the hydrophobic interior of the phospholipid bilayer. Others are peripheral, meaning they lie on either side of the membrane but are not bound to its hydrophobic interior.

- The functions that membrane proteins serve include (1) structural support; (2) cell identification, by serving as external recognition proteins that interact with immune system cells; (3) communication, by serving as external receptors for signaling molecules; and (4) transport, by providing channels for the movement of compounds into and out of the cell.

- The plasma membrane today is described by a conceptualization called the fluid-mosaic model, which views the membrane as a fluid, phospholipid bilayer that has a mosaic of proteins either fixed within it or capable of moving laterally across it.

### 5.2 Diffusion, Gradients, and Osmosis

- Diffusion is the movement of molecules or ions from a region of their higher concentration to a region of their lower concentration.

A concentration gradient defines the difference between the highest and lowest concentrations of a solute within a given medium. Through diffusion, compounds naturally move from higher to lower concentrations—meaning down their concentration gradients. Energy must be expended, however, to move compounds against their concentration gradients, meaning from a lower to a higher concentration.

- A semipermeable membrane is one that allows some compounds to pass through freely while blocking the passage of other compounds. Osmosis is the net movement of water across a semipermeable membrane from an area of lower solute concentration to an area of higher solute concentration. Because the plasma membrane is a semipermeable membrane, osmosis operates in connection with it. Osmosis is a major force in living things; it is responsible for much of the movement of fluids into and out of cells. Osmotic imbalances can cause cells either to dry out from losing too much water or, in the case of animal cells, to break from taking too much water in. Plant cells generally do not have the latter problem because their cell walls limit their uptake of water.

- Cells will gain or lose water relative to their surroundings in accordance with what the solute concentration is inside the cell as opposed to outside it. A cell will lose water to a surrounding solution that is hypertonic—a solution that has a greater concentration of solutes in it than does the cell's cytoplasm. A cell will gain water when the surrounding solution is hypotonic to the cytoplasmic fluid. Water flow will be balanced between the cell and its surroundings when the surrounding fluid and the cytoplasmic fluid are isotonic to each other—when they have the same concentration of solutes.

Web Animation 5.1: Plasma Membranes and Diffusion

### 5.3 Moving Smaller Substances In and Out

- Some compounds are able to cross the plasma membrane strictly through diffusion; others require diffusion and special protein channels; still others require protein channels and the expenditure of cellular energy. Active transport is any movement of molecules or ions across a cell membrane that requires the expenditure of energy. Passive transport is any movement of molecules or ions across a cell membrane that does not require the expenditure of energy.

- There are two forms of passive transport: simple diffusion and facilitated diffusion. For either form of transport to bring about a net movement of materials into or out of a cell, a concentration gradient must exist. Such a gradient is all that is required for simple diffusion to operate. Facilitated diffusion, however, requires both a concentration gradient and a protein channel. In facilitated diffusion, transport proteins function as channels for larger hydrophilic substances—substances that, because of their size and electrical charge, cannot diffuse through the hydrophobic portion of the plasma membrane.

- Cells cannot rely solely on passive transport to move substances across the plasma membrane. It may be that a given cell needs to maintain, on one side of its membrane, a greater concentration of a given substance than exists on the other side. Yet, passive transport serves to equalize concentrations of substances on both sides of the plasma membrane. To deal with such needs, cells employ chemical pumps to move compounds across the plasma membrane against their concentration gradients. This is active transport. One example of such transport is the pumping of glucose into cells that line the small intestines.

### 5.4 Getting the Big Stuff In and Out

- Larger materials are brought into the cell through endocytosis and moved out through exocytosis. Both mechanisms employ vesicles, the membrane-lined enclosures that alternately bud off from membranes or fuse with them. In exocytosis, a transport vesicle moves from the interior of the cell to the plasma membrane and fuses with it, at which point the contents of the vesicle are released to the environment outside the cell.

- There are two principal forms of endocytosis: (1) pinocytosis, the movement of moderate-sized molecules into a cell by means of the creation of transport vesicles produced through an infolding or "invagination" of a portion of the plasma membrane; and (2) phagocytosis, in which certain cells use pseudopodia or "false feet" to first surround and then engulf whole cells, fragments of them, or other large organic materials. As in pinocytosis, such materials are brought into the cell inside vesicles that bud off from the plasma membrane.

## KEY TERMS

| | |
|---|---|
| **active transport**  99 | **isotonic solution**  98 |
| **concentration gradient**  96 | **osmosis**  97 |
| **diffusion**  96 | **passive transport**  99 |
| **endocytosis**  101 | **peripheral protein**  93 |
| **exocytosis**  101 | **phagocytosis**  102 |
| **facilitated diffusion**  99 | **phospholipid bilayer**  93 |
| **fluid-mosaic model**  95 | **pinocytosis**  102 |
| **glycocalyx**  95 | **plasma membrane**  95 |
| **hypertonic solution**  98 | **receptor protein**  94 |
| **hypotonic solution**  98 | **simple diffusion**  99 |
| **integral protein**  93 | **transport protein**  95 |

## BRIEF REVIEW

*Answers to Brief Review questions are in the back of the book.*
*For multiple-choice quiz questions, go to* **www.aw.com/krogh4**.

1. Imagine a receptor on the surface of one of your muscle cells that responds only to glucose. Plenty of insulin molecules are passing by this receptor, but it does not respond to them. Why not?

2. You often hear how dangerous it is for athletes to take steroids to "bulk up" and how many problems steroids cause in various parts of the body. Why do these drugs affect all parts of the body—not just the muscles?

3. Explain why the phospholipid "heads" of the plasma membrane are always pointed toward the cytosol and extracellular fluid, whereas the "tails" are always oriented toward the middle of the membrane.

4. Describe the form of endocytosis, called clathrin-mediated endocytosis, that is reviewed in the text.

5. As you saw in this chapter, all cellular membranes are similar in structure. However, what differences are likely to exist between plasma membranes and the membranes that exist in the interior of the cell?

## APPLYING YOUR KNOWLEDGE

1. All life-forms on Earth exist either as single cells or as collections of cells. All such cells have a plasma membrane at their periphery (although many cells have cell walls outside their plasma membrane). Why do you think the plasma membrane exists in all living things? Why aren't there living things that don't have a membrane at their periphery?

2. Because materials move from high concentration to low concentration via simple diffusion, and because they move the same way via facilitated diffusion, why do cells make proteins to carry out facilitated diffusion? Isn't this just a waste of energy?

3. Place a number of marbles on one end of a cafeteria tray. Shake the tray gently a few times, keeping it level. What happens to the marbles? Start over and shake faster the next time. What happens now? Explain how this is like diffusion.

*All living things must have a constant supply of energy. Here, a golden stonefly (Hesperoperla pacifica) expends considerable energy getting from one place to another near the banks of Oregon's Metolius River.*

**6.1** Energy Is Central to Life    **109**

**6.2** The Nature of Energy    **110**

**6.3** How Is Energy Used by Living Things?    **112**

**6.4** The Energy Dispenser: ATP    **113**

**6.5** Efficient Energy Use in Living Things: Enzymes    **115**

**6.6** Lowering the Activation Barrier through Enzymes    **116**

**6.7** Regulating Enzymatic Activity    **118**

Energy flows through life in a never-ending pattern of acquisition, storage, and use.

# LIFE'S MAINSPRING:
# An Introduction to Energy

**W**hen my daughter Tessa was young, she had a teddy bear with a music box in it that let us listen to "Brahms's Lullaby." All I had to do was wind up the key that attached to the music box. After a few turns of the wrist, the melody would cycle through several times, then slow for a last few bars, then stop. This process could be repeated as many times as I wound the music box. As we listened, a mechanism in the music box kept the melody coming out at an even tempo. Even the slowing at the end seemed just right for a lullaby, since the audience was a sleepy child.

## 6.1 Energy Is Central to Life

The forces that determined how Tessa's teddy bear worked also turn out to condition life since, to function, both the music box and life must be supplied with the same thing: energy. Just as the music from the box depended on energy from an "outside hand" (mine) to continue, so almost all life on Earth depends on energy from an outside source: the sun, whose daily showering of energetic rays lets the green world of plants and algae flourish. These organisms then pass their bounty along to animals and other living things in the form of food to eat and oxygen to breathe.

Beyond this, the sun and the music box have something else in common: Their energy always runs downhill, from more concentrated to less. The spring in the music box spontaneously unwinds, but it will not rewind itself. The sun sends out light and heat, but this energy does not spontaneously come together to form another sun. The music from the music box eventually must come to a stop, and in a similar way, even the sun will not shine forever. With all its grandeur, its energy is winding down, its brilliance dispersing. The difference is that once the sun's showering of energy comes to a stop, nothing will restart it.

While the sun's gift lasts, however, living things can transform its energy and use it to do things like sprout leaves and swim through oceans. Such complex activities make it imperative, though, that living things be able to *control* the energy they capture. The spring within the music box unwinds at a measured rate, spinning out the melody at just the right pace; similarly, living things have elaborate mechanisms in place that allow them to make the most of the energy they receive.

Principles such as these are important enough to biology that this chapter and the two that follow are devoted to the subject of energy and life. In this chapter, you'll learn some basics about energy and some energy-saving molecules called enzymes. Then, in Chapter 7, you'll see how living things "harvest" energy, meaning how they extract energy from food. Finally, In Chapter 8, you'll look at the ways that plants, algae, and some bacteria capture the sun's energy in the process called photosynthesis. We'll start now by looking at the energy basics.

## 6.2   The Nature of Energy

What is energy? It is commonly defined as the capacity to do work, but this definition merely passes the question mark along to the word *work*. Given this, here's one possible definition of **energy**: the capacity to bring about movement against an opposing force. The thing that makes energy so tricky as a concept is that, although we can measure it with great precision, and we can experience its effects, we cannot grasp it or see it. We can see the water that drives a waterwheel, but who has seen the energy the water contains (**Figure 6.1**)?

### The Forms of Energy

Energy can be thought of as coming in different forms, many of them familiar. Mechanical energy is captured in the wound-up spring of the musical teddy bear. Chemical energy is the energy held in the chemical bonds of a peach or a lump of coal. One other way to conceptualize energy is as potential and kinetic. **Potential energy** is stored energy: the rock perched precariously at the top of the hill; the charged ions kept on one side of a cell membrane. Conversely, **kinetic energy** is energy in motion, as with the rock tumbling down the hill or charged ions rushing in through a protein channel. If we define energy as the capacity to bring about movement against an opposing force, we can see how kinetic energy fits into this scheme. A rock that tumbles off a cliff and into a lake would cause water in the lake to splash upward—against the force of gravity.

**Figure 6.1**
**Where's the Energy?**

A water wheel turns as a result of energy supplied by moving water. Though the movement of the water is plainly visible, the energy contained in the water is not.

### The Study of Energy: Thermodynamics

Given how important energy is to life, it makes sense that there is a branch of biology, called *bioenergetics*, that links biology with a scientific discipline known as **thermodynamics**, the study of energy. Although thermodynamics is a mathematical and rather abstract discipline, research into it was prompted originally by a very down-to-earth goal: the desire to build a better steam engine. British inventors had harnessed steam power in the eighteenth century, and this discovery had the awesome impact of sparking the Industrial Revolution. The word *revolution* is fitting here because before the advent of steam power, people used the stored energy in, say, wood or coal solely to produce *heat*. The steam engine was a device that allowed them to use heat to perform *work* on a scale that was previously impossible. Heat could be channeled, it was discovered, to drive an engine, which in turn could power any number of industrial processes.

With this development, new questions confronted scientists: How exactly did heat drive a steam engine? Was there a finite quantity of energy, such that it could all be used up? Research carried out in England, Germany, and France during the nineteenth century yielded the answers to these questions, and the result was the development of some insights known as the laws of thermodynamics.

### The First Law of Thermodynamics: The Transformation of Energy

The **first law of thermodynamics** states that energy is never created or destroyed, but is only transformed. The sun's energy is not used up by green plants; rather, some of this energy is *converted* by the plants into chemical form.

On first blush, this is a tricky concept to grasp. We might ask: How can something as seemingly unenergetic as, say, a peach contain energy? Well, setting aside the water in it, a peach is mostly composed of carbohydrates and, as you may recall from Chapter 3, carbohydrates are molecules made of carbon, oxygen, and hydrogen atoms that are linked together in very specific ways. You could think of the sun's rays as having moved the electrons in these atoms up an energy hill, at which point they were locked into place, at their higher level, through the process of chemical bonding. At this point, a given electron is like a rock perched at the top of a hill. Just as a rock can fall down the hill, releasing energy as it goes, so the electrons in the peach can fall down an energy hill, releasing energy as they go.

A critical point here, however, is that the tree that sprouted this peach could put only *some* of the solar energy it received into chemical form. As it turns out, a plant cannot convert all of the energy it receives into carbohydrates. Likewise, a steam engine actually converts only a fraction of the energy contained in coal into the movement of a piston. What happens to the energy that is not stored in carbohydrates or transformed into motion? It can't be destroyed; that's clear from the first law. What happens is that it is converted into the form of energy known as *heat*. For every energy "transaction" that takes place, at least some of the original energy is converted into heat. This is true for every transaction in every energy system: trees producing peaches, electric utilities burning coal, and your brain decoding this sentence. The way this concept is usually phrased, however, is that some of the original energy is *lost* to heat. At first glance, this may seem like an unfair knock on heat, which after all has intrinsic value. Why, then, is energy "lost" to heat?

Well, consider what might be called the before and after of an energy transaction involving coal. In the "before," a lump of coal is a very ordered object containing carbon atoms that exist in a precise spatial relationship to one another. Thus bonded, these atoms are not free to move. However, bringing a flame to the lump of coal causes the coal's carbon atoms to react with oxygen. Chemical bonds are broken through this process, and the energy stored in these bonds is released as heat. But heat is the random motion of molecules, and these molecules have no tendency to stay concentrated. Thus, it's clear that we have gone from the *ordered, concentrated* energy in the chemical bonds of coal to the *disordered, dispersed* energy of heat.

### The Second Law of Thermodynamics: The Natural Tendency toward Disorder

And this gets us to the critical thing. Energy transformations will spontaneously run *only* from greater order to lesser order. Some of the heat produced in a steam engine is useful; it's driving a piston. Some of it is not, however—it's simply dissipated. But all of it amounts to a form of energy that is more dispersed than the chemical bonds in coal and that will quickly disperse further. Once this happens, there is no chance that it will spontaneously convert into anything *but* heat. We expect to see a lump of coal burn to ashes and to feel the hot air that results disperse, but we do not expect to see the ashes re-form into a lump of coal or to feel the hot air concentrate itself again.

In yielding energy, matter goes from a more-ordered state to a less-ordered state. The scientific principle that speaks to this is the **second law of thermodynamics:** Energy transfer always results in a greater amount of disorder in the universe. In connection with this law, there is a term you will hear—*entropy*, which is a

measure of the amount of disorder in a system. The greater the entropy, the greater the disorder. **Figure 6.2** summarizes the principles of the first and second laws of thermodynamics. **Figure 6.3** shows you how much energy is lost to heat in several energy systems.

### The Consequences of Thermodynamics

All of this may seem like just so much . . . *hot air*, but it turns out that few things so profoundly condition the universe we live in as these thermodynamic rules. Indeed, the relentless increase in entropy that the second law entails may dictate nothing less than the fate of the universe.

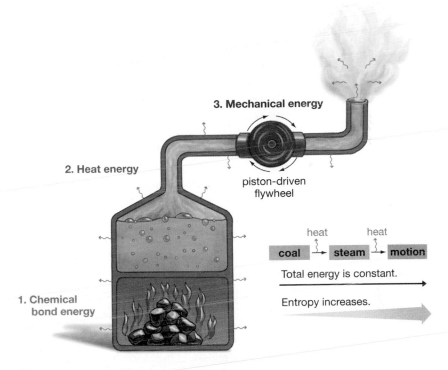

**Figure 6.2**
**The Transformations of Energy**

In a steam engine, energy locked up in the chemical bonds of coal is transformed into heat energy and mechanical energy. There is no loss of energy in this process, but energy is transformed from a more-ordered, concentrated form (the chemical bonds of coal) to a less-ordered, more dispersed form (heat). Thus, the amount of disorder—or entropy—increases in the transaction.

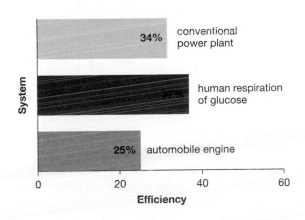

**Figure 6.3**
**Energy Efficiency**

The efficiency of several energy systems, as measured by the proportion of energy they receive relative to what they then make available to perform work. In measuring the efficiency of the car engine, the question is: How much of the energy contained in the chemical bonds of gasoline is converted by the car into the kinetic energy of wheel movement? In each system, most of the energy not available for work is lost to heat.

Allow yourself, for a second, to think of the universe as consisting solely of the sun, the Earth, and the space between them. The sun is a concentration of hydrogen that is undergoing a nuclear reaction, thus releasing light. Living things on Earth manage to capture buckets of the solar energy in this light through the activities of plants and the other photosynthesizers. But every time one of these buckets is poured into another—every time solar energy from the sun is transformed into energy stored in plants, for example—there is some spillage into heat. And once heat is generated, there is no spontaneous way it can make its way back into the orderly form of sunlight or carbohydrates. With every firing of a car's cylinder or contraction of a muscle, then, some amount of energy is lost to heat.

Against this, life as we know it is made possible by ordered concentrations of energy. The sun can sustain life, but dispersed heat will not. One day, our sun will have wound down like the spring in a music box, bringing to an end the energy source for life on Earth. One possible scenario for the universe as a whole is that all its stars eventually will darken, leaving a cold universe with little potential for life—indeed, with little activity of any sort. The laws of thermodynamics are powerful indeed.

---

### SO FAR...

1. Energy that is stored is _____ energy, while energy in motion is _____ energy.

2. The first law of thermodynamics says that energy is never _____ or _____ but is only _____. The second law of thermodynamics says that energy transfer always results in a greater amount of _____ in the universe.

3. Whenever energy is transferred—for example, from the chemical bonds in gasoline to the movement of a car's pistons—some of the original quantity of energy is lost to _____.

---

## 6.3 How Is Energy Used by Living Things?

A sprouting plant is constantly building itself up; it is making larger, more complex molecules (proteins, starches) from smaller, simpler ones (amino acids, simple sugars). This increasing organization stands in contrast to the spontaneous course of things—breakdown and disorder—and thus may seem to be a violation of the second law, but it is not. In the context of the universe as a whole, living

things are contributors to its entropy. Any given biological activity, such as building a leaf, generates heat and thus increases the disorder in the universe as a whole. But living things can bring about *local* increases in order (in themselves) by using the energy that comes ultimately from the sun.

### Up and Down the Great Energy Hill

The building up of complex molecules from simpler ones is an example of so-called synthetic work, and in it, we begin to see how living things use energy. A starchy carbohydrate is a more complex thing than the simple sugars it is made of. It is a more *ordered* arrangement of atoms than is a group of individual simple sugars. As such, the second law tells us, it should *take* energy to make a starch from simple sugars, and this is indeed the case. Going in the opposite direction, the breakdown of a starchy carbohydrate into simple sugars is an action favored by the second law because it brings about *less* order. Such a process therefore should not take energy. Indeed, this process should *release* energy, and this too is the case.

Breakdown operations, such as starches breaking down into sugars, are examples of **exergonic reactions**: reactions in which the starting set of molecules (the reactants) contains more energy than the final set of molecules (the products). Meanwhile, buildup operations are examples of **endergonic reactions**: reactions in which the products contain more energy than the reactants. This is what happens when, for example, simple glucose molecules are brought together to form glycogen, which is a storage form of carbohydrates (**Figure 6.4**).

### Coupled Reactions

With this as background, we arrive at an important insight: Endergonic and exergonic reactions are linked in living things. Why should this be so? If you want to get something uphill—into a higher energy state—you have to power that process, which means *using* some energy. Just as it takes energy to get a bicycle uphill, so it takes energy to power the endergonic reaction that takes us from glucose molecules to glycogen. Now, where is this energy going to come from? It must come from an energy-releasing *exergonic* reaction. The result is a **coupled reaction**: a chemical reaction in which an exergonic reaction powers an endergonic reaction.

With this, we reach a critical point about how living things operate: through an endless series of coupled reactions. Want to move a muscle? Two kinds of long, slender muscle protein filaments must slide past each other for this to happen. For this to work, a huge number of tiny, pivoting extensions on one kind of muscle filament must first latch onto and then pull the other kind of filament (**Figure 6.5**). Now, how are these extensions going to "cock" themselves and get ready for a pull? This cocking is an endergonic reaction—it is a movement up the energy hill. Once our muscle extensions are cocked, they will be in a higher energy state, just as the bar on a mousetrap is in a higher energetic state after it's pulled back. Following what was just said, an endergonic reaction such as this must be powered by an exergonic reaction—energy must be released from somewhere for those muscle extensions to be moved into their cocked state. Now, where does the energy for this come from? To follow the whole trail, this energy came originally from the sun, after which it was captured by a photosynthesizing organism—a peach tree, for example. Then, say we ate a peach from the tree; now, the energy that was locked up in the peach is locked up in us in the form of glucose molecules. So, does the energy to pull back the muscle extensions go directly from the glucose to the extensions? No. One more energy transfer must be made first. The energy in the glucose must be transferred into a molecule called ATP.

## 6.4 The Energy Dispenser: ATP

It is ATP that will release the energy that allows the muscle extensions to be cocked. Only with the energy dispensed by ATP can our muscles contract. You could think of ATP as a kind of "middleman"—it receives the energy from food and then doles it out for activities such as muscle contraction. Why do we need an intermediate molecule of this sort? Well, think of it this way. Power plants have considerable use for coal because, as we've seen, coal can yield energy. But it may go without saying that people don't use coal's energy *directly* in their homes. Instead, power plants must convert the energy in coal into a

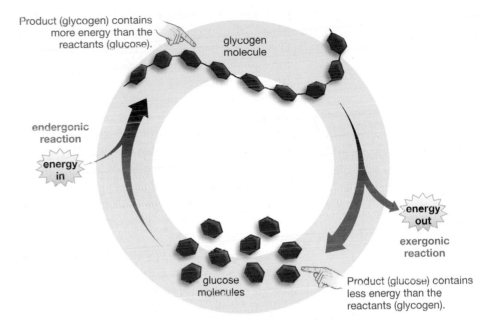

**Figure 6.4**
**Energy Stored and Released**

It takes energy to build up a more complex molecule (in this case, glycogen) from simpler molecules (in this case, glucose units). Such a buildup is thus an endergonic or uphill reaction. Conversely, energy is released in reactions in which more complex molecules are broken down into simpler ones. Such reactions are exergonic or downhill reactions.

form that people *can* use in their homes—electricity. In this same way, the energy contained in food must be converted into a form the human body can use, and this form is overwhelmingly the molecule known as ATP. Every time a muscle contracts or a nerve signal fires, ATP is doling out the energy for these operations. Every time a sperm cell whips its tail, ATP is supplying the energy. And what goes for energy use in human beings goes for energy use in all forms of life. The simplest bacterium is as dependent on ATP as we are. In sum, **ATP** can be defined as the most important energy transfer molecule in living things.

So, what's the structure of this molecule? To answer this question, let's start by thinking about where ATP's full name, adenosine triphosphate, comes from. ATP has a sugar (called ribose) and a nitrogen-containing

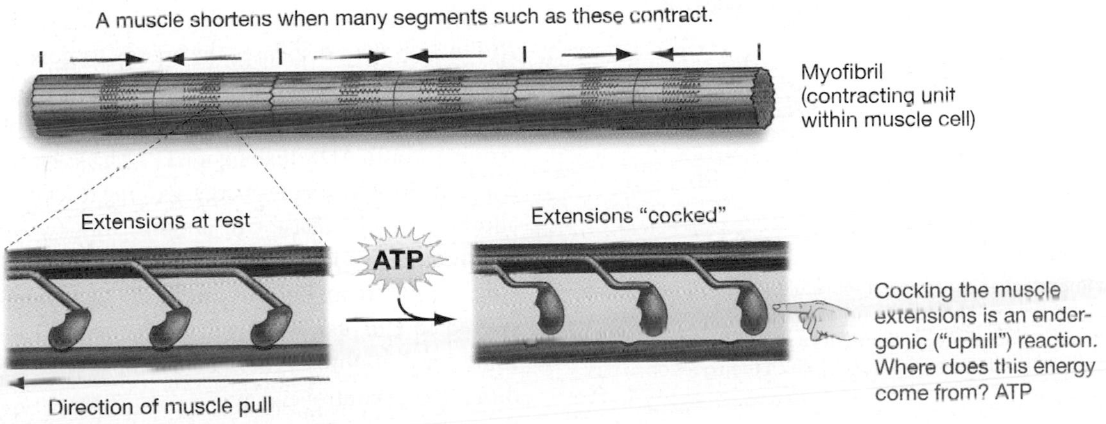

**Figure 6.5**
**Powering a Coupled Reaction**

An endergonic or energy-storing reaction takes place each time a set of muscle extensions are put in a "cocked" position that allows them to pull on a muscle filament. Energetically, this is a movement up the energy hill and thus must be powered by an energy-releasing or exergonic reaction. Together, the two reactions constitute a single reaction—a coupled reaction. This coupled reaction, like so many in living things, is powered by energy released from the molecule ATP.

A muscle shortens when many segments such as these contract.

Myofibril (contracting unit within muscle cell)

Extensions at rest

**ATP**

Extensions "cocked"

Cocking the muscle extensions is an endergonic ("uphill") reaction. Where does this energy come from? ATP

Direction of muscle pull

base (called adenine) that together make up a compound called adenosine, as you can see in **Figure 6.6.** Phosphate groups are then linked to adenosine—three phosphate groups, which gives us adenosine triphosphate. The place to look for energy storage in ATP is in the phosphate groups. Notice that they all are negatively charged. This is the key to understanding why ATP serves as such an effective energy transfer molecule. Remember that like charges repel each other, and in this case three phosphate groups are doing just that. Energy was required to put these phosphate groups together as they are, in this relatively unstable state. This linkage represents a move *up* the energy hill. With ATP, it is as though someone squeezed a jack-in-the-box back into the box and shut the lid. It should be no surprise that when the lid is opened, a good deal of energy will be released.

If you look now at **Figure 6.7**, you can see the details of how ATP does its job in the muscle contraction we just touched on. Note that when the

muscle is in its starting state, one of its extensions rests on a nearby filament, at a 45-degree angle to it. What comes next is the addition of ATP; first, a molecule of it binds with the extension and, with this, the extension detaches from the filament. Next comes the infusion of energy into the process. ATP's third phosphate group splits off from it, an action that reattaches the extension to the filament and that moves the extension to its cocked, *90*-degree angle relative to the filament. In this position, the extension is in a higher energy state: It is storing potential energy in the same way that a bar on a mousetrap does once it's pulled back. Finally, the release of the phosphate group from the extension triggers what is known as the power stroke: The extension expends the energy that was stored in it by pulling the filament to the left—contraction has taken place. The result, when carried out by a huge number of extensions and filaments, is contraction of a muscle.

## The ATP/ADP Cycle

There is another side to this ATP-driven reaction, however. Note that once ATP loses its outer phosphate group, it is not ATP anymore, but two molecules—one called ADP and another labeled P (for phosphate). ATP works by losing (or "donating") a phosphate group in this way, but once this happens, it becomes a molecule with only *two* phosphate groups: adenosine *di*phosphate or ADP. However, this is a temporary condition. Once an ADP molecule is finished with a reaction, it is free to have a third phosphate group added to it again. This is just what happens: ADP returns to being ATP, which is capable of providing energy for yet another reaction. This shuttling from ATP to ADP and back, illustrated in Figure 6.6, takes place constantly in cells.

## Between Food and ATP

The essence of our story so far is that, in animals, energy that comes originally from the sun is first stored in the food we eat and is then transferred to ATP, which then dispenses the energy to power our daily activities. But if you think about it, there's a piece missing here. How does that third phosphate group get onto ADP, making it ATP? This is an uphill reaction. So, where's the energy-*yielding* reaction that allows it to take place? Put another way, by which reactions does energy get from food to ATP? It is almost time to answer this question—it will be answered in Chapter 7—but first you need to know something about a group of proteins that living things use to control energy.

**Web Animation 6.1**
Energy and Biology

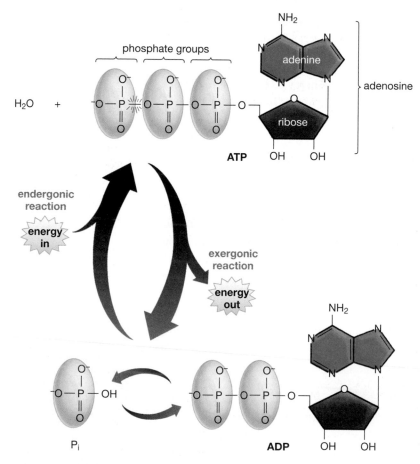

**Figure 6.6**
**Life's Most Important Energy Transfer Molecule**

ATP (adenosine triphosphate) is life's most important energy transfer molecule. It stores energy in the form of chemical bonds between its phosphate groups. When the bond between the second and outermost phosphate group is broken, the outermost phosphate separates from ATP, and energy is released. This separation transforms ATP into adenosine diphosphate or ADP, which then goes on to pick up another phosphate group, becoming ATP again.

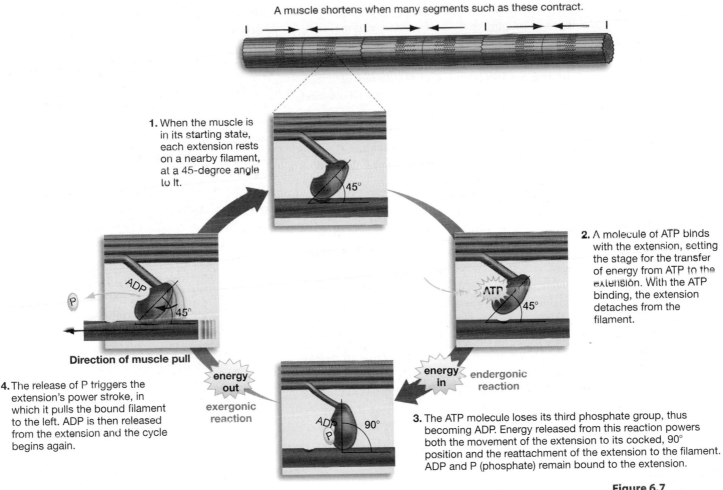

A muscle shortens when many segments such as these contract.

**1.** When the muscle is in its starting state, each extension rests on a nearby filament, at a 45-degree angle to it.

**2.** A molecule of ATP binds with the extension, setting the stage for the transfer of energy from ATP to the extension. With the ATP binding, the extension detaches from the filament.

**Direction of muscle pull**

**energy out**

exergonic reaction

**energy in**   endergonic reaction

**4.** The release of P triggers the extension's power stroke, in which it pulls the bound filament to the left. ADP is then released from the extension and the cycle begins again.

**3.** The ATP molecule loses its third phosphate group, thus becoming ADP. Energy released from this reaction powers both the movement of the extension to its cocked, 90° position and the reattachment of the extension to the filament. ADP and P (phosphate) remain bound to the extension.

**Figure 6.7**
**Muscle contraction, powered by ATP**

---

**SO FAR...**

1. Coupled reactions combine an endergonic reaction, in which the _____ have more energy than the _____, and an exergonic reaction, in which the _____ have more energy than the _____.

2. The energy for most coupled reactions in the body is supplied by the molecule _____, in particular by the splitting off of this molecule's third or outer _____.

3. After powering a reaction, ATP becomes the molecule _____, which, with the addition of another _____ becomes ATP once more.

## 6.5   Efficient Energy Use in Living Things: Enzymes

Lactose—better known as milk sugar—is composed of two simple sugars, glucose and galactose, that are linked together by a single atom of oxygen. It fol-lows from what you've seen so far that lactose will *split* into its glucose and galactose parts without any energy requirement. Indeed, a little energy will be released in this process because chemical bonds have been broken, and two smaller molecules have been produced. Such a splitting is therefore a downhill reaction.

If we actually took a small amount of lactose, however, and put it in some plain water, it would cer-tainly be hours—it might be days—before any of the lactose molecules would break down into glucose and galactose. How can this be? Most of us drink milk, and it doesn't seem to pile up in us.

### Accelerating Reactions

The secret is that when lactose is metabolized in living things, it isn't being split up in plain water. Something else is present that speeds up this process immensely, accelerating it perhaps a billionfold. That something is an **enzyme**, a type of protein that accel-erates a chemical reaction. This particular enzyme is called lactase, but it is merely one of thousands of en-zymes known to exist. Each of these compounds

**... ANSWERS**

1. products; reactants; reactants; products

2. ATP; phosphate group

3. ADP; phosphate group

works on some chemical process but not all are involved in splitting molecules. Some enzymes combine molecules; some rearrange them.

Given their numbers, it may be apparent that enzymes are involved in a lot of different activities, but this doesn't begin to give them the recognition they deserve. Enzymes facilitate nearly every chemical process that takes place in living things. No organism could survive without them. Technically, these compounds are only accelerating chemical reactions that would happen anyway, as with lactase splitting lactose. But in practical terms, they are *enabling* these reactions because no living thing could wait days or months for the milk sugar it ingests to be broken down, or for hormones to be put together, or for bleeding to stop.

To get an idea of the importance of enzymes, consider the millions of Americans who are known as lactose intolerant because, after childhood, their bodies reduced their production of lactase. When these people consume milk, the lactose in it *does* simply stay in their intestines. It is eventually digested not by them, but by the bacteria that live in the human digestive tract (something that causes bloat and gas). Troublesome as this affliction may be, it is minor compared to other conditions caused by enzyme dysfunction. All newborn babies in the United States are screened for a list of inherited diseases, three of which are Tay-Sachs disease, phenylketonuria, and congenital adrenal hyperplasia. Each of these diseases is caused by an enzyme deficiency. In the case of Tay-Sachs disease, the lack of an enzyme leads to a buildup of fat deposits within nerve cells, and the result usually is death by age 4.

**Figure 6.8**
**Metabolic Pathway: Sequence of Enzyme Action**
Most processes in living organisms are carried out through a metabolic pathway—a sequential set of enzymatically controlled reactions in which the product of one reaction serves as the substrate for the next. Enzymes perform specific tasks on specific substrates. In this example, enzyme A combines two substrates, enzyme B removes part of its substrate, and enzyme C changes the shape of its substrate.

## Specific Tasks and Metabolic Pathways

From the sheer number of enzymes that exist, you can probably get a sense of how specifically they are matched to given tasks. Although some enzymes can work on groups of similar substances, a specific enzyme usually facilitates a specific reaction: It works on one or perhaps two molecules and no others. Thus, lactase breaks down lactose—and *only* lactose—and the products of its activity are always glucose and galactose. The substance that is worked on by an enzyme is known as its **substrate**. Thus, lactose is the substrate for lactase. The -*ase* ending you see in lactase is common to most enzymes. As you can see, the substrate (lactose) is identified in the enzyme's name (lactase).

Certain activities in living things, such as blood clotting, are much more complex than the breakdown of lactose. They are multistep processes, each step of which requires its own enzyme. Most large-scale activities in living things work this way: leaf growth, digestion, hormonal balance. The result is a **metabolic pathway**: a set of enzymatically controlled steps that results in the completion of a product or process in an organism. In such a sequence, each enzyme does a particular job and then leaves the succeeding task to the next enzyme, with the product of one reaction becoming the substrate for the next (**Figure 6.8**). This is the way that, for example, cholesterol gets produced: One cholesterol precursor leads to another through a series of enzymatically enabled reactions, with the final product being cholesterol itself. The sum of all the chemical reactions that a cell or larger organism carries out is known as its **metabolism**. To put things simply, enzymes are active in all facets of the metabolism of all living things.

## 6.6 Lowering the Activation Barrier through Enzymes

All this information about enzymes is well and good, you may be saying at this point, but what does it have to do with energy? The answer is that in carrying out their tasks, enzymes are in the business of lowering the amount of energy needed to get chemical reactions going. And this means that the reactions can get going faster.

An analogy may be helpful here. Consider the now-familiar rock perched at the top of a hill. Situated in this way, it has a good deal of potential energy. If it rolled down the hill, it would release this potential energy as kinetic energy. But, here is the critical

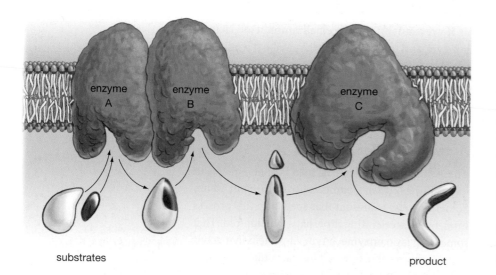

substrates                                    product

question: What would be required to get this rock going? It is perched at the top of the hill because it is *stable* at the top of the hill. To get it going would require *additional* energy in the form of, say, a push.

Now consider the lactose molecule. It is "perched" like the rock, in a sense, because it lies "uphill" from the glucose and galactose it can break into. But just as the rock is stable because of its position in the ground, so the lactose is stable because it has a strong set of chemical bonds holding its atoms in a fixed position. The question thus becomes: Is there anything that can lower the amount of energy required to break these bonds? The answer is yes—an enzyme. If you look at **Figure 6.9**, you can see this in schematic terms. Imagine that you had to push a boulder up the hill in (a) to get it to roll down the other side. Now imagine that you only had to push a boulder up the hill in (b) to get the same result. In which situation would you be able to get your desired result faster? What is at work here is lowering of **activation energy**: the energy required to initiate a chemical reaction.

### How Do Enzymes Work?

How do enzymes carry out this task of lowering activation energy? As you'll see shortly, they bind to their substrates and in so doing make these substances more vulnerable to chemical alteration. The amazing thing is that they do this without being permanently altered themselves. Enzymes are **catalysts**—substances that retain their original chemical composition while bringing about a change in a substrate. At the end of its splitting of the lactose molecule, lactase has exactly the same chemical structure as it did prior to this operation. It is thus free to pick up another lactose molecule and split *it*. All this takes place in a flash: The fastest-working enzymes can carry out 100,000 chemical transformations per second.

Enzymes generally take the form of globular or ball-like proteins with a shape that includes a kind of pocket into which the substrate fits. If you look at **Figure 6.10**, you can see a space-filling model of one enzyme, called hexokinase. You can also see, buried there in the middle, its substrate—glucose, which is better known as blood sugar. As with all proteins, enzymes are made up of amino acids, but only a few of the hundreds of amino acids in an enzyme are typically involved in actually binding with the substrate. Five or six would be common. These amino acids help form the substrate pocket, known as the **active site**: the portion of an enzyme that binds with a substrate, thus helping transform it.

In some cases, the participants at the active site include, along with amino acids, one or more accessory molecules. One variety of these molecules is the **coenzymes**: molecules other than amino acids that facilitate the work of enzymes by binding with them. If you've ever wondered what vitamins do, participation in enzyme binding is a big part of it. Once we have ingested them, many vitamins are transformed into coenzymes that sit in the active site of an enzyme and provide an added chemical attraction or repulsion that allows the enzyme to do its job.

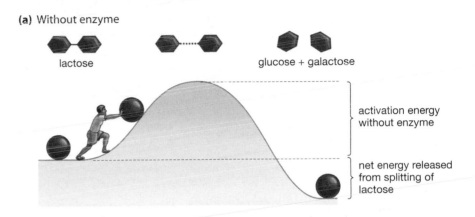

**(a)** Without enzyme

lactose          glucose + galactose

activation energy without enzyme

net energy released from splitting of lactose

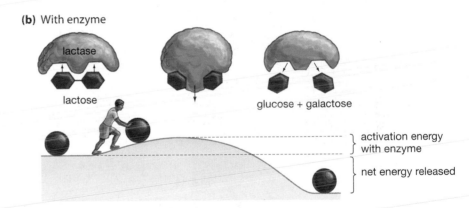

**(b)** With enzyme

lactase

lactose          glucose + galactose

activation energy with enzyme

net energy released

**Figure 6.9**
**Enzymes Accelerate Chemical Reactions**
How is lactose split into glucose and galactose? Without an enzyme, the amount of energy necessary to activate this reaction is high. In the presence of the enzyme lactase, however, a low activation energy is sufficient to get the process started. The energy released from the splitting of lactose is the same in both cases.

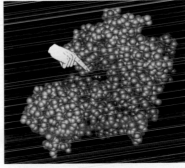

Glucose molecule binding to active site of hexokinase enzyme.

**Figure 6.10**
**Shape Is Important in Enzymes**

Substrates generally fit into small grooves or pockets in enzymes. The location at which the enzyme binds the substrate is known as the enzyme's active site. Pictured is a computer-generated model of a glucose molecule (in red) binding to the active site of an enzyme called hexokinase (in blue).

## An Enzyme in Action: Chymotrypsin

The details of how enzymes carry out their work are complex, and they vary according to enzyme. Let's consider in a general way, however, the activity of a much-studied enzyme called chymotrypsin.

Chymotrypsin is delivered from the human pancreas to the small intestine, where it works with water to break down proteins we have ingested. Its function is to snip protein chains in between their building block amino acids. It does this by breaking the single bond that binds one amino acid to the next (**Figure 6.11**). How does it work? After binding to part of a protein chain, chymotrypsin then interacts with it to create a transition-state molecule. In effect,

**Web Animation 6.2**
Enzymes

it distorts the shape of the protein—and holds it in this new shape briefly—in such a way that the protein becomes vulnerable to bonding with ionized water molecules. This allows carbon and nitrogen atoms to latch on to new partners, and the protein chain is cut in two. Chymotrypsin then returns to its original form and proceeds to a new reaction.

## 6.7    Regulating Enzymatic Activity

The activity of enzymes is regulated in several ways. The question here is: What factors influence the amount of "product" an enzyme turns out? One of these factors, logically, is the amount of substrate in the enzyme's vicinity. If there's no substrate to work on, no product can be turned out.

The work of enzymes can, however, be reduced in other ways. For example, some molecules that are not the normal substrate for an enzyme can nevertheless bind to the enzyme's active site. This happens all the time in natural ways in our bodies, but to sharpen the point, consider the well-known drug Lipitor, which millions of people take to lower their cholesterol levels. How does this drug work? Recall that there is a metabolic pathway by which cholesterol is produced—one cholesterol precursor after another is modified in a set of reactions, each one of which is facilitated by an enzyme. But, one of these enzymes has an active site that *Lipitor* can fit into. When this happens, this enzyme is occupied—it can't bind with the cholesterol precursor that is its normal substrate. And this effect is dose-dependent: The greater the number of enzyme molecules that are occupied by Lipitor, the fewer the number of enzyme molecules that can help produce cholesterol. Thus, the ultimate effect of Lipitor is to lower the amount of cholesterol produced in the body. What we see with Lipitor is an example of **competitive inhibition**: a reduction in the activity of an enzyme by means of a compound other than the enzyme's usual substrate binding with it in its active site.

### Allosteric Regulation of Enzymes

Living things routinely regulate enzyme activity by another means. Earlier, you saw a model of an enzyme called hexokinase, which works on the substrate glucose. The product of this reaction is something called glucose-6-phosphate. Although this is an enzymatic product, it can bind to hexokinase at a site *other* than its active site and in so doing change the shape of hexokinase. This change renders hexokinase less able to bind with its substrate, glucose, meaning hexokinase's work is temporarily reduced. If you look at **Figure 6.12**, you can see how this process works schematically.

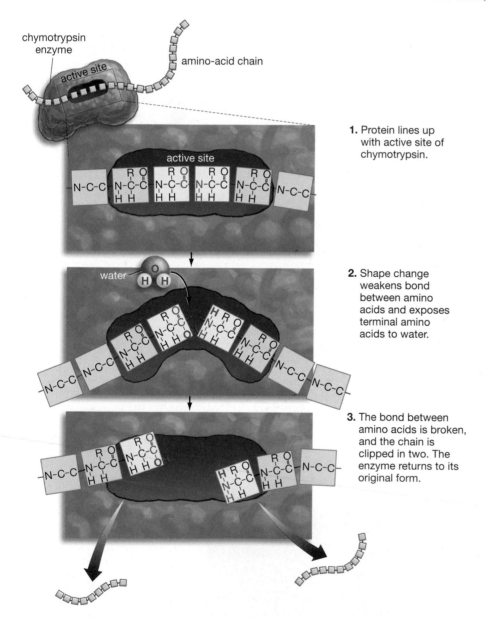

1. Protein lines up with active site of chymotrypsin.

2. Shape change weakens bond between amino acids and exposes terminal amino acids to water.

3. The bond between amino acids is broken, and the chain is clipped in two. The enzyme returns to its original form.

**Figure 6.11**
**How the Enzyme Chymotrypsin Works**
In this example, the enzyme chymotrypsin is facilitating the breakdown of a protein by changing the protein's shape. In the absence of chymotrypsin, this process would take a billion times longer.

This is but one example of the process known as **allosteric regulation**—the regulation of an enzyme's activity by means of a molecule binding to a site on the enzyme other than its active site. Many variations on this theme are possible. In the case of hexokinase, its own product (glucose-6-phosphate) bound with it and cut down on its activity, but molecules other than the enzymatic product can bind with an enzyme at one of its additional binding sites. Furthermore, the effect of such binding need not be negative; allosteric binding can increase the activity of enzymes as well as reduce it.

The importance of such regulation is that enzymes are not simply fated to turn out product as long as there is substrate to work on. If a cell has too much of a product, allosteric control can cut down on it; if there is too little, its concentration can be increased.

## On to Energy Harvesting

Our story of energy concluded in this chapter with a look at enzymes and energy. Next chapter, however, we'll pick up the story of energy with a more detailed look at a process touched on earlier—the transfer of energy from food to ATP. The question we will tackle is this: What are the steps by which our bodies extract energy from food and then channel this energy into pushing phosphate groups onto ADP, thus making it ATP? Put another way, how does our primary energy-dispensing molecule become energized?

(a) Substrate becomes product

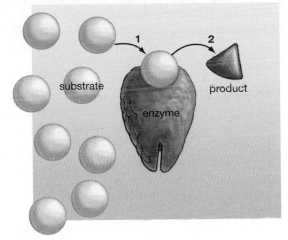

1. Substrate binds to enzyme.

2. Enzyme transforms substrate to product.

(b) Product feeds back on enzyme

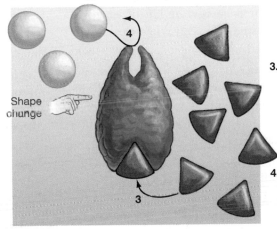

3. Product binds to a *different* site on the enzyme, causing the enzyme to change shape.

4. The new shape of the enzyme prevents it from binding to any more substrate.

**Figure 6.12**
**Allosteric Regulation of Enzymes**
The frequency of chemical reactions can be controlled by the binding of a molecule to an enzyme at a site other than its active site. In this instance, the enzyme's own product binds with the enzyme in this way, thereby reducing the enzyme's activity.

# CHAPTER 6 REVIEW

For study help, activities, and more quiz questions, go to **www.aw.com/krogh4**.

## SUMMARY

### 6.1 Energy is Central to Life

- All living things require energy. The sun is the ultimate source of energy for most living things. The sun's energy is captured on Earth by photosynthesizing organisms (such as plants), which then pass this energy on to other organisms in the form of food.

### 6.2 The Nature of Energy

- Energy can be defined as the capacity to bring about movement against an opposing force. Energy can be conceptualized as either potential energy, meaning stored energy; or kinetic energy, meaning energy in motion. The study of energy is known as thermodynamics.

- Two fundamental principles of energy are the first law of thermodynamics, which states that energy is never created or destroyed, but is only transformed; and the second law of thermodynamics, which states that energy transfer will always result in a greater amount of disorder in the universe.

- In line with the second law of thermodynamics, in every energy transaction, some energy will be lost to the most disordered form of energy, heat. Thus, in the operation of a car engine, only part of the energy released by the combustion of gasoline actually helps propel the car; the rest of the energy released in the combustion is lost to heat. Entropy is a measure of the amount of disorder in a system; the greater the entropy, the greater the disorder.

### 6.3 How is Energy Used by Living Things?

- Living things can bring about local increases in order (in themselves) through their metabolic processes. They can, for example, build up more-ordered molecules (starches, proteins) from less-ordered molecules (simple sugars, amino acids). However, it takes energy to do this.

- Energy in living things is stored away in endergonic (uphill) reactions in which the products of the reaction contain more energy than the starting substances (or reactants). Conversely, energy is released in exergonic (downhill) reactions, in which the reactants contain more energy than the products. The linkage of simple sugars to form a complex carbohydrate is an endergonic reaction. Such a reaction will not occur without an input of energy. Conversely, the breakdown of a complex carbohydrate into simple sugars is an exergonic reaction. Such a reaction releases energy. Endergonic and exergonic reactions are linked in coupled reactions—reactions in which an energy-yielding exergonic reaction powers an energy-requiring endergonic reaction. The molecule most often used in living things to power coupled reactions is ATP.

### 6.4 The Energy Dispenser: ATP

- Adenosine triphosphate (ATP) is the most important energy transfer molecule in living things. In animals, energy that is extracted from food is transferred to ATP, and this energy is then used to drive a vast array of metabolic processes. Energy supplied by ATP is used, for example, to power muscle contraction and nerve signal transmission.

- ATP's energy transfer powers stem from the fact that it contains three phosphate groups, each of which is negatively charged, meaning these groups repel each other. ATP drives chemical reactions by donating its third phosphate group to them. In the process, it becomes the two-phosphate molecule adenosine diphosphate (ADP). To again become ATP, it must have a third phosphate group attached to it. This shuttling back and forth between ATP and ADP takes place constantly in living things.

  **Web Animation 6.1:** Energy and Biology

### 6.5 Efficient Energy Use in Living Things: Enzymes

- An enzyme is a type of protein that accelerates the rate at which a chemical reaction takes place in an organism. Nearly every chemical process that takes place in living things is facilitated by an enzyme. For example, the enzyme lactase facilitates the splitting of the sugar lactose into its component sugars, glucose and galactose.

- The substance that an enzyme helps transform through chemical reaction is called its substrate. Lactose is the substrate of the enzyme lactase. Many activities in living things are controlled by metabolic pathways, in which a series of reactions is undertaken in sequence, each facilitated by its own enzyme. In such a series, the product of one reaction becomes the substrate for the next. The sum of all the chemical reactions that a cell or larger living thing carries out is its metabolism. Enzymes are active in all facets of the metabolism of all living things.

### 6.6 Lowering the Activation Barrier through Enzymes

- Enzymes work by lowering activation energy, which is the energy required to initiate a chemical reaction. Enzymes are catalysts: They bring about a change in their substrates without being chemically altered themselves.

- Enzymes generally take the form of globular or ball-like proteins whose shape includes a pocket into which the enzyme's substrate fits. This pocket is the active site—that portion of an enzyme that binds with a substrate, thus helping transform it. Only a few of the hundreds of amino acids that typically make up an enzyme will be involved in substrate binding. In some cases, substrate binding is facilitated by coenzymes: molecules other than amino acids that facilitate the work of enzymes by binding with them.

## 6.7  Regulating Enzymatic Activity

- Enzyme activity can be controlled in several ways. One of these is competitive inhibition: a reduction in the activity of an enzyme by means of a compound other than the enzyme's usual substrate binding with the enzyme in its active site. Another means of control is allosteric regulation, in which a molecule binds with the enzyme at a site other than its active site. Such binding changes the enzyme's shape, thereby decreasing or increasing the enzyme's ability to bind with its substrate. Because of such processes as allosteric regulation, enzymes are not fated to turn out product in strict accordance with the amount of substrate in their environment; rather, enzyme activity can be finely tuned in accordance with cellular needs.

Web Animation 6.2: Enzymes

## KEY TERMS

| | |
|---|---|
| activation energy   117 | enzyme   115 |
| active site   117 | exergonic reaction   112 |
| adenosine triphosphate (ATP)   113 | first law of thermodynamics   110 |
| allosteric regulation   119 | kinetic energy   110 |
| catalyst   117 | metabolic pathway   116 |
| coenzyme   117 | metabolism   116 |
| competitive inhibition   118 | potential energy   110 |
| coupled reaction   112 | second law of thermodynamics   111 |
| endergonic reaction   112 | substrate   116 |
| energy   110 | thermodynamics   110 |

## BRIEF REVIEW

*Answers to Brief Review questions are in the back of the book. For multiple-choice quiz questions, go to* **www.aw.com/krogh4**.

1. A plant leaf is more ordered and complex than are carbon dioxide and water. In photosynthesis, plants take small, relatively disordered carbon dioxide, water, and mineral molecules and make complex, ordered sugars out of them. How is it possible for this to happen without violating the second law of thermodynamics?

2. The text notes that, in connection with muscle contraction, extensions of one kind of muscle filament are "cocked" in an uphill reaction, and that ATP plays a part in this. Explain how this is a coupled reaction.

3. Where does the energy come from that is stored and released by ATP? In what way is ATP an energy transfer molecule?

4. The average person thinks of enzymes only as substances in their digestive tract that break down food. How would you explain the role of enzymes to your little sister, who has heard about digestion in grade-school biology?

5. What is meant by allosteric regulation of an enzyme?

## APPLYING YOUR KNOWLEDGE

1. In light of the laws of thermodynamics, explain why you get so much hotter when you run than when you are sitting still.

2. Which is easier to have: a messy, disordered room or a neat, ordered room? Is the second law of thermodynamics at work in this situation? What is the price of bringing order to a room?

3. Physicists say that it is impossible to create a perpetual motion machine—a machine whose own activity keeps it running perpetually. What energy principle precludes the possibility of a perpetual motion machine?

*How do living things extract energy from food? Through the process of cellular respiration. Here, a snowcap hummingbird (Microchera albocoronata) feeds from a flower in a Costa Rican rainforest.*

CHAPTER **7**

**7.1** Energizing ATP: Adding a Phosphate
Group to ADP                                    **124**

**7.2** Electrons Fall Down the Energy
Hill to Drive the Uphill Production
of ATP                                          **124**

**7.3** The Three Stages of Cellular
Respiration: Glycolysis, the Krebs
Cycle, and the Electron Transport
Chain                              *BioFlix*  **127**

**7.4** First Stage of Respiration: Glycolysis  **127**

**7.5** Second Stage of Respiration:
The Krebs Cycle                                 **131**

**7.6** Third Stage of Respiration:
The Electron Transport Chain                    **133**

**7.7** Other Foods, Other Respiratory
Pathways                                        **135**

**Essays**

When Energy Harvesting Ends at
Glycolysis, Beer Can Be the Result              **130**

Energy and Exercise                             **136**

To remain alive, living things must put the energy

they acquire into a form they can use—ATP.

# VITAL HARVEST:
# Deriving Energy from Food

**I**n May of 1996, the world's attention was riveted on news coming out of Nepal, in Southeast Asia. Disaster had struck several groups of mountain climbers who had been attempting to scale the world's highest

peak, Mt. Everest. A vicious storm had taken the members of two Everest expeditions by surprise as they were moving down from Everest's summit, toward the safety of lower altitudes. Within 36 hours, five climbers had lost their lives, and another was so badly frostbitten that one of his hands had to be amputated.

One of the expeditions was led by a highly respected guide from New Zealand, Rob Hall. Courageously remaining near the summit in an effort to bring down one of his clients (who had collapsed), Hall ended up spending a night in a howling storm on Everest without a tent. When the sun rose the next day, the factors working against him included cold and wind, which he might have encountered in lots of inhospitable places on Earth. Also bedeviling him, however, was something that human beings rarely encounter: a lack of oxygen. At his altitude—sitting in the snow at 8,700 meters (or about 28,500 feet)—the oxygen he took in with each breath was little more than a third of the amount he would have inhaled with each breath at sea level. Technology might have come to his aid, as he had two oxygen bottles with him, but his intake valve for them had become clogged with ice.

Down at Everest's lower altitudes, other climbers—some of them Hall's friends—had to endure the agony of talking to him by two-way radio while not being able to reach him physically. All of them knew that a lack of oxygen was draining not only his muscles but his mind. Here is Hall in one of his radio transmissions asking about another guide on the

**Exhausted Near the Summit**
Two members of the ill-fated expeditions that climbed Mount Everest in May 1996 approach the mountain's peak.

ill-fated expedition: "Harold was with me last night, but he doesn't seem to be with me now. Was Harold with me? Can you tell me that?" In the end, the mountain without mercy claimed Rob Hall. He never got up from the place where he spent his night on Everest.

The tragedy of the 1996 Everest expeditions drives home a point and raises a paradox: No one needs to be reminded that we need to breathe in order to live, and most people are aware that oxygen is the most important thing that comes in with breath. That said, of the next 100 people you meet, how many could tell you what oxygen is doing to sustain life? Put another way, why do we need to breathe?

The short answer is that breathing and oxygen are in the energy transfer business. They are part of a system that allows us to extract, from food, energy that is then used to put together the "energy-dispensing" molecule, ATP. As you'll recall from Chapter 6, the body uses ATP (adenosine triphosphate) to power activities that range from muscle contraction to thinking to cell repair. Living things need large amounts of ATP to live, and organisms such as ourselves use oxygen to produce most of our ATP. If we don't get enough oxygen to keep making ATP, our bodies and minds start failing.

Oxygen is not required for all the "harvesting" we do of the energy contained in food. When you lift a box, the short burst of energy that is required comes largely from a different sort of energy harvesting, which you'll learn more about shortly. But even here, the essential *product* of energy harvesting is ATP. Keep that in mind, and you won't get lost in the byways of energy transfer. How does the body produce enough ATP to allow us to read a book or climb a mountain? In this chapter, you'll find out.

## 7.1 Energizing ATP: Adding a Phosphate Group to ADP

Given the importance of ATP to this story, let us begin with a brief review of how this molecule works. Recall from Chapter 6 that each ATP molecule has three phosphate groups attached to it, and that ATP powers a given reaction by losing the outermost of these phosphate groups. In this process, ATP is transformed into a molecule with *two* phosphate groups, adenosine diphosphate or ADP. To return to its more "energized" ATP state, it must have a third phosphate group attached again. This is, however, a trip *up* the energy hill because the product (ATP) contains more energy than the reactants (ADP + a phosphate group). Thus, getting that third phosphate group onto ADP requires energy. It is like preparing a spring-loaded mousetrap for action. It takes energy to pull a mousetrap bar back, and the same is true of putting a third phosphate group onto ATP (**Figure 7.1**). Where does the energy come from? For animals such as ourselves, it comes from food; energy that is extracted from food powers the phosphate group up the energy hill—and literally onto ADP.

## 7.2 Electrons Fall Down the Energy Hill to Drive the Uphill Production of ATP

In tracking the story of how energy goes from food to ATP, we will use as an example one particular food molecule, glucose, to see how energy is harvested from it. Although the details here are complex, the essential story is simple. *Electrons* derived from glucose, which is high in energy, will be running

**Figure 7.1**
**Storing and Releasing Energy**

Adenosine triphosphate (ATP) is the most important energy-releasing molecule in our bodies. The energy it contains is used to power everything from muscle contraction to thinking.

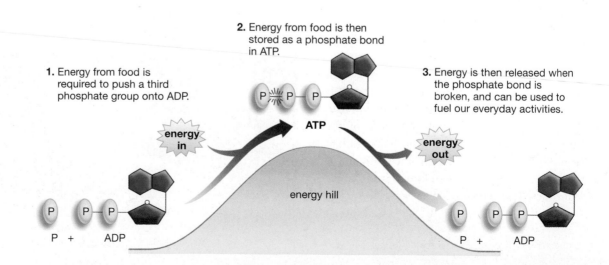

1. Energy from food is required to push a third phosphate group onto ADP.

2. Energy from food is then stored as a phosphate bond in ATP.

3. Energy is then released when the phosphate bond is broken, and can be used to fuel our everyday activities.

ATP

energy in

energy out

energy hill

P  +  ADP

P  +  ADP

downhill; they will be channeled off, a few at a time, and their downhill drop will power the *uphill* push needed to attach a phosphate group onto ADP. The glucose-derived electrons will be transferred to several intermediate molecules in their downhill journey, but the *final* molecule that will receive them at the bottom of the energy hill is oxygen. We need to breathe because we need oxygen to serve as this final electron acceptor.

As you are following this process, you may well wonder why so many steps are required to transfer energy from food to ATP. Why isn't there just one step in energy transfer—food to ATP? It turns out that the gradual transfer of energy, through many steps, allows the body to make the most of the energy it receives. Think of it this way. You could have water drop 1,000 feet straight off a cliff onto a waterwheel below. In this case, the wheel would simply be pulverized by the water crashing onto it. In the process, the potential energy that had been contained in the water at the top of the cliff would be transformed almost entirely into useless heat. Conversely, you could have a stream that ran briskly downhill for 1,000 vertical feet with waterwheels spaced periodically to channel the energy off a little at a time, thereby conserving the energy for work. The steps you are going to read about amount to this kind of controlled channeling off of energy—the energy that's contained in food.

### Redox Reactions

The basis for electron transfers down the energy hill is straightforward: Some substances more strongly attract electrons than do others. A substance that loses one or more electrons to another is said to have undergone **oxidation**. We hear the related word *oxidized* all the time—when paint on an outdoor surface has become dulled, for example, or when metal rusts (**Figure 7.2**). Meanwhile, the substance that *gains* electrons in this reaction is said to have undergone **reduction**. (This seems about as logical as saying a country lost a war by winning. The way to think about it is that, because electrons carry a negative charge, any substance that gains electrons has had a reduction in its *positive charge*.)

In cells, oxidation and reduction never occur independently. If one substance is oxidized, another must be reduced. It's like a teeter-totter; if one side goes down, the other must come up. The combined operation is known as a reduction-oxidation reaction, or simply a **redox reaction**: the process by which electrons are transferred from one molecule to another. Critically, the substance oxidized in a redox reaction has its electrons traveling energetically downhill.

**Figure 7.2**
**An Effect of Oxidation**
Rust has formed on a hook atop a corroded steel sheet.

**Web Animation 7.1**
Oxidation and Reduction

### Many Molecules Can Oxidize Other Molecules

The term *oxidation* might give you the idea that oxygen must be involved in any redox reaction, but this is not the case. Any compound that serves to pull electrons from another is a so-called oxidizing agent. In living things, a large number of molecules are involved in energy transfer, and each has a certain tendency to gain or lose electrons relative to the others.

This is how electrons can be passed down the energy hill: The starting "energetic" molecule of glucose is oxidized by another molecule, which in turn is oxidized by the *next* molecule down the hill. The whole thing might be thought of as a kind of downhill electron bucket brigade. Molecules that serve to transfer electrons from one molecule to another in ATP formation are known as **electron carriers**. The thing that makes their role a little complicated is that many of the electrons they accept are bound up originally in hydrogen atoms. You may remember from Chapter 3 that a hydrogen atom amounts to one proton and one electron. In transferring a hydrogen atom, then, a molecule is transferring a single electron (bound to a proton), which means a redox reaction has taken place.

### Redox through Intermediates: NAD

The most important electron carrier in energy transfer is a molecule known as **nicotinamide adenine dinucleotide**, or NAD, which can be thought of as a city cab. It can exist in two states: loaded with passengers or empty. And like a cab, it can switch very easily between those two states. The passengers that NAD picks up and drops off are electrons (**Figure 7.3**).

The "empty" state that NAD comes in is ionic: $NAD^+$. Remembering the discussion about ions, you can see that $NAD^+$ is positively charged, meaning that it has fewer electrons than protons. In a redox reaction, what $NAD^+$ does is pick up, in effect, one hydrogen atom (an electron and a proton) and one solo electron (from a second hydrogen atom). The isolated electron that $NAD^+$ picks up turns it from positively charged to neutral $NAD^+ \longrightarrow NAD$; the whole hydrogen atom takes it from NAD to NADH. Keeping an eye on redox reactions here, $NAD^+$ has become NADH by oxidizing a substance—by accepting electrons from it.

So much for half of NAD's role: picking up passengers. *Now*, as NADH, it is loaded with these passengers. It can proceed down the energy hill to donate them to molecules that have a greater potential to accept electrons than it does. Having dropped its passengers off with such a molecule, it returns to being the empty $NAD^+$ and is ready for another pickup. Through this process, NADH transfers energy from one molecule to another.

### How Does NAD Do Its Job?

The molecule that $NAD^+$ is oxidizing is the starting glucose molecule (or actually derivatives of it). To carry out its oxidizing role, $NAD^+$ obviously needs to be brought together with the glucose derivatives. What have you been looking at that brings substances together in this way? Enzymes! One of the things you'll be seeing, therefore, is $NAD^+$ and glucose derivatives brought together by enzymes, at

which point $NAD^+$ accepts electrons from these derivatives, later to hand them off to another molecule. Electron transfer through intermediate molecules such as $NAD^+$ provides the energy for most of the ATP produced. Thus, when looking at the diagrams in this chapter that outline respiration, you'll see:

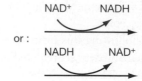

What these drawings indicate is the electron carrier shifting between its empty state ($NAD^+$) and its loaded state (NADH) or vice versa.

---

### SO FAR...

1. The essence of energy transfer in animals is that electrons derived from _____ run energetically downhill to power the uphill process by which a third _____ is put onto the body's energy-dispensing molecule _____.

2. A substance that loses one or more electrons to another is said to have undergone _____, while the substance that received the electrons in this reaction has undergone _____. The entire, coupled reaction is called a _____ reaction.

3. The electrons derived from food are transferred to the later stages of energy harvesting by molecules known as _____, the most important of which is one called _____ in its "empty" form and _____ in its "loaded" or reduced form.

---

**Figure 7.3**
**The Electron Carrier NAD⁺**

In its unloaded form NAD⁺ and its loaded form (NADH), this molecule is a critical player in energy transfer, picking up energetic electrons from food and transferring them to later stages of respiration.

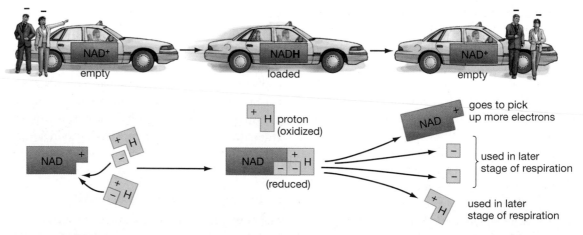

1. NAD⁺ within a cell, along with two hydrogen atoms that are part of the food that is supplying energy for the body.

2. NAD⁺ is reduced to NAD by accepting an electron from a hydrogen atom. It also picks up another hydrogen atom to become NADH.

3. NADH carries the electrons to a later stage of respiration then drops them off, becoming oxidized to its original form, NAD⁺.

## 7.3 The Three Stages of Cellular Respiration: Glycolysis, the Krebs Cycle, and the Electron Transport Chain

All the energy players are finally in place: glucose, redox reactions, electron carriers, enzymes, and the rest. In terms of a molecular formula, the big picture on respiration looks like this:

$$C_6H_{12}O_6 + 6O_2 + 36ADP + 36P \longrightarrow$$

$$6CO_2 + 6H_2O + 36ATP$$

The starting molecule on the first row is glucose, which is food, storing chemical energy. You can also see, next to it, the oxygen ($6O_2$) that is needed as the final electron acceptor. Then there is ADP, the two-phosphate molecule; and inorganic phosphate molecules (the 36P), which are pushed on ADP to make it ATP. The *products* of this reaction, on the second row, are carbon dioxide ($6CO_2$) and water ($6H_2O$) as by-products and energy in the form of ATP. How *much* ATP does the breakdown of one molecule of glucose yield? The answer is contained in the formula: a maximum of 36 molecules.

Although many separate steps are involved in actually getting energy from glucose to ATP, respiration can be divided into three main phases. These are known as *glycolysis*, the *Krebs cycle*, and the *electron transport chain*.

In overview, energy harvesting through these three phases goes like this: We get some ATP yield in glycolysis and the Krebs cycle, and glycolysis is useful in providing small amounts of ATP in quick bursts. But for most organisms, the main function of glycolysis and the Krebs cycle is the transfer of electrons to the electron carriers such as NAD$^+$. Bearing their electron cargo, these carriers then move to the third phase of energy harvesting, the electron transport chain (ETC). Here, the electron carriers are themselves oxidized, and the resulting movement of their electrons through the ETC provides the main harvest of ATP. Indeed, about 32 of the approximately 36 molecules of ATP that are netted per glucose molecule are obtained in the ETC. If you look at **Figure 7.4** on the next page, you can see two representations of the three stages of energy extraction.

### Glycolysis: First to Evolve, Less Efficient

The division among the three energy-harvesting phases is in one sense a division of evolutionary history. For most organisms, energy harvesting's first phase, glycolysis, could be likened to a nineteenth-century steam engine working side by side with a state-of-the art electric turbine. By itself, glycolysis produces only two ATP molecules per glucose molecule, which is pretty small in comparison to the 36 ATP molecules likely to come from all three phases. And glycolysis doesn't even yield many electrons compared to energy harvesting's second phase, the Krebs (or citric acid) cycle. Yet, glycolysis takes place in all living things—at the very least, as the first phase in energy harvesting—and for certain primitive one-celled organisms it can be the *only* phase in energy harvesting.

All of this points to glycolysis as a more ancient form of energy harvesting than the other two phases. It existed first, and then at some time in the evolutionary past, most organisms added to it the processes we now call the Krebs cycle and the ETC. The critical factor in this change was the use of oxygen in energy harvesting—specifically the use of oxygen in the ETC. Because the products of glycolysis feed ultimately into the ETC, the entire three-stage energy-harvesting process is referred to as oxygen-dependent or *aerobic* energy transfer. Taken as a whole, the aerobic harvesting of energy is known as **cellular respiration**.

The separation between glycolysis and the other two phases is also a physical separation. In eukaryotic cells such as ours, glycolysis takes place in the watery material that lies outside the cell's nucleus—the cytosol—while the Krebs cycle and the ETC take place within tiny organelles, called mitochondria, that are immersed in the cytosol. Let's start now to look at all three phases of cellular respiration.

## 7.4 First Stage of Respiration: Glycolysis

**Glycolysis**, the first stage of energy harvesting, means "sugar splitting," which is appropriate because our starting molecule is the simple sugar glucose. Briefly, here is what happens during glycolysis. First, this one molecule of glucose has to be prepared, in a sense, for energy release. ATP actually has to be *used*, rather than synthesized, in two of the first steps of glycolysis, so that the relatively stable glucose can be put into the form of a less-stable sugar. This sugar is then split in half. The two molecules formed have three carbons each (whereas the starting glucose had six). Once this split is accomplished, the steps of glycolysis take place in *duplicate*; it is like taking one long piece of cloth, dividing it in two, and then proceeding to do the same things to both of the resulting

*BioFlix*

Cellular Respiration

**(a)** In metaphorical terms

insert 1 glucose

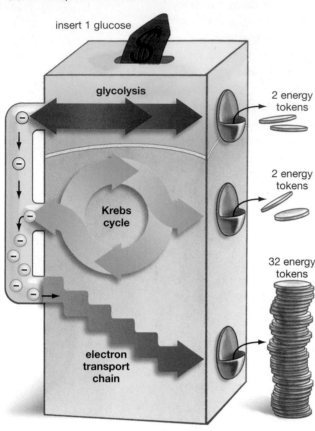

glycolysis

2 energy
tokens

Krebs
cycle

2 energy
tokens

32 energy
tokens

electron
transport
chain

**(b)** In schematic terms

reactants

products

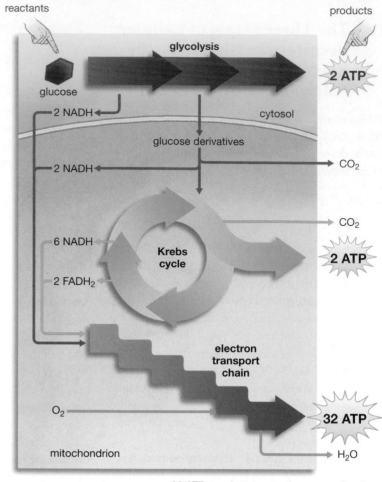

glycolysis

**2 ATP**

glucose

2 NADH

cytosol

glucose derivatives

2 NADH

$CO_2$

6 NADH

Krebs
cycle

$CO_2$

2 $FADH_2$

**2 ATP**

electron
transport
chain

$O_2$

**32 ATP**

mitochondrion

$H_2O$

**36 ATP maximum per glucose molecule**

Just as the video games in some arcades can use only tokens (rather than money) to make them function, so our bodies can use only ATP (rather than food) as a direct source of energy. The energy contained in food—glucose in the example—is transferred to ATP in three major steps: glycolysis, the Krebs cycle, and the electron transport chain. Though glycolysis and the Krebs cycle contribute only small amounts of ATP directly, they also contribute electrons (on the left of the token machine) that help bring about the large yield of ATP in the electron transport chain. Our energy-transfer mechanisms are not quite as efficient as the arcade machine makes them appear. At each stage of the conversion process, some of the original energy contained in the glucose is lost to heat.

As with the arcade machine, the starting point in this example is a single molecule of glucose, which again yields ATP in three major sets of steps: glycolysis, the Krebs cycle, and the electron transport chain (ETC). These steps can yield a maximum of about 36 molecules of ATP: 2 in glycolysis, 2 in the Krebs cycle, and 32 in the ETC. As noted, however, glycolysis and the Krebs cycle also yield electrons that move to the ETC, aiding in its ATP production. These electrons get to the ETC via the electron carriers NADH and $FADH_2$, shown on the left. Oxygen is consumed in energy harvesting, while water and carbon dioxide are produced in it. Glycolysis takes place in the cytosol of the cell, but the Krebs cycle and the ETC take place in cellular organelles, called mitochondria, that lie within the cytosol.

**Figure 7.4**
**Overview of Energy Harvesting**

pieces. **Figure 7.5** sets forth the individual steps of glycolysis.

Now, what are the results of these steps? Glycolysis accomplishes three valuable things in energy harvesting: It yields two ATP molecules, it yields two energized molecules of NADH, and it results in two molecules of pyruvic acid. These pyruvic acid molecules are the derivatives of the original glucose molecule. Now *they* are the molecules that will be oxidized—that will have electrons removed from them—in the next stage of energy harvesting. Keeping our eye on the ball here, we have some net ATP through glycolysis, some traveling electrons that will help make more ATP in the ETC, and a derivative of

the original glucose molecule (pyruvic acid) that will now move to the next stage of energy harvesting, where more energy will be extracted from it.

Before continuing on to this next stage, you may wish to read more about what might be called the consequences of glycolysis. Even oxygen-using organisms such as ourselves rely on glycolysis when short bursts of energy are needed. And, as noted, for some organisms glycolysis is the end of the line in energy harvesting. You can read about this in "When Energy Harvesting Ends at Glycolysis" on page 130. Glycolysis fits into a special place in providing energy for human exercise; you can read about this in "Energy and Exercise" on page 136.

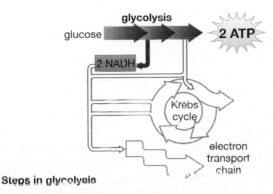

### Steps in glycolysis

1. Delivered by the bloodstream, glucose enters a cell and immediately has a phosphate group from ATP attached to it. Because this process, called phosphorylation, attaches the phosphate to the sixth carbon of glucose, it now goes under the name glucose-6-phosphate. Note that one molecule of ATP has been *used* in this step.

   **ATP ledger now reads: -1 ATP**

2. Glucose 6-phosphate is rearranged to become a molecule called fructose-6-phosphate.

3. Another molecule of ATP is used to add a second phosphate to fructose-6-phosphate, which now becomes fructose-1,6-diphosphate.

   **ATP ledger now reads: -2 ATP**

4. The single, six-carbon sugar fructose-1,6-diphosphate now becomes two molecules of a 3-carbon sugar, glyceraldehyde-3-phosphate, each with a phosphate group attached. From here on out, glycolysis happens in duplicate: What happens to one of the glyceraldehyde molecules happens to the other.

5. An enzyme brings together glyceraldehyde-3-phosphate, the electron carrier NAD$^+$, and a phosphate group. The glyceraldehyde-3-phosphate molecule is oxidized by NAD$^+$, which in its new form, NADH, moves to the electron transport chain bearing its electron cargo. The oxidation of NAD$^+$ is energetic enough that it allows the phosphate group to become attached to the main molecule, now called 1,3-diphosphoglyceric acid. Because everything is happening in duplicate, two NADH molecules are produced.

6. 1,3-diphosphoglyceric acid loses one of its phosphate groups, thus becoming 3-phosphoglyceric acid. The reaction is energetic enough to push this phosphate group onto an ADP molecule, yielding ATP. Because of duplication, two ATP are produced.

   **ATP ledger now reads: 0 ATP**

7. In two reactions, 3-phosphoglyceric acid becomes phosphoenolpyruvic acid, which generates more ATP as it transfers its phosphate group to ADP. Two more ATP molecules are produced. The phosphate transfer turns phosphoenolpyruvic acid into pyruvic acid—the derivative of the original glucose that now will enter the Krebs cycle.

   **ATP ledger now reads: +2 ATP**

**molecules in**

**molecules out**

glucose

ATP             ADP

1.

Red balls are carbons and gold ovals are phosphate groups

glucose-6-phosphate

2.

fructose-6-phosphate

ATP             ADP

3.

fructose-1,6-diphosphate

4.

glyceraldehyde-3-phosphate

$2\ NAD^+ + 2\ P$           **2 NADH** $+\ 2\ H^+$

5.

1,3-diphosphoglyceric acid

2 ADP            **2 ATP**

6.

3-phosphoglyceric acid

2 ADP            **2 ATP**

7.

pyruvic acid

**Figure 7.5**
**Glycolysis**

In glycolysis, the single glucose molecule is transformed in a series of steps into two molecules of a substance called pyruvic acid. These two molecules then move on to the next stage of cellular respiration (the Krebs cycle). Glycolysis also produces two molecules of electron-carrying NADH, which move to the electron transport chain. Although two molecules of ATP are used in the early stages of glycolysis, four are produced in the later stages, for a net production of two ATP.

# When Energy Harvesting Ends at Glycolysis, Beer Can Be the Result

For some bacteria—and even sometimes for people—energy harvesting ends with the set of steps called glycolysis rather than proceeding on through the Krebs cycle and oxygen-dependent electron transport chain (ETC). When this happens, organisms need a way to recycle their energy transfer molecules in a way that doesn't depend on oxygen. The NADH they produced in glycolysis needs to lose its added electrons and become $NAD^+$ again, so it can be reused in glycolysis. The solution to this dilemma is a kind of electron dumping, the *products* of which turn out to be some of the most familiar substances in the world.

## Fermentation in Yeast

Yeasts provide a good example of the way in which this works because yeasts are busy single-celled fungi who can live by glycolysis alone, or who can go through aerobic respiration. Say, then, that yeasts are working away on sugar, going through glycolysis, but

*For yeasts, alcohol is simply a by-product of the fermentation they carry out.*

doing so in an oxygenless environment. Recall that the final "substrate" product of glycolysis is two molecules of a substance called pyruvic acid. In yeasts, the pyruvic acid they end up with is converted to a molecule called acetaldehyde, and *it* takes on the electrons from NADH, meaning the recycling problem is now solved. Having taken on NADH's electrons, though, acetaldehyde now is converted to something very familiar: ethanol, better known as drinking alcohol.

Human beings *put* yeasts in environments in which this will happen, of course, because this is how we make wine and beer. Imagine yeast cells inside a dark, airless wine cask, working away on burgundy grape juice, harvesting energy from the grape juice sugars, but in so doing turning out pyruvic acid, which is turned into alcohol. This continues until the alcohol content of the wine reaches a level (about 14 percent) at which the yeast can no longer survive in it. This whole process is known as *alcoholic fermentation*, the process by which yeasts produce alcohol as a by-product of glycolysis they perform in an oxygenless environment.

For yeasts, alcohol is simply a waste by-product of the glycolysis they employ in *anaerobic* energy conversion, meaning conversion in an oxygenless environment. The same is true of the carbon dioxide that also results from their conversion of pyruvic acid. Here again, humans have found great use for the refuse of nature. Put the right yeast into dough, and its continuing fermentation

**Yeast Inside, Performing Fermentation**
This man is inspecting a gauge on a huge metal tank of fermenting wine. The alcohol in the wine is a by-product of the glycolysis carried out by yeast in an oxygenless environment.

produces the $CO_2$ that causes bread to rise and become "light" because of the air holes now in it. (And the alcohol? It evaporates.)

## Fermentation in Animals

*Alcoholic* fermentation is the solution that fungi (and occasionally plants) employ to sustain glycolysis in the absence of oxygen. But animals take a different tack, because in them the product of glycolysis, pyruvic acid, accepts the electrons from NADH. In this process, pyruvic acid is turned into another substance: *lactic acid*. Thus, the kind of fermentation that animals (and certain bacteria) carry out is called *lactate fermentation*. Have you ever experienced the muscle "burn" that comes with, for example, climbing several flights of stairs? If so, you have experienced a buildup of lactic acid in your muscles.

But why this should happen? Why should glycolysis ever end in lactate fermentation for us, when, unlike yeast, we always have access to oxygen? The problem turns out not to be a matter of oxygen access but a matter of oxygen *delivery*. During a quick burst of energy use, oxygen cannot be delivered into our muscle cells fast enough to accommodate the big increase in our energy expenditure. When a muscle's capacity for aerobic energy transfer has reached its limit, it turns to glycolysis and lactate fermentation to supply more ATP. Note, however, that glycolysis and lactate fermentation are limited, short-term means of supplying energy in animals such as ourselves. These processes do not supply enough energy to sustain us for long; their role is to supply enough energy to meet our needs until such time as oxygen-aided energy transfer can ramp itself up.

## 7.5 Second Stage of Respiration: The Krebs Cycle

Since glycolysis yielded only two of the 36 molecules of ATP that eventually are harvested, the glucose derivative that resulted from glycolysis, pyruvic acid, still clearly has a lot of energy left in it. Some of this energy will be harvested in the second stage of cellular respiration, the **Krebs cycle**. This stage is named for the German and English biochemist Hans Krebs, who in the 1930s used the flight muscles of pigeons to find out how aerobic respiration works. Because the first product of the Krebs cycle is citric acid, the cycle is sometimes referred to as the **citric acid cycle**. The ATP yield in this cycle is once again a paltry two molecules, but the Krebs cycle serves to set the stage for the big harvest of ATP in the ETC by stripping our molecules of pyruvic acid of their energetic electrons and sending them to the ETC, as you'll see.

### Site of Action Moves from the Cytosol to the Mitochondria

Glycolysis yielded its two molecules of pyruvic acid in the cytosol, but the site of energy harvesting quickly shifts, with the pyruvic acid, to the new site of energy transfer, the mitochondria. These organelles are where all the action takes place from here on out, so now would be a good time to look

at **Figure 7.6** and examine their structure. You can see that mitochondria have both an inner and an outer membrane. The Krebs cycle reactions take place to the interior of the inner membrane—in an area known as the inner compartment—while the reactions of the ETC take place *within* the inner membrane.

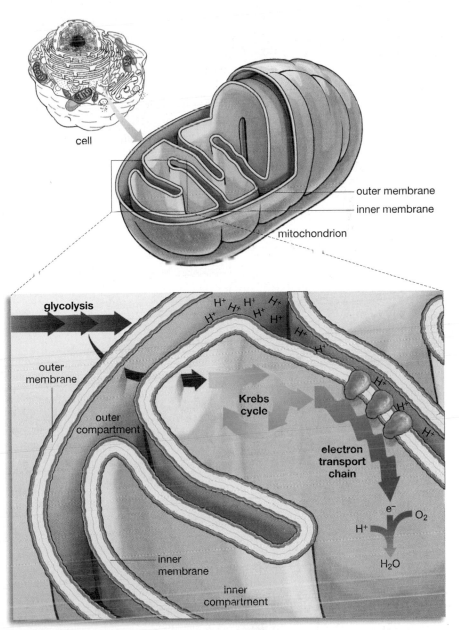

**Figure 7.6**
**Energy Transfer in the Mitochondria**
Mitochondria are organelles, or "tiny organs," that exist within cells. They are the location for the second and third sets of steps In cellular respiration, the Krebs cycle and the electron transport chain, respectively. Following a transitional step (see Figure 7.7), the products of glycolysis—the downstream products of the original glucose molecule—pass into the inner compartment of a mitochondrion, where the Krebs cycle takes place. Electrons derived from the Krebs cycle then migrate, via electron carriers, from the Krebs cycle site into the highly folded inner membrane of the mitochondrion, where the bulk of ATP is produced in the electron transport chain.

## Between Glycolysis and the Krebs Cycle, an Intermediate Step

There actually is a transition step in respiration between glycolysis and the Krebs cycle. In it, the three-carbon pyruvic acid molecule combines with a substance called coenzyme A, thus forming acetyl coenzyme A (acetyl CoA). One outcome of this reaction is that acetyl CoA enters the Krebs cycle as the derivative of the original glucose molecule. There are, however, two other products of this reaction. One is a carbon dioxide molecule that, like others you'll be seeing, eventually diffuses out of the cell and into the bloodstream. (Ever wonder where the $CO_2$ that you exhale comes from? Energy harvesting is the source.) The other product is one more molecule of NADH, which finds its way to the

ETC. This reaction takes place twice for each glucose molecule (**Figure 7.7**).

## Into the Krebs Cycle: Why is it a Cycle?

Acetyl CoA now enters the Krebs cycle. The reason this is a *cycle* becomes clear when you look at **Figure 7.8**. You can see that, in the first step of the cycle, acetyl CoA combines with a substance called oxaloacetic acid to produce citric acid. If you look over at about 11 o'clock on the circle, however, you can see that the *last* step in the cycle is the *synthesis* of oxaloacetic acid. Thus, a substance that's necessary for this chain of events to take place is itself a product of the chain.

Stroll around the circle now to see how the Krebs cycle functions in a little more detail. In essence, what's happening is that, as the entering acetyl CoA molecule is transformed into these various molecules, it is oxidized by electron carriers, with the resulting electrons moving on to the ETC. To look at this another way, the cycle starts with an energetic compound, citric acid, that is methodically stripped of electrons (and hence energy). Accordingly, ATP is also derived from this process, while $CO_2$ is a by-product. The only major player that's unfamiliar here is the $FAD - FADH_2$ that can be seen taking part in a redox reaction at about 8 o'clock. It is simply another electron carrier, similar to $NAD^+/NADH$.

In counting the Krebs cycle's yield of both ATP and electron carriers (NADHs and so on), note that *two* turns around this cycle result from the original glucose molecule (because it provided two molecules of the starting acetyl CoA). This means a total yield of 6 NADH, 2 $FADH_2$, and 2 ATP from Krebs for each glucose molecule.

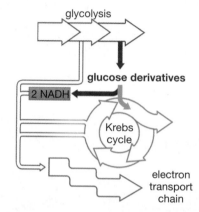

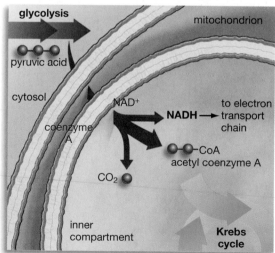

**Figure 7.7**
**Transition between Glycolysis and the Krebs Cycle**
The pyruvic acid product of glycolysis does not enter directly into the Krebs cycle. Rather, it must first be transformed into acetyl coenzyme A. The consequences of this reaction are the production of $CO_2$, which diffuses into the bloodstream, and the production of an NADH molecule, which continues onto the electron transport chain. Because one molecule of glucose produces two molecules of pyruvic acid, two molecules of NADH are produced in this step.

---

**SO FAR...**

1. In aerobic respiration, the pyruvic acid that is the product of glycolysis is converted to the molecule _____, which is the substrate that is transferred to the Krebs cycle.

2. In the Krebs cycle, the starting substrate is methodically stripped of _____, which then are transferred to the _____. Some ATP is produced in the cycle, as is the by-product _____.

3. Electrons are transferred to two different electron carriers in the Krebs cycle. One of these is the _____ carrier seen in glycolysis; the other is the carrier _____.

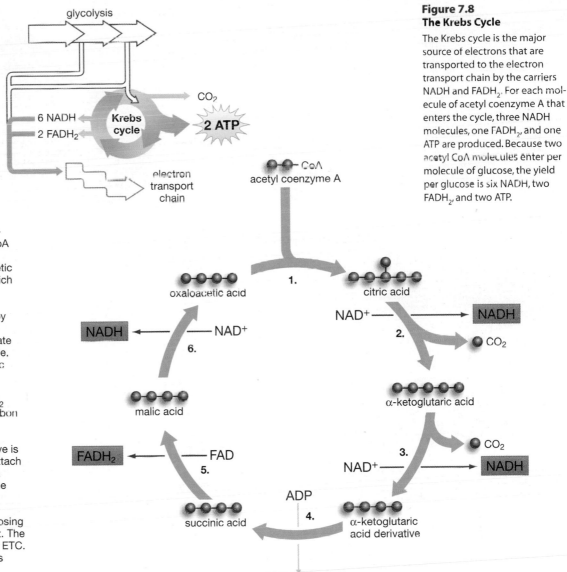

**Figure 7.8**
**The Krebs Cycle**

The Krebs cycle is the major source of electrons that are transported to the electron transport chain by the carriers NADH and $FADH_2$. For each molecule of acetyl coenzyme A that enters the cycle, three NADH molecules, one $FADH_2$, and one ATP are produced. Because two acetyl CoA molecules enter per molecule of glucose, the yield per glucose is six NADH, two $FADH_2$, and two ATP.

**Steps in the Krebs cycle**

1. Acetyl CoA combines with the four-carbon oxaloacetic acid and the CoA fragment separates from this compound. The result is the energetic six-carbon molecule citric acid, which will now be oxidized.

2. A citric acid derivative is oxidized by $NAD^+$; the resulting NADH carries electrons to the ETC. An intermediate molecule then loses a $CO_2$ molecule. Citric acid is now alpha-ketoglutaric acid.

3. Alpha-ketoglutaric acid loses a $CO_2$ molecule and the resulting four-carbon molecule is oxidized by $NAD^+$.

4. An alpha-ketoglutaric acid derivative is split, releasing enough energy to attach a phosphate to ADP, making it ATP. Alpha-ketoglutaric acid has become succinic acid.

5. Succinic acid is oxidized by FAD, losing two complete hydrogen atoms to it. The resulting $FADH_2$ then moves to the ETC. In a series of steps, succinic acid is transformed into malic acid.

6. Malic acid is oxidized by $NAD^+$. This step transforms malic acid to oxaloacetic acid—the molecule that enters the first step of the Krebs cycle.

## 7.6 Third Stage of Respiration: The Electron Transport Chain

Having moved through the Krebs cycle, you've reached the main event in cellular respiration: the production of 34 more ATP molecules through the work of the **electron transport chain (ETC)**, the third stage of aerobic energy harvesting. Glycolysis took place in the cytosol, and the Krebs cycle took place in the inner compartment of the mitochondria. Now, however, the action shifts to the mitochondrial inner membrane, the site of the ETC (Figure 7.9 on the next page). The "links" in this chain are a series of molecules. You've been looking for some time now at how NADH and $FADH_2$ carry off the elec-

trons derived from glycolysis and the Krebs cycle. This is the destination of these electron carriers and their cargo.

Upon reaching the mitochondrial inner membrane, the electron carriers donate electrons and hydrogen ions to the ETC. Again, this is a trip down the energy hill, only *NADH* is now the molecule whose electrons exist at a higher energy level. Thus, when NADH runs into the appropriate enzyme in the ETC, NADH is oxidized, donating its electrons and proton to the ETC. This process then is simply repeated down the whole ETC, each carrier donating electrons to the next electron carrier in line. A carrier is reduced by receiving these electrons; donating them to the next carrier, it then returns to an oxidized state. Each carrier thus alternates between oxidized and reduced states.

••• ANSWERS

1. acetyl CoA (acetyl coenzyme A)

2. energetic electrons; electron transport chain (ETC); $CO_2$

3. $NAD^+$/NADH; FAD/$FADH_2$

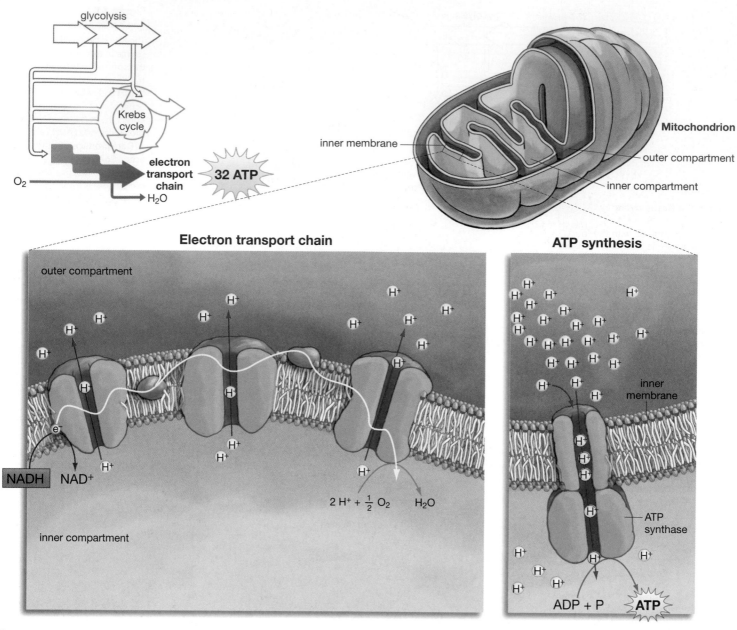

**Figure 7.9**
**The Electron Transport Chain (ETC)**

The movement of electrons through the ETC powers the process that provides the bulk of the ATP yield in respiration. The electrons carried by NADH and $FADH_2$ are released into the ETC and transported along its chain of molecules. The movement of electrons along the chain releases enough energy to power the pumping of hydrogen ions ($H^+$) across the membrane into the outer compartment of the mitochondrion. It is the subsequent energetic "fall" of the $H^+$ ions back into the inner compartment that drives the synthesis of ATP molecules by the enzyme ATP synthase.

Looking at **Figure 7.9**, you can see the carriers in the ETC: There are three large enzyme complexes with two smaller mobile molecules that link them. When NADH arrives at the inner membrane, it bumps into the ETC's first carrier and donates electrons to it; this carrier then donates these electrons to the next carrier, and so on down the line to the last electron acceptor, which is oxygen in the inner compartment (note the "$(\frac{1}{2}O_2)$," which will be explained shortly).

## Where's the ATP?

At this point, you may be tapping your foot, waiting for this promised harvest of ATP to appear. So far, all that's taken place is a transfer of electrons in the ETC. Well, note what happens in the first ETC enzyme complex. The movement of electrons through it releases enough energy to power the movement of *hydrogen ions* ($H^+$ ions) through the complex, pushing

them from the inner compartment into the outer compartment. (The fall of the electrons causes the enzyme complex to change its shape in a way that facilitates the passage of $H^+$ ions through it and into the outer compartment.) By the time the electrons have completed their movement through the ETC, this pumping of $H^+$ ions will occur with the two other large enzyme complexes. Critically, these $H^+$ ions are being pumped *against* their concentration and electrical gradients. Put another way, the ions are being pumped up the energy hill with energy supplied by the *downhill* fall of electrons through the ETC.

Here's where the ATP is produced at last. The $H^+$ ions that have been pumped into the outer compartment now move back down their concentration and energy gradients, into the inner compartment, through a special enzyme called ATP synthase. This remarkable enzyme is driven by the $H^+$ ions flowing through it, which cause part of the enzyme to rotate (as fast as 100 revolutions per second). We could think of it as being something like a waterwheel, except that its movement is driven not by water, but by the movement of $H^+$ ions that are passing through it, on their way back into the inner compartment. It is this energetic spinning that puts a third phosphate group (P) onto ADP, thus making ATP. **ATP synthase** can thus be defined as an enzyme that functions in cellular respiration by bringing together ADP and inorganic phosphate molecules to produce ATP.

The basic mechanism here—the pumping of hydrogen ions powering the synthesis of ATP—was first proposed in 1961 by British biochemist Peter Mitchell, who won the Nobel Prize in 1978 for his contribution to bioenergetics. Lest we pass by it too quickly, note that in this mechanism, we can see the essence of aerobic respiration. In the links of the ETC, the downhill drop of electrons derived from food has powered the uphill synthesis of the molecule that supplies the energy for the vast majority of our activities.

### Finally, Oxygen Is Reduced, Producing Water

The story of the breakdown of glucose—the saga of *one molecule's* transformation!—is nearly complete. The only unfinished business lies at the end of the ETC. As noted earlier, oxygen is the final acceptor of the electrons that move through the ETC. In the mitochondrial inner compartment, a single oxygen atom ($\frac{1}{2}O_2$) accepts two electrons from the last enzyme complex in the ETC and two $H^+$ ions. These then come together to produce $H_2O$: water.

This process for dealing with the electrons that made their way through respiration may sound like some minor housekeeping detail, but it's not. The electrons that are donated *to* the ETC by NADH and $FADH_2$ must be taken *from* the ETC somehow. To understand why, imagine a factory that makes, say, refrigerators. Now imagine that these refrigerators move down an assembly line on metal platforms, but that the platforms are suddenly allowed to start piling up at the end of the assembly line. Once this pile grows large enough, nothing can come off the end of the assembly line—indeed, the whole thing grinds to a halt. Just so, the body has an assembly line that results in both a product (ATP) and some materials that are used in manufacturing it (our de-energized electrons). Should these electrons not be taken away by oxygen, they would remain in place in the ETC—in its last enzyme complex. Should this happen, the complex would cease to accept electrons from the *previous* complex up the chain, and this process would continue until the whole assembly line would

be "backed up" all the way to glycolysis. Some organisms (such as yeast) can get by with just the small amount of ATP that glycolysis provides, but this is not the case for large organisms such as ourselves. We must have the larger amount of energy provided by aerobic energy transfer. And ground zero for a *lack* of energy transfer in us is the brain, which has cells that are voracious users of energy. Five minutes without oxygen is enough to put a human being into the condition known as a persistent vegetative state. We need to breathe because the energy assembly line will only keep moving if the electrons that are part of it are taken away by oxygen.

### Bountiful Harvest: ATP Accounting

Let's think for a moment about what might be called the ATP accounting in our story. In the ETC, each NADH molecule will be responsible for the production of three molecules of ATP. And how many NADH are coming? Ten: two produced in glycolysis, two from pyruvic acid conversion, and six from the Krebs cycle. Each $FADH_2$ carrier joins the ETC a little later than does NADH and as such is responsible for roughly two ATP. And how many $FADH_2$ are coming? Two, both of them produced in the Krebs cycle. When we put the $FADH_2$ and NADH output together, we find that 34 ATP are produced in the ETC as a result of the electrons brought by these carriers. Given this, think about what a mighty energizer the ETC is: 34 ATP compared to 2 in glycolysis. Imagine one car that got 34 miles to the gallon compared to another that got 2. Our final accounting on ATP yield is the 34 from the ETC, plus the 2 from glycolysis, plus 2 from the Krebs cycle, yielding 38 in all. From this, we need to subtract two ATP that are spent transporting the NADH produced in glycolysis into the mitochondria. The final result is thus 36 net ATP per molecule of glucose from all the steps of aerobic respiration.

## 7.7 Other Foods, Other Respiratory Pathways

By looking at aerobic respiration as it applies to glucose, which was used as the starting molecule, you've been able to go through the whole respiratory chain, but there are many kinds of nutrients besides glucose—for example, fats, proteins, and other sugars. All these can provide energy by being oxidized within the chain of reactions you've just gone over. However, none of them proceeds through the chain in exactly the same way as glucose. Instead, various nutrients can be channeled into the cell respiration chain in different ways.

To give one example of how this channeling of nutrients works, consider the possible fates of

ESSAY # Energy and Exercise

One minute we're sleeping, the next we're looking for our running shoes, and 10 minutes after that we're bounding down the road trying to cover three miles at a faster pace than yesterday. Coming to the steep hill at the end of our run, we go all out—charging to the top of it, slowing only after we start going downhill, breathless and spent. How can the human body cope with a range of energy demands that runs from sleeping to sprinting? Its secret is having a kind of energy trio at its service, with each member of the group specializing in delivering energy in particular situations.

The essence of any kind of exercise is the contraction of muscles, and the only energy molecule that can power the contraction of skeletal muscle is ATP. You have seen that the human body uses a three-stage process to produce ATP: glycolysis, the Krebs cycle, and the electron transport chain (ETC). As a whole, this system is dependent on oxygen, but glycolysis is not directly dependent on it. For brief periods of time, we can produce a good deal of the ATP we need through a glycolysis that is decoupled from the Krebs cycle and the ETC. Oxygen-independent glycolysis can thus be thought of as the first member of the body's energy trio, while *aerobic* or oxygen-using energy transfer can be thought of as the second member. Our cells also have a capacity to *store* small amounts of ATP, however, and they can also stockpile another molecule, phosphocreatine (PCr), that acts as a reservoir of phosphate groups that can be used to produce ATP. This stored ATP/PCr is the third member of the energy ensemble. Let's now see how all three players would work together in the course of a bike ride, starting out at a fairly brisk pace.

With the first turns of the pedals, bike riders bump up their body's energy demands tremendously. In their cycling, they could easily burn up 10 times the calories they would while in a resting state. (As a point of comparison, in doing housework a person typically burns up about three times the calories he or she does while at rest.)

So, how do muscle cells accommodate such large increases in their need for ATP? In the first few seconds, the small reservoirs of ATP/PCr take the lead role, with a proportionately smaller contribution from glycolysis and a smaller contribution yet from aerobic metabolism. These differences reflect the time it takes for each of these systems to get going. You've seen how many steps there are in glycolysis and how many *more* steps there are in the Krebs cycle and ETC. By comparison, the output of ATP from stored ATP/PCr is instantaneous. Remember, though, that the ATP/PCr reservoir is small. A person who had to depend solely on it for a full-ahead sprint would have only enough energy to last about six seconds. On a longer bike ride, glycolysis and ATP/PCr are contributing roughly equal amounts of ATP within 10 seconds. Within 30 seconds, glycolysis has greatly eclipsed PCr as an energy provider.

At about this same 30-second mark, the third player, aerobic respiration, is supplying only about 20 percent of the ATP, but its contribution is growing fast. In **Figure 1**, you can see how the aerobic contribution to ATP yield grows over the course of a long, strenuous workout. At the start, oxygen-dependent respiration is supplying less than 10 percent of our ATP, but 10 minutes into the workout it is delivering 85 percent.

**Figure 7.10**
**Many Respiratory Pathways**
Glucose is not the only starting material for cellular respiration. Other carbohydrates, proteins, and fats can also be used as fuel for cellular respiration. These reactants enter the process at different stages.

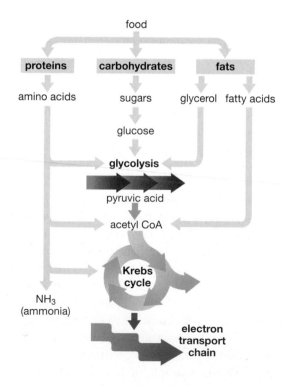

triglycerides. As you learned in Chapter 3, triglycerides are fats made of a three-carbon glycerol "head" and three long hydrocarbon chains that look like tines on a fork. Imagine that a given cell needs energy from fat. The question is, how do triglycerides go through aerobic respiration, as opposed to glucose?

The first thing that happens is that a triglyceride molecule is split by enzymes into its constituent parts of fatty acids and glycerol. Glycerol is then converted into glyceraldehyde phosphate. This formidable name may not ring a bell, but if you look back to step 4 in Figure 7.5, you'll see that it is one of the downstream derivatives of glucose in glycolysis. To put this another way, glycerol does not first get converted to glucose and then march through *all* of the steps of glycolysis. Instead, it joins the glycolytic pathway several steps down from glucose and is then converted to pyruvic acid. Then, as pyruvic acid, it goes through the

## Figure 1
### Different Contributions over Time

**Duration of maximal exercise**

|  | Seconds | | | Minutes | | | | | |
|---|---|---|---|---|---|---|---|---|---|
|  | 10 | 30 | 60 | 2 | 4 | 10 | 30 | 60 | 120 |
| Percent anaerobic | 90 | 80 | 70 | 50 | 35 | 15 | 5 | 2 | 1 |
| Percent aerobic | 10 | 20 | 30 | 50 | 65 | 85 | 95 | 98 | 99 |

Relative contributions of anaerobic and aerobic respiration to exercise during the duration of a workout. "Anaerobic" here means a combination of glycolysis and stored ATP/PCr release. At the 1-minute mark, aerobic respiration is supplying only 30 percent of the body's energy needs; at the 10-minute mark, it is supplying 85 percent. (Adapted from P. O. Astrand and K. Rodahl, *Textbook of Work Physiology*. New York: McGraw Hill Book Company, © 1977.)

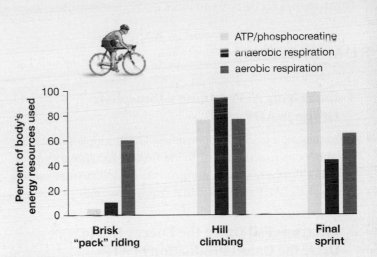

## Figure 2
### Different Activities, Different Energy Stores

Energy-harvesting processes used at different stages in a bicycle race. (Adapted from J. T. Kearney, "Training the Olympic Athletes." *Scientific American*, June 1996, p. 54.)

You might think that this would be the final division of ATP contributions among the three players, but the amounts they supply can vary over time depending on the intensity of the activity. **Figure 2** shows how this would work in an actual professional-level bicycle race. Going along at a steady state in "pack" riding, you can see the overriding contribution of aerobic metabolism. In the final sprint, however, stores of ATP/PCr—replenished during the pack riding— are fully used, with the contribution of glycolysis increasing as well. In stretches of difficult hill climbing, meanwhile, glycolysis is fully used.

Of course, lots of sports have no aerobic component to them. The energy for a sport like football is supplied almost entirely by stored ATP/PCr. Even in a competition as long as a 400-meter sprint—about once around a regulation track—about 70 percent of the energy comes from a combination of ATP/PCr and glycolysis.

Krebs cycle and ETC, producing ATP. And the fatty acids? They are converted to acetyl CoA, which you may remember is the substrate that enters the Krebs cycle, yielding energy by becoming oxidized there. In **Figure 7.10**, you can see a summary of the way foods are broken down through various respiratory pathways.

## On to Photosynthesis

This long walk through cellular respiration has illustrated how living things harvest energy from food. Recall, however, that almost all living things have one source to thank for this food: the sun's energy, which plants capture and then lock up in a sugar they produce. This capture of energy takes place through a process that has a beautiful symmetry with the respiration you've just looked at. The process in question is called photosynthesis.

### SO FAR...

1. In the electron transport chain (ETC), the electron carriers NADH and FADH _____ electrons to a series of enzyme complexes located within the inner _____ of the mitochondria. The energetic downhill drop of this electron transfer powers the movement of _____ into the outer compartment of the mitochondria.

2. The physical movement of _____ back into the inner compartment through the enzyme _____ is a downhill energetic drop that powers the process by which _____ are added to ADP, thus yielding _____.

3. The electrons picked up at various stages of energy harvesting by NADH and FADH must be accepted by _____ at the end of the ETC if energy transfer is to continue.

**••• ANSWERS**

1. lose (donate); membrane; hydrogen ions (H⁺ ions)

2. H⁺ ions; ATP synthase; phosphate groups; ATP

3. oxygen

# CHAPTER 7 REVIEW

For study help, activities, and more quiz questions, go to **www.aw.com/krogh4**.

## SUMMARY

### 7.1 Energizing ATP: Adding a Phosphate Group to ADP

- The molecule adenosine triphosphate (ATP) supplies the energy for most of the activities of living things. For ATP to be produced, a third phosphate group must be added to adenosine diphosphate (ADP), a process that requires energy.

### 7.2 Electrons Fall Down the Energy Hill to Drive the Uphill Production of ATP

- In animals, the energetic fall of electrons derived from food powers the process by which the third phosphate group is attached to ADP, making it ATP.

- Electron transfer in the production of ATP works through redox reactions, meaning reactions in which one substance loses electrons to another substance. The substance that loses electrons in a redox reaction is said to have been oxidized, while the substance that gains electrons is said to have been reduced. In energy transfer in living things, starting food molecules are oxidized, and the downhill fall of the energetic electrons they lose ultimately powers the uphill process by which ATP is produced.

- Electrons are carried between one part of the energy-harvesting process and another by electron carriers, the most important of which is nicotinamide adenine dinucleotide or NAD. In its "empty" state, this molecule exists as $NAD^+$. Through a redox reaction, it picks up one hydrogen atom and another single electron from food, thus becoming NADH, a form it will retain until it drops off its energetic electrons (and a proton) in a later stage of the energy-harvesting process.

*Web Animation 7.1:* Oxidation and Reduction

### 7.3 The Three Stages of Cellular Respiration: Glycolysis, the Krebs Cycle, and the Electron Transport Chain

- In most organisms, the harvesting of energy from food takes place in three principal stages: glycolysis, the Krebs cycle (or citric acid cycle), and the electron transport chain (ETC). Taken as a whole, this oxygen-dependent three-stage harvesting of energy is known as cellular respiration.

- Some organisms rely solely on glycolysis for energy harvesting. For most organisms, however, glycolysis is a primary process of energy extraction only in certain situations—when quick bursts of energy are required, for example—but it is a necessary first stage to the Krebs cycle and the ETC.

- Glycolysis takes place in the cell's cytosol, while the Krebs cycle and the ETC take place in cellular organelles, called mitochondria, that lie within the cytosol. Glycolysis yields two net molecules of ATP per molecule of glucose, as does the Krebs cycle. The net yield in the ETC is a maximum of about 32 ATP molecules per molecule of glucose. Glycolysis and the Krebs cycle are critical, however, in that they yield electrons that are carried to the ETC (via electron carriers such as $NAD^+$) for the final high-yield stage of energy harvesting.

*BioFlix* Cellular Respiration

### 7.4 First Stage of Respiration: Glycolysis

- When glycolysis begins with a single molecule of glucose, the ultimate products are two molecules of NADH (which move to the ETC, bearing their energetic electrons) and two molecules of ATP (which are ready to be used). Glycolysis also produces two molecules of pyruvic acid—the derivatives of the original glucose molecule—which move on to the Krebs cycle.

### 7.5 Second Stage of Respiration: The Krebs Cycle

- There is a transition step in respiration between glycolysis and the Krebs cycle. In it, each pyruvic acid molecule that was produced in glycolysis combines with coenzyme A, thus forming acetyl coenzyme A (acetyl CoA), which enters the Krebs cycle. There are also two other products of this reaction. One is a molecule of carbon dioxide, which diffuses to the bloodstream; the other is one more molecule of NADH, which moves to the ETC. For each starting molecule of glucose, two molecules of pyruvic acid go through this step; thus, the step's product per molecule of glucose is two molecules of carbon dioxide, two NADH, and two acetyl CoA.

- In the Krebs (or citric acid) cycle, the derivatives of the original glucose molecule are oxidized, with the result that more energetic electrons are transported by the electron carriers NADH and $FADH_2$ to the ETC. The net energy yield of the Krebs cycle is six molecules of NADH, two molecules of $FADH_2$, and two molecules of ATP per molecule of glucose.

### 7.6 Third Stage of Respiration: The Electron Transport Chain

- The ETC is a series of molecules located within the mitochondrial inner membrane. On reaching the ETC, the electron carriers

NADH and FADH$_2$ are oxidized by molecules in the chain. Each carrier in the chain is then reduced by accepting electrons from the carrier that came before it. The last electron acceptor in the ETC is oxygen.

- The movement of electrons through the ETC releases enough energy to power the movement of hydrogen ions (H$^+$ ions) through the three ETC protein complexes, moving them from the mitochondrion's inner compartment to its outer compartment. The movement of these ions down their concentration and charge gradients, back into the inner compartment through an enzyme called ATP synthase, drives the synthesis of ATP from ADP and phosphate. In the inner compartment, oxygen accepts the electrons from the ETC and hydrogen ions, thus forming water. If oxygen is not present to accept the ETC electrons, the entire energy-harvesting process downstream from glycolysis comes to a halt.

## 7.7 Other Foods, Other Respiratory Pathways

- Different nutrients and their derivatives can be channeled through different pathways in cellular respiration in accordance with the needs of an organism. Proteins and lipids enter the metabolic pathway for ATP production at different points than does the glucose reviewed in detail in the chapter.

## KEY TERMS

| | |
|---|---|
| **ATP synthase**   135 | **Krebs cycle**   131 |
| **cellular respiration**   127 | **nicotinamide adenine |
| **citric acid cycle**   131 | dinucleotide (NAD)**   125 |
| **electron carrier**   125 | **oxidation**   125 |
| **electron transport chain** | **redox reaction**   125 |
| **(ETC)**   133 | **reduction**   125 |
| **glycolysis**   127 | |

## BRIEF REVIEW

*Answers to Brief Review questions are in the back of the book. For multiple-choice quiz questions, go to* **www.aw.com/krogh4**.

1. Living things need ATP to power most of the processes that go on within them. In essence, how does cellular respiration yield this ATP?

2. Where do you expect people would have more mitochondria: their skin cells or their muscle cells? Why?

3. In the ETC, what is physically moved across the inner membrane—and energetically moved to a higher state—to drive the attachment of phosphate groups to ADP?

4. Where does the energy come from that powers this movement across the inner membrane in the ETC?

5. Most of the ATP we use is produced in the ETC. So, when we have to run across the street to avoid traffic, does the ATP we are using come primarily from the ETC?

6. Where does the Krebs cycle take place in this model of a mitochondrion? What is the name of this portion of the mitochondrion?

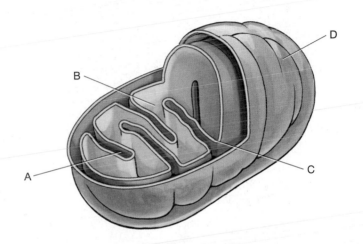

## APPLYING YOUR KNOWLEDGE

1. In Madeleine L'Engle's children's novel *A Wrinkle in Time*, the mitochondria in one of the characters start to die. Describe what would happen to people who lost their mitochondria and explain why it would happen.

2. Most of the heat in the human body is generated within mitochondria. Why should this be?

3. The food we eat is broken down in our digestive tract, but where are most of the calories in this food actually "burned"?

Energy from the sun helps produce the food that sustains nearly all living things.

Plants use the sun's energy to make their own food—and ours.

8.1    Photosynthesis and Energy    141

8.2    The Components of Photosynthesis    142

8.3    Stage 1: The Steps of the Light Reactions    146

8.4    What Makes the Light Reactions So Important?    146

8.5    Stage 2: The Calvin Cycle    *BioFlix*    148

8.6    Photorespiration and the $C_4$ Pathway    150

8.7    Another Photosynthetic Variation: CAM Plants    151

**Essay**

**THE PROCESS OF SCIENCE:** Plants Make Their Own Food, But How?    154

# THE GREEN WORLD'S GIFT:
# Photosynthesis

**O**n a summer's evening, out by a pond as the sun is going down, nature tunes up her symphony. The crickets bring their chirping to the fore, while frogs chime in with their own countermelody. Silently, something else is changing as the last light fades: The green world is shutting down. Microscopic pores that dot the undersides of leaves are closing up, ceasing to be openings for the carbon dioxide that flows in and the water vapor that flows out during the day. The green world is alive at night, but it is only fully active during the day, when a special talent that it has is set in motion by the warm rays of the sun.

make proteins. This, then, is what plants do: They *make their own food*. They use the energy provided by the sun to take a simple gas, carbon dioxide, join it to a carbohydrate (a sugar), and then *energize* that sugar, thus transforming it into food. In this activity, they are joined by a variety of other photosynthesizing organisms (**Figure 8.1**).

## 8.1 Photosynthesis and Energy

The green world's special talent is called photosynthesis, and the wonder of it can perhaps best be appreciated by imagining what life would be like if we humans possessed it. The bone and muscle within us—and the energy to make them work—come from the foods we eat: carbohydrates, fats, and proteins. Imagine, though, having the ability to flourish simply by spending time in the sun and having access to three things: water, small amounts of minerals, and carbon dioxide. In this condition, we would not *eat* carbohydrates in the usual forms (bread, sugar, potatoes). Rather, we would transform the simple gas carbon dioxide *into* carbohydrates using nothing more than water, minerals, and sunlight. Carbohydrates produced in this way are as useful as any food; they can be stored away as starches, used to provide ATP for energy needs, or transformed to

**(a)** Sunflowers (plants)  **(b)** Giant kelp (algae)  **(c)** Cyanobacteria (bacteria)

**Figure 8.1**
**Three Types of Photosynthesizers**
Photosynthesis is carried out not only by familiar plants, such as these sunflowers **(a)**, but also by algae, such as this giant kelp **(b)**, and by some bacteria, such as these cyanobacteria **(c)**, magnified ×1,025.

### From Plants, a Great Bounty for Animals

The process of photosynthesis is not only good for plants; it's good for other living things, including animals, since most of the living world ultimately depends on photosynthesis for food. Try naming something you eat that isn't a plant or doesn't itself eat plants to survive, and you will come up with a very short list indeed.

Food, however, is only half the bounty that plants provide. A by-product of photosynthesis—a kind of castoff from the work of plants—is oxygen. As part of photosynthesis, plants break water molecules apart. In doing so, they use electrons and protons from $H_2O$, but they leave behind oxygen molecules ($O_2$). This is the oxygen that exists in our atmosphere; it is the oxygen that we breathe in every minute of every day or suffer the mortal consequences.

With this, you can step back and view the largest picture of them all with respect to energy flow in living things. It is a great pathway in which energy comes from the sun and then, in photosynthesis, is stored in plants in the complex molecules we call carbohydrates. Thus stored—in such forms

as wheat grains or grass leaves—these carbohydrates can be broken down and used, either by the plants themselves or by the organisms that eat them. The breakdown of this food ends with the "cellular respiration," covered last chapter, that provides ATP.

### Up and Down the Energy Hill Again

Looked at another way, photosynthesis and cellular respiration are trips up and down the energy hill that became so familiar in Chapter 7. In respiration, the trip was down the hill, meaning from more stored energy (in food) to less, as the food was broken down to produce ATP. Photosynthesis is a trip up the energy hill. Here, electrons are removed from water, boosted to a more energetic state by the power of sunlight, and then brought together with a sugar and carbon dioxide, resulting in an *energy-rich* sugar—food in the form of a carbohydrate. The story of how this carbohydrate production is accomplished is the subject of this chapter. To encapsulate this story in a definition, **photosynthesis** is the process by which certain groups of organisms capture energy from sunlight and convert this solar energy into chemical energy that is initially stored in a carbohydrate.

## 8.2 The Components of Photosynthesis

We can think of photosynthesis as starting with the absorption of sunlight by leaves. *Absorption* here means just what it sounds like: Light is taken in by the leaves. Actually, leaves capture only a *portion* of the light that falls on them. The sunlight that makes it to the Earth's surface is composed of a spectrum of energetic rays (measured by their "wavelengths") that range from very short ultraviolet rays, through visible light rays, to the longer and less-energetic infrared rays. (In **Figure 8.2** you can see that these wavelengths are, in turn, part of a larger range of electromagnetic radiation.) Photosynthesis is driven by part of the *visible light* spectrum—mainly by blue and red light of certain wavelengths. Tune a laser to emit blue light of a certain wavelength, shine this beam on a plant, and the plant will absorb a great deal of this light. Tune the laser to emit *green* light, on the other hand, and the plant will scatter most of this light. This is why

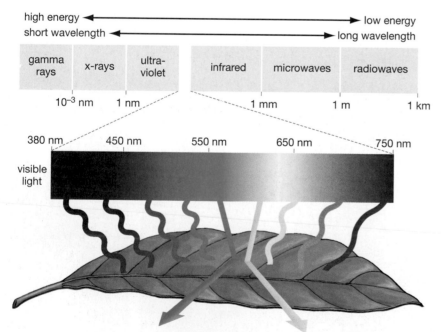

**Figure 8.2**
**The Electromagnetic Spectrum**
Sunlight is composed of rays of many different wavelengths, but we can see only those in the visible light range (represented by the color spectrum). When the light of the sun shines on a green leaf, it is primarily the sun's green rays that are scattered, while frequencies in other ranges are absorbed; frequencies in the red and blue ranges drive photosynthesis most strongly.

plants are green: They strongly scatter the green portion of the visible light spectrum.

## Where in the Plant Does Photosynthesis Occur?

So, where is this light absorption occurring within plants? To answer this question, you need to know something about the playing fields of photosynthesis, leaves.

If you look at the idealized leaf in **Figure 8.3**, you can see that it can be likened to a kind of cellular sandwich, with one layer of outer (or epidermal) cells at the top, another layer of epidermal cells at the bottom, and several layers of mesophyll cells in between. Take time now to read through Figure 8.3's drawings and captions; as you do, you'll go from the leaf's layers of cells, then into the cells themselves, and then into the tiny structures *within* the cells—the organelles called chloroplasts—that are the sites of photosynthesis.

Here is some detail on the structures seen in Figure 8.3. The **stomata** seen in the figure are the microscopic pores, mentioned at the start of the chapter, that let carbon dioxide pass into leaves and water vapor pass out of them. There is no shortage of stomata for this work; a given square centimeter

Web Animation 8.1
Properties of Light

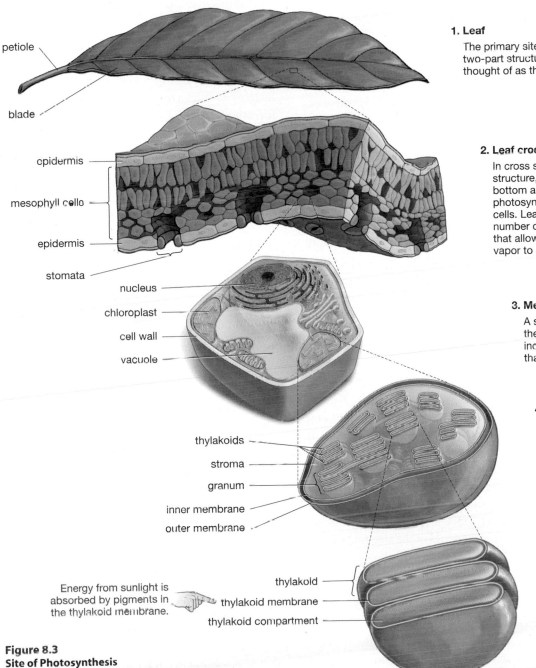

petiole

blade

epidermis

mesophyll cells

epidermis

stomata

nucleus

chloroplast

cell wall

vacuole

thylakoids

stroma

granum

inner membrane

outer membrane

Energy from sunlight is absorbed by pigments in the thylakoid membrane.

thylakoid

thylakoid membrane

thylakoid compartment

**1. Leaf**

The primary site of photosynthesis in plants, leaves have a two-part structure: a petiole (or stalk) and a blade (normally thought of as the leaf).

**2. Leaf cross section**

In cross section, leaves have a sandwich-like structure, with epidermal layers at top and bottom and mesophyll cells in between. Most photosynthesis is performed within mesophyll cells. Leaf epidermis is pocked with a large number of microscopic openings, called stomata, that allow carbon dioxide to pass in and water vapor to pass out.

**3. Mesophyll cell**

A single mesophyll cell within a leaf contains all the component parts of plant cells in general, including the organelles—called chloroplasts—that are the actual sites of photosynthesis.

**4. Chloroplast**

Each chloroplast has an outer membrane at its periphery; then an inner membrane; then a liquid material, called the stroma, that has immersed within it a network of membranes, the thylakoids. These thylakoids sometimes stack on one another to create. . .

**5. A Granum**

Electrons used in photosynthesis will come from water contained in the thylakoid compartment, and all the steps of photosynthesis will take place either within the thylakoid membrane, or in the stroma that surrounds the thylakoids.

**Figure 8.3**
**Site of Photosynthesis**

of plant leaf may contain from a thousand to a hundred thousand of them.

Going down a couple of levels, **chloroplasts** can be defined as the organelles within plant and algae cells that are the sites of photosynthesis. Anywhere from one to several hundred of these oblong structures might exist within a given leaf cell, each one capable of carrying out photosynthesis on its own. As Figure 8.3 shows, chloroplasts present another example of something you've seen before—membranes within membranes. The chloroplast has outer and inner membranes at its periphery. Moving inside them, to the interior of the chloroplast, there is a network of chloroplast membranes, active in photosynthesis, called **thylakoids**. (These thylakoids often stack on top of one another like so many pancakes, creating structures called *grana*.) Thylakoids are immersed in the liquid material of the chloroplast, the **stroma**. Keep your eye on the thylakoid membranes and on the stroma because all the steps of photosynthesis occur in one of these two places.

The thylakoid membranes are, in fact, the place where photosynthesis starts, with the absorption of sunlight we just talked about. Any compound that strongly absorbs certain visible wavelengths of sunlight is called a *pigment*. Thylakoid membranes contain a pigment, called **chlorophyll *a*,** that can be defined as the primary pigment active in plant photosynthesis. Chlorophyll *a* is then aided by several substances known as *accessory pigments*, which do just what their name implies: They aid chlorophyll *a* in absorbing energetic rays from the sun, after which they pass the absorbed energy along.

### There Are Two Essential Stages in Photosynthesis

Keeping these structures and concepts in mind, you can now begin tracing the steps of photosynthesis. It is helpful to divide photosynthesis into two main stages. In the first stage (the *photo* of photosynthesis), the power of sunlight will do two things—strip water of electrons and then boost these electrons to a higher energy level. Thus boosted, the electrons are passed along through a series of the electron carriers introduced in Chapter 7. All of this takes place in the thylakoid membranes. This first stage of photosynthesis ends when the original, energized electrons get attached to a mobile electron carrier, called $NADP^+$, that transports them to the second set of reactions. In this second stage (the *synthesis* of photosynthesis), the electrons come together with carbon dioxide and a sugar. The attachment of the electrons and $CO_2$ to

this sugar produces a high-energy sugar—meaning food. This second stage takes place in the stroma of the chloroplast.

The steps in the first stage of photosynthesis obviously depend directly on sunlight. Thus, these steps are sometimes referred to as the *light reactions*. Meanwhile, the second set of steps is referred to as the *Calvin cycle*, in honor of Melvin Calvin, the scientist who first revealed their details. You will be following both sets of steps as they occur in time—that is, from the collection of sunlight to the synthesis of carbohydrates.

### The Working Units of the Light Reactions

Looking first at the light reactions, we see that they take place within a set of working units. Each of these units is a **photosystem**—an organized complex of molecules within a thylakoid membrane that, in photosynthesis, collects solar energy and transforms it into chemical energy. If you look at **Figure 8.4**, you can see the components that make up a photosystem. First, there is a group of a few hundred pigment molecules that serve to absorb the sunlight. The majority

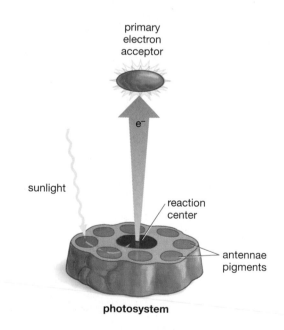

primary
electron
acceptor

$e^-$

sunlight

reaction
center

antennae
pigments

**photosystem**

**Figure 8.4**
**The Working Units of Photosynthesis**

Photosystems are multipart units that bring together electrons derived from water and energy derived from the sun. Hundreds of antennae pigments absorb sunlight and transfer solar energy to the photosystem's reaction center. This energy gives a boost to an electron within the reaction center in two ways: It physically moves the electron to the center's primary electron acceptor, and it moves the electron up the energy hill to a more energetic state.

of these molecules—most of the chlorophyll *a* and all of the accessory pigments—serve only as "antennae" that absorb energy from the sun and pass it on. At the center of the antennae system, however, there is a **reaction center**—a pair of special chlorophyll *a* molecules and associated compounds that first receive the solar energy from photosystem pigments and then transform this solar energy into chemical energy. As you can see in Figure 8.4, the reaction center includes a molecule that could be thought of as the first recipient of the absorbed solar energy, the "primary electron acceptor."

You could conceptualize photosynthesis as beginning when sunlight is absorbed by any of the hundreds of antennae pigment molecules in a photosystem unit. The absorbed energy is then passed on to the pair of chlorophyll *a* molecules in a reaction center. As a result, electrons from this pair are "moved" in a couple of ways, one physical and one metaphorical. Physically, these electrons are transferred to the primary electron acceptor. You also could think of the initial "distance traveled" by the electrons in a metaphorical way, however: as a movement up the energy hill. The energy from sunlight is pumping the electrons up the hill in photosynthesis.

## Energy Transfer in Photosynthesis Works through Redox Reactions

How can electrons move up an energy hill? Recall from Chapter 7 that some substances have a tendency to *lose* electrons to other substances. Any time one substance loses (or "donates") electrons to another, it is said to have been oxidized. Meanwhile, the substance that gained electrons is said to have been reduced. These two reactions always happen together; if one substance is oxidized, another must be reduced. This kind of reduction-oxidation reaction has a shorter name: a *redox reaction.*

Many of the photosynthetic steps you will be seeing are redox reactions. Indeed, you have already seen one. The movement of electrons within the reaction center—from the chlorophyll *a* molecules to the primary electron acceptor—is a redox reaction. The chlorophyll molecules were oxidized by the primary electron acceptor, meaning they lost electrons to it. What happens in the light reactions is a continuation of this energy transfer: Electrons are passed on by means of one electron carrier oxidizing another.

As electrons physically move through the light reactions in this way, they are also moving both up and down the energy hill—they are becoming first more and then less energetic. If you look at **Figure 8.5**, you can trace their movement up and down the energy hill as they make their journey. With energy supplied by the sun, electrons are pumped up a couple of formidable energy gradients (in photosystems II and I) only to come partway back down them as they "seek" their lowest energy state. Critically, they are releasing energy as they fall—just as a boulder releases energy by rolling downhill.

### Figure 8.5
### Collecting Solar Energy, Boosting Electrons

The solar energy gathered by photosystems II and I is used to energize electrons that exist within the photosystems' reaction centers. Energetically, the electrons are boosted up, and then move down, two energy hills. Physically, they first move to primary electron acceptors and then down electron transport chains until they are at last taken up by $NADP^+$ to form NADPH. The NADPH molecules then transfer the electrons to the Calvin cycle, in which they are used to make sugars. When an electron is boosted in photosystem II, the resulting energy imbalance drives the process by which water molecules are split. Electrons derived from the water adjacent to photosystem II then move to its reaction center, where they are the next to be boosted by the sun's energy. The energetic fall of electrons down the electron transport chain between photosystems II and I provides the energy for the synthesis of ATP, which is used to power the Calvin cycle.

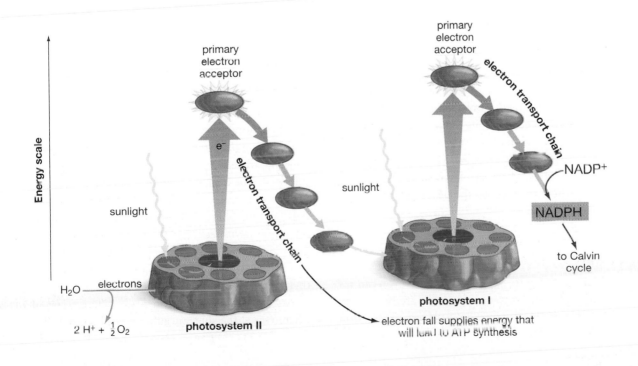

primary electron acceptor

primary electron acceptor

electron transport chain

Energy scale

e⁻

electron transport chain

electron transport chain

sunlight

NADP+

NADPH

sunlight

to Calvin cycle

H₂O — electrons

$2 H^+ + \frac{1}{2}O_2$

**photosystem II**

**photosystem I**

electron fall supplies energy that will lead to ATP synthesis

## SO FAR...

1. In photosynthesis, the power of the sun is used to boost the energy of _____ that are derived from water and bring them together with both a low-energy sugar and _____ from the atmosphere, thus producing an energy-rich _____, meaning food.

2. In plants, photosynthesis takes place within the organelles called _____, which are found most abundantly in certain cells that make up the interior portions of _____. Pigments within the membranes of these organelles, in particular a pigment called _____, serve to absorb sunlight, which provides the energy that drives photosynthesis.

3. Photosynthesis can be divided into two major sets of steps. In the first set, known as the _____, absorbed solar energy moves electrons derived from water in two ways: (1) physically to a molecule called a _____ and (2) metaphorically to a higher _____.

## 8.3   Stage 1: The Steps of the Light Reactions

With these components and processes in mind, let's trace the steps of the light reactions (see Figure 8.5). The first step is that solar energy, collected by photosystem II's antennae molecules, arrives at the reaction center. This energy then gives a boost to an electron in the reaction center in the two ways noted: The electron *physically* moves to another part of the reaction center complex, the primary electron acceptor, and is also pumped up the energy hill. With this activity, the reaction center chlorophyll has *lost* an electron. That loss leaves an energy "hole" in this chlorophyll, making it an oxidizing agent. With the energy provided by this imbalance, a special enzyme in the reaction center splits water molecules that lie within the thylakoid compartment. These water molecules are now being oxidized, which means *they* are losing electrons. The electrons travel to the reaction center, where they will be the next electrons in line for an energy boost.

### A Chain of Redox Reactions and Another Boost from the Sun

By following Figure 8.5 in a general way, you can see what happens next to electrons that are boosted in photosystem II. After arriving at the primary electron

acceptor, they fall back down the energy hill as they are transferred through a series of electron transport molecules, each oxidizing its predecessor. At the bottom of this hill, they arrive at photosystem I, which includes a slightly different kind of reaction center. This center is also receiving solar energy and uses it to boost electrons to a higher energy state. From this energetic state, electrons are transferred down the second energy hill—until they are received by the electron carrier $NADP^+$, which is very much like the electron carrier $NAD^+$ reviewed in Chapter 7. In accepting electrons, $NADP^+$ becomes reduced to NADPH, an electron carrier that ferries the electrons into the next stage of photosynthesis. This second stage is the Calvin cycle, which will yield the high-energy sugar that is the essential product of photosynthesis.

### The Physical Movement of Electrons in the Light Reactions

In this movement up and down the energy hill, the electrons have moved physically through the chloroplast. As **Figure 8.6** shows, they started out in the water of the thylakoid compartment, moved into and then through the thylakoid membrane—handed off at each step by the various electron transport molecules—and then finally ended up in the stroma, attached to NADPH.

## 8.4   What Makes the Light Reactions So Important?

The steps just reviewed are the essence of the light reactions. To make sure you don't miss the forest for the trees, let's be clear about two momentous things that have taken place within these steps: the splitting of water and the transformation of solar energy to chemical energy.

### The Splitting of Water: Electrons and Oxygen

The first notable event takes place near the start: the splitting of water. Critically, this reaction provides the traveling electrons whose path you've just followed. But it also provides something else: the oxygen that today accounts for 21 percent of the Earth's atmosphere. In the water splitting that goes on in photosynthesis, hydrogen atoms are removed from $H_2O$, while the oxygen is left behind. When two liberated oxygen atoms come together in the thylakoid compartment, the result is $O_2$—the form of oxygen that exists in our atmosphere. *Here* is where we get the substance that is the breath of life

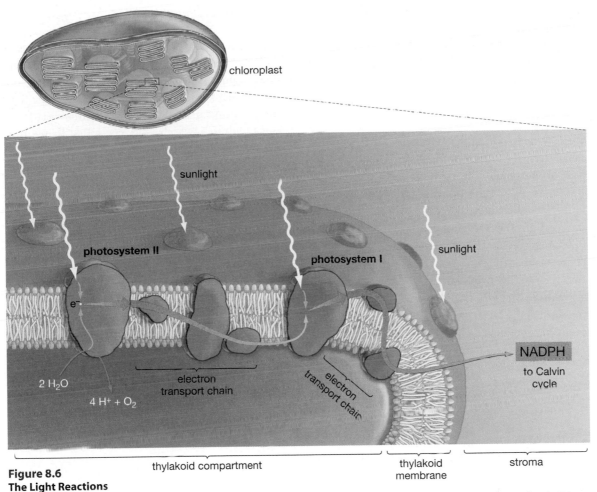

**Figure 8.6**
**The Light Reactions**

The light reactions take place within the thylakoid membranes in the chloroplasts. Electrons are donated by water molecules located in the thylakoid compartments. Powered by the sun's energy, these electrons are passed along the electron transport chain embedded in the thylakoid membrane and end up stored in NADPH in the stroma. An additional product of the splitting of water molecules is individual oxygen atoms, which quickly combine into the $O_2$ form. This is the atmospheric oxygen that we breathe.

itself for most species, all of it contributed by living things. Given its importance, it is no small irony that oxygen is, in another sense, a kind of green-world refuse —leftovers from the process by which plants strip water of the electrons they need to carry out photosynthesis.

## The Transformation of Solar Energy to Chemical Energy

The second major feat that takes place during light reactions is the transformation of solar energy to chemical energy. When the energy of sunlight moves to a chlorophyll molecule in a reaction center, it boosts an electron there from what is known as a ground state to an excited state. Thus far, this has been described as a movement up a metaphorical energy hill. In physical terms, however, this electron is moving farther out from the nucleus of an atom. To appreciate what's special about photosynthesis, consider what the *common* fates are for such

excited electrons. One fate is for the electrons to drop back down to the original (and more stable) state, in the process releasing as *heat* the energy they have absorbed. This is the process by which black objects get hot on sunny days. Another possible fate is for falling electrons to release part of their energy as light in the process known as fluorescence.

In photosynthesis, however, the energized electrons are *transferred to a different molecule*—the initial electron acceptors of photosystems II and I. They don't fall back to their ground state, releasing relatively useless heat or light in the process. They are passed on in a redox reaction. This is the bridge to life as we know it. Without this step, the energy of the sun merely makes the Earth warmer. With it, the green world grows profusely; it captures some of the sun's energy and with it builds trunks and grains and leaves. By one reckoning, photosynthesis produces up to 155 billion tons of material each year. This is an amazing

••• ANSWERS

**1.** electrons; carbon dioxide; carbohydrate (a sugar)

**2.** chloroplasts; leaves; chlorophyll *a*

**3.** light reactions; primary electron acceptor; energy state

amount—about 25 tons for every person on Earth. Electrons are taken from water, boosted to a higher energy state by the sun, and transferred to other molecules. The result is the food that sustains nearly all living things.

### Production of ATP

It's also worth noting that a primary function of the fall of electrons between photosystems II and I is a release of energy that is used for the production of ATP. Put together through a process similar to the one you looked at last chapter, this ATP will be used to power the reactions that are coming up in the second stage of photosynthesis. Thus, the end result of the light reactions is that the sun's power has provided energy that is stored in *two* forms: ATP and the energetic electrons in NADPH. Looking at the big picture, you can see that the *photo* of photosynthesis is an energy-capturing operation.

## 8.5 Stage 2: The Calvin Cycle

In the second stage of photosynthesis, the energy captured in the light reactions will be *used*. It will power a process by which carbon dioxide, taken from the atmosphere, will be joined to a sugar, with the resulting product energized, thus creating a carbohydrate—which is to say, food. To see how this happens, let's begin by taking stock of where the light reactions left off. The short answer is, in the stroma. That's where $NADP^+$ has become NADPH by accepting the energized electrons pouring out from photosystem I. There's also ATP in the stroma, another product of the light reactions.

### Energized Sugar Comes from a Cycle of Reactions

Even with the right ingredients, however, it takes more than one step to go from carbon dioxide and a sugar to a sugar that is energetic enough to serve as food. It takes a set of chemical reactions that make up a cycle. This is the **Calvin cycle** or the $C_3$ cycle, the set of steps in photosynthesis in which energetic electrons are brought together with carbon dioxide and a sugar to produce an energetic carbohydrate. The highlights of this cycle are set forth in **Figure 8.7**. This

**Figure 8.7**
**The Calvin Cycle**

$CO_2$, ATP, and electrons (contained in hydrogen atoms) from NADPH are the input into the Calvin cycle, while a sugar (G3P) is its output.

1. **Carbon fixation.** An enzyme called rubisco brings together three molecules of $CO_2$ with three molecules of the sugar RuBP. In this reaction, one carbon from each $CO_2$ molecule is being added to the five-carbon RuBP, and this is being done three times. The three resulting six-carbon molecules are immediately split into six three-carbon molecules named 3-PGA (3-phosphoglyceric acid).

2. **Energizing the sugar.** In two separate reactions, six ATP molecules react with six 3-PGA, in each case transferring a phosphate onto the 3-PGA. The six 3-PGA derivatives oxidize (gain electrons from) six NADPH molecules; in so doing, they are transformed into the energy-rich sugar G3P (glyceraldehyde 3-phosphate).

3. **Exit of product.** One molecule of G3P exits as the output of the Calvin cycle. This molecule, the product of photosynthesis, can be used for energy or transformed into materials that make up the plant.

4. **Regeneration of RuBP.** In several reactions, five molecules of G3P are transformed into three molecules of RuBP, which enter the cycle.

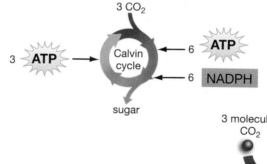

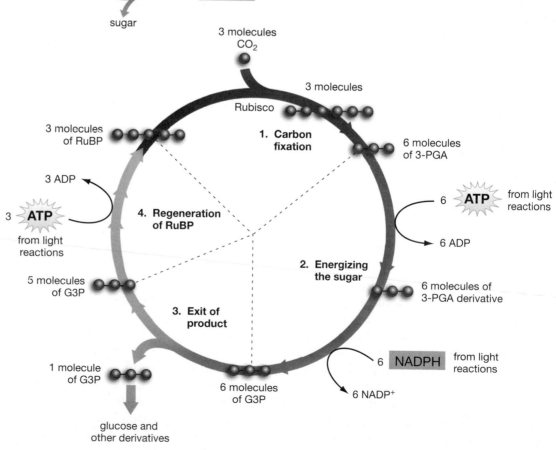

3 molecules $CO_2$

Rubisco

**1. Carbon fixation**

3 molecules

6 molecules of 3-PGA

6 ATP from light reactions

6 ADP

**2. Energizing the sugar**

6 molecules of 3-PGA derivative

6 NADPH from light reactions

6 NADP+

6 molecules of G3P

3 molecules of RuBP

3 ADP

3 ATP from light reactions

**4. Regeneration of RuBP**

5 molecules of G3P

**3. Exit of product**

1 molecule of G3P

glucose and other derivatives

is the *synthesis* of photosynthesis: a bringing together of elements.

The first steps of the cycle can be thought of as a process of **fixation**—of a gas being incorporated into an organic molecule. Specifically, carbon dioxide, which comes in through the leaves of the plant, is being fixed into the starting sugar, which is called RuBP. (From the time plants are embryos, they have small amounts of this sugar in them as a legacy from their parent plants.) This is a terrifically important step for it is the way life builds itself up with materials that lie *outside* of life. Note that carbon is taken in from the atmosphere and made part of a living thing—a plant.

The next reactions in the cycle are the energizing steps of the process (which you can see in section 2 of the cycle). Here is where our low-energy sugar receives the energetic products of the light reactions. The RuBP derivatives first interact with ATP and then receive energetic electrons from NADPH—the electrons that came originally from water and that were boosted up the energy hills by sunlight. With this, a relatively low-energy sugar has been energized—it has been moved up the energy hill—from which position it can now serve as food. The energized sugar that is produced, called G3P, is the essential product of photosynthesis. One molecule of it is netted with each turn of the Calvin cycle, as you can see in section 3. The remainder of the cycle can be thought of as a preparation of molecules for another trip around the cycle. In section 4, several reactions are carried out that result in the formation of more RuBP, which then enters the cycle again.

## The Ultimate Product of Photosynthesis

The importance of photosynthesis becomes clear only with an understanding of what G3P can give rise to once it exits the Calvin cycle. This sugar has something in common with steel coming out of a factory in that it can be turned into many things. Put two molecules of G3P together, and you get the more familiar six carbon sugar glucose. In turn, many molecules of glucose can come together to form the large storage molecules known as starches, familiar to us in such forms as potatoes or wheat grains. Beyond this, sugar is used to make proteins, which are then used as structural components of the plant or as enzymes. Given all this, if we ask what the *ultimate* product of photosynthesis is, the

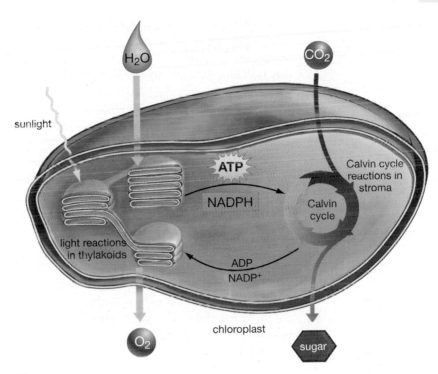

**Figure 8.8**
**Summary of Photosynthesis in the Chloroplasts of Plant Cells**

In the light reactions, solar energy is converted to chemical energy in the thylakoids, and the chemical energy is stored temporarily in the form of ATP and NADPH. Water is required for this reaction, and oxygen is a by-product. The stored chemical energy is in turn used in the Calvin cycle, taking place in the stroma, in which a high-energy sugar is made from carbon dioxide and the sugar RuBP. The sugar can be used for food or may become part of the plant's structure.

answer turns out to be: *the whole plant*. If you look at **Figure 8.8**, you can see a summary of the process of photosynthesis.

### SO FAR...

1. The energy acquired in the light reactions takes two forms: energetic _____ stored in the carrier _____ and energy stored in the energy transfer molecule _____. An important by-product of the light reactions is the _____ produced when water is split.

2. The energy from the light reactions is used to power the second major set of steps in photosynthesis, the _____ cycle.

3. In this cycle, an energy-poor _____ is brought together with carbon dioxide from the _____, and the resulting product is energized with _____ derived from the light reactions. This process is powered by the _____ produced in the light reactions.

**... ANSWERS**

1. electrons; NADPH; ATP; oxygen

2. Calvin

3. sugar; atmosphere; electrons; ATP

## 8.6 **Photorespiration and the C$_4$ Pathway**

You have now gone through the basic process of photosynthesis. It's necessary, however, to say something about what might be called a *glitch* in this process and then to note a mechanism that some plants have evolved to deal with it.

The glitch is called *photorespiration*, and it takes place within the Calvin cycle you just went over. Recall that this cycle begins with carbon dioxide (from the atmosphere) being joined to a low-energy sugar (from the plant). What is it that brings these two compounds together? Readers of the last few chapters will not be surprised to learn that it is an enzyme—a massive one called *rubisco*. Unfortunately, this molecular worker turns out to be something like the giant in "Jack and the Beanstalk": extremely powerful but slow moving and prone to making errors.

Extremely powerful? Rubisco is the bridge between the living world and the nonliving world. It is the molecule that allows plants to capture, from the atmosphere, the carbon atoms that are the building blocks for all living things. One of the problems with the way it works, however, is its speed—or lack of it. Whereas an average enzyme can catalyze 1,000 reactions per second, rubisco can manage only three per second. This snail's pace has consequences. Because rubisco works so slowly, every plant must have a lot of it; as a result, rubisco is the most abundant protein on Earth. A good motto for it might be: "Large numbers through inefficiency."

Slow pace is one thing, but it turns out that rubisco has a problem in even binding with the right substance: It can bind to *oxygen* as well as to carbon dioxide. Indeed, it does this all the time. In C$_3$ photosynthesis, rubisco may fix up to one molecule of O$_2$ for every three of the carbon dioxide it incorporates. Worse, when this happens, some of the CO$_2$ that *has* been fixed by a plant is released by it. It is as if a carpenter building a house took boards that were stored inside and began throwing them out the window.

Photorespiration is especially likely to take place when temperatures rise because heat prompts the stomata on leaves to close to preserve water. The effect of closed stomata, however, is not only that water is kept *in* but that CO$_2$ is kept *out*. With a deficit of CO$_2$, rubisco tends to bind more frequently with oxygen, prompting photorespiration—and plants that simply do not grow as much. With all this in mind, here is a definition of **photorespiration**: a process in which the enzyme rubisco undercuts carbon fixation in photosynthesis by binding with oxygen instead of with carbon dioxide. Meanwhile, **rubisco** can be defined as an enzyme that allows organisms to incorporate atmospheric carbon dioxide into their own sugars during the process of photosynthesis.

Evolution produced an adaptation that allows some warm-climate plant species to lessen the photorespiration problem. It is a mechanism that uses a special carbon-fixing "front-end" and then a carbon-dioxide shuttle into the Calvin cycle. The so-called C$_4$ plants that make use of this mechanism (**Figure 8.9**) don't initially use rubisco to fix

**Figure 8.9**
**Warm-Weather Photosynthesis**

Plants such as this sugarcane, shown here in Hawaii, use C$_4$ Photosynthesis.

carbon dioxide. Rather, they employ a different enzyme that will not bind with oxygen. The immediate product of this reaction is a four-carbon molecule (hence the $C_4$ cycle's name). A derivative of that molecule moves to a special group of cells and releases the $CO_2$, which will then bind to rubisco and go through the Calvin cycle. These special cells are wrapped around leaf veins and are hence named the bundle-sheath cells (**Figure 8.10**). What the $C_4$ system provides—even at times when little $CO_2$ may be coming in through the stomata—is a relatively high concentration of $CO_2$ where it's needed: in the cells where the Calvin cycle takes place. We can thus define **$C_4$ photosynthesis** as a form of photosynthesis in which carbon dioxide is first fixed to a four-carbon molecule and then transferred to special cells in which the Calvin cycle is undertaken, bundle-sheath cells.

### The $C_4$ Pathway Is Not Always Advantageous

The $C_4$ pathway is known to be employed in about 1,000 species of plants, most notably in some grasses, and in corn, sugarcane, and sorghum. But there are vastly more $C_3$ than $C_4$ plants. You might wonder why evolution hasn't, you might say, weeded out $C_3$ plants in favor of the $C_4$ variety since the latter seem to be so much more efficient at photosynthesis. The answer is that $C_4$ fixation does not confer an across-the-board advantage. It has a cost, which is the expenditure of ATP in shuttling carbon dioxide to the bundle-sheath cells. What this means is that $C_4$ fixation is advantageous only where the weather is warm enough to bring about a significant increase in photorespiration. Where the weather is cooler, $C_4$ plants will be outcompeted by $C_3$ species.

## 8.7 Another Photosynthetic Variation: CAM Plants

When plants live in climates that are not just warm but *dry*, a large part of their survival comes down to retaining water. Photosynthesis, however, works against water retention because, as you have seen, when $CO_2$ can pass in, water vapor can pass out.

The plants we call succulents—a group that includes most cacti—have a solution for this, which is to close their stomata during the day and open them at night. The plants then carry out $C_4$ metabolism at night, but only up to a point. They fix carbon dioxide into an initial four-carbon molecule and then stand

pat. The $CO_2$ stays "banked" in them, awaiting the energy of the next day's sun, which will power the production of the ATP they need to carry out the Calvin cycle. This is **CAM photosynthesis**, defined as a form of photosynthesis, undertaken by plants in hot, dry climates, in which carbon fixation takes place at night and the Calvin cycle occurs during the day. This process gets the job done for cactus, pineapple, orchid, and some mint family plants, among others (**Figure 8.11**). As you've no doubt guessed, CAM is an acronym; it stands for crassulacean acid metabolism. The succulent plant family Crassulaceae (which includes jade plants) was the first group of plants in which CAM metabolism was discovered. The three methods of photosynthesis

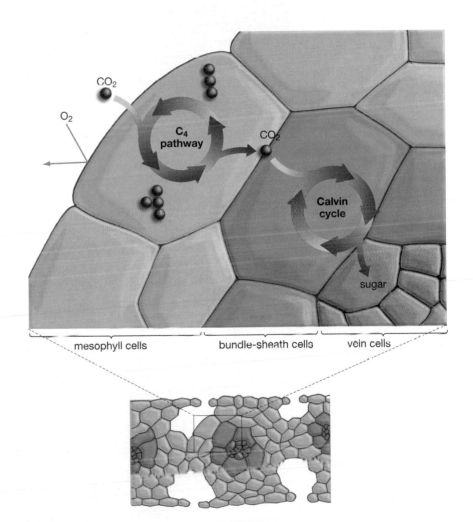

**Figure 8.10**
**$C_4$ Photosynthesis**
$C_4$ photosynthesis is an adaptation of some warm-climate plants to the problem of photorespiration. Photorespiration occurs when the $C_3$ enzyme rubisco binds to oxygen rather than carbon dioxide (thereby undercutting the production of sugar). $C_4$ plants contain a different enzyme in their mesophyll cells that binds to carbon dioxide but not to oxygen. Then, the carbon dioxide is escorted into special bundle-sheath cells, where rubisco can bind to it, thus initiating the Calvin cycle reviewed earlier, meaning the production of sugar can proceed.

**Web Animation 8.2**
Different Kinds of
Photosynthesis

we've looked at—$C_3$, $C_4$, and CAM—are summarized in **Table 8.1**.

It was not until after World War II that scientists pieced together the story of how plants join carbon dioxide to a sugar to create food. As you've seen, this happens through a set of steps that today are called the Calvin cycle. You can read about the man the cycle was named for, Melvin Calvin, and his process of discovery in "Plants Make Their Own Food, But How?" on page 154.

## Closing Thoughts on Photosynthesis and Energy

Readers who have stayed the course on this account of photosynthesis now know at least one thing very well about it: how complicated it is, with its many components and long metabolic pathways. It is also probably clear by this point why scientists continue to study photosynthesis in such detail. It is the foundation of plant growth, and on plant growth hinges nothing less than the survival of all animals—including human beings. Without an understanding of this linkage, it's easy to see plants as a set of mute fixtures

### Table 8.1 Three Modes of Photosynthesis

| | $C_3$ | $C_4$ | CAM |
|---|---|---|---|
| Used by: | Majority of plants | Corn, sugarcane (warm environments) | Cactus, pineapple, orchid (dry environments) |
| Benefits: | Efficient use of ATP | Less photorespiration | Less water loss |
| Problems: | Photorespiration | Uses up more ATP | Uses up more ATP; hard to "bank" enough $CO_2$ |
| How it works: | | | |

whose main contribution to human life is aesthetic. With this knowledge, you can begin to see the central position that plants occupy in the interconnected web of life.

Over the last three chapters, you have learned about the endless back and forth of oxygen, carbon dioxide, and energy in the living world. *Cycle* has been a recurring word in this long discussion because the only one-way trip you've encountered has been the relentless "spillage" of energy from the sun down into heat. Looked at in a cynical way, Earth and its inhabitants constitute a kind of leaky holding tank for energy that comes from the sun. Looked at another way, however, the living world has been able, through photosynthesis, to take the sun's energy and build a remarkable edifice with it. Think of life's innumerable forms and the sheer mass of living material on Earth. One of the most amazing things about this structure is that we humans get to be both a part of it and witnesses to it.

## SO FAR...

1. The enzyme rubisco brings together the _____ (from the plant) and _____ (from the atmosphere), creating the product that moves through the Calvin cycle. Unfortunately, rubisco also has an affinity for binding with _____, a process known as _____.

2. An evolutionary response to this process, found in plants that thrive in _____ climates, is _____ photosynthesis, in which carbon dioxide is first fixed to a four-carbon molecule and then transferred into cells that carry out the Calvin cycle.

3. Plants that exist in climates that are both warm and dry have evolved a variation on photosynthesis, _____ photosynthesis, in which carbon dioxide is brought into the plant only _____, while the Calvin cycle is carried out only _____.

••• ANSWERS

1. low-energy sugar; carbon; oxygen; photorespiration
2. warm-weather; $C_4$
3. CAM; at night; during the day

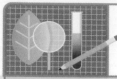

# THE PROCESS *of* SCIENCE
## Plants Make Their Own Food, But How?

**S**cience constantly is confronted with what are known as "black box" problems. In these problems, researchers know what goes into a process, and they know what comes out of it, but what happens *in between* is a mystery. At the conclusion of World War II, plant science had a doozy of a black box problem. It was clear that plants were starting with carbon dioxide from the air and ending with high-energy carbohydrates, but no one knew what came in between. What were the *intermediate* substances that carbon dioxide was being made part of on the route to food? Getting inside this black box turned out to require a three-part combination: insightful researchers, new scientific techniques, and old-fashioned hard work.

One of the peacetime benefits of World War II's atomic weapons research was that scientists could produce quantities of substances known as *radioactive isotopes*. As you saw in Chapter 2, isotopes result from variations in the number of neutrons an element has in its nucleus. Radioactivity, meanwhile, is the release of radiant energy from atoms undergoing rapid decay. The element carbon is everywhere in nature, and as it turns out, one of its isotopes, carbon-14, is radioactive. Carbon-14 occurs naturally, but the ability to *manufacture* this substance opened a host of possibilities for research in the mid-1940s.

**Web Animation 8.3**
The Calvin Experiments

Stepping up to the plate in 1946 to take advantage of these possibilities was a University of California, Berkeley, chemist named Melvin Calvin. Then just 35 years old, Calvin was encouraged by the physicist Ernest Lawrence of Berkeley to apply radiocarbon techniques to organic chemistry research. With carbon-14, Calvin realized, he could *tag* carbon dioxide to find out what it becomes part of during photosynthesis, much as a zoologist might tag a bear with a signaling device to find out where it goes during winter. The radiation that carbon-14 emits might be thought of as a signal with a constant message of, "Here I am."

The steps that Calvin and his colleagues followed were simple in conception but very difficult in execution. They first took some photosynthetic algae and put them in a flask that had light shining on it. Like any photosynthetic organism, these algae would, of course, be taking in carbon dioxide to carry out their work. In this case, however, the carbon dioxide in question contained carbon-14. The trick was to allow the algae to perform a little photosynthesis using this $^{14}CO_2$. Then, the algae were plunged into boiling alcohol, which killed them. By the time the plunge was taken, the carbon dioxide would have made it to *some point* in its transformation to carbohydrate; it would have been incorporated into one molecule in the series of intermediate molecules involved. Stopping photosynthesis at differing points—two seconds after it started, five seconds, and so on—meant stopping the progress of carbon dioxide as it marched through its changes.

All of this would have been for naught, however, without a way to figure out where the carbon was when the process was stopped. This is where the carbon-14 could do its job, but it needed to be joined to another technique that was also newly developed at that time: two-dimensional paper chromatography. If you look at **Figure 1a**, you can see how this technique works. Put a spot of ink on some filter paper and then stick one end of the paper in a solvent—water, for example. The water moves through the paper, carrying different components of the ink along with it at different rates. You can tell something about what the ink is made of by measuring how far its components move through the paper. You can tell still more by doing this in two dimensions: Let the solvent move through the spot one way, then set the paper down at a right angle in a *second* solvent and let it move through in another direction (**Figure 1b**). This is what Calvin and his colleagues did, using as their "ink" the algae that had been killed following photosynthesis.

The paper chromatography left the algae cells drawn out in two directions by the solvents. But, where was the original carbon in this material? Because it was radioactively tagged carbon, it was sending out its signal, saying, "I'm here now," but only a special "receiver" could pick up this signal: photographic film (**Figure 1c**). Setting X-ray film tightly on top of the chromatography paper for long periods—usually about two weeks—caused the radiation from the carbon-14 to expose the film. Spots would show up saying, in effect, "Carbon-14 is now here, in this concentration."

There is a wide gap, however, between seeing a black blotch on a piece of film and *identifying* that blotch as a specific substance. The difficulty the researchers encountered can be measured in time—10 years in this case. That's how long it took these workers to complete their investigation. During that time, Calvin and his principal colleagues—postdoctoral researcher Andrew Benson and graduate student James Bassham—worked extraordinarily long hours in deciphering what the downstream products of photosynthesis were and the order in which they were produced.

This long operation was a great success, however. Calvin's team dismantled the black box of carbon fixation, and the information they uncovered can be used to aid in such things as increased food production. The set of steps these workers elucidated has variously been referred to as the $C_3$ cycle, Calvin-Benson cycle, or Calvin cycle. In 1961, Melvin Calvin received the ultimate in congratulations for a piece of scientific work when he was awarded the Nobel Prize in Chemistry for shining a light on photosynthesis.

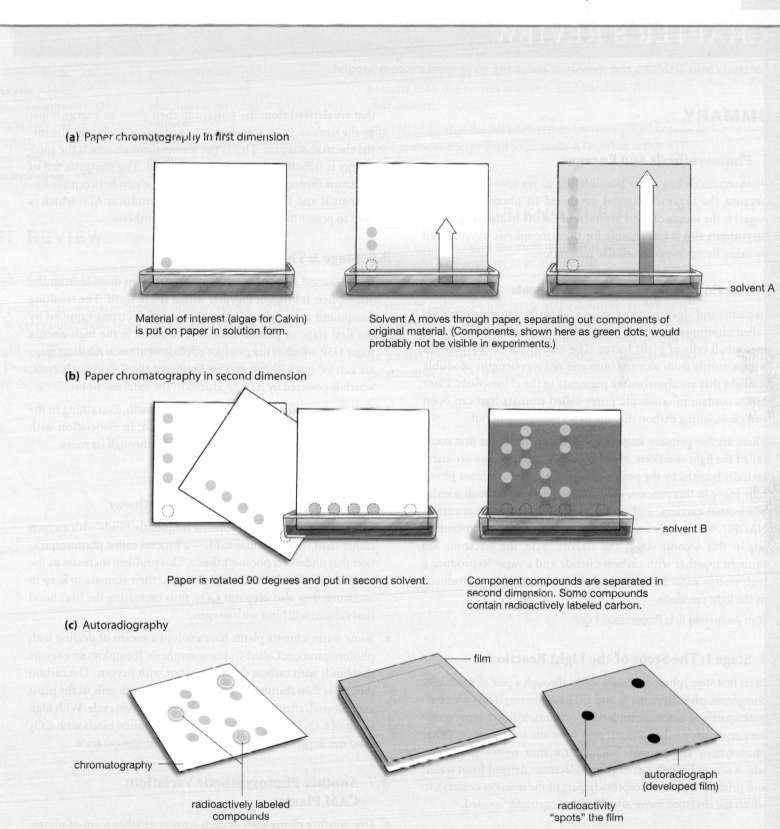

**(a)** Paper chromatography in first dimension

Material of interest (algae for Calvin) is put on paper in solution form.

Solvent A moves through paper, separating out components of original material. (Components, shown here as green dots, would probably not be visible in experiments.)

solvent A

**(b)** Paper chromatography in second dimension

Paper is rotated 90 degrees and put in second solvent.

Component compounds are separated in second dimension. Some compounds contain radioactively labeled carbon.

solvent B

**(c)** Autoradiography

chromatography

radioactively labeled compounds

film

radioactivity "spots" the film

autoradiograph (developed film)

**Figure 1**

How do living things grow and reproduce? Through cell division. Here, one human cell is about to become two through this tightly choreographed process.

Life depends on information that is stored in DNA. This information is duplicated and then apportioned every time a cell divides.

CHAPTER 9

9.1 An Introduction to Genetics 159

9.2 An Introduction to Cell Division 162

9.3 DNA Is Packaged in Chromosomes 164

9.4 Mitosis and Cytokinesis *BioFlix* 168

9.5 Variations in Cell Division 170

**Essay**
When the Cell Cycle Runs Amok: Cancer 172

158

# Genetics and Cell Division

**E**veryone knows how one generation of living things gives rise to another, at least in a general sense. But the details of how this works are not apparent just by looking at Earth's creatures. A brief sexual encounter takes place, and the next thing we know there are new kittens, chicks, or babies. Only the encounter and the newborns are obvious; what comes in between is mysterious.

The problem, of course, is that most of the component parts of reproduction are hidden away and, in any event, are so small that the process is completely invisible to the naked eye. Given this, it's no wonder that for centuries human beings wondered what it was that one generation passed on, so that the next generation could be formed.

A similar kind of puzzlement existed over a related question: What is the mechanism that controls the development of a living thing as it goes from being a microscopic cell to being a human baby in full cry? For that matter, how is it that the adult body is able to build muscle or repair a minor wound? What, in short, is the mechanism that controls day-to-day physical functioning?

## 9.1  An Introduction to Genetics

It took roughly a century of hard-won scientific advances for these questions to be answered. In the end, it turned out that a single substance is central to the reproduction of living things, their development

from single cells, and their day-to-day functioning. This substance is deoxyribonucleic acid, a long, vanishingly thin molecule known everywhere today by its acronym, DNA.

DNA contains what is known as an organism's **genome**: the complete collection of that organism's genetic information. This information exists in units called genes that lie along DNA's famous double helix.

How is it that this one molecule can undertake all the tasks just described? To the question about what it is that humans *pass on* in reproduction, the answer is: half of a father's genome and half of a mother's genome, with both these halves coming together in a fertilized egg to produce a whole, new genome (**Figure 9.1** on the next page). To the other questions about what controls the development and everyday functioning of living things, the answer is: the instructions provided by the genes in the genome.

### DNA Contains Instructions for Protein Production

To grasp how genes can control the development and functioning of living things, note that there are logically two parts to such control. If genes *contain* instructions, something needs to carry them out. As it turns out, the workers here are proteins; in

**Figure 9.1**
**Passing on Genetic Information**
Half of each person's genome comes from the father and half from the mother.

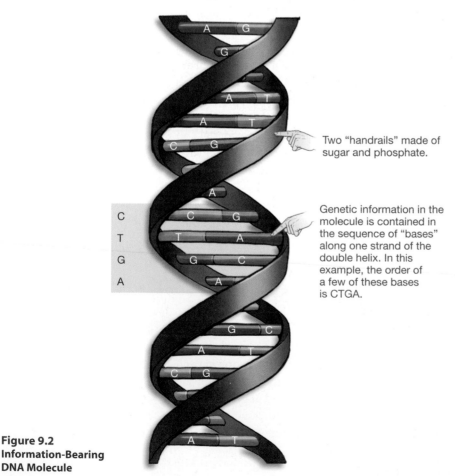

Two "handrails" made of sugar and phosphate.

Genetic information in the molecule is contained in the sequence of "bases" along one strand of the double helix. In this example, the order of a few of these bases is CTGA.

**Figure 9.2**
**Information-Bearing DNA Molecule**

tremendous diversity and number, they are the working products of the information contained in genes.

Readers who went through Chapter 3 (on biological molecules) and Chapter 6 (on energy) should have a good grasp already of what proteins are capable of. As we have seen, there are proteins that form cartilage and hair; there are communication proteins that help form the receptors on cells; there are hormonal proteins and transport proteins. And then there are **enzymes**, the chemically active proteins that speed up—or in practical terms, *enable*—chemical reactions in living things. Without these protein molecules, almost nothing could get done in the body; with them, a dazzling set of chemical reactions takes place every second in every living thing.

What most genes do, then, is contain the information for the production of proteins. In Chapter 14, you'll get the details of how protein production comes about, but for now let's just look at the essentials of the story. Imagine you are a telegraph operator in the Old West, and your "key" starts tapping frantically. You leap forward in your chair, listening for the sound of short and long bursts on the key, and you begin writing down the letters these sounds stand for. Here's the message you get:

.—.. ——— —.—. —.—    — .... .    ... .— ..—. .

which means:

L o c k t h e s a f e

You walk over and do just that. There is thus a code here, Morse code, that has resulted in some action. In DNA there is likewise a code, but instead of a code specified in short or long sounds, it is a code specified in chemical substances. There are four of these substances, and they lie along the double helix. They are the "bases" adenine, thymine, guanine, and cytosine, abbreviated A, T, G, C, respectively.

A particular sequence of these bases—usually thousands of letters long—contains information that can be acted on, just as the Morse code did. Only the action here is the production of a particular protein. One series of A's, T's, C's, and G's contains the information for the production of one protein, but a *different* sequence of A's, T's, G's, and C's specifies a different protein. These separate sequences of bases are separate genes.

You saw in Chapter 3 that proteins are composed of building blocks called amino acids. The base sequence that makes up a gene, then, is like a message that says, "Give me this amino acid, now this one, then this one, . . ." and so on for hundreds of amino acids until a protein is created that folds up and gets busy on some task. This task can be one small part of *forming* a human being—as in embryonic development—or of *maintaining* one, as in an adult. One generation gives rise to another by passing on this entire complement of information.

### The Architecture of DNA

It is the architecture of the DNA molecule that makes this protein production process possible. Look at the general scheme of it pictured in **Figure 9.2**. As you can see, DNA looks something like an open, spiral staircase, the handrails of which are a repeating series of sugar (deoxyribose) and phosphate molecules. Meanwhile, the messages you've just been reading about are contained in the steps of the staircase in the order in which the bases A, T, C, and G are placed along the helix. As you can see, these bases extend inward from both of the DNA rails. The order in which the bases occur *along the rails* is extremely varied. It is this order that "encodes" the information for the production of proteins.

### The Path of Protein Synthesis

Look now at **Figure 9.3** and see the path that is followed in protein synthesis. The DNA you have been reading about is contained in the nucleus of the cell. The first step is that a stretch of it unwinds there, and its message—the order of a string of bases—is copied onto a molecule called messenger RNA (mRNA). This length of mRNA then exits from the cell nucleus. Its destination is a molecular workbench in the cell's cytoplasm, a structure called a ribosome. It is here that both message (the mRNA sequence) and raw materials (amino acids) come together to make the product (a protein). The mRNA sequence is "read" within the ribosome, and as this happens a chain of amino acids is put together in the ribosome in the order called for by the mRNA sequence. When the chain is finished and folded up, a protein has come into existence.

### Genetics as Information Management

The story you have been reading about is that of **genetics**, meaning the study of physical inheritance among living things. In reflecting on this story, the first thing to note is that it concerns the storage, duplication, and transfer of *information*. True, it is information encoded in chemical form, but that shouldn't bother us. Information comes in lots of forms; there are the familiar words on a page, but there are also bar graphs, smoke signals, or a baby's smile.

To grasp the fact of genetics as molecular information, think back to Chapter 6, where you looked at a particular protein, an enzyme called lactase that helps

break up milk sugar (lactose) into its components of glucose and galactose. In line with what you've seen so far, there is a gene that prompts the production of lactase. This gene's message is copied onto a length of mRNA, which then migrates to a ribosome in the cytoplasm. Lactase is put together there, amino acid by amino acid, and once synthesized, it folds up and gets busy clipping lactose molecules.

Now, consider the role of the gene in this process. It didn't do the lactose clipping; it didn't even migrate to the cytoplasm to be involved in the synthesis of lactase. In this case, its role was more like that of a . . . cookbook recipe. It specified this amino acid, then that amino acid, and eventually lactase was produced.

### From One Gene to a Collection

Now, in considering not just one gene, but rather the entire *collection* of genes in a living thing—its genome, in other words—it is easy to see that we are dealing with a vast *library* of information. The size of such a collection can vary, but to give you some idea, the human genome is estimated to include 20,000–25,000

**Figure 9.3**
**The Path of Protein Synthesis**

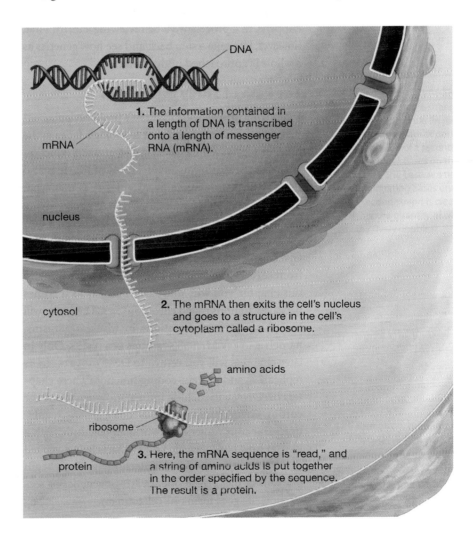

DNA

1. The information contained in a length of DNA is transcribed onto a length of messenger RNA (mRNA).

mRNA

nucleus

cytosol

2. The mRNA then exits the cell's nucleus and goes to a structure in the cell's cytoplasm called a ribosome.

amino acids

ribosome

protein

3. Here, the mRNA sequence is "read," and a string of amino acids is put together in the order specified by the sequence. The result is a protein.

Animals, fungi, plants. How did the living world become so diverse? In part through the process of meiosis.

| 10.1 | An Overview of Meiosis | | 177 |
| 10.2 | The Steps in Meiosis | *Bio**Flix*** | 178 |
| 10.3 | What Is the Significance of Meiosis? | | 182 |
| 10.4 | Meiosis and Sex Outcome | | 185 |
| 10.5 | Gamete Formation in Humans | | 186 |
| 10.6 | Life Cycles: Humans and Other Organisms | | 188 |

One special kind of cell division yields the sex cells that give rise to succeeding generations and serves as the foundation for much of the diversity seen in the living world.

# PREPARING FOR SEXUAL REPRODUCTION: Meiosis

**A** real-life couple—let's call them Jack Fennington and Jill Kent—combine their last names when they get married and thus become Jack and Jill Fennington-Kent. Now, what would happen if the

Fennington-Kent's children were to continue in this tradition? Their daughter, Susie, might marry, say, Ralph Reeson-Dodd, which would make her Susie Fennington-Kent-Reeson-Dodd. If *Susie's* daughter, Alicia, were to keep this up, she might be fated to become Alicia Fennington-Kent-Reeson-Dodd-Garcia-Lee-Minderbinder-Green, and so on.

This growing chain of names illustrates a kind of problem that living things have managed to address in carrying out reproduction. *Reproduction* in this context does not mean the mitotic cell division you looked at in Chapter 9. Reproduction here means *sexual* reproduction, in which specialized reproductive cells come together to produce offspring. The cells described in Chapter 9—the ones that undergo mitosis—are known as *somatic* cells, which in animals means all the cells in the organism except for one type. What type is this? Sex cells; cells called *gametes*. These are reproductive cells, known more commonly as eggs and sperm.

Eggs and sperm are not commonly thought of as cells, but that's just what they are, complete with plasma membranes, nuclei, chromosomes, and all the rest (though sperm eventually lose much of this material). Eggs and sperm are special cells, however, in that they are not destined to divide like somatic cells and thus make more eggs and sperm. Rather, they will *fuse*—sperm fertilizing egg—to make a zygote, which grows into a whole organism. It is sperm and egg that take part in sexual reproduction and that thus bear some relation to the Fennington-Kent problem. Here's how it plays out in humans. Human somatic cells, as you have seen, have 23 pairs of chromosomes or 46 chromosomes in all. If an egg and sperm each brought 46 chromosomes to their *union*, the result would be a zygote with 92 chromosomes. The next generation down would presumably have 184 chromosomes, the next after that 368, and so on. This would be about as functional as having Fennington-Kent-Reeson-Dodd-Garcia-Lee-Minderbinder-Green as a last name.

## 10.1 An Overview of Meiosis

How do chromosomes avoid the problem of doubling with each generation? In sexual reproduction, chromosome union is preceded by chromosome *reduction*. The reduction comes in the cells that give rise to sperm and egg. When these cells divide, the result is sperm or egg cells that have only *half* the usual, somatic number of chromosomes. Human sperm or eggs, in other words, have only 23 chromosomes in them. (Not the 23 *pairs* of chromosomes that somatic cells have, mind you; 23 chromosomes.) Each 23-chromosome sperm can then unite with a 23-chromosome egg to produce a 46-chromosome zygote that develops into a new

Three frog siblings survive to reproductive age from a clutch of eggs laid by a female and fertilized by a male. Thanks to crossing over and independent assortment, one of these frogs may end up having, say, a darker coloration than all the other frogs of its species in its area. In the formation of the eggs and sperm that led to this frog, chromosome parts were swapped (in crossing over), and chromosomes lined up (in independent assortment) in such a way that this frog was the darkest frog possible among all the frogs in its population. Further, it happens that, because of shifting rain patterns in this frog's habitat, this dark coloration is starting to be useful for evading predators. This frog will thus survive to have more offspring of its own than do other frogs around it. By being selected for survival in this way—by undergoing what is called natural selection—this frog may have started its population down the road to having a darker coloration.

Now, it's easy to see that coloration is only one of a countless number of traits that could be selected for in this way. Taller height in a tree? Diving stamina in a dolphin? Meiosis and sexual reproduction would lead the way to these traits by producing, in one generation after another, *this* tree that grew a little taller or *that* dolphin that dove a little deeper. Just as General Motors provides an array of models for consumers to choose from each year, meiosis and sexual reproduction provide an array of models for *nature* to choose from each generation.

**Web Animation 10.1**
Meiosis and Sex Outcome

To put a finer point on this, consider an alternative to meiosis and sexual reproduction. Bacteria are single-celled organisms that possess a single, circular chromosome. As you saw in Chapter 9, bacteria divide through the process of binary fission, one bacterium becoming two in the course of cell division. As it turns out, that's it for bacteria as far as reproduction goes. There is no chromosomal part swapping or independent assortment of chromosomes in reproduction. And there is no fusion of two gametes from separate organisms, as occurs with eggs and sperm. Instead, there is just an exact duplication of the bacterial chromosome and a splitting of one bacterial cell into two. In general, then, every "daughter" bacterial cell is a genetic replica of its parent cell—a clone of it. What this means is that differences among bacteria are not a built-in feature of their reproduction. Instead, such differences as come to pass are dependent on their DNA mutating (which is rare) or on a form of gene transfer they are capable of (which is intermittent). Natural selection simply does not have as much to choose from in asexually reproducing organisms such as bacteria. Where is the equivalent of our darker-colored frog in a group of bacterial cells descended in a line? How could an equivalent difference come about, given that each bacterium is a clone of its predecessor?

What evidence do we have that sexual reproduction makes a practical difference? Well, consider that bacteria and their fellow single-celled organisms, the archaea, came into existence perhaps 3.8 billion years ago. Then, for 1.8 billion years after that, the living world consisted of nothing *but* bacteria and archaea—it took that long for anything else to evolve from them. We are not sure when sexual reproduction got going, but we have some evidence that it was taking place by 1.2 billion years ago. So, in the last 1.2 billion years, all the fish and birds and trees and fungi and flowers that we see before us developed through sexual reproduction, whereas in the 1.8 billion years after bacteria and archaea got going, they remained pretty much as they were when they began. Although other factors contributed to the ramping up of diversity on Earth, meiosis and sexual reproduction were key players in it. The British geneticist J. B. S. Haldane is said to have remarked that, "Evolution could roll on fairly efficiently without sex but such a world would be a dull one to live in." How did the living world come to be such a diverse place? In part through the very small-scale process you've been looking at, meiosis.

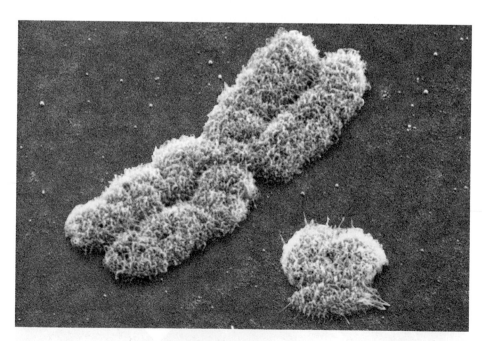

**Figure 10.4**
**The X and the Y**

The human X chromosome can be seen on the left, while the human Y chromosome is on the right. The difference in size between them is in part a reflection of the difference in the number of genes each of them contains. The Y chromosome only has only 78, while the X has about 1,500. (×10,000)

## 10.4 Meiosis and Sex Outcome

Let us think now about what meiosis means in relation to passing on sex to offspring. How do humans, in particular, come to be male or female? In humans, there is one exception to the rule that chromosomes come in homologous pairs. Human females do indeed have 23 pairs of matched chromosomes, including 22 pairs of "autosomes" and one matched pair of **sex chromosomes**, meaning the chromosomes that determine what sex an individual will be. The sex chromosomes in females are called X chromosomes, and each female possesses two of them. In males, conversely, there are 22 autosome pairs, one X sex chromosome, and then one Y sex chromosome. It is this Y chromosome that leads to the male sex. (**Figure 10.4** illustrates the differences in the X and Y sizes.)

In meiosis I in a female, the female's two X chromosomes, being homologous, line up together at the metaphase plate. Then, these chromosomes separate, each of them going to different cells (**Figure 10.5**). In males, the non-homologous X and Y chromosomes line up as if they were homologues. Then, one resulting cell gets an X chromosome, while the other gets the Y chromosome.

Sex determination is then simple. Each of the eggs produced by a female bears a single X chromosome. If this egg is fertilized by a *sperm* bearing an X chromosome, the resulting child will be a female.

If, however, the egg is fertilized by a sperm bearing a Y chromosome, the child will be a male.

Because females don't have Y chromosomes to pass on, it follows that the Y chromosome that any male has had to come from his father. Meanwhile, the single X chromosome that any male has must come from his mother. Females, conversely, carry one X chromosome from their mother and one from their father.

**SO FAR...**

1. Two features of meiosis ensure that each generation of sexually reproducing organisms will differ from its parents genetically. In one of these features, crossing over, _____ exchange _____. In the other feature, independent assortment, there is a random _____ of _____ at the metaphase plate.

2. The reason that crossing over and independent assortment ensure genetic diversity is that any two members of a pair of homologous chromosomes _____ in the genetic information they carry.

3. In human beings, any fertilized egg that carries two _____ chromosomes will be a female; fertilized eggs destined to become males, meanwhile, carry one _____ chromosome (donated by the mother) and one _____ chromosome (donated by the father).

**... ANSWERS**

1. homologous chromosomes; reciprocal portions of themselves; distribution; homologous chromosome pairs

2. differ

3. X; X; Y

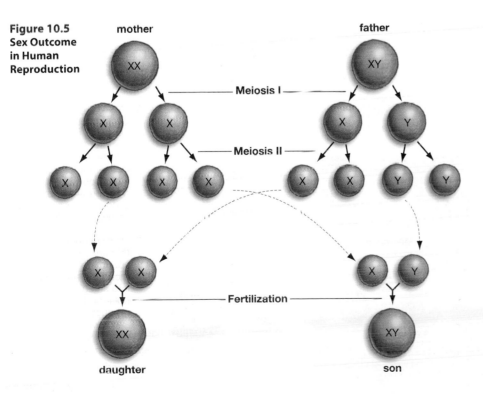

**Figure 10.5
Sex Outcome in Human Reproduction**

mother    father

Meiosis I

Meiosis II

Fertilization

daughter    son

1. Early in meiosis I, the mother's two X chromosomes line up at the metaphase plate (XX). Meanwhile, in the father's meiosis, his X and Y chromosomes line up (XY).

2. The X-X and X-Y pairs then separate into different cells.

3. The chromatids that made up the duplicated chromosomes separate, yielding individual eggs and sperm.

4. Should an X-bearing sperm from the male reach the egg first, the child will be a girl; should the male's Y-bearing sperm reach the egg first, the child will be a boy.

generation of cells, the primary oocytes, which are the cells that enter into meiosis I (Figure 10.6). Amazingly enough, meiosis I is where these oocytes *stay*—for years. Imagine a collection of oocytes in the female ovaries (which are walnut-sized structures just right and left of center in the female pelvic area). Once puberty begins, each oocyte is theoretically capable of maturing into an egg, but an average of only *one* oocyte per month actually starts this maturation process. Which oocyte is this? The one released from the ovary each month in the process called ovulation. All the other oocytes in the ovaries halt their meiosis in prophase of meiosis I. Their chromosomes are duplicated, and they have completed crossing over, but they have not yet lined up at the metaphase plate. And this is where things stay for most of the oocytes a female possesses. Only the one oocyte per month that is expelled from the ovary will complete meiosis I and thus enter into meiosis II. Even this meiosis is "arrested," however, until this oocyte is fertilized by a sperm. It is the union with a sperm that prompts the completion of meiosis II. Thus, an oocyte that came into being when a female was an embryo might remain suspended in the cell cycle for years or even decades. Of the millions of oocytes originally created, only a few hundred will mature even through metaphase of meiosis I in a woman's lifetime.

How do certain oocytes get "selected" to mature through metaphase I—how do they get selected to be ovulated, in other words? This occurs in connection with an intricate interplay that involves their surrounding tissue, hormones, and the happenstance of timing. You'll read more about this process in Chapter 32.

**Figure 10.9**
**All-Female Species**
Pictured is a whiptail lizard, *Aspidoscelis tigris*, that reproduces solely through parthenogenesis. Because the eggs these lizards produce are not fertilized by males, they contain only one set of chromosomes, passed on by the mother. Hatchlings develop from these haploid eggs, however, and as a consequence the hatchlings themselves are haploid and genetically identical to the mother. The end result is a species that is entirely female.

## One Egg, Several Polar Bodies

For the few oocytes selected, the process of going through meiosis I and II means proceeding from one diploid cell to four haploid cells. In this process, however, almost all the cytoplasm will be shunted to *one* of the four cells, the better to build up its stores of nutrients and other cytoplasmic material. This happens in two rounds: meiosis I and meiosis II. When cytokinesis comes in meiosis I, one daughter cell gets almost all the cytoplasm, while the second daughter cell gets almost none. This same thing happens in meiosis II. The result? One richly endowed haploid ovum and two, or perhaps three, very small haploid **polar bodies**: nonfunctional cells produced during meiosis in females (Figure 10.6). (Only two polar bodies and the ovum will be the end result if the polar body produced in meiosis I does not continue through meiosis II, which often happens.) The ovum has a chance at taking part in producing offspring, but the polar bodies are bound for oblivion rather than a shot at immortality. They will eventually degrade into their constituent substances.

## 10.6 Life Cycles: Humans and Other Organisms

The process of the formation of the human gametes called egg and sperm is followed, as you have seen, by the union of these gametes. In humans, the oocyte that is expelled from the ovary makes a slow trip down the female's uterine (or Fallopian) tube, during which time it may encounter a swimming sperm. If it does, the result will be a fertilized egg (or zygote), which then develops into a whole human being. One of the things that will eventually occur in this new human being is the development of another group of eggs or sperm through the process of meiosis. Thus, do we see a **life cycle**: the repeating series of steps that occur in the reproduction of an organism. You have looked at the human life cycle, but it's important to recognize that it amounts to only one among many.

### Not All Reproduction Is Sexual

As noted earlier, human reproduction is **sexual reproduction**: the union of two reproductive cells to create a new organism. In many other species, there is **asexual reproduction** or reproduction that does not involve sex. Asexual reproduction actually is seen in a vast array of organisms. In fact, the only two groupings of living things that *never* carry out asexual reproduction are birds and mammals. Admittedly, asexual reproduction is rare in creatures as complex as fish or reptiles, but it does happen. In the American southwestern desert, for example, there are a few species of whiptail lizard that are entirely female (**Figure 10.9**). The female whiptail lays haploid eggs, but they are never fertilized. Instead, they develop into haploid clones of her through regular mitotic cell division. This method of reproduction (called *parthenogenesis*) also results in "drone" male honeybees. Those eggs that a queen bee chooses not to fertilize (with stored up sperm) develop into drones through parthenogenesis.

Other forms of asexual reproduction also are possible. In fact, you've already been introduced to one of them. Recall that a bacterial cell engages in a form of asexual reproduction, called *binary fission*, in which it duplicates its single chromosome and then divides in two. Beyond this, anyone who has ever seen a "cutting" taken from a plant knows that a snipped-off stem or branch will, when properly planted, sprout its own roots and grow into an independent plant, complete with the ability to take part in sexual reproduction. This process is called *vegetative reproduction*, and it has a counterpart in the animal world. Cut the arm and part of a central disk from a sea star, for example, and, through *regeneration*, this severed limb will grow into a whole, multiarmed sea star (**Figure 10.10**).

As we go through the branching tree of life, from more complex creatures to less complex ones, asexual reproduction becomes more common. By the time we reach bacteria, it is the only form of reproduction that exists. Note that all forms of asexual reproduction produce offspring that are genetically identical to their parents. There is no mixing and matching of chromosomes from different parents in these asexual processes; just a cloning of one genetically identical individual from its parent.

### Variations in Sexual Reproduction

Within the framework of sexual reproduction, as you can imagine, there are many variations. For example, separate gametes need not come from separate organisms. Some animals, such as tapeworms, are hermaphrodites, meaning they have both male and female reproductive parts. In some cases, hermaphrodites fertilize themselves, but generally they are fertilized by another member of their species. Most plants contain both male and female parts, and some plants "self-fertilize"—their sperm can

fertilize their eggs—as with a pea plant you will be looking at in Chapter 11.

## On to Patterns of Inheritance

Our tour of cell division over the last couple of chapters has yielded not only a look at how one cell becomes two, but at how chromosomes operate within this process. Given the paired nature of chromosomes and their precisely ordered activity during meiosis, it stands to reason that there would be some predictability or pattern to the passing on of traits. This is the case, as it turns out. And the person who first recognized this pattern was a monk who worked in nineteenth-century Europe in an obscurity he did not deserve.

**Figure 10.10**
**Regenerating the Whole from a Part**

A sea star off the coast of Papua New Guinea is regenerating a body and other arms from a piece of one arm and part of its central disk.

### SO FAR...

1. Spermatogonia are regarded as _____ cells because they can produce not only a specialized variety of cell (the spermatocyte), but also more _____. Meanwhile, it is possible that the female counterpart to spermatogonia, the oogonia, are produced in females only _____.

2. The function of eggs as rich sources of nutrients and cellular materials results in oogenesis producing not four cells but one viable cell and several _____.

3. Each organism produced through asexual reproduction is genetically _____ to its parent.

••• ANSWERS

1 stem; spermatogonia; prior to birth

2. polar bodies

3. identical

# CHAPTER 10 REVIEW

For study help, activities, and more quiz questions, go to **www.aw.com/krogh4.**

## SUMMARY

### 10.1  An Overview of Meiosis

- In human beings, nearly all cells have paired sets of chromosomes, meaning these cells are diploid. Meiosis is the process by which a single diploid cell divides to produce four haploid cells—cells that contain a single set of chromosomes.

- The haploid cells produced through meiosis are called gametes. Female gametes are eggs; male gametes are sperm. They are the reproductive cells of human beings and many other organisms.

- When the haploid sperm and haploid egg fuse, a diploid fertilized egg (or zygote) is produced, setting into development a new generation of organism.

### 10.2  The Steps in Meiosis

- In meiosis, there is one round of chromosome duplication followed by two rounds of cell division. There are two primary stages to meiosis, meiosis I and meiosis II. In meiosis I, chromosome duplication is followed by a pairing of homologous chromosomes with one another, during which time they exchange reciprocal sections of themselves. Homologous chromosome pairs then line up at the metaphase plate—one member of each pair on one side of the plate, the other member on the other side. These homologous pairs are then separated, in the first round of cell division, each of them becoming part of a separate daughter cell. In meiosis II, the chromatids of the duplicated chromosomes are separated into separate daughter cells.

  *Bio***Flix**  Meiosis

### 10.3  What Is the Significance of Meiosis?

- Meiosis generates diversity by ensuring that the gametes it gives rise to will differ genetically from one another. In this, it is unlike regular cell division, or mitosis, which produces daughter cells that are exact genetic copies of parent cells.

- Meiosis generates genetic diversity in two ways. First, in prophase I of meiosis, homologous chromosomes pair with each other and, in the process called crossing over or recombination, exchange reciprocal chromosomal segments with one another. Second, in metaphase I of meiosis, there is a random alignment or independent assortment of maternal and paternal chromosomes on either side of the metaphase plate. This chance alignment determines which daughter cell each chromosome will end up in.

- The genetic diversity brought about by meiosis and sexual reproduction is responsible, to a significant extent, for the great diversity of life-forms seen in the living world today. Evolution is spurred on by differences among offspring, and meiosis and sexual reproduction ensure such differences. By contrast, asexual reproduction, as is seen in bacteria and other organisms, produces organisms that are exact genetic copies, or clones, of the parental organism.

### 10.4  Meiosis and Sex Outcome

- Human females have 23 matched pairs of chromosomes—22 pairs of autosomes and one pair of sex-determining chromosomes, which in females are X chromosomes. Human males have 22 autosome pairs, one X chromosome, and one Y chromosome. Each egg that a female produces has a single X chromosome in it. Each sperm that a male produces has either an X or a Y chromosome within it. If a sperm with a Y chromosome fertilizes an egg, the offspring will be male. If a sperm with an X chromosome fertilizes the egg, the offspring will be female.

  **Web Animation 10.1:** Meiosis and Sex Outcome

### 10.5  Gamete Formation in Humans

- The starting cells in human male gamete formation, spermatogonia, are diploid cells that are stem cells in that they give rise not only to a set of specialized cells—the diploid primary spermatocytes—but to more spermatogonia as well. Primary spermatocytes then go through meiosis, producing haploid secondary spermatocytes, which in turn give rise to the spermatids that develop into mature sperm cells.

- In adult females, there may be no counterparts to the male spermatogonia—no cells that function as reproductive stem cells, though this is currently a matter of scientific controversy. Female gamete formation begins with cells called oogonia, most or all of which are produced prior to the birth of the female. These give rise to primary oocytes—the cells that will go through meiosis I, thus producing haploid oocytes. The vast majority of primary oocytes never complete meiosis, however. It is only the single primary oocyte released each month, in the process of ovulation, that completes meiosis I. Only those ovulated oocytes that are fertilized by sperm complete meiosis II. Only one of the cells produced in meiosis will have the potential to develop into a haploid egg; the other cells receive very little cytoplasmic material during the process of cell division and thus are destined to become nonfunctional cells called polar bodies.

### 10.6  Life Cycles: Humans and Other Organisms

- Not all reproduction is sexual reproduction, meaning reproduction that works through a fusion of two reproductive cells. Most types of organisms are capable of asexual reproduction, although such reproduction is rare among more complex organisms and is never carried out by mammals or birds. Asexual reproduction can take several forms. Bacteria reproduce through one form of asexual reproduction, called binary fission, in which a given bacterial cell replicates its single chromosome and then divides in two. Plants can engage in vegetative reproduction; other organisms, such as worms and sea stars, can carry out regeneration, in which a new, complete organism can be formed from a portion of an existing one.

## KEY TERMS

| | | | |
|---|---|---|---|
| n | 178 | meiosis | 178 |
| 2n | 178 | oogonia | 186 |
| asexual reproduction | 188 | polar body | 188 |
| crossing over | 182 | primary oocyte | 186 |
| diploid | 178 | primary | |
| gamete | 178 | spermatocyte | 186 |
| haploid | 178 | sex chromosome | 185 |
| independent | | sexual reproduction | 188 |
| assortment | 183 | spermatogonia | 186 |
| life cycle | 188 | tetrad | 178 |

## BRIEF REVIEW

*Answers to Brief Review questions are in the back of the book. For multiple-choice quiz questions, go to* **www.aw.com/krogh4**.

1. Does chromosome reduction take place during meiosis I or II?

2. Is asexual reproduction confined to the plant world?

3. Why are men able to keep producing sperm throughout their lifetime, from puberty forward?

4. How does cytokinesis in the production of the eggs that females produce differ from cytokinesis in regular mitotic cell division?

5. Define and distinguish between somatic cells and gametes.

6. A diploid organism has four chromosomes. Name the phases of mitosis or meiosis shown in the following figures.

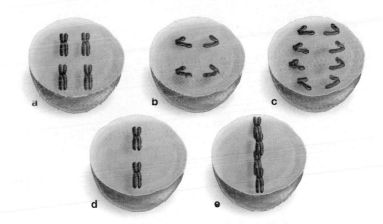

7. In what ways does meiosis ensure genetic diversity in offspring?

8. When human egg and sperm come together, how many chromosomes does each kind of gamete have? What is the genetic makeup of the resulting zygote?

## APPLYING YOUR KNOWLEDGE

1. Ultimately, is it the paternal or maternal gamete that delivers the sex-determining chromosome? Describe the process.

2. Mammals have now been cloned by scientists (think of Dolly the sheep). Suppose that half the world's human births were achieved through cloning. How would our world be different?

3. All the bananas we eat come from trees that are produced through "cuttings"—a stem from an existing tree is planted in the ground, resulting in a new tree. Thus, each tree is a clone of another. Why would growers find it advantageous to produce an enormous series of identical clones, rather than using trees that reproduce sexually?

By experimenting with generations of garden peas such as these, Gregor Mendel became the first person to understand some of the basic rules of genetics.

CHAPTER **11**

| 11.1 | Mendel and the Black Box | 193 |
| 11.2 | The Experimental Subjects: *Pisum sativum* | 194 |
| 11.3 | Starting the Experiments: Yellow and Green Peas | 196 |
| 11.4 | Another Generation for Mendel | 198 |
| 11.5 | Crosses Involving Two Characters | 202 |
| 11.6 | Reception of Mendel's Ideas | 205 |
| 11.7 | Incomplete Dominance | 205 |
| 11.8 | Lessons from Blood Types: Codominance | 206 |
| 11.9 | Multiple Alleles and Polygenic Inheritance | 207 |
| 11.10 | Genes and Environment | 209 |
| 11.11 | One Gene, Several Effects: Pleiotropy | 210 |

**Essays**

Proportions and Their Causes: The Rules of Multiplication and Addition | 201
Why So Unrecognized? | 204

A straightforward set of rules governs the way living things pass on many of their traits. Gregor Mendel was the first person to understand what those rules are.

# THE FIRST GENETICIST:
# Mendel and His Discoveries

**I**n the Czech Republic, in the oldest part of the city of Brno, stands the former Monastery of St. Thomas. Now a research center, St. Thomas has a small, fenced-in garden that sits just outside its historic living quarters. At one end of the garden stands an ivy-covered tablet, inscribed with a few simple words: "Prelate Gregor Mendel made his experiments for his law here." Visitors to the garden can cast their eyes upward and see the rooms that Gregor Johann

**Figure 11.1**
**Austrian Monk and Naturalist Gregor Mendel**

Mendel lived in while carrying out his work in the years between 1856 and 1863 (**Figure 11.1**). Mendel labored during this time on a species of common peas, *Pisum sativum*, using tweezers and an artist's paintbrush as his tools. Like many scientific discoverers, Mendel looked at ordinary objects; and as with many scientists, his work was tedious and exacting. Nevertheless, the distance from what Mendel did to what Mendel learned is the distance from the ordinary to the historic.

Coming to adulthood in the mid-nineteenth century, Mendel knew nothing of the elements of genetics we have reviewed so far. Chromosomes and the genes that lie along them had not even been discovered when he was working with his peas; little knowledge existed of meiosis or mitosis, to say nothing of DNA. And as it happened, Mendel himself played no direct role in discovering any of these things. Yet, this unassuming monk, the son of eastern European peasant farmers, generally is accorded the title of the father of genetics. What contribution earned him this honor?

## 11.1 Mendel and the Black Box

Mendel's achievement was to comprehend what was going on inside what might be called the "black box" of genetics without ever being able to look inside that box himself. Science is filled with so-called black-box problems, in which researchers know what goes *into* a given process and what comes *out*. It is what is going on in between—in the black box—that is a mystery. In the case of genetics, what lies inside the black box is DNA, chromosomes, and meiosis, and so forth—all the component parts of genetics, in other words, and the way they work together. Because of the timing of his birth, Mendel had no knowledge of any of these things. Until Mendel, however, nobody had looked carefully at even the starting and ending points of this black-box problem: at what went in and what came out. Mendel's original pea plants represented the input side to the black box of genetics, while the offspring he got from breeding these plants represented the output. By looking carefully at generations of parents and offspring—at both sides of the box, in a sense—Mendel was able to infer something about what had to be going on within. As you will see, his main inferences were correct: (1) that the basic units of genetics are material elements; (2) that these elements come in pairs; (3) that these elements (today called genes) can retain their character through many generations; and (4) that gene pairs *separate* during the formation of gametes.

These insights may sound familiar because all of them were approached from another direction in Chapters 9 and 10. Here are the lessons you went over then:

1. Genes are material elements—lengths of DNA.

2. In human beings, genes come in pairs, residing in pairs of homologous chromosomes.

3. Chromosomes make copies of themselves, thus giving the genes that lie along them the ability to be passed on intact through generations.

4. In meiosis, homologous chromosomes line up next to each other and then separate, with each member of a pair ending up in a different egg or sperm cell.

By observing generations of pea plants and applying mathematics to his observations, Mendel inferred that something like this had to be happening in reproduction. Because Mendel was the first to perceive a set of principles that govern inheritance, we date our knowledge of genetics from him.

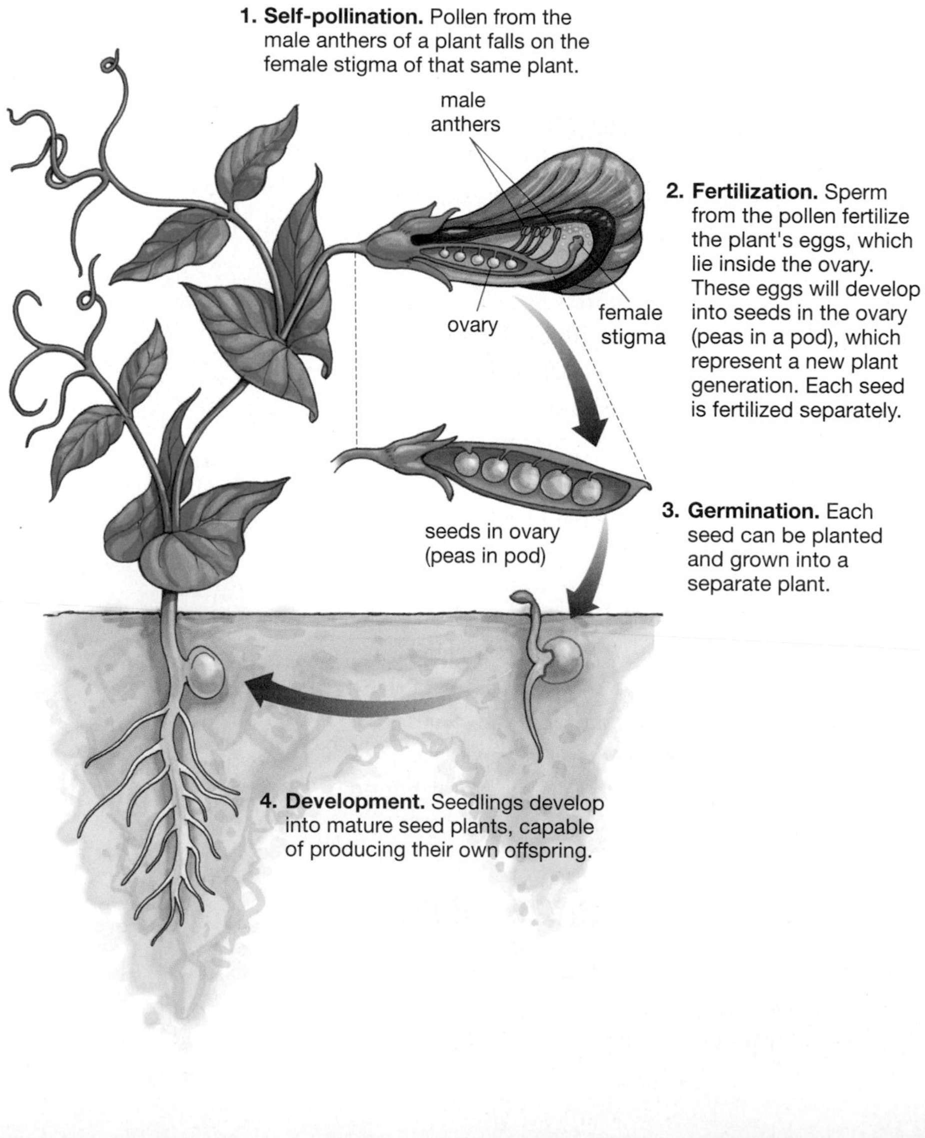

## 11.2 The Experimental Subjects: *Pisum sativum*

Beginning work in his monastery garden after a period of study in Vienna, Mendel managed to pick, in *Pisum sativum*, a nearly perfect species on which to carry out his experiments. If you look at **Figure 11.2**, you can see something of the life cycle of this garden pea. Note that what we think of as peas in a pod are seeds in this plant's ovary. Each of these seeds begins as an unfertilized egg, just as a human baby begins as a maternal egg that is unfertilized. Sperm-bearing pollen, landing on the plant's stigma, then set in motion the fertilization of the eggs.

Importantly, *Pisum* plants can *self*-pollinate. The anthers of a given flower release pollen grains that land on that flower's stigma. Each seed that develops from the resulting fertilizations can then be planted in the ground and give rise to a new generation of plant. The way to think of the seeds in a pod is as multiple offspring from multiple fertilizations— one pollen grain fertilizes this seed, another pollen grain fertilizes another. This is why, as you'll see, seeds that are in the same pod can have different characteristics.

**Figure 11.2**
**Life Cycle of the Pea Plant**

1. **Self-pollination.** Pollen from the male anthers of a plant falls on the female stigma of that same plant.

male anthers

ovary

female stigma

2. **Fertilization.** Sperm from the pollen fertilize the plant's eggs, which lie inside the ovary. These eggs will develop into seeds in the ovary (peas in a pod), which represent a new plant generation. Each seed is fertilized separately.

seeds in ovary (peas in pod)

3. **Germination.** Each seed can be planted and grown into a separate plant.

4. **Development.** Seedlings develop into mature seed plants, capable of producing their own offspring.

Although the pea plants can self-pollinate, Mendel could also **cross-pollinate** the plants at will—he could have one plant pollinate another—by going to work with his tweezers and paintbrushes, as shown in **Figure 11.3**.

In directing pollination, Mendel could control for certain attributes in his pea plants—qualities now referred to as *characters*. If you look at **Table 11.1** on the next page, you can see that there were seven characters in all—such things as stem length, seed color, and seed shape. Note that each of these characters comes in two varieties. There are yellow or green seeds, for example, and purple or white flowers. Such character variations are known as *traits*. Mendel referred to each of these variations as being either a "dominant" or a "recessive" trait, as the table shows. For now, you can think of a recessive trait as one that tends to remain hidden in certain generations of pea

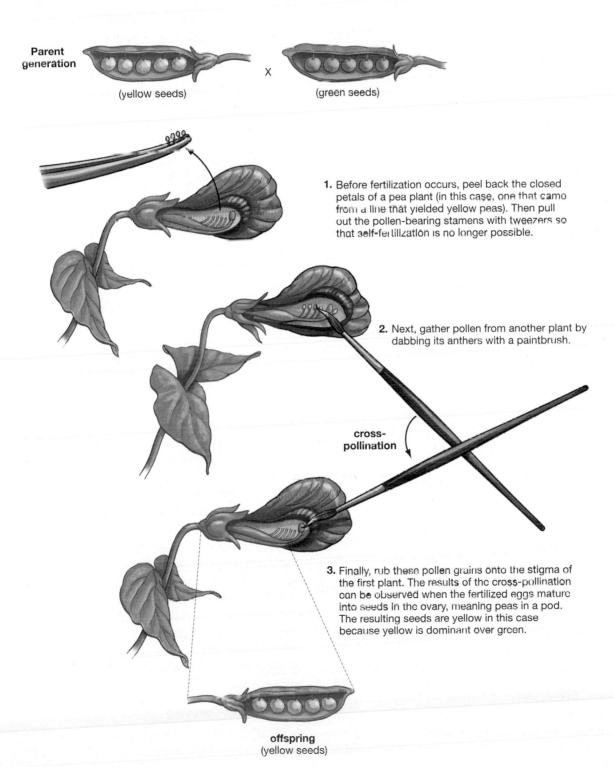

**Parent generation**

(yellow seeds)

X

(green seeds)

**Figure 11.3**
**How to Cross-Pollinate Pea Plants**

1. Before fertilization occurs, peel back the closed petals of a pea plant (in this case, one that came from a line that yielded yellow peas). Then pull out the pollen-bearing stamens with tweezers so that self-fertilization is no longer possible.

2. Next, gather pollen from another plant by dabbing its anthers with a paintbrush.

cross-pollination

3. Finally, rub these pollen grains onto the stigma of the first plant. The results of the cross-pollination can be observed when the fertilized eggs mature into seeds in the ovary, meaning peas in a pod. The resulting seeds are yellow in this case because yellow is dominant over green.

**offspring**
(yellow seeds)

## Table 11.1 Pea-Plant Characters Studied by Mendel

| Character studied | Dominant trait | Recessive trait |
|---|---|---|
| Seed shape | smooth | wrinkled |
| Seed color | yellow | green |
| Pod shape | inflated | wrinkled |
| Pod color | green | yellow |
| Flower color | purple | white |
| Flower position | on stem | at tip |
| Stem length | tall | dwarf |

plants. A dominant trait, meanwhile, can be thought of as a trait that tends to appear in those same generations. You'll get a formal definition of dominant and recessive later.

### Phenotype and Genotype

Now, let's put these traits in another context. Taken together, traits represent what are called **phenotypes**. Broadly speaking, a phenotype is a physiological feature, bodily characteristic, or behavior of an organism. In the context of Mendel, phenotype means the pea plant's visible physical characteristics. Purple flowers are one phenotype, white flowers another; yellow seeds are one phenotype, green seeds another.

Meanwhile, any phenotype is in significant part determined by an organism's underlying **genotype**, meaning its genetic makeup.

## 11.3 Starting the Experiments: Yellow and Green Peas

Mendel began by making sure that his starting plants "bred true" for the phenotypes under study. That is, he assured himself that, for example, all of his purple-flowered plants would produce nothing but generations of purple offspring if they self-pollinated.

### Parental, F₁ and F₂ Generations

The **parental generation** in such experiments is often referred to in shorthand as the **P** generation. Meanwhile, the offspring of the parental generation are known as the **first filial generation**, the word *filial* indicating a son or a daughter. The shorthand for first filial is $F_1$. The $F_x$ form can be used for any succeeding generation. You will be seeing a lot of the second filial, or $F_2$ generation, for example.

So, how did Mendel proceed with his plants? Let's start with just one example. Mendel took, for his P generation, some plants from a line that bred true for yellow seeds and other plants from a line that bred true for green seeds. Then he took pollen from one variety of the plants and fertilized the other variety, as seen in Figure 11.3. The plants fertilized in this way then produced their own seeds—every one of which was yellow. (Remember that although these seeds were developing within the pods of the parental generation, they represented the new generation—the $F_1$ generation.) Getting all yellow seeds was an interesting result in itself. Because each seed in a pea-plant pod is fertilized separately, it can be the case that a given pod will contain both yellow and green seeds. (Other pea characters can be variable within a pod as well, as you can see in **Figure 11.4**.) Yet, all the seeds in this generation were yellow, indicating that yellow was a dominant trait and green a recessive one.

Having viewed these results, Mendel then planted his $F_1$ seeds and let plants grow from them. These plants were then allowed to *self*-pollinate—sperm from a given plant fertilized the eggs from that same plant, so that instead of "crossing" one plant with another, Mendel was crossing a plant with itself. What he got from this self-pollination, in the $F_2$ generation, was 6,022 yellow seeds and 2,001 green seeds.

### The Power of Counting

In counting the seeds, Mendel was taking a giant step forward. Why? Experimenters before Mendel had

Some of these peas have a smooth texture, while others are wrinkled.

**Figure 11.4**
**Variation within a Pea Pod**
Since each garden pea is fertilized separately, individual peas within a pod can have different character traits.

gotten the kind of results he had. What they did not do, however, was undertake a careful counting of their results and then analyze the results in terms of proportions.

Looking at the specific results, two things stand out. First, green seeds disappeared in $F_1$, but came back in $F_2$. Recall that there was not a green seed to be found in the $F_1$ generation pods, but in $F_2$, there are 2,001 of them. Second, green seeds came back in $F_2$ as *one-fourth* of the seeds as a whole. The $F_2$ generation isn't divided 50/50, half yellow and half green; instead, there are roughly 3 yellow seeds for every

1 green seed—or to put it another way, a 3:1 ratio of yellow to green seeds. As it turned out, Mendel got the same result with each of the seven characters he was studying, as you can see in **Table 11.2**. Note that the 3:1 ratio is a proportion of dominant to recessive traits—more yellow seeds than green, more purple flowers than white. It was *solely* the dominant traits that appeared in the $F_1$ generation, but in the $F_2$ generation the dominants are simply appearing in greater proportion than the recessives.

## Interpreting the $F_1$ and $F_2$ Results

What did Mendel learn from these results?

### No "Blending" in Inheritance

For one thing, he saw that inheritance for his peas was not a matter of the "blending" of their characteristics. For example, the flower colors purple and white were retained as just that. No intermediate phenotypes—say, pink flowers—resulted from the cross in any generation. This finding run contrary to the notion, popular in Mendel's time, that two given traits would blend into a homogeneous third entity, as coffee and milk will blend together.

### Dominant and Recessive Elements Come in Pairs

Beyond this finding, it was apparent that, dominance notwithstanding, plants could retain the *potential* for recessive phenotypes, even though those phenotypes might not appear in a given generation. Mendel's $F_1$ plants had no green seeds, but in $F_2$ the green seeds were back. It was reasonable to assume, therefore,

| Table 11.2  Ratios of Dominant to Recessive in Mendel's Plants | | |
|---|---|---|
| **Dominant trait** | **Recessive trait** | **Ratio of dominant to recessive in $F_2$ generation** |
| Smooth seed | Wrinkled seed | 2.96:1 (5,474 smooth, 1,850 wrinkled) |
| Yellow seed | Green seed | 3.01:1 (6,022 yellow, 2,001 green) |
| Inflated pod | Wrinkled pod | 2.95:1 (882 inflated, 299 wrinkled) |
| Green pod | Yellow pod | 2.82:1 (428 green, 152 yellow) |
| Purple flower | White flower | 3.14:1 (705 purple, 224 white) |
| Flower on stem | Flower at tip | 3.14:1 (651 along stem, 207 at tip) |
| Tall stem | Dwarf stem | 2.84:1 (787 tall plants, 277 dwarfs) |
| | **Average ratio, all traits:** | **3:1** |

**I**f fame belonged to me, I could not escape her—if she did not, the longest day would pass me on the chase," wrote Emily Dickinson in 1862, in seeming ambivalence about becoming a published poet. Dickinson's words came in a letter to Thomas Higginson, who had recently advised her against seeking publication of her verses on grounds that they were not yet strong enough. Higginson was the second, and as it turned out, last person to disparage Dickinson's efforts to publish her poetry. After his advice was offered, Dickinson kept her own company as a writer; by the time she died, only a close circle around her had seen her work.

Half a world away, Dickinson had a counterpart of sorts in her contemporary, Gregor Mendel (born eight years before her and deceased two years before). Like Dickinson, Mendel had something special to offer the world, but it was an offer made to a world that was not prepared to listen. Mendel died in obscurity in 1884, his work appreciated by, so far as we can tell, no one but himself. Indeed, 34 years were to pass between publication of his research results and recognition of their significance. Yet, from the time his work was rediscovered in 1900, Mendel has been treated with great respect, such that today he is generally recognized as the founder of genetics.

*Mendel appears to have been a victim of his own originality.*

How, then, could Mendel's work have gone unrecognized in his own time? A couple of easy but unsatisfying answers are that he labored in a scientific backwater (in the hinterland of the Austrian empire), and that his results were published in an obscure scientific journal.

Even granting these points, a good number of scientists still had access to Mendel's report—if not by reading it in the journal that published it, then by receiving one of the reprints of it that Mendel himself mailed out. If fame didn't follow from such exposure, why not simple recognition by at least one person?

The answer appears to be fairly straightforward: Mendel wasn't appreciated in his lifetime by anyone because he wasn't comprehended in his lifetime by anyone. Indeed, intense exposure to his ideas seemed to make no difference in this regard. Consider that he carried on a seven-year correspondence regarding his work with a famous botanist, Carl Nägeli, during which time Nägeli,

like everyone else, seemed to miss the significance of Mendel's experiments entirely. Assuming that nineteenth-century scientists were as bright as those in the twenty-first century, how could this have happened?

First, Mendel himself did not make much of the principles of heredity he uncovered. In his long paper, he punctuates a great deal of detail about his experiments with some understated formulations of general principles. In itself, this is not surprising. Scientists often let data "speak for themselves," resting assured that their colleagues will not miss a major point, however subtly it may be presented. Most scientists can do this, however, because they are confident that their colleagues will be *looking* for major points in their papers. In Mendel's case, nobody was looking for points about how heredity worked because scarcely anybody but Mendel had a concept of pure heredity. As contemporary geneticists Daniel L. Hartl and Vitezslav Orel have observed, in the minds of nineteenth-century scientists, heredity was merely a part of development—the process by which a fertilized egg becomes a fully formed organism. Mendel, meanwhile, thought of heredity as operating under its own independent set of rules, after which he went on to say what those rules were. Mendel thus delivered an answer to a question that nobody had asked: How does heredity work?

Other barriers existed to understanding as well. Mendel used mathematics to derive his insights on genetics, an approach that was novel in the biology of his day. Beyond this, in 1859 Charles Darwin's *On the Origin of Species* had been published, an event that sent scientists scurrying to look for "continuous" or subtle variations between organisms—a phenomenon that they thought might shed some light on evolution. Conversely, Mendel's paper, coming seven years later, talked about pairs of seemingly unchangeable genetic elements that gave rise to such either/or features as smooth or wrinkled seeds. It was difficult to fit such elements into a scheme of gradual evolutionary transformation.

Looking over all these factors, Mendel appears to have been, in essence, a victim of his own originality. (This again puts him in the company of Emily Dickinson, whose language was too idiosyncratic for her early readers.) There is a saying that it doesn't pay to be more than 10 minutes ahead of your time. Gregor Mendel had the misfortune of being about 34 years ahead of his.

was correct: that characters were being transmitted independently of one another. If one character had affected another's transmission, these ratios would have been very different. This insight of Mendel's is now known as Mendel's Second Law, or the **Law of Independent Assortment**. It states that during gamete formation, gene pairs assort independently of one another.

### Independent Assortment and Chromosomes

Now recall that the underlying physical basis for this law was set forth in Chapter 10: In meiosis, pairs of homologous chromosomes *assort independently* from one another at the metaphase plate. It may be, for example, that paternal chromosome 5 will line up on side A of the metaphase plate; if so, then maternal chromosome 5 ends up on side B.

For chromosome 6, however, things could just as easily be reversed: The *maternal* chromosome might end up on side A and the paternal on side B (Figure 10.2, page 180). The genes for Mendel's seed color exist on the plant's chromosome 1, while the genes for seed shape exist on its chromosome 7. Because these are separate chromosomes, they assort independently at the metaphase plate, meaning they are passed on independently to future generations.

**SO FAR...**

1. Mendel's law of segregation states that differing traits in organisms result from two differing _____ that _____ in gamete formation, such that each gamete gets only one of the two.

2. Homozygous refers to an organism that has two _____ alleles for a character, while heterozygous refers to an organism that has two _____ alleles for the character.

3. An allele that is expressed in the heterozygous condition is said to be _____, while an allele that is not expressed in the heterozygous condition is said to be _____.

## 11.6 Reception of Mendel's Ideas

When Mendel finished his experiments, he delivered two lectures on them to a local scientific society in 1865. It would be nice to report that the society members' jaws dropped open once Mendel let them in on how inheritance worked in the living world, but no such thing happened. His findings were ultimately published in a scientific journal that was distributed in Germany, Austria, the United States, and England. Mendel even took it upon himself to get 40 reprints of his paper, thereafter sending some out to various scientists in an effort to spark some exchange on his findings—all to no avail. His work sank nearly without a trace, finally to be rediscovered in 1900, 16 years after his death. Today, the consensus within the scientific community is that nobody cared about Mendel's findings in his own time simply because nobody grasped their significance (see "Why So Unrecognized?" on page 204).

This poor early reception notwithstanding, Mendel's insights have stood the test of time. For certain kinds of phenotypes, his rules of inheritance are extremely reliable. To this day, biologists will begin an observation by noting that "Mendelian rules tell us that... " or "Mendelian inheritance operates here." Even allowing

that he had intellectual predecessors, the fact is that Mendel did not just add to the discipline of genetics; he founded it.

## 11.7 Incomplete Dominance

Mendel knew as well as anyone that the rules he discovered did not apply in all instances of inheritance. In his peas, crossing white- and purple-flowered plants resulted in an F₁ generation in which all flowers were purple. In some other species, however, the results are different. Crossing a true-breeding red-flowered *snapdragon* with a true-breeding white-flowered snapdragon produces neither red- nor white-flowered F₁ snapdragons, but *pink* snapdragons (**Figure 11.10**). This might seem to indicate that

••• ANSWERS

1. alleles; separate
2. identical; differing
3. dominant; recessive

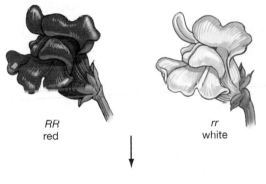

*RR*
red

*rr*
white

**P generation**

1. The starting plants are a snapdragon homozygous for red color (*RR*) and snapdragon homozygous for white color (*rr*).

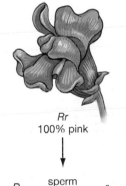

*Rr*
100% pink

**F₁ generation**

2. When these plants are crossed, the resulting *Rr* genotype yields only enough pigment to produce a flower that is pink—the only phenotype in the F₁ generation.

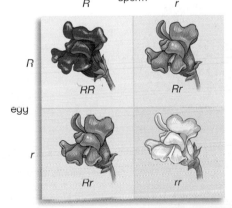

sperm
R                    r

R

egg

r

*RR*        *Rr*

*Rr*        *rr*

1  :  2  :  1
red   pink   white

**F₂ generation**

3. In the F₂ generation, alleles combine to produce red, pink, and white phenotypes.

**Figure 11.10**
**How Red and White Yield Pink: Incomplete Dominance in Snapdragons**

inheritance can indeed work through the blending of genetic traits—a notion that Mendel's work had quashed. Blending is not taking place, however. When breeding is continued through the $F_2$ generation, red and white flowers come back, along with the pink, in a ratio of 1 part red, 2 parts pink, 1 part white. This 1:2:1 ratio demonstrates that red and white alleles have not irretrievably blended into pink but rather have come together in $F_1$, only to separate

out again in $F_2$. Still, how can the pink $F_1$ snapdragons be explained?

### Genes Code for Proteins

First, think about what genes do: They contain information regarding the production of proteins. In the case of colors, such proteins can bring about the formation of pigments. This is the case with the snapdragons; they have a gene for color, and one of the alleles of this gene brings about the production of red pigment. Meanwhile, the second allele of this gene is nonfunctional—it brings about the production of no pigment at all. Two of the red alleles, then, yield a red color, while one red allele produces only enough pigment to yield a pink color. Neither the red nor the white allele, therefore, is completely dominant; each is thus said to have **incomplete dominance**, meaning a genetic condition in which the heterozygote phenotype is intermediate between either of the homozygous phenotypes.

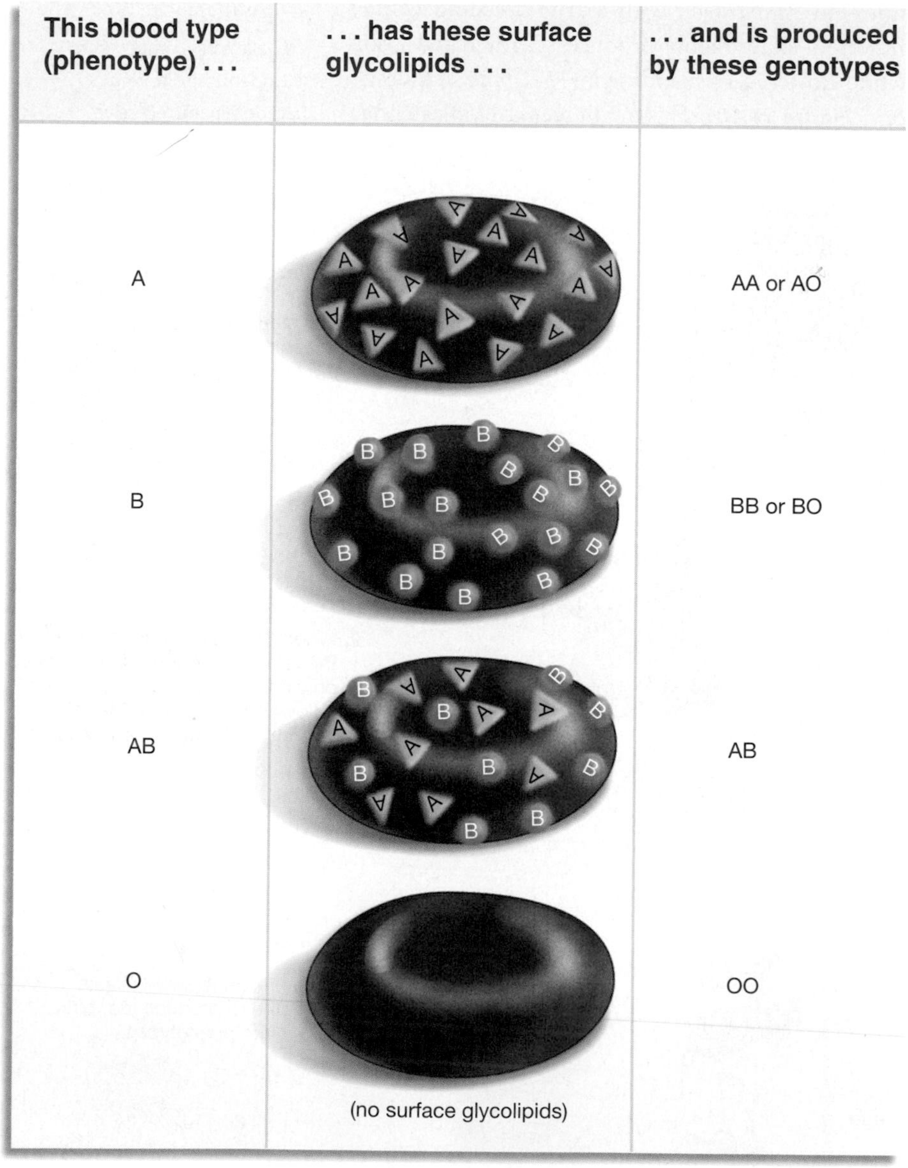

## 11.8 Lessons from Blood Types: Codominance

The notion of genes as protein-producing entities can help us steer through another variation on Mendelian inheritance, this one having great importance for variety in the living world. When people speak of blood "types," what they mean is types of "glycolipids"—lipid molecules with carbohydrate chains attached—that extend from the surface of red blood cells. These surface extensions come in many different varieties, the two most important of which are designated A and B (**Figure 11.11**).

Blood types are completely under genetic control, with a single gene that lies on chromosome 9 determining what blood type a person will have—A, B, AB, or O. There are two copies of this gene in each individual because there are two copies of chromosome 9 in each individual—one inherited from the mother, the other inherited from the father. Thus, there exists the possibility for *alleles*, or gene variants, in each individual. One of these alleles codes for a protein that modifies surface extensions into the A form, while the other codes for a protein that modifies the extensions into the type B form.

Now, a single person could have two alleles that code for type A extensions, or a person could have both alleles coding for type B extensions. In the first case, a person would have type A blood and in the second type B blood. However, a person could have

| This blood type (phenotype) . . . | . . . has these surface glycolipids . . . | . . . and is produced by these genotypes |
|---|---|---|
| A | | AA or AO |
| B | | BB or BO |
| AB | | AB |
| O | (no surface glycolipids) | OO |

**Figure 11.11**
**Human Blood Types**

The familiar ABO human blood-typing system refers to glycolipid molecules that extend from the surface of red blood cells. People whose blood is "type A" have A extensions on their blood cells. It is also possible to have only B extensions (and be type B); to have both A and B extensions (and be type AB); or to have none of these extensions (and be type O). The column on the right shows the genotypes that produce these phenotypes. Note that a person whose genotype is AO is phenotypically type A; likewise, a person whose genotype is BO is phenotypically type B.

one allele that codes for the type A extensions *and* one that codes for type B, in which case that person would have type AB blood. Finally, a person could have two alleles that are inactive—that code for neither extension—and would thus have type O blood, as Figure 11.11 shows.

## Getting Both Types of Surface Proteins

What's new in this case is, first, another aspect of the idea of dominance in alleles. In the blood types, *both* the A and B alleles produce proteins that modify the cell-surface extensions. People who are heterozygous for these alleles thus do not get a phenotype that, like pink flowers, lies halfway between A and B. They get both the A and B versions of the molecules on the surface of their blood cells. Alleles that have this kind of independent effect result in **codominance**: a condition in which two alleles of a given gene have different phenotypic effects, with both effects manifesting in organisms that are heterozygous for the gene.

## Dominant and Recessive Alleles

All this gets us to an important point, which is the relationship between dominant and recessive alleles. In common speech, the word *dominance* is used all the time to mean "bringing another under submission by means of superior power." Alleles, however, are not locked in a power struggle with one another. Alleles contain information for the production of proteins. Such proteins (or lack of them) can create dominant phenotypes (smooth peas), incompletely dominant phenotypes (pink snapdragons), or codominant phenotypes (AB blood). "Dominance," then, just refers to the ways in which a given allele is expressed. As you've seen, a *recessive* allele is not eliminated by being paired with a dominant allele—it simply isn't expressed in a given generation. When this generation gives rise to another, the recessive allele will be passed on intact, at which point it may be expressed. When the pea plant's recessive green allele was passed on, for example, it was expressed when brought together with another green allele. This has relevance for traits more familiar to us. People seem to have an idea that human blond hair, to give one example, eventually will cease to exist because the alleles for it are recessive to those for dark hair. The alleles for blond hair are, however, not disappearing. To the extent that alleles for it come together in reproduction, there will always be blond-haired people, as you can see in **Figure 11.12**.

**Figure 11.12**
**Alleles Are Retained Across Generations**
The fraternal Hodgson twins, Kian on the left and Remee on the right, were born one minute apart in 2005 to English parents Kylie Hodgson (at far left) and Remi Horder. Both parents are the product of mixed-race, black-white unions, a circumstance that set the stage for the unusual genetic outcome in their twins. Kian happened to inherit alleles, passed on from both of her black grandparents, that produced her darker skin, black hair, and brown eyes, while Remee inherited alleles, passed on from both of her white grandparents, that led to lighter skin, blond hair, and blue eyes.

### SO FAR...

1. Incomplete dominance, such as that of the pink snapdragon, exists when a _____ phenotype is intermediate between that of either of the _____ phenotypes.

2. Codominance exists when alleles have _____ effects on a phenotype, with both effects manifesting in organisms that are _____ for a codominant gene.

## 11.9  Multiple Alleles and Polygenic Inheritance

ABO blood types have one more lesson for us. In Mendel's seed shapes, there were two phenotypic variants: smooth and wrinkled. In blood types, there were four: A, B, AB, and O. As you know, no one person can have more than *two* alleles for a given gene. As a result, a person may be, say, A and B, but he or she cannot be A, B, and O. It is thus only in a *population* of humans that the full range of ABO alleles can be found. And therein lies the lesson: In a population, alleles can come not just in two, but in many variants.

••• ANSWERS
1. heterozygous, homozygous
2. independent; heterozygous

When three or more alleles of the same gene exist in a population, they are known as **multiple alleles**. Such alleles are important in and of themselves for they can bring diversity to traits that are governed by single genes (such as human blood types). Their more general significance comes, however, in connection with traits that are controlled not by one gene, but by many. Recall that seed texture in Mendel's peas was governed by a single gene (which had smooth and wrinkled allelic variants). This situation is, however, the exception rather than the rule in the living world. Most traits are governed by many genes. The human traits of height, weight, eye color, and skin color are each controlled by several genes. In the wider living world, the color of a wheat grain, the length of an ear of corn, and the amount of milk a cow gives are all controlled by many genes acting together. The term for such genetic influence is **polygenic inheritance**, meaning the inheritance of a genetic character that is determined by the interaction of multiple genes, with each gene having a small additive effect on the character.

When we have many genes contributing some small increment to a trait, the result is what you see in **Figure 11.13**. Human beings don't come in two heights, or three or four; they display what is known as a "continuous variation" in height, each person being just barely taller or shorter than the next. Contin-uous variation also holds true for human skin color. Human beings don't really have "black" or "white" skin. Instead, they have skin that comes in a range of hues, in which one color shades imperceptibly into another. Likewise, trees of a given species come in a range of heights, and the beaks of birds come in a range of lengths. Polygenic inheritance creates this fine-grained diversity.

Now, note something else about Figure 11.13a. Most of the students fall in the middle range of heights in the group. Put another way, there are fewer tall or short students than there are students of medium height. This height distribution means that the group as a whole takes on a kind of shape—a bell shape. This is, in fact, the famous **bell curve**, meaning a distribution of values that is symmetrical and largest around the average. Most biological traits manifest in this way. Look, for example, at the beak depths in a group of Darwin's finches in Figure 11.13b. Graphs like this just confirm what we know intuitively—that traits in living things cluster around what is average rather than what is extreme.

In polygenic inheritance, with its many genes and alleles, gene interactions are so complex that predictions about phenotype are a matter of *probability*, not certainty. Human height is largely under genetic control, and there are means of trying to predict the height of children based on the height of their

**(a)** Continuous variation in human height

**(b)** The bell curve

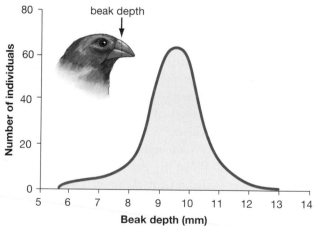

**Figure 11.13**
**Continuous Variation and the Bell Curve**

**(a)** Mendel's pea seeds may have been green or yellow, but traits such as human height do not have this either/or quality. Human heights exist in a range, with no fixed increments between heights of individuals. Such "continuous variation" results from polygenic inheritance, in which each of several genes contributes a small additive effect to a character. University of Connecticut students in the picture have been arranged by height to show how continuous variation works in this one human trait. Note that the group as a whole takes on the shape of a bell. The students' heights are distributed in a pattern that creates a bell curve.

**(b)** A bell curve can also be seen in this graph, which plots the average beak depth of a population of Darwin's finches from the Galápagos Islands. Note how most of the finches had beak depths close to the average depth of the population as a whole.

parents. (The starting point generally is to add together the height of the parents and divide the result by two.) All that such predictions can do, however, is specify that if parents are of a given height, then there is a certain probability that their children will fall into a given range of heights. Think how different this is from the situation with Mendel's peas. There, you could confidently predict that a given cross would yield, say, all green peas.

## 11.10   Genes and Environment

Scientists work awfully hard at trying to determine the probabilities for certain polygenic outcomes— those related to disease. In carrying out this work, they are essentially doing what Mendel did: observing traits in a parental generation and then observing traits in the parents' offspring. Such work is the basis for the warnings we sometimes hear about how much a person's risk for a particular cancer goes up if that person has a relative who has contracted the disease. With polygenic illnesses such as cancer, however, it is difficult to separate the genetic factors from what are called "environmental" factors. What's an environmental factor? Any external influence that is favorable or unfavorable for the development of a trait in an organism. Smoking, for example, is an environmental factor that influences development of lung cancer. Indeed, smoking is responsible for about 90 percent of all lung cancer cases. Even so, only a minority of smokers ever contract lung cancer. Why do some smokers get this disease while others are spared? A large part of the answer is genetic susceptibility: The genetic makeup of some individuals makes them more susceptible to the environmental influence of smoking. Thus, we can see genes and environment working together to create the outcome of lung cancer. This is almost always the case in nature. For polygenic traits— which is to say, most traits—it is rare for genes to be the sole determining factor.

A clear example of how genes and environment work together can be seen in **Figure 11.14**, which shows a picture of the most popular species of hydrangea flower, *Hydrangea macrophylla*. Any gardener could start with a blue hydrangea, like the one in the

**Figure 11.14**
**Flower Color Is Affected by Environment**
Both the blue hydrangeas in the background and the pink hydrangeas in the foreground are from the same species (*Hydrangea macrophylla*), yet they have different coloration because of different chemical conditions in their soil. (The soil the blue plants are in is more acidic than the soil the pink plants are in.) This is but one example of how genes and environment work together to produce the traits we see in living things. The plants are blooming along a wall in Salem, Oregon.

picture on the left, take a "cutting" from it, do a little preparatory work, plant the cutting in the ground, and get a whole new hydrangea plant as a result. For our purposes, the important point is that this second plant is a clone of the first. It is not produced by mixing eggs and sperm; it is an exact genetic replica of the first plant. It can, however, be a plant with *pink* flowers instead of blue if some garden lime is mixed into its surrounding soil. It's possible, therefore, to have two hydrangea plants with the exact same genotype but with very different phenotypes—pink or blue flowers. The message here is one of genetic limitation. Genotype specifies only so much about phenotype. Protein products that bring about a blue plant in one environment will bring about a pink plant in another. Across the natural world, things generally work this way. Most genes do not come with an unvarying ability to bring about an effect. Rather, genes and environment work together to create the phenotypes we see in living things.

### 11.11  One Gene, Several Effects: Pleiotropy

You've seen that many genes can work together to produce one effect, such as human height. It is also possible, however, for a single gene to have many effects. This is frequently the case, actually, for the simple reason that the processes of living things are so interrelated.

The phenomenon of one gene having many effects is called **pleiotropy**. Many examples of this phenomenon exist; one of the more interesting, perhaps, is a debilitation known as *fragile-X syndrome*, which is the most common cause of inherited mental retardation. The X in its name refers to the X chromosome. *Fragile* comes from the break on the long arm of the X chromosome that can be seen in people who suffer from the syndrome. Fragile-X is now thought to be caused by a defect in a single gene, dubbed *FMR-1* by the team of researchers who discovered it.

The upshot of this defect is that fragile-X victims lack a functional *FMR-1* gene, which is to say their *FMR-1* does not appear to prompt production of any protein. One clear result of this is mental retardation; people with fragile-X commonly have an IQ of about 40–70 (as opposed to the norm of 100). For our purposes, fragile-X also has effects that seemingly stand far afield from mental retardation. Those with fragile-X often have an abnormally long face; large, protuberant ears, and (when male) large testicles. This broad range of effects, seemingly stemming from the inactivation of a single gene, contains a message: Genes normally work together in an interrelated web rather than functioning under a one-gene → one-phenotype model.

## On to the Chromosome

This completes the survey of Mendel's work and the variations on his ideas of inheritance. As you have seen, Mendel's findings lay dormant for some decades in the nineteenth century. What sent scientists scurrying back to his reports were experiments on those tiny entities you have looked at in detail before: chromosomes, the subject of the next chapter.

••• ANSWERS

**1.** many; polygenic; false; polygenic inheritance tends to produce continuous variation among organisms.

**2.** environmental

**3.** pleiotropy

---

**SO FAR...**

1. The usual case in nature is that a given character in an organism is influenced by _____ genes. To put this another way, the character is produced through _____ inheritance. True or false: This form of inheritance tends to produce fixed increments of difference between organisms.

2. The hydrangea example in the book shows that the same genes can produce different phenotypes depending on _____ influences that act on an organism.

3. The phenomenon of _____ is said to occur when one gene has several effects.

# CHAPTER 11 REVIEW

For study help, activities, and more quiz questions, go to **www.aw.com/krogh4**.

## SUMMARY

### 11.1 Mendel and the Black Box

- Gregor Mendel was the first person to comprehend some of the most basic principles of genetics. He reached these understandings in the mid-nineteenth century, working in what is now the Czech Republic and using as his experimental subjects a species of garden pea, *Pisum sativum*.

### 11.2 The Experimental Subjects: *Pisum sativum*

- Mendel looked at seven characters in his plants—such attributes as seed color and texture. In his plants, each of these characters came in two varieties or traits, one of them dominant, the other recessive. His experiments involved breeding pea plants, starting with plants that had a given set of traits and then observing which of those traits showed up in succeeding generations.

- A phenotype is any physiological feature, bodily characteristic, or behavior of an organism. In Mendel's plants, purple flowers were one phenotype, and white flowers were another. Phenotypes in any organism are in significant part determined by that organism's genotype, meaning its genetic makeup. Mendel realized that the phenotypes in his plants were being controlled by what we would today call their genotypes.

### 11.3 Starting the Experiments: Yellow and Green Peas

- Mendel realized that it was possible for organisms to have identical phenotypes—for all his pea plants to have yellow seeds, for example—and yet to have differing underlying genotypes.

- One of Mendel's central insights was that the basic units of genetics are material elements that, in his pea plants, came in pairs. These elements, today called genes, come in alternative forms called alleles. One member of an allele pair resides on one chromosome, while the other allele resides on a second chromosome that is homologous to the first.

- Another of Mendel's insights was that genes retain their character through many generations rather than being "blended" together. Genes that coded for green pea color, for example, were retained in their existing form over many generations.

### 11.4 Another Generation for Mendel

- A third insight of Mendel's was that alleles separate prior to the formation of gametes. Although Mendel did not know it, the physical basis for this is that the alleles he was observing resided on homologous chromosomes, which always separate in meiosis. An organism that has two identical alleles of a gene for a given character is said to be homozygous for that character. An organism that has differing alleles for a character is said to be heterozygous for that character. Dominant means: expressed in the heterozygous condition (as with the yellow color of peas, present in the heterozygous *Yy* condition). Recessive means: not expressed in the heterozygous condition (as with the green color of peas, absent in the *Yy* condition).

**Web Animation 11.1:** Mendel's Experiments and Probability

### 11.5 Crosses Involving Two Characters

- Mendel observed that the genes for the different characters he studied were passed on independently of one another. This was so because the genes for these characters resided on separate, non-homologous chromosomes. The physical basis for what he found is the independent assortment of chromosome pairs during meiosis.

### 11.6 Reception of Mendel's Ideas

- Gregor Mendel published his work, but the significance of it was never recognized in his lifetime. It was only rediscovered 16 years after his death, in 1900.

### 11.7 Incomplete Dominance

- Not all inheritance works through the principles Mendel perceived in his peas. Incomplete dominance operates when neither allele for a given gene is completely dominant, with the result that heterozygous genotypes can yield an intermediate phenotype (such as pink snapdragons).

**Web Animation 11.2:** Variations on Mendel

### 11.8 Lessons from Blood Types: Codominance

- In some instances, differing alleles of the same gene will have independent effects in a single organism. Such is the case with the gene that codes for the type A and B glycolipids that extend from the surface of human red blood cells. An individual who has one A and one B allele will have type AB blood. In such a situation, neither allele is dominant; rather, each is having a separate phenotypic effect. When differing alleles of a single gene have independent effects on the phenotype of an individual, the alleles are said to be codominant.

### 11.9 Multiple Alleles and Polygenic Inheritance

- Human beings and many other species can have no more than two alleles for a given gene, each allele residing on a separate, homologous chromosome. Many allelic variants of a gene can, however, exist in a population. Most traits in living things are governed not by one gene but by many, with these genes often having several allelic variants. The term for such genetic influence is polygenic inheritance, meaning the inheritance of a genetic character that is determined by the interaction of multiple genes, with each gene having a small additive effect on the character.

- Polygenic inheritance tends to produce continuous variation in phenotypes, meaning variation in which there are no fixed increments of difference between individuals. Human skin, for example, comes in a range of colors in which one color shades imperceptibly into the next.

- The traits produced in polygenic inheritance tend to manifest in bell-curve distributions, in which most individuals display near-average trait values rather than extreme trait values. Gene interactions and gene–environment interactions are so complex in polygenic inheritance that predictions about phenotypes are a matter of probability, not certainty.

## 11.10 Genes and Environment

- The effects of genes can vary greatly in accordance with the environment in which the genes are expressed. An organism's genotype and environment interact to produce that organism's phenotype.

## 11.11 One Gene, Several Effects: Pleiotropy

- Genes work in an interrelated fashion, such that a single gene is likely to have multiple effects. Pleiotropy is a phenomenon in which one gene has many effects.

## KEY TERMS

| | |
|---|---|
| **allele** 198 | **Law of Independent** |
| **bell curve** 208 | **Assortment** 204 |
| **codominance** 207 | **Law of Segregation** 201 |
| **cross-pollinate** 195 | **monohybrid cross** 202 |
| **dihybrid cross** 202 | **multiple alleles** 208 |
| **dominant** 202 | **parental generation (P)** 196 |
| **first filial generation (F$_1$)** 196 | **phenotype** 196 |
| **genotype** 196 | **pleiotropy** 210 |
| **heterozygous** 202 | **polygenic inheritance** 208 |
| **homozygous** 202 | **recessive** 202 |
| **incomplete** | **rule of addition** 201 |
| **dominance** 206 | **rule of multiplication** 201 |

## BRIEF REVIEW

*Answers to Brief Review questions are in the back of the book.*
*For multiple-choice quiz questions, go to* **www.aw.com/krogh4**.

1. What are the differences between:
   a. phenotype and genotype?
   b. dominant and recessive?
   c. codominance and incomplete dominance?

2. True or false: When Mendel conducted his genetic research, neither DNA nor chromosomes had been discovered.

3. What genetic principle is demonstrated in the fragile-X syndrome? Explain how this principle differs from the relationship between genes and physical outcomes that Mendel observed in his pea plants.

4. In a case of Mendelian genetics where one gene affects one trait, how many different genotypes may bring about a dominant phenotype (for example, yellow pea color)?

5. If a heterozygous yellow pea plant is crossed with a homozygous green pea plant, approximately what percentage of the progeny peas would be expected to be yellow?

6. What term is now given to Mendel's "element of inheritance"?

7. What two laws are credited to Mendel and his research?

8. Using the symbols *A*, *a*, *B*, and *b*, what Mendelian crosses would be expected to result in the following phenotypic ratios?
   a. 1:1
   b. 3:1
   c. 1:1:1:1
   d. 9:3:3:1

## APPLYING YOUR KNOWLEDGE

1. Stands of aspen trees often are a series of genetically identical individuals, with each succeeding tree growing from the severed shoot of another tree. Using what you've learned of genetics in this chapter, would you expect one aspen tree in a stand to differ greatly from another in its phenotype? Would you expect each to look exactly like the next in terms of phenotype?

2. The ABO blood-typing system mentioned in the text has great practical significance because the human immune system will attack blood-cell surface molecules that it recognizes as "foreign"—that is, molecules that a person does not have on his or her own blood cells. Given this, what is the "universal donor" blood type? What is the "universal recipient" type?

3. If, as the text states, the effects of genes can be expected to vary in accordance with the environment in which these genes work, would you expect this phenomenon to be applicable to human genes, such as those that help produce such traits as height, weight, or introversion and extroversion?

### Genetics Problems
*Answers are in the back of the book.*

One of Mendel's many monohybrid (single-character) crosses was between a true-breeding (homozygous) parent bearing purple flowers and a true-breeding (homozygous) parent bearing white flowers. All the offspring had purple flowers. When these offspring were allowed to self-fertilize, Mendel observed that their offspring occurred in the ratio of about three purple-flowered to one white-flowered plants. The allele determining purple flowers is symbolized $P$, and the allele determining white flowers is symbolized $p$. Use this information to answer the following questions:

1. What are the genotypes of the parents?
   a. $PP; pp$
   b. $PP; Pp$
   c. $Pp; Pp$
   d. $pp; pp$

2. Which generation has only purple-flowered plants?
   a. P
   b. $F_1$
   c. $F_2$
   d. $F_3$

3. Which generation tells you which of the traits is dominant?
   a. P
   b. $F_1$
   c. $F_2$
   d. $F_3$

4. How many phenotypes are present in the $F_2$ generation?
   a. 1
   b. 2
   c. 3

5. How many different genotypes are present in the $F_2$ generation?
   a. 1
   b. 2
   c. 3

6. Mendel allowed the white-flowered $F_2$ plants to self-fertilize. What proportion of these $F_2$ plants produced only white-flowered offspring?
   a. none
   b. $\frac{1}{4}$
   c. $\frac{1}{3}$
   d. $\frac{3}{4}$
   e. all

7. Mendel allowed the purple-flowered $F_2$ plants to self-fertilize. What proportion of these $F_2$ plants produced only purple-flowered offspring?
   a. none
   b. $\frac{1}{4}$
   c. $\frac{1}{3}$
   d. $\frac{3}{4}$
   e. all

8. If a heterozygous purple-flowered pea plant is crossed with a homozygous white-flowered plant, what proportion of offspring will be white flowered?
   a. none
   b. $\frac{1}{4}$
   c. $\frac{1}{2}$

d. $\frac{3}{4}$
e. all

9. Given what you know about the alleles that determine purple and white color in pea flowers, why is it possible to have a purple-flowered heterozygote but not a white-flowered heterozygote?

10. In considering flower color in pea plants, we are dealing with _____ gene(s) and _____ allele(s).
   a. 1; 1
   b. 1; 2
   c. 1; 4
   d. 2; 1
   e. 2; 2

11. The next three questions require you to apply rules of probability.
   a. In a toss of two coins, what is the chance of obtaining two heads?
   (1) 0
   (2) 12.5 percent
   (3) 25 percent
   (4) 50 percent
   (5) 75 percent

   b. What rule of probability was used to determine the chance of obtaining two heads?
   c. Now, use this rule to solve the following problem. In a cross between two purple-flowered heterozygotes, 75 percent of their offspring are expected to be purple flowered. In a cross between two heterozygotes with yellow colored seeds, 25 percent of their offspring are expected to be green. If two purple-flowered, yellow-seeded individuals, each heterozygous for both genes, are crossed, what proportion of their offspring is expected to be purple flowered with green seeds?

12. By observing flower color in snapdragons, is it possible to unambiguously determine the genotype? Is the same true for flower color in peas? Why or why not?

13. If the alleles for snapdragon flower color were codominant instead of incompletely dominant, a possible phenotype for a heterozygous individual would be:
   a. pure red flowers
   b. pure pink flowers
   c. pure white flowers
   d. a mosaic pattern of red and white regions on flowers

14. If two individuals with AB blood type marry, what proportion of their children are expected to have AB blood type?
   a. none
   b. 0.25
   c. 0.50
   d. 0.75
   e. all

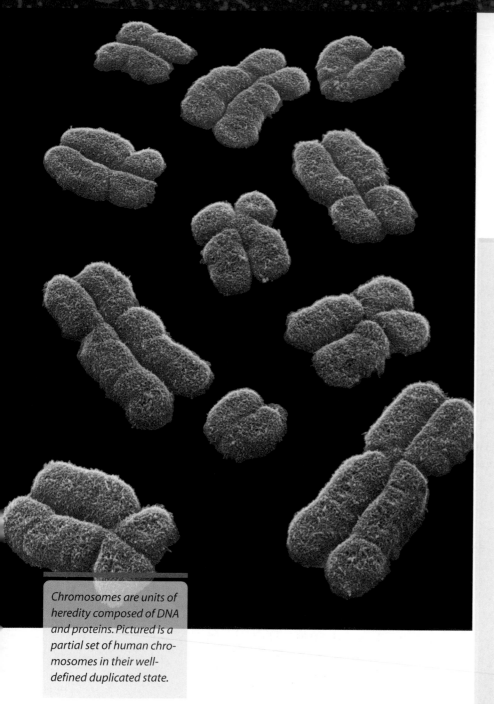

*Chromosomes are units of heredity composed of DNA and proteins. Pictured is a partial set of human chromosomes in their well-defined duplicated state.*

CHAPTER **12**

| 12.1 | X-Linked Inheritance in Humans | 216 |
| 12.2 | Autosomal Genetic Disorders | 217 |
| 12.3 | Tracking Traits with Pedigrees | 220 |
| 12.4 | Aberrations in Chromosomal Sets: Polyploidy | 220 |
| 12.5 | Incorrect Chromosome Number: Aneuploidy | 222 |
| 12.6 | Structural Aberrations in Chromosomes | 225 |

**Essays**

PGD: Screening for a Healthy Child     226

**THE PROCESS OF SCIENCE:** Thomas Hunt Morgan: Using Fruit Flies to Look More Deeply into Genetics     228

Chromosomes are critical players in the process by which living things pass on traits. Some of our worst diseases result from chromosomes that have failed to function properly.

# UNITS OF HEREDITY:
# Chromosomes and Inheritance

The common condition known as Down syndrome is almost always caused by the inheritance of an extra chromosome. But how can a parent who does not have an extra chromosome pass one on to a son or daughter? The gene that causes sickle-cell anemia also seems to offer some protection against malaria. How can it have both effects? The chromosomal makeup of women offers them a kind of protection against being color-blind. Why should this be so?

From these questions, it may be apparent that chromosomes and the genes on them are centrally involved in issues of human health. You saw last chapter that chromosomal activity lies at the root of several of the genetic principles that Gregor Mendel uncovered. Mendel's "pairs of elements," which separate in reproduction, had as their physical basis the pairs of homologous chromosomes—one inherited from the male, one from the female—that first pair up and then separate from one another in meiosis. The differing traits that Mendel investigated so thoroughly, such as yellow or green peas in a pod, had as their physical basis the pairs of *alleles* or differing forms of a gene that lie on homologous chromosomes.

But, there are aspects of chromosomal functioning that Mendel's work scarcely touched on. For one, there is the issue of sex chromosomes. As noted in Chapter 10, the sex of human beings is determined by their chromosomal makeup. Human females have two X chromosomes, while human males have one X and one Y chromosome. Apart from determining sex, what part do these special chromosomes play in

heredity? Next, how do the terms *dominant* and *recessive* that you looked at last chapter play out with respect to human disease? And how can genetically based diseases be traced through generations of a family? Finally, you saw in Chapter 10 that a critical step in meiosis is the *crossing over* (or recombination) that occurs early in it. In this process, homologous chromosomes first intertwine and then swap pieces of themselves, after which they separate in meiosis. But what happens when chromosomes fail to properly carry out this part swapping or separation? The short answer is that such mistakes are responsible for a number of the afflictions that affect human beings, including Down syndrome.

So, in this chapter you will look at sex chromosomes, dominant and recessive disease genes, tracing diseases through generations, and diseases caused by chromosomal mistakes. All the disorders that will be reviewed are transmitted through so-called Mendelian inheritance, meaning they are all disorders in which a single gene or chromosome means the difference between sickness and health. In contrast, many widespread diseases, such as heart disease and cancer, are polygenic—they are based on many genes working in concert, with a given set of genes only predisposing a person toward sickness or health, rather than ensuring it. Let's start now with sex and heredity.

Web Animation 12.1
X-Linked Recessive Traits

# 12.1 X-Linked Inheritance in Humans

Minor cuts are simply an irritation to most of us, but for people suffering from the condition known as hemophilia, such cuts can be life threatening. Hemophilia is a failure of blood to clot properly. A group of proteins interact to make blood clot, but about 80 percent of hemophiliacs lack a functioning version of just one of these proteins (called factor VIII). Genes contain the information for the production of proteins, of course, so at root hemophilia is a genetic disease—a faulty gene is causing the faulty factor VIII protein. This same process is at work in a couple of other well-known afflictions: the disease called Duchenne muscular dystrophy and a far less serious condition, red-green color blindness. These disorders have more in common than being genetically based, however. All of them are known as *X-linked* disorders because the genes that cause them lie on the X chromosome. As it turns out, X-linked disorders claim more male victims than female. Why should this be so? A look at color blindness can provide the answer.

### The Genetics of Color Vision

Despite its name, "color blindness" almost never means a *total* inability to perceive color. Rather, it most often means an inability to distinguish among certain shades of red and green (**Figure 12.1**). To understand how color vision can malfunction in this way, it's important to understand how it commonly works. When light enters our eyes, it travels to the back of them and lands on a group of cells (called cones) that have within them molecules called pigments. A pigment is simply a substance that strongly absorbs a given color of light. The pigments in cone cells turn out to be molecules composed of two main parts, one of which is a protein called opsin. Because genes code for proteins, it's easy to see how color blindness is genetically based: We have genes that are coding for the protein portions of our light-absorbing pigments. More specifically, we have a gene that codes for blue-absorbing pigment, and it lies on chromosome 7; then we have genes that code for both red- and green-absorbing pigments, and they lie very close to one another on the X chromosome. A red-green color-blind person, then, is one in whom these X-located genes fail to code for the proper pigment proteins.

### Alleles and Recessive Disorders

In the normal case, as you've seen, a given gene comes in two variant forms or *alleles* in each person. One allele for a gene might reside on, say, the chromosome 1 we inherit from our mother; if so, the second allele for this gene will reside on the chromosome 1 we inherit from our father. There is therefore a kind of backup system in human genetics: Genes come in pairs of alleles that lie on separate, homologous chromosomes. If one allele for a given trait is defective, there will usually exist a second, functional allele for that same trait on a homologous chromosome.

This genetic structure usually works fine when a condition is a so-called **recessive disorder**, meaning a genetic disorder that will not exist in the presence of a functional allele. In such a condition, a person does not have to have *two* functional alleles; a single "good" allele will do. Red-green color blindness turns out to be a recessive disorder: A person who has even one functional allele for red pigment and one functional allele for green pigment will not be color-blind. So, why do men suffer more from this condition than women? Because men have no backup: They have only one allele for red pigment and one allele for green pigment because they have only one X chromosome (and one Y). Meanwhile, women have two X chromosomes. If the pigment alleles are defective on one of their chromosomes, the alleles on their other X chromosome will still provide them with full color vision.

Given this, think about the interesting way that color blindness is passed along. Say there is a mother who is not color-blind herself, but who is heterozygous for the trait. That is, she has functional color alleles on one of her X chromosomes but dysfunctional alleles on her other X chromosome. Should her son happen to inherit this second X chromosome, he will be color-blind because the X chromosome

**Figure 12.1**
**Typical Test for Red-Green Color Blindness**
If you do not see a number inside the large circle, you may have this recessive trait.

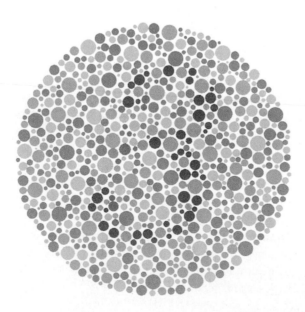

he got from his mother is his *only* X chromosome (**Figure 12.2**). Meanwhile, a daughter who inherited this chromosome would likely be protected by her second X chromosome—the one she got from her father. Not surprisingly, then, more males are color-blind than females. About 8 percent of the male population has some degree of color blindness, while for females the figure is about half a percent. Women enjoy a similar protection in connection with other X-linked disorders, such as hemophilia and muscular dystrophy. (For more on the way hemophilia can be passed down through generations, see Figure 12.6 on page 221.)

## 12.2 Autosomal Genetic Disorders

The X and Y chromosomes are, of course, only two of the chromosomes in the human collection, and any chromosome can have a malfunctioning gene on it. The chromosomes other than the X and Y chromosomes are called *autosomes*. A recessive dysfunction related to an autosome is thus known as an **autosomal recessive disorder**. (For a list of all the disorders you'll be looking at and more, see **Table 12.1**.)

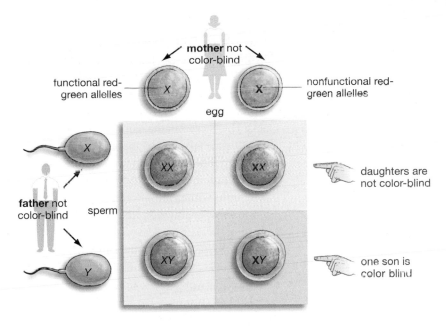

**Figure 12.2**
**The Value of Having Two X Chromosomes**

The mother in the figure is not herself color-blind but has one dysfunctional set of red-green alleles (represented by the red X), which she passes on to one son and one daughter. The daughter is not color-blind because she has inherited, from her father, a second X chromosome—one that has functional red-green alleles. Meanwhile, the son is color-blind because his only X chromosome is the flawed one he inherited from his mother. (The other son, shown in the square's lower-left cell, is not color-blind because he has inherited his mother's functional alleles.)

Web Animation 12.2
Some Human Genetic Disorders

**Table 12.1  Selected Examples of Human Genetic Disorders**

| Type | Name of condition | Effects |
|---|---|---|
| X-linked recessive disorders | Hemophilia | Faulty blood clotting |
| | Duchenne muscular dystrophy | Wasting of muscles |
| | Red-green color blindness | Inability to distinguish shades of red from green |
| Autosomal recessive disorders | Albinism | No pigmentation in skin |
| | Sickle-cell anemia | Decreased oxygen to brain and muscles |
| | Cystic fibrosis | Impaired lung function, lung infections |
| | Phenylketonuria | Mental retardation |
| | Tay-Sachs disease | Nervous system degeneration in infants |
| | Werner syndrome | Premature aging |
| Autosomal dominant disorders | Huntington disease | Brain tissue degeneration |
| | Marfan syndrome | Ruptured blood vessels |
| | Polydactyly | Extra fingers or toes |
| Aberrations in chromosome number | Down syndrome | Mental retardation, shortened life span |
| | Turner syndrome | Sterility, short stature |
| | Klinefelter syndrome | Dysfunctional testicles, feminized features |
| Aberrations in chromosome structure | Cri-du-chat syndrome | Mental retardation, malformed larynx |
| | Fragile-X syndrome | Mental retardation, facial deformities |

## Sickle-Cell Anemia

A well-known example of an autosomal recessive disorder is sickle-cell anemia, which affects populations derived from several areas on the globe, including Africa. In the United States it is, of course, most widely known as a disease affecting African Americans. The "sickle" in the name comes from the curved shape that is taken on by the red blood cells of its victims. Red blood cells carry oxygen to all parts of the body; in their normal shape, they look a little like doughnuts with incomplete holes. When they take on a sickle shape, however, red cells clog up capillaries, resulting in decreased oxygen supplies to brain and muscle (**Figure 12.3**). In the United States, the average life expectancy for men with the condition is 42 years; for women, it is 48 years.

The question is, what causes a red blood cell to take on this lethal, sickled shape? There is a protein, called hemoglobin, that carries the oxygen within red blood cells. The vast majority of people in the world have one form of hemoglobin, called hemoglobin A, but sickle-cell anemia sufferers have another form of this protein, hemoglobin S, which coalesces into crystals that distort the cell.

Now, how is sickle-cell anemia passed from parents to offspring? Because it is a recessive condition, any child who gets it must inherit *two* alleles for it—one allele from the mother, one from the father, each of them coding for the hemoglobin S protein. This means, of course, that both parents must themselves have at least one hemoglobin S allele; both parents must be at least heterozygous for the trait. In this situation, the laws governing inheritance of this disorder are simply those of Mendel's monohybrid cross: A child of these two parents has a 25 percent chance of inheriting the disorder. You can look at the Punnett square in **Figure 12.4a** to see how this works out.

It's important to note the status of the parents in this example. They are heterozygous for a recessive debilitation, meaning they do not suffer from the condition themselves because they are protected by their single "good" allele. Each of them is, however, a **carrier** for the condition: a person who does not suffer from a recessive genetic debilitation, but who carries an allele for it that can be passed along to offspring. In this respect, they are just like the mother in the color blindness example.

## Malaria Protection: Hemoglobin S as a Useful Protein

Sickle-cell anemia offers another lesson in connection with the notion of hemoglobin S as a "faulty" protein. It is in fact quite a functional protein in certain circumstances, namely, when it appears heterozygously (with hemoglobin A) in people who live in regions where malaria is common. The most severe form of malaria is caused by a single-celled parasite that is transmitted into human beings by the *Anopheles* mosquito. Traveling through the bloodstream, these parasites invade and ultimately destroy red blood cells. For reasons we still don't understand, having hemoglobin S in one's system makes red blood cells resistant to invasion. Thus, it is likely that hemoglobin S became as widespread as it is because of its value in resisting malaria. The price of this resistance, however, is that some offspring are not heterozygous for hemoglobin S. They are homozygous for it, meaning they have sickle-cell anemia.

## Dominant Disorders

Although sickle-cell anemia is an autosomal disorder and red-green color blindness an X-linked disorder, both are recessive disorders. That is, a person with even a single properly functioning allele will not suffer from them. However, there are also **dominant disorders**: genetic conditions in which a single faulty allele can cause damage, even when a second, functional allele exists. This leads to the concept of an **autosomal dominant disorder**, simply meaning a dominant genetic disorder caused by a faulty allele that lies on an autosomal chromosome. There is, for example, an autosomal dominant disorder called Huntington disease. Affecting about 30,000 Americans, it results in both mental impairment and uncontrollable spastic movements called *chorea*.

**Figure 12.3**
**Healthy Cells, Sickled Cell**
Three normal red blood cells flank a single sickled cell. (×7400)

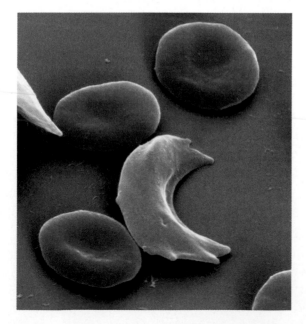

**Figure 12.4**
**Transmission of Recessive and Dominant Disorders**

**(a)** Sickle-cell anemia: transmission of a recessive disorder.

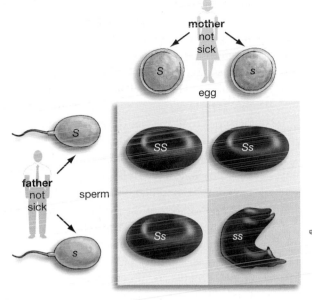

Sickle-cell anemia is a recessive autosomal disorder; both the mother and father must carry at least one allele for the trait in order for a son or a daughter to be a sickle-cell victim. When both parents have one sickle-cell allele, there is a 25 percent chance that any given offspring will inherit the condition.

👉 25% probability of inheriting the disorder

**(b)** Huntington disease: transmission of a dominant disorder.

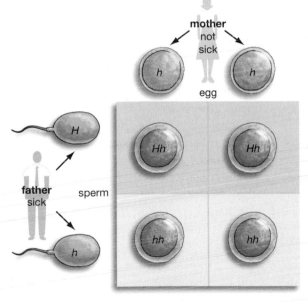

👉 50% probability of inheriting the disorder

In Huntington disease, if only a single parent has a Huntington allele there is a 50 percent chance that a son or daughter will inherit the condition.

Perhaps its most famous victim was the American folksinger Woody Guthrie. Like most Huntington sufferers, Guthrie did not begin to show symptoms of the disease until well into adulthood—*after* he had children, who then may have had the disease passed along to them. You can look at the Punnett square in **Figure 12.4b** to see how the inheritance pattern of an autosomal dominant disease such as Huntington differs from that of an autosomal recessive illness. Because a parent need only pass on a single Huntington allele for a son or daughter to suffer from the condition, the chances of any given offspring getting the disease from a single affected parent are one out of two.

**SO FAR...**

1. A recessive genetic disorder is one that will not exist in the presence of a _____, while a dominant disorder is one in which a _____ can cause trouble, even in the presence of a _____.

2. True or false: A woman who has one X chromosome containing dysfunctional alleles for red-green color vision is not color-blind.

3. To inherit an autosomal recessive disorder, both parents of an offspring must be at least _____ for the condition, meaning each parent must have at least _____.

•••ANSWERS

1. functional allele; faulty allele; functional allele

2. True; her second X chromosome contains functional alleles, which are enough to keep her from being color-blind.

3. heterozygous; one faulty allele

## 12.3 Tracking Traits with Pedigrees

Confronted with a medical condition, such as Huntington, that is running through a family, scientists find it helpful to construct a medical **pedigree**, defined as a familial history intended to track genetic conditions. Normally set forth as diagrams, medical pedigrees do more than give a family history of disease. They can be used to ascertain whether a condition is dominant or recessive, and X-linked or autosomal. This can help establish probabilities for *future* inheritance of the condition—something that can be very helpful for couples thinking of having a child.

If you look at **Figure 12.5**, you can see a simple pedigree for albinism: a lack of skin pigmentation, which is known to be an autosomal recessive condition. In the figure, you can see some of the standard symbols used in pedigrees. A circle is used for a female and a square for a male. Parents are indicated by a horizontal line connecting a male and a female, while a vertical line between the parents leads to a lower row that denotes the parents' offspring. This second row—a horizontal line of siblings—has an order to it: oldest child on the left, youngest on the right. A circle or square that is filled in indicates a family member who has the condition (in this case, albinism), while a symbol that is half filled in indicates a person known to be heterozygous for a recessive condition (a carrier). One of the strengths of a pedigree is that it can sometimes tell researchers which persons are heterozygous carriers of a recessive condition. Remember that a carrier does not display *symptoms* of the condition. So, how can a pedigree reveal this?

Take a look at Figure 12.5. The only thing that would be apparent from actually seeing any of the people in the pedigree is that two of them—a female in generation II and a male in generation III—have the condition of albinism. But a little knowledge of Mendelian genetics also allows some other deductions. If the condition had been dominant, then it would have manifested itself in at least one of the parents on the left in generation I. Because this was not the case, it is fair to deduce that this is a recessive condition. The fact that it is recessive, however, means that *both* parents in generation I had to be carriers for the allele—both have to be heterozygous for the condition, which is why their symbols can be half shaded in. Things are less clear with the parents on the right in generation I. One of their sons did not manifest the condition himself but went on to have a son who did. From this, we know that the son in generation II had to have been heterozygous for the condition. But, from which parent did this son get his albinism allele? Either, or both, of his parents in generation I could have been heterozygous for the condition and yet between them, only have passed along a single albinism allele.

This mixture of certainty and uncertainty leads to the genotype labeling you see in the figure, with *A* representing the dominant "normal" allele and lowercase *a* representing the recessive albinism allele. We know, for example, that both parents on the left in generation I had to be *Aa*, but all we can say for sure about the parents on the right in the same generation is that at least one of them was a carrier. As you can imagine, pedigrees can get much more elaborate than this one. To see how a trait can be tracked through many generations, look at **Figure 12.6**.

## 12.4 Aberrations in Chromosomal Sets: Polyploidy

All of the maladies you've looked at so far have resulted from dysfunctional genes that exist among a standard array of chromosomes. In humans, this array is 22 pairs of autosomes and either an XX combination (in females) or an XY (in males). Meanwhile, Mendel's peas had seven pairs of chromosomes, and the common fruit fly has four pairs. Whatever the *number* of chromosomes, note the similarity in all these species: Their chromosomes come in pairs, or to put it another way, they all have two *sets* of chromosomes, meaning the organisms that posses them are diploid.

In many situations, however, organisms don't end up with the two sets of chromosomes that are standard for their species. This is called **polyploidy**, a

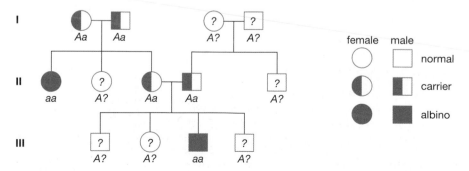

**Figure 12.5**
**A Hypothetical Pedigree for Albinism through Three Generations**

# Three Lessons from a Royal Pedigree

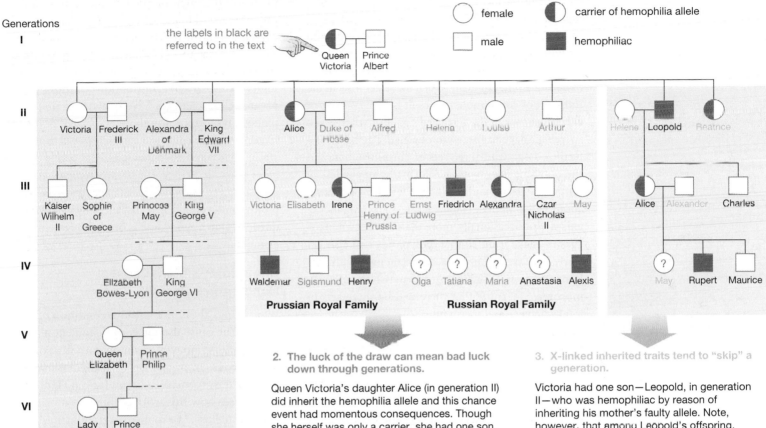

Generations

the labels in black are referred to in the text

○ female    ◐ carrier of hemophilia allele

□ male    ■ hemophiliac

**British Royal Family**

**Prussian Royal Family**    **Russian Royal Family**

## 1. The luck of the draw can mean good luck down through generations.

The hemophilia allele almost certainly arose as a mutation in either Queen Victoria or in one of her parents. Whichever the case, the Queen was undoubtedly a carrier for the allele. For Britain's royal family, the key event was that King Edward VII (toward the left in generation II) did not inherit the hemophilia allele from his mother, but instead inherited her second "good" allele. With this, the British Royal House—right down through today's princes William and Harry—avoided hemophilia, barring its introduction through marriage. Note that Edward was the second of Victoria's nine children.

## 2. The luck of the draw can mean bad luck down through generations.

Queen Victoria's daughter Alice (in generation II) did inherit the hemophilia allele and this chance event had momentous consequences. Though she herself was only a carrier, she had one son, Friedrich, who was hemophiliac (dying at age 3); and two daughters who, as carriers, went on to pass the allele along. One of these daughters, Alexandra, became the czarina of Russia after marrying Czar Nicholas II. The couple had a male child, Alexis, who inherited his mother's hemophilia allele—now passed down through at least three generations—and was thus a hemophiliac. So precarious was Alexis' condition that his mother turned to a semiliterate monk, Rasputin, as a means of trying to care for him. Anger with Rasputin among the Russian populace may have played a part in bringing down the Russian Royal House in the revolutions of 1917. Note that it is unclear whether Alexis' sisters—among them the famous Anastasia—were carriers for the hemophilia allele. We will never know because they didn't live long enough to have children. Along with their parents and brother, they were murdered by communist revolutionaries in 1918.

## 3. X-linked inherited traits tend to "skip" a generation.

Victoria had one son—Leopold, in generation II—who was hemophiliac by reason of inheriting his mother's faulty allele. Note, however, that among Leopold's offspring, there are no hemophiliacs. The next hemophiliac is found one generation further down, in Rupert, in generation IV. Why should things work this way? Remember that the gene for hemophilia is on the X chromosome and that every male must get his lone X chromosome from his mother. Thus, Leopold's son, Charles, was not at risk of being hemophiliac because Charles' single X chromosome came from his mother. However, Leopold's daughter, Alice, carried two X chromosomes—one "good" chromosome from her mother and the faulty X chromosome from Leopold. She then passed on Leopold's chromosome to one of her sons (Rupert), though not to the other (Maurice). The luck of the draw was on display again.

## Figure 12.6
### A Pedigree for Hemophilia in Europe's Royal Families

Hemophilia is a rare, recessive genetic disorder, but it managed to affect a large portion of Europe's royal family members in the nineteenth and twentieth centuries after being passed on to them from England's Queen Victoria. As the main text notes, hemophilia is nearly always caused by a particular variant form of a gene (an allele) that leads to a faulty form of a blood-clotting protein (called factor VIII). Since the gene for factor VIII resides on the X chromosome, hemophilia is a condition that is much more likely to affect men than women. The pedigree shown here for hemophilia among Europe's royal families has been greatly simplified; in most of the British royal family, for example, only the heirs to the throne and their spouses are shown.

condition in which one or more entire sets of chromosomes has been added to the genome of a diploid organism.

You may wonder how a human being could end up with more than two sets of chromosomes. After all, in meiosis, sperm and egg end up with just one set of chromosomes in them. When they later fuse, in the moment of conception, the result is then two sets of chromosomes in the fertilized egg. But what if *two* sperm manage to fuse with one egg? The result is a fertilized egg with three sets of chromosomes, and the outcome is polyploidy. Normally, an egg lets in only one sperm, and it blocks the others hovering about with a kind of chemical gate-slamming mechanism. But occasionally, more than one sperm manages to make it in. Human polyploidy can also get started before fertilization, when meiosis malfunctions and gives a single egg or sperm two sets of chromosomes, meaning that there will be three sets in total when egg and sperm fuse.

Polyploidy actually is tolerated well in some organisms. Indeed, it has come about countless times in plant species and generally gives rise to perfectly robust plants. In human beings, however, polyploidy is a disaster. Perhaps only 1 percent of human embryos with the condition will survive to birth, and none of these babies lives long. Concerns about chromosome number in living persons, then, center not on addition or deletion of whole sets of chromosomes, but rather on the gain or loss of *individual* chromosomes.

## 12.5 Incorrect Chromosome Number: Aneuploidy

The condition called **aneuploidy** is one in which an organism has either more or fewer chromosomes than normally exist in its species' full set. This usually means that an organism has gained or lost a *single* chromosome, although more complex patterns are possible, as you'll see. Among human genetic malfunctions, aneuploidy is unusual in that it occurs quite commonly and yet goes largely unrecognized. This is so because it most often occurs not in fully formed human beings, but in embryos. A would-be mother may know only that she is having a hard time getting pregnant. What she may not know is that she actually *has* been pregnant—perhaps several times—but that aneuploidy has doomed the embryo in each case. Doomed it in what way? The most common outcome of aneuploidy is a pregnancy that ends in miscarriage. Many of these miscarriages happen so early in a

pregnancy that a woman never realizes she was pregnant; the microscopic embryo may simply be part of material that is discharged, perhaps with the woman's next menstrual period. Even when a miscarriage is recognized as such, tests can seldom be run that can identify aneuploidy as the cause. But by undertaking careful examinations of thousands of pregnancies, scientists have concluded that aneuploidy is responsible for about 30 percent of miscarriages. And how common is miscarriage? In the best study conducted to date, scientists found that it takes place in 31 percent of pregnancies. A small proportion of embryos do manage to survive aneuploidy, but even in these instances there are consequences. Down syndrome is one of the outcomes of aneuploidy.

### Aneuploidy's Main Cause: Nondisjunction

What brings about the harmful condition of aneuploidy? The cause usually is a phenomenon known as **nondisjunction**, which simply means a failure of homologous chromosomes or sister chromatids to separate during meiosis. If you look at **Figure 12.7**, you can see how this works. Note that it can occur either in meiosis I or in meiosis II. In meiosis I, two homologous chromosomes, on the opposite side of the metaphase plate, can be pulled to the *same* side of a dividing cell, producing daughter cells with imbalanced numbers of chromosomes. Conversely, nondisjunction can take place in meiosis II by means of sister chromatids going to the same daughter cell after failing to "disjoin" (hence the cumbersome term *nondisjunction*). Such actions then produce an egg or sperm that has 24 or 22 chromosomes instead of the standard haploid number of 23; after union with a normal egg or sperm, this results in an embryo that has either 47 or 45 chromosomes instead of the standard 46.

The effect of this is that the embryo has a genetic imbalance—one that perhaps can be made clear by an analogy. Imagine that while baking chocolate chip cookies, you decided to increase the amount of flour called for in the recipe. This change might be of little consequence if you increased all the other ingredients in the same proportion. If, however, you increased nothing but the flour, the final product would be hard and flavorless because the ingredients in it would be unbalanced. An added chromosome has this same effect: It unbalances the proportions of *biological* ingredients (the proteins) that help produce and maintain a living thing.

## SO FAR...

1. A medical pedigree is used to identify _____ disorders that are passed along in _____.

2. Aneuploidy exists when an organism has more or fewer _____ than normally exist in its species' _____.

3. The most common cause of aneuploidy is a failure of homologous chromosomes or sister chromatids to _____ in meiosis, which leads to one daughter cell having _____, while the other daughter cell has _____.

### The Consequences of Aneuploidy

As noted, embryos with the wrong number of chromosomes are not likely to survive. Setting aside aneuploidies that affect the X and Y chromosomes, *all* aneuploidies result in miscarriage except for those that give an embryo an additional chromosome 13, 18, or 21. It is the gain of an additional chromosome 21 that causes the most well-known outcome of aneuploidy in human beings. **Down syndrome** is, in some 95 percent of cases, a condition in which a person has three copies of chromosome 21 rather than the standard two. Ninety percent of these cases come about because of nondisjunction in egg formation, with the remaining 10 percent caused by nondisjunction in sperm formation. Seen in about 0.1 percent of all live births, Down syndrome results in an array of effects: smallish, oval heads; IQs that are well below normal; infertility in males; short stature and reduced life span in both sexes.

It is well known that when women pass the age of about 35, their risk of giving birth to a Down syndrome child increases dramatically. It seems less well known that even at maternal age 40, the odds of conceiving such a child are less than 1 in 100 (**Figure 12.8**, on the next page). Scientists are not certain why the mother's age should figure so prominently in Down syndrome, although several hypotheses have been put forward to account for this. Most scientists agree that one piece of the puzzle is something you saw in Chapter 10: In egg development, the process of meiosis generally stretches on for decades. It may be that any given egg that develops in a woman starts meiosis I before the woman is born, but may not complete this meiosis until she is, say, in her thirties. In the intervening decades, this egg's cellular machinery has aged, perhaps to the point that it can no longer separate its chromosomes or chromatids properly.

### Abnormal Numbers of Sex Chromosomes

Aneuploidy can affect not only autosomes but sex chromosomes as well, meaning the two X chromosomes that females have and the X and Y chromosomes that

**... ANSWERS**

1. genetic; families

2. chromosomes; full set

3. separate; one chromosome too many; one too few

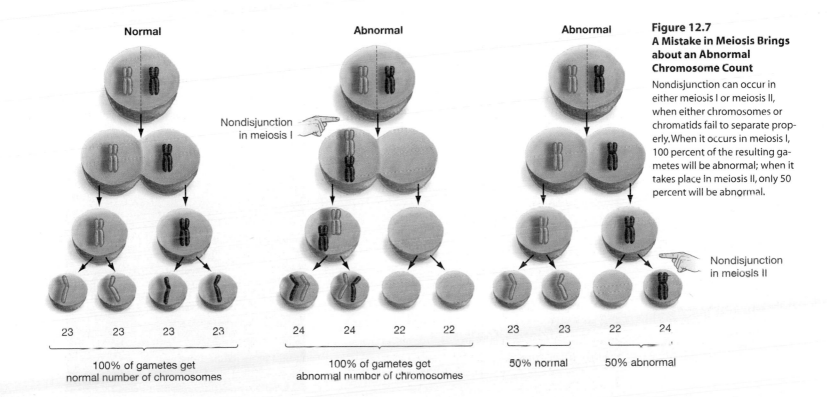

**Figure 12.7**
**A Mistake in Meiosis Brings about an Abnormal Chromosome Count**

Nondisjunction can occur in either meiosis I or meiosis II, when either chromosomes or chromatids fail to separate properly. When it occurs in meiosis I, 100 percent of the resulting gametes will be abnormal; when it takes place in meiosis II, only 50 percent will be abnormal.

males have. Embryos often survive sex chromosome aneuploidies, but the result, once again, is usually debilitations. An example is Turner syndrome, which produces people who are phenotypically female, but who have only one X chromosome. Such females, then, have only 45 chromosomes, rather than the usual 46. Their state is sometimes referred to as XO, the "O" signifying the missing X chromosome. The absence of a second sex chromosome causes a range of afflictions. Females with Turner syndrome have ovaries that don't develop properly (which causes sterility), they are generally short, and they often have brown spots (called nevi) over their bodies.

Turner syndrome results from a loss of a chromosome, but all manner of sex chromosome *additions* can also take place. There are, for example, XXY men, who while phenotypically male in most respects, tend to have a number of feminine features: some breast development, a more feminine figure, and lack of facial hair. When coupled with other characteristics (such as tall stature and dysfunctional testicles), the result is a condition called Klinefelter syndrome.

Beginning in the late 1960s, it became possible to test developing fetuses for genetic abnormalities. In recent years, however, reproductive technology has moved beyond testing. Prospective parents can now choose, from a group of embryos they have created, embryos that do not have any recognizable genetic defects. You can read more about this in "PGD: Screening for a Healthy Child" on page 226.

### Aneuploidy and Cancer

We've thus far looked at varying conditions that can result when aneuploidy takes place in meiosis, which is to say the cell divisions that produce eggs or sperm. But it's also possible for aneuploidy to take place in *mitosis*, meaning regular cell division in nonsex or "somatic" cells. As one cell divides into two, a given chromatid can fail to migrate to its proper "pole," and one resulting daughter cell ends up with one chromosome too many, while the other ends up with one chromosome too few.

Now, note that this kind of aneuploid event might take place in a child, a teenager, or an adult. If so, the only cells affected would be the "line" of cells that stem from the original two cells that underwent aneuploidy. Put another way, not every cell in the body would be affected, as is the case with aneuploidy that comes about in meiosis; only the daughter cells of the original aneuploid cells would have the wrong number of chromosomes. You might think: Well, this is bound to be less serious than aneuploidy resulting from meiosis; and in general, this is the case. Most aneuploid cells resulting from mitosis will die, but a few can survive. Indeed, aneuploidy is often seen in a destructive kind of "supercell" found in the body, the cancer cell.

Researchers have known about aneuploidy in cancer cells for decades. Today, it's clear that almost all cancer cells have undergone aneuploidy—that is, almost all cancerous cells have the wrong number of chromosomes in them. But for many years, it was assumed that such aneuploidies were the *result* of a cancer in progress rather than the *cause* of it. In recent years, however, this idea has been turned on its head. Some prominent cancer researchers believe that aneuploidy is a cause of cancer. To be sure, this idea differs from the leading hypothesis about cancer: that it results from a series of mutations to individual genes on chromosomes. (For an account of this view, see "When the Cell Cycle Runs Amok: Cancer," on page 172.) The problem with this idea is that, despite decades of research, no one has yet identified a series of mutations that will predictably cause any of the most serious forms of cancer. Meanwhile, it has been demonstrated that, at the least, aneuploidy can occur before the very earliest stages of some forms of cancer.

**Figure 12.8**
**Down Syndrome**

**(a)** The risk of giving birth to a Down syndrome child increases dramatically past maternal age 35.

**(b)** A karyotype is a visual display of an entire set of chromosomes. Pictured is a karyotype of a person with Down syndrome. Note the three chromosomes, stained blue-green, above the number 21. The karyotype also shows that this person is a female, as indicated by the two X chromosomes.

**(a)** Maternal age and Down syndrome risk

| Mother's age | Chances of giving birth to a child with Down syndrome |
| --- | --- |
| 20 | 1 in 1925 |
| 25 | 1 in 1205 |
| 30 | 1 in 885 |
| 35 | 1 in 365 |
| 40 | 1 in 110 |
| 45 | 1 in 32 |

**(b)** Karyotype of a person with Down syndrome

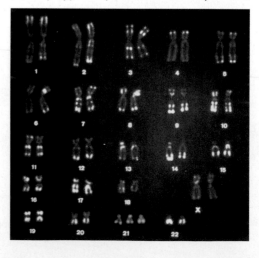

The question is whether it is the triggering event in these cancers (**Figure 12.9**). Only more research will answer this question. For now, don't be surprised if you start seeing newspaper stories about aneuploidy and cancer. We're certain that the cellular mistake of aneuploidy has devastating consequences every day. The question is whether cancer is one of them.

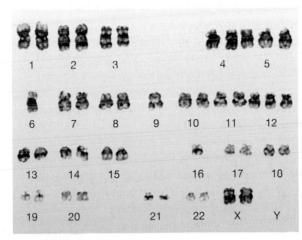

**Figure 12.9**
**Aneuploidy before Cancer**

In this karyotype of a cell taken from a young girl, aneuploidy can be seen to exist in a complex form. Note that the cell has the standard two copies of chromosomes 1 through 4, but then three copies of chromosome 5, only one copy of chromosome 6, and so on. This aneuploidy was present in the girl during embryonic development, meaning it was present prior to the time she developed a form of childhood cancer (called an embryonal rhabdomyosarcoma). Evidence of this sort has convinced some researchers that aneuploidy can be a cause of cancer. (Karyotype courtesy of Dr. Sandra Hanks and Dr. Nazneen Rahman.)

## 12.6 Structural Aberrations in Chromosomes

Aberrations can occur *within* a given chromosome, sometimes resulting from interactions *between* chromosomes. A frequent cause of such change is that pieces of chromosomes can break off from the main chromosomal body. Such a chromosomal fragment may then be lost to further genetic activity, or it may rejoin a chromosome—either the one it came from or another—often to harmful effect. These structural aberrations may come about spontaneously, but they may also be caused by exposure to such agents as radiation, viruses, and chemicals.

### Deletions

A chromosomal **deletion** occurs when a chromosome fragment breaks off and then does not rejoin any chromosome. Such an event might take place during meiosis in a healthy parent, who then could pass along a complement of 23 chromosomes, one chromosome of which would be missing a segment. If the deleted piece were large enough, the zygote would likely not survive. You can see how critical even a portion of a chromosome is by looking at children who suffer from a rare condition called *cri-du-chat* ("cry of the cat") syndrome, which results from the deletion of the far end of the "short arm" of chromosome 5. If you look at **Figure 12.10**, you can see a chromosome set that shows you this deletion. Children born with the cri-du chat condition exhibit a host

**(a)** Cri-du-chat syndrome

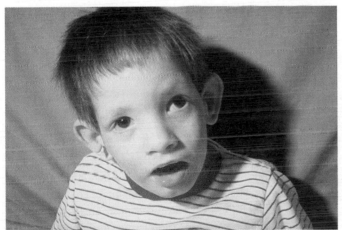

**(b)** Cri-du-chat karyotype

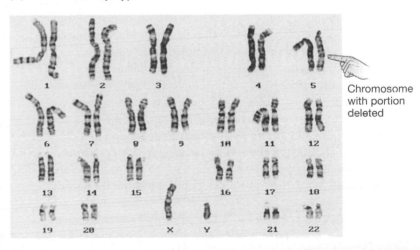

Chromosome with portion deleted

**Figure 12.10**
**Chromosomal Deletion**

**(a)** A five-year-old boy affected by cri-du-chat syndrome, which results from a chromosomal deletion. Note the small head and low-set ears.

**(b)** Karyotype of a person with cri-du-chat syndrome. Note, in chromosome 5, the abnormality in the homologous chromosome on the right. It lacks a section that is present in the left chromosome.

# PGD: Screening for a Healthy Child

"**I**s the baby OK?" In a life full of questions, it's possible that none carries more weight than this one. In most cases, the answer will be reassuring, but in some it will be devastating. Small wonder, then, that scientists constantly are trying to perfect ways to know in advance whether a newborn will be healthy. You have seen that a large number of debilitating human conditions have genetic causes. Further, you know that nearly every kind of cell in a person's body contains a complete copy of that person's genome, or set of genes. Therefore, to check on someone's genetic well-being, all that is necessary is to have access to a small collection of that person's cells. Such cells exist, of course, in embryos as well as in newborn babies and adults. Thus, it's not hard to see what the procedure would be for prenatal genetic testing: Gather embryonic cells and examine their DNA and chromosomes.

This kind of testing has been going on, in ever-more-refined forms, since the late 1960s. From that time until recently, however, most prospective parents who got bad news from a genetic test had two choices: abort the fetus or bring a child with an incurable condition into the world. Now, an ever-growing number of reproductive clinics are providing a third choice. Parents who are willing to undertake so-called in vitro fertilization—fertilization of the mother's eggs in a laboratory—can have each one of the early-stage embryos that results from this process tested for genetic trouble. Of this initial group of embryos, only those found to be genetically healthy are candidates to be inserted back into the mother. The hope is that at least one of them will implant in the mother's uterus and result in a child.

This process, called preimplantation genetic diagnosis or PGD, can be used by couples who have no known genetic risk factors but who are simply having a hard time conceiving a child. As the main text notes, a large proportion of human embryos have genetic defects (such as aneuploidy). Moreover, some couples may be particularly prone to producing such embryos, which generally are washed away in miscarriages. PGD allows couples to screen their embryos for genetic problems before implantation in the uterus, thus greatly upping their chances of bringing a child to term.

PGD can also be used, however, in connection with couples who are *likely* to produce a child with genetic defects. In an extreme example, imagine a prospective father who knows he is fated to fall victim to Huntington disease because he has been found to carry the defective Huntington allele. Since Huntington is a dominant disorder, his odds of passing the disease along to one of his children are one in two. With PGD, only those embryos without the harmful allele would be selected for implantation.

Of course, more conventional means of genetic screening, such as amniocentesis, can provide information about conditions such as Huntington, but there is a difference. In amniocentesis, cells are obtained from an embryo that has been developing in a mother's uterus for a minimum of 14 weeks. By the time the results of amniocentesis are in, more than four months of a nine-month pregnancy may have elapsed. Most prospective parents would make a distinction between aborting a four-month-old fetus on the one hand or not implanting a four-day-old embryo on the other. Of course, not everyone feels this way. For some people, all embryos represent human life; therefore, producing an embryo and then not using it amounts to taking a human life. In PGD, most embryos have two fates: Those that are not implanted are eventually destroyed (although a small number of "embryo adoptions" are now taking place across the country).

Beyond this issue, PGD also raises another ethical question: What limits should be placed on choosing from among embryos? Right now, some clinics allow parents to use PGD to choose the *sex* of their child. If this seems disturbing, consider that, with our expanding knowledge of genetics, it may be possible in the future to select for a tall or muscular child. Prospective parents who would never abort a fetus to get such an outcome might not object to choosing from among embryos to get it, particularly since more embryos must be produced than will be used. Parents may think: Since a selection process is going to take place anyway (for healthy genes), why not select for things like height as well?

*Right now, some clinics allow parents to use PGD to choose the sex of their child.*

Such questions become less abstract with a little knowledge of how PGD works. Many of its steps are those that occur with any in vitro fertilization (IVF). The starting point for IVF is to administer to the prospective mother a group of hormones that stimulate formation of a large group of eggs in her ovaries. (These eggs are the oocytes you looked at in Chapter 10.) Anywhere from a week to two weeks later, a physician uses a needle attached to an ultrasound guide to remove the mature eggs from the ovaries. Although numbers vary, it would not be uncommon for 10 or 12 eggs to be recovered in this process. The husband then provides sperm to fertilize the eggs, with this fertilization taking place in a laboratory.

Three days after fertilization, each egg has divided into an eight-cell embryo. If you look at **Figure 1a**, you can see what comes next: Using a tiny hollow tube called a pipette, a physician gently suctions one of these cells away from the others. (Embryonic cells are so undifferentiated at this point that the loss of any one of them does not interfere with development of the others into a child.) The cell that is removed provides the DNA for testing. If you look at **Figure 1b**, you can see the results of one of the tests. Recall that children born with Down syndrome generally have a third copy of chromosome 21. It is possible to *test* for a third chromosome 21 by preparing DNA "bases" that will latch onto the unique set of DNA bases that exist on chromosome 21. Moreover, these "homing" DNA bases are fluorescent—they light up. The result is what you see in the figure: The prepared bases have latched onto three chromosome 21's in these embryonic cells; they have revealed that, had this embryo been implanted in the mother, the result would have been a child with Down syndrome.

By use of this test and others, embryos produced through IVF are ranked according to their health. Parents then choose from among them, selecting perhaps three or four that will be transferred back into the mother by a physician.

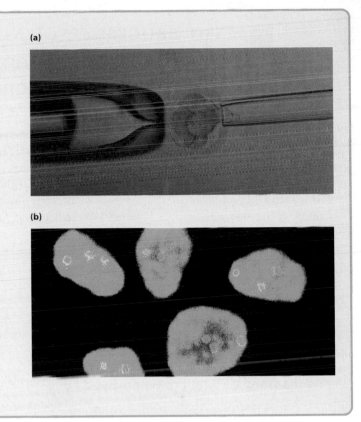

(a)

(b)

**Figure 1**

**(a) Removing a Cell for Testing**

A human embryo at the eight-cell stage (center) being manipulated to have one of its cells removed for preimplantation genetic diagnosis. A pipette at left holds the embryo, while a smaller pipette at right gently suctions off one cell from it. The cell's genetic material is then checked for abnormalities. If this embryo is determined to be normal, it may be inserted back into the mother's uterus, where it can develop into a baby.

**(b) Down Syndrome Diagnosed**

Genetic testing of this embryo has revealed that each of its cells has three copies of chromosome 21, indicated by the red areas in the nuclei of the cells. Such an embryo would develop into a child with Down syndrome. The red areas represent fluorescently labeled DNA that has bound or "hybridized" with regions of DNA specific to chromosome 21.

of maladies, among them mental retardation and an improperly constructed larynx that early in life produces sounds akin to those of a cat.

### Inversions and Translocations

When a chromosome fragment rejoins the chromosome it came from, it may do so with its orientation "flipped," so that the fragment's chemical sequence is out of order. This is an **inversion**: a chromosomal abnormality that comes about when a chromosomal fragment that rejoins a chromosome does so with an inverted orientation (**Figure 12.11**). A **translocation** is a chromosomal abnormality that occurs when two chromosomes that are not homologous exchange pieces, leaving both with improper gene sequences, which can have phenotypic effects.

### Duplications

There is, of course, a perfectly normal process of chromosomal part swapping—crossing over. But, consider what might happen if two homologous chromosomes exchanged *unequal* pieces of themselves in crossing over. One would lose genetic material, while the other would gain it. Because these are homologous (meaning paired) chromosomes, the fragment added to the latter chromosome would

**Figure 12.11**
**Chromosomal Structural Changes**

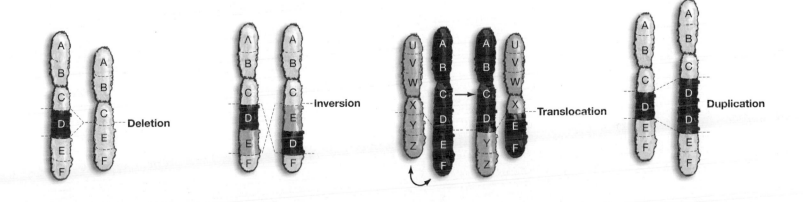

Deletion

Inversion

Translocation

Duplication

# THE PROCESS *of* SCIENCE
## Thomas Hunt Morgan: Using Fruit Flies to Look More Deeply into Genetics

Gregor Mendel used pea plants in his work, but generations of researchers after him have employed a common fruit fly, *Drosophila melanogaster*, in trying to unravel the secrets of genetics. One of the earliest *Drosophila* workers was a Kentuckian named Thomas Hunt Morgan, who greatly deepened our understanding of heredity with work he began in 1908 in his lab at Columbia University.

The *Drosophila* fly has a lot going for it as an experimental subject: It is small, it has unusually large chromosomes, and it has what is known as a short generation time. Whereas Mendel got one generation of pea plants a year, Morgan got one generation of *Drosophila* every 12 days or so. In practical terms, this meant that Morgan's group had to wait at most a month to see the results of their experiments.

## Lessons from a Mutation: White-Eyed Flies

Morgan's path to discovery began with a chance event. One day he looked into one of the empty milk bottles in which he kept his flies and saw something strange: Among numerous normal or "wild-type" *Drosophila* with red eyes was a single male with white eyes (**Figure 1**). Morgan correctly assumed that the variation he saw had come about because of a spontaneous genetic change or "mutation" in the fly. He then crossed his white-eyed male with females that he knew bred true for red eyes and got all red-eyed flies in the F₁ generation. No surprise

there; it simply looked as though red eyes were the same type of dominant trait that Mendel had seen in, for example, his yellow peas. Indeed, when Morgan went on to breed the F₁ generation—those that had mixed red and white alleles—he got a 3:1 red-to-white ratio in the F₂ generation. The strange thing was that every white-eyed fly was a *male*. Why should that be?

> *The importance of Morgan's insight was that he had linked a particular trait to a particular chromosome, which was a first.*

As it turns out, sex chromosomes exist in *Drosophila* as they do in human beings: Females have two X chromosomes, while males have one X and one Y. Doing further experiments on these flies, Morgan eventually decided that the gene for eye color had to lie on a *particular* chromosome: the X chromosome. Why? Any fly that was white-eyed had to have only white-eyed alleles (because white was recessive). Normally, of course, this would mean *two* white-eyed alleles, but there was an exception: Males could be white-eyed if they had but a single white allele—because they have only a *single X chromosome*. You can look at **Figure 2** to see how this plays out schematically. The importance of Morgan's insight was not that he had grasped something about eye color in the fly. It was that he had linked a particular trait to a particular chromosome, which was a first.

## When Genes Separate: Recombination

Continuing with *Drosophila* breeding work, Morgan and his colleagues soon were finding all kinds of mutations in the flies. There was, for example, a mutation for "miniature" wings, as opposed to full ones. This trait followed the rules described earlier for the white-eyed mutation and accordingly was deemed to be an X-linked characteristic. Not surprisingly, the white-eyed and miniature-winged mutations tended to be transmitted *together* when both of these mutations were bred for because the genes for both of them lay on the same chromosome. The question was, though, did genes on the same chromosome *always* travel together? Surprisingly, the answer was no. For example, miniature wings were usually passed

**(a)** Eye of red-eyed *Drosophila*   **(b)** Eye of white-eyed *Drosophila*

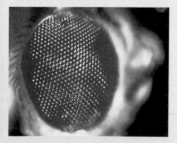 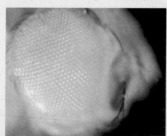

**Figure 1**
**The Mutation Morgan Saw**

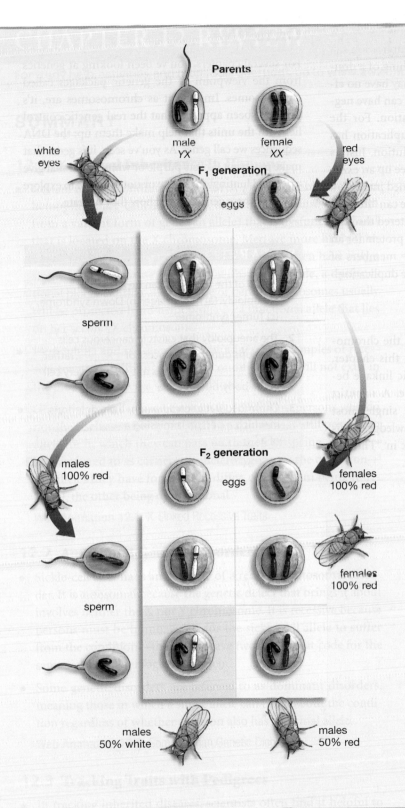

Parents

white eyes

male
YX

female
XX

red eyes

F₁ generation

eggs

sperm

males
100% red

F₂ generation

eggs

females
100% red

sperm

females
100% red

males
50% white

males
50% red

on together with white eyes, but not always. Sometimes, one trait would appear in an offspring, but the other would not. Since genes for these traits were on one chromosome, how did they get separated from one another?

A paper published in Europe in 1909 gave Morgan a clue. Viewing cells under a microscope, F. A. Janssens had observed chromosomes entwining with each other during an early phase of meiosis. Given this and his own data, Morgan made a creative leap. He suggested that chromosomes can swap parts with each other during their meiotic intertwining. Sound at all familiar? Remember, in the Chapter 10 review of meiosis, the discussion of the phenomenon known as crossing over? Here, with Morgan, is the insight that led to this now-established tenet of science. Appropriately enough, insights such as these won Morgan a Nobel Prize in 1934.

### Figure 2
### Why All Morgan's White-Eyed F₂'s Were Males

The F₂ female flies that inherited one white-eyed X-chromosome allele also inherited a second red-eyed X allele, making them red-eyed. However, the males inheriting the white-eyed allele had no second X chromosome. Instead, they inherited a Y chromosome—which has no genes on it for eye color—leaving them white-eyed.

**Web Animation 12.3**
Morgan's Flies

# BRIEF REVIEW

*Answers to Brief Review questions are in the back of the book.*
*For multiple-choice quiz questions, go to* **www.aw.com/krogh4.**

1. How can the serious genetic defect known as aneuploidy be both widespread and relatively unrecognized?

2. In sex-linked diseases, which gender is more frequently affected—male or female? Why?

3. The text notes the possibility that some forms of human cancer may result from aneuploidy. What is the fundamental difference between an aneuploidy that might lead to cancer and one that results in Down syndrome?

4. Compare and contrast the four chromosome structural aberrations discussed in this chapter: deletions, inversions, translocations, and duplications.

5. Given the following situations, answer accordingly:
   a. A human cell with 47 chromosomes has probably undergone _____.
   b. A cell duplicates its DNA, but fails to undergo cytokinesis. This would lead to _____.

6. Why would a person with Klinefelter syndrome exhibit both male and female features?

7. Certain genotypes may be advantageous in specific situations. Give an example of a heterozygous genotype that is advantageous over either homozygous genotype (hint: especially in mosquito-infested areas). Why is this the case?

# APPLYING YOUR KNOWLEDGE

1. The text notes that an alternate allele of the hemoglobin gene can cause sickle-cell anemia when a person is homozygous for this allele, but that a person who is heterozygous for the allele actually can derive a benefit from it—protection from malaria. In the United States, 8 percent of African Americans are carriers for the sickle-cell allele, while in central Africa the figure is 20 percent. What could account for this difference?

2. The text notes that the technique known as preimplantation genetic diagnosis (PGD) allows prospective parents to screen embryos not only for serious genetic defects such as Down syndrome, but for traits that may be regarded as more or less desirable, such as male or female gender. What limits, if any, should be placed on parents' ability to choose from among embryos for traits such as this?

3. Thinking back to what you learned in Chapter 10 about how sperm are produced, why is it that the cells that give rise to sperm undergo nondisjunction less frequently than the cells that give rise to eggs?

# Genetics Problems

*Answers are in the back of the book.*

1. A man who is a carrier for sickle-cell anemia, a recessive genetic disease, marries a normal, noncarrier woman. What proportion of their children are expected to be afflicted with sickle-cell anemia?

2. Hemophilia is an X-linked recessive disease that prevents blood clotting. If a woman who is a carrier for hemophilia marries a normal man, what proportion of their sons are expected to have hemophilia? What is the chance that their first child will have hemophilia? (Remember, the sex of this child is unknown until birth.)

3. Neurofibromatosis is a genetic disorder associated with an allele of one autosomal gene. Individuals with neurofibromatosis have uneven skin pigmentation and skin tumors. A man with neurofibromatosis marries a normal woman who does not carry the allele for this disorder. The couple has five children, three of whom have neurofibromatosis. The most likely explanation for this outcome is that neurofibromatosis is a _____ trait and that the man is _____.
   a. recessive; homozygous recessive
   b. recessive; heterozygous
   c. dominant; homozygous dominant
   d. dominant; heterozygous
   e. a and b are both possible explanations

Consider the following pedigree for a human autosomal trait. (Note that in pedigrees, generations are indicated by Roman numerals and individuals within a generation are numbered from left to right with Arabic numerals.)

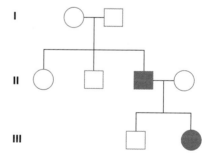

4. Is the allele that determines this trait dominant or recessive? Explain your reasoning.

5. Using symbols *A* and *a* for the dominant and recessive alleles, respectively, what is the genotype of the affected male in the second generation (individual II-3)?

6. What is the genotype of this individual's mate?

7. If female III-2 marries a heterozygous man, what proportion of their children are expected to have this trait?

8. Consider the following pedigree, which assumes a fantasy gene for ear shape, symbolized by *P* for the regular allele or *p* for a pointy-shaped allele if this is an autosomal trait; or by $X^+$ for the regular allele or $X^p$ for the pointy allele if this is an X-linked trait.

This pedigree indicates the trait is most likely inherited as an:

a. autosomal dominant

b. autosomal recessive

c. X-linked dominant

d. X-linked recessive

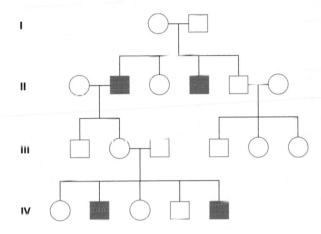

9. What is the genotype of individual I-1?

10. If a woman who is a carrier for this trait married individual III-1, what proportion of their children are expected to show this trait? What proportion of their sons are expected to show this trait?

11. Construct a pedigree to answer the following question: Red-green color blindness is an X-linked recessive trait. A woman who has a color-blind mother and a father with normal color vision marries a man with normal vision. This couple has a son. What is the chance that the son is color-blind?

12. Hyperphosphatemia, a disease that causes a form of rickets (abnormal bone growth and development), is inherited as an X-linked dominant trait. A woman who is heterozygous for the disease allele marries a normal man. What proportion of their sons will have hyperphosphatemia? What proportion of their daughters will have hyperphosphatemia? If hyperphosphatemia were an X-linked recessive trait, how would your answer differ?

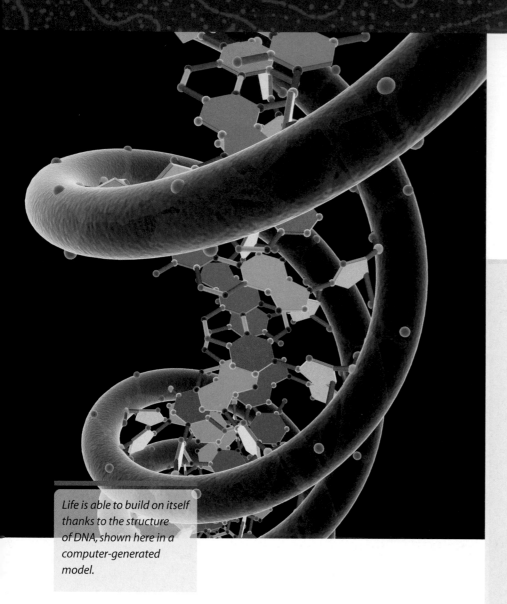

*Life is able to build on itself thanks to the structure of DNA, shown here in a computer-generated model.*

CHAPTER **13**

**13.1** **What Do Genes Do, and What Are They Made Of?** **235**

**13.2** **Watson and Crick: The Double Helix** **236**

**13.3** **The Components of DNA and Their Arrangement** **237**

**13.4** **Mutations** **240**

**Essay**

**The Process of Science:** Getting Clear about What Genes Do: Beadle and Tatum **242**

DNA serves as a storehouse of information for living things. Scientists could not understand how DNA's information could be passed on from one generation to the next until they understood how all of its component parts fit together.

# PASSING ON LIFE'S INFORMATION: DNA Structure and Replication

"**I**n research the front line is almost always in a fog," wrote Francis Crick in his 1988 memoir, *What Mad Pursuit*. Crick had first-hand knowledge of this. In 1953, with James Watson, he discovered the structure of DNA, an effort that brought him to the front line of research, which is where he remained to the end of his long, productive life. Crick's book makes clear what a tremendous difference there is between learning something about nature and discovering something about it. In learning, dozens of voices stand ready to instruct; there are books, lectures, videos, CDs—a world of well-organized information. The person who seeks to discover something about nature, meanwhile, is confronted with silence. Nature goes on: Cells divide, chromosomes condense, birds migrate. But *why* do these things operate the way they do? All the researcher has as a guide are the things themselves, working away. The trick is to devise some test (called an experiment) that can make nature's routine operations yield information. In this process of teasing out truth, knowledge is gained in very small increments. Scientists are like the first people down a darkened maze of tunnels; they must make their way exceedingly slowly, feeling each square inch of wall space as they go, after which they can only leave lights on *behind* them (in the form of their scientific papers, books, and so forth).

## 13.1 What Do Genes Do, and What Are They Made Of?

To get a feel for the process of discovery, consider the state of genetics early in the twentieth century. By 1920, it had been demonstrated beyond any doubt that genetic information resided on chromosomes. Within a decade or so, by looking at some abnormally large fly chromosomes, scientists could observe in great detail such chromosomal processes as crossing over. By viewing the banding patterns on these chromosomes, they could even identify the rough location of some genes.

Yet, what was a gene? What was the physical nature of this unit of heredity, and how did it work? No one knew. Because genes are much smaller than the chromosomes on which they reside, there was no hope of simply viewing one under a microscope. In observing chromosomes, scientists were only "looking" at genes in the way that any of us is "looking" at lunar rocks by glancing up at the moon. Through the 1920s and 1930s, then, genes could be described only in vague, functional terms, as in: A gene is an entity that lies along a chromosome and brings about a phenotypic trait in an organism (for example, the green or yellow peas of Mendel's experiments).

Things were to clear up somewhat in the ensuing years. The central achievement in genetics in the late 1930s and early 1940s was a more concrete description of what genes do: They bring about the production of proteins. The central achievement in genetics from the mid-1940s to the early 1950s was strong evidence indicating what genes are composed of: deoxyribonucleic acid, or DNA for short.

## DNA Structure and the Rise of Molecular Biology

By the early 1950s, a key question became *how* DNA carried out its genetic function. To find out, scientists had to piece together its exact chemical structure. This investigation turned out to be a watershed event in biology. Just as opening a mechanical watch and observing how its parts fit together would allow a person to understand how the watch works, so deciphering the structure of DNA allowed biologists to understand how genetics works at its most fundamental level. How could genetic information be passed on from one generation to the next? How could DNA be a versatile enough molecule to specify the wide range of proteins that are produced in living things? The structure of the DNA molecule suggested an answer to these questions, as you'll see.

Apart from what this investigation uncovered, the inquiry itself stood as a symbol of a new era in biology. When Gregor Mendel did his experiments with peas, he was looking at whole organisms. When T. H. Morgan was doing his work with *Drosophila* flies, he was interested in whole chromosomes, which he knew to be composed of several types of molecules.

In the early 1950s, however, the search turned to a single molecule. How was DNA structured, and how did this structure allow it to carry out its various functions? Biological research of this sort has grown ever more important in the decades since the 1950s. It is today known as **molecular biology**: the investigation of life at the level of its individual molecules.

## 13.2  Watson and Crick: The Double Helix

James Watson and Francis Crick may not be instantly recognized scientific names in the way that, say, Albert Einstein or Louis Pasteur are, but there's a certain public awareness that these two researchers did something important in connection with DNA (**Figure 13.1**). What they did was present to the world, in 1953, the structure of DNA. Atom by atom, bond by bond, this is how DNA fits together, they said in unveiling DNA's now-famous configuration, the double helix.

The two were a seemingly unlikely pair to make an epochal scientific discovery. Watson, an American, was a 23-year-old who was scarcely more than a year out of graduate school, and Crick, an Englishman, was a 35-year-old just then working on his doctorate when the two met in the fall of 1951 at Cambridge

**Figure 13.1**
**Young and Famous**

James Watson, on the left, and Francis Crick, with a model of the DNA double helix, shortly after they published their paper on the molecule's structure.

**(a)** Rosalind Franklin

**(b)** X-ray diffraction image of DNA

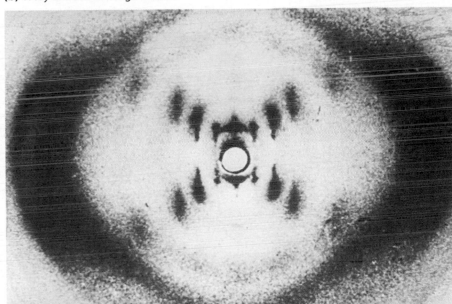

**Figure 13.2**
**(a) DNA Investigator**
Rosalind Franklin, whose work in X-ray diffraction was important in revealing the structure of the DNA molecule.
**(b) Imaging DNA**
One of Franklin's X-ray diffraction images of DNA. The "cross" formed of dark spots indicated the molecule had a helical structure.

University in England. Ending up together by coincidence at the same laboratory, they realized in short order their mutual interest in the structure of DNA, and to the neglect of projects they were supposed to be working on, they began several rounds of model building and brainstorming that resulted in their breakthrough.

Watson and Crick were greatly aided in their investigation by the work of others. Although the DNA molecule was too small to be seen by even the most powerful microscopes of the time, something about its structure could be inferred from a technique called X-ray diffraction. In this process, a purified form of a molecule is bombarded with X rays. The way these rays scatter on impact then reveals something about the structure of the molecule.

If you look at **Figure 13.2**, you can see the results of some X-ray diffraction. As you can imagine, it takes a highly trained observer to be able to deduce anything about the structure of a molecule from such an image. Fortunately for Watson and Crick, such a person was working just up the road from them. She was Rosalind Franklin, a researcher at King's College

in London and one of the handful of individuals then skilled in performing X-ray diffraction on DNA. She and her colleague, Maurice Wilkins, were themselves working on the structure of DNA at the time, as were other researchers in America. Thus did Watson, at least, regard the search for DNA structure to be a race between several teams, a fact that concentrated his efforts wonderfully. In 1962, Watson, Crick, and Wilkins were awarded the Nobel Prize in Medicine or Physiology for their work on DNA. Rosalind Franklin died of cancer in 1958 at the age of 37. Nobel Prizes are not awarded posthumously; it's unknown what would have happened had she lived.

Let's start now to look at DNA's structure, which will put you in a position to appreciate the achievement of Watson, Crick, and their fellow researchers.

## 13.3 The Components of DNA and Their Arrangement

A good way to understand the structure of DNA is to look at the building blocks that make it up. First, there is a phosphate group and second, a sugar called

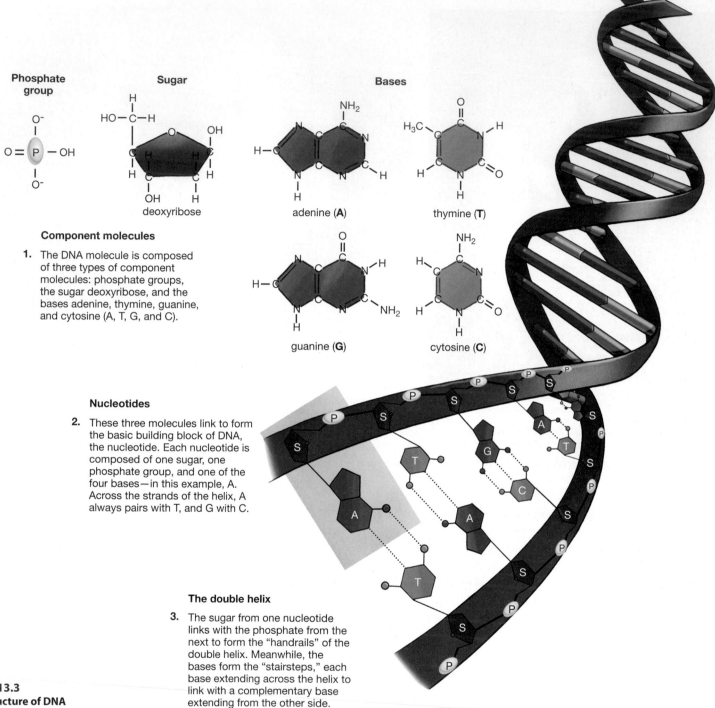

**Phosphate group**

**Sugar**

deoxyribose

**Bases**

adenine (**A**)

thymine (**T**)

guanine (**G**)

cytosine (**C**)

**Component molecules**

1. The DNA molecule is composed of three types of component molecules: phosphate groups, the sugar deoxyribose, and the bases adenine, thymine, guanine, and cytosine (A, T, G, and C).

**Nucleotides**

2. These three molecules link to form the basic building block of DNA, the nucleotide. Each nucleotide is composed of one sugar, one phosphate group, and one of the four bases—in this example, A. Across the strands of the helix, A always pairs with T, and G with C.

**The double helix**

3. The sugar from one nucleotide links with the phosphate from the next to form the "handrails" of the double helix. Meanwhile, the bases form the "stairsteps," each base extending across the helix to link with a complementary base extending from the other side.

**Figure 13.3**
**The Structure of DNA**

*deoxyribose*, both of which can be seen in **Figure 13.3**. Third, there are four possible DNA "bases": adenine, guanine, thymine, and cytosine—A, G, T, and C, for short. When you link together a phosphate group, a deoxyribose molecule, and one of the four bases, you get the basic building block of DNA, the *nucleotide*, one of which is boxed in the larger DNA molecule in the figure.

When you look at this larger figure as a whole, you can see that the double-helix looks something like a spiral staircase. By focusing on its exterior "handrails,"

you can see how sugar and phosphate fit together to form a kind of chain, with its links ordered as: sugar-phosphate-sugar-phosphate (symbolized in the figure by S and P). The "steps" lying between these handrails are composed of DNA's bases. As the figure shows, each of the bases amounts to a *half*-stairstep, if you will. Each extends inward from one handrail of the double helix and is then joined to a base extending inward from the other handrail of the DNA molecule. (The bases are linked via hydrogen bonds, symbolized by the dotted lines in the figure.)

The figure shows some examples of one base being paired with another. At the bottom, for example, an A base on the left is paired with a T base on the right. Go two sets of nucleotides up, and a G on the left is being paired with a C on the right. This turns out to be one of the fundamental rules about DNA structure. If you viewed a billion DNA *base pairs*, as they're called, you would find the same thing: A always pairing with T, and G always pairing with C, across the helix. (Any two bases that can pair together in this way are said to be *complementary*; thus, A is complementary to T, and G is complementary to C. Likewise, you can have complementary chains of DNA.)

## The Structure of DNA Gives Away the Secret of Replication

It was DNA's very structure that suggested the answer to one of the great questions of genetics: How is genetic information passed on? To put this another way, how does a cell make a copy of its own DNA? As you've seen, a full complement of our DNA is contained in nearly every cell in our body. Yet, cells divide, and each pair of daughter cells contains exactly the same complement of DNA as did the parent cell they came from. This means that genetic information is passed on by means of DNA being copied, with one copy of this molecule ending up in one daughter cell and a second copy in the other. Indeed, such copying extends to the formation of egg and sperm cells, meaning this is the way genetic information is passed on from one *generation* to the next. But how could this work?

The structure Watson and Crick discovered suggested a way. We've observed that A must always pair with T, and G with C, across the two strands of the double helix. This rule, Watson and Crick saw, meant that each single strand of DNA could serve as a *template* for the synthesis of a new single strand (**Figure 13.4**). Each A on an old strand would specify the place for a T on the new, each G on the old a place for a C on the new, and so forth. All that was required was for the two old strands to separate—splitting the stairsteps right down the middle—and for new strands to be synthesized that were complementary to the old.

## DNA Structure and Protein Production

The second thing DNA structure suggested was a partial answer to another great question of genetics: How could DNA be versatile enough to specify the dazzling array of proteins that all living things produce? The handrails of DNA's double helix are monotonous—phosphate, sugar, phosphate, sugar. But the bases can be laid out along these handrails in an extremely varied

manner. A has to pair with T and G with C *across* the helix; but *along* the handrails, the bases can come in any order. Look at the top drawing in Figure 13.4 and see the order this hypothetical group of bases comes in; starting from the bottom right, A-G-C-T-A-C. Strung together by the thousands, such bases can specify a particular protein, just as in Morse code a series of short and long clicks can specify a particular word. And, thanks to the variability of DNA's bases along the handrails, an enormous number of proteins can be specified. Next chapter, we'll look at the process by which cells start with the information contained in DNA and end up with completed proteins.

**Web Animation 13.1**
DNA

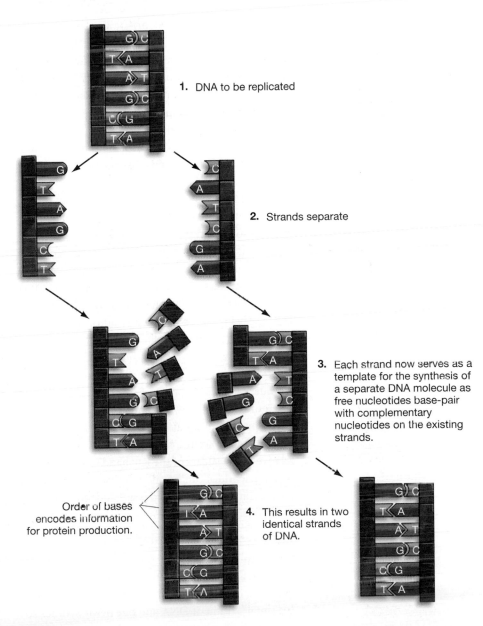

1. DNA to be replicated

2. Strands separate

3. Each strand now serves as a template for the synthesis of a separate DNA molecule as free nucleotides base-pair with complementary nucleotides on the existing strands.

Order of bases encodes information for protein production.

4. This results in two identical strands of DNA.

**Figure 13.4**
**DNA Replication**
The result of DNA replication is two identical molecules of DNA, whereas the process began with one.

**SO FAR...**

1. James Watson and Francis Crick discovered the _____.

2. The key to DNA replication is that each _____ base on an original strand of DNA serves as a template for the addition of a T base on the new strand, while each _____ on the original strand serves as a template for the addition of a G base.

3. A particular sequence of DNA _____ can specify a particular _____.

## The Building Blocks of DNA Replication

You have already looked briefly at the process by which DNA is copied or "replicated," but we now need to review some of the details of this process. The basic steps here are straightforward. The essential building block of DNA is the unit called the **nucleotide**, which can be seen in the box in Figure 13.3. Every nucleotide contains one sugar and one phosphate group. What differentiates one nucleotide from another is the base that is attached to it; in this case the base is A, but as you can see, others have T, C, or G.

Figure 13.4 shows the process of DNA replication in overview. Note that the first thing that happens is that the joined strands of the double helix unwind, separating from one another. The nucleotides on each of the single strands are then paired with free-floating nucleotides that line up in new, complementary strands. Because *both* strands of the original double helix are being paired with new strands, the end product is two double helices, where before there was one.

You can see in Figure 13.4 that the addition of nucleotides moves in differing directions on the two strands. This is so because of something you can see in Figure 13.3: the two strands of the double helix have opposite orientations. Note that the sugars of the right-hand strand, for example, are upside down relative to the sugars of the left-hand strand. Because new nucleotides can be added to only one end of a DNA strand, the nucleotides are added in different directions.

### Something Old, Something New

It's worth drawing a little finer point on one aspect of this process: Each resulting double helix is a combination of the old and the new. Each has one "parental" strand of DNA and one newly synthesized complementary strand. This combination is conceptually important because this is how life builds on itself. You can see this illustrated in **Figure 13.5**.

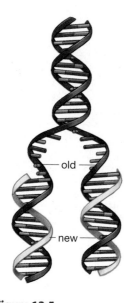

**Figure 13.5**
**How Life Builds on Itself**
Each newly synthesized DNA molecule is a combination of the old and the new. An existing DNA molecule unwinds, and each of the resulting single strands (the old, in red) serves as a template for a complementary strand that will be formed through base pairing (the new, in blue).

•••ANSWERS
1. structure of DNA
2. A; C
3. bases; protein

Eventually, the two double helices are fully formed. A little later they part company, during mitosis, with each newly formed double helix moving into a separate cell and thereafter serving as an independently functioning segment of DNA. At the chromosomal level, each newly replicated helix is, when joined by appropriate proteins, one of the sister chromatids you've been seeing so much of.

However simple this process may sound in overview, its details are enormously complicated. As you might imagine, such a process could not proceed without enzymes to catalyze it. To name just two groups of them, there are enzymes called helicases that unwind the double helix, separating its two strands to make the bases on them available for base pairing. There is also another group of enzymes, collectively known as **DNA polymerases**, that move along each strand of the double helix, joining together nucleotides as they are added—one by one—to form the new, complementary strands of DNA.

### Editing Out Mistakes

The base pairing that goes on in replication goes on millions of times in a given stretch of DNA: Free-standing A's are aligned with complementary T's, while C's are aligned with G's. Given the number of base pairs involved, the amazing thing is how few mismatches there are by the time the process is completed. The error rate in DNA replication—the rate at which the *wrong* bases have been brought together—might be only one in every billion bases by the end of replication. Yet *during* replication, such a mistake might be made once in every 100,000 bases. Obviously, to start with a given number of mistakes but to end up with far fewer, the cell's genetic machinery has to be capable of correcting its errors.

This happens partly through the services of the versatile DNA polymerases, which are able to perform a kind of DNA editing: They remove a mismatched nucleotide and replace it with a proper one.

## 13.4  Mutations

With this consideration of alterations in DNA structure, you have arrived at an important concept: that of a **mutation**, which can be defined as a permanent alteration of a DNA base sequence. Such alterations come about because the cell's various DNA error-correcting mechanisms are not foolproof; they do not correct all the mistakes that occur. For the balance of this chapter, we'll focus on these genetic mistakes, which have a powerful effect both on human health and on evolution.

DNA's makeup can be altered in many ways. A slight change in the chemical form of a base might, for example, cause a G to link up across the helix with a T (instead of with its normal partner, C), as you can see in **Figure 13.6**. Then, in a subsequent round of DNA replication, the cell might "repair" this error in such a way that a permanent mistake is introduced: An A-T pair now exists, whereas the original sequence had a G-C pair. Permanent mistakes like these are called **point mutations**, meaning a mutation of a single base pair in the genome. Such mutations stand in contrast to the kind of whole-chromosome aberrations presented in Chapter 12, although these also qualify as mutations under some definitions.

In what way are mutations "permanent"? Think of how DNA replication works. Before any cell can divide, it must first make a copy of its complement of DNA. Should this DNA contain an uncorrected mistake, it too will be copied—again and again, actually, with each succeeding cell division. Most mutations have no noticeable effect on an organism, and as you'll see, mutations are vital to the process of evolution. But the concept of mutation is quite rightly fearful to us because, in relatively rare instances, mutations can have disastrous effects.

## Cancer and Huntington Mutations

A cancerous growth is a line of cells that has undergone a special kind of mutation—one that causes the affected cells to proliferate wildly. As an example, the skin cancer known as melanoma causes skin cells called melanocytes to start dividing very rapidly. For our purposes, the question is: How does this process get going? In 2002, British scientists looked at numerous lines of melanoma cells that had been taken from cancer patients. They found that 59 percent of these cells contained a mutation in a gene called *BRAF*. This gene, like all others, amounts to a sequence of DNA bases—a sequence of C's, T's, A's, and G's. When the British scientists looked at how the *BRAF* gene had actually mutated, they found that in 92 percent of cases, there had been a single change: an A had been substituted for a T at *BRAF*'s 1,796th base. What they found, in other words, was a point mutation, nearly 2,000 bases within the sequence of bases that makes up the *BRAF* gene. And the effect of this mutation? It produces a protein that is a slightly altered version of the normal BRAF protein. This protein is altered enough, however, that it keeps melanocytes moving through the cell-division cycle, causing them to multiply wildly.

Cell proliferation is not the only kind of trouble mutations can bring about. Chapter 12 introduced Huntington disease, which causes spastic movements, severe dementia, and ultimately death among its victims. In all people, there is a gene, called *IT15*, that has within it a number of repeats of a particular "triplet" of bases, CAG. People with 34 or fewer CAG repeats in their *IT15* gene are fine; they will suffer no illness at all from the gene. People with 35–39 CAG repeats, however, may develop Huntington, although this is not a certainty. Meanwhile, people who have more than 40 repeats definitely will develop the disease, and the more repeats they have, the younger they will be when its symptoms first appear. Unlike the case with cancer, however, the Huntington mutation does not cause cells to multiply wildly. Instead, the mutated *IT15* gene creates a faulty protein called huntingtin, which cannot be broken down by nerve cells and ends up building up inside them, eventually killing them. Because Huntington is caused by a repeating group of three nucleotides, it is referred to as a "trinucleotide repeat" disease. At least eight other diseases of the nervous system fall into this category.

## Heritable and Nonheritable Mutations

Though Huntington disease and melanoma are both caused by mutations, there is an important distinction to be made between them. Most mutations come about in the body's **somatic cells**, which is to say cells that do not become eggs or sperm. This is the case with the mutations that bring about melanoma. Conversely, some mutations arise in **germ-line cells**, meaning the cells that do become eggs or sperm. This is the case with the Huntington mutation. The important point here is that germ-line cell mutations are *heritable*—they can be passed on from one generation to the next, in the ways reviewed in Chapter 12. In contrast, although a line of melanoma cells may be quite harmful, it is separate from the line of cells that

Web Animation 13.2
Mutations

**Figure 13.6**
**One Route to a Mutation**

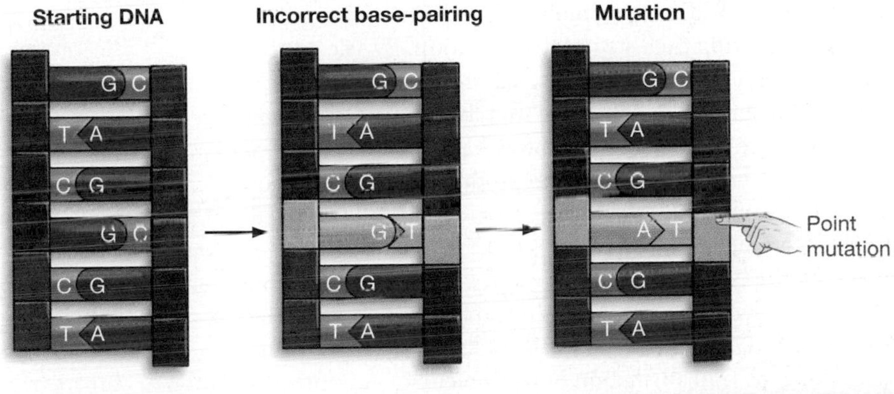

Starting DNA · Incorrect base-pairing · Mutation

Point mutation

1. In replicating a cell's DNA, mistakes are sometimes made, such that one base can be paired with another base that is not complementary to it (G with T in this case).

2. The next time a cell replicates its DNA, the replication repair mechanism may "fix" this error in such a way that a permanent alteration in the DNA sequence results. The original G will be replaced, instead of the wrongly added T. The result is an A-T base pair, whereas the cell started with a G-C base pair.

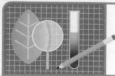

# THE PROCESS *of* SCIENCE
## Getting Clear about What Genes Do: Beadle and Tatum

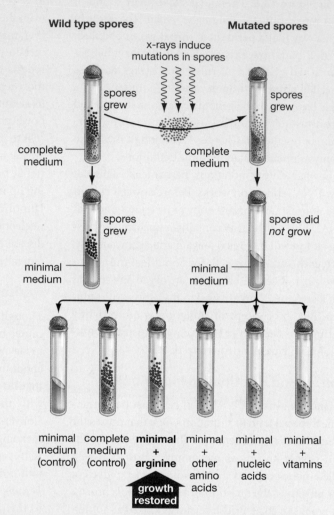

**Figure 1**
**Beadle and Tatum Experiment with Mutated Spores**
Using spores of the fungus *Neurospora,* Beadle and Tatum found that some mutated versions of the spores could not grow in a medium that required them to synthesize all their biological molecules. By selectively adding molecules to this "minimal medium," they demonstrated that the mutation was preventing the spores from producing certain compounds, among them the amino acid arginine. By tracing the metabolic pathway leading to arginine, Beadle and Tatum realized that the *Neurospora* genes that had been made dysfunctional were coding for enzymes that facilitated the steps leading from the precursors of arginine to arginine itself.

**I**n the 1930s, biologists were confused not only about what genes were made of (DNA) but also about what it was that genes did. We now know that a large part of what they do is contain the information for the production of proteins, including the reaction-hastening proteins called enzymes. But it took work over a period of years by Stanford University researchers George Beadle and Edward Tatum to make this clear. Their experimental subject was a red fungus, the bread mold *Neurospora.*

Research scientists are always looking for ways to control the confusing conditions of nature to tease out the rules it plays by. Beadle and Tatum's control mechanism was, first, to *induce* genetic mutations in *Neurospora* spores by bombarding them with X-rays. These mutated spores were then transferred one at a time into test tubes, where the second element of control came in. Natural or "wild-type" *Neurospora* can manufacture or "synthesize" most of what they need to live, given just a few basic biological molecules in their surroundings. But could the descendants of the *mutated* group of *Neurospora* do this? Could they be put in a so-called minimal medium—a liquid environment that forces the fungus to manufacture almost all its necessary biological molecules—and still multiply into a thriving colony? The answer for many of the mutant strains was no. The mutations these organisms had undergone had rendered them unable to synthesize some substance that they needed to thrive. By being selective about *adding* substances to the minimal media, Beadle and Tatum could tell what substance the mutant cells were unable to synthesize. If they added the right thing, the colony would grow, as you can see in **Figure 1**. Thus, they reasoned, whatever substance they added was the substance a mutated strain could no longer synthesize on its own. For some of the mutant strains, the missing substance turned out to be the amino acid arginine.

We now know that arginine is produced in one of the metabolic pathways reviewed in Chapter 6. In such a pathway, a precursor compound is modified several times in a chain, with a different enzyme facilitating each step of the operation. By selectively supplying different strains of mutant *Neurospora* with arginine's *precursor* substances, Beadle and Tatum could discern the step at which the arginine metabolic pathway had been blocked. One strain required only precursor A to produce arginine, for example, while another strain required precursors A and B. In the tradition of T. H. Morgan, the researchers linked these blockages to genes on three separate *Neurospora* chromosomes. The summary result then was clear. A given *Neurospora* strain had become a mutant by having its genes altered through radiation. As a result, it no longer carried out the transformation of, say, precursor A to precursor B. It took an *enzyme* to facilitate this transformation. What genes were doing, then, was bringing about the production of enzymes.

Eventually, a famous label was applied to these experimental results: The *one-gene, one-enzyme hypothesis.* What each gene does is call up production of one enzyme, which undertakes a particular metabolic task. Although the notion of what a gene is has now been broadened, the essential insight was not only correct, it was critical. With it, scientists at last understood what genes were doing. For this work, which they completed in 1941, Beadle and Tatum shared the Nobel Prize in 1958.

gives rise to eggs or sperm. For this reason, melanoma cannot be passed on from one generation to the next.

## What Causes Mutations?

A critical question, of course, is what causes DNA to mutate? One answer is so-called environmental insults. The chemicals in cigarette smoke are powerful *mutagens*, meaning substances that can mutate DNA. So is the ultraviolet light that comes from the sun. To look at the latter example in a little more detail, ultraviolet light is a form of radiation that can link adjacent T's together in a single strand of DNA; sometimes it even causes both strands of the DNA helix to break. When this latter damage takes place in a cell, one of three things can happen. First, enzymes can successfully repair the damage. Second, the cell will recognize that this damage *cannot* be repaired, in which case it may either stop dividing (in the process known as cellular senescence) or commit suicide (in the process known as apoptosis). Third, the cell will undertake none of these responses to its DNA alteration but will simply keep on dividing. In this instance, the erroneous "spelling" in the cell's DNA will be copied over and over as this original cell gives rise to two cells, which then give rise to four, and so on. In short, a mutation has occurred.

Not all mutations are caused by environmental influences. Mutations happen simply as random, spontaneous events. The collision of water molecules with DNA can remove nucleotide bases or alter them in such a way that their base-pairing properties are changed. The very process of eating and breathing produces so-called free radicals that can damage DNA. And the DNA replication machinery itself may introduce errors, irrespective of any outside influences.

Once such errors occur, it is understandable that some of them will go uncorrected simply because of the size of the replication operation. When a human cell divides, billions of DNA base pairs have to be copied. And some 25 million cells are dividing each second in human beings. The surprise, therefore, is not that mistakes happen, but that so *few* of them happen, and that fewer still become permanent. (Recall the one-in-a-billion error rate in DNA replication that was observed earlier.) After learning about the kinds of effects mutations can have on organisms, however, you can understand why this error rate is so low: Life as we know it could not exist with a high error rate, at least in the portions of genome that code for proteins. Remember that within 3 billion A's, T's, G's, and C's, the substitution of a single A for a single T can help bring about melanoma. How could life exist if such mistakes were common?

## Mutations and Evolutionary Adaptation

While it's true that, for individuals, mutations can have a negative effect, for the living world as a whole mutations have been vitally important because of a role they play in evolution. It turns out that germ-line mutations are the only means by which completely new genetic information can be added to a species' genome, in the form of new alleles (meaning variant forms of a gene). Organisms can combine *existing* alleles in myriad ways. Think of meiosis, with its chromosomal part swapping (crossing over) and reshuffling of chromosomes (independent assortment). Valuable as these processes are, no amount of genetic recombination could have produced, for example, the eyes that some living things possess. To go from no eyes to eyes, there had to have been some mutations along the line—some accidental reorderings of DNA sequences such that entirely new proteins were produced. Such adaptations are vital to living things given their struggle to get along in environments that are constantly changing. (Temperatures shift; streams dry up.) If environments change, species need to change, too, in order to survive. And the primary way major changes can come about is through mutations. You'll be looking at this topic again in the evolution unit of this book. For now, however, isn't it interesting to ponder the fact that the living world adapts partly through its mistakes?

## On to Protein Production

Back in Chapter 9, you saw that since antiquity, human beings have speculated about how one generation of living things can give rise to another—about *how* it is that life goes on. The detailed answer to this question turned out to be the something-old, something-new quality of DNA replication: parental DNA strands serve as templates for new strands, with the result that the qualities of one generation can be passed along to the next. Splendid as this function is, it is only one of the two great tasks carried out by our genetic machinery. You have just reviewed the process by which genetic information is replicated; now let's look at the process by which this information is used. How can a stretch of DNA bring about the production of a protein? You'll see in the chapter coming up.

**Web Animation 13.3**
One-Gene, One-Enzyme Hypothesis

**...ANSWERS**

1. new, complementary strands
2. DNA polymerases
3. permanent alteration

---

**SO FAR...**

1. In DNA replication, both of the strands of the double helix serve as templates for _____.

2. The _____ are a group of enzymes that add nucleotides to a replicating DNA chain.

3. A mutation is a _____ of a DNA base sequence.

# CHAPTER 13 REVIEW

For study help, activities, and more quiz questions, go to **www.aw.com/krogh4**.

## SUMMARY

### 13.1 What Do Genes Do, and What Are They Made Of?

- James Watson and Francis Crick discovered the chemical structure of DNA in 1953. This event ushered in a new era in biology because it allowed researchers to understand some of the most fundamental processes in genetics.

- In trying to decipher the structure of DNA, Watson and Crick were performing work in molecular biology. This is the investigation of life at the level of its individual molecules. Molecular biology has grown greatly in importance since the 1950s.

### 13.2 Watson and Crick: The Double Helix

- Watson and Crick met in the early 1950s at Cambridge University in England and set about to decipher the structure of DNA. Their research was aided by the work of others, including Rosalind Franklin, who was using X-ray diffraction to learn about DNA's structure.

### 13.3 The Components of DNA and Their Arrangement

- The DNA molecule is composed of building blocks called nucleotides, each of which consists of one sugar (deoxyribose), one phosphate group, and one of four bases: adenine, guanine, thymine, or cytosine (A, G, T, or C, respectively). The sugar and phosphate groups are linked together in a chain that forms the "handrails" of the DNA double helix. Bases then extend inward from the handrails, with base pairs joined to each other in the middle by hydrogen bonds. In this base pairing, A always pairs with T across the helix, while G always pairs with C.

- DNA is copied by means of each strand of DNA serving as a template for the synthesis of a new, complementary strand. The DNA double helix first divides down the middle. Each A on an original strand then specifies a place for a T in a new strand, while each G specifies a place for a C, and so forth.

- Each double helix produced in replication is a combination of one parental strand of DNA and one newly synthesized complementary strand. This is how life builds on itself. A group of enzymes known as DNA polymerases is central to DNA replication; these enzymes move along the double helix, bonding together new nucleotides in complementary DNA strands.

- DNA can encode the information for the huge number of proteins used by living things because the sequence of bases along DNA's handrails can be laid out in an extremely varied manner. A collection of bases in one order encodes the information for one protein, while a different sequence of bases encodes the information for a different protein.

Web Animation 13.1: DNA

### 13.4 Mutations

- The error rate in DNA replication is very low, partly because repair enzymes are able to correct mistakes. When such mistakes are made and then not corrected, the result is a mutation: a permanent alteration in a cell's DNA base sequence.

- Most mutations have no effect on an organism, but when they do have an effect, it is generally negative. Cancers result from a line of cells that have undergone types of mutations that cause them to proliferate wildly.

- Some mutations come about in the body's germ-line cells, meaning cells that become eggs or sperm. Such mutations are heritable: They can be passed on from one generation to another. The gene for Huntington disease, which is expressed in nerve cells, is a heritable, mutated form of a normal gene. Most mutations, however, come about in the body's somatic cells, which are cells that do not give rise to eggs or sperm. Dangerous as these mutations may be, they cannot be passed along to offspring.

- Mutations can come about through the effects of mutagens: substances, such as cigarette smoke or ultraviolet light, that can mutate DNA. Mutations can also come about as random, spontaneous events that are brought about by normal cellular processes.

- Mutations have been important to evolution because they are the only means through which completely new genetic information can be added to a species' genome. The accidental reorderings of DNA sequences that mutations bring about can, in rare instances, produce new proteins that are useful to organisms.

Web Animation 13.2: Mutations
Web Animation 13.3: One-Gene, One-Enzyme Hypothesis

## KEY TERMS

| | |
|---|---|
| **DNA polymerase** 240 | **nucleotide** 240 |
| **germ-line cell** 241 | **point mutation** 241 |
| **molecular biology** 236 | **somatic cell** 241 |
| **mutation** 240 | |

# BRIEF REVIEW

*Answers to Brief Review questions are in the back of the book.*
*For multiple-choice quiz questions, go to* **www.aw.com/krogh4**.

1. How does the phrase "something old, something new" describe the method of DNA replication employed by the cell?

2. What fundamental question of genetics did Watson and Crick's discovery help answer?

3. True or False: DNA polymerases can correct errors made in DNA replication.

4. What is a mutation? Are all mutations harmful?

5. What is the nature of the Huntington disease mutation, and what is its effect at a cellular level?

# APPLYING YOUR KNOWLEDGE

1. Given the following DNA sequence, determine the complementary strand that would be added in replication:

   ATTGCATGATAGCC

2. A gene's nucleotide bases are known to be composed of 15 percent guanine. What percentages of each of the other three bases are contained in the same gene?

3. Would you expect cancer to arise more often in types of cells that divide frequently (such as skin cells) or in types of cells that divide rarely or not at all (such as nerve cells in the brain)? Explain your reasoning.

4. The Huntington disease referred to in the main text sends all of its victims into dementia, physical immobility, and ultimately death, but usually does not start doing so until the person is at least 35. Imagine yourself as the adolescent child of a parent who has started to manifest Huntington symptoms. Your chances of inheriting this incurable condition are one out of two, but you have thus far shown no symptoms of it. The question is: Do you get yourself tested to see if you inherited the harmful Huntington allele? Do you find out for sure whether you will, or will not, have this illness, or do you live with uncertainty? Assuming you've arrived at an answer, consider it now in the context of the conclusion that most people in this situation come to. Ninety-five percent of such people decline to be tested.

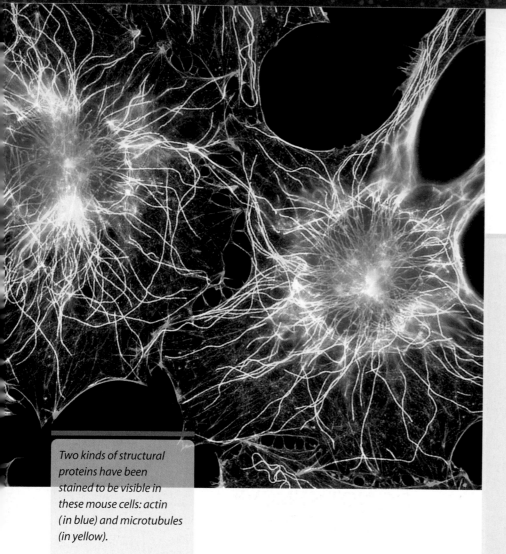

*Two kinds of structural proteins have been stained to be visible in these mouse cells: actin (in blue) and microtubules (in yellow).*

CHAPTER **14**

| 14.1 | The Structure of Proteins | 247 |
| 14.2 | Protein Synthesis in Overview | 248 |
| 14.3 | A Closer Look at Transcription | 249 |
| 14.4 | A Closer Look at Translation *BioFlix* | 252 |
| 14.5 | Genetic Regulation | 257 |
| 14.6 | Genetics and Life | 262 |

**Essays**

Cracking the Genetic Code 256

Like workers using a set of blueprints to construct a house, cells use the information contained in DNA to construct proteins. Each cell can finely tune its production of proteins in accordance with its needs.

# HOW PROTEINS ARE MADE:
# Genetic Transcription, Translation, and Regulation

**T**he scientific advance that was reviewed in Chapter 13, the discovery of the structure of DNA in 1953, bears some comparison to the birth of Napoleon: Both events had momentous consequences, but not for

a while. Only slowly did the discoveries come that, in time, were seen as hinging on James Watson and Francis Crick's work. So significant were these advances, however, that the years between 1953 and 1966 are now regarded by some scholars as a kind of golden age of genetics, a time when basic findings about this field came fast and furious, such that by the mid-1960s scientists had at last grasped the essentials of how the genetic machinery works.

These discoveries were made in connection with both of the two great tasks of genetics: on the one hand, the copying or "replication" of DNA that you reviewed in Chapter 13; on the other, the means by which DNA's genetic information brings about the production of proteins. The first task has to do with how genetic information is passed on, the second with how genetic information is used. This second task is the subject of this chapter.

## 14.1 The Structure of Proteins

Because you'll be dealing extensively with proteins here, now may be a good time to review some basic information about them. As you saw in Chapter 3, proteins fit into the "building blocks" model of biological molecules. The blocks in this case are amino acids. String a number of these together, and you have a **polypeptide** chain, which then folds up in a specific three-dimensional manner, resulting in a protein. Proteins are likely to be made of hundreds of amino acids strung together, often in several linked polypeptide chains.

### Synthesizing Many Proteins from 20 Amino Acids

Though there are hundreds of thousands of different proteins, each one of them is put together from a starting set of a mere 20 amino acids. If you wonder how such diversity can proceed from such simplicity, think of the English language, which has thousands of words, but only 26 letters in its alphabet. It is the *order* in which letters occur that determines whether, for example, "cat" or "act" is spelled out; just so, it is the order of amino acids that determines which protein is synthesized.

If you look at **Figure 14.1a** on the next page, you can see the chemical structure of two free-standing amino acids, glycine (gly) and isoleucine (ile). If you look at **Figure 14.1b**, you can see that these two amino acids are the first two that occur in one of the two polypeptide chains that make up the unusually small protein we call insulin. **Figure 14.1c** then shows a three-dimensional representation of insulin—the folded-up form this protein assumes before taking on its function of moving blood sugar into cells. A list of

glycine **(gly)**                                            isoleucine **(ile)**

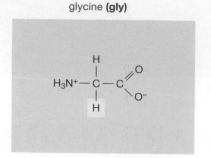     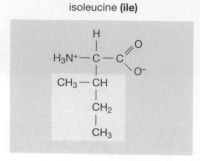

**(a)** Amino acids
The building blocks of proteins are amino
acids such as glycine and isoleucine, which
differ only in their side-chain composition
(light colored squares).

$H_3N^+$ — gly | ile | val | glu | gln | cys | cys | ala | ser | val | cys | ser | leu | tyr | gln | leu | glu | asn | tyr | cys | asn

**(b)** Polypeptide chain
These amino acids are strung together to form
polypeptide chains. Pictured is one of the two
polypeptide chains that make up the unusually
small protein insulin.

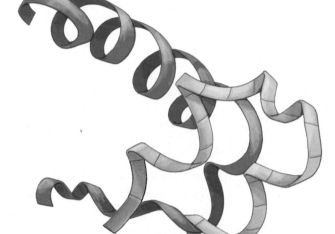

**(c)** Protein
Polypeptide chains function as proteins only
when folded into their proper three-dimensional
shape, as shown here for insulin. Note the
position of the glycine and isoleucine amino
acids in one of the insulin polypeptide chains
(colored light green).

**Figure 14.1**
**The Structure of Proteins**

**Web Animation 14.1**
Structure of Proteins

all 20 primary amino acids and their three-letter abbreviations can be found in **Table 14.1**.

For our purposes, the question is: How do the chains of amino acids that make up a protein come into being? How do gly, ile, val, and so forth come to be strung together in a specific order to create this protein? You know from what you've studied so far that genes can be thought of as "recipes" for proteins, and that these recipes are set forth as a series of chemical "bases"—the A's, T's, C's, and G's that lie along a DNA strand. How, then, do we get from *this* series of DNA bases to *that* series of amino acids—to a protein? Let's find out.

## 14.2 Protein Synthesis in Overview

In overview, the process of protein synthesis can be described fairly simply. In eukaryotic organisms such as ourselves, the DNA just referred to is contained in the nucleus of the cell. The first step is that a stretch of it unwinds there, and its message—the order of a string of A's, T's, C's and G's—is copied onto a molecule called messenger RNA (mRNA). This length of mRNA then exits from the cell nucleus (**Figure 14.2**).

The destination of this mRNA is a molecular workbench in the cell's cytoplasm, a structure called a ribosome. More formally, a **ribosome** is an organelle, located in the cell's cytoplasm, that is the site of protein synthesis. At the ribosome, both the recipe (the mRNA sequence) and the raw materials (amino acids) come together to make the product (a protein). As the mRNA sequence is "read" within the ribosome, something grows from it: a chain of amino acids that have been linked together in the ribosome in the order specified by the mRNA sequence. When the chain is finished and folded up, a protein has come into existence. And how do amino acids get to the ribosomes? They are

| Table 14.1 Amino Acids | |
|---|---|
| **Amino Acid** | **Abbreviation** |
| Alanine | ala |
| Arginine | arg |
| Asparagine | asn |
| Aspartic acid | asp |
| Cysteine | cys |
| Glutamine | gln |
| Glutamic acid | glu |
| Glycine | gly |
| Histidine | his |
| Isoleucine | ile |
| Leucine | leu |
| Lysine | lys |
| Methionine | met |
| Phenylalanine | phe |
| Proline | pro |
| Serine | ser |
| Threonine | thr |
| Tryptophan | trp |
| Tyrosine | tyr |
| Valine | val |

**Transcription**

1. In transcription, a section of DNA unwinds and nucleotides on it form base pairs with nucleotides of messenger RNA, creating an mRNA chain.

2. This segment of mRNA then leaves the cell nucleus, headed for a ribosome in the cell's cytoplasm, where translation takes place.

**Translation**

3. Joining the mRNA chain at the ribosome are amino acids, brought there by transfer RNA molecules. The length of messenger RNA is then "read" within the ribosome. The result? A chain of amino acids is linked together in the order specified by the mRNA sequence.

4. When the chain is finished and folded up, a protein has come into existence.

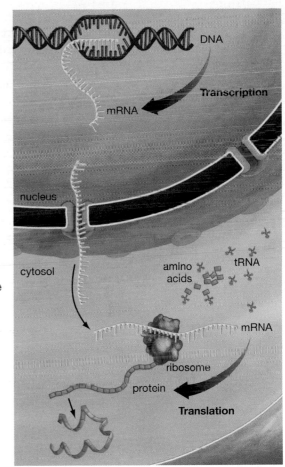

**Figure 14.2
The Two Major Stages of
Protein Synthesis**

**SO FAR...**

1. Though there are hundreds of thousands of proteins active in living things, all of them are put together from a starting set of _____ of the building blocks known as _____.

2. The information for building proteins encoded in DNA is passed on first to _____, which then migrates to an organelle called a _____, where the protein is put together.

3. Amino acids are brought to the _____ by another form of RNA, called _____.

brought there by a second type of RNA, transfer RNA (tRNA).

As may be apparent from this account, protein synthesis divides neatly into two sets of steps. The first set is called **transcription**: the process by which the genetic information encoded in DNA is copied onto messenger RNA. The second set is called **translation**: the process by which information encoded in messenger RNA is used to assemble a protein at a ribosome. Let's look in more detail now at both of these processes, starting with transcription.

## 14.3 A Closer Look at Transcription

From what's been reviewed so far, you can see that a key player in transcription (and translation) is RNA, whose full name is ribonucleic acid. As it turns out, RNA is structurally very similar to DNA. For one thing, RNA has the sugar and phosphate "handrail" components you saw in DNA last chapter. Then, in both molecules,

••• ANSWERS
1. 20; amino acids
2. messenger RNA (mRNA); ribosome
3. ribosome; transfer RNA (tRNA)

this two-part structure is joined to a third element, a base, as you can see in **Figure 14.3**.

There are, however, differences between DNA and RNA. For one, RNA is usually single stranded, whereas DNA is structured in two strands that form its famous double helix. Beyond this, recall that any given DNA building block or "nucleotide" has one of four bases: adenine, guanine, cytosine, or thymine (A, G, C, or T, respectively). RNA uses the first three of these, but then substitutes uracil (U) for the thymine (T) found in DNA.

## Passing on the Message: Base Pairing Again

Given the chemical similarity between DNA and RNA, it's not hard to see how DNA's genetic message can be passed on to messenger RNA: Base pairing is at work again. Recall from Chapter 13 how DNA is replicated. The double helix is unwound, after which bases along the now-single DNA strands are paired up with complementary DNA bases. Every T on a single DNA strand is paired with a free-floating A, and every C is linked with a G, thus yielding a complementary DNA *chain*. Because of RNA's similarity to DNA, the bases RNA has can also form base pairs with DNA. The twist to this RNA-DNA base pairing is that each A on a DNA strand links up with a U on

the RNA strand, instead of the T that would be A's partner in DNA-to-DNA base pairing.

You can see how transcription works by looking at **Figure 14.4**. It will come as no surprise to you that an enzyme is critically involved in this process. **RNA polymerase**, as this complex of enzymes is known, actually undertakes two critical tasks: It unwinds the DNA sequence and then strings together the chain of RNA nucleotides that is complementary to it, thus producing the initial RNA chain.

In organisms such as ourselves (the eukaryotes), this initial chain is not actually the finished messenger RNA we've been talking about. Instead, it is an RNA chain called a *primary transcript* that must undergo some modification or "editing" before it qualifies as messenger RNA. You can read more about this editing in section 14.5, on genetic regulation. For now, we'll set it aside and follow the fate of a completed mRNA chain. Before leaving the subject of mRNA, however, here's a more formal definition of it: **Messenger RNA (mRNA)** is a type of RNA that encodes, and carries to ribosomes, information for the synthesis of proteins.

## A Triplet Code

Thus far, we have been looking at what might be called the flow of genetic information (from DNA

---

**Figure 14.3**
**RNA and DNA Compared**

**(a)** Each building-block nucleotide of both DNA and RNA is composed of a phosphate group, a sugar—ribose in RNA, deoxyribose in DNA—and one of four bases. RNA and DNA both use the bases adenine, guanine, and cytosine (A, G, and C), but RNA uses the base uracil (U) instead of the thymine (T) that DNA uses.

**(b)** Both RNA and DNA amount to linked chains of these nucleotides. The "handrails" of both RNA and DNA are composed of the sugar molecule of one nucleotide linked to the phosphate molecule of the next nucleotide (thus the S-P-S-P labeling). The "stairsteps" stemming from the handrails are formed by the bases of the nucleotides, as with the G and C bases extending inward at the top of the DNA strand. While DNA is double stranded, RNA generally is single stranded.

**(a)** Comparison of RNA and DNA nucleotides

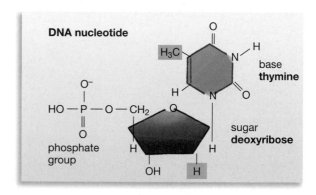

**(b)** Comparison of RNA and DNA three-dimensional structure

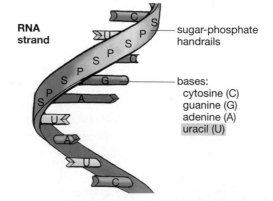

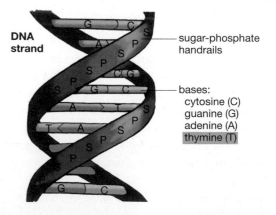

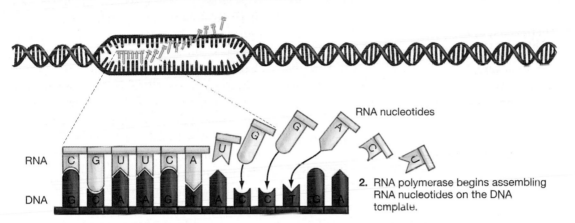

1. RNA polymerase unwinds a region of the DNA double helix.

RNA nucleotides

RNA

DNA

2. RNA polymerase begins assembling RNA nucleotides on the DNA template.

RNA

3. The completed portion of the RNA transcript separates from the DNA. Meanwhile, RNA polymerase unwinds more of the untranscribed region of the DNA.

RNA

4. The RNA transcript is released from the DNA, and the DNA is rewound into its original form. Transcription is completed.

**Figure 14.4**
**Transcription Works through Base Pairing**

Thanks to their chemical similarity, DNA and RNA can engage in base pairing, and this base pairing is how RNA transcripts are synthesized. The enzyme complex RNA polymerase undertakes two tasks in transcription: It unwinds the DNA sequence to be transcribed, and it brings together RNA nucleotides with their complementary DNA nucleotides, thus producing an RNA chain.

**Figure 14.5**
**Triplet Code**

Each triplet of DNA bases codes for a triplet of mRNA bases (a codon), but it takes a complete codon to code for a single amino acid.

to mRNA). A separate issue in protein synthesis, however, has to do with the linkage between DNA and RNA on the one hand and amino acids on the other. The question here is: How many DNA bases does it take to code for an amino acid? As **Figure 14.5** shows, the answer is three. Each three bases in a DNA sequence pair with three mRNA bases, but each group of three mRNA bases then codes for a *single* amino acid. Each coding triplet of mRNA bases is known, appropriately enough, as a **codon**.

A question that follows is this: If you have a given three mRNA bases, *which* amino acid do they specify? In Morse code, the sounds . _ . (short-long-short) code for the letter *R*. But if we know that a given

**The Triplet Code**

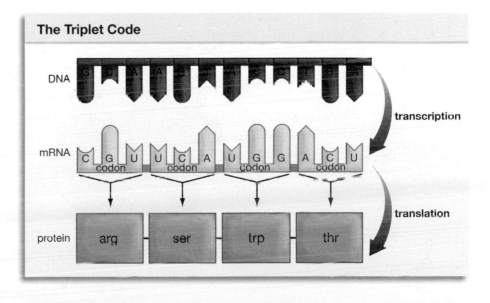

DNA

transcription

mRNA

codon    codon    codon    codon

translation

protein    arg    ser    trp    thr

mRNA codon has the base sequence UCC, what *amino acid* does this code for? Today, we know that the answer is serine (ser), but at one time this was not clear. Indeed, it took years for scientists to figure out all the linkages between codons and amino acids. When this work was completed, the result was the **genetic code**, which can be defined as the inventory of linkages between nucleotide triplets and the amino acids they code for. If you look at "Cracking the Genetic Code" on page 256, you can learn more about the importance of this inventory.

## 14.4 A Closer Look at Translation

With a length of RNA having first been transcribed from a length of DNA and having then moved to a ribosome as mRNA, many of the players are in place for translation—the second stage of protein production. What's also needed, however, are the building blocks of proteins, amino acids. As noted, they are brought to ribosomes by a second form of RNA, called transfer RNA (or tRNA).

### The Nature of tRNA

If you look at **Figure 14.6**, you can see why tRNA is aptly placed within the translation phase of pro-

tein synthesis. When, in everyday speech, we think of a *translator*, we think of someone who can communicate in *two* languages. Transfer RNA effectively does this. One end of each tRNA molecule links with a specific *amino* acid, which it finds floating free in the cytoplasm. Then, transferring this amino acid to the ribosome, this tRNA molecule binds with a *nucleic* acid—a triplet of bases on the mRNA that is moving through the ribosome. Thus, tRNA is binding with two very different kinds of molecules—amino acids on the one hand, nucleic acids on the other—thereby serving as a translator between them. **Transfer RNA (tRNA)** can be defined as a form of RNA that, in protein synthesis, binds with amino acids, transfers them to ribosomes, and then binds with messenger RNA.

**Figure 14.7a** provides a more detailed look at how transfer RNA carries out this function. You can see that it binds with an mRNA codon by means of three bases it possesses, called an **anticodon**. At its other end (the 12 o'clock position in the figure), there is an attachment site for the amino acid. The linkage between amino acids and tRNA molecules is specific enough that any tRNA that binds to an mRNA codon will be carrying an appropriate amino acid. **Figure 14.7b**

**Figure 14.6
Bridging Molecule**

Transfer RNA (tRNA) molecules link up with amino acids on the one hand and mRNA codons on the other, thus forming a chemical bridge between the two kinds of molecules in protein synthesis. They also transfer amino acids to ribosomes, as shown in the steps of the figure.

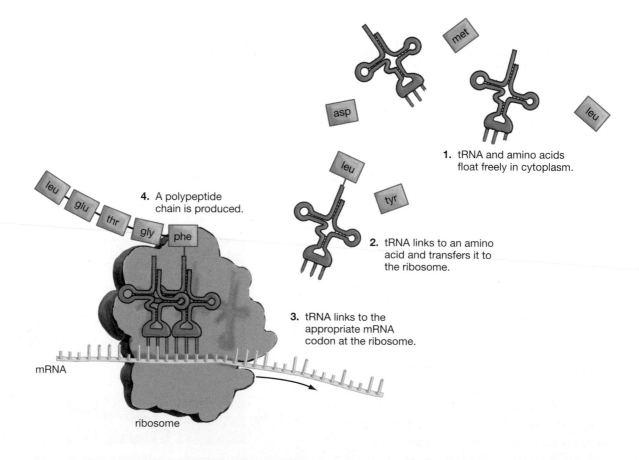

**4.** A polypeptide chain is produced.

**1.** tRNA and amino acids float freely in cytoplasm.

**2.** tRNA links to an amino acid and transfers it to the ribosome.

**3.** tRNA links to the appropriate mRNA codon at the ribosome.

mRNA

ribosome

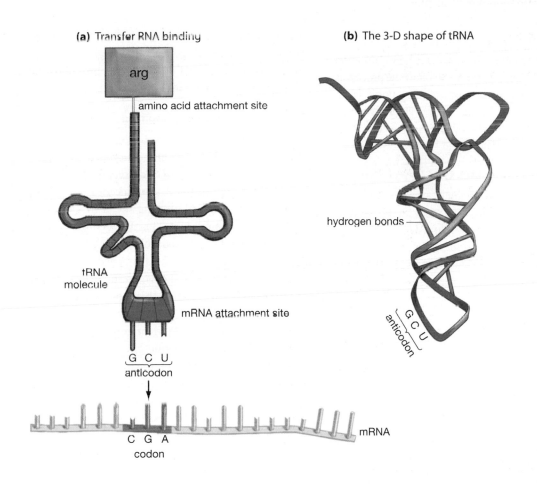

**(a)** Transfer RNA binding

arg

amino acid attachment site

tRNA molecule

mRNA attachment site

G  C  U
anticodon

C  G  A
codon

mRNA

**(b)** The 3-D shape of tRNA

hydrogen bonds

G  C  U
anticodon

**Figure 14.7**
**The Structure of Transfer RNA**

**(a)** In this two-dimensional model of transfer RNA, one end of the tRNA molecule can be seen binding to a specific amino acid (arg), while the other end binds to its counterpart mRNA codon (CGA).

**(b)** This three-dimensional model of tRNA shows the molecule in the "folded-up" form it takes when carrying out its dual-binding function.

gives you a more realistic idea of what tRNA looks like in its folded-up form.

## The Structure of Ribosomes

You've now been introduced to almost all the players that will be active in protein synthesis at the ribosome. But what is the structure of the ribosome itself? If you look at **Figure 14.8** on the next page, you can begin to get an idea. Note that ribosomes are composed of two "subunits"—one larger than the other—both made of a mixture of proteins and yet another type of RNA, **ribosomal RNA (rRNA)**. The two subunits may float apart from one another in the cytoplasm until prompted to come together by the process of translation. You can see that when the subunits have been joined, there exist three binding sites (which it is convenient to think of as slots), one of them an "E" site, the next a "P" site, and the third an "A" site. You'll be looking at their roles shortly. **Table 14.2** sets forth the three types of RNA reviewed so far that are active in protein synthesis.

**Table 14.2  Types of RNA**

| Type of RNA | Functions in | Function |
|---|---|---|
| Messenger RNA (mRNA) | Nucleus, migrates to ribosomes in cytoplasm | Carries DNA sequence information to ribosomes |
| Transfer RNA (tRNA) | Cytoplasm | Provides linkage between mRNA and amino acids; transfers amino acids to ribosomes |
| Ribosomal RNA (rRNA) | Cytoplasm | Structural component of ribosomes |

## The Steps of Translation

With all the players introduced, let's see how translation works. To keep things simple, we'll follow the process as it occurs in prokaryotes. Our starting point is an mRNA transcript that is ready to begin binding with a ribosome. Meanwhile, nearby tRNA molecules have linked to their appropriate amino acids.

### mRNA Binds to Ribosome, First tRNA Arrives

In this first step, the mRNA chain arrives at the ribosome and binds to the ribosome's small subunit (**Figure 14.9**). The mRNA codon AUG is the usual "start" codon for a polypeptide chain. Next, a tRNA molecule with the appropriate anticodon sequence (UAC) binds to this AUG codon. This tRNA arrives

bearing its appropriate amino acid, which is methionine (met). Following this, the large ribosomal subunit becomes part of the ribosome, providing the ribosome's A, P, and E binding sites.

### Polypeptide Chain Is Elongated

Next, amino acids will start to be joined together in a chain. As you can see, this elongation process begins with a second incoming tRNA molecule binding to an mRNA codon in the A site. Because it's a CUG codon that has moved into the site, a tRNA with a GAC anticodon binds to it. This tRNA comes bearing the amino acid leucine (leu). Now here's how we get the chain: The met amino acid attached to the tRNA in the P site bonds with the leu amino acid attached to the tRNA in the A site. In this process, the bond is broken between met and its original tRNA.

Once this occurs, a kind of molecular musical chairs ensues: The ribosome effectively shifts one codon to the right. With this, the tRNA that had been in the P site is relocated to the E site. What does the E stand for? Exit. The tRNA in this site bears no amino acid now; soon, it will be ejected from the ribosome altogether. Meanwhile, the tRNA that had been in the A site moves to the P site, and a new codon shifts into the now-vacated A site. You can see what comes next: A tRNA binds with this new codon in the A site. Soon, the growing polypeptide chain will bind with this tRNA's amino acid, and the process will continue.

### Termination of the Growing Chain

There are three separate codons that don't code for any amino acid but that instead act as stop signals for polypeptide synthesis. Any time one of these "termination" codons moves into the ribosome's A site, it doesn't bind with an incoming tRNA but instead brings about a severing of the linkage between the P-site tRNA and the polypeptide chain. Indeed, the whole translation apparatus comes apart at this point, with the polypeptide chain being released to fold up and be processed as a protein. Translation has been completed.

### Speed of the Process; Movement through Several Ribosomes

How fast does this process go? Very fast. An *Escherichia coli* bacterium can string together up to 40 amino acids per second, meaning that an average-sized protein of about 400 amino acids could be put together in 10 seconds. Note, however, that mRNA sequences often are read not by one ribosome, but

**Figure 14.8
The Structure of Ribosomes**

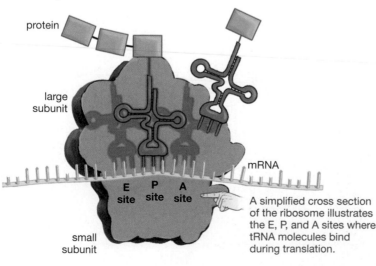

**(a)** Large and small ribosomal units

protein

large
subunit

E P A

small
subunit

mRNA

Ribosomes are composed of two subunits that come together during translation.

**(b)** Binding sites in the ribosome

protein

large
subunit

mRNA

E      P      A
site   site   site

small
subunit

A simplified cross section of the ribosome illustrates the E, P, and A sites where tRNA molecules bind during translation.

# The steps of translation

1. A messenger RNA transcript binds to the small subunit of a ribosome as the first transfer RNA is arriving. The mRNA codon AUG is the "start" sequence for most polypeptide chains. The tRNA, with its methionine (met) amino acid attached, then binds this AUG codon.

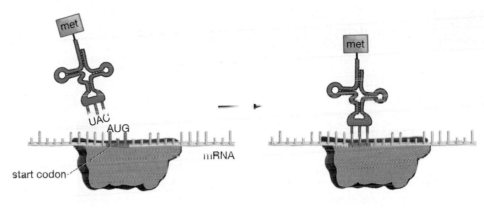

2. The large ribosomal subunit joins the ribosome, as a second tRNA arrives, bearing a leucine (leu) amino acid. The second tRNA binds to the mRNA chain, within the ribosome's A site.

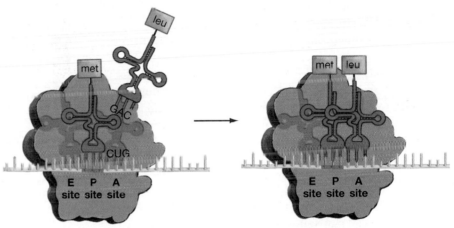

3. A bond is formed between the newly arrived leu amino acid and the met amino acid, thus forming a polypeptide chain. The ribosome now effectively shifts one codon to the right, relocating the original P site tRNA to the E site, the A site tRNA to the P site, and moving a new mRNA codon into the A site.

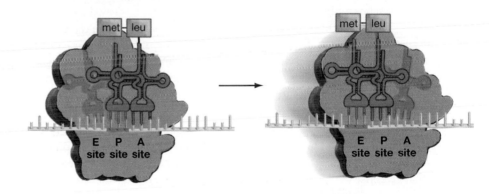

4. The E site tRNA leaves the ribosome, even as a new tRNA binds with the A site mRNA codon, and the process of elongation continues.

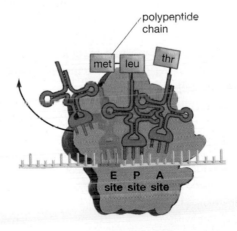

**Figure 14.9**
**The Steps of Translation**

# ESSAY Cracking the Genetic Code

Like scholars trying to extract meaning from an ancient written language, biologists in the early 1960s were trying to extract meaning from an ancient molecular language. They knew that triplets of mRNA bases were coding for amino acids, but what was the linkage? If you had a given RNA triplet, *which* amino acid did it code for? Although it took them the better part of a decade to find out, eventually they succeeded. If you look at **Figure 1**, you can see a summation of their work—the genetic code in its entirety. Let's look a little closer at this code to understand its nature and significance.

One of the notable things about the code is that not every mRNA triplet in it codes for an amino acid. Note, in Figure 1, that three mRNA codons specify "stop" codes that bring an end to the synthesis of a polypeptide chain. One triplet (AUG) specifies a special amino acid, methionine (met), which serves as the starting amino acid for most polypeptide chains. Note also that the genetic code is redundant. The amino acid phenylalanine (phe), for example, is coded for not only by the UUU triplet, but by UUC as well. Indeed, almost all the amino acids are coded for by more than one mRNA codon. Leucine (leu) and serine (ser) are coded for by no fewer than six.

What is the importance of knowing this code? For one thing, it gives scientists a powerful tool. If they know one "end" of a genetic chain, they are instantly given insight into the other. Knowing something about a DNA sequence means knowing something about a protein sequence—and vice versa. What's the practical importance of this? Well, in the 1980s, researchers wanted to produce a genetically engineered version of a protein, called erythropoietin, that is manufactured in the kidneys and that stimulates the production of red blood cells. (Kidney disease patients often suffer from anemia—a lack of red blood cells—because their kidneys can't produce enough erythropoietin.) The researchers started with a known entity, which was a portion of the amino acid sequence of erythropoietin. With this in hand, the genetic code allowed them to "work backward" to learn the DNA sequence that gives rise to this protein. If they saw a tryptophan (trp) amino acid at one point in the erythropoietin amino acid chain, that meant that a UGG triplet had to exist at a corresponding point in erythropoietin's messenger RNA chain. And each mRNA triplet has, of course, a counterpart DNA triplet. Thus, the researchers could work backward—from protein to mRNA to DNA—to piece together portions of the sequence of the erythropoietin *gene*. Doing this allowed them to find this gene within the human genome, after which they cloned it. Today, erythropoietin is manufactured not just in the human body, but also in biotech facilities.

Apart from providing the critical linkage between amino acid and DNA sequences, the cracking of the genetic code also had the effect of revealing how unified life is at the genetic level. As biologists looked among various species at the linkages between DNA sequences and amino acids, they found that, with only a few exceptions, the genetic code is universal in all living things. This means that the base triplet CAC codes for the amino acid histidine, whether this coding is going on in a bacterium or in a human being.

second base

| first base | U | C | A | G | third base |
|---|---|---|---|---|---|
| U | UUU } phe | UCU } ser | UAU } tyr | UGU } cys | U |
| | UUC | UCC | UAC | UGC | C |
| | UUA } leu | UCA | UAA Stop | UGA Stop | A |
| | UUG | UCG | UAG Stop | UGG trp | G |
| C | CUU } leu | CCU } pro | CAU } his | CGU } arg | U |
| | CUC | CCC | CAC | CGC | C |
| | CUA | CCA | CAA } gln | CGA | A |
| | CUG | CCG | CAG | CGG | G |
| A | AUU } ile | ACU } thr | AAU } asn | AGU } ser | U |
| | AUC | ACC | AAC | AGC | C |
| | AUA | ACA | AAA } lys | AGA } arg | A |
| | AUG met (start) | ACG | AAG | AGG | G |
| G | GUU } val | GCU } ala | GAU } asp | GGU } gly | U |
| | GUC | GCC | GAC | GGC | C |
| | GUA | GCA | GAA } glu | GGA | A |
| | GUG | GCG | GAG | GGG | G |

**Figure 1**
**The Genetic Code Dictionary**
If we know what a given mRNA codon is, how can we find out what amino acid it codes for? This dictionary of the genetic code offers a way. Say you want to confirm that the codon CGU codes for the amino acid arginine (arg). Looking that up here, C is the first base (go to the C row along the "first base" line), G is the second base (go to the G column under the "second base" line), and U is the third (go to the codon parallel with the U in the "third base" line).

This insight turned out to be important in two ways. First, it is evidence that all life on Earth is derived from a single ancestor. How else can we explain all of life's diverse organisms sharing this very specific code? Such a complex molecular linkage—this triplet equals that amino acid—is extremely unlikely to have evolved *more* than once. What seems likely is that it came to be employed in an ancient common ancestor, after which it was passed on to all of this organism's descendants—every creature that has subsequently lived on Earth. Note, then, that this code has been passed on from one generation to the next, and from one evolving species to the next, for billions of years. We humans share in an informational linkage that stretches back billions of years and runs through all the contemporary living world.

Apart from this, the universality of the genetic code has a very practical consequence. It means that genes from one organism can function in another. This has both good and bad consequences for human beings. First the bad: Viruses that cause diseases ranging from colds to AIDS can "hijack" the human cellular machinery for their own purposes precisely because their genes function in human cells. (The human cellular machinery will put together proteins whether these proteins are called for by human or viral DNA.) Now the good news: Using biotechnology processes you'll be learning about in Chapter 15, human beings can today use viruses and bacteria to manufacture all kinds of products, including medicines such as human insulin and human growth hormone. In these cases, human genes are being put to work inside microorganisms. This transferability of genes from one species to another has, as its basis, the universality of the genetic code.

by many. As you can see in **Figure 14.10**, several ribosomes—perhaps scores of them—might move over a given mRNA transcript, with the result that identical polypeptide chains grow out of each ribosome. This greatly increases the number of proteins that can be put together in a given period of time.

**SO FAR...**

1. DNA passes on its information to messenger RNA by means of _____.

2. Each _____ DNA bases code for _____ RNA bases, which code for _____ amino acid(s).

3. A transfer RNA molecule attaches to an _____ on one end and a _____ on the other.

## 14.5 Genetic Regulation

From what you've read so far, you might have gotten the impression that transcription never ends for genes—that all genes are constantly being transcribed, thus bringing about a constant production of the proteins they code for. But a genome that worked like this would be like a restaurant that ceaselessly turned out every dish on its menu, no matter how many customers it had. Life is complex, which means that organisms need to finely tune their production of proteins. To look at but one example of this, after we eat a meal, one of the proteins that comes into play is a hormonal protein, called CCK, that prompts our pancreas to release enzymes that break down the food we've consumed. Now, in a person who hasn't eaten in awhile, the gene for the CCK protein is transcribed only at very low levels. After a person consumes a meal high in proteins and fats, however, the story changes. Now transcription of the CCK gene proceeds at high levels, with the result that much greater quantities of the CCK protein start being produced. In short, the activity of the CCK gene is *regulated*, as is the case with all genes. Because genetic regulation is so complex, you'll only be introduced to a couple of its aspects here. Before you can learn about regulation, however, you need to learn a little more about the entity that gets regulated, the genome.

### A Genome Contains Lots More than Genes

At one time, scientists thought that genes in the genome sat right beside each other like pearls on a string. Beginning in the 1970s, however, scientists began to realize that the human genome contains large segments of DNA that do not code for proteins. Indeed, over time it became apparent that, in all eukaryotic organisms, it is the *coding* sections of the genome that are few and far between. Completion of the Human Genome Project in 2004 revealed that less than 1.2 percent of the human genome codes for proteins. To gain some perspective on this number, consider that although the DNA in any of our cells would stretch to about 6 feet in length if uncoiled, less than 1 inch of this length is protein-coding sequence.

So what is the rest of this DNA doing in the genome? Some may actually be "junk DNA," which is to say DNA that came mostly from outside our genome (from invading viruses) and has never had any function within it. Other noncoding segments, however, seem to have had enabling functions—they have enabled organisms such as ourselves to become more complex and capable, as you'll see. Then,

••• ANSWERS

1. complementary base pairing
2. three; three; one
3. amino acid; messenger RNA codon (triplet)

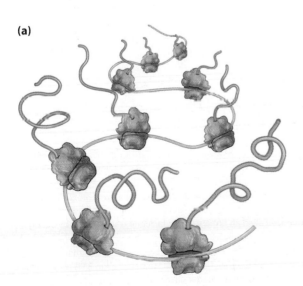

(a)

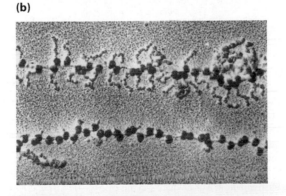

(b)

**Figure 14.10**
**Mass Production**

**(a)** An mRNA transcript can be translated by many ribosomes at once, resulting in the production of many copies of the same protein.

**(b)** A micrograph of this process in operation. The figure shows two mRNA strands with ribosomes spaced along their length. In the upper strand, translation is under way, and polypeptides can be seen emerging from the ribosomes.

beyond these two categories, we have noncoding segments that are regulatory—that are helping to control the production of proteins. Let's look now at a real-life example of genetic regulation to get a first glimpse of some regulatory DNA.

### Promoters, Enhancers, and Proteins that Bind Them

When organisms develop as embryos, there are genes in them that control the development of their midbody or "thoracic" structures, such as the vertebrae that make up their backbones. One of these genes, called *Hoxc8*, is nearly identical in creatures as diverse as chickens, mice, and snakes. Yet, as you can see in **Figure 14.11**, a chicken has 7 vertebrae, while a mouse has 13. So the same developmental gene is at work in these two animals, yet it is yielding very different outcomes in them. If we ask what accounts for

this, the answer is that the mouse *Hoxc8* is being *transcribed* more than the chicken *Hoxc8*. More transcription means the mouse has more of the protein that *Hoxc8* codes for; this in turn means a broader distribution of this protein in the mouse embryo, and the result is more vertebrae in the mouse.

But why would *Hoxc8* get transcribed more in a mouse than a chicken? You may recall that for any gene to be transcribed, an enzyme complex called RNA polymerase has to go down the line on the gene's sequence, pairing up its DNA bases with complementary RNA bases. But this is not an all-or-nothing proposition. RNA polymerase is *more or less* likely to get busy transcribing a gene depending on how it is chemically positioned on a DNA sequence that lies just "upstream" from the gene. This is the "promoter" sequence that you can see in the figure—our first example of noncoding

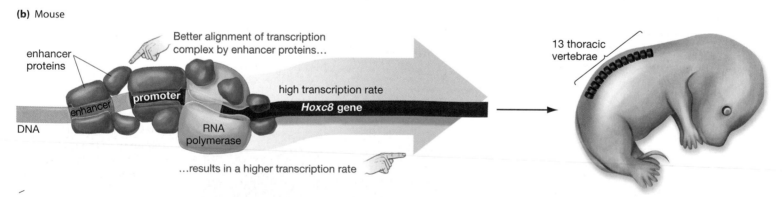

**(a)** Chicken

enhancer proteins

enhancer

promoter

low transcription rate

*Hoxc8* gene

DNA

RNA polymerase

transcription complex

7 thoracic vertebrae

**(b)** Mouse

Better alignment of transcription complex by enhancer proteins…

enhancer proteins

enhancer

promoter

high transcription rate

*Hoxc8* gene

DNA

RNA polymerase

…results in a higher transcription rate

13 thoracic vertebrae

**Figure 14.11**
**Genetic Regulation in Action**

The *Hoxc8* gene regulates chest-level or "thoracic" development in the embryos of a wide range of vertebrate animals, including the chicken and the mouse. The mouse ends up with more vertebrae in its backbone than the chicken does, however, because the mouse *Hoxc8* gene is transcribed with greater frequency.

**(a)** For transcription of any gene to begin, RNA polymerase must be aligned at a sequence of DNA called the promoter, which lies just upstream from a given gene—in this case the *Hoxc8* gene. A multipart protein complex, seen here surrounding RNA polymerase, is active in getting a base level of transcription going. Another DNA sequence, called an enhancer, can facilitate transcription by acting as a binding site for other proteins that can interact with the promoter-site protein complex.

**(b)** The enhancer in the mouse has a different DNA sequence than that of the enhancer in the chicken—a sequence that binds a different set of proteins. These proteins interact with the promoter sequence proteins in such a way as to better align RNA polymerase at the promoter. The result is increased transcription of the *Hoxc8* gene in the mouse, a broader distribution of the *Hoxc8* protein in the mouse embryo, and more vertebrae in the mouse.

DNA. For transcription to take place in any gene, a multipart protein complex must bind with the **promoter sequence** to help RNA polymerase align with it. The question then becomes: What other factors stand to facilitate or impede this alignment? If you look just upstream from the promoter in the figure, you can see a facilitating factor: a sequence called an enhancer. This, too, is a sequence of DNA, but like the promoter, it does not code for anything. Its sole role is to serve as a binding site for proteins that help get RNA polymerase positioned at the promoter. Now, remember that the *Hoxc8* gene is nearly identical in lots of animals. But this is not true of the *Hoxc8* enhancer; *its* sequence differs markedly from chickens, to mice, to snakes. Accordingly, the mouse's enhancer binds with a different set of aligning proteins than does the chicken's. And these mouse proteins more strongly induce transcription than do the chicken's. The result is more *Hoxc8* transcription in the mouse and ultimately more vertebrae in it.

So, note what's at work here. First, there are two stretches of DNA—the promoter and enhancer sequences—that do not code for protein but that do serve a regulatory function (facilitating transcription by RNA polymerase). Second, note what it is that's binding to these regulatory sequences: proteins. One multipart protein complex binds to the promoter sequence (getting a base-level of transcription going), while another complex binds to the enhancer sequence (increasing the rate of transcription). Now, where did these proteins come from? They were produced through transcription and translation, just like any other protein. Thus, DNA codes for proteins that then *feed back* on DNA itself, controlling its transcription. (Appropriately enough, these DNA-binding proteins are called transcription factors.) With this, you can begin to

see what lies at the root of genetic regulation: The whole system is *self*-regulating. DNA brings forth proteins and some of these proteins then go on to control the transcription of DNA.

Finally, notice that it is not necessary to have a change in the sequence of a gene to have a change in the effect of that gene. The mouse and the chicken have the same *Hoxc8* gene; where they differ is in their segments of regulatory DNA for this gene. Over evolutionary time, the mutation you've been reading about worked to make the *Hoxc8* regulatory segments different in the two species. So when we think of mutation working hand-in-hand with evolution to channel species in different directions, it's not necessarily a mutation of *genes* that we're talking about; it may be a mutation of regulatory DNA.

## More Regulation: Alternative Splicing

When the first stage of the Human Genome Project was completed in 2001, biologists got a surprise: The human genome was shown to contain far fewer genes than the 100,000 that had been the standard estimate for years. The teams that sequenced the genome first came up with an estimate of about 30,000 human genes and then, three years later, produced the estimate that still stands today—between 20,000 and 25,000 genes. These figures actually were surprising in two ways: We learned from them not only that we have fewer genes than previously imagined, but that some rather humble organisms have nearly as many genes as we do. You can see some comparisons in **Figure 14.12**. Note that we humans—with our vision, speech, and 10 trillion cells—have, at most, 6,000 more genes than the microscopic roundworm *Caenorhabditis elegans*, which is eyeless, speechless, and made up of 959 cells. (Indeed, we probably have

**Figure 14.12**
**Not as Much Difference as We Thought**

At one time, scientists assumed that the human genome contained about 100,000 genes. Genome sequencing revealed, however, that the human genome probably contains about 20,000–25,000 genes. This is, at most, about 6,000 more genes than the tiny roundworm *C. elegans* has and about 12,000 more than the *Drosophila* fruit fly has. Scientists can make comparisons among genomes because, although the human genome has gotten most of the attention, the genomes of many other organisms have now been sequenced as well, including those shown here.

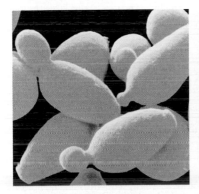

*Saccharomyces cerevisiae*
(baker's yeast)

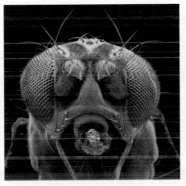

*Drosophila melanogaster*
(fruit fly)

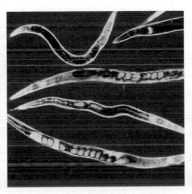

*Caenorhabditis elegans*
(roundworm)

*Arabidopsis thaliana*
(mustard plant)

| Estimated number of genes: | 6,034 | 13,061 | 19,099 | 25,000 |

*fewer* genes than a mustard plant.) So how are we able to do so much more with our 25,000 genes than the roundworm does with its 19,000? The explanation lies in part in a second form of genetic regulation.

Remember how, back in this chapter's section on transcription, we noted that the initial length of RNA that is transcribed from DNA—the primary transcript—has to undergo some editing before it becomes messenger RNA? **Figure 14.13** presents the classic view of how this editing takes place. As you can see, the primary transcript has some sequences cut out of it, after which the sequences that remain are spliced back together. The sequences that are cut out are called introns (for *int*ervening sequences), while the sequences that are retained are called exons (because most of them are *ex*pressed as proteins). It is the exons that code for protein, in other words, while the introns are not coding for protein. Given the exons' more important role, you might think that a gene would mostly be made up of them, but the reverse is true: Introns account for more than 90 percent of the length of the average human gene. (So why are these seemingly useless sequences in the genome at all? Stay tuned.)

Note that, in the classic view, all the introns are edited out, while all the exons are retained. But if you look at **Figure 14.14**, you can see that many genes are more versatile than this in that their primary transcripts can be edited in different ways. Each of these alternative edits then yields a different messenger RNA, and each of these mRNAs can yield a different protein. What's at work here is another form of genetic regulation, this one known as **alternative splicing**, a process in which a single primary transcript can be edited in different ways to yield multiple messenger RNAs. This is how we do so much with our relatively small number of genes.

Thanks to alternative splicing, our 20,000–25,000 genes don't produce 20,000–25,000 proteins; they produce more than 90,000 proteins. Under one estimate, up to 75 percent of human genes can undergo alternative splicing, whereas for an organism like the roundworm, the figure is about 10 percent.

But what is it that makes our higher level of alternative splicing possible? One factor is our higher number of introns. Over evolutionary time, some classes of introns have inserted themselves into genomes such as ours, something like genetic squatters. When such an event is followed by a mutation within an intron, the effect can be the creation of a new *exon*. (The mutation turns a noncoding DNA sequence into a coding sequence.) Then, when the genetic machinery gets busy editing a primary transcript, it can leave the new exon in (thus making protein A) or take this exon out (thus making protein B). Hence, introns have served as a kind of raw material for the creation of new exons. While this function is not strictly regulatory, it has been *enabling*, we might say—it has enabled organisms such as ourselves to produce more proteins and hence to become more complex. We'll return to this subject soon; for now, just note that we human beings have more introns per gene than any other organism.

## RNA in Genetic Regulation

The message "DNA makes RNA makes protein" has come through loud and clear in this chapter, but this message actually needs a little modification. DNA does indeed code for RNA, but not all RNA codes for proteins. Transcription can also produce noncoding RNA segments that are *regulatory*. A collective name for one variety of these segments is *micro-RNAs*—"micro" because they are only about 22 bases long. So far, about 1,500 micro-RNA transcripts have been

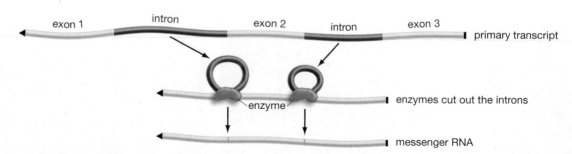

**Figure 14.13**
**Editing Out Noncoding Sequences**
In eukaryotes, lengths of DNA that are transcribed contain sequences that code for amino acids as well as sequences that do not code for them. Both kinds of sequences are copied onto primary transcripts of RNA. Once this is done, however, the noncoding sequences (called introns) are removed by enzymes through the editing process shown, while the coding sequences (called exons) are spliced back together. The result is a messenger RNA molecule.

discovered, and the betting is that many more actually exist. All micro-RNAs identified to date have the effect of reducing the production of specific proteins, and the most common means by which they do this is well known. They interfere with *messenger* RNAs, actually targeting them for destruction. There is a micro-RNA, for example, that targets insulin messenger RNAs. This micro-RNA is transcribed from a small stretch of DNA; it then migrates out of the cell's nucleus and into the cell's cytoplasm, where it sets into motion a process by which insulin mRNAs are cut in two (or otherwise interfered with). The final result is a diminished production of insulin.

Now, it may seem strange that our genome would first produce messenger RNAs that it then goes on to destroy, but remember that our genome is the product of evolution, which doesn't design perfect systems, but merely shapes working systems. (If it designed perfect systems, why would any of us need to wear glasses?) In the case of micro-RNAs, evolution has produced something that works as an additional layer of genetic control. In some instances, it may be cost-effective to "silence" genes after they have been transcribed rather than to stop their transcription altogether.

## The Importance of Regulation

Having looked at a few examples of genetic regulation in action, it may be tempting to think of these processes as a kind of fine-tuning of the protein production process. But an argument can be made that it is genetic regulation, rather than genes per se, that is the most important factor in bringing about the differences between living things—the differences between whales and bats and human beings, for example. Earlier, you saw the difference that regulatory DNA made in connection with mice and chickens. Now consider a regulatory example that is a little closer to home.

One of the things that everyone seems to know about genetics is that the genomes of chimpanzees and human beings are "nearly identical." In practice, this means that if we were to look at stretches of human and chimp DNA that are directly comparable, 98.8 percent of the A's, T's, G's, and C's that make up these stretches would line up perfectly. Given this, people understandably ask: if we humans are genetically so similar to chimps, what makes us physically so different from them? Both the human and chimp genomes have now been sequenced, and a recent comparative study of them provides some insight into this question. The authors of the study looked at 907 genes that are active in the livers of both humans and chimps. Of these genes, 14 turned out to be "expressed" more—that is, code for messenger RNAs more—in humans than in chimps. So, what are these genes doing? A surprisingly high 42 percent of them are coding for transcription factors—the proteins we talked about earlier that feed back on DNA, binding to its regulatory sequences. With this, we have a partial answer to our question. Why is it that such small differences between the chimp and human genomes can lead to such large differences between chimps and humans? Because the differences between the two genomes are *regulatory* differences to a high degree. It is not so much a variance in genes, but rather a variance in the control of genes that makes for the differences between our species.

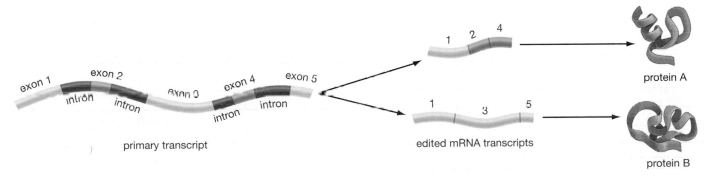

**Figure 14.11**
**Alternative Splicing**

The human genome may contain only 20,000–25,000 genes, but it codes for more than 90,000 proteins, thanks to a form of genetic regulation called alternative splicing. In the figure, a single primary transcript, containing five exons or protein-coding portions, has been transcribed from a single gene. The human genetic machinery can then splice these exons together in alternative ways to yield different proteins—A or B in the figure.

## The Importance of DNA that Doesn't Code for Proteins

If genes aren't central to diversity among living things, what about their role in complexity among them? Well, consider again the comparative gene numbers that can be seen in Figure 14.12. It's pretty clear from these figures that there is little relationship between the number of genes an organism has and the complexity of that organism. What does seem to go hand-in-hand, however, is the complexity of an organism and the proportion of its DNA that does *not* code for proteins. And here, we human beings do shine. Remember how less than 2 percent of our DNA codes for proteins? You can see this proportion in relation to the proportions that exist in some of our less complex fellow organisms in **Figure 14.15**. Perhaps what's most important in making us complex is the 98-plus percent of our genome that doesn't code for proteins.

### What Is Central in Genetics?

If the ideas set forth above about complexity and diversity are correct, then perhaps the gene—defined here as a sequence of DNA that codes for protein—is simply one player in the genetic operation rather than the central player in it. The reason it's necessary to ask whether this is true, however, is that many of the elements of regulation we've just been going over sit right at the cutting edge of contemporary genetics research. The basic process of protein production we looked at earlier—DNA makes RNA makes protein—has been understood for decades, at least in outline. We are only beginning to comprehend, however, the nature of micro-RNAs and the functions of some noncoding portions of the genome. (The term *micro-RNA* wasn't even coined until 2001.) There are researchers who feel that these component parts of genetics are important enough that we have, in essence, *mis*understood the nature of protein production for decades. While the jury is still out on this issue, what is clear is that recent discoveries about regulation and noncoding DNA are forcing biologists to ask fundamental questions about genetics. How does our genetic machinery really work? The hope is that research will soon be providing some answers.

## 14.6 Genetics and Life

The 20,000–25,000 genes that human beings have are contained in a genome that is about 3.2 billion base pairs long. To make this large number a little less abstract, consider the following exercise. If we took the base sequence of the human genome and simply

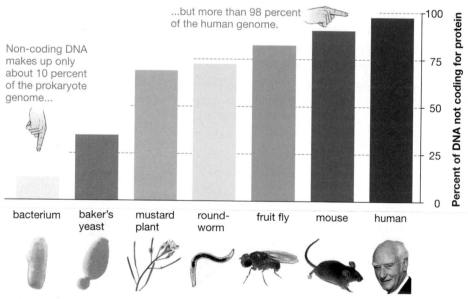

**Figure 14.15**
**What Makes for Complexity in Living Things?**
The seven categories of living things in the figure are arranged from least complex on the left to most complex on the right. Note that complexity in these organisms increases in accordance with the proportion of DNA in their genomes that does not code for proteins. Such noncoding DNA only makes up about 10 percent of the *E. coli* bacterium's genome, but more than 98 percent of the human genome. These figures support the hypothesis that noncoding DNA has been a critical factor in bringing about complexity in the living world. This hypothesis stands in contrast to the idea—at one time prevalent among scientists—that most noncoding DNA has no function and thus amounts to "junk DNA." (Adapted with permission from "Increasing Biological Complexity Is Positively Correlated with the Relative Genome-wide Expansion of Non-Protein-Coding DNA Sequences," Ryan J. Taft and John S. Mattick, *Genome Biology*, 5:1, 23, 2003.)

arrayed the single-letter symbols for the bases on a printed page, going like this:

AATCCGTTTGGAGAAACGGCCCTATTG
GCAGCAAGGCTCTCGGGTCGTCAACG
CGTATTAAACATATTTCAAGGCTCTA...

it would take about 1,000 telephone books, each of them 1,000 pages long, merely to record it all. We're talking, then, about an unbroken series of these base symbols that goes on for a million pages. That is one measure of the size of the human genome. Simply maintaining this genome—making sure that it is passed on from one cell and generation to another—requires that all of these 3.2 billion bases be faithfully copied each time a cell divides. (They have to be copied twice, actually, because we have two copies of each chromosome.) When we start thinking about *using* this genome—about producing proteins from it—all the layers of genetic regulation just sketched out come into play. Genes help bring about proteins; some of these proteins feed back on other genes and control their production of other proteins; RNA sequences get spliced in different ways, yielding different proteins from the same primary transcript, and so forth.

The obvious message here is that the human genetic operation is unimaginably complex. But it turns out that humans are only marginally special in this regard. If we were to walk through the genetic operation of even a humble bacterium, you would see that its complexity is stunning, too. What's important to recognize is that this complexity is necessary because it *enables life*. Living things must obtain energy, they must grow, they must respond constantly to their environment, they must reproduce, and they must coordinate all these activities. Only an incredibly sophisticated system of information storage and use could allow a self-contained entity to do all these things. And genetics is exactly such a system. With each twitch of a muscle or swig of a soft drink, information repositories called genes are opened or closed within us, setting in motion the delivery of the tiny working entities called proteins—or stopping delivery of them, if that's what is called for. Most of

the universe is made up of relatively simple, inanimate objects. But we living things are different. We get to inherit and then use an enormous library of information that was amassed by our ancestors over a period of 3.8 billion years. And this endowment has made us complex, capable, and conscious to greater degree than anything else we know of in the universe. Looked at one way, you might say that all living things are born rich.

## Biotechnology Is Next

If each gene in a genome is something like a book in a library, think what the effect would be if organisms could, in effect, check books out of each other's libraries. Actually, you don't have to imagine what this would be like; it's happening right now with great frequency. Soybeans in the fields of Indiana and Iowa have the genes of bacteria spliced into their genomes. Likewise, human genes have been spliced into the genomes of bacteria so that these one-celled creatures can turn out human medicines. And we are now able to read individual letters in our own genetic library with an efficiency that lets us catch criminals on the basis of flakes of dandruff they leave at crime scenes. This may sound fanciful, but it's not. It's biotechnology, the subject of Chapter 15.

### SO FAR...

1. True or false: Most of the DNA in the human genome codes for proteins.

2. In eukaryotic organisms such as ourselves, the RNA chains, called primary transcripts, that code for proteins must undergo _____ on their way to becoming _____. The segments of the primary transcripts that are removed in this process are called _____, while the segments that are retained are called _____.

3. Small lengths of RNA called _____ have the effect of "silencing" genes in the genome by means of targeting their _____ for destruction.

••• ANSWERS

1. false; less than 1.2 percent of the human genome codes for proteins

2. editing; messenger RNA; introns; exons

3. micro-RNAs; messenger RNAs

# CHAPTER 14 REVIEW

For study help, activities, and more quiz questions, go to **www.aw.com/krogh4**.

## SUMMARY

### 14.1 The Structure of Proteins

- Proteins are composed of building blocks called amino acids. A string of amino acids is called a polypeptide chain. Once such a chain has folded into its working three-dimensional shape, it is a protein. Although there are hundreds of thousands of different proteins, all of them are put together from a starting set of 20 amino acids. It is the order in which the amino acids are linked in a polypeptide chain that determines which protein will be produced. Proteins often are composed of two or more linked polypeptide chains.

**Web Animation 14.1:** Structure of Proteins

### 14.2 Protein Synthesis in Overview

- There are two principal stages in protein synthesis. The first stage is transcription, in which the information encoded in DNA is copied onto a length of messenger RNA (mRNA), which in eukaryotes moves from the cell nucleus to a structure in the cytoplasm called a ribosome. The second stage is translation, in which amino acids brought to a ribosome by transfer RNA (tRNA) molecules are linked together within the ribosome in the order specified by the mRNA sequence.

### 14.3 A Closer Look at Transcription

- The information in DNA is transferred to messenger RNA through complementary base pairing. Each "C" nucleotide in a segment of DNA being transcribed results in a "G" nucleotide being added to a segment of RNA, and so forth. The enzyme RNA polymerase unwinds the DNA sequence to be transcribed and then strings together the chain of RNA nucleotides that is complementary to it. In all eukaryotes (including humans), the initial RNA chain transcribed from a DNA sequence is not the finished messenger RNA chain, but instead is a sequence, called a primary transcript, that must undergo some editing before becoming an mRNA chain.

- Each three coding bases of DNA pair with three RNA bases, but each group of three mRNA bases then codes for a single amino acid. Each triplet of mRNA bases that codes for an amino acid is called a codon. The inventory of linkages between base triplets and the amino acids they code for is called the genetic code.

### 14.4 A Closer Look at Translation

- Transfer RNA serves as a bridging molecule in protein synthesis thanks to its ability to bind with both amino acids on the one hand and nucleic acids on the other (in the form of mRNA).

- A given tRNA molecule binds with a specific amino acid in the cell's cytoplasm and then transfers that amino acid to a ribosome in which an mRNA transcript is being "read." There, another portion of the tRNA molecule, called an anticodon, binds with the appropriate codon in the mRNA chain.

- Ribosomes, the complex "workbenches" of protein synthesis, are composed of proteins and ribosomal RNA (rRNA). Each exists as two subunits in the cytoplasm that come together only with the initiation of protein translation. Each bears A, P, and E binding sites to which tRNA molecules bind, thus facilitating the synthesis of an amino acid chain.

- Translation works by means of a succession of tRNA molecules arriving at a ribosome, bound to their appropriate amino acids, and then binding to their appropriate codon in the mRNA transcript. As this process takes place, the succession of amino acids is linked together into a polypeptide chain.

**BioFlix** Protein Synthesis

### 14.5 Genetic Regulation

- Protein production is carefully controlled or "regulated" in living things. Most genes do not simply stay "on," but instead are transcribed in accordance with the needs of an organism.

- Less than 2 percent of the DNA in the human genome codes for proteins. Some noncoding segments of DNA may be "junk" that never had a function; other segments seem to have served an enabling function—they have enabled organisms to become more complex—and still other segments are regulatory, meaning they help regulate the production of proteins.

- All gene transcription requires that RNA polymerase be properly aligned at a noncoding sequence of DNA bases, called a promoter, that lies just "upstream" from a gene sequence. There is usually a second noncoding segment of DNA, called an enhancer, that lies at some distance from the promoter sequence. Separate groups of proteins, called transcription factors, bind to both the promoter and enhancer sequences, thus facilitating the alignment of RNA polymerase at the promoter. Transcription factors are themselves produced through normal transcription and translation—they are coded for by DNA, but then feed back on it, helping control its transcription. Thus, the entire system is self-regulating.

- In all eukaryotes, the initial RNA chain produced during transcription—the primary transcript—undergoes editing by means of some sequences being cut out of it, after which the remaining sequences are spliced back together. The result is a completed messenger RNA chain. The sequences that are removed from the primary transcript are called introns, while the sequences that are retained are called exons. Introns do not code for protein, but most exons do.

- Some relatively simple organisms have nearly as many genes as human beings do (20,000–25,000). Human beings are able to be much more complex than these organisms, however, thanks in part to a form of genetic regulation called alternative splicing, in which a primary transcript can be edited in different ways. Through this process, a single primary transcript can result in different messenger RNA chains, which in turn can result in different proteins.

- DNA codes for several different forms of RNA, but of these, only messenger RNA then goes on to code for proteins. DNA also codes for several varieties of regulatory RNA that go under the collective name of micro-RNAs. Of the 1,500 micro-RNA sequences discovered to date, all have the effect of reducing the production of particular proteins, usually by targeting their messenger RNAs for destruction.

- A case can be made that it is not genes, but instead the regulation of genes that is the most important factor in bringing about the differences among organisms in the living world. In a similar vein, there seems to be little relation between the number of genes an organism has and the complexity of that organism. What does seem to be correlated is the complexity of an organism and the proportion of its DNA that does not code for protein. Thus, genetic regulation and noncoding DNA may be more important to genetics than has traditionally been assumed. We have no definitive answers about these questions, however, because research on genetic regulation and noncoding DNA lies at the cutting edge of contemporary genetic research.

## 14.6 Genetics and Life

- Life is made possible by the fantastic ability of genetic systems to store, use, and pass on information.

## KEY TERMS

| | |
|---|---|
| alternative splicing   260 | promoter sequence   259 |
| anticodon   252 | ribosomal RNA (rRNA)   253 |
| codon   251 | ribosome   248 |
| genetic code   252 | RNA polymerase   250 |
| messenger RNA | transcription   249 |
| (mRNA)   250 | transfer RNA (tRNA)   252 |
| polypeptide   247 | translation   249 |

## BRIEF REVIEW

*Answers to Brief Review questions are in the back of the book. For multiple-choice quiz questions, go to **www.aw.com/krogh4**.*

1. What two great tasks are carried out by our genetic machinery?

2. What are the three types of RNA that are used by a cell any time a protein is produced? What do their abbreviations stand for? What are their functions?

3. What is the difference between transcription and translation?

4. During transcription, DNA neither unwinds itself nor pairs its own bases with complementary RNA bases. How are these tasks accomplished?

5. How many DNA bases does it take to code for an RNA codon? How many amino acids does an RNA codon code for?

6. What is the difference between introns and exons?

7. Give two examples of mutations that would not affect protein-coding sequences of DNA but that ultimately would affect the production of proteins.

## APPLYING YOUR KNOWLEDGE

1. Whether the subject is living things, businesses, or music, do complexity and regulation always go hand in hand?

2. Why aren't genes transcribed constantly?

3. Given the following sequence of mRNA, what would the resulting amino acid sequence be?

AUGAAACGGGGACCAAUGGAUAACUAA

*Is it possible to produce an artificial spider's thread? Scientists have done it by splicing genes into bacteria—just one of the marvels made possible through biotechnology.*

CHAPTER **15**

| 15.1 | What Is Biotechnology? | 268 |
| 15.2 | Transgenic Biotechnology | 268 |
| 15.3 | Reproductive Cloning | 272 |
| 15.4 | Forensic Biotechnology | 275 |
| 15.5 | Stem Cells | 277 |
| 15.6 | Biotechnology in the Real World | 284 |

### Essays

| Reading DNA Profiles | 278 |
| **THE PROCESS OF SCIENCE:** Making Embryonic Stem Cells Work | 282 |

By manipulating the information contained in DNA, scientists are raising hopes—and some fears—on a grand scale.

# THE FUTURE ISN'T WHAT IT USED TO BE: Biotechnology

**T**hose pork chops you had recently? However good they may have tasted, didn't you wish they could be just a little *healthier*? And maybe it's the same thing with ham or with bacon? Whenever they're served,

does a side order of health-related guilt seem to come with them? If so, some news delivered in 2006 might have been tailor-made for you. In the future, you may be able to walk down the supermarket aisle and pick up pork products that are rich in healthy fats—the omega-3 fatty acids that today are found in greatest abundance in fish. In March 2006, scientists from three universities unveiled some pigs whose meat contained at least four times as many omega-3s as regular pigs. And how were these pigs produced? Through a series of procedures *each one of which* would have been regarded as science fiction not so long ago.

The essential dilemma in producing pork that's high in omega-3s is that pigs lack an enzyme that can convert one kind of fat they have in abundance (omega-6) to the healthier omega-3. On the other hand, there is an organism—a tiny roundworm called *Caenorhabditis elegans*—that does produce this enzyme, thanks to the fact that it has a gene that codes for it. So, what's the solution? Get the roundworm's gene inside pigs, working away in them. That's just what the scientists did. First, they modified this "*fat-1*" gene so that it would work in mammals. Next, they spliced this gene into the pig's genome, meaning its collection of genes. This was accomplished by taking a single pig cell and moving the *fat-1* gene past the cell's outer membrane and into its nucleus, where the gene inserted itself into the pig's DNA.

Finally, this cell was used as the basis for producing a full-grown pig—a pig was *cloned* from this original cell by use of a process that's now become commonplace in science. The result, a few months later, was one of the pigs you can see in **Figure 15.1**.

Now, what did the scientists end up with here? A pig that was seemingly ordinary in every way except that it had a gene from another organism spliced into it—a gene that allowed the pig to convert its own

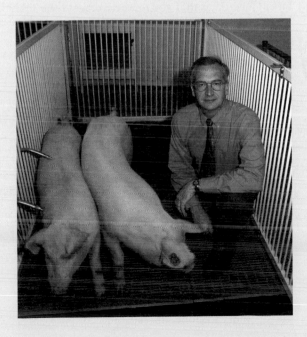

**Figure 15.1**
**Heart-Healthy Pigs through Biotech**
University of Missouri, Columbia researcher Randy Prather poses with two of the pigs who were genetically modified to produce high levels of omega-3 fatty acids. Scientists from the University of Pittsburgh and Massachusetts General Hospital also participated in the research.

omega-6 fat into omega-3 fat. And this gene was not just present in a few cells in the pig. Remember that every cell in an animal is, in a genetic sense, a copy of the original cell from which the animal is formed. In this case, this cell was the one that had the *fat-1* gene spliced into it. So as the pig developed by means of this one cell dividing into two, those two into four, and so on, every one of these cells got its own copy of the *fat-1* gene, including the pig's sperm cells. This meant that the pig could pass the *fat-1* gene on; pigs modified in this way can produce more of their own kind, in other words. The final result could be an endless supply of pigs with meat that is high in omega-3s—a supply big enough such that supermarket shelves could be stocked with their meat.

Now, what are the odds that this supermarket scenario will actually come to pass? Not great. The road to any change this big would be filled with obstacles. But this sequence of events is *possible*, and therein lies the tale. In one instance after another, scientists working today in the field known as biotechnology are upsetting our notions of what is possible and what is impossible. We may assume that all animals must have both a biological mother and father, but the cloned pig we just talked about arguably had neither. We may assume that humans and sheep are fundamentally separate entities, but scientists have been adding human "stem" cells to sheep fetuses, thus producing sheep that have mostly human liver tissues. Even the humble bacterium routinely has human genes inserted into it, which allows it to turn out human proteins for medical use. Meanwhile, if you have eaten anything made from soybeans recently, you probably have eaten food from a plant that was bioengineered. And if you're worried about the survival of the panda, you may take some comfort in knowing that the Chinese government is investigating ways to clone it as a means of staving off its extinction.

Taken together, achievements such as these amount to a revolution; but like all revolutions, this one seems to be inspiring both hope and fear in equal measure. For many people, the production of a cloned pig that is carrying a roundworm gene doesn't mean we're living in a world that is better controlled; it means we're living in a world that is out of control. Forget about the "promise" of biotechnology, this view goes; just put the genie back in the bottle. On the other hand, who wouldn't want healthier foods at lower cost or more medicines for people who are sick?

In thinking about such questions, it's helpful to view biotechnology not as a monolithic entity, but rather as a collection of separate *capabilities*. There is the capability to clone, the capability to splice genes into organisms, and so forth. The purpose of this chapter is to provide you with a basic understanding of some of these capabilities. Once you see what's at work in them, you'll be in a better position to make judgments about the degree to which biotechnology holds out promise or peril.

## 15.1 What Is Biotechnology?

The term *biotechnology* is familiar to most of us, but what does it really mean? **Biotechnology** can be defined as the use of technology to control biological processes as a means of meeting societal needs. One biological process that can be controlled is the way a cell copies its DNA every time it divides. In the 1980s, a California scientist realized there was a way to use technology to exploit this natural copying, such that a tiny DNA sample could rapidly be copied over and over, thus yielding a relatively large DNA sample. One result of this *polymerase chain reaction* or PCR process is that police can now recover, from the tiniest specks of blood—or flakes of dandruff, even—enough DNA to get a "match" in a criminal case. Like other biotech processes we'll be looking at, PCR is a marriage of both biology and technology. By the time you finish this chapter, you'll see why biotechnology is such an appropriate term.

Because biotechnology is so varied, any account of it has to be limited to a few examples chosen from among thousands that exist. In this chapter, we will consider four aspects of biotechnology:

- Transgenic biotechnology—the splicing of DNA from one species into another

- Reproductive cloning—the production of mammals through cloning

- Forensic biotechnology—the use of biotechnology to establish identities (of criminals, of crime victims, and so forth)

- Stem cells—the use of special cells to produce new human tissues

Let's start now by reviewing an area of biotechnology in which genes are moved from one organism to another.

## 15.2 Transgenic Biotechnology

Human beings grow to their full height under the influence of human growth hormone (HGH), which normally is secreted by the human pituitary gland. HGH's role in promoting growth is, of course, most important during childhood and adolescence. A faulty pituitary gland can greatly reduce the amount of HGH young people have in their system, leaving them abnormally short. For years, the only way to get HGH was to laboriously extract it from the pituitaries of dead human beings, a practice that not only yielded too little HGH to go around but that also turned out to be unsafe.

Enter biotechnology, which in the mid-1980s produced synthetic HGH in the following way. Using collections of human cells, the gene for HGH was snipped out of the human genome and inserted into the *Escherichia coli* bacterium. Each bacterium that took on the HGH gene began transcribing and then translating this gene, which is to say turning out a small quantity of HGH protein. These bacteria were then grown in vats by the billions. The result? Collectible quantities of HGH, clinically indistinguishable from that produced in human pituitary glands, manufactured by a biotech firm, and shipped to pharmacies worldwide.

Note that in this process, a gene was taken from one species (a human being) and spliced into another species (a bacterium). With this, the bacterium became a **transgenic organism**: an organism whose genome has stably incorporated one or more genes from another species. This same phenomenon is repeated time and time again in biotechnology. The pigs in Figure 15.1 are transgenic—they incorporated a roundworm gene into their genome. Likewise the goats you can see in **Figure 15.2** are transgenic—they have incorporated a human gene into their genome. But how is this possible? How do you cut DNA out of one genome and paste it into another? To get an answer, we'll take a walking tour of some basic biotech processes. As a starting point, imagine that, as with

human growth hormone, the goal is to produce quantities of a hypothetical human protein through the use of a living organism.

## A Biotech Tool: Restriction Enzymes

In the early 1970s, it became possible for scientists to cut genomes at particular places, with the discovery of **restriction enzymes.** These are enzymes, occurring naturally in bacteria, that are used in biotechnology to cut DNA into desired fragments. (In nature, bacteria use them to cut up the DNA of invading viruses.) In isolating restriction enzymes, scientists found that many of them had a wonderful property: They didn't just cut DNA randomly; they cut it at very specific places. Here is how it works with an actual restriction enzyme called *Bam*HI. The two strands of DNA's double helix are complementary, as we've seen, and they also happen to run in opposite directions. Thus, the sequence GGATCC would look like **Figure 15.3** if we were viewing both strands of the helix

**Figure 15.2**
**Goats, with Human Genes Spliced In**

A firm called GTC Biotherapeutics has specialized in splicing DNA into the genomes of goats. The result is goat's milk that contains therapeutic human proteins, among them a protein called antithrombin that prevents the formation of blood clots.

1. A portion of a DNA strand has the highlighted recognition sequence GGATCC.

"sticky ends"

2. A restriction enzyme moves along the DNA strand until it reaches the recognition sequence and makes a cut between adjacent G nucleotides.

**DNA fragment**

3. A second restriction enzyme makes another cut in the strand at the same recognition sequence, resulting in a DNA fragment.

**Figure 15.3**
**The Work of Restriction Enzymes**

Now, the *Bam*HI restriction enzyme will move along the double helix, leaving the DNA alone until it comes to this series of six bases, known as its recognition sequence, and here it will make identical cuts on both strands of the DNA molecule, always between adjacent G nucleotides. When another *Bam*HI molecule encounters another GGATCC sequence, it also makes cuts, which effectively is like making a second cut in a piece of rope, giving us rope *fragments*.

*Bam*HI's recognition sequence may be GGATCC, but another restriction enzyme will have a different recognition sequence and will make its cuts between a different pair of bases. Indeed, several thousand restriction enzymes have been identified so far, which cut in hundreds of different places. The fact that they make cuts at so many specific locations has given scientists a terrific ability to cut up DNA in myriad ways.

Note that with *Bam*HI each of the resulting DNA fragments has one strand that protrudes. Restriction enzymes that make this kind of cut are particularly valuable for they produce "sticky ends" of DNA, so named because they have the potential to *stick to* other complementary DNA sequences. In the fragment on the lower left in Figure 15.3, for example, the protruding sequence CTAG could now easily form a base pair with any piece of DNA whose sequence is the complementary GATC. So great is the need for restriction enzymes that they can now be ordered from biochemical suppliers, much as a person might order a set of socket wrenches from a hardware store.

## Another Tool of Biotech: Plasmids

From our look at manufacturing human growth hormone, recall that the human gene for HGH was inserted into *E. coli* bacteria, which then started turning

**Web Animation 15.1**
Action of Restriction Enzymes and Plasmids

out quantities of this protein. The question is, how did this human gene get into a bacterium? Several methods of transfer are available, but for now let's focus on a specific kind of DNA delivery vehicle. As it turns out, bacteria have small DNA-bearing units that lie *outside* their single chromosome. These are the **plasmids,** extrachromosomal rings of bacterial DNA that can be as little as 1,000 base pairs in length (**Figure 15.4**). Plasmids can replicate independently of the bacterial chromosome, but just as important for biotech's purposes, they can *move into* bacterial cells.

How do plasmids do this? Bacteria are capable of taking up DNA from their surroundings, after which this DNA will function—that is, code for proteins—inside the bacterial cells. Appropriately enough, this process is known as **transformation**: a cell's incorporation of genetic material from outside its boundary. Some bacterial cells are naturally adept at transformation, while others, such as *E. coli*, can be induced to perform it by means of chemical treatment. Critically, plasmid DNA can be taken in via transformation and continue to function, as plasmid DNA, inside the bacteria.

## Using Biotech's Tools: Getting Human Genes into Plasmids

At this point, you know about a couple of tools in the biotech tool kit: restriction enzymes and the transformation process involving plasmids. Let's now see how they work together. As you can see in **Figure 15.5**, the process starts with a gene of interest in the human genome. The first step is to use restriction enzymes on this human DNA. Knowing, say, the starting and ending sequence of the gene of interest, a restriction enzyme is selected that allows part of the genome to be cut into a manageable fragment including this sequence of interest, preferably in a sticky-ended form.

Here's where the beauty of restriction enzymes really comes into play. If the same restriction enzyme is now used on the DNA of isolated *plasmids*, the result is complementary sticky ends of plasmid and human DNA. In other words, these segments of human and plasmid DNA, through sticky-ended base pairing, fit together like puzzle pieces.

When the DNA fragments are mixed with the "cut" plasmids, that's just what happens: Human and plasmid DNA form base pairs, and the human DNA is incorporated into the plasmid circle. With this, a segment of DNA that was once part of the human genome has now been *re-combined* with a different stretch of DNA (the plasmid sequence). This process gives us the term **recombinant DNA,** defined as two or more segments of DNA that have

**Figure 15.4**
**Transfer Agent**

Plasmids are small rings of bacterial DNA that are not a part of the bacterial chromosome. They can exist outside bacterial cells and then be taken up by these cells. This artificially colored micrograph shows a type of plasmid, from *E. coli* bacteria, that is commonly used in genetic engineering.

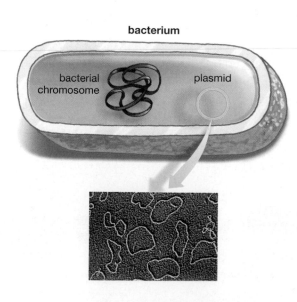

**bacterium**

bacterial chromosome

plasmid

Figure 15.5
How to Use Bacteria to
Produce a Needed Human
Protein

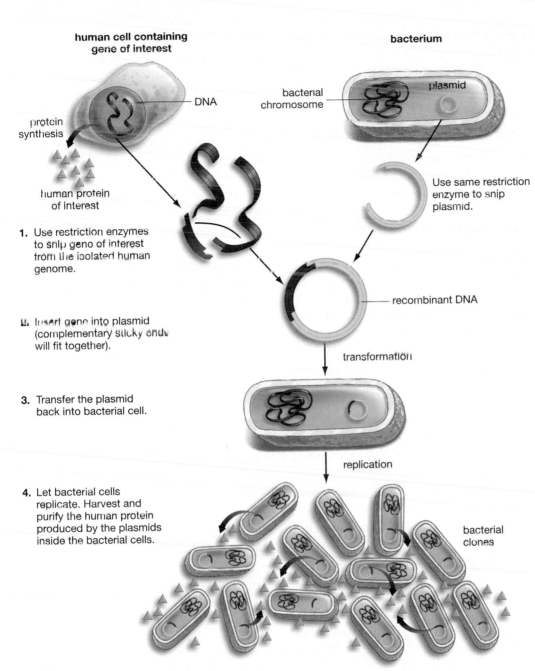

human cell containing
gene of interest

bacterium

DNA

bacterial
chromosome

plasmid

protein
synthesis

human protein
of interest

Use same restriction
enzyme to snip
plasmid.

1. Use restriction enzymes
to snip gene of interest
from the isolated human
genome.

recombinant DNA

2. Insert gene into plasmid
(complementary sticky ends
will fit together).

transformation

3. Transfer the plasmid
back into bacterial cell.

replication

4. Let bacterial cells
replicate. Harvest and
purify the human protein
produced by the plasmids
inside the bacterial cells.

bacterial
clones

been combined by humans into a sequence that does not exist in nature.

## Getting the Plasmids Back inside Cells, Turning out Protein

To this point, what has been done is to produce a collection of independent plasmids having a human gene as part of their makeup. Remember, though, the goal is to turn out quantities of whatever protein the human gene is coding for. To do this requires a vast quantity of plasmids working away, which means working away *back inside* bacterial cells, and it's here that transformation comes into play. If the plasmids are put into a medium containing specially treated bacterial cells, a few of these cells will take up plasmids

through transformation (at which point these cells have become transgenic, as noted earlier). Once this happens, the plasmids start *replicating* along with the bacterial cells themselves. As one bacterial cell becomes two, two become four, and so on, the plasmids are replicating away as well. And as the cell count reaches into the billions, collectible quantities of the protein begin to be turned out via instructions from the human gene inserted into the plasmid DNA.

## A Plasmid Is One Kind of Cloning Vector

In this example, plasmids were the vehicle that served to take on some foreign DNA and then to ferry it into working bacterial cells. Note, however, that plasmids are only one vehicle among several that can be used.

Collectively, such vehicles are known as **cloning vectors**, meaning self-replicating agents that serve to transfer and replicate genetic material. Next to plasmids, the most common cloning vector is a type of virus that infects bacteria—a bacteriophage.

### Real-World Transgenic Biotechnology

The example we just went over concerned a hypothetical medicine, but lots of actual medicines are being produced through biotechnology today. By one industry count, more than 250 biotechnology drugs and vaccines have been approved by the U.S. Food and Drug Administration. We've already touched on one biotech medicine: the human growth hormone that is produced inside the *E. coli* bacterium. As it turns out, *E. coli* is something of a workhorse in medical biotech. Human insulin and several kinds of cancer-fighting compounds are also produced through it.

*E. coli* and its fellow bacteria are not the only transgenic "bio-factories" used in medical biotechnology, however. Yeast, hamster cells, and even mammals are used in this way as well. Goats of the type pictured in Figure 15.2 produce several kinds of human proteins in their milk, including one called antithrombin that prevents the formation of blood clots. You might wonder why large, expensive goats would be used to turn out proteins when small, inexpensive bacteria are available. The answer is that lactating goats naturally produce a tremendous amount of protein: A single goat can produce over a kilogram (2.2 pounds) of needed human protein per year. You also might wonder how a human gene can be spliced into a goat genome. The answer is, through the splicing of a human gene into a single goat cell and then a cloning of an adult goat from that cell—the same process we looked at earlier with pigs.

### Genetically Modified Food Crops

Animals are not the only kind of transgenic organisms. You may have heard of genetically modified or "GM" food crops. It won't surprise you to hear that all of these crops are transgenic. If you look at **Figure 15.6**, you can see a transgenic rice that has been dubbed "golden rice" because of its ability to produce its own beta carotene, which the human body converts to vitamin A. Conventional rice does not contain beta carotene, and this fact has health consequences. The World Health Organization estimates that half a million children go blind each year from vitamin A deficiency, and that one to two million children die from it. The hope is that golden rice can help alleviate this problem (though, as you'll see, this is a hope that has existed for a long time now). Golden rice actually is transgenic in more ways than one: Genes from both a bacterium and a daffodil have been spliced into it, thus enabling it to produce the beta carotene.

Though the public seems scarcely aware of it, several other transgenic food crops are in full commercial production right now. Indeed, in the United States, 80 percent of all soybeans, 79 percent of all cotton, and 52 percent of all corn crops currently are genetically modified, though only for a couple of limited purposes. One of these is to allow the crops to resist insects; the other is to allow them to tolerate the chemical weed killers known as herbicides. So, why is biotechnology's role in agriculture so limited? Stay tuned.

---

**SO FAR...**

1. A transgenic organism is one whose genome has stably incorporated one or more _____ from _____.

2. DNA is cut into useful fragments through use of the molecular scissors known as _____.

3. The plasmids noted in the text had _____ DNA spliced into them as a first step in using them to turn out a human _____.

---

## 15.3 Reproductive Cloning

The concept of *cloning* has come up a couple of times so far. Thanks mostly to movies, this word has taken on some sinister implications, as if biological cloning were inherently a Frankenstein-like procedure. But it need be no more threatening than the process of making copies of a single gene. To **clone** simply means "to make an exact genetic copy of." An

**Figure 15.6**
**A More Nutritious Rice**
Grains of the genetically engineered "golden rice" stand out next to grains of ordinary rice. The rice has a golden color because it is able to produce its own beta carotene, which the human body converts into Vitamin A.

individual clone is one of these exact genetic copies. Thus, in the biotech world, the gene for human growth hormone is cloned by means of being snipped out of the human genome and then copied within bacterial cells. A collection of such cells can be thought of as a single clone because all of them are genetically identical. (One bacterium splits into two, and both "daughter" cells are genetic replicas of the parent cell.)

Cloning can involve not just bacteria, however, but larger organisms as well. Human beings actually have been making clones for centuries, although by low-tech rather than high-tech means. A "cutting" taken from a plant and put into soil can sometimes grow into a whole new plant. When this happens, there is no mixing of eggs and sperm from two different plants. The new plant comes entirely from one individual—the original plant—and thus meets the definition of a clone in that it is an exact genetic copy of another entity.

All this said, in recent years biotechnology has greatly expanded the range of what can be cloned. The potential for such cloning appears to be very broad indeed—something that has inspired both great optimism and great concern. Biotechnology has now joined with various reproductive technologies to produce clones not just of genes or cells or plants, but of mammals, with the starting cells for these mammals coming from adult mammals. Though we've mentioned a couple of these cloned animals already, the most famous one by far is Dolly the sheep, cloned by researcher Ian Wilmut and his colleagues in Scotland in 1997 (**Figure 15.7**). Dolly and similar animals are a product of **reproductive cloning:** cloning intended to produce adult mammals of a defined genotype. Powerful in its own right, this procedure gains added potential when combined with the basic biotech processes you've been reviewing.

### Reproductive Cloning: How Dolly Was Cloned

Dolly, who died in 2003, was a clone as that term is defined: She was, to a first approximation, an exact genetic replica—in this case, of another sheep. Here's how she was produced (**Figure 15.8** on the next page). A cell was taken from the udder of a six-year-old adult sheep and then grown in culture in the laboratory, meaning that the original cell divided into many "daughter" cells. While this was going on, researchers took an *egg* from a second sheep and removed its nucleus, meaning they removed all its nuclear DNA. Then, they placed the udder cell (which had DNA) next to the egg cell (which did not have DNA) and applied a small electric current to the egg. This had two effects: It caused the two cells to fuse into one, and it mimicked the stimulation normally provided when a *sperm* cell fuses with an egg. With this, the udder cell DNA began to be reprogrammed. As a result of this reprogramming, the fused cell started to develop as an embryo. (Though the egg cell had its DNA removed, it still contained all kinds of egg-cell proteins whose normal function is to trigger development of an embryo. It was these factors still in the egg that reprogrammed the donor-cell DNA.) After the embryo had developed to a certain point, the researchers implanted it in a third sheep, which served as a surrogate mother. The result, 21 weeks later, was the lamb Dolly, who went on to give birth to two sets of her own lambs.

Every cell in Dolly had DNA in its nucleus that was an exact copy of the DNA in the six-year-old donor sheep. Thus, Dolly was a clone of that sheep. There was no second parent contributing chromosomes to Dolly, no mixing of genetic material at the moment of conception—just a copying of one individual from another. The procedure through which she was cloned is known today as **somatic cell nuclear transfer (SCNT):** a means of cloning mammals through fusion of one somatic (non-sex) cell with an egg cell whose nucleus has been removed (an "enucleated" egg cell).

### Cloning and Recombinant DNA

Once Dolly was born, science was off to the races in terms of producing clones of mammals. To date, horses, mules, cows, pigs, cats, mice, goats, and a

**Figure 15.7**
**Revolutionary Sheep**

Dolly, the first mammal ever cloned from an adult mammal, is shown here as a young sheep with her surrogate mother. Note that Dolly was white-faced, like the sheep she was cloned from, while her surrogate mother was black-faced.

**•••  ANSWERS**

1. genes; another species
2. restriction enzymes
3. human; protein

**Web Animation 15.3**
Copying DNA through PCR

The essence of "forensic DNA typing," as it is known, is to get two sets of physical samples—first, a sample left by a crime's perpetrator in such forms as blood or semen; and second, a blood sample from a suspect. By comparing the DNA patterns of both samples, forensic scientists can establish odds of whether the suspect was present at the crime scene. If he or she was, but had no good reason to be there, this is strong evidence that the suspect committed the crime. Another way of gathering DNA evidence is to use a victim's blood sample and test it against, for example, blood found on a suspect's clothing.

**Figure 15.9**
**DNA Copying Machine**
The polymerase chain reaction (PCR) makes copies of a given length of DNA very quickly.

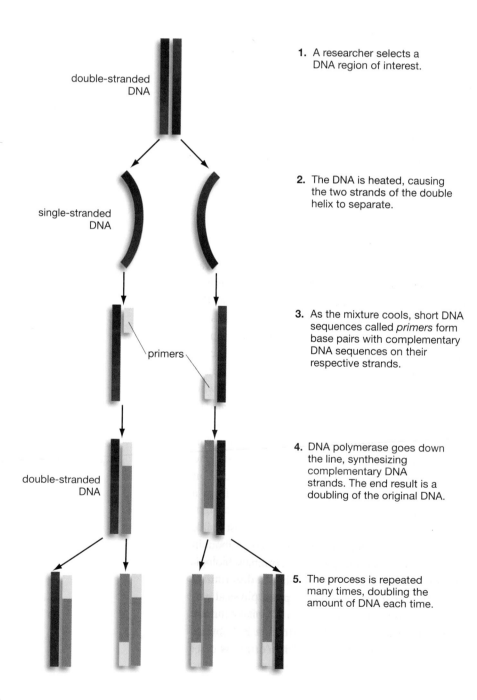

double-stranded DNA

single-stranded DNA

primers

double-stranded DNA

1. A researcher selects a DNA region of interest.

2. The DNA is heated, causing the two strands of the double helix to separate.

3. As the mixture cools, short DNA sequences called *primers* form base pairs with complementary DNA sequences on their respective strands.

4. DNA polymerase goes down the line, synthesizing complementary DNA strands. The end result is a doubling of the original DNA.

5. The process is repeated many times, doubling the amount of DNA each time.

## The Use of PCR

When forensic DNA typing first came into use in the 1980s, a significant limitation on it was that relatively *large* samples of human tissue had to be obtained to get enough DNA to work with. Eventually, however, this problem was all but eliminated by something called the **polymerase chain reaction (PCR):** a technique for quickly producing many copies of a specific segment of DNA. Thanks to PCR, if a criminal so much as licks an envelope and leaves it behind, police can retrieve enough DNA to get a match.

The PCR process is very simple in outline. Four kinds of materials are mixed together to carry it out. The first of these is a quantity of DNA itself. Then, there is a collection of DNA nucleotides and DNA polymerase (which, remember, goes down the line on single-stranded DNA, affixing nucleotides to available bases). Finally, there are two DNA "primer" sequences—short sequences of single-stranded DNA that act as signals to DNA polymerase, saying "start adding nucleotides here."

The first step in the process is that the starting quantity of DNA is heated until the two strands of DNA's double helix separate, resulting in two single strands of DNA (**Figure 15.9**). As the heated DNA mixture cools, primers will attach to each of the now-separate strands of DNA. Then, the DNA polymerases go down the line, starting from the primers, linking nucleotides to the template DNA and thus producing strands that are complementary to the original strands. The result is two double-stranded lengths of DNA, both identical to the original double strand. In short, the DNA sample has been doubled in one copying "cycle." Then the entire process of heating and cooling is repeated. Since each copying cycle takes only 1–3 minutes, by the time 90 minutes have passed, millions of copies can be created. Moreover, thanks to the specificity of primers, the sequence that is copied can be limited to a small sequence of interest within the larger genome. So useful is PCR that it is employed not just in forensic biotechnology, but in countless facets of biological research. We are focusing here on one specific use for PCR, but it has so many applications that in 1993, its inventor, Kary Mullis, was awarded the Nobel Prize in Chemistry for his achievement.

## Finding Individual Patterns

Once PCR has done its job, the police have enough DNA to begin looking for the individual patterns within it—one set of patterns found in the crime-scene DNA, one set of patterns seen in the suspect's DNA. But what are these patterns that each of us carries? Human genomes are filled with short sequences of DNA that are repeated over and over, from 3 to 50 times, like this:

TCATTCATTCATTCATTCAT

This TCAT example is not hypothetical. It is an actual short tandem repeat (or STR) that police departments use in today's DNA typing. The usefulness of STRs is that, at a given location in the genome, one person will have one number of tandem repeats, while another person is likely to have a different number of repeats, like so:

Individual 1: TCAT TCAT TCAT TCAT

Individual 2: TCAT TCAT TCAT TCAT TCAT TCAT

Now, the chances are small that two unrelated individuals will have an identical number of repeats at even one location in the genome. But modern forensic DNA typing looks at STR repeats at 13 different locations. Imagine yourself, then, as a forensic investigator who is reviewing both DNA from a crime scene and DNA from a suspect in the crime. At genomic location 1, the STR patterns are:

Crime scene: TCAT TCAT TCAT TCAT

Suspect: TCAT TCAT TCAT TCAT

So, you have a match at this one location. If the STR patterns continue to line up this way at the other 12 locations then what once was just probable becomes certain. Your suspect was certainly at the crime scene because the odds of any two people having identical STR patterns at 13 locations are astronomically small.

It's important to recognize that this kind of DNA evidence can be used to prove innocence as well as guilt. DNA testing regularly exonerates not only suspects, but people who have been *convicted* of crimes. In the past 17 years, more than 200 prisoners who were jailed for various offenses have been freed on the basis of DNA testing. If you look at "Reading DNA Profiles" on page 278, you can see the means by which STR patterns are visualized in today's criminal cases.

Beyond matching a known suspect to a crime scene or victim, forensic DNA typing has another power. Increasingly, it is allowing police to make so-called *cold hits*. All 50 states now require criminals convicted of violent crimes to provide DNA samples. Capitalizing on this, the FBI has put together a national database of the DNA profiles of these criminals. Given this, imagine a situation in which a crime is committed, and no suspect was identified, but in which a DNA sample was retrieved. A DNA profile will be produced from this sample—a profile that can be matched against all those that are stored in a DNA database. If the profile from the crime scene matches a criminal's stored profile, that in itself is sufficient evidence to make an arrest. And this is happening all the time—cold hits result in hundreds of arrests across the nation each year.

---

**SO FAR...**

1. To "clone" means to make an exact _____.

2. True or false: Dolly was a transgenic organism.

3. PCR has been described as a copying machine for _____.

---

## 15.5 Stem Cells

In the United States, it would have been hard to avoid the term *stem cells* over the past few years, given the national political fight over them and several state ballot initiatives regarding them. But what are stem cells, and why are they so controversial? Let's see.

### Cell Fates: Committed or Not

In one way, human cells are like human beings: They start off with the potential to be lots of things, but eventually most of them commit themselves to being just one thing. This course of events takes place during human "development," meaning the process by which an embryo becomes a fully formed human being. Early in this process, cells need to be adaptable enough to go down different developmental pathways. Why? Because the embryo is developing in so many basic ways. For example, it needs to create both muscle cells and bone cells. Fortunately, there is a type of embryonic cell that is flexible enough to give rise to either type of cell. As time goes on, however, more and more tissues become fully specialized, which means there is less and less need for cells that are developmentally flexible. By the time humans reach adulthood, specialization has almost completely won out over flexibility. The vast majority of cells in our bodies have undergone what is known as *commitment*: a developmental process that results in cells with roles that are completely determined. Almost all your muscle cells, for example, are committed in that they can be nothing but muscle cells and can give rise to nothing but muscle cells.

## ESSAY   Reading DNA Profiles

**W**hat does a forensic DNA profile look like? Law enforcement agencies use the kind of imaging you can see below. The upper profile was produced from DNA taken from a crime scene, while the lower profile was produced from DNA obtained from a suspect in the crime. The short tandem repeat or STR patterns that differ from one person to another are represented by the peaks that run across the rows in both figures. Each peak, or set of two peaks, represents one location in the genome—a location whose STR pattern was examined. In the upper figure's first row, the peak labeled 16 indicates that the person who left DNA at the crime scene had 16 repeats at this location. The nearby 17 then indicates that this person had 17 repeats at this same location, but on his second, homologous chromosome. Meanwhile, the 24 that can be seen further to the right represents the number of repeats at a different location in the genome. In this case, the person had 24 repeats on both chromosomes—he was homozygous for repeats at this location.

The scale running across the top of the figure indicates the number of DNA base pairs in the lengths of DNA that were examined at each location. Thus, the 16 repeats at the first peak were found in a section of DNA that was about 125 base pairs long. The height of each peak has no bearing on identity but simply reflects the quantity of DNA that was derived, through the PCR process, for each location. The X and Y peaks on the second row represent a section of the genome that differs between men and women. Had a woman's DNA been found at the crime scene, there would have been no Y peak.

A quick look at both the upper and lower figures reveals that the STR patterns in them match up perfectly. The obvious implication is that the suspect was at the crime scene. Indeed, the odds that the crime-scene DNA came from anyone *other* than the suspect are infinitesimally small. What are the chances that a person selected at random from a general population would have the same STR profile as the one found in the crime-scene DNA? The odds of this occurring range from 1 in 4.6 trillion for the U.S. Caucasian population to 1 in 36 trillion for the U.S. southwestern Hispanic population.

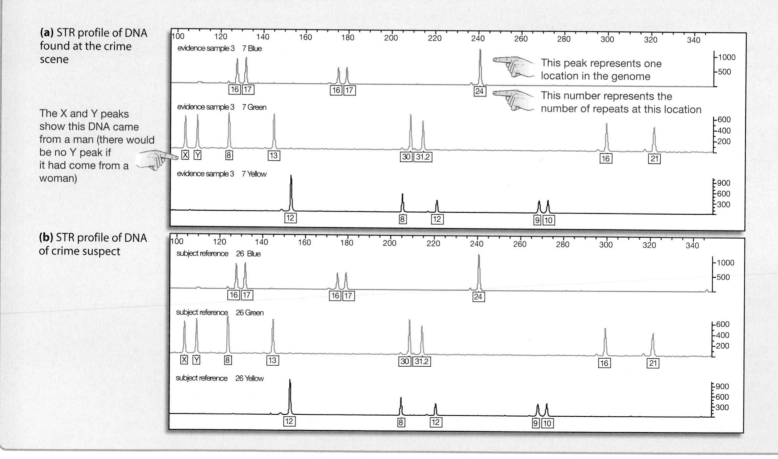

**(a)** STR profile of DNA found at the crime scene

This peak represents one location in the genome

This number represents the number of repeats at this location

The X and Y peaks show this DNA came from a man (there would be no Y peak if it had come from a woman)

**(b)** STR profile of DNA of crime suspect

From a human perspective, the problem with this sequence of events is that the tissues of fully developed human beings can become damaged. They break down or become diseased; they are injured in car accidents or fires. And many damaged tissues will not regenerate themselves. A person whose spinal cord is damaged does not generate a new spinal cord. Thus, for medicine, the question for decades has been: Could we use developmental processes to generate new tissues? One obvious way to go about this would be to harness the power of the cells that, in nature, generate new kinds of tissue—early embryonic cells. It turns out, however, that there is also another possibility. A small proportion of *adult* cells retain the ability to produce specialized cells. Both the adult and the embryonic cells that have this generative capacity are known as *stem cells*: cells with a continuing capacity to produce more cells of their own type, along with at least one type of specialized daughter cell.

## Stem Cells from Embryos

So, what is the nature of this first kind of stem cell, the type derived from an embryo? When human sperm and egg come together, they create a fertilized egg (an embryo) that has begun to undergo cell division. If we go forward about five days from the first division, the embryo has developed into a hollowed-out, fluid-filled ball of 100–150 cells—a **blastocyst**. One section of the blastocyst's cells becomes the placenta that will help nurture a growing embryo, while another section, called the inner cell mass, constitutes the embryo itself (**Figure 15.10**). It is these latter cells that interest us, for among the flexible, early embryonic cells we talked about, these cells are *very* flexible. Indeed, these are the cells you've heard so much about in the news, **embryonic stem cells**. How flexible are they? We can define them as cells from the blastocyst stage of a human embryo that are capable of giving rise to all the types of cells in the adult body.

## Adult Stem Cells

Our second variety of stem cells, adult stem cells, differs from embryonic stem cells in some fundamental ways. First, while there is only one type of embryonic stem cell, adult stem cells come in many varieties—there are muscle stem cells, blood stem cells, nervous system stem cells, and more. Second, while there is only one source for embryonic stem cells (human embryos), adult stem cells are found, in small numbers, in various tissues in the body. Nervous system stem cells are found in neural tissue, for example, while muscle stem cells are found in muscle tissue. Finally, whereas embryonic stem cells can give rise to any type of cell in the adult body, adult stem cells are more limited in their ability to differentiate, although scientists do not yet know what the full range of their ability is. Recent research has shown that a type of adult stem cell found in our bone marrow—a cell that normally gives rise to fat or bone cells—can be coaxed into differentiating into a type of *nerve* cell. Moreover, this same variety of adult cell has also been steered toward becoming a type of cell that lines the passageways of our lungs.

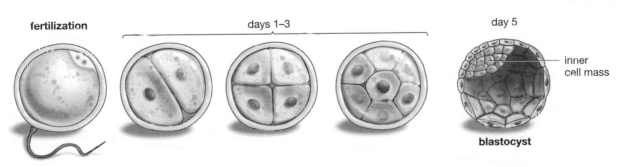

fertilization    days 1–3    day 5

inner cell mass

blastocyst

**Figure 15.10**
**Cells with a Great Ability to Specialize**

Human development starts with the fertilization of egg by sperm and then proceeds through several days of cell division, such that by day 3 the embryo has become a tiny, solid ball of cells. By day 5, however, it has developed into a hollow ball of 100–150 cells called a blastocyst. In nature, one segment of the blastocyst, the inner cell mass, develops into the baby. Each collection of embryonic stem cells starts as the inner cell mass of an embryo produced through *in vitro* fertilization, meaning fertilization outside a woman's body. Scientists remove these cells from a blastocyst and transfer them to a laboratory dish, where they are "cultured," or grown into a large collection of cells. Then cells from the culture—now referred to as embryonic stem cells —can be chemically treated in ways that make them differentiate into one of the specialized types of cell found in the human body. Pictured on the right is a human blastocyst on day 6 of development. It has recently undergone the process of "hatching" or breaking out of the protective covering that surrounds early embryos, which is why it no longer has the spherical shape of the day 5 blastocyst.

Thus, some adult stem cells do have the capacity to differentiate fairly broadly, and scientists have been able to harness this capacity to some degree. It shouldn't be surprising, however, that adult cells don't have the embryonic cells' ability to become any cell type since all adult cells have undergone some level of commitment—all of them have gone down the path of specialization to some degree, which seems to put a limit on their ability to differentiate. Beyond this, adult cells are weaker than embryonic cells in their ability to *keep on* giving rise to new cells, generation after generation (although strides have been made to strengthen this ability in them). This is important because the kind of tissue regeneration we talked about requires huge numbers of cells from a single source.

Given limitations such as these, you might ask: why would scientists work with adult stem cells rather than embryonic cells? One reason involves a scientific consideration we'll touch on shortly; the other has to do with ethics and societal values.

## The Fight over Embryonic Stem Cells

To state the obvious, human embryonic stem cells come from human embryos—from the collections of cells that have the potential to become fully formed human beings. Where do researchers get the embryos that are the source of embryonic stem cells? In most cases, from fertility clinics, which each year discard thousands of excess embryos that have been produced for in vitro fertilizations. (A couple undergoing fertility treatments generally will produce many more embryos than they need to bring about a pregnancy.) With the couples' consent, excess embryos that would have been discarded are used instead in scientific research. When scientists extract stem cells from these early-stage embryos, however, the embryos are destroyed, and this is a matter of great concern to many people. For them, the destruction of an embryo is a destruction of human life.

In contrast, many people feel that an embryo in the blastocyst stage represents only a *potential* human life. And others hold that, no matter what one's view of this issue, society has a greater responsibility toward impaired adults and children than it does toward embryos. Given the superior ability that embryonic cells have to differentiate and multiply, there is a consensus in the scientific community that, if stem cell research is to proceed as rapidly as possible, then this research cannot be limited to adult stem cells; embryonic stem cells must be included as well.

One practical consequence of this controversy concerns research funding. In 2001, President Bush decided that the federal government—the largest single source of biological research funds—would fund research on embryonic stem cells but only in a limited way. Federal funding is now provided to researchers who work with cells from any of the 71 stem-cell "lines" that were in existence before the date of the president's announcement. Four years later, however, only 22 or so of the "presidential lines" were available to researchers, and some of those were not robust or were feared to have been contaminated by non-human viruses. Meanwhile, countries like Great Britain have imposed almost no restrictions on stem cell research funding. In the view of many American experts, it is as if the United States is trying to develop the latest computer technology with the stipulation that no microchips produced after 2001 be used in the effort. In response, private institutions and state governments have started to take matters into their own hands. Harvard University has announced a $100 million stem cell initiative, Stanford University has announced something similar, and—using private money—university researchers are developing new stem cell lines that can be distributed to the research community. In the biggest response of all, Californians voted in 2004 to spend $3 billion of state money over the succeeding 10 years on stem cell research, with the vast majority of these funds going to embryonic stem cell research.

## The Potential of Embryonic Stem Cells

So, why is there so much optimism about embryonic cells (ESCs)? No human disease has yet been cured with them, and so far they have demonstrated therapeutic potential mostly in mice and rats. Nevertheless, most experts seem to agree that they hold enormous promise. Consider an affliction such as Parkinson's disease, which has symptoms of shaking and stiffness that are brought about because of the death of neurons in the brain that produce a substance called dopamine. The question is: Can we get ESCs to differentiate into useful, dopamine-producing neurons? As it turns out, researchers have already done this in rats. They took stem cells, differentiated them, injected them into the brains of rats who had half their dopamine-making neurons destroyed, and then watched as the injected cells were transformed into functional neurons, complete with the ability to secrete dopamine. Another disease of the nervous system, multiple sclerosis, does its damage by destroying a kind of insulation, called myelin, that wraps around nerves. In 2004, researchers from California announced that in the laboratory, they coaxed human ESCs into becoming myelin-producing nerve cells. Better yet, when cells of the same type were later injected into the spinal cords of rats with spinal cords that had been damaged, the new cells started producing a myelin covering around spinal cord neurons in the damaged area. The result? Rats who had been paraplegic—who had lost control of their hind legs—regained enough function in them to do a kind of impaired walking, rather than dragging themselves along with their front legs. (For more on how this research feat was accomplished, see "Making Embryonic Stem Cells Work" on page 282.)

Encouraging as these results are, they don't guarantee anything about the ability of ESCs to regenerate damaged tissue inside human beings. Researchers have some daunting hurdles to get over before stem cells can even be tested in connection with most human illnesses. Can ESCs be developed that will differentiate *only* into the desired type of cell? Can implanted cells be counted on to stay where they are needed, rather than "migrating" to other areas? Can we rest assured that implanted cells will be free of any harmful viruses? Given these issues, most researchers think that it may be 5 or even 10 years before a large number of human ESC trials can get under way. This is not to say, however, that all human trials are this far off. The California researchers who did the work with the spinal cords of rats say their procedure is ready to be tested in human clinical trials right now.

## Stem Cells Meet Cloning: Therapeutic Cloning

One of the additional challenges to stem cell therapy is the same challenge posed by organ transplants: Cells that are derived from an embryo, and that are then put into the body of a patient, are likely to be regarded as invading *foreign* cells by the patient's immune system, meaning these cells can set off an immune system attack. Thus, transplanted cells could be "rejected" by a body in the same way a transplanted kidney can be. So, how can this hurdle be overcome? Well, imagine that each of us could supply physicians with our *own* stem cells in the form of *adult* stem cells from our bodies. The idea is that these cells could then be coaxed into giving rise to the kind of specialized cells we would need in the event of injury. In this scenario, there would be no risk of rejection because the transplanted cells would be our own—they would be regarded as "self" by our immune system. This is the theoretical scientific advantage, mentioned earlier, that adult stem cells have over ESCs, and it is one of the motivations for continued research into adult cells.

It turns out, however, that adult cells might not have a monopoly on solving the tissue rejection problem. What if each of us could produce our own *embryonic* stem cells, which would then generate the type of specialized cells we need for a transplant? At first, this may sound impossible. ESCs exist only in early-stage embryos, after all, which none of us has been for a long time. But remember the cloning procedure we talked about earlier? The idea scientists have is to combine it with stem cell therapy. Here's how it would work. A standard cell from our body would be used as the donor-DNA cell in the cloning process. This cell would then be fused with an egg cell whose nucleus has been removed. The result, as you've seen, would be an embryo. This embryo, however, would only be allowed to divide to the blastocyst stage. At this point, ESCs would be harvested from this blastocyst, coaxed into specialization, and the resulting cells would be used to repair our own body.

This may seem too far-fetched to be believed, but scientists are working on it right now. In fact, scientists around the world thought that the first steps in this process *had* been achieved, in 2004 and 2005, by a research team from South Korea. Unfortunately, the team's leader, Woo Suk Hwang, was lying about most of his group's work (**Figure 15.11**). By manipulating images and making false claims, Hwang managed to carry out one of the most damaging instances of scientific fraud in recent decades. We now know that his team did derive some human embryos through cloning (although all of these embryos were in "poor condition" according to a Korean panel who investigated his work). Completely bogus, however, were his claims to have derived 12 stem cell lines from these embryos—he maintained he had removed ESCs from the embryos and then managed to get these cells to keep dividing in the lab.

To date, one other scientist has succeeded in producing a human embryo through cloning, but no

**Figure 15.11**
**Disgraced Researcher**

In March 2006, South Korean scientist Woo Suk Hwang was surrounded by reporters at the office of state prosecutors in Seoul. Two months later, Hwang was charged with criminal fraud and embezzlement in connection with one of the worst instances of scientific fraud in recent history. Hwang manipulated research reports on therapeutic cloning to make it appear that he had derived lines of embryonic stem cells from cloned human embryos when in fact no such cell lines existed.

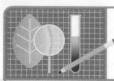

# THE PROCESS *of* SCIENCE
## Making Embryonic Stem Cells Work

In a world hoping for the best from embryonic stem cells, a scientific report published in 2005 contained the best news ever to that point: Scientists had restored function to animals through the use of stem cells. Specifically, rats with spinal cords that had been damaged regained some ability to walk on all fours, whereas previously they could only drag themselves along by their front paws. The treatment that made the difference was the injection of cells into the spinal cords of the rats. All these cells were a type of nervous system cell, and all of them had been derived from embryonic stem cells (ESCs). A young University of California researcher and his team had started out with ESCs, had coaxed them chemically into becoming the nervous system cells, had injected the rats with these cells, and then had watched as the treated rats went from crawling to unsteady walking.

The scientist who led this effort was Hans Keirstead, a Canadian-born biologist now at the University of California, Irvine (UCI). Keirstead's findings, which appeared in the *Journal of Neuroscience*, were the culmination of years of strategic planning, corporate partnership, and detailed scientific work by himself and his fellow researchers—mostly graduate students and postdoctoral scholars in his lab.

Speaking by telephone from UCI, Keirstead noted that the 2005 paper had its roots in thinking he had done six years earlier about ways to simplify the treatment of spinal cord injuries. Historically, he said, most thinking has involved puzzling over ways to "address the cavity"—ways to reconstruct all the damaged tissue that results from a spinal cord injury. His idea, conversely, was to work on just one aspect of the damage: a phenomenon called myelination. The cells in the spinal cord that transmit nervous system signals are called neurons. But the work of neurons is enabled by another group of cells, called oligodendrocytes, which produce a material called myelin that wraps around the neurons. Following spinal

**Stem Cell Researcher**
Hans Keirstead of the University of California, Irvine.

cord injuries, there are neurons that have been "spared," as spinal cord researchers phrase it, but that have ceased to function because they have lost their myelination. This happens because the

one has yet derived any stem cell lines from such embryos, and it's not clear what the prospects are for this taking place. Nevertheless, scientists worldwide are continuing to work on what is known as **therapeutic cloning**: the use of cloning to produce human embryonic stem cells that can be used to treat disease and injury. We need to be clear that the objective of this work is not the *reproductive* cloning we talked about earlier. Nobody working in therapeutic cloning is contemplating letting a cloned embryo develop into a baby. The idea instead is that each of us could help generate an early-stage embryonic clone of ourselves—one whose cells would be collected in the blastocyst stage and then used in our own repair.

## A Stunning Breakthrough in 2007

Late in 2007, two teams of scientists announced a stem-cell advance whose potential was so enormous that it made news around the world. Shinya Yamanaka of Kyoto University in Japan and James Thomson of the University of Wisconsin, Madison announced that, by inserting just four genes into ordinary skin cells, they were able to produce cells that appear to have all the developmental potential of embryonic stem cells. These so-called induced pluripotent cells (iPCs) could thus be used, like embryonic stem cells, to generate needed human tissues. But they would not raise any of the ethical concerns associated with

oligodendrocytes that produce the myelination have died following the injury.

Keirstead's plan was to try to bring about a *remyelination* of the neurons by means of introducing new oligodendrocytes to the site of injury. And where would these oligodendrocytes come from? Keirstead saw that embryonic stem cells held out the best hope of producing the millions of cells that would be needed. The fantastic potential of embryonic stem cells lies in their ability to give rise to quantities of any type of cell in the adult body. Keirstead wanted to channel this potential in one direction: He wanted to coax ESCs into becoming immature oligodendrocytes and then see what these cells could do in injured animals.

> *The fantastic potential of embryonic stem cells lies in their ability to give rise to quantities of any type of cell in the adult body.*

With this long-term strategy in mind, Keirstead approached a California biotech firm, the Geron Corporation, and proposed that it provide him with both the ESCs and the funding he needed for his project. (What was in it for Geron? It owns or licenses several basic stem cell patents; to the extent that stem cell therapy comes into use, Geron stands to profit.) Geron said yes to the proposal, and Keirstead began the difficult task of trying to get ESCs to differentiate into oligodendrocytes. Though the effort took years, it succeeded. Indeed, its full fruits can be seen in Keirstead's 2005 paper on remyelination. Within this report, readers can find a "protocol"— a kind of scientific recipe—for deriving oligodendrocytes from ESCs. Developing this protocol was, in and of itself, a crucial step. With it, Keirstead demonstrated that it was possible to derive a pure collection of a specific kind of nervous system cell from ESCs.

With the means of producing oligodendrocytes in hand, Keirstead set out to put these cells to the test. Could they bring about a change in animals with damaged spinal cords? The path to finding out was exacting and complicated. First, rats had to be videotaped before injury—one at a time, with each of them rated on several kinds of movements. Then, the rats were anesthetized, and each of them was given an identical spinal cord injury. Seven days after the injury, Keirstead operated on them. At the site of their injuries, some rats got injections of immature oligodendrocytes. (Each rat got 1.5 million cells, an amount that would occupy a volume about half the size of a pea.) To keep the experiment carefully controlled, other rats got injections of another kind of cell, while still others got injections of nothing but the liquid medium the cells were immersed in. Then, the waiting and watching began. Over time, it became apparent that some of the rats were regaining function in their hind legs. But were these the rats that had received the oligodendrocytes? None of the research team members could be certain since, under the design of the experiment, none of them were allowed to know which rats fell into which category. But they had their hopes, and in the end these hopes were realized: It was the rats who got the oligodendrocytes that regained function in their legs.

Once these behavioral results were in, Keirstead and his team needed to confirm that remyelination had indeed occurred, and that it—rather than some other factor—was responsible for returning function to the animals' hind legs. So, the rats were killed, and their tissues were examined under a microscope. When all the possibilities were considered, remyelination emerged as the factor that made the difference.

Once Keirstead's paper appeared, it took on added significance following an announcement by Geron that it intended to move quickly into *human* trials of the remyelination procedure. This is a controversial decision—many stem cell experts think more animal testing should be done before human trials are attempted. But Geron thinks the time is right, and Keirstead supports the company's position. The human trials will be overseen by medical doctors rather than by biologists such as Keirstead. But his work has provided the building blocks for the trials. No one knows how they will come out, but in a world still hoping for the best from stem cells, everyone's hopes are running high.

ESCs since they are derived from ordinary cells that could easily be obtained from children or adults.

Moreover, iPCs appear to be capable of solving a second major issue associated with ESCs—that of tissue rejection—and they would do so in a way that no one would find morally objectionable. Since iPCs are derived from regular cells, the donors of these original cells could be anyone in need of therapy. Thus, patients could be the source of the stem cells used in their own treatment. Prior to this advance, most scientists believed that the most promising route to producing patient-specific stem cells was therapeutic cloning, which is technically unproven and which, even if it successful, would always entail the destruction of human embryos.

Working separately, the Yamanaka and Thomson teams began their research by picking out genes within the human genome that are active in early-stage embryos. Using a virus as a vector to ferry these genes into the starting skin cells, the researchers then watched as the genes reprogrammed the skin cells, effectively turning them into ESC-like stem cells. Scientists around the world agreed that this achievement holds potential for ushering in a new era in stem-cell research. With announcement of the breakthrough, other groups of scientists began efforts to duplicate and broaden the findings of the teams from Japan and Wisconsin.

## 15.6 Biotechnology in the Real World

Thus far, we've taken a tour though some notable successes in biotechnology—the cloning of Dolly the sheep, the development of DNA fingerprinting, and so forth. What has been missing from this account, however, is a notion of how *difficult* it is to make advances of this sort. To get a sense of this, consider the following three stories, all of which provide some perspective on biotech examples we looked at earlier.

**Omega-3-rich pigs**—The university researchers we talked about who worked on this project were able to produce three pigs whose tissues produced high levels of omega-3 fatty acids. But, how many embryos had to be first cloned and then transferred into surrogate mothers to get these three pigs? The answer is 1,633.

**Golden rice**—This transgenic rice, which produces its own beta-carotene, was first developed in 1999. Field tests of it began in 2004, and an improved version of it was developed, with corporate sponsorship, in 2005. So, how long will it be before golden rice is actually in the hands of farmers? Its chief inventor, Ingo Potykus, says 2010 at the earliest.

**Transgenic goats**—The Massachusetts company that developed these medicine-yielding goats started work on them in 1989. So, how many goat-derived medicines have been approved for human use? One, in Europe. Thus far, not a single drug from any transgenic animal has been approved for use in the United States (though, as you've seen, numerous drugs produced by bacteria have been approved for use).

Stories such as these provide two general lessons about biotechnology as it exists in the real world. First, the road to any success in biotech is almost always long and expensive. To be sure, costs and difficulties are part of any large-scale commercial endeavor. But biotech faces especially high hurdles in this regard because the processes it is developing are fundamentally new. The researchers who worked on the omega-3 pigs were not trying to better an existing process; they were trying to do something that had never been done before. And this quality of newness is the reason that regulators, such as the Food and Drug Administration and the U.S. Department of Agriculture (USDA), are obliged to provide a high level of scrutiny to most biotech inventions. But there is another reason that progress in biotech tends to come slowly: The processes it is developing are sometimes not just new, but are controversial in ways we'll look at now.

### Controversies in Biotechnology

You may recall the biotech crops we looked at earlier—the ones that are transgenic, in that they have one or more genes from another species spliced into them. Worldwide, tens of millions of acres have been planted with these genetically modified or GM crops. It turns out, however, that almost all this acreage is devoted to just *four* crops: cotton, corn, soybeans, and canola. Moreover, GM crops have not been modified to make their ultimate products—food or clothing—preferable to consumers in some way. Instead, all four GM crops have been modified to be preferable to *farmers*, either by being resistant to pests or tolerant of herbicides. (One of the latter crops is "Roundup-ready" soybeans, whose seeds contain a bacterial gene that allows them tolerate the herbicide Roundup.) You might think that the list of GM crops will certainly be getting longer in the future, but this may not be the case. Consider that in 1999 about 120 GM fruits and vegetables were being field tested in the United States, but that by 2003 this number had dropped to about 20.

So, why are there so few GM crops, either in the fields or in the works? A large part of the answer can be summed up in two words: consumer resistance. Worldwide, a significant number of consumers have become persuaded that biotech foods are dangerous, and the effect of this has been a continuing shelving of plans for biotech crops. In 2004, for example, the Monsanto corporation gave up on a program it had started seven years earlier aimed at producing GM wheat. The company worried that few American or Canadian farmers would plant the wheat because they feared resistance to it from European and Japanese consumers. Such resistance, the farmers thought, might effectively put a "do not import" label on all

U.S. and Canadian wheat, whether genetically modified or not. This same chain of reasoning is operating in connection with other GM foods.

And what is the basis for consumer fears about GM foods? Nothing that has occurred so far. In 2004, the National Academies of Sciences published a report whose central conclusion is still valid: "To date, no adverse health effects attributed to genetic engineering have been documented in the human population." The report went on to note that food crops actually have been "genetically modified" for centuries, although this has come about through the low-tech means of breeding different plant strains with each other. This form of genetic mixing, which is relatively uncontrolled, stands in contrast to the biotech variety, in which a small number of identifiable genes are inserted into (or removed from) a plant's genome. Yet, it is biotech crops that have encountered resistance as "frankenfoods," while plants produced through conventional means can be changed without notice.

Resistance to GM crops is based on more than the fear of their health effects, however. The other great concern about them has to do with environmental effects. If you look at **Figure 15.12**, you can see a case in point in this battle. Pictured are two sets of cotton plants, one set genetically modified, the other not. The modified plants contain genes from a bacterium, called *Bacillus thuringiensis*, that is found naturally in the soil and that produces proteins toxic to a number of insects. Collectively, these proteins constitute a natural insecticide known as Bt, which has been sprayed on crops for years, mostly by "organic" farmers, meaning those who do not use human-made fertilizers, pesticides, or herbicides. Enter biotechnology firms, which in the 1990s spliced Bt genes into crop plants, with the result that these plants—corn, cotton, and potatoes—now produce their own insecticide. The results have in some ways been an environmentalist's dream. In one survey conducted in the American Southeast, farmers who planted Bt cotton reduced the amount of chemical insecticides they applied to their fields by 72 percent. They did this, moreover, while increasing cotton yields by more than 11 percent.

Despite such benefits, we can also see, in the use of Bt seeds, the qualities that make environmentalists uneasy about GM crops in general. Those few insects that survive in Bt-enhanced fields are likely to give rise to populations resistant to the natural toxin. This raises the prospect of a valued natural insecticide losing its effectiveness against insects over the long run. So alarming is this possibility that the U.S. Environmental Protection Agency requires that at least 20 percent of any given farmer's crops must be non-Bt plants. The idea is to create non-Bt "refuges" near the Bt fields that will harbor bugs that have not built up a resistance to Bt. These bugs will then mate with Bt-resistant bugs, thus helping to ensure that Bt resistance does not spread. And

**Figure 15.12**
**Built-in Resistance to Pests**
The cotton plants on the left have had genes spliced into them from the bacterium *Bacillus thuringiensis*. These genes code for proteins that function as a pesticide. Meanwhile, the cotton plants on the right are not transgenic in this way—they have not had genes from another organism spliced into them. Both stands of cotton were under equal attack from insects, but the Bt-enhanced plants fared much better.

how has this strategy worked so far? Seemingly very well. A USDA survey completed in 2003 found that pests in or near Bt fields had developed little or no resistance to the insecticide.

### Ethical Controversies in Biotechnology

Disputes over GM crops involve mostly practical questions—do GM foods pose a health hazard or not?—but some biotech controversies are essentially ethical in nature. We've touched on two of these ethical controversies already: the disputes over stem cells and human cloning. To be clear, scarcely any controversy exists over the idea of human *reproductive* cloning since nearly everyone—scientists and laypersons alike—thinks it is a repulsive idea. (No one could ensure that a cloned person would be perfectly healthy, mentally or physically. Who among us would be willing to accept the idea of producing "test" human beings in cloning experiments?)

But then there is the issue of therapeutic cloning, which despite its differences with reproductive cloning, would still result in the production—and probable destruction—of human embryos. Moreover, the embryos used in this process would be derived from a living person and then used to heal that person in the event of injury or illness. So, suppose this came to pass. What would it represent: a triumph of medical science or an act of societal narcissism? (Under the latter view, members of society would be creating and then eliminating potential lives as a means of benefiting their own.)

Beyond these issues, there is a more general ethical objection to biotech research that has to do with the question of altering living things—of altering nature, if you like. Of course, human beings have been altering nature for centuries through such means as the plant breeding we talked about. But for an example that is a little more relevant, consider one of our favorite pets, the dog. The fantastic variety in dog breeds that we see today has mostly come about in the past 400 years, and this has happened almost entirely as a result of the human practice of selectively breeding one kind of dog with another. So is this inherently unethical? If your answer is no, then what about producing a sheep with liver tissue that is mostly made of human cells? Is there some ethical line that is crossed when we go from mixing different breeds of an animal to mixing different species of animal? (The animal that results from this mixing is "chimeric" in scientific parlance.) If the answer to this is no, then what about injecting immature human brain cells into the brain of a monkey? This has been done already, as a means of conducting Parkinson's disease research, and the monkeys have, to outside observers eyes, remained "all monkey." But is there a point at which this becomes unethical? Perhaps the point at which a monkey might develop the first glimmer of human consciousness?

The possibilities that biotech has opened up for chimeric animals drives home the point that we have entered a new era in altering nature. Given biotech's power, however, it isn't necessary to consider chimeras to confront the kind of "altering" dilemmas that might soon be upon us. Suppose you could alter a chicken's own genes so that you could create an animal that could lay eggs but that had no eyes or legs. (This would make these chickens easier to control, which would in turn bring the price of eggs down.) Suppose further that these animals suffered no apparent discomfort as a result of this manipulation. If you could do such a thing, would you?

## On to Evolution

In touring the strands of DNA's double helix, you've had many occasions to see how this microscopic molecule can profoundly affect our macroscopic world. DNA replicates, it mutates, it is passed on from one generation to the next. And some organisms will have more *success* in passing on their DNA—in reproducing, in other words—than will others. This last fact has been critical in bringing about the huge variety of life-forms that Earth houses, from bacteria to bats to trees. Over billions of years, genetics has interacted with Earth's myriad environments to produce the grand story of life. That story is called evolution, and it is the subject of the next four chapters.

# CHAPTER 15 REVIEW

For study help, activities, and more quiz questions, go to **www.aw.com/krogh4**.

## SUMMARY

### 15.1  What Is Biotechnology?

- Biotechnology can be defined as the use of technology to control biological processes as a means of meeting societal needs.

### 15.2  Transgenic Biotechnology

- A transgenic organism is an organism whose genome has stably incorporated one or more genes from another species. Many biotechnology products are produced within transgenic organisms. Human growth hormone, for example, is produced within a bacterium that has been made transgenic by means of incorporating a human gene.

- Restriction enzymes are proteins derived from bacteria that can cut DNA in specific places. Plasmids are small, extrachromosomal rings of bacterial DNA that can exist outside of bacterial cells and that can move into these cells through the process of transformation.

- Human DNA can be inserted into plasmid rings. Scientists use the same restriction enzyme on both the human DNA of interest and the plasmids. Complementary "sticky ends" of the fragmented human and plasmid DNA will then bond together, splicing the human DNA into the plasmid. This produces recombinant DNA: two or more segments of DNA that have been combined by humans into a sequence that does not exist in nature.

- Once plasmids have had human DNA spliced into them, the plasmids can then be taken up into bacterial cells through transformation. As these cells replicate, producing many cells, the plasmid DNA inside them replicates as well. These plasmids are producing the protein coded for by the human DNA that has been spliced into them. The result is a quantity of the human protein of interest.

- Plasmids are one type of cloning vector, meaning a self-replicating agent that functions in the transfer of genetic material. Viruses known as bacteriophages are another common cloning vector.

- A large number of medicines and vaccines are produced today through transgenic biotechnology. Transgenic organisms that are used for this purpose include not only bacteria but also yeast, hamster cells, and mammals such as goats. Transgenic food crops are planted in abundance today in the United States.

Web Animation 15.1: Action of Restriction Enzymes and Plasmids
Web Animation 15.2: Recombinant DNA

### 15.3  Reproductive Cloning

- A clone is a genetically identical copy of a biological entity. Genes can be cloned, as can cells and plants. Reproductive cloning is the process of making adult clones of mammals of a defined genotype. Dolly the sheep was a reproductive clone. Today, reproductive cloning of mammals is carried out through variants of the process that was used with Dolly: An egg cell has its nucleus removed and is fused with an adult cell containing a nucleus and, therefore, DNA. The fused cell then starts to develop as an embryo and is implanted in a surrogate mother. This process is called somatic cell nuclear transfer (SCNT).

- Reproductive cloning can work in tandem with various recombinant DNA processes to produce adult mammals possessing special traits. A cell can be made transgenic for such a trait and then used as the starting cell (the donor-DNA cell) in producing an adult mammal with the trait.

- It is possible that a human being will be cloned in the future, although anyone attempting to undertake the process would face technical, legal, and ethical obstacles. A human clone would be a genetic replica of the person who provided the donor-DNA cell. The donor and his or her clone would be genetically identical in the same way that identical twins are.

### 15.4  Forensic Biotechnology

- Identities of criminals, biological fathers, and disaster victims often are established today through the use of DNA. Forensic DNA typing is the use of DNA to establish identities in connection with legal matters, such as crimes.

- The polymerase chain reaction (PCR) is a technique for quickly producing many copies of a segment of DNA. It is useful in situations, such as crime investigations, in which a large amount of DNA is needed for analysis, yet the starting quantity of DNA is small.

- Forensic DNA typing usually works through comparisons of short tandem repeat (STR) patterns that are found in all human genomes. Police will, for example, compare the STR pattern in a suspect's DNA with the STR pattern in DNA that has been extracted from a crime scene.

Web Animation 15.3: Copying DNA through PCR

### 15.5  Stem Cells

- Most cells in the adult human body have undergone commitment, a developmental process that results in cells whose roles are completely determined. Most muscle cells have undergone this process, for example, and hence can be nothing but muscle cells and give rise to nothing but muscle cells.

- Cells exist in the early embryo that have not yet undergone commitment and can give rise to various kinds of cells. A relatively small number of cells in the adult body have a similar capability. Both the embryonic and adult cells that have this capability are called stem cells: cells that have a continuing capacity to produce more cells of their own type, along with at least one type of specialized daughter cell.

- One source of stem cells is the early embryonic structure known as the blastocyst. Cells from the blastocyst's inner cell mass, known as embryonic stem cells (ESCs), can give rise to all the different cell types in the adult human body. ESCs stand in contrast to adult stem cells, which are found, in small numbers, in various types of tissues in the adult body. These adult cells have demonstrated some ability to differentiate into various types of specialized cells and hence are the subject of continued research interest. However, they do not have the differentiation potential of ESCs nor the ESC's ability to continue to produce specialized cells generation after generation.

- There is an ethical controversy connected to the use of ESCs for research purposes. To harvest ESCs, researchers must destroy the embryos the cells are a part of, something many people regard as a destruction of human life. Others hold that embryos are potential human life, and that society has a greater responsibility toward impaired children and adults than it does toward embryos.

- Federal funding of embryonic stem cell research has been sharply limited by order of President Bush. In consequence, an increasing portion of this research is being funded by states and by private sources.

- Most experts believe embryonic stem cells hold out great potential, but this potential has yet to be realized in human beings. Most human trials involving embryonic stem cells are likely to be 5 to 10 years away.

- Cells derived from ESCs and then introduced into a human body may set off an immune system attack. Such an attack could be avoided if individuals could produce their own stem cells—either adult or embryonic—which could then be used to derive the therapeutic cells introduced into the body.

- A possible means of doing this through ESCs is known as therapeutic cloning: the use of cloning to produce human embryonic stem cells that can be used to treat disease and injury. In this process, a standard cell from an adult's body would be used as the donor-DNA cell in the SCNT cloning process. The resulting embryo would then be allowed to divide to the blastocyst stage, at which point ESCs from the embryo would be harvested, coaxed into specialization, and the resulting cells used to repair the body of the individual who provided the donor-DNA cell. Therapeutic cloning suffered a setback in 2005 when it was revealed that a

leading researcher in the field, Woo Suk Hwang of South Korea, had faked many of his results.

- In 2007, separate teams of scientists from Japan and the United States announced that, by inserting four genes into normal skin cells, they had managed to produce cells that seemingly have the developmental potential of embryonic stem cells. Because these induced pluripotent cells (iPCs) are derived from ordinary cells, there is no moral controversy surrounding their use. Moreover, they hold the potential of solving the dilemma of tissue rejection in stem-cell therapy: any given patient could theoretically provide the cells that would be used to derive iPCs. Thus, patients could be the source of the stem cells used in their own therapy.

## 15.6 Biotechnology in the Real World

- Advances in biotechnology tend to be costly and to come slowly. This is so partly because the biotech processes being developed are fundamentally new and hence require considerable regulatory scrutiny.

- Biotech progress also comes slowly because so many of the processes it is developing are not just new, but controversial. A notable biotech controversy concerns genetically modified (GM) crops. Opponents of these crops are concerned about their effect on human health and the environment. There is no evidence so far that GM crops have had detrimental effects in either area, but consumer resistance to the crops has sharply limited both the types being planted and the types being put into development.

- Some biotech controversies are essentially ethical in nature. Among these are the controversies concerning embryonic stem cells and therapeutic cloning. A more general controversy has to do with the question of what level of constraint society ought to impose on the modification of living things.

## KEY TERMS

| | |
|---|---|
| **biotechnology**   268 | **recombinant DNA**   270 |
| **blastocyst**   279 | **reproductive cloning**   273 |
| **clone**   272 | **restriction enzyme**   269 |
| **cloning vector**   272 | **somatic cell nuclear transfer** |
| **embryonic stem cells**   279 | **(SCNT)**   273 |
| **plasmid**   270 | **therapeutic cloning**   282 |
| **polymerase chain reaction** | **transformation**   270 |
| **(PCR)**   276 | **transgenic organism**   269 |

## BRIEF REVIEW

*Answers to Brief Review questions are in the back of the book.*
*For multiple-choice quiz questions, go to* **www.aw.com/krogh4**.

1. Why is it useful to have organisms such as bacteria turning out human proteins?

2. Why are restriction enzymes that cut DNA to produce sticky ends useful? If the following piece of DNA were cut with *Bam*HI, what would the fragment sequences be?

ATCGGATCCTCCG
TAGCCTAGGAGGC

3. Why did Dolly the sheep end up looking like the sheep that "donated" the udder cell rather than the sheep that donated the egg?

4. Why is the polymerase chain reaction (PCR) so valuable?

5. If embryonic stem cells derived from fertility clinic embryos work as hoped in fighting disease, what is the point of using therapeutic cloning to produce embryonic stem cells?

## APPLYING YOUR KNOWLEDGE

1. One of the motives put forth for human cloning is that people want to replace children or other loved ones who have died. To what extent could a clone of a loved one be a replacement for that person? If the technique had then been available, should doctors in the nineteenth century have preserved the DNA of Abraham Lincoln for cloning?

2. Should society demand that there be no risks to genetically modified foods before it allows them to be developed? Does any significant technology have no risks associated with it?

3. What limits, if any, should be put on the ability of human beings to modify the genomes of other living things? Should human beings be free to carry out any modification that does not harm an organism?

*The theory of evolution explains countless features of the natural world, including the dazzling plumage of the peacock (Pavo cristatus).*

CHAPTER **16**

16.1   Evolution and Its Core Principles   **292**

16.2   Charles Darwin and the Theory of Evolution   **293**

16.3   Evolutionary Thinking before Darwin   **294**

16.4   Darwin's Insights Following the *Beagle*'s Voyage   **296**

16.5   Alfred Russel Wallace   **297**

16.6   Descent with Modification Is Accepted   **298**

16.7   Darwin Doubted: The Controversy over Natural Selection   **299**

16.8   Opposition to the Theory of Evolution   **300**

16.9   The Evidence for Evolution   **301**

**Essay**

The Evolution of Human Skin Color   **304**

Charles Darwin is the person most responsible for deepening our understanding of how life evolved on Earth. Evidence uncovered from his time to ours has confirmed his insights.

# AN INTRODUCTION TO EVOLUTION: Charles Darwin, Evolutionary Thought, and the Evidence for Evolution

**S**tanding in a coastal redwood grove in California, we see trees stretching hundreds of feet into the air and wonder: Why are they so tall? Watching a nature program on television, we observe the exquisite plumage of the male peacock but are puzzled that such a *cumbersome* display could exist on a creature that has to be wary of predators. Leafing through a book, we discover that the color vision of honeybees is most sensitive to the colors that exist in the very flowering plants they pollinate. We wonder: Can this be an accident?

For as long as humans have existed, people have asked questions like these. Yet for most of human history, even the most informed individuals had no way to make sense of the variations that nature presented. Why does the redwood have its height or the peacock its plumage? Through the middle of the nineteenth century, the answer was likely to be either a shrug of the shoulders or a statement such as: "The creator gave unique forms to living things for reasons we can't understand."

Jumping ahead to our own time, modern-day biologists would give different answers to these questions. Their replies would go something like this:

- Redwoods (and trees in general) are so tall because, over millions of years, they have been in competition with one another for the resource of sunlight. Individual trees that had the genetic capacity to grow tall got the sunlight, flourished, and thus left relatively many offspring, which made for taller trees in successive generations.

- Extravagant plumage exists on male peacocks because it is attractive to *female* peacocks when they are choosing mating partners. Such finery is indeed cumbersome and carries a price (in lack of mobility), but this price is outweighed by the value of the feathers. Over time, the peacock ancestors with more attractive plumage mated more often and thus sired more offspring than peacocks with less-impressive plumage. This made for successive generations of peacocks with more elaborate plumage.

- It is not an accident that honeybee eyesight is most sensitive to the colors that exist in the flowering plants they pollinate. Both bees and flowering plants flourish because of their interactions with one another—bees get food, and flowers get pollinated. It is likely that flower color, or bee vision, or both, were modified over time to maximize this interaction.

Comparing contemporary answers such as these to that single nineteenth-century answer, you might well ask: What is it that came *in between,* so that we no longer must simply stand in puzzled wonderment before the diversity of nature?

## 16.1 Evolution and Its Core Principles

What intervened was a set of intellectual breakthroughs in the mid-nineteenth century that today go under the heading of the *theory of evolution*. Two principles lie at the core of this theory.

### Common Descent with Modification

One of these principles has been labeled **common descent with modification**. It holds that particular groups, or species, of living things can undergo modification in successive generations, with such change sometimes resulting in the formation of new, separate species—one species separating into two, these two then separating further. If you had a videotape of this "branching" of species and then ran it *backward,* what you'd eventually see is a reduction of all these branches into one, meaning that all living things on Earth ultimately are descended from a single, ancient ancestor. Evolution itself can have several definitions. In this chapter, we will think of it as synonymous with common descent with modification. In Chapter 17, you'll get a more technical definition of it.

### Natural Selection

Descent with modification is joined to a second evolutionary principle, **natural selection**: a process in which the differential adaptation of individual organisms to their environment selects those traits that will be passed on with greater frequency from one generation to the next. In the redwood example, trees that grew taller were better adapted to their environment than shorter trees, in that the taller trees got more sunlight. Accordingly, the taller trees left more offspring than the shorter trees. These offspring would likewise tend to inherit the genetic capacity to grow tall, meaning they, too, would have more offspring than other trees of *their* generation. In this way, the trait of tallness is being *selected* for transmission to ever-greater proportions of the tree population. As a result, the population evolves in the direction of greater tallness. As you will see, there are other processes that shape evolution, but none is as important as natural selection.

### The Importance of Evolution as a Concept

Evolution is not just important to biology; it is central to it. If biology were a house, evolution's principles would be the mortar binding its stones together. Why? Because all life on Earth has been shaped by evolu-

tion's key principles: natural selection and common descent with modification. This is easy to see once you think about any given organism. Take your pick: a fish, a tree, a human being? *All* of them evolved from ancestors, all of them ultimately evolved from a common ancestor, and natural selection was the primary process that shaped their evolution.

Given this, the theory of evolution provides a means for us to understand nature in all its buzzing, blooming complexity. It allows us to understand not just things that are familiar to us, such as the forms of the redwoods and peacocks, but all manner of new natural phenomena that come our way. When physicians observe that strains of infectious bacteria are becoming resistant to antibiotics, they don't have to ask: How could this happen? General evolutionary principles tell them that organisms evolve; that bacteria are capable of evolving very quickly (because they can produce many generations a day); and that antibiotic resistance is simply an evolutionary adaptation to the human use of antibiotics.

### Evolution Affects Human Perspectives Regarding Life

So far-reaching is the theory of evolution that its importance stretches beyond the domain of biology and into the realm of basic human assumptions about the world. Once persuasive evidence for the theory had been amassed, it meant that human beings could no longer view themselves as something *separate* from all other living things. Common descent with modification tells us that we are descended from ancestors, just like other organisms. Thus, we sit not on a pedestal, viewing the rest of nature beneath us, but rather on one tiny branch of an immense evolutionary tree.

Beyond this, evolution meant the end of the idea of a *fixed* living world in which, for example, birds have always been birds and whales have always been whales. Birds are the descendants of dinosaurs, while whales are the descendants of medium-sized mammals that walked on land. Life-forms only appear to be fixed to us because human life is so short relative to the time frames in which species undergo change.

Finally, in evolution human beings are confronted with the fact that, through natural selection, life has been shaped by a process that has no mind and no goals. Like a river that digs a canyon, natural selection shapes, but it does not design; it has no more "intentions" than does the wind. Trees that received more sunlight left more offspring, and thus trees grew taller over time. This is not a "decision" on the part of nature; it is an outcome of an impersonal process. And if evolution has no consciousness, it follows that it can have no morals. This force that has so powerfully shaped the natural world is neither cruel nor kind, but simply indifferent.

## 16.2 Charles Darwin and the Theory of Evolution

How did such a far-reaching theory come into existence? Thinking back to this book's unit on genetics, you may recall that the basic principles of that field were developed over the course of about a century by a large number of scientists. In contrast, a single person—nineteenth-century British naturalist Charles Darwin—is credited with bringing together the essentials of the theory of evolution (**Figure 16.1**).

### Darwin's Contribution

Darwin's contribution was twofold. First, he developed existing ideas about descent with modification while providing a large body of evidence in support of them. Second, he was the first person to perceive natural selection as the primary driving force behind descent with modification.

To be sure, the study of evolution has been refined and expanded since Darwin. And it should be noted that before Darwin had ever published anything on his theory, a contemporary of his, fellow Englishman Alfred Russel Wallace, independently arrived at his insight about the importance of natural selection to evolution. (This makes Wallace the co-discoverer of this principle.) Yet, there is universal agreement that it is Charles Darwin who deserves primary credit for providing us with the core ideas that exist in evolutionary biology even today.

### Darwin's Journey of Discovery

Darwin was born on February 12, 1809, in the country town of Shrewsbury, England. He was the son of a prosperous physician, Robert Darwin, and his wife, Susannah Wedgwood Darwin, who died when Charles was eight. Young Charles seemed destined to follow in his father's footsteps as a doctor, being sent away to the University of Edinburgh at age 16 for medical training. But he found medical school boring, and his medical career came to a halt when, in the days before anesthesia, he found it unbearable to watch surgery being performed on children. His father then decided that he should study for the ministry. At the age of 20, Darwin set off for Cambridge University to spend three years that he later recalled as "the most joyous in my happy life." Darwin's happiness came in part from the fact that theology at Cambridge took a backseat to what had been his true passion since childhood: the study of nature. From his early years, he had collected rock,

**Figure 16.1**
**Scientist of Great Insight**
Charles Darwin, late in his life. (Painting by John Collier, 1883. London, National Portrait Gallery. ©Archiv/Photo Researchers.)

animal, and plant specimens and was an avid reader of nature books. His studies at Cambridge did yield a divinity degree, but they also gave Darwin a solid background in what we would today call life science and Earth science.

### An Offer to Join the Voyage of the *Beagle*

Darwin's training, and the contacts he made at school, came together in one of the most fateful first-job offers ever extended to a recent college graduate. One of Darwin's Cambridge professors arranged to have him be the resident naturalist aboard the *HMS Beagle*, a ship that was to undertake a survey of coastal areas around the world (**Figure 16.2**). If you look at **Figure 16.3**, you can trace the *Beagle*'s journey. In addition to numerous stops on the east coast of South America (the ship's primary survey site), the *Beagle* also stopped briefly at the remote Galapagos Islands, about 970 kilometers or 600 miles west of Ecuador—a visit you'll learn more about later.

Thus did Darwin spend time on a research vessel—five years in all—beginning in England two days after Christmas 1831 and ending back there in October 1836. Just 22 when he left, he was prone to seasickness; he had to share a 10-by-15-foot room with two other officers; he was not a traveler by nature; and the journey was dangerous (three of the *Beagle*'s officers died of illness during it). Yet Darwin was happy because of the work he was doing: looking, listening, collecting, and thinking about it all. Here he is writing about a part of coastal Brazil:

> A most paradoxical mixture of sound and silence pervades the shady parts of the wood. The noise from the insects is so loud that it may be heard even in a vessel anchored several hundred feet from the shore; yet within the recesses of the forest a universal silence appears to reign. To a person fond of natural history such a day as this brings with it a deeper pleasure than he can ever hope to experience again.

## 16.3 Evolutionary Thinking before Darwin

Darwin beheld these sights and sounds during a period in the nineteenth century in which change was in the air with respect to ideas about life on Earth.

**Figure 16.2**
**What the *Beagle* Looked Like**

A reconstruction of the *HMS Beagle*, sailing off the coast of Tierra del Fuego in South America.

**Figure 16.3**
**Journey into History**

The main mission of *HMS Beagle*'s 5-year voyage of 1831–1836 was to chart some of the commercially promising waters of South America. Charles Darwin served on board the ship as a resident naturalist and companion to ship's captain Robert FitzRoy. Darwin observed nature and collected specimens throughout the ship's journey, but for purposes of the theory of evolution, the ship's most important stops came in 1835 on the Galapagos Islands, west of South America. The *Beagle* had a complex itinerary; for example, it stopped twice at the Falkland Islands off the southeast coast of South America, once in 1833 and again in 1834.

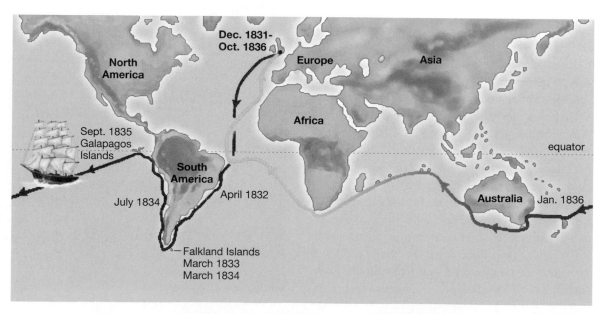

**(a)** Strata of sedimentary rock with fossils embedded

**(b)** Fossilized sea urchin, at least 65 million years old

**Figure 16.4**
**Geologic Strata and Their Contents**

## Charles Lyell and Geology

The single book Darwin took with him when he boarded the *Beagle* was the first volume of Charles Lyell's *Principles of Geology*, published in 1830 and bearing a message that even Lyell's fellow scientists found hard to accept: that geological forces *still operating* could account for the changes geologists could see in the Earth's surface. (As Darwin put it, "that long lines of inland cliffs had been formed, the great valleys excavated, by the agencies which we still see at work.") Under this view, Earth had not been put into final form at a moment of creation but rather was steadily undergoing change. If such a thing were possible for the Earth itself, why not for the creatures that lived on it (**Figure 16.4**)?

## Jean-Baptiste de Lamarck and Evolution

An idea along this line *had* been proposed, by the French naturalist Jean Baptiste de Lamarck, in a book published in 1809. Lamarck believed that organisms changed form over generations through what has been termed the inheritance of acquired characteristics. Ducks or frogs did not originally have webbed feet, he said, but in the act of swimming they stretched out their toes to move more rapidly through

the water. In time, membranes grew between their toes, effectively becoming webbing; critically, this characteristic was passed along to their offspring (**Figure 16.5**). Over time, he believed, an animal would acquire enough changes that one species would diverge into two, with this branching extending all the way to human beings. In his work, Lamarck lent support to a *means* of evolution that we know today is false. (Animals don't pass along traits in accordance with the activities they carry out.) Yet, note what he got right: that organisms can evolve; that one kind of organism can be ancestral to a different kind of organism.

**Figure 16.5**
**The Duck's Feet**
Jean-Baptiste de Lamarck proposed that new species evolve through a branching evolution, which is correct. He also proposed, however, that one of the forces driving this evolution is the inheritance of acquired characteristics, which does not take place. Under Lamarck's view, common at the time, a duck got its webbed feet because of activities it carried out during its lifetime, and this change was then passed on to its offspring.

### Georges Cuvier and Extinction

Another French scientist, Georges Cuvier, in examining the fossil-laden rocks of the Paris Basin early in the nineteenth century, provided conclusive evidence of the extinction of species on Earth. This was a radical notion at the time because many Christians believed that the creator would never allow one of his creatures to perish. (One amateur fossil collector who believed this was Thomas Jefferson.) This much about extinction Cuvier got right. However, seeing in his rock layers seeming "breaks" in the sequence of animal forms—a layer of simple forms would be followed by a layer of more complex forms as he went from older to newer layers—he held that there had been a series of catastrophes, such as floods, that wiped out life in given areas, after which the creator had carried out *new* acts of creation, bringing more complex life-forms into being with each act.

If you were to survey a broad swath of scientific thought leading up to Darwin, you'd repeatedly find what you've seen with Lamarck and Cuvier: a given scientist getting things partly right and partly wrong (which is almost always the way in science). The result was a rich mix of scientific findings and fanciful speculation that existed in Darwin's world as he boarded the *Beagle*. He himself believed at the time that species were fixed entities; that they did not change over time.

**Figure 16.6**
**The Galapagos Islands**

*Galapagos* is Spanish for *tortoise*. A part of Ecuador, the islands are still an important site of evolutionary research. Here a giant tortoise moves through the Alcedo Volcano area on one of the islands, Isabela.

## 16.4 Darwin's Insights Following the *Beagle*'s Voyage

Darwin observed and collected wherever he went, but the most important stop on his journey took place nearly four years into it, when the *Beagle* stopped for a scant five weeks at the remote series of volcanic outcroppings called the Galapagos Islands (**Figure 16.6**). There, in the dry landscape, amid broken pieces of black lava, he saw strange iguanas and tortoises and mockingbirds that varied from one island to another. He shot and preserved several of these mockingbird "varieties" for transport back home, along with a number of small birds that he took to be blackbirds, wrens, warblers, and finches.

### Perceiving Common Descent with Modification

Arriving back in England in October of 1836, Darwin soon donated a good deal of his *Beagle* collection to the Zoological Society of London. If there was a single most important flash of insight for him regarding descent with modification, it came when one of the society's bird experts gave him an initial report on the birds he had collected on the Galapagos. Three of the mockingbirds were not just the "varieties" that Darwin had thought; they were separate species, as judged by the standards of the day. Beyond that, the small birds he had believed to be blackbirds, finches, wrens, and warblers were all finches—separate species of finches, each of them found nowhere but the Galapagos (**Figure 16.7**).

With this, Darwin began to see a pattern: The Galapagos finches were related to an ancestral species that could be found on the mainland of South America, hundreds of miles to the east. Members of that ancestral species had come by air to the Galapagos and then, fanning out to separate islands, had diverged over time into separate species. Thus it was with Galapagos tortoises and iguanas and cactus plants as well; they had common mainland ancestors, but on these islands, had diverged into separate species. Darwin began to perceive the infinite branching that exists in evolution.

### Perceiving Natural Selection

But what drives this branching? In England and set on a career as gentleman naturalist, Darwin married and settled in London for a time before moving to a small village south of London, where he would live out his days (**Figure 16.8**). Two years after his homecoming, inspiration on evolution's key process came to him

**Figure 16.7**
**Three of Darwin's Finches**
Thirteen species of finch evolved on the Galapagos Islands, all of them descendants of a single species of finch native to South America. Shown, left to right, are *Geospiza fuliginosa*, *G. fortis*, and *G. magnirostris*.

not from looking at nature, but from reading a book on human population and food supply—*An Essay on the Principle of Population*, by T. R. Malthus. As Darwin later wrote in his autobiography:

> I happened to read for amusement Malthus on Population, and being well prepared to appreciate the struggle for existence which everywhere goes on . . . it at once struck me that under these circumstances favourable variations would tend to be preserved and unfavourable ones to be destroyed.

Thus did natural selection occur to Darwin as the driving force behind evolution. Organisms that had "favorable variations" were preserved (they lived and left many offspring), while those that had unfavorable variations were destroyed (they left fewer offspring or none at all). As a result, over generations, organisms evolved in the direction of the favorable variations.

You might think that, with two major insights in place—common descent with modification and natural selection—Darwin would have rushed to inform the world of them. In fact, more than 20 years were to elapse between his reading of Malthus and the 1859 publication of his great book, *On the Origin of Species by Means of Natural Selection*. In between, though incapacitated much of the time with a mysterious illness, Darwin published on geology, bred pigeons, and spent eight years studying the variations in barnacles—activities that each had relevance to his theory and that yet constituted a kind of holding pattern for him as he contemplated informing the world of a theory that he knew would be controversial. Darwin finally got to work on what he thought would be his "big species book" in 1856.

**Figure 16.8**
**Where *On the Origin of Species* Was Written**

In this study in his house in rural England, Charles Darwin conducted scientific research and wrote his groundbreaking work, *On the Origin of Species by Means of Natural Selection*.

## 16.5 Alfred Russel Wallace

Two years after Darwin began this labor, half a world away and unbeknown to him, a fellow English naturalist lay in a malaria-induced delirium on an Indonesian island. Alfred Russel Wallace made a living collecting bird and butterfly specimens from the then-exotic lands of South America and Southeast Asia, selling his finds to museums and collectors. He had thought long and hard (and even published) about how species originate. Now, shivering with malarial fever in a hut in the tropical heat, with the question on his mind again, Wallace came on the very insight that had come to Darwin 20 years earlier: Natural selection is the process that shapes evolution. Recovering from his illness, Wallace wrote out his ideas over the next few days and sent them to a scientific hero of his in England—Charles Darwin! He asked Darwin to read his manuscript and submit it to a journal if Darwin thought it worthy.

Darwin was stunned as he faced the prospect of another man being the first to bring to the world an insight he thought was his alone. Nevertheless, he would not allow himself to be underhanded with the younger scientist. He informed some of his scientist friends about his plight, and (without informing Wallace) they arranged for both Wallace's paper and some of Darwin's letters, sketching out his ideas, to be presented at a meeting of a scientific society in London on July 1, 1858. The readings before the scientific society turned out to have little immediate effect, but they did prod Darwin into working on what would become *On the Origin of Species*, published some 16 months later. This event had a great effect. Indeed, it set off a thunderclap whose reverberations can still be heard today.

## 16.6 Descent with Modification Is Accepted

Sparking both scorn and praise, Darwin's ideas were fiercely debated in the years after 1859. At least within scientific circles, however, it did not take long for common descent with modification to be accepted. Fifteen years or so after *On the Origin of Species* was published, almost all naturalists had become convinced of it, and it's not hard to see why. With evolution as a framework, so many things that had previously seemed curious, or even bizarre, now made sense.

For example, years before *Origin* was published, scientists had known that, at a certain point in their embryonic development, organisms as diverse as fish, chickens, and humans all have structures known as pharyngeal slits (**Figure 16.9**). In fish, these structures go on to be *gill* slits; in humans, they develop into the eustachian tubes, among other things. The question was: Why would land-dwelling and sea-dwelling animals share this common structure in embryonic life? For that matter, why would organisms as different as a human and a chicken share this structure? With evolution, the answer became clear. All of them shared a common vertebrate *ancestor*, which had the slits. The vertebrate ancestral line had branched out into various species, yet elements of the common ancestor persisted in these different species in their embryonic stage.

You might think that this would leave open the question of why organisms should differ more as adults than as embryos, but Darwin's theory made sense of this, too. Remember that natural selection is the primary process that brings about the differences that we see among species. Now consider one trait that natural selection stands to operate on in all animal species—the trait of movement or "locomotion." If you think about it, *adult* animals must deal with the issue of movement all the time, but embryonic animals never deal with it (because they don't move). Hence, for millions of years, natural selection was channeling the adult forms of vertebrate species in different ways with respect to their means of movement (fins, legs, wings). But this channeling was going on *only* in the adult forms. Likewise, natural selection was only affecting adults with respect to such traits as mating and the means of obtaining food. (Embryos don't mate, and food is delivered to them.) Hence, natural selection was driving differences in the adult forms of animals

**Figure 16.9
Different Embryos, Same Structure**

(Adapted from M. K. Richardson, 1997.)

Pharyngeal slits exist in these five vertebrate animals . . .

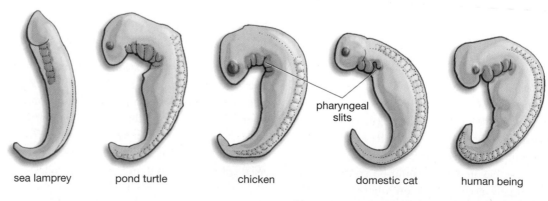

pharyngeal slits

sea lamprey    pond turtle    chicken    domestic cat    human being

. . . evidence that all five evolved from a common ancestor.

more than it was driving differences in embryonic forms. The result? Early embryos of various species could retain the ancestral vertebrate form more than adults could.

In bringing together countless loose ends such as this, evolution became the mortar that unified the study of the living world. (It's interesting to think of biologists from this period as readers who start going over a detective novel a *second* time, saying, "Of course!" and "Why didn't I see that?") Given this, scientists had little trouble accepting the *fact* of evolution—that is, the occurrence of descent with modification. But the primary process that Darwin said was driving evolution, natural selection, was not embraced so quickly. Indeed, the twentieth century would be almost halfway over before it was generally accepted.

## 16.7 Darwin Doubted: The Controversy over Natural Selection

The essential stumbling block for the theory of evolution by natural selection was this: It asserted that traits can vary in ways that confer reproductive advantages on given individuals and that these variations can be passed on from one generation to the next. With the trees considered earlier, the trait was height, and the variation that conferred a reproductive advantage was greater height. In a given environment, therefore, added height represents a *difference* that allows some members of the species to leave more offspring than others, leading eventually to a taller species over all. But in the middle of the nineteenth century, no one could imagine how such differences could reliably be passed down over many generations. Even scientists of the time could not understand this because they had no grasp of the field that we today call genetics.

### Coming to an Understanding of Genetics

Thinking back to the genetics section of this book, recall, from Chapter 11, that a trait such as tree height is likely to be governed by many genes working in tandem, with each of these genes coding for a protein that has some small, additive effect on height. Indeed, such multigene or "polygenic" inheritance is at work with most of the traits in living things. (See "Multiple Alleles and Polygenic Inheritance" in Chapter 11, page 207.) A tree's "height genes" will be shuffled in countless ways when tree egg and sperm are formed and

when they fuse in fertilization. As a result, gene combinations that yield particularly tall trees will come together in certain instances. In addition, outright mutations—alterations in genetic information—can take place, showing up as new physical traits, one of them being a taller tree. Most important for our purposes, whatever the genetic information is, it *persists;* it is passed on largely intact from one generation to another in the small informational packets we call genes. The result is small physical differences that can be reliably passed on from one generation to the next.

Darwin could not have imagined a better mechanism for carrying out natural selection as he had envisioned it. The problem was that neither he nor anyone else in his time knew that things worked like this. Most nineteenth-century scientists believed that inheritance worked through a "blending" process akin to mixing different-colored paints. If you blend red and blue paint, for example, you get something that lies halfway between them, purple. Under such a system, differences in traits would not be preserved over time; they would be averaged into oblivion. Blending would work to make groups of organisms the same, not to allow differences among them. How could natural selection work if there were no persistent differences to select from?

### Vindicating Natural Selection's Role in Evolution

The vindication of natural selection required, first, a demonstration (by Gregor Mendel) that inheritance doesn't work by blending. Rather, it works through the transmission of the stable genetic units we now call genes. Then, in the early part of the twentieth century, came the demonstration that, through polygenic inheritance, genes could account for very small physical differences that arise within a given population. Trees come in a *range* of heights, with the differences in this range governed by the combinations of individual genes described earlier.

### Darwin Triumphant: The Modern Synthesis

Advances in genetics such as this were later joined to other types of evolutionary research, such that in the period roughly from 1937 to 1950 there took place what is known as the **modern synthesis**: the convergence of several lines of biological research into a unified evolutionary theory. Taxonomists reported on how species are distributed throughout the world. Mathematical geneticists provided guidance on how evolution *could* work, given the shifting around of genes among organisms. Paleontologists studied evolution in the fossil record. Critically, these different types of scientists were involved in a process of mutual education with one another; the findings of geneticists, for example, were now known to taxonomists, and vice versa. The upshot was a greatly deepened understanding of how evolution works—an understanding that provided conclusive evidence for Darwin's assertions about the importance of natural selection. The architect of evolutionary theory had been vindicated nearly 100 years after publishing his great work. (For another example of evolution's explanatory power, see "The Evolution of Human Skin Color" on page 304.)

---

**SO FAR...**

1. True or false: Charles Darwin was the first person to perceive common descent with modification.

2. The co-discoverer of natural selection was _____.

3. In his voyage around the world, Darwin gained the most insight from the time he spent on the _____ Islands, which lie off the coast of _____.

---

## 16.8 Opposition to the Theory of Evolution

So, why are Darwin's essential insights so widely believed to be correct? We'll now review some of the evidence for these insights, but first a word about why a look at this evidence is desirable. When you studied cells in this book, you did not review the "evidence for cells." Nor did you go over the "evidence for energy transfer" when studying cellular respiration nor the "evidence for lipids" when studying biological molecules. What makes evolution different from these topics?

### The False Notion of a Scientific Controversy

Evolution has an unusual status among major biological subdisciplines in that, almost alone among them, its findings are regularly challenged as being unproven or simply wrong. For the average person, who can't be bothered with the details of such a controversy, these attacks make it appear that the theory of evolution is not a body of knowledge, solidly grounded in evidence, but rather a kind of scientific guess about the history of life on Earth.

### What Is a Theory?

Several factors are critical in allowing the continued appearance of a "scientific debate" on the validity of the theory of evolution, when in fact none exists. The first of these has to do with a simple misunderstanding regarding terminology. You saw in Chapter 1 that, to the average person, the word *theory* implies an idea that certainly is unproven and that may be pure speculation. In science, however, a theory is a general set of principles, supported by evidence, that explains some aspect of the natural world. Accordingly, we have Isaac Newton's theory of gravitation, Albert Einstein's general theory of relativity, and many others. Lots of principles go into making up the theory of evolution, some of them on surer footing than others. Over time, however, some of these principles have achieved the status of established fact—we are as sure of them as we are sure that the Earth is round. Among these principles are the fact that evolution has indeed taken place and the fact that this has occurred over billions of years.

### The Nature of Historical Evidence

A second factor that provides an opening for the opponents of evolution is the nature of the evidence for it. If you were called on to provide the "evidence for cells" referred to earlier, a look through a microscope at some actual cells might be enough to stop this "debate" in its tracks. Conversely, one of evolution's most important manifestations—radical transformations of life-forms—can never be observed in this way because these transformations take place over vast expanses of time. Asking to "see" the equivalent of a dinosaur evolving into a bird before you'll believe it is like asking to see the European colonization of North America before you'll acknowledge it.

Evolution is taking place right now, all around us, but the evidence for evolution is historical evidence to a degree that is not true of, say, the study of genetics or of photosynthesis. Any ancient historical record is fragmentary, and evolution's historical record is ancient indeed. The fossils that scientists analyze are what remain from hundreds of millions of years of weathering and decay. Such an incomplete record leaves room for a great deal of interpretation among scientists—far more than is the case in purely experimental science. This interpretation has to do, however, with the *details* of evolution, not with its core principles. It has to do with what group descended from what other; with the rate at which evolution has proceeded, or with the role that pure chance has played in evolution. It does not have to do with *whether* evolution has occurred.

# 16.9 The Evidence for Evolution

So, what makes scientists so sure about the core principles of evolution? In any search for truth, we are more comfortable when lines of evidence are internally consistent and then go on to agree with *other* lines of evidence. The evidence for evolution satisfies both criteria. If we were to find even a single glaring inconsistency within or between lines of evidence, the whole body of evolutionary theory would be called into question. But no such inconsistencies have turned up yet, and scientists have been looking for them for about 150 years.

## Radiometric Dating

One claim of evolutionary theory, as you have seen, is that evolution proceeds at a leisurely pace, with billions of years having elapsed between the appearance of life and the present. Yet, how do we know that Earth isn't, say, 46,000 years old, as opposed to the 4.6 billion that scientists believe it to be? The conceptual bedrock on which we have determined the age of the Earth and its organisms is **radiometric dating**, a technique for determining the age of objects by measuring the decay of the radioactive elements they contain. As volcanic rocks are formed, they incorporate into themselves various elements that are in their surroundings. Some of these elements are radioactive, meaning they emit energetic rays or particles and "decay" in the process.

With such decay, one element can be transformed into another, the most famous example being uranium-238, which becomes lead-206 through a long series of transformations. The critical thing is that this transformation proceeds at a fixed *rate*; it is as steady as the most accurate clock imaginable. It takes 4.5 billion years for half a given amount of uranium-238 to decay into lead-206 (hence the term half-life). When such a transformation takes place in a cooled rock, the original or "parent" element is trapped within the rock, as is the "daughter" element, and the atoms of both elements can be counted. Therefore, comparing the proportion of a parent element to the daughter element in a rock sample provides a date for the rock with a fair amount of precision. There are now more than 40 different radiometric dating techniques used, each of them employing a different radioactive isotope. (See Chapter 2, page 21, for information on isotopes.) Some of these isotopes have very long half-lives (as with uranium-238), but some have relatively short half-lives. Given the range of such "radiometric clocks," scientists have been able to date objects from nearly the formation of the Earth to the present, although there are some gaps in the picture.

## Fossils

Today, one line of evidence for evolution is the similarity of fossil types by sedimentary layers. Looking at the same geologic layers of sediment worldwide, scientists find similar types of fossils, with a general movement toward more complex organisms as they go up through the newer strata. In Chapter 1, it was noted that scientific claims must be *falsifiable*; they must be open to being proved false with the discovery of new evidence. The fossil record presents a falsifiable claim. For example, creatures called trilobites (**Figure 16.10**) had a long run on Earth, existing in the ancient oceans from about 500 million years ago until some 245 million years ago, when they became extinct. By contrast, the evolutionary lineage that humans are part of, the primates, *began* no more than 65 million years ago. If scientists were to find, in a single fossil bed, fossils of trilobites existing side by side with those of early primates, our whole notion of evolutionary sequences would be called into question. No such incompatible pairing has happened, however, and the strong betting is it won't. Given this, the line of fossil evidence is internally consistent, but there's more. When we compare fossil placement with the dates we get from radiometric dating, we get excellent agreement *between* these two lines of evidence. We don't find trilobites embedded in sediments that turn out to be 60 million years old.

## Comparative Morphology and Embryology

**Morphology** is the study of the physical forms that organisms take, while **embryology** is the study of how animals develop, from fertilization to birth. As it turns out, the evidence provided for evolution by comparative embryology was touched on earlier, in

**••• ANSWERS**

1. False; common descent with modification was an idea that pre-dated Darwin. Darwin was, however, the first person to perceive natural selection as an evolutionary mechanism.
2. Alfred Russel Wallace
3. Galapagos; South America

**Figure 16.10
Ancient Organism, Now Extinct**

Shown is a fossilized trilobite, a kind of arthropod that became extinct long before the animals known as primates came to exist. This fossilized trilobite dates from about 370 million years ago and was found in present-day Morocco.

noting the pharyngeal slits that exist in the embryos of creatures as diverse as fish and humans.

In comparative morphology, some classic evidence for evolution is seen in the similar forelimb structures found in a very diverse group of mammals—in a whale, a cat, a bat, and a gorilla, as seen in **Figure 16.11**. Such features are said to be **homologous**, meaning the same in structure owing to inheritance from a common ancestor. Look at what exists in each case: one upper bone, joined to two intermediate bones, joined to five digits. Evolutionary biologists postulate that the four mammals evolved from a common ancestor, adapting this 1-2-5 structure over time in accordance with their varying needs.

## Evidence from Biogeography

The Hawaiian islands are places of such lush natural diversity that it can be surprising to learn that they harbor only one native variety of land-dwelling mammal: the bat. Then again, this same thing is true of the island of Mauritius, in the Indian Ocean; of the island of New Zealand, in the south Pacific; and of the Bonin islands, which are part of Japan. In every case, the only native terrestrial mammal is the bat. So, what accounts for this seemingly odd distribution of mammals? One guess ventured in the past was that a benevolent creator thought that larger mammals might not do well on islands and hence did not put them there. But it turns out that mammals such as pigs and goats have done just fine on islands when *human beings* have put them there, so this reasoning doesn't seem to hold up. An alternative explanation is that bats are the only mammals who can fly, and that flying was the only way that

mammals could reach isolated islands such as these. This seems like a common-sense explanation, but note that it contains a profound implication: that living things have been distributed around the world through *natural* processes (such as their means of movement). This principle stands in contrast to the idea just mentioned, which holds that a creator *placed* Earth's creatures where they are.

This story then deepens when we look at islands in a little more detail. New Zealand's only land mammal may be the bat, but prior to the arrival of human beings, it housed 14 species of giant, flightless birds (the moas; **Figure 16.12**). Hawaii may have just one native species of land mammal, but it is home to 800 species of the fruit flies known as drosophilids—a number so large that it represents one-fourth of all the drosophilid species found in the world. Meanwhile, the island of Madagascar, off the coast of Africa, has no native members of the cat family, no dogs, no antelopes, no rabbits or monkeys or apes, but it houses nearly 100 percent of the world's 49 species of the primates known as lemurs (**Figure 16.13**).

Now, what can explain this seemingly odd pattern of distribution—so few varieties of some animals on these islands, but so many varieties of others? Evolution explains it. We know from geology that Madagascar broke off from a giant supercontinent (that included Africa) about 165 million years ago. With this, Madagascar became isolated from all other land masses in the world. Only a few large land mammals could migrate to Madagascar, so this accounts for its lack of many of the species commonly seen in mainland Africa. The absence of these mammals, however, had another consequence:

**Figure 16.11**
**Four Animals, One Forelimb Structure**

Whales, cats, bats, and gorillas are all descendants of a common ancestor. As a result, the bones in the forelimbs of these diverse organisms are very similar despite wide differences in function. Four sets of homologous bones are color-coded for comparison. Note that in each case there is one upper bone, joined to two intermediate bones, joined to five digits.

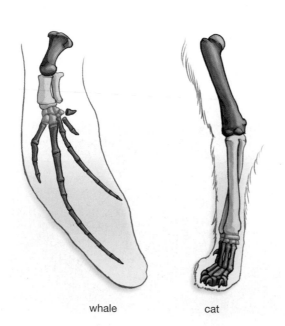

whale          cat

bat          gorilla

It left creatures on Madagascar free to *evolve* in ways that animals on mainland Africa could not. Why are there 49 living species of lemurs on Madagascar? Because Madagascar's lemurs didn't have to deal with competition from the primates found in abundance in Africa—the monkeys and the apes.

In a similar vein, why are there 800 species of drosophilid flies on Hawaii? The Hawaiian islands are volcanic in origin, with the oldest of the islands (Kauai) created about 5 million years ago. This means that when the islands were first formed, they were barren rock, with no life-forms on them. In time, however, life did come—on the wind and on the water—with bacteria, fungi, and plants tending to get established first and animals later. When the ancestors of the drosophilid flies arrived, they found a kind of natural paradise awaiting them: a land rich in food and ecological diversity, but a land lacking any flies of the drosophilid variety. As "early arrivers" among flies, the drosophilids moved into Hawaii's myriad environments (which range from tropical to desert-like). On the mainland, a drosophilid group moving into a new environment would have found plenty of similar flies already occupying each habitat. But not in Hawaii; there the drosophilid species could move into different environments without competition. And as they did so, they adapted to these environments, which is to say they *evolved* in these environments. Natural selection shaped one population of drosophilid flies after another to one kind of environment after another. The ultimate result was 800 species of the flies.

Finally, how could 14 species of giant moas even exist in New Zealand given that these birds were *flightless*? The special circumstance that allowed their survival was the absence of large land mammals on New Zealand. And the birds' species numbers? Like the lemurs of Madagascar and the drosophilids of Hawaii, the moas could move safely into lots of environments, and they did so over a very long period of time (about 80 million years). Put these elements together and what you get is the appearance of many species through evolution.

The geographic distribution of living things is called **biogeography**. To understand the power of evolutionary insight, imagine yourself for a moment trying to account for the biogeography of islands like New Zealand or Hawaii *without* reference to evolution. You might get away with saying that lopsided species numbers are an accident on one island, but when such imbalances are seen on one island after another, you've got some explaining to do. Would a creator engage in a type of special "island creation"—lots

**Figure 16.12**
**Really Big, and Flightless**

In 1877, British naturalist Richard Owen stood beside the remains of one of New Zealand's extinct, flightless moa birds. In 1839, Owen had deduced, from his inspection of a single moa thighbone fragment, that birds of the moa's size must have once roamed New Zealand. His claim was met with skepticism at the time—no complete moa skeletons had been found to that point, and people found it hard to believe that a flightless bird of such size could ever have existed. Subsequent excavations of moa remains proved Owen right, however. In his right hand, he is holding the moa thigh bone fragment that led to his famous deduction.

**Figure 16.13**
**Rare outside Madagascar**

Two ringtailed lemurs (*Lemur catta*) on the island of Madagascar, which is home to nearly 100 percent of the world's species of lemurs.

| ESSAY | The Evolution of Human Skin Color |

**D**ark-skinned Kenyans, lighter-skinned Algerians, and fair-skinned Norwegians. Everyone knows that as you travel from the equator to the poles, human skin color tends to go from darker to lighter. But why should this be so? Two California researchers have made a case that human skin color results from a precarious position we humans are in with respect to sun exposure: If we get too much sun, we suffer one kind of damage; if we get too little, we suffer another. These competing influences, the researchers say, have worked through natural selection to make human skin lighter in some populations and darker in others.

Nina Jablonski and George Chaplin of the California Academy of Sciences believe that all human beings are involved in a balancing act with respect to sun exposure. On the one hand, people who get too little sun may have a hard time making enough vitamin D, meaning their bones cannot grow properly, and their immune system starts to break down. On the other, people who get too much sun deplete their body's supply of a B vitamin called folic acid. When this happens in grown men, the result may be low sperm counts. When it happens in pregnant women, the result may be babies that have devastating "neural tube" defects—nervous system damage caused by the failure of the fetal spine to close properly during development.

The vitamin D part of this story has been understood in outline for decades. Vitamin D allows calcium to be absorbed by the intestines, and calcium, in turn, allows bones to grow strong. Human beings can get vitamin D from food, but they can also get it from sun exposure; the ultraviolet portion of sunlight prompts the body to turn cholesterol into vitamin D.

Since the 1960s, researchers have hypothesized that vitamin D's connection with sunlight has pushed human beings toward developing *lighter* skin. Why? The pigment that turns human skin dark, melanin, has the effect of blocking UV light absorption. Therefore, in a place like Scandinavia, where there is less sunlight during the year, it would be advantageous to have less melanin in the skin as this would mean lighter skin, more UV absorption, and more vitamin D production. If skin color were based solely on this factor, however, we would expect all human beings to have the fair skin of Scandinavians. They clearly don't, so what other factors might be at work in pushing human beings toward *darker* coloration?

Nina Jablonski was prompted to think about this in the early 1990s because of a fortuitous set of circumstances. She was then at the University of Western Australia; researchers there were among the first to discover the link between a lack of folic acid in expectant mothers and neural tube defects in their children. (So strong is this connection that, by the late 1990s, the U.S. and Canadian governments ordered that flour and other grain products be "fortified" with folic acid.) While at the university, Jablonski was asked to give a lecture on the biology of human skin color. In doing research for

her talk, she ran across a 1978 *Science* magazine paper that noted that fair-skinned people who are exposed to sun-like levels of ultraviolet light can suffer drastic reductions in the amount of folic acid in their system. Making a mental note of this, she went on, a few weeks later, to attend a seminar at the university on the link between folic acid and neural tube defects. Now it all came together. For some years, scientists had speculated that folic acid levels might have something to do with darker skin, but *how* could these two things be linked? Now, Jablonski thought she knew: Sunlight could lower folic acid levels, and low folic acid levels could bring about neural tube defects. As an evolutionary anthropologist, she realized that these two factors could easily work with a third to bring about different human skin colors. What was the third factor? Natural selection.

Imagine two women—human mothers on the sunny African savannah, say, 50,000 years ago, both of them pregnant. One of them has enough melanin in her skin to block the sun's UV rays such that her supply of folic acid remains high. As a result, she delivers a normal, healthy child. The other mother, however, has insufficient pigment for the UV blockage. As a result, she delivers a child with a neural tube defect. This second child will not live long enough to have children of his or her own. The first child, however, *will* go on to have children; this child has been *selected*, if you will, to have offspring, who themselves are likely to be dark-skinned. When this process is repeated with many generations of children, the result is a population of dark-skinned people.

Jablonski published a short paper on the possible link between folic acid and dark skin in 1992, and there the matter stayed until 1996, when she saw a paper by an Argentine pediatrician describing how three healthy women under his care had given birth to children with neural tube defects after using tanning beds in the early weeks of their pregnancy. Spurred on by this information, Jablonski and her research colleague (and husband) George Chaplin set out to find evidence that would confirm or disprove a hypothesis they had arrived at: that skin color is a result of a push-pull between retaining enough folic acid (for nervous system development) and making enough vitamin D (for proper bone development). From that time to this, they say, all the evidence they have uncovered supports this idea.

On the vitamin D side, a question the two researchers wanted to answer was: How much sun do you have to get for your body to start making this vitamin? Fortunately, scientists in Boston had answered that question for their area by exposing human skin, day after day, to sunlight on the roof of a hospital, and then measuring its vitamin D production. Combining information from this work with satellite data on the amount of UV radiation that falls around the world allowed Jablonski and Chaplin to construct what you see in **Figure 1**—a map dividing the world into three kinds of areas: those in which enough sunlight falls to allow people to synthesize

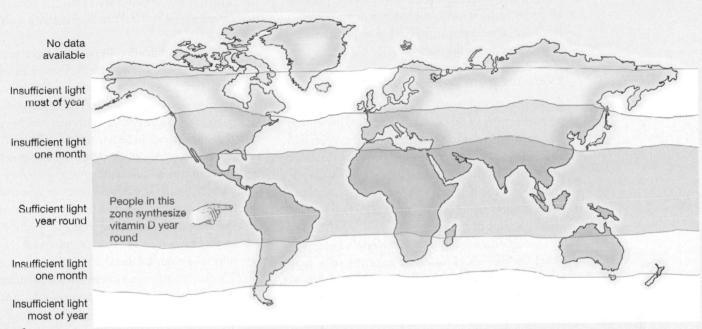

**Figure 1**
**Ultraviolet Light and Vitamin D**

Does the sun provide enough ultraviolet light to produce the vitamin D human beings need? In some parts of the world it does, but in others it does not. Note that in northern Europe there is insufficient UV light for vitamin D production most of the year, but that near the equator UV light is sufficient all year round.

vitamin D all year long; those in which vitamin D cannot be synthesized at least one month out of the year; and those in which there is not enough UV light, averaged throughout the year, to satisfy human vitamin D needs.

With this, the researchers were able to make predictions about how dark the skin of various peoples should be in given areas based on the amount of UV light that falls there. Then, these predictions were checked against *actual* skin colors for the areas. The result? The deeper the ancestral roots of a population in a given area, the more likely it was that this population's skin color would match the color predicted for it. The ancestors of the !Kung tribesman on the left in **Figure 2** came to southern Africa thousands of years ago, while the ancestors of the Zulu woman on the right came to the area about 1,000 years ago from equatorial Africa. Thus, natural selection has had longer to work on the !Kung population than on the Zulu population and has given the !Kung a lighter skin color that is a better fit with the relatively cool climate of the region.

Beyond these findings, the researchers believe that their hypothesis may help explain another aspect of human skin color: why, in all cultures, men tend to have slightly darker skin than women. This difference, of perhaps 3 to 4 percent, may exist because men need slightly higher folic acid levels than women. Lack of folic acid in mice and rats causes their sperm production to drop so much that they become infertile. Meanwhile, human males who are given folic acid experience a rise in sperm count. "Men need to protect their

**Figure 2**
**Adaptation to UV Levels**

The !Kung tribesman on the left is a member of the Khoisan ethnic group, which came to southern Africa thousands of years ago. By contrast, the woman on the right is part of the Zulu ethnic group, who came to southern Africa from equatorial Africa about 1,000 A.D. The Zulus have thus had less time to adapt to the lower levels of UV radiation that prevail in southern Africa. As a result, their skin is still the darker color common to the indigenous peoples of equatorial Africa.

folic acid levels year-round," Jablonski says, whereas women are most at risk for low levels only during pregnancy. If so, then in the slight differences between male and female skin color, we can see the results of an ancient push-pull. Standing in the middle of this, acting as the judge of what works and what doesn't, is the no-nonsense process of natural selection.

of species of a few types but almost no species of other types? If not, what accounts for this pattern? The evolutionary account is coherent and consistent. Against this, what other framework even begins to make sense of who lives on Earth's islands?

## Evidence from Gene Modification

In the genetics unit, you saw that every living thing on Earth employs DNA and uses an almost identical "genetic code" (*this* triplet of DNA bases specifies *that* amino acid). At the very least, this means there is a unity running through all earthly life; it is also consistent with the idea that all life on Earth ultimately had a single starting point—the single common ancestor mentioned earlier.

In recent decades, molecular biology has also provided another check on what scientists have long believed to be true about evolutionary relationships—about which species are more closely related and about how long it's been since they've shared a common ancestor. All genes amount to a sequence of A, T, G, and C bases along DNA's double helix. And there are some genes that do very similar things in different organisms. There is, for example, a gene called *hedgehog* that helps regulate embryonic development in the *Drosophila* fruit fly and a gene called *sonic hedgehog* that helps regulate embryonic development in mice.

Though the *hedgehog* and *sonic hedgehog* genes are similar, we would not expect them to be identical. This is so because gene base sequences *change* over time through the process of mutation. If the *rate* of this change is constant, then mutations are like a molecular clock that's ticking; with the passage of a given amount of time, you get a set number of mutations. Given this, the longer it has been since two organisms shared a common ancestor, the greater the number of base differences we should see in the sequence of genes like *hedgehog* and *sonic hedgehog*.

To better understand this last point, it may be helpful to think in terms of something more familiar: natural language. The European settlers of Australia were predominantly English. Once the Australians and English became fundamentally separate, however, their manner of speaking began to evolve independently. For example, we could think of the word *outback* as a "mutation" in Australian speech. Meanwhile, England would be developing its own "mutations," meaning words not used in Australia. And the longer the Australians went on being separate from the English, the greater the number of such differences we would see between the two languages. Just so, the longer it has been since two species went down

separate evolutionary lines, the greater the number of base differences we would see in similar genes they have, as measured by the different mutations each has acquired.

This gene-modification hypothesis has been put to the test in connection with a gene that codes for an enzyme called cytochrome *c* oxidase, which exists in organisms as different as humans, moths, and yeasts. Going into this experiment, evolutionary theory predicted that there should be fewer DNA base-pair differences between the cytochrome *c* oxidase genes of, say, a human and a duck than between a human and a moth because all the *other* evidence we had told us that humans and ducks share a more recent common ancestor than do humans and moths. If you look at **Figure 16.14**, you can see that evolutionary theory was fully borne out by the DNA sequencing. There are 17 sequence differences between a human and a duck, but 36 between a human and a moth. Indeed, all the differences between the species pictured fall into line with evolutionary theory. Thus, we have another confirmation for evolution between lines of evidence—between DNA sequencing *and* the fossil record *and* radiometric dating *and* comparative morphology.

## Experimental Evidence

Finally, much evidence for evolution has been provided in recent years by experiment and observation. At first glance, this may seem rather improbable because, as you've seen, evolutionary change can often be perceived only over long stretches of time. Scientists have devised clever ways to catch evolution in the act, however.

Consider the experiments of John Endler, who believed that the male guppies he was studying were being pulled in two directions by natural selection. On the one hand, those who were larger and had brighter coloration were chosen more often by females for mating. On the other, these very characteristics made the males more vulnerable to predators. Endler saw that he could test natural selection by putting some guppies in a predator-free environment. In only a few generations, the males evolved brighter coloration and larger tails. When Endler then reintroduced predators into this population, things went the other way: Over several generations, the males evolved smaller tails and less-brilliant colors. Both of these outcomes are precisely what evolutionary theory would predict. In both instances, the guppies evolved in a direction that would maximize their reproductive success—more mates with the brighter colors in the predator-free environment

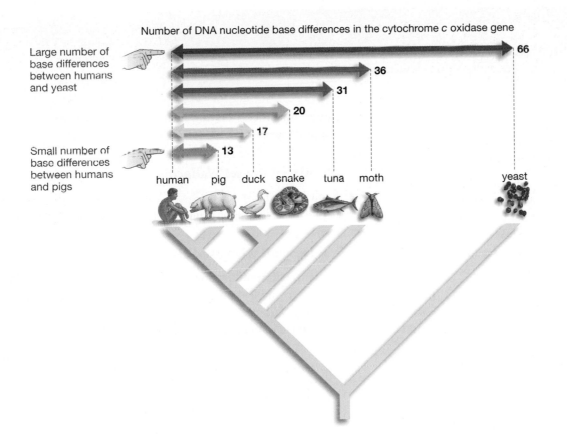

Number of DNA nucleotide base differences in the cytochrome *c* oxidase gene

Large number of base differences between humans and yeast

Small number of base differences between humans and pigs

66

36

31

20

17

13

human    pig    duck    snake    tuna    moth                    yeast

**Figure 16.14**
**Using Molecules to Track Evolution**

Diverse organisms—such as yeast, moths, and pigs—all have a gene that codes for an enzyme called cytochrome *c* oxidase. These organisms inherited cytochrome *c* oxidase genes from a common ancestor many millions of years ago. Over time, however, the cytochrome *c* oxidase genes have undergone mutations that have altered the sequence of their DNA "building blocks," called bases. The longer it has been since any two species shared a common ancestor, the more differences there should be in their cytochrome *c* oxidase bases. There are 13 differences between the bases found in the human cytochrome *c* oxidase gene and the one found in pigs, but there are 66 differences between the human and yeast genes. (Data from Whitfield, Philip. 1993. From *So Simple a Beginning: The Book of Evolution*. New York: Macmillan: Maxwell Macmillan International.)

and a longer life (and hence more reproduction) with the drab coloration they evolved in the predator-laden environment.

## On to How Evolution Works

Given the abundance of evidence for evolution, biologists long ago stopped asking whether it occurred. The really interesting questions for decades have been: Through what means has evolution proceeded? At what pace? In what direction, if any? These are the questions we'll look at in the next three chapters.

**SO FAR...**

1. The age of the Earth has been established through the use of _____, in which scientists test for the _____ of parent and daughter elements in a given sample of rock or other material.

2. _____ features are features in different species that are the same in structure owing to inheritance from a _____.

3. The term *biogeography* refers to the _____ of Earth's organisms.

••• ANSWERS

1. radiometric dating; proportions

2. homologous; common ancestor

3. geographic distribution

# CHAPTER 16 REVIEW

For study help, activities, and more quiz questions, go to **www.aw.com/krogh4**.

## SUMMARY

### 16.1 Evolution and Its Core Principles

- Within the theory of evolution, a key principle is that of descent with modification. This principle describes the process by which species of living things can undergo modification over time, with such change sometimes resulting in the formation of new, separate species. All species on Earth have descended from other species, and a single, common ancestor lies at the base of the evolutionary tree.

- A second key principle in the theory of evolution concerns natural selection: a process in which the differential adaptation of individual organisms to their environment selects those traits that will be passed on with greater frequency from one generation to the next.

- The theory of evolution has an importance beyond the domain of biology. Through it, human beings have become aware that (1) they are descended from other varieties of living things, and (2) the organisms that populate the living world are not fixed entities, but instead are constantly undergoing modification.

### 16.2 Charles Darwin and the Theory of Evolution

- Charles Darwin deserves primary credit for the theory of evolution. He developed existing ideas about descent with modification while providing a large body of evidence in support of them. And he was the first to perceive natural selection as the primary process that drives evolution.

- Darwin's insights were inspired by the research he carried out during a five-year voyage he took around the world on the ship *HMS Beagle*, beginning in 1831.

### 16.3 Evolutionary Thinking before Darwin

- Some of Darwin's ideas can be traced to the work of Charles Lyell, Jean-Baptiste de Lamarck, and Georges Cuvier, who respectively noted the dynamic geological nature of the Earth, the possibility of descent with modification, and the extinction of some species on Earth and the appearance of others within different time-frames.

### 16.4 Darwin's Insights Following the *Beagle's Voyage*

- Darwin understood descent with modification for several years before he comprehended that natural selection was the most important process driving it. It was his reading of a work by Malthus that sparked his realization about natural selection.

### 16.5 Alfred Russel Wallace

- English naturalist Alfred Russel Wallace is the co-discoverer of natural selection as the principal process underlying evolution.

### 16.6 Descent with Modification Is Accepted

- Descent with modification was accepted by most scientists not long after publication of Darwin's *On the Origin of Species by Means of Natural Selection* in 1859. Scientists accepted it because it explained so many facets of the living world.

### 16.7 Darwin Doubted: The Controversy over Natural Selection

- The hypothesis that natural selection is the most important process underlying evolution was not generally accepted until the middle of the twentieth century. Its acceptance hinged on a modern synthesis in the theory of evolution that brought together lines of evidence from genetics, the fossil record, and the distribution of organisms throughout the world.

### 16.8 Opposition to the Theory of Evolution

- Even today, the theory of evolution is regularly challenged as being unproven or simply wrong. One factor leading to the appearance of a "scientific debate" over evolution is confusion about the meaning of the word theory. Though the average person may equate "theory" with speculation, in science, a theory is a general set of principles, supported by evidence, that explains some aspect of the natural world.

### 16.9 The Evidence for Evolution

- Six lines of evidence are consistent with the theory of evolution. First, radiometric dating has confirmed the immense age of the Earth—an age that is consistent with the long periods of time scientists believe it has taken species to evolve. Second, around the globe, fossils from the same evolutionary periods are consistently found together in geologic strata; moreover, there is excellent agreement between the relative ages assigned to fossils by evolutionary theory and the absolute ages assigned to them by radiometric dating. Third, the theory of evolution explains the common occurrence of homologous physical structures in different organisms. Fourth, island biogeography—the geographic distribution of species on Earth's islands—is explained by the theory of evolution.

Fifth, variations found in the DNA sequences of various organisms are consistent with evolutionary theory. And sixth, experimental demonstrations of evolution have been carried out in the laboratory and in nature.

Web Animation 16.1: Principles of Evolution

## KEY TERMS

| | |
|---|---|
| biogeography  303 | modern synthesis  299 |
| common descent with modification  292 | morphology  301 |
| | natural selection  292 |
| embryology  301 | radiometric dating  301 |
| homologous  302 | |

## BRIEF REVIEW

*Answers to Brief Review questions are in the back of the book. For multiple-choice quiz questions, go to* **www.aw.com/krogh4**.

1. What is homology, and why does it provide evidence for evolution?

2. How does the evidence presented in Figure 16.14 support the conclusion that humans are more closely related to pigs than to yeast?

3. Describe two examples in which agreement among different lines of evidence provides evidence for evolution.

4. What is one observation that Darwin made in the Galapagos that influenced his thinking about evolution, and why was it important? Would he have been as likely to formulate his theory had he not had the opportunity to sail on the *Beagle*? Why or why not?

5. What major tenet of Darwin's theory of evolution was accepted by almost all scientists within Darwin's lifetime? What major tenet of his theory did not gain acceptance until midway through the twentieth century?

## APPLYING YOUR KNOWLEDGE

1. Using evolutionary principles, explain why large ears might be expected to evolve in a terrestrial plant eater such as a rabbit or deer but not in an aquatic mammal such as a seal.

2. Explain the evolutionary steps by which bacteria may become resistant to antibiotics, using the core requirements for evolution by natural selection.

3. Critics of the theory of evolution say it leaves no room for human purpose since the theory asserts that human beings evolved through a process—natural selection—that has no goals or intentions. Does the theory undercut the idea of purpose in human life?

*What is it that evolves? The essential unit that does so is the population. Pictured is part of a migrating population of snow geese (Anser caerulescens).*

CHAPTER **17**

17.1    **What Is It That Evolves?**    **311**

17.2    **Evolution as a Change in the Frequency of Alleles**    **313**

17.3    **Five Agents of Microevolution**    **313**

17.4    **Natural Selection and Evolutionary Fitness**    **320**

17.5    **Three Modes of Natural Selection**    **322**

**Essays**

Lessons from the Cocker Spaniel: The Price of Inbreeding    **318**

Detecting Evolution: The Hardy-Weinberg Principle    **325**

Evolution at its most fundamental level is called microevolution.

# THE MEANS OF EVOLUTION:
# Microevolution

**W**hen people hear the word *evolution*, what generally comes to mind is the grand sweep of evolution: the story of how microorganisms were the first life-forms; of how dinosaurs came and went; of how human

beings arose as newcomers on the planet. It is quite a story, to be sure, but to fully appreciate it we need to look first not at these *outcomes* of evolution, but rather at evolution's *processes*. How do generations of organisms become modified over time? That's the subject of this chapter. When you understand it, you'll be in a position to comprehend evolution on its grander scale.

## 17.1  What Is It That Evolves?

In approaching the topic of how organisms become modified over generations, the first question is: What is it that evolves? If you think about it, it's pretty clear that it is not individual organisms. A tree may inherit a mix of genes that makes it slightly taller than other trees, but if this tree is considered in isolation, it is simply one slightly taller tree. If it were to die without leaving any offspring, nothing could be said to have evolved because no persistent quality (added height) has been passed on to any group of organisms.

In Chapter 16 you were introduced to the idea of a **species**, which can briefly be defined as a group of organisms who can successfully interbreed with one another in nature, but who don't successfully interbreed with members of other such groups. You might think that it is species—such as horses or American elm trees—that evolve. Indeed, scientists often speak

this way when they refer, for example, to a species of amphibious mammal having evolved into today's whales. But this is a kind of shorthand whose inaccuracy becomes apparent when species are considered as they live in the real world.

### Populations Are the Essential Units That Evolve

Think of a hypothetical species of frogs living in, say, equatorial Africa in a single expanse of tropical forest. Suppose that a drought persists for years, drying up the forest such that this single lush range is now broken up into two ranges separated by an expanse of barren terrain (**Figure 17.1** on the next page). When the separation occurs, there is still only a single species of frog, but that species is now divided into two *populations* that are geographically isolated and hence no longer interbreeding. Each population now faces the natural selection pressures of its *own* environment—environments that may differ greatly even though they are nearly adjacent. One area may get more sun, while the other has a larger population of frog predators, for example. Thus, each population stands to be modified individually over time—to evolve. What is it that evolves? The essential unit that does so is a **population**, which can be defined as all the members of a species that live in a defined geographic region at a given time.

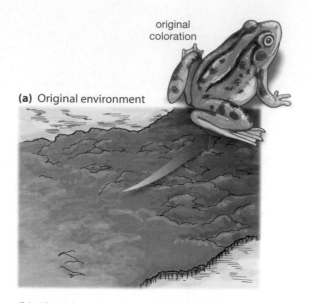

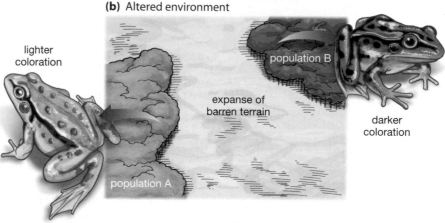

**Figure 17.1**
**Evolution within Populations**

**(a)** A hypothetical species of frog lives and interbreeds in one expanse of tropical forest.

**(b)** After several years of drought, the forest has been divided by an expanse of barren terrain. The single frog population has thus been divided into two populations, separated by the barren terrain. The two frog populations now have different environmental pressures, such as the kinds of predators each faces, and they no longer interbreed; after many generations their coloration diverges as they adapt to the different pressures.

The question then becomes, *how* do populations evolve? The population of frogs that has more predators may, over many generations, evolve a coloration that makes them less visible to predators. But how does this slightly different coloration come about? Through genes. A given frog may be unable to change its spots, but *generations* of frogs can have theirs changed through the shuffling, addition, and deletion of genetic material.

## Genes Are the Raw Material of Evolution

Recall from Chapter 11 that the genetic makeup of any organism is its **genotype**—and that a genotype provides an underlying basis for an organism's **phenotype**, meaning any observable traits that an organism has, including its physical characteristics and behavior. Many genes are likely to be involved in producing a phenotype such as coloration. In sexually reproducing organisms, *each* of these coloration genes will likely come in variant forms, called **alleles**, with offspring inheriting one allele from their father and one from their mother (**Figure 17.2**). Although both alleles help code for coloration, one may result in slightly lighter or darker coloration than the other.

Genes don't just come in two allelic variants, however; they can come in many. In most species, no one organism can possess more than two alleles of a given gene, but a *population* of organisms can possess many allelic variants of the same gene (which is one reason human beings don't just come in one or two colors, but in a continuous *range* of colors).

When the concept of a population is put together with that of genes and their alleles, the result is the concept of the **gene pool**: all the alleles that exist in a population. This gene pool is the raw material that evolution works with. If evolution were a card game, the gene pool would be its deck of cards. Individual cards (alleles) are endlessly shuffled and dealt into different "hands" (the genotypes that individuals inherit), with the strength of any given hand dependent on what game is being played. (Survival in a drought? Survival against a new set of predators?)

## 17.2 Evolution as a Change in the Frequency of Alleles

Thinking of genes in terms of a gene pool gives us a new perspective on what evolution is at root: a change in the frequency of alleles in a population. This may sound a little abstract until you consider the frog coloration example. A frog inherits, from its mother and father, a coloration that allows it to evade predators slightly more successfully than other frogs of the same generation. It thus lives longer and leaves more offspring than the other frogs. It is successful in this way because of the advantageous set of alleles it inherited. Because this frog is more successful at breeding, its alleles are being passed to the *next* generation of frogs in relatively greater numbers than the alternative alleles carried by less-successful frogs. So this frog's alleles are increasing in frequency in the frog population. In looking at any example of evolution in a population over time, you would find this kind of change in allele frequency as its basis, assuming a stable environment.

With this perspective in mind, you're ready for a definition of **microevolution**: A change of allele frequencies in a population over a relatively short period of time. Why the *micro* in microevolution? Because evolution *within* a population is evolution at its smallest scale. This concept of microevolution allows you to understand the formal definition of evolution promised to you last chapter. **Evolution** can be defined as any genetically based phenotypic change in a population of organisms over successive generations. Taking a step back, the large-scale patterns produced by microevolution eventually become visible, as with the evolution of, say, mammals from reptiles. This is **macroevolution**, defined as evolution that results in the formation of new species or other large groupings of living things.

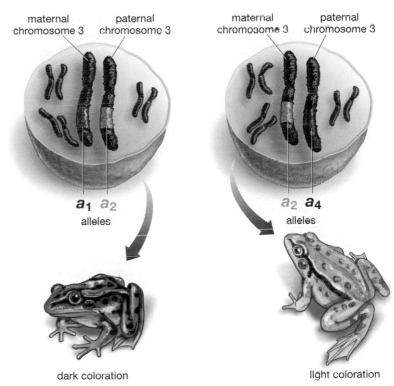

**Figure 17.2**
**Genetic Basis of Evolution**

Many genes can produce a trait such as body coloration, and each gene often has many alleles or variants. Each individual in a population, however, can possess only two alleles for each gene, one allele inherited from its mother and one inherited from its father. The two frogs in the figure both have maternal and paternal copies of chromosome 3 that house genes for coloration. The chromosomes of the two frogs may differ, however, in the allelic variants they have of these genes. The frog with dark coloration may possess alleles $a_1$ and $a_2$, while the light-colored frog may possess alleles $a_2$ and $a_4$ of this same gene.

---

**SO FAR...**

1. The essential unit that evolves in nature is the _____, which can be defined as all the members of a _____ living in a defined geographic region at a given time.

2. Genes come in variant forms called _____. A gene pool is defined as all the _____ that exist in a _____.

3. Microevolution is defined as a _____ in _____ in a population over a relatively short period of time.

Web Animation 17.1
Evolution and Genetics

## 17.3 Five Agents of Microevolution

So, what is it that causes microevolution? Put another way, what causes the frequency of alleles to change in a population? In the frog-coloration example, a familiar process was at work: natural selection. The frog's coloration allowed the frog to

... **ANSWERS**

1. population; single species

2. alleles; alleles; population

3. change; allele frequencies

evade predators, thus helping the frog to produce more offspring than did other frogs. This particular coloration was thus *selected* for greater transmission to future generations. With this selection,

allele frequencies began to be altered in the frog population. Over generations, the alleles for protective coloration increased relative to other sets of alleles.

Natural selection is not the only process that can change allele frequencies, however. There are five "agents" of microevolution that can alter allele frequencies in populations. These are mutation, gene flow, genetic drift, sexual selection, and natural selection. You can see these processes summarized in **Table 17.1**. (For a review of how scientists tell whether one of these processes is actually at work changing allele frequencies, see "Detecting Evolution: The Hardy-Weinberg Principle," on page 325.) We'll now look at each of these agents of change in turn.

## Mutations: Alterations in the Makeup of DNA

As you saw in the genetics unit, a mutation is any permanent alteration in an organism's DNA. Some of these alterations are *heritable*, meaning they can be passed on to future generations. Mutations can be as small as a change in a single base pair in the DNA chain (a point mutation) or as large as the addition or deletion of a whole chromosome or parts of it. Whatever the case, a mutation is a change in the informational set an organism possesses (**Figure 17.3**). Looked at one way, it is a change in one or more alleles.

### Table 17.1 Agents of Change: Five Forces That Can Bring about Change in Allele Frequencies in a Population

| Agent | Description |
|---|---|
| Mutation | Alteration in an organism's DNA; generally has no effect or a harmful effect. But beneficial or "adaptive" mutations are indispensable to evolution. |
| Gene flow | The movement of alleles from one population to another. Occurs when individuals move between populations or when one population of a species joins another, assuming the second population has different allele frequencies than the first. |
| Genetic drift | Chance alteration of gene frequencies in a population. Most strongly affects small populations. Can occur when populations are reduced to small numbers (the bottleneck effect) or when a few individuals from a population migrate to a new, isolated location and start a new population (the founder effect). |
| Sexual selection | Occurs when some members of a population mate more often than other members. |
| Natural selection | Some individuals will be more successful than others in surviving and hence reproducing, owing to traits that give them a better "fit" with their environment. The alleles of those who reproduce more will increase in frequency in a population. |

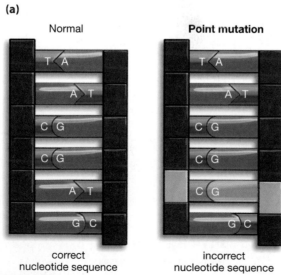

**(a)**

Normal     **Point mutation**

correct nucleotide sequence     incorrect nucleotide sequence

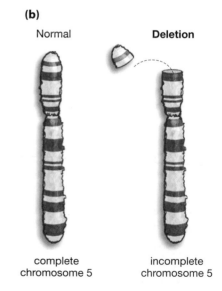

**(b)**

Normal     **Deletion**

complete chromosome 5     incomplete chromosome 5

**Figure 17.3**
**Basis of New Genetic Information**
A mutation is any permanent alteration in an organism's DNA. Examples of mutations include:
**(a)** point mutations, in which the nucleotide sequence is incorrect, and
**(b)** deletions, in which part of a chromosome is missing.

The rate of mutation is very low in most organisms; during cell division in humans, it might be just one DNA base pair per billion. And of the mutations that do arise, very few are beneficial or "adaptive." Most do nothing, and many are harmful to organisms. So, mutations usually are not working to *further* survival and reproduction, as the frog-coloration alleles did. Given this, mutations generally are not likely to appear with greater frequency in successive generations. The upshot is that mutations are not likely to account for much of the change in allele frequency that is observed in any population.

But a few mutations occur that are adaptive. These genetic alterations are something like creative thinkers in a society: They are rare but very important. Such mutations are the only means by which new genetic information comes into being—by which new proteins are produced that can modify the form or capabilities of an organism. The evolution of eyes or wings had to involve mutations. No amount of shuffling of existing genes could get the living world from no eyes to eyes. Of course, no mutation can bring about a feature such as eyes in a single step; such changes are the result of many mutations, followed by rounds of genetic shuffling and natural selection, generally over millions of years.

## Gene Flow: When One Population Joins Another

Allele frequencies in a population can also change with the mating that can occur after the arrival of members from a *different* population. This is the second microevolutionary agent, **gene flow**, meaning the movement of genes from one population to another. Such movement takes place through **migration**, which is the movement of individuals from one population into the territory of another. Some populations of a species may truly be isolated, such as those on remote islands, but migration and the gene flow that goes with it are the rule rather than the exception in nature. It may seem at first glance that migration would be limited to animal species, but this isn't so. Mature plants may not move, but plant seeds and pollen do; they are carried to often-distant locations by wind, sea-currents, and animals (**Figure 17.4**). Of course, for a migrating population to alter allele frequencies of another population, its gene pool must be different from that of the population it is joining.

## Genetic Drift: The Instability of Small Populations

To an extent that may surprise you, evolution turns out to be a matter of chance. You can almost see the dice rolling in the third microevolutionary agent, genetic drift. Imagine a hypothetical population of 10,000 individuals. An allele in this gene pool is carried by 1 out of 10 of them, meaning that 1,000 individuals carry it. Now imagine that some disease sweeps over the population, killing half of it. Say that this allele had nothing to do with the disease, so the illness might be expected to decimate the allele carriers in rough accordance with their proportion in the population. If this were the case, 5,000 individuals in the population would survive, and 1 in 10—or about 500 of them—could be expected to be

**(a)** Hawaiian silversword      **(b)** Tarweeds in California

**Figure 17.4**
**Migration in Plants**
Brought about by volcanic eruptions, the Hawaiian Islands have always been surrounded by the Pacific Ocean. Therefore, all the plant species existing on the islands today are descended from species that were introduced to the islands through one means or another—human activity, wind currents, water currents, or animal dispersal of seeds and pollen.
**(a)** Hawaiian silverswords are derived from
**(b)** a lineage of California plants commonly known as tarweeds.

carriers of this allele. Let us say, however, that just by chance, *550* of the allele carriers were killed, thus leaving the surviving population of 5,000 with only 450 allele carriers. In that scenario, the frequency of the allele in this population would drop from 10 percent to 9 percent (**Figure 17.5a**).

Now for the critical step. Imagine the same allele, with the same 1-in-10 frequency, only now in a population of *10*. There is now but a single carrier of the allele, and that individual may not be one of the 5 members of the population to survive the disease. In this case, the frequency of this allele drops from 10 percent to zero (**Figure 17.5b**). It can be replaced only by a mutation (which is extremely unlikely) or by migration from another population. Assuming that neither happens, no matter how this population grows in the future, in genetic terms it will be a different population than the original in that it lost this allele. The allele might be helpful or harmful, but its adaptive value doesn't matter. It has been eliminated strictly through chance. This is an example of **genetic drift**: the chance alteration of allele frequencies in a population, with such alterations having the greatest impact on small populations. It is true that some genetic drift has taken place in the larger population, but consider how small the effect is. The allele simply went from a 10 percent frequency to a 9 percent, with no loss of allele. Chance events can have much greater effects on small populations than on large ones.

Genetic drift can have large effects on small populations through two common scenarios: Populations can be greatly reduced through disease or natural catastrophe, or a small subset of a population can migrate elsewhere and start a new population. The first of these scenarios is called the *bottleneck effect*; the second is called the *founder effect*. Let's have a look at both of them in turn.

## The Bottleneck Effect and Genetic Drift

The **bottleneck effect** can be defined as a change in allele frequencies in a population due to chance following a sharp reduction in the population's size. Real populations, or even species, can go through dramatic reductions in numbers. For example, northern elephant seals, which can be found off the Pacific coast of North America, were prized for the oil their blubber yielded and thus were hunted so heavily that, by the 1890s, fewer than 50 animals remained. Thanks to species protection measures, the seals' numbers have rebounded somewhat in recent decades. But genetically, all the members of this species are very similar today because they all descended from the few seals that made it through the nineteenth-century bottleneck. What occurs in these reductions is a "sampling" of the original population—the "sample" being those who survived the devastation (**Figure 17.6**).

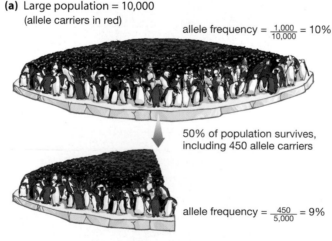

**(a)** Large population = 10,000
(allele carriers in red)

allele frequency = $\frac{1,000}{10,000}$ = 10%

**(b)** Small population = 10
(allele carriers in red)

allele frequency = $\frac{1}{10}$ = 10%

50% of population survives, including 450 allele carriers

50% of population survives, with no allele carrier among them

allele frequency = $\frac{450}{5,000}$ = 9%

allele frequency = $\frac{0}{5}$ = 0%

**little change in allele frequency**
**(no alleles lost)**

**dramatic change in allele frequency**
**(potential to lose one allele)**

**Figure 17.5**
**Genetic Drift**

**(a)** In a hypothetical population of 10,000 individuals, 1 in 10 carries a given allele. The population loses half its members to a disease, including 550 individuals who carried the allele. The frequency of the allele in the population thus drops from 10 percent to 9 percent.

**(b)** A population of 10 with the same allele frequency likewise loses half its members to a disease. Because the one member of the population who carried the allele is not a survivor, the frequency of the allele in the population drops from 10 percent to zero.

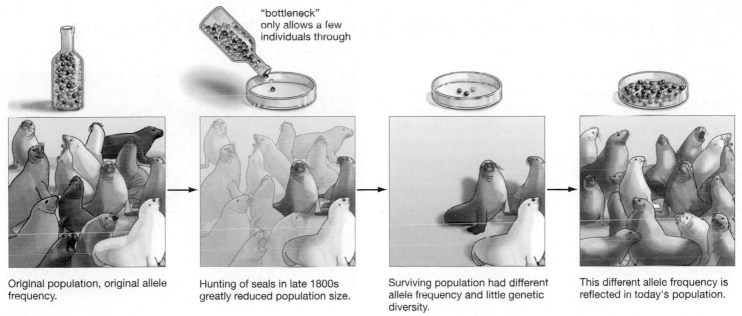

"bottleneck" only allows a few individuals through

Original population, original allele frequency.

Hunting of seals in late 1800s greatly reduced population size.

Surviving population had different allele frequency and little genetic diversity.

This different allele frequency is reflected in today's population.

**Figure 17.6**
**The Bottleneck Effect**
Northern elephant seals were hunted so heavily by humans that in the 1890s fewer than 50 animals remained. The population has grown from the few survivors (represented here by the three seals in the second frame), but the resulting genetic diversity of this population is very low. (Seal coloration for illustrative purposes only.)

The reason allele frequencies change in such an event has to do with the nature of probability and sample size. Imagine a box filled with M&Ms candies, with equal numbers of red, green, and yellow M&Ms inside. You close your eyes and grab a handful of M&Ms and pull out 12. With such a small sample, you might get, say, six reds, four greens, and two yellows, rather than the four of each kind that would be expected from probability. If you pulled out *120* M&Ms, conversely, your reds, greens, and yellows would be much more likely to approach the 40 of each kind that would be expected. In just such a way, a small sample of *alleles* is likely to yield a gene pool that's different from the distribution found in the larger population.

### The Founder Effect and Genetic Drift

Genetic drift can also result from the **founder effect**, which is simply a way of stating that when a small subpopulation *migrates* to a new area to start a new population, it is likely to bring with it only a portion of the original population's gene pool. This is another kind of sampling of the gene pool, in other words; but in this case it's caused by the migration of a few individuals rather than the survival of only a few. This sample of the gene pool now becomes the founding gene pool of a new population. As such, it can have a great effect; whichever genes exist in it become the genetic set that is passed on to all future generations as long as this population stays isolated.

The genetic drift that comes with the founder effect can be seen most clearly when a founding population brings with it the alleles for rare genetic diseases. There is, for example, a very rare genetic affliction of the eyes called *cornea plana* that results in a misshapen cornea—the first structure of the eye through which light passes. The result can be impaired close-range vision and a general clouding of eyesight. Cornea plana is known to affect only 113 people worldwide. The strange thing is that 78 of these people are in Finland, most of them in an area in northern Finland. Current research indicates that about 400 years ago, a small population arrived in this isolated area, with at least one member of this population carrying the recessive allele that causes this affliction. Since then, this allele has continued to profoundly affect subsequent generations in the area. This will always be the case if the descendants of a founder population stay relatively isolated over time—that is, if the descendents breed mostly among themselves.

From this and the earlier examples, you may have perceived by now that there is an inherent value in genetic diversity; you can read about this in "Lessons from the Cocker Spaniel," on page 318.

## Lessons from the Cocker Spaniel: The Price of Inbreeding

"Marry afar" is advice given by the pygmy men of Africa, by which they mean, "marry a woman who lives far from your home." The pygmies point to the hunting rights a man acquires in distant territory by marrying a woman from a remote region, but there is another benefit to this practice as well: It helps ensure genetic diversity among the pygmies. When pygmies from remote locations marry, it cuts down on the chances of *inbreeding*, meaning mating in which close relatives produce offspring. Inbreeding can have harmful effects not only in humans, but in any species.

*In the mid-1990s it was estimated that up to one-fourth of all U.S. purebred dogs had some sort of serious genetic defect.*

What are these effects? Well, consider what has happened in the United States with "purebred" dogs (**Figure 1**), which is to say dogs that over many generations have been bred solely with members of their own breed (cocker spaniels mating only with cocker spaniels, and so forth). In the mid-1990s it was estimated that up to one-fourth of all U.S. purebred dogs had some sort of serious genetic defect—ranging from improper joint formation, to heart defects, to deafness. We also have a good deal of experimental evidence from other animals on the effect of inbreeding. In one instance, a group of white-

**Figure 1**
**Result of Generations of Controlled Breeding**

footed mice that had been inbred and then let out into the natural environment survived at only 56 percent the rate of a group of genetically diverse mice. As already noted, negative biological outcomes can appear in humans, too; for that reason, many Western societies have made it illegal for first cousins to marry.

Why does inbreeding cause such trouble? As you've seen, every organism has a whole series of genes, almost all of which come

### Sexual Selection: When Mating Is Uneven across a Population

You've now looked at mutation, gene flow, and genetic drift as agents of change with respect to allele frequencies in a population. Now let's look at a fourth agent, **sexual selection**, defined as a form of natural selection that produces differential reproductive success based on differential success in obtaining mating partners. In practice, this is mating based on *phenotype*, which you may recall is any observable trait in an organism, including differences in appearance and behavior. It is the appearance of particularly nice plumage on a male peacock that makes female peacocks choose it for mating, rather than another male, while it is the behavior of strutting and chest-swelling in a male sage grouse that causes a succession of female grouse to mate with it rather than nearby competitors (**Figure 17.7**). It's easy to see that if one male mates four times as much as the average male of his generation, his alleles stand to increase proportionately in the next generation.

Differential mating success among members of one sex in a species often is based on choices made by members of the opposite sex in that species. In general, it is females who are doing the choosing in these situations. Differential mating success can also, however, be based on differences in combative abilities that give individuals of one sex (generally males) greater access to members of the opposite sex. Sexual selection is a form of natural selection in that some alleles—those that help bring about attractiveness, for example—are being preferentially selected for transmission to future generations. But while natural selection has to do with differential survival and reproductive capacity (and hence reproduction), sexual selection has to do with differential *mating* (and hence reproduction).

### Natural Selection: Evolution's Adaptive Mechanism

You've already gone over a good deal about natural selection in this review of evolution, but it's time now

in variant forms, or alleles. In any individual, one allele will be inherited from the father and one from the mother. In all organisms, a very small proportion of these alleles is potentially harmful or even lethal. But in most instances such alleles are recessive. *One defective allele is not enough to cause trouble* because the second differs from it and provides sufficient information for the organism to function properly. Only an individual who has two copies of the defective allele—an individual who is homozygous for it— would suffer from the condition. Meanwhile, an individual who has one "good" and one harmful allele would have no trouble.

So how do these biological facts play out in inbreeding? Imagine two seals, brother and sister, that have mated together because their population has been so reduced. Imagine that their mother had a recessive allele that brings about a heart defect, and that this allele occurs in only 1 of every 100 seals. There is a one-in-two chance that the mother will pass this allele on to any one of her offspring, and let's say it happened with the male and female here: Both brother and sister have inherited a single defective allele. When *they* mate, the chance that either of them will pass along the defective allele to their offspring is likewise one in two. More important, however, the chance that *both* of them will pass the allele along—producing an offspring with two defective alleles—is one in four. If, on the other hand, the brother had mated with an unrelated member of the population, the chances of this union passing along two copies of this defective allele are 1 in 400.* Why the difference? Because the unrelated female had only a 1-in-100 chance of having the allele in the first place.

The essence of this lesson, as you may have realized, is that inbreeding has brought together rare, identical alleles—the recessive alleles that are required for many genetic diseases. Random breeding, meanwhile, tends to keep such alleles apart. It tends to preserve genetic diversity, to put it another way.

You might think from this that there are no uses for genetic *similarity,* but consider what human beings have been doing for centuries with dog breeding. Cocker spaniels did not come about by accident. Dogs that had cocker spaniel features were bred together for many generations, ultimately giving us today's dog. It was the *novel features* of the spaniel that people wanted, and novel features are often the result of recessive alleles coming together. While the breeding done to produce American cocker spaniels has resulted in dogs with cute, floppy ears, the unintended effect of this practice is a breed that is also prone to ear infections and a "rage syndrome" that can result in unprovoked aggression, even against familiar humans. In short, inbreeding produces a range of "novel features," some of which are desirable, but others of which are more aptly referred to as genetic defects.

---

*Remembering the rule of multiplication from Chapter 11, to calculate the odds of passing on two defective alleles in the random mating example, take the son's chances of passing along the allele (1 in 2, or 50 percent, or 0.5) $\times$ 0.01 (the chance of his female partner *having* the allele) $\times$ 0.5 (the chance of her passing the allele along). This works out to 0.0025, which equals 1 in 400. For the brother-and-sister mating, the figures are 0.5 (the chance of the brother passing it along) $\times$ 0.5 (the chance of the sister passing it along), which equals 0.25—or 1 in 4.

---

to look at it as the last agent of change in microevolution. First, what is meant by natural selection? Here is a short definition: **Natural selection** is a process in which the differential adaptation of individual organisms to their environment selects those traits that will be passed on with greater frequency from one generation to the next. What does this mean in practice? Here is what biologist Julian Huxley had to say about it almost 60 years ago:

> Since there is a struggle for existence among individuals, and since these individuals are not all alike, some of the variations among them will be advantageous in the struggle for survival, others unfavorable. Consequently, a higher proportion of individuals with favorable variations will on the average survive, a higher proportion of those with unfavorable variations will die or fail to reproduce themselves.

By this process, the traits of those who are more successful in reproducing will become more widespread in a population—as the alleles that bring about these traits increase in frequency from one generation to the next.

**Figure 17.7**
**Sexual Selection**
Individuals in some species choose their mates based on appearance or behavior. Female sage grouses prefer to mate with males who put on superior "displays," which include sounds, a kind of strutting, and a puffing up of their chests. Here, two males display before females on a sage grouse breeding ground.

A key concept here is that of **adaptation**, meaning a modification in the form, physical functioning, or behavior of organisms in a population over generations in response to environmental change. Environmental change may come to a population (through streams drying up, for example), or a population may come to environmental change (through migration). Either way, natural selection is the process that pushes populations to adapt to new environmental conditions. The frog population noted at the outset of the chapter adapted to a new predator, for example, by evolving a darker coloration.

Among the five agents of microevolution, natural selection is the only one that consistently works to adapt organisms to their environment. Genetic drift is random; it could as easily work against an adaptive trait as for it. Although mutations can have an adaptive effect, they more often have no effect or even a negative effect. Gene flow doesn't necessarily bring in genes that are better suited to a given environment. And sexual selection has to do with the ways mating partners are selected, not with matching individuals to environment.

Natural selection is, however, constantly working to modify organisms in accordance with the environment around them. As such, it is generally regarded as the most important agent in having shaped the natural world—in having given zebras their stripes and flowers their fragrance. Because it is so important, we'll now look at it in a little more detail.

---

### SO FAR...

1. What the agents of microevolution all have in common is that each of them can _____ in a population.

2. Genetic drift is a _____ alteration of _____ in a population that has its greatest effect on _____ populations.

3. In sexual selection, differential reproductive success comes about because of differential success in obtaining _____.

---

## 17.4 Natural Selection and Evolutionary Fitness

Even the strongest supporters of natural selection as a shaping force would not maintain that it is working to produce perfect organisms. The concept of "fitness" is helpful here, if only to clear up some

misconceptions. To a biologist, **fitness** means the success of an organism in passing on its genes to offspring *relative to* other members of its population at a particular time. An organism cannot be deemed "fit," even if it has 1,000 offspring. It can only have *more or less* fitness relative to other members of its population (who might have 900 or 1,100 offspring). This has to do, once again, with allele frequencies in a gene pool. No matter how many offspring an individual has, its allele frequencies will increase in a population only if it has *more* offspring than other members of its generation. Thus, fitness is a measure of impact on allele frequencies in a population.

This concept then gets us to the notion of "survival of the fittest" and the misunderstandings that arise from it. The phrase can be taken to imply the existence of superior beings; that is, organisms that are simply "better" than their counterparts, with images of being faster, more muscular, or smarter coming to mind. In fact, however, evolutionary fitness tells us nothing about organisms being *generally* superior, and it certainly tells us nothing about the value of any particular capacity, be it brawn or brain. All it tells us about are organisms who are better than others in their population at passing along their genes in a given environment at a given time. An accurate phrase, as others have pointed out, would not be "survival of the fittest," but rather "survival of those who fit—for now." Let's look at a real-life example of natural selection to see why this is true.

### Galapagos Finches: The Studies of Peter and Rosemary Grant

When Charles Darwin stopped at the Galapagos Islands in 1835, some of the animal varieties he collected were various species of finches. Over the years, biologists kept coming back to "Darwin's finches" because of the very qualities Darwin found in them: They seemed to present a textbook case of evolution, with their 13 species having evolved from a single ancestral species on the South American mainland. Yet for more than 100 years after Darwin, it was a puzzle for scientists to figure out how Darwin's posited mechanism of evolution, natural selection, could have been at work with the finches. This changed beginning in the 1970s, when the husband-and-wife team of Peter and Rosemary Grant began a painstakingly detailed study of the birds.

Natural selection in the finches came into sharp focus in 1977, when a tiny Galapagos island, Daphne Major, suffered a severe drought. Rain that normally begins in January and lasts through July scarcely came at all that year. This was a disaster for the island's two species of finches; in January 1977 there were 1,300 of them, but by December the number had plunged to fewer than 300. Daphne's medium-size ground finch, *Geospiza fortis*, lost 85 percent of its population in this calamity. The staple of this bird's diet is plant seeds. When times get tough, as in the drought, the size and shape of *G. fortis* beaks—their beak "morphology"—begins to define what one bird can eat as opposed to another (**Figure 17.8**). In *G. fortis*, larger body size and deeper beaks turned out to make all the difference between life and death in the drought of 1977. Measuring the beaks of *G. fortis* that survived the drought, the Grant team found they were larger than the beaks of the population before the drought by an average of some 6 percent. This was a difference of about half a millimeter, or roughly two-hundredths of an inch; by such a difference were the survivors able to get into large, tough seeds and make it through the catastrophe, eventually to reproduce (**Figure 17.9**).

This is natural selection in action, but there is more to be learned from the Grant study, which is to say *evolution* made visible. The Grant team knew that beak depth had a high "heritability" in the finches, meaning that beak depth is largely under genetic control. As it turned out, the offspring of the drought survivors had beaks that were 4 to 5 percent deeper than the average of the population before the drought. In other words, the drought had preferentially preserved those alleles from the starting population that brought about deeper beaks, and the result was a population that evolved in this direction.

But the Grant study yielded one more lesson. In 1984–1985 there was pressure in the opposite direction: Few large seeds and an abundance of small seeds provided an advantage to smaller birds, and it was these birds that survived this event in disproportionate numbers.

### Lessons from *G. fortis*

So, where is the "fittest" bird in all this? There isn't any. Evolution among the finches was not marching toward some generally superior bird. Different traits were simply favored under different environmental conditions. Second, there is no evolutionary movement toward combativeness or general intelligence here. Survival had to do with size—and not necessarily *larger* size at that. Looking around in nature, it's true that some showcase species, such as lions and mountain gorillas, gain success in reproduction

**Figure 17.8**
**A Product of Evolution on the Galapagos**

A male of the finch species *Geospiza fortis*, which is native to the Galapagos Islands.

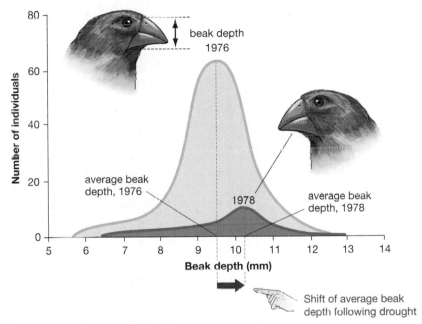

**Figure 17.9**
**Who Survives in a Drought?**

A large percentage of the population of *Geospiza fortis* died on a Galapagos Island, Daphne Major, during a drought in 1977. Peter and Rosemary Grant observed in 1978 that individuals that survived the drought had a greater average beak depth than average individuals surveyed before the drought, in 1976. Individuals with larger beaks were better able to crack open the large, tough seeds that were available during the drought. The offspring of the survivors likewise had larger average beak size than did the population before the drought. Thus, evolution through natural selection was observed in just a few years on the island.

by being aggressive. And it's true that our own species owes such success as it has had to intelligence. But in most instances, it is not brawn or brain that make the difference; it is something as seemingly benign as beak depth and its fit with the environment.

**⋯ ANSWERS**

1. alter allele frequencies
2. chance; allele frequencies; small
3. mating partners

**Web Animation 17.2**
Three Modes of Natural
Selection

Finally, consider how imperfect natural selection is at the genetic level. Suppose the smaller *G. fortis* that disproportionately died off in 1977 also disproportionately carried an allele that would have aided just slightly in, say, long-distance flying. In the long run this might have been an adaptive trait, but it wouldn't matter. The flight allele would have been reduced in frequency in the population (as the smaller birds died off) because flight distance didn't matter in 1977; beak depth did. Evolution operates on the phenotypes of whole organisms, not individual genes. As such, it does not work to spread *all* adaptive traits more broadly. Instead, the destiny of each trait is tied to the constellation of traits the organism possesses. Genes are "team players," in other words, that can only do as well as the team they came in on.

## 17.5 Three Modes of Natural Selection

In what directions can natural selection push evolution? As noted in Chapter 11, a character such as human height is under the control of many different genes and is thus **polygenic**. Such polygenic characters tend to be "continuously variable." There are not one or two or three human heights, but an innumerable number of them in a range. (See "Multiple Alleles and Polygenic Inheritance" on p. 207.) When natural selection operates on characters that are polygenic and continuously variable, it can proceed in any of three ways. The essential question here is: Does natural selection favor what is average in a given character, or what is extreme (**Figure 17.10**)?

**Figure 17.10**
**Three Modes of Natural Selection**

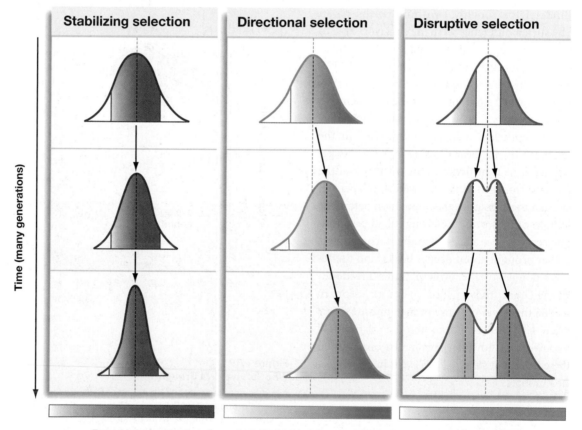

**Range of a particular characteristic** (in this instance, lightness or darkness of coloration)

In stabilizing selection, individuals that possess extreme values of a characteristic—here, both the lightest and the darkest colors—are selected against and die or fail to reproduce. Over succeeding generations, an increasing proportion of the population becomes average in coloration.

In directional selection, one of the extremes of a characteristic is better suited to the environment, meaning that individuals at the other extreme are selected against. Over succeeding generations, the coloration of the population moves in a direction—in this case toward darker coloration.

In disruptive selection, individuals with average coloration are selected against and die. Over succeeding generations, part of the population becomes lighter, while part becomes darker, meaning the range of color variation in the population has increased.

## Stabilizing Selection

In **stabilizing selection**, intermediate forms of a given character are favored over extreme forms. An example of this is human birth weights. In **Figure 17.11**, you can see, first, the weights that human babies tend to be. Notice that there are not only relatively few 3-pound babies, but relatively few 9-pound babies as well. A great proportion of birth weights falls at the average or mean of a little less than 7 pounds. Now look at the infant mortality curve. Infant deaths are highest at both extremes of birth weight; low-birth-weight babies *and* high-birth-weight babies are more at risk than are average-birth-weight babies (although low birth weight poses the greater risk). Put another way, the children most likely to survive (and reproduce) are those carrying alleles for intermediate birth weights. Thus, natural selection is working to make intermediate weights even more common. It is not working to move birth weights toward the extremes of higher or lower weights. Stabilizing selection is assumed to be the most common type of selection operating in the natural world. This should not be surprising because most organisms are well adapted to their environments.

## Directional Selection

When natural selection moves a character toward *one* of its extremes, **directional selection** is in operation—the mode in which we most commonly think of evolution operating. If you look at **Figure 17.12**, you can see an example of directional selection that took place over a very long period of time—about 4 million years—involving evolution toward larger brain size in

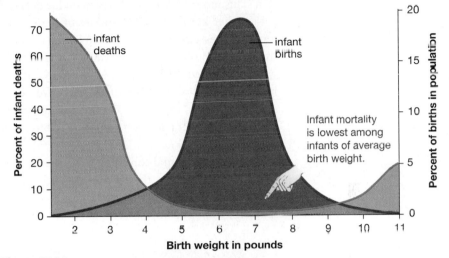

**Figure 17.11**
**Stabilizing Selection: Human Birth Weights and Infant Mortality**
Note that infant deaths are more prevalent at the upper and lower extremes of infant birth weights.

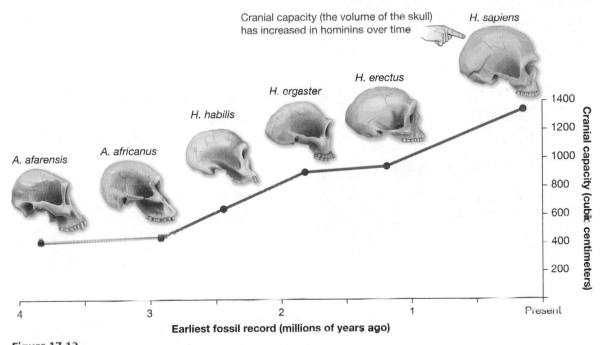

**Figure 17.12**
**Directional Selection: Evolving toward Increased Brain Size**
Modern humans are part of an evolutionary group, called the hominins, that includes all the upright-walking or "bipedal" primate species. Cranial capacity is a measure of the volume of brain tissue that can fit inside a skull. Directional selection for cranial capacity can clearly be seen in the ever-increasing capacities of this group of hominins (all of whom are now extinct, except for ourselves). The line in the figure does not trace ancestry, but merely shows the trend toward increased brain size. Going from left to right, the full names of the species shown are *Australopithecus afarensis* (the "Lucy" species), *Australopithecus africanus, Homo habilis, Homo ergaster, Homo erectus,* and *Homo sapiens* (ourselves).

some close, but now extinct, relatives of ours, a group called the hominins.

### Disruptive Selection

When natural selection moves a character toward *both* of its extremes, the result is **disruptive selection**, which appears to occur much less frequently in nature than the other two modes of natural selection. This mode of selection is visible in the beaks of yet another kind of finch. A species of these birds found in West Africa (*Pyrenestes ostrinus*) has a beak that comes in only two sizes. Thomas Bates Smith, who has studied the birds, has observed that if human height followed this pattern, there would be some Americans who are 4 to 5 feet tall and some who are 6 to 7 feet tall, but no one who is 5 to 6 feet tall (**Figure 17.13**).

The environmental condition that leads to this mode of selection in the finches is, once again, diet. When food gets scarce, large-billed birds specialize in cracking a very hard seed, while small-billed birds begin feeding on several soft varieties of seed, with each type of bird being able to outcompete the other for its special variety. Given how these birds have evolved, it's probably safe to assume that a bird with an intermediate-size bill would get less food than one with a bill of either extreme type. You might think that what really exists here are two separate species of bird, but large- and small-billed birds mate with one another. Bill size seems to be under the control of a single genetic factor, so that bills are able to come out either large or small. This genotype was presumably shaped by natural selection over generations, such that any alleles for intermediate-size bills were weeded out.

## On to the Origin of Species

Stabilizing selection does what it says: It stabilizes given traits of a population, thereby keeping it a single entity. However, both disruptive and directional selection can serve as the basis for speciation—for bringing about the transformation of a single species into one or more *different* species. How is it, though, that such speciation works? And how do we classify the huge number of species that are the outcome of evolution's myriad branchings? These are the subjects of the next chapter.

**Figure 17.13**
**Disruptive Selection: Evolution that Favors Extremes in Two Directions**
Finches of the species *Pyrenestes ostrinus*, found in West Africa. Birds of this species have beaks that come in two distinct sizes—large and small.

••• ANSWERS

1. False: Evolutionary fitness exists only in a relative sense; in a population, one organism is more or less fit than another depending on whether it has more or fewer offspring than another. So, having 1 offspring or 1,000 does nothing to qualify an organism as fit or not.
2. intermediate; extreme
3. one extreme

**SO FAR...**

1. True or false: If an insect has 1,000 offspring, we can label it as "fit" in an evolutionary sense.

2. In stabilizing selection, forms that are _____ are favored over forms that are _____.

3. We normally think of evolution working through directional selection, in which a character moves toward _____.

# Detecting Evolution: The Hardy-Weinberg Principle

How can scientists tell when evolution is taking place? As you've seen, evolution always entails a change in allele frequencies in a population. And there are five primary "agents" of allele frequency change—processes such as genetic drift and natural selection. So, how can scientists tell when one of these agents actually is at work? They can employ a procedure—a kind of test—devised early in the twentieth century by a British mathematician, Godfrey Hardy, and a German physician, Wilhelm Weinberg. Working independently, Hardy and Weinberg arrived at an insight, now called the *Hardy-Weinberg principle*, that today is used routinely not just in evolutionary biology, but in fields like public health as well.

The starting point for the Hardy-Weinberg principle is its assertion that, in a sufficiently large population, allele frequencies will *not* change unless one of the five agents is at work. A population with alleles that are not being altered by one of the agents is said to be at *Hardy-Weinberg equilibrium*—a state in which the population's allele frequencies remain constant, generation after generation. Let's walk through an example now that shows how a population at Hardy-Weinberg equilibrium stays the same. Along the way, you'll see the kinds of insights that are made possible by the Hardy-Weinberg principle.

Imagine a hypothetical species of plant whose flowers come in two colors, red and white, and say that we are looking at a population of 100 of these individuals. Further, we know from breeding these plants that red results from a dominant allele for a gene, while white results from a recessive allele for that gene. In the tradition of Mendel, we signify the red allele as *R* and white as *r*. Finally, we have sampled this population's DNA and found that *R* has a frequency of 78 percent, while *r* has a frequency of 22 percent. What does "frequency" mean in this context? It means that if we looked at the gametes of these plants—their eggs and sperm—78 percent of them would contain a dominant *R* allele while 22 percent would contain a recessive *r* allele. Now, keep your eye on this 78:22 allele frequency because that is what we are primarily interested in. The essential question is: When we look at the *next* generation of plants, will the allele frequencies in it remain 78:22? Or do allele frequencies change merely by virtue of reproduction? Hardy and Weinberg say no; in a large population, an agent of change is required to alter allele frequencies. Let's find out.

To see what happens in the next generation, we need to calculate the *genotypes* of our starting generation, which the Hardy-Weinberg principle allows us to do. What is a genotype in this context? Each egg or sperm may contain only *one* allele for a gene—in this case *R*

or *r*—but in diploid organisms such as flowering plants, egg and sperm fuse in fertilization, thus bringing alleles together to create an organism that has *two* alleles for each gene—one from the male, one from the female. In this context, then, genotype means the combination of *R* and *r* alleles that an individual possesses. Now, what possible combinations could a given individual have? An easy way to visualize them is to return once again to the Punnett squares that were used in Chapter 11. Imagine, in the parental generation, a male and female that are both heterozygous for the color gene—each of them has one *R* allele and one *r* allele. In this case, our Punnett square looks like this:

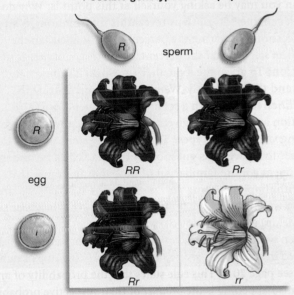

**Possible genotypes in the population**

So, the possible genotypes in the next generation are *RR*, *Rr*, or *rr*. (Further, note that because the *r* allele is recessive, only the *rr* genotype yields white flowers, signified by the white flower in the square.) But how do we calculate the genotype frequencies that will come about from this range of possibilities? How do we calculate what percentage of the population will have the *RR* genotype, for example, based on the allele frequencies we started with? The heart of the Hardy-Weinberg principle, the Hardy-Weinberg equation, allows us to do this. To generalize beyond the case of our red- or white-flowered plants, let's use a couple of conventional symbols in the equation: *p* will stand for the frequency of the more common *R* allele, and *q* will stand for the frequency of the less common *r* allele. The Hardy-Weinberg equation then looks like this:

| The Hardy-Weinberg equation . . . | $p^2$ | + | $2pq$ | + | $q^2$ |
|---|---|---|---|---|---|
| Applied to this example means . . . | $R^2$ | + | $2Rr$ | + | $r^2$ |
| And this yields the . . . | frequency of the *RR* genotype | + | frequency of the *Rr* genotype | + | frequency of the *rr* genotype |

# CHAPTER 17 REVIEW

For study help, activities, and more quiz questions, go to **www.aw.com/krogh4**.

## SUMMARY

### 17.1 What Is It That Evolves?

- A population is defined as all the members of a single species living in a defined geographical area at a given time. The smallest unit that evolves is a population.

### 17.2 Evolution as a Change in the Frequency of Alleles

- Genes exist in variant forms called alleles. In most species, no individual will possess more than two alleles for a given gene—one allele coming from the individual's father, the other allele coming from the individual's mother. A population, however, is likely to possess many alleles for a given gene. The sum total of alleles in a population is referred to as that population's gene pool.

- The basis of evolution is a change in the frequency of alleles in a gene pool. To the extent that a given set of alleles increases in frequency from one generation to the next within a population, the phenotypes, or observable characteristics, produced by those alleles will be exhibited to a greater extent within the population. With such a change, a population can be said to have evolved. Evolution at this level is referred to as microevolution: a change of allele frequencies within a population over a relatively short period of time. Conversely, macroevolution, a product of microevolution, is evolution on a larger scale; it is evolution that results in the formation of new species or other large groupings of living things.

Web Animation 17.1: Evolution and Genetics

### 17.3 Five Agents of Microevolution

- Five evolutionary forces can result in changes in allele frequencies within a population. These agents of microevolution are mutation, gene flow, genetic drift, sexual selection, and natural selection.

- A mutation is any permanent alteration in an organism's DNA, and some mutations are heritable, meaning they can be passed on from one generation to the next. Mutation happens fairly infrequently, and most mutations either have no effect or are harmful; yet rare adaptive mutations are vital to evolution in that they are the only means by which entirely new genetic information comes into being.

- Gene flow, the movement of genes from one population to another, takes place through migration, meaning the movement of individuals from one population into the territory of another.

- Genetic drift, the chance alteration of allele frequencies in a population, has its greatest effects on small populations. Genetic drift can have large effects on small populations through two common scenarios. The first of these is the bottleneck effect, defined as a change in allele frequencies due to chance during a sharp reduction in a population's size. The second is the founder effect: the fact that when a small subpopulation migrates to a new area to start a new population, it is likely to bring with it only a portion of the original population's gene pool.

- Sexual selection is a form of natural selection that can affect the frequency of alleles in a gene pool. It occurs when differences in reproductive success arise because of differential success in mating. A given male in a population may, for example, sire many more offspring than the average male in the population. If so, this male's alleles will increase in frequency in the next generation of the population.

- In a population, some individuals will be more successful than others in surviving, and hence reproducing, owing to traits that better adapt them to their environment. This phenomenon is known as natural selection. Natural selection is the only agent of microevolution that consistently acts to adapt organisms to their environments. As such, it is generally regarded as the most powerful force underlying evolution.

### 17.4 Natural Selection and Evolutionary Fitness

- The phrase "survival of the fittest" is misleading because it implies that evolution works to produce generally superior beings who would be successful competitors in any environment. Evolutionary fitness, however, has to do only with the relative reproductive success of individuals in a given environment at a given time. One individual is said to be more fit than another to the extent that it has more offspring than another. But individuals are not born with invariable levels of fitness; instead, fitness can change in accordance with changes in the surrounding environment.

### 17.5 Three Modes of Natural Selection

- Natural selection has three modes: stabilizing selection, directional selection, and disruptive selection. Stabilizing selection moves a given character in a population toward intermediate forms and hence tends to preserve the status quo; directional selection moves a given character toward one of its extreme forms; disruptive selection moves a given character toward two extreme forms.

Web Animation 17.2: Three Modes of Natural Selection

## KEY TERMS

| | |
|---|---|
| adaptation 320 | genotype 312 |
| allele 312 | macroevolution 313 |
| bottleneck effect 316 | microevolution 313 |
| directional selection 323 | migration 315 |
| disruptive selection 324 | natural selection 319 |
| evolution 313 | phenotype 312 |
| fitness 320 | polygenic 322 |
| founder effect 317 | population 311 |
| gene flow 315 | sexual selection 318 |
| gene pool 313 | species 311 |
| genetic drift 316 | stabilizing selection 323 |

## BRIEF REVIEW

*Answers to Brief Review questions are in the back of the book.*
*For multiple-choice quiz questions, go to* **www.aw.com/krogh4**.

1. Explain the statement, "Individuals are selected, populations evolve."

2. What is a gene pool, and why do changes in gene pools lie at the root of evolution?

3. What are the five causes of allele frequency changes (microevolution), and how does each work?

4. Why is evolutionary fitness always a measure of relative fitness in a population?

5. Why is natural selection the only agent of evolution that consistently produces adaptation?

6. Is the evolutionary fitness of an individual expected to be the same or different in different environments? How did the Grant's study of beak depth of finches after the drought and then again in 1984–1985 provide a real example of this?

## APPLYING YOUR KNOWLEDGE

1. Cheetahs have long legs relative to other large cats. However, leg length in a cheetah population is more likely to be considered stabilizing rather than directional selection. Why?

2. The text notes that natural selection is the process that pushes populations to adapt to environmental change. Is natural selection still going on in human populations? Why or why not?

3. Explain why genetic drift may be important when captive populations of animals or plants are started with just a few individuals.

4. Two moth populations of the same species use different host plants. One rests on leaves and has evolved a green color that allows it to escape predation by blending in with the leaves. The other population rests on tree trunks and is brown in color. The colors of these moths are genetically determined by different alleles of the same gene. What would be some consequences of migration of moths leading to gene flow between these two populations?

*Male and female little green bee-eaters* (Merops orientalis) *engage in a courtship ritual in which the male feeds the female. Such rituals are one of the mechanisms that help produce new species.*

| 18.1 | What Is a Species? | 331 |
| 18.2 | How Do New Species Arise? | 333 |
| 18.3 | Many New Species from One: Adaptive Radiation | 337 |
| 18.4 | The Pace of Speciation | 340 |
| 18.5 | The Categorization of Earth's Living Things | 341 |
| 18.6 | Classical Taxonomy and Cladistics | 345 |

**Essay**

New Species through Genetic Accidents: Polyploidy     338

New species come about when populations of existing species cease to breed with one another. Each new species that arises constitutes another branch on life's family tree. Scientists are trying hard to figure out what this tree looks like.

# THE OUTCOMES OF EVOLUTION:
# Macroevolution

**H**ow many types of living things are there on Earth? How many varieties of life-forms are there that we can recognize as being fundamentally different from one another? It may surprise you to learn that we haven't

the foggiest idea. The lowest estimate is about 4 million species, but higher-end estimates often come in at 10 to 15 million, and the highest of them all is 100 million. Scientists simply have been unable to catalogue the vast diversity that exists on our planet. In 250 years of watching, digging, netting, and bagging, they have identified about 1.8 million species—a large number, to be sure, but only a fraction of the species that actually exist, even if our lowest estimates are correct.

Since no one knows how many species there are, let's suppose that there are "only" 4 million on Earth. Isn't this number staggering? This seems particularly true given its starting point. A single type of organism arose perhaps 3.8 billion years ago, branched into two types of organisms, and then the process continued—branches forming on branches until at least 4 million different types of living things came to exist on Earth. Now here's the real surprise: These are just the *survivors*. The fossil record indicates that more than 99 percent of all species that have ever lived on Earth are now extinct. Branching indeed.

The questions to be answered in this chapter are: How does this branching work? How do we go from the microevolutionary mechanisms explored in Chapter 17 to the actual divergence of one species into two? And how do we classify the creatures that result from this branching?

## 18.1 What Is a Species?

The category of life that's been mentioned so far, the species, has a great importance not only to biology, but to society in general. If we allow that some bacteria are harmful to human beings while others are helpful, that some varieties of mosquitoes transmit malaria while others do not, that some species of birds are endangered while others are not, then it follows that there must be some means of distinguishing one of these groupings from another (**Figure 18.1** on the next page). The whole notion of knowing something about the natural world begins to break down if we can't say *which* grouping it is that is endangered or harmful.

Over the centuries, human beings have devised various ways of defining the groups now called species. But in science today, the most commonly accepted definition stems from what is known as the **biological species concept**, which uses breeding behavior to make classifications. We actually looked at a version of the biological species concept in Chapter 17. Here it is again, as formulated by the evolutionary biologist Ernst Mayr:

> Species are groups of actually or potentially interbreeding natural populations which are reproductively isolated from other such groups.

**Figure 18.1**
**Separate Entities**
The concept of a species can have very practical effects.

**(a)** This bird is a northern spotted owl (*Strix occidentalis*), which is an endangered species, protected by federal law.

**(b)** This bird is a barred owl (*Strix varia*), which is not endangered.

**(a)** Endangered species

**(b)** Not endangered

**Figure 18.2**
**Hybrid Animal**
Pictured is a liger—a big cat whose father was a lion and whose mother was a tiger. These animals are produced only in captivity, never in nature.

a lion (**Figure 18.2**). In natural surroundings, however, they apparently have never interbred, even when their ranges overlapped centuries ago.

You might think that, with the biological species concept in hand, scientists would be able to study any organism and, by discerning its breeding behavior, pronounce it to be a member of this or that species. Nature is so vast and varied, however, that this doesn't always work. We can't look at the breeding activity of the single-celled bacteria or archaea, for example, because they don't *have* any breeding behavior; they multiply instead by simple cell division. (Microbiologists define bacterial and archaeal groupings by sequencing their DNA or RNA and looking for identifying patterns among these sequences.) Then there are separate species that sometimes interbreed in nature, producing so-called hybrid offspring as a result. Such mixing between species happens more in the plant world than the animal, but it does take place in both. If species are supposed to be "reproductively isolated" from one another, what are we to make of these crossings? (Mayr pointed out that it is *populations* in his definition that are reproductively isolated, not individuals, and that whole populations do tend to stay within their species confines.)

Note that the breeding behavior Mayr talks about can be real or potential. Two populations of finch may be separated from one another by geography, but if, on being reunited in the wild, they began breeding again, they are a single species. Note also that Mayr stipulates that species are groups of *natural* populations. This is important because breeding may take place in captivity that would not in nature. No one doubts that lions and tigers are separate species, yet they will mate in zoos, producing little tiglons or ligers, depending on whether the father was a tiger or

Despite these difficulties, the biological species concept provides a useful way of defining the basic category of Earth's living things. Bacteria and archaea notwithstanding, most multicellular species carry out sexual reproduction at least part of the time, and relatively few species outside the plant kingdom regularly produce hybrids. Moreover, this species concept is rooted in a critical behavior of organisms themselves—mating, which controls the flow of genes. And as you saw in Chapter 17, it is the change in the genetic makeup of a population (a change in its allele frequencies) that lies at the root of evolution.

## 18.2 How Do New Species Arise?

Having defined a species, let's now see how one variety of them can be transformed into another in the process called **speciation**: the development of new species through evolution. As you've observed, a single species can diverge into two species, the "parent" species continuing while another branches off from it. A key question for scientists is: What brings this branching evolution about? The answer has to do with the flow of genes reviewed in Chapter 17. As you saw then, evolution within a population means a change in that population's allele frequencies. Now imagine *two* populations of a single species of bird, with one being separated from the other—say, by one of them having flown to a nearby location. To the extent that they continue to breed with one another, with individuals moving between the locations, each population will *share* in whatever allele frequency changes are going on with the other population. Hence, the two populations will evolve together, remaining a single species. Now, however, imagine that the migration stops between the two populations. Through the various agents of change we looked at last chapter—natural selection, genetic drift, and so forth—each population continues to undergo allele frequency changes. But now neither population is sharing these changes with the other. And, over time, such genetic changes may bring about alterations in form and behavior. Coloration or bill lengths may change; feeding habits may be transformed; mating preferences may be modified. After enough time, should the two populations find themselves geographically reunited, they may no longer freely interbreed. At that point, they are separate species. Speciation has occurred.

The critical change here came when the two populations quit interbreeding. For scientists, the key question thus becomes: Why would this happen? What could drastically reduce interbreeding, and hence gene flow, between two populations of the same species?

### The Role of Geographic Separation: Allopatric Speciation

In the preceding example, a very clear factor reduced gene flow between the bird populations: They were separated geographically. And geographical separation turns out to be the most important starting point in speciation.

Populations can become separated in lots of ways. On a large scale, glaciers can move into new territory, cutting a previously undivided population into two. Rivers can change course, with the result that what was a single population on one side of the river may now be two populations on different sides of it. On a smaller scale, a pond may partially dry up, leaving a strip of exposed land between what are now two ponds with separate populations. Such environmental changes are not the only ways that populations can be separated, however. Part of a population might *migrate* to a remote area, as did the bird population, and in time be cut off from its larger population. For a real-life example of how migration can bring about speciation, see **Figure 18.3**.

**Figure 18.3**
**Speciation in Action**

Millions of years ago, the salamander *Ensatina eschscholtzii* began migrating southward from the Pacific Northwest. When it reached the Central Valley of California—an uninhabitable territory for it—populations branched west (to the coastal range) and east (to the foothills of the Sierra Nevada). Over time, the populations took on different colorations as they moved southward. By the time the two populations were united in Southern California, they differed enough, genetically and physically, that they either did not interbreed or produced infertile hybrid offspring when they did. (Salamanders are falsely colored in drawing for illustrative purposes.)

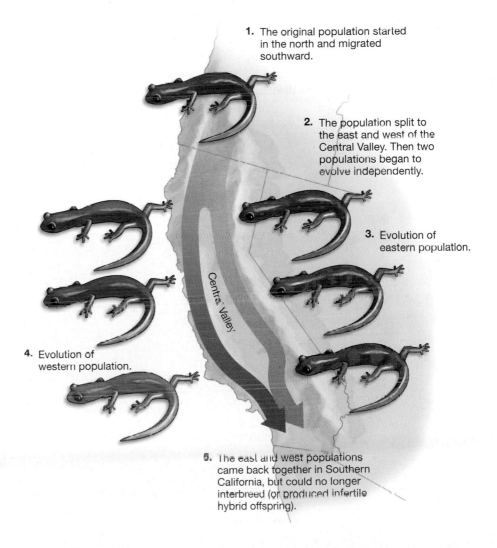

1. The original population started in the north and migrated southward.

2. The population split to the east and west of the Central Valley. Then two populations began to evolve independently.

3. Evolution of eastern population.

4. Evolution of western population.

Central Valley

5. The east and west populations came back together in Southern California, but could no longer interbreed (or produced infertile hybrid offspring).

When geographical barriers divide a population and the resulting populations then go on to become separate species, what has occurred is **allopatric speciation.** *Allopatric* literally means "of other countries" (**Figure 18.4**). Look along the banks on either side of the Rio Juruá in western Brazil, and you will find small monkeys, called tamarins, that differ genetically from one another in accordance with how wide the river is at any given point. Where it is widest, the members of this species do not interbreed, while at the narrow headwaters they do. Are the nonbreeding populations on their way to becoming separate species? Perhaps, but only if gene flow is drastically reduced between them for a very long time.

### Reproductive Isolating Mechanisms Are Central to Speciation

While geographical separation is the most important factor in getting speciation going, it cannot bring about speciation by itself. Following geographical separation, two populations of the same species must then undergo physical or behavioral changes that will *keep* them from interbreeding should they ever be reunited. Allopatric speciation thus operates through a one-two process: first the geographic separation, then the development of differing characteristics in the two resulting populations. These are characteristics that will *isolate* them from each other in terms of reproduction.

Thus arises the concept of **reproductive isolating mechanisms**, which can be defined as any factor that, in nature, prevents interbreeding between individuals of the same species or of closely related species.

Geographic separation is itself a reproductive isolating mechanism because it is a factor that prevents interbreeding. But, because the mountains or rivers that are the actual barriers to interbreeding are outside of or *extrinsic* to the organisms in question, geographic separation is called an **extrinsic isolating mechanism.** For allopatric speciation to take place, what also must occur is the second in the one-two series of events: the evolution of *internal* characteristics that keep organisms from interbreeding. Such characteristics are referred to as **intrinsic isolating mechanisms:** evolved differences in anatomy, physiology, or behavior that prevent interbreeding between individuals of the same species or of closely related species.

---

**SO FAR...**

1. Under the biological species concept, species are defined by their _____ behavior.

2. The critical factor leading to speciation is reduced _____ between two _____ of the same species.

3. Two factors always play a part in allopatric speciation: first, the development of _____ and second, the development of one or more _____.

---

### Six Intrinsic Reproductive Isolating Mechanisms

So, what are these intrinsic mechanisms? A list of the most important ones is presented in **Table 18.1.** An easy way to remember them is to think about what

**(a)** Abert squirrel, south rim of Grand Canyon

**(b)** Kaibab squirrel, north rim of Grand Canyon

**Figure 18.4**
**Geographical Separation—Leading to Speciation?**
These two varieties of squirrel had a common ancestor that at one time lived in a single range of territory. Then the Grand Canyon was carved out of land in northern Arizona, leaving populations of this species separated from one another, about 10,000 years ago. In the area of the Grand Canyon, **(a)** the Abert squirrel lives only on the canyon's south rim, while **(b)** the Kaibab squirrel lives on the north rim. It's unclear whether these two varieties of squirrel would be reproductively isolated if reunited today—that is, whether they are now separate species—but the geographical separation they have experienced is the first step on the way to allopatric speciation.

sequence of events would be required for fertile offspring to be produced in any sexually reproducing organism. First, organisms that live in the same area must encounter one another; if they don't, the "ecological isolation" mechanism is in place. If they do encounter one another, they must then mate in the same time frame; if they don't, then temporal isolation is in place, and so on.

## Ecological Isolation

Two closely related species of animals may overlap in their ranges and yet feed, mate, and grow in separate areas, which are called *habitats*. If they use different habitats, this means they may rarely meet up. If so, gene flow will be greatly restricted between them. Lions and tigers *can* interbreed, but they never have in nature, even when their ranges overlapped in the past. One reason for this is their largely separate habitats: Lions prefer the open grasslands, tigers the deep forests.

## Temporal Isolation

Even when two populations share the same habitat, if they do not mate within the same time frame, gene flow will be limited between them. Two populations of the same species of flowering plant may begin releasing pollen at slightly different times of the year. Should their reproductive periods cease to overlap altogether, gene flow would be cut off between them.

## Behavioral Isolation

Even when populations are in contact and breed at the same time, they must choose to mate with one another for interbreeding to occur. Such choice is often based on specific courtship and mating displays, which can be thought of as passwords between members of the same species. Birds must hear the proper song, spiders must perform the proper dance, and fiddler crabs must wave their claws in the proper way for mating to occur.

## Mechanical Isolation

Reproductive organs may come to differ in size, shape, or some other feature, such that organisms of the same or closely related species can no longer mate. Different species of alpine butterfly look very similar, but their genital organs are different enough that one species cannot mate with another.

## Gametic Isolation

Even if mating occurs, offspring may not result if there are incompatibilities between sperm and egg or between sperm and the female reproductive tract. In plants, the sperm borne by pollen may be unable to

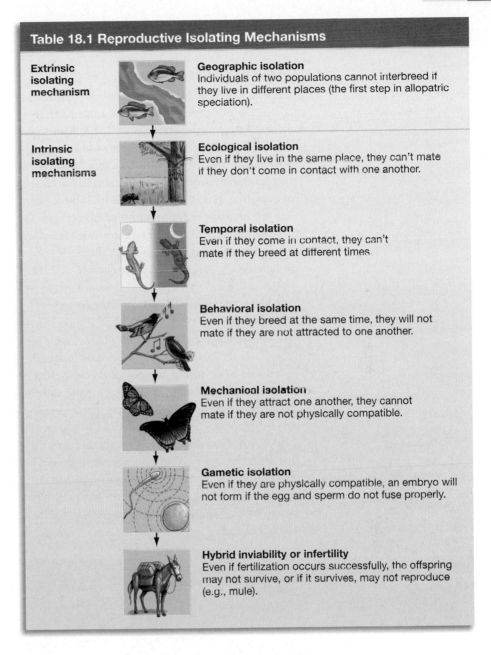

**Table 18.1 Reproductive Isolating Mechanisms**

| | |
|---|---|
| **Extrinsic isolating mechanism** | **Geographic isolation** Individuals of two populations cannot interbreed if they live in different places (the first step in allopatric speciation). |
| **Intrinsic isolating mechanisms** | **Ecological isolation** Even if they live in the same place, they can't mate if they don't come in contact with one another. |
| | **Temporal isolation** Even if they come in contact, they can't mate if they breed at different times. |
| | **Behavioral isolation** Even if they breed at the same time, they will not mate if they are not attracted to one another. |
| | **Mechanical isolation** Even if they attract one another, they cannot mate if they are not physically compatible. |
| | **Gametic isolation** Even if they are physically compatible, an embryo will not form if the egg and sperm do not fuse properly. |
| | **Hybrid inviability or infertility** Even if fertilization occurs successfully, the offspring may not survive, or if it survives, may not reproduce (e.g., mule). |

reach the egg lying within the plant's ovary. In animals, sperm may be killed by the chemical nature of a given reproductive tract or may be unable to bind with receptors on the egg.

There is one form of gametic isolation, called *polyploidy*, that is very important in plants and that may have played a part in giving rise to the vertebrate evolutionary line that we humans are part of. You can read about it in "New Species through Genetic Accidents: Polyploidy" on page 338.

## Hybrid Inviability or Infertility

Even if offspring result, they may develop poorly—they may be stunted or malformed in some way—or they may be *infertile*, meaning unable to bear offspring of their own. A well-known example of such

**··· ANSWERS**
1. breeding
2. gene flow; populations
3. geographic separation; intrinsic isolating mechanisms

infertility is the mule, which is the infertile offspring of a female horse and a male donkey (**Figure 18.5**).

## Sympatric Speciation

Thus far, you have looked at the development of intrinsic reproductive isolating mechanisms strictly as a second step in speciation—one that follows a geographic separation of populations. As it happens, however, intrinsic reproductive isolating mechanisms can develop between two populations in the *absence* of any geographic separation of them. If these isolating mechanisms reduce interbreeding between the populations sufficiently, speciation can take place. What occurs in this situation is not allopatric speciation, however; it is **sympatric speciation**, which can be defined as any speciation that does not involve geographic separation. (*Sympatric* literally means "of the same country.")

Sympatric speciation has been a contentious subject in biology for decades. Scientists have long recognized one of its forms: the polyploidy reviewed in "New Species through Genetic Accidents." But for years, most biologists were skeptical that it existed in any other form. Over the past decade, however, new evidence has convinced most experts that sympatric speciation has operated in a number of instances. No one claims that it has the importance of allopatric speciation—clearly, most speciation events involve a geographic separation of populations. But for most biologists today, the key question about sympatric

**Figure 18.6**
**A Species Undergoing Sympatric Speciation?**
The fruit-fly species *Rhagoletis pomonella*, pictured here on the skin of a green apple, may be undergoing speciation.

speciation is not whether it has taken place, but how frequently it has occurred.

## Sympatric Speciation in a Fruit Fly

To see how sympatric speciation works, consider one of its best-studied examples, a species of fruit fly named *Rhagoletis pomonella*. Prior to the European colonization of North America, *R. pomonella* existed solely on the small, red fruit of hawthorn trees (**Figure 18.6**). The Europeans brought *apple* trees with them, however, and by 1862 some *R. pomonella* had moved over to them. It seemed to at least one mid-nineteenth-century observer, however, that with the introduction of apples there had arisen separate varieties of the flies—what are sometimes called apple flies and haw flies—with each variety courting, mating, and laying eggs almost exclusively on its own type of tree. A modern researcher named Guy Bush led the way, beginning in the early 1960s, in investigating whether this was the case, and the answer turned out to be yes. The two varieties of flies are separated from each other in all these ways. In one study undertaken on them, only 6 percent of the apple and haw flies interbred with one another. The apple and haw *R. pomonella* are not separate species yet, but they certainly give indication of being in transition to that status.

For our purposes, the first thing to note is that this separation has not come about because of *geographic* division. The apple and hawthorn trees that the two

**Figure 18.5**
**Mules Are Infertile Hybrids**
Even though mules cannot themselves reproduce, humans frequently cross horses and donkeys to produce mules to take advantage of their exceptional strength and endurance. Chromosomal incompatibilities between horses and donkeys leave the mules sterile.

varieties of flies live on may scarcely be separated in space at all. Given this, how did this single species move toward becoming two separate species? Bush offers a likely scenario, based on a critical difference between the hosts the two species live on. Apples tend to ripen in August and September, while hawthorn fruit ripens in September and October. In the summer, all fruit flies emerge as adults after wintering underground as larvae, after which they fly to their host tree to mate and lay eggs in the fruit.

Bush believes that about 150 years ago (when hawthorns were hosts to all the flies), some individuals in a population of *R. pomonella* experienced either a mutation or perhaps a chance combination of rare existing alleles. In either event, this change did two things: It caused these flies to emerge slightly *earlier* from their underground state than did most flies, and it drew these flies to the smell of *apples* as well as hawthorns. Because apples mature slightly earlier than the hawthorn fruit, these flies had a suitable host waiting for them. More important, these flies would have bred *among themselves* to a high degree. Recall that the other flies were emerging later, and the adult fly only lives for about a month. Thus, the variant alleles were passed on to a selected population in the next generation. Today, the apple *R. pomonella* flies indisputably do emerge earlier than the hawthorn flies. This is one of the things that ensures reproductive isolation between the two groups; their periods of mating don't fully overlap.

Such a lack of overlapping mating periods may sound familiar because it is one of the intrinsic reproductive isolating mechanisms you looked at earlier. (It is temporal isolation.) You also can see ecological isolation in operation with these flies. Although the two types of flies live in the same area, they meet up with relatively little frequency because they have different habitats (hawthorn vs. apple trees). In short, these populations have developed intrinsic reproductive isolating mechanisms without ever having been separated from one another geographically. They are headed toward speciation, but in this case it is sympatric speciation.

To sum up, the most common form of speciation, allopatric speciation, takes place when a geographic separation is followed by the development of intrinsic reproductive isolating mechanisms. A much less common form of speciation, sympatric speciation, occurs when these isolating mechanisms develop in the absence of geographic separation. One special means of sympatric speciation is speciation through polyploidy.

> **SO FAR...**
>
> 1. Name as many intrinsic isolating mechanisms as you can.
>
> 2. Sympatric speciation is speciation that does not involve _____.
>
> 3. Polyploidy is a special form of sympatric speciation that is most important in _____.

## 18.3 Many New Species from One: Adaptive Radiation

In Chapter 16, you saw that, although the Hawaiian Islands have only one native variety of land mammal (the bat), they do have numerous varieties of one type of fruit fly, called the drosophilid flies (**Figure 18.7**). Some 800 species of drosophilids exist in Hawaii, and genetic analysis has revealed that

**(a)**

**(b)**

**Figure 18.7**
**Two of Many**

The Hawaiian islands are home to an estimated 800 species of the flies known as drosophilids. Pictured are two varieties of these flies, *Drosophila plantibia*, on the left, and *D. cyrtoloma* on the right. Both flies live in rainforest habitats on the island of Maui.

... **ANSWERS**

1. ecological, temporal, behavioral, mechanical, and gametic isolation and hybrid inviability or infertility

2. geographic separation

3. plants

## ESSAY    New Species through Genetic Accidents: Polyploidy

In this chapter so far, speciation has been portrayed as something that takes place over many generations. But plants have a means of speciating in a single generation. It is called *polyploidy*, and it is very important in the plant world; about 35 percent of all flowering plant species are thought to have come about through it. Beyond this, polyploidy may have had great importance in the evolution of animals. Evolutionary biologists currently are debating whether a polyploidy "event" helped speed the rise of animals with backbones—the vertebrates, such as ourselves. As you'll see, whenever polyploidy comes about, it helps speed up evolution.

How does polyploidy work? Here is one of the ways. From the genetics unit, you may recall that human beings have 23 pairs of chromosomes, *Drosophila* flies 4 pairs, and Mendel's peas 7. Whatever the *number* of chromosomes, the commonality among all these species is that their chromosomes come in *pairs*, which is the general rule for species that reproduce sexually.

As noted at the start of the chapter, plants are more adept than animals at producing "hybrids," with the gametes from two separate species coming together to create an offspring. Often these offspring are sterile, however, because when it comes time for them to produce *their* gametes (eggs and sperm) the sets of chromosomes they inherited from their different parental species may not "pair up" correctly in meiosis owing to differences in chromosome number or structure.

Now comes the accident. Suppose that, back when it was first created as a single-celled zygote, the hybrid offspring carried out the usual practice of doubling its chromosome number in preparing for its initial cell division. Now, however, the cell fails to actually divide; whereas it was supposed to put half its complement of chromosomes into one daughter cell and half into another, it doesn't do this. It doubles the chromosomes but keeps them all in one cell. Equipped with *twice* the usual number of chromosomes, this single cell then proceeds to undergo regular cell division, meaning that every cell in the plant that follows will have this doubled number of chromosomes. Critically, this plant will have

doubled both its *sets* of chromosomes—the set it got from parental species A and the set it got from species B. If you double a set, by definition every member of that set now has a partner to pair with in gamete formation. Thus, the roadblock to a hybrid producing offspring has been removed; the chromosomes can all pair up.

This pairing up yields eggs and sperm that can then come together and fuse thanks to another capability of plants. Recall that many plants can *self-fertilize*. A single plant contains male gametes that can fertilize that same plant's female gametes. Thus, the plant with the doubled set of chromosomes can fertilize itself, theoretically beginning an unending line of fertile offspring. This line of organisms is "reproductively isolated" from either of its parental species because it has a different number of chromosomes than either parental species. With reproductive isolation, we have a *separate* species, and with self-fertilization, we have a species that can perpetuate itself.

A multiplication of the normal two sets of chromosomes to some other set number is known as **polyploidy**; here we have speciation by polyploidy, which is one type of sympatric speciation. The importance of this in the plant world is immense. Many of our most important food crops are polyploid, including oats, wheat, cotton, potatoes, and coffee. Indeed, the type of polyploid speciation we've looked at—which begins with a hybrid offspring— often produces bigger, healthier plants. As a result, breeders have developed ways to artificially induce polyploidy.

*Did polyploidy speed the evolution of our own vertebrate line?*

But what about animals? The self-fertilization that some plants are capable of does not exist among more complex animals. As a result, polyploidy that can be maintained over generations is extremely rare in the animal kingdom. But in the course of hundreds of millions of years of evolution, it clearly has been maintained several times. An extremely important polyploidy event may have occurred about 500 million years ago, just as vertebrate

every one of these species evolved from a single drosophilid species that made it to the islands millions of years ago. This plot thickens with the knowledge that Hawaii's 800 species of drosophilids represent a quarter of all the drosophilid species found in the world. Hence we have mystery: A small geographic area has been home to a tremendous amount of speciation within one group of living things. What can explain this kind of supercharged speciation? A phenomenon known as adaptive radiation.

### New Environments and Speciation

As you may know, Hawaii's islands were formed by volcanic activity. The oldest of the islands, Kauai, arose through this activity about 5 million years ago, while the youngest island, Hawaii, formed about 375,000 years ago. If you've ever seen a landscape recently coated by lava, you know the meaning of the word *barren*. Each Hawaiian island originally was barren in this way, which meant that, for a brief time, each island was sterile, with no

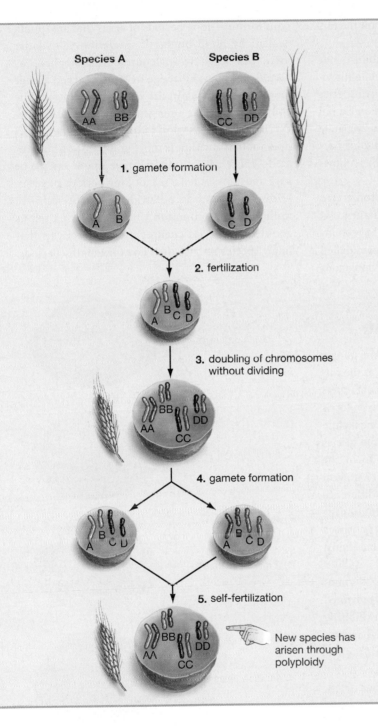

**Species A** **Species B**

1. gamete formation

2. fertilization

3. doubling of chromosomes
   without dividing

4. gamete formation

5. self-fertilization

New species has
arisen through
polyploidy

animals were evolving from invertebrates. A controversial hypothesis holds that this event doubled the chromosomal set of our invertebrate ancestors, thus greatly accelerating the pace of their evolution.

How can polyploidy speed up evolution? Think of it this way: If you added an extra room to a house, that would give you the flexibility to transform, say, an existing bedroom into a den. Now, if you add genes to a genome, as polyploidy does, that provides the *genome* with a kind of flexibility: the flexibility to have one kind of gene transformed into another. Where once an organism would have but two copies of a given gene—one allele from each parent—with polyploidy it has four. With this change, members of one pair of these alleles can mutate without causing harm to the organism because the *other* pair of alleles can carry out their function all by themselves. And in a few cases, alleles that mutate will produce new proteins that can help transform an organism—that can help it evolve down different lines. Polyploidy, then, provides for a tremendous increase in the genetic capacity to evolve. And it provides this capacity in a single generation.

### Polyploidy in Wheat

Two different species of wheat exist in nature, with slightly different "genomes" or complements of DNA.

**1.** Gametes (eggs and sperm) are formed in the different species.

**2.** These gametes fuse, in fertilization, to form a zygote—a single cell that will develop into a new plant. Such a mixed-species or "hybrid" zygote generally develops into a sterile plant because its chromosomes cannot pair up correctly when the plant produces its own gametes during meiosis.

**3.** In this case, however, the zygote doubles its chromosomes in preparation for cell division, but then fails to divide. With this doubling, each chromosome now has a compatible homologous chromosome to pair with during meiosis. This is the polyploidy event.

**4.** Gamete formation then takes place in the plant.

**5.** These gametes from the same plant then fuse because this is a self-fertilizing plant. With this, a new generation of wheat plant has been produced—one that is a different species from either parent generation because each of the parent species has two pairs of chromosomes, while this new hybrid species has four pairs.

life on it. Very quickly, however, life did come to the islands, with bacteria, fungi, plant seeds, and tiny animals landing on them, all being borne by ocean currents, wind, or migrating animals. In time, numerous large plant species became established. By the time the drosophilid flies arrived, it's likely they encountered two things that were critical to their speciation: an environment that *had* plenty of food and habitat, but that *lacked* any flies of the drosophilid variety. To understand the importance of these factors, imagine that you are a graphic artist, working

in a big city with lots of other graphic artists, most of whom specialize in this or that (magazines, Internet websites, etc.). Now you and a few other graphic artists move to a new city in which there are few graphic artists but a good number of *possibilities* for graphic arts work. You would thus be able to specialize fairly easily—filling a *niche* or working role in this new environment—because many of these niches would not yet have been taken. Just so did the drosophilids rush in to fill previously unoccupied niches on the Hawaiian Islands—this mating environment, this food source, this place on the forest floor for laying eggs.

Such a situation is ripe with possibilities for change (meaning speciation) because, while niches are in flux, there is a good deal of shaping of species to environment. But more of this occurred on Hawaii because this niche-filling was taking place on separate *islands*. These islands represented a geographic barrier to interbreeding, and you know what follows from this: allopatric speciation.

The drosophilid species of Hawaii proliferated through **adaptive radiation**: the rapid emergence of many species from a single species that has been introduced to a new environment. The flies radiated out to fill new niches on the islands, with populations adapting to the environments over time.

## 18.4  The Pace of Speciation

If you look at **Figure 18.8**, you can see some photos of *Macrocallista maculata*—better known as the calico clam—which is found in the warm coastal waters of the Atlantic Ocean. Note that the picture on top is of a living *M. maculata*, but that if you go one picture down, you are looking at a fossil of this same species from a million years ago. Proceeding on down, the clams are 2 million, 4 million, and then 17 million years old. If you don't see much of a difference among these specimens, don't worry, evolutionary biologists don't either. To put things simply, what's exhibited here is a species that has undergone almost no outward change over the course of 17 million years of evolution. *Macrocallista maculata* is somewhat extreme in the degree to which it has remained the same over time, but it is not alone in exhibiting this quality. Indeed, to judge by the fossil record, most species spend enormous lengths of time in what is called *stasis*—periods in which they stay exactly the same or undergo some minute modifications. Then, after this long stasis, they appear either to give rise to new species very quickly or simply to die out altogether, without heirs.

So, is this the mode in which evolution operates—long periods of stasis interspersed with short periods of change? If so, then evolution proceeds differently from the way Charles Darwin imagined it. He envisioned it as a series of infinitesimally small changes that accumulated in a slow, steady way in populations until such time as one species diverged into two. If you look at **Figure 18.9**, you can see this gradualist view of Darwin's as contrasted to a conceptualization known as punctuated

equilibrium—long periods of stasis that are "punctuated" by rapid bursts of speciation. This latter view, put forth by paleontologists Stephen Jay Gould and Niles Eldredge in 1972, sparked a huge amount of research in the years that followed. The consensus that has emerged from this work, however, is that we just don't know yet whether, in its operation, speciation is more gradualist or punctuated—more smooth or jerky, we might say. To be sure, most fossils tell a tale of long periods of stasis. (How long? Among ocean-dwelling species 5–10 million years is common.) But this doesn't tell us about how species change; it merely tells us that many species remain the same for lengthy periods.

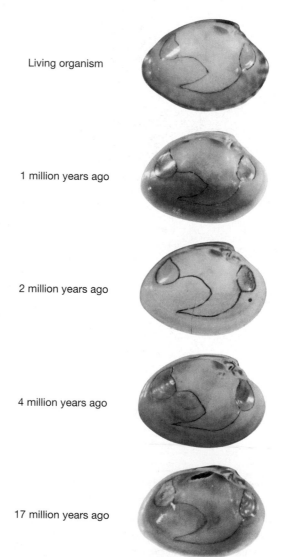

Living organism

1 million years ago

2 million years ago

4 million years ago

17 million years ago

**Figure 18.8**
**Over Millions of Years, Not Much Change**

The calico clam, *Macrocallista maculata*, has undergone very little change in form in 17 million years, as shown in this series of living and fossilized clams unearthed in Florida by researcher Steven Stanley.

**Figure 18.9
Is Speciation Smooth or
Jerky?**

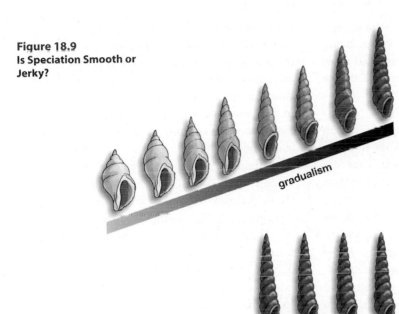

According to the gradualism model, speciation occurs through gradual change over long periods of time.

gradualism

punctuated equilibria

According to the punctuated equilibria model, long periods of stasis may periodically be "punctuated" by rapid bursts of speciation.

It turns out that when scientists have been able to look at fairly complete fossil records—those in which species have left frequent fossil "snapshots" of themselves in transition—the evidence sometimes points to gradual change and other times points to punctuated equilibrium.

So, how long does speciation take? It can occur in as little as a few thousand years or as many as several million. If you look at **Figure 18.10**, you can see an example from the speedy side of this spectrum. Pictured are two species of fish from a small lake in Nicaragua, Lake Apoyo. The species on the right (*Amphilophus zaliosus*) evolved from the species on the left (*Amphilophus citrinellus*) in a sympatric speciation that certainly took no more than 10,000 years and may have taken as few as 2,000. (One of the ways we know who evolved from whom is that *A. citrinellus* is widespread in the region, while *A. zaliosus* exists nowhere in the world except Lake Apoyo.) On the other end of the spectrum, species within a group of freshwater fish called the characins—which include piranhas and many aquarium fish—have been shown to take up to 9 million years to speciate. In between these extremes, there are species that appear to have taken hundreds of thousands of years to give rise to sister species.

*Amphilophus citrinellus*

*Amphilophus zaliosus*

**Figure 18.10
Speedy Speciation**

Through sympatric speciation, the fish on the right, *Amphilophus zaliosus*, evolved from the fish on the left, *A. citrinellus*, within less than 10,000 years in a relatively small lake in Nicaragua. The two species are kept separate by both ecological and behavioral isolating mechanisms. Ecologically, *A. zaliosus* is a surface feeder, while *A. citrinellus* is a bottom-feeder. Behaviorally, differences in courtship signals keep the two species from mating, even when males and females from the two species come together in nature or in the laboratory.

## 18.5 The Categorization of Earth's Living Things

This chapter began by noting the importance of the species concept—of being able to say that this organism is fundamentally separate from that. To have a species concept means that there has to be some means of *naming* separate species. You have probably noticed that most of the species considered in this chapter have been referred to by their scientific names; you didn't just look at a fruit fly, but rather at *Rhagoletis pomonella*. To the average person, such names may be regarded as evidence that scientists are awfully exacting—or just plain fussy. Why the two names? Why the Latin? Can't we just say fruit fly?

To take these questions one at a time, scientists can't just say "fruit fly" for the same reason a person can't say "the guy in the white shirt" while trying to identify a player at a tennis tournament. There are lots of guys in white shirts at tennis tournaments, and there are lots of different kinds of flies in the world. It is important to be able to say *which* fly we are talking about. The importance of this can be seen with *Rhagoletis pomonella* itself, which is a major pest in the apple industry. Tell some apple growers that you have spotted flies on their apples, and you may get a shrug of the shoulders; tell them you have spotted *R. pomonella,* and you may get a very different reaction.

The Latin that is used stems from the fact that this naming convention was standardized by the Swedish scientist Carl von Linné in the eighteenth century, at a time when Latin was still used in the Western world in scientific naming. (In fact, Linné is better known by the Latinized form of his name—Carolus Linnaeus.)

Linnaeus recognized the confusion that can result from having several common names for the same creature and thus devoted himself to giving specific names to some 4,200 species of animals and 7,700 species of plants—all that were known to exist in his time. Many of the names he conferred are still in use today.

The fact that there are two parts to scientific names—a **binomial nomenclature**—points to the central question of groupings of organisms on Earth. Consider the domestic cat, which has the scientific name *Felis domestica*. The first part of its name is its genus, *Felis*, which designates a group of closely related but still separate species. It turns out that, worldwide, there are five other species of small cat within the *Felis* genus, such as the small cat *Felis nigripes* ("black-footed cat") found in southern Africa.

## Taxonomic Classification and the Degree of Relatedness

The practical importance of the genus classification is that, if we know that two organisms are part of the same genus, then to know something about one of them is to know a good deal about the other. But what is the basis for placing organisms in such a category? Modern science classifies organisms largely—although not entirely—in accordance

with how closely they are *related*. In this context, "related" has the same meaning as it does when the subject is extended families. If people have the same mother, they are more closely related than if they shared only a common *grandmother*, which in turn makes them more closely related than if they had only a common great-great grandmother, and so on.

In the same way, species can be thought of as being related. Domestic dogs (*Canis familiaris*) are very closely related to gray wolves (*Canis lupus*), with all dogs being descended from these wolves. Indeed, by some reckonings dogs and gray wolves are the same species (*Canis lupus*) because they will interbreed in nature. Such differences as exist between them are the product of a mere 15,000 years or so of the human domestication of dogs. It may be, then, that 15,000 years ago there was but one species (the gray wolf) that gave rise to both today's wolf and today's dog—a close relation indeed.

Domestic dogs are also related, however, to domestic cats; if we look far back enough in time, we can find a single group of animals that gave rise to both the dog and cat lines. This does not mean going back 10,000 years, however; it means going back perhaps 60 million years. So there is a big difference in how closely related dogs and wolves are as opposed to dogs and cats. Establishing such *degrees* of relatedness is the most important task of the scientific classification system. There is a field of biology, called **systematics**, that is concerned with the diversity and relatedness of organisms; part of what systematists do is try to establish the truth about who is more closely related to whom. They study the evolutionary history of groups of organisms.

Setting aside for the moment the difficulty in *determining* what such evolutionary histories are, there obviously is a tremendous cataloguing job here given the number of living and extinct species mentioned earlier. Given this diversity, a method of classifying organisms, called a *taxonomic system*, is employed to classify every species of living thing on Earth. There are eight basic categories in use in the modern taxonomic system: **species, genus, family, order, class, phylum, kingdom,** and **domain.** The organisms in any of these categories make up a grouping of living things, a **taxon.**

## A Taxonomic Example: The Common House Cat

If you look at **Figure 18.11,** you can see how this taxonomic system works in connection with the domestic cat. As noted, this cat is only one species in a genus (*Felis*) that has five other living species in it. The genus then is a small part of a family (Felidae) that has 17 other genera in it (panthers, snow leopards, and others). The family is then part of an order (Carnivora) that includes not only big and small "cats," but other carnivores, such as bears and dogs. On up the taxa we go, with each taxon being more inclusive than the one beneath it until we get to the highest category in this figure, the kingdom Animalia, which includes all animals. (Later you'll look at the "supercategory" above kingdom, called domain, but for simplicity's sake, kingdom is the highest-level taxon considered here.)

## Constructing Evolutionary Histories

So how do systematists go about putting organisms into these various groups? What evidence do they use in constructing their evolutionary histories? They rely on radiometric dating, the fossil record, DNA sequence comparisons—all the things reviewed in Chapter 16 that are used to chart the history of life on Earth. In practice, there has been a revolution in this field in the last 20 years or so in that most systematics work today is molecular work—scientists are comparing DNA, RNA, and protein sequences among modern organisms and using patterns among these sequences to make judgments about lines of descent.

If you look at **Figure 18.12**, you can see the outcome of some of this work, an evolutionary "tree" for one group of organisms, in this case one of the major groups of mammalian carnivores. You can see that the tree is "rooted," about 60 million years ago, with an ancestral carnivore whose lineage split two ways: to dogs on the one hand and everything else on the other. One interesting facet of this evolutionary history is how closely related bears are to the aquatic carnivores. All such histories are hypotheses about evolutionary relationships, with each such hypothesis known as a **phylogeny**.

### Obscuring the Trail of Evidence: Convergent Evolution

One of the things that makes the interpretation of evolutionary evidence difficult is that similar features may arise *independently* in several evolutionary lines. A bedrock of systematic classification is the existence of **homologies**, which can be defined as common structures in different species that result from a shared ancestry. In Chapter 16, a strong homology was noted in forelimb structure that exists in organisms as different as a gorilla, a bat, and a whale. If you

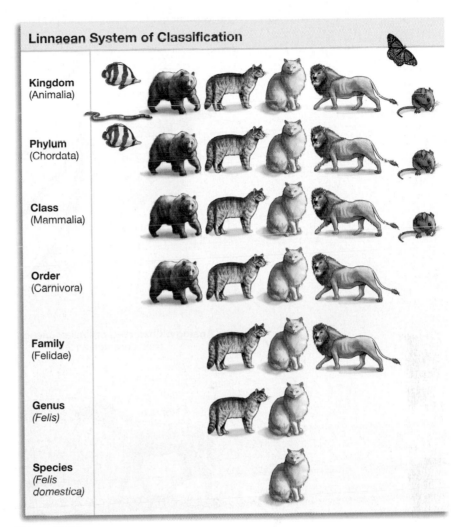

**Linnaean System of Classification**

Kingdom (Animalia)

Phylum (Chordata)

Class (Mammalia)

Order (Carnivora)

Family (Felidae)

Genus (Felis)

Species (*Felis domestica*)

**Figure 18.11**
**Classifying Living Things**
The classification of the house cat, *Felis domestica*, based on the Linnaean system.

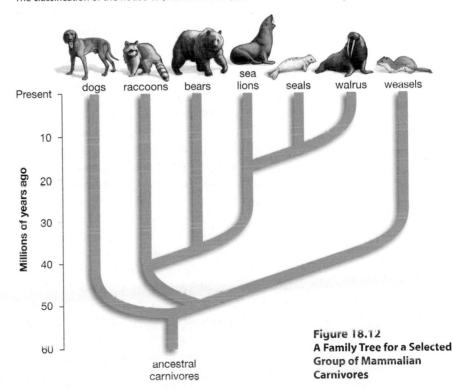

Present

dogs  raccoons  bears  sea lions  seals  walrus  weasels

Millions of years ago

10

20

30

40

50

60

ancestral carnivores

**Figure 18.12**
**A Family Tree for a Selected Group of Mammalian Carnivores**

**(a) Homology:** Common structures in different organisms that result from common ancestry

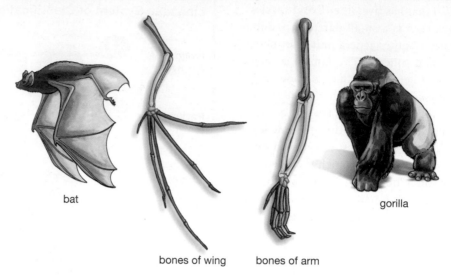

bat

bones of wing    bones of arm

gorilla

**(b) Analogy:** Characters of similar function and superficial structure that have *not* arisen from common ancestry

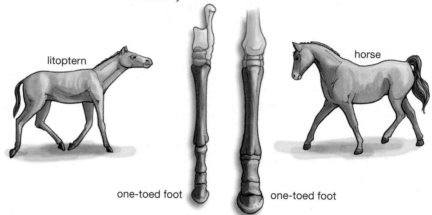

litoptern

horse

one-toed foot    one-toed foot

**Figure 18.13**
**(a)** Bats and gorillas have the same bone composition in their forelimbs—one upper bone, joined to two intermediate bones, joined to five digits—because they share a common ancestor. Of course, these forelimbs are used today for very different functions.
**(b)** Horses and the extinct litopterns have long legs and one-toed feet that serve the same function, but these features were derived independently in the two creatures—in separate lines of descent.

look at **Figure 18.13**, you can see the gorilla and the bat again.

Now, however, consider an extinct group, the litopterns, that once lived in what is now Argentina. Like the modern-day horse, they roamed on open grasslands. Over the course of evolutionary time, they developed legs that were extraordinarily similar to the horse's, right down to having an undivided hoof. In fact, litopterns are only very distantly related to the horse, but the leg similarities were enough to convince one 19th-century expert that he had found in litopterns the ancestor to the modern horse.

What is exhibited in the legs of the litoptern and horse is an **analogy**: a feature in different organisms that is the same in function and superficial appear-

ance. When nature has shaped two separate evolutionary lines in analogous ways, what has occurred is **convergent evolution**. (Both the litoptern and horse lines converged in their development of similar legs.) But analogous features have nothing to do with common descent; they merely show that the same kinds of environmental pressures lead to the same kinds of designs. (There are only so many leg structures that work well for grass-eating mammals that roam over long distances, so it's not surprising that natural selection would have pushed two separate evolutionary lines toward a similar leg shape.) The problem is that analogy can be confused with homology; analogies can make us think that species share common ancestry when in fact they don't.

## 18.6 Classical Taxonomy and Cladistics

Given difficulties such as convergent evolution and a less-than-perfect fossil record, systematists over the years have been driven to devise several methodologies for producing phylogenies. The essential question they have asked is: What means of *interpreting* the evidence will give us the most accurate picture of how Earth's living and extinct organisms are related?

One of the methodologies for interpreting evidence actually predates the field of systematics and most of the techniques used in it. Handed down from the time of Darwin, this methodology is sometimes called *classical taxonomy* (or descriptive taxonomy). In thinking about animals in Earth's family tree, classical taxonomists would look first and foremost at the physical form or morphology of the animals in question, as preserved in the fossil record, and compare it to the morphology of modern animals. Skull shape, "dentition" (teeth patterns), limb structure, and much more would be considered. They would also look at *where* the ancient forms existed compared to modern forms. More recently, they have been using molecular techniques, such as comparing DNA or protein sequences in different living species to determine the relatedness among them. In essence, classical taxonomists would look at how many similarities one group has with another and, on that basis, try to judge relatedness among them. The word *judge* is used advisedly here because, at the end of the day in the classical system, subjective judgments usually need to be made about who is more closely related to whom. It is a matter of weighing one piece of evidence against another, but who's to say which piece of evidence is more important? In addition, as you'll see, classical taxonomy sometimes makes a distinction between the phylogenies it establishes and the taxonomic categories into which it puts different organisms.

### Another System for Interpreting the Evidence: Cladistics

The subjectivity inherent in classical taxonomy helped bring about the formation of another system for establishing relatedness. Developed in the 1950s by the German biologist Willi Hennig, it is called **cladistics** (from the Greek *klados*, meaning "branch"). In practice, cladistics has become the core of most phylogenetic work going on today.

In **Figure 18.14** you can see a **cladogram**, which is an evolutionary tree constructed within the cladistic system. Note that there is a very simple branched line, and that no time scale is attached to it, as was the case with the evolutionary tree of carnivores. Cladistics concerns itself first with lines of descent—with the *order* of branching events. Once this order has been established, efforts can be made to fix events in time, but that is a secondary concern. First and foremost, this cladogram is a proposed answer to the question: Among the animals lizard, deer, lion, and seal, which two groups of animals have the most recent common ancestor? Then, which other two have the *next* most recent common ancestor, and so on. By extension, the

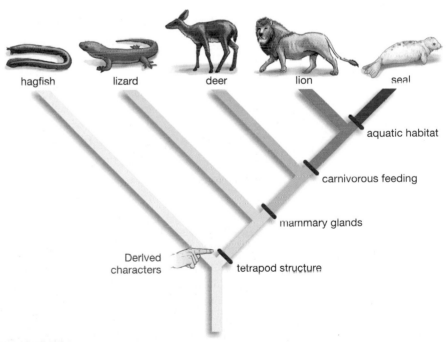

hagfish    lizard    deer    lion    seal

aquatic habitat

carnivorous feeding

mammary glands

Derived characters    tetrapod structure

**Figure 18.14**
**A Simple Cladogram**
Among the lizard, deer, lion and seal, which two are most closely related? The tentative answer displayed in this cladogram is the lion and the seal. The basis for this assertion is the number of unique or "shared derived" characters these two creatures have compared to any other two organisms under consideration. (They share three characters: tetrapod structure, mammary glands, and carnivorous feeding.) Such a system pays no attention to superficial similarities among organisms. Like fish, seals swim in the water, but they share far more derived characters with lions than they do with fish—enough to persuade scientists that seals have returned to the water in relatively recent times, following a period in which their ancestors lived on land.

question is: Who is more closely related to whom? Note that these questions are being asked about only four of the five animals listed. You'll get to the role of the hagfish in a moment.

### Cladistics Employs Shared Ancestral and Derived Characters

The starting point for cladistic analysis is the difference between so-called ancestral characters and derived characters. If you look at any group of species (any taxon), its **ancestral characters** are those that existed in an ancestor common to them all. There are about 50,000 species of animals on Earth that possess a dorsal vertebral column (better known as a backbone). The ancestor to all fishes, reptiles, and mammals had such a feature, which makes the vertebral column an ancestral character for all these groups. On the other hand, only *some* vertebrates then went on to become animals that have four limbs (and are thus tetrapods). The tetrapod feature is thus a **derived character** of the vertebrate state; it is a character *unique* to taxa descended from a common ancestor.

In Figure 18.14, you can see a continuation of this process of ever-more selective grouping in that only some mammals are carnivorous feeders, and only some members of this taxon are aquatic mammalian carnivores. The presumption in cladistics is that shared derived characters are evidence of common ancestry, and that the *more* derived characters any two organisms share, the more recently they will have shared a common ancestor compared to other organisms under study. At its most fundamental level, cladistics becomes a matter of counting shared derived characters. Seals and lions share three derived characters (tetrapod structure, mammary glands, and carnivorous feeding), while the lizard has only one derived character that it shares with all the other animals (tetrapod structure). Cladistics infers from this that the lion and the seal have a more recent common ancestor than do any other two groups in the cladogram—that their branching from this common line of descent came later than any other branching in the line. And the hagfish? It is a starting-point organism for the mammalian carnivores, known as an *out-group*. If we are to say that the animals under consideration have derived characters, they must be derived *compared to* some other organism—in this case, the hagfish.

The examples here employ visible, physical traits, but in practice modern cladistics relies heavily on comparisons among molecular sequences, such as the number of shared, derived base pairs in lengths of DNA or RNA. Furthermore, since cladistics often compares many species, its phylogenies cannot be produced through simple counting; instead, they are the result of multiple rounds of computer-aided calculations.

Taking a step back, the important thing to note about the cladistic method is that it has done away with one of the problems of classical taxonomy: It has removed the element of subjectivity. There is a firm rule in cladistics for inferring relatedness—the number of shared derived characters—which means that judgment need not play a part in its analyses.

### Should Anything but Relatedness Matter in Classification?

Strictly speaking, cladistic analysis is not about the classification of Earth's living things. Rather, cladistics is concerned solely with establishing lines of descent. Once these are established, categories can be overlaid on them. Having looked at various cladograms, we could say that this group constitutes a class called Reptilia, while this other group constitutes a class called Amphibia. But cladistics per se is not concerned with this. If the question, however, is not just one of phylogeny, but how to *categorize* Earth's organisms, then it's not clear that cladistic analysis always provides the most sensible categories.

To appreciate this point, consider the following question. Who is more closely related: dinosaurs and lizards or dinosaurs and birds? It may surprise you to learn that it is dinosaurs and birds. Birds split off from a dinosaur line within the last 200 million years or so, while the split between the lizard and dinosaur lines came much earlier. Despite this, conventional taxonomy says that dinosaurs and lizards belong in one class (Reptilia), while birds belong in another (Aves; **Figure 18.15**). The assumption here is that, in classifying creatures, something should count *besides* their relatedness—in this case, the special qualities of birds (feathers, flight).

To some scientists, a categorization such as this makes no sense. Why should relatedness dictate taxonomy in some instances but not in others? Shouldn't we consistently categorize organisms on the basis of relatedness? On the other hand, aren't birds different enough from lizards that they deserve to be in a separate grouping? If you think that both sides in this debate are making sensible arguments, then you understand why it is so difficult to come up with a universally accepted means of classifying Earth's living things.

**(a) Classical** view of relationships among tetrapods

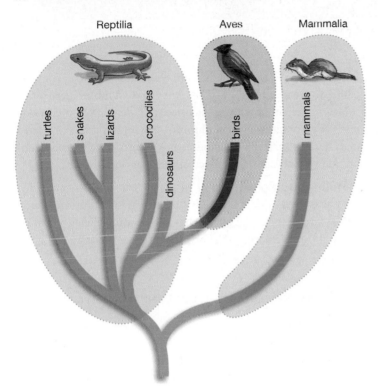

**(b) Cladistic** view of relationships among tetrapods

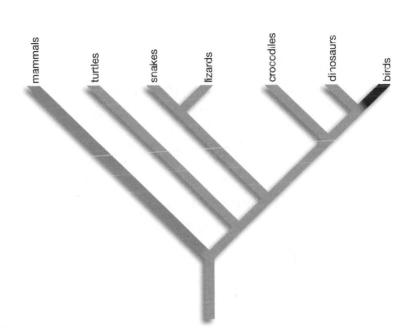

**Figure 18.15**
**Are Birds Reptiles?**

**(a)** Classical taxonomy employs three classes, Aves, Mammalia, and Reptilia, for birds, mammals, and reptiles, respectively. Thus, dinosaurs and birds are in separate classes—a recognition on the part of classical taxonomists of the unique features of birds, such as feathers.

**(b)** In cladistic analysis, by contrast, birds are grouped very closely with dinosaurs in recognition of the recent shared ancestry of the two taxa. The only consideration in this system of classification is who is more closely related to whom.

## On to the History of Life

The millions of species that exist on Earth today took billions of years to evolve. During that time, life went from being exclusively microscopic and single-celled to being the staggeringly diverse entity it is today. How did this happen over time? And what is the order of appearance for the Earth's living things? These are the subjects of Chapter 19.

**SO FAR...**

1. Cladistics determines relatedness among organisms by counting shared, derived characters: characters that are _____ to taxa descended from a _____.

2. One of the key features of cladistics is that it does away with the element of _____ in proposing phylogenies.

••• ANSWERS
1. unique; common ancestor
2. subjectivity

# CHAPTER 18 REVIEW

For study help, activities, and more quiz questions, go to **www.aw.com/krogh4**.

## SUMMARY

### 18.1 What Is a Species?

- The most accepted definition of a species is derived from the biological species concept: "Species are groups of actually or potentially interbreeding natural populations that are reproductively isolated from other such groups." Earth's organisms are so varied that this definition does not apply to all of them.

### 18.2 How Do New Species Arise?

- Branching evolution occurs when a single "parent" species diverges into two species, the parent species continuing while a second species arises from it. Branching evolution is based on reductions in gene flow between populations of the same species. To the extent that gene flow is maintained between two geographically separate populations, they will experience similar allele frequency changes, meaning they will evolve together. When gene flow is drastically reduced for a long period, however, the populations will evolve separately. When evolution has altered the physical or behavioral characteristics of the populations enough that they will no longer interbreed, even if reunited, speciation has occurred.

- Geographical separation is the most important factor in reducing gene flow between populations of the same species. Such separation can come about because of a change in environment—a river may change course, for example—or because of the migration of part of a population. When geographic separation plays a part in the evolution of a species, allopatric speciation is said to have occurred.

- Geographical separation cannot bring about speciation by itself. Following geographical separation, populations of the same species must develop internal characteristics that will keep them from interbreeding should they be reunited. Such characteristics are referred to as intrinsic isolating mechanisms. The six most important mechanisms are ecological, temporal, behavioral, mechanical, and gametic isolation and hybrid inviability or infertility.

- Speciation can occur in the absence of geographical separation, but such sympatric speciation is thought to occur much less frequently than allopatric speciation. The key to sympatric speciation is that populations of a single species need not be separated geographically in order to become reproductively isolated from one another.

- Polyploidy is a special form of sympatric speciation that can give rise to a new species in a single generation. One of its forms is initiated when eggs and sperm from two different species come together to create a hybrid zygote. This zygote doubles its chromosomes, in preparation for cell division, but then fails to divide. The result is an organism whose chromosome set has doubled relative to its starting set. If this organism is able to carry out self-fertilization (as many plants can), then a new species has been formed.

- Polyploidy is most important in plants, but it has occurred in animal species as well. It speeds the process of evolution by providing a redundancy in alleles that allows some alleles to mutate without causing harm to the organism.

  **Web Animation 18.1:** Speciation and Polyploidy

### 18.3 Many New Species from One: Adaptive Radiation

- Speciation is more likely to occur when a species is introduced to an environment in which few other species of its kind exist. Such an environment has many available working roles or niches available to the new species—a situation that leads to rapid specialization, which in turn leads to speciation. This series of events is known as adaptive radiation: the rapid emergence of many species from a single species that has been introduced to a new environment.

### 18.4 The Pace of Speciation

- There is considerable debate among evolutionary biologists regarding the pace of evolution. Under one view, called the gradualist model, evolution proceeds at a slow, steady rate; under another view, called the punctuated equilibrium model, organisms undergo long periods of stasis or little change, followed by relatively brief episodes of rapid speciation. Speciation can occur in as little as a few thousand years or as many as several million.

### 18.5 The Categorization of Earth's Living Things

- In science, a two-name or binomial nomenclature is used for each of Earth's species. The first name designates the genus, or group of closely related organisms, that the species is a member of; the second name is specific to the species. Genus and species categories fit into a larger framework of taxonomy, meaning the classification of species. Going from least to most inclusive, the eight most important categories in the taxonomy of Earth's organisms are species, genus, family, order, class, phylum, kingdom, and domain.

- Species are put into these categories largely on the basis of relatedness—species that are closely related are in the same genus while species that are distantly related may only be in the same phylum or kingdom. The biological discipline of systematics is concerned with establishing degrees of relatedness among both living and extinct species.

- Systematists establish evolutionary family trees or "phylogenies" by reviewing various kinds of evidence, including radiometric dating, the fossil record, and DNA sequence comparisons. Based on this evidence, they determine the evolutionary relationships among species. In practice, most of the work being done in systematics today is molecular work: Scientists compare DNA, RNA, and protein sequences, looking for identifying phylogenetic patterns among them.

- In establishing phylogenies, one of the things systematists look for is homologous structures: common structures in different species that result from a shared ancestry. One problem with the use of these structures is that they can be confused with analogous structures: similar features that developed independently in separate lines of organisms (as with the legs of modern horses and extinct litopterns).

## 18.6 Classical Taxonomy and Cladistics

- One method of determining phylogeny, classical taxonomy, establishes evolutionary relationships among living and extinct organisms in accordance with such factors as physical form, distribution, and molecular similarities. But classical taxonomy employs subjective judgments in deciding what weight to give one piece of evidence as opposed to another. The phylogenetic system known as cladistics does away with this element of subjectivity by following a firm rule for inferring relatedness. It counts the number of shared derived characters two organisms have, meaning the features they uniquely share that are derived from another group of organisms.

- Cladistics is centrally concerned with establishing lines of descent, although species can be put into taxonomic categories that are derived from cladistic analysis. Classical taxonomy holds that, in putting organisms into various categories, factors other than phylogeny ought to be taken into account. Cladistics holds that the only criterion for taxonomic placement is phylogeny—a group must contain all descendants stemming from a given branch point in a family tree.

## KEY TERMS

| | |
|---|---|
| adaptive radiation 340 | biological species |
| allopatric speciation 334 | concept 331 |
| analogy 344 | cladistics 345 |
| ancestral character 346 | cladogram 345 |
| binomial nomenclature 342 | class 342 |

| | |
|---|---|
| convergent evolution 344 | order 342 |
| derived character 346 | phylogeny 343 |
| domain 342 | phylum 342 |
| extrinsic isolating | polyploidy 338 |
| mechanism 334 | reproductive isolating |
| family 342 | mechanisms 334 |
| genus 342 | speciation 333 |
| homology 343 | species 342 |
| intrinsic isolating | sympatric speciation 336 |
| mechanism 334 | systematics 342 |
| kingdom 342 | taxon 342 |

## BRIEF REVIEW

*Answers to Brief Review questions are in the back of the book. For multiple-choice quiz questions, go to* **www.aw.com/krogh4**.

1. State the biological species concept.

2. Why is the evolution of intrinsic reproductive isolation mechanisms required for two groups to be called separate species? Why isn't simple geographical isolation sufficient?

3. Describe one major difference between allopatric and sympatric speciation. What is one thing that they have in common?

4. Contrast the gradual and punctuated equilibrium models of long-term evolution.

5. What is convergent evolution, and why is it a problem for phylogenetic analyses?

## APPLYING YOUR KNOWLEDGE

1. Living things are classified into a set of hierarchical categories. Why is this more useful to biologists than simply giving everything a one-word name (or a number) and eliminating some of the categories?

2. Populations of some kinds of organisms may be isolated by a barrier as small as a roadway, while other organisms require a much larger physical barrier to block gene flow. What features of organisms might determine how readily they become geographically isolated? What impact might these features have on how often or how rapidly speciation is likely to occur in these groups?

3. In an adaptive radiation, one species may colonize a previously empty environment, such as an island, and diversify evolutionarily into a set of closely related species that occupy a very wide range of habitats. For example, with Darwin's finches on the Galapagos Islands, one species fills the role of woodpecker by using a stick to tap on trees and dig out insects. Do you think that such a species would have been likely to evolve in an area that already had woodpeckers? Why or why not?

The remains of a group of ocean-dwelling animals called crinoids have been preserved in this piece of rock dating from Earth's Carboniferous Period, which began 359 million years ago.

CHAPTER **19**

| 19.1 | The Geological Timescale: Life Marks Earth's Ages | 352 |
| 19.2 | How Did Life Begin? | 354 |
| 19.3 | The Tree of Life | 360 |
| 19.4 | A Long First Era: The Precambrian | 360 |
| 19.5 | The Cambrian Explosion | 363 |
| 19.6 | The Movement onto the Land: Plants First | 365 |
| 19.7 | Animals Follow Plants onto the Land | 367 |

**Essays**

| Physical Forces and Evolution | 356 |
| Who Gets a Kingdom? Evolution within Life's Categories | 362 |
| **THE PROCESS OF SCIENCE:** Going after the Fossils | 370 |

Life started out small perhaps 3.8 billion years ago and stayed that way for a long time. Eventually it exploded into a multitude of forms that included human beings.

# A SLOW UNFOLDING:
# The History of Life on Earth

**W**ithin 5 minutes of its birth, a newborn wildebeest on the plains of central Africa can be off and running with its herd (**Figure 19.1**). By contrast, human infants probably will not be able to take a step by themselves

until they are about a year old. Female lions reach sexual maturity at about 3 years, and killer whales at about 7, but in human hunter-gatherer societies, reproduction generally begins at age 18 or 19. Even in comparison to our fellow mammals, we humans mature at a notably slow rate.

**Figure 19.1**
**No Time to Lose**

A newborn wildebeest struggles to its feet immediately after its birth on the plains of Africa. Wildebeest predators, such as hyenas, often concentrate their efforts on the newborn.

At first glance, it might seem puzzling that our evolution took this turn. From what we have learned of natural selection, wouldn't it make sense that survival would *decrease* for offspring that developed more slowly? What is the fitness payoff in having young who are essentially helpless for several years and then immature for many more? Here is one possible answer: As a species, we have survived not so much because of our physical capabilities but because of our wits. Our success is the product of our remarkable capacity to learn. But learning takes time, and it is arguably best undertaken by minds that remain for a long time in what might be called a state of flexible immaturity. (Think of how easily young children acquire a second language in comparison with adults.) Among our human ancestors, it may be that those who survived tended to be those who could learn the most. And who could learn the most? Those who took the longest to mature.

Under this view, we carry our evolutionary heritage with us in the form of delayed maturity. We did not leave this heritage behind in the African savanna, where human beings first evolved; we can see it every time we look at a 2-year-old. In the same way, we can see evolutionary vestiges in us from a much earlier time as well. Note the way we walk, our left arm swinging forward as our right leg does the same. Almost all four-limbed animals walk this way. One view

**Figure 19.2**
**Our Evolutionary Heritage**

The gait exhibited by almost all four-limbed animals, including human beings, stems from the reversing S-pattern that our evolutionary ancestors used in getting around when they were making the transition from aquatic to terrestrial life. Here a salamander employs this same motion in getting from one place to another.

is that we inherited this from the elongated, four-finned fish we evolved from. They came onto the muddy land by slithering in an S-pattern, the left-front fin going back as the left-rear fin went forward, then the reverse (**Figure 19.2**).

Note the sweep of evolution in all this. Maturation is with us from one period, walking style from another. If we look at the genetic code—this DNA sequence specifying that amino acid—we have within us a heritage that stretches back to the beginning of life on Earth because this code is shared by nearly every living thing. Under this view, evolution is not some abstract process that lies separate from us; on the contrary, it is with us in every step we take. But how has evolution proceeded in its long sweep down the ages? Starting with the spark of life itself, how did it produce the range of creatures that inhabit the Earth? In this chapter and the one that follows, you'll begin to find out.

**Web Animation 19.1**
Evolutionary Timescales

## 19.1 The Geological Timescale: Life Marks Earth's Ages

We can start this inquiry by getting a sense of life's timeline. The Earth, which is home to all the life we know of, came into being about 4.6 billion years ago. Given the human life span of 80-plus years, 4.6 billion years is an unimaginably long time. Indeed, it's a long period of time relative to any time frame we know of. The universe as a whole is thought to be 13.7 billion years old, so the Earth's history stretches back a third of the way to the beginning of time.

In measuring something as long as 4.6 billion years, it obviously won't do to think in terms of individual years, so scientists use something called the geological

timescale, which you can see in **Figure 19.3**. It divides earthly history into broad eras and shorter periods. The scale begins with the formation of the Earth (at the bottom of the figure) and runs to "historic time," meaning time in the last 10,000 years.

But what do the demarcations in this timescale indicate? What, for example, is the distinction between the Permian period you can see at the middle-left of Figure 19.3 and the Triassic period that followed it? To understand what separates them, it's important to know *when* the timescale was developed: in the late eighteenth and early nineteenth centuries, at which point there was there was no radiometric dating of materials—no using the decay of uranium, for example—to provide absolute dates for any fossils that might be found. Hence, no one could say whether a fossil was 10 million or 100 million years old. Geologists of the time did know, however, that sedimentary rocks had been put down in layers (or strata), with the oldest strata on the bottom and the youngest strata on top (**Figure 19.4**). And as it turned out, each stratum tended to have within it a group (or "assemblage") of fossils that

**Figure 19.4**
**Revealing Layers**

The history of life can be traced through fossilized life-forms found in layers of sediment, seen here in America's Grand Canyon.

was unique to it. Thus, the strata could be divided up in accordance with the transitions the scientists saw in the life forms. The Permian period is consid-ered different from the Triassic period because, with movement from one to the other, a different set of life-forms appears, as recorded in the fossil record.

## Features of the Timescale

Now a couple of notes on the timescale. First, it may surprise you to learn what the basis is for many of the transitions in it: death on a grand scale, in the form of "major extinction events." Five of these events are noted in Figure 19.3 (over on its right), with all of them defining the end of an era or period. The most famous extinction event, the **Cretaceous Extinction**, occurred at the boundary between the Cretaceous

88% of Earth's history was spent in the Precambrian era

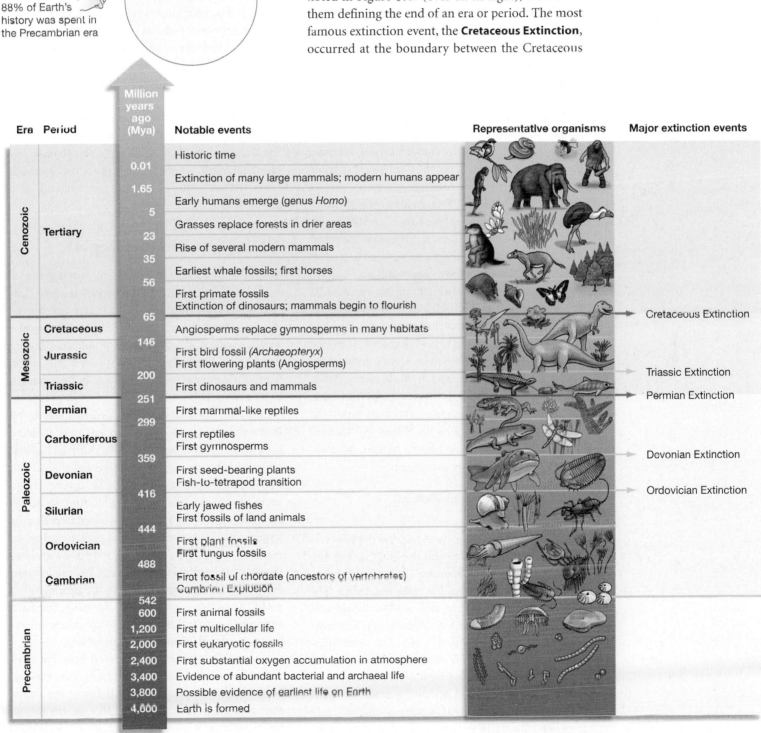

**Figure 19.3**
**Earth's Geologic Timescale**

and Tertiary periods. This was the extinction, aided by the impact of a giant asteroid, that brought about the end of the dinosaurs along with many other life-forms. The greatest extinction event of all, however, occurred earlier, in the boundary between the Permian and Triassic periods. This was the **Permian Extinction**, in which as many as 96 percent of all species on Earth were wiped out.

This is not to say, however, that all Earth's eras or periods are marked off by extinctions. The transition between the Precambrian and Paleozoic eras marks not a major extinction, but a seeming explosion in the diversity of animal forms. Nevertheless, this so-called Cambrian Explosion, which you'll read about later, is just a different kind of transition in living things.

Now, when did some of these transitions come about? You might think that a transition such as a sudden increase in animal forms would have occurred very early in life's history, but the Cambrian Explosion took place 542 million years ago. (Another way of writing this is 542 Mya.) This is a long time ago, to be sure, but life is thought to have begun 3.8 *billion* years ago. There may not have been an animal on the planet for 3.2 billion years after life got going—nor any plants either. Life was single-celled to begin with, and 2 billion years later, it still consisted of nothing but microbes. As you'll see, these tiny organisms were undergoing important evolutionary changes during Earth's long first era, the Precambrian. But to find changes in the *forms* of life's organisms, we have to go very far forward indeed. If the fossil record is correct, all the birds, reptiles, fish, plants, and mammals that exist today only came about in the last 16 percent of evolutionary time—the last 600 million years of life's 3.8-billion-year span. As you'll see, our own species, *Homo sapiens,* is an extreme latecomer in this series of events.

### What Is Notable in Evolution Hinges on Values

One final word about the timescale and the path you will follow in this chapter and the one that follows. Picking out the "notable events" in evolution is inevitably an exercise in making value judgments. The notable events covered in this book will lead you along a line that begins with microscopic sea creatures and ends with human beings. The value judgment guiding this path is that students have a great interest in knowing about their own evolution, and that the path to humans presents a broad sweep of evolutionary development. Such a course of study may leave the impression, however, that all of evolu-

tion amounts to a march toward the development of human beings, who then get to occupy the highest branch of an evolutionary tree.

In reality, we occupy one ordinary branch among a multitude of branches. Under notable events in the Cenozoic era, Figure 19.3 could just as easily have listed milestones in the evolution of fish or birds or the emergence of social insects such as bees. These creatures have continued evolving in the modern era, along with our own species, but in lines separate from us. Focusing on a line that leads to humans is like focusing on a railroad route that leads from, say, Baltimore to Denver, while ignoring the multitude of lines that are separate from it. We regard the Baltimore-Denver route as special because we are *interested* in it, but that does not mean that it is fundamentally *different* from any other line. Some special qualities have evolved in human beings, but the same could be said for almost any species. In phylogenetic terms, we simply lie at the tip of one evolutionary branch, while bees lie at another, and cactus plants lie at still another.

## 19.2 How Did Life Begin?

Taking a historical approach to tracing life on Earth, the first question that arises is one of the toughest: How did life begin? Darwin himself thought about this and imagined that life began in what he called a "warm little pond" in the early Earth. Well, maybe; but the *very* early Earth had no warm little ponds. What it had was an environment more akin to hell on Earth.

You have seen that Earth was formed about 4.6 billion years ago. This took place in a process of "accretion," in which ever-larger particles clumped together: cosmic dust to gravel, gravel to larger balls, larger balls to objects the size of tiny planets. One consequence of such development was that the accretion now called Earth was periodically being slammed into by *other* large accretions—meteorites and comets and protoplanets. One of these objects was as large as Mars, and the result of its impact on the Earth, 100 million years after Earth formed, may have been the creation of our moon. There were no oceans at this time, and Earth was so hot—up to 1,800 degrees Fahrenheit—that it was covered with a layer of molten rock.

At one time, scientists believed that Earth remained in this state until about 3.8 billion years ago, but recent evidence indicates it may have cooled enough by 4.2 billion years ago that oceans formed on its surface. With this change, we are left with a large question about what the planet was like. Scientists believe that the sun shone some 30 percent less brightly then than it does now. So one of two things is possible. If Earth's atmosphere contained a sufficient concentration of "greenhouse" gases, such as methane and carbon dioxide, then it would have remained a warm place—perhaps even a broilingly hot place. Conversely, if the concentration of these gases was low, then the Earth could have been frozen over in a global glaciation, at least for a time.

These two views lead to varying ideas about where life could have arisen. It may have originated in a *cold* little pond—one that was under a glacier—but then again, it may have gotten started in just the opposite environment: a hot-water environment, as we'll see. A third possibility is that something like Darwin's warm little pond did exist, in the form of tide pools at the edge of an ancient ocean. As different as all of these environments are, note that all of them are *aqueous* environments. Life got going in water.

## From the Simple to the Complex

But how did this happen? The question of the origin of life really is a question of chemistry. Life had to get going through a kind of chemical bootstrapping: Simple, available chemical compounds came together in reactions that facilitated the production of molecules that were more complex. These then came together to bring about the production of molecules that were more complex yet. At some point, a group of these molecules became capable of *self-replication:* They could make copies of themselves in ways we'll be looking at. With this step, life began.

But how did this chain of events get started? Are we sure it's possible to go from the simple compounds available on the early Earth to life's more complex molecules? You saw in Chapter 3 that life's informational molecules (DNA and RNA) are composed of certain kinds of building blocks (nucleotides), while its proteins are composed of other kinds of building blocks (amino acids). How difficult would it have been for these sorts of molecules to have been produced from the starting chemical ingredients available on the early Earth?

In 1953, a young graduate student named Stanley Miller was asking himself this very question. In seeking to answer it, he performed what is undoubtedly the most famous experiment in origin-of-life studies. In essence, he tried to re-create the conditions of the young Earth in the set of glass flasks and tubes you can see in **Figure 19.5**. Working with his mentor, Harold Urey, Miller put together an "atmosphere"

inside the glassware that he thought would approximate the atmosphere of the early Earth—water vapor combined with the gases methane, ammonia, and hydrogen. He then ran an electric current through the system, thus simulating the lightning he believed would have been flashing through Earth's ancient skies. After a couple of days, he got a brown goo in the apparatus that, on analysis, turned out to contain . . . amino acids! Some of the prime building blocks of life had been created out of simple compounds in just a few days! This result had an enormous impact on origin-of-life research for it meant that it was *not* difficult to go from some very simple substances to life's building blocks under the conditions believed to exist on the young Earth. With this experiment, orgin-of-life studies were turned into an experimental science.

Since the time of Miller's work, generations of researchers have essentially been asking the same kinds of questions he did: What was Earth's atmosphere like at the time life got going? What

**Figure 19.5**
**Origins-of-Life Researcher**

In 1994, Stanley Miller posed with part of the laboratory apparatus he had used 41 years earlier in attempting to re-create the atmosphere that surrounded the young Earth. Miller ran an electric current through the gases in the apparatus and discovered that, in only a few days time, the system generated amino acids, some of the key building blocks of life.

# Physical Forces and Evolution

## Climate Change

The living world is very much affected by the physical forces operating on the Earth. You can see this in connection with the mass extinctions discussed in the main text. What has caused these extinctions? Evidence links at least two of those listed in Figure 19.3 to episodes of global climate change, specifically global cooling. (The two are the Ordovician and Devonian Extinctions.) All kinds of things happen with global cooling. Glaciers develop, wiping out everything in their path and tying up huge amounts of water in ice. This reduces sea levels, thereby killing many organisms that live on the continental shelves. The oceans themselves get colder, particularly at greater depths, which is its own kind of killing event.

## Extraterrestrial Objects

In the public mind, the Cretaceous Extinction that finished off the dinosaurs had a single, nearly instantaneous cause: the explosive impact of an enormous asteroid, 10 kilometers (6.5 miles) in diameter, that struck off the coast of Mexico's Yucatan Peninsula 65 million years ago (Mya). There is no doubt that this catastrophic event happened; what is in doubt is that it was the sole or even predominant cause of the Cretaceous Extinction. For one thing, fossil evidence indicates that many Cretaceous life-forms, including the dinosaurs, were in severe decline *before* the asteroid struck. Second, we have evidence of massive volcanic activity, in what is present-day India, beginning 66 Mya. This activity is likely to have caused a "greenhouse" warming of the Earth (because it put carbon dioxide into the atmosphere) and to have put high levels of poisonous gases into the air. If there is a consensus view on the Cretaceous Extinction, it is that one or more asteroids probably were responsible for the nearly instantaneous demise of some species, but that in other cases these objects were merely the final blow in extinctions that already were under way.

At the very least, however, the Cretaceous Extinction shows that evolution can be affected by sudden, catastrophic events that have nothing to do with Earth's usual physical processes. A fascinating point is raised by this: Fitness, in the evolutionary sense, can change very rapidly. The qualities that made the dinosaurs so dominant in the Triassic period were of no use at the end of the Cretaceous when the game of survival had changed. Now *other* qualities came to the fore in survival. Among these may have been small size, which mammals had, perhaps to their great benefit.

## Continental Drift

As noted in Chapter 18, the primary engine driving speciation is the geographic separation of populations. Geographic separation was

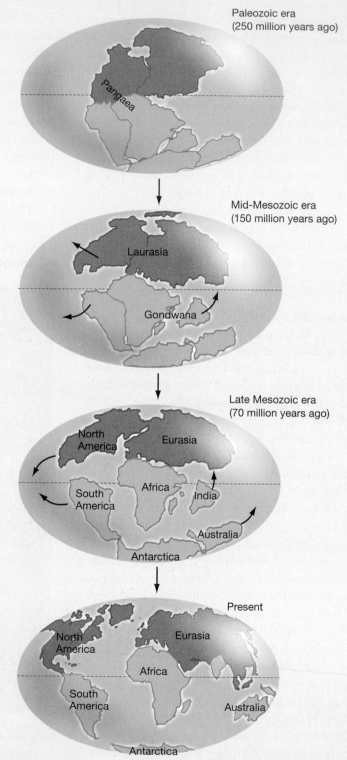

**Figure 1**
**Continental Drift**
During Earth's history, the tectonic plates that make up the surface layer of Earth have moved with respect to one another, causing the continents to spread apart and slam together.

described as taking place in scales that were both small (a stream changing course) and large (a glacier's movement). It turns out that geographic separation can take place on a much larger scale yet, in that whole *continents* have come together and separated again during Earth's history.

If you look at **Figure 1**, you can see the landmasses of the world as they have existed at four separate times: 250 Mya, 150 Mya, 70 Mya, and the present. In the first of these, nearly all the Earth's continental mass was part of a single supercontinent, Pangaea, which stretched from the South to North Pole. By the Early Cretaceous, Pangaea was separating along the lines you see in the figure. This in turn gave way to our current division, which like its predecessors, is a configuration in flux.

These maps are only four "snapshots" in a process of **continental drift**, or the movement of continents in geological time, that has produced lots of other alignments. How does such movement come about? In essence, the Earth's crust and part of its upper mantle are divided into a series of plates that move laterally over the globe. The movement of these plates can be thought of in terms of a conveyor belt. Halfway between North America and Europe there is a long underwater mountain chain, the Atlantic Ridge. This ridge has in its middle a deep valley that forms the border between the North American and Eurasian Plates. Hot material in the Earth's interior is constantly being moved up into the valley from either plate, after which, as new crust, it moves eastward on one side of the ridge but westward on the other. What we have, then, is a spreading of the ocean floor in opposite directions. *Continents* sit atop this spreading material, and they are being moved in opposite directions—Europe to the east, North America to the west—as if they were objects on conveyor belts (see **Figure 2**).

## The Impact of Continental Drift on Evolution: Climate and the Separation of Populations

Continental drift has had profound effects on evolution. As noted, global cooling has been the most significant force behind at least two mass extinctions. But what causes global cooling? One factor is the development of glaciers, which as it turns out can grow only when continents are located at or near the Earth's poles. This *was* the case 300 Mya, then it was *not* the case 200 Mya, and now it is again. By moving landmasses on and off the poles, continental drift has helped change the climate.

Continental drift has had a second profound effect on evolution in that it has brought about the separation and mixing of living *populations*. For example, the breakup of the supercontinent Pangaea eventually meant the creation of island continents such as Australia and Antarctica. Until the coming of humans, Australia had scarcely any placental mammals—mammals with embryos that are nourished by a placenta (see Chapter 23). Since prehistoric times, however, Australia has had an abundance of marsupial (or "pouched") mammals, such as kangaroos and koalas. We can read evolutionary history in this distribution. As

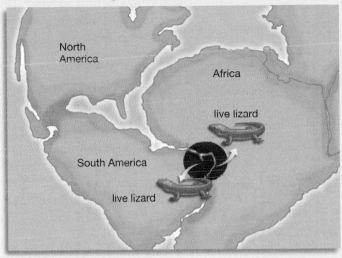

**(a)** Paleozoic era Pangaea (250 million years ago)

North America

Africa

live lizard

South America

live lizard

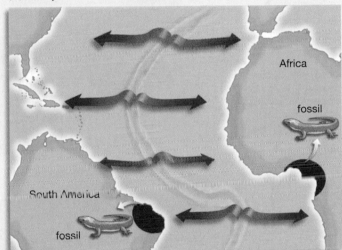

**(b)** Today

Africa

fossil

South America

fossil

**Figure 2**
Similar fossils are found on the east coast of South America and the west coast of Africa. How can this be?

**(a)** 250 million years ago, the two continents were contiguous and formed part of the giant continent Pangaea.

**(b)** Animals that were alive then are now found as fossils on opposite sides of the ocean because the ocean floor has been spreading westward and eastward along the central Atlantic Ridge. The ocean floor is spreading because hot material from the Earth's interior is constantly being moved up into the central Atlantic Ridge, even today.

traditional theory has it, marsupials developed in Pangaea before its breakup, in what is now North America and western Europe, and spread to Australia across land bridges to what is now Antarctica. (See the first and second drawings, Figure 1.) Meanwhile, placental mammals were spreading, too, but few were able to reach the Australia-Antarctic landmass before it broke away from Pangaea. In effect, placental mammals missed the Antarctic-Australian boat.

was the climate like? What compounds might have been made available through the action of ocean waves or volcanic activity? The idea is to test plausible scenarios for Earth's early conditions against the likelihood of life being produced in those conditions.

### The Source of Life's Raw Materials

We've already touched on one of the key questions in this research: Where do you get the raw materials to produce the more complex building blocks of life? As one origin-of-life researcher has put it, these materials could have come from three possible sites: above, beyond, or below. Stanley Miller assumed an "above" site—he saw methane, hydrogen, and

ammonia being provided by the *atmosphere* of the young Earth. This hypothesis went out of favor in the 1960s because it appeared that hydrogen would have dissipated from Earth's early atmosphere, but just recently it has been made more plausible by some calculations indicating that the hydrogen would have been retained.

The "beyond" hypothesis for life's component molecules means beyond Earth—outer space is proposed as the source of fully formed building blocks, such as amino acids. The idea is that these molecules were delivered by the meteorites and comets that smashed into the young Earth. It's clear, from meteorites that have landed on Earth in recent decades, that such objects can carry organic molecules (**Figure 19.6**). It appears unlikely, however, that such accidental "seeding" from outer space could have brought a sufficient *quantity* of organic materials to get life going.

**Figure 19.6**
**Space Travelers**

Fragments of the Murchison meteorite, which landed on the township of Murchison in Australia in 1969. The meteorite contained water and organic molecules such as amino acids. Some scientists hypothesize that organic materials like these could have seeded life on Earth.

### Life May Have Begun in Very Hot Water

The third possibility for the supply of organic materials—the "below" possibility—is that they came from the methane and hydrogen sulfide that gush out even today from deep-sea vents on the floors of the world's oceans. Interestingly, there are creatures living near these vents today. Some of these are microscopic archaea and bacteria that thrive at temperatures of 105 degrees Celsius (247 degrees Fahrenheit), which is above the temperature at which water boils (**Figure 19.7**). Some molecular sequencing work indicates that the very oldest organisms on Earth may have been heat-tolerant organisms such as these. This work has recently been challenged, but various other lines of evidence continue to converge around the idea that life began in the "prebiotic soup" of hot-water systems—in the deep-sea vents or in the kind of hot-spring pools that exist in Yellowstone Park (**Figure 19.8**).

On the other hand, there are respected researchers who feel that, whatever the deep-sea vents may have had in the way of life-inducing qualities, their scalding temperatures would have obliterated any early self-replicating molecules, which were bound to be fragile. A much more likely first home for life, these researchers feel, are sheltered stretches of ancient ocean beaches, where the combination of tides and the composition of shoreline rocks would have "sorted" organic materials in accordance with their adhesiveness and other qualities. In this environment, the right initial combination of materials could have come together and then been joined by fresh materials, delivered continuously by the ocean waves.

**Figure 19.7**
**Hot Habitat**

Material pours forth from a hot-water vent on the floor of the Atlantic Ocean. The fluid being emitted is mineral-rich enough to support bacteria and archaea, which in turn form the basis for a local food web.

## The RNA World

Wherever living things first appeared, they had to have at least one critical feature: the ability to reproduce themselves. Today, a key player in reproduction is DNA. In Chapter 14, you saw that our genome, or inventory of DNA, can be thought of as a kind of cookbook containing recipes (DNA sequences) that lead to products (proteins) that undertake a wide variety of tasks. Yet, while DNA is the cookbook, it is not the cook. When it comes time for DNA to be copied—a necessary step in reproduction—the double helix cannot unwind itself, pair its bases up, or do anything else. Instead, all these tasks are carried out by the *proteins* known as enzymes. Hence, there is a chicken-and-egg dilemma: Enzymes can't be produced without the information contained in DNA, but DNA can't copy itself without the activity of enzymes. So, how did life get going if neither thing can function without the other?

In the late 1960s, researchers hit on a solution that has come to have a good deal of support. In early life-forms, a single molecule performed both the DNA and enzyme roles. This molecule was RNA (ribonucleic acid), which was portrayed back in Chapter 14 as a kind of genetic middleman, ferrying DNA's information to the sites where proteins are put together. A critical piece of evidence about RNA's role in early life came in 1983, when researchers discovered the existence of enzymes that are composed of RNA instead of protein: **ribozymes**. These molecules can encode information *and* act as enzymes (by, for example, facilitating bonds that bind RNA units together).

This evidence has led researchers to the presumed existence of an "RNA world"—an early living world that consisted solely of self-replicating RNA molecules. But there are questions about this RNA world. One of them is how it came into being in the first place. RNA is so complex, and so fragile, that many researchers believe it had to have taken over the role of a simpler, sturdier precursor.

### The Step at Which Life Begins

The critical step in life's development comes in imagining RNA's precursor beginning to carry out what might be called variable self-replication. *This* is the step at which life can be said to have begun. Given a molecule that can make copies of itself, a line of such molecules can come into existence. But, it's important that these molecules be able to *vary* from one another. Why? Only then could we have one molecule that would differ from oth-

**Figure 19.8**
**Only Seemingly Inhospitable**
The hot-water Grand Prismatic Spring in Yellowstone National Park, shown here in an aerial view, gets its blue color from several species of heat-tolerant cyanobacteria. The colors at the edge of the pool come from mineral deposits. The "road" at the top of the pool is a walkway for visitors.

ers in being, say, more resistant to the elements. A molecule that had such an advantage would produce a greater number of daughter molecules, which we might just as well call offspring. What would happen, in short, is natural selection among self-replicating molecules. This selection is the basis for evolution among these molecules, and with evolution, life begins to diversify into the innumerable forms we see today.

Once replication got going, we were still a long way from life as we know it today. Even the most primitive contemporary living things are encased in protective linings, and they can get rid of wastes, react to their environments, and so forth. Whatever the sequence of events, once life's elaborations are joined to replication, we move from simple molecules to the cellular ancestors of today's organisms.

---

**SO FAR...**

1. Each of the eras or periods within the geological timescale is characterized by a particular grouping of _____ preserved today as _____.

2. True or false: The first animals and plants evolved early in life's history.

3. Life is generally regarded as having begun with the development of _____ molecules.

---

••• ANSWERS

1. life-forms; fossils

2. False. To judge by the fossil record, animals and plants don't appear until the last 16 percent of life's history.

3. self-replicating

**Web Animation 19.2**
RNA-like Self-Replication

**Figure 19.9**
**Earth's Organisms and How They Evolved**

Scientists divide Earth's living organisms into three major categories: the Domains Bacteria, Archaea, and Eukarya. There are then four kingdoms within Domain Eukarya: Protista, Plantae, Animalia, and Fungi, which represent the plants, animals, and fungi of the world, along with a group of mostly microscopic, water-dwelling organisms, the protists. The lower drawing shows the evolutionary relationships among life's three domains. The universal ancestor at the base of the tree gave rise to all three domains, with an early evolutionary split resulting in Domain Bacteria on the one hand and an evolutionary line leading to Domains Archaea and Eukarya on the other. Different evolutionary lines of protists gave rise to all of the planet's plants, animals, and fungi.

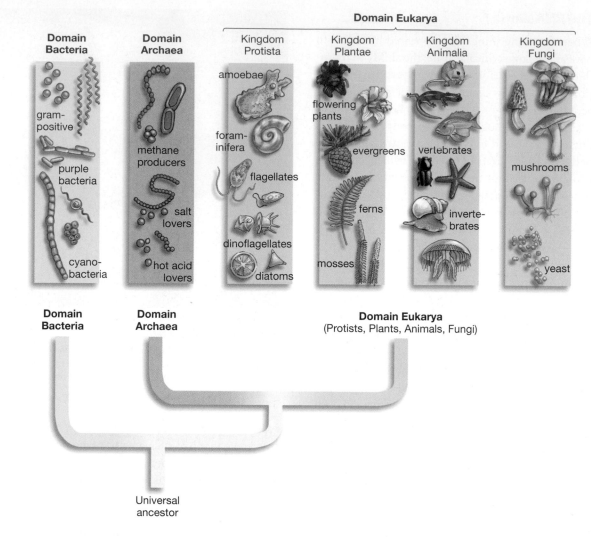

## 19.3 The Tree of Life

Once life was established, how did it evolve? If you look at **Figure 19.9**, you can see the biggest picture of all. In the upper drawing, you can see life as it currently exists on Earth, divided up into its major categories. As noted last chapter, the broadest of these categories are the three "domains" of life—**Bacteria**, **Archaea**, and **Eukarya**. Then there are four "kingdoms" within Eukarya: Protista, Plantae, Animalia, and Fungi. The figure gives you some idea of the kinds of organisms that fall into each of these categories. Domains Bacteria and Archaea are made up strictly of single-celled microbes, while Domain Eukarya has within it all the organisms that are most familiar to us—plants, animals, and fungi—along with a group that is less familiar, the protists, which are mostly microscopic and water-dwelling.

If you then look at the lower drawing, you can see how these groups evolved. At the bottom you see a "universal ancestor," which is the organism—or perhaps group of organisms—that gave rise to all current life. The evolutionary line that leads from the universal ancestor then branches out to yield, on the one hand, Domain Bacteria and, on the other, a line that leads to both Domain Archaea and Domain Eukarya. The archael microbes were long thought of as simply a different kind of bacteria, but we now know that these organisms are distinct from bacteria in fundamental ways. Nevertheless, there are some important similarities. All bacteria and archaea are essentially single-celled, and no archael or bacterial cell has a nucleus. This latter fact makes the archaea and bacteria **prokaryotes**. Conversely, all the organisms in Domain Eukarya have nucleated cells and are thus **eukaryotes**. For some detail on how life evolved within its three domains, see "Who Gets a Kingdom?" on page 362. In subsequent chapters, we will be taking walks through all of these major categories of life.

## 19.4 A Long First Era: The Precambrian

With this picture in mind, let's start tracing life from its cellular beginnings in the first of Earth's eras—the era called the Precambrian. Recall that the Earth was formed about 4.6 billion years ago. The earliest evidence we have of life is not a fossil but instead a kind of chemical signature of life left in the barren, frozen Isua rocks of Greenland (**Figure 19.10**). Carbon trapped in these 3.8-billion-year-old rocks is of a type produced only by living things. The evidence here is sparse, but most authorities think we can date life from this point. Shift forward another 400 million years, and we don't have to wonder whether life existed; we have abundant evidence of it, much of it found in the rock deposits of Western Australia. These signs of life are not fossils but rather evidence of living things having altered the rock deposits we see today. If you look at **Figure 19.11**, you can see a 3.4-billion-year-old rock deposit from Australia containing structures, called stromatolites, whose internal patterns are so complex it's likely they could only have been created by living organisms.

What were these organisms? All of them were microbes. To put a finer point on it, all of them were microbes that were either bacteria or archaea. For perhaps 1.8 billion years, these simple, single-celled creatures had Earth to themselves. During all this time, there were no fish, no trees, no insects, no mushrooms. Indeed, there were no eukaryotes of any kind—no organisms composed of cells that have a nucleus. The oldest eukaryote fossils we have date from a mere 2 billion years ago. Since today's eukaryotes *do* include fish, trees, and the like, you might think that the early eukaryotes would have had some size to them, but they were single-celled, too—and they stayed that way for a long time. The important feature of multicellularity didn't evolve in eukaryotes until about 1.2 billion years ago.

Taking a step back from these various dates, note the big picture in early evolution. It took about 800 million years for life to appear after the Earth formed, but then it took another 2.6 billion years to get from life's earliest single-celled life-forms to any multicelled life forms. Some authorities say this is evidence that the real hurdle in evolution was not the initiation of life but rather the initiation of complex, multicellular life. What is certain is that, in Earth's long first era, evolution took its time. The years rolled by in the millions, the land was barren, and the oceans were populated mostly by creatures too small to see with the naked eye.

## Notable Precambrian Events

None of this is to say, however, that "nothing happened" in the Precambrian era. Chapter 8 showed how critical photosynthesis is for life on Earth. This capability first came about in the Precambrian era—and fairly early in the era at that.

Photosynthesis began in bacteria by at least 3.4 billion years ago. This event was a turning point. Had this capacity not been developed, evolution would have been severely limited for the simple reason that there would have been so little to eat on Earth. The earliest organisms subsisted mostly on organic material in their surroundings, but the supply of this material was limited. In photosynthesis, the sun's energy is used to *produce* organic material—in our own era, the leaves and grasses and grains on which all animal life depends. Large organisms require more energy, and through photosynthesis, a massive quantity of energy-rich food was made available.

One particular kind of photosynthesis, again beginning in the Precambrian, had a second dramatic impact on evolution. A single type of bacteria, the cyanobacteria, were the first organisms to produce *oxygen* as a by-product of photosynthesis, beginning no later than 2.7 billion years ago. For us, the word *oxygen* seems nearly synonymous with the word life, but until 2.4 billion years ago, very little of it existed

**Figure 19.10**
**Looking at the Oldest Rocks**
The rocks at Isua, Greenland, are believed to be the oldest in the world. In examining them, scientists have found possible chemical signatures of life that have been dated to nearly 3.8 billion years ago.

**Figure 19.11**
**Traces of Ancient Organisms**
A geologic structure called a stromatolite in the Strelley Pool Chert of Western Australia shows evidence of microbes that flourished in a shallow-water ocean environment there some 3.4 billion years ago. Note the cone-shaped structures in the formation; scientists believe that such features could only have been produced by living organisms.

Who Gets a Kingdom? Evolution within Life's Categories

**H**ow did life evolve within its largest categories—within the three domains of Bacteria, Archaea, and Eukarya? If you look at **Figure 1**, you can get some idea. A few of the main branches within each domain have been labeled, such that one important variety of bacteria—the cyanobacteria—can be seen as a branch within Domain Bacteria, for example, while Methanococcus archaea can be seen as a branch within Domain Archaea.

It is when we turn to Domain Eukarya, however, that there may be a few surprises. In our daily lives, it is plants and animals that come front and center when we think of the living world. And fungi may cross our minds fairly frequently—when we see toadstools or bread mold, for example. Yet, note the place the plant, animal, and fungal kingdoms have within Domain Eukarya: Each is an ordinary branch that is rooted within the kingdom of protists. There is one protist ancestor, for example, that gave rise to the protists called choanoflagellates on the one hand and to the entire kingdom of animals on the other. (See the branch point in the tree that leads to both choanoflagellates and animals? That's the common ancestor.)

Of course, it's likely you never heard of the microscopic water dwellers called choanoflagellates. They didn't branch out in the myriad ways that animals did. Yet in evolutionary terms, they stand as a "sister" group to the animals, having shared a common ancestor with them. If you could be transported back to the time of the first animals and choanoflagellates, nothing you could see about either group (while using your microscope) would indicate their evolutionary futures. Nothing would tell you that the animal you're looking at would eventually give rise to birds, bats, and insects, while the choanoflagellate would give rise to . . . well, to about 150 species of single-celled water dwellers that live in cooperative groups.

The question of kingdoms within Domain Eukarya is a tricky one because, as it turns out, there are probably good reasons to call the choanoflagellates a kingdom—either that or give plants, animals, and fungi a somewhat less-exalted title. In essence, there is a problem in our current system for classifying living things, and the nub of the problem lies with one of the kingdoms within the system, Kingdom Protista, which is made up of organisms as diverse as one-celled amoebae and multicelled algae.

To understand what's wrong with this "kingdom," consider animals, which justifiably constitute their own kingdom. The reason animals have this status is that all of them are closely related enough that they deserve to be put into a single category (Kingdom Animalia). And this same thing is true of fungi and plants. For an organism to be a part of Kingdom Plantae, it has to be more closely related to every other plant than it is to any animal or fungus.

This sensible system of kingdoms breaks down, however, with "kingdom" Protista. To get a notion of what's wrong, look at Figure 1 again. Note that the choanoflagellates in it are as distantly related from, say, the red algae as animals are from plants. Yet animals and plants are classified as separate kingdoms, while choanoflagellates and red algae are lumped together in the single "kingdom" of Protista. It is as if we had multiple family lines—the Garcias, the Lees, the Beesons—and then decided to call all of them the Johnsons.

This naming fiction does serve a purpose: It holds down the number of kingdoms within Domain Eukarya, which makes for a manageable number of "major" categories of life. (Imagine if each of the branch terminals within Domain Eukarya were elevated to the status of kingdom.) Yet, all authorities are agreed that the term *protist* is meaningless in an evolutionary sense. As long as it remains in wide use, the important thing to remember is that it refers not to one evolutionary line, but to many.

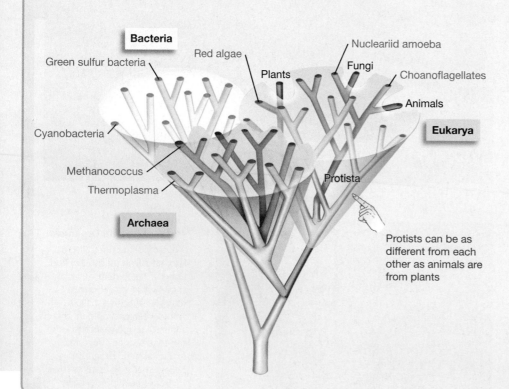

Bacteria

Green sulfur bacteria

Red algae

Plants

Nucleariid amoeba

Fungi

Choanoflagellates

Animals

Eukarya

Cyanobacteria

Methanococcus

Thermoplasma

Protista

Archaea

Protists can be as different from each other as animals are from plants

**Figure 1**
**Evolutionary relationships within life's three domains.**

in the atmosphere. When it did arrive, through the work of the cyanobacteria, it had the effect of greatly increasing the capacity of Earth to support a diverse group of organisms because it allowed for an additional infusion of energy into the living world. As Chapter 7 made clear, organisms that obtain their energy through oxygen-aided or "aerobic" means can extract a great deal more energy from a given quantity of food than can organisms that lack this aerobic capacity. Just as cars need oxygen to burn gasoline efficiently, so living things need oxygen to burn food efficiently. As such, the appearance of atmospheric oxygen set the stage for the emergence of life-forms larger and more complex than the early microbes. One group of microscopic organisms, the early eukaryotes, benefited from the oxygenic change through what sounds like a rather unpromising event: They were invaded.

Those who read Chapter 7 will recall that most of the energy that eukaryotes get from food is extracted within special cellular structures called *mitochondria.* Although mitochondria are tiny organelles within eukaryotic cells, they have characteristics of free-living *cells,* specifically free-living bacterial cells. As it turns out, that is what they once were. Ancient bacteria that could metabolize oxygen took up residence in early eukaryotic cells and eventually struck up a mutually beneficial relationship with them. The bacteria benefited from the eukaryotic cellular machinery, and the eukaryotes got to survive in an oxygenated world (**Figure 19.12**). A later bacterial invasion (by cyanobacteria) allowed one group of eukaryotes, the algae, to start carrying out photosynthesis.

### The Malnourished Earth

Surviving in an oxygenated world is not the same thing as thriving in it, however, and the early eukaryotes didn't seem to manage much more than survival. For the first billion years after they appeared, most of them remained single-celled and fairly simple. Eukaryote evolution was not quite dead in the water, but it was uneventful. Given the astounding variety of forms that eukaryotes eventually assumed, we might wonder why they experienced such a long holding pattern early in their evolution. A fascinating idea recently put forward is that they essentially were starving. According to what's called the Malnourished Earth hypothesis, the oxygen that started building up in Earth's atmosphere 2.4 billion years ago had the effect of weathering sulfur off the land and dumping it into the ancient oceans. This was, in essence, a global toxic spill that effectively removed elements from the oceans that the photosynthesizing eukaryotes—the

algae—needed in order to take up nutrients. The end result was a group of organisms so malnourished that they could not diversify very far through the kinds of evolutionary processes we looked at last chapter. The first big increase in eukaryote diversity did not arrive until about 600 million years ago (600 Mya), and when it did come, it was not algae that were in the vanguard; it was animals.

## 19.5 The Cambrian Explosion

As the Precambrian was coming to a close, life consisted of bacteria, archaea, and the eukaryote protists, almost all of which lived in the oceans. By about 575 Mya, however, we get the first fossil evidence of *animals* or animal-like creatures in the seas. The term "animal" here fits our common sense definition: They all are multicelled organisms, and they get their nutrition from other organisms or organic material.

There currently is a debate about whether the earliest animal fossils represent a line of creatures that died out or are instead related to today's animals. Whatever the case, a classic view of animal evolution was that these early animals were a kind of early stirring, a first blip on the screen bringing news of something big to come. To judge by the fossil record, some 30 million years after the emergence of these animals, we get it: a tidal wave of new animal forms, the like of which has never been seen before or since in

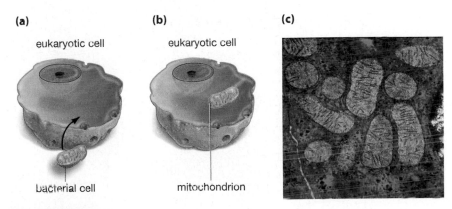

(a)            (b)            (c)

eukaryotic cell        eukaryotic cell

bacterial cell        mitochondrion

**Figure 19.12**
**Adapting to Oxygen**
The "powerhouses" of our cells—the organelles called mitochondria—have several characteristics of free-standing bacteria.

**(a)** There is general agreement among scientists that our mitochondria originally were bacteria that invaded eukaryotic cells many millions of years ago. These bacteria benefited from the eukaryote's cellular machinery, and the cell benefited from the bacteria's ability to metabolize oxygen.

**(b)** Over time, the bacteria came to be integrated into the host cells, replicating along with them. Two symbiotic organisms had now become a single organism.

**(c)** Today, the mitochondria in your cells turn the energy from food into a form that allows you to run, think, and turn the pages of this book. These artificially colored mitochondria functioned in a liver cell of a mouse.

evolution. This is the **Cambrian Explosion**: an alleged sudden appearance of the ancestors of modern animals beginning about 542 Mya.

Recalling the taxonomic system you went over in Chapter 18, you may remember that just below the domain and kingdom categories there is the **phylum**, which can be defined informally as a group of organisms that share the same body plan. There are at least 36 phyla in the animal kingdom today, some of these being Porifera (sponges), Arthropoda (insects, crabs), and Chordata (ourselves, salamanders). With one exception, the fossil record indicates, every single one of these came into being in the Cambrian Explosion. This seeming riot of sea-floor evolution was actually more extensive than even this implies in that a good number of phyla that *don't* exist today—phyla that have become extinct—also seem to have first appeared in the Cambrian. Some of these creatures are so bizarre it's surprising that Hollywood hasn't mined the Cambrian archives for new ideas on monsters (**Figure 19.13**). As if all this weren't enough, the Cambrian Explosion appears to have taken place in a very short period of time relative to evolution's normal pace—a few tens of millions of years at most.

But did it? The Cambrian Explosion is so extreme that one of the primary challenges to evolutionary biology has been to explain why such a thing came about—or why it came about only once. Note the series of events: only a few modern animal forms before the Cambrian, then the explosion, then almost no fundamentally new animal forms since. In the 1990s, several lines of evidence suggested something else: that today's animal forms actually began to emerge well back in the *Pre*cambrian. Under this view, all the Cambrian Explosion amounted to was an explosion of forms big and hard enough to leave fossils. The forms themselves appeared earlier and more gradually, this idea goes, but were too small and fragile to leave imprints behind.

Where does the truth lie? As things stand, the reality of the Cambrian Explosion is a matter of great debate. Most molecular studies indicate that modern animal phyla emerged much earlier than the supposed explosion—perhaps a billion years ago. Yet, a couple of recent molecular studies essentially side with the evidence from the fossils: They indicate that the modern phyla really did develop later and quite suddenly. In 2004, something new was added to this controversy. Microfossils found in China that were dated to about 600 Mya appeared to be not just of animals, but of the later-evolving "bilateral" animals, which is to say animals that have right and left sides that are symmetrical. (Think of your own body: Your chest is very different from your back, but your left side is very similar to your right.) These complex forms didn't just suddenly appear 600 Mya, the argument goes, so it's clear that the roots of today's complex animals stretch back much further in time.

Pending the resolution of this controversy, the most that can be said is that modern animal phyla may have began to appear as much as a billion years ago, but that large specimens of these phyla only appear in the fossil record beginning about 542 Mya. Note, however, that even if the Cambrian Explosion was only an explosion of size, this was not a trivial event. With this change, animals developed the capacity to *eat* one another, and this ability helped bring about the "arms race" that has been spurring on animal evolution ever since. (One animal's predatory adaptation spurs another's defensive adaptation and vice versa.)

We might ask, however: Why should animals have experienced a sudden increase in size beginning 542 Mya? A factor that we've talked about before may have played a part. Recall that significant quantities of oxygen first appeared in Earth's atmosphere about 2.4 billion years ago, thanks to the

**Figure 19.13**
**Life on the Ancient Sea Floor**

The Cambrian Explosion may have resulted in a multitude of new animal forms. In this artist's interpretation of the fossil evidence, a number of now-extinct Cambrian animals are shown in their ancient sea-floor habitat. Two of these are Anomalocaris, raising up with the hooked claws, and Hallucigenia, on the sea floor with the spikes extending from it.

photosynthesis carried out by bacteria. It turns out, however, that even with this increase, atmospheric oxygen levels remained relatively low until about 600 Mya when, for reasons unknown to us, they climbed to something approaching current levels. This second oxygen increase occurred right at the point at which we get the first large animals of any sort—the variety, noted earlier, that may have been an evolutionary dead end—and it came about 60 million years before the Cambrian explosion. Large organisms need large amounts of energy, and only oxygen-driven or "aerobic" respiration is capable of delivering this kind of energy. It may be that the oxygen that had done so much to change life early on had one more card to play as animals began to evolve.

---

**SO FAR...**

1. Life has three domains, and one of these domains has four kingdoms within it. Name both the domains and the kingdoms.

2. Life is thought to have begun as early as _____ years ago. Name some important evolutionary events that took place during life's first era, the Precambrian era.

3. The Cambrian Explosion is a seeming explosion in the number of _____ life-forms.

---

## 19.6 The Movement onto the Land: Plants First

The teeming seas of the early Cambrian stand in sharp contrast to what existed on land at the time, which was no life at all except for some hardy bacteria. Earth was simply barren—no greenery, no birds, no insects. When multicelled life did come to the land, the first intrepid travelers were plant-fungi combinations. (Even today, about 80 percent of plants have a symbiotic relationship with fungi, the plants providing fungi with food through photosynthesis, the fungi providing plants with water and mineral absorption through their filament extensions.) Exactly when this transition onto land took place is a matter of debate. The earliest undisputed fossil evidence we have of plants and fungi on land dates from 460 Mya, meaning these organisms had become abundant by that time. But when did they first appear? Our best estimates are sometime before 500 Mya. The general transition here began with marine algae in

the ocean, then continued with freshwater algae, then freshwater algae that came to exist in *shallow* water, living partially above the waterline. It was one variety of this algae, the green algae, that gave rise to all of Earth's plants. When these plants made the transition to full-time living on land, it was to damp environments.

### Adaptations of Plants to the Land

Such a change required a lot of adaptation. Aquatic algae did not have to deal with water loss or the crushing effects of gravity, but land plants did. One of the plants' responses to the water problem was to evolve a waxy outer covering, called a *cuticle*, that could retain moisture. Meanwhile, an initial response to gravity was to stay low. Some of the most primitive land plants are mosses that often hug the ground like so much green carpet (**Figure 19.14**).

Then there was the problem of reproduction. When green algae reproduce sexually, their gametes (eggs and sperm) float off from them as individual cells, after which they are brought together by ocean currents or the cells' own movement. Once this happens, the resulting embryos develop completely on their own. An embryo left to mature by itself on *land*, however, would dry out and perish. Plants adapted to the land by developing a protective housing for embryos; in a plant, embryos mature *within* a parent.

**Figure 19.14**
**Lying Low**
Moss covering rocks along a stream in Tennessee's Great Smoky Mountains National Park. Mosses are members of the earliest-evolving group of plants still alive, the bryophytes.

## Another Plant Innovation: A Vascular System

If you look at **Figure 19.15**, you can see how the major divisions of plants evolved. The most primitive land plants that are still with us, the **bryophytes** (represented by today's mosses), had no vascular structure—meaning a system of tubes that transports water and nutrients. Plants without such a system have very limited structural possibilities. They can grow out, but they lack ability to grow *up* very far, against gravity. By contrast, the ancestors of today's ferns developed a vascular system; with this, over the ensuing 100 million years, a variety of **seedless vascular plants** evolved, including huge seedless trees such as club mosses, some of them 40 meters tall, which is about 130 feet.

## Plants with Seeds: The Gymnosperms and Angiosperms

Even as the seedless vascular plants were reaching their apex, a revolution in plants was well under way as some of them developed *seeds*. With the first seed plants, offspring no longer had to develop within a delicate plant on the forest floor. Instead, they developed within a dispersible seed, which can be thought of as a reproductive package that includes not only an embryo (brought about when sperm fertilized egg), but food for this embryo and a tough outer coat.

The seed plants split into two main evolutionary lines that are still with us today. The earliest of these lines to develop was the **gymnosperms**—today's pine and fir trees, for example—with seeds that are visible in the well-known pinecone. Gymnosperms also represent the final liberation of plants from their aquatic past. Mosses and ferns had sperm that could make the journey to eggs only through water. Not much water was needed; a thin layer would do, as with the last of a morning's dew on a fern leaf. With seed plants, however, the water requirement is gone entirely. Sperm are encased inside pollen grains, which can be carried by the *wind*. Think how far a windblown pollen grain could disperse compared to a sperm moving across a watery leaf. Several types of gymnosperms developed, beginning from earlier seed plants sometime before 300 Mya. Today, gymnosperm trees cover huge stretches of the Northern Hemisphere.

## The Last Plant Revolution So Far: The Angiosperms

At the time the dinosaurs reigned supreme among land animals, the first flowering on Earth occurred with the development of flowering plants, also known as **angiosperms**. Evolving between 180 and 140 Mya, the angiosperms eventually succeeded the gymnosperms as the most dominant plants on Earth. Today, there are about 700 gymnosperm species, but some 260,000 angiosperm species, with more being identified all the time. Angiosperms are not just more numerous than gymnosperms; they are vastly more diverse as well. They include not only magnolias and roses, but oak trees and cactus, wheat and rice, lima beans and sunflowers.

One of the reasons for the angiosperms' success was that they developed a new way for their pollen to be transported. Whereas the gymnosperms used the

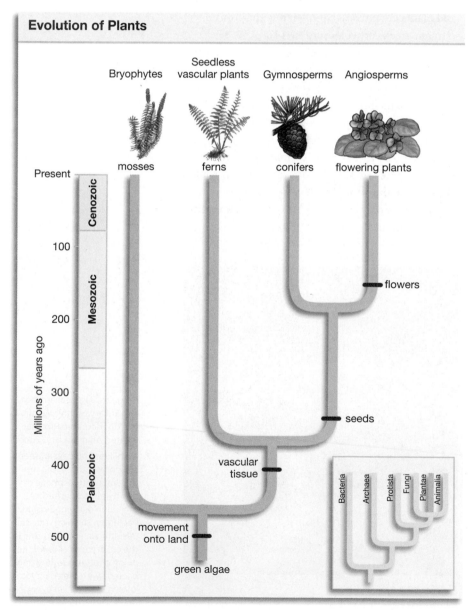

**Evolution of Plants**

Bryophytes — Seedless vascular plants — Gymnosperms — Angiosperms

mosses — ferns — conifers — flowering plants

Present

Cenozoic

Mesozoic

Paleozoic

Millions of years ago

100

200

300

400

500

flowers

seeds

vascular tissue

movement onto land

green algae

Bacteria | Archaea | Protista | Fungi | Plantae | Animalia

**Figure 19.15
How Plants Evolved**

The tree of life for bryophytes, seedless vascular plants, gymnosperms, and angiosperms.

**Figure 19.16**
**A Possible Early Flowering Plant**

In 2002, a team of Chinese and American scientists announced the discovery of *Archaefructus sinensis*; Its fossilized, 125-million-year-old impression can be seen at left. The plant clearly had angiosperm-like traits—for example, the male reproductive structures known as anthers. Yet, it had no petals. One hypothesis is that it represents a transitional form in the evolution of angiosperms from lake-dwelling herbaceous plants. On the right is an artist's conception of what *A. sinensis* looked like.

wind, the angiosperms used animals—bees, birds, and bats—that travel from one angiosperm to another, picking up nectar and inadvertently delivering pollen. If you look at **Figure 19.16**, you can see the fossilized impression of what may be an early flowering plant, dating from 125 Mya, along with an artist's conception of what it looked like.

## 19.7 Animals Follow Plants onto the Land

The movement of plants onto the land made it possible for animals to follow. With plants came food and shelter from the sun's rays. Recalling the division of the animal kingdom into various major groups called phyla, it was the phylum of arthropods that first moved to land. About 440 Mya—20 million years after the first plant fossils were laid down—a creature similar to a modern centipede laid down the oldest set of terrestrial animal tracks we know of. It makes sense that arthropods were the first animals to come onto land because the hallmark of all of them is a tough external skeleton, or *exoskeleton*, which can prevent water loss and guard against the sun's rays.

Other kinds of arthropods, the insects, came on land some 50 million years after the first arthropod immigrants. The insect pioneers were wingless, perhaps being something like modern silverfish. Insect flight would not develop for tens of millions of years more, but when it did insects had the skies to themselves for 100 million years. The more primitive flying insects are represented today by dragonflies, whose wings cannot be folded back on their body (**Figure 19.17**). The compactness that came with foldable wings gave later insects the advantage of being able to get into small spaces. (Think of a common housefly.) In both winged and wingless forms, the insects took to the land with a vengeance. About 1.8 million species of living things have been identified on Earth to date, of which *half* are insects. For millions of years, the insects and their fellow arthropods were the only animals on land—but this would change.

### Vertebrates Move onto Land

You'll recall that back at the time of the Cambrian Explosion, almost all animals existed on the ancient ocean floor. Within just a few million years, however, animals with backbones, known as *vertebrates*, were moving throughout the ocean; several of them were primitive, now extinct, orders of fishes. About 450 Mya, we get the rise of the family of fishes, the gnathostomes, that are the ancestors of nearly all the fish species still alive today, from sharks to gars to goldfish. Critically, gnathostomes were the first creatures to have *jaws*. With the development of jaws, these vertebrates and their descendants (one of whom is us) could securely grasp all kinds of things, big chunks of food, to be sure, but also offspring and inanimate objects. The new foods that became available to those with jaws allowed the gnathostomes to outgrow their jawless vertebrate relatives.

**Figure 19.17**
**The First to Take Flight**

Dragonflies, such as this dew-covered individual, are descendants of the insects that first developed the ability to fly. Flying insects that evolved later could fold their wings back, as with today's housefly.

**Figure 19.18**
**Vertebrates onto Land**

**(a)** Our ray-finned fish ancestors lacked bones at the end of their fins.

**(b)** The lobe-finned fish that evolved from ray-finned fish had small bones in their fins that enabled them to pull out of the water and support their weight on tidal mudflats and sandbars.

**(c)** Finally, the early tetrapods lost ray fins altogether and had limbs like ours.

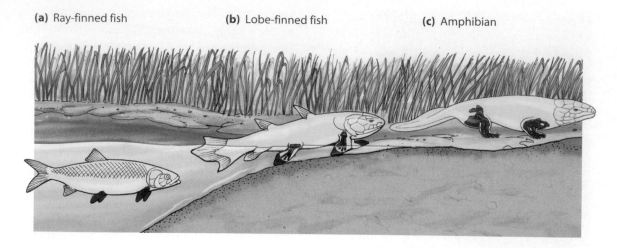

**(a)** Ray-finned fish     **(b)** Lobe-finned fish     **(c)** Amphibian

**Figure 19.19**
**Transitional Animal**

An artist's conception of *Ichthyostega,* one of the earliest known amphibians, seen at upper left climbing on the log and at center. Sometimes thought of as a "four-legged fish," *Ichthyostega* spent most of its time in water and probably used its back legs mostly for paddling. All land creatures with backbones, including human beings, ultimately trace their ancestry to tetrapod fishes such as *Ichthyostega.*

### Fins to Limbs

A particular, later-arriving type of gnathostome, the lobe-finned fish, had an anatomical feature that turned out to be critical in vertebrate evolution. The lobed *fins* these fish possessed were the precursors to the four *limbs* that are found in all tetrapods, including ourselves. With fins, the fish swam in water, but with limbs the tetrapods moved onto land (**Figure 19.18**).

This momentous transition is known to have occurred in a period of about 10 million years, starting 380 Mya and ending 370 Mya. How can we be sure of the dates? We now have fossils that span the entire period. By searching from nearly the North Pole to Australia, scientists have retrieved the fossils of nine species whose forms are intermediate between fully aquatic fish and fully terrestrial tetrapods. If you look at **Figure 19.19**, you can see an artist's interpretation of the appearance of a creature, called *Ichthyostega,* that lay closer to the fully terrestrial end of this spectrum. And if you turn to "Going after the Fossils,"

on page 370, you can get a view of a recently discovered creature, named Tiktaalik, that lay closer to the fish end. It's worth noting that one of the arguments sometimes advanced against the theory of evolution is that there are no "intermediate forms" in the fossil record—no forms that show us the "in-between" organisms in an evolutionary transition. We now have so many intermediate forms in the fish-to-tetrapod transition that even experts are hard-pressed to say where fish end and tetrapods begin.

### Evolutionary Lines of Land Vertebrates

It is several lines of tetrapods—the amphibians, the reptiles, and the mammals—that you'll follow for the rest of your walk through evolution. Here is the order of emergence among these forms. The early tetrapods gave rise to amphibians, and early, salamander-like amphibians gave rise to reptiles. Early reptiles then diverged into several lines, one leading to the dinosaurs and another leading to mammals. Dinosaurs eventually died out, of course, but before they did, a lineage that lived on—the birds—diverged from them (**Figure 19.20**).

### Amphibians and Reptiles

*Amphibian* literally means "double life," which is appropriate because in these creatures we can see the pull of both the watery world from which they came and the land onto which they moved. Modern-day frogs can serve as an example here. Many species of frog spend their adult lives as air-breathing land dwellers but must return to the water to reproduce. Female frogs deposit eggs directly into the water, and the males then deposit sperm on top of them. The young that survive from this clutch go on to live as swimming tadpoles, complete with gills and a tail. After a time, however, both these features disappear, even as lungs start to develop, along with ears for hearing and legs for hopping. The tadpole has become a frog—a double life indeed.

### A Critical Reptilian Innovation: The Amniotic Egg

In the transition from amphibian to *reptile*, there is a severing of the amphibian ties to the water through a new kind of protection for the unborn. Amphibian eggs require a watery environment; lacking any sort of outer shell, they dry out if taken out of the water, thus killing the embryo inside. The early reptiles solved this problem with the remarkable adaptation of the **amniotic egg**: an egg that has not only a hard outer casing, but an inner padding in the form of egg "whites" and a series of membranes around the growing embryo. These membranes help supply nutrients and get rid of waste for the embryonic reptile. With this hardy egg, the tie to the water had been broken; reptiles could move *inland*.

Amphibians were the sole terrestrial vertebrates for about 30 million years, beginning about 350 Mya. We think of these creatures now in terms of small frogs and salamanders, but the amphibians that roamed the Earth in the Carboniferous period were often the size of modern-day pigs or crocodiles. The carnivores among them probably fed on small early reptiles, but reptiles ultimately became the hunters rather than the hunted. Thriving everywhere from water to dry highlands, reptiles grew in number and diversity.

One line of reptiles evolved, about 225 Mya, into the most fearsome creatures ever to walk the Earth—the dinosaurs. For 160 million years, no other land creatures dared challenge them for food or habitat. Not all dinosaurs were huge, and many did not eat meat. Yet, even a leaf eater such as *Apatosaurus* must have commanded a good deal of respect because it measured more than 26 meters, or 85 feet, from head to tail and weighed more than 30 tons. By comparison, a modern African elephant may weigh about 7 tons.

### From Reptiles to Mammals

Scurrying about in the underbrush when the dinosaurs reigned supreme were numerous species of insect-eating animals, few of them bigger than a rat. These animals fed their young on milk derived from special female mammary glands, however. In addition, they probably possessed fur coats over their skins—an important feature for animals restricted to feeding at night, when the dinosaurs were less active. These were the mammals, an evolutionary line of organisms that split off from reptiles 220 Mya.

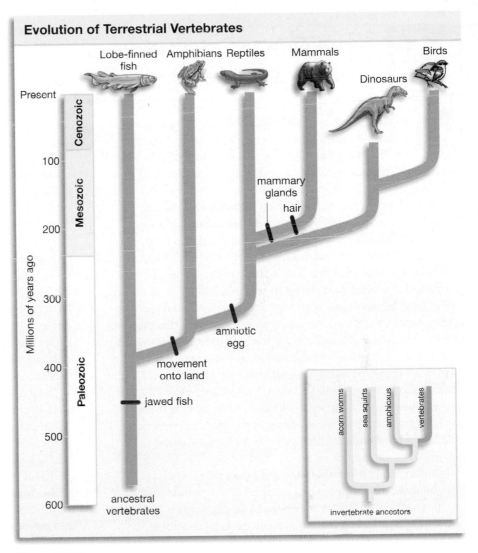

**Figure 19.20**
**How Terrestrial Vertebrates Evolved**
A tree of life for lobe-finned fish, amphibians, reptiles, mammals, and birds.

The conventional wisdom has long been that it took the disappearance of the dinosaurs, in the Cretaceous Extinction, to bring the early mammals into more diverse environments, thus allowing them to evolve into the forms seen today. A major analysis of fossil and molecular evidence completed in 2007 told a different tale, however. According to it, all the basic groups of modern mammals (all of today's mammalian "orders") had appeared by 75 Mya, which is some 10 million years *before* the dinosaurs disappeared. A second round of speciation, this one within the mammalian orders, did take place after the demise of the dinosaurs, but it took place so *long* after the dinosaurs died out—from 55 to 10 Mya—that it's clear the extinction of the dinosaurs had nothing to do with the evolution of modern mammals. An obvious question here is: Why did the mammals survive the Cretaceous Extinction while the dinosaurs didn't?

# THE PROCESS *of* SCIENCE
## Going after the Fossils

For the University of Chicago's Neil Shubin and his colleagues, it would have been nice if the ancient animal they dubbed *Tiktaalik roseae* had made its home in what is now Tahiti or Hawaii, but that's not the way things happened. Instead, this startling creature made its home on what is now Canada's Ellesmere Island. Back when Tiktaalik was sloshing around in streambeds, Ellesmere actually was a balmy place. But that was 375 million years ago, when Canada's Hudson's Bay sat on the equator. In the eons that followed, North America moved north, and this movement put Ellesmere Island *way* north—you have to travel 600 miles south from it just to reach the Arctic Circle.

But fossil hunters have to go where the fossils are, and the evidence told Shubin and his colleague Ted Daeschler that the Ellesmere region might well contain the kind of fossils they were looking for: those of ancient animals—you could probably call them fish—that lived mostly in water but were starting to make the transition to living on land. So, beginning in the summer of 1999, the two researchers, along with colleague Farish Jenkins Jr. and a changing group of graduate students, began spending summers walking the barren tundra of northern Canada looking for 375-million-year-old fish encased in rock.

In undertaking this task, the researchers were practicing a time-honored discipline known as expeditionary paleontology. Scientists who study evolution by examining fossils are called paleontologists, and *finding* these fossils often means mounting full-scale expeditions. Certainly this is what Shubin's team has had to do in connection with the Ellesmere fossils. For starters, the sites at which they're found lie 240 miles from the nearest civilization. As such, the researchers and all their supplies have to be helicoptered in to these sites, after which the researchers are on their own. Work can basically only take place during the month of July—in June winds can rip up tents, and by August it's snowing. But even July is no picnic. Sleeping can be difficult because the sun never sets.

Then there are the polar bears; their presence means the researchers must look for fossils while toting shotguns (**Figure 1**).

Despite these difficulties, the prize Shubin and his colleagues sought was important enough that they spent the summers of 1999, 2000, and 2002 trekking across Ellesmere and another island to its west, eyes to the ground, digging in one spot and then another, only to come away empty handed. All told, they walked across some 400 miles of tundra before finally hitting the jackpot in the summer of 2004 (**Figure 2**).

"It was getting to the point where we were ready to chalk it up in the loss column," Shubin said by telephone, days after returning from another Ellesmere expedition. "The 2004 year was going to be our last." And although that year ended wonderfully, it had started badly. The researchers' arrival had been delayed for two days by foul weather. And once they did get to work, a morning rain washed out their second day of digging on a fossil site.

"So, I said 'Well, let's just walk for the second half of the day,'" Shubin recalls. "And on that walk I was sitting down having a snack. And next to where I was sitting was the side of a Tiktaalik, exposed—sitting right out on the surface. And I was like 'Holy cow! That rings a bell.'" The researchers started digging right away, then returned to the site the next day and found a second exposed side of a Tiktaalik. Then on the third day, they made the biggest discovery of all.

"My colleague Steve Gatesy was at the top of the hill by the site, and he's flipping rocks, and he says 'Hey, what's this?' So we go over there and look. And it was the snout of a Tiktaalik sticking out of the hill."

This was the kind of specimen every paleontologist dreams of finding: an important creature in evolutionary history whose

**Figure 1**
**Fossil Hunters**
Paleontologists Neil Shubin, on the left, and Ted Daeschler in the field in the Canadian north in 2006. Daeschler is carrying one of the shotguns research team members use to guard against attacks by polar bears.

**Figure 2**
**A Long, Difficult Search**
Research team members trek across the valley in Ellesmere Island where they made their discovery.

**Figure 3**
**Transitional Creature**
Fossilized remains of *Tiktaalik roseae*, which the research team unearthed in Ellesmere Island in the summer of 2004. The creature represents a transitional form between water-dwelling fish and land-dwelling tetrapods as it possesses features of both kinds of animals.

most important body part—its head—is well preserved. (See **Figure 3** for a picture of it.) In time, biologists around the world would agree it was a terrific find, for it revealed, in the words of paleontologist Jennifer Clack, "What our ancestors looked like when they began to leave the water." Tiktaalik had the scales of the fish it evolved from, but the neck and flexible forelimbs of the land dwellers that came later. It had the gills of a fish, but also had lungs that it probably filled by gulping air the way frogs do. Overall, it sat midway in the evolutionary transition between fish and the four-limbed land-walkers called tetrapods. This was an important transition because tetrapods are the ancestors of all land vertebrates, including ourselves. Tiktaalik is thus a candidate

for what *human* ancestors looked like when they started moving onto land.

None of this was lost on the researchers in the summer of 2004 as they looked at the creature—soon to be named Tiktaalik—with its snout sticking out of the Ellesmere rock formation. Shubin and his colleagues couldn't wait to find out what the whole fossil would reveal. Their problem was that, having discovered this specimen and the others, they then had to get them out of the rocks and safely home. And they had to do this quickly, before the August snows set in. The first challenge facing them was to leave the right amount of rock around the fossils: enough so that they wouldn't be damaged, but not so much as to make them too heavy to transport. The second trick was to figure out how to get a protective layer of plaster to dry around them, even as rains were washing over Ellesmere every day.

In the end, all the specimens made it safely to the United States. Over the course of the next several months, as museum experts carefully separated rock from fossil, it became apparent that the researchers had discovered a creature rich in evolutionary information. They decided to give it a name based in the Inuktitut language used on Ellesmere. They called it Tiktaalik (tik-TA-lik), which means large, shallow-water fish. In April of 2006, the wider world got its first look at Tiktaalik when Shubin and his colleagues published two scientific papers on the creature in the journal *Nature*. Newspaper, radio, and television stories followed. It was a busy time for Shubin because he was simultaneously doing two things: working away in Chicago, fielding press calls all the while, and laying plans to be 3,000 miles away in two months' time—in Ellesmere, looking for fossils.

Forty-three modern mammalian lineages made it through this asteroid-aided extinction, while not one group of dinosaurs did. It's not clear, however, which factors account for the difference. Pure size may have had something to do with it; small creatures seem to survive extinction events better than large ones.

## The Primate Mammals

A species of placental mammals gave rise to the order of mammals called *primates*, although exactly when this happened is not clear. The oldest primate fossils we have date from 55 Mya, but it will not surprise you to learn that molecular work places their origins much further back—at about 90 Mya. If you look at **Figure 19.21**, you can see, in the lemur of Madagascar, a modern-day descendant of these earliest primates. Three things characteristic of most primates are apparent in this picture: large, front-facing eyes that allow for binocular vision (which enhances depth perception); limbs that have an opposable first digit, like our thumb (which makes grasping possible); and a tree-dwelling existence. There are about 230 species of

**Figure 19.21**
**Primate Characteristics**

These modern-day lemurs in Madagascar exhibit several characteristics common to most primates: opposable digits that enable grasping, front-facing eyes that allow binocular vision, and a tree-dwelling existence.

**(a)** A prosimian

**(b)** A New World monkey

**(c)** A great ape

**Figure 19.22**
**Several Types of Primates**

**(a)** A loris from India (*Loris tardigradus*) is representative of an early primate lineage, the prosimians, that includes lemurs.

**(b)** A woolly spider monkey (*Brachyteles arachnoids*) from the Atlantic rain forest in Brazil. Note the tail, which is capable of grasping objects. Only New World monkeys have such tails.

**(c)** A lowland silverback gorilla (*Gorilla gorilla*), native to Africa.

primates living today (**Figure 19.22**), which is a small portion of the 4,554 species of mammals. It's a tiny number of species indeed compared to, say, the 60,000 species of molluscs or the 750,000 species of insects. **Figure 19.23** sets forth a primate family tree.

## On to Human Evolution

If you look to the upper right in the primate tree, you can see a fork that leads to chimpanzees on the one hand and human beings on the other. The simplicity of the fork might give you the impression that a couple of speciation events, occurring about 6 Mya, led to today's chimps and today's humans. But looking at an evolutionary tree of this sort is something like looking at a river from an airplane: It's valuable in that you get the big picture, but it's misleading in that you miss the details. In reality, the evolution of human beings was a complicated thing—a process that produced many species of primates that were somewhat like us. Where are they now? They're all gone except for us. Among the primates, we're the last two-legged walkers still standing. The story of how we evolved is coming up.

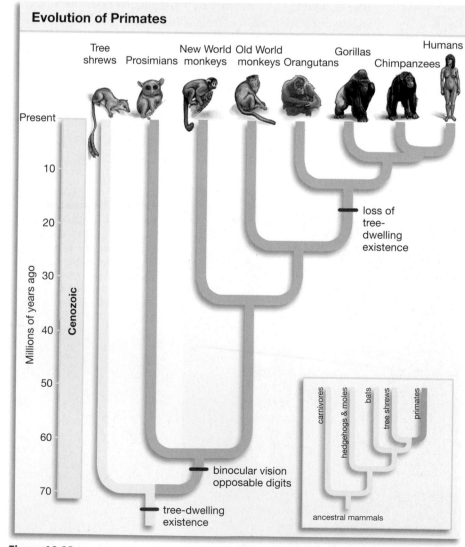

**Figure 19.23**
**How Primates Evolved**

# CHAPTER 19 REVIEW

For study help, activities, and more quiz questions, go to **www.aw.com/krogh4**.

## SUMMARY

### 19.1 The Geological Timescale: Life Marks Earth's Ages

- Earth's 4.6 billion years of history are measured in the geological timescale, which is divided into broad eras and shorter periods. These time frames are defined in large part by the different life-forms that evolved within them, as reflected in the fossil record. Many of the major transitions in the fossil record have come about because of extinction events.

- Life is thought to have begun on Earth about 3.8 billion years ago; from that time until perhaps 1.2 billion years ago, all life was strictly microscopic and single-celled. To judge by the fossil record, all the animals and plants that exist today came about in the last 16 percent of life's 3.8-billion-year history.

- Evolution can be conceptualized as a branching tree, with humans sitting on one ordinary branch. It is not shaped as a pyramid with humans at the apex. The story of evolution recounted in this chapter and the next leads to human beings, but it could just as easily have led to any other modern organism.

**Web Animation 19.1:** Evolutionary Timescales

### 19.2 How Did Life Begin?

- Life arose through a chemical process: Simple elements and compounds available on the early Earth came together to produce more complex molecules. Ultimately, a group of these molecules became capable of self-replication.

- In a famous experiment conducted in 1953, researcher Stanley Miller attempted to re-create, in a set of tubes and flasks, the environmental conditions that he believed existed on the early Earth. Miller found that one variety of life's chemical building blocks, amino acids, could be produced in these conditions in a short period of time.

- There currently is a debate about the kind of environment in which life is likely to have begun. One hypothesis is that life began in a "prebiotic soup" of hot-water systems—in deep-sea vents or in the kind of hot-spring pools that exist in Yellowstone Park today. Some researchers feel, however, that early replicating molecules could not have withstood the high temperatures of such environments. A more likely site for life's origins, they believe, are the tide pools of Earth's early oceans.

- A critical question for origin-of-life researchers has been to account for the existence in living things of both the worker molecules called enzymes and the information-bearing molecules

DNA and RNA, since neither kind of molecule can be synthesized without the other. The discovery of RNA molecules, called ribozymes, that have enzymatic capabilities helped solve this dilemma. Ribozymes provided evidence for an ancient "RNA world" in which the only living things were simple RNA molecules that could bring about their own replication.

- Life could be said to begin with the origin of molecules capable of variable self-replication. It was not enough for molecules to make copies of themselves; these copies had to differ one from another, such that some would be better suited to their environment than others. With this, Darwinian natural selection could have acted on the molecules, thus setting in motion the evolution that would ultimately result in today's diverse living world.

**Web Animation 19.2:** RNA-like Self-Replication

### 19.3 The Tree of Life

- Every living thing on Earth belongs to one of three domains of life: Bacteria, Archaea, or Eukarya. All members of Domains Bacteria and Archaea are single-celled microbes, and no bacterial or archael cell has a nucleus, which makes the bacteria and the archaea prokaryotes. In contrast, all organisms in Domain Eukarya are composed of one or more cells that have a nucleus, which means that all members of this domain are eukaryotes. Domain Eukarya is composed of four kingdoms: plants, animals, fungi, and protists.

- Life today can be conceptualized as beginning with a universal ancestor that produced an evolutionary line leading to Domain Bacteria on the one hand and to Domains Archaea and Eukarya on the other. Separate evolutionary lines within Domain Eukarya—lines collectively referred to as protists—gave rise to the plant, animal, and fungal kingdoms.

### 19.4 A Long First Era: The Precambrian

- The earliest evidence we have for life is a chemical signature of it: carbon utilization, as recorded in rocks found in Greenland dated to about 3.8 billion years ago. The internal patterns of sedimentary rock formations known as stromatolites, dating from 3.4 billion years ago, tell us that life was abundant by that point. All of this life was bacterial or archael, and all of it appears to have existed in the oceans.

- Photosynthesis was first performed by bacteria beginning no later than 3.4 billion years ago. This was a critical event because photosynthesis provides the food that supports almost all life on Earth. Oxygen came to exist in significant quantity in Earth's atmosphere through the activity of the cyanobacteria, which produce oxygen as a by-product of the photosynthesis they carry out. This increase in atmospheric oxygen levels was important because organisms that metabolize food through oxygen-aided means can extract much more energy from a given quantity of food than can organisms that lack this "aerobic" capacity. Early eukaryotes benefited from the rise in atmospheric oxygen after being invaded by, and then continuing to live in symbiosis with, bacteria that were able to metabolize oxygen. A later bacterial invasion allowed one variety of eukaryotes, the algae, to perform photosynthesis.

- Eukaryotes experienced only a limited amount of diversification during the first billion years after they appeared. One possible explanation for this is that they were malnourished. The oxygen that appeared in Earth's atmosphere may have had the effect of weathering sulfur into the oceans. This toxin would have prevented the photosynthesizing eukaryotes—the algae—from taking up the nutrients they needed.

### 19.5 The Cambrian Explosion

- The fossil record indicates a tremendous, rapid expansion in the number of animal forms in a "Cambrian Explosion" that began about 542 Mya, but this evidence has been challenged. It may be that the Cambrian Explosion was merely an explosion in the number of animal forms big and hard enough to leave fossils, and that the animal forms we see today are the product of evolution that took place over a relatively long period of time, beginning as much as a billion years ago.

### 19.6 The Movement onto the Land: Plants First

- Plants, which evolved from green algae, made a gradual transition to land in tandem with their symbiotic partners, the fungi. Plant–fungi fossils exist from about 460 Mya, but they are thought to have come onto land sometime before 500 Mya. Major transitions in plant life came with the development of a water-retaining covering (the cuticle), embryos that matured inside parent plants, and a "vascular" or fluid-transport system.

- Later plants went on to develop seeds, which can be thought of as packages containing an embryo and food for it, encased in a tough outer covering. The living descendants of the first seed plants are the gymnosperms, represented by today's conifers. The flowering plants, called angiosperms, developed between 180 and 140 Mya and eventually succeeded the gymnosperms as the most dominant plants on Earth. Today, angiosperms include many food crops, cactus, and tree varieties.

### 19.7 Animals Follow Plants onto the Land

- The first land animals were arthropods; a centipede-like creature laid down the oldest terrestrial animal tracks we know of. Insects soon followed in great abundance. The arthropods were the only land animals for millions of years thereafter.

- One group of fish, the lobe-finned fishes, gave rise to the four-limbed vertebrates, called tetrapods, that moved onto land between 380 and 370 Mya.

- Early tetrapods gave rise to amphibians—represented by today's frogs and salamanders—that in turn gave rise to reptiles. All mammals then evolved from reptiles. Birds are the living descendents of the reptile branch that included the dinosaurs.

- Amphibians inhabit two worlds, water and land, in keeping with their close evolutionary relationship with animals that lived strictly in water. Reptiles evolved a protective amniotic egg that allowed their offspring to develop away from water, freeing reptiles to migrate inland.

- Mammals appeared as a group of small, insect-eating animals about 220 Mya. Contemporary research indicates that all of today's basic forms of animals—all of the modern mammalian orders—had evolved by 75 Mya, some 10 million years prior to the demise of the dinosaurs in the Cretaceous Extinction. Another round of mammalian speciation began about 10 million years after the dinosaurs died out. Thus, contrary to a long-standing assumption, the demise of the dinosaurs seemed to have no effect on the evolution of modern mammals.

- The order of mammals called primates, whose fossils date from 55 Mya, is characterized by large, front-facing eyes, limbs with an opposable first digit (thumbs in today's human beings), and a tree-dwelling existence (which some later-arriving primates lost).

## KEY TERMS

| | |
|---|---|
| **amniotic egg** 369 | **Eukarya** 360 |
| **angiosperm** 366 | **eukaryote** 360 |
| **Archaea** 360 | **gymnosperm** 366 |
| **Bacteria** 360 | **Permian Extinction** 354 |
| **bryophytes** 366 | **prokaryote** 360 |
| **Cambrian Explosion** 364 | **phylum** 364 |
| **continental drift** 357 | **ribozyme** 359 |
| **Cretaceous Extinction** 353 | **seedless vascular plant** 366 |

## BRIEF REVIEW

*Answers to Brief Review questions are in the back of the book.*
*For multiple-choice quiz questions, go to* **www.aw.com/krogh4**.

1. What are two differences between the organisms in Domain Bacteria and those in Domain Eukarya?

2. What is the best date we have for the origination of life? What forms of life existed for the next 1.8 billion years thereafter?

3. What group of organisms did all land plants evolve from?

4. In what order did the following vertebrate animals evolve: reptiles, amphibians, mammals, lobed-finned fish, early tetrapods?

5. What is the amniotic egg, and why was it a breakthrough for terrestrial animals? In which vertebrate class did the amniotic egg evolve?

6. Describe two characteristics that evolved in mammals that had not been present in earlier organisms.

## APPLYING YOUR KNOWLEDGE

1. How would Earth be different if photosynthesis had never developed in any organism?

2. Why does it make sense that life started so small? Why does it make sense that it began in water?

3. In what way are the bryophyte plants and the amphibian animals alike?

*A skull of* Homo habilis *or "handy man," surrounded by additional fragments of its fossilized skeleton. This close human relative lived about 1.8 million years ago in East Africa.*

| 20.1 | The Human Family Tree | 377 |
| 20.2 | Human Evolution in Overview | 378 |
| 20.3 | Interpreting the Fossil Evidence | 381 |
| 20.4 | Snapshots from the Past: Three Hominins | 382 |
| 20.5 | The Appearance of Modern Human Beings | 385 |
| 20.6 | Next-to-Last Standing? The Hobbit People | 387 |

**Essay**

Sequencing the Neanderthal Genome | 386 |

Human beings evolved as one species within an evolutionary group called the hominins whose origins stretch back 6 or 7 million years.

# ARRIVING LATE, TRAVELING FAR:
# The Evolution of Human Beings

**I**n August of 1856, workers at a limestone quarry in Germany's Neander Valley blasted away the entrance to a cave and discovered within it some bones they recognized as unusual. Thinking that they had perhaps come upon the bones of a cave bear, they set some aside and asked a local schoolteacher and naturalist, Johann Fuhlrott, to come take a look at them. On seeing them, Fuhlrott knew immediately that these were not the bones of a bear. But what were they? Although they were much like the bones of a human being, they were not *exactly* human in form. There was a partial, human-like skull, but its browridge was strangely thick and prominent. And there were limb bones, but they were too curved and massive to belong to an ordinary human.

Together with Hermann Schaaffhausen, a professor of anatomy at the University of Bonn, Fuhlrott presented both the bones and drawings of them to a natural history society in 1857. News about the presentation followed, and with this, humanity got a wake-up call about itself. Although the notion was resisted at first, over time it became clear that there once had been beings on Earth who were much more human in form than, say, chimpanzees or gorillas, but who nevertheless were not *fully* human. So, how should these beings be classified? Schaaffhausen's view was that the Neander individual was a member of a "savage race" of Europeans who had been supplanted by modern Germans. Although it took a while, scientists eventually rejected this view in favor of another: that there once had been separate *species* of human-like beings on the Earth. The bones of the Neander individual belonged to one of these species, which in time came to be called *Homo neanderthalensis* or Neanderthal Man.

## 20.1 The Human Family Tree

Like a child who has matured to the point that he or she can start to recognize family members in photos, humanity had matured enough by the nineteenth century that it could start to recognize its ancient family members in fossils. In 1891, Dutch physician Eugène Dubois found, in present-day Indonesia, a partial skull of a being who had both ape- and human-like qualities. Today, we know this 400,000-year-old individual as a member of the relatively modern species *Homo erectus*. In 1924, Raymond Dart, an Australian physician then working in South Africa, identified the skull of a child who came to be known as Taung baby. Today, we know this child as a member of the relatively primitive species *Australopithecus africanus*, who lived between 2 and 3 million years ago (Mya). By the early 1960s, at least three more human-family species members had been found, but as it turned out, things were just getting started in the 1960s. Discoveries kept coming in the 1970s, and a trove of them were added in the 1990s and early 2000s. Today, there are at least two dozen separate species who are candidates for membership in the taxonomic category known as **Hominini**, which is to say the taxon of human-like primates.

The Neander find that started this chain of discovery came 3 years before Charles Darwin published his path-breaking work, *On the Origin of Species*. This work, and the research on evolution that followed, provided researchers with a framework for making sense of the human-family fossils. Over time, a growing understanding of evolution converged with the fossil evidence to produce a single, startling realization: Some members of Hominini were not just human relatives, but human ancestors—we humans had evolved from them. All the fossil evidence told us that human beings were extreme *latecomers* among the Hominini. At the time Taung baby lived, there were no modern human beings. And the fossil evidence told us that, with movement forward in time from species like the Taung baby's, hominin species became progressively more human-like in form. The obvious implication was that we humans formed a single, late-budding branch on a Hominini family tree.

Throughout these years of discovery, scientists were asking a question that is still with us today: What does this Hominini tree look like? **Figure 20.1** presents one attempt to answer this question. What's pictured is a proposed phylogeny, or tentative family tree, for hominins—species that are members of Hominini. You'll notice that it contains not only a convoluted line that leads to our own species, **Homo sapiens**, but several lines that turned out to be evolutionary dead ends. Note also that our own genus (*Homo*) is only one of several within Hominini. At the center of the figure, for example, you can see members of the genus *Australopithecus*, whose species are, in general, more ape-like than any of the *Homo* species.

If you look at the smaller tree in the figure's lower right, you'll also get an idea of who's *not* included in the Hominini. Note that there is a primate ancestor who gave rise to all the primate lines you can see—monkey, gorilla, chimpanzee—but that the final split at right comes between the chimpanzee and hominin lines. By working backward through a set of primate molecular clocks, scientists have determined that, sometime between 6 and 7 Mya, there was a single primate "common ancestor" that gave rise to the chimpanzee line on the one hand and the hominin line on the other. Figure 20.1's large tree is simply one idea of what the hominin line looks like in detail.

So, who was the common ancestor of the chimps and the hominins? No one knows. We know that

it was not the equivalent of a modern-day chimpanzee; today's chimpanzees did not evolve until about the same time modern humans did. All we can say is that it was an ape-like creature whose closest living relatives include not only ourselves and chimpanzees but a group of chimp-like primates known as bonobos.

## Connecting the Dots in the Tree

Figure 20.1 was drawn up by a respected expert in human evolution, Ian Tattersall, but if we asked six experts of Tattersall's stature to provide hominin family trees, we would almost surely get six different trees. Why the uncertainty? We just don't have enough evidence to *connect* the species with confidence. We don't have enough fossils to be sure about who evolved from whom. This is the case even though our inventory of fossils is now quite large—we have at least one good fossil from every species you see on the tree and scores of fossils from some species. Moreover, we have good dates for most of these species. See the species called *Australopithecus anamensis* toward the center? We're certain that the fossils we have for it date to between 3.9 and 4.2 Mya. Meanwhile, the fossils of the *Ardipithecus ramidus* species you can see just to the lower left of the main tree date from about 4.4 Mya. But did *Ar. ramidus* evolve into *Au. anamensis*? A number of experts think the answer is yes; they think, in other words, that the two *Ardipithecus* species in the tree ought to be moved over to the right, and that a line from them ought to lead straight up to *Au. anamensis*. Tattersall isn't convinced, however, so he's left *Ardipithecus* with no known evolutionary heirs and *Australopithecus* with no known evolutionary predecessors. If we were to walk through the tree, we'd find many such points of disagreement. Even with such uncertainty about linkage, however, we still know a great deal about human evolution, as a little more analysis of the tree will confirm.

## 20.2 Human Evolution in Overview

One of the notable features of the hominin tree is something we've touched on already. If you look right at the top of it, you'll notice that, among all the hominins, only one species made it to the line marked "present." Numerous hominin species arose over the past 6 or 7 million years, but all of them are extinct now except for us. If you could take a time machine back 50,000 years, you would find, in Africa and elsewhere, fully modern human beings, which is to say hominins who physically are indistinguishable

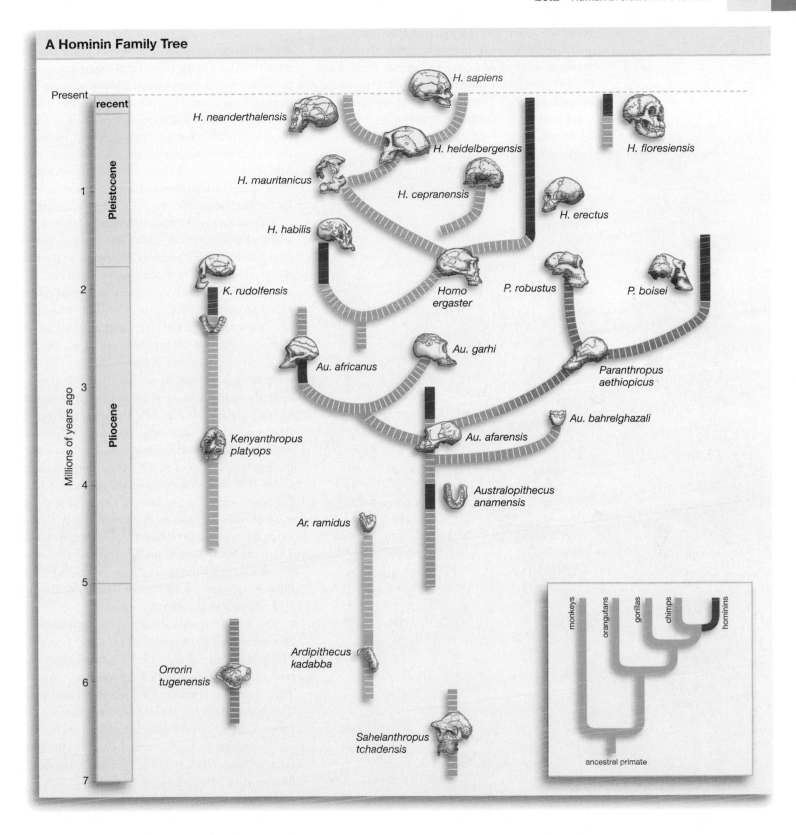

A Hominin Family Tree

from us. But in Europe you would also find stocky Neanderthals, and over in the Far East you would find the dwindling remnants of the *Homo erectus* species. Jump forward another 10,000 years, and all the *H. erectus* are gone; jump forward another 12,000, and all the Neanderthals are gone. Today, only we *Homo sapiens* remain among the hominins.

Another way to think of this tree is in terms of relative time frames. An "ancient" civilization, such as that of Egypt, stretches back perhaps 5,000 years.

**Figure 20.1
A Possible Hominin Family Tree**

Yet the Neanderthals disappeared from the face of the Earth 23,000 years before Egyptian civilization began, and they had been in existence for at least 130,000 years by the time they died out. Even so, the Neanderthals were latecomers among the hominins. Recall that the hominin line stretches back as far as 7 million years. Thus, human evolution took a great deal of time, even when dated just from the last primate ancestor. On the other hand, when we consider the entire sweep of evolution, all hominins are latecomers. If we assume that life on Earth began 3.8 billion years ago and think of all evolution as taking place in one 24-hour day, then the most ancient of the hominins evolved in the last minutes of that day. And modern human beings? We arose about 200,000 years ago, which means we have only been around for the last 5 seconds of Earth's evolutionary day.

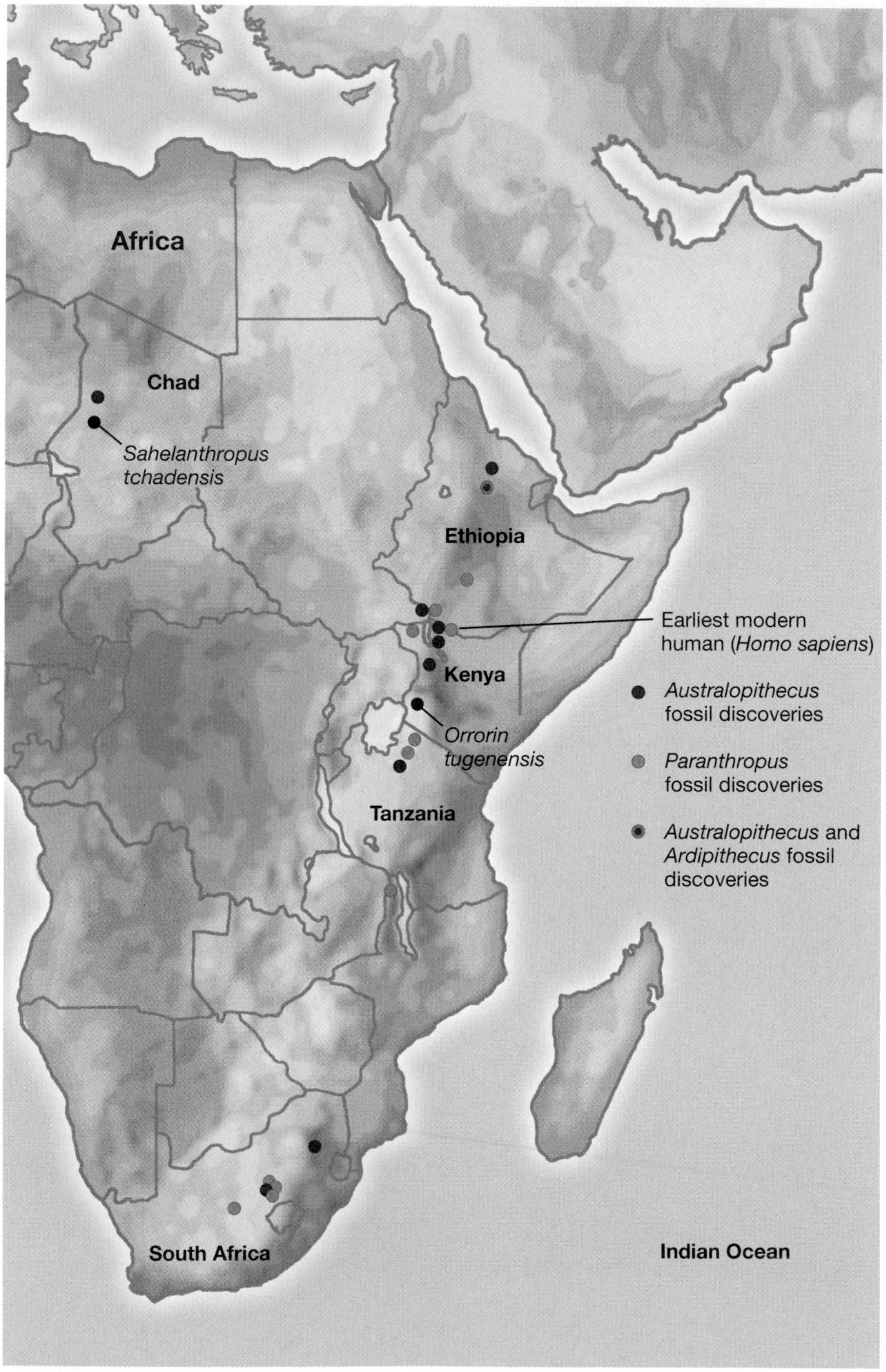

Yet another way to think of human evolution is in terms of *where* it took place. The message here is straightforward: For perhaps 70 percent of the time we hominins have been around, we lived and died solely in Africa. This is clear because the earliest hominin fossils found *outside* of Africa date to only 1.8 Mya. Meanwhile, all the early- and mid-period hominin fossils have been unearthed on the African continent, as you can see in **Figure 20.2**.

As little as 10 years ago, a finer point could have been put on this: Human evolution largely took place not just in Africa, but in *East* Africa. Until the mid-1990s, nearly all early- and mid-period hominin fossils were found in a swath of Africa that runs from the eastern portion of South Africa up through the Great Rift Valley that traverses modern Kenya, Tanzania, and Ethiopia. But the range of likely hominin fossils has now been expanded some 2,500 kilometers west of the Great Rift Valley, to Chad. There, in 2001, Michel Brunet and his colleagues found a very primitive hominin, a creature they dubbed *Sahelanthropus tchadensis* but whom they informally refer to as Toumaï.

The Toumaï fossils actually were remarkable in another way. They capped a period of 2 years in which the root of the hominin family tree was pushed back by more than 2 million years. Prior to 2000, the oldest hominin fossil was dated at 4.4 million years old. Then, in rapid succession, fossil discoveries for all three of the earliest species you can see in Figure 20.1 were announced by different teams of researchers. With the Toumaï find, hominin fossils were pushed back to between 6 and 7 Mya. As you may recall, the hominin and chimpanzee lines only split at about 7 Mya at most. One possibility, then, is that Brunet's Toumaï is a creature that lies right at the base of the hominin family tree. Another possibility, however, is that Toumaï is not a hominin but belongs in another line of primates, specifically the apes. How do scientists determine placements like these? Let's see.

**Figure 20.2**
**Hominin Fossil Sites in Africa**

All of the oldest hominin fossils have been found on the African continent. In addition, the oldest known remains of fully modern human beings (*Homo sapiens*) have been found in an area known as the Omo Kibish site, in southern Ethiopia, and dated to 195,000 years ago.

## 20.3 Interpreting the Fossil Evidence

If you look at **Figure 20.3a**, you can see part of the *S. tchadensis*/Toumaï fossils that Brunet's team unearthed in the Chadian desert (**Figure 20.4**). This remarkably complete cranium of Toumaï was complemented by additional artifacts the researchers found: two lower jaw segments and some individual teeth. Fragmentary as this evidence may seem, it actually is more extensive than most hominin fossil finds. We tend to think that the term *fossils* means complete skeletons, but this is rarely the case. The claim for the primitive *Orrorin tugenensis* species seen in Figure 20.1 originally was based mostly on three thigh bones and several teeth.

The researchers who look at this evidence have a professional name that is a mouthful. Any scientist who studies evolution by analyzing fossils is a paleontologist. Meanwhile, a scientist who studies human culture or development is an anthropologist. Putting these two terms together, a scientist who studies fossils in the human lineage is a **paleoanthropologist.** These researchers are, to be sure, not the only variety of scientist who investigates human evolution. As noted, the dates we have for the hominin-ape split do not come from fossil evidence, but from *molecular* evidence—from analyses of primate DNA sequences. The importance of molecular evidence has increased greatly in recent years (see "Sequencing the Neanderthal Genome" on p. 386). Nevertheless, the central evidence in human origins research has always been fossils, and this is as true today as it was when the first hominin fossils were being dug up in Germany's Neander Valley in the nineteenth century.

So, what are paleoanthropologists looking for in such fossils? What features help them decide, for example, whether a given species should be placed in the hominin family tree or be put over with the apes? When Brunet first looked at Toumaï, three features stood as evidence that it was a hominin. One was its canine teeth; they are smaller and less sharp than those of the apes. Another was the enamel on its teeth—it is thicker than the enamel apes have. A third was what might be called the slope of the face; the lower face doesn't jut out beyond the eyes in the way ape faces do. Against this, Toumaï clearly had some

**(a)** Cranium of Toumaï.

**(b)** Computer-generated model of Toumaï.

**Figure 20.3**
**An Ancient Human Ancestor?**

**(a)** This cranium of *Sahelanthropus tchadensis*, or Toumaï, was discovered in the deserts of Chad and has been determined to be 6 to 7 million years old. Toumaï's discoverers believe it is a very ancient hominin, but other researchers believe it should be grouped with apes.

**(b)** A procedure called computed tomography yielded this clay reconstruction of the appearance of Toumaï.

**Figure 20.4**
**The Hard Work of Finding Fossils**

Hominin fossils can be located in some inhospitable places. Here Michel Brunet (in the foreground) looks for hominin fossils with his crew in Chad's Djurab Desert in 2006. It was Brunet's team that discovered the Toumaï fossils in Chad in 2001.

ape-like features. Note its widely spaced eyes and massive brow, for example. Moreover, Brunet's fellow paleoanthropologist, Brigitte Senut, interpreted Toumaï's canine teeth not as evidence of a male hominin but as evidence of a female ape.

These various arguments were set forth at the time Brunet unveiled Toumaï, in 2002. In 2005, however, two experts reported on a painstaking computer analysis they did of Toumaï's skull that yielded an additional, important piece of information: Toumaï appears to have walked upright, on two legs. This is critical because, along with tooth structure, upright walking, or **bipedalism**, is the single most important defining feature of a hominin. Our closest living relatives, the chimpanzees, are "knuckle walkers" who only occasionally rise up to walk on two legs. But all the hominins you see in Figure 20.1 are either known to have walked on two legs or arguably did so. The evidence that Toumaï did is one of the factors paleoanthropologists would take into account in deciding whether to place it among the hominins or the apes. The computer analysis of Toumaï's skull that yielded the evidence of its upright walking also yielded another benefit, which is an idea of what this creature may have looked like in the flesh (**Figure 20.3b**).

## 20.4 Snapshots from the Past: Three Hominins

It shouldn't surprise us that a controversy exists over whether Toumaï belongs in the hominin or the ape family tree. After all, Toumaï lived right about the time that the hominin and ape lines were parting company. If we looked at each of the species noted in Figure 20.1, moving from most ancient to most recent, what we would see is a transition that reflects this split—a transition from creatures that are very much like apes to those that look very much like ourselves. We don't need to examine all the species noted in the figure to get a sense of this transition, however. Here are three snapshots from the past that will provide some idea of the nature of our hominin relatives over time. An artist's interpretation of the appearance of three hominin species can be seen in **Figure 20.5**.

*Australopithecus afarensis:* **Lucy and Her Kin.** If we jump forward about 3 million years from Toumaï, we find a hominin whose features illustrate a partial transition from ape-like to human-like. The species, *Australopithecus afarensis*, is best known through its most famous individual, dubbed Lucy, whose remains were found in Ethiopia in the 1970s. We have a very

**(a)** *Paranthropus boisei*

**(b)** *Homo ergaster*

**(c)** *Homo neanderthalensis*

**Figure 20.5**
**Three Hominins**

What did now-extinct members of the hominin group look like? Working from fossil and other evidence, artist Jay Matternes has produced drawings that provide some idea of their appearance.

**(a)** *Paranthropus boisei* lived about 2 million years ago. Although it was not a member of our genus (*Homo*) and not an ancestor of modern humans, it was a hominin.

**(b)** *Homo ergaster* is the species represented by Turkana Boy. Its fossils date from a time only slightly later than *P. boisei*, but its physical form clearly is more human-like—not surprising since it probably was a human ancestor.

**(c)** *Homo neanderthalensis* lived in Europe and elsewhere in a period running from about 130,000 years ago to 28,000 years ago. These were the Neanderthals, a species that lived in proximity to modern humans for thousands of years before becoming extinct.

good idea of when Lucy lived; her fossils have been dated to 3.18 Mya.

There is no doubt that, like us, Lucy was bipedal; the structure of her pelvis makes this a certainty. Nevertheless, she possessed long arms, short legs, and feet that were built for grasping—features consistent with a species that no longer lived exclusively in the trees but that probably still spent considerable time in them. Meanwhile, our own species is thought not to have been "arboreal" or tree-dwelling at all. Lucy also differed from us in the size of her brain; she had a cranial volume of about 450 cubic centimeters—

about that of a chimpanzee—while ours is about 1,400 cubic centimeters.

When we think about these various features together, we can see that hominin features evolved in what is called a *mosaic pattern*—different features developed at different points in time in different species. Lucy didn't have upright walking, a big brain, and modern limb lengths; she only had upright walking. To get some idea of Lucy's anatomy compared with ours, see **Figure 20.6**.

One of the notable things about Lucy's species is that there is good agreement about its location in

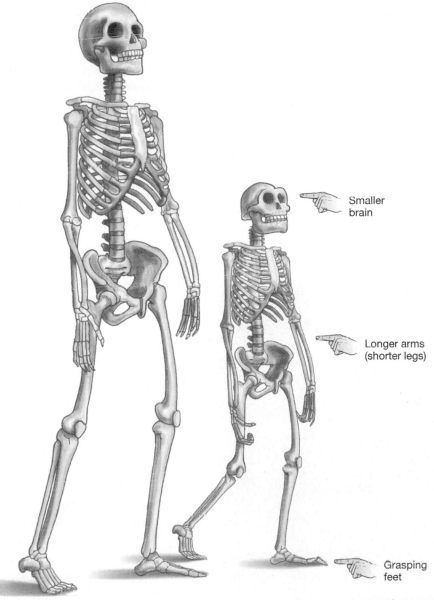

Smaller brain

Longer arms (shorter legs)

Grasping feet

**Figure 20.6**
**"Lucy" and Modern Humans Compared**
What was the hominin Lucy like compared to a modern human female? The figure gives an idea. Lucy stood about three-foot-seven and had a much smaller brain than modern humans, even allowing for her smaller stature. Lucy's hip and pelvic bones make clear that she was bipedal, but note the longer arms and grasping feet common to tree-dwelling primates. (Adapted from Boyd, R. and Silk, J.B., *How Humans Evolved*. New York: W.W. Norton & Company, 2000, p. 329.)

the hominin family tree. Most authorities would give *A. afarensis* the central place you can see in Figure 20.1—an important place because, as you can see, *A. afarensis* is thought to have given rise to our own genus (*Homo*). In Lucy, then, we probably are seeing not just a relative, but an ancestor.

*Homo ergaster:* **More Like Us.** Changes to physical forms that are more like ours come at several steps in the hominin line, but the most dramatic change comes with the rise of *Homo ergaster*, whose best-preserved remains come from a boy who died 1.6 Mya on the shores of Lake Turkana in Kenya. Experts estimate that, had this 9-year-old "Turkana Boy" grown to maturity, he would have reached a height of 6 feet. His brain was more than half the size of the average modern *Homo sapiens* brain. He had a much more modern face, long limbs typical of those humans who dwell today in arid climates in Africa, and advanced tool technology. If you look at **Figure 20.7**, you can see Turkana Boy's skeleton, which was found in amazingly complete form.

*Homo ergaster* is regarded by many researchers as being ancestral not only to our own species, but to another species—*Homo erectus*—that was perhaps the first to migrate out of Africa, beginning at least 1.8 Mya. *Homo erectus* ended up very far from Africa indeed as its remains have been found in both China and in Indonesia, where the last remaining *H. erectus* individuals died out about 40,000 years ago. Keep *H. erectus* and its extinction date in mind as this will be important for what's coming up.

*Homo neanderthalensis:* **Modern and Far-flung.** In the Neanderthals, we see a modern species of hominin—one that nobody ever confused with an ape. Various dates are assigned to their first appearance; we have signs of them as much as 350,000 years ago, but they didn't evolve into their most modern form until about 130,000 years ago. Either way, the Neanderthals appeared recently indeed compared to the 1.6 Mya when *H. ergaster* flourished, to say nothing of the 3.18 Mya when Lucy was alive. The Neanderthals evolved in Europe, but they migrated great distances; their fossils have been recovered as far away as Uzbekistan and the Arabian Peninsula.

So, what were these hominins like? Figure 20.5c shows you an artist's interpretation of their physical stature: short, stout bodies—the average man was about five foot six—powerfully built, with a heavy "double-arched" brow and a receding chin; big-boned, with large joints to match. It's easy to see how such features were turned into a *caricature* that has been with us since the nineteenth century: that of the caveman. It is the Neanderthals who provided us with the image of dumb, prehuman brutes, grunting their way through the Stone Age. But does this image fit with the reality? Well, consider that the Neanderthals had a cranial capacity slightly *larger* than ours; that they took the trouble to bury their dead; and that they appear to have taken care of those within their groups who were old or sick.

This is not to say, however, that Neanderthals were "just like us." Whenever direct comparisons are made, the Neanderthals do indeed look primitive in comparison with ancient *Homo sapiens*. Neanderthals thrust spears at prey, but humans developed the valuable technique of *throwing* spears from a distance. The Neanderthals used nothing but stone for their relatively primitive tools, while *H. sapiens* used bone and antler as well as stone in constructing their much finer implements. And the Neanderthals were "foragers," while ancient humans were "collectors." What's the difference? Humans monitored their environments and used "forward planning" by, for example, placing their campsites near animal migration paths. By contrast, Neanderthals do not appear to have timed their migrations in this way. Such differences may have been important to the fate of the Neanderthals, as we'll now see.

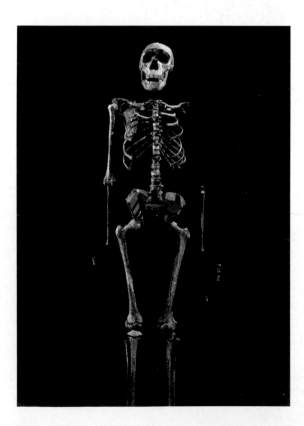

**Figure 20.7**
**Much More Like Us**

In the remarkably complete skeleton of Turkana Boy we can see the evolution of hominins to a form more like our own. This member of the species *Homo ergaster* was tall and had a much larger brain capacity than the earlier hominin Lucy. His remains, dating from 1.6 million years ago, were found near Kenya's Lake Turkana.

**SO FAR...**

1. The two most important features in distinguishing hominins from apes are _____ walking and _____ structure.

2. The "Lucy" hominin was like modern human beings in that she was _____, but unlike them with respect to _____ size.

3. Lucy shows us that hominin features evolved in a _____ pattern, which is to say that different features evolved at different points in time and in different species.

## 20.5 The Appearance of Modern Human Beings

The Neanderthals had Europe to themselves for tens of thousands of years, but beginning about 40,000 years ago, they had company, in the form of ourselves—modern *Homo sapiens*. Where had we come from? The short answer is Africa. Beginning sometime between 55,000 and 85,000 years ago, we human beings left our ancestral African home and began a migration that eventually took us all over the globe. Our travels initially took us not into Europe, however, but north and east, into Asia, Indonesia, and then into Australia, through one of the routes you can see in **Figure 20.8**. The oldest human fossils found outside of Africa and its immediate environs are 46,000 years old and come from the Lake Mungo region in Australia. Of course, to reach Australia by this date, we would had to have been in Asia much earlier.

Critically, our global migration seems to have begun after we humans evolved into our present anatomical form. Put another way, *H. sapiens* fully evolved *in Africa* before migrating out to the rest of the world. This "out-of-Africa" hypothesis stands in contrast to another long-standing school of thought, called multiregionalism, which holds that several late-stage hominins left Africa at different times and then interbred in Europe and Asia. Over time, the hypothesis goes, these species evolved into both modern *H. sapiens* and the late-stage Neanderthals. So, why does this idea now have so little support? Scientists have actually been able to analyze the DNA of several Neanderthals—in their first try, they recovered DNA from the original Neander Valley fossils and the results of this work indicate that human beings and Neanderthals did not interbreed. The

genetic makeup of the two species is too different for this to have occurred. We humans shared a common ancestor with the Neanderthals (probably the *H. heidelbergensis* you can see in Figure 20.1), but this was at least 500,000 years ago. Thus, the Neanderthals are not our ancestors but instead are more like cousins who happened to live in proximity to us for an extended period of time.

So, when did modern human beings first appear in Africa? Molecular evidence indicates that we evolved no earlier than 200,000 years ago, and we now have fossil evidence that dovetails nicely with this date. The oldest known *H. sapiens* fossils, found in present-day Ethiopia, have been dated to 195,000 years ago. (You can see the location of these remains in Figure 20.2 on p. 380.)

### Who Lives, Who Doesn't?

After human beings left Africa, it's possible that they encountered not only Neanderthals, but individuals from the species *Homo erectus*, which you may remember had arrived in the Far East by at least 1.8 Mya. Now, let's think about extinction and survival among these three species. Recall that about 40,000 years ago, *H. sapiens* arrived in Europe, where *H. neanderthalensis* was living, but that by 28,000 years ago, all the Neanderthals were gone. Then recall that we humans had arrived in Australia by 46,000 years ago, but that by 40,000 years

••• ANSWERS

1. bipedal; tooth
2. bipedal; brain
3. mosaic

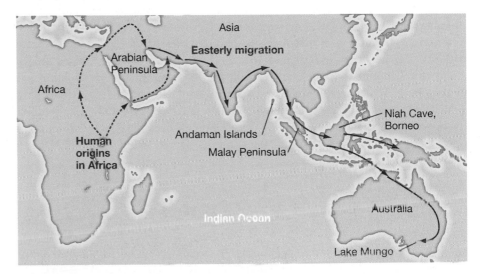

**Figure 20.8**
**Out of Africa, Headed East**

The first long-distance migration of human beings out of Africa went in an easterly direction. Examination of the mitochondrial DNA of indigenous, contemporary people living in the Andaman Islands and the Malay Peninsula (see map) indicates that early humans reached these areas by about 65,000 years ago. The Lake Mungo site (in Australia) and the Niah Cave site (in Borneo) are locations for which we have human fossils dating from earlier than 45,000 years ago. The two dotted lines leading from Africa into the Arabian Peninsula represent alternative routes that our ancestors may have taken in their migration to Asia. Adapted with permission from Peter Forster and Shuichi Matsumura, "Did Early Humans Go North or South?" *Science* 13 May 2005 308; 965–966.

# Sequencing the Neanderthal Genome

Sequence the genome of a living species? It's done all the time. Since 2003, when the Human Genome Project was completed, scientists have managed to get complete DNA sequences for chimpanzees, sheep, fruit flies, a mustard plant, several fungi, and many bacteria—the genomes of hundreds of species are now in hand. But how about sequencing the genome of an *extinct* species? This is much more difficult because of what might be called the DNA access problem. DNA degrades over time, and of course, many extinct species have been gone a *long* time. Moreover, any DNA that remains in ancient bones is likely to be not only fragmented but also contaminated: it will be mixed with the DNA of both bacteria and any scientists who have handled the bones it came from.

Given such obstacles, the scientific world cheered in 2006 when a team of German scientists announced that they intended to sequence the genome of a long-extinct relative of ours, *Homo neanderthalensis* or **Neanderthal** Man. Led by geneticist Svante Pääbo of the Max Planck Institute for Evolutionary Anthropology, the team decided to begin its work by extracting DNA from the bones of a Neanderthal man who lived 45,000 years ago in what is now Croatia.

Pääbo and other scientists actually have sequenced Neanderthal DNA several times before. But in each case, the starting genetic material was so-called mitochondrial DNA, meaning DNA that comes from the tiny organelles within cells (the mitochondria) that carry their own small complement of DNA. In the new project, Pääbo will be sequencing the Neanderthal's "genomic DNA," meaning its primary complement of DNA, derived from all its chromosomes.

Proof that useful genomic DNA can be extracted from an organism the age of a Neanderthal came in 2005 when Max Planck scientists collaborated with researchers from the U.S. Department of Energy's Joint Genome Institute in sequencing the DNA of two 40,000-year-old cave bears. As expected, the DNA retrieved from the bones and teeth of these bears was overwhelmingly bacterial— only 6 percent of it came from bears. But scientists can recognize bacterial genetic sequences when they see them, and high-powered computers can be programmed to discard unwanted sequences while retaining those that are useful.

It's one thing to identify a species' DNA, however, and another to assemble fragments of this DNA in the right *order*. Since the Neanderthal DNA will exist in fragments that are about 100 base pairs long, the challenge is akin to having a puzzle that is broken up into 30 million pieces and then asking yourself how they fit together. To assemble the fragments, Pääbo's team is collaborating with a Connecticut company that has developed a new means of sequencing DNA. The so-called 454 machines this company has invented process DNA in segments that are just the right size for the Neanderthal project—about 100 base pairs per segment.

So, what do we stand to learn from sequencing the Neanderthal genome? Well, how many Neanderthals were there over time? Did their numbers shrink or grow dramatically at any one period? What color was their skin? Did they interbreed with human beings? (Thus far, the mitochondrial DNA evidence says no, but the genomic evidence could settle things.) And how about this: Were the Neanderthals capable of producing spoken language in the way we do? There is a gene, called *FOXP2* that, in today's human beings, controls the ability to manipulate the tongue and lips in speech. (How do we know? Researchers in England found a family that has members who cannot produce intelligible speech. All of these family members have a mutation in their *FOXP2* gene.) It turns out that chimpanzees have a *FOXP2* gene, too, but that it differs slightly from the human version. The human form underwent mutations that altered it, and these mutations probably became universal in the human population only within the last 120,000 years. So, where did the Neanderthals stand with respect to *FOXP2*? If they had the chimpanzee form of the gene, then it's highly unlikely that they could have spoken in the way modern humans beings do. Information on such questions stands to be forthcoming if Pääbo and his team can manage to make a picture emerge from the puzzle pieces of Neanderthal DNA.

ago the last *H. erectus* were gone, having died out in Indonesia.

You can probably see the implication in this chain of events. If there is a consensus account of why we humans are the last surviving hominins, it is that we "replaced" our fellow hominins throughout the world once we encountered them. This is not to say that we simply killed them outright. We have no evidence of violent encounters between ourselves and the Neanderthals or *H. erectus*. But similar species are bound to compete for the same kinds of resources, such as food and shelter. And we have every reason to believe that modern humans could *out*-compete both the Neanderthals and *H. erectus*, thanks to our superior brain power. (Remember the comparisons noted earlier between Neanderthals and modern humans?) So, one possibility is that the hominin family is now a family of one—ourselves—in part because our success brought about the extinction of our near relatives.

## 20.6  Next-to-Last Standing? The Hobbit People

In 2004, a fascinating twist was added to this story. Researchers working on the Indonesian island of Flores reported finding fossils of a hominin species whose members stood just 3 feet tall and who lived on Flores as recently as 18,000 years ago (**Figure 20.9**). This discovery was stunning for three reasons. First, anthropologists have long believed that the last hominins other than ourselves died out not 18,000 years ago, but 28,000 years ago, with the passing of the Neanderthals. Second, if modern humans really were spanning the globe replacing other hominin species 50,000 years ago, how did the Flores people, dubbed *Homo floresiensis*, manage to avoid this fate for some 32,000 years? Third, the Flores people used fire and had sophisticated tools. Yet, they had a cranial capacity of only 380 cubic centimeters—less than that of a chimpanzee! How could they do so much in the way of culture with so little in the way of brain size?

The answers to these questions may be a while in coming. One possibility is that much of the hoopla about *H. floresiensis* has been undeserved. Two teams of researchers who have analyzed the Flores fossils believe *H. floresiensis* represents not a new hominin species, but a modern human who had a medical condition—microcephaly, which results in a very small head and brain. The researchers who made the Flores discovery have rejected this assertion, however, and their position has been supported by other research teams. Indeed, one examination of the Flores fossils suggested that *H. floresiensis* had a brain structure that would allow it to do more with less—that would allow it to build sophisticated tools, for example, even with a very small brain.

Assuming that these Flores "Hobbit people" really are a new species, how might they fit into the greater hominin family? No one is sure, but some recent evidence suggests that they were a later-evolving offshoot of *Homo erectus* (which you'll recall lived in Indonesia). The researchers who made the Flores find believe, however, that *H. floresiensis* didn't evolve into its small stature on the island, but instead was small upon arrival there. More research will hopefully sort out the truth about the Flores people. For now, we're left with a fascinating question: Did a hominin species other than our own—a tiny one at that—survive until just 18,000 years ago?

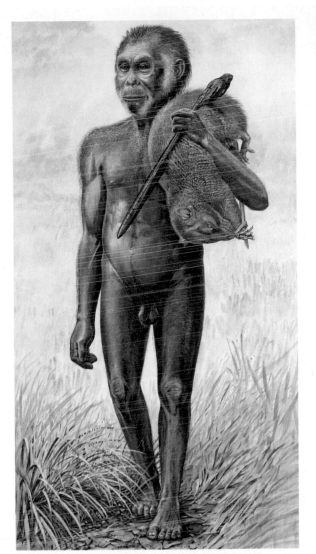

**Figure 20.9**
**Small Hominin, Possibly Big Discovery**
An artist's interpretation of the appearance of the 3-foot tall hominin dubbed *Homo floresiensis*, whose skeletal remains were found on the Indonesian island of Flores.

••• ANSWERS
1. Africa
2. No; the Neanderthals lived in Europe long before human beings migrated there, and the species *Homo erectus* reached Southeast Asia by at least 1.8 million years ago.
3. 18,000; Indonesia

## On to the Diversity of Life

In reviewing the history of life in the last two chapters, you necessarily looked, in some small way, at almost all of Earth's major life-forms—at primates in this chapter and at other animals, bacteria, plants, and so forth in Chapter 19. In the three chapters coming up, you'll take a more detailed look at each of these life-forms. This is a tour that will make plain the astounding diversity that exists in the living world.

### SO FAR...

1. Human beings appear to have evolved into their present anatomical form in_____.

2. Were human beings the first hominin species to reach Europe and Asia?

3. The "Hobbit" people, *Homo floresiensis*, are thought to have survived until as recently as _____ years ago on an island that is part of present-day_____.

# CHAPTER 20 REVIEW

## SUMMARY

### 20.1 The Human Family Tree

- Human evolution is the study of the taxonomic grouping called the Hominini or human-like primates. Every member of this group is referred to as a hominin, including human beings.

- A common primate ancestor is believed to have given rise to both the chimpanzee and the human family evolutionary lines between 6 and 7 Mya. The structure of the hominin tree is a matter of considerable debate among researchers. We do not have enough fossil evidence to say with certainty which species were ancestral to which others.

### 20.2 Human Evolution in Overview

- All the members of the hominin grouping are extinct except for *Homo sapiens*, the human species. Human beings were late in arriving among the hominins, but all hominins are late arrivers when the entire sweep of evolution is considered. All early and mid-period hominin evolution took place in Africa.

- Until the past 10 years, evidence indicated that most early- and mid-period hominin evolution took place in east Africa. But in 2002, the range of likely hominin fossils was expanded 2,500 kilometers to the west, to Chad, with the discovery of the 6- to 7-Mya remains of a primitive hominin named *Sahelanthropus tchadensis* or Toumaï.

### 20.3 Interpreting the Fossil Evidence

- In human evolution studies, molecular evidence—usually the sequencing of DNA—has increased in importance in recent years. But the primary evidence in the field remains fossil evidence. The researchers who find and analyze hominin fossils are known as paleoanthropologists.

- These scientists interpret features of fossils in order to make judgments about where a given fossil form lies in the hominin family tree. Interpretations can differ, however. The two most important defining characteristics of a hominin are tooth structure and upright or "bipedal" walking.

### 20.4 Snapshots from the Past: Three Hominins

- *Australopithecus afarensis*—The bipedalism seen in our own species clearly existed in the hominin species *Australopithecus afarensis*, whose most famous individual, Lucy, lived 3.18 Mya in what is now Ethiopia. Lucy had a much smaller brain than modern humans do, however, and she probably was partly arboreal,

or tree-dwelling. Lucy's set of features demonstrate that hominin features developed in a mosaic pattern—different features evolved at different points in time and in different species. Lucy's species is generally regarded as being ancestral to the *Homo* genus that human beings are a part of. Thus, Lucy probably is a human ancestor.

- *Homo ergaster*—A change to a physical form and mental capacity much closer to ours comes with the evolution of *Homo ergaster*, exemplified by Turkana Boy, who lived 1.6 Mya.

- *Homo neanderthalensis*—In their modern form, the Neanderthals (*Homo neanderthalensis*) populated Europe as well as parts of Asia for about 130,000 years. The last of them died 28,000 years ago in Europe. The modern caricature of the caveman comes from the Neanderthals, but this image is difficult to reconcile with facets of Neanderthal life such as the burying of the dead. Even given these facets, however, it is clear that the Neanderthals were a primitive species in comparison with *H. sapiens*.

### 20.5 The Appearance of Modern Human Beings

- Human beings appear to have evolved into their modern anatomical form in Africa prior to the time they began to migrate into the wider world. This "out-of-Africa" hypothesis stands in contrast to the less-accepted "multiregional" hypothesis, which holds that several species of hominins migrated from Africa at different times, interbred, and evolved into *H. neanderthalensis* and *H. sapiens*.

- The initial wave of human migration out of Africa was initiated sometime between 85,000 and 55,000 years ago. The earliest fossils we have of modern human beings outside Africa and its immediate environs date from 46,000 years ago and were found in Australia.

- Molecular evidence indicates that modern human beings evolved no earlier than 200,000 years ago. The earliest human fossils we have date from 195,000 years ago and were found in Ethiopia.

- The arrival of modern human beings in Europe 40,000 years ago was followed by the extinction of the Neanderthals 12,000 years later, while the arrival of human beings in the Far East, at least 46,000 years ago, was followed by the extinction of the *Homo erectus* species 6,000 years later. If there is a consensus about why we human beings are the only living species of hominin, it is that we "replaced" such species as *H. erectus* and *H. neanderthalensis* by out-competing them after having migrated to Asia and Europe.

## 20.6 Next-to-Last Standing? The Hobbit People

- In 2004, researchers reported finding, on the Indonesian island of Flores, fossils of a previously unknown hominin, *Homo floresiensis*, who stood only 3 feet tall and seemed to have survived until 18,000 years ago. This interpretation of the Flores fossils has been challenged, however, by two teams of researchers who have concluded that *H. floresiensis* actually was a small, modern human being suffering from the medical condition of microcephaly. If *H. floresiensis* is a new species of hominin, one possibility is that it evolved from *H. erectus*.

## KEY TERMS

| | |
|---|---|
| **bipedalism**  382 | **Neanderthal**  386 |
| **Hominini**  378 | **paleoanthropologist**  381 |
| ***Homo sapiens*** **(*H. sapiens*)**  378 | |

## BRIEF REVIEW

*Answers to Brief Review questions are in the back of the book. For multiple-choice quiz questions, go to* **www.aw.com/krogh4**.

1. Did human beings evolve from chimpanzees?

2. All phylogenies or family trees are tentative, but why is the hominin family tree particularly tentative?

3. East Africa has sometimes been called the cradle of humanity. Why has it gotten this title?

4. Were human beings the first hominins to leave Africa?

5. Should the Neanderthals be thought of as human ancestors or human cousins?

## APPLYING YOUR KNOWLEDGE

1. In his 1903 essay "Was the World Made for Man?" Mark Twain wrote, "If the Eiffel tower were now representing the world's age, the skin of paint on the pinnacle-knob at its summit would represent man's share of that age; and anybody would perceive that that skin was what the tower was built for. I reckon they would. I dunno." After having learned something about the history of life on Earth, do Twain's words seem insightful, more than a century after he wrote them?

2. Suppose some isolated populations of Neanderthals had survived into modern times. How do you think society would treat them? As a species more akin to chimpanzees, which is to say subject to being kept in zoos? Or as regular, if somewhat different, human beings?

3. Is connecting the dots in the human family tree important? Is it important that we know who our ancestors were and what they were like?

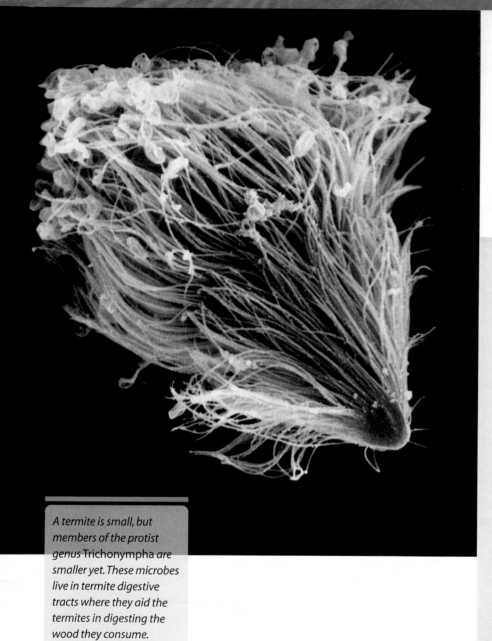

A termite is small, but members of the protist genus Trichonympha *are smaller yet. These microbes live in termite digestive tracts where they aid the termites in digesting the wood they consume.*

CHAPTER **21**

| 21.1 | Life's Categories and the Importance of Microbes | 392 |
| 21.2 | Viruses: Making a Living by Hijacking Cells | 394 |
| 21.3 | Bacteria: Masters of Every Environment | 397 |
| 21.4 | Intimate Strangers: Humans and Bacteria | 400 |
| 21.5 | Bacteria and Human Disease | 401 |
| 21.6 | Archaea: From Marginal Player to Center Stage | 403 |
| 21.7 | Protists: Pioneers in Diversifying Life | 405 |
| 21.8 | Protists and Sexual Reproduction | 406 |
| 21.9 | Photosynthesizing Protists | 407 |
| 21.10 | Heterotrophic Protists | 408 |

**Essays**

| Unwanted Guest: The Persistence of Herpes | 398 |
| Modes of Nutrition: How Organisms Get What They Need to Survive | 400 |
| **THE PROCESS OF SCIENCE:** The Discovery of Penicillin | 410 |

Across its categories, life is stunningly diverse. Some of the most interesting living things are microscopic.

# VIRUSES, BACTERIA, ARCHAEA, AND PROTISTS: The Diversity of Life 1

The smallest living thing discovered to date is a type of bacteria that measures about two-tenths of a micrometer in diameter—that is, two-tenths of one-millionth of a meter in diameter. When we turn to the *biggest* living thing, elephants or whales might come to mind, but it turns out that animals aren't even in the running for this title. At a maximum length of 27 meters (or about 89 feet), the blue whale is the largest animal that has ever existed, but it's small compared to California's coastal redwood trees, which might reach a height of 100 meters or about 330 feet. This is large indeed, but consider that a giant sea kelp once was found that was 274 meters in length, meaning it was about 900 feet long. It may be, though, that in terms of size, both kelp and redwoods have to take a backseat to a life-form normally thought of as quite small. In 1992, researchers reported finding a fungus growing in Washington State that runs underground through an area of 1,500 *acres* and that arguably is a single organism.

Now, how about the *deepest*-living things? The current record holders are bacteria that were found in the early 1990s—in a combination of oil drilling and scientific exploration—living almost 3 kilometers beneath the Earth's surface, which is about 1.9 miles down. Food and water are so scarce at such depths that these bacterial cells may live in a kind of suspended animation, dividing perhaps once a year or even once a century. Turning to the highest living things, 12 species of bacteria have been found living in the Himalayas at 8,300 meters or 27,000 feet above sea level, which is just a couple thousand feet lower than the peak of Mount Everest. A species of chickweed survives in the Himalayas at over 20,000 feet.

All kinds of life-forms—clams, tube worms, crabs—live kilometers beneath the surface of the oceans, near the "hot-water vents" that spew out water and minerals from the Earth's interior onto the ocean floor. Microbes from a group called archaea, living at these deep-sea vents, set the pace for life in a hot environment. An average, midday temperature in the Sahara Desert would be so *cold* it would end the reproduction of the archaeon *Pyrolobus fumarii*, which lives within hot-water vents at temperatures of up to 113°C, which is hotter than the temperature at which water boils. At the other extreme, the well-named bacterium *Phormidium frigidum* carries out photosynthesis in Antarctic lakes with temperatures that sit right at the freezing point of water.

Finally, what about the oldest living things? Humans have been known to live to 120, but this is an eyeblink compared to the lives of some plants. In California, there is a sequoia tree that was growing 3,500 years ago, when the pharaohs ruled Egypt, and there is a bristlecone pine tree (*Pinus longaeva*) that has been living in the dry slopes of eastern California for 4,900 years. These trees are mere youngsters, however, compared to some microbes, which can survive in a form called spores for immensely long periods of time. In the 1990s, bacterial spores were found living within a bee that had been encased in amber for 25 to 40 million years. Even this eye-popping number did not prepare scientists for what came next, however. In 2000, researchers announced that they had found, encased within salt crystals at the bottom of a New Mexico air shaft, archaeal spores that date from 250 million years ago. These spores were revived, and billions of their offspring now exist. This claim is controversial—one possibility is that the crystal sample the researchers were looking at was contaminated with modern DNA. But if this finding is borne out, it raises the question of whether some microbes are "effectively immortal," as one researcher has put it.

Looking at life like this—at its biggest or deepest or oldest extremes—is one way to get at its incredible diversity. But, it does not begin to do justice to life's variety because life is diverse in so many ways. Honeybees do a "dance" to let their hive-mates know the location of a food supply. Corn plants, when attacked by army worms, can call in an air force: The plants release an airborne substance that attracts parasitic wasps, which then prey on the worms. Pacific salmon live for years in the ocean, only to make one arduous, upstream journey to spawn in the waters where they were born, after which they die. Look closely at almost any creature, and you are likely to find something only slightly less dramatic (**Figure 21.1**).

## 21.1 Life's Categories and the Importance of Microbes

The purpose of this chapter and the two that follow is to introduce you to life as it exists across all its large-scale categories. As you may recall from Chapter 19, all living things can be placed into one of three *domains* of life. Two of these domains are Archaea and Bacteria, whose members are all single-celled and microscopic. The third domain, Eukarya, encompasses such incredible diversity that it is further divided into four separate *kingdoms*. These are

**Figure 21.1**
**An Amazing Diversity in the Living World**

**(a)** Mouse beneath the foot of an elephant

**(b)** Aquatic snails moving across kelp

**(c)** Fungus growing on a fallen log

**(d)** Water flea swimming near green algae

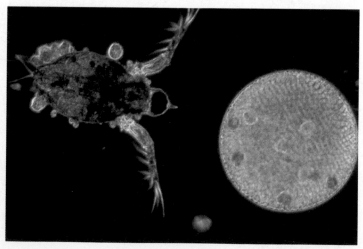

the kingdoms of plants, animals, fungi, and a group of mostly microscopic organisms called protists. We will begin our tour of the living world by starting small, looking in this chapter at Domains Bacteria and Archaea and at the protists within Domain Eukarya. Then, in Chapter 22, fungi and plants will be covered; in Chapter 23 animals will be reviewed. **Figure 21.2** shows the evolutionary relationships among all three domains of life and the kinds of organisms that are part of each domain.

Because this chapter is devoted to mostly unfamiliar *microbes*— meaning living things so small they can't be seen with the naked eye—it might be helpful to say a word at the start about their significance. Microbes are something like the foundation of a house: seldom thought of but critically important. To the extent that they cross our minds at all, we are aware of them mostly for the diseases they cause. This is an important topic—one we'll go into extensively—but thinking of microbes strictly as disease-causing organisms is like thinking of cars strictly as wreck-causing machines. Where does the oxygen we breathe

come from? If your answer is "plants," that's partly correct. But, the photosynthesis carried out by plants supplies less than half our oxygen; the balance comes from microscopic algae and bacteria—mostly drifters on the ocean's surface. How about the nitrogen that is indispensable to all plants and animals? Among living things, only bacteria and archaea are capable of taking nitrogen that exists in the air and transforming it into a form that plants can use. If there were no bacteria and archaea, there would be no plants—and consequently no land animals. And how is it that dead tree branches, orange peels, or the bones of animals are broken down and recycled into the Earth? Almost entirely through the work of bacteria and fungi. Without these organisms, Earth would long ago have become a garbage heap of dead, organic matter. Animals cannot decompose a dead tree branch, but bacteria and fungi can.

The upshot of all this is that life on Earth would grind to a halt without microbes; they are an essential underpinning to all forms of life. They come to this status in part because of the environments they

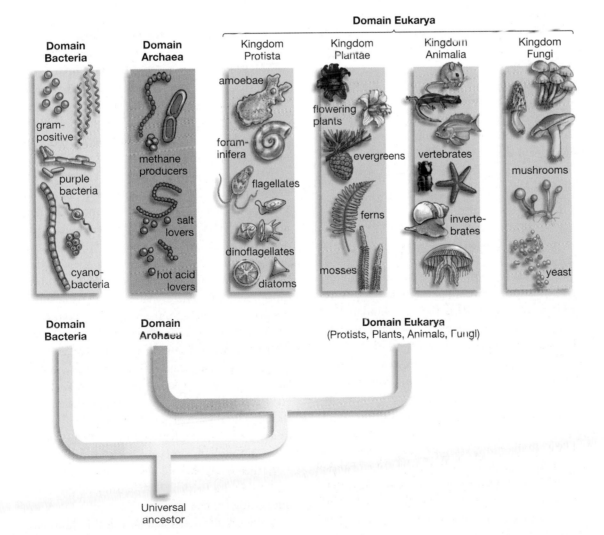

**Figure 21.2**
**The Tree of Life**

Life is presumed to have developed from a single common ancestor (represented by the base of the tree), which gave rise to the three domains of the living world: Bacteria, Archaea, and Eukarya. The boxes above the tree show the kinds of organisms that are members of each domain.

live in, which is to say all environments on Earth that support life of any kind. It probably didn't escape your attention that the coldest-, hottest-, highest-, and deepest-living things were all bacteria or archaea. Wherever a plant or animal can live, microbes can be found, too. Just as impressive is the sheer number and tonnage of these tiny organisms. A common denominator for measuring the amount of living material is something called *biomass*, defined as the total dry weight of material produced by a given set of organisms. Each bacterium or microscopic alga has an infinitesimally small weight, of course—but then again, there are so many of them! It has been estimated that there are more bacteria living in your mouth now than the number of people who have ever lived, and there are far more bacterial cells in the human body than there are human cells (a topic we'll return to). A typical gram of fertile soil—an amount about the size of a sugar cube—has about 100 million bacteria living in it, and a mere drop of water near the ocean's surface contains thousands of the tiny algae called phytoplankton. All this adds up. When the calculations are done, we find that microbes probably make up more than half of Earth's biomass. Put another way, the microbes of the world outweigh all the plants, animals, and visible fungi of the world combined.

In sum, microbes matter, and not just in ways that are harmful. We'll now start to look at the portion of the living world that is made up of them. Our first stop on this tour, however, will be a category of tiny replicating entities that most biologists would say lie just outside of life, although they have a great effect on living things.

## 21.2 Viruses: Making a Living by Hijacking Cells

If a living thing is too small to be seen and can cause disease, most people would put it in the vague category known as "germs." But, there is a fundamental distinction to be made between *two* kinds of infectious organisms. On the one hand there are bacteria, which may be very small but nevertheless are *cells*, complete with a protein-producing apparatus, a mechanism for extracting energy from the environment, a means of getting rid of waste—in short, complete with everything it takes to be a self-contained living thing. On the other hand there are *viruses*, which by themselves possess none of these features. Indeed, viruses can be likened to a thief who arrives at a factory he intends to rob possessing only two things: the tools to get inside and some

software that will make the factory turn out items he can use. The factory being broken into is a living cell, and the software that the virus brings is its DNA or RNA, which it puts inside this "host" cell. Once this is accomplished, viruses employ different tactics, as you'll see, but in most cases the result is that viruses make more copies of themselves. Viruses are an integral part of the living world, and not all of them cause harm. But unlike the case with bacteria, it's right to think of them first and foremost in terms of the diseases they cause, not only in humans, but in every variety of living thing. **Viruses** are noncellular replicating entities that must invade living cells to carry out their replication.

### HIV: The AIDS Virus

If you look at **Figure 21.3a**, you can see a simplified rendering of a particular virus, the human immunodeficiency virus (HIV), which is the cause of AIDS. All viruses have two of the three large-scale structures you can see in HIV: the genetic material at the core (in HIV's case, two strands of RNA) and a protein coat, called a **capsid**, that surrounds the viral genetic material. Many viruses then go on to have a third major element—a fatty membrane, called an *envelope*, which you can see surrounding the HIV capsid. Protruding from the HIV envelope are a series of receptors, often referred to as "spikes," which are proteins capped with carbohydrate chains. These serve the critical role of binding with receptors that protrude from the target cell, thus giving the virus a way to get in.

If you look at **Figure 21.3b**, you can see how the life cycle of HIV proceeds from this initial binding. The target cell in this case is the one most commonly invaded by HIV, an immune system cell called a helper T-cell. As you can see, once HIV binds with two receptors on the T-cell's surface, the viral envelope fuses with the T-cell membrane. With this, HIV's capsid, with the genetic material enclosed, is inside the cell. Once there, the capsid disintegrates. Now two HIV enzymes that had been enclosed in the capsid get to work. If you look at Figure 21.3a, you can see these enzymes, integrase and reverse transcriptase. The reverse transcriptase gets busy first, using HIV's RNA strands as a template to produce double-stranded viral DNA. Then, integrase does just what its name implies: It integrates the viral DNA into the *cell's* DNA through a cut-and-paste operation.

At this point, the viral DNA might simply stay integrated in the T-cell's DNA, causing no great harm but getting copied, along with the cell's DNA, each time the cell divides. The effect of this, however, is

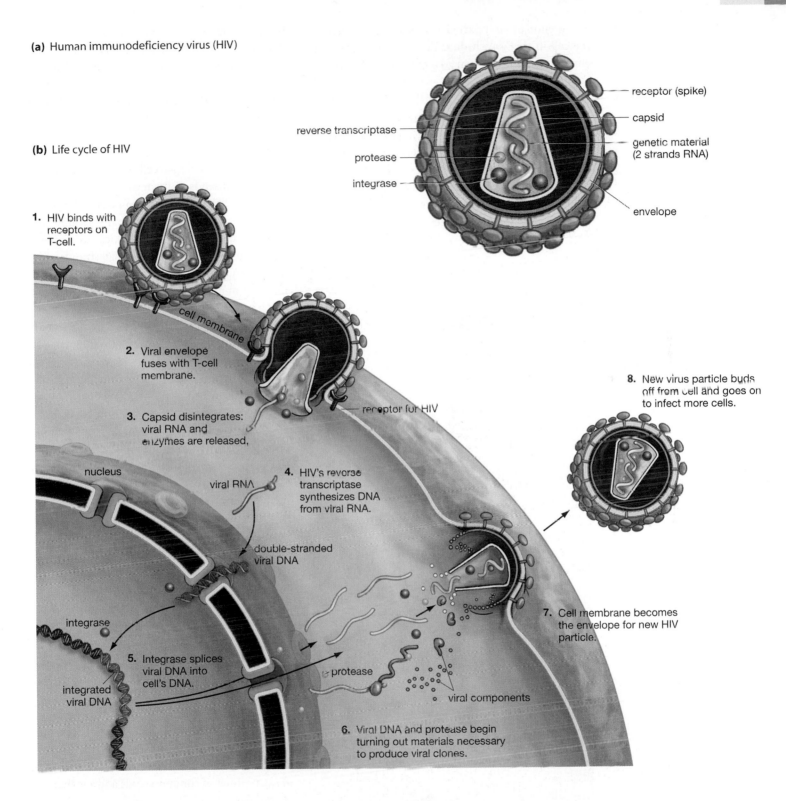

**(a)** Human immunodeficiency virus (HIV)

receptor (spike)

capsid

reverse transcriptase

genetic material
(2 strands RNA)

protease

integrase

envelope

**(b)** Life cycle of HIV

**1.** HIV binds with receptors on T-cell.

cell membrane

**2.** Viral envelope fuses with T-cell membrane.

**8.** New virus particle buds off from cell and goes on to infect more cells.

receptor for HIV

**3.** Capsid disintegrates: viral RNA and enzymes are released,

nucleus

viral RNA

**4.** HIV's reverse transcriptase synthesizes DNA from viral RNA.

double-stranded viral DNA

integrase

**7.** Cell membrane becomes the envelope for new HIV particle.

**5.** Integrase splices viral DNA into cell's DNA.

integrated viral DNA

protease

viral components

**6.** Viral DNA and protease begin turning out materials necessary to produce viral clones.

that every "daughter" cell of this infected cell will be infected too; each new cell will have viral DNA spliced into its own DNA. Each of these cells then has a time bomb ticking within it because at some point the HIV DNA will change course. It will no longer simply go on replicating within the cell's DNA. Instead, it will start turning out the materials necessary to make whole new copies of the virus. As you can see

in the figure, virus construction takes place just inside the cell's outer membrane (helped along by a third enzyme that HIV brought with it, protease). This location is important because the cell's membrane ends up serving as the final structural part of each new viral particle. When the capsid is completed, it buds off from the cell, taking with it part of the cell's membrane, which now becomes the capsid's envelope

**Figure 21.3**
**The Virus that Causes AIDS**

The anatomy and life cycle of HIV, the human immunodeficiency virus.

When enough viral particles have budded off from the cell in this way, it dies. The viral particles that emerge from the cell then go on to infect other cells.

## Viral Diversity

In the HIV life cycle, you can see four steps that are common to almost all viruses: get genetic material inside the host cell, turn out viral component parts, construct new particles from these parts, and move these particles out of the cell. HIV uses RNA as its genetic material, but many viruses use DNA. Viruses that lack an envelope don't get their genetic material

inside the cell by fusing with the cell's membrane in the way HIV does. If you look at **Figure 21.4**, you can see a spacecraft-shaped virus called T4 that infects bacteria through an alternate means. T4 keeps its capsid outside the target cell but gets its DNA inside by injecting it through the bacterium's cell wall and membrane.

As you might be able to tell from the size of T4 compared to its target cell, viruses are very small things. Thousands of them can typically fit within a bacterial cell. Some amount to nothing but their genetic material and the capsid that surrounds it, and there are even viruslike entities, called *viroids*, that lack a capsid; they are simply small strands of infectious RNA. What viruses lack in size, however, they make up in numbers. One researcher has calculated that a typical milliliter—a thousandth of a liter—of ocean water has 10 million virus particles in it. If so, then ocean-going viruses contain as much as 270 million metric tons of carbon, which is more than 20 times the amount of carbon contained in all the whales in the world.

In every case, viruses are not only small, but *simple* things compared to the organisms they infect. First, they lack the usual machinery that cells have—a protein assembly line, a waste disposal system, and so forth. Even when we look at the one thing all of them do have, which is genetic material, they are still very simple entities. Human beings are thought to have between 20,000 and 25,000 genes, and even a primitive organism like baker's yeast has 6,000 genes. But the AIDS virus? It has nine genes. To be sure, some viruses are more complex than HIV, but in no case is a virus complex enough to replicate by itself. Because viruses can carry on scarcely any of life's basic functions by themselves, most scientists don't classify them as living things.

## The Trouble Viruses Cause: Avian Flu

But, what trouble is caused by something that isn't even alive! Apart from AIDS, viruses are responsible for chickenpox, measles, rabies, polio, herpes, rubella, and some forms of cancer, hepatitis, and pneumonia, to say nothing of common colds and the many varieties of flu.

As if this weren't enough, public health officials around the world are now on high alert in connection with a viral disease merely because of its *potential* to cause harm. Avian flu, as the disease is commonly known, has claimed only a few hundred human lives thus far. Yet, the threat it poses is so severe that human populations are constantly being monitored for signs of it, and millions of domesticated birds have been

**Figure 21.4**
**Another Variety of Virus**

**(a)** The T4 virus looks like a spacecraft that has landed on the surface of a bacterium.

**(b)** A viral invader lands. An artificially colored T4 virus injects its DNA into an *Escherichia coli* bacterium (in blue).

**(a)**

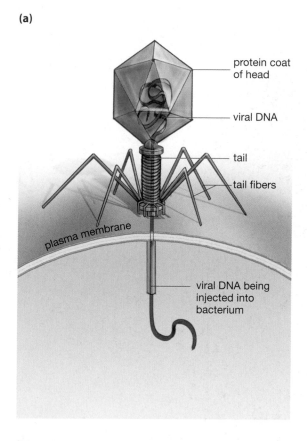

protein coat of head

viral DNA

tail

tail fibers

plasma membrane

viral DNA being injected into bacterium

**(b)**

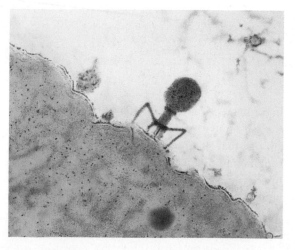

slaughtered or vaccinated worldwide to keep it from spreading (**Figure 21.5**).

Why have birds been slaughtered in connection with a human disease? There is an entire class of viruses—the influenza type A viruses—that exist naturally in wild ducks and shore birds. Avian flu is caused by one of these viruses, a microbe named A (H5N1). Although wild birds are unharmed by A (H5N1), they can pass it on to domesticated birds, such as chickens and turkeys, and these birds *are* affected by it, often fatally. The real problem, however, is that A (H5N1) has now spread from domesticated birds to humans. Such infection is rare and has almost always involved prolonged contact between birds and humans. Nevertheless, of the humans who have been infected by A (H5N1), more than half have died as a result. This alone is reason enough for countries to engage in the mass slaughter of infected domestic birds, but there's more. Any contact between infected birds and humans increases the likelihood of the virus undergoing a critical genetic change, such that it can be transmitted not just from birds to humans, but from *humans* to humans. (Only a single instance of such transmission has been confirmed so far, although it can be difficult to judge infection routes in cases of human infection.)

A change of this sort is possible because viruses are capable of undergoing genetic alteration through two different routes. One is that they evolve into new "strains" through mutations to their DNA or RNA. The other is that two different viruses—say, A (H5N1) and a human virus—can infect a single cell and then exchange genetic sequences while within the cell, thus producing a hybrid virus that is different from either of the parental varieties.

What are the chances that A (H5N1) could be altered through one of these routes, such that it could easily move from person to person? On the positive side, it has been around for more than 10 years without this occurring. On the other hand, no one can take any chances that this state of affairs will continue. The twentieth century saw three *pandemics*, or worldwide outbreaks of fatal, type A influenzas. One of them, in 1957, killed between 1 and 4 million people, while another, in 1968, killed about a million people. Both paled in comparison, however, to an outbreak that occurred in 1918–1919. That pandemic is estimated to have infected a third of the people then living in the world and to have killed 50 million of them. (By way of comparison, AIDS took a quarter century to kill 25 million people.) For the foreseeable future, therefore, you are likely to be seeing frequent news reports on the worldwide state of avian flu. Where has it affected domesticated birds? Where has it been transmitted from one human

**Figure 21.5**
**Trying to Prevent a Pandemic**

A health worker injects avian flu vaccine into a chicken at a poultry farm in Yichang, in the Hubei province of central China.

to another? What are the prospects of developing a vaccine for it? The monitoring will be constant because the stakes are so high.

---

### SO FAR...

1. The three domains of life are _____, _____, and _____. The four kingdoms within one of these domains are composed of _____, _____, _____, and _____.

2. Microbes perform several activities that are critical to the maintenance of other life-forms on Earth. Name three of these activities.

3. In order to replicate, viruses must invade _____.

---

## 21.3 Bacteria: Masters of Every Environment

Domain Bacteria is made up entirely of the microbes called bacteria. Keeping in mind that almost all the organisms we're looking at are microbes, what separates bacteria from any of the others? Recall from Chapter 4 that if we set a bacterial cell and, say, a fungal cell side by side, anyone peering through a microscope could see a fundamental difference between the two. The fungal cell would have a large, circular-shaped structure near its center, while the bacterial cell would not. The circular structure is the fungal cell's nucleus, which contains almost all of its DNA. But bacteria are **prokaryotes**: organisms whose DNA is not contained within a cell nucleus. Only bacteria and archaea lack a

## ESSAY    Unwanted Guest: The Persistence of Herpes

The first outbreak usually is the worst, but the real problem is that the first outbreak usually isn't the last. Herpes comes back, as herpes sufferers well know. A few weeks or months after the initial outbreak, everything seems fine, but then there may be a tingling—nothing that can be seen yet, but instead just a feeling. Not much later, however, it's back: the same unsightly "cold sore" on the mouth or lesions on the genitals, in the same place that the first outbreak occurred.

This series of events is familiar to lots of people. An estimated one in five Americans over the age of 12 is infected with the herpes virus, although this statistic overstates the trouble caused by it. The vast majority of herpes carriers do not even realize they harbor the disease because such outbreaks as they have are so mild as to go unnoticed. This very thing facilitates the spread of herpes, however. People who have no symptoms can nevertheless be "silent shedders": They can pass herpes along to others, who may get the full-blown form of the affliction.

But, why does herpes come back in the way it does? Where does it go between outbreaks, and why can't we just get rid of it permanently? The term "herpes" actually is shorthand for *herpes simplex viruses*, a pair of viruses that belong to a group of 100 known viruses in the herpes family. (Other viruses in the family cause chicken pox, shingles, and the Epstein-Barr disease.) Herpes simplex viruses come in type 1 and type 2 varieties. The type 1 variety supposedly is "oral," while the type 2 is "genital." But, either variety can end up in either location because of oral sex.

The herpes sores that can be so distressing are the result of newly formed virus particles breaking out of skin cells, thereby destroying them. Fever, swollen glands, and other symptoms often accompany these lesions, particularly in an initial infection. Within a couple of weeks or so, however, the immune system has killed all the virus particles in and around the skin cells, and the symptoms are gone. The problem is that, by this time, the virus has also infected *nerve* cells in the area. Such nerve cells (or neurons) have long extensions of themselves called axons. With the initial infection, herpes virus particles start making a slow migration up these axons to the *body* of the nerve cells, specifically to the nucleus inside the cell body. In the case of genital herpes, this means a journey from the vagina or penis to cell bodies that sit right at the base of the spinal cord. And these cell bodies are where the virus stays—permanently. Once there, it does next to nothing; it's not causing any nerve damage or replicating. The nerve cells it is in never divide, so it isn't passed on to a larger group of cells. But, neither is it eliminated. It just resides in the nerve cell nuclei until one day—often a stressful day—it gets activated by an unknown mechanism and begins a journey back up the very axons it came in on. Once it reaches the original nerve cell endings, it spreads to skin cells, and the result is new lesions in the same old place.

As noted, some people with herpes can shed virus particles from time to time even when they have no lesions, meaning they are infectious during these periods. This, of course, puts these carriers at terrible risk of passing the virus on to their uninfected partners. The good news about herpes is that there is a drug, called valacyclovir, that has shown significant ability to suppress the virus at all times. In a large clinical trial, herpes carriers who took one valacyclovir tablet a day transmitted the virus to their partners at less than half the rate of carriers who took only a placebo.

---

**Figure 21.6**
**Small Is a Relative Thing**

The bacteria, protist, and viral particles shown in the figure are all microscopic. Nevertheless, they differ greatly among themselves in terms of size. A typical human cell might be about a third the size of the *Paramecium*.

viral clones (T4)
0.2 μm long

bacteria (*E. coli*)
2 μm long

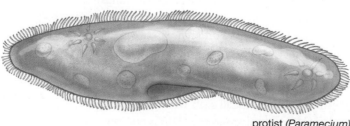
protist (*Paramecium*)
75 μm long

nucleus. All other organisms—all plants, animals, fungi, and protists—are composed of cells that have a nucleus, which makes them **eukaryotes**. The lack of a cell nucleus is, however, just one defining feature of bacteria. Here is a list of others.

- **No membrane-bound organelles.** A nucleus is an example of an *organelle*: a highly organized structure within a cell that carries out specific cellular functions. You may remember from Chapter 4 that, in addition to a nucleus, eukaryotic cells have such organelles as mitochondria and lysosomes, which are surrounded by a membrane. Bacteria have only a single kind of organelle, the ribosome, which differs from other organelles in that it is not surrounded by a membrane.

- **Single-celled organisms with small cell size.** Bacteria often come together in groupings that have an organization to them, but each bacterial cell is a self-contained living thing—each one can live and reproduce without the aid of other cells. Most bacterial cells are also very small, relative to eukaryotic cells; thousands of them could typically fit into a single eukaryotic cell. (If you look at **Figure 21.6**, you can see a size comparison of bacteria, a T4 virus, and a protist you'll be looking at called a paramecium.)

- **Asexual reproduction.** Bacteria reproduce by a simple cell splitting or **binary fission**: one cell splits into two, with both "daughter" cells exact replicas of the parental cell. In contrast, most eukaryotes are capable of sexual reproduction, in which the genetic material of two separate organisms produces offspring, often through the fusion of egg and sperm.

If you look at **Figure 21.7**, you can see micrographs of three forms that bacterial cells commonly take: round, rod shaped, and spiral shaped. A round bacterium is known as a *coccus*; a rod-shaped bacterium is a *bacillus*; and a spiral-shaped bacterium goes under a variety of names, the best known of which is a *spirochete*. Different species of bacteria fit into all three groups.

Almost all bacteria have a structure called a cell wall at their periphery that is largely composed of a material called peptidoglycan. This may seem like a technical detail, but the fact that bacteria have cell walls is of great importance to humans because it represents a *difference* between bacterial cells and human cells. As you'll see, we humans have exploited this difference in our development of the medical drugs known as antibiotics.

## Bacterial Numbers and Diversity

In 2006, scientists reported on a bacterial census they undertook by means of looking for bacterial DNA sequences within tiny quantities of soil from Alaska. The result? Each gram of soil contained about 4,000 different species of bacteria. Meanwhile, a comparative analysis indicated that a gram of Minnesota farm soil contained about 10,000 bacterial species. Fascinatingly, only about 20 percent of the species the researchers identified were present in the soils of *both* states. You might think that surveys such as these could give us a handle on how many bacterial species there are in total, but so far that's not the case. We really have no idea how many different kinds of bacteria there are in the world; the most we can say is that the number appears to be very large.

When we begin to examine these various types of bacteria, we find that an incredible diversity exists among them. It may seem strange to think of bacteria as "diverse," given that all of them are microscopic and single celled. But physical form is only one kind of diversity. Imagine that only some human beings got their nutrition in the conventional way—by ingesting food—while others could make their own food by carrying out photosynthesis. Bacteria do things both ways. Imagine that only some human beings needed oxygen to live, while others could take it or leave it, and still others were poisoned by it. Bacteria fit into all three categories. Animals and plants have diverse forms, but bacteria have diverse metabolisms. If you

**(a)** Round bacteria (cocci)

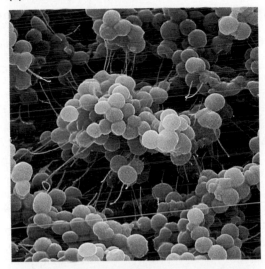

**(b)** Rod-shaped bacteria (bacilli)

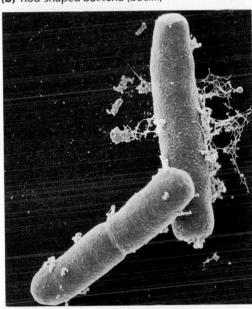

**(c)** Spiral-shaped bacterium (spirochete)

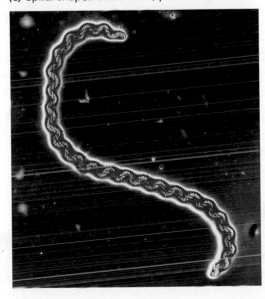

**Figure 21.7**
**Different Shapes of Bacteria**

**(a)** The round bacterium *Staphylococcus epidermis* is a normal part of the bacteria that cover our skin.

**(b)** The rod-shaped bacterium *Bacillus anthracis* is the cause of the deadly disease anthrax.

**(c)** The spiral-shaped bacterium *Leptospira interrogans* often infects rodents, sometimes infects dogs, and occasionally infects human beings.

# Modes of Nutrition: How Organisms Get What They Need to Survive

A ll living things need energy and nutrients. The question is, how do they get them? What is their "nutritional mode," as biologists would put it?

The most fundamental distinction in nutritional mode separates groups known as autotrophs from those called heterotrophs. "Autotroph" means "self-feeding." **Autotrophs** are organisms that can manufacture their own food, defined as some form of organic (carbon-containing) molecule that can be broken down to yield energy. All organisms that are not autotrophs are **heterotrophs**, meaning "other feeders"; they cannot manufacture their own food but must get it from elsewhere, as animals do.

The idea of heterotrophs is not strange to us—it's what we are, after all—but the notion of autotrophs may be a little more exotic. Chapter 8, on photosynthesis, explained that almost all plants are autotrophs in that they manufacture their own food, initially a sugar. To do this, they need a carbon source, because carbon is the "backbone" of the organic molecules we know as food, and they need an energy source that can drive the complex process of photosynthesis. The carbon source for plants turns out to be carbon dioxide ($CO_2$), obtained from the atmosphere, while the energy source is sunlight. Here, we see the *two-part* requirement for nutrition that holds for all organisms: a source of carbon on the one hand and an energy source on the other.

Within this framework, there are four nutritional modes, which you can review in **Table 1**. The two most important of these modes are *photoautotrophy*, which is the nutritional mode of plants, and *chemoheterotrophy*, in which organic materials (better known as food) act as both carbon supplier *and* energy source. This is another way of saying that the cereal you ate this morning supplied you, as a chemoheterotroph, with both carbon and high-energy electrons.

There is also *chemoautotrophy*, in which organisms—some bacteria and archaea—get their carbon from carbon dioxide, like plants do, but power the production of food from this carbon by oxidizing (pulling electrons from) such inorganic materials as hydrogen sulfide and ammonia. A small number of bacteria and archaea practice *photoheterotrophy*, in which the sun supplies the energy but the carbon comes from surrounding organic material.

| Table 1 | | | |
|---|---|---|---|
| **Nutritional mode** | **Carbon source** | **Energy source** | **Practiced by** |
| Autotrophy | | | |
|    Photoautotrophy | Carbon dioxide ($CO_2$) | The sun's rays | Almost all plants, some bacteria, and many protists |
|    Chemoautotrophy | Carbon dioxide | Inorganic compounds such as hydrogen sulfide and ammonia | Some bacteria and archaea |
| Heterotrophy | | | |
|    Photoheterotrophy | Organic material | The sun's rays | A few bacteria and archaea |
|    Chemoheterotrophy | Organic material | Organic material | Almost all animals, all fungi, most bacteria, many protists, and a few plants |

look at "Modes of Nutrition," above, you can see the various ways that life's creatures get the nutrition they need. Note that bacteria utilize every one of these modes, while humans utilize only one.

## 21.4 Intimate Strangers: Humans and Bacteria

In the chapter introduction, you saw how bacteria are indispensable to life in general because they can capture atmospheric nitrogen, decompose dead organic material, and so forth. As it turns out, however, bacteria may be indispensable to *human* life in an even more direct way. From the time we travel down our mother's birth canal, we enter into a lifelong interdependence with these microorganisms. If you look at **Figure 21.8**, you can see the places bacteria are found in quantity outside and inside the human body. Outside, we are covered in bacteria from head to toe, with heavy concentrations in the armpits and scalp. Inside, they exist in our nasal passages, in the vaginas of females, and most especially, in our digestive tracts.

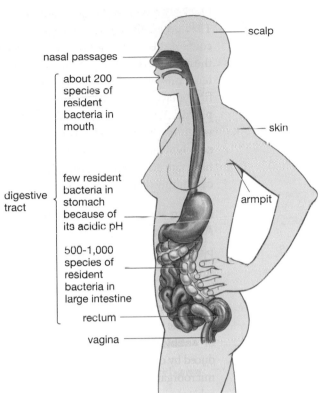

All told, perhaps 100 trillion bacterial cells live in or on the human body. Meanwhile, the number of *human* cells in a human body is only a tenth of this number—about 10 trillion cells. (So why aren't we bloated with bacterial cells? They're so small that their collective weight is only about three pounds.) This is not to say that bacteria are everywhere within us; most of our tissues are kept bacteria free, except for occasional invaders. But, about half the contents of our colon are bacteria, as are perhaps a fourth of our feces by weight. These are not transient bacteria that come for brief periods and then are gone. They are so-called resident bacteria whose permanent home is the human body.

This account may make you feel like you've been colonized, but the relationship between us and many of the bacteria we harbor actually is one of **mutualism**, which is to say a relationship between two organisms that benefits both of them. If we could shrink ourselves down to the size of a bacterium and enter, say, the human large intestine, we would see a community of competing but *interdependent* organisms, most of which are bacteria, but one of which is us. What bacteria get from participating in this community is food and habitat. What humans get is a fully functional digestive system. How do we know that bacteria are helping maintain this system? Consider that rats whose bacteria have been wiped out through laboratory techniques require 30 percent more calories to maintain their body weight than do normal rats. Without bacteria, the structure of these rats' intestines is altered in a way that's unhealthy: They have low numbers of the cells that move nutrients from the digestive tract to the bloodstream. Researchers have made mice "germ free," then reintroduced bacteria into them and watched as mouse genes were turned on that help metabolize sugars and fats. Bacteria even metabolize sugars that we cannot digest, and they produce some vitamins as well. The essential message here is that, within the digestive tract, mammals and bacteria have had such a long working relationship that each has become dependent on the other. We humans don't just tolerate the bacteria in our gut; we need them.

## 21.5 Bacteria and Human Disease

It goes without saying that not all bacteria are beneficial to human beings. Even within our digestive tract, there are bacteria that are *commensal* with us—they benefit from the relationship, while we are unaffected by it. And, of course, there are bacteria that are **pathogenic**, which is to say disease-causing. These

**Figure 21.8**
**At Home in Us**
We human beings live with trillions of "resident" bacteria, meaning bacteria that are always present on us or within us. (Bacteria on the scalp or skin might temporarily be washed away, but they quickly return.) The greatest concentration of these bacteria is found in the digestive tract, particularly in the large intestine.

are not bacteria that routinely inhabit any part of our body in large numbers. Instead, they are opportunists that invade specific tissues. Among the diseases they cause are tuberculosis, syphilis, gonorrhea, cholera, tetanus, leprosy, typhoid fever, and diphtheria, not to mention all manner of food poisonings and blood-borne infections. In the fourteenth century, the bacterially caused bubonic plague wiped out a *third* of Europe's population in just 4 years. Closer to home, you may recall a substance referred to as *anthrax* that was maliciously mailed to various offices after the September 11 attacks. Anthrax actually is a bacterium (*Bacillus anthracis*) that, when not in the hands of terrorists, occasionally infects farm animals.

How do pathogenic bacteria cause illness? A few invade human cells and reproduce there. More often, however, bacterial damage comes from substances bacteria either secrete or leave behind when they die. These compounds are *toxins*, which is to say substances that harmfully alter living tissue or interfere with biological processes. In the case of respiratory anthrax, *Bacillus anthracis* secretes a trio of toxins that cause blood vessels in the lungs and brain to "hemorrhage" or leak. The botulism bacterium, *Clostridium botulinum*, secretes a toxin that

sequenced the first genome of an archaeon, they were stunned to find that 56 percent of its genes were completely unknown to science—they were unlike anything seen in either bacteria or in eukaryotes. As for the remaining 44 percent of the genome, some genes worked like those of eukaryotes, while others worked like those of bacteria.

Findings such as these produced the picture you can see in **Figure 21.10**; they convinced scientists that life evolved along three principal evolutionary lines, one of which is the archaea. Nevertheless, all three domains of organisms are linked in that they share the common ancestor at the base of life's tree—something consistent with archaea having some genes in common with bacteria and others in common with eukaryotes.

### Archaea and Their Habitats

So, what about the second confusion we talked about—the misunderstanding about where archaea live? A large number of archaea species do in fact live in "extreme" environments, as we'll see. But archaea also live in large numbers in a variety of common habitats. Scientists divide the ocean into a series of vertical zones, one of which—starting at 200 meters down—is informally dubbed the "twilight zone" because it lies just beneath the level at which sunlight penetrates strongly. An eye-opening survey done in the 1990s found that archaea account for 40 percent of the microbial life in this zone over vast stretches of the ocean. This alone makes the archaea "one of the most abundant organisms on the planet," in the words of researchers Stephen Giovannoni and Ulrich Stingl. Beyond this, archaea have recently been found to exist in abundance in common soil; there actually are more

archaea than bacteria taking part in one phase of the "nitrogen-fixing" process we talked about earlier.

### Extremophiles

All this said, many species of archaea do live in environments that are harsh or forbidding. We can thus think of some archaea as being **extremophiles**: organisms that grow optimally in one or more conditions that would kill most other organisms. Archaea are not the only extremophiles; many bacteria, some fungi, and even some organisms such as lichens thrive under tough conditions as well. Some of these organisms can withstand high pressure, while others can resist radiation; some flourish in cold temperatures, while others can get by with very little water. Here are three important categories of extremophiles, each of which has many species of archaea within it:

- **Thermophiles.** These are microbes that make their homes in extremely hot environments, such as the hot-water vents on the ocean floor or hot-water pools such as exist in Yellowstone National Park. At the start of the chapter, we noted that the archaeon *Pyrolobus fumarii* lives in hot-water vents at temperatures of up to 113° Celsius, but even it may not be the record holder in this regard. In 2003, researchers captured an archaeon from an undersea vent and then proceeded to "culture" it—to grow it in the lab—at a temperature of 121° Celsius, which is 250° Fahrenheit.

- **Halophiles.** These are microbes that live in environments so *salty* that few other types of organisms can survive in them. Utah's Great Salt Lake and the Dead Sea that separates Israel and Jordan are examples of *hypersaline* environments, meaning environments with salt concentrations greater than those found in seawater. Since all life needs water, the challenge for halophiles becomes how to preserve water internally while living in environments in which osmosis tends to pull water out of their cells.

- **Anaerobes.** These are microbes that either can live without oxygen or that actually are poisoned by it and thus can only live in oxygen-free environments. You might think it would be impossible to get away from oxygen on Earth, but oxygenless environments are surprisingly widespread. They exist in ocean sediments, swamps, rice fields, and the digestive tracts of many animals, including ourselves. (The microbial menagerie within the human gut is mostly made up of anaerobic bacteria, but it includes

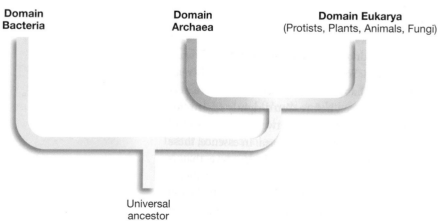

**Figure 21.10**
**Archaea and the Universal Tree of Life**
Archaea form one of three main branches on the universal tree of life. Note that they appear to be more closely related to eukaryotes (Domain Eukarya) than to bacteria (Domain Bacteria). The archaea and eukaryotes shared a common ancestor that was not shared by bacteria.

archaea as well.) One group of anaerobic microbes—all of them archaea—produce the gas methane in the course of their metabolism and thus are *methanogens* (**Figure 21.11**). Their production of this gas can be substantial; the amount diffusing from rice fields alone has been estimated at a minimum of 20 million tons per year. The problem with this? Methane is a potent greenhouse gas—it contributes to global warming.

### Prospecting for Extremophiles

Various species of extremophiles have in recent years become the target of a kind of new-age prospecting, with "miners" being scientists from chemical and biotechnology firms (**Figure 21.12**). What's the commercial attraction of extremophiles? Organisms that live in extreme environments must have *enzymes* that function in these environments. And, through modern biotechnology processes, enzymes can be isolated from an organism and then turned out in quantity in factories. These enzymes can then be put to use in *human-made* extreme environments, such as the inside of a washing machine or the confines of a laboratory container. If you've ever watched one of the *CSI* TV shows, the PCR (polymerase chain reaction) process used in sequencing DNA depends on an enzyme that came originally from a hot-spring-dwelling bacterium, *Thermus aquaticus*. Meanwhile, one U.S. company has introduced a clothing detergent containing a cleaning additive that is simply an enzyme from a heat-loving extremophile, while another produces an extremophile enzyme that gives jeans a softer feel and a "stonewashed" look.

**Figure 21.11**
**An Extreme Archaeon**

*Methanopyrus kandleri* is an archaeon that lives in the ocean at temperatures near those of boiling water. It is thus a thermophile, but that's not all. It's also an anaerobe—it can only live in oxygenless environments—and it is a methanogen in that it produces the gas methane as part of its metabolism. It has been found in geothermal vents on the sea floor at depths of 2,000 meters or about 1.2 miles down.

**Figure 21.12**
**Prospecting for Extremophiles**
Researchers looking for archaeon extremophiles in the Obsidian Pool in Yellowstone National Park.

### SO FAR...

1. The relationship between human beings and bacteria that are resident in their digestive tract can be characterized as one of mutualism, which is to say a relationship in which _____.

2. Antibiotics are chemical compounds produced by _____ that are _____.

3. Name three ways that bacteria and archaea are similar life-forms.

## 21.7 Protists: Pioneers in Diversifying Life

We've so far looked at two of life's domains, Bacteria and Archaea. As noted, the third domain, Eukarya, is so diverse we will look at only one of its four kingdoms in this chapter, the mostly microscopic organisms known as protists.

What's a protist? No one can provide a satisfactory definition. Domain Eukarya has within it three well-defined kingdoms—plants, animals, and fungi—but the protists within Eukarya are defined in terms of what they are *not* relative to these other life-forms: **Protists** are eukaryotic organisms that do not have all the defining characteristics of a plant, an animal, or a fungus. Why do we have such an unsatisfactory definition for this one group of organisms? The root of the problem is something first mentioned in Chapter 19: The term *protist* is a label of convenience rather than a label that reflects evolutionary reality. If you look

••• ANSWERS
1. both parties benefit
2. one microorganism; toxic to another microorganism
3. Both are strictly microscopic, single-celled, and prokaryotic (their cells do not have nuclei).

person. The catchall name for them is **algae**. If you look at **Figure 21.15**, you can see three varieties we'll be reviewing. Figure 21.15a shows a microscopic golden alga called *Synura scenedesmus* that can be found swimming in freshwater ponds in the United States. In this species, we can see a transition between the single-celled life carried out by bacteria and the multicelled life carried out by more complex organisms. All varieties of golden algae come in species that exhibit **colonial multicellularity**: a form of life in which individual cells form stable associations with one another but do not take on specialized roles. Each of the three objects you see in the picture of *Synura* is a cluster of *Synura* cells—a colony of them. Each member of the colony keeps its two whiplike flagella pointed to the outside of the group, the flagella beat back and forth, and the colony literally rolls through the water. None of these cells, however, performs a function different from any other cell, which is why this *Synura* is referred to as colonial and not truly multicellular.

If you look at Figure 21.15b, you can see a form of *green* algae, called *Volvox*, that represents an evolutionary step beyond *Synura*. The beautiful, translucent circles that outline *Volvox* are a gelatinous material that initially encloses a single layer of anywhere from 500 to 60,000 *Volvox* cells, all of them spaced around the periphery of the sphere. Almost all of these cells have flagella that they move back and forth in unison, so that *Volvox* moves slowly through the water in a given direction, spinning along an axis as it goes (something like a torpedo). For our purposes, the key thing about *Volvox* is that not all of the cells in this colony will reproduce. Only a select group of cells without flagella, generally at the rear pole of the axis, will divide to produce new colonies. These colonies in turn give rise to new spheres that exist inside the original parent sphere. Eventually, the gelatinous material around the parent sphere ruptures, and the daughter spheres emerge, each of them then continuing the process. The important point is that *Volvox* has cells that specialize.

As such, *Volvox* has arguably achieved **true multicellularity**: a form of life in which individual cells exist in stable groups, with different cells in a group specializing in different functions. In human beings, of course, we can see a great specialization in, for example, nerve and muscle cells. *Volvox* shows us specialization in a rudimentary form.

Finally, if you look at Figure 21.15c, you can see some representative brown algae—the giants of the protist world. One variety of brown algae, the sea kelp, has individuals that may be 60 meters long and that often grow in enormous kelp forests. For our purposes, brown algae are notable because their leaflike "blades" and stemlike "stalks" represent another leap up in complexity: They contain organized groups of cells that conduct food throughout the organism.

One other thing is important about the algae. Note that the golden algae and *Volvox* share three characteristics: They are microscopic, aquatic, and carry out photosynthesis. If we were to go through all the algae, we would find these same qualities in countless species. Creatures in this category, including some bacteria, are known as **phytoplankton**: small photosynthesizing organisms that float near the surface of water. Such organisms are extremely important to life on Earth for two reasons. First, they produce most of the Earth's oxygen. Second, they sit right at the base of most aquatic food chains. All large life-forms on Earth ultimately depend on photosynthesizers for their food. On land, lions may eat zebras, but zebras eat grass. On the ocean, whales may eat the shrimp-like creatures called krill, but krill eat phytoplankton.

## 21.10 Heterotrophic Protists

Lots of species of protists do not get their nutrients from photosynthesis, as algae do, but instead get them from consuming other organisms or bits of organic matter that they find. This makes them heterotrophs (or "other-eaters").

**Figure 21.15**
**Transitions Visible in Algae**

**(a)** Three colonies of the golden algae *Synura*, each of which is composed of a cluster of individual cells. *Synura* is a colonial protist in that none of the cells in a colony takes on any specialized function.

**(b)** Colonies of the green algae *Volvox* with daughter colonies growing inside. *Volvox* colonies can be regarded as truly multicellular organisms in that they have specialized reproductive cells. The largest *Volvox* colonies can be seen with the naked eye.

**(c)** A frond of the brown algae known as sea kelp. These unusually large protists have assemblages of cells that conduct food throughout the organism.

**(a)** Golden algae

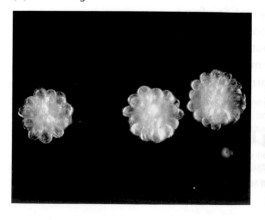

**(b)** Green algae

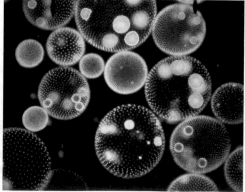

**(c)** Brown algae

**(a)** A *Paramecium*

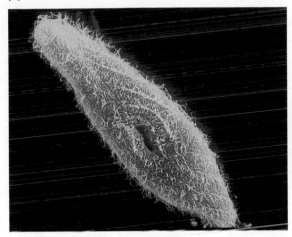

**(b)** The *Giardia* parasite

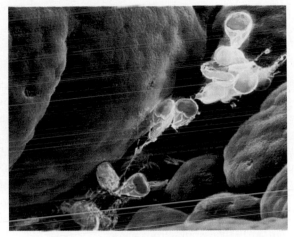

**Figure 21.16**
**Movement through Cellular Extensions**

**(a)** A green *Paramecium*; its many cilia both help it move and bring food to its gullet, visible in the center.

**(b)** The parasite *Giardia lamblia* within the human small intestine. Seen here as the pear-shaped green and white objects, *Giardia* use sucking disks to attach themselves to fingerlike extensions of the intestine's inner lining.

## Heterotrophs with Locomotor Extensions

Some of these protists have evolved slender extensions, called cilia and flagella, that allow them to move toward their prey or away from danger. If you look at **Figure 21.16**, you can see two varieties of these highly mobile heterotrophic protists.

Figure 21.16a shows a *Paramecium*, which is known as a ciliate because of all the hairlike cilia that cover its exterior. These cilia beat in unison, allowing paramecia to move forward or back in an exacting fashion—toward or away from objects in its environment. A *Paramecium* may be single-celled, but it is complex. It takes food in through a mouth-like passageway called a gullet, digests it in a special organelle called a food vacuole, and then empties the resulting waste into the outside world through a special pore on its exterior. Some species even have tiny dartlike structures that they can shoot out when threatened. Most animals require different *groups* of cells to do all these things, but a *Paramecium* does them within a single cell. Given this, it may be that paramecia represent the most complex single cells in all of nature.

Figure 21.16b shows a famous human parasite that enters human beings through contaminated water. So widespread is this "camper's parasite," *Giardia lamblia*, that health authorities have advised Americans never to drink stream or lake water without purifying it. *Giardia* lives in the small intestines of humans and other animals, where it causes symptoms that include nausea, diarrhea, and vomiting. It gets around not by cilia like the *Paramecia*, however, but by pairs of whiplike flagella, as we saw in the algae protists.

## Heterotrophs with Limited Mobility

**Figure 21.17a** shows a so-called amoeboid protist, meaning a protist that gets around by means of a pseudopod or "false foot." The amoeba sends out a slender extension of itself (the foot), and then the rest of its body flows into this extension. It takes in bits of food by encircling them with extensions and then fusing these extensions together. What once had been

**(a)** Amoeba

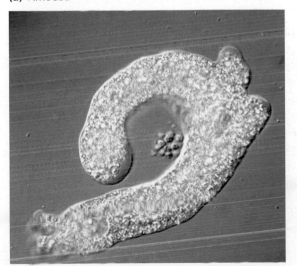

**(b)** Slime mold

**Figure 21.17**
**Life as a Blob**

**(a)** An amoeba bends itself around food, which it will soon ingest.

**(b)** This plasmodial slime mold, *Physarium polycephulum*, is growing on and around a tree trunk in Pennsylvania.

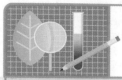

# THE PROCESS *of* SCIENCE
## The Discovery of Penicillin

In wartime England, late in 1940, as the German bombings raged, a 43-year-old policeman lay dying in the Radcliffe Infirmary in Oxford. Albert Alexander was, however, not a victim of the German air raids. His enemy was smaller and more subtle. A few months earlier, he had scratched his cheek on a rosebush thorn. This minor wound had then become infected, and from that point he steadily went downhill as bacteria worked away inside him. By December his face was a mess of festering wounds, one of which cost him an eye.

Incomprehensible as it may seem, in 1940 a scratch from a rosebush thorn could do this to a person. The notable thing about Albert Alexander was not that he became deathly ill from such a trivial injury, but that he came close to beating his tiny enemies. He happened to live near an Oxford University laboratory where a handful of researchers were hurriedly trying to extract, from a common fungus, a product that the world eventually came to know as penicillin. When Albert Alexander became the first human being to receive a therapeutic dose of penicillin in February 1941, the results were startling. His fever dropped, his infection began to clear; even his appetite returned. But there was to be no happy ending for him, for the researchers had only a 5-day supply of the drug. Trying everything they could think of, they even extracted precious micrograms of penicillin from Alexander's own urine to put back into him. But it was not enough; he died days later.

Nevertheless, Alexander's case had shown that penicillin could work in human beings, and clear-cut successes with the drug were not long in coming. A 15-year-old boy with an infected hip underwent a remarkable turnaround when injected with it; children had their eyesight saved when blindness would have been the expected outcome. These early trials made clear that, if ever there was a "miracle drug," penicillin was it. Prior to its advent, doctors could only watch and hope for the best as their patients were wracked by infections that ranged from pneumonia to meningitis to diphtheria to blood poisoning. With the advent of penicillin, at last doctors could do something.

How did it happen that a small band of researchers in wartime England ended up realizing the wondrous potential of this substance? The story begins not in the 1940s, but in 1928, when another scientist in England, Alexander Fleming, discovered the existence of penicillin through a combination of perceptiveness and luck. Returning to his hospital laboratory from a summer vacation, he noticed a spot of mold (a type of fungus) in one of the Petri dishes in which he was growing or "culturing" bacteria. Such bacterial cultures can easily become contaminated; this dish probably became tainted by means of fungal spores wafting in from another lab in the building. Something about this particular contamination caught Fleming's eye, however: The bacteria in the vicinity of the mold had clearly burst—they were dead. Realizing that chance had brought to him a substance that had the power to kill bacteria,

Fleming grew more of the fungal mold, prepared a broth from it, tested its power against several kinds of deadly bacteria, and even went so far as to inject it into healthy mice to see if it had adverse effects (which it didn't). This last finding was crucial. There are lots of substances that can kill bacteria; the trick is to find one that can kill bacteria while leaving human beings unharmed.

Having advanced his work with penicillin to this point, however, Fleming then abandoned it. He had clearly recognized its therapeutic potential right from the start, but his subsequent research seemed to convince him that penicillin was just one bacteria killer among many that had no future in medicine. He found that it was unstable; that it appeared very difficult to purify from the stew of fungus chemicals it existed in; and that it quickly lost its power to kill bacteria once they had been exposed to it. After 1931, Fleming didn't publish anything on penicillin or encourage his students to work with it. He seemed to regard it as a substance that was useful mostly for killing off unwanted bacteria in Petri dishes. A great deal of credit goes to him for being perceptive enough to recognize what had been laid in front of him. But had research on penicillin ended with him, there never would have been a drug called penicillin. By 1935, as Trevor Williams has written, "There was not one person in the world who believed in penicillin as a practical aid to medicine."

> *The team had to brew up liters of fungal fluid to get mere millionths of a gram of active penicillin.*

But this was soon to change, thanks to the work of two immigrants to Britain. One was Howard Florey, an Australian trained as a medical doctor who was then head of a school of pathology at Oxford University. The other was Ernst Chain, a refugee from Hitler's Germany who had been hired by Florey to develop a biochemistry unit in the school. The critical decision the two made in 1938 was to investigate substances in the same class as penicillin—antibiotics, which is to say substances made by one microorganism that are toxic to another. Critically, one of the compounds Florey and Chain decided to investigate was penicillin; they knew of its existence from reading the earlier papers of Fleming and others.

From the beginning, the team's research on penicillin proceeded on two tracks. One was simply seeing if the substance worked. The other was finding out what the substance *was* and trying to get enough of it to work with. Remember that penicillin is the product of a living organism—originally the fungus *Penicillium notatum*—and as such it existed within a mixture of thousands of other compounds the fungus produces. The trick was to isolate and characterize the germ-killing substance within the fungal soup. This biochemical work was the job of Chain. While this was going on, the team had to brew up liters of fungal fluid to get mere millionths of a gram of active penicillin.

Finally armed with enough of the drug to put it to the test, the group carried out the critical experiment. On the morning of May 25, 1940, they injected eight mice with deadly doses of streptococcal bacteria, and an hour later they injected four of the mice with penicillin. Then, a round-the-clock watch began. Normally, the dose given would have killed all the mice within 24 hours, and indeed by 3:30 a.m. the next day the four unprotected mice were dead. But the four mice given the penicillin were all fine! A long road lay ahead with human subjects such as Albert Alexander, but penicillin had done in mice what had so long been sought: It killed bacterial invaders while leaving their "host" unharmed.

This effort was being carried out against the backdrop of the outbreak of World War II. By the time the mouse experiments were started, much of Europe had fallen to Hitler, and it was feared that England might be next. (The wartime importance of a drug such as penicillin was clear to the Oxford team, which had contingency plans to destroy all its lab equipment and notes in the case of a successful German invasion. Only the precious *P. notatum* spores would be retained, by means of the researchers rubbing them into their clothing.) In this chaotic environment, the scientists were forced to go outside England to get the help they needed to ramp up the production of penicillin to an industrial scale. Howard Florey thus went to America to enlist aid there. Oxford researchers eventually found themselves working with chemical engineers from the U.S. Department of Agriculture in the unlikely location of Peoria, Illinois. Where once the English scientists had fermented penicillin in milk bottles, eventually they and their American colleagues were brewing it up in 25,000-gallon tanks.

Getting to this point meant coming up not only with better fermenting methods but also with better starting ingredients. To this end, military pilots collected soil samples from around the world in an effort to find a *Penicillium* species that would yield more of the precious drug. As luck would have it, a species that did the trick was found on the project's doorstep. In 1943, lab worker Mary Hunt bought a moldy cantaloupe at a Peoria market. The melon contained a fungus, *Penicillium chrysogenum*, that doubled the amount of penicillin produced; its daughter spores are the ones used to this day to produce penicillin.

Despite some detours, this crash program was a smashing success. The drug that once was restricted to mice, then to a few human subjects, was put into commercial production early in 1944. By the time of the Allied invasion of Europe, in June 1944, enough was being produced that every soldier who needed it could have it. Not much later, all restrictions on it were lifted; penicillin was available by prescription from the corner drugstore.

In 1945, Alexander Fleming, Howard Florey, and Ernst Chain were awarded the Nobel Prize in Physiology or Medicine for their work in discovering this wonder drug. Bestowing the prize just after the war ended, the Nobel committee noted:

In a time when annihilation and destruction through the inventions of man have been greater than ever before in history, the introduction of penicillin is a brilliant demonstration that human genius is just as well able to save life and combat disease.

**Figure 1**
**Penicillin's Discoverers**
The three scientists who were awarded the Nobel Prize for their work on the development of penicillin. From left to right, Alexander Fleming, Howard Florey, and Ernst Chain.

outside the amoeba is now inside, being digested. Many amoeboid protists are parasites, and one of them, *Entamoeba histolytica*, is a serious human parasite that causes an estimated 100,000 deaths per year, mostly in the tropics, by entering the human digestive system through contaminated water. When a parasite invades the human intestinal wall, the resulting condition is known as *dysentery*. When that parasite is an amoeba, the result is amoebic dysentery.

**Figure 21.17b** shows a protist known as a plasmodial slime mold. This organism assumes several forms during its life cycle. In the form you see in the picture, its name is appropriate because it is essentially moving slime. If you looked at this organism under a microscope, you would see lots of cell nuclei, but almost none of the *boundaries* (the plasma membranes) that normally separate one cell from another. The result is a free-flowing, but slowly moving mass of cytoplasm—the interior of a single cell spread out over a large area, with only one plasma membrane at its periphery. This material moves by the same "cytoplasmic streaming" you saw in the amoeboids. As it rolls over rocks or forest floor, it consumes underlying bacteria, fungi, and bits of organic material. Should an object stand in its path, it simply flows around it, or even through it, if the object is porous enough.

By now, you may be getting the impression that protists could be star contestants in a show called "Can You Top This for Strangeness?" If you look at **Figure 21.18**, you can see another contestant. Pictured is a much-studied organism, *Dictyostelium discoideum*, a so-called cellular slime mold that has a multiple personality. At one point in its life cycle, it exists in the form of individual amoeba cells that crawl along the forest floor feeding on bacteria. Each "Dicty" is on its own at this stage, but if the bacterial food runs low, these individual cells begin "aggregating," or coming together in an organized group. Eventually, they develop into a barely visible "slug" of up to 100,000 cells that has front and back ends and that migrates to a more fertile area of the forest. Then, at a certain point, these cells change form again. They arrange themselves into a tower-like reproductive structure that has, at its top, spores that will be dispersed and develop into new individual cells that once again crawl along the forest floor. The whole aggregation process will not take place unless starvation is imminent, however; without the danger of starvation as a motivator, the individual cells stay separate.

Now, once you get past the shock of realizing that such a thing could happen—that individual cells could just collect themselves and create a larger, moving organism—you can begin to see what interesting questions it raises. What's the signaling mechanism that brings 100,000 separate cells together? How do they begin to cooperate to form these various structures? (To put this another way, how does multicellularity begin to work?) And, perhaps most intriguingly, why should some cells sacrifice themselves by serving on the *stalk* of the reproductive structure—from which location they will not reproduce themselves—while others get to be at the top and disperse to a new life in better hunting grounds? If evolution is all about passing on genes, why should some cells give up the chance to pass on theirs? Scientists study Dicty in hopes of learning more about all these questions.

## On to Fungi and Plants

The bacteria, archaea, and protists reviewed in this chapter represent only a portion of the living world. Coming up next are two life-forms that most people think of as related and similar, although in fact they are only distantly related and are quite dissimilar. These are fungi and plants.

**Figure 21.18**
**Changing Forms**

**(a)** A *Dictyostelium* in its slug stage, composed of individual cells that have come together to produce an organism capable of moving across the forest floor. Although the slug is composed of tens of thousands of cells, it is no bigger than a grain of sand.

**(b)** Cells in the slug eventually arrange themselves into this tower-like reproductive structure. Reproductive cells called spores will be dispersed from the top of the structure to the forest floor. There they will begin life as individual *Dictyostelium* cells.

**(a)** *Dictyostelium* slug

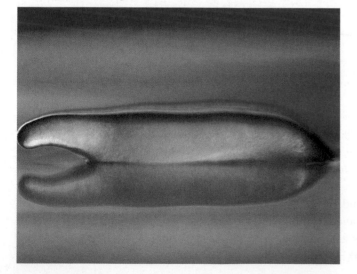

**(b)** *Dictyostelium* reproductive structure

# CHAPTER 21 REVIEW

For study help, activities, and more quiz questions, go to **www.aw.com/krogh4.**

## SUMMARY

### 21.1 Life's Categories and the Importance of Microbes

- All living things on Earth can be classified as falling into one of three domains of life: Bacteria, Archaea, or Eukarya. All the members of Domains Bacteria and Archaea are single-celled and microscopic. Domain Eukarya is further divided into four kingdoms: plants, animals, fungi, and protists.

- Microbes—living things so small they cannot be seen with the naked eye—are indispensable to all life on Earth. They produce more than half of Earth's atmospheric oxygen; the bacteria and archaea among them are responsible for putting atmospheric nitrogen into a form plants can use; and bacteria and fungi are the most important decomposers of the natural world—they break down dead organic matter, such as tree branches, and recycle the resulting elements back into the Earth.

- Microbes live in all environments in which larger life-forms exist. They are present in numbers so immense that the weight or biomass of all microbes on Earth exceeds the biomass of all larger life-forms.

### 21.2 Viruses: Making a Living by Hijacking Cells

- Viruses are noncellular replicating entities that must invade living cells to carry out their replication. Because viruses can carry out so few of life's basic processes on their own, most scientists do not classify them as living things.

- The human immunodeficiency virus (HIV), which causes AIDS, has two structures common to all viruses: genetic material and a protein coat, called a capsid, surrounding this material. HIV also has one other structural element that many viruses possess: a fatty membrane, called an envelope, which surrounds the capsid.

- Most viruses carry out four steps in their life cycle. They get their genetic material inside a "host" cell; turn out viral component parts; construct new virus particles from these parts; and move the new particles out of the cell, at which point the particles go on to infect more cells.

- Viruses cause a host of human illnesses. Health officials worldwide are now on high alert merely because of the potential harm that stands to come from one viral illness, avian flu. Scientists are watching to see if the virus that causes this illness, A (H5N1), will undergo a genetic transformation that will allow it to move easily between one human being and another. Viruses can undergo such transformations through two different means. One of them is mutation to their DNA or RNA; the other is that two different viruses can infect a single cell and then exchange genetic sequences while within the cell, thus producing a virus with different properties.

### 21.3 Bacteria: Masters of Every Environment

- Bacteria are microscopic, single-celled organisms that are prokaryotes: organisms whose genetic material is not contained within a nucleus. Other defining features of bacteria are that they have only a single organelle (the ribosome) and reproduce asexually through a simple cell splitting called binary fission. Millions of species of bacteria exist. Bacteria are metabolically far more diverse than plants or animals.

### 21.4 Intimate Strangers: Humans and Bacteria

- Bacteria live on and in human beings in great numbers. In the digestive tract, the relationship between humans and many bacteria is one of mutualism: a relationship between two organisms that benefits both of them. Bacteria get food and habitat from this relationship; human beings get an efficiently functioning digestive system.

### 21.5 Bacteria and Human Disease

- Only a small proportion of bacteria are pathogenic or disease causing, but these bacteria are responsible for some of humanity's worst diseases. A few pathogenic bacteria cause harm by invading human cells, but bacteria generally do their damage by releasing or leaving behind harmful substances called toxins.

- The primary human defense against pathogenic bacteria is the class of drugs known as antibiotics, defined as substances produced by one microorganism that are toxic to another. The first antibiotic, penicillin, was developed in the 1940s. Antibiotics work by exploiting the differences between bacterial and human cells, such that they kill bacteria while leaving human cells unharmed.

- The power of antibiotics is being threatened by the emergence of antibiotic-resistant strains of bacteria. These bacteria are evolving in greater numbers because of an overuse of antibiotics in medicine and agriculture. One antibiotic-resistant bacterium, methicillin-resistant *Staphylococcus aureus*, is being seen with increasing frequency in the general public, in particular among high school and college athletes.

### 21.6 Archaea: From Marginal Player to Center Stage

- Archaea were once thought to be a form of bacteria but are now known to constitute their own domain of life, standing beside Domains Bacteria and Eukarya. Archaea are superficially similar to bacteria in that they are single-celled prokaryotes that reproduce through simple cell splitting. However, archaea are unique in the living world at the level of the chemical structure of their cells. This structural uniqueness is based on a genetic uniqueness; many of the genes found in archaea are unlike the genes found in either bacteria or eukaryotes.

- Archaea exist in large numbers in some common environments. They make up 40 percent of the microbial life in large portions of the world's oceans, and they are seen in large numbers in common soil, where they join bacteria in carrying out one phase of the nitrogen-fixing process.

- Many species of archaea live in extreme environments and thus are extremophiles: organisms that grow optimally in environments whose conditions would kill most other organisms. Three large classes of extremophiles are thermophiles—organisms that live in extremely hot environments; halophiles—organisms that live in extremely salty environments; and anaerobes—organisms that can either do without oxygen or that actually are poisoned by it.

- Today, pharmaceutical and biotechnology firms are "prospecting" for novel extremophiles in their home environments, with the intent of developing commercial products from the enzymes these organisms produce.

## 21.7 Protists: Pioneers in Diversifying Life

- A protist is a eukaryotic organism that does not have all the defining features of a plant, an animal, or a fungus. This unsatisfactory definition stems from the fact that the term protist doesn't refer to a single evolutionary grouping but instead is used as a label for several different evolutionary lines of organisms, many of which are only distantly related.

- Protists are mostly microscopic, and all of them live in environments that are at least moist, if not aquatic. About 100,000 species are known to exist. The small portion of these that are pathogenic include *Plasmodium falciparum*, the cause of malaria, and the intestinal parasite *Giardia*, which contaminates water.

## 21.8 Protists and Sexual Reproduction

- For nearly the first 2 billion years after life appeared, it consisted solely of bacteria and archaea. Protists were the first life-form to evolve other than bacteria or archaea; they were the organisms that made transitions to many of the capabilities and forms seen in larger organisms today. Among these transitions was the change to sexual reproduction, which protists were the first to practice.

## 21.9 Photosynthesizing Protists

- Protists that get their nutrition by performing photosynthesis are known as algae. Some algal species provide examples of colonial multicellularity, defined as a form of life in which individual cells form stable associations with one another but do not take on specialized roles. Other algal protists provide exam-

ples of true multicellularity: a form of life in which individual cells exist in stable groups, with different cells specializing in different functions.

- Microscopic algae are important members of the group of organisms known as phytoplankton: small photosynthesizing organisms that float near the surface of water. Phytoplankton are very important to life in general because they produce most of Earth's oxygen, and because they form the base of so many aquatic food chains. All phytoplankton are either algae or bacteria.

## 21.10 Heterotrophic Protists

- Heterotrophic protists do not get their nutrients by performing photosynthesis but instead get them from consuming either other organisms or bits of organic matter. Some have evolved tiny slender extensions, cilia and flagella, with which they move toward prey or away from danger.

- The protists called amoeba move through use of pseudopods or "false feet"—slender extensions of the amoeba into which the rest of the body flows. Likewise, the protists called plasmodial slime molds and cellular slime molds move by means of this "cytoplasmic streaming." The cellular slime mold called *Dictyostelium discoideum* exists as a collection of individual amoeboid cells that come together to form a tiny "slug" during times of little food. Subsequently, these cells form a tower-like reproductive structure possessing cells at the top that will disperse to begin life as individual cells on the forest floor. "Dicty" raises questions about cell signaling and the origins of multicellular life.

## KEY TERMS

| | |
|---|---|
| **algae**   408 | **heterotroph**   400 |
| **antibiotics**   402 | **mutualism**   401 |
| **autotroph**   400 | **pathogenic**   401 |
| **binary fission**   398 | **phytoplankton**   408 |
| **capsid**   394 | **prokaryote**   397 |
| **colonial multicellularity**   408 | **protist**   405 |
| **eukaryote**   398 | **true multicellularity**   408 |
| **extremophile**   404 | **virus**   394 |

## BRIEF REVIEW

*Answers to Brief Review questions are in the back of the book.*
*For multiple-choice quiz questions, go to* **www.aw.com/krogh4**.

1. How do plants get the nitrogen they need?

2. Why do most biologists classify viruses as nonliving?

3. Antibiotics used in humans must have at least two qualities. What are they?

4. How can we say that archaea are superficially like bacteria and yet fundamentally different from them?

5. Name three features of more complex organisms that first appeared in protists.

6. What is an amoeba? What class of diseases is associated with them?

## APPLYING YOUR KNOWLEDGE

1. What does it mean to be a "successful" organism? Bacteria are certainly the most numerous organisms on Earth, they live in nearly all environments, and they are the least susceptible to being eliminated through environmental catastrophe. Are they Earth's most successful organisms?

2. What does it mean to be an "organism"? The cellular slime mold *Dictyostelium discoideum* exists initially as a collection of separate cells moving over the forest floor. In times of starvation, these cells come together to form a single "slug" that moves to more fertile territory. Eventually, it will change into a new structure that will throw off separate cells again. Is "Dicty" an organism or a group of temporarily cooperating cells?

3. What would life on Earth be like if there were no decomposers?

*Mushrooms of the fungus Amanita muscaria sit side by side with several species of plants in a Russian forest.*

Two of life's kingdoms seem alike but are very different. These are the fungi, which grow toward their food, and plants, which make their own.

| | | |
|---|---|---|
| 22.1 | **The Fungi: Life as a Web of Slender Threads** | **417** |
| 22.2 | **Roles of Fungi in Society and Nature** | **419** |
| 22.3 | **Structure and Reproduction in Fungi** | **420** |
| 22.4 | **Categories of Fungi** | **422** |
| 22.5 | **Fungal Associations: Lichens and Mycorrhizae** | **426** |
| 22.6 | **Plants: The Foundation for Much of Life** | **427** |
| 22.7 | **Types of Plants** | **429** |
| 22.8 | **Angiosperm–Animal Interactions** | **434** |

**Essay**

A Psychedelic Drug from an Ancient Source     **425**

# FUNGI AND PLANTS:
# The Diversity of Life 2

**C**ommon houseflies can suffer a hostile takeover by a fungus called *Entomophthora muscae*. It's not unusual for one organism to live off another, but in the case of *E. muscae*, the story has a bizarre twist. This

fungus gets inside flies either by drilling through their tough outer skeleton or by squeezing between some cracks in it. The strange thing is that, on entry, the first thing *E. muscae* does is send out extensions of itself to the fly's brain—specifically to parts of the brain that control crawling behavior. Eventually, the fly begins to crawl *upward*. This usually means upward on a plant stem or leaf, but it may mean upward on the wall of a house, toward the ceiling. Like any fungus, *E. muscae* lives by digesting organic material, and in this case the organic material is the fly's body. By the time the fly begins making its upward trek, it is nearly dead from the internal damage *E. muscae* has done.

If you look at **Figure 22.1**, you can see the end result of all this: a fly that has crawled up on a conifer needle and, in one of its last acts, has put its mouth parts on the needle, to which they are now glued by the fungus. The white growths you see around the fly's abdomen are fungal spores. Soon they will be showered over everything *beneath* the fly—including, perhaps, some more flies, which then can be parasitized as well. Fungi have various ways of getting their spores up off the ground for dispersal. The structure we call a mushroom is essentially a spore dispersal tower. But *Entomophthora muscae* has taken another tack. It gets its *prey* to lift it to higher ground, and sticks its prey in place; then it uses the prey as a launching pad for a shower of fungal

spores. *Entomophthora muscae* shows us once again that, while nature is certainly diverse, it is not always pretty.

## 22.1 The Fungi: Life as a Web of Slender Threads

*Entomophthora muscae* is a member of one of the "kingdoms" of the living world, the fungi. There are lots of fungi that make their living as parasites, as *E. muscae* does, but then again there are fungi that live in a beneficial association with plants and fungi

**Figure 22.1**
**Doing a Parasite's Bidding**
Pictured is a root maggot fly (genus *Delia*) that has been killed by an *Entomophthora* fungus. The white spores of the fungus can be seen bursting out from the fly's abdomen. Just before death, the fly made its way up to the conifer needles pictured and stuck its feeding organ to them. From the height the fly reached, the *Entomophthora* spores showered down on any additional flies that may have been below. *Entomophthora* brings about a behavioral change in its prey that causes them to climb or fly to heights from which its spores can be dispersed.

417

**(a)**

**(b)**

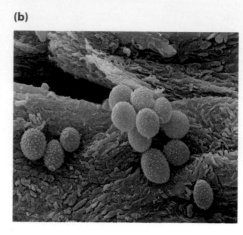

**(c)**

**Figure 22.2**
**Fungal Diversity**

**(a)** What happens to dead branches and leaves on the forest floor? They are decomposed, mostly by fungi. The two mushrooms sprouting from this leaf in Guyana are outgrowths of a network of fungal fibers that are growing within the leaf. Such fibers digest and absorb nutrients from the leaf, a process that eventually breaks the leaf down into its chemical component parts, which then are recycled back into the soil.

**(b)** Fungi are also human pests. Pictured is a human nail with fungal spores on it that are the cause of athlete's foot.

**(c)** Fungi distribute their spores in various ways. Pictured are spores being expelled from some puffballs in Costa Rica. What brings about the expulsion? Drops of rain falling on the puffballs.

that do us the favor of decomposing fallen trees (**Figure 22.2**). Fungi are a large and diverse kingdom, and hence the subject of the first half of this chapter. Once we've reviewed them, we'll look at a better-known kingdom, the plants.

At first glance, plants and fungi may seem to be alike. After all, they're fixed in one spot and tend to grow in the ground. But as you'll see, the main

connection between plants and fungi is that many of them have struck up underground partnerships. When it comes to evolution, the surprising thing is that fungi and *animals* are more closely related than fungi and plants. A single, common ancestor seems to have given rise to animals on the one hand and fungi on the other, with plants only distantly related to the members of either kingdom.

But if a fungus is not a plant, what is it? As the fly example shows, fungi do not make their own food as plants do. Instead, fungi are heterotrophs—they consume existing organic material in order to live. They make a living by sending out webs of slender, tube-like threads to a food source. This food source could be a fly, as in *E. muscae*'s case, but it could just as easily be bits of organic matter below ground or a fallen log in a forest. Whatever the case, fungi cannot immediately take in or "ingest" their food in the way that animals do. Fungal cells have cell walls, which means that only relatively small molecules can pass into the fungal filaments. The fungus' solution to this problem is to digest its food *externally*, through digestive enzymes it releases. When the food has been broken down sufficiently by these enzymes, the resulting molecules are taken up into the filaments. Almost all fungi obtain nutrition in this manner. Whether we are talking about a microscopic bread mold or extensive webs of fungi that spread underneath the forest floor, this is a defining characteristic of fungi. This characteristic and others can be stated in a more formal way.

**Figure 22.3**
**Structure of a Fungus**

Fungi are composed of tiny slender tubes called *hyphae*. The hyphae form an elaborate network called a *mycelium*. The same hyphae also form a reproductive structure called a fruiting body, which in this case is the familiar mushroom.

fruiting body (mushroom)

mycelium

spore

hyphae

- Fungi largely consist of slender, tube-like filaments called **hyphae**.

- Fungi get their nutrition by dissolving their food externally and then absorbing it into their hyphae.

- Collectively, the hyphae make up a branching web, called a **mycelium**. If you look at **Figure 22.3**, you can see the nature of this structure. You may have thought of a mushroom as the main part of a fungus, but in reality it is merely a reproductive structure, called a *fruiting body*, that sprouts up from the larger mycelium below. A fruiting body amounts to a large, folded-up collection of hyphae.

- Most fungi are **sessile** or fixed in one spot, but their hyphae grow. The fungal mycelium does not move, but its individual hyphae grow toward a food source. This growth comes, however, only at the tips of the hyphae, not throughout their length.

- Fungi are almost always multicellular. The primary exception to this is the yeasts, which are unicellular and thus do not form mycelia.

This last point leads to a more general concept about the living world. You may recall from last chapter that all bacteria and archaea are single-celled, and that protists are usually single celled. But protists are the limit of single-celled life. With fungi, plants, and animals, life becomes almost completely multicellular. There is no such thing as a single-celled plant or animal, and the one large group of single-celled fungi, the yeasts, seem to have evolved into this state from an earlier multicelled form. This is not to say that all life becomes *big* once we get past the protists. As you'll see, there are plenty of microscopic fungi (and even some microscopic animals). But from here on out, all the life-forms we'll be looking at are made up of more than one cell.

## 22.2 Roles of Fungi in Society and Nature

Like bacteria, fungi are mostly hidden from us—and mostly known to us by the trouble they cause. This trouble is considerable as fungi are responsible for mildew, dry rot, vaginal yeast infections, bread molds, general food spoilage, toenail and fingernail infections, and a variety of related skin afflictions, among them athlete's foot, "jock itch," and ringworm. The agricultural blights of corn smut and wheat rust are fungal infections. Indeed, fungi probably are the single worst destroyers of crop plants. One survey in Ohio indicated that only 50 plant diseases there were caused by bacteria and 100 by viruses, but 1,000 by fungi. If you look at **Figure 22.4**, you can see an example of what happened to American elm trees in cities across the United States beginning in the 1930s because of a fungus called *Ophiostoma ulmi*—the cause of Dutch elm disease. In recent years, there has been great concern about the fungi commonly called molds and the effects they have on both houses and the people who live in them.

Even given all this fungal pestilence, however, we could say of the fungi what we said of bacteria: While

**(a)** Gillet Avenue, Waukegan Illinois, 1962

**(b)** Gillet Avenue, Waukegan Illinois, 1972

**Figure 22.4 The Effects of Dutch Elm Disease**

we could do without some of them, we could not live without others. Mold in houses may be a huge problem, but there is another mold that is the source of penicillin. Fungal infections may be a problem, but organ transplants are made possible today by the immune-system-suppressing drug cyclosporin, which is derived from a fungus. If you know someone who has high cholesterol, he or she may control it with the drug lovastatin, whose source is also a fungus. In the realm of food production, brewer's yeast makes bread rise and beer ferment, and blue cheese is blue because of the veins of mold within it. Most soft drinks contain citric acid that is produced from a fungus, and we simply eat fungi outright in the form of mushrooms. Out in nature, fungi join bacteria as the major decomposers of the living world, breaking down organic material such as garbage and fallen logs and turning it into inorganic compounds that are recycled into the soil. In fact, the final breakdown of woody material is almost entirely the work of fungi. And, as noted, fungi are involved in a critical association with plants, about which you'll learn more later.

Some 70,000 species of fungi have been identified so far, but this is just a fraction of the number that actually exist. Because fungi tend to grow in inaccessible places, most of them are unknown to us. By one estimate, the total number of fungal species may be about 1.6 million.

**ANSWERS**

1. hyphae; mycelium
2. (b)
3. breaking down and recycling organic matter

### SO FAR...

1. The essential structure of most fungi is a set of slender filaments called _____ that collectively make up a structure called a _____.

2. Which of the following is not a characteristic of fungi? (a) usually fixed in one spot; (b) make their own food; (c) tend to be multicellular.

3. Fungi are indispensable to the natural world because of their role in _____.

## 22.3 Structure and Reproduction in Fungi

Let's now look at one type of fungus as a means of learning something about fungi in general. In **Figure 22.5**, you can see a so-called club fungus that we know as the common mushroom. If you look at the figure's lower-right portion, you can see that the mushroom's hyphae amount to thin lines of cells—something that is true of all hyphae. In many cases, individual cells in the hyphae are separated from one another by dividers called septa. But as it happens, septa are somewhat porous dividers; they have tiny openings that allow for a fairly free flow of cellular material between one cell and the next. This allows for rapid movement of cellular resources right to the tip of the hyphae—the site of hyphal growth. The result is that hyphae can grow very quickly, whether they are growing toward a food source or

organizing themselves into a mushroom cap. Did it ever seem to you that a mushroom sprouted in your yard overnight? It probably did.

If you look at the upper left of the figure, you can see how the mushroom cap is structured. The underside of each cap is made up of a profusion of accordion-like folds called gills. Each gill is simply a collection of hyphae. And, right at the tip of some of these hyphae is a reproductive structure called a basidium. Reproductive cells called spores are produced within the basidium and then ejected from it. Caught by the wind, these spores are then carried away to new ground on which they can "germinate" or sprout new hyphae. Fungal spores are, however, something like lottery tickets: A multitude are made, but only a few will be winners. Such a tiny fraction of spores end up germinating that huge numbers must be released to ensure that all nearby environments receive some. Tens of millions will be

**Figure 22.5
The Life Cycle of a Fungus**

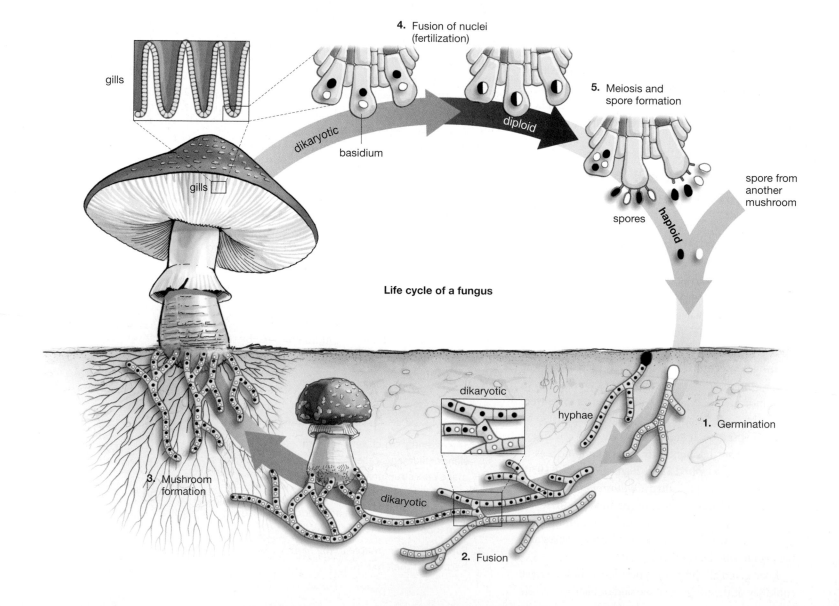

ejected each hour from the underside of a fairly small mushroom cap.

The mushroom cap is one example of a large fungal structure, but most fungi don't have this kind of size. Think of a flat circle of blue-gray mold on some spoiled bread. It has its own mycelium, with reproductive structures topped by spores, but magnification is necessary to make out any of this detail, as you can see in **Figure 22.6**. To a great extent, fungi can be regarded as microbes.

## The Life Cycle of a Fungus

We've noted the role that mushroom caps play in releasing mushroom spores, but how do these spores get produced in the first place? For that matter, what's a spore? These questions can best be understood within the context of the life cycle of one type of fungus, which is set forth in Figure 22.5. To understand it, we need to go over three tricky concepts related to fungal reproduction. One of these can be called the dikaryotic phase of life; the others are the nature of sex in fungi and the nature of spores.

### The Dikaryotic Phase of Life

Taking the life cycle in Figure 22.5 from the beginning, over at step 1, you can see that it starts with spores that have been released from fruiting bodies. These spores germinate, generate hyphae, and in step 2, two of these hyphae fuse—two hyphal cells come together and become one cell, from which a single hypha grows. Note, however, that the cells in this new hypha have *two* nuclei each (symbolized by the little black and white dots within them). This is an unusual condition for a cell; in the normal case, one cell has one nucleus. This condition has come about because, while two cells fused in step 2, their nuclei remained separate. Thus begins a phase of life that is unique to fungi: the **dikaryotic phase**, in which cells in a fungal mycelium have two nuclei. And, what is the nature of the genetic material in these nuclei? Each nucleus exists in what is known as a haploid state—there is but a *single set* of chromosomes in each one. This contrasts with the diploid state in almost all human cells. Our nuclei have *paired* sets of chromosomes within them. The haploid state of these nuclei may seem like a technical detail, but it's important for what's coming up. Once a mushroom mycelium enters dikaryotic phase it can stay there for a long time. The phase can go on for years, in fact, and ends only when the most visible manifestation of the mycelium, the mushroom cap, sprouts above ground as seen in step 3.

**Figure 22.6**
**Bread Mold Up Close**
The dark spots on moldy bread are composed of a mycelium that forms stalks tipped by spores, visible in this micrograph as tiny black balls. The spores will be dispersed to produce new mycelia elsewhere—perhaps on more bread. These are spores of the mold *Rhizopus nigricans*, growing on the surface of some bread.

Web Animation 22.1
Life Cycle of Fungi

### Sex in Fungi

The dikaryotic state initially exists in all the cells in the mushroom—even the cells that make up the spore-releasing basidia that line the gills. But in some of these cells, right at the tips of the basidia, this will change. The two nuclei in these cells will fuse into one nucleus as seen in step 4. This change means these cells have become diploid; they now have a single nucleus that contains *two* sets of chromosomes. Now, think about where each of these sets of chromosomes came from. One of them came originally from the black spore over at step 1, while the other came from the white spore. They existed separately in a single cell for a while, but now they have fused. With this, we have had a fusion of genetic material from two separate organisms—sex has taken place. It's not the merger of egg and sperm we're used to in human reproduction, but it's sex nevertheless.

Next, as you can see in step 5, each of the cells in which this fusion has taken place undergoes the meiosis that was reviewed in Chapter 10. This means a mixing of chromosomes from the two parents and two rounds of cell division. The result is four haploid spores attached like little bags to the tip of a basidium; these will be among the millions of spores released from the mushroom cap. Each of these spores will be of one "mating type" or another—symbolized by the black and white colors in the figure. Just as only opposite sexes can come together to create offspring in humans, so only opposite mating types can come together (in hyphae fusion) to create offspring in fungi when they reproduce sexually.

### The Nature of Spores

This gets us to the nature of a **spore**, which can be defined as a reproductive cell that can develop into a new organism without fusing with another reproductive cell. Note the difference between the spores of the mushroom and the *gametes* of a human being—eggs or sperm. A human being cannot develop from an egg or a sperm; a fusion of both cells is required. Meanwhile, the spores that our mushroom developed from did not undergo fusion. A fungal spore was produced in meiosis, and once it came into being, it was dispersed and developed into a whole new mycelium by itself.

### Reproduction in Other Types of Fungi

Complicated as all this may seem, it actually amounts to only one means of fungal reproduction among many that exist. Under some conditions, fungi that are capable of sexual reproduction will bypass it altogether in favor of *asexual* reproduction. The little black balls you can see on top of the bread mold in Figure 22.6 are spores that were produced through mitosis, or simple cell division, as carried out by one organism. There was no fusion of the cells and nuclei from *two* organisms to produce these spores—there was no sex, in other words. Spores such as the ones on the bread mold develop through cell division, after which they disperse to form new mycelia; then these mycelia develop new spores through cell division and the process continues. Note that the *constant* within fungal reproduction is spores—they are employed no matter what. But spores can be produced either sexually or asexually. And the sexual processes that fungi employ can vary considerably. In short, fungi exhibit great diversity in the means by which they reproduce themselves.

## 22.4 Categories of Fungi

We're used to dividing the animal world up into various categories, such as insects, mammals, and reptiles. So, what are the major categories of fungi? It turns out there are four principal groups or "phyla" of fungi,

and that three of these phyla are defined by their reproductive structures. As you've seen, toadstool-shaped mushroom fungi have a reproductive structure called a basidium; accordingly, these mushrooms are in the fungal group known as the basidiomycetes. Likewise, there is a group called the ascomycetes and one called the zygomycetes—both named for the reproductive structures they have. Finally, there is a group called the chytrids whose members give us an idea of how close the evolutionary relationship is between fungi and animals. Let's look briefly now at each of these four fungal phyla.

- **Basidiomycetes**—Toadstool-shaped mushrooms are the best-known members of Phylum Basidiomycota, but this group also includes two banes of the agricultural world—the rusts and the smuts—along with the familiar "shelf" fungi that often are seen sprouting from trees (**Figure 22.7**). You may recall that the spore-releasing basidium on a mushroom is shaped something like a club. It is this shape that gives the basidiomycetes the common name of the club fungi. Most of the living world benefits from the work of basidiomycetes as they serve as a major group of decomposers. This is small comfort to farmers worldwide, however, who must live with the incredible agricultural damage caused by the basidiomycete smuts and rusts. Wheat, corn, barley, oats, soybeans, coffee—all these crops and more are infected by these fungi, which can sweep through crops in epidemic waves. In 1993, leaf rust destroyed over 40 million bushels of wheat in Kansas and Nebraska alone.

- **Ascomycetes**—Morel mushrooms are members of Phylum Ascomycota, as are the highly prized truffles that are used in cooking. On the other

**(a)** Shelf fungus on beech tree trunk

**(b)** Brown loose smut on barley

**Figure 22.7**
**Basidiomycete Fungi**

**(a)** The basidiomycetes include the well-named shelf fungi, such as this *Ganoderma* fungus, seen growing from the trunk of a beech tree in England.

**(b)** The agricultural scourges called smuts are also basidiomycetes. Note the difference between healthy heads of barley and those infected with the basidiomycete smut *Ustilago nuda*.

hand, the cause of Dutch elm disease is the ascomycete *Ophiostoma ulmi*, and if you see some brown rot on a peach or other stone fruit, it is likely to be the ascomycete *Monilinia fructicola*. All members of this phylum reproduce by using tiny spore-releasing structures whose sac-like shape gives the ascomycetes their common name: the sac fungi. In ascomycete mushrooms, these sacs can be contained in a large fruiting body that is shaped like a cup (**Figure 22.8**).

- **Zygomycetes—** When you see a patch of blue-gray mold on bread, you're probably looking at a fungus that's part of Phylum Zygomycota. This is so much the case that a common name for the zygomycetes

is the bread molds. There are only about 1,050 known species of zygomycetes, which is a small number compared to the more than 22,000 known basidiomycetes or the 32,000 ascomycetes. Nevertheless, the zygomycetes are diverse. Many of them form underground associations with plant roots, but one group of them makes a living by parasitizing insects. (Remember the *Entomophthora* species from the start of the chapter that parasitized the fly? It was a zygomycete.) What all zygomycetes have in common is the use of a reproductive structure whose slender extensions are topped by spherical sacs filled with spores; you can get a close-up view of this structure in **Figure 22.9**.

**(a)** Common morel mushroom (*Morchella esculenta*)

**(b)** Orange peel fungus (*Aleuria aurantia*)

**Figure 22.8**
**Ascomycete Fungi**

**(a)** The ascomycetes include the morel mushrooms, such as these specimens (*Morchella esculenta*), growing in Pennsylvania.

**(b)** Another group of ascomycetes are "cup" fungi, such as this orange peel fungus (*Aleuria aurantia*) growing in Scotland.

**(a)** Bread mold (*Mucor mucedo*) growing on bread

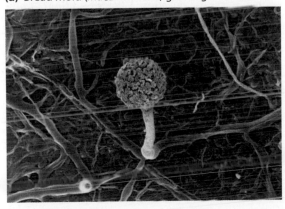

**(b)** Close up of spore-laden bread mold sporangium

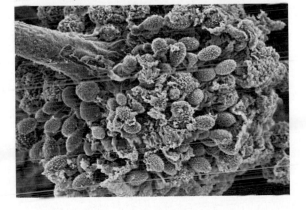

**Figure 22.9**
**Zygomycete Fungi**

**(a)** When they reproduce asexually, zygomycetes employ variations on the lollipop-like reproductive structure, called a sporangium, you can see in the photo on the left. In this zygomycete (*Mucor mucedo*), seen here growing on a piece of bread, the nutrient-absorbing filaments of the mycelium can be seen forming a network below the sporangium.

**(b)** A closer view of a spore-laden zygomycete sporangium.

**Figure 22.10**
**The Chytrid Fungi**

**(a)** In this highly magnified micrograph, a Chytrid fungus can be seen releasing a cloud of spores even as it feeds on a pollen grain, visible as the pale brown circle at the photo's center.

**(b)** Like many of its fellow amphibians, this great barred frog has has been infected by the Chytrid fungus *Batrachochytrium dendrobatidis*. Such infections have caused steep, continent-wide declines in some frog species and have brought about the actual extinction of other species. *B. dendrobatidis* invades the skin of frogs, altering it in ways that can be lethal.

**(a)** Chytrid fungus growing on a pollen grain

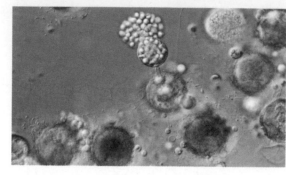

**(b)** Frog infected with a chytrid fungus

- **Chytrids**—The fungi of Phylum Chytridiomycota are very different from any of the others we've looked at. For one thing, most of them live in water. But beyond this, all of them are *mobile* at one point in their life cycle. Indeed, these fungi move through the water using the kind of tail-like flagellum that we saw last chapter in various species of protists. (To picture a flagellum, just think of the "tail" of a human sperm.) And, some chytrids don't develop a mycelium but instead are like the yeasts in that they are single-celled. Given all this, you might wonder why the chytrids are classified as fungi at all. The answer is that they duplicate their DNA in a manner used by all fungi; their cell walls are partly composed of a material (chitin) found in most fungal cell walls; and contemporary sequencing of their DNA shows that their closest relatives are fungi rather than any other life-form. The fascinating thing is that the chytrids probably give us an idea of what fungi were like *originally*—back when fungi first evolved from protists. The sessile fungi we've been looking at seem to have evolved from a group of mobile fungi that resembled today's chytrids.

Moreover, the chytrids show us how close the association is between fungi and animals. Think of it: DNA sequencing tells us how close fungi and animals are to each other; the chytrids show us what the early, mobile fungi were like; and a good deal of evidence indicates that early animals had the flagellated form of the chytrids. Given all this, the message is that primitive animals probably were not all that different from primitive fungi. Like most of the fungi we've looked at, the chytrids take a toll on other life-forms. One species of chytrid, *Batrachochytrium dendrobatidis*, has been decimating populations of frogs worldwide in recent years (**Figure 22.10**).

## Yeasts: *Saccharomyces cerevisiae*

Several times so far, we've taken note of the single-celled fungi called yeasts. Where do they fit within the four-phylum structure of the fungi? The answer is that, with the exception of the chytrids, each of the fungal phyla has yeast species within it—there are basidiomycete yeast, ascomycete yeast, and zygomycete yeast, in other words. Thus, "yeast" is not a taxonomic category; it merely describes a form that fungi can take. Most yeasts are ascomycetes, but the basidiomycetes are well represented in this category as well. We can define a **yeast** as any single-celled fungus that tends to reproduce by budding. This, of course, raises the question: What's budding? **Figure 22.11** shows it in action. As you can see, budding is not a splitting of one cell down the middle; it is a means of fungal reproduction in which daughter cells are produced as outgrowths of parental cells.

The particular yeast that's shown in the picture deserves special mention since it has been tremendously useful to people for as long as human civilizations have existed. *Saccharomyces cerevisiae* is the most important yeast by far in making bread rise and in making wine and beer ferment. The alcohol in wine is simply a product of *S. cerevisiae*—it feeds on

**Figure 22.11**
**Our Partner in Food and Drink**

The fungal yeast *Saccharomyces cerevisiae* has been used by human beings for thousands of years in baking and brewing. The *S. cerevisiae* cell at the forefront of the picture is giving rise to a daughter cell through the process of budding. The circular areas on this cell and the one behind it are bud scars—areas from which previous daughter cells have broken away.

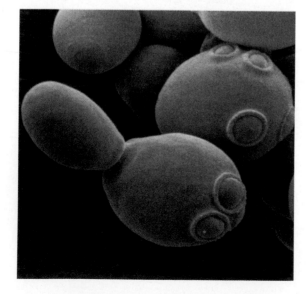

## ESSAY | A Psychedelic Drug from an Ancient Source

In 1943, chemist Albert Hoffman of Switzerland's Sandoz pharmaceutical company found himself working with a substance that had a long history, most of it tragic. The substance came from a fungus in the genus *Claviceps*, which infects several grain crops, among them rye. The visible manifestation of a *Claviceps* infection is little sickle-shaped growths of black hyphae that pop out among the rye grains (**Figure 1**). Centuries ago, the French perceived these growths to look like an *ergot*—a little spur that some birds have on their legs—and the name stuck.

In nature, when wild grasses get ergot infections, the effect is not all that notable. The fungus takes some plant nutrition away, the ergot growths fall to the ground in autumn, and spores from it infect the next generation of grass in the spring. The problem for humans comes when the fungus infects *food* crops such as rye, and the ergot are ground up along with the rye grains. The result is that ergot becomes a flour ingredient. This is serious because ergot contains a powerful alkaloid; its active ingredient is a member of the family of alkaline substances that also includes cocaine, nicotine, and morphine, along with such poisons as hemlock. In numerous instances, recorded in Europe since the Middle Ages, ergot made its way into the bread that people ate, and the results were disastrous. France in particular had repeated instances of "ergotism," with two of these episodes alone resulting in 50,000 deaths. What did ergot do? Here is a description by Oswald Tippo and William Louis Stern. Ergot's effects included:

> Abortions in pregnant females; burning sensation in extremities; feelings of being consumed by fire; constriction of blood vessels leading to the blockage of circulation and resulting in the appearance of blue discolored areas on the body; onset of gangrene; shriveling and falling off of hands, arms, and legs; experiencing of hallucinations such as wild beasts in pursuit of the victim.

Horrible as these effects were, medical people hearing of them recognized that ergot had the ability to do good as well as harm. Unwanted abortions are a terrible thing, but labor sometimes needs to be *induced* when a fetus is overdue; likewise, the constriction of blood vessels has its uses when properly controlled. Alkaloids often have such double-edged capabilities. Cocaine was first isolated from the coca plant in 1860, but in the 1880s it became a miracle drug for eye surgery when it served as the first *local* anesthetic. Morphine, which is derived from poppy plants, is used in medicine as a painkiller. Thus, it's not surprising that drugs were also isolated from ergot over time. In the 1930s, for example, a drug called ergonovine—still in use today—was developed that helps stop bleeding after childbirth.

Thus it was that Albert Hoffman found himself working with ergot in his Basel lab in 1943. He was looking for other useful alkaloids that the fungus might yield. Years before, he had isolated from ergot a chemical core—a compound called lysergic acid—and added to it a manufactured compound called diethylamine, but the results seemed uninteresting to him. This time, however, he apparently absorbed a tiny quantity of the substance through his skin. The outcome, as he later wrote, was that:

> I was seized by a peculiar sensation of vertigo and restlessness. Objects appeared to undergo optical changes and I was unable to concentrate on my work. . . . With my eyes closed, fantastic pictures of extraordinary plasticity and intensive color seemed to surge toward me.

Although he had not intended to, Albert Hoffman had discovered one of the world's most powerful psychoactive drugs: lysergic acid diethylamide, which the world would come to know by its initials, LSD.

**Figure 1**
**Damaging Intruder**
Black ergots of the *Claviceps purpurea* fungus sprout from among the grains of a wheat plant.

the sugars in grape juice and produces alcohol in the course of its metabolism. In a similar vein, the rise that we see in yeasted breads almost always comes from the carbon dioxide given off by *S. cerevisiae* as it feeds on other bread ingredients. As if all this weren't enough, *S. cerevisiae* is a mainstay of scientific research: an entire field of genetics has been built around the study of it.

**SO FAR...**

1. A spore is a reproductive cell that can develop into a new organism without _____.

2. Most fungi are capable of reproducing in cycles that are carried out entirely through simple cell division or mitosis. When fungi reproduce through this means, they are carrying out _____ reproduction.

3. Most fungi are multicelled organisms that take the form of a web of filaments called hyphae. However, three of the four fungal phyla also have single-celled species that collectively are referred to as _____.

**Figure 22.12**
**Life on the Rocks**
Lichens growing on sandstone rocks near the Montana-Wyoming border.

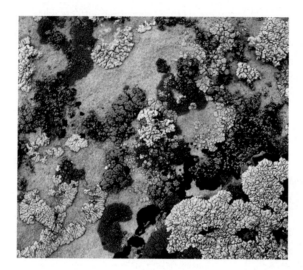

**Figure 22.13**
**The Structure of a Lichen**
A lichen is not a single organism. It is composed of a fungus and an alga (or sometimes a photosynthesizing bacterium) living in a mutually beneficial arrangement. The fungal hyphae form a dense layer on the top and the bottom and a loose layer in the middle, within which the alga is nestled. The fungus provides water, carbon dioxide, and a protective environment for the alga, and the alga provides food for the fungus through the photosynthesis it performs.

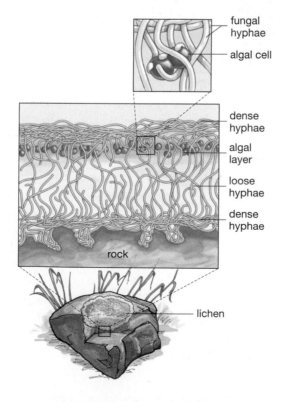

fungal hyphae

algal cell

dense hyphae

algal layer

loose hyphae

dense hyphae

rock

lichen

**••• ANSWERS**

1. fusing with another reproductive cell
2. asexual
3. yeasts

## 22.5 Fungal Associations: Lichens and Mycorrhizae

### Lichens

Everyone has seen the thin, sometimes colorful coverings called *lichens* that seemingly can grow anywhere—on rocks as well as on trees, in Antarctica as well as in a lush forest (**Figure 22.12**). As it turns out, lichens are not a single organism. A **lichen** is a composite organism composed of a fungus and either algae or photosynthesizing bacteria. If you look at **Figure 22.13**, you can see that this association can be structured as a kind of sandwich, with an upper layer of densely packed fungal hyphae on top, then a zone of less-dense hyphae that includes a layer of algal or bacterial cells, then another layer of densely packed hyphae on the bottom that sprout extensions down into the material the lichen is growing on (such as a rock). The example shown is an association between a fungus and an alga, which is the case in about 90 percent of lichens. In these cases, the fungal hyphae either wrap tightly around the algal cells or actually extend into them.

A lichen is mostly a fungus, then, but it is a fungus that could not grow without its algal partner. Why? Because the alga makes its own food through photosynthesis and then proceeds to supply the fungus with most of the nutrients it needs. A debate existed for years about whether the fungus was in turn doing something for the alga, but this now seems to have been settled. The relationship between a fungus and an alga is *mutualistic*—both parties are benefiting from it. The alga may be supplying food *to* the fungus, but it is getting several things *from* the fungus in return: water, carbon dioxide, minerals, and protection from the elements. This last factor explains why lichens are often so colorful. Their fungi produce brightly colored pigments that can screen out harsh sunlight—something that's important in the exposed environments that lichens favor.

In nature, the general significance of lichens comes from their role as pioneers in barren habitats where life is just getting established. Because the algae in lichens make their own food, and because the fungi in them can dig into rock itself, lichens can make a living in these tough habitats, whereas most other organisms cannot. Beyond this, lichen fungi produce acids that can help break down rock and build up soil—activities that create habitat for *other* organisms.

Lichens have an ability not only to establish themselves in tough environments but also to persist in them for a long time. If left undisturbed, they can survive for thousands of years in colder climates. The lives of lichens, however, could be characterized not only as long, but as slow. At most, they spread out at a rate of about 20 millimeters per year, which is about three-fourths of an inch. Indeed, they have been observed to grow as little as 0.2 millimeters per year, at which pace they would grow by about three-fourths of an inch per century. What accounts for this sluggish rate of expansion? In times of little water, lichen algae stop performing photosynthesis. This puts the entire organism into a state of dormancy—a kind of suspended animation—that persists until water becomes available again.

### Mycorrhizae

By some estimates, 90 percent of seed plants live in a cooperative relationship with fungi. The linkage here is one of plant roots and fungal hyphae, with both the plants and the fungi benefiting. What the fungus gets from the association is food, in the form of carbohydrates that come from the photosynthesizing plant; what the plant gets is minerals and water, absorbed by the fungal hyphae. So important is this relationship to plants that some species of trees, such as pine and oak, cannot grow without fungal partners. Other plants will have stunted growth without a fungal partner.

Fungal hyphae generally grow into the plant roots, but in some instances they wrap around the root without penetrating it. In either case, the root–hyphae associations of plants and fungi are known as **mycorrhizae** (**Figure 22.14**). The mycorrhizal relationship is one of the oldest and most important in nature. It appears that at least 460 million years ago, when the very first plants made the transition to life on land, they did so with fungal partners.

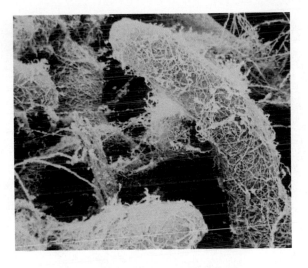

**Figure 22.14**
**Underground Partners**
Mycorrhizae are associations between plant roots and fungal hyphae that benefit both the plant and the fungus. Shown is an association between Aspen tree roots and threadlike mushroom hyphae that are wrapped around them. The hyphae help bring water and minerals to the trees, while the tree provides the fungus with food.

---

**SO FAR...**

1. A lichen is a composite organism, always composed of a _____ and either _____ or _____.

2. In a lichen, the _____ is providing the _____ with water, carbon dioxide, minerals, and protection from the elements.

3. The cooperative associations called mycorrhizae provide the _____ partner with sugars, while providing the _____ partner with water and minerals.

---

## 22.6 Plants: The Foundation for Much of Life

In your review of the living world so far, you've seen that human beings have reason to feel ambivalent about bacteria, protists, and fungi. While there is much to appreciate in these organisms, there is a good deal to fear as well. When it comes to plants, however, it's hard to find much to fear—but easy to find much to appreciate. For starters, we humans are utterly dependent on plants for food. Try naming something you eat that isn't a plant or that didn't itself eat plants, and you will end up with a very short list. Then there is the atmospheric oxygen that plants produce as a byproduct of photosynthesis. Beyond this, plants stabilize soil, provide habitat for animals, and lock away the "greenhouse" gas carbon dioxide. They yield lumber and medicines, and certain varieties of them are so beautiful that human beings spend hours tending them. Against all this, what's the worst thing we could say of them? That some of them grow in a greater abundance than we'd like? What other form of life helps us so much and harms us so little?

### The Characteristics of Plants

We all have an intuitive sense of what plants are, although not all plants match our intuitions. A few are parasites (on other plants) and some, like the Venus flytrap, consume animals. Allowing for such exceptions, plants share the following characteristics:

- **Plants are photosynthesizing organisms that are fixed in place, multicelled, and mostly land-dwelling.** We know that plants are fixed in one spot (sessile), and that they make their own food through photosynthesis. That they are multicelled follows from the fact that they develop from embryos, which are multicelled by definition. During the course of their evolution from green algae, plants made a transition from water to land, and land is where most of them are found today—although some have made the transition *back* to water, as with water lilies.

- **Plant cells have cell walls.** Both plant and animal cells are enclosed by a plasma membrane, but plant

... ANSWERS
1. fungus; algae; bacteria
2. fungus; algae or bacteria
3. fungal; plant

**Figure 22.15**
**Characteristics of Plants**

All cells have an outer membrane, but plant cells have a wall external to this membrane. The compounds cellulose and lignin, which help make up the cell wall, impart strength to it. Plant cells have a higher proportion of water in them than do animal cells, with much of this water located in an organelle called a central vacuole. The sites of photosynthesis in plants are the organelles called chloroplasts.

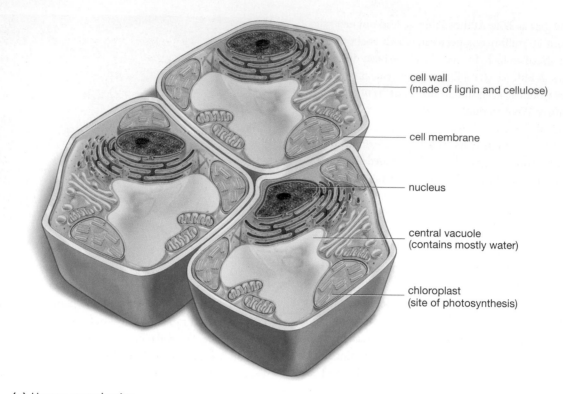

cell wall
(made of lignin and cellulose)

cell membrane

nucleus

central vacuole
(contains mostly water)

chloroplast
(site of photosynthesis)

**Figure 22.16**
**The Alternation of Generations in Plants, Compared to the Human Life Cycle**

**(a) Humans.** Almost all cells in human beings are diploid or 2n, meaning they have paired sets of chromosomes in them. The exception to this is human gametes (eggs and sperm), produced through meiosis, which are haploid or 1n, meaning they have a single set of chromosomes. In the moment of conception, a haploid sperm fuses with a haploid egg to produce a diploid zygote that grows into a complete human being through mitosis.

**(b) Plants.** In plants, conversely, diploid (2n) plants—the multicellular sporophyte fern in the figure—go through meiosis and produce individual haploid (1n) spores that, without fusing with any other cells, develop into a separate generation of the plant. This is the multicellular gametophyte shown. This gametophyte-generation plant then produces its own gametes, which are eggs and sperm. Sperm from one plant fertilizes an egg from another, and the result is a diploid zygote that develops into the mature sporophyte generation. The alternation between the sporophyte and gametophyte forms is called the alternation of generations.

**(a)** Human reproduction

multicellular
diploid
adults

meiosis

egg

sperm
1n

fertilization

2n

mitosis and
development

zygote

**(b)** Plant alternation of generations

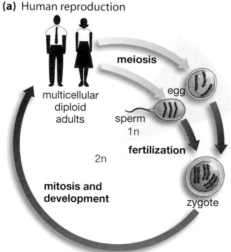

mitosis

mitosis

multicellular
gametophyte

spores

gametes

1n

meiosis

fertilization

2n

multicellular
sporophyte

zygote

mitosis

haploid (1n)

diploid (2n)

cells have something just outside the membrane: a **cell wall**, defined as a relatively thick layer of material that forms the periphery of plant, bacterial, and fungal cells (**Figure 22.15**). The plant cell wall is composed in large part of a tough, complex compound called cellulose. In woody plants, cellulose is joined by a strengthening compound called lignin. Together, cellulose and lignin allow plant cells to support the massive weight of trees.

- **Plant (and algae) cells have organelles called chloroplasts that are the sites of photosynthesis.** If there is one feature that has fundamentally shaped plant evolution, it is the chloroplast. Why are plants *able* to be fixed in one spot? Because they make their own food in chloroplasts and thus don't need to move toward food, as animals do.

- **Successive generations of plants go through what is known as an alternation of generations.** One way to conceptualize this characteristic is to compare plant reproduction with human reproduction. The eggs and sperm that human beings produce are so-called gametes that are *haploid* cells, meaning they have but a single set of chromosomes in them. When these haploid cells fuse in the moment of conception, what's produced is a *diploid* fertilized egg, meaning an egg with two sets of chromosomes—one set from the father and one from the mother. This egg gives rise to more diploid cells through cell division, and the result is a whole new human being (**Figure 22.16a**).

A typical *plant* in its diploid phase will, like human beings, produce a specialized set of haploid reproductive cells. Instead of these cells being egg or sperm, however, what's produced is a spore: a reproductive cell that can develop into a new organism without fusing with another reproductive cell. Like the spores we saw in fungi, plant spores have the ability to grow into a new generation of plant all on their own. In some plants, this spore lands on the ground, starts dividing, and after a time a mature plant exists. When this plant reproduces, however, it does not do so through more spores. It produces gametes—eggs and sperm. These go on to fuse, and the result is a diploid plant just like the one we started with.

The upshot of all this, as you can see in **Figure 22.16b**, is that plants move back and forth between two different kinds of generations, one of them spore-producing, the other gamete-producing. In this cycle, the generation of plant that produces spores is known as the **sporophyte generation**, while the generation of plant that produces the gametes is the **gametophyte generation**. The fern shown in Figure 22.16h gives you some idea of how physically different these generations can be, but the disparities actually can be much greater than this. In a massive organism such as a redwood tree, the tree that is familiar to us is the sporophyte generation. But, what do its spores develop into? On the male side, barely visible pollen grains. On the female side, small collections of cells, hidden deep within a redwood "pinecone." The fact that these gametophyte individuals are tiny, however, doesn't mean that they are unnecessary. The eggs and sperm they produce are required to bring about a

whole new redwood tree. In sum, all plants go through an **alternation of generations**: a life cycle practiced by plants in which successive plant generations alternate between the diploid sporophyte condition and the haploid gametophyte condition.

---

**SO FAR...**

1. Which of the following traits is *not* characteristic of plants? Fixed-in-place, no cell wall, photosynthesizing, land-dwelling.

2. All plants go through an alternation of generations in which one generation produces the reproductive cells called _____, while the alternate generation produces the reproductive cells _____ and _____.

---

## 22.7 Types of Plants

For our purposes, it is convenient to separate the members of the plant kingdom into a mere four types (**Figure 22.17**). These are the *bryophytes*, which include mosses; the *seedless vascular plants*, which include ferns; the *gymnosperms*, which include coniferous ("cone-bearing") trees; and the *angiosperms*, a vast division of flowering plants—by far the most dominant on Earth today—that includes not only flowers such as orchids, but also oak trees, rice, and cactus. We'll look at all four types briefly here. Chapters 24 and 25 provide detailed descriptions of the workings of flowering plants.

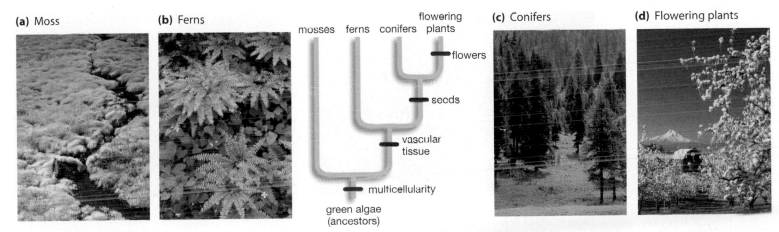

**(a)** Moss **(b)** Ferns **(c)** Conifers **(d)** Flowering plants

mosses ferns conifers flowering plants
flowers
seeds
vascular tissue
multicellularity
green algae (ancestors)

**Figure 22.17**
**Four Main Varieties of Plants**
The most primitive type of plant **(a)** the bryophytes (moss in the figure) lack a fluid-transporting vascular system. **(b)** Ferns, representing the seedless vascular plants, have a vascular system but do not have seeds. **(c)** Gymnosperms (conifers in the picture) do utilize seeds. **(d)** Flowering plants (blossoming pear trees in the picture) produce seeds and are responsible for several other plant innovations, among them fruit.

## Bryophytes: Amphibians of the Plant World

If you look in low-lying, wet terrain, you are likely to see a carpet-like covering of moss. In countries such as Ireland, whole fields of peat moss grow and are harvested to be used as fuel for heating and cooking. Mosses are the most familiar example of a primitive type of plant called a **bryophyte**, which can be defined as a type of plant lacking a true vascular system (**Figure 22.18**). What's a vascular system? A network of tubes within an organism that serves to transport fluids. We can get a sense of what the bryophytes are like by looking at the mosses.

Like all bryophytes, mosses are close living relatives of the earliest plants that made the transition from living in water to living on land. As such, mosses can be conceptualized as plants that have made only a partial break with the aquatic living of the evolutionary ancestors of all plants, the algae. In living on land, the mosses have to deal with something the algae don't, which is the effect of gravity. Lacking a vascular system, they cannot transport water and other substances very *far* against the force of gravity; instead, they must lie low, hugging the surface to which they are attached while spreading out horizontally to maximize their exposure to sunlight. Nevertheless, mosses are hearty competitors in some tough environments, such as the arctic tundra and the cracks in sidewalks.

In keeping with their aquatic origins, mosses have sperm that usually get to eggs by swimming through water, although not much water is necessary. (A thin film left over from the morning dew will suffice.) For decades, scientists believed that this was the only means by which moss sperm got to eggs, but in 2006 scientists reported that the bodies

**(a)** Mosses

**(b)** Liverworts

**(c)** Hornworts

**Figure 22.18**
**Three Kinds of Bryophytes**

**(a)** Mosses, **(b)** liverworts, and **(c)** hornworts. These plants have no vascular tissue and thus tend to be small. The sperm of bryophytes usually travels through water to get to eggs, although mites and tiny insects can serve this function. Bryophytes are found most often in moist environments.

of animals also carry out this transport function. Mites and tiny insects called springtails move moss sperm from one kind of moss "tuft" to another. This was a revelation because, up until this point, scientists thought that only one kind of plant used animals as sperm couriers—the flowering plants we'll be looking at shortly.

Given the close ties that mosses have to water, it's not surprising that they can usually be found in moist environments. Like all bryophytes, mosses have no roots at all but instead use single-celled extensions called rhizoids to anchor themselves to their underlying material, which may be soil or rock or wood. You might expect that rhizoids would serve to absorb water, as roots do, but this is true only to a very limited extent. Bryophytes take in water almost entirely through their above-ground exterior surface.

## Seedless Vascular Plants: Ferns and Their Relatives

As noted, all plants except the bryophytes have a *vascular system*, or network of fluid-conducting tubes that transport both food and water. When we begin to look at vascular plants, we find that the most primitive variety of them are the so-called **seedless vascular plants**: plants that have a vascular system, but that do not produce seeds as part of reproduction.

Easily the most familiar representatives of the seedless vascular plants are ferns, with their often beautifully shaped leaves, called fronds (**Figure 22.19**). These plants have moved a step further in the direction of separation from an aquatic environment. Their vascular system allows them to grow up as well as out, and it allows for roots that extend into the

**(a)** Ferns

**(b)** Horsetails

**(c)** Club mosses

**Figure 22.19**
**Three Kinds of Seedless Vascular Plants**

**(a)** Fall-colored ferns in New Hampshire, **(b)** horsetails, and **(c)** club mosses. Because these plants have vascular tissue, they are able to grow taller than most bryophytes. Like bryophytes, however, they do not produce seeds and are tied to moist environments.

ground, where they serve their absorptive function. Despite this evolutionary innovation, the sperm of the seedless vascular plants is like that of most of the bryophytes: It needs to move through water to fertilize eggs.

**(a)** Spruce tree

**(b)** *Ginkgo biloba*

**Figure 22.20**
**Gymnosperms**

**(a)** This spruce tree is a member of the grouping of gymnosperms known as conifers, which account for about three-fourths of all gymnosperm species. The conifers also include redwood, pine, juniper, and cypress trees.

**(b)** There are other types of gymnosperms as well. Pictured are the leaves and seeds of the maidenhair tree *Ginkgo biloba*, which are used today in herbal preparations.

## The First Seed Plants: The Gymnosperms

There are two kinds of seed plants, the gymnosperms and the angiosperms, but putting things this way makes it sound as though these plants are on a kind of equal footing with the bryophytes and seedless vascular plants. In reality, the gymnosperms replaced seedless vascular plants as the most dominant plants on Earth, although this event took place a long time ago—about the time the dinosaurs came to dominance. Then the gymnosperms lost this status to the angiosperms, which began flourishing about 80 million years ago. Today, there are only about 700 gymnosperm species compared to 260,000 angiosperm species. Nevertheless, the gymnosperm presence is considerable, especially in the northern latitudes, where they exist as vast bands of coniferous trees, including pine, fir, and spruce. Conifers such as these provide most of the world's lumber (**Figure 22.20**).

## Reproduction through Pollen and Seeds

Why were the gymnosperms able to outcompete so many of the seedless vascular plants back in the dinosaur days? Part of the answer lies in sperm transport. In gymnosperms, sperm are contained in tiny spheres called pollen grains. Critically, these grains are transported to the female eggs through the *air*, rather than being limited to *swimming* to them or being carried on the bodies of tiny, crawling animals. Given this innovation, the gymnosperms could propagate over great distances, which gave them a competitive advantage over the seedless plants.

Another advantage the gymnosperms had were their seeds, which can be thought of as tiny packages of food and protection. If you look at **Figure 22.21**, you can see one example of a gymnosperm seed, this

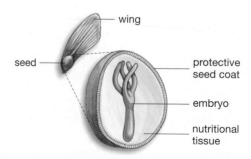

**Figure 22.21**
**A Gymnosperm Seed**

Seeds are tiny packages of food and protection. They come in many shapes and sizes, but they all contain an embryo, some food, and a protective seed coat.

one from a pine tree. **Figure 22.22** then shows you how this seed fits into the life cycle of the pine. As you can see, pollen grains are carried by the wind to the female reproductive structure, which resides in the familiar pinecone. Then the sperm in the pollen grain comes together with an egg in the reproductive structure, and the result is a fertilized egg that begins to develop as an embryo. This is no different in principle from human egg and sperm coming together to create a human embryo. In the pine tree, however, the embryo is developing inside a seed, a structure that is unique to plants. Seeds have a tough coat, inside of which is not only the embryo but also a food supply for it —a kind of sack lunch of stored carbohydrates, proteins, and fats. In sum, a **seed** is a plant structure that includes a plant embryo, its food supply, and a tough, protective casing. When seeds reach the ground, they contain all the starting materials necessary to bring about germination of a new generation of plant.

Gymnosperm seeds turn out to differ from the seeds produced by angiosperms. Angiosperm seeds come wrapped in a layer of tissue—called fruit—that gymnosperm seeds do not have. The details of this anatomical feature are presented in Chapter 25. For now, just be mindful that a **gymnosperm** can be defined as a seed plant whose seeds are not surrounded by fruit. The very name *gymnosperm* comes from the Greek words *gymnos*, meaning "naked," and *sperma*, meaning "seed."

### Angiosperms: Nature's Grand Win–Win Invention

As noted, there are about 260,000 known species of the flowering plants or angiosperms. The term *flowering plants* may bring to mind roses or tulips, and indeed, these flowers are angiosperms. But the angiosperm grouping includes all manner of other plants as well—almost all trees except for the conifers, all our important food crops, cactus, shrubs, common

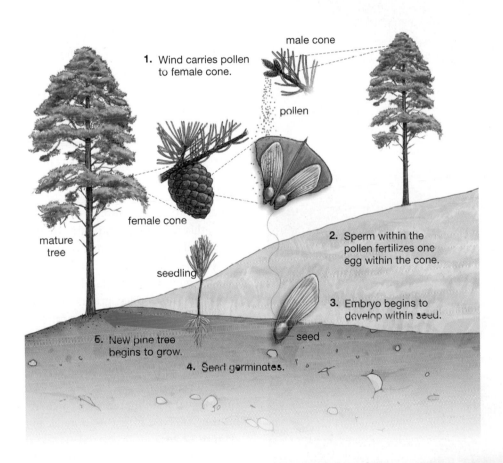

**1.** Wind carries pollen to female cone.

male cone

pollen

mature tree

female cone

seedling

**2.** Sperm within the pollen fertilizes one egg within the cone.

**3.** Embryo begins to develop within seed.

**5.** New pine tree begins to grow.

seed

**4.** Seed germinates.

**Figure 22.22**
**The Life Cycle of a Gymnosperm**
Pine trees have two kinds of cones. Pollen is produced within the smaller male cones, while eggs are produced within the larger female cones. When the wind carries pollen onto the female cone, the sperm within the pollen fertilizes one of the eggs within the cone. An embryo then begins to develop inside a seed, which falls to the ground. Once conditions are suitable, the seed germinates, and a whole new pine tree begins to grow.

**(a)** Calla lilies on the California coast

**(b)** Cholla cactus in Arizona

**(c)** Corn in a field

**Figure 22.23**
**Angiosperm Variety**

The incredible diversity of angiosperms can be seen in the differences between these three types of angiosperm plants.

**(a)** Pollen-covered honeybee on a dandelion

**(b)** Carib martinique pollinating a flower

**(c)** Lesser long-nosed bat pollinating a cactus

**Figure 22.24**
**Animals Help Pollen Get from Here to There**

Flowering plants take advantage of the mobility of animals to transport pollen from one plant to another. Relationships between flowers and pollinators can be species specific, which helps to ensure that the pollen will be delivered to the right address.

grass; the list is very long (**Figure 22.23**). As noted, **angiosperms** are defined by an aspect of their anatomy; their seeds are surrounded by the tissue called fruit. The details of angiosperm anatomy and physiology are covered in Chapters 24 and 25. In this chapter, we will look only at why angiosperms and animals are so important to each other.

## 22.8 Angiosperm–Animal Interactions

For an angiosperm such as a new honeysuckle plant to come into existence, a two-part process must take place. First, sperm—developing once again inside a pollen grain—must get to an egg and fertilize it, which produces the embryo encased in a seed coat. Next, this seed must land on a patch of soil somewhere and begin to germinate or sprout.

Now, in the angiosperms, how is it that sperm gets to egg in the first place? Remember that with gymnosperms, pollen grains were carried by the wind. By contrast, the angiosperms have induced *animals* to carry pollen from one plant to another—to pollinate them. Not all angiosperms are pollinated in this way; some are pollinated by wind, and a few aquatic species rely on water currents. But most are pollinated with the help of animals, be they insects, birds, mammals, or even snails (**Figure 22.24**).

Angiosperms use various attractants to encourage animal pollination. The most important of these is nectar, which is essentially sugar-water, but angiosperms don't stop there. So important is pollination to them that they have developed a host of pollination marketing strategies. How about a

**(a)** This is what we see (normal sunlight).

**(b)** This is what the bee sees (ultraviolet light).

**(c)** This is what the bee "thinks."

**Figure 22.25**
**Food Lies This Way**
Animals may be attracted to a particular flower by its fragrance, color, or pattern of visual elements. Insects are guided into the nutrients in the center of the flower by color patterns called nectar guides, visible to an insect—because it perceives ultraviolet light—but not to us.

fragrance that, say, bees are sensitive to? How about colors that they're attracted to? Insect-pollinated flowers generally have both sweet fragrances and coloration patterns—nature's equivalent of homing signals and landing lights. They serve to get an animal to the flower in the first place and then into the right position for feeding and pollination (**Figure 22.25**).

Why do plants spend so much energy attracting animal pollinators while gymnosperms let the wind do the work? Windborne pollination in gymnosperms can be thought of as a kind of scattershot approach: If the wind happens to blow a pollen grain onto the female reproductive structure, then pollination occurs. Think how much more directed things are with animal pollination. It is essentially door-to-door service. Animal pollination is one of the reasons that angiosperms are by far the most dominant plants on Earth.

### Seed Endosperm: More Animal Food from Angiosperms

Once pollination has occurred, the resulting seed, with the embryo inside, has to reach the ground and then germinate in it. All seeds contain food reserves for the growing embryo, but angiosperm seeds develop a special kind of nutritive tissue, called **endosperm**, that often surrounds the embryo (**Figure 22.26**). This endosperm supplies much of the food for human beings throughout the world. Rice and wheat grains are seeds consisting in large part of the endosperm meant to sustain the plant embryo. It's worth noting that *whole-grain* wheat is wheat that still has its embryo (usually called "wheat germ") and its seed coat (usually called "bran"). By contrast, white bread, regular pasta,

and so forth are made from wheat whose wheat germ and bran have been removed through processing.

### Fruit: An Inducement for Seed Dispersal

Some flowering plants also add an inducement for the *dispersal* of their seeds: the tissue called fruit. This is not fruit in the technical sense, but fruit as that term is commonly understood—the flesh of an apricot or cherry, for example. As such, it represents one more piece of bounty that plants provide to animals. To understand how fruit benefits plants as well, imagine a bear that consumes a wild berry, which consists not only of the fruit flesh, but of the seeds inside this flesh. The fruit is digested by the bear as food, but the *seeds* are tough; they are passed through its system intact to be deposited, with bear feces as fertilizer, at a location that may be very remote from the place

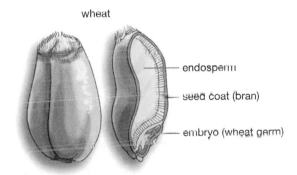

wheat

— endosperm

— seed coat (bran)

— embryo (wheat germ)

**Figure 22.26**
**Food for the Seedling, Food for Humans**
Endosperm is a food reserve in angiosperm seeds. In nature, its function is to feed a growing plant embryo in the period before the embryo germinates into a grown plant and starts making its own food through photosynthesis. It is this same endosperm that nourishes humans throughout the world in the form of wheat, corn, rice, and even coconut.

**Figure 22.27**
**Seed Carriers**
Angiosperms have taken advantage of the mobility of animals not only to transfer pollen, but also to disperse seeds.

**(a)** Some seeds are wrapped in tasty fruit that is consumed by animals, such as bears.

**(b)** Other seeds come wrapped in burrs and spines that stick to the fur of animals and are carried away.

**(a)** Wild berries induce seed dispersal

**(b)** Burrs help seeds travel

where the bear *ate* the berry. The result is seed dispersal at what may be a promising new location for a berry plant (**Figure 22.27**).

Given pollination and the varieties of food that angiosperms produce, it's easy to see that, with the rise of the flowering plants, animals and plants entered into a much more interdependent relationship than had existed before, one that is still evolving today.

## On to a Look at Animals

In this chapter and the one that preceded it, you've looked at bacteria and archaea, and at three of the kingdoms within the eukaryotic domain—protists, fungi, and plants. This leaves one more eukaryotic kingdom to go. It is a kingdom that is arguably the most familiar of all because we human beings are part of it. Yet it is a kingdom so diverse that it holds a good many surprises as well. It is the kingdom of animals.

**SO FAR...**

1. Of the following plants, which produce(s) seeds? moss, fern, spruce tree, orchid.

2. Pollen grains are tiny objects that contain plant _____.

3. The pollen of gymnosperms generally is transported by _____, while the pollen of angiosperms generally is transported by _____.

••• ANSWERS
1. spruce tree, orchid
2. sperm
3. wind; animals

# CHAPTER 22 REVIEW

For study help, activities, and more quiz questions, go to **www.aw.com/krogh4**.

## SUMMARY

### 22.1  The Fungi: Life as a Web of Slender Threads

- Fungi do not make their own food, as plants do, but instead are heterotrophs—they consume existing organic material in order to live. Fungi consist largely of webs of slender tubes, called hyphae, that grow toward food sources. Fungi cannot immediately ingest the food the hyphae reach since only small molecules can pass through the cell walls of fungal cells. Fungi thus digest their food externally, through release of digestive enzymes, and then bring the resulting small molecules into the hyphae.

- Collectively, hyphae make up a branching web, called a mycelium, that forms the bulk of most fungi. Many fungi have a reproductive structure called a fruiting body that produces and releases reproductive cells called spores. All fruiting bodies are organized collections of hyphae.

- Most fungi are sessile or fixed in one spot, but their hyphae grow toward food sources. Such growth comes only at the tips of the hyphae, not throughout their length.

- Most fungi are multicellular. The primary exception to this, the single-celled yeasts, are thought to have evolved to the single-celled state from an earlier multicelled condition.

### 22.2  Roles of Fungi in Society and Nature

- Fungi are the cause of many crop diseases, human infections, and forms of damage to buildings and homes. They are also sources of medicines and are used extensively in food processing. In nature, fungi join bacteria as major decomposers of the living world, breaking down the organic material in such objects as fallen trees and recycling the inorganic products that result back into the soil.

### 22.3  Structure and Reproduction in Fungi

- Individual cells that form hyphae often are separated from one another by dividers called septa. These septa are porous enough, however, that they allow for a fairly free flow of cellular material between one hyphal cell and the next—so much so that hyphae can bring resources very quickly to the point of growth at the hyphal tips. As a result, fungi can grow very quickly.

- The fruiting body of the toadstool mushroom fungi, the mushroom cap, is made up on its underside of accordion-like gills that both produce spores and release them into the wind, which carries them to new locations. Only a tiny fraction of fungal spores will successfully germinate, or sprout new mycelia. Because of this, fruiting bodies release huge quantities of spores as a means of ensuring new fungal growth.

- Within their life cycle, many fungi have a so-called dikaryotic phase of life in which the fusion of cells from different parental hyphae produces a single hypha whose cells each have two nuclei—one nucleus coming from one parental hypha, the other nucleus coming from the second parental hypha. Each of these nuclei is haploid, meaning each contains only a single set of chromosomes.

- When the below-ground mycelium in a mushroom fungus sprouts into the above-ground mushroom cap, all the cells in the cap initially are dikaryotic, including those in the reproductive structure called the basidium. The nuclei in certain cells of the basidia then undergo fusion, which is a step in carrying out sexual reproduction. These cells are now diploid—they have a single nucleus that contains a paired set of chromosomes. These same cells then undergo meiosis, and the result is four haploid spores attached to the tip of the basidium. These spores are then released and blown by the wind to new locations, where they may germinate and result in new mycelia.

- A spore is a reproductive cell that can develop into a new organism without fusing with another reproductive cell. Unlike human eggs and sperm, which fuse to make a fertilized egg, fungal spores do not need to fuse with any other cell to give rise to a new generation of fungi.

- Most fungi can reproduce by asexual as well as sexual means. In asexual reproduction, fungal spores are produced through mitosis (or simple cell division) that is carried out by one organism. In sexual reproduction, fungal spores are produced through a fusion of nuclei from two separate fungi.

**Web Animation 22.1:** Life Cycle of Fungi

### 22.4  Categories of Fungi

- There are four major categories or "phyla" of fungi: the basidiomycetes, the ascomycetes, the zygomycetes, and the chytrids. The first three of these phyla are defined by reproductive structures. Any fungus that reproduces through use of the reproductive structure called a basidium is a basidiomycete, for example, while the ascomycetes and zygomycetes are defined by their own reproductive structures.

- The basidiomycetes, also known as the club fungi, include "toadstool" mushrooms, "shelf" fungi that often can be seen sprouting from trees, and the agricultural pathogens called smuts and rusts. The ascomycetes, also known as the sac fungi, include truffles, morel mushrooms, and fungi that attack stone fruits. The zygomycetes, also known as the bread molds, include many fungi that form associations with plant roots, but also include one group of fungi that live by parasitizing insects. The chytrids are a primitive group of mostly aquatic fungi, all of which are mobile at one point in their life cycle. The chytrids probably give us an idea of what the earliest fungi were like, as the sessile fungi that are so common today seem to have evolved from chytrid-like ancestors. Early fungi were probably very similar to early animals.

- A yeast is any single-celled fungus that tends to reproduce by the process known as budding. In this process, fungal daughter cells are produced as outgrowths of parental cells. Yeast species exist within three of the four fungal phyla—the ascomycetes, the basidiomycetes, and the zygomycetes. The majority of yeasts are ascomycetes, but many are basidiomycetes. The yeast *Saccharomyces cerevisiae* has been very important to human beings because of the critical roles it has played in the production of beer, wine, and bread and because of the important role it has played in scientific research.

## 22.5  Fungal Associations: Lichens and Mycorrhizae

- Lichens are composite organisms made up usually of fungi and algae but sometimes of fungi and bacteria. Within a lichen, the photosynthesizing algae are supplying the fungi with food, while the fungi are supplying the algae with water, minerals, carbon dioxide, and protection from the elements. Lichens can establish themselves in barren environments, and some lichens have been alive for thousands of years. Lichens grow at an extremely slow pace, however, thanks to their practice of entering a state of dormancy during periods when water is scarce.

- Up to 90 percent of seed plants live in a cooperative association with fungi that links plant roots with fungal hyphae. In this relationship, plants supply fungi with food produced in photosynthesis, while the fungi supply plants with minerals and water, gathered by the web of fungal hyphae. Associations of plant roots and fungal hyphae are called mycorrhizae.

## 22.6  Plants: The Foundation for Much of Life

- Plants are the foundation for much of life on Earth because they are responsible for much of the living world's production of food and oxygen. In addition, they stabilize soil, provide habitat for animals, and lock up carbon dioxide. All plants are multicelled, and almost all are fixed in one spot and carry out photosynthesis. All plant cells have a cell wall and contain organelles called chloroplasts, which are the sites of photosynthesis. Plants reproduce through an alternation of generations: a life cycle in which successive plant generations produce either spores (the sporophyte generation) or gametes (the gametophyte generation). Within a given species, these two generations can differ greatly in size and structure.

## 22.7  Types of Plants

- The four principal categories of plants are bryophytes, seedless vascular plants, gymnosperms, and angiosperms. Bryophytes include mosses; seedless vascular plants include ferns; gymnosperms include coniferous trees; and angiosperms include a wide array of plants, such as orchids, oak trees, rice, and cactus.

- Bryophytes are close living relatives of the earliest plants that made the transition from living in water to living on land. They lack a fluid transport or vascular system and thus tend to be low lying. Bryophyte sperm get to eggs primarily by swimming through water, although it's recently been discovered that tiny animals sometimes transport sperm among mosses. Bryophytes tend to inhabit damp environments.

- Seedless vascular plants have a vascular system but do not produce seeds in reproduction. Their sperm must move through water to fertilize eggs.

- Gymnosperms are seed-bearing plants whose seeds are not encased in tissue called fruit. There are only about 700 gymnosperm species, but their presence is considerable, particularly in northern latitudes, where gymnosperm trees, such as pine and spruce, often dominate landscapes. The sperm of gymnosperms is encased in pollen grains that are carried to female reproductive structures by the wind. Gymnosperms produce seeds in carrying out reproduction. Seeds are structures that include a plant embryo, its food supply, and a tough, protective casing.

- Angiosperms, or flowering plants, are seed plants whose seeds are encased in tissue called fruit. Angiosperms are easily the most dominant group of plants on Earth, with some 260,000 species having been identified to date. Angiosperm species include not only plants with flowers, such as roses, but almost all trees except for the conifers, all important food crops, cactus, shrubs, and common grass.

## 22.8  Angiosperm–Animal Interactions

- Angiosperm pollen grains generally are transferred from one plant to another by animals, such as insects and birds. To induce animals to carry out this pollination, angiosperm flowers produce nectar and have developed attractive colorations and fragrances.

- Angiosperm seeds contain tissue called endosperm, which functions as food for the growing embryo. Endosperm supplies much of the food that human beings eat. Rice and wheat grains consist largely of endosperm.

- Angiosperm seeds are unique in the plant world in being wrapped in a layer of tissue called fruit. Fruit that is edible functions in angiosperm seed dispersal because animals will eat and digest the fruit but then excrete the tough seeds inside, often in a different location.

# KEY TERMS

| | |
|---|---|
| alternation of generations   429 | gymnosperm    433 |
| angiosperm   434 | hyphae   418 |
| bryophyte   430 | lichen   426 |
| cell wall   428 | mycelium   418 |
| chloroplast   428 | mycorrhizae   427 |
| dikaryotic phase   421 | seed   433 |
| endosperm   435 | seedless vascular plant   431 |
| gametophyte generation   429 | sessile   419 |
| | spore   421 |
| | sporophyte generation   429 |

# BRIEF REVIEW

*Answers to Brief Review questions are in the back of the book.*
*For multiple-choice quiz questions, go to* **www.aw.com/krogh4**.

1. Explain the phrase "animals ingest then digest, while fungi digest then ingest."

2. Fungi could be said to take a "lottery" approach to reproduction. In what respect?

3. How does a spore differ from a gamete?

4. How does a seed differ from a gamete?

5. Describe two means by which angiosperm reproduction is aided by animals.

# APPLYING YOUR KNOWLEDGE

1. Animals move because they need to—they have to get to prey and they must evade predators. These dual needs led to the evolution of the animal nervous system, which in turn led to the development of varying levels of intelligence in animals. Meanwhile, plants never had a need to move because they made their own food through photosynthesis. Do the separate evolutionary paths of animals and plants mean that there is an inherent value to struggle in the living world—in the animals' case, the struggle to obtain food?

2. Like all living things, fungi require water to begin growing. The recent scourge of mold fungi in American homes has primarily affected new homes. Why should they be affected more than old homes?

3. The *Entomophthora muscae* fungal parasite that invades flies has no consciousness, yet it manages to undertake actions that affect the behavior of the flies it attacks. (It sends hyphae into the flies' brains, and as a result, the flies crawl upward.) Almost all parasitism is disturbing to human sensibilities, but this brand is particularly unsettling. Why should this be so?

*The blue darner dragonfly (Aeshna cyanea). The darner's name seems to have come from a folk belief that they use their slender bodies to sew up or "darn" the lips of misbehaving children.*

The diversity of animals is one of the wonders of the natural world.

23.1   What Is an Animal?                                      441

23.2   Animal Types: The Family Tree                           442

23.3   Phylum Porifera: The Sponges                            446

23.4   Phylum Cnidaria: Jellyfish and Others                   447

23.5   Phylum Platyhelminthes: Flatworms                       450

23.6   Phylum Annelida: Segmented Worms                        451

23.7   Phylum Mollusca: Snails, Oysters, Squid, and More       453

23.8   Phylum Nematoda: Roundworms                             455

23.9   Phylum Arthropoda: Insects, Lobsters, Spiders, and More 456

23.10  Phylum Echinodermata: Sea Stars, Sea Urchins, and More  460

23.11  Phylum Chordata: Mostly Animals with Backbones          462

# ANIMALS:
# The Diversity of Life 3

**W**e human beings are, in a sense, intimate strangers with the bacteria and fungi that were reviewed in the last two chapters. We are intimate with these organisms in that we carry out our lives right along beside them: We walk through clouds of them, wash off layers of them, and have trillions of bacteria living inside us. That said, except for the occasional outbreak of athlete's foot or case of upset stomach, these organisms never cross our minds. So small are bacteria and most fungi that, while they may be everywhere, for us they are effectively nowhere. It's a different story with the animals that are the subject of this chapter. Our cats and dogs and neighborhood birds are part of our conscious lives. We pay attention to a new spider web on our porch or to the lizard on our hiking path. We wonder about the circling hawk overhead.

Animals come from the long evolutionary line of *other* feeders: the heterotrophs, who cannot manufacture their own food but must get their nutrition from outside themselves. Many bacteria fit into this category, as do lots of protists and all of the fungi. But animals, in their quest for food, added something that is unique to them: a nervous system; a system for transmitting complex messages rapidly over long pathways in the body. Once this happened, the living world was off to the races in terms of adding novel abilities. Think of sight, hearing, smell, flight, walking, singing, and reading. Then there are such things as the echolocation of bats and the waggle dance of honey bees. For the average person, to be alive means to sense, to investigate, to respond, to move. In all these areas, animals reign supreme in the living world.

This chapter is about the broad diversity that exists in the animal kingdom (**Figure 23.1** on the next page). Given that animals are part of our daily lives, some of what follows will be familiar, but there are likely to be many surprises as well.

## 23.1 What Is an Animal?

In coming to an understanding of animals, we first have to understand what separates animals from other varieties of living things. And to do this, the question that needs to be answered is: What characteristics do *all* animals have that other organisms *don't* have? It turns out there is a single, rather technical feature that is sufficient to set animals apart from all other living things.

- All animals pass through something called a blastula stage in their embryonic development. A *blastula* is a hollow, fluid-filled ball of cells that forms soon after an egg is fertilized by sperm (see Chapter 31 for details). All animals go through a blastula stage, but no other living things do.

**(a)** Octopus

**(b)** Feather star

**(c)** Chimpanzee

**Figure 23.1**
**Animal Diversity**

An octopus, a feather star, and a chimpanzee are very different creatures, yet all are members of the animal kingdom.

**(a)** Octopuses often are found crawling over ocean rocks, but they swim as well by expelling a jet of water from a hoselike siphon, often propelling themselves "backward" as with this one.

**(b)** Although they look like plants, ocean-dwelling feather stars such as this one are animals that are part of the same phylum as sea stars. They extend their arms to catch bits of food drifting in the ocean currents.

**(c)** Chimpanzees are the closest living relatives of human beings.

Three other characteristics are found in all animals, but they're found in other kinds of organisms as well. All animals:

- Are multicelled; there are no single-celled animals.

- Are heterotrophs; they must get their nutrition from outside themselves.

- Are composed of cells that do not have cell walls. (The outer lining of animal cells is the plasma membrane. By contrast, all plants and most other creatures have a relatively thick additional lining—the cell wall—outside their plasma membrane.)

Apart from these universals, animals *usually* go on to share other characteristics. Animals tend to move, although there are animals that, for at least part of their lives, are as *sessile*, or fixed in one spot, as any plant. Except for the animal group that includes sponges, animal bodies are organized into **tissues**, each tissue being a collection of similar cells that serves a common function. (In our own bodies, for example, we have muscle tissue, which is composed of a group of similar cells that serve the common function of contracting.)

Then, there are characteristics of animals that don't fit with our preconceived notions. While it's true that animals generally are large relative to bacteria or protists, some animals are microscopic. We often think of animals as creatures with vertebral columns or "backbones," but there are far more **invertebrates**, or animals without vertebral columns, than animals with them. Indeed, 99 percent of all animal species are invertebrates. We generally think of animals as land creatures, and with a huge assist from insects, there certainly are a lot of

land animals. However, in terms of animal *diversity*, as measured by basic body plans, there are more kinds of ocean-going animals than land animals, as you'll see. This makes sense because animals existed in the sea for at least 100 million years before any of them came onto land.

**SO FAR...**

1. A feature common only to animals is that all animals pass through a _____ stage in their _____ development.

2. Three other features that all animals share is that they are _____, _____, and their cells do not have _____.

3. True or false: vertebrates make up a small portion of the animal kingdom.

## 23.2 Animal Types: The Family Tree

One way to get a handle on the range of animal life is to look at the large-scale categories of animals and see how the creatures in these categories differ from one another. The animal kingdom's broadest categories are known as phyla, with each **phylum** informally defined as a group of organisms that shares a basic body structure. Depending on who is counting, there are between 36 and 41 animal phyla. So, in what ways are the members of these various phyla related to one another? To put this another way, what does the animal family tree look like? If you look at **Figure 23.2**, you can see one version of it. While not all of the 36–41 animal phyla are shown in this tree, those phyla that are

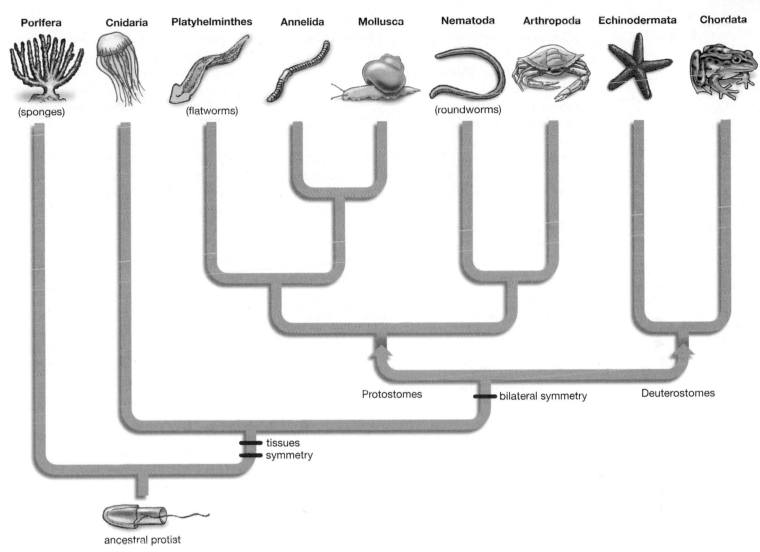

Porifera (sponges)  Cnidaria  Platyhelminthes (flatworms)  Annelida  Mollusca  Nematoda (roundworms)  Arthropoda  Echinodermata  Chordata

Protostomes — bilateral symmetry — Deuterostomes

tissues
symmetry

ancestral protist

**Figure 23.2**
**A Family Tree for Animals**
The animal kingdom is divided into groups called phyla. The members of each phylum have in common physical features that are evidence of shared ancestry. Nine of the estimated 36–41 animal phyla are shown here. All these phyla are regarded as having evolved from an ancestral species of protist, shown at lower left in the family tree. The red horizontal lines on the tree mark the points in animal evolution at which certain structural innovations appeared, such as tissues and bilateral symmetry. Human beings are members of Phylum Chordata, at upper right.

included will give you a good sense of the scope of animal diversity.

One way to conceptualize this tree is that, over time, animals became more complex through a series of *additions* to the characteristics found in more primitive animals. The twist is that only some varieties of animals evolved to get these additions, while others retained the primitive, "ancestral" condition. Let's take a tour of the tree now, focusing on features that were added to the animal kingdom through evolution.

## Additions 1 and 2:
## Tissues and Symmetry

Looking down at the trunk of the tree in Figure 23.2, you can see that all animals have a common ancestor—

probably a protist similar to the modern-day protists called choanoflagellates. Now, notice that the tree splits above this common ancestor and yields, over to the left, a phylum called Porifera, which are sponges. Porifera are truly the outliers of the animal world in that, almost alone among animals, they lack tissues and a quality known as symmetry (which will be defined shortly). If you look again at the split above the common ancestor, this time going to the right, you can see that the animals on this branch did get the additions of tissues and symmetry. Note also that this right branch leads to all the other animal phyla. Thus, all the animals that stem from this branch will have tissues and symmetry (which is what makes Porifera the outlier).

••• ANSWERS
1. blastula; embryonic
2. multicelled; heterotrophic; cell walls
3. True; 99 percent of all animal species are invertebrates

We talked about tissues earlier, but what is this addition called symmetry? If you look at **Figure 23.3a**, you can see that the jellyfish that's pictured can be thought of as being divided by imaginary planes, thus yielding body sections. These sections are mirror images of one another; put another way, they have **symmetry**, meaning an equivalence of size, shape, and relative position of parts across a dividing line or around a central point. The jellyfish, part of Phylum Cnidaria, actually has a particular kind of symmetry. It has **radial symmetry**, meaning a symmetry in which body parts are distributed evenly around a central point. In simpler terms, the symmetry of jellyfish is the symmetry of pie sections.

To appreciate symmetry, consider what a lack of it is like. Look at the sponge in **Figure 23.3b**. Where is its symmetry? There isn't any because there is no section of the sponge that is a mirror image of any other.

Symmetry can, however, come in several forms. The dog in **Figure 23.3c** is divided by what is known as an imaginary "sagittal" plane. The dog obviously has symmetry, but it is a different kind of symmetry from that of the jellyfish. It has *bilateral* symmetry. We'll take a moment to consider this form of symmetry since it constitutes the third addition to the animal kingdom.

## Addition 3: Bilateral Symmetry

Animals usually are different front-to-back and top-to-bottom but symmetrical side-to-side. Put another way, your head is different than your feet, and your chest is different than your back, but your left *side* is very similar to your right. This is what **bilateral symmetry** means: a bodily symmetry in which opposite sides of a sagittal plane are mirror images of one another.

In looking at the animal family tree, you can see that bilateral symmetry was an evolutionary innovation that affected all modern-day animals except for the sponges and the Phylum Cnidaria, which includes the jellyfish. Thus, bilateral symmetry is the general rule in the animal world.

The bilateral symmetry/radial symmetry divergence is in one sense the divergence between having a head and not having one. Jellyfish and sponges don't have one, but most other animals have at least a "cephalization," or concentration of nerve cells at one end of their bodies. Animals more complex than the cnidarian jellyfish evolved to sense their world primarily at one end of themselves—their heads—which is the end with which they move into their worlds. This is understandable because it is more advantageous to know something about the future (where you're going) than the past (where you've been).

## Addition 4: A Body Cavity

Your stomach expands when you've just had a big meal but then contracts when you're busy for a few hours. Your heart expands and contracts perhaps 70 times a minute. You bend over to tie your shoes, and unbeknown to you, many of your internal organs slide out of the way. What allows you to do all this? The answer is an internal space you have, a centrally placed, fluid-filled body cavity called a **coelom** ("SEE-lome"). We human beings are not alone in having such a cavity. There are only three animal phyla in Figure 23.2 that *don't* have this cavity: the sponges, the cnidarians (jellyfish, etc.), and the members of Phylum Platyhelminthes, which is made up of flatworms.

What's the value of such a cavity? Well first, an expandable stomach has the same value as a gas tank: It allows you to go for a while without refueling. It also allows reproductive organs to expand to accommodate eggs or offspring. Then, there is the fact that if a heart couldn't expand and contract, it couldn't work at all. Finally, a body cavity protects organs from bodily blows and provides a large part of the body with flexibility.

In most instances, the coelom surrounds another physical structure, the *digestive tract*—meaning the tube, functioning in digestion, that runs from the mouth to the anus. We can therefore think of the coelom as one tube that encircles another; the coelom

**Figure 23.3**
**Symmetry Is an Equivalence in Body Sections**

If an imaginary plane drawn through an animal can divide that animal into sections that are mirror images of one another, then that animal has symmetry.

**(a) Radial symmetry** The imaginary planes drawn through the jellyfish show that it has radial symmetry: a symmetry in which body sections are distributed evenly around a central point. Radial symmetry is characteristic of the Phylum Cnidaria, which includes jellyfish.

**(b) Asymmetry** The sponge has no symmetry—no plane drawn through the sponge body would yield sections that are mirror images of one another.

**(c) Bilateral symmetry** The dog has a kind of symmetry common to most animals: bilateral symmetry, in which the sides of an animal are mirror images of one another. Notice also the terms for different parts of the dog—dorsal and ventral, meaning "back" side and "belly" side; and anterior and posterior, meaning "head" end and "tail" end.

**(a)** Radial symmetry:
Symmetry around a central point

**(b)** Asymmetry:
No planes of symmetry

**(c)** Bilateral symmetry:
Symmetry across the sagittal plane

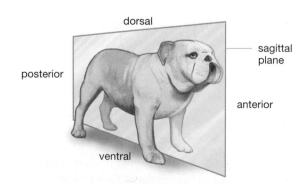

dorsal

sagittal plane

posterior

anterior

ventral

is generally tube shaped, and it surrounds the tube that is the digestive tract. (There are, however, lots of variations on this general principle.) **Figure 23.4** displays what it means to have no coelom, as in flatworms; and a *true coelom*, as in earthworms. (Not pictured is another variation on this theme, a pseudocoel, which roundworms have.) You may wonder why there are no red lines on Figure 23.2 showing where the coelom was added in animal evolution. So valuable is this internal space that it seems to have evolved independently several times among animals—at least once each among the two large groupings you can see toward the base of the tree, the protostomes and the deuterostomes. But, what are these two large groupings? Let's see.

## A Split in the Animal Kingdom: Protostomes and Deuterostomes

The protostome/deuterostome categories you can see toward the bottom of the animal family tree simply represent an early split among animals along two evolutionary paths. The difference between the two types of animals is grounded in how each type develops in its early embryonic stages. Remember how you saw earlier that the chief defining characteristic of animals is that they all go through something called the blastula stage in embryonic development? Well, in **Figure 23.5**, you

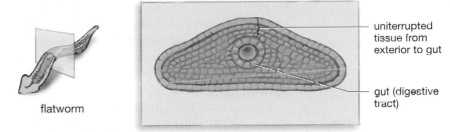

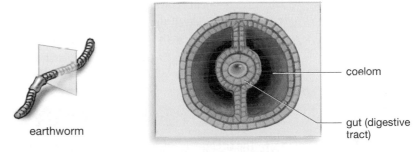

**No coelom.** Phylum Platyhelminthes, composed of flatworms, is one of the phyla that has no coelom. Note that from its gut to its exterior, the flatworm is composed of uninterrupted tissue.

**Coelom.** The earthworm is one of many animals that has a coelom – a fluid filled central cavity that usually surrounds the gut.

**Figure 23.4**
**An Important Space**

Most animal bodies have an enclosed cavity or coelom—an internal space that surrounds their digestive tract or other internal structures. A coelom gives an animal flexibility, protects its organs from external blows, and provides space for the expansion of such organs as the stomach. Only three of the phyla covered in the chapter lack a coelom.

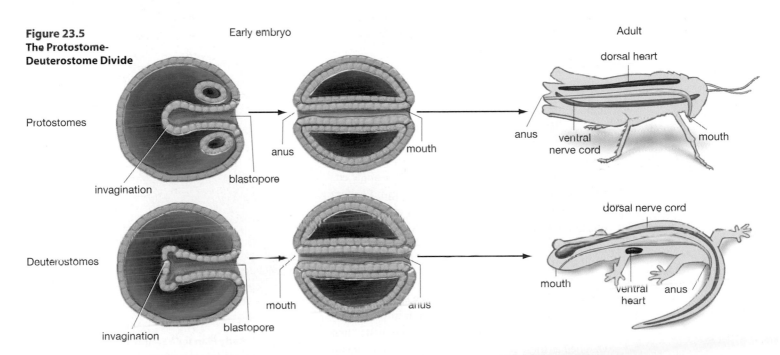

**Figure 23.5**
**The Protostome-Deuterostome Divide**

All animals pass through a blastula stage of development. In most animals, the blastula invaginates to form a structure that develops into the animal's gut. The opening to this invagination is called a blastopore. In protostomes, the blastopore becomes the mouth. In deuterostomes, by contrast, the blastopore becomes the anus.

Developmental differences such as these have consequences in the adult. Protostomes, such as the insect shown, have their heart on their top or dorsal side and nerve cord on their bottom or ventral side. In deuterostomes, such as the reptile shown, this placement is reversed.

can see what happens to the hollow ball of cells called a blastula: It develops an invagination whose opening is called a blastopore. In protostomes, this blastopore becomes the mouth, but in deuterostomes it becomes the anus. (Protostome means "mouth first," while deuterostome means "mouth second.")

If you look at the right side of **Figure 23.5**, you can see the effect that this developmental difference and others have on adult protostomes and deuterostomes. Note that the protostome has a dorsal heart and a ventral nerve cord, while in the deuterostome this arrangement is reversed.

Having reviewed the broad-scale features of the animal kingdom, we'll now take a brief tour of it by walking through each of the nine phyla shown in Figure 23.2.

---

### SO FAR...

1. The sponges in Phylum Porifera are "outliers" in the animal kingdom in that, almost alone among animals, their bodies are not composed of _____ , and their body structure lacks the quality of _____.

2. All animals except the sponges and those in Phylum Cnidaria exhibit the quality of _____ symmetry, in which an animal's _____ are mirror images of each other.

3. A coelom is a _____ that is possessed by almost all animals except sponges, jellyfish, and flatworms.

---

## 23.3 Phylum Porifera: The Sponges

Back in the 1780s, the United States was just that: a group of states that had just united into a single entity. The sponges that make up the Phylum Porifera bear some comparison to this. Each sponge is a single, unified entity, but not by much. Sponges have no stomach, no heart; no organs of any sort. They don't really even have tissues. Each of their cells acquires its own oxygen and eliminates its own wastes. In experiments, scientists have strained some sponge species through a filter, "disaggregating" them into the individual cells they are made of. The scientists then watched as these cells came back together to make up a single sponge again. This is possible only in an organism in which each cell functions with a great deal of independence. If you were disaggregated into your individual cells, there is no chance they would come back together to re-form you. On the other hand, the

cells in sponges clearly work together in a coordinated way. Sponges may not show much organization relative to fish, but they show a great deal of organization relative to, say, bacteria.

So, what manner of organism is a sponge? The simplest varieties have a layer of outer cells that is pockmarked with thousands of microscopic pores. Water flows in through these pores in the outer-layer cells and then is expelled out through a large opening at the top of the sponge (**Figure 23.6**). Because sponges are fixed in one spot (generally on the sea floor), everything they need in the way of food and oxygen must come to them, while everything they need to get rid of must be washed away from them. The water that flows through them takes care of both needs. Food and oxygen wash in through the microscopic pores, and wastes wash out through the large opening (called an *osculum*). Most sponges have multiple osculi, but in these cases the concept remains the same: Move the water in, capture the food that comes with it, and move the water out, expelling wastes in the process. Millions of cells, called *collar cells*, wave tiny, whiplike flagella that keep the water flowing through.

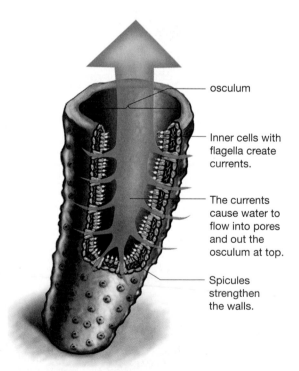

osculum

Inner cells with flagella create currents.

The currents cause water to flow into pores and out the osculum at top.

Spicules strengthen the walls.

**Figure 23.6**
**Sponges: A Body Plan for a Simple Lifestyle**
Sponges filter water through themselves in order to live. Cells on their interior surface have hairlike extensions, called flagella, with rapid back-and-forth movement that draws water in through pores in the sponge's side walls. After filtering this water for nutrients, the sponge then expels it through a large opening at the top of its body, called an osculum. Sponges lack true tissue layers but possess many specialized cell types.

Most sponges have skeletons composed of small, barbed structures called *spicules* that keep the sponge pores open and generally give the sponge structure. Some sponges have no spicules, but instead have skeletons made of a special variety of the protein collagen, which can be a soft, pliable material. (Think of the collagen *injections* that people sometimes get to improve their looks.) Sponges with collagen skeletons are the sponges from which we get our everyday notion of a sponge, meaning a soft, absorbent object used in cleaning. Put another way, the sponges that people once used were the skeletal remains of ocean-dwelling animals (which were harvested by divers). Today, our sponges generally are made of synthetic materials.

Depending on whose count you accept, between 5,000 and 10,000 species of sponge have been recognized so far, with probably lots more remaining to be identified. Almost all these species dwell in ocean water, although there are a few freshwater varieties (**Figure 23.7**).

Sponges can reproduce merely by budding. A group of cells breaks off from an existing sponge and then develops into a new sponge. Sexual reproduction also exists, however, and here again the sponge puts water currents to use. Sperm are released from one sponge and make their way via water currents to another sponge, where they are taken up and transported to egg cells. An offspring that comes from sexual reproduction has a moving or *motile* stage. It swims for a time as a larva before landing on a solid surface—perhaps a rock or the body of a living animal—whereupon its outer layer of cells secretes a substance that will allow it to attach to the material beneath it. Then, it begins its life of filtering water.

## 23.4 Phylum Cnidaria: Jellyfish and Others

If the signature activity of sponges is filtering water to get food, the signature activity of cnidarians is stinging prey to get it. In the main, cnidarians (knee-DAR-ce-uns) harpoon their prey with extensions that are not only barbed, but that may release poisons. These extensions are tiny since each one springs from inside a single cell, but they are numerous enough that a single jellyfish may be able to immobilize and eat animals the size of small fish. Only a few species have a sting potent enough to cause real harm to human beings, but it makes sense to give all jellyfish a wide berth.

Jellyfish are undoubtedly the best-known cnidarians, but the phylum also includes corals, whose calcified remains make up coral reefs; the delicate sea anemones, whose petal-like tentacles close up when touched; and hydrozoans, which can look more like sea plants than animals. One type of hydrozoan is the hydra that you may see in your biology lab. There are

**(a)** A leaf sponge

**(b)** A yellow, warm-water tube sponge

**Figure 23.7**
**Sponge Diversity**
Sponges differ greatly in size and shape, as you can see from these examples.

••• ANSWERS
1. tissues; symmetry
2. bilateral; sides
3. fluid-filled internal cavity

**(a)** Jellyfish

**(b)** Sea anemone

**(c)** Coral polyps

**Figure 23.8**
**Cnidarian Diversity**

One characteristic that unifies cnidarians as diverse as jellyfish, sea anemones, and corals is their ability to immobilize prey with harpoon-like stinging cells.

**(a)** Pictured are some Pacific Ocean jellyfish, *Chrysaora fuscescens;* their sting, although painful, generally is not harmful to human beings. One of its relatives, *Chrysaora quinquecirrha*, is a major irritant during the summer in the Chesapeake Bay.

**(b)** The large sea anemone *Urticina piscivora* is found off the shores of British Columbia in Canada. At about 20 centimeters or 8 inches across, it is so big it can eat small fish.

**(c)** What builds a coral reef? Coral polyps, such as these seen in a coral reef in Western Australia. Note the mouths and extended tentacles of the polyps, which help form the living outer layer of coral reefs.

about 9,000 identified species of cnidarians, nearly all of them ocean dwellers, with a few living in freshwater (**Figure 23.8**).

The basic body plan of a cnidarian is that of a sack, although it often is a sack that is turned upside down. If you look at **Figure 23.9**, you can see how this plays out in connection with a hypothetical jellyfish. In the familiar adult or medusa stage of a jellyfish, the sack is upside down; note that the mouth is on the *underside* of the body in this stage. Conversely, look at this jellyfish's immature "polyp" stage, when it is fixed to the ocean floor; now the mouth is on top. Medusa and polyp stages are present in many cnidarians, although corals, sea anemones, and hydra have only the polyp stage. The medusa stage predominates in jellyfish—misnamed, because they are not fish— while the polyp stage predominates in most hydrozoans. In all medusa and in many polyps, however, the basic body plan is the same: a single opening to the outside, serving as both mouth and anus, is surrounded by tentacles that both sting prey and bring the prey to the mouth. Note also the material that lies between the jellyfish's gastrovascular cavity and its exterior. This is the *mesoglea*, a secreted, gelatinous material that makes up most of the medusa-stage jellyfish. Its consistency accounts for the name jellyfish. The gastrovascular cavity functions in both digestion (*gastro-*) and in circulation (*vascular*).

**Figure 23.10** shows you the life cycle of a cnidarian, this one a hydrozoan known as *Obelia*. Medusastage male *Obelia* release sperm into the water, while females release eggs; these unite, eventually producing a larval stage offspring that settles to the oceanfloor bottom, now becoming a polyp. This starting

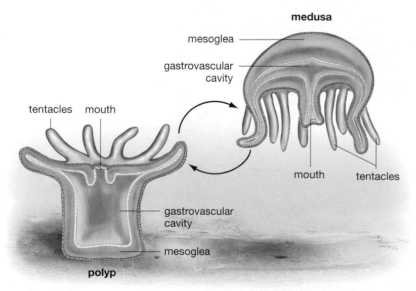

**Figure 23.9**
**Two Stages of Life for Many Cnidarians**

Many cnidarians go through both a polyp stage, in which they are attached to a solid surface beneath them, and a medusa stage, in which they swim freely through the water. Note that in the polyp stage, the cnidarian's mouth and tentacles face upward, while in the medusa stage, they face downward. The gastrovascular cavity of the cnidarian functions in both digestion and circulation. The mesoglea noted in the figure is a secreted, gelatinous material that makes up the bulk of the medusa stage in jellyfish cnidarians. All cnidarians use stinging tentacles to capture prey.

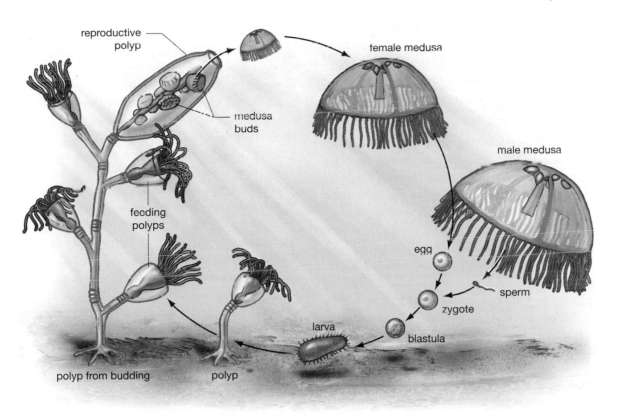

reproductive
polyp

female medusa

medusa
buds

male medusa

feeding
polyps

egg

sperm

zygote

larva

blastula

polyp from budding

polyp

**Figure 23.10**
**Cnidarian Reproduction**

This hydrozoan cnidarian,
known as *Obelia*, reproduces
through both sexual and asexual means. Sexual reproduction
can be observed at the right of
the figure, where male and female medusae release sperm
and eggs, respectively, that fuse
to form the new generation of
*Obelia*. The polyp that results
from this fusion, however, can
reproduce by asexual means.
Cells bud off from the first polyp
(center) and grow into another
polyp, shown at left, which may
be the first of many that will
develop this way. Polyps within
these larger structures specialize. Some are involved in feeding, while others function in
reproduction. The reproductive
polyps have medusa developing
inside them (the medusa buds),
which are released to the surrounding water, starting the life
cycle over again. The different
stages of the life cycle are drawn
at different scales.

polyp can give rise to an entire colony of *Obelia* polyps through the process of *budding*: Groups of cells break off from one polyp and form another polyp. Note that in this polyp stage, some of the branchlike outgrowths are "feeding polyps" that have the familiar sacklike appearance of cnidaria (tenta-cles, mouth), while other outgrowths are entirely given over to reproduction, containing "medusa buds." These buds are immature medusa-stage *Obelia*, which eventually will be released to become free-swimming individual medusae. The *Obelia* life cycle is one among many that exist in cnidarians.

Cnidarians don't have any true organs, but they are a long way from sponges in that they have a ner-vous system and cells that function like muscle cells. This gives medusa-stage cnidarians the ability to swim. (Polyp-stage cnidarians sometimes can slide slowly along the ocean floor.) The typical jellyfish medusa contracts cells that ring its mesoglea, thus ex-pelling a jet of water from its underside. This action both moves the animal in the opposite direction from the jet and pushes in the mesoglea. Like a com-pressed cylinder of foam rubber, the mesoglea then snaps back into shape, ready for the next contraction.

Special mention should be made of the reef-building corals, which have no medusa stage but live almost entirely as polyp-stage cnidarians. The fantastic ocean cities known as coral reefs take their names from the coral animals, which actually are close relatives of sea anemones. Coral reefs have a rock-like appearance, and indeed, they are composed mostly of calcium car-bonate, better known as limestone. But this limestone comes from living organisms. The corals secrete it, a practice they share with a type of red algae with whom they share their habitat. The limestone forms an external skeleton for the corals, and when each coral polyp dies, its limestone skeleton remains. Thus, coral reefs are composed largely of the stacked-up re-mains of countless generations of coral polyps, along with the remains of their neighboring red algae. But the thin outer veneer of each reef is composed of *living* algae and the latest generation of pinhead-sized polyps. If you looked up close at any one of these coral animals, you would see the basic features of a cnidar-ian polyp: stinging tentacles pointed upward, sur-rounding a mouth.

---

**SO FAR...**

1. Sponges obtain nutrition and oxygen, and get rid of wastes, by _____ through themselves.

2. All cnidarians _____ prey to obtain food.

3. Cnidarians tend to exhibit both a mobile _____ stage in which their mouths are on _____ of their bodies, and a sessile _____ stage, in which their mouths are _____.

••• ANSWERS

1. moving water

2. sting

3. medusa; the undersides; polyp; on top

THE PROTOSTOMES

## 23.5  Phylum Platyhelminthes: Flatworms

With the flatworms, you've arrived at animals that fit our everyday notions of what animals look like in that they have something like a head and have the same side-to-side (bilateral) symmetry as, say, a mouse. In addition, flatworms go beyond the tissues of the cnidarians and have **organs**, meaning highly organized structures formed of several kinds of tissues. (Your stomach is an organ that includes nerve and muscle tissues.)

Despite this complexity, flatworms lack a good many features that exist in almost all the phyla further up the complexity ladder. They have no coelom or enclosed internal cavity. Thus, except for the tube of their digestive tract, they have very little internal space in them. This may sound like a good thing, but it actually limits their flexibility. Further, they have no system of blood circulation. Therefore, every cell in a flatworm must get its oxygen directly from the environment around it, through diffusion from water or air outside the body. This is why flatworms are flat—they have to maximize the number of cells that are near an exterior surface, as opposed to being buried deep inside. (How flat are they? About the width of a sheet of typing paper.) Those flatworms that are on land tend to be found under stones or rotting wood; many live at the bottom of the ocean, and a few live in freshwater (**Figure 23.11**). Some 25,000 species of flatworm have been identified.

Humans are unlikely to notice flatworms because, of the minority that live around us, most are very small. When they do come into our consciousness, it is often in the unpleasant role of parasites. Two of the three traditional classes of flatworms live strictly as

**parasites**, defined as organisms that feed off their prey but do not kill them, at least immediately. Some of these worms can infect human beings. There are, for example, the tapeworms that can enter the human body in undercooked meat or fish. Tapeworms can be small, like most flatworms, but the beef tapeworms that infect human beings can exceed 7 meters in length, which is about 23 feet. Then, there are several species of another variety of flatworm, the flukes, that cause the disease schistosomiasis, which affects more than 200 million people in tropical areas around the world, often causing serious damage to the human bladder, liver, or spleen. With flukes, the means of entry is not food but unprotected human skin, as with people who are standing in fishing areas or flooded rice paddies.

The flukes are in one class of flatworms (Trematoda), while the tapeworms are in another (Cestoidea). There is then a third class of flatworms, the Turbellaria, whose members generally are not parasitic but free-living. If you look at **Figure 23.12**, you can see something of the anatomy of one variety of turbellarian, the flatworm *Dugesia*, which is likely to be found crawling under rocks at the bottom of freshwater streams. About 1 centimeter or less than half an inch long, *Dugesia* has a "head," in that one end of it has not only a pair of primitive eyes (called *ocelli*) but also a concentration of nerve cells (the cerebral ganglia) that are connected to nerve trunks that run the length of the body.

Given this concentration of features in *Dugesia's* head, the location of its mouth may come as a surprise. It's located midway down its body, on its ventral (bottom) side. To feed, it shoots a muscular structure called a pharynx outside its mouth, coats the prey with enzymes to break down its tissues, penetrates the prey with the pharynx, and then sucks the resulting semiliquid material back in. (Your pharynx is the area at the back of your throat where your windpipe and digestive tract meet. Imagine being able to shoot it outside your mouth.) Food goes from the pharynx into an intestine, where it is digested and then simply diffuses into nearby tissues. Remember that flatworms have no blood circulation system. Thus, *Dugesia's* intestine has to be highly branched because it serves as both a food digestion and a food delivery system. Flatworms have no anus; their mouth is the single opening to their digestive tract, and waste material is expelled out of it.

Like most flatworms, each *Dugesia* has both ovaries and testes (the organs noted earlier). It is thus **hermaphroditic,** meaning a state in which one animal possesses both male and female sex organs. When two

**Figure 23.11
Flatworms**

The flatworm in this picture, *Pseudoceros ferrugineus,* was photographed in the oceans off the coast of the Philippines. Most flatworms are free-living like this one, but there are also many parasitic flatworms that live for at least part of their lives inside a host.

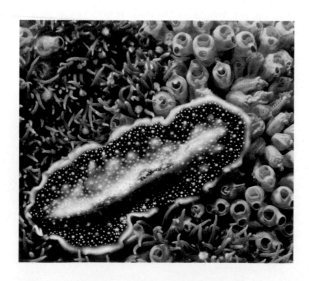

*Dugesia* copulate, each projects its penis and inserts it in the genital pore of the other. Most turbellarians can also reproduce asexually by a straightforward means: They break themselves in half. The posterior ("tail") part of the worm grasps the material beneath it while the anterior portion moves forward. The separated halves then grow whatever tissues they need to make themselves complete worms.

## 23.6 Phylum Annelida: Segmented Worms

More worms? Actually, three of the nine phyla covered in this chapter are worms. (Still to come are the roundworms in Nematoda.) Evolution resulted in a lot of worm or wormlike phyla—perhaps a dozen altogether—but the reason so many are covered here is that they occupy interesting places in the spectrum of animal diversity.

Annelid worms occupy such a position in that they provide the clearest example of a feature that exists in a lot of animals, including ourselves: **body segmentation**, which can be simply defined as a repetition of body parts in an animal. Just as a set of identical Lego blocks can be snapped together to make a larger structure, so can a repeated set of segments be connected to make a larger body. Think of the vertebrae that make up your backbone; 24 of them run in a line from your neck to the base of your back, each of them having a similar basic structure and a similar function (support and protection).

Segmentation is so useful in animals that it evolved independently in them several times. It happened at least once in the protostome lineage that you've just been going over and then once again in the deuterostome lineage—more specifically in our own phylum, Chordata.

Although segmentation is widespread in the animal kingdom, it is most visible in the annelid worms. Indeed, the name Annelida comes from the Latin word *anellus*, or "little ring." That pretty much describes what we see in the best-known annelid, the common earthworm; its body seems to be composed of a series of little rings that have been joined together (**Figure 23.13** on the next page).

Segmentation is valuable in earthworms for the same reason that individual rooms are valuable in a house: The activities carried out in one room can *differ* from those carried out in another. Each of the earthworm's 100–150 segments is separated from adjoining segments by a partition, and each segment can function independently of the others to some degree. The muscles in individual segments (or groups of seg-

### (a) Flatworm anatomy

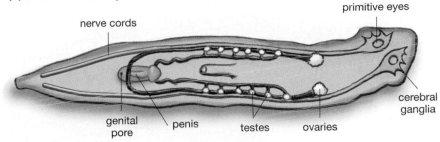

*Dugesia*'s nervous system includes primitive eyes and two collections of nerve cells, the cerebral ganglia, that connect to nerve cords that run the length of the animal. In reproduction, *Dugesia* is hermaphroditic, meaning it possesses male sex organs (testes and penis) as well as female sex organs (ovaries and other structures). When two *Dugesia* copulate, each projects its penis and inserts it in the genital pore of the other.

### (b) Feeding and digestion

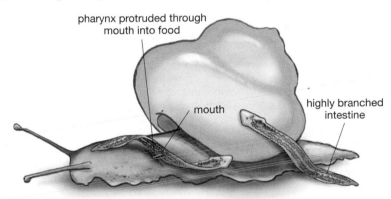

*Dugesia* feeds by turning its muscular pharynx inside out and projecting it into the food source. Food taken up through the pharynx is distributed through the body by *Dugesia*'s highly branched intestine. Flatworms lack a true circulatory system and thus depend on their digestive system for the distribution of nutrients. Their digestive system lacks an anus, meaning *Dugesia* has only one digestive-tract opening, its mouth.

**Figure 23.12
The Flatworm *Dugesia***

ments) can contract independently. This, in turn, makes for independent control of needle-like bristles that the worm can extend from each segment.

These features account for the mesmerizing way in which earthworms move. One set of muscles toward the rear of the earthworm contracts, making the worm thick and compact in the area, even as the bristles in these segments are being pushed into the ground like so many stakes. Now, with its hindquarters fixed in place, the worm uses a different set of muscles to make the midportion segments lengthen, thinning out as they do, meaning this part of the worm has moved forward. Then, the worm plants bristles in the front, anchors itself, and literally brings up its rear. Note that this movement is possible only because of segmentation. If the *whole worm*

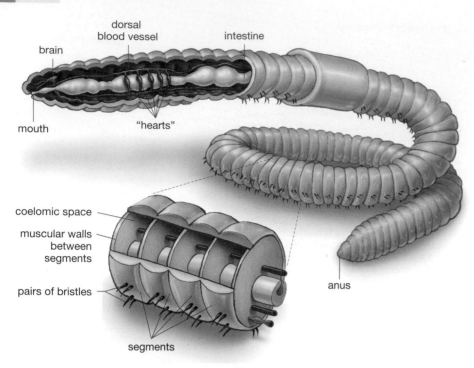

brain

dorsal
blood vessel

intestine

mouth

"hearts"

coelomic space

muscular walls
between
segments

pairs of bristles

anus

segments

**Figure 23.13**
**The Body Plan of an Earthworm**

The segmentation of body parts—a widespread feature in the animal kingdom—is clearly seen in earthworms, whose segments are repeated hundreds of times. In general, such segmentation provides flexibility in combination with strength. In the earthworm, it also allows the actions being carried out in one segment to differ from those being carried out in another. Note, in the blowup of the earthworm's body, the muscular walls between segments; these give the earthworm independent control of the segments and the bristles that can be extended from them. The earthworm also expands our notion of what a "heart" amounts to. It has a dorsal blood vessel that contracts, as do our human hearts, giving the main impetus to the circulation of red blood that moves through the worm's system. But, it also has five pulsating vessels on each side of its intestine that serve as accessory hearts.

lengthened, without part of it being compacted and fixed in one spot, it would get nowhere.

Segmentation is important in annelids, then, and is thought to have evolved in them precisely because of its power to help them move. (Earthworms use this power extensively because they burrow in the dirt for hours at a time, swallowing soil for the nutrients it contains and eating leaves or other decaying vegetation.) In more general terms, segmentation is valuable to living things because it can confer flexibility while maintaining strength. We can see this in our own "backbones." Imagine how inflexible you would be if, instead of 24 movable vertebrae, you had a single bone running from your skull to your posterior.

Earthworms generally are helpful to human beings—and not just as fish bait. They churn up soil, aerating it and depositing their cylindrical "worm casting" waste, which enriches the soil. Earthworms have no lungs; they take in oxygen right through their skin. Their skin needs to be moist for this to happen, however. The great threat to an earthworm is drying out, which is why these "nightcrawlers" generally stay underground during the daytime. Like the flatworms, earthworms are hermaphrodites. Each member of a copulating pair inseminates the other.

Although we've focused on earthworms, they actually represent only one small part of Phylum Annelida, which contains about 17,000 named species. Most annelids actually are ocean dwelling, and some of these colorful creatures, the polychaetes, do not fit in with our idea of drab worms (**Figure 23.14**). There

**(a)** Hawaiian Christmas tree worm

**(b)** Medicinal leech

**Figure 23.14**
**Annelid Diversity**

Not all annelid worms are drab, and not all live underground.

**(a)** Polychaetes are ocean-dwelling worms. Pictured is a Hawaiian Christmas tree worm in the genus *Spirobranchus*. These polychaetes stay in one spot, burrowed into tubes they themselves have constructed, generally within a coral. The bushy "Christmas trees" that extend from the worm's head are structures called radioles that it uses to filter-feed from the water.

**(b)** Leeches are a smaller class of mostly parasitic annelid worms. Pictured is a leech that is feeding from a human arm. This particular leech, *Hirudo medicinalis*, serves a therapeutic function. These leeches are used today in connection with human reattachment and plastic surgery. Enzymes they secrete serve as anticoagulants—they keep blood from clotting—and help reduce pain at the site of surgery. A leech might be attached at the site of surgery for half an hour or so and then be removed, after which its enzymes will continue working for hours longer.

**(b)** Cephalopod

**(a)** Gastropod

**(c)** Bivalve

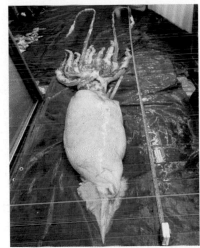

**Figure 23.15**
**Mollusc Diversity**
Pictured are species in three of the best-known classes of molluscs:

**(a)** Snails are gastropods. Pictured is a colorful terrestrial snail in the *Liguus* genus. These snails live in trees in southern Florida and through much of the Caribbean, although their numbers are dwindling in Florida.

**(b)** The largest cephalopod of all is the giant squid *Architeuthis dux*, shown here, which can reach a length of about 20 meters or 60 feet. This squid, prepared for an exhibit at New York's American Museum of Natural History, measured a mere 25 feet in length and weighed 250 pounds. It was netted off the coast of New Zealand in 1997.

**(c)** The bivalve *Argopecten irradians* is better known as the bay scallop—a part of many seafood dishes. Note the small blue dots around the edge of the scallop. These are a few of the animal's many primitive eyes.

also are freshwater annelids, which are part of the same class as the earthworms but tend to be much smaller. Then there is the smallest class of the annelids, which is composed entirely of leeches. Most leeches live up to their reputation of being bloodsucking parasites. Although most live in freshwater, some live in the ocean, and others dwell in damp-land habitats.

**SO FAR...**

**1.** The shape of flatworms is dictated by the fact that they have no system of _____. And unlike almost all other bilateral animals, they have no _____ or fluid-filled internal cavity.

**2.** Flatworms tend to be hermaphroditic, meaning that an individual worm will possess _____.

**3.** Annelid worms such as the earthworm provide one of nature's clearest examples of _____, a quality seen in human vertebrae.

## 23.7 Phylum Mollusca: Snails, Oysters, Squid, and More

With the molluscs, we reach a phylum that has lots of members familiar to us. Everyone has seen a snail, many of us have eaten clams or mussels, and squid and octopus are probably known to us from television or books. But molluscs are even more varied than these examples would suggest. Some molluscs are slugs and others wormlike; some have highly developed brains and others no brains at all; some are filter feeders and others fierce ocean predators. Some are very small, but one is a behemoth that is one of the enduring mysteries of the animal world. The giant squid, *Architeuthis dux*, reaches a length of about 20 meters or 60 feet and weighs up to a ton (**Figure 23.15b**).

Depending on the expert consulted, there are between 50,000 and 100,000 described species of molluscs. They live in freshwater, saltwater, and on land. There actually are eight classes of molluscs, but here we will look only at the three most familiar classes, the **gastropods** (snails, slugs), the **bivalves** (oysters, clams, mussels), and the **cephalopods** (octopus, squid, nautilus).

**•••** ANSWERS

**1.** blood circulation; coelom

**2.** both female and male sex organs

**3.** body segmentation

**(a) Aquatic snail**

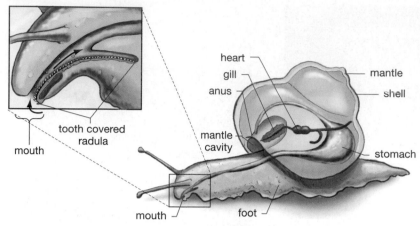

Aquatic snails such as this one have features that are common to many molluscs. All molluscs possess a skin-like tissue called a mantle, which secretes material that can form a shell, whether it is an external shell (as in the snail) or an internal shell (as in a squid). Snails also possess a tooth-covered rasping structure for feeding, called a radula, and they are complex enough to have evolved some specialized organs that are familiar to us—the heart and the stomach. The unitary "foot" of the snail, which works through wave-like contractions, is another common mollusc feature.

**(b) Squid**

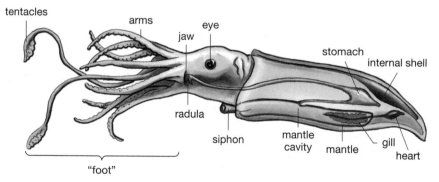

In the squid, the mollusc's unitary foot has evolved into a multisegmented appendage of eight arms and two tentacles. Note how other mollusc features have been modified in the squid. The mantle cavity houses not only gills, but a siphon the squid jets water out of, thus propelling itself. The radula now exists near jaws that function like a parrot's beak.

**Figure 23.16**
**The Body Plans of Molluscs**

What features unite this huge, diverse group as a single phylum? Outside the realm of technical anatomical details, there really is only one feature common to all molluscs. If you look at **Figure 23.16a**, you can see an aquatic snail that displays this feature along with others. Note a layer near the dorsal (top) surface of the snail called a *mantle*. This is a fold of skin-like tissue that surrounds the upper body of the mollusc and that usually secretes material that forms a shell (although there are lots of molluscs without shells).

Past this universal feature, the snail displays a number of structures that are common to most molluscs. The mantle generally drapes over part of the mollusc body like a tablecloth, meaning there is a mantle cavity—a space underneath the mantle. This cavity generally is the site of gills, by which all aquatic molluscs obtain oxygen and by which the filter-feeding bivalves (clams, mussels, oysters) obtain not only oxygen but food. In land-dwelling molluscs, such as garden snails, much of the cavity is a primitive lung that evolved from the gill.

Then, there is the single or "unitary" foot, as shown on the snail's ventral (bottom) side. Through wave-like muscular contractions, this foot can propel the mollusc along the surface underneath it, be it a leaf or the ocean bottom. Some sessile bivalves, such as oysters, have lost the foot altogether, however; and as you'll see, in squid and octopus it has evolved into something else entirely. Note, in the blowup of the snail's mouth, there is a tooth-covered membrane called the *radula*. Present in most molluscs except the bivalves, this organ can be extended outside the mouth, where it serves as a kind of rasping file. Molluscs use it for scraping a surface or for cutting small prey, then they retract it back into the mouth, thus moving food into their digestive tracts.

The snail has some other features worth pointing to now (although we could have looked at variations on them in the annelids as well). These are a heart and a stomach. These organs obviously aren't unique to snails (or even molluscs), and that's just the point. Along with gills and lungs, these organs are signs of the molluscs' advanced position in terms of structural specialization.

Why does something like a heart exist? Because creatures as large as molluscs need a *delivery system* for food and oxygen. Simple diffusion of food from a gut won't work anymore because some tissues are buried so deep they would never be served. The delivery system's vehicle is blood, but that blood must be propelled by some force so that it can make a set of rounds through the animal, picking up oxygen from the gills and food from the digestive tract. The propelling force for this work is the beating of the heart.

The circulation of the cephalopods (squid, octopus) is like ours in that it is a **closed circulation system**—a circulation system in which the blood *stays within* the blood vessels, while the oxygen and nutrients diffuse out of the vessels to the surrounding tissues. Most molluscs, however, have an **open circulation system**: a circulation system in which blood flows out of blood vessels altogether, into spaces or "sinuses" where it *bathes* the surrounding tissues. Veins collect the blood (actually "hemolymph") in these sinuses for a return trip to the gills and the heart. Gastropods such as snails can be either single sex or hermaphroditic, and this is the case for the bivalves as well. All cephalopods, however, are single sex.

**Figure 23.16b** gives you some idea of one set of variations evolution has wrought with the basic body plan of the mollusc. Remember how the unitary foot is a usual mollusc feature? Look at what has become of it in the squid; it has been divided into eight arms (with suckers) and two longer tentacles. The radula is still there, but now it lies next to a set of jaws that operate something like a parrot's beak. Meanwhile, the mantle cavity now has another function; the siphon you can see at one end of it acts like a jetting water hose that not only helps propel squid movement—they expel water out of it—but also directs this movement by being aimed first one direction and then another.

Cephalopods deserve mention for another reason, which is that they are not only the biggest invertebrates in the world, they are the smartest as well. One researcher who worked for decades with octopus said that they had the intelligence of dogs, which is very smart indeed in the animal world. Many researchers would go further, however, saying that octopus and squid seem to plan and calculate in very sophisticated ways. These animals also are fast swimmers, have keen eyesight, can change color in an instant (to hide from predators), and as a last resort in defending themselves, have the ability to shoot out black "ink" from a special sac. Given their remarkable capabilities, the curious thing about these molluscs is that they live only a year or two; most animals this sophisticated live much longer lives.

## 23.8 Phylum Nematoda: Roundworms

Nematodes, commonly referred to as roundworms, exist in such enormous numbers that it is impossible to imagine life without them. They fit into numerous ecological roles, such as that of "detritivore," meaning an organism that feeds on dead or cast-off organic material. Nevertheless, many farmers in the United States would *like* to imagine life without them because several species of these animals are a major cause of damage to such crops as soybeans and corn. Farmers and research scientists have been united for decades in an entrenched battle against them. In

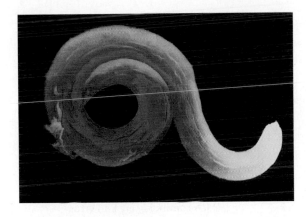

**Figure 23.17**
**Roundworms**

The tiny worms called nematodes are a normal part of many ecosystems, including soil ecosystems. However, a small number of roundworm species are harmful to humans, either directly as parasites or indirectly as crop pests. Pictured is an artificially colored micrograph of the nematode *Toxocara canis*, a parasite that infects dogs. Humans can become infected with *T. canis* eggs if they have contact with an infected dog or contaminated soil. The worm is magnified × 450.

addition, some are human parasites. The disease trichinosis, usually brought on by eating undercooked pork, is caused by a roundworm; likewise, there is hookworm, which affects hundreds of millions of people in warmer climates throughout the world (**Figure 23.17**).

An acre of prime farmland might contain several hundred billion roundworms, but you don't have to go to farmland to find these animals. They exist in desert sands, polar ice, ocean bottoms, lakes—name the place where life is going on, and the nematode worms are likely to be among the living. Indeed, scientists have only the vaguest idea of how many *species* there are; about 15,000 have been catalogued, but 100,000 seems to be the minimum number that actually exist.

Most roundworm species are very small—the crop pests, which feed on roots, generally are microscopic—but some parasitic roundworms are several meters long. The roundworms that have been studied most intensely are those that cause disease and crop damage, and these have a characteristic appearance: transparent, smooth, C-shaped, and cylindrical with tapered ends (hence the term *roundworm*). Across the breadth of roundworm species, however, there is a great diversity in form.

Roundworms generally are of one sex or the other, meaning they are not hermaphroditic, as the flatworms are. Reproduction is always sexual; unlike flatworms, roundworms can't reproduce by pulling themselves apart. And, you may remember that roundworms do not have a "true" coelom or central cavity, but instead have a "pseudocoel." Roundworms are the first of two *molting* animal phyla we will look at; they have an external skeleton, called a cuticle, that they shed or molt four separate times during their lifetime. We'll take a more detailed look at molting, however, in the animals coming right up, the arthropods.

## 23.9 Phylum Arthropoda: Insects, Lobsters, Spiders, and More

Phylum Arthropoda is large and varied indeed. It is so large that there are more described species in one arthropod group, the insects, than there are in all the other animal phyla combined. It is so varied that its members range from legless barnacles to hundred-legged millipedes (**Figure 23.18**). There are lots of ways to group the animals in this phylum, but for our purposes, it will be convenient to think about them batched into three subphyla:

- Subphylum Uniramia, which includes insects, millipedes, and centipedes, among other organisms

- Subphylum Crustacea, which includes shrimp, lobsters, crabs, and barnacles, among other organisms

- Subphylum Chelicerata (chel-is-er-AH-ta), which includes spiders, ticks, mites, and horseshoe crabs, among other organisms

Before getting to these subphyla, it will be useful to review the characteristics that all arthropods share. First, all arthropod animals have an **exoskeleton**, defined as an external material covering the body, providing support and protection. You can see an example of an exoskeleton in the grasshopper pictured in **Figure 23.19**. Second, all arthropods have what are known as **paired, jointed appendages**, which are just what they sound like: appendages, such as legs, that come in pairs and that have joints. (The word *arthropod* actually means "jointed leg.")

Taking the exoskeleton first, it is made of a tough carbohydrate called chitin that comes embedded in protein. The crustacean arthropods (crabs, lobsters,

**(a)** Subphylum Chelicerata

**(b)** Subphylum Crustacea

**(c)** Subphylum Uniramia

**Figure 23.18**
**Big Numbers, Much Diversity**

The arthropods represent the largest animal phylum, in sheer numbers as well as in diversity. Shown are representatives of the three arthropod subphyla:

**(a)** A web-building *Argiope* spider, this one found on the Snake River at the Idaho-Oregon border. The Chelicerata grouping to which it belongs is the only arthropod subphylum whose members do not have antennae.

**(b)** A flame lobster from Hawaii. Note the well-defined arthropod characteristics in this crustacean: an exoskeleton and the legs that exist as paired, jointed appendages.

**(c)** The arthropod subphylum Uniramia includes insects. Pictured is a short-horned grasshopper native to western North America.

and so forth) have, in addition, a calcified component to their skeleton that makes it very hard, as anyone who has ever "cracked" a lobster knows. Such a skeleton is not only protective but also serves to anchor arthropod muscles very securely. Your own muscles work by pulling against your *endo*skeleton—your internal skeleton. Arthropod muscles pull against a skeleton, too, but it's a skeleton that surrounds their bodies.

Strong though it may be, the exoskeleton only works because it is also flexible. Like a knight's armor, the exoskeleton has plates that overlap and thin, flexible sections between these plates that can bend as necessary. Even with this flexibility, however, the exoskeleton presents a problem: The arthropod body can only grow so much before it expands right into it.

So, how does the arthropod solve this problem? Its solution is **molting**: a periodic shedding of an old skeleton followed by growth of a new one. A crustacean such as a crab will retreat to a relatively safe place and begin to slip out of its old exoskeleton, as if it were Houdini trying to get out of a straightjacket (**Figure 23.20**). Through chemical processes, the old skeleton has begun to split by this time. Since the new, developing exoskeleton lies just underneath the old one, you might think this process would leave the crab just as hemmed in as before. But, at this stage the new skeleton can be *stretched*, and the crab proceeds to do just that by ingesting a large amount of water, thus inflating both its body and the skeleton to a new size. The new skeleton then hardens in its inflated size, the crab loses the extra water, and shrinks to its previous size. The result? It now has room to grow. Insects pull off this same feat, but stretch themselves with air instead of water.

The jointed appendages that arthropods possess come in many forms. There are legs for walking, claws for predation and defense, and wings for flying. These appendages are flexible and often come with sensory attachments, such as sensory "hairs," that keep the arthropod in touch with its environment.

Together, jointed appendages and an exoskeleton make for an animal that is often nimble and well defended—a combination that is one of the reasons arthropods are so numerous. In addition, arthropods often have keen senses, such as sight and chemical perception (as with ants following a "pheromone," or chemical scent trail).

## Subphylum Uniramia: Insects First

Recall that the arthropod phylum is divided into three subphyla. The first of these subphyla, Uniramia, has within it a class of animals we all have some familiarity with, the insects.

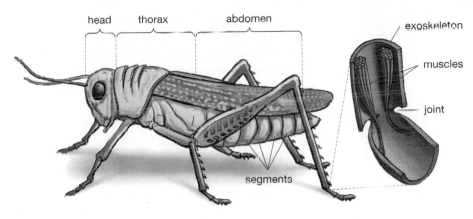

**Figure 23.19**
**Arthropod Features**
Arthropods have paired, jointed appendages, as with the legs of this grasshopper. The insects among the arthropods have bodies made up of three large-scale sections: head, thorax, and abdomen. Like the annelid worms, arthropods are segmented. This segmentation is clearly visible in the abdomen region of the grasshopper but is less pronounced in the head and thorax because there the segments have fused. The entire arthropod body is covered in a rigid exoskeleton that is jointed to allow movement.

Who can comprehend how many insects there are in the world? There are more species of them—about a million have been counted so far—than there are species in all the other groupings of animals combined. If we could get an accurate count of their numbers, we would probably find that there are tens of millions of insect species. To look at their numbers another way, for every person on Earth, there may be 200 million insects.

We human beings are completely unaffected by most of these creatures, of course, but then again, we have been involved in a long struggle with a few varieties of them. Think, for example, of the everyday troubles caused by ants, lice, fleas, and cockroaches.

**Figure 23.20**
**Old and New**
Arthropods have an outer or exoskeleton of a fixed size. They are able to grow by periodically shedding, or molting, this stiff skeleton. The crab in the photograph has just finished molting; its old exoskeleton lies nearby.

**••• ANSWERS**
1. snails or slugs; oyster, clams, or mussels; octopus, squid, or nautilus
2. open; closed; blood vessels
3. nematodes; roundworms

Then, there are more serious predations, such as the malaria transmitted by mosquitoes and the crop destruction caused by locusts. Conversely, we have insects to *thank* for much of the plant growth we see in the world, in that insects pollinate flowering plants. In addition, insects help break down organic matter, and many of them are beneficial to humans in that they feed on the very insect pests that are so troublesome to humans.

So, how are these creatures structured? If you look again at Figure 23.19, you can see the body parts of a typical insect. Note that it has three principal body regions—head, thorax, and abdomen—and that each region has its own pair of legs, which gets us to the famous six legs possessed by all insects. In the abdomen, you can see a segmentation reminiscent of that you saw with the annelid worms. Most insects have two pairs of wings, but there are many exceptions to this rule. (The flies with which we're most familiar have only one pair of wings, and worker ants have no wings at all.) Sexes are separate in the insects, and reproduction is generally sexual. We human beings take in oxygen through our lungs but then rely on circulating blood to carry this oxygen to all the tissues in our body. Not so with the insects. Their oxygen comes in through a set of small openings on each side of the body, and each opening is then linked, via a small tube, to a network of tubes that is so extensive that every part of the body receives oxygen through it.

The well-known transformation of a caterpillar into a butterfly is representative of a phenomenon common to most insects: **metamorphosis**, meaning a change in form in an organism as it develops from an embryo into an adult. The caterpillar-to-butterfly transformation exemplifies the most common variety of insect metamorphosis in that *three* developmental stages are involved following the hatching of an insect from an egg. An initial, wormlike form (a larva) gives rise to an intermediate, hibernating form (a pupa), which in turn gives rise to a fully grown, reproductive form (an adult). Caterpillars are larvae, as are the maggots that develop into flies and the grubs that develop into beetles. The pupae that larvae develop into do not feed during their hibernation but instead are locked away in a protective casing—called a cocoon in moths and a chrysalis in butterflies—in which they may spend cold winter months. The adults that pupae give rise to are the only insect form that reproduces; when the eggs that an adult female produces are fertilized, they develop into larvae, and the cycle continues.

## Other Uniramians: Millipedes and Centipedes

In addition to insects, the uniramian subphylum also includes millipedes and centipedes (**Figure 23.21**). Millipedes don't really have a thousand legs, despite their name. But hundreds of legs are possible because these creatures have up to 100 body segments, each of which has two pairs of legs. This structure makes for movement that is slow but relatively powerful as millipedes make their way across the damp, dark environments they favor, which often means the forest floor.

Centipedes look much like millipedes, but are a different class of arthropod. They have only one pair of legs per segment and a pair of more prominent antennae extending from their head. Moreover, centipedes are carnivores, whereas millipedes feed largely on decaying plants. Centipedes come equipped for their predatory work, as their front two legs have become modified into venom-injecting fangs. This

**Figure 23.21**
**Millipedes and Centipedes**

Insects constitute only one of the three classes of arthropods in the arthropod subphylum Uniramia. Millipedes and centipedes make up the other two classes.

**(a)** Millipedes have two pairs of legs per segment and live mostly on decaying plant matter.

**(b)** Centipedes have only one pair of legs per segment and are carnivores. Shown is the Texas giant centipede, *Scolopendra heros*.

**(a)** Millipede

**(b)** Centipede

may sound ominous, but only a few tropical species pose a threat to human beings.

## Subphylum Crustacea: Shrimp, Lobsters, Crabs, Barnacles, and More

The second arthropod subphylum, Crustacea, includes several kinds of animals that don't seem to have much in common at first glance. How are a crab and a barnacle alike, for example? But, remember that in putting living things together in a category, biologists are first and foremost concerned with shared *ancestry*, not with whether one creature has the same general appearance as another. At a crime scene, convincing evidence is likely to be found in small details, such as hair samples and fingerprints. The same thing holds true for evidence about the relatedness of groups of animals: the telling details are likely to involve small features.

This turns out to be the case with crustaceans. What all of them have in common is five pairs of appendages extending from their heads—two pairs of antennae and three pairs of feeding appendages. If you look at **Figure 23.22**, however, you can see the fantastic diversity that evolution wrought in crustaceans.

One of the crustaceans pictured in Figure 23.22, the water flea in the genus *Daphnia*, certainly is not familiar in the way that crabs or barnacles are, but it is representative of a variety of small, aquatic crustaceans that exist in huge numbers, mostly in the oceans, but also in freshwater bodies. As a whole, the crustaceans are mostly ocean dwelling, but many live in freshwater, while a few live in damp-land environments. Crustaceans are known to humans mostly as a food source, such as shrimp and lobsters, or as a marine pest, such as the barnacles that attach themselves to ships. It is not just humans who use crustaceans as a food source, however. Closely related to shrimp are the crustaceans known as krill. Generally no more than an inch long, these animals feed on the plant life that floats in the icy ocean waters off Antarctica. Traveling in huge, dense swarms, they then become an important part of the diet of whales, seals, and penguins.

## Subphylum Chelicerata: Spiders, Ticks, Mites, Horseshoe Crabs, and More

The third arthropod subphylum, Chelicerata, provides another example of small physical features yielding critical clues about relatedness. You might think that horseshoe crabs should be categorized as crustaceans, along with the other crabs, but it turns out that horseshoe crabs aren't really crabs at all.

**(a)** Barnacles

**(b)** Water flea

**(c)** Shrimp

**Figure 23.22**
**Crustacean Diversity**
Diversity among the arthropod subphylum Crustacea is stunning.

**(a)** One class of crustaceans includes barnacles. When seen on a pier or ship, barnacles may seem lifeless to us, but notice the feathery extensions coming from these barnacles in the south Pacific. These are cirri, which barnacles extend from their shell and then use to catch drifting plankton.

**(b)** Water fleas in the genus *Daphnia*, such as the one shown here, live in huge numbers in ponds and lakes. Although barely visible with the naked eye, they are complex animals. Under microscopes, their tiny hearts can be seen beating a furious rate. As they are not insects, they are not really fleas. They get their name from the jumping, jerky motions they make in the water. Note the clearly visible digestive tract in this water flea.

**(c)** The crustacean class that includes shrimp also includes lobsters, crabs, and the pill bugs so often seen when we turn over rocks. Shown is the colorful fire shrimp, *Lysmata debelius*.

**(a)** Horseshoe crabs

**(b)** Tarantula spider

**(c)** Dust mite

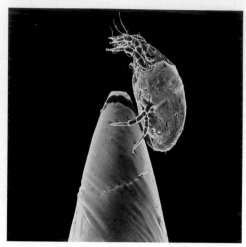

**Figure 23.23**
**Chelicerate Diversity**

All members of the arthropod subphylum Chelicerata possess paired chelicerae, which are pincer-like appendages used for feeding. These appendages stand in contrast to the mandibles used by the other two arthropod subphyla.

**(a)** In the springtime, East Coast beaches often have lots of horseshoe crabs such as these. In this picture, a group of males is competing for access to a female.

**(b)** The much-feared but generally harmless tarantula, this one a dry-weather Chilean rose tarantula, *Grammostola spatulata*, that is often kept as a pet. Like all spiders, the tarantula is a member of the Chelicerata class Arachnida, from which we get the term *arachnophobia*.

**(c)** Dust mites actually are eight-legged arachnids as well. The dust mite shown is standing on the head of a sewing needle. Such mites exist in almost all homes in the millions. Body fragments and excrement from them actually make up a large portion of the "dust" that is visible when sunlight shines into a room. This material sets off allergic reactions in many people.

If you looked at the underside of a horseshoe crab, near its mouth, you would see two small appendages called chelicerae, which have pincers and which the crab uses to bring food to its mouth. If you looked near the mouth of a common spider, you would again see two chelicerae, only now each chelicera has become modified to become a venom-injecting *puncturing* appendage, rather than a grasping appendage. Ticks, mites, and all the other chelicerates likewise have paired chelicerae, from which this subphylum gets its name (**Figure 23.23**).

Horseshoe crabs may be familiar to anyone who has spent time on the beaches of the Atlantic seaboard in May or June. Fearsome though they may look, they are harmless creatures that feed on worms, clams, or whatever else they can scavenge from the ocean bottom.

It is another grouping of chelicerates, however, that really inspires human fear: the arachnids, a class that includes mites, ticks, and scorpions, along with the more familiar spiders. Why "arachnophobia" should be so widespread in human beings is something of a mystery since, of the 34,000 identified species of spiders, only a handful cause any real harm to people. Part of the problem may be that spiders are blamed for wounds actually inflicted by other creatures. In one study of 600 suspected "spider bite" occurrences in southern California, 80 percent turned out to be the work of other arthropods—mostly ticks and some of the true bugs of the insect world (meaning bedbugs and other members of their order). Spiders evolved one of the world's great prey-snaring abilities, which is the spinning of their remarkable silk webs.

If we want to look for true causes of human misery, a better place to start would be with another arachnid, the microscopic dust mite, which is now known to be a major cause of asthma and other

human allergic reactions. In the last decade, the mite's fellow arachnid, the *Ixodes* tick, has become feared because of its role as a carrier of the bacterium that spreads Lyme disease.

**SO FAR...**

1. All arthropods share the features of an _____ or external material that provides support and protection, and _____, _____ appendages. All arthropods also go through the process of _____, in which an old _____ is shed and replaced by a new one.

2. Within the arthropod phylum, Subphylum Uniramia includes not only insects but _____ and _____. Subphylum Crustacea includes not only shrimp but _____, _____, and _____. Subphylum Chelicerata includes not only spiders but _____, _____, and _____.

3. There are more species of one group of arthropods, the _____, than there are species of all other groupings of animals combined.

THE DEUTEROSTOMES

## 23.10 Phylum Echinodermata: Sea Stars, Sea Urchins, and More

With the echinoderms, we cross into the animal grouping noted earlier, the deuterostomes, that also includes our own phylum, Chordata. You might expect that, as our relatives, the echinoderms would have the kind of sophisticated features we've recently been seeing—brains, well-developed sense organs, strictly

**(a)** Sea star

**(b)** Sea cucumber

**(c)** Sea urchins

**Figure 23.24**
**Echinoderm Diversity**
There are five classes in the echinoderm phylum, three of which are represented here.
**(a)** Sea stars such as this one are fierce predators. Here a sea star is managing to open the shell of a bivalve known as a rock cockle.
**(b)** True to its name, this sea cucumber does look something like the vegetable, but sea cucumbers come in many shapes and colors. Like most echinoderms, sea cucumbers move slowly along the ocean floor.
**(c)** Kelp is a favorite food of many sea urchins. Here, a group of sea urchins is shown foraging on some kelp.

sexual reproduction—but none of this is true in the best-known echinoderm, the sea star (**Figure 23.24a**). It has no brain, its only sensory organs are the primitive eyespots found on the tips of its arms, and a whole new sea star can be regenerated from the arm and part of a central disk of an existing sea star. Even the bilateral symmetry that you've seen in every animal since the jellyfish is absent in the adult sea star: It has radial symmetry, like the jellyfish.

This is not to say, however, that the sea star and its echinoderm relatives are completely primitive. Most have a remarkable system for moving that involves forcing water into a series of suction-tipped tube feet

that then extend, like so many inflated water balloons, from the underside of the sea star's arms (**Figure 23.25**). Moreover, the sea star evolved into radial symmetry *from* bilateral symmetry. How do we know? Sea star larvae have bilateral symmetry, a quality that is lost when the larvae develop into adults. Looked at together, these echinoderm features add up to a phylum that is something of a mystery to scientists—a mixture of characteristics from all over the animal map.

Sea stars aren't the only familiar echinoderms. Sea urchins, with their spine-covered spherical bodies, are echinoderms as well, as are sand dollars. Then, there are the less familiar but well-named sea cucumbers

**••• ANSWERS**

1. exoskeleton; paired, jointed; molting; exoskeleton

2. millipedes, centipedes; lobsters, crabs, barnacles; ticks, mites, horseshoe crabs

3. insects

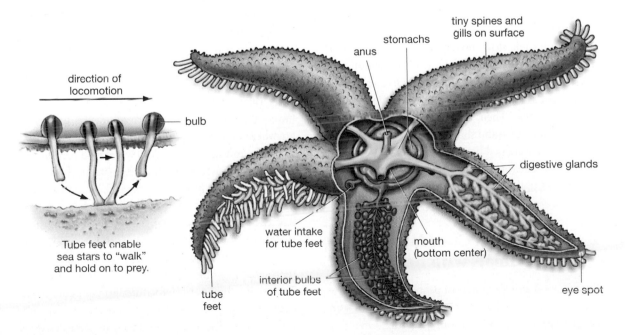

direction of locomotion

bulb

Tube feet enable sea stars to "walk" and hold on to prey.

tube feet

interior bulbs of tube feet

water intake for tube feet

mouth (bottom center)

anus

stomachs

tiny spines and gills on surface

digestive glands

eye spot

**Figure 23.25**
**The Body Plan of a Sea Star**

The sea star is radially symmetrical in its adult form, and it has no brain. Its only sensory organs are the primitive eyespots found on the tips of its arms. It has, however, evolved a remarkable system for moving along the sea floor and grasping prey. It first takes in water from the surrounding sea and then channels this water into a series of water bulbs. When inflated with the water, these bulbs lengthen, functioning as suction-tipped "tube feet."

**(a)** Subphylum Cephalochordata

**(b)** Subphylum Vertebrata

**(c)** Subphylum Urochordata

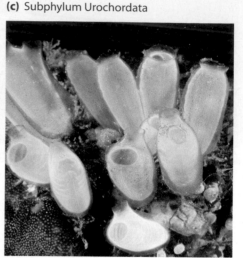

**Figure 23.26**
**Chordate Diversity**

**(a)** A single type of animal, the lancelet, is the sole representative of one of the three subphyla of chordates, Cephalochordata. Lancelets spend much of their time burrowed into the sand or mud of the shallow ocean floor, filter-feeding on small organisms or passing food particles. Only a few centimeters long, the diminutive lancelets are chordates, but not vertebrates.

**(b)** The African cheetah shown here is a vertebrate and thus a representative of another of the chordate subphyla Vertebrata.

**(c)** Urochordates, representing the third chordate subphylum Urochordata, come in about 3,000 species, most of them similar to the vase-like tunicates shown here. These animals are filter feeders that spend their adult lives attached to rocks. Their common name, sea squirts, comes from their practice of suddenly ejecting a spout of water from an opening near the top of their bodies. This group was photographed in the ocean off the coast of Indonesia.

(Figure 23.24b). There are about 7,000 identified species of echinoderms, all of them ocean dwelling and nearly all of them inhabiting the ocean floor, although "floor" in this case may just mean the bottom of a tide pool. Some are sessile, but most move across the ocean floor at a slow pace. Their shared habitat and measured pace make sense when we realize that the entire phylum evolved from a group of sessile filter feeders.

Despite their slow movement, echinoderms can be formidable predators. Sea stars often feed on molluscs, such as oysters and mussels. In some cases, they use the suction tubes on their feet to pry open a mollusc's shell just a little. Then, they "evert" their stomach—turn it inside out—and slide it in through the narrow gap they have created between shell halves. Juices secreted by their stomach then start to digest the prey's tissues. Sea urchins, meanwhile, are some of the animal kingdom's most voracious algae eaters.

## 23.11 Phylum Chordata: Mostly Animals with Backbones

Say the word *animal*, and most people think of a vertebrate animal—be it tiger, elephant, deer, lizard, or some other large and probably toothy creature. One thing

that ought to be clear by now, however, is what a small portion of the animal world vertebrates make up. Nevertheless, there are a lot of vertebrate species. The 50,000 or so that have been identified are nothing compared to the 1 million species of insects, but there are more species of vertebrates than there are of sponges, or cnidarians (jellyfish, etc.), or flatworms, and the vertebrate numbers may match those of molluscs.

In other ways, too, the vertebrates have been a big success. All the largest and swiftest animals are vertebrates, whether in water or in air or on land. And, the sheer variety of sensory capabilities that vertebrates have—from echolocation to reading—is unrivaled anywhere in the living world.

Large and important as they are, the vertebrates don't even constitute a phylum; instead, they make up only a subphylum, Vertebrata. The larger phylum into which vertebrates fit, called Chordata, actually has two additional subphyla in it. One of these (Cephalochordata) is made up of only a single kind of animal, a small, eel-like creature called a lancelet, represented by only 25 or so species (**Figure 23.26a**). The other subphylum (Urochordata) has about 3,000 ocean-going species in it but is represented by only three classes of animals. The largest of these classes is made up of the tunicates, whose practice of expelling water earns them their alternative name of sea squirt.

## What Is a Vertebrate?

What distinguishes vertebrates from these other chordates? Only the vertebrates have just what you'd expect: a **vertebral column**. Better known as a backbone, this flexible column of bones extends from the anterior to posterior end of an animal. Conversely, what unites all the chordates? Four features, which you can see pictured in a lancelet in **Figure 23.27**. At some point in their lives, all chordates have:

- A **notochord**: a stiff, rod-shaped support structure composed of cells and fluid and surrounded by a lining of fibrous tissue, running from the chordate's head to its tail

- A **dorsal nerve cord**: a rod-shaped structure consisting of nerve cells, running from the chordate's head to its tail

- A series of **pharyngeal slits**: openings to the pharyngeal cavity

- A **post-anal tail**: a tail located posterior to the anus

At this point, as a chordate, you may be wondering where your *own* post-anal tail or pharyngeal slits went, so a little explanation of these features is in order. Although the word *notochord* sounds like "nerve cord," the notochord is a flexible but tough *support* structure, not a nervous system feature. A lancelet retains its notochord throughout life, but most vertebrates lose theirs in embryonic life. (Its outer portion develops into the backbone.)

The dorsal nerve cord may not seem like a feature unique to chordates—you've seen nerve cords in lots

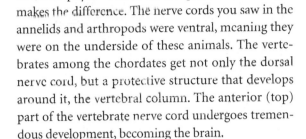

of other phyla—but it is the qualifier *dorsal* that makes the difference. The nerve cords you saw in the annelids and arthropods were ventral, meaning they were on the underside of these animals. The vertebrates among the chordates get not only the dorsal nerve cord, but a protective structure that develops around it, the vertebral column. The anterior (top) part of the vertebrate nerve cord undergoes tremendous development, becoming the brain.

The pharyngeal slits you see in the lancelet in Figure 23.27 take on more meaning when you remember that, in both humans and lancelets, the pharynx is a passageway just behind the mouth that can be thought of as the first chamber of the digestive tract. Pharyngeal *slits*, then, are perforations of the pharynx, and these serve different functions in different chordates. In the filter-feeding lancelets, they serve to trap food. Meanwhile, in embryonic fish, the slits *develop into* gills, which fish use to obtain oxygen. And in people? If you look at a month-old human embryo, you can see a series of pharyngeal clefts that for a time have openings in them. These pharyngeal slits, however, develop not into gills but into various other structures, including the space inside the middle ear.

Finally, what about that tail? If you look at **Figure 23.28**, you can see an 8-week-old human embryo with a very clearly developed tail. By the

### Figure 23.27
**Four Universal Chordate Features**

All chordates possess four structures at some point in their lives: pharyngeal slits, a support structure called a notochord, a dorsal nerve cord, and a post-anal tail. In the lancelet, all these features are present in the adult. In humans, only the dorsal nerve cord is clearly evident in adults.

**(a)** Human embryo tail at eight weeks

**(b)** Human embryo tail at ten weeks

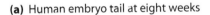
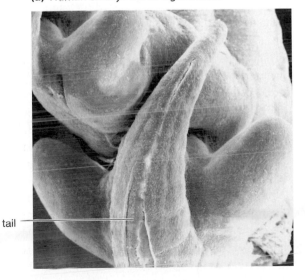

tail

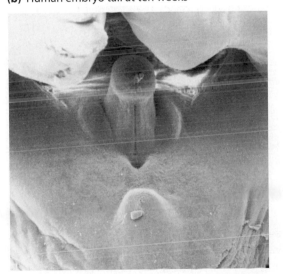

### Figure 23.28
**The Human Tail**

**(a)** Eight weeks after conception, the human embryo has a very clearly developed tail.

**(b)** Two weeks later, the tail has almost disappeared. In adult humans the only vestige of our embryonic tail is a small bone at the base of the vertebral column, the coccyx.

time the embryo is 10 weeks old, however, the tail has almost disappeared. All chordates have a tail at some point. In our oceangoing ancestors, it was used to help move them through the water, as is the case with modern fish. In human adults, though, all that remains of our embryonic tails is a small bone, called the coccyx, located at the base of our vertebral column.

---

**SO FAR...**

1. The best-known echinoderm is the _____, which is unusual in that, like a jellyfish, its adult form has _____ symmetry.

2. Human beings are vertebrates and are thus members of one of three subphyla within Phylum _____.

3. Which of the following traits are not shared by all chordates at some point in their lives: notochord, dorsal nerve cord, exoskeleton, pharyngeal slits, sessile living, post-anal tail?

---

## Diversity among the Vertebrates

For the remainder of the chapter, we'll concentrate on the vertebrates among the chordates. The evolutionary relationships among these vertebrates were covered in Chapter 19. For a summary of which of them gave rise to which others, see Figure 19.20 on page 369.

A signal shift in vertebrate evolution came with the development of a very familiar feature: jaws. There is general agreement that ancient, jawless vertebrates gave rise to the jawed variety. The development of jaws is considered perhaps the single most important event in vertebrate history for it gave the vertebrates a great power to capture prey, carry offspring, and generally interact with the world in more effective ways. So useful was this feature that today all true vertebrates have jaws—with one exception.

### A Jawless Vertebrate

The lamprey is a long, thin animal that looks something like an eel, but it should not be confused with one. Eels are fish, complete with jaws; lampreys, by contrast, have a sucking disk at their anterior end with which they attach themselves to their prey, which are usually fish (**Figure 23.29**). The lamprey's oral disk has numerous teeth around its edge. It combines these teeth with a rasping tongue to pierce the flesh of its prey. It then hangs on to the victim through suction, continuing to rasp away, ingesting the fluids and flesh that it obtains in this way.

Lampreys are best known in the United States as a destructive force in the Great Lakes. Once confined to the eastern Great Lakes, parasitic lampreys were able to move west by the mid-twentieth century thanks to human alteration of a canal around Niagara Falls. By the 1950s, they had decimated a once-thriving trout fishing industry in the western Great Lakes. Human efforts to kill off lampreys while reintroducing trout have had only limited success. The main culprit seems to be human pollution, which is keeping stocked trout from reproducing.

### Fish: Cartilaginous and Bony

All the early jawed vertebrates were fish, which today are a big success story in the vertebrate world. Of the 50,000 vertebrate species, more than half are fish. And of the 25,000 or so fish species, more than 24,000 are so-called bony fishes whose name tells their story: With a couple of exceptions, their skeletons are made of bone. Another 750 fish species are cartilaginous fishes, meaning their skeletons are made of the more pliable connective tissue, called cartilage, that also gives shape to human ears.

By far the best-known cartilaginous fish are sharks, but not far behind are the graceful rays that seem to fly through the ocean (**Figure 23.30**). The sharks' reputation for being an ancient life-form is well deserved. Their lineage dates back more than 350 million years; hence, they had been around for 130 million years by the time the first dinosaurs appeared. Unlike most of their bony fish relatives, sharks have no **swim bladder**, which is an inflatable

**Figure 23.29**
**No Jaws, Powerful Predator**
Lampreys, such as the one shown here, are jawless fish. Note the lamprey's sucking oral disk, which it uses to latch onto and then parasitize prey. This sea lamprey, *Petromyzon marinus*, is the species that, thanks to human intervention, was able to invade the western Great Lakes in the mid-twentieth century. It remains there today, although in smaller numbers than before.

**Figure 23.30**
**Cartilaginous Fish**

Sharks and rays, such as the giant manta ray shown here, have cartilaginous, rather than bony, skeletons. The ancient cartilaginous fishes represent only about 3 percent of all fish species. This ray and its accompanying diver were photographed off the coast of Mexico.

**Figure 23.31**
**Bony, Ray-finned Fish**

About 97 percent of fish species have bony (rather than cartilaginous) skeletons, and of the bony fish, almost all are "ray-finned," meaning that their dorsal fins are supported by stiff rays such as the ones visible on this yellowtail parrotfish, *Sparisoma rubripinne*.

organ in a fish that the fish can fill with gas for optimum buoyancy. Fish that do have swim bladders can inflate them as needed to maintain a neutral buoyancy—that is, to float in the water at a given depth without expending any energy. By contrast, a shark must keep swimming or it will sink.

Sharks, rays, and their close kin aside, all the rest of the fish world is made up of bony fish, from goldfish to tuna to herring (**Figure 23.31**). What accounts for their great numbers today? Their basic body plan seems to have allowed them to adapt to every kind of underwater environment. Some have evolved a slender shape (eels), others a fantastic camouflage (anglerfish), and still others a biological "antifreeze" for frigid waters (the Antarctic icefish).

Almost all bony fish are so-called ray-finned fishes, so named because their dorsal fins are supported by straight-line structures that look like rays (see Figure 23.31). Four surviving species of bony fishes belong to another category, however: the lobed-finned fishes. These fish are important because it is their ancestors who brought vertebrates onto the land by struggling out of the swamps some 375 million years ago, using their muscular, lobed fins to propel themselves. Their lobed *fins* evolved into the four *limbs* of the **tetrapods**, meaning the four-limbed vertebrates (see Figure 19.18 on page 368 for this transition). All land vertebrates are tetrapods: amphibians, reptiles, birds, and mammals. Even snakes and whales

**Figure 23.32**
**Lobe-finned Fish**

One of the few living lobe-finned (rather than ray-finned) fish is the coelacanth, which was first discovered in modern times in the Indian Ocean near Madagascar. A fish similar to this one is thought to be the ancestor of all land vertebrates. Note its four fleshy, lobed fins. Fins such as these evolved into the four limbs of land vertebrates.

are tetrapods; these animals simply lost the limbs their ancestors had. If you look at **Figure 23.32**, you can see a picture of one of the rare, living lobe-finned fish, the coelacanth (SEE-la-kanth).

### Amphibians: At Home in Two Worlds

Amphibians were the first truly terrestrial vertebrates, having evolved from early tetrapods, but this statement needs qualification. Are amphibians really land animals? Most of them actually live in two worlds—land

and water. (Their very name means "double life.") If you look at **Figure 23.33**, you can see the life cycle of the best-known amphibian, the frog.

Even adult frogs must live in environments that are at least moist, because water can evaporate right through their skin. But frog offspring are fully aquatic for a time. They hatch from eggs that must be laid in water lest they dry out, and most young frogs spend the first part of their lives as swimming tadpoles. There are exceptions to the rule that amphibians must remain near standing water. Some frogs have become fully terrestrial, but even they must inhabit environments that are moist.

Apart from frogs, there are two other varieties of amphibians: salamanders and newts, collectively known as the tailed amphibians; and some wormlike, tropical amphibians known as caecilians. Salamanders may look like lizards, but their watery amphibian heritage betrays them. With some exceptions, they must live where it is moist, a fact that often keeps them hidden from human view.

Amphibians once were the dominant land animals; indeed, they were the only land vertebrates for some 75 million years. But, about 250 million years ago they ceded their dominant status to the very group that evolved from them, the reptiles, who in turn gave rise to birds.

## Reptiles and Birds

It may seem strange to lump reptiles and birds together, but remember that scientific classification has mostly to do with shared ancestry. There is general agreement that birds are the direct descendants of the reptiles we know as dinosaurs. If so, then the separate categories of "bird" and "reptile" make sense mostly in terms of describing features of the two kinds of animals.

One seemingly innocuous feature unites all reptiles and birds, but it is a feature that had far-reaching consequences. If you look at **Figure 23.34**, you can see an illustration of something known as the **amniotic egg**, named after a membrane (called an amnion) that aids the embryo inside of the egg. Recall that most amphibians must lay their eggs in water, lest the eggs dry out and the embryos inside them perish. The amniotic egg of reptiles and birds is different; it can be laid in environments ranging from moist forest floor to desert. Its tough shell keeps the

**Figure 23.33**
**Amphibian Life Cycle**
Amphibians (such as frogs, salamanders, and newts) require an aquatic, or at least moist, environment for at least part of their life cycle. Note the female frog in the figure is laying her eggs into the water. The male then releases sperm that falls on the eggs, fertilizing them. Most frogs dwell solely in the water through much of their development, first as embryos, then as tadpoles.

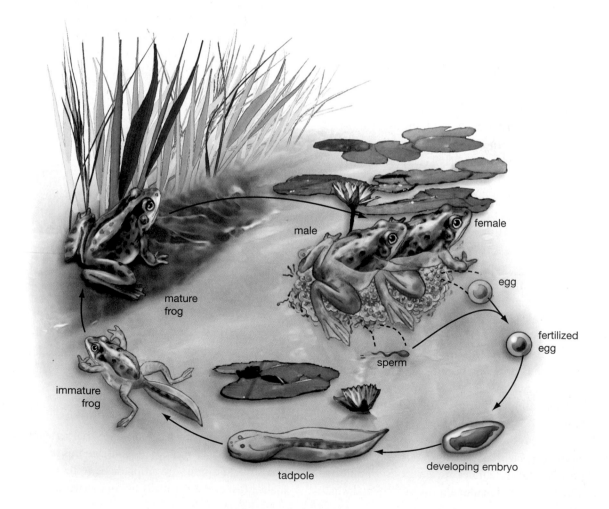

egg from drying out, and the various membranes within it protect the embryo. The amniotic egg therefore gave reptiles a tremendous advantage over the amphibians in the kinds of locations the reptiles could call home; unlike the amphibians, reptiles could settle away from the water.

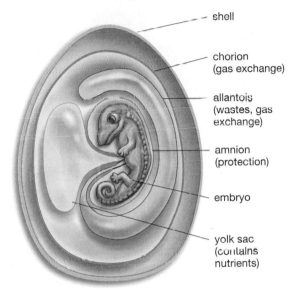

**Figure 23.34**
**An Egg for Many Environments**

The amniotic egg was an evolutionary development that allowed vertebrates to move inland from the water. The egg's features meant that, in dry environments, the embryo would not "desiccate" or perish through a loss of fluids. The egg's tough shell is the first line of defense; it limits evaporation of fluids. The amnion is one of several membranes that provide some cushioning protection for the embryo. The allantois serves as a repository site for the embryo's waste products. The chorion provides cushioning protection and works with the allantois in gas exchange, meaning the movement of oxygen to the embryo and the movement of carbon dioxide away from it. The yolk sac provides the nutrients the embryo will need as it develops.

## Characteristics of Reptiles

Several other things distinguish reptiles from amphibians. Looking back at the frog life cycle in Figure 23.33, you can see that the female frog lays her eggs in the water, after which the male spreads his sperm on top of them. This is external fertilization. By contrast, all reptiles employ internal fertilization—eggs are fertilized inside the female's body. A requirement for this, of course, is that males possess a copulatory organ that allows sperm to be deposited inside the female. Another difference between amphibians and reptiles is that reptiles have a tough, scaly skin that conserves water, as opposed to the thin amphibian skin that allows water to escape. Reptiles also have a stronger skeleton than amphibians, more efficient lungs, and a better-developed nervous system.

There are three main varieties of modern-day reptiles: lizards and snakes; crocodiles and alligators; and turtles (**Figure 23.35**). Then, there is an ancient, extinct variety of reptiles that is as well known as the living examples. The dinosaurs first appeared about 220 million years ago and were certainly the most fearsome creatures ever to have walked on land. The last of them died out 65 million years ago.

## Characteristics of Birds

Surprising as it may seem, "dinosaurs," as properly defined, never did die out, but live on in their descendants, the birds. So secure is this conclusion now that many evolutionary biologists routinely refer to birds as "avian dinosaurs." Two lines of evidence have convinced almost all scientists that birds descended from a line of bipedal (two-footed), meat-eating dinosaurs. One line concerns bone similarity. Long, S-shaped necks, types of hip bones, relative leg sizes. Only dinosaurs and birds share this list of characteristics.

**Figure 23.35**
**Reptile Diversity**

The three main groups of modern-day reptiles are the turtles, the lizards and snakes, and the crocodiles and alligators. Although some of these animals live in water, their scaly skin and amniotic eggs free them from dependency on it.

**(a)** This green turtle (*Chelonia mydas*) is swimming in the Red Sea.

**(b)** A newborn green tree python (*Chondropython viridis*) in New Guinea. It will change into its adult color in 6 to 8 months.

**(c)** American crocodiles such as this one (*Crocodylus acutus*) are found in far-southern Florida, through the Caribbean, Central America, and northern South America.

**(a)** Green turtle

**(b)** Tree python

**(c)** American crocodile

**Figure 23.36**
**Dinosaur and Bird**

Scientists believe that this fossil, called *Archaeopteryx*, represents a transitional form between dinosaurs and birds. It has the teeth and claws of a dinosaur, but the feathers of a bird. The drawing at right is an artist's interpretation of the fossil at left.

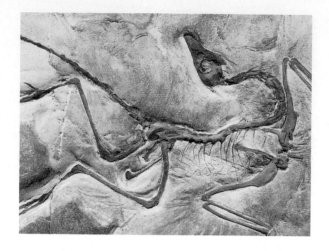

Second, there is the famous crow-sized fossil *Archaeopteryx*—a clear transitional form between dinosaurs and birds that had the dinosaur's teeth and claws, but the bird's feathers (**Figure 23.36**). In the 1990s, scientists began filling in the transitional gaps on the other side, unearthing fossils of dinosaurs that were flightless but nevertheless had feathers.

Birds are a vertebrate success. They exist all over the globe in great numbers, and their 9,100 recognized species exceeds the number of reptile species (7,000) or amphibian species (4,300). Although there are obvious differences between an eagle and a hummingbird, all flying birds have a very similar appearance. By contrast, think how different snakes are from crocodiles among the reptiles or frogs are from salamanders among the amphibians. The similarity among birds is a consequence of the strict requirements of vertebrate flight: light, hollow bones, wings, and powerful flight muscles that attach to a breastbone (**Figure 23.37**). All birds use their own metabolism to maintain a relatively stable internal temperature—a quality called *endothermy* that we'll review shortly in connection with our last major grouping of animals.

## Mammals: Small Numbers, Big Impact

The story of mammals could be titled "Big Effects from a Small Group." There are only about 4,400 species of mammals—not even half the number of bird species and scarcely more than the number of amphibian species. The total number of individual mammals is infinitesimal compared to the number of insects or roundworms. Yet, it's the mammals that are the biggest creatures not only on land (the elephants) but in the sea (the whales). It's the mammals that are the fiercest creatures from the savanna (lions) to the arctic ice (polar bears). And, of course, it's the mammals, in the form of humans, who control the fate of so much of the rest of the living world.

But what is a mammal? Here are two mammal universals and two near-universals. All mammals have mammary glands and maintain a near-constant internal temperature. Most mammals have hair and

**Figure 23.37**
**Bird Similarity**

Despite the obvious differences between the heron on the left and the bald eagle on the right, there are great similarities in the birds as well. Indeed, all flying birds display a similarity in body form. This is so because animals as large as birds must meet a strict set of requirements in order to fly. All birds have thin, hollow bones, tails to stabilize their flight, and powerful flight muscles that originate on their breastbone or sternum.

**(a)** Heron

**(b)** Bald eagle

have eggs that develop inside their mother's body. Here's some detail on each of these characteristics.

Mammals are named for the **mammary glands** that all of them possess: a set of glands that, in females, provide milk for the young. Next, almost all mammals have hair, while no other animals have it. This feature is related to another universal mammal characteristic, which is that mammals are **endothermic**: Their internal body temperature is relatively stable, and their body heat is generated internally—by their own metabolism. Conversely, amphibians and reptiles are **ectothermic**, meaning their internal temperature is controlled largely by the temperature of their environment.

Endothermy, sometimes misleadingly called warm-bloodedness, is rare in the animal world. In fact, other than mammals, the only creatures to possess it are birds. Endothermy and ectothermy are very important in determining which creature will live where in the animal world. After a cold night in the desert, the muscles of a lizard cannot function at full capacity until that lizard has warmed itself in the sun. By contrast, a desert rat is as functional at sunrise as it is at sunset. But, there is a price to be paid for its readiness, and that price is energy expenditure. It takes a great deal of energy to maintain the constant, relatively high body temperatures of mammals and birds. Pound per pound, the "basal," or resting metabolic, rate of a rat may be four times that of a lizard. And, it takes food to fuel this energy expenditure. The upshot is that, pound per pound, mammals and birds simply have to eat more than reptiles or amphibians.

The very quality of being able to maintain a stable body temperature has, however, allowed mammals (and birds) to live where amphibians and reptiles cannot—in cold climates. Consider the fact that there are arctic hares, arctic foxes, and arctic birds, but there are no arctic lizards or frogs. Past a certain point, north or south on the globe, there is no place a lizard can go to warm its body enough to live. Mammals have their limits as well, but the range of climates they can adapt to is greater. Their hair is a great aid in maintaining temperature, as is the thick layer of fat that lies just beneath the skin of such creatures as whales and seals.

With two important exceptions, mammals are **viviparous**: a condition in which fertilized eggs develop inside a mother's body. Contrast this with the **oviparous** condition, seen in all birds and most reptiles, in which fertilized eggs are laid outside the mother's body and then develop there. All embryos need nutrients and protection. In egg-laying (oviparous) animals, both things are provided by the egg itself (though parents also may provide protection). Conversely, a human egg (and then embryo) is protected by the mother's body and draws its nutrients directly from the mother in ways we'll look at shortly.

Reproduction turns out to be the defining feature of the three principal evolutionary lines of mammals: the monotremes, the marsupials, and the placental mammals (**Figure 23.38**). **Monotremes** are egg-laying mammals, represented by the duck-billed platypus and spiny anteaters found in Australia. The monotremes do nourish their young on milk from mammary glands, but they have no nipples; thus, the young must lap up the milk that diffuses onto the hair of the mother.

**Figure 23.38**
**Mammal Diversity**

All mammals have mammary glands that provide milk for their young. However, mammals differ in their reproductive strategies.

**(a)** This duck-billed platypus is a monotreme—a mammal that lays eggs.

**(b)** Kangaroos are marsupials—they give birth to physically immature young, which then develop further within the mother's pouch. Pictured are a mother and her fairly mature "joey" in eastern Australia.

**(c)** The young of placental mammals, such as this grizzly bear (*Ursus arctos*), develop to a relatively advanced state inside the mother, with nutrients supplied not by an egg, but by a network of blood vessels and membranes, the placenta.

**(b)** Marsupial

**(a)** Monotreme

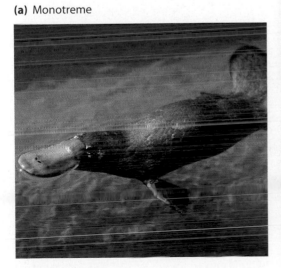

**(c)** Placental mammal

## Table 23.1  Animal Phyla

| Phylum | Members include | Live in | Characteristics |
|---|---|---|---|
| Porifera | Sponges | Ocean water (marine), with a few living in freshwater | Sessile (fixed in one spot) as adults, no symmetry. System of pores through which water flows serves to capture nutrients. Bodies do not have organs or tissues. Sexual and asexual reproduction. |
| Cnidaria | Jellyfish, hydrozoans, sea anemones, coral animals | Almost all marine; a few in freshwater | Radial symmetry, medusa (swimming) and polyp (largely sessile) stages of life. Use stinging tentacles to capture prey. Have tissues. Can reproduce sexually or by budding. |
| Platyhelminthes | Flatworms (flukes, tapeworms, turbellarians) | Marine, freshwater, or moist terrestrial environments | Bilateral symmetry, hermaphroditic, possess organs, no central cavity, most primitive animals with triploblastic tissue structure. Many are parasites. Can reproduce by dividing themselves. |
| Annelida | Segmented worms | Most are marine, some freshwater and moist terrestrial | Distinct body segmentation in varieties such as earthworms. Leeches are annelids. |
| Mollusca | Gastropods (snails, slugs), bivalves (oysters, clams, mussels), cephalopods (octopus, squid, nautilus) | Marine, freshwater, moist terrestrial | Mantle that can secrete material that becomes shell; mantle cavity housing gills or lungs; unitary foot (lost in some sessile species, evolved into arms and tentacles in cephalopods). |
| Nematoda | Roundworms | Almost all environments | Small size in most, but some are several meters long. Many are crop pests, others animal parasites. Possess pseudocoelom. Reproduction always sexual. |
| Arthropoda | Vast group that includes insects, spiders, mites, crabs, shrimp, and centipedes | Many aquatic and terrestrial environments | External skeleton made of chitin; jointed, bilateral appendages; molting in many species; segmented bodies; generally an open circulation system. |
| Echinodermata | Sea stars, sea urchins, sea cucumbers | Always marine, most on ocean floor | Generally have radial symmetry as adults; tube feet in some species, slow ocean floor movement in most. |
| Chordata | Vertebrates, lancelets, sea squirts | Great variety of aquatic and terrestrial environments | At some point, all possess notochord, dorsal nerve cord, pharyngeal slits, post-anal tail. Vertebrates have vertebral column. |

**Marsupials** are mammals in which the young develop within the mother to a limited extent, inside an egg that has a membranous shell. Marsupials are represented by several Australian animals, including the kangaroo, and by several animals in the Western Hemisphere, including the North American opossum. Early in development, the egg's membrane disappears, and in a few days time, a developmentally immature but active marsupial is delivered from the mother. In the case of a kangaroo, the tiny youngster has just enough capacity to climb up its mother's body, into her "pouch," and begin suckling from a nipple there. We can think of the kangaroo as developing partly inside its mother's reproductive tract and partly inside her pouch.

All other mammals are known as **placental mammals**. They are mammals nurtured before birth by the **placenta**, a network of maternal and embryonic blood vessels and membranes that allows nutrients and oxygen to diffuse *to* the embryo from the mother while allowing embryonic wastes to diffuse *from* the embryo to the mother. (For more on the placenta, see Chapter 32, page 681.) Thus, in placental mammals, the embryonic young derive their nutrition not from food stored in an egg but directly from the circulation of the mother. Further, the embryos of placental mammals can develop for a very long time inside the mother. The human gestation period—the time from conception to birth—is about 9 months, and in elephants it is almost 2 years. By contrast, the gestation time for a kangaroo is about a month, after which its "pouch life" might last 6 months.

The diversity of the placental mammals is impressive. There are flying mammals (bats); swimming ocean behemoths (whales, dolphins, and so forth); all manner of large herbivores (horses, deer, cows); and smaller herbivores (rodents, which make up more than half of all mammal species). Then, of course, there are the primates such as ourselves, defined by our forward-facing eyes, grasping hands, and arboreal or tree-dwelling habitat. Several species of primates, including our own, have come down from the trees to dwell on the ground (see Chapter 19, page 372). **Table 23.1** summarizes all of the animal phyla reviewed in this chapter.

## On to Plants

In life's family tree, the animals you've just looked at are far removed from plants. Yet animals and plants are alike in one sense: They sit atop two great evolutionary lines. Plants occupy the higher branches of the line of "self-feeders": the autotrophs that make their own food by harnessing the power of the sun. The first creatures who performed this feat were bacteria; then came the algae (which are protists). And then, evolving from the algae, came the plants, moving onto land and constructing their green edifice wherever they took root, thereby providing food and shade for all animals that came after them. In a sense, this sequence of events describes the relationship that animals and plants have: plants first, then animals. In a world without plants, there would be no land animals, but in a world without land animals, there would still be plants. The two chapters coming up describe the nature of the indispensable, green members of the living world, the plants.

---

**SO FAR...**

1. With the exception of lampreys, all true vertebrates have the important anatomical feature of _____.

2. All land-dwelling vertebrates are tetrapods, meaning species with four _____. All tetrapods evolved from _____.

3. The four principal varieties of tetrapods are _____, _____, _____, and _____.

••• ANSWERS

1. jaws

2. limbs; fish

3. amphibians; reptiles; birds; mammals

# CHAPTER 23 REVIEW

For study help, activities, and more quiz questions, go to **www.aw.com/krogh4**.

## SUMMARY

### 23.1  What Is an Animal?

- A single physical feature is sufficient to define animals. All animals pass through a blastula stage in embryonic development, but no other living things do. A blastula is a hollow, fluid-filled ball of cells that forms once an egg is fertilized by sperm.

- Three other features are characteristic of all animals, but also are shared, to varying degrees, by members of other kingdoms in the living world. All animals are multicelled, are heterotrophs (they cannot make their own food) and are composed of cells that do not have cell walls.

### 23.2  Animal Types: The Family Tree

- The animal kingdom is divided into large-scale categories called phyla. There are between 36 and 41 animal phyla, each phylum being a group of organisms that share basic body structure.

- Animals can be thought of as having evolved from simpler to more complex forms through a series of additions to the characteristics found in more primitive animals. These additions are tissues, radial symmetry, bilateral symmetry, and an enclosed body cavity or coelom.

- A fundamental split in animal evolution came with the divergence of the protostome and deuterostome animal lines. Protostome animals include flatworms, roundworms, molluscs, annelid worms, and arthropods; deuterostome animals include echinoderms and chordates.

### 23.3  Phylum Porifera: The Sponges

- Sponges are simple animals, lacking organs, tissues, or symmetry. Nearly all sponges live in marine environments, although there are a few freshwater varieties.

- Sponges live by drawing water into themselves through a series of tiny pores on their exterior and then filtering this water for food.

- Sponges can reproduce sexually, with eggs and sperm carried from one sponge to another by water currents, or they can produce asexually, through budding.

### 23.4  Phylum Cnidaria: Jellyfish and Others

- The defining characteristic of members of Phylum Cnidaria is their use of stinging tentacles to capture prey.

- Cnidarians include jellyfish, sea anemones, hydras, and the reef-building coral animals, which have skeletons that make up the bulk of coral reefs.

- Many cnidarians have both an adult, medusa stage of life, which swims in the water, and an immature, polyp stage of life, which generally remains fixed to rocks, animals, or other solid surfaces. Some cnidarians have only the polyp stage. Cnidarians have no organs, but they do have tissues.

- Cnidarians have radial symmetry, and reproduction can be sexual or asexual.

### 23.5  Phylum Platyhelminthes: Flatworms

- The flatworms of Phylum Platyhelminthes are mostly small creatures, dwelling either in aquatic or moist terrestrial environments. Some parasitic flatworm species can, however, reach enormous lengths.

- Flatworms have bilateral symmetry but have no coelom or system of blood circulation. They have no anus, meaning the flatworm's mouth is the single opening to its digestive system.

- Most flatworms are hermaphroditic—a single flatworm is likely to possess both male and female sex organs. Many flatworms can reproduce asexually by breaking themselves in half.

- Parasitic tapeworms, which can infect human beings, are flatworms, as are the flukes that cause the disease schistosomiasis.

### 23.6  Phylum Annelida: Segmented Worms

- The worms of Phylum Annelida provide a clear example of a physical feature that is widespread in the animal kingdom, body segmentation, meaning a repetition of body parts in an animal. Segmentation is generally valuable because it produces a body that is both strong and flexible.

- Earthworms are representative of the annelid or segmented worms; earthworm bodies have a segmentation that allows the worm to exercise a degree of independent control over each segment or group of segments.

- Most annelids are marine, and some dwell in freshwater. All annelids exist in environments that are at least moist. Leeches are parasitic annelid worms.

### 23.7  Phylum Mollusca: Snails, Oysters, Squid, and More

- Phylum Mollusca is an extremely varied group of animals whose members range from sessile, brainless mussels to agile, intelligent squid. Mollusc habitats range from marine and freshwater to moist terrestrial.

- Three important classes of molluscs are the gastropods, which include snails and slugs; the bivalves, which include oysters, clams, and mussels; and the cephalopods, which include octopus, squid, and nautilus.

- All molluscs possess a fold of skin-like tissue, called a mantle, that usually secretes material that forms a shell, although many molluscs do not have shells. Molluscs tend to have a mantle cavity that houses either gills (if the molluscs are aquatic) or lungs (if they are terrestrial); many have a unitary foot; the foot's wave-like contractions allow movement. Many molluscs have a tooth-lined membrane called a radula that can be extended outside the body for scraping and retrieving food.

- Cephalopods have a closed circulation system in which blood stays within blood vessels, while oxygen and nutrients diffuse out of the vessels to the surrounding tissues. Most molluscs, however, have an open circulation system, in which blood flows out of blood vessels into openings called sinuses, where the blood bathes tissues.

## 23.8 Phylum Nematoda: Roundworms

- The roundworms of Phylum Nematoda exist in enormous numbers in all kinds of habitats on Earth. Most are microscopic, although some are large.

- A number of roundworms are agricultural pests, and some are human parasites. The disease trichinosis is caused by roundworms, and the parasites known as hookworms are roundworms.

- Roundworms are the least-complex animals to possess an enclosed cavity—technically a pseudocoel instead of a true coelom. Their reproduction is always sexual, and they are not hermaphroditic.

## 23.9 Phylum Arthropoda: Insects, Lobsters, Spiders, and More

- Phylum Arthropoda is enormous and extremely varied. There are more species in one class of arthropods, the insects, than there are in all the other groupings of animals combined.

- Arthropods can be divided into three subphyla: Uniramia, which includes millipedes and centipedes along with insects; Chelicerata, which includes spiders, ticks, mites, and horseshoe crabs; and Crustacea, which includes shrimp, lobsters, crabs, and barnacles. All arthropods have an external skeleton or exoskeleton, and paired, jointed appendages. Arthropods go through molting, meaning the periodic shedding of an old exoskeleton.

- Members of Subphylum Uniramia include the insects, which have bodies divided into three sections: head, thorax, and abdomen. Sexes are separate among the insects, reproduction is generally sexual, and the life cycle usually involves metamorphosis: a change in form in an organism as it develops from an embryo into an adult. Centipedes and millipedes make up the two other classes of Uniramians.

- All members of Subphylum Crustacea have five pairs of appendages extending from their heads—two pairs of antennae and three pairs of feeding appendages. Crustaceans range from large lobsters to microscopic water fleas.

- All members of Subphylum Chelicerata have in common a pair of appendages, called chelicerae, that serve various feeding functions—grasping in horseshoe crabs, for example, and puncturing predation in spiders.

## 23.10 Phylum Echinodermata: Sea Stars, Sea Urchins, and More

- Echinoderms include sea stars, sea urchins, sand dollars, and sea cucumbers, among others. All members of Phylum Echinodermata are marine, and most inhabit the ocean floor.

- Echinoderms have a radial symmetry as adults. The sea stars among them evolved to this condition from bilateral symmetry; sea star larvae have a bilateral symmetry that is lost as they develop into adults. Sea stars have no brain and only one sensory organ, the primitive eyespots on the tips of their arms. A new sea star can be generated asexually from the arm and part of the central disk of an existing sea star.

- Echinoderms tend to move slowly across the sea floor, but many are formidable predators, often feeding on such molluscs as oysters or mussels.

## 23.11 Phylum Chordata: Mostly Animals with Backbones

- Phylum Chordata is made up of three subphyla: Vertebrata, which includes all the vertebrates, including human beings; Cephalochordata, made up of creatures called lancelets; and Urochordata, made up of three classes of animals, including the tunicates or sea squirts. Only the vertebrates have a vertebral column, meaning a flexible column of bones running from the anterior to posterior ends of an animal.

- All chordates possess, at some point in their lives, a rod-shaped support structure called a notochord; a nerve cord on their dorsal side; a post-anal tail; and a series of pharyngeal slits that develop into various structures, depending on the type of chordate.

- With the exception of lampreys, all true vertebrates have jaws. Today's jawed vertebrates developed from ancient jawless vertebrates.

- Fish account for more than half the 50,000 species of vertebrates, with the vast majority of fish being bony fish, as opposed to the more ancient line of cartilaginous fishes, such as sharks, which have skeletons that are made of the connective tissue cartilage.

- Most bony fish are ray-finned fishes, named after the raylike structures in their dorsal fins. Four species of fish belong to another category of fish, the lobe-finned fishes. These fish are important because an ancient variety of them is thought to have given rise to all tetrapods, meaning four-limbed vertebrates. All land vertebrates are tetrapods: amphibians, reptiles, birds, and mammals.

- Amphibians were the first terrestrial vertebrates. Today, amphibians include frogs; the tailed amphibians (salamanders and newts); and some wormlike amphibians known as caecilians. All amphibians must live in environments that are at least moist, most employ external fertilization of eggs, and most live in aquatic environments for part of their lives.

- Reptiles evolved from amphibians. An important feature in their evolution was the development of the amniotic egg, which can keep embryos moist even in dry environments. The amniotic egg allowed the reptiles to move inland from water sources. Reptiles employ internal fertilization and have a tough, scaly skin that conserves water. There are three main varieties of modern-day reptiles: turtles, lizards and snakes, and crocodiles and alligators. Dinosaurs were reptiles.

- Birds, the descendants of dinosaurs, exist all over the globe in great numbers. The similar appearance of all flying birds stems from the strict requirements of flight for creatures as large as vertebrates: light bones, wings, and powerful flight muscles that attach to a breastbone. All birds are endothermic: Their internal body temperature is relatively stable, and body heat is generated internally, by their own metabolism. Conversely, amphibians and reptiles are ectothermic: Their internal temperature is controlled largely by the temperature of their environment. Birds and mammals are the only true endothermic animals.

- Although there are relatively few mammal species, mammals have had a great impact on the natural world. They account for the largest creatures on land and sea, and one species of mammal, human beings, controls the fate of much of the living world.

- All mammals have mammary glands, meaning glands that deliver milk to the young. Most mammals have hair, and their fertilized eggs develop inside the mother's body (making them viviparous). There are three principal evolutionary lines of mammals: the monotremes, which are egg-laying (or oviparous) mammals, represented by the platypus; marsupials, in which the young develop within the mother only to a limited extent, represented by kangaroos; and placentals, in which young are nourished within the mother through a placenta, represented by most of world's mammals.

## KEY TERMS

| | | | |
|---|---|---|---|
| amniotic egg | 466 | dorsal nerve cord | 463 |
| bilateral symmetry | 444 | ectothermic | 469 |
| bivalves | 453 | endothermic | 469 |
| body segmentation | 451 | exoskeleton | 456 |
| cephalopods | 453 | gastropods | 453 |
| closed circulation system | 454 | hermaphroditic | 450 |
| | | invertebrate | 442 |
| coelom | 444 | mammary glands | 469 |

| | | | |
|---|---|---|---|
| marsupial | 471 | phylum | 442 |
| metamorphosis | 458 | placenta | 471 |
| molting | 457 | placental mammal | 471 |
| monotreme | 469 | | |
| notochord | 463 | post-anal tail | 463 |
| organ | 450 | radial symmetry | 444 |
| open circulation system | 454 | swim bladder | 464 |
| oviparous | 469 | symmetry | 444 |
| paired, jointed appendages | 456 | tetrapods | 465 |
| | | tissue | 442 |
| parasite | 450 | vertebral column | 463 |
| pharyngeal slits | 463 | viviparous | 469 |

## BRIEF REVIEW

*Answers to Brief Review questions are in the back of the book.*
*For multiple-choice quiz questions, go to* **www.aw.com/krogh4**.

1. A coelom amounts to internal space in an animal. Why is such a feature valuable?

2. Do all animals have a single ancestor, or do the different phyla of animals have different ancestors? What evidence supports your answer?

3. What are the benefits and the costs of endothermy? What animals are endothermic?

4. Why is it fair to characterize sponges as "primitive" animals?

5. What is a coral reef chiefly composed of?

6. What feature common to molluscs was modified through evolution to become the arms and tentacles of the squid?

7. Which animal is a spider more closely related to, a dust mite or a grasshopper?

8. A mother provides milk for her offspring, and yet those offspring are hatched from eggs. What kind of animal is the mother?

## APPLYING YOUR KNOWLEDGE

1. The text notes the tremendous number of insects in the world—both the number of insect species and the number of individual insects. One of the most spectacular examples of insect numbers can be seen in the so-called periodical cicadas, which emerge in the eastern and central United States in groups called *broods* after completing an underground larval stage that lasts either 13 or 17 years. One 17-year brood that emerged in 1956 near Chicago was found in densities of 1.5 million per acre in lowland forests—about 533 tons of cicadas per square mile over their entire range. Biologists often talk of the reproductive "strategies" of various species. These are the

traits of a species that help ensure its continued reproduction, generation after generation. What is the reproductive strategy of the periodical cicadas? Does their long underground immature stage aid in this strategy?

2. One of the distinguishing features of animals is that their cells do not have the thick outer lining known as a cell wall. Plant cells have a cell wall, as do fungi cells. Why would such a feature be advantageous for plants but disadvantageous for animals?

3. Plants can have elaborate defense systems, and they can respond in sophisticated ways to their environment—for example, in preserving resources, they can go into a dormant state in winter. Given this, can plants have intelligence? Or, is it only animals that have this trait, to one degree or another?

*Earth's flowering plants or angiosperms are a vast and varied group that includes not only all wildflowers, but all cacti, including the familiar saguaro cactus (Carnegiea gigantea), seen here in Arizona's Sonoran Desert.*

CHAPTER **24**

| 24.1 | The Importance of Plants | 477 |
| 24.2 | The Structure of Flowering Plants | 479 |
| 24.3 | Basic Functions in Flowering Plants | 485 |
| 24.4 | Responding to External Signals | 492 |

**Essays**

| What Is Plant Food? | 484 |
| Keeping Cut Flowers Fresh | 488 |
| Ripening Fruit Is a Gas | 490 |

Plants are indispensable to most other living things, and are not as passive and defenseless as they may seem. The flowering plants known as angiosperms are the most important plants of all.

# THE ANGIOSPERMS:
# An Introduction to Flowering Plants

**T**he optometrist has almost completed the examination but would like a better look at the interior of the patient's eyes. Would she agree to have her pupils dilated with the application of a couple of drops to

each eye? The patient tilts her head back, and in her pose, it's possible to imagine her as a lady of the Italian Renaissance, lifting her eyes upward and applying to them a few drops of a substance she believes will make her a *bella donna*—a beautiful lady. And what does this substance do? It dilates her eyes just slightly, making her pupils large and entrancing. Spin back further in time to the eleventh century in Scotland where, as legend has it, the Scottish leader Earl Macbeth poisons an invading army of Danish soldiers by lacing their drinks with juice from a plant known as sleepy nightshade. Only later will the English-speaking world call this plant by the name used today: *Atropa belladonna* (**Figure 24.1**). Skip forward to

1831, when scientists isolate from the belladonna plant one of its active ingredients, a compound called atropine. This is the substance that actually dilated the eyes of the Italian ladies and that, at one time, dilated eyes in optometry clinics. Then, come forward to the summer of 2004, when the federal government announces that it will soon be supplying states with an assortment of antidotes to chemical weapons. As a response to the threat posed by terrorists, the government says, it will distribute sets of "chem-packs" that include the drug atropine. This makes sense because U.S. armed forces units have for years been outfitted with "auto-injectors" that can pump a dose of atropine into the thigh of any soldier who has been gassed. Now step back from the government officials making their announcement and observe, just a few miles away, some paramedics hard at work. They're making use of atropine, too, although not in connection with a terrorist assault. In ambulances and emergency rooms, atropine is used to revive heart attack victims. An injection of it can bring a pulse back to a person whose heart is failing.

**Figure 24.1**
**A Plant that Has Served Many Purposes**
The flowering plant *Atropa belladonna.*

## 24.1 The Importance of Plants

Like a talented employee, belladonna keeps getting called back to work by human beings. If you were keeping track, we humans have derived, from just this one plant, a poison, an antidote to poison, a cosmetic, an aid to visual diagnosis, and a heart medicine. But that's the way with plants. The number of things they do for human beings is enormous. The best-known thing they do, of course, is provide us with food. But even in this role, their significance is not fully appreciated. In the world today, up to 90 percent of the calories that human beings consume come directly from plants. Indeed, as some botanists have noted, there are about a dozen plants that stand between humanity and starvation. (The most important of these plants is rice, with wheat coming in second.) Plants are able to serve this function because they make their own food through photosynthesis. The bounty they produce through this process has made Earth a planet of surplus—a planet of grains and leaves and roots that are available for the taking. Along these same lines, plants also produce much of the oxygen that most living things require. Then there are the products that human beings have learned to derive from plants, among them lumber, medicines, fabrics, fragrances, and dyes (**Figure 24.2**).

Apart from these uses, plants are a kind of anchoring environmental force. Their roots prevent soil erosion, while their leaves absorb such pollutants as sulfur dioxide and ozone. One of the main greenhouse gases warming Earth is carbon dioxide, and plants absorb carbon dioxide when they perform photosynthesis. Thus, there is not only a *production* side to atmospheric carbon dioxide—in significant part the pollution human beings produce—there is also a *consumption* side to it in the amount plants absorb. One proposed way to fight greenhouse warming, therefore, is to plant trees. A single mature maple tree will absorb about 450 kilograms (about 1,000 pounds) of carbon dioxide from the atmosphere over the course of a single summer.

## A Focus on Flowering Plants

Although the plant kingdom is vast and varied, it turns out that there are just four principal types of plants in it. These are the bryophytes, represented by mosses; the seedless vascular plants, represented by ferns; the gymnosperms, represented by coniferous (evergreen) trees; and the flowering plants, also known as angiosperms. Chapter 22, which introduced plants, contained information on the bryophytes, the seedless vascular plants, and the gymnosperms. In this chapter, we focus strictly on flowering plants. Why pay so much attention to just one type of plant? Because of the overwhelming dominance of flowering plants within the plant kingdom. There are about 16,000 species of bryophytes, 13,000 species of seedless vascular plants, and 700 species of gymnosperms, but there are an estimated 260,000 species of flowering plants. The term *flowering*

**Figure 24.2**
**Uses Galore**

Plants are used by human beings in innumerable ways.

**(a)** Food, here in the form of Concord grapes.

**(b)** Wood, being harvested from a forest in British Columbia.

**(c)** A foxglove plant, which yields the heart medicine digitalis.

**(a)**

**(b)**

**(c)**

*plant* may bring to mind roses or orchids, and these plants are indeed angiosperms. But food crops such as rice and wheat are flowering plants, as are all cacti, almost all the leafy trees, innumerable bushes, pineapple plants, cotton plants, ice plant—the list is very long (**Figure 24.3**). In this chapter, you'll get an overview of how flowering plants are structured, of how they function internally, and of how they respond to signals from the outside world. Chapter 25 goes into greater detail on angiosperm structure and function. A formal definition of angiosperms will be provided once you've learned a little about them.

## 24.2 The Structure of Flowering Plants

Let's begin our tour of the angiosperms by looking at their component parts.

### The Basic Division: Roots and Shoots

We'll first look at the larger-scale structures of the angiosperms, starting with a simple, two-part division that rhymes: roots and shoots. Plants live in two worlds, air and soil, with their root system

below ground and their shoot system above it (**Figure 24.4**).

The function of the root system is straightforward: Grow to reach water and minerals, absorb

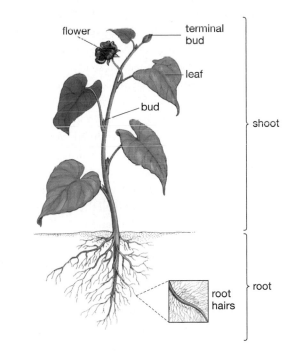

**Figure 24.4**
**A Life-Form in Two Parts**
Flowering plants live in two worlds, with their shoots in the air and their roots in the soil. Although flowering plants differ enormously in size and shape, they generally possess all the external features shown.

**(a)**

**(b)**

**(c)**

**Figure 24.3**
**Angiosperm Variety**
**(a)** A trumpet vine (*Clytostoma callistegiodes*) spreading on a bush.
**(b)** A Senita cactus, on the left, stands next to a saguaro cactus in the Arizona desert.
**(c)** The cereal grain rye (*Secale*) growing in a field.

them from the soil, and then begin transporting them up through the rest of the plant. (For more on the minerals plants absorb, see "What Is Plant Food?" on page 484.) Most roots also serve as anchoring devices for plants, and some act as storage sites for food reserves.

The function of the shoot system is more complex. Photosynthesis takes place in this system, primarily in leaves, which means leaves must be positioned to absorb sunlight. Plants are in competition with one another for sunlight, which is a primary reason so many plants are tall. Once food is produced in photosynthesis, it must be distributed throughout the plant, starting with the shoots. But it is not just plant food production that is centered in the shoot system; it is plant reproduction as well. Within flowers and their derivatives, we find all the components of reproduction—seeds, pollen, and so forth. Now, let's look in more detail at the components of the root and shoot systems.

## Roots: Absorbing the Vital Water

If you look at **Figure 24.5**, you can see pictures of the two basic types of plant root systems, one a **taproot system** consisting of a large central root and a number of smaller lateral roots, and the second a **fibrous**

**Figure 24.6**
**Root Hairs**

Root hairs enormously increase the surface area of a root. The greater the surface area, the more fluid absorption can occur. Shown here are root hairs on the taproot of a sweet-corn plant.

**root system** consisting of many roots that are all about the same size. **Figure 24.6** then shows you not only the taproot of a young sweet-corn plant but also another feature common to root systems. These are **root hairs**: threadlike extensions of roots that greatly increase their absorptive surface area. Each root hair actually is an elongation of a single outer cell of the root. Between roots and root hairs, the root structure of a given plant can be extensive. One famous analysis, conducted in the 1930s on a rye plant, concluded that its taproot and lateral roots alone, if laid end to end, would have totaled some 622 kilometers, which is about 386 miles. Meanwhile, the root hairs collectively were 10,620 kilometers long, meaning the plant had almost 6,600 miles of root hairs! This from a plant whose *shoot* stood 8 centimeters (about 3 inches) off the ground.

Why do plants lavish such resources on their roots? Essentially because plants have such a great need to take up water, and roots are the structures that allow them to do this. To perform photosynthesis, plants must have a constant supply of carbon dioxide, which enters the plants through microscopic pores, called *stomata*, that exist mostly on the underside of plant leaves. Open stomata, however, don't just let carbon dioxide in; they let the plant's water *out*, as water vapor. To accommodate this loss while still keeping vital tissues moist, plants continually pass water through themselves, from roots, up through the stem, and into the leaves (at which point it exits as water vapor). The evaporation of water from a plant's shoot is known as **transpiration**; through it,

**(a)** Taproot system

**(b)** Fibrous root system

**Figure 24.5**
**Two Root Strategies**

**(a)** Dandelions such as this one employ a taproot system—a large central root and a number of smaller lateral roots.

**(b)** The French marigold employs a fibrous root system—a collection of roots that are all about the same size.

more than 90 percent of the water that enters a plant evaporates into the atmosphere. Because the roots of a single tall maple tree can absorb about 220 liters (or nearly 60 gallons) of water per *hour* on a hot summer day, the scale of transpiration is immense. Thus do we see the importance of root development for plant metabolism. And the nutrient storage function of roots that was mentioned earlier? We human beings benefit from it along with plants—sweet potatoes and carrots are roots.

## Shoots: Leaves, Stems, and Flowers

Now let's turn to the shoot system, looking first at the leaves within it.

### Leaves: Sites of Food Production

The primary business of leaves is to absorb the sunlight that drives photosynthesis, which is why most leaves are thin and flat. This leaf shape maximizes the surface area that can be devoted to absorbing sunlight while minimizing the number of cells in leaves that are irrelevant to photosynthesis. Beyond this, through their tiny stomata, leaves serve as the plant's primary entry and exit points for gases. As noted, the most important gas that's entering is carbon dioxide, which is one of the starting ingredients for photosynthesis. What's exiting is the by-product of photosynthesis, oxygen, and the water vapor.

The broad, flat leaves that are so common in nature have in essence a two-part structure—a **blade** (which we usually think of as the leaf itself) and a **petiole**, more commonly referred to as the leaf stalk. If you look at the idealized cross section of this leaf in **Figure 24.7b**, you can see that the blade can be likened to a kind of cellular sandwich, with layers of cuticle, or waxy outer covering, on the outside, and a layer of epidermal cells just inside them. In the leaf's interior, there are vascular bundles, which bear some relation to animal veins in that they are part of a transport system. Then, there are several layers of mesophyll cells. It is these cells that are the sites of most photosynthesis.

Now note, on the underside of the blade, the openings called **stomata**. These are the pores, noted earlier, that let water vapor out and carbon dioxide in—but for most plants only during the day. When the sun goes down, photosynthesis can no longer be performed, and it is not cost-effective for a plant to lose water without gaining carbon dioxide. As such, in most plants the stomata close up until photosynthesis

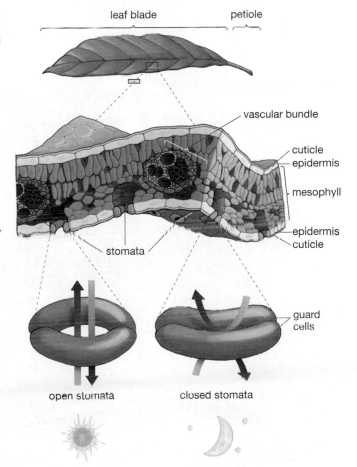

**(a)** Leaves tend to be broad and flat to increase the surface area exposed to sunlight.

**(b)** Photosynthesis occurs primarily within the mesophyll cells in the interior of the leaf. The vascular bundles carry water to the leaves and carry the product of photosynthesis, sugar, to other parts of the plant. Pores called stomata, mostly on the underside of leaves, allow for the passage of gases in and out of the leaf.

**(c)** Guard cells of the stomata control the opening and closing of the stomata. When sunlight shines on the leaf during the day, the guard cells engorge with water that makes them bow apart, thus opening the stomata. When sunlight is reduced, water flows out of the guard cells, causing the stomata to close.

leaf blade    petiole

vascular bundle
cuticle
epidermis
mesophyll
epidermis
cuticle
stomata

guard cells

open stomata    closed stomata

**Figure 24.7**
**Site of Photosynthesis**

begins again the following day. If you look at **Figure 24.7c**, you can see how this opening and closing of stomata is achieved: Two "guard cells," juxtaposed like the sides of a coin purse, are arranged around the stomata. When sunlight strikes the leaf, these cells engorge with water, which makes them bow apart. Then, in the absence of light, water flows out and the door closes again. A given square centimeter of a plant leaf may contain from 1,000 to 100,000 stomata.

**Figure 24.8** gives you some idea of the varied forms of leaves. In addition to the so-called simple leaf we've been looking at, there are compound leaves in which individual blades are divided into a series of "leaflets." Pine needles and cactus spines are likewise actually leaves. The cactus, in particular, has reduced the amount of leaf surface it has as a means of conserving water. So, how does it perform enough

photosynthesis to get along? It carries out most of its photosynthesis in its stem.

## Stems: Structure and Storage

We all generally understand what the stem of a plant is, although it is less generally appreciated that the trunk of a tree is simply one kind of plant stem. The main functions of stems are to give structure to the plant as a whole and to act as storage sites for food reserves. In addition, water, minerals, hormones, and food are constantly shuttling through (and to) the stem, with water on its way up from the roots and food on its way down from the leaves to the rest of the plant.

If you look at **Figure 24.9**, you can see a cross section of one type of plant stem, showing the vascular bundles or veins you first saw in the leaves and the outer, or epidermal, tissue just as the leaves had. You

**Figure 24.8**
**Leaves Come in Many Sizes, Shapes, and Colors**

Simple leaves have just one blade. In compound leaves, the blade is divided into little leaflets. Some leaves are so modified that the average person can hardly recognize them as leaves—for example, the spines of a cactus.

compound leaves

simple leaves

leaves modified as spines

highly reduced leaves (needles)

leaves modified as tendrils

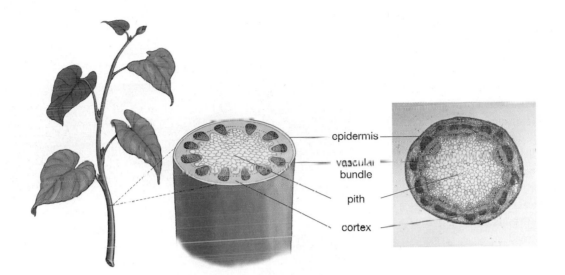

**Figure 24.9**
**The Stem and Its Parts**
Stems provide support to the rest of the plant, act as nutrient storage sites, and conduct fluids. Vascular plants have a "plumbing" system that transports food, water, minerals, and hormones. The vascular bundles in the figure are groups of tubes, running in parallel, that serve this transport function.

can also see so-called ground tissue of two types: an outer cortex and an inner pith, both of which can play a part in food storage and wound repair and provide structural strength to the plant.

## Flowers: Many Parts in Service of Reproduction

Flowers are the reproductive structures of plants. A single flower generally has both male and female reproductive structures on it, which might make you think that a given plant would fertilize *itself*. This is indeed the case with some plants, such as Gregor Mendel's pea plants, reviewed in Chapter 11. However, because the evolutionary benefit of sexual reproduction is to get the genetic diversity that comes with *mixing* genetic material from different individual organisms, natural selection has worked against self-fertilization by endowing many flowering plants with ways to reduce the incidence of it. For example, the male pollen of a given plant might be genetically incompatible with that plant's female reproductive structures. Some of the illustrations you'll be seeing show a plant fertilizing itself, but this is done only for visual simplicity.

If you look at **Figure 24.10**, you can see the components of a typical flower. Taking things from the bottom, there is a modified stem, called a pedicel, which widens into a base called a receptacle, from which the flowers emerge. Flowers themselves can be thought of as consisting of four parts: sepals, petals, stamens, and a carpel. The **sepals** are the leaflike structures that protect the flower before it opens. (Drying out is a problem, as are hungry animals.) The function of the colorful **petals** is to announce "food here" to pollinating animals.

The heart of the flower's reproductive structures consists of the stamens and the carpel. If you look at Figure 24.10, you can see that the **stamens** consist of a long, slender **filament** topped by an **anther**. These anthers contain cells that ultimately will yield sperm-bearing pollen grains. Thus, the anthers are the place in the flower where *male* reproductive cells are produced. Pollen grains ultimately will be released from the anther and then carried—perhaps by a pollinating bee or bird—to the carpel of another plant. As Figure 24.10 shows, a **carpel** is a composite structure

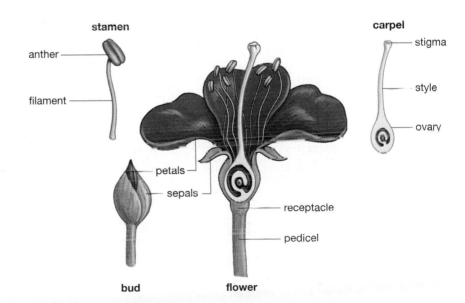

**Figure 24.10**
**Parts of the Flower**
Flowers are composed of four main parts: sepals, petals, stamens, and a carpel. The sepals protect the young bud until it is ready to bloom. The petals attract pollinators. The stamens are the reproductive structures that produce pollen grains (which contain sperm cells). The carpel is the female reproductive structure; it includes an ovary that contains one or more eggs.

## ESSAY What Is Plant Food?

**P**lants are able to make their own food through photosynthesis if they have just four things: water, sunlight, carbon dioxide, and a few nutrients. Water, sunlight, and probably even carbon dioxide are no mystery to you by now, but what exactly are these nutrients you've been reading about?

A **nutrient** is simply a chemical element that is used by living things to sustain life. Recall from Chapter 2 that "elements" are the most basic building blocks of the chemical world. Silver is an element, as is uranium, but these are not nutrients because living things do not need them to live. In the broadest sense, carbon, oxygen, and hydrogen are nutrients. And, it turns out that almost 96 percent of the weight of the average plant is accounted for by these three elements, which come to plants primarily from water and air.

Plants need at least another 13 elements to live, however, and not all of these are supplied by water and air. When we think of common "plant food," we are generally referring to those nutrients that plants can *use up* in their soil and that thus must be replenished to ensure continued growth and reproduction. Of these, the most important three are nitrogen, phosphorus, and potassium, which have the chemical symbols N, P, and K, respectively. When you look at a package of an average plant fertilizer, you will see three numbers in sequence on it (for example, 10-20-10). What these refer to are the percentage of the fertilizer's weight accounted for by these nutrients (**Figure 1**). The growth of plants usually is enhanced by a fertilizer that has equal ratios of the three elements, but if increased *blooming* is your goal, an increased phosphorus ratio generally is recommended, as with the 10-20-10 example.

It is not just houseplants or lawns that benefit from fertilizer, but farm crops as well. Indeed, the planting of a crop on a parcel of land generally requires a large investment in fertilizer because a crop such as wheat removes a great deal of N, P, and K from the soil in a single season. Historically, fertilizers used on farms were *organic*, meaning that they came from decayed living things, such as the fish that Native Americans taught the Pilgrims to use when planting corn. Now, however, commercial fertilizers generally are *inorganic*, meaning they are mixtures of pure elements within binding materials, produced by chemical processes.

**Figure 1**
**Needed Nutrients**

Plants make their own food from sunlight, water, carbon dioxide, and a few nutrients, some of which come from the soil. The most important of these nutrients are nitrogen (N), phosphorus (P), and potassium (K). Because garden plants and houseplants are isolated from natural ecosystems, they must often get these nutrients in the form of fertilizer.

Nature is, of course, perfectly capable of producing a green bounty in such places as forests and marshes without the aid of any human-made fertilizer. Decaying plant and animal matter put nutrients back into the soil, and bacteria are able to fix nitrogen, meaning to absorb it from the air and transform it into a form that plants can take up. Looked at one way, however, houseplants and farm crops are *isolated* plants with very limited participation in this web of life, and thus they require a helping hand from humans to remain robust.

composed of three main parts: the **stigma**, which is the tip end of the carpel, on which pollen grains are deposited; the **style**, a slender tube that raises the stigma to such a prominent height that it can easily catch the pollen; and the **ovary**, the area in which fertilization of the female egg and then early development of the plant embryo take place. Thus, the ovary is the structure in the flower that houses the *female* reproductive cells, and it is the structure in which the male and female reproductive cells come together.

### SO FAR...

1. The primary function of roots is to _____.

2. The primary function of leaves is to _____.

3. Flowers are the _____ structures of plants.

4. The male reproductive cells of plants are produced in their _____, while the female reproductive cells are produced in their _____.

## 24.3 Basic Functions in Flowering Plants

Having looked at the structure, or anatomy, of the angiosperms, let's now go over the activities of some of these components. We'll think in terms of systems here: the reproductive system, the transport system, the hormonal system, the communication system, and the defense system. In addition, you'll look briefly at the nature of plant growth. Because you've just reviewed the parts of a flower, let's start with reproduction.

### Reproduction in Angiosperms

A key event in angiosperm reproduction is that a plant sperm must fertilize a plant egg. As noted, the sperm in flowering plants is contained inside pollen grains. But how do sperm and pollen grains get produced? In the anthers of a flowering plant, there are cells that undergo the type of cell division reviewed in Chapter 10, meiosis, thereby producing cells called **microspores**. Each pollen grain develops from one of these microspores—a process you can see pictured in **Figure 24.11**. In the type of plant shown there, the pollen grain will consist, by the time it leaves the anther, of a tough outer coat and three cells: one tube cell and two sperm cells.

You can also see in Figure 24.11 what happens next. When the pollen grain leaves the anther, it is bound for the stigma of another plant. But, for a grain to merely *land* on a stigma doesn't mean that anything has been fertilized. The **tube cell** in the pollen grain must then begin to *germinate* on the stigma, sprouting a **pollen tube** that grows down through the style. Once this has taken place, one of the **sperm cells** in the grain travels through the tube, gets to the female egg, and fertilizes it. (The other sperm moves down along with the first, but then spurs the growth of food for the fertilized egg, a process we'll review in Chapter 25.)

So, how does the egg that gets fertilized come into being? Inside the plant's ovary, there is a type of cell that also goes through meiosis, thus producing a cell known as a **megaspore** (*mega* because it's bigger than the male microspore). It in turn gives rise to a cluster of cells, one of which is the egg that the sperm from the pollen grain will fertilize. Once this happens—once sperm has fertilized egg—we're on our way to a new generation of plant. Many a step remains before arriving at something that *looks* like the original plant. The fertilized egg (technically a zygote) must develop into an embryo that has a tough covering around it. The combination of embryo, its surrounding food supply, and the covering is called

a **seed**. This seed must be released and then land on a suitable patch of earth, there to germinate and grow to a full flowering plant. But the fertilization of egg by sperm sets all this in motion.

The microspores and megaspores that were produced by our original flowering plant may appear to be nothing more than component parts of that plant, but it's important to recognize that these cells actually are the start of an alternate *generation* of plant. After all, it wasn't the original plant that produced the eggs and sperm that were so critical. It was the microspores (on the male side) and megaspores (on the female side) that produced these gametes, as they're called. Thus, the microspores and megaspores represent the **gametophyte generation** or gamete-producing generation of plant, while our starting plant was a **sporophyte generation** or spore-producing generation of plant. (Remember how it

**... ANSWERS**

1. absorb water and minerals and begin transportation of them to the rest of the plant
2. serve as the sites of photo-synthesis
3. reproductive
4. anthers; ovaries

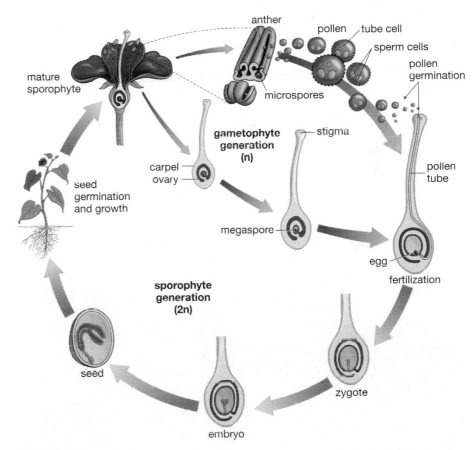

**Figure 24.11**
**The Angiosperm Life Cycle**
The mature sporophyte flower produces many male microspores within the anther and a female megaspore within the ovary. The male microspores then develop into pollen grains that contain the male gamete, sperm. Meanwhile, the female megaspore produces the female gamete, the egg. Pollen grains, with their sperm inside, move to the stigma of a plant. There, the tube cell of the pollen grain sprouts a pollen tube that grows down toward the egg. The sperm cells then move down through the pollen tube, and one of them fertilizes the egg. This results in a zygote that develops into an embryo protected inside a seed. This seed leaves the parent plant and sprouts or "germinates" in the earth, growing eventually into the type of sporophyte plant that began the cycle.

produced the microspores and a megaspore?) You can read more about the alternation of generations in plants on page 428 of Chapter 22.

As noted, the angiosperm egg is fertilized and develops inside the structure called the ovary (Figure 24.10). And, remember that as the fertilized egg develops into an embryo, it becomes enclosed within a seed. As this is taking place, the ovary that surrounds the seed is also developing into something: a tissue called fruit. We are all familiar with fruit, of course, and sometimes the tissue surrounding the seed is fruit as we commonly understand that term—the flesh of an apricot, for example. But ovaries can mature into fruit in other forms. The pod that surrounds peas is fruit in this scientific sense, as is the outer covering of a kernel of corn. A **fruit** is simply the mature ovary of a flowering plant. All angiosperms have fruit in this sense, and this provides us with our definition of the flowering plants. An **angiosperm** is a flowering seed plant whose seeds are enclosed within the tissue called fruit.

### Asexual Reproduction in Angiosperms: Vegetative Reproduction

We are used to the human mode of reproduction, which—pending the arrival of cloned human beings—is always *sexual* reproduction: At some point, sperm must fertilize egg to produce a new embryo. But plants, including angiosperms, have another option—**vegetative reproduction**, which is *asexual* reproduction. This is familiar to us in the form of "cuttings" that can be taken from one house-

plant to start another. A growing cutting represents a new plant that is an exact genetic replica of the original plant. No alternate generation or fertilization is required; one sporophyte plant has been grown from another. Vegetative reproduction is also common in nature. The roots of aspen trees, for example, produce a form of shoot known as a sucker that, if physically separated from the parent plant, will grow into a new aspen tree. Often, then, a stand of aspen trees amounts to a group of clones of one another (**Figure 24.12**).

**SO FAR...**

1. A pollen grain develops from a cell called a _____ and is composed, at maturity, of an outer coat, one _____, and two _____ cells.

2. The egg that is fertilized is one of a cluster of cells that develops from a cell called a _____ that is located in the plant's _____.

3. The entity composed of the fertilized egg, its food supply, and a tough outer coat is the plant's _____, which in angiosperms is surrounded by the tissue called _____.

## Plant Plumbing: The Transport System

In looking at leaves and stems earlier, you saw that they have something called *vascular bundles* running through them. This term refers to collections of tubes through which fluid materials move from one part of the plant to another. This transport system bears obvious similarities with the circulation systems of animals, but there are differences as well. Many animals employ blood as a transport medium, and animals such as humans have a transport pumping device, called a heart. Plants do not have blood, and they have no pumping system. (In fact, plants have almost no "moving parts" at all.) Yet, think of the transport job plants have to do. Water that is transpired from the top of a redwood tree may have made a journey of more than 100 meters straight up—more length than a football field! What kind of power is at work here?

There are two essential components to the plant transport system. First there is **xylem**, which can be defined as the tissue through which water and dissolved minerals flow in vascular plants. (**Tissue** means a group of cells that perform a common function.) Second, there is **phloem**, the tissue through which the *food* produced in photosynthesis—mostly sucrose—is conducted, along with some hormones and other compounds. As noted before, water is

**Figure 24.12**
**Reproduction without Sex**
Individual aspen trees often are clones of one another, reproducing without the fusion of eggs and sperm from different individuals. This stand of aspen trees is in Colorado.

making a directional journey from root through leaf, but the plant must be able to transport food and hormones everywhere within itself.

If you look at **Figure 24.13**, you can see an idealized view of a plant's transport system as it would appear within a stem. You'll also see that the vascular bundles noted earlier are composed of bundles of linked xylem and phloem tubes running in parallel.

Xylem is composed of two types of fluid-conducting cells, *vessel elements* and *tracheids*, which have different shapes, as you can see. On reaching maturity, these cells do the rest of the plant the favor of dying. The cells' content is then cleared out, leaving strands of empty cells stacked one on top of another—leaving tubes, in other words. The walls of these cells remain, however, and indeed are reinforced with the materials cellulose and lignin, which provide the *load-bearing* capacity that allows something as massive as a redwood tree to be so tall. (For more on the role of xylem in everyday houseplants, see "Keeping Cut Flowers Fresh," on page 488.)

Phloem is also composed of two types of cells, but the arrangement here is very different. One type of phloem cell—a *sieve element*, which does the actual nutrient conducting—doesn't undergo cell death on maturity, but it does lose its cell nucleus, meaning nearly all its DNA. Those of you who went through the genetics unit may wonder how any cell could function without its DNA information center, but there is an answer. Each sieve element cell has associated with it one or more *companion cells* that retain their DNA and that seem to take care of all the housekeeping needs of their sister sieve elements. So, why do sieve elements lose their nuclei? Apparently to make room for the rapid flow of food through them. In Chapter 25, you'll look at the means by which phloem and xylem conduct their respective materials.

## Communication: Hormones Affect Many Aspects of Plant Functioning

We're so used to associating hormones with animal functioning that it may come as a surprise to learn that plants also have hormones. Taken as a whole, plant hormones do many of the same things that animal hormones do. Their most important roles are to regulate growth and development and to integrate the functioning of the various plant parts. In less abstract terms, hormones help buds grow and leaves fall and fruit ripen. This goes hand in hand with helping plants respond to their environments—to heat and cold, munching goats, and voracious insects.

Although most people have an intuitive sense of what hormones are, a definition might be helpful. **Hormones** are chemical messengers; they are substances that, when released in one part of an organism, go on to prompt physiological activity in another part of that organism. In animals, hormones are generally synthesized in well-defined organs, called *glands*, whose main function is to produce hormones. (For example, the thyroid or adrenal glands.) In plants, however, hormone production is a more diffuse process, taking place not in glands, but in collections of cells that carry out a range of functions.

If you look at **Table 24.1** on the next page, you can see a list of five of the most important hormones that function in plants. Three of these actually are *classes* of hormones, and two are individual hormones. You can read about the hormone ethylene in "Ripening Fruit Is a Gas." Let's go over just one other plant hormone, a member of the family of hormones known as *auxins*, to give you some idea of what plant hormones do.

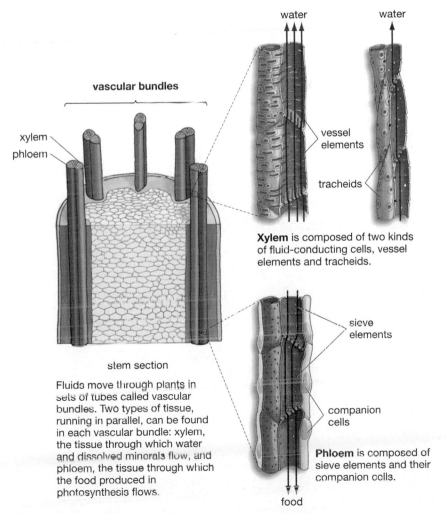

water

water

**vascular bundles**

xylem

phloem

vessel elements

tracheids

**Xylem** is composed of two kinds of fluid-conducting cells, vessel elements and tracheids.

stem section

Fluids move through plants in sets of tubes called vascular bundles. Two types of tissue, running in parallel, can be found in each vascular bundle: xylem, the tissue through which water and dissolved minerals flow, and phloem, the tissue through which the food produced in photosynthesis flows.

sieve elements

companion cells

**Phloem** is composed of sieve elements and their companion cells.

food

**Figure 24.13**
**Fluid-Transport Structure of Plants**

## ESSAY Keeping Cut Flowers Fresh

A s already noted, xylem is the plant tissue through which water moves *up*, from roots through leaves. The flowers we put in vases in our homes have lost their roots, of course, but they haven't lost their xylem, which continues to function long after the flower has been picked. Given this, many flowers can last a long time indoors, but we can maximize their stay if we follow a few simple rules.

First, realize that the liquid in the xylem is under negative pressure—its natural tendency is to move up *into* the stem, not to flow out of it. As such, if the stems are cut when they are out of water, *air* gets sucked up into the cut ends, creating air bubbles that can then get trapped in the xylem and keep water from rising up through it. When this happens, flowers can wilt, even when their stems are submerged in clean water. Recutting the stem under water (or under a steady stream from the faucet) can remove this blockage. Better yet, cut the stems under water the first time.

Beyond this, acidic sugar water, such as can be found in citrus-flavored soft drinks, will prolong the life of some flowers by keeping bacterial growth down; changing water frequently is a good idea, particularly when it starts to look gummy or discolored. Keep your arrangement out of direct light or heat and remove dead and dying flowers in the arrangement because the hormone ethylene is given off by dying flowers and in many cases will hasten the demise of the healthy ones in the bunch. (For another role of ethylene, see "Ripening Fruit Is a Gas," on page 490.)

**Preserving Beauty**
Following a few simple rules can prolong the life of cut flowers.

### Table 24.1 Plant Hormones

| Hormone | Major functions | Where found or produced in plant |
|---|---|---|
| Auxins | Suppression of lateral buds; elongation of stems; growth and abscission (falling off) of leaves; differentiation of xylem and phloem tissue | Root and shoot tips; young leaves |
| Cytokinins | Stimulate cell division; active in the development of plant tissues from undifferentiated cells | Roots |
| Gibberellins | Stem elongation, growth of fruit, promotion of seed germination | Seeds, apical meristem tissue, young leaves |
| Ethylene | Ripening of fruit, retardation of lateral bud growth, promotion of leaf abscission | Nearly all plant tissue |
| Abscisic acid | Induces closing of leaf pores (stomata) in drought; promotes dormancy in seeds; counteracts growth hormones | Young fruit; leaves, roots |

### An Auxin Gives the "A" Shape to Trees

Why do Christmas trees have their characteristic "A" shape (**Figure 24.14**)? They develop in this way thanks in large part to the effects of an auxin known as IAA. Unlike animals, plants do not grow globally, over their whole surface, but instead confine most of their vertical growth to special regions, called apical meristems, that exist at the *tips* of the roots and shoots. **Figure 24.15** shows you how this works with the shoots. The shoot apical meristem gives rise to a series of growth modules: more stem, one or more leaves, and a lateral bud that forms in tandem with each leaf. These lateral buds have meristem tissue in them and can grow into new branches, but they normally don't *when they lie close to the apical meristem.* Why? The apical meristem is producing IAA, which works in tandem with other hormones to suppress the growth of these buds. But, the farther away from the apical meristem, the smaller the concentration of

**Figure 24.14**
**Uniform "A" Shapes**
The effects of apical dominance are clearly visible in this stand of blue spruce trees at a Michigan Christmas tree farm. The apical meristem tissue at the tip of the trees produces a hormone called IAA that inhibits the growth of lateral branches. Because the concentration of IAA is highest at the top near the apical meristem and lowest at the bottom, the branches are very short near the top and longest at the bottom.

IAA that is available to the lateral buds. The result is greater budding of branches at the base of a tree and tapering of this growth going from the base up through the tree's apex. This phenomenon is called **apical dominance**: a suppression of the growth of lateral branches through the activity of apical meristems. It is most pronounced in certain conifers, such as the Douglas firs often used for Christmas trees. Meanwhile, trees that don't have such a strict IAA gradient don't get the strict "A" shape. This maintenance of apical dominance is just one of the many things that IAA does. To give you some idea of the complexity of its function, IAA may serve to suppress the emergence of lateral buds, as noted, but it works to *promote* the cellular elongation that is important in plant growth, as we'll see later.

## Plant Growth: Indeterminate and at the Tips

Having looked a little at how plant growth is affected by hormones, we can note some basic characteristics of plant growth. To visualize how it works, imagine a boy about 8 years old who decides to drive a long nail into the trunk of a secluded tree. He doesn't drive the nail in completely, but lets it stick out a bit. After 10 years pass, the boy happens to walk by the same tree; on seeing it, a couple of things occur to him (**Figure 24.16**). First, he has grown, but the nail

is in the same place; he would have to bend down now to pull the nail out, whereas when he drove it in he was standing up straight. Second, he would have a hard time actually pulling the nail out now because not much of it is visible; the tree has grown around it. Third, if the nail's height is any indication, the trunk of the tree may not have moved up, but the *top* certainly has; it's much taller than he remembered it.

The lessons here are that, in their vertical growth, plants do not grow the way people do—throughout their entire length—but instead grow only at their

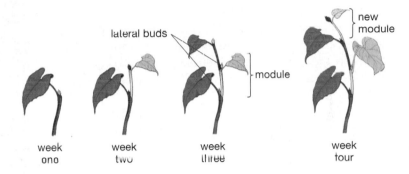

**Figure 24.15**
**Modular Growth**
Vertical growth in plants occurs in "modules." Each module includes a new portion of stem, at least one leaf, and a lateral bud that can give rise to a branch or a flower. Here, alternating dark-green and light-green segments indicate modules.

**Figure 24.16**
**The Basic Characteristics of Plant Growth**
If a young boy drives a long nail into a tree and returns 10 years later, two things will be apparent: The nail is at the same height it used to be despite the increased height of the tree, and it is almost completely buried in the tree. What does this say about tree growth? Trees do not elongate throughout their whole length, but only at their tips in "primary growth." Further, they widen laterally in "secondary growth."

# Ripening Fruit Is a Gas

That banana you had? The one you put in that plastic bag, thinking you'd make it part of your lunch? It already had a few black spots on it by the time it went in the bag, but then you forgot about it for lunch, and by the time you looked at it again, it was more black than yellow and something you were no longer interested in eating. The question is: Why did it go downhill so *fast*?

The answer is that those little black spots on the banana result from concentrations of a gas called *ethylene*—a plant hormone that can ripen lots of fruits. The bag you put the banana in became a kind of gas chamber, with the ethylene concentrations inside very high. Moreover, ethylene stimulates its own production; the more ethylene there is, the more that gets produced. Given this contagious quality to ethylene, you can see why it is true that one bad apple can spoil the bunch.

Ethylene doesn't work on all fruits; grapes, strawberries, and cherries, for example, are unaffected by it. But apples, avocados, and tomatoes join bananas in being very sensitive to it. If you wish to ripen one of these latter fruits fast, putting it in a plastic bag will give you an edible fruit faster than putting it on the window sill.

But, what does it mean for a fruit to be ripe? Let's match up our intuitive sense of this with the underlying physical changes. Compared to unripened fruits, ripe fruits tend to be sweeter (starches are converted to sugars); softer (a secretion of enzymes softens cell walls); and more fragrant (ripening releases compounds that our nose detects as sweet smelling). Ethylene helps bring about all these changes, basically by upping the tempo at which a plant carries out its metabolic processes. The reason a green banana turns to yellow is that ethylene hastens the breakdown of the green chlorophyll in the banana, which allows us to see yellow pigments that had been there all along.

For years, commercial fruit growers have employed ethylene, or test-tube compounds that help release it, to *time* the ripening of their fruit. This is why we can pick green tomatoes at the source and then ripen them when they get closer to market. It's also possible to *suppress* ethylene production by storing fruit in a room that has high concentrations of carbon dioxide and low concentrations of oxygen. In this way, a fruit such as apples can be kept on hold for months before being brought to ripeness by simply being exposed to normal air. Ever wonder why you can have fresh apples in the spring when they generally are picked in the fall? Ethylene control is part of the answer.

**Spurring Their Own Ripening**
These bananas ripen under the influence of a gas they produce, ethylene. Concentrations of ethylene produce the black spots on bananas.

tips (of both shoots and roots, although the boy couldn't see the roots). In addition, as the half-buried nail attests, this tree has carried out the lateral growth that botanists refer to as secondary growth, which you'll look at next chapter.

Beyond these things, it turns out that a plant's growth is indeterminate. In most plants, it can go on indefinitely at the tips of roots and shoots. By contrast, animal growth is generally determinate: It comes to an end at a certain point in development. Imagine if, say, the tips of your fingers just kept grow-

ing long after your trunk and arms had reached their full extension.

## Defense and Cooperation

Plants may be immobile and toothless, but they are not defenseless. Indeed, they have developed a formidable defense arsenal (**Figure 24.17**). One variety of plant defenses, the structural defenses, are familiar to us because they're so visible—think, for example, of the spines on a cactus or the thorns on a rose. A second variety of defenses, the chemical defenses, is not so visible but is potent nevertheless. The chomping a beetle does on a potato plant induces the plant to produce substances (called *protease inhibitors*) that are believed to give the insect an ache in its gut that it won't forget. Other plants, when attacked by a fungus, release antifungal compounds that spread not only at the site of infection, but throughout the plant.

Beyond these self-defense mechanisms, plants can warn *other* plants of dangers in their midst. Airborne signaling compounds can be produced from the leaf surface of a plant under attack, waft to a neighboring plant, and prompt it to start producing compounds that are noxious or harmful to the would-be plant eater. Some plants can even call in animal allies. Corn plants that are attacked by army worms can call in an air force: The plants release a substance that attracts parasitic wasps, which then devour the worms.

Plants also enter into cooperative relationships with other organisms. In Chapter 22, you saw that most plant roots are linked to the thin, underground tubes, called *hyphae*, that extend from fungi. What the fungi get from this is nutrition from the photosynthesizing plant; what the plants get is a greatly expanded water and mineral absorption network. The combined root–hyphae associations are known as **mycorrhizae**. Other species, notably the legumes, such as soybeans, have developed partnerships with nitrogen-fixing bacteria, which is to say bacteria that are able to take nitrogen out of the atmosphere and put it into a form that living things can use. The bacterium infects the plant through a root hair, directs the construction of its own lodgings (called a *root nodule*), and then begins converting atmospheric nitrogen, while obtaining nutrition from the plant. This process must take place in the absence of oxygen, which the plant sees to by eliminating the available oxygen in the nodule. Lots of crops require applications of human-manufactured nitrogen, but legumes such as beans and alfalfa can do fine on their own (**Figure 24.18**).

**Figure 24.17**
**Silent and Fixed in One Spot, but Not Defenseless**
The spiny bark of this Costa Rican tree, a member of the genus *Zanthoxylum*, protects it from plant predators.

**Figure 24.18**
**Working Arrangement**
Plants need nitrogen to grow. Some get it through a cooperative relationship with bacteria that are nitrogen fixing, meaning they can convert atmospheric nitrogen into a form living things can use. Shown are root nodules (the small spheres) on the roots of a soybean plant. Bacteria make a home in the nodules, where they fix nitrogen that is then used by the plant. In turn, the plant supplies the bacteria with nutrients, which it produces through photosynthesis.

**SO FAR...**

1. Xylem is the plant tissue that conducts _____, while phloem is the plant tissue that conducts _____.

2. Plant growth is largely controlled by the chemical messengers called _____.

3. Vertical plant growth occurs only at the _____ of shoots and roots, and can go on _____.

••• ANSWERS

1. water and minerals; the food produced in photosynthesis

2. hormones

3. tips; indefinitely

## 24.4 **Responding to External Signals**

Plants may lack a nervous system, but this doesn't mean they can't respond to their environment. They must do so, actually, because their life depends on it. They sense the march of the seasons by measuring available light and the surrounding temperature; they use internal clocks to synchronize their development and physical functioning; they respond to light, gravity, touch, heat, cold, and pH, among other things. Let's conclude the chapter by reviewing the plant's interactions with gravity, light, touch, and the seasons as examples of these environmental responses.

### Responding to Gravity: Gravitropism

When a seed first starts to sprout, or "germinate," its roots and shoots must go in very specific directions: The root must go down, toward the water and minerals the plant needs, and the shoot must go up, toward the sunlight. Imagine a seed that has simply fallen on the ground; it would not do for either root or shoot to grow *horizontally*; each must grow vertically. If you look at **Figure 24.19**, you can see dramatic evidence of a plant's ability to sense which way is up and which way down. The impatiens in Figure 24.19a was placed on its side, and within 16 hours, the shoot began curving upward, as you can see.

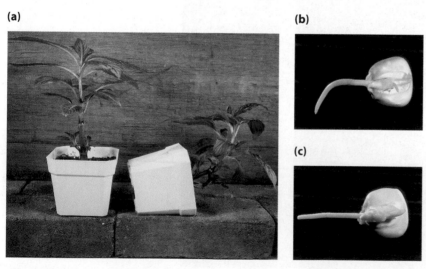

**(a)**

**(b)**

**(c)**

**Figure 24.19**
**Plants Respond to Gravity**

**(a)** Potted impatiens demonstrate the effects of gravitropism. The plant on the right was laid on its side 16 hours before the photo was taken, by which time the shoot had curved, beginning to grow upward.

**(b)** Similarly, the root of a germinating sweet-corn seed begins to curve downward only hours after being placed on its side. How do these plants sense which way is up and which way down?

**(c)** One hint is that if the very tip or "root cap" of the root is removed, the root no longer bends downward. Researchers believe that organelle "sedimentation" in the root cap—the effect of the movement of certain organelles in response to gravity—is the sensing mechanism in gravitropism.

Figure 24.19b shows the *root* of a germinating sweet-corn seed, likewise oriented horizontally (by researchers), but quickly beginning to bend downward. What's being exemplified here is **gravitropism**, meaning a bending of a plant's root or shoot in response to gravity. Figure 24.19c again shows the root of a sweet-corn plant, only this one has had its very tip, or root cap, snipped off. Here, instead of bending in accordance with gravity, the root continues its straight, horizontal growth. Conclusion? Cells or substances in the root cap are essential for root gravitropism.

But, how do roots "know" which way is down, and how do they orchestrate a course correction when they need one? Logically, there must be at least two elements at work here: first, a gravitational *sensing* mechanism and then a means of *responding* to gravitational cues, such that corrective bending points the root toward the center of Earth.

The consensus among botanists today is that the sensing mechanism in gravitropism is the "sedimentation" that various plant cell organelles perform in response to gravity. Recall that organelles are small but highly organized structures inside cells. (Mitochondria and ribosomes are organelles, for example.) Like the bubble in a carpenter's level, some plant-cell organelles can be seen changing position in response to gravity. The most important of these organelles is a group of starch-storing structures called *amyloplasts* (although some plants seem to employ alternate organelles). When these organelles move in response to gravity, it sets in motion the responding mechanism in gravitropism. The organelles "land" on other structures inside the cell (such as the cytoskeleton), and the resulting impact triggers a redistribution of a substance you were recently introduced to: the hormone IAA. The result of this redistribution is differential growth within the plant—one *side* of the plant stem or root will grow more than the other side. With this growth, a root that is oriented horizontally will start bending toward the ground, while a stem that is oriented horizontally will start bending toward the sky.

### Responding to Light: Phototropism

Many plants will bend toward a source of light, with the value of this perhaps being obvious: Plants produce their own food, and that production depends on light. Thus, a plant needs to respond when its sunlight becomes blocked by another plant or some other physical object in its surroundings. As it turns out, it is once again IAA—this time produced in the shoot tips of growing plants—that controls this

**phototropism**, defined as a curvature of shoots in response to light. When light strikes one side of the shoot, it causes IAA to migrate to the other side, where the IAA acts to promote the *elongation* of cells on this far side. The effect is to make the shoot curve toward the light (**Figure 24.20**).

## Responding to Contact: Thigmotropism

We've all seen plants that manage to climb upward by encircling the stem of another plant. Whole stems can undertake such encircling, but it is often a thin, modified leaflet called a tendril that does so (**Figure 24.21**). But, how is a tendril able to wrap around another object? Once again, through differential growth—more rapid growth on one side of a tendril than on the other. Contact with the object is perceived by outer, or epidermal, cells in the tendril. This sets into motion the differential growth, which probably is controlled by the hormones IAA and ethylene. This process is called **thigmotropism**, meaning growth of a plant in response to touch. Plants with this capability can piggyback on other plants to get more access to sunlight.

## Responding to the Passage of the Seasons

The profusion of brightly colored leaves in the fall is one of the great seasonal markers in temperate climates such as those in the American Midwest and most of the East Coast. As green leaves turn red, gold, and purple, we know that autumn has arrived.

### Dormancy in Winter

Trees that exhibit a coordinated, seasonal loss of leaves are called **deciduous** trees. But, why should these broad-leafed trees lose their leaves, while the evergreen pines and firs keep the modified leaves known as needles?

First and foremost, there is relatively little water available in winter in cold climates (most of it being frozen in the ground), and flat-leafed trees transpire more than do evergreens. In addition, because of the way wind flows over flat leaves, they lose more heat than do pine needles. Thus, the deciduous trees have evolved a strategy that is a matter of straight economics. Such winter photosynthesis as they might perform would not be worth the water loss that would result. The strategy thus becomes to lose the leaves, grow new ones next spring, and in the meantime perform no photosynthesis but exist instead completely on food reserves. This state, in which growth is suspended and metabolic activity is low, is called

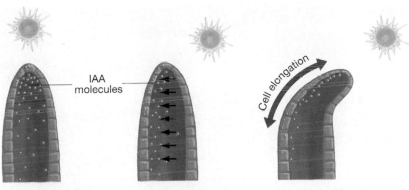

**(a)** When sunlight is overhead, the IAA molecules produced by the apical meristem are distributed evenly in the shoot.

**(b)** Once the sunlight shines on the shoot at an angle, the IAA molecules move to the far side and induce the elongation of cells on that side.

**(c)** Cell elongation results in the bending of the shoot toward the light.

**Figure 24.20**
**Plants Respond to Light**

dormancy. Evergreens exhibit dormancy, too, in low temperatures, but deciduous trees are locked into it until the coming of spring.

### Limitation of Loss in the Deciduous Strategy

The deciduous strategy, of first losing leaves and then growing them back, may at first seem very wasteful, akin to a company destroying much of its machinery each year only to rebuild it months later.

**Figure 24.21**
**Moving Up by Curling Around**
This dodder plant is able to curl around its host plant because of thigmotropism—a plant's ability to grow in response to touch.

**Figure 24.22**
**Seasonal Marker**
Brightly colored fall leaves in Michigan's Ottawa National Forest. Plants have a number of mechanisms that allow them to respond to the passage of the seasons.

But, consider how economical trees are about this. They *reclaim* the nutrients they have stored in their leaves. Proteins are broken down by enzymes and shipped, along with the nutrients, to storage cells in the stem and roots. By the time a leaf is ready for detachment from the tree, it is little more than an empty shell of cell walls.

Before this happens, however, the change in coloration takes place. It may surprise you to learn there is less adding of new colors here than an unmasking of colors that existed in the leaves all along. With the approach of fall, green chlorophyll, which is the main pigment in photosynthesis, begins to break down in the leaf. With it gone, the yellow and orange colors provided by the carotenoid pigments become visible (**Figure 24.22**). New colors are added as well; blue and red come with the synthesis of pigments called anthocyanins. (These pigments are largely responsible for the stunning red color of many maples in fall.) The combination of carotenoids and anthocyanins gives us the final result, which is multihued leaves.

### Photoperiodism

So, how does a deciduous tree sense when it's time to begin preparing for cold weather? In general, it is an interaction between two factors. First, there is cold itself, which brings about changes in metabo-

lism. In addition, there is a phenomenon known as **photoperiodism**, which is the ability of a plant to respond to changes it is experiencing in the daily duration of darkness, relative to light. This is another example of the plant's ability to respond to its environment, but it should not be confused with the "phototropism" or *bending* toward the light that was discussed earlier. In photoperiodism, plants are making seasonal adaptations based on the length of the nights they are experiencing. Most plants in Earth's temperate regions and many in the subtropics exhibit photoperiodism, which can affect such processes as the timing of flowering as well as the onset of dormancy. The value of such a sensing mechanism becomes apparent when we think about the kind of trouble plants could get into if they relied solely on temperature signals to provide them with seasonal cues. It's entirely possible to have, say, an unusual cold snap in September—one that, on its own, would be a signal to start losing leaves. A tree that acted on such a false signal would, however, miss out weeks of growth and food production. In contrast, seasonal dark/light cues are much more reliable—there is no such thing as an unusual *light* condition for a given month. Given this, it makes sense that photoperiodism has an important place in providing plants with a sense of the seasons as they pass.

To get a feel for how photoperiodism works, consider ragweed, a plant that is the bane of many hay-fever sufferers. After sprouting in the spring and reaching a certain level of maturity, ragweed flowers only when it is in darkness for 10 hours *or more* per day. When would this threshold be crossed? It occurs as the nights begin to get longer. Ragweed thus joins a group of plants, called long-night plants, whose flowering comes only with an *increased* amount of darkness. As you can imagine, most of these plants are like ragweed in that they flower only in late summer or early fall. Their counterparts are the so-called short-night plants, which flower only when the nights get short enough—that is to say, when nighttime hours are *decreasing* as a proportion of the day. Predictably, these plants tend to flower in early to midsummer.

The mechanism by which photoperiodism works is now fairly well understood in one short-night plant, a member of the mustard family called *Arabidopsis*. Flowering in *Arabidopsis* is controlled by a protein, known as CO, that can switch genes on by binding to DNA. When enough CO accumulates in *Arabidopsis*, the genes that set flowering in motion are switched on. So, what can bring about CO levels that are high enough to get the flowering started? Shorter nights. Substances called *photoreceptors*, which are sensitive to light, act mostly to allow CO to accumulate during the day, but then predominantly act to destroy it at night. As the nights grow shorter, the balance is shifted toward CO accumulation. When this accumulation reaches a certain threshold, the relevant genes are switched on and flowering begins.

## On to a More Detailed Picture of Plants

This chapter has served to introduce you to the basic features of plants—their varieties, structures, and the most important systems that operate in them. There is more to say, however, about many of the topics that we haven't gone over, such as how trees grow *out* as well as up, or what that second sperm cell from a pollen grain fertilizes when it reaches an ovary. Chapter 25 goes into more detail on plant structure and functioning, including the function of reproduction.

---

**SO FAR...**

1. Plants respond to gravity, touch, and the direction of sunlight through the _____ of roots or stems.

2. In temperate climates, the coming of fall puts _____ trees into a state called dormancy, in which their growth is _____ and their metabolic rate is _____.

3. Plants in temperate climates make seasonal adjustments based on the length of the _____ they are experiencing.

**••• ANSWERS**

1. differential growth
2. deciduous; suspended; lowered
3. nights

# CHAPTER 24 REVIEW

For study help, activities, and more quiz questions, go to **www.aw.com/krogh4**.

## SUMMARY

### 24.1  The Importance of Plants

- Plants are vital to many types of living things on Earth. The photosynthesis they carry out indirectly feeds many other life-forms. The oxygen they produce as a by-product of photosynthesis is vital to many organisms. The lumber and paper that trees provide is important to human beings. Plants act as an anchoring environmental force by preventing soil erosion and absorbing carbon dioxide and pollutants.

- There are four principal varieties of plants: bryophytes, represented by mosses; seedless vascular plants, represented by ferns; gymnosperms, represented by coniferous (evergreen) trees; and the flowering plants or angiosperms. Of the four varieties, angiosperms are by far the most dominant on Earth.

### 24.2  The Structure of Flowering Plants

- Plants live in two worlds, above the ground and below it. As such, their anatomy can be conceptualized as consisting of the above-ground shoots and the below-ground roots. Roots absorb water and nutrients, anchor the plant, and often act as nutrient storage sites. Shoots include the plant's leaves, stems, and flowers.

- Leaves serve as the primary sites of photosynthesis in most plants. Leaves have a profusion of tiny pores, called stomata, that open and close in response to the presence or absence of light. In this way, the stomata control the flow of carbon dioxide into the plant and the flow of oxygen and water vapor out of the plant.

- Stems give structure to plants and act as storage sites for food reserves.

- Flowers are the reproductive structures of plants, with most flowers containing both male and female reproductive parts. The male reproductive structure, called a stamen, consists of a slender filament topped by an anther. The anther's chambers contain the cells that will develop into sperm-containing pollen grains. The female reproductive structure, the carpel, is composed of a stigma, on which pollen grains are deposited; a tube called a style, which raises the stigma to such a height that it can catch pollen; and a structure called an ovary, where fertilization of the female egg and early development of the resulting embryo take place.

### 24.3  Basic Functions in Flowering Plants

- *Reproduction*  Pollen grains develop from cells called microspores located inside the plant's anthers. At maturity, each pollen grain consists of two sperm cells, one tube cell, and an outer coat. Plant eggs develop from a cell called a megaspore within the plant's ovary. When a pollen grain from one plant lands on the stigma of a second plant, the pollen grain germinates, developing a pollen tube that grows down through the second plant's style. Sperm cells from the pollen grain travel through the pollen tube, with one of the sperm cells reaching the egg in the ovary of the second plant and fertilizing it. Once it is fertilized by the sperm cell, the egg (or zygote) develops into an embryo that eventually will be surrounded by a tough outer covering. The combination of embryo, its food supply, and the outer covering is called a seed, which is capable of being implanted in the ground and growing into a new generation of plant. Angiosperms can be defined as plants whose seeds are surrounded by a layer of tissue called fruit. Fruit is the mature ovary of a flowering plant.

- Plants can reproduce asexually through such means as grafting. This is known as vegetative reproduction.

- *Fluid transport* in plants is handled through two kinds of tissue: xylem, through which water and dissolved minerals flow; and phloem, through which the food the plant produces flows, along with hormones and other compounds. Xylem is composed of two types of fluid-conducting cells—vessel elements and tracheids—while phloem is composed of cells called sieve elements and their related companion cells.

- *Plant hormones* regulate plant growth and development and integrate the functioning of various plant structures. Many fruits ripen under the influence of the plant hormone ethylene, while the hormone IAA is important in controlling plant growth.

- *Plant growth*  Plants do not grow vertically throughout their length but instead grow almost entirely at the tips of both their roots and shoots. Some plants, such as trees, thicken through lateral or "secondary" growth. The growth of most plants is indeterminate, meaning it can go on indefinitely.

- *Defense*  Plants have formidable defenses, both structural (such as cactus spines) and chemical (such as antifungal compounds).

- *Cooperative relationships*  Plants enter into cooperative relationships with other organisms. Most plant roots are linked to underground fungal extensions called hyphae. This relationship brings added water and nutrients to the plant (from the hyphae) and food to the fungi (from the photosynthesis the plant performs). The combined root–hyphae associations are known as mycorrhizae. Some plants form cooperative relationships with nitrogen-fixing bacteria, with the bacteria taking in atmospheric nitrogen and transforming it into a form the plants can use, and the plants providing the bacteria with nutrients.

**Web Animation 24.1:** Leaves, Reproduction, and Fluid Transport

## 24.4  Responding to External Signals

- Plants are able to sense their orientation with respect to the Earth and direct the growth of their roots and shoots accordingly—roots into the Earth, shoots toward the sky. This ability is called gravitropism.

- Plants will bend toward a source of light through the process of phototropism, meaning a curvature of shoots in response to light.

- Some plants can climb upward on other objects by making contact with them and then encircling them in growth. This is thigmotropism, defined as the growth of a plant in response to touch.

- Differential growth on one side of the root or stem makes possible phototropism, gravitropism, and thigmotropism.

- In temperate climates, deciduous trees exhibit a coordinated, seasonal loss of leaves and enter into a state of dormancy, existing on stored nutrient reserves in colder months.

- Plants can sense the passage of seasons and time their metabolic and reproductive activities accordingly. One mechanism that assists in this process is photoperiodism, which is the ability of a plant to respond to changes it is experiencing in the daily duration of darkness relative to light. Some plants that exhibit photoperiodism are long-night plants, meaning those whose flowering comes only with an increased amount of darkness—in late summer or early fall. Others are short-night plants, meaning those whose flowering comes only with a decreased amount of darkness—in early to midsummer.

Web Animation 24.2:  Phototropism

## KEY TERMS

| | |
|---|---|
| angiosperm   486 | microspore   485 |
| anther   483 | mycorrhizae   491 |
| apical dominance   489 | nutrient   484 |
| blade   481 | ovary   484 |
| carpel   483 | petal   483 |
| deciduous   493 | petiole   481 |
| dormancy   493 | phloem   486 |
| fibrous root system   480 | photoperiodism   494 |
| filament   483 | phototropism   493 |
| fruit   486 | pollen tube   485 |
| gametophyte generation   485 | root hair   480 |
| gravitropism   492 | seed   485 |
| hormone   487 | sepal   483 |
| megaspore   485 | sperm cell   485 |

| | |
|---|---|
| sporophyte generation   485 | thigmotropism   493 |
| stamen   483 | tissue   486 |
| stigma   484 | transpiration   480 |
| stomata   481 | tube cell   485 |
| style   484 | vegetative reproduction   486 |
| taproot system   480 | xylem   486 |

## BRIEF REVIEW

*Answers to Brief Review questions are in the back of the book.*
*For multiple-choice quiz questions, go to* **www.aw.com/krogh4**.

1. The stomata that exist mostly on leaves allow important gas exchanges to take place in plants, but these exchanges also present plants with a problem they must solve. What gases are exchanged through the stomata and in which direction? What problem must plants solve because of these exchanges, and how do they solve it?

2. What functions do roots perform? What are the main types of roots produced by flowering plants?

3. List the main parts of a flower and describe the functions performed by each part.

4. Contrast the way people and plants grow over their lifetimes.

5. Give examples of mutually beneficial relationships between plants and bacteria and between plants and fungi.

6. Distinguish between the gametophyte and sporophyte plant generations.

## APPLYING YOUR KNOWLEDGE

1. Gardeners often "pinch off" the terminal shoot apex to stimulate bushiness in young plants. Explain the physiological basis for this common horticultural practice.

2. The text makes clear that plants respond to their environments in sophisticated ways, even to the point of signaling one another during attacks by predators. Given this, can plants be said to be conscious beings, or are animals the only conscious beings?

3. The "race for sunlight" in plants led to the evolution of the tallest living things in existence, trees. But sunlight is not the only thing that stands to make a tree successful or not in reproducing. Think about how plants function and then answer this question: What are the costs of being taller, as opposed to shorter?

The flowering plant sometimes called the chincherinchee (Ornithogalum thyrsoides) is native to South Africa. Its flowers grow on a stem that can be up to three feet long.

CHAPTER **25**

25.1    Two Ways of Categorizing
        Flowering Plants                          500

25.2    There Are Three Fundamental
        Types of Plant Cells                      502

25.3    The Plant Body and Its Tissue Types       503

25.4    How a Plant Grows: Apical Meristems
        Give Rise to the Entire Plant             506

25.5    Secondary Growth Comes from
        a Thickening of Two Types of Tissue       509

25.6    How the Plant's Vascular
        System Functions        *BioFlix*         513

25.7    Sexual Reproduction in
        Flowering Plants                          517

25.8    Embryo, Seed, and Fruit:
        The Developing Plant                      522

**Essays**

A Tree's History Can Be Seen in Its Wood         512

The Syrup for Your Pancakes Comes
from Xylem                                        516

Plants are fixed in one spot and have almost no moving parts, yet they manage to reproduce, transport fluids inside themselves, and keep growing for as long as they live.

# THE ANGIOSPERMS:
# Form and Function in Flowering Plants

**I**nsects that eat plants are not much of a surprise, but plants that eat insects? About 500 species of flowering plants are carnivorous, with at least one variety, the tropical pitcher plant *Nepenthes*, able to capture animals up to the size of a small bird. But the most famous plant predator of them all undoubtedly is the Venus flytrap (*Dionaea muscipula*), which is native to the wetlands of North and South Carolina. So, how does this plant, which has no muscles and is fixed in one spot, end up consuming insects?

Each trap is a single leaf folded in two, with the interlocking "fingers" of the two halves able to snap together in less than half a second, imprisoning any small wanderer (**Figure 25.1**). But what keeps the leaf halves from wasting their energy by snapping together in pursuit of wind-borne blades of grass or other objects? The chemical reaction that puts the trap in motion is set off by a series of trigger hairs on the inside of each of the leaf's two sections. The trick is that *two* trigger hairs must be tripped for the halves of the leaf to snap shut. Insects wandering around inside the leaf are likely to trip two hairs, but blades of grass are not. And, how does the plant slam its "jaws" shut without muscles? When the trigger hairs are tripped, they set off a chemical reaction that engorges cells at the base of the leaf with water. This rapid movement causes the leaf halves to come together. Enzymes are then released that begin to digest the struggling insect.

It may surprise you to learn that Venus flytraps make their own food through photosynthesis, just like any other plant. So, why do they need these side orders of insects? For nutritional supplements. Like many other carnivorous plants, Venus flytraps grow naturally in mineral-poor soil, and the insects they catch are a good source of nitrogen and phosphate.

**Figure 25.1**
**Fatal Entry**
A fly enters the leaf of a Venus flytrap. Once the fly pulls on two trigger hairs within the leaf, its opposing halves will snap shut, trapping the fly. Then the plant will begin to digest its prey.

The world of plants is filled with species as unique in their own way as the Venus flytrap is. Queen Victoria water lilies (*Victoria regia*) have circular leaves that can be nearly 2 meters (about 6 feet) in diameter and that can serve as a floating platform for a person if pressed into service (**Figure 25.2**). The seeds of orchids are so tiny that they resemble dust more than seeds, while the seeds of the double coconut come wrapped in a fruit that might be 0.6 meter or about 2 feet across.

Yet there is a unity to plant life, particularly in the case of the flowering plants, or angiosperms, that you'll be looking at in this chapter. All angiosperms have a common set of cell types, for example, and all of them reproduce in a similar way. The goal in this chapter is to cover common angiosperm features in four areas: cell and tissue types, growth, fluid transport, and reproduction. Before we begin, it might be helpful to revisit the representation of the whole plant that you first saw in Chapter 24 so that you can "find your place" when reading about a plant's component parts. If you look at **Figure 25.3**, you can see a diagram of a typical plant with its two-system division—roots and shoots—and the various parts of the plant that lie within these two systems. The function of each of these parts will become clear to you as you go through the details on them.

**Figure 25.2**
**Floating Platforms**
A woman plays a violin while standing atop a Victoria water lily at the Missouri Botanical Garden in St. Louis at the turn of the twentieth century. The lilies are native to South America. Some South American Indians call the lilies *Yrupe*, which can be translated as "big water tray."

**Figure 25.3**
**Anatomy of a Flowering Plant (an Angiosperm)**

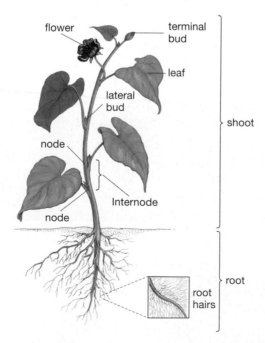

## 25.1 Two Ways of Categorizing Flowering Plants

The constituent parts of plants are often organized one way in a given kind of plant, but another way in a different type. So, what are these types? Here are a couple of ways of *categorizing* plants, the significance of which will become apparent as you go along.

### The Life Spans of Angiosperms: Annuals, Biennials, and Perennials

One important question that might be asked about any plant is: How long does it take for it to go through its entire life cycle—from being a seed germinating in the ground, through growth, flowering, seed dispersal, and then death? Plants that go through this cycle in 1 year or less are known as **annuals**. Some of these are food crops, such as tomatoes and the commercial grains. Plants that go through the cycle in about two years are **biennials**, which include carrots and cabbage. (Have you ever seen the flowers of a carrot plant? Probably not; we *pick* the taproots called carrots after 1 year, so their shoots don't get a chance to flower.) Plants that live for many years are known as **perennials**, a category that includes trees, woody shrubs such as roses, and many grasses. **Figure 25.4** shows some representative annuals, biennials, and perennials.

### A Basic Difference among Flowering Plants: Monocotyledons and Dicotyledons

A second distinction among the flowering plants has to do with their anatomy, meaning the arrangement of the structures that make them up. With respect to anatomy, there are two broad classes of flowering plants: monocotyledons and dicotyledons, which are almost always referred to as monocots and dicots. A cotyledon is an embryonic leaf, present in the seed, as you can see in **Figure 25.5**. **Monocotyledons** are plants that have one embryonic leaf; **dicotyledons** are plants that have two.

This distinction in the embryos is only the start of how monocots and dicots differ. Their roots are different, their leaves are different, and their transport tubes, or vascular bundles, are arranged differently (Figure 25.5). More than 75 percent of all flowering plants are the broad-leafed dicots. But, given the 260,000 known species of flowering plants, this still leaves more than 50,000 species of narrow-leafed monocots, which include most of the important food

**(a)** One-year life span

**(b)** Two-year life span

**(c)** Variable life span

**Figure 25.4**
**Categorizing Plants by Life Span**

**(a)** Annuals such as tomato plants live for only 1 year.

**(b)** Biennials such as carrot plants have a 2-year life span.

**(c)** Perennials such as this white oak tree live for many years, some for many hundreds of years. Plants with long life spans tend to be woody and larger than plants with short life spans.

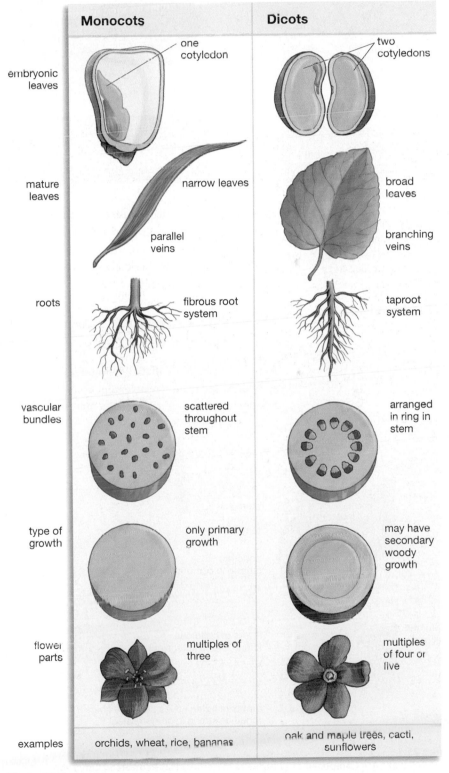

| | Monocots | Dicots |
|---|---|---|
| embryonic leaves | one cotyledon | two cotyledons |
| mature leaves | narrow leaves / parallel veins | broad leaves / branching veins |
| roots | fibrous root system | taproot system |
| vascular bundles | scattered throughout stem | arranged in ring in stem |
| type of growth | only primary growth | may have secondary woody growth |
| flower parts | multiples of three | multiples of four or five |
| examples | orchids, wheat, rice, bananas | oak and maple trees, cacti, sunflowers |

**Figure 25.5**
**A Cotyledon Is an Embryonic Leaf**

Plants with one embryonic leaf (the monocots) are structured differently from plants with two embryonic leaves (the dicots), as shown in the figure.

**(a)** Monocot plant

**(b)** Dicot plant

**Figure 25.6**
**Categorizing Plants by Physical Features**

**(a)** Corn is one of several monocot plants that are food crops.

**(b)** Geraniums are an example of a dicot. See Figure 25.5 for a list of the anatomical differences between monocots and dicots.

crops (corn, wheat, rice). **Figure 25.6** shows you an example of both a monocot and a dicot.

With these distinctions in mind, you're now ready to start looking at the basic elements that go into making up flowering plants.

## 25.2 There Are Three Fundamental Types of Plant Cells

You saw in Chapter 4 that all living things are made up of cells, and plants are no exception to this rule. As it happens, there are three fundamental types of plant

cells that, alone or in combination with each other, go on to make up most of the plant's **tissues**, which are *groups* of cells that carry out a common function.

### Parenchyma Cells

First, there are **parenchyma cells**, which have thin cell walls and are easily the most abundant type of cell in the plant, being found almost everywhere in it. Accordingly, these cells have a lot of different functions. They form the flesh of many fruits, the outer surface of most young plants, parts of leaves that are active in photosynthesis, and much of the "ground" tissue you'll be hearing about shortly. Further, parenchyma are alive at maturity, a quality that might hardly seem worth mentioning except that some plant cells are dead in their mature, functioning state.

### Sclerenchyma Cells

This dead-at-maturity condition characterizes a second type of plant cell, **sclerenchyma cells**, which have thick cell walls that help the plant return to its original shape when it has been deformed by some force, such as the push of wind or animals. Sclerenchyma cells are found in parts of the plant that are mature—that will not be growing further, and that need the support that sclerenchyma can provide. Like all plant cells, sclerenchyma have, running through their cell walls, a compound called **cellulose**, which can be likened to the steel bars in reinforced concrete. Most sclerenchyma cell walls also are infused with a tough compound called *lignin*. The combination of lignin and cellulose results in cells that can bear the crushing weight of massive trees (**Figure 25.7**).

### Collenchyma Cells

The third type of plant cell, **collenchyma cells**, can be thought of as support cells, like sclerenchyma cells, but with the function of stretching or elongating in parts of the plant that are growing, such as young leaves and stems. Collenchyma cells combine the properties of parenchyma and sclerenchyma cells, performing some of the functions of each of these cell types. They provide mechanical strength in areas that are actively growing and can also participate to some extent in photosynthesis and wound repair.

### Parenchyma as Starting-State Cells

Both collenchyma and sclerenchyma are derived from parenchyma cells; that is, some parenchyma cells differentiate *into* collenchyma or sclerenchyma cells. Parenchyma cells are thus a kind of starting-state cell, but their capabilities are greater than this. They can be transformed "backward," in a sense, into

reinforced concrete

steel bar (strong in tension)

concrete (strong in compression)

reinforced cell walls

layers of **cellulose** fibers (strong in tension)

matrix of **lignin** (glues cellulose fibers together, strong in compression)

**Figure 25.7**
**Plant Cell Walls Are Composed of Composite Materials, as Is Reinforced Concrete**
The cellulose fibers of plant cell walls are very strong in tension, as are the steel bars in reinforced concrete. Lignin is similar to the matrix of concrete in that it binds the fibers together and resists compression. Both plant cell walls and reinforced concrete are important in providing support for structures that are tall and heavy.

the type of embryonic cells that give rise to the whole plant. You may have wondered in Chapter 24 how we can get a new plant by taking a cutting from an existing plant and placing it in the ground. The answer is that when a plant is cut, parenchyma cells lying close to the cut are transformed into *growth* cells of a type you'll look at shortly, and these can sprout a new root. This obviously gives plants a flexibility not possessed by animals such as ourselves. If one of our fingers is cut off, we do not grow a new finger, to say nothing of a whole new person.

### SO FAR...

1. Plants that go from germination in the ground to death in 1 year or less are referred to as _____, while those that live for many years are referred to as _____.

2. Most plants fall into the category of the broad-leafed _____, but many of our most important food crops are narrow-leafed _____.

3. _____ cells can be thought of as a starting-state plant cell that can differentiate into the other two main plant cell types _____, which are dead in their mature state, and _____, which are support cells that are alive in their mature state.

## 25.3 The Plant Body and Its Tissue Types

You saw earlier that tissues are groups of cells that carry out a common function. So, what kinds of tissues are there in angiosperms? The short answer is dermal, ground, vascular, and meristematic, with each of these four types of tissue generally composed of one or more of the cell types you just looked at. Dermal tissue can be thought of as the plant's outer covering, vascular tissue as its transport or "plumbing" tissue, meristematic tissue as its growth tissue, and ground tissue as almost everything else in the plant.

### First: A Distinction between Primary and Secondary Growth Tissue

Before proceeding further with this topic, we need to take note of a basic distinction in plant tissue that has to do with the way plants grow. Some plants are capable only of what is called **primary growth**, meaning growth at the tips of their roots and shoots that principally increases their *length*. In contrast, other plants exhibit not only primary growth but also

something known as **secondary growth**, which can be thought of as the *lateral* growth, or thickening, that occurs in **woody plants**. Secondary growth occurs in trees, for example, but not in orchids or strawberry plants. Non-woody plants such as an orchid are known as **herbaceous plants**, meaning those that never develop wood (or bark) and that thus contain only primary tissue. It is this *primary* tissue that you will look at first, with a review of secondary tissue to come shortly.

If you look at **Figure 25.8**, you can see the location of all four kinds of primary tissue in a dicot. Looking at the stem in cross section, you can see that ground tissue is well named because visually it forms a kind of background against which you can see the tubes of the vascular tissue. Dermal tissue then is at the periphery of the plant, forming a "skin" layer around the ground tissue. The placement of the meristematic tissue is discussed shortly. First, let's look in a little more detail at dermal, ground, and vascular tissue.

### Dermal Tissue Is the Plant's Interface with the Outside World

Plants can't move, they can't bite, and as you saw in Chapter 24, they must carefully control their water supply. All this makes their **dermal tissue**, or

••• ANSWERS

1. annuals; perennials
2. dicotyledons (dicots); monocotyledons (monocots)
3. Parenchyma; sclerenchyma; collenchyma

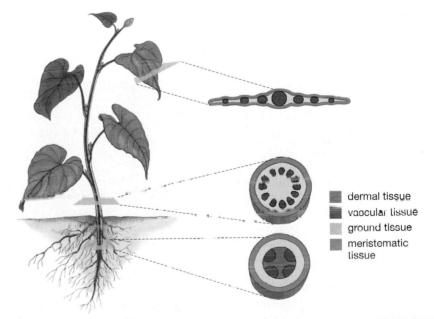

**Figure 25.8**
**Four Types of Tissue in Primary Plant Growth**
Dermal tissue (dark green) is found on the outside of the plant and is the interface between the plant and its environment. Ground tissue (pale green) is found throughout the plant and gives the plant its shape, stores food, and is active in photosynthesis. Vascular tissue (purple) transports water, nutrients, and sugar throughout the plant. Meristematic tissue (blue) occurs at the tips of shoots and roots and at leaf bases and gives rise to the entire plant, through the production of cells that develop into the other three tissue types.

dermal tissue
vascular tissue
ground tissue
meristematic tissue

*epidermis*, very important to them. This outer coat is generally only one layer of cells thick, but it serves to protect the plant and control its interaction with the outside world (**Figure 25.9**). A plant's first line of defense against outside predators is a waxy coating called the *cuticle*, which covers the epidermis of the plant's above-ground parts (its shoot). The cuticle is a kind of waterproofing, serving to keep water in, and an invader-proofing, helping to keep infecting bacteria and fungi out.

Plants need to exchange gases with their environments, however. They take in carbon dioxide and emit oxygen and water vapor through the microscopic pores called stomata, found mostly on the underside of their leaves. (You went over the functioning of stomata in Chapter 24, starting on page 481.) Stomata also represent a kind of weak point for plants, however, because they provide a passageway for microbial invaders and a conduit for excessive loss of moisture.

Dermal tissue also forms several kinds of extensions, called trichomes, that protrude from the plant's surface. You learned about one variety of them in Chapter 24: root hairs, meaning the thread-like outgrowths of individual epidermal root cells. As we noted there (on page 480), these wispy outgrowths greatly increase the plant's ability to absorb water and minerals. Other types of trichomes serve not to absorb substances, but to secrete them. In some plants, trichomes secrete toxic chemicals that ward off planteating animals.

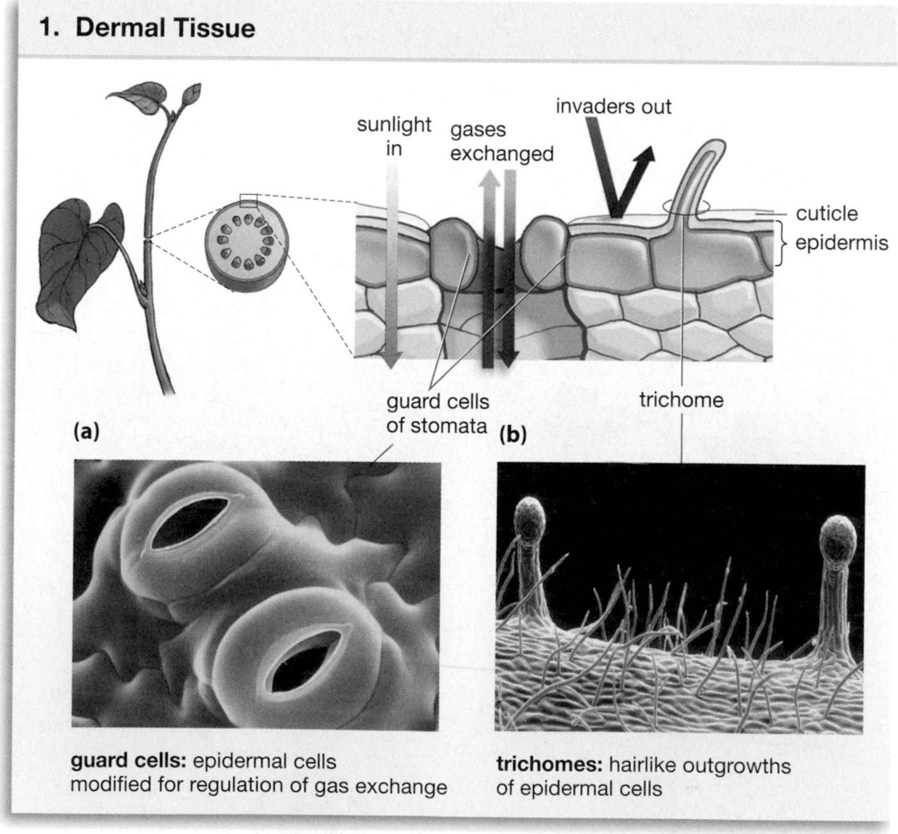

## Ground Tissue Forms the Bulk of the Primary Plant

Most of the primary plant is ground tissue, which can play a role in photosynthesis, storage, and structure (**Figure 25.10**). Large parts of ground tissue may be simple parenchyma, but other parts are combinations of cell types.

## Vascular Tissue Forms the Plant's Transport System

As you saw in Chapter 24, the plant's vascular system has two main functions. First, it must transport water and minerals; second, it must transport the food made in photosynthesis. In line with this, there are two main components to the plant vascular system. First, there is **xylem**, the tissue through which water and dissolved minerals flow. Second, there is **phloem**, the tissue that conducts the food produced in photosynthesis, along with some hormones and other compounds (**Figure 25.11**). The movement of water through xylem is directional—from root up through stem and then out through leaves as water vapor. Meanwhile, the food made in photosynthesis must travel through phloem to every part of the plant, although as a practical matter, in temperate climates the net flow is downward in summer (from foodproducing leaves) but upward in early spring (from root and stem storage sites).

Xylem and phloem are arranged in *vascular bundles*, which is to say collections of xylem and phloem tubes that run together in parallel in the stem—xylem tubes toward the inside of the stem, phloem tubes toward the outside. The bundles are arranged in different ways within a stem, depending on whether the plant is a monocot or a dicot; the dicot bundles are configured in a circle, the monocot in an irregular pattern. This is why dicot ground tissue often is conceptualized as existing in two

**1. Dermal Tissue**

sunlight in — gases exchanged — invaders out

cuticle
epidermis

guard cells of stomata

trichome

(a)

(b)

**guard cells:** epidermal cells modified for regulation of gas exchange

**trichomes:** hairlike outgrowths of epidermal cells

**Figure 25.9**
**Where Plant Meets Environment**

Dermal tissue serves as the interface between a plant and the environment around it. The waxy cuticle and the single layer of cells in the epidermis work together to protect the plant and to control interactions with the outside world.

**(a)** Specialized epidermal cells called guard cells regulate the opening of the plant pores called stomata, thus controlling gas exchange between the plant and its environment. Here, guard cells have opened two stomata. In most plants, the majority of stomata are found on the leaves.

**(b)** Some epidermal cells have hairlike projections, called trichomes, that serve various functions. Two kinds of trichomes are visible in this rose plant. The larger of the two varieties, with the bulbous tips, help secrete chemicals that guard against plant-eating predators.

## 2. Ground Tissue

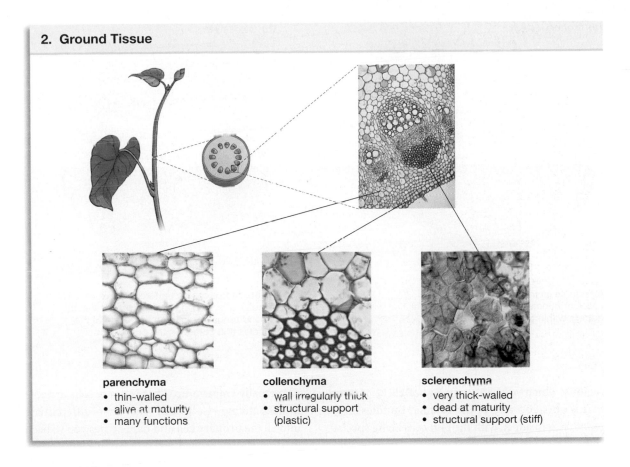

**parenchyma**
- thin-walled
- alive at maturity
- many functions

**collenchyma**
- wall irregularly thick
- structural support (plastic)

**sclerenchyma**
- very thick-walled
- dead at maturity
- structural support (stiff)

**Figure 25.10**
**Ground Tissue**

Most of a plant is made of ground tissue, which is composed of parenchyma, collenchyma, and sclerenchyma cells. Examples of these cell types shown here come from several different plants. In a cross section of a stem, thin-walled parenchyma can be found in most of the interior. Collenchyma—at the bottom of the collenchyma picture, surrounded by parenchyma—are support cells that can elongate in growing parts of the plant. Very thick-walled sclerenchyma, such as these from a pear, can provide strength to tissues.

## 3. Vascular Tissue

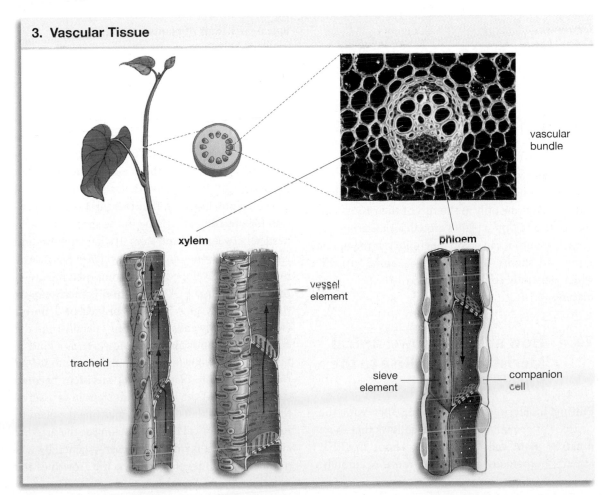

vascular bundle

**xylem**

vessel element

tracheid

**phloem**

sieve element

companion cell

**Figure 25.11**
**Materials Movers**

Vascular tissue transports fluids throughout the plant. Xylem is composed of cells called tracheids and vessel elements that transport water and dissolved minerals. Phloem is composed of sieve elements (and their companion cells), which transport the food produced during photosynthesis. These cell types are stacked on one another to form long tubes.

**Figure 25.12**
**Two Types of Flowering Plants**

These cross sections show the organization of primary tissues in a monocot and a dicot.

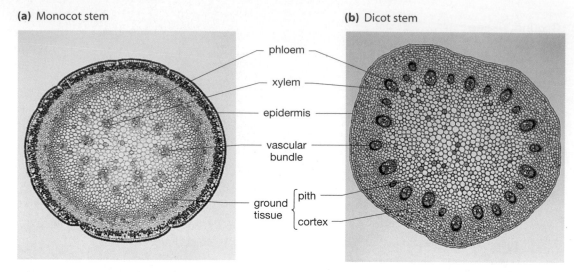

**(a)** Monocot stem

**(b)** Dicot stem

phloem

xylem

epidermis

vascular bundle

ground tissue { pith

cortex

Monocots (in this case an onion) have vascular bundles arranged in an irregular pattern within ground tissue.

Dicots (in this case a buttercup) have vascular bundles arranged in a ring that separates the ground tissue into pith on the inside of the bundles and cortex on the outside.

regions, which you can see in **Figure 25.12**: a *cortex* that is outside a ring of the vascular bundles and a *pith* that is inside it, with the pith cells being specialized for storage.

### Meristematic Tissue and Primary Plant Growth

So, you've seen three types of primary plant tissues—dermal, ground, and vascular. The question now is: How do these arise? How is it that a plant gets its vertical growth? The short answer is that the fourth type of plant tissue, meristematic tissue, gives rise to all the other tissue types. In considering the role of meristematic tissue, recall from Chapter 24 that, in their vertical growth, plants do not grow the way people do—throughout their entire length—but instead grow only at the tips of their roots and shoots. In addition, a plant's growth is *indeterminate*; in most plants, it can go on indefinitely at the tips of roots and shoots, as opposed to animal growth, which generally comes to an end when the animal matures.

### 25.4 How a Plant Grows: Apical Meristems Give Rise to the Entire Plant

Putting indeterminate growth together with the concept of growth at the tips, it follows that there must be plant cells at root and shoot tips that remain perpetually young—or more accurately,

perpetually embryonic. That is, these cells are able to keep giving rise to cells that then differentiate into all the primary cell and tissue types you've been reading about. Indeed, everything in plants (both their primary and secondary tissues) develops ultimately from these cells, which collectively form the **apical meristems** of plants.

If you look at **Figure 25.13**, you can see the apical meristem locations in a typical plant. Note that in this plant with a taproot, each lateral root tip has its own apical meristematic tissue that is capable of giving rise to yet more roots. In the shoot, plants confine their growth to the shoot apices that lie at the tip of each stem. Why? The better to compete for precious sunlight. It is the **shoot apical meristem** that gives rise to all the cells that allow this vertical growth. Note, however, that there is a second location for meristematic tissue in the shoot—the area nestled between leaf and stem. It's here that we find lateral buds (sometimes called *axillary buds*). Any **bud** is an undeveloped shoot, composed mostly of meristematic tissue. A **lateral bud** is meristematic tissue that may give rise to a branch or a flower (which obviously ends up growing laterally from the stem). Lateral buds also serve, however, as a kind of insurance policy for the plant in that they can switch roles. Should the plant's apical meristem become damaged, one of the lateral buds steps in to assume the role of the shoot apex, thus allowing the plant to maintain its vertical growth. As long as the original apical meristem is intact, however, it generally will produce hormones that *suppress* the growth of the

**4. Meristematic Tissue**

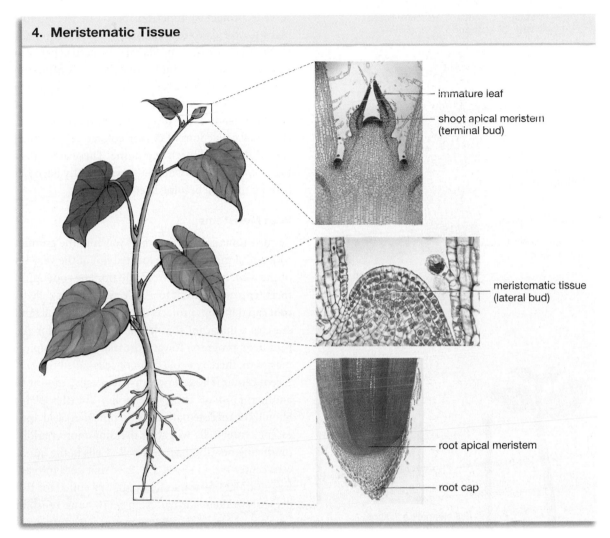

immature leaf

shoot apical meristem (terminal bud)

meristematic tissue (lateral bud)

root apical meristem

root cap

**Figure 25.13**
**Meristematic Tissue**
Growth originates from meristematic tissue. The shoot and root elongate at the apical meristems, while new branches originate at the lateral buds. Meristematic tissue gives rise to more dermal, ground, and vascular tissue, as well as to more of itself.

lateral buds near it, leaving them dormant. The shoot apex is itself sometimes referred to as a **terminal bud**, particularly when *it* is dormant, as with trees in winter.

We can think of plant growth in terms of growth modules. Each module consists of an internode (the stem between leaves), a node (the area where a leaf attaches to stem), and one or more leaves (**Figure 25.14**).

### A Closer Look at Root and Shoot Apical Meristems

Now, let's look a little closer at these important collections of cells from which all else comes, the shoot and root apical meristems. The right-hand pictures in Figure 25.13 show you what shoot apical meristem, lateral bud, and root apical meristem look like up close.

### Shoot Meristems

As you can see in the topmost picture, the shoot apical meristem lies atop a dome-shaped collection of cells. Overlapping it are two budding leaves that

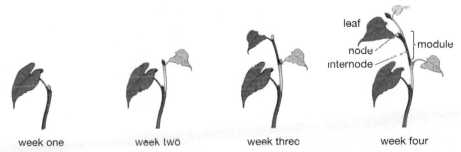

leaf

node

internode

module

week one          week two          week three          week four

**Figure 25.14**
**Vertical Growth in Plants Occurs in Modules**
Each growth module (indicated here by alternating dark green and light green segments) consists of an internode, a node, and one or more leaves. This plant progresses from one module in week 1 to four modules in week 4.

**Figure 25.15**
**Tissue Development from Apical Meristems**

Apical meristems give rise to the primary tissues. In development, cells first are mostly engaged in division, just above the apical meristem, then elongation, and finally differentiation. The result is three fully formed kinds of tissue.

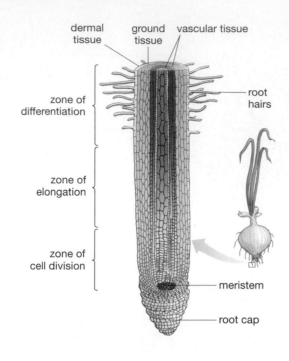

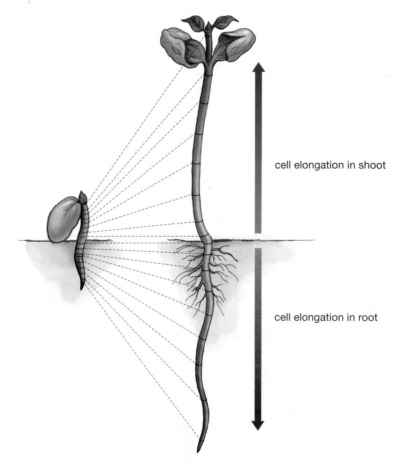

**Figure 25.16**
**Growth by Cell Elongation**

Much of the elongation of the shoot and root in a young plant results from cell elongation rather than from cell division. Thus, it is cells just above the root tip, in the zone of elongation, that provide most of the root's growth, as shown here. The tip of the root (where cell division occurs) has elongated very little, and the cells in the upper root have completed most of their elongation. This same pattern holds true for the shoot.

serve a protective function. Next, in the middle picture, you can see a second shoot location for meristematic tissue—the lateral bud, nestled between leaf and stem. These buds are collections of meristematic tissue that are left behind, in a sense, by the apical meristem tissue as it continues its growth. (If you look again at the topmost picture, you can see two small collections of darker-colored cells on each side of the apical meristem dome. These are collections of meristem tissue that have recently been left behind; they will become lateral buds.)

## Root Meristems

Because roots are pushing their way into the ground, root apical meristems are located not at the very tip of the root, but just up from the tip. The root apical meristems are adjacent to a collection of cells called a **root cap** that both protects the meristematic cells and secretes a lubricant that helps ease the way for the root in its progress. Toward the middle of the apical meristem, there is a collection of cells called the quiescent center. These slowly dividing cells amount to another kind of insurance policy for the plant. Should the meristematic cells become damaged, quiescent center cells will start dividing more rapidly, producing new meristematic cells. Cells in the quiescent center are, in short, a reserve that goes into action in times of trouble. They are well suited for this because, in their dormancy, they are more resistant than apical meristem cells are to injury from such influences as toxic chemicals.

## How the Primary Tissue Types Develop

You've looked so far at ground, dermal, and vascular tissues. And, you know that the fourth type of primary tissue, meristematic tissue, gives rise to all these. So, how do plants get from meristem to the others? Not in a single step, it turns out—not directly from apical meristem cells to, say, fully functioning vascular tissue. Instead, there is a transition involving three main changes: *More* cells are produced through cell division; cells *elongate* through cell growth; and cells *differentiate* into the three kinds of tissue. These three changes are overlapping—cells are elongating even as they are differentiating—but events do take place in roughly this order. You can see this by looking at **Figure 25.15**. In a gradual transition with no sharp boundaries, the apical meristem gives way to a **zone of cell division**, then there is a **zone of elongation**, and finally a **zone of differentiation**, which is where the ground, dermal, and vascular tissues fully take shape. Note that root hairs don't appear until the zone of differentiation. Although the three kinds of tissues *finish* taking shape in the zone of differentiation, they actually begin this process just above the apical meristem (in the zone of cell division). It's accurate to think of the tissues in the zone of cell division as being precursors to the three tissue types.

If you look at **Figure 25.16**, you can see the practical effect of a zone of elongation: It is where most of the primary plant growth occurs. There is some growth with the addition of more cells down near the apical meristem, but most of the vertical expansion of the plant takes place because of cell elongation. Such elongation has the effect of extending the roots into the soil and the stem into the air.

### Intercalary Meristems Keep Grasses Growing

There is one important variation on primary growth through apical meristems. What about plants such as grasses that are constantly losing their tops through such means as grazing or mowing? Grasses get their growth from tissue known as **intercalary meristems**, which are found at the base of *each node*. There thus exists a series of vertical growth tissues that are intercalated, or interspersed, between regions of nondividing cells as we go up the plant. One practical effect of this structure is that when we mow the top half of a blade of grass, it grows back. A second is that, because these plants are growing at *many* points along their length (rather than just their apex), they can manifest very rapid growth. We can easily see nodes in the giant grass known as bamboo, which exhibits intercalary growth (**Figure 25.17**).

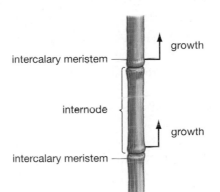

**Figure 25.17**
**Intercalary Meristems Keep Grasses Growing**

One reason bamboo grows so rapidly is that growth occurs at several points along its length at once. The intercalary meristems responsible for this growth are located at the base of each internode.

## 25.5 Secondary Growth Comes from a Thickening of Two Types of Tissue

So far, you've been looking at a plant's vertical or primary growth. In the herbaceous plants, that's all the growth there is. But you've seen that some plants—the woody plants—grow out as well as up and down; they have primary *and* secondary growth, in other words (**Figure 25.18**). So, how does this secondary growth come about? Once again, it is through cells that are meristematic. In essence, secondary growth yields three new kinds of tissue: a different type of xylem and phloem—secondary xylem and secondary phloem—and several layers of outer tissue that collectively go under the familiar name of *bark*.

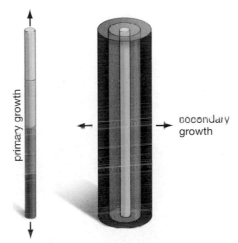

**Figure 25.18**
**Growing Up versus Growing Out**

Herbaceous plants grow at their tips, mostly vertically, through primary growth. Woody plants, however, grow not only vertically but also laterally, through what is called secondary growth. Dicots can be either woody or herbaceous, but almost all monocots are herbaceous.

---

### SO FAR...

1. An orchid is capable only of primary growth, which principally increases a plant's _____, while a tree is capable not only of primary growth, but of secondary growth, which can be thought of as _____ growth.

2. Dermal tissue is tissue located at the _____ of a plant, while vascular tissue forms a plant's _____ system.

3. Xylem is the tissue through which _____ and _____ flow, while phloem is the tissue that conducts the _____.

4. Everything in the plant ultimately develops from _____ tissue in structures known as _____ _____.

**••• ANSWERS**

1. length; lateral

2. periphery; fluid transport

3. water; minerals; food produced in photosynthesis

4. meristematic; apical meristems

At the start, it's worth noting that secondary growth in plants is related to two of the plant characteristics that we reviewed at the start of the chapter. First, secondary growth almost always takes place in perennials, rather than annuals (although plenty of perennials exhibit only primary growth, as with grasses). Second, with very few exceptions, the monocots described earlier have only primary growth. Meanwhile, dicots are more often woody, although many are herbaceous.

### Secondary Growth through the Vascular Cambium: Secondary Xylem and Phloem

You saw earlier that the plant's vascular tissues are arranged into bundles of tube-like vessels, with the food-carrying phloem tubes toward the outside of the stem and the water-carrying xylem tubes toward the inside. In a cross section of a dicot plant stem, these tubes form concentric rings. If you look at the plant another way—through its entire *length*—you can see that these rings take the shape of cylinders that are nested, as if these layers of tissue were a series of open-ended cans, one inside the next (**Figure 25.19**).

Now, notice that woody plants develop a thin layer of tissue *between* the primary xylem and primary phloem cylinders. This is the **vascular cambium**, one of the additional types of meristematic tissue that brings about secondary growth. The vascular cambium has one group of cells in it, called *ray initials*, that produce exactly what their name indicates: *rays* of parenchyma cells that carry water

and dissolved compounds through the *width* of the plant. (If you've ever looked at the stump of a tree, you've probably noticed fine lines that seem to radiate out, from near the center to the periphery of the trunk. These are the tree's transport rays, made up of parenchyma cells that form streaks of living cells in an expanse of dead xylem. You can see an example of them in Figure 25.22.)

More important for our purposes, the vascular cambium is something like a cell factory that is continually pushing out two different kinds of cells to either side of itself. The result is two new cylinders of tissue: **secondary xylem** to the inside of the vascular cambium and **secondary phloem** to the outside. Secondary xylem and phloem initially do just what primary xylem and phloem do: transport, through the *length* of the plant, water through xylem and food through phloem. **Figure 25.20** shows you how the vascular cambium pulls off its feat of two-way xylem/phloem production. Note in the figure that the cells of the vascular cambium divide, producing one cell that differentiates into either xylem or phloem and another cell that remains vascular cambium. This process continues over and over again, causing the stem of the plant to thicken.

### Secondary Xylem Is Responsible for Most of a Plant's Widening

It is the secondary xylem that is responsible for most of the widening of trees and other woody

**Figure 25.19**
**Vascular Cambium and Secondary Growth**

Vascular cambium is meristematic tissue that produces secondary xylem tissue to its interior and secondary phloem tissue to its exterior. Over the years, the thickening of a tree comes about primarily because of the growth of secondary xylem.

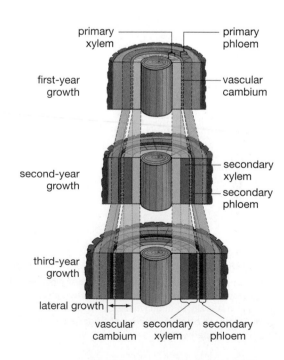

primary xylem — primary phloem

first-year growth — vascular cambium

second-year growth — secondary xylem / secondary phloem

third-year growth

lateral growth ⊢—⊣

vascular cambium · secondary xylem · secondary phloem

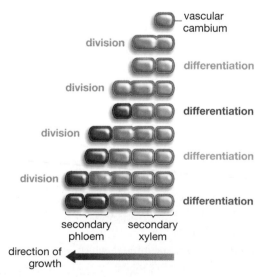

vascular cambium

division
differentiation

division
**differentiation**

division
differentiation

division
**differentiation**

secondary phloem · secondary xylem

direction of growth

**Figure 25.20**
**How Vascular Cambium Thickens the Plant**

The cells of the vascular cambium divide, producing one cell that differentiates into either xylem or phloem and another cell that remains vascular cambium. This process repeats, causing the stem of the plant to thicken.

plants. Indeed, another term for secondary xylem is **wood,** a material that comprises about 90 percent of an average tree. After an initial period of carrying water, the secondary xylem cells will cease doing this but are still valuable to the tree because of their strength.

Almost everyone is familiar with the annual rings of trees in temperate climates such as ours. But, what is each visible ring? **Figure 25.21** tells the tale. Each ring represents the abrupt change between the secondary xylem of late summer and the secondary xylem of the following spring. In late summer, when plants are getting little water, they produce xylem cells with smaller interiors and proportionately thicker cell walls. Because cell walls are *darker* than the cell interiors, the summer xylem cells are relatively dark. Come next spring, with its plentiful water, the cells that are produced are much larger and thus have lighter color. Thus, the "line" of any given ring is made up of darker, late-summer cells. We can look just to the outside of any line and say "spring began here." To learn more about what we can tell by looking at a cross section of a felled tree, see "A Tree's History Can Be Seen in Its Wood" on the next page.

## Secondary Growth through the Cork Cambium: The Plant's Periphery

With the secondary xylem cells pushing out from the interior of the tree or shrub, the plant's original outer covering (its epidermis) can't last. It splits apart, eventually to be shed altogether, usually within a year. Before it goes, however, the plant develops a new kind of outer covering, one that has the ability to renew itself constantly as the plant continues to grow. This quality may sound familiar by now, and indeed **cork cambium** represents the final type of meristematic tissue you'll be encountering—one that gives rise to the plant's outer tissues (**Figure 25.22**).

The cork cambium itself is a product of other tissue in the stem—in a mature tree, the secondary phloem tissue you just looked at. Cork cambium is once again secondary meristematic tissue, which is like a cell factory that is continually pushing out different kinds of cells to either side of itself. The main product of this activity goes to the *outside* of the cambium in the form of **cork** cells. As these cells mature, they go on to infuse their cell walls with a waxy substance that acts as nature's own waterproofing and invader-proofing. But cork cells are born to die, in the sense that their genetic blueprint brings about a so-called programmed cell death. It is in this dead,

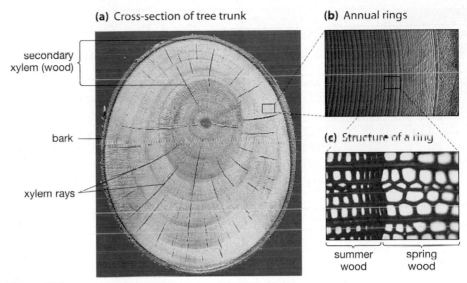

**(a)** Cross-section of tree trunk

secondary xylem (wood)

bark

xylem rays

**(b)** Annual rings

**(c)** Structure of a ring

summer wood  spring wood

**Figure 25.21**
**Annual Rings Tell Time and More**

**(a)** The thick secondary xylem (the wood) of a tree contains many concentric rings, with the dark line of each ring representing the end of 1 year's growth. The xylem rays transport water and nutrients laterally through the trunk.

**(b)** This close-up of the annual rings from an actual tree shows how growth can vary with environmental conditions. The earlier growth rings on the left are thicker than the rings on the right. The tree grew less in each of the later years because it was being affected by acid rain.

**(c)** This micrograph of a pine tree shows the structural basis for annual rings. On the left—toward the inside of the trunk—are smaller cells that grew in late summer and early fall, when water was less plentiful. The very narrow cells at left-center create the dark line of an annual ring. Immediately to the right of these densely packed cells are the larger cells that grew the following spring, when water was again plentiful. These larger cells are visible as the lighter wood in the trunk.

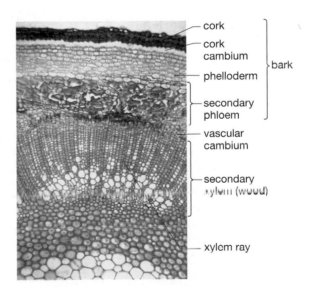

cork

cork cambium

phelloderm

bark

secondary phloem

vascular cambium

secondary xylem (wood)

xylem ray

**Figure 25.22**
**Secondary Growth Tissues**

The secondary xylem, the wood with which we are all familiar, is responsible for water transport and the structural support of the plant. Xylem rays are lateral conduits for water. Bark is composed of four layers of tissue: secondary phloem, phelloderm, cork cambium, and cork.

mature state that they serve their protective function. In a single growing season, the cork cambium can produce layer on layer of cork cells, resulting in a well-protected tree. To its interior, the cork cambium can produce, in a few species, a layer of parenchyma cells called phelloderm.

Almost everyone knows about looking at the cross section of a felled tree trunk and counting its "rings" to gauge the age of the tree. But what other stories does this kind of tree wood have to tell?

First, are the rings wider in some years than others? To the extent that the tree grew more in a given year, it will have a wider ring. The wider the ring, the more likely it is that optimal conditions for growth existed during that year: plentiful rainfall, enough sun, a lack of pests. A series of unusually narrow rings probably means just the opposite (**Figure 1**).

Next, are the rings perfectly concentric, or are they *eccentric* at some point in the tree's growth? Is a given set of circles wide on, say, the left as you look at the trunk, but narrower on the right, as in Figure 1? If so, it may be because some force—for example, pressure from another tree—was bending the tree trunk toward the left at that point in its history. Because trees prosper by being as tall as they can, they will resist being pushed over. They do so by growing extra wood on one *side* of themselves. The wider areas in a given set of circles represent this extra wood, known as *reaction wood*.

Look at the center of any large log, and the wood you see there is almost always darker than the surrounding wood. Why? The inner wood—which like all wood is secondary xylem—has long since closed down with respect to its water-transport function. The cells that make up the water columns were always dead in their mature, water-transporting state, but eventually each column was severed altogether, meaning it quit conducting water. The problem is that an empty xylem column is a *dangerous* column as far as the tree is concerned because a fungus could grow right up through it. Therefore, trees have developed a plugging mechanism: Surrounding living cells push material into the columns, sealing them off. Over time, these cells produce oils, lignin, and microbe-fighting substances and then die themselves. The result is a center of a tree that is completely dead—and dark in coloration—but strong in its load-bearing strength and resistant to bacteria and fungi. This is the tree's heartwood. Its counterpart is the tree's sapwood—the lighter, outer wood whose xylem tubes that are still conducting xylem sap.

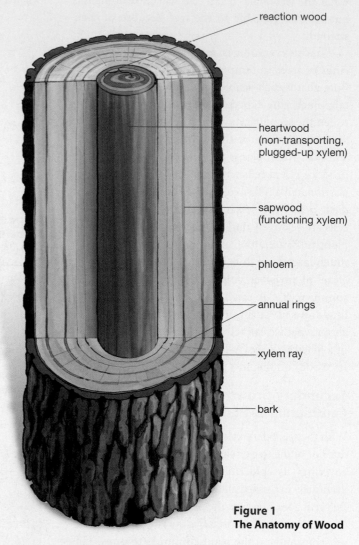

reaction wood

heartwood
(non-transporting,
plugged-up xylem)

sapwood
(functioning xylem)

phloem

annual rings

xylem ray

bark

**Figure 1**
**The Anatomy of Wood**

### What Is Bark?

So, woody plants have cork cambium and its two products (cork and phelloderm). Then, *inside* this three-part structure there is the secondary phloem. Put together, all four parts constitute a region of a woody plant that goes by a familiar name, **bark**. Put another way, bark is everything outside the vascular cambium. Figure 25.22 shows all the layers of tissues in secondary growth.

If a tree is "ringed" by having a strip removed around its entire circumference—as hungry animals might do in winter—it might live without its cork, the cork cambium interior to it, and the phelloderm inside that. But if this cutting continues into the younger, food-conducting secondary phloem, that would be the end of the tree because it could no longer move its energy stores to where they're needed. Indeed, this kind of cutting describes the "girdling" that is a common way to kill unwanted trees (**Figure 25.23**).

To conclude this section on plant growth, **Figure 25.24** shows you how the entire plant grows, with the starting point for both secondary and primary growth being apical meristem tissue.

**Figure 25.23**
**Death from Lack of Nutrients**
This tree has been girdled by someone who intends to kill it. The bark was removed all the way down to the secondary phloem, the tissue layer that transports food from one part of the plant to another.

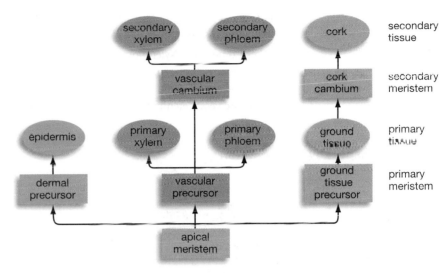

**Figure 25.24**
**Summary of Primary and Secondary Growth in the Stem of a Vascular Plant**
The "precursor" tissues that arise from the apical meristem are those that take shape in the zones of cell division, elongation, and differentiation noted earlier in the chapter.

**Figure 25.25**
**Water Carriers**
The fluid-conducting cells of the xylem are tracheids and vessel elements. Both kinds of cells stack to form tubes.

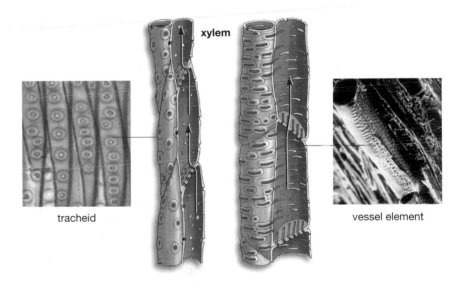

xylem

tracheid

vessel element

## SO FAR...

1. Most of the widening in trees and other woody plants comes about as a result of the growth of secondary xylem, a tissue commonly known as _____. Secondary xylem and secondary phloem are produced by the meristematic tissue known as _____.

2. A woody plant's outer tissues are produced by the meristematic tissue known as _____.

3. Bark is another name for all the secondary growth tissue outside the _____.

## 25.6 How the Plant's Vascular System Functions

You have had ample opportunity already to look at the plant's vascular or "plumbing" system as it relates to plant growth, but now it's time to take a closer look at how the vascular system itself works, starting with the water-conducting xylem and then continuing to the food-conducting phloem.

### How the Xylem Conducts Water

Remember the statement at the beginning of the chapter that a good many plant cells are dead in their mature, functioning state? You saw this once already with cork cells; here is another variety of these cells, which form most of the xylem. Two types of cells make up the water-conducting portions of both the primary and secondary xylem. These are **tracheids** and **vessel**

**elements**. On reaching maturity, these cells die, and their contents are then cleared out. What's left is a strand of empty, thick-walled cells stacked one on top of another—tubes, in other words (**Figure 25.25**). The walls of these xylem cells remain, however; indeed, *these* are some of the sclerenchyma cells noted earlier that have the strong lignin reinforcement in their cell walls. The upshot is not only a transport system but a *load-bearing* system, which is why something as massive as a tree can grow so tall. As noted earlier, this load-bearing function remains in a given group of xylem cells long after they have ceased to act as a conduit for the water and dissolved nutrients that are sometimes called

Web Animation 25.3
Plant Reproduction

**Figure 25.39
Development of a Plant
Embryo, Germination of a
Plant Seed**

Egg and sperm have come
together in this dicot to produce a single-celled zygote,
which divides repeatedly in
developing as the plant embryo.
The zygote is surrounded by the
plant ovule, which develops into
a seed.

### Development of the Embryo and Germination of the Seed

While fruit and seed coat are maturing, the embryo inside them is also developing. **Figure 25.39** shows you the outline of the development of a plant embryo, this one a dicot plant. Recall that the process starts with a single cell, the zygote, encased within the structure that will become the seed, the ovule. The zygote divides, yielding two cells, one of which becomes the embryo. The other cell gives rise to a paddle-shaped structure that pushes the growing embryo into the endosperm.

Eventually, the embryonic tissue develops two **cotyledons**, or embryonic leaves. When the embryo is still tightly encased in the seed coat, the cotyledons can take on an important role you've seen before, which is to provide nutrients to the growing embryo. You may say: But isn't this the function of the endosperm? In most monocots (for example, the grains), the endosperm retains this function, but in most dicots the nutrients stored in endosperm tissue are *taken up* by the cotyledons as they develop. The foods we know as beans are comprised of embryos with large cotyledons.

The seed that encases the embryo eventually separates from the sporophyte parent plant and makes its way into a suitable patch of earth, there to germinate or to sprout both roots and shoots. First to emerge from the seed is the root structure or radicle, which is attached to **hypocotyl**, which can be thought of as all the plant tissue below the cotyledons. Then, there emerges the **epicotyl**, meaning the tissue above the cotyledons. The epicotyl gives rise to the first true leaves of the plant. The seed has sprouted into a new sporophyte plant.

### Seed Dormancy Can Be Used to a Plant's Advantage

One of the survival strategies of plants is the use of dormancy in seeds, meaning a prolonged low level of metabolism in them. Seeds can *postpone* germinating until they have favorable conditions around them, such as the proper temperature or amount of light. What generally sets seed germination in motion is an uptake of water from the surrounding environment. It is, however, the triggering mechanisms for this water uptake that tell the tale in dormancy. Some seeds require a scraping or abrasion before they will take up water; others simply need to dry out once they've shed the fruit from around them; still others require the action of enzymes from an animal's gut or even charring by fire.

Once the plant has germinated and matured a little, you could take a cross section of its stem and

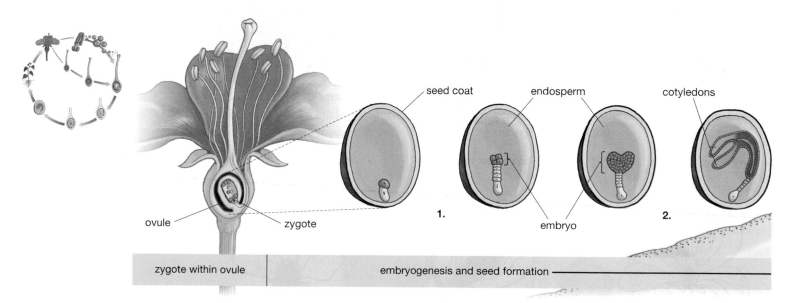

seed coat    endosperm    cotyledons

ovule    zygote

**1.**    embryo    **2.**

zygote within ovule    embryogenesis and seed formation

**1.** The zygote divides into two cells, one of which is the embryo (green in the figure). The other cell (yellow in the figure) develops into a structure that will push the embryo into the endosperm.

**2.** The embryo starts to develop its cotyledons or embryonic leaves. Most of the endosperm nutrients eventually are taken up by these cotyledons. (Because this plant is a dicot, it has two cotyledons.)

find concentric circles of the tissue types mentioned near the start of this chapter: dermal, vascular, and ground. Meanwhile, meristematic tissue could be found at the tips of the root and shoot, allowing the plant to grow further. With this, we have come full cycle: Another generation of plants is maturing.

## On to Animals

In Chapters 24 and 25, you have looked at a group of living things, the plants, whose members are silent, fixed in one spot, and have almost no moving parts. Now, it's time to shift gears. There are animals that are silent and animals that are fixed in place, but you'll also find in animals nature's only sound-makers and its supreme travelers. The next seven chapters are concerned to a small degree with the physical functioning of animals in general, but they focus on one animal that is very familiar to us—the human animal.

### SO FAR...

1. The plants with which we're familiar, such as maple trees, do not produce gametes (eggs and sperm), but instead produce _____, making them the _____ generation of plants. This generation alternates with the _____ generation, which on the male side consists of _____ and on the female side consists of small collections of cells, each one of which is known as an _____.

2. One of the fertilizations in double fertilization produces a _____ which is the original cell in the new generation of _____ plant. The other fertilization produces _____ in the form of the tissue called _____.

3. In all angiosperms, a growing embryo is first surrounded by a _____, which in turn becomes surrounded by the tissue called _____. This tissue can be defined as the mature _____ of an angiosperm.

••• ANSWERS

1. spores; sporophyte; gametophyte; pollen grains; embryo sac

2. zygote; sporophyte; food; endosperm

3. seed coat; fruit; ovary

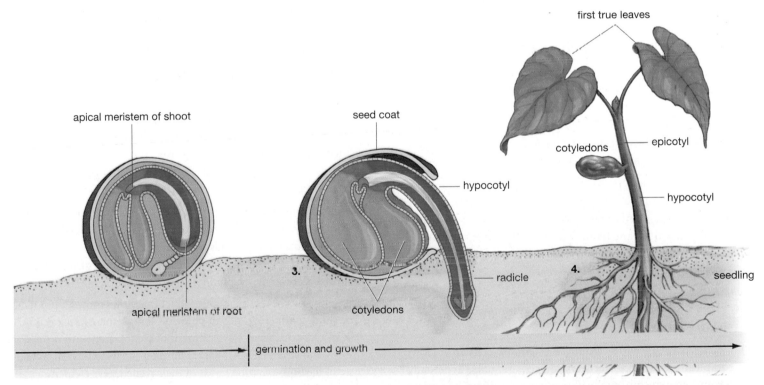

3. Having made its way to the ground, the seed is ready to "germinate" or start sprouting. The seed coat splits, allowing the emergence of the hypocotyl (tissue below the cotyledons), which includes the plant's radicle, or early root structure.

4. Root and shoot have both sprouted. The plant's cotyledons, having completed their function, will wither away. The plant sprouts its first true leaves and begins the process of photosynthesis. The reproductive cycle is ready to begin again.

# CHAPTER 25 REVIEW

For study help, activities, and more quiz questions, go to **www.aw.com/krogh4**.

## SUMMARY

### 25.1 Two Ways of Categorizing Flowering Plants

- Plants can be categorized by how long it takes them to go through a cycle that runs from germination to death. Those that go through this cycle in a year or less are annuals; those that go through it in about 2 years are biennials; those that live for many years are perennials.

- A cotyledon is an embryonic leaf, present in the seed. Angiosperms are classified according to how many cotyledons they have—one in the case of the narrow-leafed monocotyledons, two in the case of the broad-leafed dicotyledons. Monocots and dicots differ in structure in many ways.

### 25.2 There Are Three Fundamental Types of Plant Cells

- There are three fundamental types of cells in plants that, alone or in combination, make up most of the plant's tissues, meaning groups of cells that carry out a common function. These three cell types are parenchyma, sclerenchyma, and collenchyma.

### 25.3 The Plant Body and Its Tissue Types

- Some plants are capable only of primary growth, meaning growth at the tips of their roots and shoots that primarily increases their length. Plants that exhibit only primary growth are herbaceous plants, composed solely of primary tissue. Other plants exhibit both vertical growth and lateral, or secondary, growth. These are the woody plants, composed of primary and secondary tissue.

- There are four tissue types in the primary plant body: dermal, vascular, meristematic, and ground. Dermal tissue can be thought of as the plant's outer covering, vascular tissue as its "plumbing," meristematic tissue as its growth tissue, and ground tissue as almost everything else in the plant.

Web Animation 25.1: Plant Tissue and Growth

### 25.4 How a Plant Grows: Apical Meristems Give Rise to the Entire Plant

- The entire plant develops from meristematic cells in regions called apical meristems. Meristematic cells remain perpetually embryonic, able to continually give rise to cells that differentiate into all the plant's tissue types.

- Shoot apical meristems give rise to the entire shoot of the plant. In addition to providing for vertical growth, shoot apical meristems produce meristematic tissue called lateral buds at the base of leaves that can give rise to a branch or flower. Root apical meristems are located just behind a collection of cells at the very tip of the root, called the root cap.

- The plant's tissue types develop in stages from meristematic cells. This development takes place in a series of regions adjacent to the apical meristem. In a gradual transition, the apical meristem gives way to a zone of cell division, followed by a zone of elongation (in which developing cells lengthen), then followed by a zone of differentiation (in which cells fully differentiate into different tissue types).

### 25.5 Secondary Growth Comes from a Thickening of Two Types of Tissue

- Secondary growth in plants takes place through the division of cells in two varieties of meristematic tissue that develop only in woody plants: the vascular cambium, which continually produces secondary phloem and secondary xylem tissue layers to either side of itself; and the cork cambium, which gives rise to the outer tissues of woody plants. Secondary xylem, also known as wood, is responsible for most of a tree's widening.

- Looking at a tree from the secondary phloem outward to the tree's periphery, four tissues constitute the tree's bark: secondary phloem, phelloderm, cork cambium, and cork. The cork cells are dead in their mature state and provide layers of protection for the tree.

### 25.6 How the Plant's Vascular System Functions

- Two types of cells make up the water-conducting portions of xylem tissue. These are tracheids and vessel elements, both of which are dead in their mature, working state. Vessel elements, which exist almost solely in angiosperms, conduct more water than tracheids. Their existence in angiosperms is one of the reasons for the angiosperms' dominance in the plant world.

- Water movement through xylem is driven by transpiration, meaning the loss of water from a plant, mostly through the leaves. As water evaporates into the air, it pulls a continuous column of water upward through the plant. The energy for this process comes from the sun, whose rays power the evaporation of water at the leaf surface.

- The sugar sucrose is the main product that flows through phloem. The fluid-conducting cells in phloem, sieve elements, lack cell nuclei in maturity. Each sieve element has associated with it one or more companion cells, which retain their nuclei and seem to take care of the housekeeping needs of their related sieve elements.

- Plants expend their own energy to load the sucrose they produce into the phloem's sieve element cells. Once this takes place, there is a greater concentration of solutes inside the cells than outside them—a condition that brings about a flow of water into the cells through osmosis. The pressure that results from the increased water inside the cells is sufficient to move the solution of water and dissolved sucrose through the sieve element cells, from "source" (the cells into which the sugar was loaded) to "sink" (the cells in which the sugar is stored or used). The name given to this hypothesis about phloem fluid movement is the pressure-flow model.

Web Animation 25.2: The Vascular System for Plants

**BioFlix** Water Transport in Plants

## 25.7 Sexual Reproduction in Flowering Plants

- All plants, including angiosperms, reproduce through an alternation of generations. A sporophyte generation (the familiar tree or flower) produces haploid spores that develop into their own generation of plant, the gametophyte generation. In angiosperms, the male gametophyte is the pollen grain, consisting in maturity of an outer coat, two sperm cells, and one tube cell. The female angiosperm gametophyte consists at maturity of an embryo sac composed of seven cells, one of which is the egg. The female gametophyte is housed inside a structure of the parent sporophyte plant called an ovule.

- Fertilization of the egg by sperm requires that a pollen grain land on the stigma of a plant. The tube cell of the pollen grain then germinates, sprouting a pollen tube that grows down through the sporophyte plant's stigma and style, eventually reaching the female reproductive cells inside the ovule. The sperm cells inside the pollen grain travel down through the pollen tube, and one of the sperm cells fertilizes the egg in the ovule, producing a zygote. With this, the new sporophyte generation of plant has come into being.

- The second sperm cell in the pollen grain enters the central cell in the embryo sac, setting in motion the development of food for the embryo, endosperm tissue. This second fertilization completes the process of double fertilization—a fusion of gametes on the one hand and of cells producing nutritive tissue on the other.

## 25.8 Embryo, Seed, and Fruit: The Developing Plant

- With fertilization, the ovule integuments that surrounded the embryo sac begin to develop into the seed coat that will surround the growing sporophyte embryo. The ovary that surrounded the ovule then starts to develop into a layer of tissue that will surround the seed—fruit, which is defined as the mature ovary of a flowering plant. Under this definition the pod of a pea plant is fruit, as is the flesh of an apricot.

- The seed with its fruit covering eventually separates from the sporophyte parent plant and then germinates in the Earth.

Web Animation 25.3: Plant Reproduction

## KEY TERMS

angiosperm 523
annual 500
apical meristem 506
bark 512
biennial 500
bud 506
cellulose 502
collenchyma cell 502
companion cell 515
cork 511
cork cambium 511
cotyledon 524
dermal tissue 503
dicotyledon (dicot) 501
double fertilization 522
embryo sac 521
epicotyl 524
fruit 523
herbaceous plant 503
hypocotyl 524
intercalary meristem 509
lateral bud 506
megaspore mother cell 519
microspore mother cell 519
monocotyledon (monocot) 501
parenchyma cell 502
perennial 500
phloem 504
pollination 521
pressure-flow model 516
primary growth 503
root cap 508
sclerenchyma cell 502
secondary growth 503
secondary phloem 510
secondary xylem 510
seed coat 522
shoot apical meristem 506
sieve element 515
sieve tube 515
spore 518
terminal bud 507
tissue 502
tracheid 513
transpiration 514
vascular cambium 510
vessel element 513
wood 511
woody plant 503
xylem 504
xylem sap 514
zone of cell division 508
zone of differentiation 508
zone of elongation 508

## BRIEF REVIEW

*Answers to Brief Review questions are in the back of the book.*
*For multiple-choice quiz questions, go to* **www.aw.com/krogh4**.

1. Compare monocots and dicots with respect to their anatomy, morphology, and seed structure.

2. What are the three main cell types in a plant, and what distinct functions are performed by each?

3. What is meristematic tissue? Describe its organization.

4. Compare the structure and function of the water-conducting (xylem) and food-conducting (phloem) cells in a plant.

5. What does a growth ring on a tree trunk represent? Why can you tell the age of a tree by counting its growth rings?

6. Describe the process of double fertilization in plants.

## APPLYING YOUR KNOWLEDGE

1. If you were to regularly shear a lilac bush down to a stub, you would probably kill it. Yet, mowing after mowing, the grass in your lawn continues to grow. Is there a fundamental difference in the location of growing points in the lilac and the grass? Explain.

2. Beavers often kill young trees by girdling them—that is, by removing a strip of bark from the circumference of a tree trunk. Why does this bark removal kill a tree?

3. Gardeners often speak of outdoor plants having become "established," meaning the plants have reached a point at which they need less human care. What threshold does a plant have to cross to become established?

How is the human body structured and how does it work? The disciplines of human anatomy and physiology are dedicated to finding out.

| 26.1 | The Disciplines of Anatomy and Physiology | 529 |
| 26.2 | How Does the Body Regulate Itself? | 530 |
| 26.3 | Levels of Physical Organization | 531 |
| 26.4 | The Human Body Has Four Basic Tissue Types | 532 |
| 26.5 | Organs Are Made of Several Kinds of Tissues | 534 |
| 26.6 | Organs and Tissues Make Up Organ Systems | 534 |
| 26.7 | The Integumentary System: Skin and Its Accessories | 537 |
| 26.8 | The Skeletal System | 541 |
| 26.9 | The Muscular System *BioFlix* | 546 |

**Essays**

| Why Fat Matters and Where It Matters Most | 540 |
| There Is No Such Thing as a Fabulous Tan | 542 |

Just as several individuals may make up an office, and several offices a department, so our bodies are organized into a number of working units that go on to make up a functioning human being. Three of the larger units involve skin, bone, and muscle.

# INTRODUCTION TO HUMAN ANATOMY AND PHYSIOLOGY: The Integumentary, Skeletal, and Muscular Systems

**T**here is an old country saying that goes, "You don't miss your water until your well runs dry." Doesn't this seem applicable to the way we think of our bodies? We go along taking for granted our ability to eat or talk or run—until the day comes when we can't do one of these things. If you've ever, say, broken a leg, and ended up with a cast and crutches, you know that just walking around with ease can look like the most wonderful thing in the world. Broken bones heal, of course, and that reassures us—until we run up against a condition that won't heal. While still in his 20s, the actor Michael J. Fox woke up one morning and noticed, as he later remembered, that "the pinky in my left hand was twitching. It was a curiosity as much as anything because I couldn't stop it, no matter how I tried to." About a year later, after many doctor's visits, Fox got definitive word that he had Parkinson's disease, which is incurable. By then, it was not just his little finger that was twitching; it was both hands. Imagine if being able to hold your hands steady seemed as beautiful as the moon and just as far away.

We take our bodies for granted partly because they work so well most of the time, but partly because what goes on inside them is hidden from us and so complicated that our physical functioning just seems mysterious. The next seven chapters of this book are aimed at clearing up some of this mystery. At their conclusion, you will have a basic understanding of how your heart beats, for example, of how your immune system fights off invaders, and of how your eyes and brain provide you with a picture of the world.

If you think about just the first of these things—the workings of the heart—you can get a sense of what's in store. The human heart beats about 100,000 times *each day*, pumping about 8,800 quarts of blood through the body in the process. Moreover, it keeps this rhythmic contraction up without a second's rest from the third week of our embryonic life to the day we die. As if this weren't enough, it is carrying out this feat mostly with its original equipment. Very few of the muscle cells in the heart ever divide; as a result, the cells that are working for us at age 8 are pretty much the cells that are working for us at age 80.

An engineer who could invent such a machine would be considered a genius. But the truly remarkable thing is that there are lots of organs and systems in the body that are just as impressive as the heart. Once you've learned how these things work, it's hard to take your body for granted in the same way because you've come to understand what an incredible machine the human body is.

## 26.1 The Disciplines of Anatomy and Physiology

When we study the makeup or functioning of any complex organism, we have entered the realms of two overlapping disciplines in biology: anatomy and physiology. All organisms have an anatomy and a physiology. But in these chapters, our primary concern is with human beings. Within this framework, human anatomy is the study of the body's structure and the relationships between the body's constituent parts. Human physiology is the study of how these parts work. Thus, anatomy tells us where the heart is, for example, how it is structured, and how it is situated in relation to other organs. Physiology describes to us *how* the heart works.

## 26.2 How Does the Body Regulate Itself?

Before we get to specifics about the human body, it will be helpful to go over a couple of general concepts about its structure and functioning. One of these concepts concerns the question: Who's in charge here? We are in charge of our conscious actions, of course; human beings get to decide what they are going to do from one moment to the next. But what about the huge number of unconscious actions that the body undertakes each second—getting oxygen to cells, getting rid of waste, being prompted to breathe while asleep? You might think the brain would be in charge of all this, and indeed messages from the brain do prompt a lot of physical activity. But there are plenty of physical processes that don't involve brain signals, and the brain *gets* lots of messages that say, in effect, "Get this process going now." Human society is hierarchical to a great degree; we're

used to bosses and employees who work in a straight line of authority. But the body doesn't work this way. Instead, it functions through a process of *circular causation* in which A prompts B and B prompts C, but in which C may then prompt A. It is a system of self-regulation in which all parts work together to influence each other.

In understanding self-regulation, it's useful to think about what kind of internal environment the body needs to function properly. The answer is a *stable* environment. The human body must guard against being too hot or too cold; too dehydrated or too hydrated; too stimulated or too relaxed. If it is to exist at all, it must avoid extremes. To put this another way, it must seek **homeostasis**, meaning the maintenance of a relatively stable internal environment. And it turns out that a single process is almost solely responsible for bringing about homeostasis. Let's look at this process by way of an analogy.

Most people are aware of how a home heating system works. Falling temperature causes a thermostat to turn on a furnace. To look at this another way, there is a stimulus (cold air) that brings about a response (furnace operation). The *product* of this response is hot air. When enough hot air circulates to raise the temperature, the thermostat senses this and shuts the furnace down.

Now, let's think of your body, which has, as an example, a certain amount of calcium circulating through it in the bloodstream. When levels of calcium in the blood fall too low (stimulus), organs called the parathyroid glands sense this and secrete a hormone that causes your bones to release calcium (response). Thus, the product of *this* stimulus-response chain is released calcium; when enough of it is circulating, the parathyroid glands sense this and stop releasing their calcium-liberating hormone (**Figure 26.1**).

With both the thermostat and the parathyroid glands, we can see that their responses to a stimulus bring about a decrease in their own activity. In other words, their responses feed back on their activity in a negative way—they reduce it. This is **negative feedback**, which can be defined as a process in which the elements that bring about a response have their activity reduced by that response. The calcium example is only one of the countless negative-feedback loops that exist in the body. In it, we can see not only how the body maintains homeostasis, but how the body is self-regulating. There was no central boss giving commands in this process. Instead two elements (calcium levels and the parathyroid glands)

**Stimulus:** *low* calcium levels in blood

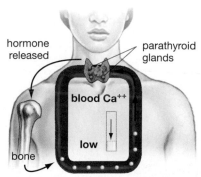

hormone released

parathyroid glands

blood Ca⁺⁺

low

bone

**Response:** release of calcium into blood

**Stimulus:** *high* calcium levels in blood

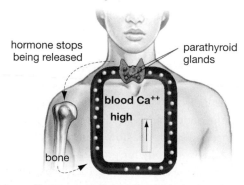

hormone stops being released

parathyroid glands

blood Ca⁺⁺ high

bone

**Response:** release of calcium is shut down

**Figure 26.1**
**Calcium Levels: An Example of Negative Feedback**
Calcium levels in the bloodstream are regulated in part by the parathyroid glands. These glands can sense when calcium levels are too low (the stimulus in the drawing at left). In response, the parathyroids secrete a hormone that causes the bones to release calcium. This brings about high calcium levels, which the glands also can sense (the stimulus in the drawing at right). High calcium levels cause the parathyroids to halt their production of the calcium-liberating hormone.

influenced each other to bring about stability. This is the way the body generally works.

## Large-Scale Features of the Body

Many of the large-scale features of a human being—four limbs, one trunk, and so forth—are obvious, but a couple of features are hiding in plain sight, we might say. The first of these is an internal body cavity. All animals more complex than a flatworm have such a cavity. Humans and other mammals actually have two primary body cavities. The first of these is a dorsal (back) cavity that is further divided into cranial and spinal cavities, into which our brain and spinal cord fit (**Figure 26.2**). The second is a ventral (front) cavity that includes a thoracic or "chest" portion containing the lungs and heart as well as a lower abdominopelvic cavity containing such organs as the stomach, liver, and intestines. A muscular sheet called the diaphragm separates the thoracic and abdominopelvic cavities. Cavities provide flexibility and protection. With the space provided by a cavity, an organ like the stomach can expand as we eat and is protected from external blows to some extent.

Humans also have an internal skeleton that supports the body. We take for granted a skeleton that exists inside our skin, but it stands in contrast to what most animals have, which is an *external* skeleton (as with crabs and grasshoppers) or no skeleton at all (as with jellyfish).

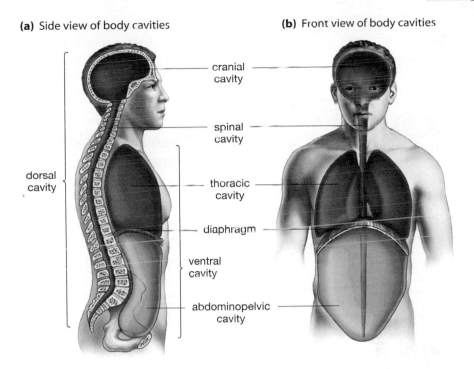

**(a)** Side view of body cavities   **(b)** Front view of body cavities

cranial cavity

spinal cavity

dorsal cavity

thoracic cavity

diaphragm

ventral cavity

abdominopelvic cavity

**Figure 26.2**
**Body Cavities**

**(a)** A side or "lateral" view of the two main cavities in the human body—the dorsal (back) and ventral (front) cavities—and the smaller cavities within them. The dorsal body cavity is bounded by the bones of the skull and vertebral column and includes both the cranial and spinal cavities. The ventral cavity includes a thoracic (chest) cavity and an abdominopelvic cavity, which is separated from the thoracic cavity by a muscular diaphragm.

**(b)** A front or "ventral" view of the body's cavities.

## 26.3 Levels of Physical Organization

Features of the human body, such as a skeleton, fit into a larger organizational framework that is common to all living things. In our everyday world, several individuals may make up an office, several offices a department, and several departments a working company. In a similar fashion, each living thing is built up from a series of individual units that go on to make up a working organism. Here's how these units are organized if we consider just the animals among life's creatures (one of which is us). Recall from Chapter 4 that the basic unit of all life is the cell. Cells and cellular products that work together to perform a common function are known as **tissues**. There is, for example, muscle tissue—a group of cells that perform the common function of contracting. Tissues in turn are arranged into various types of **organs**, meaning complexes of several kinds of tissues that perform a special bodily function. The stomach is an organ that has both muscle and nerve tissue within it. Organs then go on to form **organ systems**, which are groups of interrelated organs and tissues that serve a particular function. Contractions of the heart push blood into a network of blood vessels. The heart, blood, and

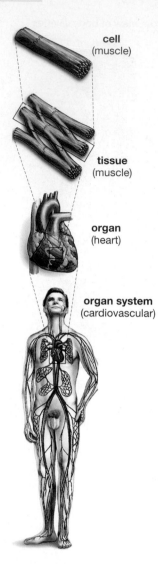

cell
(muscle)

tissue
(muscle)

organ
(heart)

organ system
(cardiovascular)

**Figure 26.3**
**Levels of Organization in the Human Body**

A group of cells that performs a common function is a tissue; here, muscle cells have formed heart muscle tissue (whose common function is heart muscle contraction). Two or more tissues combine to form an organ such as the heart. (The heart contains both muscle and connective tissue, for example.) The heart is one component of the cardiovascular system; other parts of the system are blood and blood vessels. All the organ systems combine to create an organism, in this case a human being.

blood vessels form the *cardiovascular system*, which is one of 11 systems in the human body. **Figure 26.3** shows the various levels of structural organization within the human body, using the cardiovascular system as an example.

## 26.4 The Human Body Has Four Basic Tissue Types

Looking at the smallest level of organization we're interested in, the tissue, we can then ask: What kinds of tissues are there in humans and other animals? It turns out there are only four fundamental tissue types: epithelial, connective, muscle, and nervous. All other tissue varieties are subsets of these fundamental types. Let's look briefly at the characteristics of these four varieties.

### Epithelial Tissue

The key word regarding epithelial tissue is *surfaces* because **epithelial tissue** is tissue that *covers* surfaces exposed to an external environment. The outside of our body is exposed to the external environment, and as such, the outer layer of our skin is composed of epithelial tissue. But there are other surfaces as well. Think of a pipe; it has not only an external surface but also an internal surface that comes into contact with something that is external to it—the fluid that runs through it. Similarly, your body has arteries and veins that have the fluid called blood running through them, and the tissues that line the internal blood vessel surfaces are epithelial tissues. Likewise, the lining of your stomach and the lining of your lungs are surfaces that come into contact with something from outside you; again, this contact is made by epithelial tissues. Given their location, epithelial tissues often serve as a barrier, as in the case of our skin; but many epithelial tissues are "transport" tissues that materials are moved across. (The food that enters your bloodstream does so by being transported across the epithelial tissue that lines your small intestines.) Epithelial tissues can be several layers of cells thick or as thin as a single layer, depending on their function (**Figure 26.4**).

Epithelial tissues often produce substances that aid in various bodily processes. The skin has epithelial tissue that produces the water-resistant protein keratin, for example, while the stomach has epithelial tissue that produces digestive juices. This substance-producing function can be carried out by isolated epithelial cells, but it is sometimes undertaken by concentrations of cells. Such concentrations are known as **glands**: organs or groups of cells that are specialized to secrete one or more substances. Later, we'll look at two basic kinds of glands.

### Connective Tissue

**Connective tissue**, defined as tissue that stabilizes and supports other tissue, is very different from epithelial tissue. Whereas epithelial tissue is always in contact with an external environment, connective tissue never is. Whereas epithelial tissue is almost completely composed of cells, connective tissue usually is composed of cells that are separated from each other by an extracellular material—a material that is secreted by the connective tissue cells themselves. Indeed, the prime function of many connective tissues is to produce this kind of material.

Bone and cartilage are clear examples of connective tissue. They are connecting something, after all (bodily structures), and their support and stabilization role is clear. In addition, they lie within an extracellular material, which is properly referred to as a *ground substance*. The ground substance of bone, which the bone cells themselves secrete, is a mix of closely packed protein fibers and calcified material. The hard bones we're familiar with are largely made up of these substances.

But ground substances need not be hard. Adipose or fat tissue is connective tissue, and its ground substance is soft. Indeed, a ground substance can even be fluid, as is the case with blood and a bodily fluid called lymph. How can a ground tissue be a fluid? Well, to take blood as an example, the ground substance that surrounds blood cells is mostly water, with electrolytes, proteins, and other materials in the mix.

### Muscle Tissue

**Muscle tissue** is tissue that is specialized in its ability to contract, or shorten. There are three kinds of muscle tissue—skeletal, smooth, and cardiac—that differ in their structure and in the way they are prompted to contract. **Skeletal muscle** is the ordinary muscle that is attached to bone and is contained in, for example, our biceps. It is under our conscious control and has a striped or "striated" appearance when looked at under a microscope owing to the parallel orientation of some long, fibrous units that make it up. **Cardiac muscle** exists only in the heart and likewise is striated, but it contracts under the influence of its own pacemaker cells. **Smooth muscle**, which gets its name

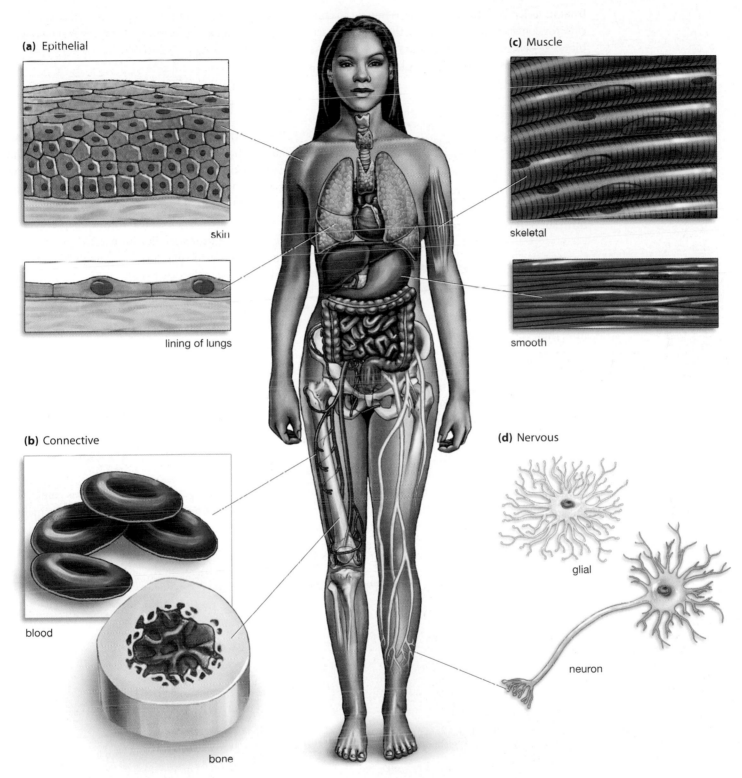

**(a)** Epithelial
skin
lining of lungs

**(b)** Connective
blood
bone

**(c)** Muscle
skeletal
smooth

**(d)** Nervous
glial
neuron

**Figure 26.4**
**Four Types of Tissue**

There are four basic tissue types in the human body: epithelial, muscle, nervous, and connective. Pictured are examples of each of the four types in some representative locations.

**(a)** Epithelial tissue, which always covers a surface exposed to an external environment, can take the form of many layers of cells (as with the skin) or a single layer of cells (as with the lining of lungs).

**(b)** Connective tissue, which supports and stabilizes other tissue, can take such diverse forms as bone and blood.

**(c)** Muscle tissue, which is specialized for contraction, comes in a striped or "striated" form (in both skeletal and cardiac muscle) and in a smooth form that lines the digestive tract, blood vessels, and other structures.

**(d)** Nervous tissue is specialized for electrical signal transmission. The two types of nervous tissue cells are neurons, which conduct nervous system signals, and glial cells, which support the neurons.

from its lack of striated appearance, is muscle responsible for contractions of the uterus, the digestive tract, blood vessels, and the passageways of the lungs. As you might guess from this list of tasks, smooth muscle is not under voluntary control.

### Nervous Tissue

**Nervous tissue** is tissue that is specialized for the rapid conduction of messages, which take the form of electrical impulses. There are two basic types of nervous tissue cells: neurons, which actually carry the nervous system messages; and glial cells, which perform support functions for the neurons.

## 26.5 Organs Are Made of Several Kinds of Tissues

Having looked at the four tissue types, let's now see how they combine to form an organ. **Figure 26.5** shows you the anatomy of the human stomach, with an emphasis on the kinds of tissues that go into making it up. The muscle tissue noted in the figure is one of several layers of smooth muscle that bring about contractions that help digest food and move it into the small intestines. Likewise, there is nervous tissue, which controls the release of digestive juices and the contractions of the stomach's muscles. Supportive connective tissue exists as well. And, of course, since the inner lining of the stomach is a surface coming into contact with an external environment, there is epithelial tissue. True to what

you saw earlier, some of this epithelial tissue is secretory—that is, it secretes substances that help with digestion.

## 26.6 Organs and Tissues Make Up Organ Systems

You've now seen how cells make up tissues and how tissues make up organs. Let's take the final step and see how tissues and organs together make up organ systems. What follows is a brief look at each of the body's 11 organ systems. Once you have finished reviewing them, the balance of this chapter will be devoted to a more detailed look at the first three of these systems. In Chapters 27 through 32, you'll be taking a look at the other eight.

### Organ Systems 1: Body Support and Movement—The Integumentary, Skeletal, and Muscular Systems

- The **integumentary system** (**Figure 26.6a**) is sometimes thought of simply as skin, but a number of structures are associated with skin, including hair and nails and the glands we noted earlier. In addition to serving as an outer, protective wrapping, skin also helps keep body temperature stable.

- The **skeletal system** (**Figure 26.6b**) is composed not only of bones, but of cartilages and ligaments. This is a complex framework, as human beings have 206 bones in their body, along with a large

**Figure 26.5
Organs Are Composed of Tissues**

The stomach is a typical organ in that it is composed of various tissue types. Nervous tissue controls the release of digestive juices into the stomach and also controls the contractions of a second type of tissue found in the stomach, its muscle tissue. Connective tissue helps support and stabilize the other tissues around it. Epithelial tissue lines the stomach's surface, which comes into contact with the food we eat. The individual epithelial cells pictured secrete digestive acids and enzymes.

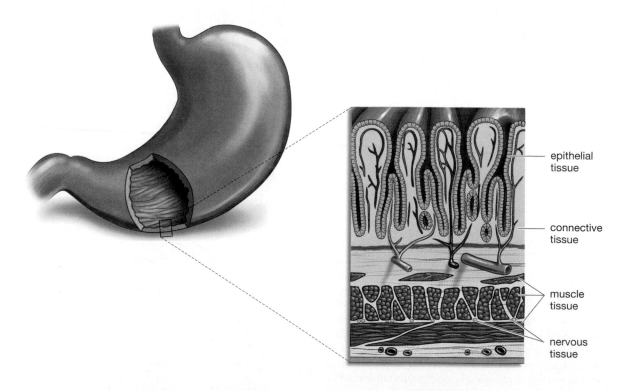

epithelial tissue

connective tissue

muscle tissue

nervous tissue

number of cartilages. Bones provide support and protection, and they can store fat and minerals.

- The **muscular system** (**Figure 26.6c**) includes all the skeletal muscles of the body—about 600 of them. These muscles provide for movement, posture, and support. They are also vital players in homeostasis in that our bodies are largely warmed by the heat given off from muscle contraction. The smooth and cardiac muscle tissues noted earlier are not part of the muscular system. Cardiac muscle is part of the cardiovascular system, and smooth muscle is part of several systems—in the stomach, for example, it is part of the digestive system.

## Organ Systems 2: Coordination, Regulation, and Defense—The Nervous, Endocrine, and Immune Systems

- The nervous tissue reviewed earlier goes on to make up the **nervous system** (**Figure 26.6d**), a rapid communication system in the body that includes the brain, spinal cord, all the nerves outside the brain and spinal cord, and sense organs such as the eye and ear.

- The endocrine system (**Figure 26.6e**) bears comparison with the nervous system in that both systems are in the business of sending signals throughout the body. The **endocrine system**, however, is a communication system that works more slowly than the nervous system through chemical messengers called **hormones**: substances that, when released in one part of an organism go on to prompt physiological activity in another part of the organism. Many human hormones are produced in specialized glands (such as the parathyroid gland we looked at). Remember the earlier remark that there are two basic varieties of glands in the body? One of these varieties is named for the endocrine system. **Endocrine glands** are glands that release their materials directly into surrounding extracellular fluid or into the bloodstream. This is what the thyroid gland does, for example. Endocrine glands stand in contrast to **exocrine glands**, which are glands that secrete their materials through tubes or "ducts." We'll see examples of these glands later.

- The **immune system** consists of a collection of cells and proteins whose central function is to rid the body of invading microbes. In **Figure 26.6f**, you can see a key component of the immune system, the body's **lymphatic network,** which serves a dual function.

**Figure 26.6a–f**
**The Organ Systems of the Human Body**

**(a)** The integumentary system

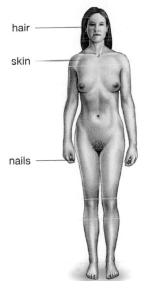

hair
skin
nails

**(b)** The skeletal system

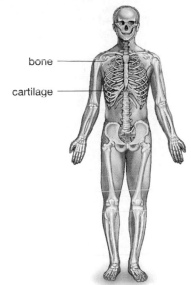

bone
cartilage

**(c)** The muscular system

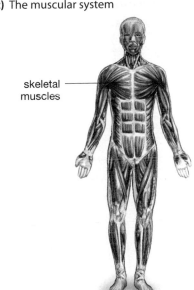

skeletal muscles

**(d)** The nervous system

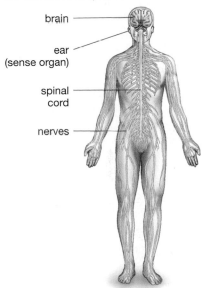

brain
ear (sense organ)
spinal cord
nerves

**(e)** The endocrine system

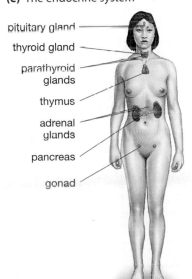

pituitary gland
thyroid gland
parathyroid glands
thymus
adrenal glands
pancreas
gonad

**(f)** The immune system (lymphatic network)

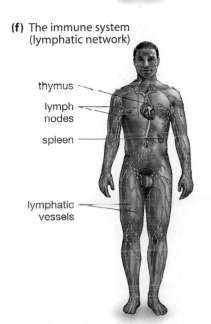

thymus
lymph nodes
spleen
lymphatic vessels

**(g)** The cardiovascular system

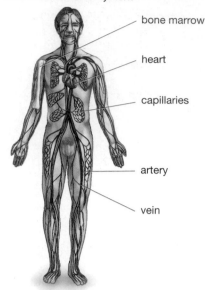

- bone marrow
- heart
- capillaries
- artery
- vein

**(h)** The respiratory system

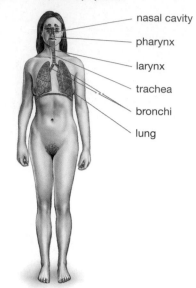

- nasal cavity
- pharynx
- larynx
- trachea
- bronchi
- lung

**(i)** The digestive system

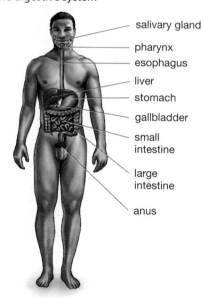

- salivary gland
- pharynx
- esophagus
- liver
- stomach
- gallbladder
- small intestine
- large intestine
- anus

**(j)** The urinary system

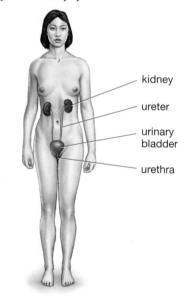

- kidney
- ureter
- urinary bladder
- urethra

**(k)** The male reproductive system

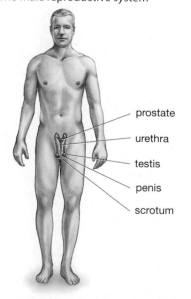

- prostate
- urethra
- testis
- penis
- scrotum

**(l)** The female reproductive system

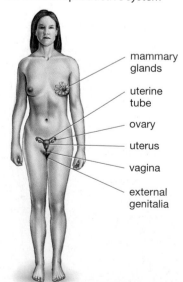

- mammary glands
- uterine tube
- ovary
- uterus
- vagina
- external genitalia

First, it acts as a kind of drainage system, collecting fluid that lies outside cells and routing it back to the bloodstream. Second, it acts as an *inspection* system. Lymphatic organs contain heavy concentrations of the body's defenders—its immune system cells—and these cells monitor lymphatic fluid as it passes by for signs of invading microbes. Lymphoid organs include the well-known lymph nodes located under the arms and elsewhere, along with the thymus gland, tonsils, and spleen. Immune system cells and proteins don't reside solely within the lymphatic network, however, but exist throughout the body.

## Organ Systems 3: Transport and Exchange with the Environment—The Cardiovascular, Respiratory, Digestive, and Urinary Systems

- The **cardiovascular system** (**Figure 26.6g**) consists of the heart, blood, and blood vessels and an inner, "marrow" portion of bones where red blood cells are formed. The cardiovascular system is the body's mass transit system. It carries nutrients, dissolved gases, and hormones *to* tissues throughout the body, and it carries waste products *from* tissues to the sites where they are removed: the kidneys and the lungs.

- The **respiratory system** (**Figure 26.6h**) includes the lungs and the passageways that carry air to and from the lungs. Through this system, oxygen comes into the body, and carbon dioxide is expelled from it.

- The central feature of the **digestive system** (**Figure 26.6i**) is the digestive tract, a long tube that begins at the mouth and ends at the anus. Along its course, there are the digestive organs of the system—the stomach, small intestines, and large intestines—and accessory glands, such as the pancreas and liver.

- The **urinary system** (**Figure 26.6j**) has two major functions, one of them well known, the other not so much appreciated. The well-known function is the elimination of waste products from the blood through the production of urine. The underappreciated function is conservation. The system is very good at retaining what is useful to the body—water, proteins, and so forth—even as it eliminates what is useless. The primary organs that carry this selection process out are the kidneys, but the system also includes various other parts, such as the bladder.

- Males and females have different **reproductive systems** (**Figures 26.6k** and **26.6l**). The two are linked, of course, in that together they produce offspring. Both systems are discussed in detail in Chapter 32.

**Figure 26.6g-l**

With this brief review of the organ systems completed, you're ready to look in more detail at the first three organ systems noted earlier—the integumentary, skeletal, and muscular systems. They are batched together in this chapter because each has to do with body support and movement. We'll start with the integumentary system.

## 26.7 The Integumentary System: Skin and Its Accessories

The primary component of the integumentary system is the organ known as skin. It may seem strange to think of skin as an organ since it is not a compact, clearly defined structure like the heart or the liver. But remember the definition of an organ: a complex of several kinds of tissues that performs a special bodily function. Clearly, that's what skin is. Joining skin to make up the integumentary system are the related structures of hair, nails, and a variety of exocrine glands.

The primary function of skin is easy to see: It covers the body and seals it off from the outside world. Skin also protects underlying tissues and organs, however, and helps regulate our temperature. In addition, it stores fat and makes vitamin D. Meanwhile, exocrine glands that are accessories to the skin excrete materials such as sweat, water, oils, and milk, and specialized nerve endings in the skin detect such sensations as pressure, pain, and temperature.

### The Structure of Skin    • • •

**Skin** is an organ that is organized in two parts: a thin outer covering, the epidermis, and a thicker underlying layer, the dermis, which is composed mostly of connective tissue. Beneath the dermis—and not part of the skin proper—is another layer, called the hypodermis, made up of connective tissue that attaches the skin to deeper structures such as muscles or bones. You can see a cross section of human skin in **Figure 26.7**.

### The Outermost Layer of Skin, the Epidermis

The outer layer of our skin, the epithelial tissue called the **epidermis**, would seem to be facing an impossible task. On the one hand, it has to serve as a permanent, protective barrier, keeping out everything from water to invading microorganisms; on the other it has to

**Figure 26.7**
**Our Outer Wrap: What Is Skin Made Of?**

**(a)** A micrograph of human skin, showing the epidermal and dermal layers. Note how the cells of the epidermis flatten out about halfway toward the top.

**(b)** A cross section of the skin's epidermis and dermis layers and the hypodermis layer beneath them (which is not part of the skin).

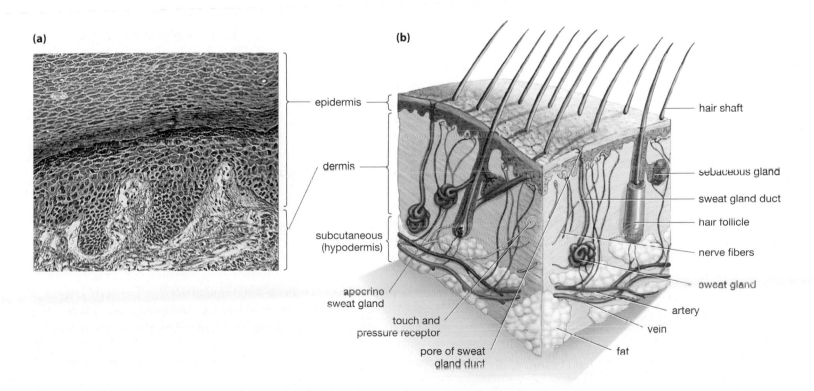

(a)

(b)

epidermis

dermis

subcutaneous (hypodermis)

apocrine sweat gland

touch and pressure receptor

pore of sweat gland duct

hair shaft

sebaceous gland

sweat gland duct

hair follicle

nerve fibers

sweat gland

artery

vein

fat

continually renew itself, given that it is cut, scraped, and simply worn away each day. How does it manage to do both things? The answer lies in the many *layers* of cells the epidermis is organized into. If we could take a microelevator down to the innermost layer of the epidermis, we would see a group of rounded cells rapidly dividing and, in the process, pushing the cells on top of them *up*, toward the surface of the skin. As we move toward the surface, we notice that the epidermal cells are flattening out. Furthermore, they're getting less active; at a certain point, they cease dividing altogether. What is happening is that they are now so far from the underlying blood vessels (down in the dermis) that they are losing their blood supply. They are still carrying out a task, however, which is the production of **keratin**—a flexible, water-resistant protein, abundant in the outer layers of skin, that also makes up hair and fingernails. By the time epidermal cells reach the surface, they are scarcely cells at all anymore. They are dead and don't even have the remnants of internal organelles within them. Instead, they have become a series of tightly interlocked keratin sacs. In another 2 weeks, they will be scraped or washed away, replaced by the next cells that have pushed up from the inner epidermis.

### Beneath the Epidermis: The Dermis and Hypodermis

Whereas the epidermis is epithelial tissue, the inner layer of skin, the **dermis,** is mostly connective tissue. But if you look again at Figure 26.7, you can see that nerves and muscles exist in the dermis as well, along with sweat glands, hair follicles, and more. This is just another example of how all the various kinds of tissues tend to exist side by side in a typical organ.

Many of the structures in the dermis are serving the epidermis. For example, the nervous system touch and pressure receptors shown in Figure 26.7 allow us to know when something makes contact with our skin, and the blood vessels pictured provide a blood supply to the dividing lower cells of the epidermis. But the dermal blood supply takes on another role as well. It plays a big part in controlling our temperature. If you get too cold, the dermal blood vessels will constrict, thus moving blood away from your surface and retaining the heat that's in the blood. If you get too hot, the dermal blood vessels expand, thereby moving blood toward your surface so that the heat that's in blood can radiate to the outside world (making your skin feel "flushed"). One more thing to note about the dermis is that its cells are relatively fixed, unlike those of the outer epidermis. Ever wonder why a tattoo must be made with a needle

sunk into the skin? The tattoo's pigment must be deposited into the fixed cells in the dermis rather than the ever-changing cells of the epidermis.

As noted earlier, the "hypodermis," the layer beneath the dermis, is not actually part of the skin. But it deserves special mention here because of one of the aspects of it you can see in Figure 26.7—its abundant adipose or fat tissue. This tissue makes up the body's *subcutaneous fat*, which can be thought of as a layer of fat located just beneath the skin that varies in thickness throughout the body. The body also has what is known as *visceral fat*, however, which is fat that lies deep within the abdomen (inside the abdominopelvic cavity we talked about). This "belly fat" doesn't exist as a single layer but instead surrounds organs such as the liver. When it exists in excess amounts, it is regarded as a more serious health risk than is an excess of subcutaneous fat. You can read more about this in "Why Fat Matters and Where It Matters Most," on page 540.

The term *hypodermis* may ring a bell because it sounds like the more familiar term hypoder*mic*, as in needles. The deeper tissue of the hypodermis doesn't have much of a blood supply and thus is a good place to inject medicines that need a relatively slow, steady entry into circulation. Here's one final thing about the skin. If you want to make it age fast, one sure way to do it is to spend a lot of time in the sun. You can read more about this in "There Is No Such Thing as a Fabulous Tan" on page 542.

### Accessory Structures of the Integumentary System

As noted, skin has a number of accessory structures associated with it, including hair, glands, and nails.

#### Hair

Five million or so hairs exist on the human body, although only about 100,000 of them are on the human head. Hair clearly serves a major cosmetic role in modern society, but why does it exist on human beings in the first place? You might think that heat retention for our ancestors would have been its main function, but maybe not. Rub your hand with just the lightest stroke over the hair on your head and you can get a sense of how exquisitely sensitive hair is to our sense of *touch*. Like a cat's whiskers, hair may have served primarily to fine-tune our sense of contact with the world around us.

How can hair have this quality since it is not alive but merely a shaft of the protein keratin, which is a product of underlying cells? If you look at Figure 26.7, you can see that each hair grows out of a tube-like structure called a *hair follicle* that begins fairly deep in the dermis. Each follicle is a tissue complex that has not only its own blood vessels and muscles, but its own nerve endings as well. Thus, when you touch a hair, you stimulate a hair follicle's nerve endings.

True to what we noted at the start of the chapter, the body's hair is perhaps more missed when it is gone than appreciated when it is here. Many men start losing the hair on their heads in their early 20s, a trend that accelerates as they get older. Women can lose hair, too, although their loss tends to come later in life and usually involves a general thinning of the hair rather than receding hairlines or bald spots. What causes hair loss?

Follicles go through growth and resting phases, with a given hair on our heads growing for anywhere from 2 to 6 years until a short resting phase is reached. At that point, the hair drops out, and its follicle's growth phase starts again. Hair loss generally is a condition in which an ever-growing proportion of follicles start to shrink, resulting in a briefer growing phase and hair that comes out either short and fine (like a baby's) or doesn't come out at all. Genes clearly play a part in this process, although contrary to popular belief, it is not just genes from the mother's side of the family that are involved. Hormones called *androgens* are involved as well; men have higher circulating levels of such androgens as testosterone and thus experience more hair loss.

Both men and women are, of course, subject to a *graying* of their hair with increasing age. What causes hair to lose its color? About midway down through the dermal portion of the hair follicle you can see in Figure 26.7b there are cells called melanocyte stem cells. When we are young, these cells produce a steady stream of cells that first migrate to near the bottom of the hair follicle and then differentiate into cells called mature melanocytes. These are the cells that produce the pigments that give hair its color. We thus have a kind of pigment-producing assembly line in our hair follicles, but it's one that aging tends to disrupt in two ways. One is that our melanocyte stem cells can die off altogether; the other is that our mature melanocytes can start maturing too quickly—they differentiate *before* they have migrated to the deepest part of the follicle, which is the only location in which they can function properly. So, does this knowledge leave us with any hope of stopping the graying process? Perhaps. The scientists who revealed this color-loss mechanism have identified two genes that appear to be vital in maintaining a healthy stem cell population.

### Glands

You may remember that when we went over the endocrine system, you saw that glands come in two varieties: those that secrete their substances directly into the bloodstream or surrounding tissue (endocrine glands) and those that secrete their substances through the tubes called ducts (exocrine glands). Skin contains two types of exocrine glands. The first of these are known as **sebaceous glands**: glands that produce an oily substance called sebum that is secreted into hair follicles and that then moves through them onto the skin (see Figure 26.7b). Once on the skin, sebum lubricates the hair and inhibits bacterial growth in the surrounding area.

The problem with this process is that, under hormonal influence, sebaceous glands can be *too* active—they can produce too much sebum, particularly in teenagers. When this excess sebum mixes with skin cells that are being shed from the sebaceous follicle, the result can be a blocked follicle that swells with sebum—a whitehead. Once this blockage occurs, bacteria normally found on the skin begin to multiply in the follicle, producing irritating substances that can cause general inflammation and pimples. Taken together, these various skin afflictions are known as *acne*. No matter what you may have heard, acne is not caused by eating fatty foods, stress, or a lack of face scrubbing. As with baldness, the hormones called androgens are the culprits. They prompt the increased sebaceous activity; since males at puberty have more androgens in them than females, it is males who suffer more from acne.

The second type of exocrine gland in the skin is the **sweat gland**, simply defined as a gland that produces the fluid sweat. This can mean a couple of different things, however, because sweat glands come in two varieties. One type is a *merocrine* gland that produces the clear, saltwater-like fluid we normally think of as sweat. We have, at a minimum, 2 million merocrine sweat glands on our bodies. They exist on almost all skin surfaces, but we have an especially dense concentration of them on our feet, forehead, and palms (hence "sweaty palms"). The primary function of the merocrine sweat glands is to bring sweat to the surface of the skin so that it can cool us by evaporating. These glands have their own ducts and follicles, which begin deep in the dermis.

The other sweat glands in the body, called *apocrine* glands, are very different from the merocrine glands. First, they only exist in a few places—notably the groin, the anal area, and the armpits. As such, they have little effect on body cooling. Second,

## ESSAY | Why Fat Matters and Where It Matters Most

A t one time, scientists thought of the body's billions of fat cells as little more than fuel depots—as storage sites for the calories we ingest but don't burn up. This model began to fall apart in 1994, however, with the discovery of a hormone called leptin that plays a role in regulating hunger and that is *produced* by fat cells. In the years that followed, scientists identified 11 more hormones whose source is fat cells, and the bet is that there are more of these substances yet to be discovered. The upshot is that fat cells are no longer regarded as passive warehouses; instead, they are thought of as active chemical factories. Indeed, many scientists now look at

*Fat cells are no longer regarded as passive warehouses.*

the body's collection of fat cells as a single hormonal *organ*, on a par with, say, the pancreas (see **Figure 1**). The problem is that the products of this organ are mostly bad news when a person has an excess of fat cells, which is to say, when a person is overweight. And who is overweight? Perhaps two-thirds of the adults in the United States. This proportion is so high—and the afflictions linked to fat cells so serious—that research into these cells is now a top public health priority.

Consider type 2 diabetes, which comes about not because the body has stopped producing insulin but because the body has become resistant to its effects. Insulin helps move glucose from the bloodstream into many cells in the body. Thus, when cells become resistant to insulin, glucose accumulates in the bloodstream. This is where the trouble starts because elevated levels of glucose can result in nerve damage, heart disease, blindness, and more. Now, why is it

that being overweight is the most important risk factor for developing type 2 diabetes? Well, it turns out that, of the dozen known hormones produced by fat cells, seven have the effect of increasing insulin resistance. One of these hormones, the well-named *resistin*,

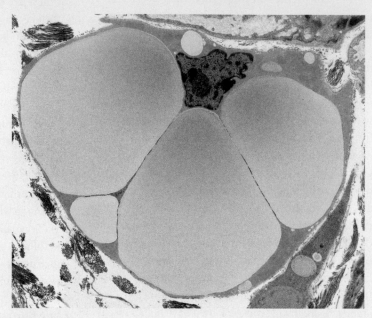

**Figure 1**
**Where Excess Calories Go**

Fat cells, such as the one shown here, often contain several lipid droplets whose size dwarfs the cell's other components. In this case, the three large central droplets do just this. The cell's nucleus is the purple object near the top.

they secrete their products into hair follicles, not up through their own ducts. Most important, apocrine gland secretions contain not only regular sweat, but a fatty, milky fluid whose function in humans is unknown. The *effect* of this fluid is very well known, however. Bacteria feed on it, and the resulting bacterial by-products are the cause of body odor. Apocrine glands only begin functioning at puberty, which is why children seldom have body odor.

### Nails

Our final accessory structure of the integumentary system is the nails found on our fingers and toes. Like hair, nails are made of a variety of the protein keratin. They are clearly useful for protection and for prying, scratching, and picking up objects. Nails allow us to look, in a sense, at the blood vessels underlying the dermis because it's the reddish color of these vessels that gives nails their pink hue.

**SO FAR...**

1. The primary component of the integumentary system is an organ called _____, which is composed of two primary layers, an outer layer called the_____ and an inner layer called the _____.

2. The outermost cells of our skin are essentially a set of water-resistant sacs composed mostly of the protein _____.

3. Fat exists in the human body in two primary locations. First there is subcutaneous fat, which can be thought of as a layer of fat located _____; second there is visceral fat, which is fat that surrounds _____ deep within the _____.

4. Each hair that we have grows out of a tissue complex called a _____.

actually is produced in greater quantity in a type of immune system cell, called a macrophage, but macrophage numbers turn out to be linked to . . . fat cell numbers. Macrophages seem drawn to collections of excess fat cells, as if these cells were foreign invaders. When we become overweight, therefore, the result is not just more (and larger) fat cells but more macrophages as well, which means increased production of resistin.

The connection between fat cells and the immune system then carries over into an equally serious health threat, heart disease. The old conception of heart disease was essentially one of a steady buildup of fatty tissue in one of the coronary arteries of the heart. When a sufficient amount of fatty tissue accumulated, this view held, the result was a complete blockage of a coronary artery—a heart attack. We now know, however, that most heart attacks come about through a fatty buildup coupled with the immune system response known as *inflammation*. And compounds that fat cells produce help bring about inflammation. Indeed, fat cells contribute to heart disease in several ways as they also produce a compound (called angiotensinogen) that raises blood pressure.

The conception of fat cells as chemical factories has seemingly answered the question of why being overweight has so many serious health consequences. It's not just the weight itself—the strain it puts on joints, for example, or the extra work it requires the cardiovascular system to do. It's that fat cells change character when they grow too large and too numerous. The set of chemical reactions they engender—their metabolism—goes from being benign to being harmful.

All fat cells are not equally harmful, however. A critical question to ask about them is where they are located in the body. As the main text notes, human beings have concentrations of fat cells in two separate locations. On the one hand, we have subcutaneous fat, which is the layer of fat that exists just beneath the skin throughout the body. On the other, we have visceral fat, which is fat that surrounds our internal organs deep within the abdomen. As it turns out, this latter "belly fat" has a much stronger linkage to both diabetes and cardiovascular disease than does subcutaneous fat.

A famous demonstration of this fact came in 2004, when researchers got to look at the health effects of losing subcutaneous fat while retaining visceral fat. How could a person lose one kind of fat without losing the other? Through the procedure of liposuction, which removes only subcutaneous fat. The researchers first measured 15 very obese women for a set of risk factors associated with diabetes and heart disease—such things as blood pressure, triglyceride levels, and circulating levels of inflammatory compounds. Next, they performed liposuctions on the women, removing an average of 10 kilograms (or 22 pounds) of fatty tissue from each one. Finally, they measured the women again for the risk factors. The result? None of their factors had changed at all. The critical thing, the researchers noted, was that, had these women lost weight through diet and exercise, *all* their risk factors would have been expected to improve. Why should this be so? Diet and exercise remove both subcutaneous and visceral fat, while liposuction leaves visceral fat untouched. The implication was that it was solely the women's' visceral fat that was putting them at greater risk of disease.

The linkage between visceral fat and health is now well-enough established that doctors have been advised to look not just at their patients' weight, but at their patients' *shape* in determining their risks for diabetes and heart disease. The idea is to act on the message being delivered by today's research: Health problems are more likely to be caused by a big belly than by a big backside or big hips.

## 26.8 The Skeletal System

We now shift gears, going from a consideration of our outer covering to a consideration of the framework that underlies it—our skeletal system. This system has only three structural elements to it: bone, ligaments, and cartilage. Most of us are quite familiar with **bone**, which can be defined as a connective tissue that provides support and storage capacity and that is the site of blood cell production. Ligaments are fairly inaccessible to us; but if you look at **Figure 26.8**, you can see some typical ligaments carrying out a typical task—linking bones together. You can also see the third element of the skeletal system, cartilage. Note that cartilage is located *in between* bones in the foot, and this turns out to be one of its common functions. **Cartilage** is a connective tissue that serves as padding in most joints. It is more familiar to us, however, in other parts of the body. It forms our larynx (voice box), the C-shaped "rings"

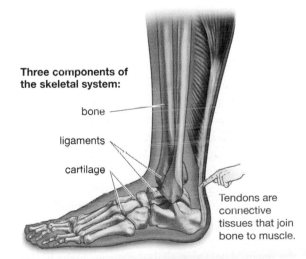

**Three components of the skeletal system:**

bone

ligaments

cartilage

Tendons are connective tissues that join bone to muscle.

**Figure 26.8**
**Elements of the Skeletal System**

Bones are the basic element of the skeletal system. Ligaments link one bone to another, and cartilage often serves as padding between bones. Tendons, while not part of the skeletal system, always link bone to muscle.

••• ANSWERS

1. skin; epidermis; dermis

2. keratin

3. just beneath the skin; organs; abdomen

4. follicle

## There Is No Such Thing as a Fabulous Tan

The French clothing designer Coco Chanel once remarked that "fashion changes, but style remains." Decades ago, however, Chanel herself helped launch a fashion that seems stubbornly resistant to change. The fashion is that of getting a suntan, but this practice might as well be called the fashion of damaging your skin so that in a couple of decades it will be saggy, wrinkled, and spotted.

Throughout the early part of the twentieth century, it was chic to have pale skin, rather than a tan, as a tan was a sign of someone who *had* to be outdoors—working a job. According to one popular account, Chanel (of Chanel No. 5 fame) got some sun on an ocean cruise in the 1920s, stepped off the boat tanned, and conveyed the message that it was chic to look that way. People in industrial countries were ready to hear this message because it came at a moment when their leisure time was increasing, meaning they had more opportunities to be out in the sun playing, rather than working.

*Sun exposure is just a faster route to most of the skin imperfections that come with aging.*

These days, the constant message from health professionals is to avoid the sun, but this is a message society has not really taken to heart. Consider that there are an estimated 25,000 businesses in the United States devoted to *giving* people tans through exposure to ultraviolet light—the very thing the sun delivers. (On an average day, more than a million Americans will visit tanning salons, with more than 70 percent of these people being women, according to the American Academy of Dermatology.)

But in what ways does sun exposure harm the skin? The most serious thing it does is help cause skin cancer—three varieties of it, actually—but let's set this aside and focus only on what it does to the appearance of the skin. Look at the forearm of a fair-skinned 50-year-old who has spent time in the sun, and you're likely to see imperfections commonly known as "liver spots." These don't actually have anything to do with the liver, however, but instead are collections of pigment-producing cells, called melanocytes, that have clustered together as a result of exposure to the sun. In a similar vein, what makes skin wrinkle? In 2000, a set of researchers reported that extreme age and spending time in the sun have the same effect. Both prompt a decrease in the production of the skin protein known as collagen, and both bring about an increase in levels of enzymes that degrade collagen—an action that puts this protein into a misshapen form. Since collagen is the major structural protein in the skin, the ultimate result of these changes is wrinkled skin. Along similar lines, if we look at moles, "spider" veins, and sagging skin, we find that all these imperfections are seen in elderly persons who have stayed out of the sun and in *middle-aged* persons who have spent time in it. Dermatologists understand the similarities between aging and sun exposure so well that they refer to any sun-caused skin damage as "photo-aging."

But what about sunscreens? They are indeed recommended by dermatologists, with the understanding that they need to be SPF 15 or higher, applied in nice thick layers at least 15 minutes before going out in the sun and reapplied after activities such as swimming. The problem is that it's difficult to be conscientious about all these factors. The general advice of dermatologists is simply to avoid lengthy exposure to sunlight, particularly between the hours of 10 a.m. and 2 p.m. If you limit your sun exposure, you may not end up with the most fashionable looking skin over the next two weeks, but over the next two decades you'll be the clear winner.

on our trachea (windpipe), and our nose tip and outer ears, and it links each of our ribs to our breastbone. In all its capacities, cartilage is a very flexible, resilient tissue, which is not surprising given that it is mostly made up of water.

It's a good idea to look at Figure 26.8 to get a sense not only of the difference between bone, ligaments, and cartilage but also of the difference between all of these skeletal elements and another type of tissue that is pictured, tendons. Although not part of the skeletal system, tendons are always associated with the bones of the system. Looking at all four elements in the figure, here's one way to think of them. Bones are the basic framework; cartilage often serves as padding between bones; **ligaments** are tissues that join bone to bone; and **tendons** are tissues that join bone to muscle.

## Function and Structure of Bones

It's tempting to think of bones as nothing more than support beams, but this conception is wide of the mark in two ways. First, support beams are not alive, but bone is very much a living, dynamic tissue, as you'll see. Second, bones do more than just support us. Once we are past infancy, all of our blood cells (both red and white) are created in the interior of our bones. Some bones store fat, in the form of yellow marrow, as a kind of energy reserve, and bones serve as the storage sites for important minerals such as calcium and phosphate. Beyond this, the only reason we can move at all is that our muscles are attached to bones. When we contract our bicep, for example, our forearm is raised by means of the bicep pulling on the bones of the forearm.

## The Structure of Bone

Each bone in our body actually is considered a separate organ because each bone is composed of several kinds of tissue. There is the calcified or "osseous" tissue we think of as bone, to be sure, but each bone also has its own blood vessels and nerves. (Cartilage, meanwhile, has neither blood vessels nor nerves.) Bones are a great example of connective tissue in that the *cells* in bones may account for as little as 2 percent of bone mass. What is the other 98 percent of bone made of? It is overwhelmingly a ground substance, secreted by bone cells, that is composed of hard, calcium-containing crystals and tough, yet flexible collagen fibers. This combination means that bones are hard without being brittle—they can support our weight and yet can bend or twist a little without breaking.

### Large-Scale Features of Bone

The typical features of a long bone such as the humerus (in the arm) are shown in **Figure 26.9**. A long bone has a central shaft, or diaphysis, and expanded ends, or epiphyses (singular, epiphysis). There are two types of bone. **Compact bone**, which is relatively solid, forms the outer portion of the bone (all the way around it, as you can see in Figure 26.9b). **Spongy bone** then fills the epiphyses and is well named. It is porous enough that it contains another type of tissue, **red marrow**, which is the tissue in which all the blood cells of the adult body are produced. (These cells include not only oxygen-carrying red blood cells, but the white blood cells that are the central cells of the immune system.) Meanwhile, **yellow marrow**, a tissue largely made up of energy-storing fat cells, is found in the *marrow cavity* of the type of long bones seen in Figure 26.9.

### Small-Scale Features of Bone

Dropping down to the microscopic level and looking at bone *cells*, we find that three different types are involved in the growth and maintenance of bone. **Osteoblasts** are immature bone cells that are responsible for the production of new bone. They secrete the material that becomes the bone ground substance. Once osteoblasts are surrounded by this material, they reduce their production of it, and this marks their transition to osteocytes, the second type of bone cell. **Osteocytes** are mature bone cells; they *maintain* the structure and density of normal bone by continually recycling the calcium compounds around themselves. The third type of bone cell, the osteoclasts, could be thought of as the demolition team of bone tissue. **Osteoclasts** are cells that move along the outside of bones, releasing enzymes that eat away at bone tissue, thus liberating minerals stored

**Figure 26.9**
**Large- and Small-Scale Features of a Typical Bone**
Note that in **(c)**, the building units of compact bone, the osteons, run parallel to the long axis of the bone. Spongy bone does not contain osteons but instead is composed of an open network of calcified rods or plates.

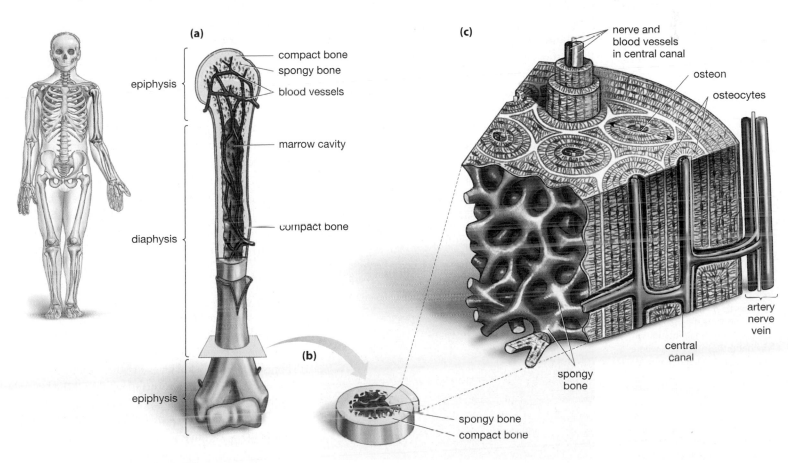

in the bone. This may sound harmful, but it actually is essential for bone growth and for the regulation of blood levels of calcium. (This is the calcium-releasing function we reviewed at the start of the chapter.) The upshot of all this is that, at any given moment, osteoblasts are adding to bone, osteocytes are maintaining it, and osteoclasts are removing it.

Within compact bone, the basic functional unit is the osteon seen in Figure 26.9c (which is also known as the Haversian system). The essence of bone growth is that, within each osteon, osteoblasts produce layer after layer of concentric cylinders of bone—much like layers of insulation wrapped around a pipe. These layers surround what could be thought of as the hole in the pipe, which is a central canal (also known as a Haversian canal). These canals parallel the long axis of the bone and have blood vessels and nerves running through them—a kind of support system for the bone.

Unlike compact bone, spongy bone has no osteons. It consists of an open network of interconnecting calcified rods or plates. The red marrow fills the spaces within this network.

## Practical Consequences of Bone Dynamics

Now, let's think about the practical consequences of this structure. Remember the earlier observation that bone is a living tissue? With its three kinds of cells constantly adding, maintaining, and breaking down the tissue, you can see the truth of this. Indeed, the bone "remodeling" that these cells undertake is so extensive that our entire skeleton is replaced through it every 10 years.

But, bone is more dynamic than even this would indicate in that bones are responsive to our activities. As physiologist Frederic Martini has noted, when you undertake an exercise like jogging, which stresses your bones, the calcium-containing crystals of the affected bones create small electrical fields that apparently attract osteoblasts. On arrival, these osteoblasts begin producing bone. The result? If you take up jogging, you get denser leg bones. But there's a flip side to this coin: People who are using crutches to take the weight off a broken leg bone may temporarily lose up to a third of the mass in this bone because it is now scarcely being stressed at all.

Interestingly, although we develop almost all our height by late adolescence, we are still gaining in bone density up until about the age of 30. Then, beginning in middle age, the body may start removing more bone than it adds. This phenomenon particularly afflicts women, in the form of a condition known as *osteoporosis*, meaning a thinning of bone tissue that can result in bone breakage and deformation (**Figure 26.10**). Why should osteoporosis affect more women than men? Hormonal changes that come with menopause can increase the activity of the bone-depleting osteoclasts relative to the activity of the bone-building osteoblasts. One way to guard against osteoporosis is to reach the age of 30 with as much bone density as possible—the equivalent of putting money in the bank as a hedge against a later withdrawal. And, there are a couple of proven ways to get more bone density before 30: Take up moderate programs of "weight-bearing" exercise such as jogging or walking and modestly increase calcium intake.

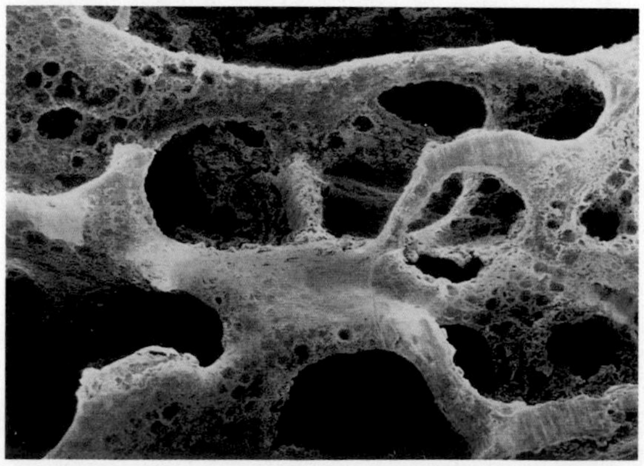

## The Human Skeleton

Taking a step back from the microscopic world, you can see in **Figure 26.11** that there are a lot of bones in the skeletal system—206 of them in all. (Only the major bones are visible in the figure.) There are two main divisions to the skeletal system: the **axial skeleton**, whose 80 bones include the skull, the vertebral column, and the rib cage that attaches to it; and the **appendicular skeleton**, whose 126 bones include those of our paired appendages—the arms and the legs, along with the pelvic and pectoral "girdles" to which they are attached. Also note the blue-colored tissue joining ribs to our breastbone and lying in between each of the vertebrae. These are but a few of the cartilages in the body.

## Joints

Joints, or articulations, exist wherever two bones meet. What qualities are required of a joint? Well, some only need strength. We think of our skull as being a single bone, but in fact, it's composed of several bones that interlock with each other in joints that are immovable. For other joints, flexibility is the key requirement. The ball-and-socket joint at our shoulder permits a range of motion so extensive that movement of our upper arms is limited more by our shoulder muscles than by our bones.

**Figure 26.10
Bones Weakened by Osteoporosis**

The damaging effects of osteoporosis can be seen in this micrograph of a spinal vertebra that has been "thinned" by the condition. Note the many small, circular holes or indentations in the bone; these are areas where bone tissue has been removed at a greater rate than it has been added. The result is a bone that is more susceptible to fractures. The micrograph is of the spongy bone portion of the vertebra; the larger circular holes are thus part of the bone's normal structure.

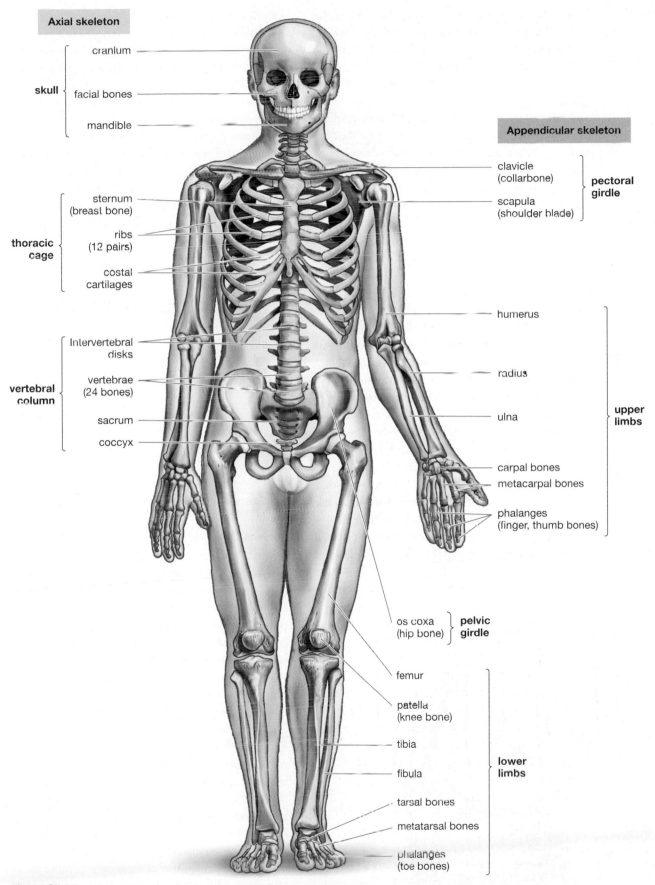

**Axial skeleton**

skull
- cranium
- facial bones
- mandible

thoracic cage
- sternum (breast bone)
- ribs (12 pairs)
- costal cartilages

vertebral column
- Intervertebral disks
- vertebrae (24 bones)
- sacrum
- coccyx

**Appendicular skeleton**

pectoral girdle
- clavicle (collarbone)
- scapula (shoulder blade)

upper limbs
- humerus
- radius
- ulna
- carpal bones
- metacarpal bones
- phalanges (finger, thumb bones)

pelvic girdle
- os coxa (hip bone)

lower limbs
- femur
- patella (knee bone)
- tibia
- fibula
- tarsal bones
- metatarsal bones
- phalanges (toe bones)

**Figure 26.11**
**Solid Framework**

A view of the human skeleton, which has two divisions, axial and appendicular. The major bones of the axial division are listed on the left, and the major bones of the appendicular division on the right.

Highly movable joints are typically found at the ends of long bones, such as those in the legs as well as the shoulder. You can see a view of such a joint—the human knee joint—in **Figure 26.12**. A joint such as this must serve two competing functions. On the one hand, it must keep the massive femur (thigh bone) and large tibia (shin bone) in a stable arrangement. On the other, it serves as a hinge that allows the tibia to swing as free as a screen door off the femur. How does it do both things? Well, the two bones are partly held in place by four main ligaments, two of which are visible in the figure. Recall that ligaments link bone to bone, but there are also tendons, which attach bone to muscle; the knee joint has these in place as well.

With respect to the motion that such a joint allows, a critical element is that one bone does not actually touch another. Instead, an extensive network of padding and lubricating tissues in between the bones performs the same function as the "gel" found in running shoes. Note that the top of the tibia and bottom of the femur are surrounded by cartilage, and that more cartilage—in the form of the meniscus tissue—lies on each side of the joint. The yellow tissue shown is fat padding that fills in spaces that are created when the bones move, and the space in between the bones (the joint cavity) is filled with a thick, viscous fluid that not only lubricates joints, but aids in their metabolism.

Joints can get injured, of course, and as we age they can wear down altogether. If you follow sports, the name of one of the ligaments shown in Figure 26.12 may ring a bell. The anterior cruciate ligament (ACL), which runs from the back of the femur to the front of the tibia, is a frequent site of injury for athletes who participate in sports that involve rapid "cutting"

movements, such as volleyball, soccer, and basketball. An athlete will come to a quick stop, hear a pop, and be down with a torn or ruptured ACL that usually will have to be replaced by grafting tissue onto it from another part of the body. For older people, the usual problem is not a sudden injury, but rather a slow, steady degeneration of hip, knuckle, or knee joints in the disease known as osteoarthritis. Such "wear-and-tear" arthritis results in a breakdown of the cartilage in between bones, such that joints become swollen, with bone sometimes grinding against bone. Although osteoarthritis generally appears later in life, young people can take steps to avoid it. Risk factors for the condition include being overweight and injuring the joints through the traumatic collisions that can take place in sports like football and soccer.

---

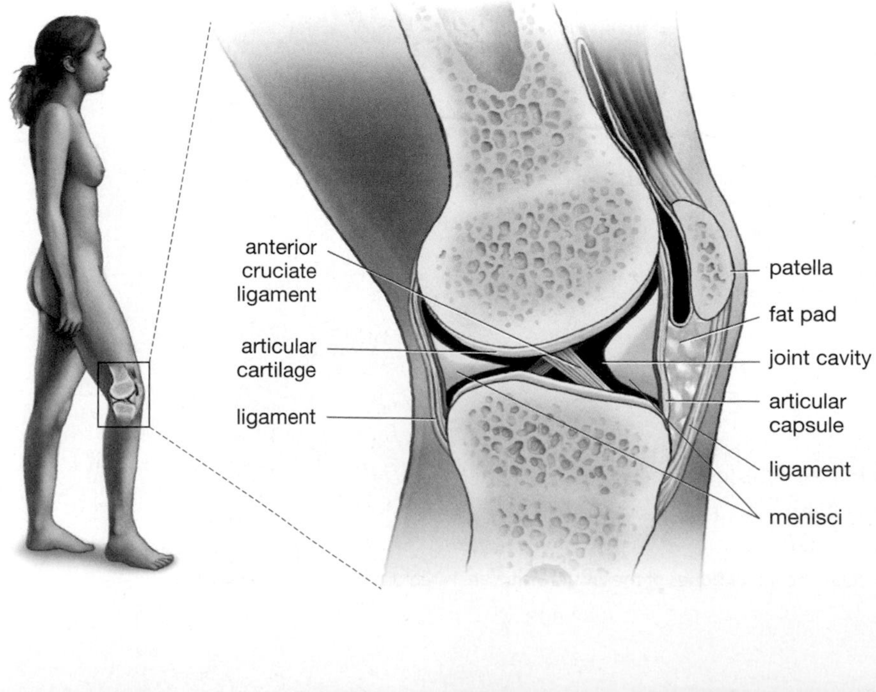

### SO FAR...

1. The three elements of the skeletal system are _____, _____, and _____.

2. Name two functions of bone other than support.

3. _____ are the cells that produce new bone, _____ maintain the structure and density of bone, and _____ release enzymes that eat away at bone tissue.

---

## 26.9 The Muscular System

Bones may provide a scaffolding for the body, but how is it that this scaffolding is capable of movement? The answer is muscles, specifically the skeletal muscles that were introduced earlier, which are numerous and large enough that they account for about 40 percent of our body weight. (When we talk about the "meat" in, say, cattle, we are mostly talking about skeletal muscle.) Like bones, skeletal muscles are organs in that they contain a variety of tissues. At each end of a muscle, fibers of the outer muscle layer come together to form the tendons that attach muscle to bone.

### The Makeup of Muscle

As you can see in **Figure 26.13**, each muscle has within it a number of oval-shaped bundles called fascicles. Inside each fascicle there is a collection of muscle cells, but the term *cell* here may be misleading because any one of these cells can be as long as the muscle itself—a gigantic length relative to the strictly microscopic dimensions of most cells. Because skeletal muscle cells are so elongated, they are referred to as **muscle fibers**. Along their length, muscle fibers are divided into a set of about 10,000 repeating units,

**Figure 26.12**
**A Highly Moveable Joint**

A simplified view of the human knee joint, which links the femur (thigh bone) to the tibia (shin bone). Note that the two bones never touch each other because they are covered with cartilage. The menisci that you can see are additional cartilage tissue that serve an important shock-absorbing function while also helping stabilize the joint. The anterior cruciate ligament is the "ACL" we hear about so much in sports injuries. It is one of four main ligaments that keep the femur and tibia in a stable arrangement.

anterior cruciate ligament
articular cartilage
ligament
patella
fat pad
joint cavity
articular capsule
ligament
menisci

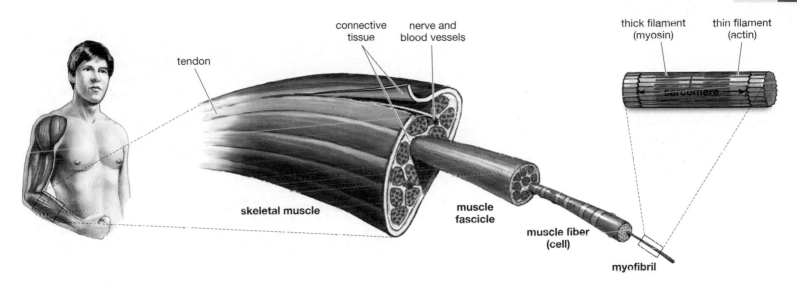

connective tissue
nerve and blood vessels
thick filament (myosin)
thin filament (actin)
tendon
sarcomere
skeletal muscle
muscle fascicle
muscle fiber (cell)
myofibril

called **sarcomeres**, that are the fundamental units of muscle contraction. But, what is it that's contracting in a muscle? A look inside a single muscle fiber reveals that it is composed of more strands—perhaps a thousand long, thin structures called *myofibrils* that run the length of the cell. Each myofibril, in turn, has a large collection of two kinds of strands inside it that alternate with one another; these are thin filaments made of the protein *actin* and thick filaments made of the protein *myosin*. It is these filaments that contract, as you'll now see.

## How Muscles Work

To understand contraction, note first that the thick myosin filaments lie in the *center* of a given sarcomere. Meanwhile, the thin actin filaments are attached to either *end* of the sarcomere and extend toward the center, where they overlap with the thick filaments (**Figure 26.14**, on the next page). In contraction, the thin filaments slide toward the center, causing the unit to shorten, which is to say, causing the muscle to contract. To understand this, place just the tips of the fingers of your right hand in between the tips the fingers of your left hand. Now slide your right hand as far as it will go to the left. Note that neither set of *fingers* shortened in length, but that the total distance made up by both sets of fingers did shorten—exactly what happens within a sarcomere.

But what enables the thin filaments to slide? The thick myosin filaments have numerous club-shaped "heads" extending from them. These myosin heads are capable of alternately binding with or detaching from the adjacent thin filaments, creating so-called cross-bridges. The myosin heads pivot at their base, as if they were on a hinge. Here's the order of their activity: Attach to the actin filament, pull it toward the center of the sarcomere, detach from the actin, reattach to it,

and pull again. In actual muscle contraction, this process happens many times in the blink of an eye; at any one point in a contraction, some myosin heads will be pulling an actin filament while others are detaching from it and still others binding to it.

This whole process gets going only when a signal arrives from a nerve telling a muscle to contract—often when a person has decided to move. This is why we speak of skeletal muscles as being under voluntary control, although it's easy to find exceptions to this rule. (The diaphragm that allows you to breathe is a skeletal muscle, but normally there is little that's voluntary in using it.)

When a nerve signal arrives, the actual message doesn't go directly from nerve to muscle. Instead, a kind of ferryboat operation takes place in which the nerve end secretes a chemical, called a *neurotransmitter*, that travels across a tiny gap that lies between nerve and muscle. Once the neurotransmitter arrives at the muscle, it binds with it, setting in motion an internal chemical reaction that results in calcium ions ($Ca^{2+}$) being released inside a sarcomere. This release of these ions has the effect of allowing the myosin heads to bind to the actin filaments, which of course has to happen before the myosin heads can *pull* the actin filaments.

Now, what about the practical consequences of this physiology? The cycle that runs from myosin binding, through pulling, and then detaching is known as a *twitch*. The muscle fibers that go through this cycle come in two varieties—fast twitch and slow twitch—and these names describe how long it takes them to complete a cycle. Fast-twitch fibers are not just faster than slow-twitch fibers, however; they also can generate more power. But there is a catch to this in that fast-twitch fibers can generate power only for brief periods. In contrast, slow-twitch fibers generate less power but can do it over a sustained period of

**Figure 26.13**
**Structure of Skeletal Muscle**

Skeletal muscles have oval-shaped bundles within them, called fascicles, that are composed of a number of muscle cells, each one of which is so elongated it is referred to as a fiber. Each fiber is composed of many thin myofibrils, and each myofibril has within it a collection of alternating thick and thin protein strands called filaments. The thin filaments are made of the protein actin, the thick filaments of the protein myosin. Each myofibril is divided lengthwise into a series of repeating functional units called sarcomeres.

••• ANSWERS
1. bones; cartilage; ligaments
2. production of blood cells; storage of energy reserves as fat
3. osteoblasts; osteocytes; osteoclasts

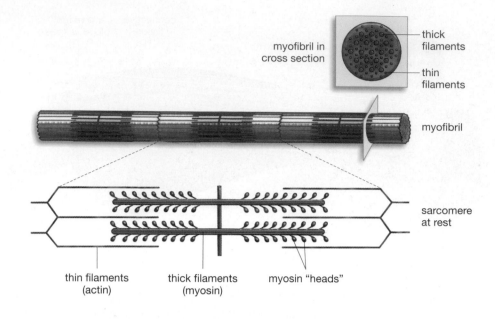

**(a)** Muscle at rest

The myofibril contains several sarcomeres, one of which is shown at rest. The thin filaments within the sarcomere, made of actin, overlap with adjacent thick filaments, made of myosin. The thick filaments have club-shaped outgrowths, called heads. The myofibril is also shown in cross section, at a location where thick and thin filaments overlap.

myofibril in cross section

thick filaments

thin filaments

myofibril

sarcomere at rest

thin filaments (actin)

thick filaments (myosin)

myosin "heads"

**(b)** Muscle contraction

Muscle contraction is a matter of the thin actin filaments sliding together within a sarcomere. This comes about when the myosin heads attach to the actin filaments and pull them toward the center of the sarcomere. The myosin heads pivot at their base and are capable of alternately binding with, or detaching from, the thin filaments.

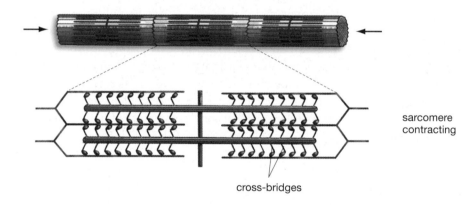

sarcomere contracting

cross-bridges

**Figure 26.14**
**How a Skeletal Muscle Contracts**

**Bio*Flix***

Muscle Contraction

••• ANSWERS

1. fibers; length

2. sliding; sarcomere

3. myosin; actin; center; sarcomere

time. Thus, fast- and slow-twitch fibers are something like the hare and tortoise of muscle tissue.

Because slow-twitch fibers are infused with an oxygen-binding molecule called *myoglobin*, they have a red coloration that true fast-twitch fibers lack. The dark meat of a chicken leg is dark because it is primarily made of myoglobin-infused slow-twitch fibers. The chicken constantly uses its legs but doesn't generate much power in doing so—the very thing that slow-twitch fibers are good at. The chicken will occasionally engage its breast muscles in a flurry of contraction so that it can flap its wings. Not surprisingly, chicken breasts are white meat, which is to say muscle that is primarily composed of fast-twitch fibers. Most muscles contain a mixture of fast- and slow-twitch fibers. We simply use different proportions of each kind of fiber, depending on what we are doing. As you might guess, a long-distance runner primarily uses slow-twitch fibers, while a sprinter primarily uses fast-twitch fibers.

# On to the Nervous and Endocrine Systems

Having looked at the basic characteristics of the body and at three of its organ systems, we're now ready to move on to two other systems, both of which function in communication. These are the nervous and endocrine systems.

**SO FAR...**

1. Muscle cells are referred to as muscle _____ because of their unusual _____.

2. Muscle contraction works through the process of a _____ of two kinds of muscle filaments past each other within muscle unit, known as a _____.

3. In this process, relatively thick _____ filaments pull relatively thin _____ filaments toward the _____ of the _____.

# CHAPTER 26 REVIEW

## SUMMARY

### 26.1 The Disciplines of Anatomy and Physiology

- Anatomy is the study of the body's internal and external structure and the physical relationships between the body's constituent parts. Physiology is the study of how these parts work.

### 26.2 How Does the Body Regulate Itself?

- The body regulates itself not through a hierarchical command structure but rather through a system of circular control, in which A prompts B and B prompts C, but in which C may then prompt A. Through use of such circular systems, the body maintains homeostasis, defined as a relatively stable internal environment. One mechanism is largely responsible for preserving homeostasis. This mechanism is negative feedback: a process in which the elements that bring about a response have their activity reduced by that response.

- Humans have two primary body cavities: a dorsal cavity that is further divided into cranial and spinal cavities; and a ventral cavity that is further divided into a thoracic cavity that contains the lungs and the heart, and an abdominopelvic cavity that contains such organs as the stomach and liver.

### 26.3 Levels of Physical Organization

- Animal bodies function through a series of smaller-to-larger working units whose order runs as follows: cells, tissues, organs, and organ systems. Tissues are groups of cells that have a common function. (Muscle tissue, for example, is made up of a group of cells that have the common function of contracting.) Organs are complexes of several kinds of tissues that perform a special function. (The heart is made up of several kinds of tissues, including muscle and connective tissue.) Organ systems are groups of organs and tissues that perform a given function. (The heart, blood, and blood vessels form the cardiovascular system.)

### 26.4 The Human Body Has Four Basic Tissue Types

- There are four fundamental tissue types in the human body: epithelial, connective, muscle, and nervous. Epithelial tissue covers all body surfaces exposed to an external environment and forms glands, which are organs that secrete one or more substances.

- Connective tissue supports and protects other tissues. Connective tissue cells are embedded in a "ground" material, which they themselves generally secrete. A prime function of most connective tissue cells is to produce this extracellular material, which varies in consistency from a fluid to a solid. For example, bone

cells secrete a connective ground material that largely forms the solid bones we are familiar with.

- Muscle tissue is specialized in its ability to contract. There are three primary types of muscle tissue: skeletal, cardiac, and smooth. Skeletal muscle is always attached to bone and is under voluntary control. Cardiac muscle exists only in the heart and beats under the control of its own pacemaker cells. Smooth muscle is "smooth" because it lacks the striations that skeletal muscle has; it is not under voluntary control and is responsible for contractions in blood vessels, the digestive tract, air passages, and hollow organs such as the uterus.

- Nervous tissue is specialized for the rapid conduction of electrical impulses. It is made up of two basic types of cells: neurons, which actually transmit the impulses; and glial cells, which are support cells for neurons.

### 26.5 Organs Are Made of Several Kinds of Tissues

- The stomach provides a case in point for how various tissue types come together to form a working organ. A typical cross section of the stomach is made up of muscle, nervous, connective, and epithelial tissues.

### 26.6 Organs and Tissues Make Up Organ Systems

- There are 11 organ systems in the body. The first three of these—the integumentary, skeletal, and muscular systems—function in body support and movement. The integumentary system is made up of skin and such associated structures as glands, hair, and nails. The skeletal system is made up of bones, ligaments, and cartilages. The muscular system includes all the skeletal muscles of the body but not smooth or cardiac muscles.

- The nervous, endocrine, and immune systems function in communication, regulation, and defense of the body. The nervous system is a rapid communication system that includes the brain, spinal cord, all the nerves, and sense organs such as the eye and ear. The endocrine system is a communication system that works more slowly, through the substances called hormones, many of which are produced in specialized glands. (Glands that release their substances directly into the bloodstream or extracellular fluid are endocrine glands. Glands that release their substances through the tubes known as ducts are exocrine glands.) The immune system is a network of cells and proteins whose central function is the elimination of microbial invaders. The immune system works in part through the body's lymphatic network, whose vessels collect extracellular fluid and deliver it to the blood vessels. Immune cells within lymph glands inspect lymph as it passes by to identify invading microbes.

- The cardiovascular, respiratory, digestive, and urinary organ systems function in transport and in exchange with the environment. The cardiovascular system consists of the heart, blood, blood vessels, and the inner "marrow" portion of the bones where red blood cells are formed. The respiratory system includes the lungs and the passageways that carry air to them. The central feature of the digestive system is the digestive tract, a tube that begins at the mouth and ends at the anus. Along its course, there are digestive organs of the system: the stomach, small intestines, and large intestines and such accessory glands as the liver and pancreas. The urinary system functions to eliminate waste products from the blood, through the formation of urine, and to retain substances that are useful to the body. It is made up of the kidneys and various other structures, such as the bladder.

- Humans have separate reproductive systems for females and males.

## 26.7 The Integumentary System: Skin and Its Accessories

- The primary component of the integumentary system is the organ called skin. Integumentary structures associated with skin are hair, nails, and a variety of exocrine glands.

- In addition to covering the body, skin protects underlying tissues and organs, controls the evaporation of body fluids, regulates heat loss, stores fat, and makes vitamin D. Exocrine glands associated with skin excrete such materials as water, oils, and milk. Specialized nerve endings in the skin detect touch, pressure, pain, and temperature.

- Skin has a thin outer epithelial covering, the epidermis; and a thicker underlying dermis, composed mostly of connective tissue. Beneath the dermis—and not part of the skin—is a third layer of tissues, the hypodermis. Epidermal skin cells are constantly worn away and replaced by new epidermal cells being pushed up from the inner epidermis. As they move toward the surface, these cells are transformed into a series of dead, tightly linked sacs made of the protein keratin. The dermis is filled with accessory structures, such as hair follicles and sweat glands, and contains blood vessels and nerves that support the surface of the skin. The blood supply of the dermis is important in controlling body temperature.

- The hypodermis is the location for the body's subcutaneous fat: a layer of fat, located just beneath the skin, that varies in thickness throughout the body. Separate from this tissue is the body's visceral fat, which is fat that lies deep within the abdomen and that surrounds organs such as the liver. An excess of visceral fat is more harmful to health than is an excess of subcutaneous fat.

- Hairs originate in tiny organs called hair follicles and may have been important for our evolutionary ancestors in both heat retention and touch sensitivity.

- The skin contains two types of exocrine glands: sebaceous glands, which produce a material called sebum, which lubricates hair and limits bacterial growth on the skin; and sweat glands, which produce sweat that cools the body by evaporating. Overactive sebaceous glands are the cause of acne. Sweat glands come in two varieties: the more-numerous merocrine glands, which produce the perspiration; and the less-numerous apocrine glands, which produce sweat mixed with a thick, cloudy secretion that moves through hair follicles onto the skin. Bacteria feed on this secretion, and the resulting bacterial by-products are the cause of body odor.

- Nails are made of the protein keratin and form over the tips of the fingers and toes, where they are useful in protection and in prying, scratching, and picking up objects.

## 26.8 The Skeletal System

- The human skeletal system is composed of bone, ligaments, and cartilage. Bone is a connective tissue that provides support and storage capacity and that is the site of blood cell production. Cartilage serves as padding in most joints, forms our larynx (voice box) and trachea (windpipe), and links each rib to the breastbone. Ligaments are tissues that join bone to bone. Tendons link bone to muscle but are not part of the skeletal system.

- Each bone of the skeleton is an organ that contains not only connective tissue—mostly the supporting connective tissue called osseous tissue—but blood vessels and nervous tissue as well.

- The typical long bone features a central shaft, called a diaphysis, and expanded ends, called epiphyses. There are two types of bone within this structure: compact bone, which is relatively solid and forms the outer portion of the bone; and spongy bone, which is less dense and fills the epiphyses. In long bones, spongy bone is filled with red marrow, which is the site of the production of all the blood cells in the adult body. Long bones have a central marrow cavity that is filled with yellow marrow, which is composed largely of energy-storing fat cells.

- Three different types of cells are involved in bone growth and maintenance. Osteoblasts are responsible for the production of new bone; osteocytes are mature bone cells that maintain the structure and density of normal bone; and osteoclasts release enzymes that break down bone, thus releasing stored substances, such as calcium.

- In compact bone, the basic functional unit is the osteon. Osteoblasts produce layers of concentric cylinders of bone around a central canal of the osteon. The central canal, which parallels the long axis of the bone, has blood vessels and nerves running through it.

- Bone is responsive to our activities. The density of a bone will decrease to the extent that the bone is not used and will increase to the extent that the bone is stressed, as in some forms of exercise.

- The human skeleton has 206 bones in all, in two main divisions: an axial skeleton, whose 80 bones include the skull, the vertebral column, and the rib cage; and the appendicular skeleton, whose 126 bones include those of the arms and legs and the pelvic and pectoral girdles to which they are attached.

- Joints or articulations exist wherever two bones meet. Some joints provide a strong linkage between bones, while others provide great flexibility. Ligaments (joining bone to bone) and tendons (joining bone to muscle) help provide the stability that large-bone joints need. Normally, the bony surfaces of highly movable joints do not meet directly but are instead covered with special cartilages and pads that provide shock absorption and lubrication.

## 26.9 The Muscular System

- A given skeletal muscle cell can be as long as the muscle itself; because of this elongation, these cells are called muscle fibers. Along their length, muscle fibers are divided into a set of repeating units called sarcomeres, which are the basic units of muscle contraction. Each muscle fiber is composed of thin structures, called myofibrils, which are in turn composed of assemblies of two kinds of protein strands that alternate with one another: thin filaments made of the protein actin and thick filaments made of the protein myosin.

- Actin filaments attached to the end of each sarcomere overlap with the myosin filaments that lie in the center of each sarcomere. The myosin filaments bring about contraction by attaching to the actin filaments, pulling them toward the center of the sarcomeres, detaching, and then pulling again. As the actin filaments slide toward the center of the sarcomere, the sarcomere shortens.

- Skeletal muscles contract only when prompted to by the arrival of a signal from a nerve. A nerve signal initiates a chemical cascade inside muscle tissue that allows the myosin filaments to bind to the actin filaments.

- Muscle fibers differ in how long it takes them to go through the cycle of myosin binding, pulling, and detaching—a cycle referred to as a twitch. Muscle fibers can be categorized as slow twitch or fast twitch. Slow-twitch fibers contract with less force but can do so in a sustained manner; fast-twitch fibers contract with relatively more force but can do so only for brief periods of time.

**BioFlix** Muscle Contraction

## KEY TERMS

| | | | |
|---|---|---|---|
| appendicular skeleton | 544 | hormone | 535 |
| axial skeleton | 544 | immune system | 535 |
| bone | 541 | integumentary system | 534 |
| cardiac muscle | 532 | keratin | 538 |
| cardiovascular system | 536 | ligaments | 542 |
| cartilage | 541 | lymphatic network | 535 |
| compact bone | 543 | muscle fiber | 546 |
| connective tissue | 532 | muscle tissue | 532 |
| dermis | 538 | muscular system | 535 |
| digestive system | 536 | negative feedback | 530 |
| endocrine gland | 535 | nervous system | 535 |
| endocrine system | 535 | nervous tissue | 534 |
| epidermis | 537 | organ | 531 |
| epithelial tissue | 532 | organ system | 531 |
| exocrine gland | 535 | osteoblast | 543 |
| gland | 532 | osteoclast | 543 |
| homeostasis | 530 | osteocyte | 543 |

| | | | |
|---|---|---|---|
| red marrow | 543 | smooth muscle | 532 |
| reproductive system | 536 | spongy bone | 543 |
| respiratory system | 536 | sweat gland | 539 |
| sarcomere | 547 | tendons | 542 |
| sebaceous glands | 539 | tissue | 531 |
| skeletal muscle | 532 | urinary system | 536 |
| skeletal system | 534 | yellow marrow | 543 |
| skin | 537 | | |

## BRIEF REVIEW

*Answers to Brief Review questions are in the back of the book. For multiple-choice quiz questions, go to* **www.aw.com/krogh4**.

1. What is negative about negative feedback?

2. What role does the lymphatic network play in immune function?

3. Why is our skin water resistant?

4. What are the two principal locations for fat tissue in the body?

5. List three functions of the skeletal system.

6. What is the general mechanism at work in hair loss?

7. A sample of bone shows an interconnecting network of calcified rods and plates filled in with red marrow. Is this sample from the shaft (diaphysis) or the end (epiphysis) of a long bone?

8. People who have torn their anterior cruciate ligament report that their knee feels "wobbly" until they get it repaired. Why would this be?

## APPLYING YOUR KNOWLEDGE

1. Cells called melanocytes that lie deep in the epidermis produce a pigment, called melanin, that gives skin its color. Exposure to ultraviolet light—from the sun or a tanning lamp—causes melanocytes to produce more melanin, which they pass along to the skin cells above them, yielding a suntan. But why would sun exposure prompt this increased melanin production? What is the function of a tan, in other words?

2. The text notes that bone density increases in response to the stresses on bone that are provided by exercises such as running. Imagine two fairly inactive people who decide to start exercising. One becomes a runner, the other becomes a swimmer. Who experiences the greater increase in bone density: the swimmer (in the arm bones) or the runner (in the leg bones)?

3. Acids secreted by the stomach give the materials within it a pH level that lies between that of lemon juice and battery acid. Yet our small intestines, which the stomach feeds into, have chemical buffers flowing into them that raise the pH level of the material coming from the stomach. Why should the stomach need such an extreme pH level while the small intestines do not?

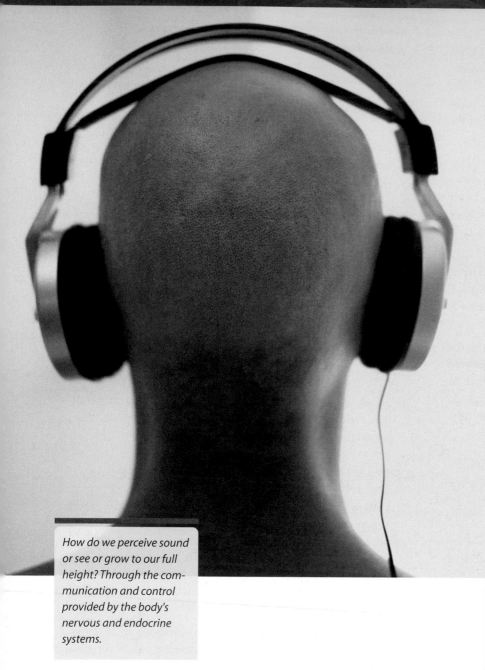

*How do we perceive sound or see or grow to our full height? Through the communication and control provided by the body's nervous and endocrine systems.*

To get through a day, our bodies must undertake a huge number of tasks each second, many of them coordinated with each other. How do we control such an enormous set of activities? Through our nervous and endocrine systems.

**552**

27.1    Structure of the Nervous System    554

27.2    Cells of the Nervous System    555

27.3    How Nervous System Communication Works    *BioFlix*    558

27.4    The Spinal Cord    561

27.5    The Autonomic Nervous System    563

27.6    The Human Brain    564

27.7    The Nervous System in Action: Our Senses    566

27.8    Our Senses of Touch    567

27.9    Our Sense of Smell    568

27.10   Our Sense of Taste    570

27.11   Our Sense of Hearing    571

27.12   Our Sense of Vision    572

27.13   The Endocrine System    576

27.14   Types of Hormones    577

27.15   How Is Hormone Secretion Controlled?    578

27.16   Hormones in Action: Four Examples    580

**Essays**

Spinal Cord Injuries    565

Too Loud: Hair Cell Loss and Hearing    571

When Blood Sugar Stays in the Blood: Diabetes    584

# COMMUNICATION AND CONTROL:
# The Nervous and Endocrine Systems

**W**ith its depiction of a man who can neither remember his present nor forget his past, the film *Memento* was a big hit in 2001. In it, we watch as the fictional Leonard Shelby relentlessly tracks down the murderer of his wife. He does so, however, while taking Polaroid pictures and scribbling notes to himself anywhere he can—some on his own body as tattoos. He does this because he has a big handicap to deal with. Having sustained a head injury in the same break-in in which his wife was killed, he can remember everything in his life up to the moment of his injury but has lost the ability to create *new* long-term memories. Now, anything that happens more than 15 minutes in the past is lost to him. Hence the constant notes and Polaroids: Where am I staying? Who did I talk to this morning? What did he look like?

Although Leonard Shelby is a movie character, the condition he has actually exists in some people. A version of it was brilliantly described by the physician-writer Oliver Sacks in his book, *The Man Who Mistook His Wife for a Hat*. Sacks introduces us to "Jimmy G.," who was admitted to Sacks' care in the mid-1970s. Although Jimmy G. is a gray-haired, 49-year-old ex-sailor, he perpetually believes himself to be the 19-year-old sailor he was back in 1945. About 1970, he apparently drank so heavily that he damaged his brain, losing not only his memory back to 1945, but his ability to make new memories as well. Hence, in his mind he is always a young man, and it is always the end of World War II. This can lead to brief, horrible moments of disorientation, as Sacks describes:

"Here," I said, and thrust a mirror toward him. "Look in the mirror and tell me what you see. Is that a 19-year-old looking out from the mirror?" He suddenly turned ashen and gripped the sides of the chair. "Jesus Christ," he whispered. "Christ, what's going on? What's happened to me?"

Sacks would not have done this but for his certainty about what would happen next. Two minutes later, Jimmy G. has no memory of the mirror or of Sacks. When Sacks briefly leaves the room and then reenters it, Jimmy G. greets him as a total stranger.

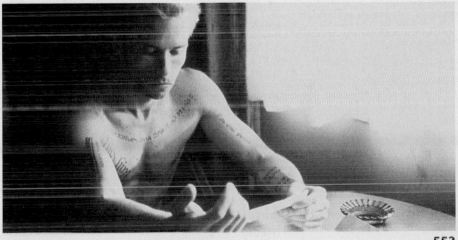

In the film Memento, the actor Guy Pearce played Leonard Shelby, a man who has lost his ability to create new long-term memories.

As Jimmy G. shows, the "self" that we take for granted largely amounts to a set of memories, which exist in the small physical space of the brain. We can thus think of memory as an agent of communication and control—one that allows us to act in the present based on our experience of the past. The human nervous system, which includes the brain, can be thought of as a wider-yet communication-and-control operation—one that constantly monitors our present, along with our past, and issues commands that allow us to function. Only some of these commands are conscious orders, however; most are actions that the brain takes without our knowledge.

Beyond the nervous system, the body has an additional communication-and-control operation in its hormonal or "endocrine" system. This system works ceaselessly, mostly outside our conscious awareness, to handle such things as our growth when we are children and our bodies' water balance when we are adults.

Our goal in this chapter is to review both these communication-and-control operations: the nervous and endocrine systems. We'll start with the nervous system, which is such a wonder when healthy and, as Jimmy G. shows, such a heartbreaker when not.

## 27.1 Structure of the Nervous System

Taking the broadest view, the nervous system includes all the cells in the body that can be defined as nervous tissue plus the sense organs we have, such as the eye and ear. Recall from Chapter 26 that two types of cells make up nervous tissue. These are neurons, which actually transmit nervous system messages; and glial cells, which support neurons and appear to modify neuronal communication.

One helpful way to think about the nervous system is to consider three essential tasks it has to perform. It first has to *receive* information, both from outside and inside our bodies. (For example, what taste sensations are coming in?) It next has to *process* this information. (Do I like this?) Then, it has to *send* information out that allows our body to deal with this input. ("Lift your hand for another spoonful.")

The nervous system in which these activities take place can be divided into two fundamental parts. The first of these is the **central nervous system**: that portion of the nervous system consisting of the brain and the spinal cord. The second is the **peripheral nervous system**: that portion of the nervous system outside the brain and spinal cord, plus the sensory organs (**Figure 27.1a**).

The central nervous system's brain and spinal cord are not physically separate entities; the spinal cord simply expands greatly at its top, and we call this flowering the brain. Because the brain and spinal cord are so different in terms of structure and function, however, they are thought of as individual units.

Meanwhile, the peripheral nervous system, or PNS, can be pictured as a group of nerves and related nerve cells that fan out from either the brain or spinal cord. One way to think about these peripheral nerves and related cells is to consider the direction of the messages they carry. Any nerves that help carry messages *to* the brain or spinal cord are said to be part of the **afferent division** of the peripheral nervous system. Any nerves that help carry messages *from* the brain or spinal cord are said to be part of the **efferent division** of the PNS. An easy way to remember the difference between these two terms is to remember that "efferent" sounds like "effect." And, the efferent division *effects change* in various organs in the body.

What kinds of change? Many of the activities the nervous system controls are voluntary activities, which often means movement. You decide to lift your finger; you decide to stand up, and so forth. You may remember from Chapter 26 that all movement is handled by skeletal muscles—muscles, like the biceps, that are attached to bones and that are under voluntary control. So, one part of the PNS's efferent division is the **somatic nervous system**: that portion of the peripheral nervous system's efferent division that provides voluntary control over skeletal muscle.

But, there are lots of processes in the body that are not under voluntary control. When you walk from bright sunshine into a darkened movie theater, your pupils dilate to let in more light, but you have no control over this dilation. Certain muscles allow our pupils to open up, but these are not skeletal muscles; they are the involuntary "smooth muscles." Likewise, the cardiac muscle of your heart beats in a rhythm that is largely out of your conscious control. And your glands release hormones in a way you have little control over. It is a *second* part of the PNS's efferent division that controls these operations, the **autonomic nervous system**: that part of the peripheral nervous system's efferent division that provides involuntary regulation of smooth muscle, cardiac muscle, and glands (**Figure 27.1b**).

Now let's go down just one more level, this time strictly within the autonomic system. It turns out that the autonomic system is divided into two divisions, the *sympathetic* and the *parasympathetic*. These terms are formally defined later; for now, just be aware of

**(a)** The nervous system has two components

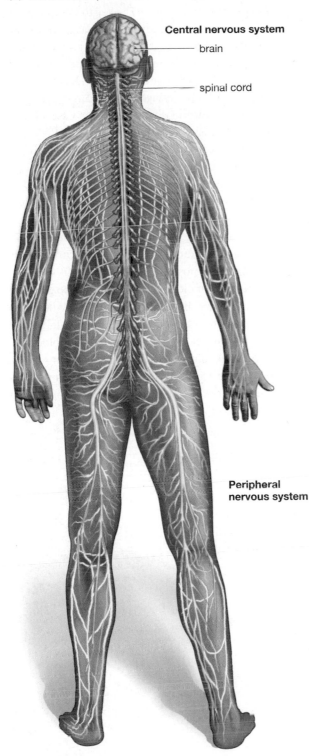

Central nervous system
— brain
— spinal cord

Peripheral nervous system

**(b)** How these two components interact

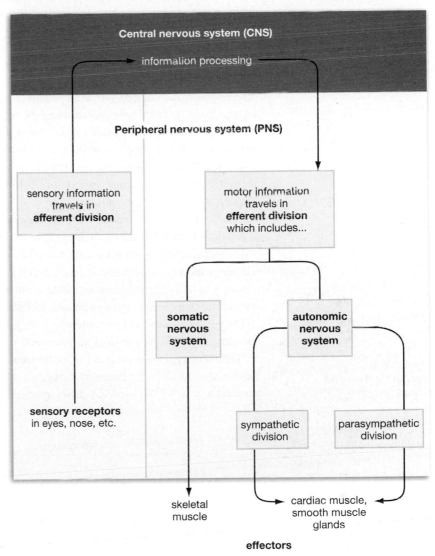

**Figure 27.1**
**Divisions of the Nervous System**

**(a)** The two central branches of the nervous system are the central nervous system (CNS), which consists of the brain and the spinal cord; and the peripheral nervous system (PNS), which consists of all the nervous tissue outside the brain and spinal cord plus the sensory organs.

**(b)** Information about the body and its environment comes to the nervous system from its sensory receptors—for example, cells in the eyes—which are part of the peripheral nervous system. This information then goes through the afferent division of the peripheral nervous system to the brain and spinal cord. After processing this information, the brain and spinal cord issue motor commands through the peripheral nervous system's efferent division. These commands go to "effectors" such as skeletal muscles and glands. The peripheral nervous system's efferent division has two systems within it: the somatic nervous system, which provides voluntary control over skeletal muscles; and the autonomic system, which provides involuntary regulation of smooth muscle, cardiac muscle, and glands. The autonomic system is further divided into sympathetic ("fight-or-flight") and parasympathetic ("rest-and-digest") divisions.

two characteristics they have. First, these divisions differ in that the nerves that make them up stem from different locations in the brain and spinal cord. Second, the sympathetic division generally has stimulatory effects on us—we get adrenaline going through this division, for example—while the parasympathetic division generally has relaxing effects.

## 27.2 Cells of the Nervous System

So, what about the cells that make up these systems? Remember first that the nervous-system cells that transmit signals are called neurons. These cells come in three varieties that neatly parallel the idea of a nervous system that receives, processes, and sends information (**Figure 27.2a**). The first type of neuron is the **sensory neuron**, which does just what its name implies: It senses conditions both inside and outside the body and brings this received information to the central nervous system. (Given the direction of this information, sensory neurons are afferent neurons.) When someone brushes the top of your hand, sensory neurons just beneath the skin sense lots of things about this touch—what direction it came from, what shape the touching object was—and then convey a message containing this information that goes from your hand, into your spinal cord, then up into your brain.

The second type of neuron is the interneuron (or association neuron). Located solely in the central nervous system, **interneurons** interconnect other neurons. These can be very simple connections, but they can be complex as well. The memory we talked about at the start of the chapter amounts to a massive mobilization of interneurons. How do we remember? We process information in complex webs of interneurons.

The third type of neuron is the **motor neuron**: a neuron that sends instructions from the central nervous system (CNS) to such structures as muscles or glands. (Given the direction of this transmission, these are efferent neurons.) The key to understanding motor neurons is to think of them as neurons that transmit messages to organs or tissues *outside* the nervous system, thus prompting some kind of action. If you look at the right side of Figure 27.2a, you can see how this works. A message comes from the cell body of a motor neuron (which lies in the CNS), travels down an extension of it (which lies in the PNS), and then gets transferred to a muscle—causing the muscle to contract. Any organ or group of cells that responds to this kind of nervous system signal is called an *effector*. Thus, a gland prompted to release hormones through a motor neuron signal is also an effector.

To sharpen the point about the difference between sensory and motor neurons, consider the popular drug Botox. Physicians inject this drug into the foreheads of older people who want to get rid of "age lines" there. Botox works by blocking the signal between motor neurons and the muscles they stimulate. The linkage works as follows: Botox blocks the nerve signal going to forehead muscles; this means there will be no forehead muscle contraction, which means there will be no age lines. Now, one of the

**Figure 27.2**
**Types of Neurons and Neuron Anatomy**

**(a)** Three types of neurons

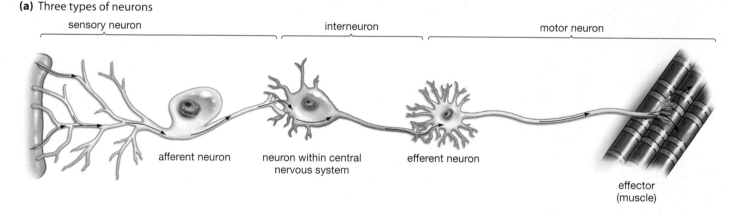

sensory neuron — interneuron — motor neuron

afferent neuron — neuron within central nervous system — efferent neuron — effector (muscle)

**(b)** Anatomy of a neuron

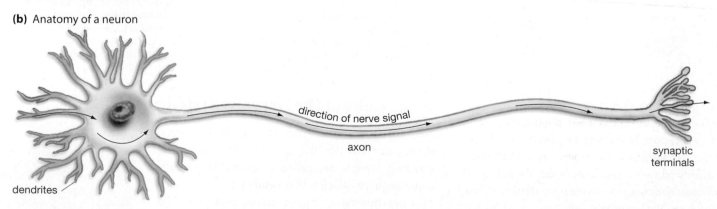

direction of nerve signal

axon

synaptic terminals

dendrites

cell body

frequently asked questions about Botox is: Will it make my forehead numb? The answer is no because Botox is blocking *motor* neuron transmission, not sensory neuron transmission. The motor signals are going out, the sensory signals are coming in, and only the motor neurons are responsive to Botox.

## Anatomy of a Neuron

Now let's look at how a neuron is structured. Like any other cell, a neuron has a nucleus, it makes proteins, it gets rid of waste, and so forth. If you look at **Figure 27.2b**, you can see, however, that the neuron has some extensions sprouting from it that clearly separate it from other cells. Projecting from the cell body are a variable number of **dendrites**: extensions of neurons that carry signals *toward* the neuronal cell body. There is also a single, large **axon**: an extension of the neuron that carries signals *away* from the neuronal cell body. Although you can't see it, the other thing that separates the neuron from other cells is its outer or "plasma" membrane, which has an amazing ability to respond to various kinds of stimulation, as you'll see.

## The Nature of Glial Cells

Having looked a bit at neurons, let's now consider the second kind of nervous system cell, the glial cell (also known as glia). Found in both the CNS and PNS, glia usually are described as "support" cells for neurons, and this is true as far as it goes. Some glia, for example, wrap their cell membranes around the axons of neurons in the CNS and PNS. The membranous covering that glia provide to neurons is called **myelin**, and an axon wrapped in this way is said to be myelinated. The importance of this is that axons that are myelinated carry nerve impulses faster than those that are not. (If this sounds like some technical detail, consider that the disease multiple sclerosis results from a dismantling of the myelin covering CNS axons by the body's own immune system.) Because myelin is fat-rich, areas of the brain and spinal cord containing myelinated axons are glossy white. Thus do we get the term *white matter* of the CNS, which contains mostly axons. By contrast, areas containing mostly neuron cell bodies are *gray matter*. If you look at **Figure 27.3a**, you can see what a myelinated axon looks like in the peripheral nervous system.

The support that glial cells provide to neurons goes beyond myelination, however. Some glial cells provide a framework for CNS neurons, acting as a kind of scaffolding for them, and still others serve as a cleanup crew for CNS neurons by taking in cellular debris—from dead or damaged cells—and disposing of it. Recent research has indicated, however, that glial cells

**(a)** A myelinated axon

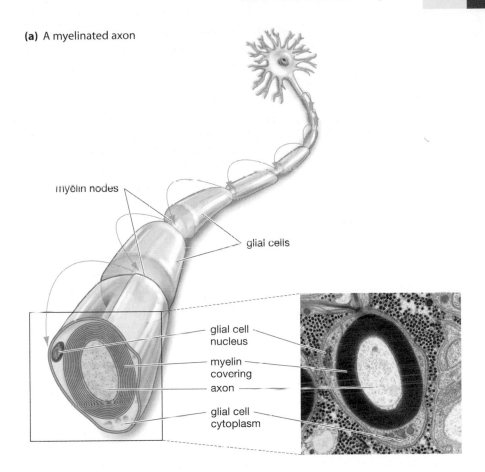

myelin nodes

glial cells

glial cell nucleus

myelin covering

axon

glial cell cytoplasm

**(b)** Anatomy of a nerve

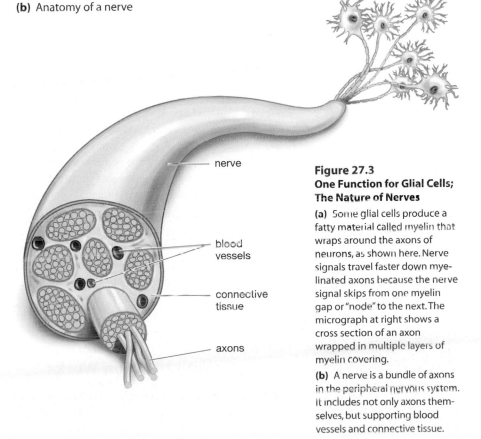

nerve

blood vessels

connective tissue

axons

**Figure 27.3**
**One Function for Glial Cells; The Nature of Nerves**

**(a)** Some glial cells produce a fatty material called myelin that wraps around the axons of neurons, as shown here. Nerve signals travel faster down myelinated axons because the nerve signal skips from one myelin gap or "node" to the next. The micrograph at right shows a cross section of an axon wrapped in multiple layers of myelin covering.

**(b)** A nerve is a bundle of axons in the peripheral nervous system. It includes not only axons themselves, but supporting blood vessels and connective tissue.

are active not just in neuron support, but in neuron communication. Glia communicate, by chemical means, both among themselves and with neurons. Indeed, communication between glia and neurons seems to be coordinated in such a systematic way that it appears that glia are modifying neural communication by, for example, increasing the level of signaling that goes on between differing sets of neurons. Another possible role for glia is that they are important in establishing the connections, called synapses, that allow one neuron to communicate with another.

### Nerves

All this information about nervous system cells and the environment in which they function provides a way to understand a term that's been used extensively thus far but that has not been defined. What is a **nerve**? It is a bundle of axons in the PNS that transmits information to or from the CNS. If you look at **Figure 27.3b**, you can see a representation of a nerve. Note that nerves have support tissue in the form of blood vessels and connective tissue.

---

### SO FAR...

1. The brain and spinal cord make up the _____ nervous system, while all the nervous tissue outside the _____ plus the sensory organs make up the _____ nervous system.

2. The efferent division of the peripheral nervous system is made up of nerves that carry messages _____ the brain or spinal cord, and it is divided up into a portion that provides voluntary control over skeletal muscles—the _____ nervous system—and a portion that provides involuntary regulation over smooth muscle, cardiac muscle, and glands—the _____ nervous system. The afferent division of the peripheral nervous system is made up of nerves that carry messages _____ the brain or spinal cord.

3. The cells of the nervous system that carry nervous system messages are called _____, each of which has a variable number of _____ that carry signals toward the cell body, and a single _____ that carries signals away from the cell body.

---

••• **ANSWERS**

1. central; brain and spinal cord; peripheral

2. from; somatic; autonomic; to

3. neurons; dendrites; axon

---

**Web Animation 27.1**
The Nervous System

---

## 27.3 How Nervous System Communication Works

With this anatomy under your belt, you're ready to see how nervous system signaling works. We'll look at a two-step process, which we can think of as (1) signal

movement down a cell's axon and (2) signal movement from this axon over to a second cell, across a synapse.

### Communication within an Axon

All the action in this first step is going to take place on either side of the thin plasma membrane that constitutes the outer border of any animal cell. Inside the plasma membrane is the cell and all its contents; outside is extracellular fluid (**Figure 27.4**). As was noted in Chapter 5, the plasma membrane regulates the passage of substances in and out of the cell. You may remember that some of these substances are the electrically charged particles called ions, meaning atoms that have gained or lost one or more electrons. Because ions are charged, they get the + or − signs after their chemical names, as with $Na^+$ for the positively charged sodium ion. Such ions cannot simply pass through the plasma membrane; rather, they need to move through special passageways called protein channels.

It turns out that a "resting" neural cell—one not transmitting a signal—has a greater positive charge outside itself than inside. This is so mostly because there are more proteins inside the cell than outside, and proteins carry a *negative* charge. The charge difference also exists, however, because there are more of the positively charged $Na^+$ ions outside the membrane than there are positively charged potassium ions ($K^+$) inside, as you can see in Figure 27.4. As you know, in electricity, opposites attract, meaning the positively charged $Na^+$ ions outside the plasma membrane are attracted to the negatively charged interior of the cell. This amounts to a form of potential energy in the same way that a rock perched at the top of a hill does. The rock would roll down the hill (releasing energy) if given a push, and the $Na^+$ ions would rush into the cell (releasing energy) if the proper channels were opened to them. The charge difference that exists from one side of the neuronal plasma membrane to the other is known as the **membrane potential**.

This potential exists in part because of what you've already seen: the greater abundance of $Na^+$ ions outside the cell. It's important to note that this abundance doesn't exist by accident. Every cell keeps this imbalance in place by constantly pumping three $Na^+$ ions out for every two $K^+$ ions it pumps in. It takes a lot of energy to maintain this *sodium–potassium pump*, but the cell does it for the same reason that you charge your cell phone battery: to have ready access to a capability. In your case, the capability is talking on the phone; in the cell's case, the capability is that of having $Na^+$ ions available to rush into it. Indeed, the action of the

### (a) Resting potential

Electrical energy is stored across the plasma membrane of a resting neuron. There are more negatively charged compounds just inside the membrane than outside of it. As a result, the inside of the cell is negatively charged relative to the outside. The charge difference creates a form of stored energy called a membrane potential. Protein channels (shown in green) that can allow the movement of electrically charged ions across the membrane remain closed in a resting cell, thus maintaining the membrane potential.

### (b) Action potential

1. Nerve signal transmission begins when, upon stimulation, some protein channels open up, allowing a movement of positively charged sodium ions (Na$^+$) into the cell. For a brief time, the interior of the cell becomes positively charged at this location. On either side of this location, however, the interior of the cell remains negatively charged. Attracted by this negative charge, the Na$^+$ ions move laterally in both directions from their point of entry.

2. The Na$^+$ gates close and the gates for positively charged potassium ions (K$^+$) open up, allowing a movement of K$^+$ out of the cell. With this, there is once again a net positive charge outside the membrane. Meanwhile, the arrival of the charged Na$^+$ ions "downstream" from their original point of entry triggers an influx of Na$^+$ ions at the next Na$^+$ channel.

3. Through repetition of this process, the nerve signal is then propagated one way along the axon. Given that Na$^+$ ions move laterally in both directions from their original point of entry, why isn't the signal propagated in both directions? Because once an Na$^+$ channel has opened, it enters a brief period in which cannot respond to any additional stimulus. Thus, each "upstream" Na$^+$ channel remains briefly closed, while each "downstream" channel is opened in succession.

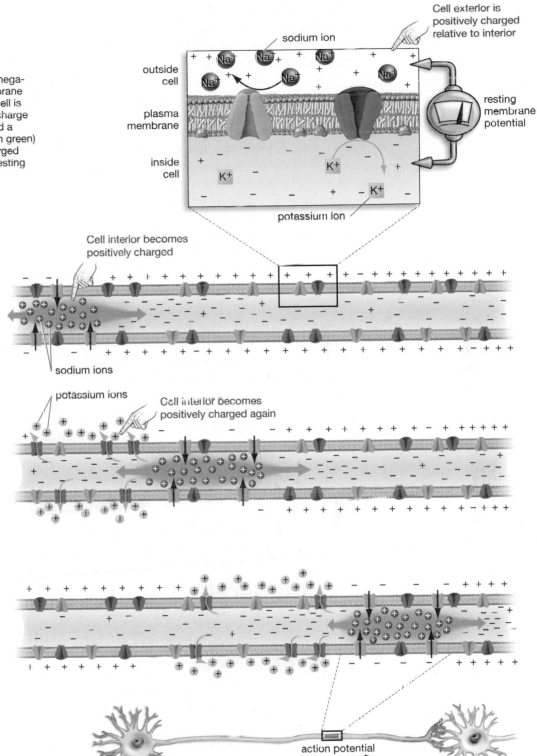

Figure 27.4
**Nerve Signal Transmission within a Neuron**

pump readies a neuron for activity in two ways. In addition to the charge difference it maintains, the pump maintains a *concentration* difference of the ions. The fact that there are more Na$^+$ ions outside the cell than inside means that the Na$^+$ ions will have a natural tendency to move down their "concentration gradient"—they will naturally move from where they are more concentrated to where they are less concentrated. Since they are more concentrated outside the cell than inside, there will be a natural net movement of these ions into the cell. (For more on concentration gradients, see

Chapter 5.) In sum, the sodium–potassium pump maintains both an electrical charge difference and a concentration difference that can get a nerve signal going, as you'll now see.

When a neuron is stimulated by getting a chemical signal from another neuron, the effect of this stimulation is that, up near where the neuron's axon meets its cell body, $Na^+$ ion gates open up. Now the $Na^+$ ions can act on their electrical attraction and concentration difference, and they do so, rushing into the cell. After a very brief time, this influx is great enough that the inside of the cell is more positive than the outside. This very state causes the potassium gates to open up more fully, however. Recall that potassium ions exist in more abundance inside the cell than outside it. With the opening of their gates, they act on their electrical and concentration differences, rushing out of the cell, and the positive charge that has briefly existed on the inside of the cell begins to decline. Eventually, enough ions have moved out that the cell returns to its resting state, with a negative charge inside the membrane. This entire process takes place in a few milliseconds, or thousandths of a second.

### Movement Down the Axon

*BioFlix*

How Neurons Work

All this alters the membrane potential at *one spot* along the neuron's axon. But, how does a nerve signal get transmitted down the entire length of the axon? The key to this transmission is that an influx of $Na^+$ ions at an initial location on the membrane triggers reactions that cause the *neighboring* portion of the membrane to begin an $Na^+$ influx (see Figure 27.4). What occurs, in other words, is a chain reaction that moves down the entire length of the axon membrane. How does this work? Keep in mind that the initial $Na^+$ ions that come into a cell are moving into an environment that is negatively charged. On entry, these positively charged $Na^+$ ions are pulled in both directions along the interior of the membrane by the negative charge they encounter (Figure 27.4). When the $Na^+$ ions reach the first "downstream" $Na^+$ channel in the axon, their charge causes it to open, letting in $Na^+$ ions from outside the cell, and getting the whole process going again. Then this same set of steps is repeated all the way down the axon. You may wonder, however: If the initial $Na^+$ ions that enter the cell are pulled in both directions by the negative charge they encounter, then why isn't the *nerve signal* propagated in both directions from this point of entry? The answer is that an $Na^+$ gate that has just opened goes through a brief "refractory" period in which it cannot open again. Thus, the nerve signal can be propagated in only one direction—down the axon to its terminal. The entire process of axon signal transmission is referred to as an **action potential**: a temporary reversal of cell membrane potential that results in a conducted nerve impulse down an axon. (Temporary reversal? Remember the inside of the cell goes from negative to briefly positive, then back.) Action potentials have been compared to what happens to a lighted fuse: The heat of the spark causes the neighboring section of the fuse to catch fire, thus moving the spark along.

All action potentials are of the same strength. Once the original signal on the cell body/axon border reaches sufficient strength to allow $Na^+$ ions to rush in, the action potential will get going. Thanks to its fuse-like quality, this potential will be the same strength all the way down the axon. How fast can this signal go? At best, about 120 meters per second—very fast indeed, but much slower than the electricity that comes from a wall socket.

## Communication between Cells: The Synapse

Once it has traveled the length of an axon, an action potential then reaches a tip of the axon. How does the signal then get to the *next* neuron (or muscle, or gland cell)? This question was once very intriguing to biologists because it was clear that, with a few exceptions, an axon does not touch the downstream cell. It comes extremely close, but there is almost always a small intervening space between the two cells. There are thus three entities involved in cell-to-cell transmission: the sending neuron, the receiving cell, and the gap between them. The area where all three come together is called a **synapse** (**Figure 27.5**).

The action potential arrives at a branch of the axon, called a synaptic terminal, which has stored within it small sacs (or vesicles) containing a chemical called a neurotransmitter. The arrival of the action potential causes the synaptic terminal to release neurotransmitter molecules into the gap in the synapse, which is called a **synaptic cleft**. With this, the molecules diffuse over to the receiving cell and bind to receptors in that cell's outer membrane. This binding stimulates the opening of sodium gates there, allowing the now-familiar influx of $Na^+$ ions in the receiving neuron, and this keeps the signal transmission going.

But why doesn't the released neurotransmitter just keep on stimulating the receiving cell? Serving as one means of control are the proteins called enzymes. Released into the synaptic cleft, they break down the neurotransmitter, thus inactivating it. The sending cell may also be capable of taking the neurotransmitter back into itself by a process known as "reuptake." In a given nerve impulse, a sending cell typically releases tens of thousands of neurotransmitter molecules, which are stored in hundreds of sacs. With all this in mind, you can understand the definition of a **neurotransmitter**: a chemical, secreted into a synaptic cleft by a neuron, that affects another neuron or an effector by binding with receptors on it.

## The Importance of Neurotransmitters

It is difficult to overstate the importance of neurotransmitters. You may recall from Chapter 26 that a neurotransmitter must travel from nerve cell terminals to muscles for any skeletal muscle to work. If this neurotransmitter (called acetylcholine) doesn't make it out of a motor neuron, we can't contract the muscles that allow us to breathe. Likewise, the shaking and stiffness of Parkinson's disease are caused by a lack of the neurotransmitter dopamine, which in turn is brought on by the death of dopamine-producing cells in the brain.

On a different level, remember the reuptake process we talked about, by which a releasing cell will take back a neurotransmitter that it has secreted into a synaptic cleft? Well, the antidepressant drugs Prozac, Zoloft, and Paxil are all called SSRIs, which stands for selective serotonin reuptake inhibitors. Serotonin is a neurotransmitter found in the brain, and SSRIs are aimed at reducing serotonin reuptake. The result is an increased amount of serotonin in synaptic clefts—and perhaps less depression. By the same token, drugs of abuse such as cocaine and the amphetamine-like ecstasy work by altering neurotransmitter release or reuptake. What makes us feel bad, or OK, or ecstatic? A big part of the answer is: the levels of neurotransmitters in our brains.

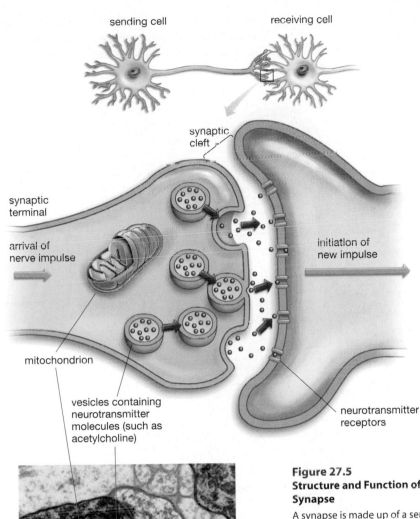

**Figure 27.5
Structure and Function of a Synapse**

A synapse is made up of a sending neuron, a receiving cell, and the gap between them, called a synaptic cleft. As a nerve impulse reaches the synapse, molecules of a neurotransmitter (such as acetylcholine) are released into the synaptic cleft from synaptic terminals of the sending neuron. This occurs when small membrane-bound vesicles—each filled with thousands of neurotransmitter molecules—fuse with the outer membrane of the synaptic terminal. The neurotransmitter molecules then move across the cleft and stimulate receptors in the receiving cell's membrane, thus initiating a nerve impulse within it. The lower figure is a color-enhanced micrograph of a synapse.

---

### SO FAR...

1. Transmission of a nerve signal down an axon takes advantage of something called a membrane potential, which can be defined as a difference in _____ that exists between the inside and the outside of a neuron.

2. The key to signal transmission down the length of an axon is that an influx of Na⁺ ions at an initial location on the axon membrane triggers an _____ on a _____ of the membrane.

3. Communication between neurons takes place by means of substances called _____ moving across a _____ from a sending cell to a receiving cell.

## 27.4 The Spinal Cord

Now that you've looked at how nerve impulses are transmitted at the cellular level, it's a good time to take a step back and look at the nervous system in its larger dimensions. Let's start by examining the spinal cord, which serves two key functions. First, it can act as a communication center on its own, receiving input from sensory neurons and directing motor neurons in response, with no input from the brain. This is what a reflex amounts to. Second, most sensory impulses that go to the brain don't go *directly* to the brain; they are channeled first through the spinal cord.

Extending from the base of the brain to an area just below the lowest rib, the spinal cord is about 14 millimeters or a half inch wide and consists of

31 segments, each having both a left and right spinal nerve stemming from it (**Figure 27.6a**). These nerves are grouped into the classes you see named in the figure, which correspond to the region of the body they

**(a)**

brain

cervical spinal nerves

thoracic spinal nerves

lumbar spinal nerves

sacral spinal nerves

tip of spinal cord

**(b)**  white matter  ventral root
gray matter  spinal nerve
dorsal root ganglion  central canal  dorsal root

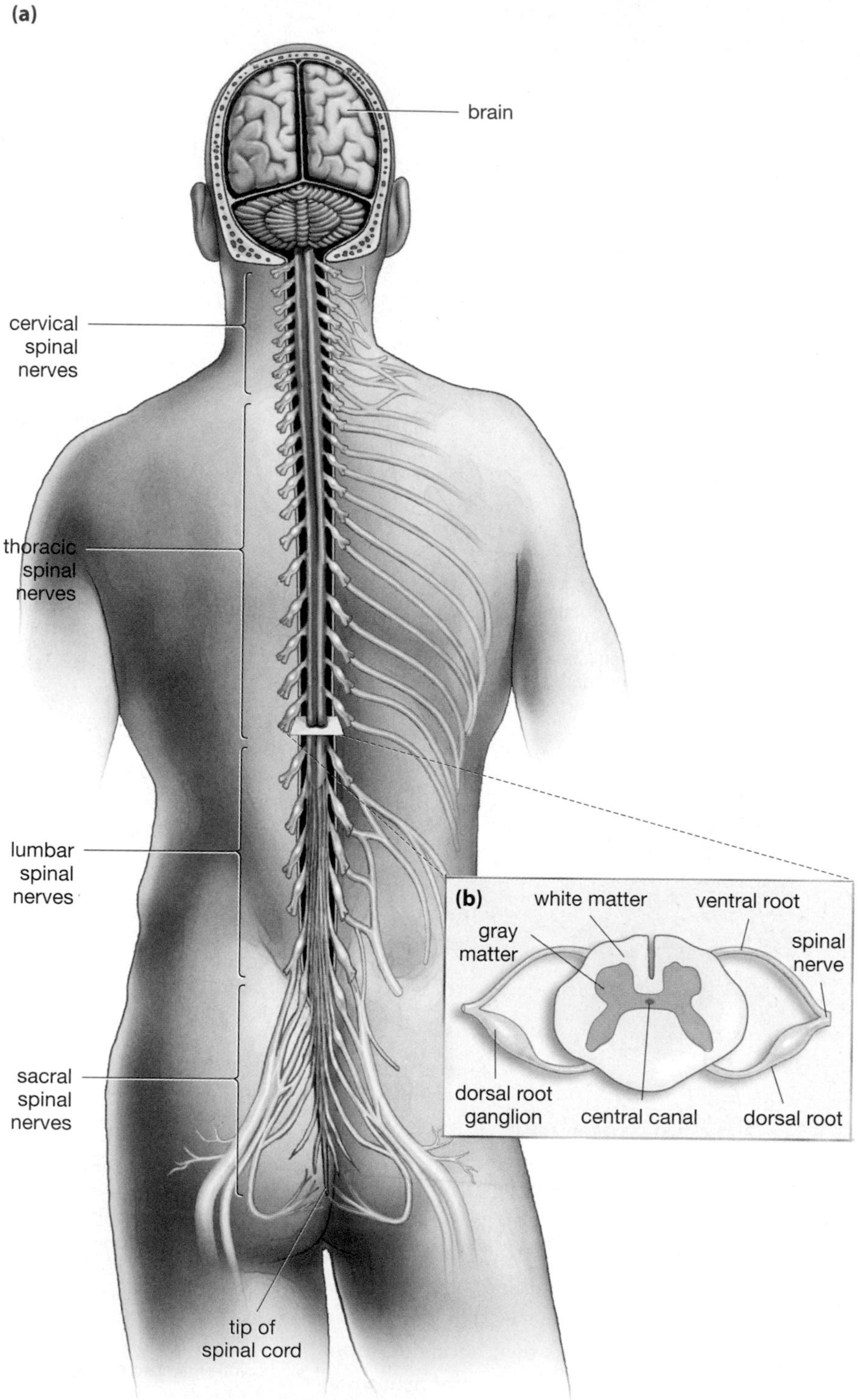

**Figure 27.6**
**Structure of the Spinal Cord**

**(a)** Thirty-one pairs of spinal nerves extend from the spinal cord. These nerves are grouped according to the body part they control. Eight cervical nerves control the head, neck, diaphragm, and arms; twelve thoracic nerves control the chest and abdominal muscles; five lumbar nerves control the legs; and five sacral nerves control the bladder, bowels, sexual organs, and feet.

**(b)** A cross section through the lower thoracic region of the spinal cord showing the arrangement of gray and white matter, the central canal, the ventral roots, the dorsal roots, and the dorsal root ganglia.

serve. ("Thoracic" is the chest area; "lumbar" the lower back; and so forth.) Remember how we noted that the peripheral nervous system could be thought of as a group of nerves and related nerve cells that fan out from the brain and spinal cord? In Figure 27.6a, you can see the start of a lot of this fanning, in the form of the spinal nerves. (The other major set of nerves that do this are the so-called cranial nerves, which fan out from the brain.)

The most striking feature of a cross section of the spinal cord, seen in **Figure 27.6b**, is a rough H- or butterfly-shaped area around the narrow central canal. This H is the gray matter of the spinal cord, composed mostly of the cell bodies of neurons. Surrounding the spinal cord's gray matter is its white matter, which, as you've seen, is largely myelin-coated axons. The central canal of the spinal cord is a space filled with **cerebrospinal fluid**, a fluid that circulates both in the spine and the brain, supplying nutrients, hormones, and immune system cells and providing the brain with protection against any jarring motion.

### The Spinal Cord and the Processing of Information

Sensory input comes *to* the spinal cord from various sensory receptors in the body, while motor commands must go *from* the spinal cord to effectors such as muscles. Where are the spinal cord cells that take care of this business? The motor neurons that send the motor commands lie in the spinal cord's gray matter (as do the interneurons discussed earlier). The axons of these motor neurons leave the spinal cord through the ventral roots you can see in Figure 27.6b. Meanwhile, the cell bodies of sensory neurons lie outside the spinal cord in the dorsal root ganglia you can see in Figure 27.6b. (A **ganglion** is any collection of nerve cell bodies in the PNS.) These sensory neurons have dendrites whose endings lie in, say, the hand. When we touch something, these dendrites' terminals sense it and send a message back to the cell body in the dorsal root ganglion. From there, this signal proceeds via the cell's axon into the spinal cord. In sum, sensory information comes to the spinal cord through a dorsal root, while motor commands leave the spinal cord through a ventral root. Note that both dorsal and ventral roots come together, like fibers being joined in a single cable, to form a given spinal nerve.

### Quick, Unconscious Action: Reflexes

The sensory/motor division of labor is clearly reflected in simple body **reflexes**, which can be defined as automatic nervous system responses that help us avoid danger or preserve a stable physical state. If you accidentally touch a hot stove, you automatically pull back your hand. No conscious thought of, "Oh, I've touched a hot stove; I'd better pull my hand back," is necessary; the reaction is automatic. The neural wiring of a single reflex is called a *reflex arc*. A reflex arc begins at a sensory receptor, runs through the spinal cord, and ends at an effector, such as a muscle or gland. **Figure 27.7** shows one of the best-known examples, the knee jerk, or patellar reflex, in which a properly placed sharp rap on the knee produces a noticeable kick. Note what's at work: A sensory receptor is stimulated, thus prompting a signal to move into the spinal cord. There, the sensory neuron is linked via a synapse to a motor neuron, which issues a command that is carried out by an effector—in this case a set of skeletal muscles. This is an example of the simplest possible reflex arc in that the sensory neuron is linked directly to the motor neuron. Because only one synapse is involved, this kind of simple reflex controls the most rapid motor responses of the nervous system. Many other reflexes have at least one interneuron placed between the sensory receptor and the motor neuron.

## 27.5 The Autonomic Nervous System

As noted earlier, the part of the peripheral nervous system's efferent or "outgoing" division over which we have no conscious control is called the autonomic nervous system. You can grasp its importance by imagining having to consciously control the digestion of the food you eat. Recall that this system controls the involuntary regulation of smooth muscle, cardiac muscle, and glands.

### Sympathetic and Parasympathetic Divisions

Also recall that, within the autonomic nervous system, there are two "divisions," the sympathetic and parasympathetic. The **sympathetic division**, often

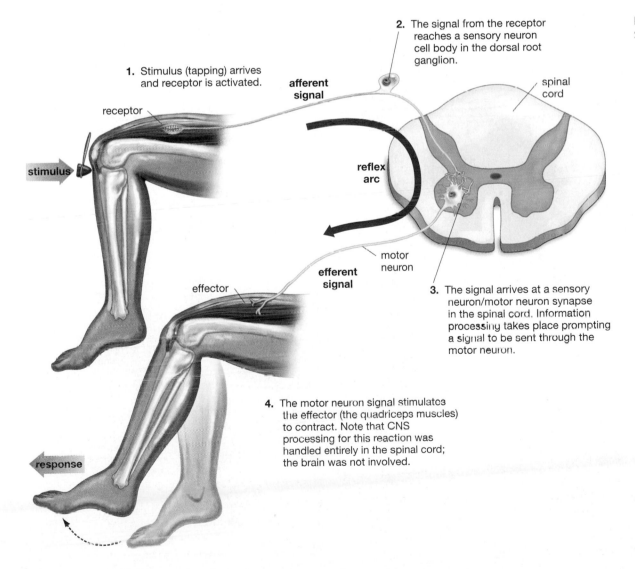

**1.** Stimulus (tapping) arrives and receptor is activated.

**2.** The signal from the receptor reaches a sensory neuron cell body in the dorsal root ganglion.

**3.** The signal arrives at a sensory neuron/motor neuron synapse in the spinal cord. Information processing takes place prompting a signal to be sent through the motor neuron.

**4.** The motor neuron signal stimulates the effector (the quadriceps muscles) to contract. Note that CNS processing for this reaction was handled entirely in the spinal cord; the brain was not involved.

afferent signal

spinal cord

reflex arc

efferent signal

motor neuron

receptor

stimulus

effector

response

**Figure 27.7**
**Steps in a Reflex Arc**

**Figure 27.8**
**Involuntary Control of Bodily Functions**

The autonomic nervous system has two divisions, the sympathetic and the parasympathetic, which exercise automatic control over the body's organs, generally in opposing ways. Note how the parasympathetic aids in digestion, while the sympathetic controls "fight-or-flight" responses, such as an increase in heart rate. Axons of the parasympathetic division emerge not only from the spinal cord but from the brain as well.

called the "fight-or-flight" division of the autonomic nervous system, usually stimulates tissue metabolism, increases alertness, and generally prepares the body to deal with emergencies. What does this mean in practice? Among other things, sympathetic stimulation puts adrenaline into circulation, it diverts blood from our digestive tract to our skeletal muscles, and it opens our lung passages further—a useful set of responses when we're faced with danger. In contrast, the autonomic system's **parasympathetic division** generally conserves energy and promotes what might be called routine maintenance activities. Parasympathetic stimulation slows our heart rate, constricts the pupils in our eyes (thus reducing the amount of light that comes in), and increases muscular activity along our digestive tract. Given effects such as these, the parasympathetic is sometimes called the "rest-and-digest" division of the autonomic nervous system.

In **Figure 27.8**, you can see which spinal and cranial nerves are part of both the parasympathetic and sympathetic divisions. You can also see a list of the effects each division has on various bodily functions. Although some organs are connected to only one division or the other, most vital organs receive both sympathetic and parasympathetic signals. Where such dual signaling exists, the two divisions often have opposing effects, keeping the body's stability mechanisms working in balance.

## 27.6 The Human Brain

Now that we've reviewed the spinal cord, it's time to look at the second major part of the central nervous system. The adult human brain is far larger and more complex than the spinal cord. Its 100 billion neurons constitute about 98 percent of the neural tissue in the

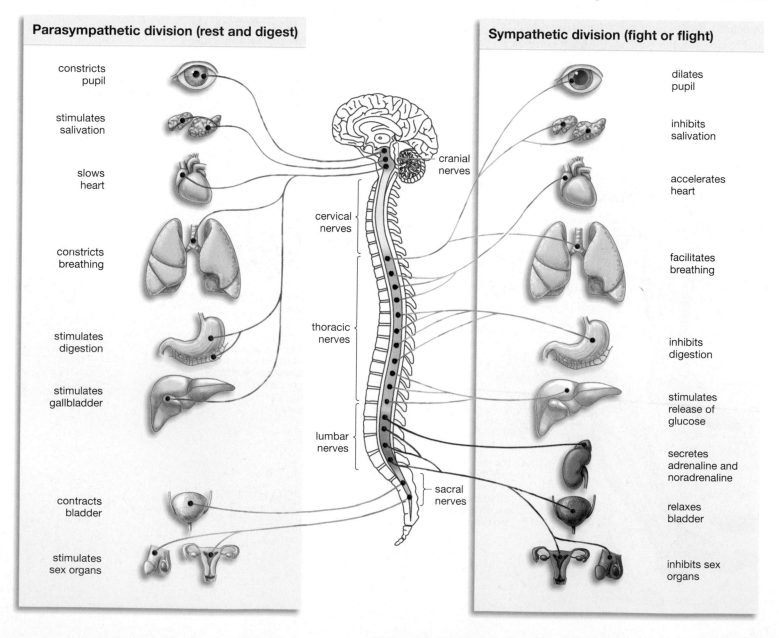

**Parasympathetic division (rest and digest)**

constricts pupil

stimulates salivation

slows heart

constricts breathing

stimulates digestion

stimulates gallbladder

contracts bladder

stimulates sex organs

cranial nerves

cervical nerves

thoracic nerves

lumbar nerves

sacral nerves

**Sympathetic division (fight or flight)**

dilates pupil

inhibits salivation

accelerates heart

facilitates breathing

inhibits digestion

stimulates release of glucose

secretes adrenaline and noradrenaline

relaxes bladder

inhibits sex organs

## Spinal Cord Injuries

I n the movie *Million Dollar Baby*, Maggie, the boxer played by Hilary Swank (**Figure 1**), sustains an injury that she says has made her "a complete C1–C2." What she means is that her spinal cord was severed at the level of cervical spinal nerves 1 and 2—C1 and C2. If you look at Figure 27.6 on page 562, you can see the cervical spinal nerves bracketed. There are eight of these nerves, the topmost of them being C1, the next C2, and so forth. (The thoracic spinal nerves are then called T1, T2, etc.) Maggie's condition was so terrible because she sustained an injury so high up in her spinal cord.

People whose spinal cords are severed between the T1 and upper lumbar segments become paraplegic—they lose at least partial use of their legs. People whose injury comes between C5 and T1 become quadriplegic—they lose at least partial use of both their legs and arms. But people whose injury comes above C3 lose all this functionality *and* the ability to breathe on their own. Why? The nerve signals that raise our diaphragm—thus allowing us to inhale—travel through the C3–C5 spinal nerves. Since Maggie's injury was above C3, she could be kept alive only by being connected to a ventilator.

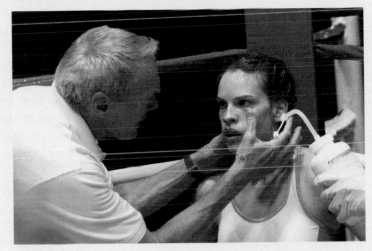

**Figure 1**
**In the Ring**

Clint Eastwood and Hilary Swank in a scene from *Million Dollar Baby*.

body. An average adult brain weighs 1.4 kilograms (a little over 3 pounds), is about the size of a grapefruit, and has the consistency of cream cheese. Like the spinal cord with its spinal nerves, the brain has nerves extending from it that allow it to communicate directly with other body tissues and organs. These are the cranial nerves and, as you might expect, some of them go to nearby organs (the eyes, the nose, the face muscles), but some also go to organs that are farther away (the heart and lungs). If you look at **Figure 27.9**, you can see the major structures that make up the brain. Let's look briefly now at each of them.

### Six Major Regions of the Brain

- The **cerebrum**, the largest region of the human brain, is responsible for much of our higher mental functioning. With its valley-like fissures and fatty appearance, the roundish cerebrum fills up most of our skulls and is effectively draped over many other portions of the brain. It is divided into left and right *cerebral hemispheres*. The outer layer of the cerebrum is the **cerebral cortex**, the site of

our highest thinking and processing. It covers the entire cerebrum, but it amounts to a *thin* covering—only 2 or 3 millimeters thick, which means it is

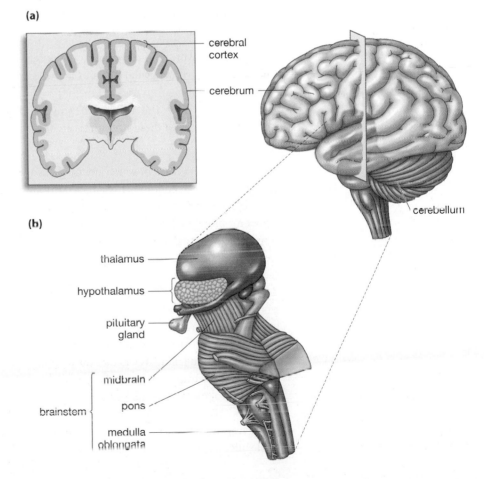

**(a)**

**(b)**

**Figure 27.9**
**The Brain**

**(a)** View of a cross section of the brain showing its cerebrum and cerebral cortex and of the left surface of the brain.

**(b)** Structures of the brain outside the cerebrum and cerebellum. The pituitary gland is connected to the hypothalamus but is a part of the body's endocrine system. The brainstem is made up of structures called the midbrain, pons, and medulla oblongata.

little more than a 10th of an inch deep at most. Its billions of neurons are arranged in functional columns within this area, and they are extremely varied in their activities. The cerebral cortex has, for example, areas called the visual cortex and the olfactory cortex, which are the primary processing centers for sight and smell, respectively. Likewise, there is a sensory cortex that gets touch information from our skin. Then, there is a "prefrontal" association area, up toward our foreheads, that is involved in "executive" functioning, meaning it helps prioritize our behaviors. As may be apparent, these cortex areas are regions of higher processing. Out in the body, there are neurons that sense various forms of stimulation, but the cerebral cortex *makes* sense of the information these neurons provide. The cerebrum is more than the cerebral cortex, however. It also contains an area called the hippocampus, which plays a critical role in forming long-term memories; and the amygdala, which functions in the fear response and our memory of it.

- The *cerebellum* is a portion of the brain that refines our movements based on our "sense memory" of them. People who have damaged their cerebellum may find that their movements have become jerky, or that they reach too far (or not far enough) for everyday objects. The cerebellum also helps us maintain balance and equilibrium.

In looking at the brain, if you take away the cerebrum and the cerebellum, what remains are (a) the thalamus and hypothalamus and (b) the brainstem, which is composed of the midbrain, pons, and medulla oblongata. All of these structures can be seen in Figure 27.9b.

- When we see, hear, taste, or touch anything, the resulting sensory perceptions are channeled through the **thalamus** before moving on to the cerebral cortex for more processing. Lying below the thalamus is the hypothalamus, which is small but extremely important in several ways. Since we'll be seeing it several times in this chapter, a formal definition might be helpful. The **hypothalamus** is a portion of the brain that is important in regulating drives and in maintaining homeostasis in the body. With respect to drives, cells in the hypothalamus are critical in telling us whether we are hungry or thirsty, for example. Likewise, the hypothalamus helps control our sleep-wake cycle, or "circadian rhythms." And the hypothalamus is important in preserving homeostasis—the maintenance of a stable internal environment—because it controls a

good number of the hormones that are released from our glands, as you'll see later.

- The *midbrain* helps us maintain muscle tone and posture through control of involuntary motor responses. With the amygdala, the midbrain is involved in our fear responses. Also located in the midbrain are the dopamine-producing cells noted earlier that keep our hands steady and our movements fluid.

- The Latin term *pons* refers to a bridge, and the pons serves as one. Its primary function is to relay messages between the cerebrum and the cerebellum. It also helps control involuntary breathing.

- Located next to the spinal cord, the *medulla oblongata* contains major centers concerned with the regulation of unconscious functions such as breathing, blood pressure, and digestion. The medulla oblongata actually has a connection to a well-known term. What does it mean to be "brain dead"? Under one definition, it means that all the centers of the brain *except* the medulla oblongata have permanently ceased to function. As long as the medulla is still working, a person continues to breathe despite the loss of all conscious ability.

The brain and spinal cord contain internal cavities filled with the cerebrospinal fluid noted in connection with the spinal cord. In the brain, this fluid provides cushioning for delicate neural structures—the brain essentially floats in the cerebrospinal fluid.

---

**SO FAR...**

1. The spinal cord can receive input from _____ and direct _____ in response, without input from the brain. This entire process is known as a _____.

2. The autonomic nervous system is divided into the sympathetic or "_____" division and the parasympathetic or "_____" division.

3. The portion of the brain that is the site of our highest thinking and processing, the _____, is the outer portion of the brain's _____.

---

## 27.7 The Nervous System in Action: Our Senses

Having learned some of the basics of the nervous system, we can see how it works in one of its most interesting areas, sensory perception.

**••• ANSWERS**

1. sensory neurons; motor neurons; reflex

2. fight-or-flight; rest-and-digest

3. cerebral cortex; cerebrum

It may come as a surprise to learn that we actually have more sensory capabilities than the famous "five senses" of vision, touch, smell, taste, and hearing. We have a sense of balance, for example, but it isn't one of these five senses. We have another little-recognized sensory capability, related to touch, that helps us do such things as dance and simply sit up straight. To demonstrate this capability to yourself, extend one arm out, parallel to the ground, and close your eyes. Now, use the same arm and touch your index finger to the tip of your nose. How were you able to do this? You have sensory neurons that monitor such things as the position of your joints and the tension of the tendons in your body. These neurons provide a sense, called *proprioception*, that in this instance helped guide your finger to your nose. (It sent out signals that said "on the right course" or "correction needed.") How do you know if you're sitting up straight? Proprioception is part of the answer.

Taking these and other sensory capabilities into account, it becomes apparent that our "five senses" really amount to the senses we are consciously aware of. It is understandable that these senses command our attention, however, because they alone make us not just sensory beings, but sensual beings. (If you wanted to evoke the richness of life in as short a space as possible, you could do worse than to write down just five words: seeing, touching, hearing, smelling, and tasting.) As such, we'll briefly review how each of these senses works. But first, it's worth noting what all the senses have in common.

Each of our senses employs the sensory receptor cells noted earlier—cells that can respond to stimulation, such as vibration (for sound) or reflected light (for vision). In addition, all sensory receptor cells *transform* their responses into the language of the nervous system, meaning electrical signals that work through action potentials. Think of this task as similar to the one undertaken by a so-called transducer clipped across the sound hole of an acoustic guitar. It has to turn the vibration of the guitar strings into an electric signal that goes to an amplifier. In a similar way, sensory receptors in your ear must turn the vibrations of your eardrum into an electric signal that goes to your brain. Fittingly enough, sensory receptors are sometimes called *transducers*. All the senses except one also have another commonality to them. If you look at **Figure 27.10**, you can see that, except for smell, all sensory perceptions are routed through the brain structure noted earlier called the thalamus, and that all these signals then go on to various parts of the cerebral cortex for additional processing. Now, let's look at the five best known senses, taking touch first.

## 27.8 **Our Senses of Touch**

Humans actually have several senses of touch. One of these is *thermal reception*, which means sensing hot and cold. Another is *tactile reception*, which includes our senses of pressure and the proprioception we talked about. A third is *pain reception*, which means just what it sounds like.

Within our skin, we have several kinds of sensory receptors that contribute to these senses. Thermal and pain sensing operate through simple nerve endings near the surface of the skin. But the tactile sense has five special types of receptors in the skin alone. Why so many? Well, pressure can be light or heavy, and we want to know the difference. Likewise, contact can be brief or ongoing, and we need to assess the difference in a sophisticated way. We want the message "your elbows have reached the table" to come to our attention front and center, but we want the message "your elbows are still on the table" to be a *low-level* message, so that we can focus on other

**Figure 27.10**
**The Routing of Sensory Information**

Nerve signals for all the senses except smell get routed through the brain's thalamus and then continue on to differing parts of the cerebral cortex for further processing. Signals for smell are routed through the olfactory cortex, the amygdala, and the hypothalamus.

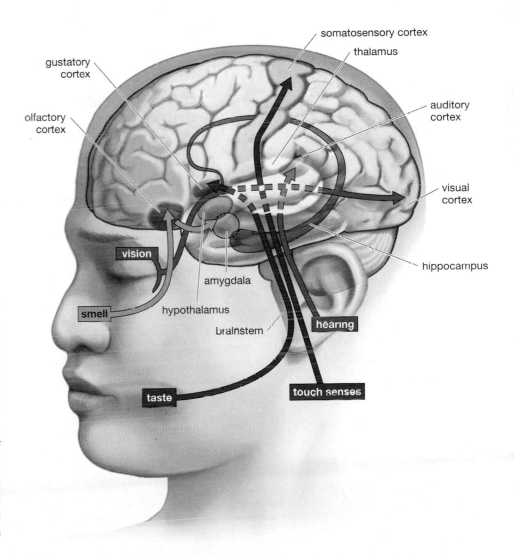

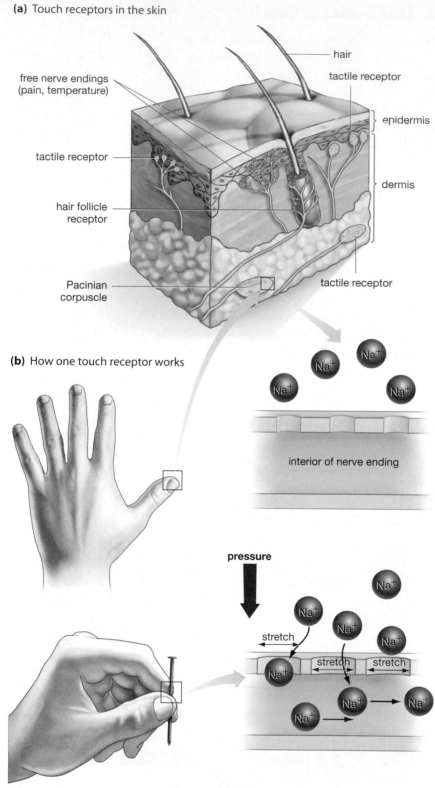

**(a)** Touch receptors in the skin

hair

free nerve endings
(pain, temperature)

tactile receptor

epidermis

tactile receptor

dermis

hair follicle
receptor

Pacinian
corpuscle

tactile receptor

**(b)** How one touch receptor works

Na⁺  Na⁺  Na⁺  Na⁺  Na⁺

interior of nerve ending

**pressure**

Na⁺  Na⁺  Na⁺  Na⁺

stretch

Na⁺  stretch  stretch

Na⁺  Na⁺  Na⁺

Na⁺

**Figure 27.11**
**Our Senses of Touch**

**(a)** Free nerve endings near the surface of the skin sense pain and temperature, while five different types of sensory receptors in the skin sense other sorts of tactile sensations.

**(b)** One of the skin's sensory receptors is the Pacinian corpuscle. When the dendrite within the corpuscle is distorted by touch, its ion channels are stretched in such a way that it allows an influx of sodium ions ($Na^+$). This instigates a nerve signal conveying the message "Touch has occurred here."

things. These various receptors allow for all this and much more.

If you look at **Figure 27.11a**, you can see several varieties of touch receptors, and if you look at **27.11b**, you can see how one particular receptor works. The Pacinian corpuscle receptor consists of a single dendrite that is wrapped within onion-like layers of fluid and supportive tissue (which make up the corpuscle). As a result, this receptor is not likely to respond to gentle pressure. Instead, it responds to somewhat deeper pressure, usually involving vibrations (like pulling your finger across a page). Undisturbed by touch, the ion channels in this receptor's outer membrane are not shaped correctly to let through the sodium ($Na^+$) ions that get a nerve signal going. When stretched, however, the ion channels open in such a way that the $Na^+$ ions rush in, and a touch message gets going. Note that the Pacinian receptor does the double duty of first responding to a stimulus (pressure) and then transforming this stimulus into the language of the nervous system.

Once the Pacinian's nerve signal is on its way, it comes into the spinal cord through one of the spinal nerves. It then moves up through the spinal cord and brainstem and into the brain's thalamus, from which point it is sent to the brain's somatosensory cortex. It's worth noting that sensory messages such as these always cross over a midpoint in the spinal cord, so that anything we feel on the left side of our body is processed on the right side of our brain and vice versa.

## 27.9 Our Sense of Smell

As noted earlier, our sense of smell stands alone in not routing its messages up through the thalamus. If you look at Figure 27.10, you can see where these messages go: to a part of the cerebral cortex (called the olfactory cortex) and to the amygdala and hypothalamus. These last two structures are active in both emotion and memory. Some scientists believe this linkage is the reason that our sense of smell—or **olfaction**, as it is properly known—is able to evoke such strong memories. People, places, and events can all come back to us in a rush if we happen to encounter the right smell. Olfaction is also thought to be our "oldest" sense in that some of the olfaction mechanisms at work in us are also at work in fruit flies, even though our evolutionary line diverged from theirs about 500 million years ago.

If you look at **Figure 27.12**, you can see how olfaction works. Air that bears fragrant molecules comes into our nasal passages. Right up at the top of these passages, back behind the plane of our forehead and just beneath the level of our eyes, we have a group of

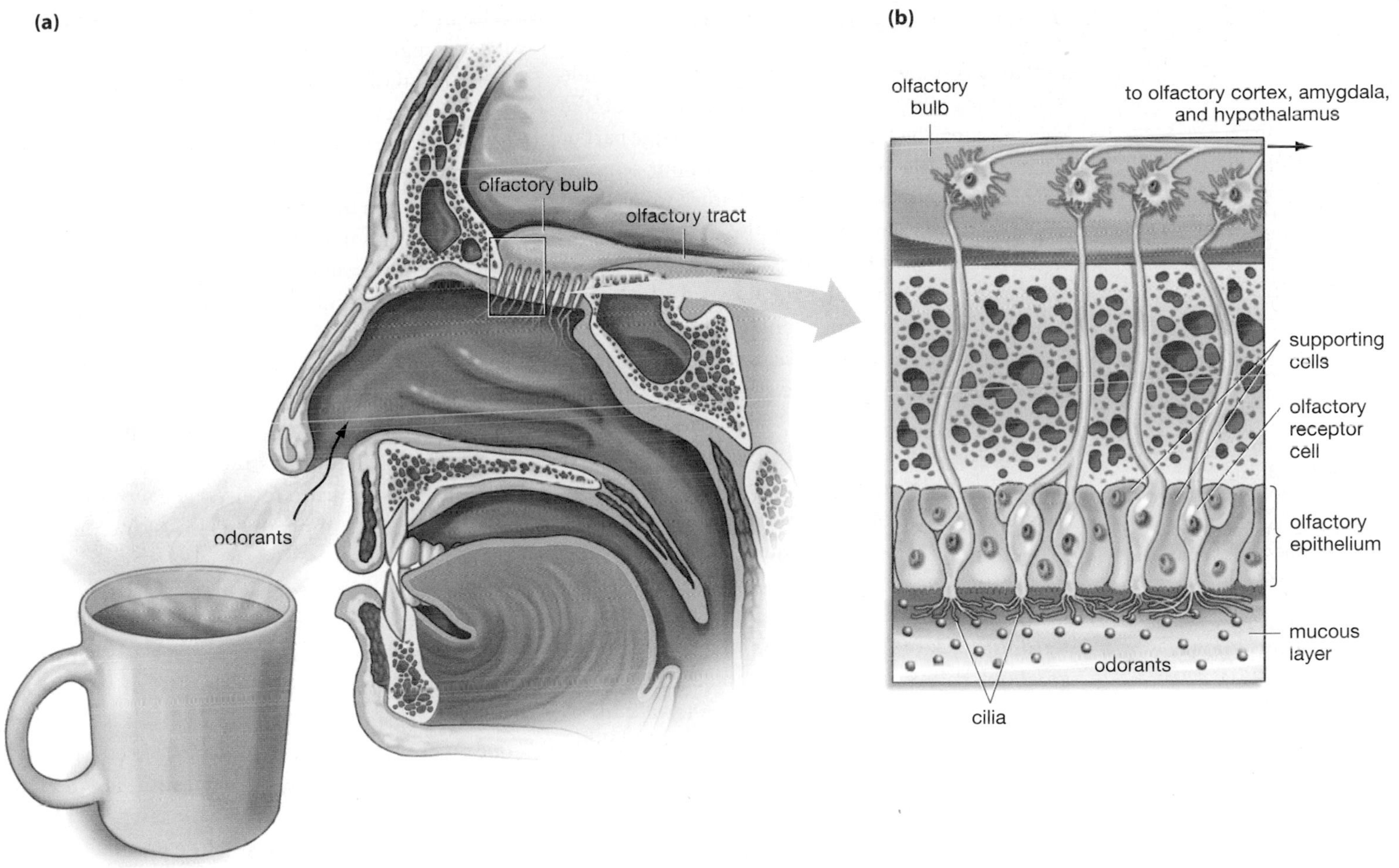

**(a)**

olfactory bulb

olfactory tract

odorants

**(b)**

olfactory bulb

to olfactory cortex, amygdala, and hypothalamus

supporting cells

olfactory receptor cell

olfactory epithelium

mucous layer

odorants

cilia

neurons, called olfactory receptor cells, whose dendrites extend down into the passages. These dendrites have some hairlike extensions, called cilia, sprouting from their tips. Each of these cilia is covered with receptors capable of binding with the fragrance molecules. These fragrance molecules, which are called "odorants," come in, get trapped in a mucus layer that lines the nasal passages, and soon bind with receptors on the cilia. This binding puts into motion a complicated set of reactions inside the dendrite that results in $Na^+$ channels being opened up in the dendrite membrane. The result is a neural signal that goes up through the dendrite and then the cell as a whole, which synapses with neurons in the olfactory bulb—a nerve tract that is part of the brain. From there, the signal gets distributed in the pattern we talked about (to the amygdala, hypothalamus, etc.).

The receptors lying on the surface of the olfactory receptor cilia seem to come in about 350 different varieties—at most, there may be a thousand varieties. This sounds like a lot, but consider that human beings are thought to be capable of discriminating among about 10,000 different smells. How can 350 to 1,000 receptor types yield 10,000 smells? Well, as researcher Richard Axel has noted, a mere 26 letters in

the English language yield a huge number of words because the letters can be *combined* in different ways. In olfaction, each receptor adds nothing more than a message saying "receptor here stimulated"—one letter in a word. But the brain is receiving input from all these receptors. When a given combination of receptor signals comes in, the brain senses a "word," meaning the smell of, say, chocolate or lemon. The brain is essentially saying, "When receptors 4, 32, 87, and 220 fire together, that's chocolate."

**SO FAR...**

1. Each of our senses employs the cells known as _____, which can both respond to stimulation and transform their responses into _____.

2. All senses except _____ route their signals through the brain's _____, after which these signals go to differing parts of the _____ for additional processing.

3. The 350 to 1,000 types of olfactory receptors we have allow us to discriminate among an estimated 10,000 smells because the various receptor types can send signals in different _____.

**Figure 27.12**
**Our Sense of Smell**

Smell or "olfaction" begins when fragrance molecules, called odorants, bind with receptors that cover cilia extending from olfactory receptor cells. This binding causes an influx of $Na^+$ ions into the cells, resulting in a nerve signal that moves from the cells to the olfactory bulb—a part of the brain.

••• **ANSWERS**

1. sensory receptors; electrical signals

2. smell (olfaction); thalamus; cerebral cortex

3. combinations

## 27.10 Our Sense of Taste

Pity the person who eats a piece of chocolate while also having a stuffed-up nose. The flavor of the chocolate may be lost entirely because a large part of what we call "flavor" comes not from our sense of taste but from our sense of smell. Earlier, we pictured smells as coming in from the nose to the nasal passages, but remember that these same nasal passages empty out into the top of the throat. Every bite of food you take sends a huge concentration of odorant molecules wafting up the *other* direction—from the throat into the nasal passages.

When we consider pure "taste," as opposed to "flavor," the action does take place on our tongues, within several kinds of bumps that lie on the tongue's surface (some of which are visible after you drink milk). If you look at **Figure 27.13**, you can see the locations of some of these "papillae," and some detail on one of the larger varieties, which exist toward the back of the tongue. As you can see, if you go about halfway down these papillae, you get to something that sounds familiar: the taste bud, so named because each one actually looks something like a plant bud. Each taste bud is a collection of 50 to 100 cells, some of these being touch receptors and some supporting cells. But taste buds also contain *taste cells*, which is what we are interested in. Taste

cells are not neurons, but they do have the ability to convey signals. Like the olfactory receptors we looked at, they have tiny hairlike projections extending from them—in this case, projecting out into the cavity that surrounds each papilla. These projections are capable of interacting with "tastants," meaning food molecules that elicit different tastes.

With all this in mind, here's how taste works. You eat some food, and the resulting tastants dissolve in your saliva. Molecules or ions from the tastants then make contact with the projections (called microvilli) sprouting from the taste cells. Through one of several chemical reaction routes, this contact results in a change in the membrane potential of the taste cells, and this change prompts them to release a neurotransmitter. Waiting to receive the neurotransmitter are the dendrites of nearby neurons, and our nerve signal is on its way.

You may have heard that humans have four basic taste sensations: sweet, sour, salty, and bitter. What makes these "basic" is that each one works through a different chemical reaction route in the taste cell. Salty foods spur neurotransmitter release through one set of reactions; sweet foods spur release through another. In recent years, some scientists have become persuaded that there is a fifth basic taste—a "meaty" one called *umami* that is close to the taste of monosodium glutamate (MSG)—but this is still a matter of contention. What is clear is that every taste cell is capable of responding to any of the basic taste sensations. And, contrary to any "tongue maps" you may have seen, no part of the tongue is particularly responsive to any of the basic tastes. Bitter taste is not sensed primarily toward the back of the tongue or sweet taste toward the front.

At this point, you may be wondering: How can there be just four or five *basic* tastes, but a huge number of tastes we actually experience? The answer is that the neurons that receive the signals from the taste cells vary in their response to specific tastants. If you eat some table sugar, one group of neurons will respond by firing perhaps 10 times over a few seconds, but another group of neurons will fire more than 100 times. Most researchers now agree that it is the *pattern* of firing activity across all taste neurons that the brain uses to tell us whether it is just basil in the soup or a particularly sharp basil. You've seen this pattern processing before, with smell, and it is at work elsewhere. We might say that in pattern processing, the brain is something like a talented news writer: It gets input from lots of individuals and turns these reports into a coherent, highly detailed story—in this instance, our sense of what we are tasting.

**Figure 27.13**
**Our Sense of Taste**

Small bumps in the tongue called papillae contain taste buds—collections of tissues and various cells, including taste cells. Microvilli extending from taste cells are capable of binding with food molecules or "tastants," which they come in contact with through taste pores. This contact causes the taste cells to release a neurotransmitter that stimulates the dendrites of adjacent nerve cells. With this, a nerve signal regarding taste is on its way to the brain.

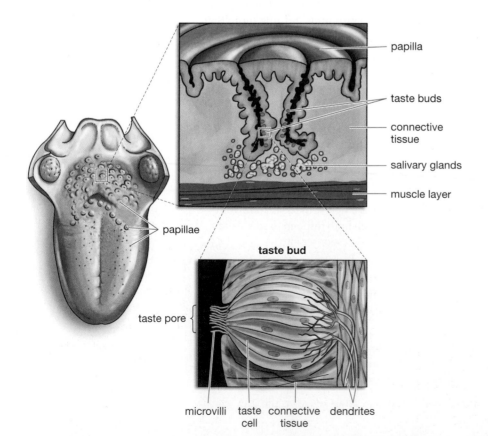

papilla

taste buds

connective tissue

salivary glands

muscle layer

papillae

**taste bud**

taste pore

microvilli  taste cell  connective tissue  dendrites

| ESSAY | Too Loud: Hair Cell Loss and Hearing |
|---|---|

Our sense of hearing can easily be damaged. Eighty percent of the hearing loss in the United States—affecting 28 million people—involves damage to either the auditory hair cells (**Figure 1**) referred to in the main text or to the large nerve into which these cells feed. This type of hearing loss often comes as a consequence of old age, but it can come as a consequence of *young* age, which is to say the tendency of young people to listen to loud music. Damage caused by loud noise may be limited only to the hair cells' cilia—their roots may break off or the links between them may be destroyed—and in most of these instances, these hair cells can recover. When noise is loud or long-lasting enough, however, it can kill hair cells outright. When this happens, the effect may be not only a permanent loss of hearing, but a so-called tinnitus, meaning a high-pitched ringing in the ears.

So, how loud is too loud? The rule of thumb is, if you have to shout to be heard, you're in the danger zone. The amount of time a person is exposed to loud noise matters, however. You could be around a typical lawn mower for several hours without harm, but after 15 minutes at a loud rock concert, you are damaging your hair cells. In terms of absolute sound levels, 85 decibels is the limit for indefinite exposure—above that and you could damage hearing if you're exposed for long enough. If you look at **Figure 2**, you can see where various everyday sounds lie on the decibel scale. (Keep in mind that the scale is logarithmic; a 110-decibel sound is 10 times as loud as a 100-decibel sound.) Note that the popular iPod MP3 player can, at maximum volume, deliver up to 120 decibels of sound—an amount equivalent to that produced by a jet airplane when it's taking off.

| Source | Decibel level |
|---|---|
| Firearm | 140+ |
| Jet engine | 140 |
| Jackhammer | 130 |
| Sporting event | 127 |
| Live music concert | 120+ |
| Jet plane takeoff | 120 |
| Band practice | 120 |
| iPods and other MP3 players at maximum volume | 120 |
| Health club and aerobics studio | 120 |
| Movie theatre | 118 |
| Motorcycle | 95-120 |
| Chain saw or pneumatic drill | 100 |
| Lawnmower | 90 |
| Subway | 90 |
| Busy street | 80 |
| Alarm clock | 80 |
| Vacuum cleaner | 70 |
| Conversation | 60 |
| Moderate rainfall | 50 |
| Quiet room | 40 |
| Whisper, quiet library | 30 |

**Figure 1**
**Vulnerable to Damage by Loud Noise**
This color-enhanced micrograph shows the ear's hair cells, in yellow. Loud noise can damage the hair like cilia visible on the cells or can even kill the cells. Two types of hair cells are shown: outer cells, with a V shape at right, and inner cells at left.

**Figure 2**
**How Loud Is It?**
The decibel levels of some everyday sounds. In general, anything above the 85-decibel level can damage hearing if exposure to the sound continues for long enough.
*Source:* American Speech-Language-Hearing Association.

## 27.11   **Our Sense of Hearing**

Sound is always depicted as coming in "waves," but it does not physically exist as a series of tiny, invisible lines that run in peaks and troughs. Pluck a single string on a harp and then slow the videotape of this action way down. The first thing that happens is that the string moves in the opposite direction from the way it was pulled. This movement *compresses the air molecules* ahead of the string, making this air more dense than that of the ambient air around it. Now the string swings back the other way, and this *lowers* the air pressure ahead of the string—below that of atmospheric air pressure. So what we perceive in hearing are "waves" of moving air molecules that are, by turns, more and less compressed than those of ambient air. This cycle of high and low compression happens very rapidly. When you hit the A above middle C on a piano, the waves of high and low compression are bumping up against your ear drum 440 times per second—meaning the piano string is going back and forth 440 times per second. It is these cycles that give sound its frequency. The faster the frequency, the higher the sound. The best human ears (usually children's ears) can perceive sounds ranging from about 20 to 20,000 cycles per second.

So, how do we go from a vibrating eardrum to the perception of a sound? If you look at **Figure 27.14a**, you can see the basic anatomy of the ear, and in **Figure 27.14b**, you can track the steps involved in hearing. (The elongated structure on the right of Figure 27.14b is the cochlea of the ear, uncoiled for illustrative purposes.)

Note that hearing begins when vibrations of the eardrum or *tympanic membrane* in turn vibrate the three smallest bones in the body: the malleus, incus, and stapes (or "hammer, anvil, and stirrup"), which lie in the middle ear. In the same way that a lens serves to take light coming from a large area and focus it on a small one, these three bones take the vibrations coming from the relatively large tympanic membrane and focus them on the smaller oval window. This is important because it *amplifies* the vibration signal for the next step. The oval window's vibrations now shake fluid that lies within the pea-sized structure called the **cochlea**: the coiled, membranous portion of the inner ear in which vibrations are transformed into the nervous system signals perceived as sound.

In the cross section of the cochlea pictured in **Figure 27.14c**, you can see how vibrations of fluid in the cochlea are transformed into the sensation of sound. The cochlea can be seen to have a couple of ducts (the vestibular and tympanic), which are filled with fluid. In the middle, there is the cochlear duct, likewise filled with fluid, which has at its bottom a "basilar" membrane on which sit the elements that will produce sound. Note that, supported by the basilar membrane, there are a group of so-called hair cells, so named because each of them sprouts a group of hairlike cilia that come close to touching the tectorial membrane that folds over them in the cochlea.

Now, how does all of this yield sound? Think of it this way: If you rested your left hand on a water balloon and then pushed rhythmically on this same balloon with your right hand, your left hand would move up and down slightly. If there were something directly above your left hand, then your hand might make contact with it. In the same way, the vibrations of the fluid in the cochlea move the basilar membrane up and down, and this movement can push the hair cells up against the tectorial membrane that lies above them. If you look at the blowup of Figure 27.14c, you can see what happens next. The cilia on the hair cells have microscopic "trapdoor" channels on them, and when these cilia bend from touching the tectorial membrane, the trap doors open. With this, potassium ($K^+$) ions flow in, leading to an influx of calcium ($Ca^{2+}$) ions at the base of the hair cell, and this leads the hair cells to release a neurotransmitter. Waiting to receive the neurotransmitter are adjacent dendrites, and with this, vibration has been turned into a nerve signal.

The details of the hearing process actually are much more complicated than this account would indicate. While we need not go over these details, it's worthwhile to note just one of the ways this system is able to discriminate one sound from another. To locate where a sound is *coming from*, the brain in effect calculates the difference between how intense a sound is in one ear as opposed to the other, and what the time difference is between when sound arrives at one ear and then the other.

Finally, it is easy for hearing to be damaged, and this damage often is to the hair cells we've been looking at, as you can see in "Too Loud" on page 571.

---

**SO FAR...**

1. Taste cells are part of organized collections of 50 to 100 cells known as _____. Taste works by means of "tastants" interacting with hair-like projections that extend from the _____.

2. Sound waves are waves of moving air molecules that are either more or less _____ than molecules of ambient air.

3. The inner ear's cochlea is filled with _____; its vibrations open and shut "trapdoor" channels on _____ cells, thus resulting in the transmission of a nerve signal.

---

## 27.12 Our Sense of Vision

Our visual system has to accomplish three central tasks. First, it has to capture light from the outside world and focus it at a very precise location within our eyes. Second, it has to take this focused light and convert it into a nervous system signal. Third, it has to make sense of the visual information it receives. If you

**(a)** Anatomy of the ear

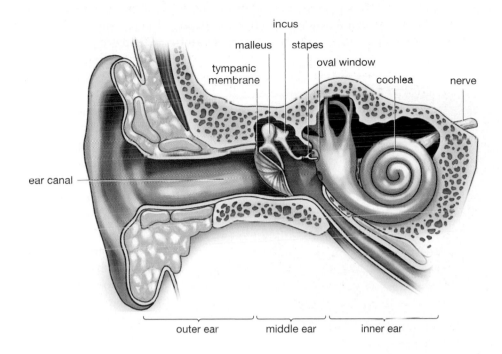

incus

malleus  stapes

tympanic
membrane

oval window

cochlea  nerve

ear canal

outer ear  middle ear  inner ear

**(b)** From air vibration to nerve signal

2. The tympanic membrane
vibrates the three bones of
the middle ear; the malleus,
incus and stapes.

3. The vibration of the stapes
focuses the sound-wave
vibration on the membrane
of the oval window.

4. The oval window's vibrations
cause fluid vibrations within the
coiled, tubular cochlea (shown
elongated here for illustrative
purposes).

1. Sound waves enter
through the ear
canal and vibrate the
tympanic membrane.

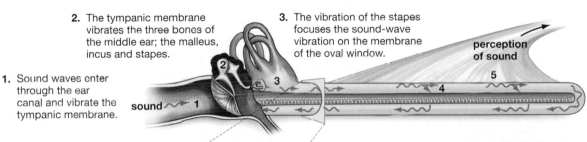

perception
of sound

5. These fluid vibrations cause
cells within the cochlea to
release a neurotransmitter
which triggers a nerve signal
to the brain.

sound  1

**(c)** How fluid triggers nerve signal

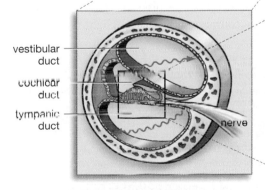

vestibular
duct

cochlear
duct

tympanic
duct

nerve

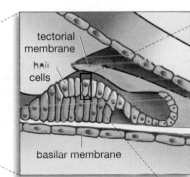

tectorial
membrane

hair
cells

basilar membrane

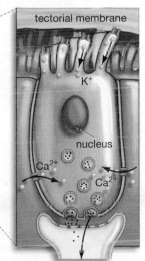

tectorial membrane

K⁺

nucleus

$Ca^{2+}$

$Ca^{2+}$

3. As the hair cells contact the
tectorial membrane, cilia on
them bend. This change in
position causes "trap door"
channels in the hair cells to
open, which allows
potassium ions ($K^+$) to flow
into them.

4. This influx triggers an influx
of calcium ions ($Ca^{2+}$) at the
base of the hair cells, which
in turn causes the cells to
release a neurotransmitter.

5. The neurotransmitter is
received by adjacent
dendrites and a nerve
signal is sent to the brain.

1. Seen in cross-section, the
cochlea has vestibular
and tympanic ducts, in
which fluid is vibrating.

2. This vibration shakes the
basilar membrane, pushing
hair cells on it up against the
overlying tectorial membrane.

**Figure 27.14**
**Our Sense of Hearing**

look at **Figure 27.15** you can see some of the structures of the eye that function in the first two steps. Light comes into our eyes through the transparent

**Figure 27.15
Anatomy of the
Human Eye**

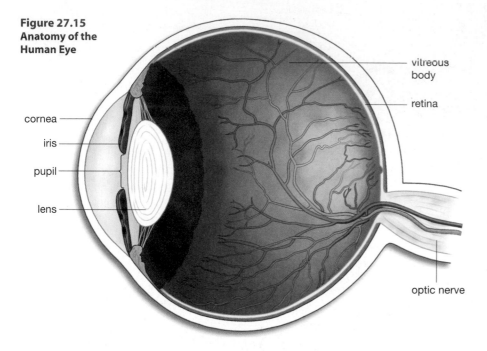

- cornea
- iris
- pupil
- lens
- vitreous body
- retina
- optic nerve

**(a)** Normal vision

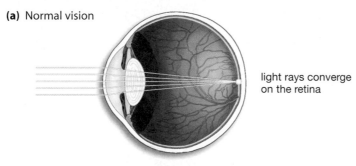

light rays converge
on the retina

**(b)** Farsighted vision

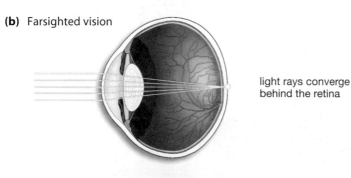

light rays converge
behind the retina

**(c)** Nearsighted vision

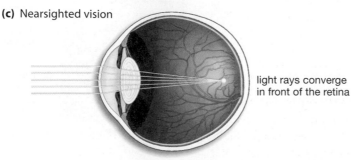

light rays converge
in front of the retina

cornea and passes through the opening known as the pupil. Surrounding the pupil is the iris, which is composed partly of smooth muscle and thus is capable of contracting or dilating, meaning it can let in less light in a bright environment or more light in a darkened one. The incoming light is bound for the layer of cells at the back of the eye called the **retina**: an inner layer of tissue in the eye containing cells that transform light into nervous system signals. But as the light rays come in, they are bent or *refracted*—first by the cornea and then by the lens that lies behind it—such that they converge on a small area of the retina, as you can see in **Figure 27.16**. This is no different in principle from what happens with a camera. We take a picture of a large object (such as a statue) that ends up as a small image on a frame of film. In a similar way, thanks to refraction, large objects we view with our eyes end up as *very* small images on our retina. How small? At fractions of a millimeter, some of them are microscopic. The eyes do not always focus light rays in such a way that they converge properly on the retina, however. You can see what it means to be nearsighted or farsighted in Figure 27.16.

Now, how do retinal cells convert rays of light into electrical signals? If you look at **Figure 27.17**, you can see that there are three layers of cells in the retina: ganglion cells, connecting cells, and photoreceptors. The **photoreceptors** are the sensory receptors for vision, and they come in two varieties: rods and cones. **Rods** are photoreceptors that function in low-light situations but that provide only black-and-white vision. **Cones** are photoreceptors that respond best to bright light and that provide color vision.

The blowup of these cells shows that their posterior ends—the ends toward the back of the head—are filled with pigments, which lie embedded in a series of flattened, membranous disks. A pigment is simply a chemical compound that absorbs light. So, our light

**Figure 27.16
Focused and Unfocused**

Sharp vision requires that incoming light rays converge just as they reach the retina.

**(a)** In normal vision, the eye's cornea and lens bring about this convergence; they refract the incoming light at an angle that matches the length of the eye.

**(b)** People who are farsighted have a mismatch between the length of their eye and the amount of refraction provided by their cornea and lens: Their eye is too short for the amount of refraction provided. Thus, light rays converge behind their retina. As a result, these people cannot see nearby objects clearly.

**(c)** People who are nearsighted have the opposite problem: Their eye is too long for the amount of refraction provided, with the result that light rays converge in front of their retina. These people cannot see distant objects clearly. Both nearsighted and farsighted vision can be corrected.

**(a)**

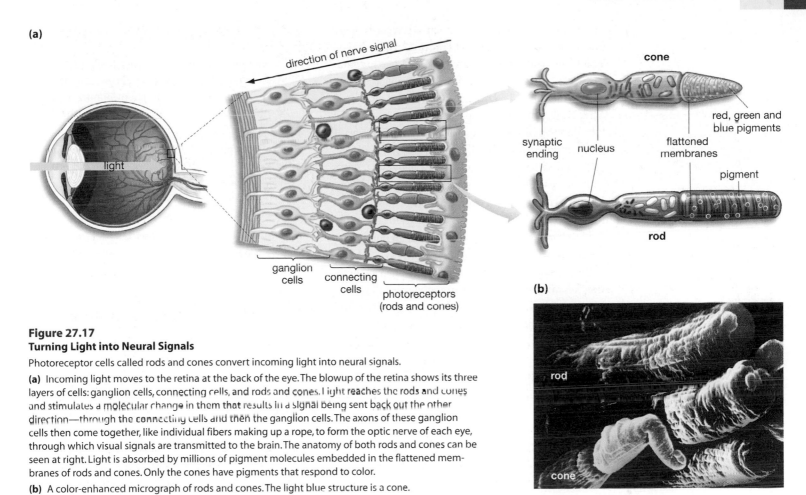

direction of nerve signal

**cone**

synaptic ending

nucleus

flattened membranes

red, green and blue pigments

pigment

**rod**

ganglion cells

connecting cells

photoreceptors (rods and cones)

light

**Figure 27.17**
**Turning Light into Neural Signals**

Photoreceptor cells called rods and cones convert incoming light into neural signals.

**(a)** Incoming light moves to the retina at the back of the eye. The blowup of the retina shows its three layers of cells: ganglion cells, connecting cells, and rods and cones. Light reaches the rods and cones and stimulates a molecular change in them that results in a signal being sent back out the other direction—through the connecting cells and then the ganglion cells. The axons of these ganglion cells then come together, like individual fibers making up a rope, to form the optic nerve of each eye, through which visual signals are transmitted to the brain. The anatomy of both rods and cones can be seen at right. Light is absorbed by millions of pigment molecules embedded in the flattened membranes of rods and cones. Only the cones have pigments that respond to color.

**(b)** A color-enhanced micrograph of rods and cones. The light blue structure is a cone.

**(b)**

rod

cone

comes in and moves to, say, a rod in the retina. There it is absorbed by a pigment, and this absorption causes a normally bent part of the pigment molecule to … straighten out. That's it. Our entire sense of vision hinges on this tiny initial effect. Yet, as a shouted "hello" might cause an avalanche, so this small step puts big things in motion. In his book *What Makes You Tick?* vision researcher Thomas Czerner put it this way:

> The infinitesimal energy of a photon [of light] changes the shape of a *single* molecule of pigment, which causes the release of *dozens* of molecules of an enzyme, which cause the rapid breakdown of *hundreds* of molecules of a chemical messenger, which cause the gates of *thousands* of ion channels in the cell membrane to slam shut, which blocks *millions* of sodium ions from entering the cell, which produces the grand finale, a local change in the resting membrane potential.

You might expect that this change in photoreceptor membrane potential would, like the others you've seen, result in the release of a neurotransmitter to an adjacent neuron. The twist in vision, however, is that in their *resting* state rods and cones are releasing

neurotransmitter molecules in abundance to their connecting cells. When light strikes the rods and cones, it has the effect of reducing this neurotransmitter release. Thus, the lack of neurotransmitter release from a rod or cone transmits the message "light stimulation here."

The signal that begins with a rod or cone goes first to connecting neurons and then to the ganglion cells you can see in Figure 27.17. These cells have axons that are like tributaries coming together to form a river, only this river is the optic nerve, composed of about a million axons (see Figure 27.15). Because we have two eyes, we thus have a pair of million-fiber optic nerves carrying visual information from the eyes to the brain. The optic nerve takes the visual information to the thalamus, and from there it is transmitted, on other nerve fibers, to an area at the very back surface of the brain, the *primary visual cortex*, where most vision processing is done.

The details of this processing are much too complicated to go into here. But a central lesson of this complexity is that the human visual system does not work like a camera; it does not passively record bits of light and register a collection of these bits as images. Rather, the brain *constructs* images as much as it records them. To get an idea of this, look at **Figure 27.18**. In it, you see a V, formed out of bumps and a dot-like indentation. Now turn the book upside down and look again. The bumps have become indentations and vice versa. Why? Note that when the book is right side up, most dots in the V are shaded in such a way that light seems to be coming from above them—there are "shadows" on the lower part of the dots. Light generally comes from above us, whether from a lamp or from the sun. This is so common that evolution has produced a genetically based "rule" in our visual system that says: "When shadows are in the lower part of an object, that object is raised above a surface."

Why would evolution produce rules such as these? After all, this rule is yielding an *inaccurate* interpretation of the dots on the page—they aren't really raised above the surface at all. Our visual system has such rules because what mattered in the evolution of our ancestors was not whether they perceived objects "correctly." What mattered was whether they perceived objects in ways that helped them survive and reproduce. Over the ages, individuals who could perceive, say, a snake more quickly than other individuals survived longer and thus left more offspring. The many visual rules we have amount to rules for survival. They provide a series of "best bets" for quickly telling us what we are looking at. (If an object's lower half is shaded, the best bet is that it's raised above a surface.) The idea that our visual system is often creating perceptions, rather than simply recording them, may seem strange, but that's the reality of visual perception.

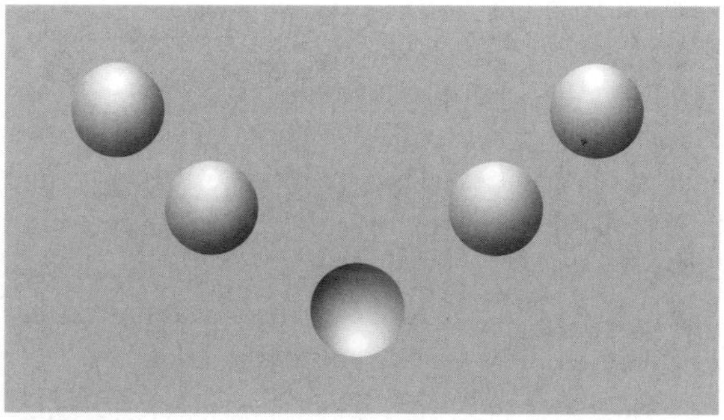

••• ANSWERS
1. retina; cornea; lens
2. rods; black-and-white; cones; color
3. optic nerves

**Figure 27.18**
**Seeing Is Believing**

Look at the figure and then turn the book upside down and look at it again. With the shift, dots that once appeared to be raised above the surface of the drawing now appear to be indented, and vice versa. Our visual system has a rule that causes us to perceive objects that are shaded at the bottom as protruding from a surface, while objects that are shaded at the top appear to be indented. This is but one example of how our visual system creates visual images rather than simply recording them.

**SO FAR...**

1. Light coming in through the eye is focused on a layer of tissue at the back of the eye called the _____ by means of being bent, or refracted, by both the eye's _____ and the _____ that lies behind it.

2. The eye's photoreceptor cells come in two varieties: _____ , which function in low-light situations but provide only _____ vision; and _____ , which respond best to bright light but that provide _____ vision.

3. Visual information is passed from the eyes to the brain via our two _____ .

## 27.13  The Endocrine System

We now leave the nervous system to focus on a different system in the body, the endocrine system. Like the nervous system, the endocrine system is in the communication-and-control business. And, like the nervous system with its neurotransmitters, the endocrine system works through a group of chemical messengers. The endocrine messengers are not neurotransmitters, however, but instead are substances called hormones. In this text, we will define **hormones** as substances secreted by one set of cells that travel through the bloodstream and affect the activities of other cells. Such a definition sets hormones apart from other kinds of signaling molecules in the body. There are signaling molecules, called *paracrines*, that do not travel through the bloodstream but instead diffuse from one or more cells to a nearby group of cells, causing a metabolic change in them. Likewise, a cell can be affected by its own secreted chemical messenger, called an *autocrine*. Note that both paracrines and autocrines carry out strictly *local* communication between cells. The endocrine hormones we'll be looking at can have their effects over short distances, but they also have the ability to carry chemical signals throughout the body given their transportation through the bloodstream.

This very means of distribution provides a key for getting to the heart of what separates the endocrine and nervous systems. In the nervous system, signals go from neurons A to B to C in well-defined lines of transmission—rather like a telephone call going through relay stations. By contrast, a typical hormone is "broadcast," in a sense, as it moves through the bloodstream. Like a television signal, it can be "picked up" by any cell that has the proper "receiver." What are these receivers? They are receptors that are shaped in such a way that they can latch onto the hormone. With this, we get to the concept of **target cells**: those cell types that can be affected by a given hormone (**Figure 27.19**). The target cells for a hormone called antidiuretic hormone are located primarily in the kidneys, while the primary target cells for the hormone insulin include not only liver cells, but skeletal muscle and fat cells that are located throughout the body. Thus, hormones differ greatly with respect to the number of target cells they affect and the location of these cells.

Hormonal production takes place to a significant extent within specialized organs called endocrine glands, and this represents another contrast with the nervous system. You may remember that a gland is any localized group of cells that work together to secrete a substance. **Endocrine glands** are glands that release their materials directly into the bloodstream or into surrounding tissues, without using ducts. Thus, the endocrine system has a set of large organs whose primary function to secrete chemical messengers. In contrast, the nervous system has no such organs—only its individual neurons, secreting neurotransmitters. The major endocrine glands are shown in **Figure 27.20** on the next page. While most hormones are secreted by these glands, it's important to note that not *all* hormones are. The heart, kidneys, stomach, liver, small intestine, placenta, and fatty tissue all secrete hormones related to their functioning, yet they are not glands.

In a final point of comparison, the endocrine system tends to work more slowly than the nervous system, but its effects tend to be more long lasting. The fastest-acting hormones take several seconds to work, while the slowest-acting ones may take several hours. Contrast this with the almost instantaneous effects of nervous system messages. The opposite side of this coin is that the longest-lasting hormones can keep exerting their effects for hours after they have been released. In contrast, nervous system signals disappear as fast as they arise. Given the time scales in which hormones act, it's not surprising that they tend to regulate processes that unfold over minutes, hours, or even years: whether we have enough water in our systems, how we grow, and so forth. If you need to swerve to avoid a collision, look to the nervous system for a response; for the maintenance of water balance and bone density, look to the endocrine system.

Despite their differences, the endocrine and nervous systems are tightly linked in that signals from one system often affect the other. Beyond this, modern research actually is blurring the distinction between the two systems. Scientists have a growing list of substances that once were thought to work strictly as hormones but that now turn out to be both hormones and neurotransmitters. For example, the hormone adrenaline works as a neurotransmitter

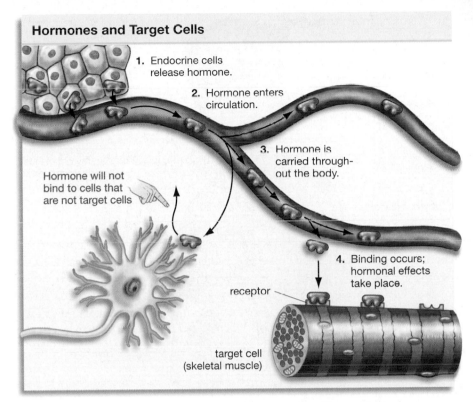

**Hormones and Target Cells**

1. Endocrine cells release hormone.
2. Hormone enters circulation.
3. Hormone is carried throughout the body.

Hormone will not bind to cells that are not target cells

4. Binding occurs; hormonal effects take place.

receptor

target cell (skeletal muscle)

**Figure 27.19**
**Hormones and Their Target Cells**

For a hormone to affect a target cell, that cell must have receptors that can bind the hormone. The binding of hormone to receptor then initiates a change in the target cell's activity. The figure shows a peptide hormone that affects skeletal muscle tissue. The hormone does not affect the nerve cell, also shown in the figure, because this cell does not have the appropriate receptors to bind with the hormone.

within the brain, shuttling across synapses; but then it also works as a hormone—it is secreted into circulation by the body's adrenal glands.

## 27.14 Types of Hormones

Hormones come in three primary classes. The first of these is the **amino-acid-based hormones**: hormones that are derived from modification of a single amino acid. The hormone adrenaline, just mentioned, is produced through the modification of the amino acid tyrosine, for example, while the hormone melatonin is produced through a modification of the amino acid tryptophan. The second class of hormones, the **peptide hormones**, are hormones composed of *chains* of amino acids. Such chains can vary greatly in length. Human growth hormone, for example, is composed of 191 amino acids, while antidiuretic hormone is composed of only 9. Insulin is a peptide hormone, as are most of the hormones secreted by an important endocrine gland we'll be looking at, the pituitary.

With a few exceptions, peptide and amino-acid-based hormones have target cells whose hormone receptors are located on their outer membranes. These receptors are the "broadcast receivers" that stand ready to bind with these hormones when they pass by, as depicted in Figure 27.19. The effects of this binding are very diverse. Ion channels may be opened in a target cell as a result, or enzymes may be activated. In all cases, the cell will change its activities as a result of the binding.

**Web Animation 27.2**
The Endocrine System

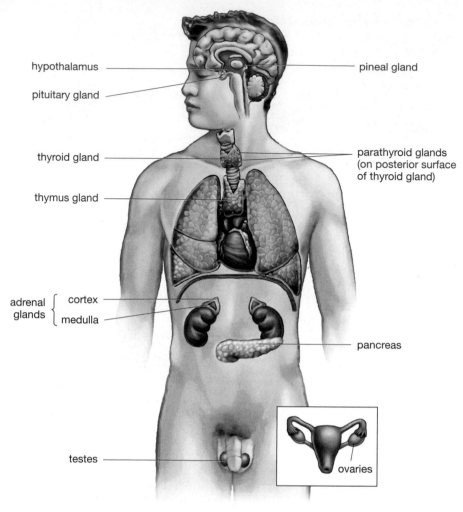

hypothalamus

pituitary gland

pineal gland

thyroid gland

parathyroid glands
(on posterior surface
of thyroid gland)

thymus gland

adrenal { cortex
glands { medulla

pancreas

testes

ovaries

**Figure 27.20**
**The Major Hormone-Secreting Glands of the Body**
Although it is part of the brain (and not an endocrine gland), the hypothalamus is shown because it plays a central role in hormonal regulation.

The third major class of hormones, the **steroid hormones**, are all constructed around the chemical framework of the cholesterol molecule. Most of the steroid hormones are released from just a few glands—notably the male and female reproductive glands and the adrenal glands (which sit atop the kidneys). Unlike amino-acid-based and peptide hormones, most steroid hormones pass *through* a cell's plasma membrane and then bind with a receptor protein inside the cell. This combined hormone/receptor molecule then enters the nucleus of the cell and binds with the cell's DNA there. This binding turns on one or more of the cell's genes, thus bringing about the production of one or more proteins. The complexity of this series of events makes steroids the snails of the hormone world; hours might pass between the time a steroid hormone is produced and the time it has an effect. While testosterone is undoubtedly the best-known steroid hormone, estrogen is one as well. From this, it's easy to see that the term *steroid*, as

commonly used, actually means one kind of steroid—the muscle-building or "anabolic" steroid hormones.

**SO FAR...**

1. Hormones are substances secreted by one set of _____ that move through the _____ and affect the activities of other _____.

2. Hormone production takes place to a significant extent within _____ glands, defined as glands that release their materials directly into the _____ or surrounding tissue, without the use of ducts.

3. Hormones come in three principal classes: _____ based, _____ , and _____ hormones. Of these, the slowest acting are the _____ hormones.

## 27.15 How Is Hormone Secretion Controlled?

Almost all hormone secretion is ultimately controlled by the negative-feedback mechanism that was reviewed in Chapter 26 (see page 530). As a reminder, this is the same mechanism that operates in a home heating system. A thermostat senses that the home temperature is too cold and thus switches on a furnace. The product of this action is hot air, and this very product then feeds back on the thermostat's activity in a negative way—it shuts it down. In hormone secretion, a given gland or set of neurons will sense that more of a certain hormone is needed and then send a signal out, prompting release of that hormone. The hormonal product that results then feeds back on the sensing mechanisms in a negative way, causing them to reduce or shut down production of the hormone. When we consider all the negative-feedback loops in operation in the endocrine system, we find that they are big players in helping to preserve **homeostasis**: an organism's tendency to maintain a relatively stable internal environment.

### Hormonal Hierarchy: The Hypothalamus

Although negative feedback ultimately controls most hormone secretion, it's accurate to view a good part of the endocrine system as a hierarchy that has, sitting on top of it, one of the brain structures we looked at earlier, the hypothalamus. The hypothalamus is prompted to act based on input it gets from other sources—sensory nerves feed into it, for example—so it is part of larger negative-feedback loops. Nevertheless, as you can see in **Figure 27.21**, the hypothalamus

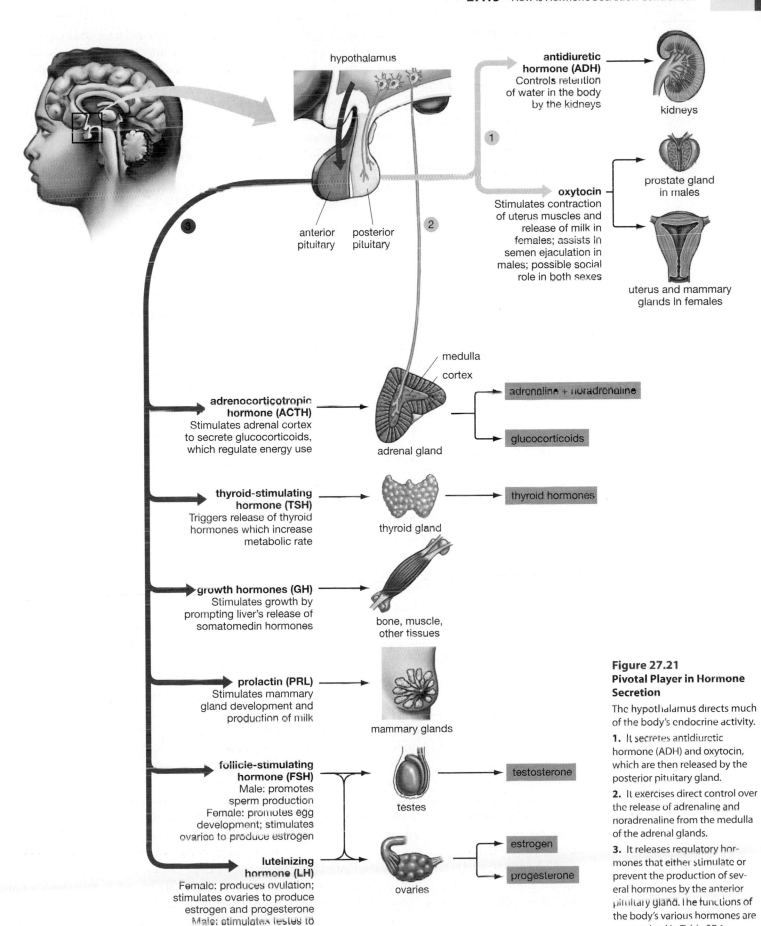

hypothalamus

**antidiuretic hormone (ADH)**
Controls retention of water in the body by the kidneys

kidneys

1

2

3

anterior pituitary    posterior pituitary

**oxytocin**
Stimulates contraction of uterus muscles and release of milk in females; assists in semen ejaculation in males; possible social role in both sexes

prostate gland in males

uterus and mammary glands in females

medulla
cortex

**adrenocorticotropic hormone (ACTH)**
Stimulates adrenal cortex to secrete glucocorticoids, which regulate energy use

adrenal gland

adrenaline + noradrenaline

glucocorticoids

**thyroid-stimulating hormone (TSH)**
Triggers release of thyroid hormones which increase metabolic rate

thyroid gland

thyroid hormones

**growth hormones (GH)**
Stimulates growth by prompting liver's release of somatomedin hormones

bone, muscle, other tissues

**prolactin (PRL)**
Stimulates mammary gland development and production of milk

mammary glands

**follicle-stimulating hormone (FSH)**
Male: promotes sperm production
Female: promotes egg development; stimulates ovaries to produce estrogen

testes

testosterone

**luteinizing hormone (LH)**
Female: produces ovulation; stimulates ovaries to produce estrogen and progesterone
Male: stimulates testes to produce androgens

ovaries

estrogen

progesterone

**Figure 27.21**
**Pivotal Player in Hormone Secretion**

The hypothalamus directs much of the body's endocrine activity.

**1.** It secretes antidiuretic hormone (ADH) and oxytocin, which are then released by the posterior pituitary gland.

**2.** It exercises direct control over the release of adrenaline and noradrenaline from the medulla of the adrenal glands.

**3.** It releases regulatory hormones that either stimulate or prevent the production of several hormones by the anterior pituitary gland. The functions of the body's various hormones are summarized in Table 27.1.

controls a lot of the body's hormone secretion. Note all that it does. First, it acts as an endocrine organ itself, releasing, through one part of the pituitary gland—the posterior pituitary—two hormones that it produces itself. Second, it exercises direct control over the endocrine cells of the inner portion (or medulla) of the adrenal glands. Upon nervous system stimulation from the hypothalamus, the adrenal medulla releases into the bloodstream the hormones adrenaline and noradrenaline. Third, the hypothalamus exercises control over a second part of the pituitary gland—the anterior pituitary—which is referred to as a "master gland" in that several of the hormones it releases go on to control *other* endocrine glands. Although lots of hormones are produced outside the control of the hypothalamus, you can see from this list that it is a central structure in the endocrine system. Note that the hypothalamus is a part of the brain, yet it is releasing some hormones and controlling the release of others. This is an example of how extensive the overlap is between the nervous and endocrine systems.

### The Pituitary Gland

Let's look a little closer now at one of the prime links in the hypothalamic hierarchy, the pituitary gland. This structure is a small, lima-bean-sized organ that looks something like a punching bag. Suspended from the hypothalamus by a slender stalk (see Figure 27.21), the pituitary is a single entity in a sense. Yet, it has anterior and posterior regions that differ greatly in function—so greatly that it's tempting to think of these regions as two neighbors who happen to live in the same apartment building. The posterior pituitary is really an extension of the hypothalamus and as such simply acts as a storage-and-release site for two hormones produced by the hypothalamus. Meanwhile, the anterior pituitary synthesizes its *own* hormones— six of them in all—although it releases them under control of the hypothalamus. Here are a couple of more formal definitions of these regions. The **posterior pituitary** is a region of the pituitary gland that stores and releases two hormones, oxytocin and antidiuretic hormone, that are produced by the brain's hypothalamus. The **anterior pituitary** is a region of the pituitary gland that, under control of the brain's hypothalamus, produces and releases hormones that work directly on target cells or on other endocrine glands.

The hypothalamus exercises its control over the anterior pituitary in an interesting way. The anterior pituitary's endocrine cells are surrounded by a network of the tiny blood vessels called capillaries. Move up to the hypothalamus, and it too has a capillary bed in its lower portion. Running between these two capillary beds are a small number of blood vessels. Nerve cells in the hypothalamus synthesize extremely small amounts of hormones that travel down their axons and are released into the hypothalamic capillaries. From there, they travel down the linking blood vessels to the anterior pituitary capillaries, and from there, they move into the pituitary gland itself. These can be so-called *releasing hormones* that stimulate pituitary cells to produce their own hormones. Conversely, they may be *inhibiting hormones* that serve to *reduce* the production of given hormones by the pituitary. The anterior pituitary itself is composed of five types of cells that release the gland's six different hormones. Once these hormones are made, they move out into the bloodstream for distribution throughout the body.

## 27.16 Hormones in Action: Four Examples

What kinds of effects do hormones have? **Table 27.1** provides a list of the hormones produced by most of the major endocrine glands in the body along with the effects of these hormones. To give you some idea of the range of the endocrine system, we'll now consider four of these hormones in detail.

### Insulin and Glucagon: Keeping a Tight Rein on Glucose

The simple sugar glucose is the single most important source of energy for the human body. All human cells use it to power their activities, and it is the sole source of energy for human brain cells. Most of the carbohydrates we eat are broken down (in the digestive tract) into glucose, after which this "blood sugar" passes to the liver and then to all the tissues of the body via the bloodstream.

Glucose is, however, a double-edged sword: Our cells require it, but it is harmful when it exists in excess amounts in our bloodstream. The body is thus constantly performing a balancing act; while it must keep some glucose in circulation—to supply cells with energy as needed—it must guard against having too *much* in circulation. So, how does it walk this fine line? It uses two hormones that have opposite effects. One of these hormones is **insulin**: a hormone that brings about a decrease in blood levels of glucose. The other hormone is **glucagon**: a hormone that brings about an increase in blood levels of glucose.

## Table 27.1  Hormones of the endocrine system: Their sources and effects

| Gland/Hormone | Effects |
| --- | --- |
| **Hypothalamus** | |
| Releasing hormones | Stimulate hormone production in anterior pituitary |
| Inhibiting hormones | Reduce hormone production in anterior pituitary |
| **Anterior pituitary** | |
| Thyroid-stimulating hormone (TSH) | Triggers release of thyroid hormones |
| Adrenocorticotropic hormone (ACTH) | Stimulates adrenal cortex cells to secrete glucocorticoids |
| Follicle-stimulating hormone | Female: promotes egg development; stimulates ovaries to produce estrogen<br>Male: promotes sperm production |
| Luteinizing hormone (LH) | Female: produces ovulation (egg release); stimulates ovaries to produce estrogen and progesterone<br>Male: stimulates testes to produce androgens (e.g., testosterone) |
| Prolactin (PRL) | Stimulates mammary gland development and production of milk |
| Growth hormone (GH) | Stimulates growth by prompting liver's release of somatomedin hormones |
| **Posterior pituitary** | |
| Antidiuretic hormone (ADH) | Controls retention of water in the body by the kidneys |
| Oxytocin | Stimulates contraction of uterus muscles and release of milk in females; assists in semen ejaculation in males; possible social role in both sexes |
| **Thyroid** | |
| Thyroxine | Increases general rate of body metabolism |
| Calcitonin | Reduces calcium ion levels in blood |
| **Parathyroid** | |
| Parathyroid hormone (PTH) | Increases calcium ion levels in blood |
| **Thymus** | |
| Thymosins | Stimulate development of white blood cells (lymphocytes) in early life |
| **Adrenal cortex** | |
| Glucocorticoids | Includes cortisol, which stimulates glucose production and breakdown of fats. A stress-response hormone. |
| Mineralocorticoids | Cause the kidneys to retain sodium ions and water and excrete potassium ions |
| **Adrenal medulla** | |
| Adrenaline | Also known as epinephrine; stimulates release of energy stores; increases heart rate and blood pressure |
| Noradrenaline | Also known as norepinephrine; effects similar to adrenaline |
| **Pancreas** | |
| Insulin | Decreases glucose levels in blood |
| Glucagon | Increases glucose levels in blood |
| **Testes** | |
| Testosterone | Promotes production of sperm and development of male sex characteristics |
| **Ovaries** | |
| Estrogens | Support egg development, growth of uterine lining, and development of female sex characteristics |
| Progesterones | Prepare uterus for arrival of developing embryo and support of further embryonic development |
| **Pineal gland** | |
| Melatonin | Establishes day/night cycle |

**Figure 27.22**
**Where Insulin and Glucagon Are Produced**

**(a)** The pancreas takes on two very different roles in the body. Most of its cells produce enzymes that help digest the food we eat. (These enzymes move into the small intestine, seen on the left of the figure, through the tube known as the pancreatic duct.) The pancreas is also home, however, to the cells that produce insulin and glucagon—the hormones that control the levels of glucose in the bloodstream. Insulin is produced by beta cells in the pancreas, while glucagon is produced by alpha cells.

**(b)** Both alpha and beta cells exist in small, roundish clusters of cells, called islets of Langerhans, that are scattered throughout the pancreas' more abundant digestive cells. Each islet has an abundant set of the blood vessels, called capillaries, that move insulin and glucagon into general circulation.

If you look at **Figure 27.22**, you can see the source of both of these hormones: the organ known as the pancreas, which lies just behind the bottom portion of our stomach. You might think that most of the pancreas would be given over to insulin and glucagon production, since these two hormones are so widely used in the body, but in fact only 1 percent of the volume of the pancreas is devoted to hormone synthesis. (The other 99 percent produces digestive enzymes that empty into the small intestine via the ducts you can see in the figure.) Insulin is secreted by "beta" cells in the pancreas, while glucagon (GLOO-ka-gon) is secreted by "alpha" cells. These two kinds of cells sit side by side within small clusters or "islets" of cells in the pancreas called Islets of Langerhans. A profusion of tiny capillaries exists in the islets as well; these blood vessels take up the insulin and glucagon as they are produced, which is the first step in moving them into general circulation.

How do these hormones do their jobs? Here's a simplified account. Insulin binds with its target cells, and this binding prompts these cells to create channels, called transport proteins, that allow glucose to move from the bloodstream into the interior of the cells. As noted earlier, the primary target cells for insulin are liver, skeletal muscle, and fat cells. (Brain cells, meanwhile, are capable of taking up glucose on their own, without the help of insulin.) Liver and skeletal muscle cells may make quick use of some of the glucose they take in, but these two kinds of cells also are capable of *storing* glucose in a form called glycogen. Looking at the other side of control, glucagon does its job by *unpacking* the storage that insulin helps brings about. In muscle cells, glucagon promotes the transformation of glycogen back into glucose, and this glucose is then

used by the muscle cells themselves. (When we're hard at work exercising, much of the energy we're expending comes from glucose that was originally stored in the muscles as glycogen.) Glucagon also promotes the transformation of glycogen in the *liver*, but in this case, much of the resulting glucose is released into general circulation. This circulating glucose can then move into cells throughout the body as needed. As you might imagine, the body produces more glucagon than insulin when blood levels of glucose are running low—typically when we haven't eaten for awhile. After we've had a meal, however, blood glucose levels surge, and the body now produces more insulin than glucagon, so that glucose can be moved out of the bloodstream and into cells (**Figure 27.23**).

All of this sounds like an efficient way for the body to keep a tight rein on its circulating levels of glucose, but as many people know in a personal way, the body's system of glucose control can break down. In a large number of children and adults, the body *fails* to move glucose from the bloodstream into cells, and the end result of this failure is the disease diabetes, which you can read more about in "When Blood Sugar Stays in the Blood" on page 584.

## Oxytocin: Wide-Ranging Roles for a Single Hormone

Say the word *hormone* to the average person, and he or she will think of a substance that affects human *mood* rather than human physical processes. We'll now look at a hormone, called oxytocin, that does have effects on mood, but that also manages to have effects on purely physical processes. The odd thing is that oxytocin's physical effects have been understood for decades, while its effects on

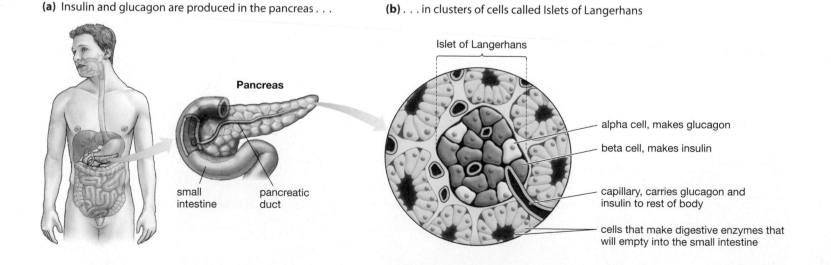

**(a)** Insulin and glucagon are produced in the pancreas . . .

**(b)** . . . in clusters of cells called Islets of Langerhans

Pancreas

small intestine

pancreatic duct

Islet of Langerhans

alpha cell, makes glucagon

beta cell, makes insulin

capillary, carries glucagon and insulin to rest of body

cells that make digestive enzymes that will empty into the small intestine

**(a)** After a meal, the role of insulin

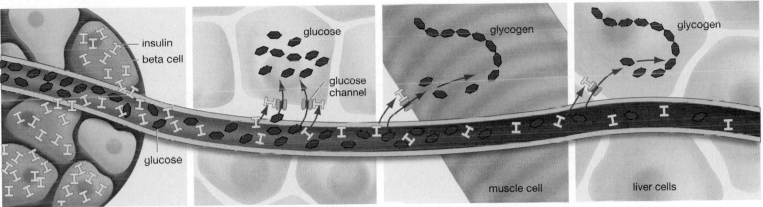

1. **Pancreas:** Stimulated by high levels of glucose in the bloodstream, beta cells in the Islets of Langerhans produce insulin.

2. **Other cells throughout the body:** Insulin enables glucose to move from the bloodstream into cells by triggering the formation of channels in the cell membranes.

3. **Skeletal muscle cells and liver cells:** With insulin's help, glucose can move into these cells and either be used right away or stored in the form of glycogen molecules.

**(b)** In between meals, the role of glucagon

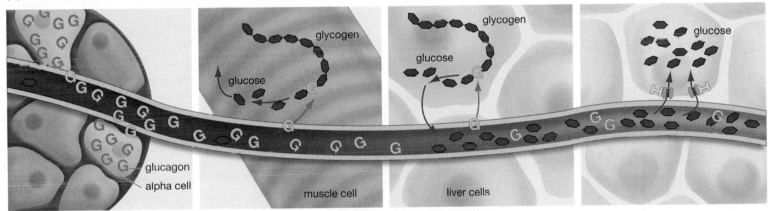

1. **Pancreas:** Stimulated by low levels of glucose in the bloodstream, alpha cells in the Islets of Langerhans produce glucagon.

2. **Skeletal muscle cells and liver cells:** With glucagon's help, glycogen is broken down into glucose. Muscle cells retain all the glucose they derive from this process, using it to power their own activities. Liver cells, meanwhile, move much of the glucose they liberate into general circulation.

3. **Other cells throughout the body:** Glucose released by the liver moves from the bloodstream into cells, supplying them with energy.

mood only came to light in the 1990s. For scientists, it was like learning that a car mechanic they had known for years had been holding down a second job all along as an opera composer. Oxytocin (ox-ee-TOE-sin) is one of the two hormones noted earlier that is produced by the hypothalamus and that is then released by the posterior pituitary gland. Its well-understood physical roles have to do with the birth and nurturing of babies. When childbirth begins, oxytocin is released (from the pituitary gland and elsewhere) and stimulates muscle contractions in the mother that push the fetus down further into the uterus, thus stretching it. This stretching then stimulates more oxytocin release, which stimulates more muscle contractions and the cycle continues. What is in process

here is one of the body's few *positive*-feedback loops. Oxytocin and stretching both prompt *more* of each other in a process that ends only when the baby is born. Oxytocin's role then continues after birth as it triggers the release of milk from the mother's breasts.

You might think that functions as diverse as these two would be plenty for any one hormone, but in recent years, researchers have found that oxytocin's role is much broader yet. For one thing, we might say that it is involved in reproduction from start to finish, although most of the evidence in this regard comes from animals other than humans. In mice, oxytocin seems to enhance the desire for physical contact, while in a mammal called the prairie vole, it makes females more receptive to sex. In human

**Figure 27.23**
**The Control of Glucose Levels: Insulin and Glucagon**

## ESSAY | When Blood Sugar Stays in the Blood: Diabetes

The paradox of the disease diabetes is that it amounts to starvation in the midst of plenty. At any given moment in the body of an untreated diabetic, billions of cells are being deprived of the life-sustaining blood sugar known as glucose. Meanwhile, less than a hair's breadth away from each of these cells, ample quantities of glucose are passing by—within the body's blood vessels. In diabetics, then, glucose moves past the body's cells but does not move into them, a state of affairs that deprives the cells of energy. When this happens, cells begin breaking down their own proteins and lipids as a means of obtaining energy. All kinds of physiological problems can follow, many of them circulatory. In the most serious cases, the result can be blindness, kidney failure, a stroke or heart attack, or a partial loss of limbs. Given such consequences, medical doctors are quick to intervene at the first signs of elevated glucose levels in any patient.

But what brings this elevation about? What causes diabetes, in other words? There actually are two principal varieties of this disease. The first variety, known as type 1 diabetes, is caused by the death of the body's beta cells—the cells in the pancreas that secrete insulin. Sadly, these cells are victims of the body turning on itself. For reasons we don't understand, the body's immune system can attack beta cells as if they were foreign invaders. With the death of these cells, type 1 diabetics have no natural source of insulin; as a consequence, they must get insulin from outside themselves, in the form of insulin injections or the newer insulin inhalations.

The second variety of diabetes is caused not by a shortage of insulin, but by a resistance to it. In type 2 diabetes, the body's beta cells are still producing insulin, but the body's biggest users of insulin—muscle, liver, and fat cells—become less responsive to it over time. As a result, even with normal levels of insulin in circulation, glucose increasingly stays within the bloodstream. Beta cells respond to this change by producing ever-greater amounts of insulin, but beyond a certain point, these cells begin to lose function. When they can no longer produce enough insulin to keep blood levels of glucose within safe limits, the body has moved from insulin resistance to type 2 diabetes.

While type 2 diabetes has always been with us, in recent years the prevalence of this disease has skyrocketed. In 2005, an estimated 7 percent of the U.S. population—20.8 million people—had diabetes, with at least 90 percent of these cases being type 2 diabetes. It appears, however, that worse is yet to come. In 2003, researchers from the federal Centers for Disease Control and Prevention estimated that one in three Americans born in the year 2000 will develop diabetes in their lifetime.

What accounts for figures such as these? The single most important risk factor for developing type 2 diabetes is being overweight. Indeed, more than 80 percent of type 2 diabetes patients are overweight. And more Americans are either now overweight or in the still-heavier category of "obese" than at any time in our history. (In 1980, 15 percent of American adults were obese; by 2004, the figure had risen to 33 percent.) Of course, weight can be controlled and therein lies the good news: Most people who are at risk for developing type 2 diabetes can avoid the disease through diet and exercise. Drugs also exist that are aimed at either preventing or controlling type 2 diabetes, but diet and exercise are the first lines of defense against this disease. So, why is weight such a critical factor in its development? For information on this topic, see "Why Fat Matters and Where It Matters Most" in Chapter 26 on page 540.

---

males, it facilitates ejaculations of some of the substances that help make up semen. And in both men and women, blood levels of oxytocin rise during sex and surge following orgasm, leading to speculations that it may be linked to making sex pleasurable.

Beyond these effects, research in recent years has generally indicated that oxytocin plays an "affiliative" role in several species—it stimulates all kinds of social contact, whereas its absence lessens this contact. Then, in 2005, our view of oxytocin's *human* affiliative role underwent a dramatic change with the publication of a study indicating that this hormone can affect human trust. Researchers in Switzerland had a group of male university students play a game of financial rewards in which some students were "investors" while others were "trustees" of the money provided by the investors. The investors stood to increase their money by turning it over to trustees, but there was an element of risk involved: The trustee could violate the investor's trust by simply keeping the money, secure in the knowledge that, under the rules of the game, trustees did not have to interact with investors more than once. Each investor thus had a decision to make before deciding how much money to turn over: How trustworthy was the trustee? The critical variable in the experiment was that, prior to playing the game, some investors were given a single dose of oxytocin, via nasal spray, while others received only a placebo spray. The result? Those who received the oxytocin exhibited much more trust than those who didn't. Those who were dosed with the hormone were, for example, more than twice as likely to invest the maximum amount with the trustee than those who received the placebo.

Findings such as these have sparked a wave of interest in the ways human behavior may be affected by oxytocin. Think, however, how wide ranging the roles are for this one hormone when we only consider what we know about it already. Its effects range from something as purely physical as muscle contractions in childbirth to something as deeply psychological as the degree to which we trust other human beings.

### Cortisol: Stress and Illness

One of the hormones that the anterior pituitary gland releases, ACTH, prompts the adrenal glands to release several so-called glucocorticoid hormones, the most important of which is cortisol. In our everyday lives, cortisol is critical in energy management. It helps bring about the production of our most important energy molecule, glucose, in a couple of ways—it provides the raw materials for glucose by facilitating the breakdown of muscle proteins, and it promotes the synthesis of glucose in the liver. Along similar lines, it promotes the breakdown of fats so that their component-part fatty acids can be used to power our activities.

In general, then, cortisol helps transform energy molecules; it takes them out of storage forms and puts them into forms that can be used. Like oxytocin, however, cortisol plays more exotic roles as well. Cortisol is a stress hormone. It is part of a group of hormones that are released as part of the body's general "stress response." When you feel the pressure building in the days before a big exam and you feel yourself being uncomfortably tense and stimulated, that's cortisol at work. (Its role overlaps that of the more famous stress hormone adrenaline, which can be thought of as functioning more in times of panic, rather than ongoing stress.) During prolonged stress, cortisol carries out its normal function, in a sense, in that it's helping to provide energy. But it is also having more sinister effects during these times: It is making you vulnerable to illnesses and temporarily destroying cell connections in the part of the brain called the hippocampus, which is critical in making long-term memories. These effects stem partly from the fact that cortisol weakens the immune system. (Among other things, it actually brings about the death of some immune cells.) But cortisol also plays a part in the narrowing of the arteries that leads to heart attacks, and it makes cells less sensitive to insulin, leaving people more at risk for developing type 2 diabetes.

All this raises the question: Why should the endocrine system work to make stressed people more vulnerable to illness? Isn't this a kind of "poor-get-poorer" syndrome? In his fine book on the stress response, *Why Zebras Don't Get Ulcers*, researcher Robert Sapolsky has suggested that human beings are stuck with a hormonal stress response that was built by evolution for beings, such as our mammalian ancestors, whose stresses were all *short-term*. Such ancestors were not stressed about what would happen next week, but human beings are. Because of our ability to envision the future, our stress responses can go on for weeks, months, or years. And it is long-term stress responses that cause all the problems noted above. In cortisol, as with oxytocin, then, we can see that the effects of hormones can be both far-reaching and not at all straightforward.

## On to the Immune System

Having looked at the capabilities of the nervous and endocrine systems, it's time to move on to another of the body's systems. This system is not involved in communication and control but instead has the job of defense. Indeed, we might say it is involved in a never-ending war of defense. On one side in this combat are all kinds of microscopic organisms—bacteria, viruses, fungi—that use every opportunity to enter the body and start reproducing in it. On the other side is the body's chief defender, the immune system, whose story is the subject of Chapter 28.

### SO FAR...

1. Hormone secretion is a big part of homeostasis, meaning an organism's tendency to maintain a relatively stable _____.

2. The part of the brain called the _____ is a central player in hormone production in that it not only controls release of several hormones it produces but also controls release of hormones produced by the body's "master gland," the _____.

3. _____ is a hormone that promotes the transport of glucose from the bloodstream into cells, while _____ is a hormone that does the reverse. Both hormones are secreted by cells in the _____.

••• ANSWERS
1. internal environment
2. hypothalamus; anterior pituitary
3. Insulin; glucagon; pancreas

# CHAPTER 27 REVIEW

For study help, activities, and more quiz questions, go to **www.aw.com/krogh4**.

## SUMMARY

### 27.1 Structure of the Nervous System

- The nervous system includes all the nervous tissue in the body plus the body's sensory organs, such as the eyes and ears. Nervous tissue is composed of two kinds of cells: neurons, which transmit nervous system messages; and glial cells, which support neurons and modify their signaling.

- The two major divisions of the human nervous system are the central nervous system (CNS), consisting of the brain and spinal cord; and the peripheral nervous system (PNS), which includes all the neural tissue outside the CNS plus the sensory organs.

- The PNS has an afferent division, which brings sensory information to the CNS; and an efferent division, which carries action (motor) commands to the body's "effectors"—muscles and glands. Within the PNS's efferent division are two subsystems: the somatic nervous system, which provides voluntary control over skeletal muscles; and the autonomic nervous system, which provides involuntary regulation of smooth muscle, cardiac muscle, and glands.

- The autonomic system is further divided into the sympathetic division, which generally has stimulatory effects; and the parasympathetic division, which generally has relaxing effects.

### 27.2 Cells of the Nervous System

- There are three types of neurons: sensory neurons, which sense conditions inside and outside the body and convey information about these conditions to neurons inside the CNS; motor neurons, which carry instructions from the CNS to such structures as muscles or glands; and interneurons, which are located entirely within the CNS and which interconnect other neurons.

- Each neuron has extensions called dendrites that receive signals coming to the neuron cell body and a single large extension, called an axon, that carries signals away from the cell body. The support that glial cells provides for neurons includes the production of a fat-rich wrapping, called myelin, that can surround neuronal axons and that increases the speed of neural signals. Recent research indicates that glial cells also modify communication among neurons by, for example, increasing the level of signaling that goes on between differing sets of neurons. A nerve is a bundle of axons in the PNS that transmits information to or from the CNS.

### 27.3 How Nervous System Communication Works

- Nervous system communication can be conceptualized as working through a two-step process: signal movement down a neuron's axon and signal movement from this axon to a second cell across a structure known as a synapse.

- An electrical charge difference, called a membrane potential, exists across the plasma membrane of neurons because the inside of the neuron is negatively charged relative to the outside. This represents a form of potential energy that is put to use when special protein channels in the neuron's membrane open up on stimulation, thereby allowing charged particles called ions to flow into the neuron. This influx of ions at an initial point on the axon triggers reactions that cause the adjacent portion of the axonal membrane to initiate the same influx of ions. Thus, a conducted nerve impulse, called an action potential, moves down the entire axon in a set of linked reactions.

- A nerve signal moves from one neuron to another (or from a neuron to a muscle or gland cell) across a synapse, which includes a "sending" neuron, a "receiving" cell, and a tiny gap between the two cells, called a synaptic cleft. A chemical called a neurotransmitter diffuses across the synaptic cleft from the sending neuron to the receiving neuron. It then binds with receptors on the receiving neuron, thus keeping the signal going. Neurotransmitters can be degraded in synaptic clefts by enzymes or taken back into a sending cell in the process called reuptake.

**Web Animation 27.1:** The Nervous System

*BioFlix* How Neurons Work

### 27.4 The Spinal Cord

- The spinal cord can act as a nervous system communication center, receiving input from sensory neurons and directing motor neurons with no input from the brain. It also channels sensory impulses to the brain. In cross section, the spinal cord has a darker, H-shaped central area, composed mostly of the cell bodies of neurons; and a lighter peripheral area, composed mostly of axons. These two areas are the gray matter and white matter of the spinal cord, respectively. The central canal of the spinal cord is filled with cerebrospinal fluid, which provides the spinal cord with nutrients. Spinal nerves extend from the spinal cord to most areas of the body.

- Spinal cord motor neurons, which are sending motor commands to muscles and other effectors, have cell bodies that lie within the gray matter of the spinal cord. The axons of these neurons leave the spinal cord through its ventral roots. Sensory neurons, which transmit information to the spinal cord, have their cell bodies outside the spinal cord, in the dorsal root ganglia. Thus, sensory information comes to the spinal cord through a dorsal root, while motor commands leave the spinal cord through a ventral root. Dorsal and ventral roots come together, like fibers being joined in a single cable, to form a given spinal nerve.

- Reflexes are automatic nervous system responses, triggered by specific stimuli, that help us avoid danger or preserve a stable physical state. The neural wiring of a single reflex, called a reflex arc, begins with a sensory receptor, runs through the spinal cord to a motor neuron, and proceeds back out to an effector such as a muscle or gland.

### 27.5 The Autonomic Nervous System

- The sympathetic division of the autonomic nervous system is often called the fight-or-flight system because it generally activates bodily functions. The parasympathetic division is often called the rest-and-digest system because it conserves energy and promotes digestive activities. Most organs receive input from both systems.

### 27.6 The Human Brain

- The brain contains almost 98 percent of the human body's neural tissue. There are six major regions in the adult brain: the cerebrum, thalamus and hypothalamus, midbrain, pons, cerebellum, and medulla oblongata. The cerebrum is divided into right and left cerebral hemispheres and is the seat of our higher thinking and processing. The cerebrum also has a thin outer layer, the cerebral cortex, that is the site of our highest thinking. The brainstem is a collective term for three brain areas—the midbrain, pons, and medulla oblongata. These brainstem structures are active in controlling involuntary bodily activities (such as breathing and digesting), in relaying information, and in processing sensory information. Most of the body's sensory perceptions are channeled through the thalamus before going to the cerebral cortex. The hypothalamus is important in sensing internal conditions and in maintaining stability or homeostasis in the body, largely through its control of many of the body's hormones.

### 27.7 The Nervous System in Action: Our Senses

- Human beings have more sensory capabilities than the famous five of vision, touch, smell, taste, and hearing. Each sense employs cells called sensory receptors that do two things: respond to stimuli (such as vibration in sound) and transform these responses into the language of the nervous system—electrical signals that travel through action potentials. Signals from every sense except smell are routed through the brain's thalamus and then to specific areas of the cerebral cortex.

### 27.8 Our Senses of Touch

- Touch works through a variety of sensory receptors that distinguish among such qualities as light or heavy pressure and new or ongoing contact. In some sensory cells, the stretching of their

outer membrane prompts an influx of ions that results in the initiation of a nerve signal.

### 27.9 Our Sense of Smell

- Our sense of smell, or olfaction, works through a set of sensory receptors whose dendrites extend into the nasal passages. "Odorants," which are molecules that have identifiable smells, bind with hair-like extensions of these dendrites, resulting in a nerve signal to the brain. The higher processing centers of the brain distinguish one odorant from another by sensing unique groups of neurons that fire in connection with given odorants.

### 27.10 Our Sense of Taste

- Our sense of taste works through a group of taste cells, located in taste buds near the surface of the tongue, which have receptors that bind to "tastants," or molecules of food that elicit different tastes. A given taste cell can respond through any of four (or perhaps five) chemical signaling routes that correspond to the four or five basic tastes of sweet, sour, salty, and bitter and a possible fifth taste of umami. The neurons that receive input from taste cells vary in their response to different tastants. The brain makes sense of the pattern of input it gets from these neurons, thus yielding the large number of tastes we experience.

### 27.11 Our Sense of Hearing

- Our sense of hearing is based on the fact that vibrations result in "waves" of air molecules that are, by turns, more and less compressed than the ambient air around them. These waves of compression bump up against our eardrums (or tympanic membranes), which in turn vibrate, initiating a chain of vibration that ends in the fluid-filled cochlea of the inner ear. "Hair cells" in the cochlea have ion channels that open and close in response to this vibration, resulting in nerve signals to the brain.

### 27.12 Our Sense of Vision

- The human visual system must accomplish three central tasks. It has to gather and focus light reflected by objects in the outside world; it must convert light signals into nervous system signals, and it must make sense of the visual information it has received. Light first enters the eye through the cornea and then passes through the lens and various materials on its way to a layer of tissue called the retina at the back of the eye. Light is bent or refracted by the cornea and the lens in such a way that it ends up as a tiny, sharply focused image on the retina.

- Light signals are converted to nervous system signals by cells in the retina called photoreceptors, which come in two varieties: rods and cones. Rods function in dim light but are not sensitive to color; cones function best in bright light but are sensitive to color. These photoreceptors have pigments, or light-absorbing molecules, embedded in membranes within them. When light strikes a pigment, it changes shape in a way that prompts a cascade of chemical reactions that results in neurotransmitter release being inhibited between the rod or cone and its adjoining connecting cell. This lack of release sends the signal, "Photoreceptor stimulated here." Vision signals travel from photoreceptors through two sets of adjoining cells, the latter of which have axons that come together to form the body's optic nerves.

- The brain does not passively record visual information. Rather, it constructs images as much as it records them. The visual system operates through a series of genetically based "rules" that allow us to quickly make sense of what we perceive. Evolution shaped our vision in this way to be maximally useful to us in survival and reproduction.

### 27.13 The Endocrine System

- The endocrine system functions in the control and regulation of bodily processes. It works through a group of chemical messengers called hormones: substances secreted by one group of cells that travel through the bloodstream and affect the activities of other cells. Hormones stand in distinction to other signaling molecules the body uses that do not travel through the bloodstream. Molecules called paracrines diffuse from one or more cells to a nearby group of cells, causing a metabolic change in them. Likewise, a cell can be affected by its own secreted chemical messenger, called an autocrine.

- Each hormone works only on specific cells—the hormone's target cells. Hormones bind to their target cells via receptors on or in the target cells. This binding then spurs chemical reactions within the target cells.

- Hormonal production and secretion take place to a significant extent within endocrine glands, meaning glands that secrete materials directly into the bloodstream or into surrounding tissues. Some hormones are secreted, however, not by specialized glands, but by organs such as the heart or kidneys.

- Once secreted, hormones can take from several seconds to several hours to work, but then can continue to have effects for extended periods of time. Given these time frames, hormones tend to regulate processes that unfold over minutes, hours, or even years.

### 27.14 Types of Hormones

- There are three principal classes of hormones: amino-acid-based hormones, peptide hormones, and steroid hormones. Each amino-acid-based hormone is derived from a chemical modification of a single amino acid. Peptide hormones are composed of amino acid chains whose lengths can vary greatly. Steroid hormones are all constructed around the chemical framework of the cholesterol molecule. Amino-acid-based and peptide hormones generally link to their target cells via receptors that protrude from the target cells' outer membranes. Some steroid hormones can bind in this manner, but most pass through a cell's plasma membrane and bind with a receptor protein inside the cell. The combined steroid hor-

mone/receptor molecule then binds with the cell's DNA, thus turning on one or more cell genes, which results in the production of one or more proteins.

**Web Animation 27.2:** The Endocrine System

### 27.15 How Is Hormone Secretion Controlled?

- Almost all hormone secretion is controlled by negative feedback. With its many negative-feedback loops, the endocrine system is important in preserving homeostasis, meaning an organism's tendency to maintain a relatively stable internal environment.

- A part of the endocrine system can be viewed as a hierarchy that has the brain's hypothalamus at the top. Although it ultimately is prompted to act via negative feedback, the hypothalamus (1) acts as an endocrine organ, producing two hormones that are released by the posterior pituitary gland; (2) exercises control, via the nervous system, over the release of two hormones—adrenaline and noradrenaline—that are produced by the adrenal glands; and (3) controls release of six hormones secreted by the anterior pituitary gland. The anterior pituitary is known as the body's "master gland" because four of the hormones it releases go on to affect the release of hormones in other endocrine glands.

- The hypothalamus exercises control over the anterior pituitary through a set of releasing and inhibiting hormones that it sends to the anterior pituitary via a tiny set of blood vessels.

### 27.16 Hormones in Action: Four Examples

- The human body controls its blood levels of the sugar glucose through the use of two hormones secreted in the pancreas. These are insulin, a hormone that reduces blood levels of glucose; and glucagon, a hormone that increases blood levels of glucose. Insulin is produced by the pancreas' beta cells while glucagon is produced by its alpha cells. After moving into circulation, insulin prompts cells to create channels through which they take in circulating glucose. The primary target cells for insulin are liver, skeletal muscle, and fat cells. In liver and skeletal muscle cells, some of the glucose that is taken in may be stored in the form of a molecule called glycogen. Glucagon prompts the transformation of glycogen back into glucose in skeletal muscle and liver cells. Muscle cells use all of the glucose produced within them through this process to power their own activities. In contrast, much of the glucose produced in the liver through this process is released into general circulation. The disease diabetes results from a failure of the body to move glucose into cells.

- The hormone oxytocin, released by the posterior pituitary gland, prompts labor contractions in childbirth and the release of milk from nursing mothers. It also appears to stimulate various forms of social contact among mammals and to affect the levels of trust that human beings are willing to place in one another.

- The hormone cortisol, released by the adrenal glands, helps bring energy stores into use but also is a "stress" hormone that can have negative effects if stress goes on for extended periods.

## KEY TERMS

| | |
|---|---|
| action potential 560 | membrane potential 558 |
| afferent division 554 | motor neuron 556 |
| amino-acid-based hormones 577 | myelin 557 |
| anterior pituitary 581 | nerve 558 |
| autonomic nervous system 554 | neurotransmitter 561 |
| axon 557 | olfaction 568 |
| central nervous system (CNS) 554 | parasympathetic division 564 |
| cerebral cortex 565 | peptide hormones 577 |
| cerebrospinal fluid 562 | peripheral nervous system (PNS) 554 |
| cerebrum 565 | photoreceptor 574 |
| cochlea 572 | posterior pituitary 581 |
| cones 574 | reflex 562 |
| dendrites 557 | retina 574 |
| efferent division 554 | rods 574 |
| endocrine gland 577 | sensory neuron 556 |
| ganglion 562 | somatic nervous system 554 |
| glucagon 580 | steroid hormones 578 |
| homeostasis 578 | sympathetic division 563 |
| hormones 576 | synapse 560 |
| hypothalamus 566 | synaptic cleft 560 |
| insulin 580 | target cells 577 |
| interneuron 556 | thalamus 566 |

## BRIEF REVIEW

*Answers to Brief Review questions are in the back of the book.*
*For multiple-choice quiz questions, go to* **www.aw.com/krogh4**.

1. Three functional types of neurons are found in the nervous system. What are they, and what role does each play?

2. In what way is the neuron a cell that is specialized for rapid communication?

3. Describe how a simple reflex arc works.

4. Most of the senses reviewed in the book had these things in common: sensory receptor stimulation, transduction, neurotransmitters, receiving neurons, thalamus, and cerebral cortex processing. Describe what happened in general at each of these steps.

5. Why is the endocrine system better at controlling something like fluid levels in the body than our ability to shoot a basketball?

6. Do hormones choose their target cells?

7. What are the differences between amino-acid-based, peptide, and steroid hormones?

## APPLYING YOUR KNOWLEDGE

1. In the section on the senses, you saw that the body can respond to many different kinds of stimulation—vibrations, smells, pressure, light, and taste among them. Can you think of any other kinds of stimulation that our bodies did not evolve to respond to but might have? Could the senses we have be extended in any way? (Hint: Think of the olfactory abilities of dogs.)

2. It is often said that seeing is believing. Is it reasonable to put this much faith in the human visual system?

3. When most of us hear the word *hormones*, we think of substances that cause us to act in unusual ways. Is this the normal function of most hormones? Why do they have this reputation?

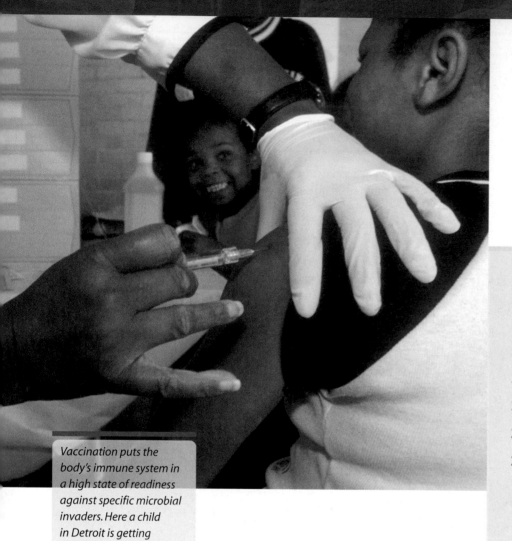

Vaccination puts the body's immune system in a high state of readiness against specific microbial invaders. Here a child in Detroit is getting vaccinated just prior to the first day of school.

| 28.1 | Two Types of Immune Defense | 592 |
| 28.2 | Nonspecific Defenses | 592 |
| 28.3 | Specific Defenses | 594 |
| 28.4 | Antibody-Mediated Immunity | 597 |
| 28.5 | Cell-Mediated Immunity | 598 |
| 28.6 | AIDS: Attacking the Defenders | 600 |
| 28.7 | The Immune System Can Cause Trouble | 602 |

Infectious microbes enter the human body every day. Waiting to meet them is a highly organized network of human cells and proteins. Collectively, these defenders of the body are known as the immune system.

# DEFENDING THE BODY:
# The Immune System

**T**he word "vaccine" comes from the word "vacca," which is Latin for "cow." What do cows have to do with vaccines? In 1796, an English country doctor, Edward Jenner, decided to test a bit of folk wisdom, which held that farm laborers who caught the mild disease cowpox from cattle could never catch the dreaded disease smallpox from humans. Accordingly, Jenner took cowpox pus from a sore on the hand of a dairymaid and rubbed it into scratches he made on the arms of an 8-year-old boy, James Phipps. The next step—a chilling one from a modern perspective—was that Jenner applied pus from *smallpox* sores onto Phipps. The boy sailed through this and several subsequent smallpox applications, however, and Jenner's belief was proved correct: A person could be made immune to smallpox by being exposed to cowpox.

With Jenner's demonstration, a revolution was launched—the vaccine revolution. Because of it, polio, diphtheria, whooping cough, and tetanus have been nearly eradicated from developed countries. And the smallpox that Jenner worked against has been eliminated from the face of the Earth as a natural disease. (The last naturally transmitted case of it occurred in 1977.)

But how does vaccination work? To take James Phipps as an example, why should his exposure to cowpox have protected him from smallpox? Two centuries of hard-won scientific advances since Jenner's time have provided the answer. Vaccination activates one part of the *immune system*, a complex network of cells and proteins whose most important job is finding invading microbes and killing them. The immune system is tenacious and has an arsenal of weapons at its disposal. Then again, it needs these very qualities because it faces such a formidable lineup of invaders. In the chapter that follows, you'll learn how the immune system rises to the challenge of protecting the human body.

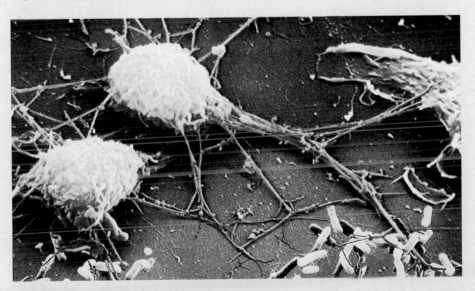

**Defenders and Invaders**

The large, spherical cells on the left are immune system macrophages that are in the process of attacking invading *E. coli* bacterial cells, at the bottom of the photo. The macrophages will engulf the bacterial cells, ingest them, and dissolve them. (Micrograph × 2150)

591

## 28.1 Two Types of Immune Defense

The immune system has two basic defensive strategies. First, it employs something called **nonspecific defenses**: immune system defenses that do not discriminate between one invader and the next. Skin cells do not discriminate between bacteria and fungi, for example; they keep both kinds of invaders out in the same way. Likewise, some immune cells will attack any cell perceived to be foreign.

The immune system's second main strategy, its **specific defenses**, provides protection against *particular* invaders. You have immune cells in your body that will latch onto bacterium A, but not to bacterium B. In a sense, these specific defenses *remember* invaders they have faced, which enables them to generate a rapid response to any subsequent invasion by the same microorganism. The result is **immunity**: a state of long-lasting protection that the immune system develops against specific microorganisms. (Here, you can begin to see what was at work with smallpox.)

Before going into details about nonspecific and specific defenses, you need to be introduced to a term you'll be seeing repeatedly. An **antigen** is any foreign substance that elicits an immune system response. A bacterial cell in the lining of our lungs is, of course, foreign to our bodies. But it is certain proteins or carbohydrates on the surface of the bacterial cell that are likely to set off the human immune response. In general, antigens are component parts of living things, but an inorganic material such as asbestos can also serve as an antigen (in that it can trigger an immune response).

**Table 28.1** lists just a few of the many players that take part in the immune process. All the entities on the lists do one of three things: They destroy antigens, they mark antigens for destruction, or they act as communication molecules. Note that all of the immune cells listed in Table 28.1a are different kinds of white blood cells. Such cells are the key players in immunity; any cell that is active in the immune system is likely to be a variety of white blood cell.

## 28.2 Nonspecific Defenses

The body's first line of defense actually is not a part of the immune system per se but instead is a set of barriers. To cause trouble, a disease-causing

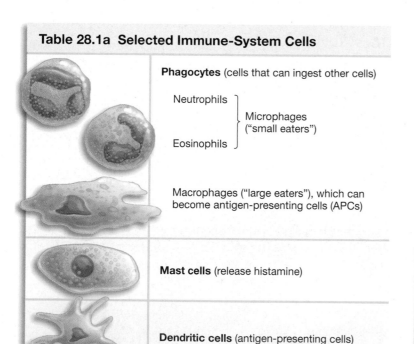

**Table 28.1a  Selected Immune-System Cells**

**Phagocytes** (cells that can ingest other cells)

Neutrophils
}  Microphages ("small eaters")
Eosinophils

Macrophages ("large eaters"), which can become antigen-presenting cells (APCs)

**Mast cells** (release histamine)

**Dendritic cells** (antigen-presenting cells)

**Lymphocytes**

Natural killer cells

B cells, which differentiate into:
Plasma cells
Memory cells

T cells, which differentiate into:
Helper T (CD4) cells
Cytotoxic (killer) T cells
Regulatory T cells

**Table 28.1b  Selected Immune-System Proteins**

**Lysozymes** (protective enzymes found in tears and saliva)

**Complement proteins** (kill some invaders, bind with others to aid phagocytes)

**Antibodies** (receptors on surface of B cells; later released by plasma cells derived from B cells)

**Interleukins** (diverse group of signaling molecules)

organism—known as a *pathogen*—must enter the body's tissues, and that often means crossing an *epithelium*, or tissue that covers an exposed body surface. The epithelial covering of the skin, described in Chapter 26, has multiple layers, a waterproof keratin coating, and a network of tight seams that lock adjacent cells together. These specializations create a very effective barrier that protects underlying tissues.

The exterior surface of the body is also protected by hairs and glandular secretions. Secretions from sebaceous and sweat glands flush the surface of the skin, washing away microorganisms and chemical agents. Some secretions also contain bacteria-killing chemicals and protective enzymes, called *lysozymes*, which can be found in tears and saliva.

From Chapter 26, recall that some of our epithelial tissue lines *interior* surfaces that are exposed to an *exterior* environment. For example, our lungs are interior to us, but they come into contact with air from the outside world. Interior epithelial tissue is delicate, but it is well defended. The lungs have a set of hair-like cilia that sweep their passageways clean like so many tiny brooms. The stomach contains a powerful acid that can destroy pathogens. Urine, which has a pathogen-killing low pH, flushes the urinary passageways. And the female reproductive tract has a low pH thanks to lactic acid produced by *helpful* resident bacteria, the lactobacilli.

### Nonspecific Cells and Proteins

But what happens if invaders get past these barriers? We have a line of nonspecific defense waiting to meet them. Let's look at some of the members of this team of defenders.

There is a type of white blood cell, called a **phagocyte**, that can be defined as a cell capable of ingesting another cell, parts of cells, or other materials. As you might guess from this description, phagocytes can take on a warrior role—they essentially surround invading microorganisms with cellular extensions, take them in, and then dissolve them in a set of acid baths. But phagocytes also take on a less-exalted role, which is that of janitor. They engulf and haul away the body's *own* cells and tissue fragments when they have become worn out or damaged. As you can see in Table 28.1a, phagocytes come in several varieties, two of them smaller (the neutrophils and eosinophils) and one larger (the macrophages). All three have the capacity to kill foreign invaders outright. Later, you'll be seeing the macrophage in a different role. It not only kills

invaders, but then presents fragments of these invaders to other immune cells, something like a soldier who is holding an enemy's uniform out for the rest of the troops to see.

When an invader is a virus, rather than a bacterium, another key player in nonspecific defense is the *natural killer* (NK) cell, so named because it will attack any cell that it perceives as "foreign" in the sense of having unusual surface proteins. Cells infected by viruses are altered in this way; when an NK cell encounters one of these infected cells, it can kill it in one of two ways. First, it can use proteins to create so many pores in the membranes of infected cells that these cells break up, like a ship that's been riddled with holes. Second, it can cause these cells to commit suicide through a process called *apoptosis*.

Table 28.1a also lists an immune system cell called the mast cell. This cell releases a substance called histamine, whose role we'll get to shortly. Finally, it is not just cells that are involved in nonspecific defense; it is proteins as well. A group of them listed in Table 28.1b, the complement proteins, will make an appearance soon. Let's now see how all the players work together to mount a type of well-orchestrated nonspecific defense, the *inflammatory response*.

### Nonspecific Defense and the Inflammatory Response

Let's say you accidentally puncture your skin with a nail that's been sitting outdoors. This means you not only have a puncture, you have lots of microscopic invaders that have gotten past the barrier of your skin. How does the immune system deal with the bacteria among them? In outline, three types of responses take place. First, blood vessels near the site of injury dilate and become more permeable—they increase in diameter, and they undergo a change that allows cells and proteins to pass out of them more freely. Second, these cells and proteins then leave the blood vessels and move to the site of injury, where they kill the bacteria. Third, various compounds are released that wall off the site of injury, thus limiting the spread of the infection.

The mast cells noted earlier help get the first step going by releasing a substance called **histamine**, a compound that, in the inflammatory response, brings about blood vessel dilation and increased blood vessel permeability. Mast cells have significant quantities of histamine to release because they store it inside themselves in tiny granules. The blood vessel dilation that histamine prompts is important because it results in increased blood flow near the injury site—an efficient means of bringing in more immune system cells and proteins. The increased permeability histamine causes is important because it allows these cells and proteins to move out of the blood vessels so that they can get to the site of the infection. Normally, blood vessel cells are locked tightly together; histamine has the effect of opening up spaces between them.

The first immune system cells to squeeze through these spaces probably will be one of the smaller phagocytes noted earlier, the neutrophils. Slower in arriving, but longer lived, are the macrophages. Both varieties of cells are guided to the site of injury by a form of chemical signaling: Injured cells release compounds that produce a kind of trail that leads to them. Once the neutrophils and macrophages arrive, they begin ingesting bacterial invaders; later, the macrophages will clean up the debris.

Flowing through the bloodstream at this time, and also activated by the attack, are the complement proteins noted earlier—about 20 of them in all. They are called complement proteins because their actions are complementary to those taken by other parts of the immune system. In this instance, one of their roles is to cut holes in the cell membranes of the bacteria. The site of infection is likewise being limited in size at this time. The area is sealed off partly through creation of a network of fibers composed of a blood protein that, appropriately enough, is called fibrin.

In looking at this simplified account of the inflammatory response, you can see that it brings more blood near a site of injury and more cells and proteins to the site. Blood is warm and red, of course, and this explains why the area around an injury site takes on these same qualities. Meanwhile, the added cells and proteins at the site cause the swelling and pain that we associate with injuries.

It's important to note that all these responses are nonspecific: Any group of invaders that entered through this route would get pretty much this same treatment by this same lineup of immune system players. But what about instances when the body's nonspecific defenses can't contain an infection? What happens when, say, a chicken pox virus comes into the nasal passages and starts to spread from there throughout the body? If this invader can't be defeated by nonspecific defenses, the body has another line of defense at the ready—the specific defenses, which have a marvelous ability to target specific invaders.

**SO FAR...**

1. The human body's _____ defenses provide protection against particular invaders. The memory of these defenses allows the body to generate a rapid response to any subsequent invasion by the same microbe. This long-lasting state of protection is called _____.

2. An antigen is any substance that _____.

3. All of the cells most important to immune system function are varieties of _____ cells. One variety of these cells, the phagocyte, is a cell that is capable of _____.

## 28.3  Specific Defenses

In contrast to the barriers and nonspecific defenses we've looked at so far, specific defenses work through a process known as *acquired immunity*, which means the immunity we acquire by virtue of what we come into contact with during our lifetime. One type of acquired immunity is provided to us by others. **Passively acquired immunity** is immunity gained by the administration of disease-fighting substances, called antibodies, that have been produced by another individual. Mothers pass antibodies to their children in the womb and in breast milk, for example. Adults might be the beneficiaries of passively acquired immunity too, when they receive antibodies through the gamma globulin shots that people sometimes get prior to taking a trip abroad (**Figure 28.1**). Invader-fighting antibodies made by others are degraded rather rapidly in our system, however, which means that the immunity these antibodies provide lasts a few weeks or months. Of more interest to us is the long-lasting immunity that our bodies generate themselves. This is **actively acquired immunity**: immunity developed as a result of accidental or deliberate exposure to an antigen.

Note the critical word there: exposure to an *antigen*—a substance the body recognizes as foreign. Think about this in relation to one well-known form of actively acquired immunity, the vaccinations mentioned at the start of the chapter. In vaccinations, a person is not being treated with a substance that kills an invader. Rather, the person is being injected with at least a part of the invader itself. This is deliberate exposure to an antigen, and it results in our body mounting an attack on the antigen in ways that we'll review shortly. Of course this "invader" has been made harmless back in the laboratory. It may be a poliovirus, for example, that has been killed, so that it can no longer cause disease. But the immune system's attack on it is

**Figure 28.1**
**Types of Acquired Immunity**

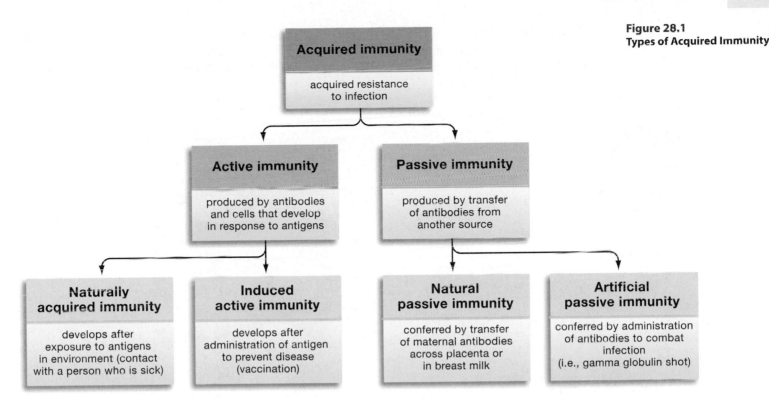

**Acquired immunity**

acquired resistance to infection

**Active immunity**

produced by antibodies and cells that develop in response to antigens

**Passive immunity**

produced by transfer of antibodies from another source

**Naturally acquired immunity**

develops after exposure to antigens in environment (contact with a person who is sick)

**Induced active immunity**

develops after administration of antigen to prevent disease (vaccination)

**Natural passive immunity**

conferred by transfer of maternal antibodies across placenta or in breast milk

**Artificial passive immunity**

conferred by administration of antibodies to combat infection (i.e., gamma globulin shot)

full blown, and having mounted it, the system enters a high state of readiness for any subsequent attack by the *real* invader—the live poliovirus. What vaccines do, then, is elicit readiness on the part of the immune system. Getting a disease like chicken pox does this as well —it elicits protection that keeps us from getting chicken pox again. But how does this trick of "invader memory" work? For that matter, how does the immune system defeat specific invaders the first time they attack? These are the subjects to which we now turn.

### Antibody-Mediated and Cell-Mediated Immunity

Actively acquired immunity has two major arms. The first arm is called **antibody-mediated immunity**: an immune system capability that works through the production of proteins called antibodies. The second arm is called **cell-mediated immunity**: an immune system capability that works through the production of cells that destroy other cells in the body —those that have been infected by an invader. Cells that are central to both arms originate in the same place, which is the marrow of our bones. It is there that all white blood cells begin development in adults. The white blood cells that are key to specific immunity are called *lymphocytes*. Some lymphocytes go on to migrate to the body's thymus (located at the base of the throat), and there they specialize. They develop into the main cells of cell-mediated immunity, the **T-lymphocyte cells** or just **T cells** (with the

T standing for "thymus" cell). Meanwhile, lymphocytes that remain in the bone marrow can develop into the central cells of *antibody*-mediated immunity, the **B-lymphocyte cells** or **B cells**. The development of T and B cells is tracked in **Figure 28.2**.

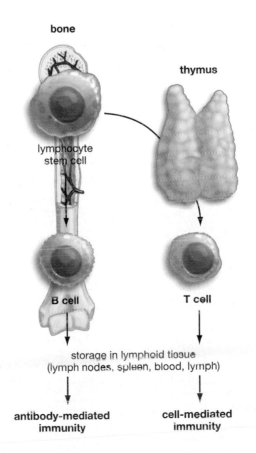

bone

thymus

lymphocyte stem cell

B cell

T cell

storage in lymphoid tissue (lymph nodes, spleen, blood, lymph)

antibody-mediated immunity

cell-mediated immunity

**Figure 28.2**
**Development of T Cells and B Cells**

Key players in the body's specific defense system are T cells and B cells, both of which start out as a type of white blood cell, called a lymphocyte stem cell, in the bone marrow. Some of these lymphocytes migrate to the thymus gland, where they differentiate into T cells (thymus cells). Others that remain in the bone marrow develop into B cells. Most T and B cells then migrate to lymphoid tissues, such as lymph nodes and the spleen, where they will serve their immune system function.

••• ANSWERS

1. specific; immunity

2. elicits an immune system response

3. white blood; ingesting another cell, parts of cells, or other materials

Eventually, both types of cells will migrate to the body's lymphatic system organs—the spleen and the lymph nodes, for example—where they are prepared to fight invaders. The lymphatic system is laid out for you in **Figure 28.3**. Think of it as a system of vessels that picks up fluids (called interstitial fluids) that

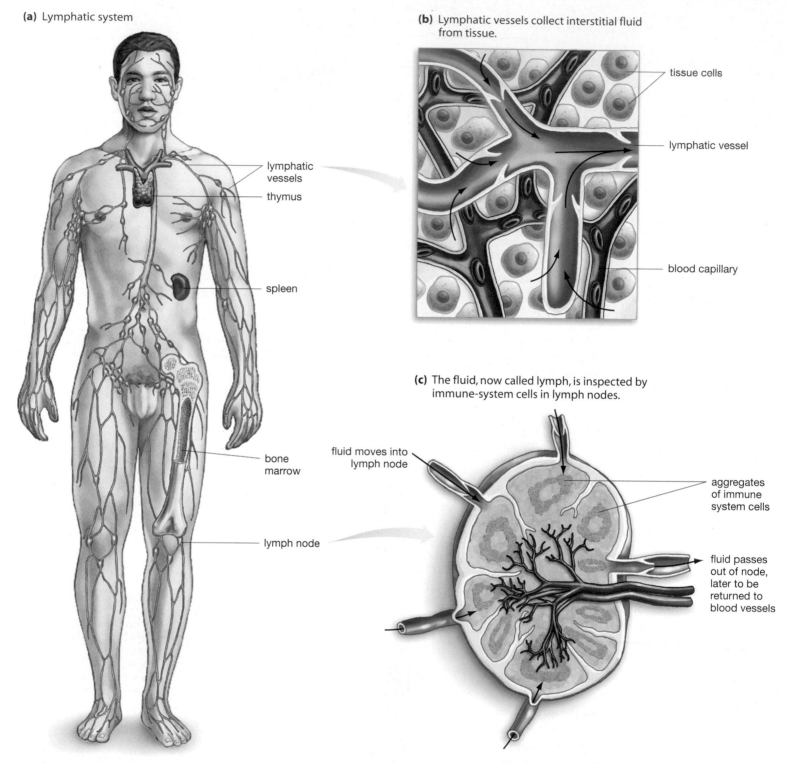

**(a)** Lymphatic system

lymphatic vessels
thymus
spleen
bone marrow
lymph node

**(b)** Lymphatic vessels collect interstitial fluid from tissue.

tissue cells
lymphatic vessel
blood capillary

**(c)** The fluid, now called lymph, is inspected by immune-system cells in lymph nodes.

fluid moves into lymph node
aggregates of immune system cells
fluid passes out of node, later to be returned to blood vessels

**Figure 28.3**
**The Human Lymphatic System**

**(a)** Fluid moving through lymphatic vessels passes through lymphatic organs, such as lymph nodes or the spleen, where the fluid is inspected by an array of immune system cells.

**(b)** Detail of interstitial fluid moving into lymphatic vessels.

**(c)** Detail of fluid moving through a lymph node.

have leaked from blood vessels. Eventually, these fluids are channeled back into the circulatory system, but while they are in the lymphatic system, they pass through the lymphoid organs, which are packed with immune system cells. This is why, when you get sick, your "lymph glands" are swollen—they are filled with an abnormally large number of cells that are busy fighting off the infection. If you look back at Table 28.1, you can see the various types of lymphocytes listed.

From this point forward, the antibody-mediated and cell-mediated immune operations are treated separately. The two systems work together, but for now we'll look at them in isolation, starting with antibody-mediated immunity.

## 28.4 Antibody-Mediated Immunity

As you might imagine, a key player in antibody-mediated immunity is the **antibody**, which can be defined as a circulating immune system protein that binds to a particular antigen. From this definition, it may be apparent that antibodies and antigens are closely linked. (They are so closely linked, in fact, that one is named for the other; the word *antigen* is short for "antibody generating.") In general, a given antibody will bind to a specific antigen and to no other antigen. In their binding capacity, you will see antibodies in two roles: first as *receptors* for antigens on the surface of B cells, extending in a generalized Y shape from the cells' outer membranes; second as *free-standing* molecules that are produced by B cells and then stream away from them in great numbers, moving through the bloodstream.

### The Fantastic Diversity of Antibodies

It's common knowledge that it's not just one virus that causes the common cold; rather, there are hundreds of viruses that bring on the general set of symptoms we call a cold. Then there are all the other potential invaders, such as bacteria, fungi, and so forth. How can the immune system cope with such a variety of foes? One of its key strengths is that it produces B cells that differ from each other with respect to the type of antigen receptors they have on their surfaces. Look at one B cell, and it will have receptors that latch only onto an antigen on *this* virus; look at a *different* B cell, and its receptors bind with an antigen of a different virus. This mind-boggling complexity is made possible by DNA arranging that takes place back in the bone marrow, when the precursors of B cells are being formed. The result is B cells with

millions of variations in antigen receptors. Now, how do these B cells do their job?

### The Cloning and Differentiation of B Cells

Lying in readiness in the lymph nodes or circulating through the bloodstream, a B cell encounters, say, a bacterium—one that has an antigen that can be bound by this B cell's antigen receptors. If you look at **Figure 28.4**, you can see what happens next. This very binding causes the B cell to start dividing very rapidly. Each

**Figure 28.4**
**B Cells and Antibody-Mediated Immunity**

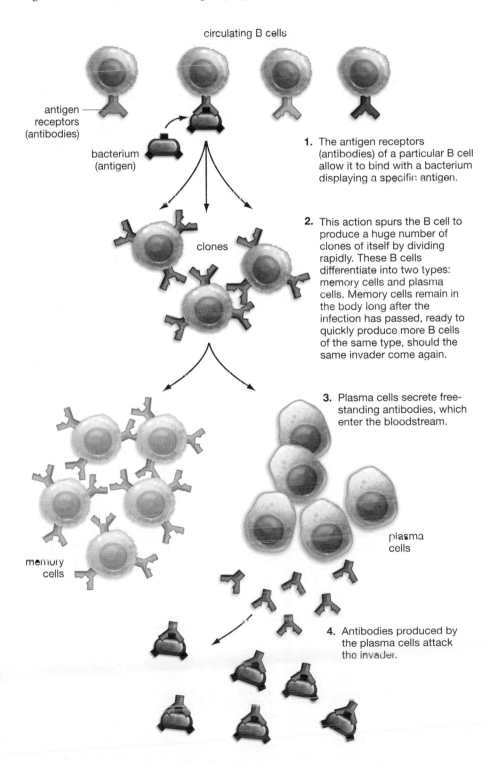

circulating B cells

antigen receptors (antibodies)

bacterium (antigen)

clones

memory cells

plasma cells

1. The antigen receptors (antibodies) of a particular B cell allow it to bind with a bacterium displaying a specific antigen.

2. This action spurs the B cell to produce a huge number of clones of itself by dividing rapidly. These B cells differentiate into two types: memory cells and plasma cells. Memory cells remain in the body long after the infection has passed, ready to quickly produce more B cells of the same type, should the same invader come again.

3. Plasma cells secrete free-standing antibodies, which enter the bloodstream.

4. Antibodies produced by the plasma cells attack the invader.

new B cell of this type gives rise to additional identical B cells, creating a "clone," or a huge number of identical cells of this type. Such numbers are necessary because the *bacteria* will be multiplying rapidly as well.

After a time, these B cells then differentiate into two types of cells. One variety is called a **plasma cell**: an immune system B cell that is specialized to produce antibodies. These antibodies are not the cell surface receptors you've been looking at, however. They are free-standing antibodies that leave the plasma cells by the millions—they are *exported* from the plasma cells—after which they move through the bloodstream and fight the invader directly.

The other type of B cell is called the **memory B cell**, and it is the body's permanent sentry. This is one part of the long-lasting invader readiness noted earlier. After the initial war with an invader is over, this specific variety of memory B cell remains in the body. Should this invader come in a *second* time, these memory cells will be ready to divide and produce more plasma cells to mount a very quick defense. This is why being inoculated with poliovirus antigens protects us permanently from this crippler. (And this is why cowpox antigens protected Edward Jenner's patients from smallpox; the antigens of the two viruses were similar enough that exposure to cowpox put the body in a state of readiness for smallpox.) If you look at **Figure 28.5**, you can see the typical time course of exposure to antigen and production of antibodies. What is known as the "primary immune response" has reached a substantial level 14 days out from first exposure. Look at the second exposure, however, to see what a difference this first exposure has made. Now, there is a much greater response in a much shorter time.

••• ANSWERS

1. exposure to an antigen; readiness
2. lymphatic system; immune system cells; circulatory
3. antigen; B; exported

### The Action of the Antibodies

But what is it that the circulating antibodies do to combat the invader? First, because they are binding to the invader's antigens, the antibodies can sometimes prevent the invader from attaching to anything *else*, which stops its spread. Antibodies can also come together with antigens in a clumping or "agglutination" that renders the invaders inactive. Third, antigens work in concert with an arm of the immune system you've already reviewed—the nonspecific defenses. Remember the cell-ingesting phagocytes and the complement proteins that you saw in nonspecific defense? Many invading bacterial cells have a slick outer coating that keeps phagocytes from latching onto them. Working together, antibodies and complement proteins often *can* bind to many of these invaders, and this gives the phagocytes something to hold onto so that they can begin ingesting the enemy.

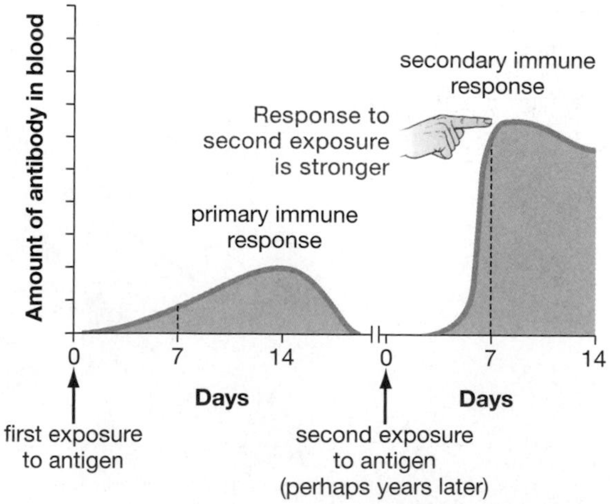


**Figure 28.5**
**Prepared for an Invasion**
The memory cells produced by the body during a first attack by an invader allow it to mount a faster, more vigorous defense should the same invader attack a second time.

**SO FAR...**

1. Vaccination, a deliberate ________, elicits a high state of ________ on the part of the immune system to any second attack by a given microbe.
2. The spleen and lymph nodes, parts of the body's ________, are packed with ________ that inspect lymphatic fluid as it is being returned to the ________ system.
3. An antibody is a circulating immune system protein that can bind to a particular ________. Antibodies are seen in two forms: as receptors on the surface of ________ cells and as free-standing proteins that are ________ by these cells.

## 28.5 Cell-Mediated Immunity

Although antibody-mediated immunity is quite spectacular, it also turns out to be limited in that it works strictly on free-standing foreign organisms. The problem with this is that many viruses (and some bacteria) are successful in invading the body's *own cells*—they exist and multiply inside these cells, which makes the viruses inaccessible to antibodies. To deal with this threat, the body has developed another arm of acquired immunity, the cell-mediated immunity mentioned earlier.

You may recall that the central player in cell-mediated immunity is the T-lymphocyte or T cell, which comes in three main varieties. Let's now go through the actions of the cell-mediated immune system, taking as a starting point a body that has been infected with a virus.

## Cells Bearing Invaders: Antigen-Presenting Cells

Any cell that has been infected by a virus puts protein fragments from the virus on its surface. Certain immune system cells, however, don't merely display such fragments; they present them, as antigens, to other immune cells as part of the immune system response. The most important of these **antigen-presenting cells** or **APCs** is called a dendritic cell (which gets its name from the spiky extensions it has that look something like the dendrites that extend from nervous system cells). Another APC is the macrophage you've already been introduced to, and still another is a B cell. Dendritic cells begin their work by engulfing and killing some of the invaders that are free-standing and then displaying the invader's fragments. Outfitted with these antigens on display, dendritic cells then migrate to lymph nodes or the spleen, where, you'll recall, immune system cells exist in abundance. If you look at **Figure 28.6**, you can see how the next interaction plays out. Helper T cells now lock onto the dendritic cells, but there is great specificity in this action. A given dendritic cell is displaying the invader's protein fragment within one of its own membrane proteins, meaning the helper T cell's binding is specific to both the dendritic and the invader proteins. Once this interaction has taken place, the helper T cell has been "activated," and this initiates the creation of a helper T cell clone. The first activated helper T cell divides into two cells, these two cells divide into four, and so forth. Critically, each cell in this clone is specific to this invader. These T cells then leave the lymph nodes or spleen and begin circulating throughout the body.

## Helper T Cells, Cytotoxic T Cells, and Regulatory T Cells

The helper T cells that have been activated now take on a central role in specific immunity. Beyond creating a helper T cell clone, they produce a signaling molecule, called interleukin 2, that puts several other processes in motion. First, this secretion spurs production not just of more helper T cells, but of the variety of T cell that actually will eliminate infected cells in the body. This is the **cytotoxic** (or killer) **T cell**: a type of T lymphocyte that binds to and kills the body's own cells when they have become infected. Critically, both cytotoxic and helper T cells are being produced not only in active types, but in "memory" types for use in future invasions. The helper T cell's interleukin 2 secretion also spurs the activation of a third type of T cell, the **regulatory T cell**: a type of immune system cell that

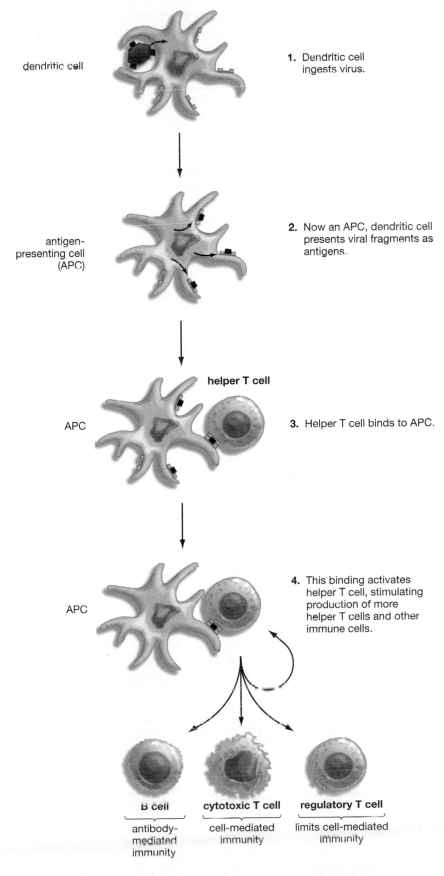

1. Dendritic cell ingests virus.

2. Now an APC, dendritic cell presents viral fragments as antigens.

3. Helper T cell binds to APC.

4. This binding activates helper T cell, stimulating production of more helper T cells and other immune cells.

| B cell | cytotoxic T cell | regulatory T cell |
| --- | --- | --- |
| antibody-mediated immunity | cell-mediated immunity | limits cell-mediated immunity |

acts to limit the body's immune system response, thus protecting the body's own tissues from attack. As if this weren't enough, other interleukins secreted by helper

**Figure 28.6
T Cells and Cell-Mediated Immunity**

T cells activate B cell immunity by promoting the production and differentiation of B cells. Indeed, the entire process of B cell development is dependent on signals that come from helper T cells.

Once this production process has ramped up, the system's cytotoxic T cells are primed to recognize any cell that is displaying the fragments of the viral invader. Recall that this means any cell in the body that's been infected by the invader because all infected cells will put some invader fragments on their surface. Using their own special receptors, the cytotoxic T cells now latch onto these infected cells and kill them in one of two ways. First, they puncture the outer membranes of infected cells, causing them to *lyse* or burst—the same kind of lethal action that natural killer cells undertake, but now carried out by a huge number of cells that are targeting this specific invader. Second, they secrete signaling molecules that cause the infected cells to commit suicide through apoptosis.

Note that cytotoxic T cells are binding with the body's own infected cells. From this, it may be apparent why this is called *cell-mediated* immunity. In the end, it is an attack involving not the antibodies described before, but several sets of cells.

Taking a step back from cell- and antibody-mediated immunity, in **Figure 28.7** you can see how the two systems work together, along with nonspecific defenses. In specific defense, however, note the central role of the **helper T cell**. We can define it as an immune system cell that helps activate both cell- and antibody-mediated immunity, since it is necessary for the production of both T and B cells. This diverse set of helper-T functions is important in what's coming up.

## 28.6 AIDS: Attacking the Defenders

The picture of the immune system just painted is one of a sophisticated operation capable of dealing with all kinds of invaders. Yet, everyone knows of at least one invader that the system seems powerless to deal with: the virus that causes AIDS.

The virus in question is **HIV**, the human immunodeficiency virus, and it is devastating for three reasons. First, it attacks the very system that is supposed to protect us from disease—the immune system we've just been reviewing. Second, its infection process gives it an immediate foothold in the body, as you'll see. Third, this virus is literally a shape-shifter: Its genetic material mutates rapidly, and this brings about a rapid change in HIV's structure—in particular, a change in the proteins that protrude from its surface. As you've seen, the antibodies and cells of the immune system work by binding to such proteins. But with HIV, these binding sites are changing all the time. Thus, HIV particles are like soldiers who are constantly "morphing" into shapes that the immune system's soldiers cannot recognize.

HIV works by attaching itself, most importantly, to helper T cells—specifically to so-called CD4 and CCR5 receptors on the surface of helper T cells. Once this binding is complete, the outer HIV envelope

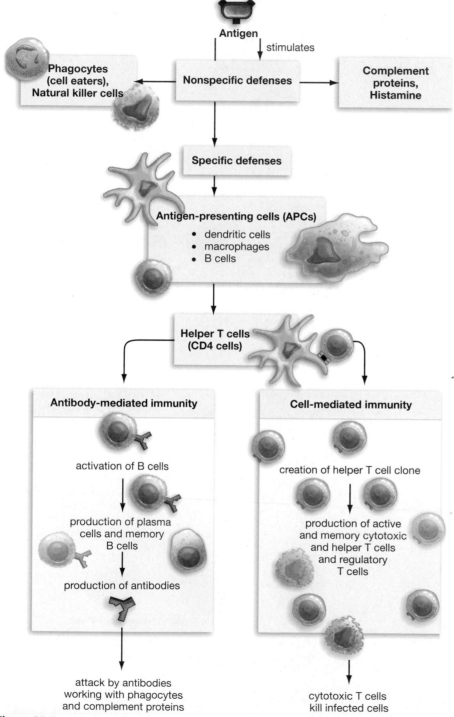

**Figure 28.7**
**An Overview of the Immune System Process**

then fuses with the helper T cell membrane, much as two soap bubbles might fuse together to become one (**Figure 28.8**). With this, HIV's proteins and its RNA genetic material are inside the T cell. Next, the virus uses these materials to create a double strand of viral DNA that it inserts into the infected cell's own length of DNA. Thus situated within a cell, the viral DNA may remain inactive or "latent" for years, during which time it is hidden from immune surveillance carried out by the body. But every time the T cell copies its own DNA in cell division, the inserted viral DNA is copied along with it and is passed along to the resulting "daughter cells."

In many infected helper T cells, however, the virus is far from latent. Remember, it now amounts to a length of DNA inserted into the cell's own DNA. In

**Figure 28.8**
**How the AIDS Virus Attacks the Immune System**

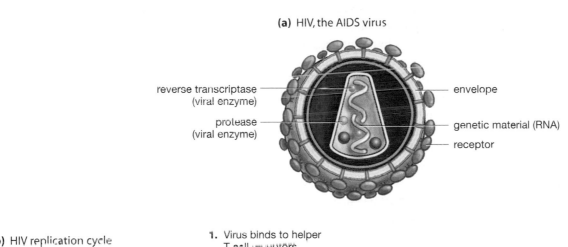

**(a)** HIV, the AIDS virus

reverse transcriptase (viral enzyme)

protease (viral enzyme)

envelope

genetic material (RNA)

receptor

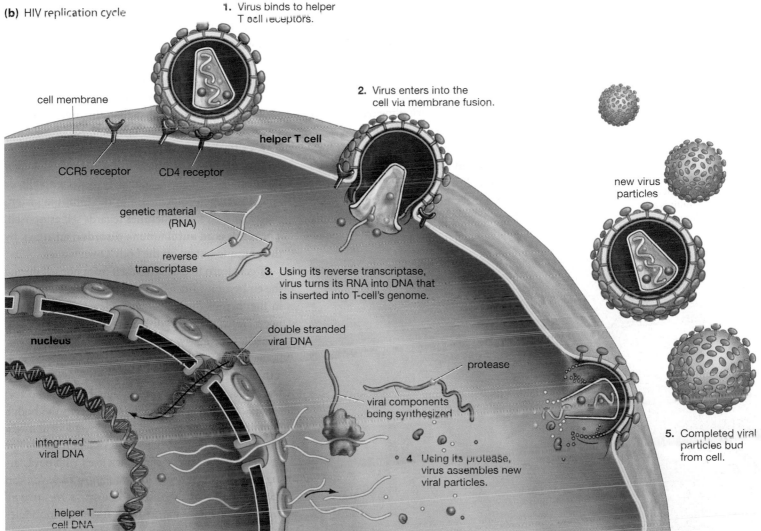

**(b)** HIV replication cycle

1. Virus binds to helper T cell receptors.

cell membrane

helper T cell

CCR5 receptor    CD4 receptor

2. Virus enters into the cell via membrane fusion.

new virus particles

genetic material (RNA)

reverse transcriptase

3. Using its reverse transcriptase, virus turns its RNA into DNA that is inserted into T-cell's genome.

nucleus

double stranded viral DNA

protease

viral components being synthesized

integrated viral DNA

5. Completed viral particles bud from cell.

4 Using its protease, virus assembles new viral particles.

helper T cell DNA

# CHAPTER 28 REVIEW

For study help, activities, and more quiz questions, go to **www.aw.com/krogh4.**

## SUMMARY

### 28.1 Two Types of Immune Defense

- The immune system has two primary defensive strategies: non-specific defenses, which do not discriminate between one invader and the next; and specific defenses, which provide protection against particular invaders.

- An antigen is any foreign substance that elicits an immune system response. These substances generally are proteins or carbohydrates that are part of an invading organism. Almost all of the cells active in immune responses are different varieties of white blood cells.

### 28.2 Nonspecific Defenses

- The body's first line of nonspecific defense is a set of barriers, such as tightly locked skin cells, and the secretions that protect these barriers, such as the acid secretions of the stomach and the low pH of urine.

- Nonspecific defenses mount a coordinated action called the inflammatory response. It involves not only cells and proteins that kill invaders, but substances that increase blood flow and blood vessel permeability near the site of an injury.

### 28.3 Specific Defenses

- Acquired immunity is the immunity that results from what humans come into contact with during their lifetime. It exists in two forms. Passively acquired immunity is achieved through the transmission of disease-fighting antibodies produced by another individual—for example, the antibodies transmitted to an infant through its mother's milk. Actively acquired immunity develops as a result of natural or deliberate exposure to an antigen. Actively acquired immunity can be produced by getting an immunization shot or by catching a disease such as chicken pox.

- Actively acquired immunity has two major arms. The first arm is antibody-mediated immunity, which works through the production of proteins called antibodies. The second arm is cell-mediated immunity, which works through the production of cells that destroy infected cells in the body.

- White blood cells called lymphocytes are fundamental to both antibody-mediated and cell-mediated immunity. One variety of lymphocyte, the B-lymphocyte cell (or B cell) is central to antibody-mediated immunity. Another type of lymphocyte, the T-lymphocyte (or T cell) is central to cell-mediated immunity and helps initiate antibody-mediated immunity.

- Mature B and T cells are most often found in organs of the body's lymphatic system, which is a system of vessels that captures fluids leaked from the bloodstream; that then subjects these fluids to scrutiny by immune system cells; and that then returns the fluids to the circulatory system.

### 28.4 Antibody-Mediated Immunity

- An antibody is a circulating immune system protein that binds to a particular antigen. Antibodies exist (1) as receptors on the surface of B cells that bind to specific antigens and (2) as free-standing invader-fighting proteins that are secreted by B cells and move through the bloodstream.

- Antibody-mediated immunity begins with the binding of a B cell antigen receptor to an antigen. This binding produces a cascade of effects, including the production of a B cell clone—a huge number of identical copies of the B cell that bound originally to the antigen. Some of these cells are plasma B cells that export free-standing antibodies that fight the invader; others are memory B cells that stay in the body to provide a quick response to a subsequent attack by the same invader.

- Antibodies limit infectious attacks in several ways. Their binding with antigens may keep the antigen from binding with anything else, and in concert with complement proteins, antibodies bind to invaders in such a way that macrophages can bind with the invaders and kill them.

### 28.5 Cell-Mediated Immunity

- Cell-mediated immunity provides protection in instances when the body's own cells have become infected by an invader. Viruses and some bacteria can infect the body's cells and remain inside them, in which case antibodies will have no access to these invaders. The central player in cell-mediated immunity is the T cell, which comes in three main varieties: helper T cells, cytotoxic (or killer) T cells, and regulatory T cells.

- Cell-mediated immunity begins with several classes of white blood cells killing invaders and displaying protein fragments from them on their cell surfaces. These fragments now serve as antigens. Any cell that plays this role is known as an antigen-presenting cell, or APC. The most important APC is called the dendritic cell, although macrophages and B cells carry out the antigen-presenting function as well. Once dendritic cells display fragments from an invader on their surface, they migrate to lymphatic organs and bind with helper T cells that have receptors specific to this invader. This "activation" of helper T cells prompts the production of a helper T cell clone: An activated helper T cell divides into two, these two into four, and so forth. Critically, all the members of this clone are specific to this invader.

- Helper T cells from the clone then migrate out of the lymphatic organs to encounter infected cells wherever they may be in the body.

A protein called interleukin 2, secreted by helper T cells, prompts the creation of a cytotoxic T cell clone. Both this clone and the helper T cell clone are being produced in active and "memory" varieties. In addition, helper T cell activity stimulates regulatory T cell activation and B cell clone development. Cytotoxic T cells bind with infected cells and bring about their destruction in two ways: They lyse or puncture the cells' membranes, and they initiate suicide or apoptosis in the infected cells. Regulatory T cells limit immune system attacks, thus sparing the body's own tissues from damage.

## 28.6  AIDS: Attacking the Defenders

- AIDS, or acquired immune deficiency syndrome, is caused by HIV, the human immunodeficiency virus. This virus is devastating for three reasons: It attacks the immune system itself; its invasion process leads to a harmful infection very quickly; and its genetic material mutates rapidly, meaning the virus itself changes shape at a rapid rate, which allows it to evade immune system detection.

- HIV attaches itself, most importantly, to helper T cells and thus gains entry to these cells. Once inside, the virus can insert a DNA copy of its own genetic material into the T cell's DNA and remain inactive for extended periods of time, hidden from the immune system. Conversely, it can use the infected cell's metabolic machinery to turn out its own proteins and genetic material—the building blocks of new virus particles. These complete viruses then leave the cell to infect more cells; when this cycle is repeated often enough, the cell dies. Because helper T cells are central to immunity, the result can be a great vulnerability to common microorganisms.

- It will be very difficult to develop a vaccine that can prevent AIDS because any such vaccine would have to prevent the infection of all T cells. Perhaps an AIDS vaccine will only be able to slow the replication of HIV, which in itself could be of great practical value.

**Web Animation 28.1:** HIV Replication in Detail

## 28.7  The Immune System Can Cause Trouble

- The immune system can attack the body's own healthy tissues, essentially mistaking the body's own tissues ("self") for foreign substances ("nonself"). These attacks produce such autoimmune disorders as rheumatoid arthritis, multiple sclerosis, and type 1 diabetes.

- An allergy is an immune system overreaction to a foreign substance (an allergen) that causes the body to release the infection-fighting compound histamine. The result is an inflammatory response that can range from small and harmless (sneezing and sniffling) to large and life-threatening (asthma or anaphylactic shock).

- The immune system is the most important problem in organ transplantation. The immune system regards any transplanted organ as nonself and thus initiates an attack on it.

## KEY TERMS

| | | | |
|---|---|---|---|
| **actively acquired immunity** | 594 | **histamine** | 593 |
| **allergen** | 603 | **HIV** | 600 |
| **allergy** | 603 | **immunity** | 592 |
| **antibody** | 597 | **memory B cell** | 598 |
| **antibody-mediated immunity** | 595 | **nonspecific defenses** | 592 |
| **antigen** | 592 | **passively acquired immunity** | 594 |
| **antigen-presenting cell (APC)** | 599 | **phagocyte** | 593 |
| **autoimmune disorder** | 602 | **plasma cell** | 598 |
| **B-lymphocyte cell (B cell)** | 595 | **regulatory T cell** | 599 |
| **cell-mediated immunity** | 595 | **specific defenses** | 592 |
| **cytotoxic T cell (killer T cell)** | 599 | **T-lymphocyte cell (T cell)** | 595 |
| **helper T cell** | 600 | | |

## BRIEF REVIEW

*Answers to Brief Review questions are in the back of the book. For multiple-choice quiz questions, go to* **www.aw.com/krogh4**.

1. Name the main actions that take place in the inflammatory response.

2. What primary immune function does a dendritic cell carry out?

3. HIV infection can lead to several forms of cancer, including Kaposi's sarcoma and a type of lymphoma; it can lead to a kind of dementia and a physical wasting; it makes its victims prone to several kinds of fungal infections that are rarely seen in the general population. How can one virus have this many effects?

4. Why is the immune system an obstacle to organ transplantation?

5. In what way does the rapid mutation of HIV make the development of an AIDS vaccine more difficult?

## APPLYING YOUR KNOWLEDGE

1. When smallpox raged in England in the eighteenth century, most of its victims were young children. Meanwhile, the people least likely to die from it were senior citizens. Why should this have been so?

2. Public health officials agree that, since the 1960s, asthma has been increasing among children in developed countries such as the United States. Among the tentative explanations for this increase is the "hygiene hypothesis," which holds that the problem is that many children are growing up in environments that are "pristine"—very clean and devoid of household pets. What could the linkage be between an immune system condition such as asthma and growing up in a pristine environment?

3. One of the key treatments for children who lack an immune system is a bone marrow transplant. Why would this procedure help to establish an immune system in a person who does not have one?

To carry out any activity, human beings must first acquire oxygen and then move that oxygen to cells throughout the body—tasks that are handled by the human respiratory and cardiovascular systems.

CHAPTER **29**

| 29.1 | The Cardiovascular System | 608 |
| 29.2 | The Composition of Blood | 608 |
| 29.3 | Blood Vessels | 609 |
| 29.4 | The Heart and Blood Circulation | 611 |
| 29.5 | What Is a Heart Attack? | 613 |
| 29.6 | Distributing the Goods: The Capillary Beds | 615 |
| 29.7 | The Respiratory System | 616 |
| 29.8 | Steps in Respiration | 618 |

**Essays**

| Blood Doping: A Dangerous Way to Cheat | 610 |
| Listening in on Blood Pressure | 614 |

Our bodies have a respiratory system that acquires oxygen and gets rid of carbon dioxide, a digestive system that breaks down food, and a urinary system that gets rid of some materials while recycling others. Tying all these systems together is another system, the circulatory system.

# TRANSPORT AND EXCHANGE 1:
# Blood and Breath

**T**he cells that make up our bodies may be tiny things, but they are still living things, which means they have their own ongoing set of needs. They have to obtain energy and get rid of wastes; they have to obtain

oxygen; they have to obtain nutrients such as nitrogen and phosphorus. When life consisted of nothing but single-celled creatures, every cell had access to the world outside itself, which meant that food could enter, waste could be expelled, and gases could be "exchanged" in a fairly simple manner. In a human being, conversely, many of our 10 trillion cells spend their entire existence wrapped deep inside layers of tissue.

During the course of evolution, this access problem, as it might be termed, was solved by the development of physiological *systems*: a special bodily apparatus for extracting nutrients from food, another for delivering oxygen, and another for removing waste. We call the system that extracts nutrients the digestive system, the one that delivers oxygen the respiratory system, and the one that removes waste the urinary system. If you think about it, what all these systems have in common is that they carry out various exchanges of materials between us and our surroundings. They do this, however, only with the help of one additional system, the cardiovascular system, which uses the fluid called blood to pick up gases, nutrients, and wastes from one location in the body and transport them to another. The cardiovascular system is thus central to our other transport and exchange systems in that it serves as a kind of mass-transit system for them. (How does oxygen get from our lungs to, say, our feet? It travels

through the cardiovascular system in blood cells.) Given the linking role of the cardiovascular system, we'll begin our review of transport and exchange by looking at it, after which we'll review the respiratory system. Then, in Chapter 30, we'll review the digestive and urinary systems. While we're looking at the digestive system, we'll review human nutrition. **Figure 29.1** summarizes the basic functions of the four systems we'll be looking at.

**Figure 29.1**
**The Transport and Exchange Systems of the Body**

The human body has four systems that are active in transport and exchange: the cardiovascular, digestive, urinary, and respiratory systems. The cardiovascular system is central to all the others in that it transports materials to and from them.

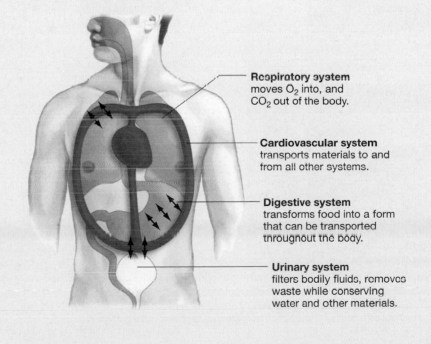

**Respiratory system**
moves $O_2$ into, and $CO_2$ out of the body.

**Cardiovascular system**
transports materials to and from all other systems.

**Digestive system**
transforms food into a form that can be transported throughout the body.

**Urinary system**
filters bodily fluids, removes waste while conserving water and other materials.

## 29.1 The Cardiovascular System

The human **cardiovascular system** is a fluid transport system that consists of the heart, all the blood vessels in the body, the blood that flows through these vessels, and the bone marrow tissue in which red blood cells are formed. This system bears a rough comparison with a car's cooling system. A car uses antifreeze and water as its circulating fluids; the cardiovascular system uses blood. A car may use a water pump to circulate its fluids; the cardiovascular system uses a pump called the heart. A car has an assortment of conducting pipes; the cardiovascular system has blood vessels. The similarities end there, however, because, while a car's cooling system transports only a couple of things, the cardiovascular system transports lots of things. It moves nutrients, vitamins, waste products, hormones, and immune system cells and proteins, not to mention the oxygen we breathe in and the carbon dioxide we breathe out. All these materials are transported in the substance we'll look at first within the cardiovascular system—blood.

## 29.2 The Composition of Blood

Collect a sample of blood from a human being, prevent it from clotting, and spin it in a centrifuge, and it will separate into two layers. The lower layer, accounting for some 45 percent of its volume, consists of blood cells and cell fragments, which collectively are known as the **formed elements** of blood (**Figure 29.2**). The yellowish, upper layer, accounting for the rest of the volume, is called **plasma**, meaning the fluid portion of blood. Within an actual blood vessel, the formed elements are suspended in the flowing plasma just as tiny bits of sediment might be suspended within a moving stream.

### Formed Elements

It turns out there are three kinds of formed elements in the plasma. The most numerous by far are the **red blood cells**, which transport oxygen to, and carbon dioxide from, every part of the body. Then there are **white blood cells**, which are critical players in the immune system reviewed last chapter. Finally, there are **platelets**—small fragments of cells that are important in the blood-clotting process. Let's now look at each of these elements in more detail.

Red blood cells, or *erythrocytes*, are red because within them is the iron-containing protein **hemoglobin** (**Figure 29.3**) Molecules of hemoglobin that lie within a red blood cell (RBC) bind with oxygen as it comes in from the lungs and then hold onto it, releasing it only when the cell reaches a part of the body whose oxygen concentration is relatively low. RBCs then work with blood plasma in picking up the carbon dioxide that cells have produced through their metabolism; they then carry this $CO_2$ back to the lungs, where it will be exhaled.

Blood is deep red because RBCs constitute 99.9 percent of all the formed elements in it. Indeed, the *number* of RBCs in an average person staggers the imagination. One cubic millimeter of blood—an amount that might scarcely be visible—contains 4.8 to 5.4 million RBCs. So numerous are these cells that they make up about a third of all the cells in the human body.

Apart from their numbers, RBCs are also notable for another quality, which is their odd structure. In their mature state, they essentially consist of a cell membrane that surrounds a mass of hemoglobin.

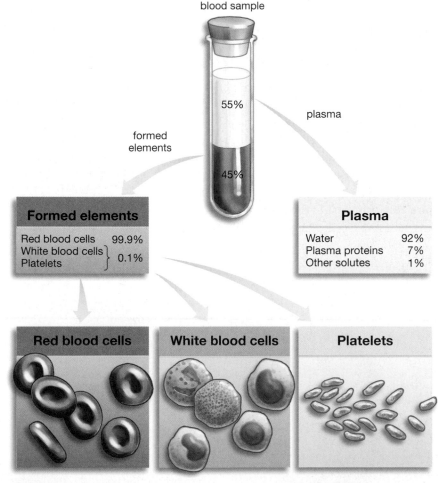

**Figure 29.2**
**The Composition of Blood**

Blood that is spun in a centrifuge separates into its two primary constituent parts: formed elements and plasma. Formed elements are cells and cell fragments that help make up the blood—red blood cells, white blood cells, and platelets. Blood plasma, the liquid in which the formed elements are suspended, is composed mostly of water but also contains proteins and other materials.

Viewed another way, an RBC has almost none of the standard cellular equipment—a nucleus, other organelles, and so forth. This lack of cellular machinery means that the average RBC life span is only 120 days, and this means that about 180 million new RBCs have to enter the circulation *each minute* for us to have enough of them in our system. (Elite athletes have, unfortunately, learned that *boosting* their number of red blood cells through artificial means can provide them with a competitive advantage—something you can read more about in "Blood Doping: A Dangerous Way to Cheat" on page 610.)

The second formed element in blood, the white blood cells or *leukocytes*, have a completely different function from red blood cells. They have standard cellular equipment, but they are not in the oxygen or $CO_2$ transport business. Instead, in their many varieties, they are central to the body's immune system operation. A typical cubic millimeter of blood contains 6,000 to 9,000 white blood cells. This is a tiny number relative to the millions of red blood cells in such a volume, but the comparison is somewhat misleading because most white blood cells (WBCs) are found in the body's tissues rather than in the bloodstream. WBCs use the bloodstream primarily to *get to* sites of invasion or injury.

The third formed element, platelets, can be thought of as enzyme-bearing packets. They are not cells but fragments of cells that have broken away from a set of unusually large cells that exist in the bone marrow. The enzymes in them aid in blood clotting at sites of injury. Apart from releasing these enzymes, the platelets clump together at the site of an injury, forming a temporary plug that slows bleeding while the process of clotting continues.

### Blood's Other Major Component: Plasma

The formed elements you've just read about are suspended in blood's other major component, plasma, which is 92 percent water. Proteins and a mixture of other materials are dissolved in this fluid as well. Why aren't these proteins and other materials "formed elements"? Well, note that they are not cells, or even cell fragments, but are instead much smaller molecules that go into making up a living thing.

There are three primary kinds of proteins in the plasma. First, there are the albumins, made in the liver, which function in the transport of both hormones and the components of fats known as fatty acids. Second, there is fibrinogen, also made in the liver, which aids in blood clotting. Third, there are globulins, which come in two forms. One is the immunoglobulins or anti-

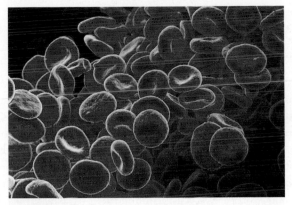

**Figure 29.3**
**Transport Cells**
A collection of erythrocytes or mature red blood cells, which carry oxygen to the body's tissues and carry carbon dioxide away from them.

bodies that you saw so much of in Chapter 28, which attack foreign proteins and disease-causing organisms. The other is the transport proteins that carry small ions, hormones, or other compounds.

Looking strictly at transport proteins, we find that one variety of them has a real importance to our hearts. These *lipoproteins* essentially are capsules of protein that surround globules of fatty material—usually cholesterol. Lipoproteins are best known by the names of their two varieties: **low-density lipoproteins** (or **LDLs**), which carry cholesterol *to* outlying tissues from the liver and small intestine; and **high-density lipoproteins** (or **HDLs**), which carry lipids *from* these tissues to the liver. Keep both LDLs and HDLs in mind because you'll be reading about them later in connection with heart attacks.

Taken together, the plasma proteins plus water make up 99 percent of blood plasma. The other 1 percent is a mixture of hormones, nutrients, wastes, and ions (which are better known as electrolytes). As you can imagine, wastes that are dissolved in the plasma are on their way out of the body altogether. For example, a substance called urea, produced by the liver's breakdown of proteins, is carried within the plasma to the kidneys, is filtered out by them, and is then urinated out of the body.

## 29.3 Blood Vessels

Having looked at what moves through the cardiovascular system, let's now look at the plumbing that carries these materials—our network of blood vessels.

Blood vessels are tubes, in a sense. To get a real idea of them, however, imagine a regular tube, such as a straw, that could change its internal diameter from second to second, that could take in some

Blood Doping: A Dangerous Way to Cheat

In 2006, the Tour de France got off to its worst start ever when 13 riders were forced to withdraw from the race before it even began (**Figure 1**). All these riders were alleged to have had ties to a doping ring based in Spain whose facilities had been raided in May by Spanish police. Among the items police seized were nearly 100 bags of frozen blood and quantities of a hormone known as EPO. These items were evidence that, despite increased testing, lifetime bans, and even accidental deaths, world-class cyclists continue to engage in a sorry form of cheating known as blood doping.

Although there actually are two forms of blood doping, both of them have the same goal: to increase the number of red blood cells athletes have in circulation during a race. What's the point? Remember that the way oxygen gets to all the tissues in the body is through red blood cells. And, in any "endurance" sport, oxygen is critical to performance. Increase the oxygen that is delivered to muscles and you increase the capacity of the muscles to continue working. In the late 1960s, some endurance athletes starting training at high altitude in the belief they could boost their red blood cell count by doing so. (The idea was that the body would respond to the decreased oxygen concentrations at high altitude by producing more red blood cells. It does seem to do this, but whether athletic performance is improved as a result is an open question.) By the 1970s, however, athletes and their trainers had hit on another, more rapid means of boosting red blood cell counts, which was blood doping.

In its original form, blood doping usually was achieved by means of athletes drawing out one or two pints of their own blood, whose cellular portion they would then freeze and store. In the ensuing weeks, their blood levels would then return to normal through the natural process of blood production. But normal blood levels were not what these athletes were seeking. In the period just before a race, they would then unfreeze the blood cells they had withdrawn from themselves weeks earlier and transfuse these cells back into their bodies. In each case, the result was a body that had an *oversupply* of red blood cells.

Although this technique is still used today, beginning in the early 1990s a more efficient means of doping became available. The key to it was a hormone called erythropoietin (or EPO for short), which is naturally produced in the kidneys and which prompts the body to generate red blood cells. Healthy individuals produce enough EPO to remain well supplied with red blood cells, but some kidney and cancer patients have a reduced ability to produce EPO and thus end up anemic—they end up having too few red blood cells in circulation to meet their oxygen needs. In the late 1980s, scientists came to the rescue of these patients by developing a genetically engineered or "recombinant" form of EPO. This boon to kidney and cancer patients quickly proved to be a bane to cycling and other endurance sports, however, as a sleazy, international black market for EPO developed among competitive athletes. (In 2003, the director of Italy's antidoping efforts noted that enough EPO was being sold in his country to supply the needs of every Italian EPO patient six times over.) EPO quickly became a banned substance in competitive sports, but a cat-and-mouse game of EPO concealment and detection has existed in many sports from the 1990s to the present, with cycling being the worst-offending sport.

The reason athletes inject EPO, or engage in old-fashioned blood doping, is simply that these techniques work. Experts are agreed that the extra blood cells that result from these procedures can increase athletic performance. But at what cost? As noted in the main text, human blood consists of two essential components: its plasma or liquid portion and its formed element or blood cell portion. Athletes who boost only their red blood cell counts are upsetting this balance. In essence, they are *thickening* their blood by adding cells without adding plasma. This apparently can have fatal consequences. To look at just one series of incidents, during a period of about a year in 2003 and 2004, eight elite cyclists died in Europe from cardiac arrest. Given the secretive nature of blood doping, it was difficult to definitively link these deaths with EPO use. But experts say that athletes who use EPO are putting their blood in a sludge-like state that can easily clog up in blood vessels. When these vessels are in the heart, the result can be a heart attack. Most heart attacks occur in people who are middle aged or older, of course, but this was not the case with the European cyclists who died in 2003–2004. Their average age was 27.

**Figure 1**
**The Tour in Tatters**
Allegations of links to a Spanish blood-doping ring forced 13 riders to withdraw from cycling's most prestigious event, the Tour de France, just before the start of its 2006 race. Spain's entire Astana-Wurth team, training here days before the race began, withdrew after five of its members were implicated in the Spanish investigation, leaving the team with too few members to compete.

materials along its length while keeping others out, and that could repair itself when damaged. Because blood vessels are composed mostly of cells, they are dynamic entities that change their activities in accordance with the body's needs. Blood vessels carrying blood *away* from the heart are the **arteries**, while the vessels carrying blood back *to* the heart are the **veins**. The farther from the heart these vessels are, the smaller they tend to be. Connecting the arteries with the veins are the smallest blood vessels of all, the **capillaries**.

Arteries and veins are always are made up of three distinct layers of tissue (**Figure 29.4**). The innermost layer is composed of a group of flat cells, which you may remember from Chapter 26 are *epithelial cells*—cells that come into contact with materials from outside the body, such as oxygen and nutrients. The middle layer of arteries and veins contains smooth muscle. These are muscles that allow the blood vessel to constrict or expand in diameter. This very ability has health consequences because the dangerous condition called hypertension or high blood pressure is brought about by a *persistent* constriction of the body's smaller arteries. (As the diameter of these arteries decreases, the pressure within them increases.) The outer layer of arteries and veins is a stabilizing sheath of connective tissue. Taken as a whole, these three layers give arteries and veins a great dual capability: strength combined with flexibility. The largest arteries have diameters of up to 2.5 centimeters or about an inch, but farther out from the heart, the body's smallest arteries, the arterioles, have diameters

that average 30 micrometers (μm), which is about a thousandth of an inch.

Tiny as this diameter is, things get a lot smaller yet with the body's other main variety of blood vessel, the capillaries. Their diameters average about three *ten*-thousandths of an inch. Why so minuscule? Their central function is to have substances pass into them and out of them along their length. As such, their walls need to be very thin. Accordingly, each capillary is made of a single layer of cells. So small are capillaries that tiny red blood cells must move through them in single file.

## 29.4 The Heart and Blood Circulation

Web Animation 29.1
The Cardiovascular System

What makes blood flow through these vessels? The answer is the muscular pump we call the heart. A small organ, roughly the size of a clenched fist, the heart actually is shaped something like a valentine heart in its major outline. It lies near the back of the chest wall, directly behind the bone known as the sternum.

The heart sends blood out, and then gets it back, through two circulation loops. As noted earlier, one major task of the blood transport system is to get oxygen to all the various bodily tissues. For blood to carry oxygen to tissues, it first has to *take up* the oxygen that is brought into the body through the lungs. So imposing is this task of oxygenating blood that one of the heart's circulation loops is devoted solely to getting blood to the lungs and then

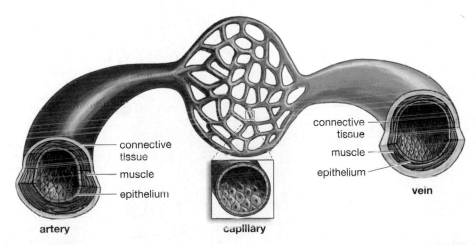

**Figure 29.4**
**The Structure of Blood Vessels**
Both arteries and veins have three layers to them: an inner epithelium, a layer of smooth muscle, and a layer of connective tissue. By contracting or relaxing, the smooth muscle can change the diameter of the vessels. Arteries and veins are linked by intervening capillary beds. Each capillary consists of a single layer of cells.

back to the heart. The oxygenated blood that comes to the heart in this loop is then pumped out to the rest of the body in the heart's second major loop (**Figure 29.5a**).

To put these things more formally, there are two general networks of blood vessels in the cardiovascular system. One of these is the **pulmonary circulation**, in which the blood flows between the heart and the lungs. Then there is the **systemic circulation**, in which the blood is moved between the heart and the rest of the body. Let's see how this two-network system works.

## Following the Path of Circulation

The human heart contains four muscular chambers, two associated with the pulmonary circulation and two associated with the systemic circulation. These are the right atrium and right ventricle (pulmonary circulation) and the left atrium and left ventricle (systemic circulation). Let's take as the starting point the blood entering the right atrium from the veins called the superior and inferior venae cavae (on the *left* side of **Figure 29.5b**). These veins are returning blood to the heart from the upper and lower portions of the body, respectively. This is blood that is coming back after distributing oxygen and other blood-borne

materials throughout the body—to our legs, our hands, our trunk, and so forth. As such, it is deoxygenated blood, which actually is dark red in color, but which is indicated by the blue color of the veins it flows through. After the right atrium receives this blood, it pumps it the short distance to the right ventricle, which contracts, pushing the blood into the pulmonary arteries. (These vessels are arteries instead of veins because blood moves *away* from the heart through them.) This blood is now bound for the lungs, where it will pick up oxygen. Once it does this, it returns to the heart through the two pulmonary veins, which empty into the left atrium. The blood then is pumped down into the left ventricle, which contracts and sends the blood coursing up into the enormous artery called the **aorta**. This vessel has branches stemming from it that will carry the blood to all the tissues of the body.

Taking a step back from this, you can see that the right side of the heart pumps blood only to the lungs, while the left side of the heart pumps blood to all the body's tissues, as shown in Figure 29.5a. All four chambers of the heart pump blood, but the right and left *atria* pump blood only into another portion of the heart—their adjacent ventricles. The right ventricle is pumping blood to the lungs, so there is some

**Figure 29.5**
**Two Circulatory Networks and How the Heart Supports Them**

(a) Veins in the systemic circulation bring blood into the right side of the heart; it is pumped out via the pulmonary circulation to the lungs, where it is oxygenated, and then returns to the left side of the heart. It is then pumped out from the left side of the heart, through the systemic circulation, to the rest of the body.

(b) The path of the blood's circulation through the four chambers of the heart.

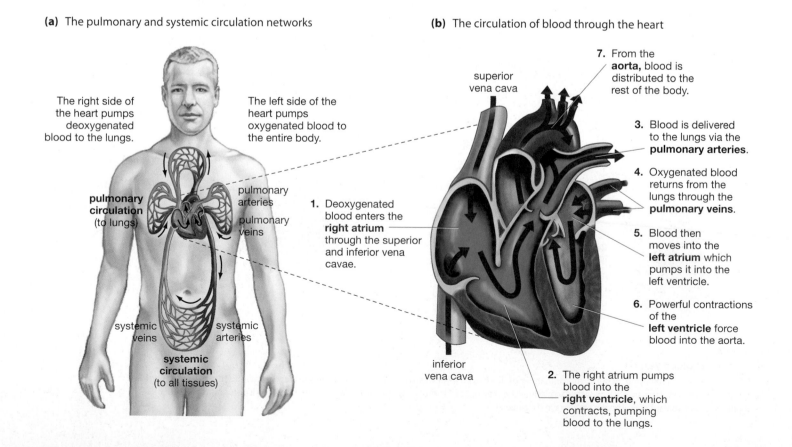

**(a)** The pulmonary and systemic circulation networks

The right side of the heart pumps deoxygenated blood to the lungs.

The left side of the heart pumps oxygenated blood to the entire body.

**pulmonary circulation** (to lungs)

pulmonary arteries

pulmonary veins

systemic veins

systemic arteries

**systemic circulation** (to all tissues)

**(b)** The circulation of blood through the heart

superior vena cava

inferior vena cava

**7.** From the **aorta,** blood is distributed to the rest of the body.

**3.** Blood is delivered to the lungs via the **pulmonary arteries**.

**4.** Oxygenated blood returns from the lungs through the **pulmonary veins**.

**5.** Blood then moves into the **left atrium** which pumps it into the left ventricle.

**6.** Powerful contractions of the **left ventricle** force blood into the aorta.

**1.** Deoxygenated blood enters the **right atrium** through the superior and inferior vena cavae.

**2.** The right atrium pumps blood into the **right ventricle**, which contracts, pumping blood to the lungs.

greater contraction power needed there. But, the *left* ventricle is pumping to the whole body, and its contractions are by far the most powerful. When we "take our pulse," what we are feeling is the surge of blood produced by the contraction of the left ventricle. (For an account of how our pulse figures in the process of measuring blood pressure, see "Listening in on Blood Pressure" on the next page.)

### Valves Control the Flow of Blood

Because both ventricles contract with such force, you may wonder why blood doesn't go back up into their respective atria with each beat. The answer is that there are valves between the atria and ventricles that allow the blood to flow only one way—from atrium to ventricle. Occasionally, because of disease or genetic predisposition, people do have this kind of backflow into the atria. The fluid turbulence that results can be heard as a kind of gurgling sound, known as a "heart murmur" (which is not necessarily a sign of a serious heart problem). Valves also exist to prevent backflow between the ventricles and the arteries they pump blood into. The heart's two sets of valves are responsible for the familiar "lub-dub" sound of the heartbeat. The "lub" is the sound of the valves closing between the atria and ventricles, while the "dub" is the sound of the valves closing between the ventricles and the arteries.

user of these materials because it essentially is a set of muscles that never stops working. Blood does not come to the muscles of the heart through any of the chambers you've been reviewing. (Think about the two ventricles; one is sending blood to the lungs, the other to the rest of the body, but not directly to the heart itself.) The way blood does come to the heart is through two arteries that branch off from the aorta just after it emerges from the left ventricle. Because these arteries encircle or "crown" the heart before they start branching, they are known as the **coronary arteries** (**Figure 29.6**).

It is difficult to overstate the importance of the coronary arteries because about half of all deaths in the United States today are caused by a blockage of one or more of them. Such blockages generally have, as their starting point, the LDLs or low-density lipoproteins, noted earlier, that carry cholesterol molecules away from the liver and small intestine to varying locations in the body. One place these LDLs go is the coronary arteries. There, they can lodge within one of the three layers of tissue that make these arteries up—the innermost layer that comes into contact with the flowing blood. Once present in large numbers, these LDLs can become damaged through oxidation (the same process that causes metal to rust), and this is where the real trouble starts. The immune system now regards these damaged LDLs as *foreign*

### SO FAR...

1. Blood is composed of formed elements, which are _____, and plasma, which is the _____ portion of blood. There are three kinds of formed elements: _____, _____, and _____. Plasma, meanwhile, is overwhelmingly composed of _____.

2. Arteries carry blood _____ the heart, while veins carry blood _____ the heart. The tiny blood vessels called _____ connect arteries to veins.

3. There are two networks of blood vessels in the body. In one of these, the pulmonary circulation, blood flows between the _____. In the other, the systemic circulation, blood flows between the _____.

## 29.5 What Is a Heart Attack?

The heart is a unique organ, but like any other organ, it needs a supply of blood and the oxygen and nutrients that come with it. Indeed, the heart is a voracious

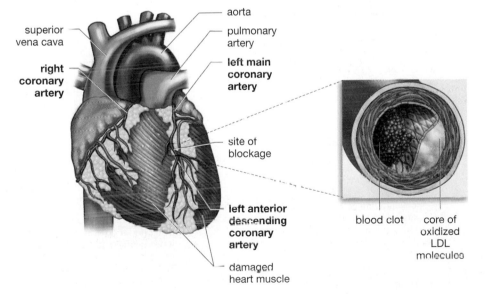

**Figure 29.6**
**Critical Vessels**

The coronary arteries begin with a right and left coronary artery that branch off from the aorta. These arteries then undergo branching themselves, thus supplying blood to the whole heart. Blockage of the left anterior descending coronary artery figures in almost half of all heart attacks. The magnification of this artery shows the factors that lead to most heart attacks. A buildup of oxidized LDL molecules within the inner artery tissue layer and a resulting blood clot together bring about a complete blockage of the artery.

Listening in on Blood Pressure

**B**ecause high blood pressure manages to be both harmful and stealthy, it has often been called a "silent killer." For a time, at least, there are no outward symptoms of this condition. Yet, as high blood pressure progresses, it is making its victims vulnerable to some very serious illnesses, among them heart attacks and strokes. Is it any wonder, then, that we can expect to get our blood pressure checked almost any time we see a physician?

You may have wondered, however, what physicians are listening for when they take our blood pressure. Here's the story. The tightening of the cuff around our upper arm collapses a major artery—the brachial artery—whose blood pressure is being measured. This ensures that no blood is getting through the artery, which means that there is no sound of a pulse coming through the stethoscope. After the physician starts to release pressure on the cuff and blood starts to flow, the point at which the physician can first hear a pulse marks our systolic blood pressure—the highest

pressure we have, brought about by the contraction of our heart's left ventricle. The point at which the sound of the pulse disappears marks our diastolic blood pressure—our lowest blood pressure. Physicians watch the gauge on the blood pressure device and note the blood pressure levels at the two sound points, as you can see in **Figure 1**.

The readings from these two points give us the well-known, two-figure blood pressure levels (as in 120/80). High blood pressure—or hypertension as it's officially known—can mean that either figure is elevated above a level regarded as healthy. What level is this? Normal systolic pressure for adults is anything below 120, while normal diastolic pressure is anything below 80. People are then regarded as being at risk for high blood pressure if their systolic level stands between 120 and 139, or if their diastolic level stands between 80 and 89. Actual high blood pressure is any systolic reading of 140 or higher or any diastolic reading of 90 or higher.

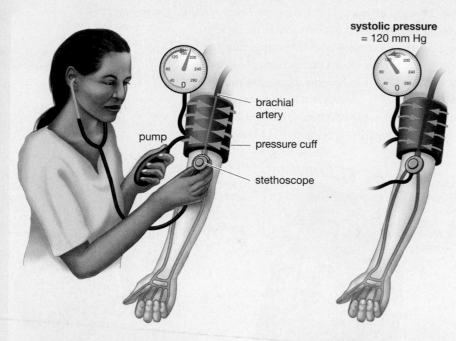

systolic pressure
= 120 mm Hg

diastolic pressure
= 80mm Hg

brachial artery

pump

pressure cuff

stethoscope

**1.** While listening with a stethoscope, the physician inflates the cuff around the patient's arm, tightening it to a level that prevents blood from moving through the patient's brachial artery.

**2.** While slowly releasing pressure on the cuff, the physician listens for the sound of a pulse. The point at which a pulse can first be heard marks the patient's systolic or highest blood pressure. This is the point at which the patient's blood pressure has overcome the pressure being applied by the cuff.

**3.** As pressure on the cuff continues to drop, the sound of a pulse begins to fade. The point at which a pulse can no longer be heard marks the patient's diastolic or lowest blood pressure.

**Figure 1**
**How Physicians Measure Blood Pressure**

*invaders* and thus initiates a complex attack against them. What follows is the inflammatory response that readers of Chapter 28 are familiar with. Inflammation that would take place with, say, a puncture wound causes tissue to become flushed and swollen.

The same thing is happening here, only this tissue is swelling into the space through which blood flows in a coronary artery, thus beginning a blockage of it.

Even so, this swelling is rarely extensive enough to completely block an artery. Eighty-five percent of all heart attacks are set off by what happens next. The "cap" of swollen tissue ruptures, letting in some of the circulating blood, and the immune cells react with this blood to make it clot. Now there is a blood clot combined with a swollen cap, and the result can be a *complete* blockage of a coronary artery—a **heart attack**. The effect of this blockage is that groups of heart muscle cells no longer have a blood supply, which kills them if it goes on for long. This can bring death to the heart attack victim in several ways. The heart muscles may no longer be able to contract with sufficient force to pump blood through the body, or a wild, irregular rhythm, called a ventricular fibrillation, can be set off. Even if the victim survives, the damaged heart muscle tissue will not regenerate; this person will have a weakened heart for life.

The good news about all this is that there are things we can do to lessen the likelihood that this sequence of events will get started. A key concept is to lower the number of LDLs we have circulating within us while raising the number of the HDLs or high-density lipoproteins mentioned earlier. The LDLs, as you've seen, lodge in the coronary arteries, setting off the immune response. The HDLs, conversely, not only remove these LDLs (bringing them back to the liver) but may also carry enzymes that break down the LDLs that have become damaged through oxidation—the LDLs that set off the immune response. So, how do we lower LDLs and raise HDLs? Exercise does both things, and losing excess weight does both things. Meanwhile, eating saturated (animal) fat raises LDL levels, and smoking may damage the LDLs in ways that set off the immune response. It's worth noting that fatty LDL deposits can begin forming in coronary arteries from childhood forward. Heart attacks may be an affliction of the middle-aged and elderly, but heart disease prevention can be practiced at any age.

## 29.6 Distributing the Goods: The Capillary Beds

Having looked at the players involved, you're now ready to see how the cardiovascular system actually carries out its central functions of bringing materials to tissues and taking materials away from them. Propelled by the heart through the aorta, blood flows through the tube-like arteries. These large-diameter arteries fork repeatedly, gradually decreasing in size until they become the smaller-diameter arterioles. These then branch into the delivery vehicles of the cardiovascular system, the capillary beds. If you look at **Figure 29.7a**, you can see a representation of one of these beds. Note that at one end, the bed extends

from the body's arterial system, but that at the other it feeds into the body's system of veins—the venous system that brings blood *back* to the heart. Looking at the venous system as it begins with the capillaries, you can see that it consists of its smallest vessels, the tiny venules, which then merge into veins. Smaller veins eventually merge into larger ones, which finally come together into the largest veins of all, the two vena cavae that feed directly into the heart.

Thus, blood comes "out" through the arteries and "back" through the veins, but the real action in the circulatory system takes place in the capillaries that lie *between* the arteries and the veins. As you've seen, capillaries are very thin. Think now about a line of oxygenated red blood cells, suspended in plasma and moving in single file through a capillary along with hormones, glucose, ions, and other materials needed by the cells *outside* the capillary (**Figure 29.7c**). The

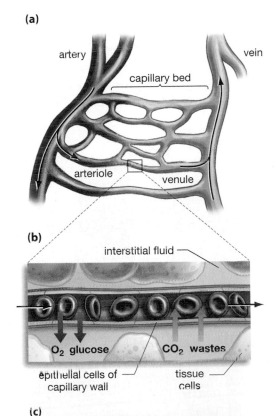

**Figure 29.7**
**Cardiovascular System Delivery Vehicle**

Capillary beds are the sites of the exchange of materials between blood vessels and the body's tissues.

**(a)** An idealized view of how blood flows from arteries to arterioles, through capillary beds, and then into the venules and veins that return it to the heart.

**(b)** Blood pressure moves materials out of the capillaries near the arterial end of capillary beds, while osmosis moves materials into the capillaries near the venous end of the beds.

**(c)** Micrograph of a cross section of a capillary, with red blood cells moving through in single file.

oxygen, glucose, and all the rest move out through the capillary wall and into the interstitial fluid—the liquid in which both the capillary and the cells around it are immersed. Once within this fluid, these needed materials will make their way to nearby cells. The distance they must travel to the cells is not far, however. So extensive is the body's capillary network that no cell is farther than two or three cells away from a capillary.

### Forces That Work on Exchange Through Capillaries

Red blood cells themselves are too big to pass out of the capillaries, but the oxygen within them diffuses across the capillary lining into the interstitial fluid. Nutrients such as glucose and small ions such as sodium likewise are distributed in this manner. Meanwhile, carbon dioxide and wastes are diffusing *into* the capillaries. The movement of all these substances is aided by concentration gradients, meaning these substances are moving from an area of their higher concentration to an area of their lower concentration. If you take oxygen, for example, there is more of it inside a capillary than outside of it. As a result, it will move "down" its concentration gradient

and leave the capillary. (For more on concentration gradients and diffusion, see Chapter 5, page 95.)

Water flows in and out of capillaries as well, propelled by a couple of forces that oppose each other. It tends to flow out of capillary beds at their arterial end, but into the beds at their venous end. The force at work at the arterial end is blood pressure, which results from the heart's contractions. This pressure is relatively strong at the arterial end—strong enough to push water out of the capillary, in a flow that brings with it some suspended materials. But the narrowness of the capillaries means that blood pressure is largely spent by the time blood approaches the venous end of a capillary bed. At that point, another factor, osmotic pressure, tends to move materials back into the capillary. (Proteins too big to leave the capillary result in an osmotic pressure that "pulls" on fluids. You can read more about osmosis in Chapter 5, on page 97.) At the venous end of a capillary, then, osmosis overcomes the force of blood pressure. Most of the water that leaves the capillary beds at the arterial end is returned to them at the venous end. Some remains as interstitial fluid, however, to be picked up by the body's other group of fluid vessels, the lymphatic vessels.

### Muscles and Valves Work to Return Blood to the Heart

Because blood pressure has dropped to such low levels by the time the blood gets to the venous system, you may wonder how it can get back to the heart. The answer is that our skeletal muscles do double duty. In contracting, they squeeze the veins in a way that moves the venous blood along. Blood can only move toward the heart in this system because the veins have a series of valves in them that block movement of blood away from the heart (**Figure 29.8**).

**Valves allow blood to go forward . . .**

valve open

muscles contracted

valve closed

**. . . but not backward**

valve closed

muscles relaxed

valve open

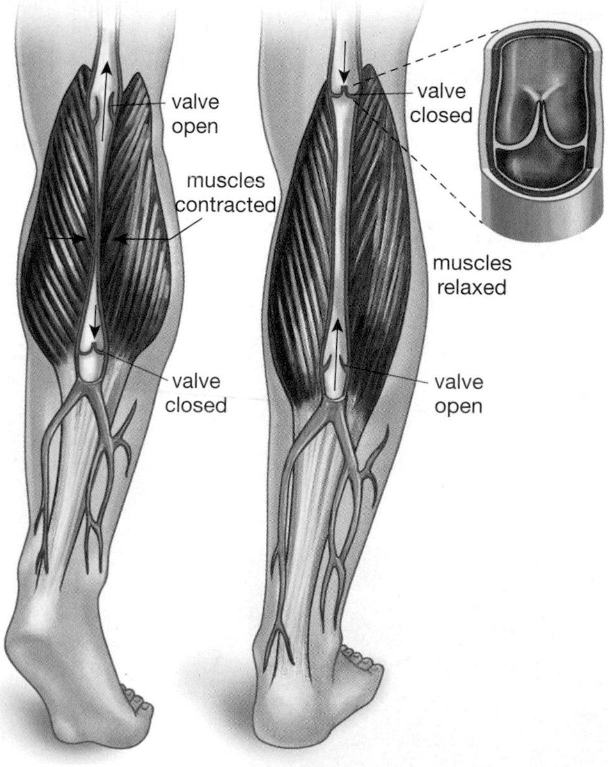

**Figure 29.8**
**One-Way Flow to the Heart**
Skeletal muscle contraction is the primary force that drives blood back to the heart through the venous circulation. A system of valves guards against the backflow of this blood.

**SO FAR...**

1. A heart attack can be defined as a complete _____ of a _____.

2. High-density lipoproteins are regarded as healthy because they carry cholesterol _____.

3. _____ are the delivery vessels of the cardiovascular system.

## 29.7 The Respiratory System

As you've seen, one of the things the cardiovascular system transports is oxygen and another is carbon dioxide. Now it's time to look at the bigger picture of how these substances are exchanged between our cells and the air around us.

People can live perhaps a couple of months without food and generally a few days without water. But if they go 5 or 6 minutes without breathing, death is usually the result. This is because oxygen is in the energy transfer business; without oxygen, our cells simply don't have enough energy to function. When oxygen is present, however, cells carry out their work and *generate* carbon dioxide as a result, and this $CO_2$ has to be disposed of. Thus do we see the two central functions of breathing: The first is capturing and distributing oxygen; the second is disposing of carbon dioxide. Apart from these tasks, breathing also helps balance the pH level in our bloodstream and allows us to talk by providing the air that passes over the vocal cords. All of these tasks are technically part of **respiration**, meaning the exchange of gases between the atmosphere outside the body and the cells within it.

## Structure of the Respiratory System

The lungs are but one part of the respiratory system, which you can see in **Figure 29.9a**. This system also includes the nose, nasal cavity, and sinuses, along with the pharynx (upper throat), the larynx (voice box), and the trachea (windpipe). In addition, there are air-conducting passageways: the left bronchus and right bronchus and the many small

**(a)** Anatomy of the lungs

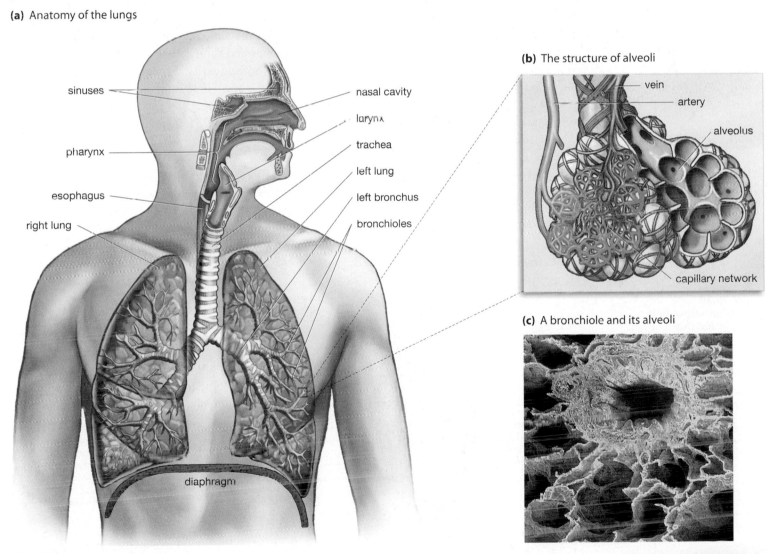

**(b)** The structure of alveoli

**(c)** A bronchiole and its alveoli

**Figure 29.9**
**Anatomy of the Respiratory System**

**(a)** The large-scale components of the human respiratory system.

**(b)** In this magnification of the alveoli, the capillary network that all alveoli are surrounded by has been cut away, on the right, to reveal the structure of the alveolar sacs.

**(c)** A micrograph of single bronchiole surrounded by many alveoli—essentially the "cutaway" view of this tissue as seen in the illustration above.

air passageways they branch into, the **bronchioles**, which deliver air to the lungs.

If we ask what the lungs themselves are, they are made up mostly of tiny, hollow sacs that sprout in grapelike clusters from the end of each bronchiole. These sacs are the **alveoli**, the air-exchange chambers of the lungs. There are about 300 million of them in a typical set of lungs—a number that gives you an idea of just how tiny these sacs must be. The dual quality of being small and numerous gives alveoli what they need to get their job done, which is *surface area*. What does being small and numerous have to do with surface area? Imagine a cube of Jell-O, 1 inch on each side. Its surface area is 6 square inches. Divide this cube in half, however, and the Jell-O's surface area goes to 8 square inches even though its volume—the amount of space it takes up—stays the same. Repeat this kind of dividing millions of times, as with our alveoli, and the results are startling. The alveoli in a typical set of lungs have a surface area about the size of a tennis court. If you look at **Figure 29.9b**, you can see how this surface area is used: to give the alveoli and the capillaries that cover them a huge area in which to exchange oxygen and carbon dioxide.

**Web Animation 29.2**
The Respiratory System

## 29.8 Steps in Respiration

### First Step: Ventilation

The first step in respiration is **ventilation**, meaning the physical movement of air into and out of the lungs. To understand ventilation, look at the drawing, in **Figure 29.10**, of the old-fashioned "accordion bellows" that can be used to get a fire going. Pull the bellows handles apart, and the volume inside the bellows increases. This lowers the air pressure inside the bellows. Because gases always flow from areas of higher to lower pressure, the air flows into the bellows from the outside. Reverse this process—squeeze the bellows handles together—and the air pressure inside becomes higher than the pressure outside, meaning the air flows out of the bellows. Similarly, we have a space inside our chests (called the thoracic cavity) whose volume can increase or decrease. The contraction of a muscular sheet called the diaphragm causes the ribs to rise and, with this, the thoracic cavity expands like the inside of the bellows. Air then rushes in. Exhaling in this kind of "quiet" ventilation is a more passive process. When the diaphragm relaxes, the ribs drop due to the combined force of gravity and natural elasticity of the lungs. As a result, the thoracic cavity shrinks, and air is expelled. During the heavy breathing that takes place in exercise, exhalation is aided by a second set of chest muscles whose contractions help shrink the thoracic cavity.

### Next Steps: Exchange of Gases

Having gotten into our lungs, oxygen now moves into the bloodstream. How does it get there? Remember that the sac-like alveoli have capillaries covering them. Think of the interface between a capillary and an alveolus as a border that can be easily crossed. An oxygen molecule is inhaled into an alveolus and then

**Figure 29.10
Ventilation—How Air Is
Moved In and Out**

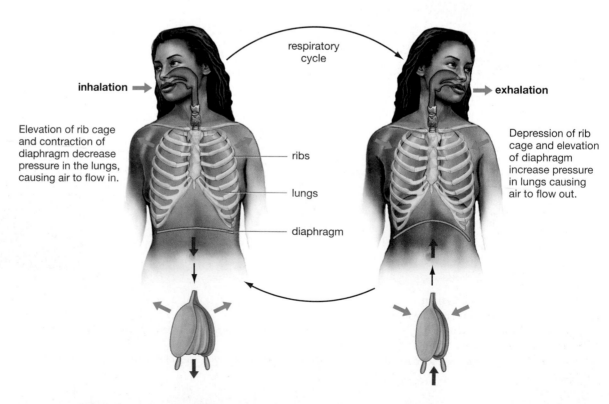

respiratory cycle

inhalation →

Elevation of rib cage and contraction of diaphragm decrease pressure in the lungs, causing air to flow in.

→ exhalation

Depression of rib cage and elevation of diaphragm increase pressure in lungs causing air to flow out.

ribs

lungs

diaphragm

diffuses across the thin alveolar tissue into an adjacent capillary. On entry, this oxygen molecule encounters a passing red blood cell, which it diffuses into, binding with one of the hemoglobin molecules inside it. Locked up by this binding, the oxygen molecule moves with the red blood cell to the heart. In its ceaseless pumping, the heart is taking oxygenated blood cells like this one and sending them up through the aorta and from there out to the rest of the body. Moving through this route to a tissue somewhere in the body, the red blood cell drops off the oxygen molecule to be used.

This drop-off occurs once again by way of diffusion: The oxygen flows out of a capillary and into the interstitial fluid that surrounds it. Other body cells are immersed nearby in this interstitial fluid, of course,

and they can now receive the oxygen. These same cells are, as noted, producing carbon dioxide, which now diffuses from them into the interstitial fluid, and then diffuses again into a capillary. This $CO_2$ is returned to the lungs via the venous system. Then, at the same capillary–alveoli interface noted earlier, the $CO_2$ moves into the alveoli, from which it is expelled into the environment outside the body (**Figure 29.11**).

As noted, all the oxygen that is loaded into red blood cells binds initially with hemoglobin molecules that exist in these cells. In contrast to this simplicity, the carbon dioxide that is being transferred to the lungs is transported by the blood in three different ways. We need only note here that, in a kind of chemical packing and unpacking, some $CO_2$ ends up traveling within the hemoglobin in red blood cells, while most of it travels outside of red blood cells, in the blood plasma.

## On to the Digestive and Urinary Systems

Oxygen and carbon dioxide are just a couple of the substances that the human body "exchanges" with the outside world. We take in food, of course, and we accumulate—and later rid ourselves of—waste products. Two different transport and exchange systems deal with the digestion of food and the removal of liquid waste. These are the digestive and urinary systems, the subjects of Chapter 30.

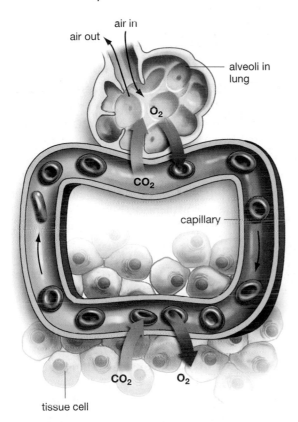

air in
air out
alveoli in lung
$O_2$
$CO_2$
capillary
$CO_2$   $O_2$
tissue cell

**Figure 29.11**
**Gas Exchange in the Body**

**SO FAR...**

1. The two central functions of respiration are capturing and distributing _____ and disposing of _____.

2. The air-exchange chambers of the lungs are tiny sacs called _____, which are surrounded by _____.

3. Oxygen binds with the molecule called _____ inside _____.

**ANSWERS**

1. oxygen; carbon dioxide
2. alveoli; capillaries
3. hemoglobin; red blood cells

# CHAPTER 29 REVIEW

For study help, activities, and more quiz questions, go to **www.aw.com/krogh4**.

## SUMMARY

### 29.1 The Cardiovascular System

- The human cardiovascular system is a fluid transport system that consists of the heart, all the body's blood vessels, the blood, and the bone marrow tissue in which red blood cells are formed. This system transports substances both to and from the body's cells. Such substances include oxygen, carbon dioxide, nutrients, vitamins, hormones, waste products, and immune system cells and proteins.

### 29.2 The Composition of Blood

- Blood has two primary components: formed elements and blood plasma. Formed elements are blood cells and cell fragments; blood plasma is the fluid portion of blood in which the formed elements are suspended. There are three kinds of formed elements: red blood cells, white blood cells, and platelets. Red blood cells carry oxygen to, and carbon dioxide from, every part of the body; white blood cells are central to the immune system; and platelets are small fragments of cells that are important in the blood-clotting process.

- Blood plasma is 92 percent water, but it also contains other materials, including proteins. There are three primary classes of plasma proteins: albumins, which transport hormones and fatty acids; fibrinogen, which aids in blood clotting; and globulins, which aid the immune system and serve as transport proteins. Two transport proteins are important in the health of the heart. Low-density lipoproteins (LDLs) carry lipids to bodily tissues from the liver and small intestines; high-density lipoproteins (HDLs) carry lipids from these tissues to the liver. Other plasma compounds include nutrients, wastes, hormones, and electrolytes.

### 29.3 Blood Vessels

- Blood vessels carrying blood away from the heart are arteries; blood vessels returning blood to the heart are veins. The smallest blood vessels, the capillaries, connect the arteries with the veins.

- Arteries and veins are always made of three distinct layers of tissue, the middle layer of which is muscle that allows arteries and veins to widen or constrict in diameter. Capillaries, conversely, are composed of only a single layer of cells; this allows the movement of blood-borne materials into and out of them along their length.

### 29.4 The Heart and Blood Circulation

- The heart's contractions propel blood out to the various tissues of the body. Two blood circulation loops exist in the body. The first loop is the pulmonary circulation, in which blood circulates between the heart and the lungs (with the result that blood is oxygenated). The second loop is the systemic circulation, in which blood circulates between the heart and the rest of the body (with the result that needed materials are transported to and from all parts of the body).

- The human heart contains four muscular chambers—two for pulmonary circulation (the right atrium and right ventricle) and two for systemic circulation (the left atrium and left ventricle). A series of valves that open and close ensures that blood flows only one way through the heart.

  **Web Animation 29.1:** The Cardiovascular System

### 29.5 What Is a Heart Attack?

- About half of all deaths in the United States today are caused by the blockage of one or more of the coronary arteries that supply heart tissue with blood. Such blockages generally are caused by a buildup of low-density lipoprotein (LDL) molecules in a coronary artery, followed by an immune system reaction to these LDLs and formation of a blood clot in the artery. A heart attack occurs when this process results in the complete blockage of a coronary artery, which cuts off the blood supply to groups of cells within the heart, thus killing them.

### 29.6 Distributing the Goods: The Capillary Beds

- Arteries near the heart branch into smaller arterioles, which feed into the delivery vehicles of the cardiovascular system, the capillary beds. The capillary beds then feed back into the body's system of veins that returns blood to the heart.

- Materials needed by the body's tissues move out of the capillaries and into the interstitial fluid that surrounds both the capillaries and nearby cells. Meanwhile, carbon dioxide and wastes from these cells flow into capillaries from the interstitial fluid. The movement of all these substances is aided by their concentration gradients—they are moving from areas of their higher concentration to areas of their lower concentration. The movement of water is driven in two directions by two opposing forces. At the arterial end of the capillary beds, blood pressure tends to drive water out of the capillaries, but at the venous end of the beds, osmosis overcomes the force of blood pressure and pulls most of this water back into the capillaries.

- Blood pressure is at low levels by the time blood has moved through the capillaries. Blood returns to the heart through the contraction of skeletal muscles, which squeeze the veins in a way that moves the venous blood toward the heart. A system of valves in the veins ensures that this movement is one way—toward the heart.

## 29.7 The Respiratory System

- The central function of the respiratory system is to capture oxygen and to dispose of carbon dioxide. It also aids in controlling pH balance in the bloodstream and in producing sounds for speaking. Respiration can be defined as the exchange of gases between the atmosphere outside the body and the cells within it.

- The respiratory system includes the lungs; the nose, nasal cavity, and sinuses; the pharynx (upper throat); the larynx (voice box); the trachea (windpipe); and the conducting passageways, called bronchi and bronchioles, that lead to the lungs. The lungs themselves are largely composed of the tiny hollow sacs, called alveoli, that lie at the end of each bronchiole and that are the air-exchange chambers of the body. The enormous surface area of the alveoli and their associated capillaries is used for the exchange of oxygen and carbon dioxide.

## 29.8 Steps in Respiration

- The first step in respiration is breathing or ventilation, meaning the physical movement of air into and out of the lungs.

- Once in the lungs, oxygen diffuses across the thin wall of an alveolus into an adjacent capillary and binds with hemoglobin protein in red blood cells. Oxygen then moves with the blood cells to the heart, which pumps the blood to body tissues, where the oxygen diffuses into the interstitial fluid and then into nearby cells. The carbon dioxide produced in the body's cells moves into nearby capillaries, to be carried to the lungs. All the oxygen loaded into red blood cells binds initially with the hemoglobin in them. Carbon dioxide, however, is transported both within red blood cells and in blood plasma.

**Web Animation 29.2:** The Respiratory System

## KEY TERMS

| | | | | |
|---|---|---|---|---|
| alveoli | 618 | formed elements | 608 | |
| aorta | 612 | heart attack | 615 | |
| artery | 611 | hemoglobin | 608 | |
| bronchiole | 618 | high-density lipoprotein (HDL) | 609 | |
| capillary | 611 | | | |
| cardiovascular system | 608 | low-density lipoprotein (LDL) | 609 | |
| coronary artery | 613 | | | |

| | | | |
|---|---|---|---|
| plasma | 608 | systemic circulation | 612 |
| platelet | 608 | vein | 611 |
| pulmonary circulation | 612 | ventilation | 618 |
| red blood cell (RBC) | 608 | white blood cell (WBC) | 608 |
| respiration | 617 | | |

## BRIEF REVIEW

*Answers to Brief Review questions are in the back of the book.*
*For multiple-choice quiz questions, go to* **www.aw.com/krogh4**.

1. What are the primary functions of the three types of formed elements found in blood?

2. Describe the essential difference between the heart's atrial chambers and its ventricular chambers.

3. Why can substances diffuse in and out of capillaries but not arteries or veins?

4. What arteries in the body carry deoxygenated blood?

5. What do eating animal fats, smoking, and exercise have to do with heart attacks?

6. Name as many anatomical structures as you can that are part of the respiratory system.

7. Describe the process of ventilation, in which air is moved into and out of the lungs.

8. Where, specifically, does oxygen enter the bloodstream?

## APPLYING YOUR KNOWLEDGE

1. Heart attacks have sometimes been called a disease of modern civilization because their incidence is much higher in affluent, more developed societies than in poorer, less-developed ones. Why should this be so?

2. Why does the term *circulation* so aptly describe the flow of blood in the body?

3. The walls of one of the heart's four chambers are thicker and more muscular than the walls of any of the other three. Given the differing functions of the heart's chambers, which chamber do you think is likely to be more muscular in this way?

The human digestive system puts the food we eat into a form we can use.

CHAPTER **30**

| 30.1 | The Digestive System | 623 |
| 30.2 | Structure of the Digestive System | 624 |
| 30.3 | Steps in Digestion | 625 |
| 30.4 | Human Nutrition | 629 |
| 30.5 | Water, Minerals, and Vitamins | 629 |
| 30.6 | Calories and the Energy-Yielding Nutrients | 633 |
| 30.7 | Proteins | 634 |
| 30.8 | Carbohydrates | 636 |
| 30.9 | Lipids | 639 |
| 30.10 | Elements of a Healthy Diet | 642 |
| 30.11 | The Urinary System in Overview | 643 |
| 30.12 | Structure of the Urinary System | 643 |
| 30.13 | How the Kidneys Function | 645 |
| 30.14 | Urine Storage and Excretion | 646 |

Our digestive system must break down the food we eat in order for it to be useful to us. Different varieties of foods must be consumed in order for us to get the nutrients we need. The body conserves what it needs and eliminates what it does not need in large part through the workings of the urinary system.

# TRANSPORT AND EXCHANGE 2:
# Digestion, Nutrition, and Elimination

**I**nnovations that work well tend to stay around for a while, but a structure that animals use to digest food has been around for a very long while. The structure is essentially a long tube through which food moves once animals ingest it. This "digestive tract," as it's called in humans, is in widespread use in the animal kingdom. Among major animal groups, only the primitive sponges and jellyfish lack such a tract. Meanwhile, nearly all other animal groups have one. As such, this structure can be viewed as having stood the test of time: Over the course of some 500 million years of evolution, nature has seemingly been unable to come up with anything that works better. Fish may swim and bats may fly, roundworms may burrow and snails may crawl, but in all these creatures, food is taken care of in the same way. To be sure, the digestive tract has had some pieces of auxiliary equipment added to it over time. Flatworms have nothing but a simple tube in a branched form, but more complex animals, such as snails, have the tube plus the digestive organs of a liver and a stomach. Human beings have all this and more—a liver and a stomach plus a gallbladder and a pancreas and no fewer than three sets of salivary glands. Even so, the digestive process that operates in humans is essentially the one that operates in snails: Take food in through a mouth, digest it in the tube, move what's useful *out* of the tube and into circulation, and move the waste out of the body altogether. In this chapter, we'll follow this whole sequence of events as it occurs in human beings. What happens to the food we eat? In the account that's coming up, you'll find out. Next, we'll look at food from another angle by asking the question: What is it that food does for us, and *which* foods stand to keep us the healthiest for the longest? The subject in this part of the chapter is nutrition. Finally, we'll look at a filtering operation the body runs that allows it to get rid of blood-borne waste materials while hanging on to important nutrients and fluids. This third subject in the chapter is the urinary system. Let's begin now by following the progress of food as it moves through the human body.

## 30.1 The Digestive System

The central structure of the human digestive system, the **digestive tract** we've been talking about, can be defined as a muscular passageway for food and food waste that runs through the human body from the mouth to the anus. Along its length, this tube receives input from the various accessory organs we've noted, such as the gallbladder and the pancreas. You saw last chapter that nutrients are delivered to tissues throughout the body by the circulatory system. So the real question for the digestive system is: What does it need to do to get food into a form that can *move into* the circulatory system? For carbohydrates and proteins, the pathway is straightforward. Partway through a portion of the digestive tract called the small intestine, carbohydrates and proteins have been broken down enough that they move out from the inner lining of the intestine into adjacent capillaries. This network of capillaries then feeds into a common vein that goes to the liver. Most nutrients, then, are carried straight from the small intestine to the liver. In a controlled release, the liver then sends these nutrients out (through another vein) to the heart, which pumps them out to all the tissues of the body.

The digestion of fats takes a somewhat more complicated route, as you'll see, but the principle remains the same. Fats must be broken down to such a point that they can leave the digestive tract for the rest of the body. For all foods, "breakdown" means just what it sounds like: a transition from large to small. The transition goes from big, visible bites of food to smaller bits and eventually from big *molecules* of food to the

building blocks that make them up. It is only these smaller molecules—simple sugars, amino acids, and fatty acids—that can leave the digestive tract for the rest of the body. Of course, not everything we eat makes this transition. A good deal of the food we ingest is retained in the digestive tract as a waste product that ultimately is eliminated from the body altogether. It's the digestive system that handles these tasks of waste retention and elimination, even as it's carrying out its other functions of food digestion and nutrient transfer. Now, let's take a look at the components of the digestive system, after which you'll follow the path of digestion.

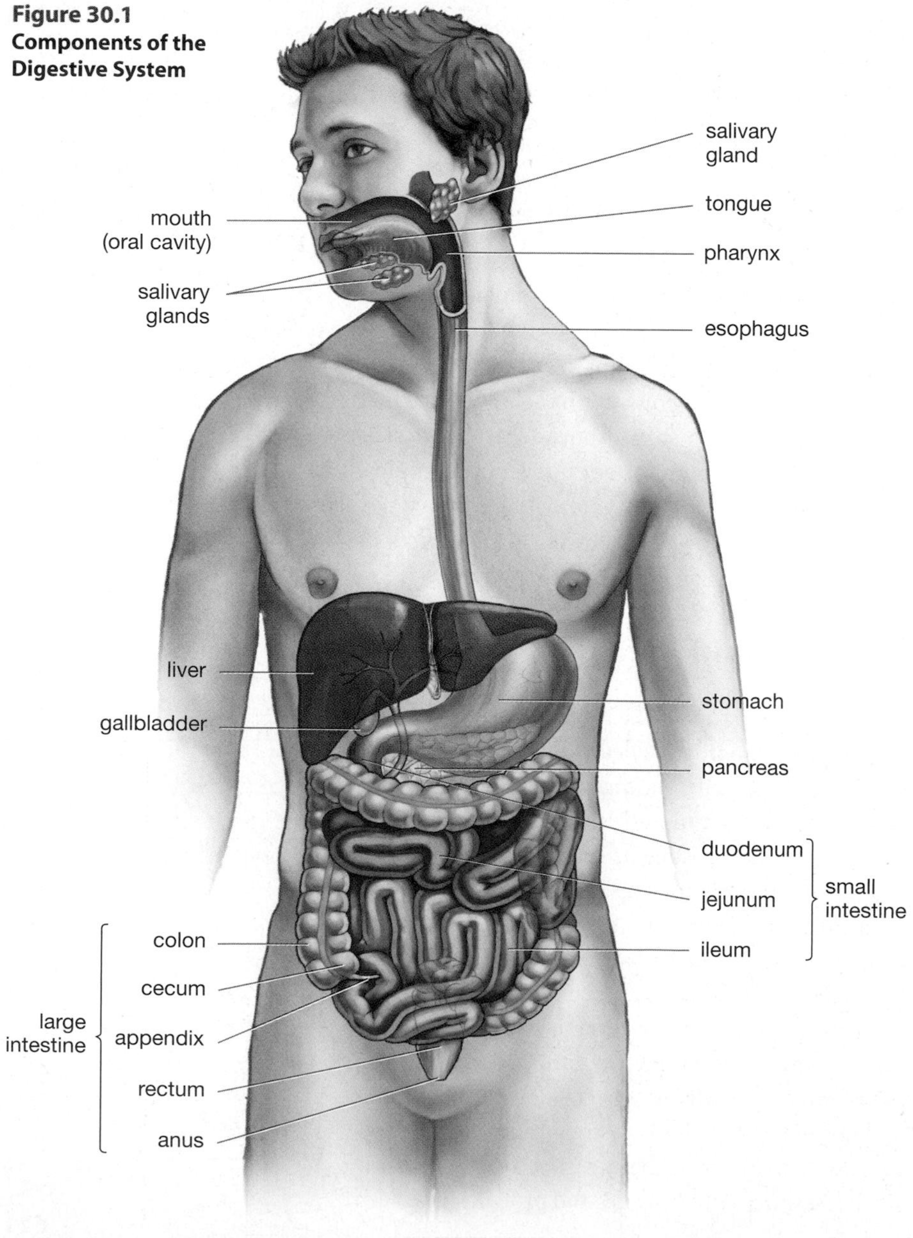

## 30.2 Structure of the Digestive System

**Figure 30.1** shows the locations and functions of the accessory glands and different subdivisions of the digestive tract. The digestive tract begins with the mouth (or oral cavity) and continues through the pharynx, esophagus, stomach, small intestine, and large intestine before ending at the rectum and anus. The digestive tract is sometimes called the alimentary canal or the gastrointestinal tract.

### The Digestive Tract in Cross Section

Thinking again of the digestive tract as a muscular tube, let's look at a cross section of it as it would appear within the small intestine. As you can see in **Figure 30.2**, the digestive tract is a four-layered tube. The outermost layer is connective tissue known as the serosa. Inside of it, there is muscle: two sets of smooth muscles, actually, which together are known as the muscularis externa. In a process called **peristalsis**, these muscles take turns contracting, which produces waves of contraction that push materials along the length of the digestive tract. Inside the muscularis externa there is the submucosa, a layer of tissue that contains large blood vessels and a network of nerve tissue that both coordinates the contractions of the smooth muscle layers and regulates the secretions of digestive juices. Finally, there is the epithelial layer that makes up the internal lining of the tract. It is called the mucosa because it is constantly bathed in mucous secretions.

Along most of the length of the digestive tract, the mucosa has a large number of folds that increase the surface area available for absorption of digested food; these folds also work like accordion folds in that they permit expansion, which is helpful after a large meal. The mucosa also has on it fingerlike projections

**Figure 30.1
Components of the
Digestive System**

mouth
(oral cavity)

salivary
glands

liver

gallbladder

colon

cecum

appendix

rectum

anus

large
intestine

salivary
gland

tongue

pharynx

esophagus

stomach

pancreas

duodenum

jejunum

ileum

small
intestine

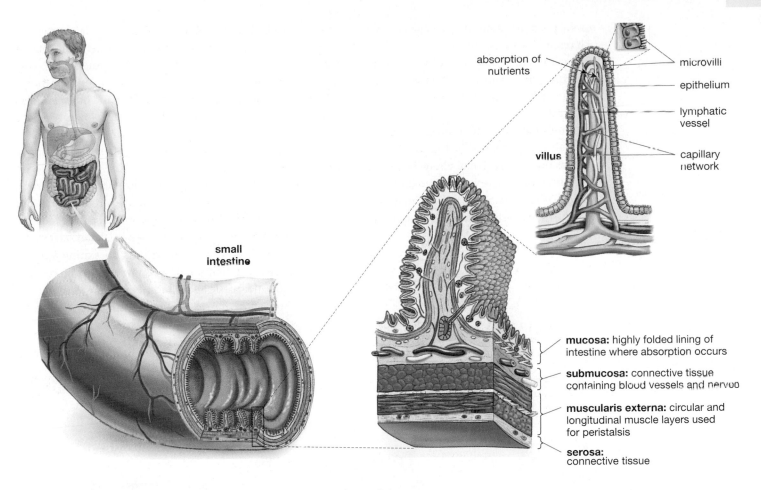

absorption of
nutrients

microvilli

epithelium

lymphatic
vessel

villus

capillary
network

**mucosa:** highly folded lining of
intestine where absorption occurs

**submucosa:** connective tissue
containing blood vessels and nervos

**muscularis externa:** circular and
longitudinal muscle layers used
for peristalsis

**serosa:**
connective tissue

small
intestine

called *villi* that further increase the area for absorption. These villi are the place where most of our digested food actually makes the transition out of the digestive system. If you look at the blowup of the villus in Figure 30.2, you can see where nutrients leave the system. Just outside it, they enter either an adjacent capillary or, in the case of fats, a lymphatic vessel whose connecting vessels will move the nutrients out of the lymphatic system and into the bloodstream.

## 30.3  Steps in Digestion

Let's now look at the process of digestion from start to finish, which means taking it from the top, so to speak. Our mouths are the entry place for food, of course, and we immediately start breaking food down in them by the mechanical means of chewing. Food starts to get digested there by chemical means as well because enzymes in the saliva—secreted by the three pairs of salivary glands noted earlier—have the ability to break down carbohydrates.

### The Pharynx and Esophagus

Food begins to move out of the mouth when the tongue pushes it back toward the pharynx, which we might think of as the upper throat (see Figure 30.1).

This action is tricky because the **pharynx** is a passageway that links the mouth with both the food-transporting esophagus and the *air*-transporting trachea. Once the tongue begins pushing food back, the esophagus opens while tissue called the epiglottis folds down over the larynx to keep the food from blocking it. (Sometimes, food does go down this "wrong way," of course, at which point we may simply start coughing or, in the worst case, need the Heimlich maneuver to keep us from choking.) The pharynx has a trio of muscles wrapping partway around it that get activated in a 1-2-3 sequence to push the food into the esophagus. The esophagus in turn has its own muscles, which work with gravity to send food on a 6 second trip to the next stop in the digestive tract.

### The Stomach

The stomach digests food partly by the mechanical means of churning it, thus breaking it into small bits. It also carries out some chemical breakdown of foods, with proteins the primary target of its digestive juices. In addition to performing digestive work, the **stomach** is an organ that has a less-appreciated second role: It serves as a temporary, expandable storage site for food. This capacity is the reason we can go

**Figure 30.2**
**Structure of the Digestive Tract**

The digestive tract is a tube formed of several layers of tissue, each performing a different function. Note that the absorption of nutrients takes place when food that has been sufficiently digested moves across the epithelium and is taken up by either a capillary or a lymphatic vessel inside a villus.

several hours without eating. When empty, the stomach resembles a tube with a narrow cavity, but it can expand to contain up to 1.5 liters or almost half a gallon of material. This expansion is possible because the stomach contains numerous folds, called rugae (**Figure 30.3**).

Glands feed into the stomach by way of millions of small depressions, called gastric pits, that open onto its interior. Each day, about 1,500 milliliters or 45 ounces of gastric juice pour into the stomach through these pits. This gastric juice is *caustic*—a mixture of hydrochloric acid and the enzyme pepsin.

So acidic is the hydrochloric acid that the contents of the stomach have a pH that lies between that of lemon juice and battery acid. Why is such an extreme environment needed? It serves not only to break down food but also to kill bacteria and other invaders that ride in on the food. You may wonder why the stomach itself wouldn't be eaten away in this kind of environment. The answer is that the cells lining it produce a protective mucus.

A few substances, such as alcohol and some drugs, can pass directly from the stomach into circulation, but in general the stomach is only preparing food for transfer within the digestive system. Food stays in the stomach for 2 to 4 hours, but on exiting it is no longer just plain food; it is now a soupy mixture of food and gastric juices known as **chyme**. A circular or "sphincter" muscle—the pyloric sphincter—regulates the flow of chyme between the stomach and the next structure in the digestive system.

**Figure 30.3
External and Internal Views
of the Stomach**

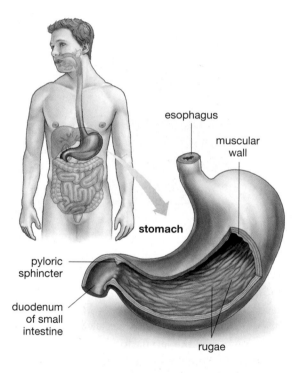

esophagus

muscular
wall

**stomach**

pyloric
sphincter

duodenum
of small
intestine

rugae

**SO FAR...**

1. The central structure of the digestive system is a muscular tube called the _____ that passes through the body from the _____ to the _____.

2. A central function of the digestive system is to get food into a form that can move into the _____.

3. Food is moved along the length of the digestive tract through the action of two sets of _____ in a process called _____.

**Figure 30.4
Structure of the Small
Intestine**

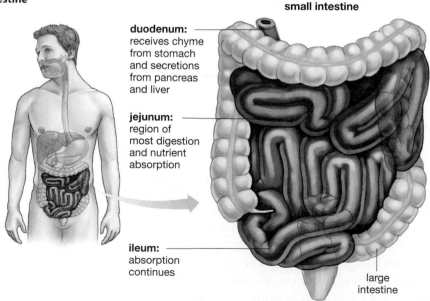

**small intestine**

**duodenum:**
receives chyme
from stomach
and secretions
from pancreas
and liver

**jejunum:**
region of
most digestion
and nutrient
absorption

**ileum:**
absorption
continues

large
intestine

## The Small Intestine

The only thing small about the small intestine is its diameter, which is about 4 centimeters or 1.6 inches up where it starts (at the stomach) and about 2.5 centimeters, or an inch, where it ends (at the large intestine). In between these two points, the "small" intestine winds on for about 6 meters, or almost 20 feet. Eighty percent of our nutrients make the transition out of the digestive system along the length of the **small intestine**, which can be defined as the portion of the digestive tract that runs between the stomach and the large intestine. (About 10 percent of our nutrients are transferred from the stomach, and the remaining 10 percent are transferred from the large intestine.)

As you can see in **Figure 30.4**, the small intestine is made up of three regions. The first of these, the *duodenum*, is the 25 centimeters or 10 inches that are closest to the stomach. Think of the duodenum as that portion of the small intestine that receives not only chyme from the stomach, but digestive secretions from three of the specialized digestive organs noted earlier: the pancreas, the liver, and the gallbladder. Next, there is the *jejunum*, which goes on for about 2.5 meters or 8 feet. This is the site within the small intestine where most chemical digestion and nutrient absorption occurs. Absorption continues to some extent in the third segment, the *ileum*, which is the longest, averaging 3.5 meters or 12 feet in length. The ileum ends at a sphincter muscle that controls the flow of chyme into the large intestine. On average, it takes about 5 hours for chyme to pass through the small intestine. Thus, material that began its digestive journey in the mouth takes perhaps 9 hours to reach the end of the small intestine. Now, let's look at what happens at the start of the small intestine, where juices from the digestive glands come into the picture.

## The Pancreas

Feeding into the small intestine is the pancreas, which lies behind the stomach and is an elongated, pinkish-gray organ about 15 centimeters or 6 inches long. The pancreas is best known as the organ that produces insulin, which it secretes directly into the bloodstream. But our concern is with materials the pancreas secretes through a tube or duct. The duct in question, as you can see in **Figure 30.5**, is the pancreatic duct, which comes together with the common bile duct (stemming from the liver and gallbladder) to empty into the small intestine's duodenum. In digestion, the **pancreas** can be defined as a gland that secretes, into the small intestine, digestive enzymes and chemical buffers that help raise the pH of the chyme coming from the stomach. (Remember, this chyme is very acidic; when its pH is being "raised," that means it is becoming less acidic.) Pancreatic digestive enzymes are classified according to the kind of food they work on. Lipases break down lipids, proteases break proteins apart, and carbohydrases digest sugars and starches.

## The Gallbladder

As Figure 30.5 shows, lodged within a recess under the right lobe of the liver is a muscular sac called the **gallbladder**: an organ that stores and concentrates a digestive material called bile. Note the operative terms about the gallbladder's role: It stores and concentrates bile. But bile is *produced* by the liver. This is why people can have their gallbladders removed and still function perfectly well. **Bile** can be defined as a

substance produced by the liver that facilitates the digestion of fats. It does two things that allow the digestive lipase enzymes to go to work on fats: It breaks up clusters of fat molecules, and it coats these molecules with a material that gives the lipases a greater ability to bind with them.

Bile comes to the small intestine through either of two routes. It always travels through passageways within the liver called hepatic ducts and then leaves the liver through the common hepatic duct. This bile may then (1) flow into the small intestine's duodenum through the common bile duct or (2) enter the cystic duct that leads to the gallbladder. If it goes to the gallbladder, only later will it flow into the duodenum through the common bile duct. Why would it take one course or the other? A sphincter at the intestinal end of the common bile duct opens only at mealtimes. At other times, bile backs up through the cystic duct for storage and concentration within the gallbladder before moving to the small intestine.

Bile can become *too* concentrated and harden into the troublesome crystals known as gallstones. Most gallstones are small enough to be flushed away into the duodenum, but they can reach 2 centimeters or more in diameter—about three-quarters of an inch, which is large enough to block either the common bile duct or the cystic duct. The pain that results is so excruciating that gallstones have to be taken care of in one way or another. One modern way of getting rid of them is to pulverize them with high-energy sound waves.

## The Liver

What about the organ that lies in front of the gallbladder, the liver? If we don't count the skin, the liver is the largest organ in the body. It weighs about 1.5 kilograms or 3.3 pounds (Figure 30.5). The **liver**

**••• ANSWERS**

**1.** digestive tract; mouth; anus
**2.** circulatory system
**3.** muscles; peristalsis

**Figure 30.5
Accessory Glands in Digestion: The Liver, Pancreas, and Gallbladder**

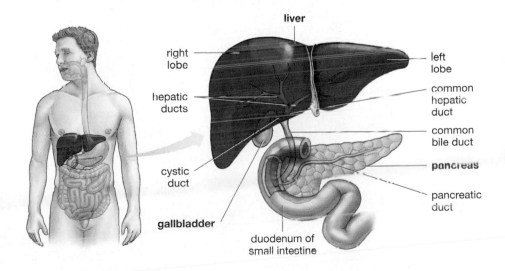

is a reddish-brown organ that is central to the body's metabolism of nutrients and that serves as a major storage site for blood. This organ actually does dozens of things for us, but here we will concentrate on a few of its digestive functions.

The liver clearly plays a central role in digestion because, as you may recall, all blood carrying nutrients from the digestive tract is channeled into a single blood vessel, called the hepatic portal vein, that flows into the liver. We could thus think of a series of blood vessel "streams"—stemming from the stomach, the small intestine, and so forth—all of them carrying digested nutrients, and all of them flowing like tributaries into the common river of the hepatic portal vein that goes to the liver. This arrangement makes the liver a first stop for most nutrients. From this position, the liver controls which nutrients it will send to the rest of the body and which nutrients it will store. The liver is, however, not only the first stop for nutrients, but the first stop for toxins such as alcohol. This is why people who have abused their bodies by drinking too much can end up with damaged livers. In addition to transferring and storing nutrients, the

liver produces the bile noted earlier, and it packages waste products for removal by the kidneys.

## The Large Intestine

If most nutrients have been transferred out of the digestive system by the time the small intestine ends, what is there left for the large intestine to do? Its main task is to serve as a kind of trash compactor: It holds and compacts material that by now is largely refuse, turning it into the solid waste we know as feces.

If you look at the upper drawing in **Figure 30.6**, you can see that the large intestine neatly frames the small intestine. The small intestine's last section, the ileum, feeds into the large intestine's first section by means of a sphincter (called the ileocecal valve) that you can see at about 7 o'clock in the lower drawing. This sphincter can relax or contract, thus controlling movement of chyme into the large intestine. The large intestine continues for about 1.5 meters or 5 feet from this point before ending at the anus. Compared to the 20-foot small intestine, the large intestine is large only in diameter.

The **large intestine** can be defined as that part of the digestive tract that begins at the small intestine and ends at the anus. It is divided into three major regions: the cecum, the colon, and the rectum. Note that the cecum has a pouch called the *appendix* attached to it; the walls of this pouch can become infected, resulting in appendicitis. (When healthy, the appendix is an arm of the immune system, although obviously not an essential one, as it can be removed.) The colon is the large intestine's second section and is by far its longest. The individual pouches that you can see in it allow it to expand and contract, depending on how full it is. The colon then empties into the expandable rectum, which serves as a storage site for feces. The rectal chamber is usually empty except when peristaltic contractions force fecal materials into it from the lower part of the colon. Stretching of the rectal wall then triggers the urge to defecate. Material takes anywhere from 12 to 24 hours to move through the large intestine. Adding this to the other transport times we've noted, you can see that a complete trip through the digestive system takes anywhere from 19 to 33 hours.

Apart from waste disposal, the large colon also functions in water conservation. Roughly 1,500 milliliters or 45 ounces of watery material arrives in the colon each day, but some 40.5 ounces move back out, leaving very little to be ejected with the feces. The large intestine also absorbs vitamin K that is produced by the abundant bacteria that live within the colon. Carbohydrates that cannot be digested arrive in the colon intact and serve as a nutrient source

**Figure 30.6**
**The Large Intestine**

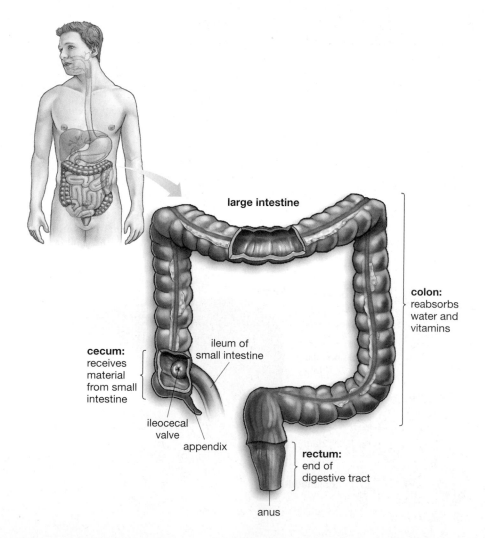

large intestine

**colon:** reabsorbs water and vitamins

**cecum:** receives material from small intestine

ileum of small intestine

ileocecal valve

appendix

**rectum:** end of digestive tract

anus

for these bacteria. Their resulting metabolic activities are responsible for intestinal gas. Why do beans often trigger such gas? Because they contain a high concentration of carbohydrates that we cannot digest but that the bacteria can.

---

**SO FAR...**

1. Eighty percent of our nutrients make the transition out of the digestive system along the length of the _____.

2. Bile is a substance secreted by the _____ and stored and concentrated by the _____, which facilitates the digestion of _____. The pancreas secretes _____ and _____ into the _____.

3. Arrange the following parts of the digestive system in the order in which food passes through them: stomach, pharynx, large intestine, esophagus, small intestine.

---

## 30.4 Human Nutrition

Having learned something about how the human body processes food, you're now in a better position to understand what kinds of foods the body needs. The subject here is **nutrition**, which can be defined as the study of the relationship between food and health.

What do the foods we eat need to contain in order for us to have a healthy body? Scientists think of foods in terms of substances they contain called *nutrients*. To be a **nutrient**, a substance must do at least one of three things: It must provide us with energy, it must provide us with one of the structural building blocks the body needs, or it must help regulate the body's physical processes. Many of the foods we eat are nutritious in all three senses. While a piece of whole wheat bread is loaded with energy, it also contains some structural building blocks (amino acids) and some substances that help regulate bodily processes (potassium, sodium).

Nutrients can play these diverse roles because nutrients themselves are a diverse group of substances. Every nutrient, however, is a member of a single *class* of nutrients, and it turns out there are only six such classes: water, minerals, and vitamins; and carbohydrates, proteins, and lipids. Given that everything we eat falls into one of these classes, we'll structure our tour of nutrition largely as a walk through these groups. Along the way, you'll see how important it is to choose carefully among foods as a means of getting the nutrients you need.

## 30.5 Water, Minerals, and Vitamins

Water makes up about 66 percent of our body by weight and thus serves not just as one substance among many within our body but rather as an environment in which our body tissues are immersed. We must have water to live, but in most instances, we do not need to be concerned about whether we are getting too little or too much of it in our diet. Recent research has shown that thirst is the best indicator of whether our bodies are sufficiently "hydrated." Healthy women consume an average of about 2.7 liters (or 91 ounces) of water per day, while men consume about 3.7 liters (or 125 ounces). Surprisingly, perhaps, about 20 percent of this water comes from the food we eat rather than the fluids we drink.

**Minerals** are chemical elements needed by the body either to help form bodily structures or to facilitate chemical reactions. You may remember from Chapter 2 that a chemical element is a substance that is "pure" in the sense that it cannot be broken down into any set of simpler substances through chemical means. All minerals are elements in this sense. They are not *combinations* of substances; they are elemental substances themselves.

Looking at the functions of minerals in nutrition, we might ask first: What does it mean for a mineral to help form a bodily structure? The mineral calcium provides a good case in point. About 1.5 percent of our body is made up of calcium. Where does this mineral exist? As you can imagine, overwhelmingly in our bones: 99 percent of the calcium in our bodies is found in bone. The 1 percent of calcium *not* found in bone is also critical, but it functions in the second role played by minerals—that of facilitating chemical reactions. Any time we perceive a sound, for example, calcium ions must flow into the cells in our ears that turn sounds into nerve signals. Any time we decide to move a muscle, calcium ions must be released from one variety of muscle fiber in order for another variety of fiber to latch on to it. Calcium thus serves both as a chemical building block and as a partner in chemical reactions. Not all minerals serve both functions—some function only as facilitators of reactions— but as a class of nutrients, minerals are important in both roles.

If you look at **Table 30.1**, on the next page, you can see a list of 16 of the most important dietary minerals. Note that there are two varieties of them. *Major minerals* are minerals needed in large amounts by the body, while *trace minerals* are minerals needed in small amounts. What's the dividing line? If we need 100 milligrams or more of a mineral per day, it's a

## Table 30.1 Minerals Required by the Human Body

### Major minerals

| Mineral | Sources | Functions | Symptoms of deficiency |
|---------|---------|-----------|------------------------|
| Calcium (Ca) | Dairy products, leafy green vegetables | Maintenance of bones, muscle contraction, nerve signal transmission | Retarded growth in children; osteoporosis in adults |
| Chloride (Cl) | Table salt | Water balance, digestion | Very rare |
| Magnesium (Mg) | Broccoli, leafy green vegetables, whole-grain wheat products | Component of bones, teeth; nerve signaling, muscle contraction | Irregular heartbeat, muscle spasms, seizures |
| Phosphorus (P) | Dairy products, meat, whole grains | Component of bone, nucleic acids, ATP | Bone loss in older women |
| Potassium (K) | Numerous fruits and vegetables, dairy products | Nerve signal transmission, fluid balance | Muscle cramps, irregular heartbeat, paralysis |
| Sodium (Na) | Table salt | Nerve signal transmission, fluid balance | In rare instances, muscle cramps, shock |
| Sulfur (S) | Protein-rich foods (eggs, meat) | Detoxification of drugs; component of some amino acids, vitamins | Unknown when dietary protein is adequate |

### Trace minerals

| Mineral | Sources | Functions | Symptoms of deficiency |
|---------|---------|-----------|------------------------|
| Chromium (Cr) | Liver, nuts, whole grains | Helps activate insulin | Elevated glucose, insulin levels |
| Copper (Cu) | Liver, seafood, nuts, legumes | Enzyme activation, hemoglobin synthesis | Rickets-like skeletal problems, anemia |
| Fluorine (F) | Fluoridated tap water, toothpaste | Resistance to tooth decay | Increased tooth decay |
| Iodine (I) | Iodized table salt, seafood, dairy products | Synthesis of thyroid hormones | Goiter, mental retardation in children of iodine-deficient mothers |
| Iron (Fe) | Whole grains, beef, fish, beans, nuts | Synthesis of hemoglobin | Iron deficiency anemia |
| Manganese (Mn) | Whole grains, nuts, leafy green vegetables | Enzyme activation | Never observed in humans |
| Molybdenum (Mo) | Dairy products, grains, legumes | Enzyme activation | Increased heart rate, night blindness, edema |
| Selenium (Se) | Seafood, liver, eggs | Enzyme activation | Muscle pain, weakness, vascular deterioration |
| Zinc (Zn) | Meat, poultry, fish, eggs, beans | Enzyme activation in many processes | Stunted growth, diarrhea |

major mineral; if we need less than 100 mg per day, it's a trace mineral. What do these differences mean in practice? To take an extreme example, an average adult needs at least 1,000 milligrams (meaning a gram) of dietary calcium per day. But this same person needs only 150 *micrograms*—that is, millionths of a gram—of dietary iodine per day. This is not to say, however, iodine and other trace minerals are unimportant. If you look at **Figure 30.7**, you can see what can happen over time when the body does not get its 150 micrograms of iodine per day. Both major and trace minerals are critical to our functioning;

the difference is that trace minerals are needed in smaller quantities.

**Figure 30.7**
**Iodine Deficiency**

The swelling in this woman's neck is a goiter—an enlargement of the thyroid gland brought about by a lack of iodine in the diet. Goiters of this size are a rare occurrence, but iodine deficiency is a major health problem in the world's less developed countries, where women who are iodine-deficient are at high risk of giving birth to mentally retarded children. In recent years, great progress has been made in combating this problem by means of adding iodine to table salt, a strategy adopted decades ago by the world's more developed countries. The strategy is cost effective: A mere two ounces of a substance called potassium iodate can iodize a ton of salt at a cost of about $1.15.

Our third category of nutrients, **vitamins**, can be defined as chemical compounds that are needed in small amounts in the diet to facilitate chemical reactions in the body. At first glance, this definition may make vitamins sound just like minerals but note the differences between these two kinds of nutrients. First, vitamins are compounds—they are made up of more than one element—whereas each mineral is a *single* element. Thus, a molecule of vitamin $B_2$ (riboflavin) is made up of the elements carbon, hydrogen, oxygen, and nitrogen. Second, unlike minerals, vitamins serve no structural role in the body; their function is purely to facilitate chemical reactions. As such, vitamins are needed strictly in *small* quantities—sometimes in milligram quantities, but often in microgram quantities. In consequence, the daily diet of a typical American contains only about 300 milligrams of vitamins compared to 20 grams of minerals.

So, in what ways do vitamins facilitate chemical reactions? If you look at **Figure 30.8** you can see an example. From Chapter 6, on energy, you may recall that nearly every chemical reaction that takes place in the body is speeded up—or, in practical terms, *enabled*—by a special kind of protein called an enzyme. Thus, when we drink milk, the lactose (or milk sugar) in it is broken down in a chemical reaction that uses an enzyme—an enzyme called lactase. Like all enzymes, lactase can only do its job by binding with its "substrate," meaning the substance it works on. Now, how do enzymes do this binding? In some cases, they come equipped, we might say, to bind on their own. A few of the units that make up an enzyme will bind to the substrate and help alter it. In other cases, however, enzymes need a chemical partner. In order for them to bind to their substrate, they must first bind with a *coenzyme*. This is the role played by all eight B vitamins: They serve as coenzymes (or parts of coenzymes). To put this another way, the B

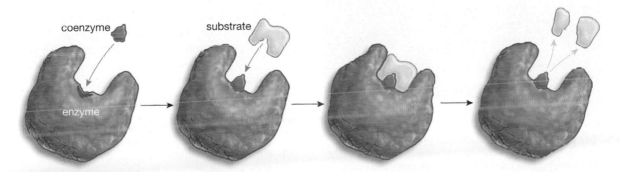

**Figure 30.8**
**One Role for Vitamins**

Many of the vitamins we consume serve as coenzymes—substances that bind to enzymes, thus allowing the enzymes to bind to their substrates, meaning the substances they help modify.

vitamins are molecules that bind to enzymes, thus allowing them to function.

If we looked at other vitamins—C and D and so forth—we'd find them playing roles other than coenzymes in the facilitation of chemical processes. In each case, however, a vitamin is pivotal in allowing some metabolic process to go forward. Thousands of bodily compounds are important in this same way, but almost all these compounds can be *manufactured* by the body. The reason vitamins are classified as nutrients is because, in most cases, our only source for them is the food we eat or the vitamin supplements we take. (The principal exceptions to this are vitamin D, which can be manufactured by our skin if it's exposed to enough sunlight, and vitamin K, which is produced by bacteria in our gut.) If you look at **Figure 30.9**, you can see examples of what can happen to the body if it's deprived of vitamin D over an extended period of time. **Table 30.2** sets forth a list of all 13 vitamins that the human body must have in order to function properly.

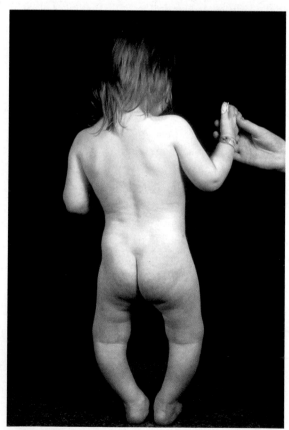

**Figure 30.9**
**The Effects of Vitamin D Deficiency**
The bowed legs of this child are a symptom of rickets, a condition in which a child's bones are too weak to support its body. In more developed countries, rickets is seen in a small number of children who have both little exposure to sunlight and an inadequate intake of dietary vitamin D. These factors decrease the body's ability to absorb dietary calcium, a condition that leads to weakened bones.

Note that vitamins are divided into two classes: water soluble and fat soluble. What's the difference? Water-soluble vitamins dissolve in water, while fat-soluble vitamins dissolve in fat. The practical importance of this distinction is that excess amounts of water-soluble vitamins are simply excreted in urine, with the result that these vitamins do not build up in us. Fat-soluble vitamins, meanwhile, *can* build up in us (particularly in the liver), which means we need to guard against taking excess quantities of them.

## Vitamin and Mineral Sources: Do We Need Supplements?

The foods we eat can be rich sources of both vitamins and minerals. Given this, the question is whether we need to take vitamin and mineral *supplements*—usually pills or capsules—to augment the quantities of vitamins and minerals we get through food. The consensus among nutritionists seems to be that, while a well-balanced diet can supply all the vitamins and minerals the average American needs, a simple one-a-day supplement can't hurt and might be helpful. When it comes, however, to very large doses of particular vitamins—E or C, for example—their advice is different. There is no reason to think that "megadoses" of any vitamin will be beneficial to health, and there is even reason to worry that such doses might be harmful.

Diet alone tends to supply all the vitamins we need in part because so many of the foods we eat are now fortified with vitamins and minerals. Many dairy products have vitamins A and D added to them, flour and other grain products are now fortified with folic acid, and breakfast cereals often come loaded with an array of vitamins. Even with these additions, though, many nutritionists believe that the average American diet is sometimes deficient in two vitamins, A and C, and in two minerals, calcium and iron.

### SO FAR...

1. To be a classified as a nutrient, a substance we ingest must do at least one of three things: It must provide us with _____, it must serve as a _____, or it must help _____ the body's physical processes.

2. The six classes of nutrients are _____, _____, and _____; and _____, _____, and _____.

3. Minerals are chemical _____, while vitamins are chemical _____. Because, unlike minerals, vitamins do not help _____, every vitamin is needed in strictly _____.

## Table 30.2 Vitamins in the Human Diet

### Water-soluble vitamins

| Vitamin | Sources | Functions | Symptoms of deficiency |
|---|---|---|---|
| Thiamin ($B_1$) | Whole grains, legumes, nuts | Coenzyme in cellular respiration | Beriberi |
| Riboflavin ($B_2$) | Dairy products, leafy green vegetables | Coenzyme in cellular respiration | Skin lesions, blurred vision |
| Niacin | Peanuts, poultry, beans | Part of cellular respiration coenzyme | Pellagra, mental disorders |
| Vitamin $B_6$ | Whole grains, dairy products, nuts, leafy green vegetables | Part of coenzyme in amino acid synthesis | Anemia, twitching, skin disorders |
| Pantothenic acid | Nuts, beans, dairy products, eggs | Coenzyme in fat synthesis | Numbness, fatigue |
| Folic acid | Leafy green vegetables, beans, orange juice | Coenzyme in hemoglobin production, DNA formation | Spina bifida in children of mothers deficient in folic acid |
| Vitamin $B_{12}$ | Dairy products, eggs, meat | Coenzyme in nucleic acid formation, red blood cell maturation | Pernicious anemia |
| Biotin | Eggs, fish, poultry | Coenzyme in synthesis of fat, amino acids | Scaly, inflamed skin, poor growth |
| Vitamin C | Citrus fruits, green peppers, cauliflower, broccoli | Promotes collagen synthesis in bone, cartilage; antioxidant | Scurvy, poor wound healing |

### Fat-soluble vitamins

| Vitamin | Sources | Functions | Symptoms of deficiency |
|---|---|---|---|
| Vitamin A | Liver, fish, eggs, fortified dairy products | Critical to vision | Night blindness, complete blindness in severe cases |
| Vitamin D | Can be produced by skin exposed to sunlight; fatty fish, fortified dairy products | Regulates bone metabolism | Rickets |
| Vitamin E | Vegetable oils, asparagus, eggs, almonds | Antioxidant; helps protect cell membranes | In infants, hemolytic anemia (rare) |
| Vitamin K | Liver, leafy green vegetables; some production by bacteria in gut | Production of compounds used in blood clotting | Blood-clotting problems in newborns |

## 30.6 Calories and the Energy-Yielding Nutrients

Thus far, we've looked at three classes of nutrients—water, minerals, and vitamins—that are critical parts of a healthy diet but that have nothing to do with the aspect of food consumption that most people are concerned about: calories. Water contains no calories at all, and minerals and vitamins contain next to none. This is not the case, however, with the carbohydrates, lipids, and proteins we'll look at now. If you want to know how many calories you're consuming per day, just add up your daily consumption of these three classes of nutrients.

A common assumption, however, is that adding up calories is an exercise in adding up units of *weight*, but this isn't the case. Calories actually are units of *energy*. A single calorie is defined as the amount of energy it takes to raise the temperature of 1 gram of water by 1 degree Celsius. It takes such a small amount of energy to do this, however, that the caloric content of food is never measured in individual

••• ANSWERS

1. energy; structural building block; regulate

2. water, minerals, vitamins; proteins, carbohydrates, lipids

3. elements; compounds; form bodily structures; small amounts

calories. Instead, each *thousand* calories is referred to as a single calorie (sometimes written as Calorie, with a capital C). This is what is meant by the term "calorie" as we see it in everyday life. Each nutritional **calorie** is the amount of energy it takes to raise the temperature of 1,000 grams of water by 1° C.

Thinking of calories in these terms gives us some insight into the nature of carbohydrates, proteins, and lipids. To say that these nutrients contain calories is the same thing as saying they are capable of yielding energy, which is just what they do. Students who read Chapter 7, on cellular respiration, are familiar with the means by which these nutrients do this: They yield electrons that are high in energy and that thus can run energetically "downhill," in the same way that the water in a stream can run downhill, yielding energy as it goes. It is the energy given up by carbohydrates, lipids, and proteins that powers every activity we carry out, from moving our muscles to reading words on a page. As such, these three classes of nutrients are sometimes referred to as the *energy-yielding nutrients*. To put a little finer point on this, however, the vast bulk of our everyday energy needs actually are met through carbohydrates and lipids alone; proteins yield only about 5 percent of the energy we use.

Given the importance of energy to our functioning, there's some irony in the fact that the calories used to measure energy generally are thought of as something to be *avoided*. We must have energy to live, and energy is measured in calories. In this sense, calories are a good thing. The problem, of course, is that in modern society it's easy to have too much of a good thing: We can easily consume more calories in, say, a day than we will burn up during that day. When this happens, the grams of food that are not burned up within us are not excreted or dissipated to any significant extent; nearly all of them are *stored* within us, mostly as fat tissue. This is why we end up thinking of calories as a measure of weight. To see calories as units of energy, however, is the first step in understanding what it means for one food to be more "caloric" than another. Why is it that lipids (more commonly known as fats) are more caloric than either carbohydrates or proteins? The answer has to do with the *density* of energy in these nutrients—with how much energy is contained in a given portion of them. If we look at carbohydrates and proteins, we find that each gram of them yields 4 calories of energy. Meanwhile, each gram of lipids yields 9 calories of energy. Notice, in this accounting, that lipids are not *heavier* than carbohydrates or proteins. (We're comparing 1-gram quantities of each kind of nutrient, after all.) Instead, gram per gram, lipids contain

**Table 30.3  Daily Calorie Requirements**

| Groups, by age | Calorie Range Sedentary ———▶ Active | |
|---|---|---|
| **Children** | | |
| 2–3 years | 1,000 | 1,400 |
| **Females** | | |
| 4–8 years | 1,200 | 1,800 |
| 9–13 | 1,600 | 2,200 |
| 14–18 | 1,800 | 2,400 |
| 19–30 | 2,000 | 2,400 |
| 31–50 | 1,800 | 2,200 |
| 51+ | 1,600 | 2,200 |
| **Males** | | |
| 4–8 years | 1,400 | 2,000 |
| 9–13 | 1,800 | 2,600 |
| 14–18 | 2,200 | 3,200 |
| 19–30 | 2,400 | 3,000 |
| 31–50 | 2,200 | 3,000 |
| 51+ | 2,000 | 2,800 |

Sedentary means a lifestyle that includes only the light physical activity associated with typical day-to-day life.

Active means a lifestyle that includes physical activity equivalent to walking more than 3 miles per day at 3 to 4 miles per hour in addition to the light physical activity associated with typical day-to-day life.

more energy than do carbohydrates or proteins. This is why lipids have such power to put pounds on us—they're so packed with energy.

So, how many calories do we need per day to satisfy our energy needs? If you look at **Table 30.3**, developed by the U.S. Department of Agriculture, you can see that there is no single answer to this question; calorie requirements vary according to age, gender, and level of physical activity. If you then look at **Table 30.4**, you can see the number of calories contained in servings of some common foods.

We've thus far thought of carbohydrates, lipids, and proteins strictly in connection with their role in supplying calories, but we need to think more broadly about these three classes of nutrients. We'll now begin looking at them individually, with an eye toward answering the question: What kinds of proteins, carbohydrates, and lipids should we be consuming to ensure that our diet is as healthy as possible?

## 30.7  Proteins

Like all nutrients, proteins are defined by their chemical structure (which you can read about in detail in Chapter 3). The particular structure of proteins is that of a set of linked building blocks called amino acids. In a typical human protein, hundreds of amino acids are strung together, one by one, to create a chain that then folds up into a three-dimensional shape that gives each protein its working ability. Interestingly, although tens of thousands of human proteins are put together through this means, all of them are produced from a starting set of a mere 20 amino acids. How can things work like this? String amino acids together in one order, and you have protein A; string them together in a different order, and you have protein B. For purposes of nutrition, the important thing to note is that amino acids are the raw materials of proteins.

When we start to look at the roles that proteins take on in the body, we find that no group of molecules is more diverse in its functions. You saw above that almost every chemical reaction in the body is facilitated by the proteins called enzymes. We might marvel at proteins for this regulatory role alone, but as it turns out, proteins play structural roles as well. They form the solid portion of our muscles, for example, and form most of the connective tissue that holds other tissues together. Beyond this, they are active players in the immune system; they serve as signaling molecules and hormones and transport capsules—the list of their functions is very long.

This versatility notwithstanding, you may remember from the discussion above that proteins are *not* very good at one function of nutrients, which is supplying energy. This is so in part because the body has no reservoir of stored proteins. Lipids can be stored in great quantities within us (think of a big belly), and carbohydrates can be stored to a lesser extent (in muscles and in the liver). But this isn't the case with proteins. All of them we have either are being used already or are being transported toward use somewhere within us. Therefore, for the body to use proteins as an energy source, an *existing* protein has to be broken down somewhere—say, in our muscles. This is an inefficient system for extracting energy, and as such, the body employs it mostly when it has nowhere else to turn—that is, when available supplies of lipids and carbohydrates have started to run low. In well-fed societies such as ours, this is likely to take place for extended periods of time only in two situations: when people are dieting or when they are engaging in prolonged bouts of endurance exercise (marathon races, long bike rides). In sum, proteins are critically important for two of the nutritional roles we talked about earlier—helping form bodily structures and regulating bodily processes—but not so important for the third nutritional role of providing energy.

## Table 30.4 Calories in Portions of Some Selected Foods

| Food | Calories |
| --- | --- |
| 8 oz. mug unsweetened green tea | 2 |
| 1 large slice whole wheat bread | 79 |
| 1/2 cup cooked white rice | 102 |
| 1/2 cup cooked brown rice | 107 |
| 12 oz. glass nonfat milk | 120 |
| 12 oz. cola (Coca-Cola®) | 140 |
| 12 oz. glass of 2% milk | 180 |
| 1 Krispy Kreme Original Glazed® doughnut | 200 |
| 1 piece thick-crust, 12" cheese pizza | 256 |
| 1 regular (2.07 oz.) Snickers® bar | 280 |
| 1 Starbucks grande Mocha Frappuccino® (with whipped cream) | 380 |
| 1 McDonald's Big Mac® | 540 |
| 1 Burger King Whopper® | 670 |

*Source:* USDA and corporate websites and products.

The proteins we eat cannot enter the bloodstream "as is" but instead must first be broken down in the digestive tract into their amino acid building blocks—molecules that are small enough to pass out of the digestive tract and into circulation. Thus, protein-laden foods don't supply our bodies with proteins per se but rather with a collection of amino acids that can be turned *back* into proteins through the body's protein-synthesizing machinery.

### Protein Sources and Requirements

Dietary proteins can come to us from both animal and plant sources. Three ounces of salmon contain about 18 grams of protein, but then again 2 tablespoons of peanut butter contain almost 10 grams. But not all proteins are equally nutritious; what counts in this regard is the amino acid makeup a protein has. Recall that a collection of 20 amino acids is needed to produce every protein in the body. Of these 20, however, 11 can be produced *by* the body and so are called **nonessential amino acids**. Meanwhile, there are nine amino acids that cannot be produced by the body that must be obtained from foods; these are the **essential amino acids**. For purposes of nutrition, an important point is that nearly every source of *animal* protein supplies us with a complete, balanced set of essential amino acids. This means that every time we eat fish, poultry, eggs, red meat, or dairy products, we are consuming this set. Meanwhile, among plant proteins, only soy protein contains all nine essential amino acids in their proper proportions. The upshot of this is that meat eaters—or vegetarians who eat dairy products—needn't worry about their sources of protein. Vegetarians who don't eat dairy products, however, may need to pay attention to their diet to make sure they are getting all their essential amino acids. One means of achieving this is to practice what is sometimes called *protein complementation*, in which the amino acids provided by several foods combine to make up a complete set. How would this work in practice? To cite the most famous example, eating both rice and beans over the course of a day would provide a person with all nine essential amino acids (**Figure 30.10**). Leave the beans out, however, and the day's diet would be deficient in the amino acid lysine; leave the rice out, and the diet would be deficient in the amino acid methionine.

About 70 percent of the protein in the average American diet comes from animal sources, with beef, poultry, and milk being big suppliers of this class of nutrients. Americans tend to consume much more dietary protein than they need. A 154-pound man needs about 56 grams of protein per day but gets about 100 grams. Meanwhile, a 125-pound woman needs about 46 grams per day but gets about 65 grams.

---

**SO FAR...**

1. In nutrition, each calorie is defined as the amount of _____ required to raise the temperature of a _____ of water by _____.

2. Each gram of lipids yields _____ of _____, while each gram of carbohydrates or proteins yields _____.

3. Proteins move from the digestive system into circulation in the form of their building blocks, _____. Nearly every _____ source of protein supplies us with a complete, balanced set of _____, but this usually is not the case with proteins that come from _____ sources.

---

## 30.8 Carbohydrates

If proteins are most important nutritionally in areas other than supplying energy, carbohydrates are the reverse: Their primary importance in the diet is that they furnish the energy that helps power our activities. Like proteins, carbohydrates follow a building-blocks model of chemical organization. Whereas the building blocks for proteins are amino acids, however, the building blocks for carbohydrates are known as monosaccharides. If we take one well-known monosaccharide, glucose, and put it together with a second monosaccharide, fructose, we get a slightly more complex *di*saccharide—sucrose, which is better known as table sugar. As it turns out, monosaccharides and disaccharides represent one of three principal classes of dietary carbohydrates, the **simple sugars**. When we

**Figure 30.10**
**Protein Complementation**

Eating both rice and beans in a single day is one way vegetarians can make sure their diet is providing them with all nine essential amino acids.

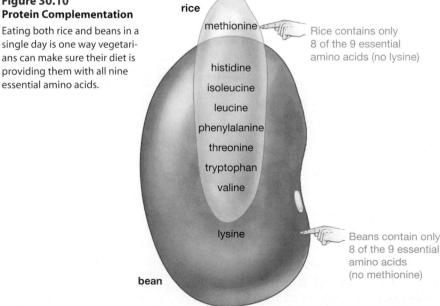

rice

methionine — Rice contains only 8 of the 9 essential amino acids (no lysine)

histidine
isoleucine
leucine
phenylalanine
threonine
tryptophan
valine

lysine — Beans contain only 8 of the 9 essential amino acids (no methionine)

bean

**Figure 30.11**
**Processed and Unprocessed**

Foods that are heavily processed, such as those on the left, tend to be less healthy than those that are fresh and unprocessed, such as those on the right.

look at the other two classes of dietary carbohydrates, we find that they are composed not of one or two monosaccharide units, but of thousands of linked units of the monosaccharide glucose. These are the so-called complex carbohydrates, which exist in two forms. One of these is the **starches**, which can be defined as complex carbohydrates that are digestible. The other is **fibers**, which can be defined as complex carbohydrates that are indigestible.

To get a sense of what these abstract classes mean in practice, imagine yourself at a picnic with a plate of food in front of you that includes a slice of whole wheat bread on your left and a can of cola over to your right. You eat some of the bread and then wash it down with the cola. What have you consumed? Apart from the water in the cola, what you have consumed is overwhelmingly carbohydrate—in all three of its dietary forms. Two of these forms are found solely in the bread you ate. These are starch and fiber. The starch is a complex carbohydrate in the bread that will be digested; it will be broken down in the digestive system and then put into circulation. The fiber in the bread is different; it is a complex carbohydrate that the body *cannot* break down. As such, it stays within the digestive tract and eventually is excreted. (This actually is a good thing, as you'll see.) Meanwhile, the third form of carbohydrate you consumed at your meal was found primarily in the soft drink. Look on most cans of regular soft drink today, and you are likely to see "high-fructose corn syrup" listed among the ingredients. Your cola probably contained about 40 grams of this simple sugar, which as the name implies, was derived originally from corn. How did this transformation work? Corn from the fields was processed into cornstarch, which was then broken down to yield a mixture of the two simple sugars fructose and glucose, the final ingredients in high-fructose corn syrup.

## Carbohydrate Choices: Think Fresh and Whole Grain

Simple sugars, starches, and fibers all have nutritional uses. But this is not to say that these three forms of carbohydrate are all equally healthy. Nutritionists today uniformly recommend a diet that is not only low in simple sugars that have been added to products—as was the case with our high-fructose corn syrup—but high in particular sorts of *complex* carbohydrates: those that come from fresh fruits, vegetables, and whole-grained cereals. (In a nutritional sense, *cereals* refers to cereal grain crops such as wheat, oats, rice, barley, and corn.) What this advice often means in practice is eating carbohydrates that have been *processed* as little as possible (**Figure 30.11**). To understand this concept, consider the whole wheat bread and the cola that we just looked at in our picnic meal. What made the wheat in the picnic bread "whole wheat" was the fact that it had undergone relatively little processing; each wheat grain used in this bread retained both its bran (outer coat) and its wheat germ (embryonic tissue), whereas the wheat used to make white bread has had both of these components stripped away (**Figure 30.12**). On the other side of the spectrum, if we look at our cola, we can see that a starting ingredient for it was a complex carbohydrate

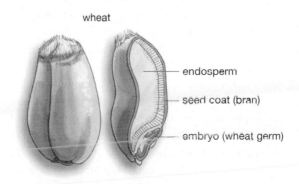

wheat

endosperm

seed coat (bran)

embryo (wheat germ)

**Figure 30.12**
**What "Whole Wheat" Means**

Each grain of wheat used in a whole wheat product retains both its seed coat (also known as its bran) and its embryo (also known as its wheat germ). In contrast, grains of wheat used to make white bread or regular pasta have both of these elements removed during processing.

(the starch in whole corn) that, through a good deal of processing, ended up yielding two simple sugars.

The question then becomes: In nutritional terms, how does this cola, with its simple sugars, compare to our slice of whole wheat bread, with its whole grains? When we look at a typical, large slice of whole wheat bread, we find that it contains about 80 calories. Meanwhile, a 12-ounce can of cola might contain 140 calories, all of which come from the corn syrup. Now, *apart* from the calories—which have a value in that they're yielding energy—what do we get from the cola in the way of nutrition? Almost nothing: water and a little sodium, but that's about it. Meanwhile, if we look at the bread, what do we get? Calcium, iron, magnesium, potassium, small quantities of several B vitamins—in short, an array of vitamins and minerals. Comparisons such as these are the reason nutritionists have long referred to soft drinks as foods that provide "empty calories," meaning little in the way of nutrition *except* calories. But the nutritional gap between our simple sugar and our whole-grained bread involves more than just vitamins and minerals. To understand a second factor that's at work, let's think for a moment about how these carbohydrates are processed by the body.

### Slower Digestion and Glycemic Load

As you saw earlier in the chapter, only small, simple molecules can pass from the digestive tract into the bloodstream. And as you've seen, the starch that helps make up whole wheat is neither small nor simple. It is a complex carbohydrate whose long-chains must be broken down into simple carbohydrates—into the individual glucose units that can move into circulation. It takes time to do this, which means that the glucose that results from the digestion of whole wheat moves *slowly* into circulation. Contrast this pace, however, with what happens with the carbohydrate from our cola. This carbohydrate is *ingested* as a set of simple sugars—it comes to us in a simple form that requires relatively little further digestion. Consequently, concentrations of simple sugars we ingest can result in a "spike" of glucose in the bloodstream, and this in turn results in a spike of insulin, released from the pancreas, whose function is to store the glucose away in cells. One problem with these insulin surges is that they seem to bring on hunger sooner than do slow, steady rises in insulin levels. Indeed, many experts believe that, to the extent that carbohydrates cause weight gain, it is not *all* carbohydrates that do so, but instead just those that are rapidly absorbed into the bloodstream. Beyond the issue of hunger, however, persistently high blood levels of insulin can raise levels

of lipids in circulation, while lowering levels of "good" cholesterol (or HDLs). One large, long-standing study of diet and health found that a high intake of calories from refined grains and potatoes is associated with a higher risk of developing both type 2 diabetes and heart disease. Given findings such as these, the takeaway message is that the faster a given carbohydrate moves from the digestive tract into circulation, the less healthy it seems to be.

This conclusion is important enough that nutritionists have come up with a couple of ways of measuring the effects of different carbohydrates on blood glucose levels. The best accepted of these is called a food's **glycemic load**: a measure of how blood glucose levels are affected by defined portions of given carbohydrates. Glycemic loads of less than 15 are considered low (which is good), while those of 15 to 20 are considered intermediate, and those of more than 20 are considered high. To give you some perspective on these figures, the single slice of whole wheat bread we've been looking at has a glycemic load of about 9, while a typical 12-ounce cola would have a glycemic load of about 25. There can be some surprises in glycemic load figures, however. A baked potato is largely composed of starch, but it's of a type that breaks down so quickly that the glycemic load of 8 ounces of baked potato is 48.

### Fiber: The Value of Not Being Digested

Beyond added nutrients and glycemic load, there turns out to be a third reason to prefer unprocessed, fresh, and whole-grained carbohydrates over any other variety. Recall that in addition to the two classes of carbohydrates we've been looking so much at—simple sugars and starches—there is then a third class, which are the various fiber carbohydrates that go undigested when we eat them. You might think that any substance that fails to move into circulation couldn't provide health benefits at all, but this isn't the case with fibers. Among other things, these tough, complex carbohydrates bind in the digestive tract with cholesterol and substances made from it, which has the effect of lowering *blood* levels of cholesterol. In addition, fiber increases the bulk of our stool, in part by absorbing water, and this effect softens our stool, making it easier to pass. (Excessive straining during defecation can produce small pouches, called diverticula, that pop out from the large intestine and that can become sites of inflammation or even infection.) Moreover, any starch that is bound up with fiber will move more slowly into circulation; thus, fiber helps with the glycemic load problem we just talked about. And fiber adds a bulk to the food we eat that helps us feel full, which may help us cut down on the calories we consume. Beyond these factors, to judge by the long-standing study noted earlier, a high-fiber diet may lower the risk of acquiring type 2 diabetes and heart disease, while as we've seen, a diet high in refined grains and potatoes may raise this risk.

So, where do different foods stand with respect to their fiber content? By now you can probably make some informed guesses. Looking once again at our cola, with its high-fructose corn syrup, we find that it has a fiber content of zero. If we then look not at our slice of whole wheat bread, but instead at bread made from more heavily processed *white* flour, we find that a slice of it contains 0.7 grams of fiber. And our whole wheat bread? One slice of it contains 2.2 grams of fiber.

### Carbohydrate Sources: Phytochemicals

You may have noticed that all the carbohydrates we've been considering have come from plant sources. This turns out to be the case generally with carbohydrates; indeed, the overwhelming preponderance of our carbohydrate calories come from plant sources—grains, fruits, vegetables, beans. The principal exception to this is the carbohydrates that exist in dairy products. The fact that carbohydrates come from plants actually is a good reason to eat carbohydrates, providing that, once again, they come from fresh, unprocessed sources. Plants contain thousands of compounds that have been dubbed **phytochemicals**: nonnutritive substances found in plants that promote health. Why are phytochemicals defined as nonnutritive? Because they don't do any of the things that we've defined nutrients as doing. As such, an *absence* of them in the diet doesn't lead to a calorie deficiency or to the breakdown of a metabolic pathway. However, the consistent *presence* of them in the diet does seem to protect against a variety of diseases, including heart disease, some forms of cancer, and age-related macular degeneration (the leading cause of blindness in older Americans).

There are at least nine classes of phytochemicals, a fact that might make us despair at ever getting a handle on what we should be eating to get enough of them in our diets. But, it turns out that following a few simple rules will lead to an abundance of dietary phytochemicals. First, these compounds are found in greatest abundance in fruits and vegetables, although beans, nuts, whole grains, and some spices can contain them as well. Second, in choosing among fruits and vegetables, a diversity of *color* is a pretty good guide to ensuring that you're getting what you need (**Figure 30.13**). The best-studied phytochemicals—a group called the carotenoids—is most abundant in fruits and vegetables that have a red or yellow-orange

**Figure 30.13**
**Bright Color as a Guide to Health**
The healthy compounds called phytochemicals are found in abundance in brightly colored fruits and vegetables.

color, among them tomatoes, carrots, acorn squash, apricots, mangoes, and watermelon. (Carotenoids can also be found in broccoli, spinach, and other leafy green vegetables.) Meanwhile, a group of phytochemicals called flavonoids are found in abundance in fruits and vegetables whose color lies toward the blue, purple, and red side of the color spectrum: blueberries, raspberries, red cabbage, and purple grapes among them. Overall, nutritionists recommend choosing from a "rainbow" spectrum of fruits and vegetables to ensure an adequate intake of these healthy plant compounds.

### SO FAR...

1. There are three principal classes of dietary carbohydrates: _____, _____, and _____.

2. To say that a carbohydrate is a source of "empty calories" is to say that it provides very few _____ or _____.

3. Glycemic load is a measure of how _____ levels are affected by defined portions of given carbohydrates. The _____ the glycemic load, the better, in terms of health. Fruits and vegetables are rich sources of _____, defined as nonnutritive substances found in plants that promote health.

## 30.9 Lipids

Our final class of nutrients, the lipids, are a terrific source of energy, as we saw earlier. In the body, lipids serve more roles than just providing energy, however. Students who read Chapter 4, on cell structure, may recall that every human cell has an outer or "plasma" membrane that is largely composed of lipids (the phospholipids and cholesterol). Beyond this, lipids are used to make hormones such as estrogen and testosterone, and the lipids in our fat cells provide us with insulation and perform a shock-absorbing function.

••• ANSWERS
1. simple sugars; starches; fibers
2. vitamins; minerals
3. blood glucose; lower; phytochemicals

The building blocks of most dietary lipids are molecules called triglycerides, one of which can be seen in **Figure 30.14**. Note that this is a molecule that has two essential parts. One is a "head," composed of a compound called glycerol; the other is a set of three *fatty acids* that are attached to the glycerol and that stem from it like tines on a fork. When our bodies need energy, they can get it by breaking triglycerides apart: The fatty acids are clipped from the glycerol, and then each fatty acid is broken down further. When, however, our bodies have plenty of energy, this process goes the other way: Glycerol and fatty acid units are brought together, and the resulting triglycerides are then stored in cells called adipocytes, which are better known as fat cells. A key point in nutrition, however, is that it doesn't *take* lipids to *make* lipids. Should we have more carbohydrates or proteins in circulation than are necessary to meet our energy needs, these nutrients will go to the liver and be transformed into triglycerides, which will end up being stored in fat cells. In short, we could get fat eating nothing but carbohydrates and proteins.

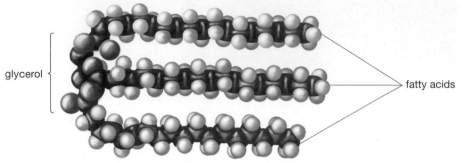

**Figure 30.14**
**Building Block of Dietary Lipids**
Most of the lipids we consume are composed of building blocks called triglycerides. Every triglyceride is composed of a single glycerol molecule linked to three fatty acids. This particular triglyceride, called tristearin, has three identical fatty acids linked to its glycerol "head," but in many cases several different fatty acids will be part of a single triglyceride.

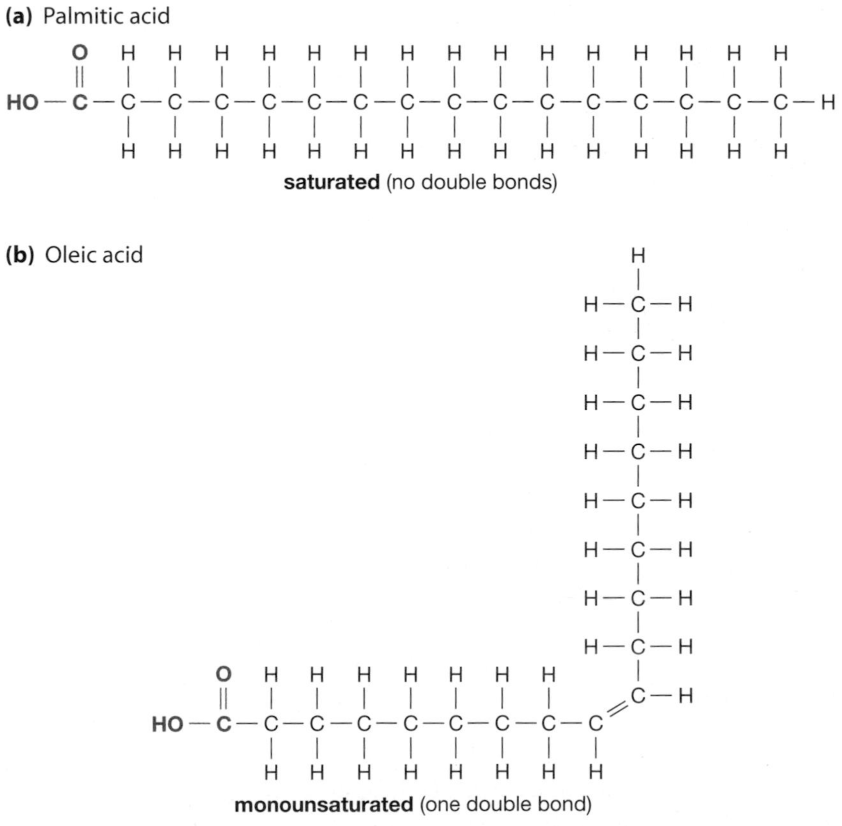

**(a)** Palmitic acid

**saturated** (no double bonds)

**(b)** Oleic acid

**monounsaturated** (one double bond)

**Figure 30.15**
**Saturated and Unsaturated Fatty Acids**
The degree to which fatty acid chains are "saturated" with hydrogen atoms has consequences both for the form these lipids take and for human health.

**(a)** The hydrocarbon chain in palmitic acid is formed by an unbroken line of carbon atoms, each with a single bond to the next.

**(b)** In oleic acid, however, a double bond exists at one point between two carbon atoms. An additional hydrogen atom could link to each of these carbon atoms instead, which would make this a saturated fatty acid. As things stand, this is a monounsaturated fatty acid. If two or more carbons in the chain had double bonds between them, it would be a polyunsaturated fatty acid.

## Oils and Fats

In everyday speech, lipids are invariably referred to as fats, but this common term obscures an important distinction. We actually consume dietary lipids in two different forms. One of these is **oils**, which can be defined as dietary lipids that are liquid at room temperature (olive oil, canola oil). The other is then **fats**, which can be defined as dietary lipids that are solid at room temperature (butter, the fat in a piece of bacon). As we'll see, these two terms will be helpful in making distinctions among different types of lipids with respect to their effects on health. And, what is the health concern connected to lipids? As most people know, the central issue is heart disease. Although recent research results have been mixed, a preponderance of evidence indicates that, as our consumption of some lipids goes up, so does our long-term risk for developing heart disease. Meanwhile, other lipids appear to be at least neutral with respect to heart disease, while still others actually seem to protect against it. So, which lipids fall into which category? Let's take a look.

## Lipid Choices: Think Unsaturated

In discussions of lipids, you've probably heard the terms *saturated* and *unsaturated* fats. What makes a "fat"—meaning a fat or oil—saturated or unsaturated is its fatty acid makeup. About 80 percent of the fatty acids in olive oil, for example, are unsaturated, while only 20 percent of the fatty acids in it are saturated. Hence, olive oil is referred to as an unsaturated fat. Meanwhile, if we look at a stick of butter, we find that 60 percent of the fatty acids in it are saturated, while 40 percent are unsaturated; hence, butter is referred to as a saturated fat.

But how do the saturated fatty acids found in butter differ from the unsaturated fatty acids found in it? To put this another way, what makes a fatty acid saturated or unsaturated? In **Figure 30.15**, you can get an idea.

Pictured are structural formulas for two individual fatty acids, palmitic acid and oleic acid. Note that the palmitic acid is a straight chain of carbon atoms (the Cs) and their associated hydrogen atoms (the Hs). Meanwhile, the oleic acid is bent; there's a kink in it, and if you look closely, you can see where this kink occurs: right at the place where two carbon atoms have a double bond with each other. As a result, these carbons are linked to only one hydrogen atom each, whereas all the other carbons in the chain are linked to two hydrogens. Now, it would be possible to replace this lone carbon double bond with a single bond, but to do so, we would have to add one more hydrogen atom to each of the carbon atoms involved. Once that happened, however, there would be no space for any *additional* hydrogen atoms in the chain. We could thus say that this fatty acid chain had become *saturated* with hydrogen atoms, and this, indeed, is where we get the term saturated fatty acid. If we left our oleic acid alone, however, it would remain an *un*saturated fatty acid since there is a carbon double bond within it where hydrogen atoms could be added. The definitions of various kinds of fatty acids turn out to be based on the existence of these double bonds. A **saturated fatty acid** is a fatty acid with no double bonds between the carbon atoms of its hydrocarbon chain. Meanwhile, a **monounsaturated fatty acid** is a fatty acid with one double bond between carbon atoms, and a **polyunsaturated fatty acid** is a fatty acid with two or more double bonds between carbon atoms.

On one level, the importance of the unsaturated-to-saturated spectrum simply has to do with the *form* of lipids we consume: Lipids that contain a high proportion of unsaturated fatty acids tend to be oils (as in our olive oil example), while those that contain a high proportion of saturated fatty acids tend to be fats (as in our butter example). On another level, however, the transition between unsaturated and saturated lipids is the single most important factor in how *healthy* various lipids are, as defined by their effect on heart disease. Here is a hierarchy of different "fats"—meaning lipids—that is based on their health effects. Running from least healthy to most healthy, we have:

- Trans fats
- Saturated fats
- Polyunsaturated and monounsaturated fats
- Polyunsaturated fats containing omega-3 fatty acids

To put this within the framework we used earlier, consumption of trans fats and saturated fats increases our long-term risk of heart disease. Meanwhile, monounsaturated and polyunsaturated fats are at least neutral with respect to heart disease, and the polyunsaturated omega-3s actually seem to protect against it.

But, how do we know which of these kinds of fats we're consuming? We can judge pretty well by the kind of *food* we're consuming. If we look at omega-3 fatty acids, for example, we find that the best source for them is certain kinds of fatty fish—notably salmon and albacore tuna—although they can also be found in lesser amounts in plant-based foods such as walnuts and the soybean-based product tofu. If we then look at the monounsaturated and other polyunsaturated fats, we find that, by and large, our sources for them are *oils*: olive, canola, and peanut oil for the monounsaturated fats and safflower and corn oil for the polyunsaturated fats. Moving to the less-healthy saturated fats, we find that our usual source for them is animal fat, primarily fatty meats and dairy products such as cheese and butter. Finally, the least healthy fats of all, the trans fats, are found in a significant, but decreasing, number of packaged and fast foods: cookies, French fries, cakes, crackers, doughnuts, popcorn, and candy, to name a few foods.

Taking a step back from this list, it's easy to see that a good rule of thumb is that if a lipid is an oil, it is at least neutral with respect to health and may actually be healthy. Which oils fall into the healthy category? A prime candidate is olive oil. Several studies have shown that residents of Mediterranean countries have significantly lower levels of heart disease than people in other parts of the developed world. What these residents have in common, in terms of diet, is consumption of large amounts of this monounsaturated oil.

On the unhealthy side of the spectrum, the fats known as trans fats occupy a special place among lipids. As noted, as we move from unsaturated to saturated fats, we also move in a general way from oils to solid fats. Trans fats are created in an industrial process in which naturally occurring oils are turned *into* fats by partially saturating them—by bubbling hydrogen through the oils in a process called hydrogenation. The upside of this is the production of fats that have a longer shelf life, or that have a taste or creamy texture that consumers like. The downside is the production of fats that raise levels of "bad" cholesterol (LDLs) while lowering levels of "good cholesterol" (HDLs). In addition, trans fats appear to boost fat levels in general in the blood while impairing the ability of blood vessels to open or "dilate." Given these effects, American food producers and fast-food chains have been moving in recent years to remove trans fats from their products. Beginning in 2006, the federal Food and Drug Administration required packaged food producers to list trans fats separately in the ubiquitous "Nutrition Facts" labels. It's thus become easy to know whether you're consuming trans fats in packaged foods: Simply look for the "trans fats" line within the Nutrition Facts label. These days, it's likely that the number of grams of trans fats you'll see there will be zero.

On the other side of the health spectrum, what makes the polyunsaturated omega-3 fats so *good* for us? These fats guard against blood clot formation, they reduce fat levels generally in the bloodstream, and they reduce the growth of the fatty deposits that clog heart arteries; in consequence, they appear to reduce both overall levels of cardiovascular disease and sudden cardiac death.

## 30.10  Elements of a Healthy Diet

We've been looking at the six classes of dietary nutrients in isolation from one another, but any real-world diet will, of course, be a mixture of different foods that contain these nutrients. If we think strictly along the lines of nutrition, what would a good combination of foods look like in a diet? More specifically, what *proportions* of different kinds of foods would go into making up a healthy diet? Nutritionists have for years found the image of a pyramid useful in making recommendations about

food proportions. If you look at **Figure 30.16**, you can see one such pyramid—in this case a pyramid constructed by researchers at Harvard School of Public Health. Many of the recommendations in this figure will by now be familiar. Note that sweets and other refined carbohydrates have been consigned to the narrow, "use sparingly" portion of the pyramid, while whole-grained foods are at its base and recommended for "most meals." Meanwhile, fruits and vegetables should make up a substantial portion of a healthy diet, while red meat and butter—both high in saturated fats—are recommended for infrequent consumption.

Although these ideas about food proportions are shared by most nutritionists, they represent only one way of looking at the elements of a healthy diet. In 2005, the U.S. Department of Agriculture abandoned its traditional food pyramid after deciding that no one set of food recommendations could take all Americans into account. The USDA still employs a food pyramid, but only as one element within an interactive website that allows individuals to enter information about themselves as a means of receiving individualized dietary recommendations. You can construct your own set of USDA dietary recommendations by visiting www.MyPyramid.gov.

**Figure 30.16**
**Foods in the Proper Proportions**

This food pyramid, constructed by researchers at the Harvard School of Public Health, provides one view of the proportions that foods ought to be consumed in to ensure a healthy diet. Note the prominent role recommended not only for whole grains, but for plant oils (canola, olive, etc.), which can be thought of as "good" fats. The recommendation for "alcohol in moderation (if appropriate)" stems from evidence indicating that drinking small amounts of alcohol as part of a healthy diet can reduce the risk of heart disease.

## SO FAR...

1. Most of the lipids we consume are composed of building blocks called _____, which in turn are composed of a molecule called glycerol and three _____.

2. Saturated fatty acids are saturated with _____. As we move from unsaturated to saturated fatty acids, we move in a general way from _____, which are dietary lipids that are liquid at _____, to _____, which are dietary lipids that are solid at _____.

3. Most saturated fats have a(n) _____ source, while most monounsaturated and polyunsaturated fats have a(n) _____ source. Saturated fats are _____ healthy than monounsaturated or polyunsaturated fats.

## 30.11 The Urinary System in Overview

If nutrition and digestion largely concern what moves into blood circulation, the subject we'll take up now is the reverse: What gets filtered out of the blood and then eliminated from the body altogether? Our subject is the urinary system, which filters waste from the blood and then passes this waste on to the bladder for elimination.

Such waste removal obviously is important, but it turns out to be only one of several critical functions the urinary system undertakes. Equally important is the system's regulation of blood volume—how much blood we have within us—a task whose necessity becomes obvious when we think about consuming lots of liquids. Suppose the body simply incorporated all the water in these liquids into circulating blood? We would eventually swell and burst like a water balloon. Instead, the urinary system *regulates* how much water the body retains and how much it excretes, with the kidneys being a critical player in this process. The urinary system also controls ion concentration and maintains pH balance in the body. In short, it is important in preserving the stable state known as homeostasis. We almost never think of the urinary system in this regard, essentially because it works so well. But a person who suffers kidney failure and must rely on a kidney dialysis machine to filter wastes knows the value of the urinary system in a personal way.

In carrying out these activities, the urinary system is very much involved in the *conservation* of bodily resources. Indeed, the kidneys are as much in the conservation business as the waste removal business. They conserve water, amino acids, ions, sugars—all kinds of things that the body needs. Consider that the kidneys process about 180 liters (or almost 48 gallons) of fluid every day. Yet our urination is only about 1.8 liters, or about *half* a gallon daily. More than 47 gallons recycled back into the body each day! Now let's see how the kidneys and the larger urination system manage this feat.

••• ANSWERS

1. triglycerides; fatty acids
2. hydrogen atoms; oils; room temperature; fats; room temperature
3. animal; plant; less

## 30.12 Structure of the Urinary System

The pivotal structures in the urinary system, shown in **Figure 30.17**, are its two filtering organs, the **kidneys**, which produce urine while conserving useful blood-borne materials. Urine leaving the kidneys travels along two tubes, the left and right **ureters**, to the **urinary bladder**, the hollow, muscular organ that acts as a temporary, expandable storage site for urine. When urination occurs, contraction of the muscular bladder forces the urine through the conducting tube called the **urethra** and out of the body.

**Figure 30.17
The Urinary System in Overview**

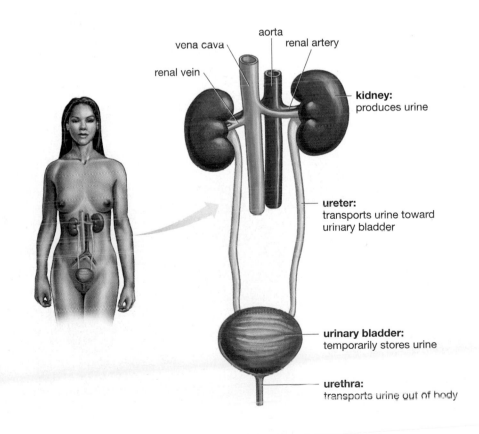

vena cava
aorta
renal artery
renal vein

**kidney:**
produces urine

**ureter:**
transports urine toward urinary bladder

**urinary bladder:**
temporarily stores urine

**urethra:**
transports urine out of body

The kidneys are located on either side of the vertebral column, at about the level of the eleventh and twelfth ribs. They are shaped like kidney beans, but since they are each about 10 centimeters or 4 inches long, their size is more like that of a small pear.

In considering how the kidneys work, it's helpful to think first in terms of input and output. What comes into the kidney in the way of fluids? Blood from the circulatory system arrives by way of the renal arteries. As you can see in Figure 30.17, these branch off from the aorta, one renal artery going to the left kidney, one going to the right. (The word *renal* comes from *renes*, which means "kidneys"; you'll be seeing a lot of "renal" this and that.) Now, what comes *out* of the kidney? Two things come out: Blood flows out of the kidneys through the renal

*veins* that are pictured, and urine exits each kidney through its ureter. So, each kidney has one input (the renal artery) but *two* outputs. This is a clue to what's happening inside. Some of the material coming into the kidney through the renal arteries is being channeled out of the blood vessels and into a *separate* system of vessels whose output tubes are the ureters. Now let's look more closely at the kidneys.

Seen in cross section in **Figure 30.18a**, each kidney has within it a chamber called the renal pelvis, whose numerous branches collect the urine produced by the kidneys and channel it into the ureter. If you look at **Figure 30.18b**, you can see a blow-up of the basic working unit of the kidney, a structure called a nephron. Each kidney has within it about 1.25 million nephrons, each serviced by a single arteriole

**Figure 30.18**
**Structure of the Kidney and Its Working Unit, the Nephron**

**(a)** A kidney in cross section.

**(b)** A nephron in the kidney, composed of a nephron tubule and its associated blood vessels, which are immersed in interstitial fluid.

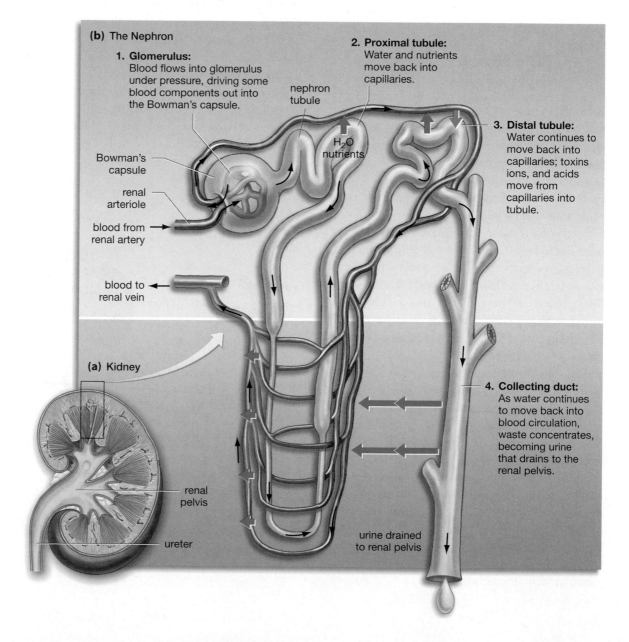

**(b) The Nephron**

**1. Glomerulus:**
Blood flows into glomerulus under pressure, driving some blood components out into the Bowman's capsule.

Bowman's capsule

renal arteriole

blood from renal artery

blood to renal vein

**2. Proximal tubule:**
Water and nutrients move back into capillaries.

nephron tubule

$H_2O$ nutrients

**3. Distal tubule:**
Water continues to move back into capillaries; toxins ions, and acids move from capillaries into tubule.

**4. Collecting duct:**
As water continues to move back into blood circulation, waste concentrates, becoming urine that drains to the renal pelvis.

**(a) Kidney**

renal pelvis

ureter

urine drained to renal pelvis

which is part of a network of vessels whose blood comes from the entering renal artery.

In the figure, you can see how a nephron begins: An arteriole enters a hollow chamber, called the Bowman's capsule, and then promptly expands into a knotted network of capillaries, called the glomerulus, that is enclosed by the capsule. *Emerging* from this network, there is again an arteriole that goes on to surround something called a nephron tubule. (In reality, this arteriole branches into a network of capillaries that surrounds the tubule, but for simplicity's sake these capillaries are shown here as a single blood vessel.) Thus, we have two sets of tubes—the nephron tubule and the blood vessel—and these will become the two *outputs* just mentioned: The nephron tubule will transmit what the body doesn't need (waste sent to the bladder), while the blood vessel will first lose, and then take back, what the body does need (nutrients, etc.). All this put together adds up to a **nephron**: the functional unit of the kidneys, composed of a nephron tubule, its associated blood vessels, and the interstitial fluid in which both are immersed. At the "output" end of the nephron, you can see that its tubule is only one of several that are emptying into a common *collecting duct*. Eventually, many of these ducts will themselves merge to feed into the renal pelvis, which feeds into the ureter. Now, let's look at what takes place within a nephron.

## 30.13  How the Kidneys Function

Imagine holding a washcloth, baglike, underneath a running kitchen faucet. What happens? The water slowly flows out of the washcloth. What if this is a faucet that hasn't been turned on in a while, however, and the water thus contains some sediment? Now the washcloth acts to retain these larger particles while letting smaller substances (the water molecules) flow through. In short, the washcloth acts as a filter. The knotted network of nephron capillaries called the **glomerulus** takes on this same filtering role. Receiving blood that is under pressure, it is porous enough to let some smaller blood-borne materials pass out of it—and into the surrounding Bowman's capsule (see Figure 30.18b). What materials pass out? Water, first of all, which makes up such a large portion of the blood, and then many of the materials suspended in the water, such as vitamins, nutrients, and waste materials. All these substances now move from the glomerulus into the Bowman's capsule. What stays in the glomerulus (and hence in the bloodstream) are plasma proteins and blood cells, which are too big to pass out.

We should note at this point the nature of the waste products that are moving out of the glomerulus. Every time you use a muscle, you produce a waste product, called creatinine, that has to be eliminated from the body. Likewise, proteins are broken down mostly in the liver, and a by-product of this breakdown is a substance called urea. Both urea and creatinine are picked up by the bloodstream and move to a glomerulus, where they flow out, along with the water and other substances.

All the materials that have entered the Bowman's capsule constitute a fluid, called filtrate, that flows from the capsule into the nephron's next structure, its proximal tubule. With this tubule, we get to the essence of how the kidneys work as a filtering mechanism—getting rid of waste but retaining what is valuable. The retention process begins immediately after the filtrate moves into the proximal tubule. Some active pumping of materials out of the tubule is coupled to the process of osmosis (reviewed in Chapter 5). The result of both actions is that about two-thirds of the water lost in filtration, as well as almost all the nutrients lost there, move back into blood circulation at the proximal tubule—back into the network of capillaries represented in Figure 30.18b by the single blood vessel. While this "reabsorption" is going on, however, the urea and other waste products have remained in the nephron tubule because, thanks to their chemical makeup, they cannot easily pass out of it.

If you just follow the nephron tubule on around, past the proximal tubule, you can see what comes next: Water continues to flow out of the tubule and back into circulation (blue arrows), while up at the distal tubule, toxins and other waste move the other way: from circulation into the tubule (green arrow). The end result is that, with movement through the tubule, the filtrate becomes an ever-more-concentrated waste product. Water and nutrients are removed from it and sent back to circulation, while urea, creatinine, and other wastes remain in it. By the time the filtrate reaches the collecting duct, it contains such a high concentration of waste that it goes by another name: urine.

### Hormonal Control of Water Retention

We noted earlier that the kidneys are important players in the body's control of the volume of fluids we have within us. How does this control work? The key to it is a substance, produced by the brain's hypothalamus, called **antidiuretic hormone (ADH)**. If we get dehydrated, the hypothalamus releases ADH. When this hormone reaches the kidneys, it has the effect of allowing water to flow more freely out of the distal nephron tubule and the collecting duct. What happens next, of course, is that this water then moves back into blood circulation. Thus, ADH prompts the kidneys to recycle more water back into circulation rather than allowing them to send this water to the bladder as urine. Of course, there are times when we have too much water in the body. In this event, the hypothalamus shuts down its production of ADH, and this prompts the distal tubule and collecting duct to hold on to more water, which means more of it is sent to the bladder. The end result of all this is that the body can tightly regulate the amount of water it retains. Along these lines, if you've ever wondered why 12 ounces of beer makes a person have to urinate more than does 12 ounces of lemonade, it's because the alcohol in the beer suppresses the production of ADH. Reduced ADH levels mean that more of the water that enters our systems will be sent to the bladder rather than being moved back into circulation.

**Web Animation 30.1**
The Urinary System

## 30.14   Urine Storage and Excretion

Having seen how the kidneys produce urine, let's now look at the steps by which urine is then excreted from the body. Recall that urine is transported to the urinary bladder by the pair of tubes called the ureters—one stemming from each kidney. The ureters are muscular tubes that extend for a distance of about 30 centimeters, or 12 inches. Starting at the kidney, about every 30 seconds a peristaltic wave of muscle contraction squeezes urine out of the renal pelvis, along the ureter, and into the urinary bladder. Occasionally, solids develop within the collecting tubules, collecting ducts, or ureters. These are kidney stones, and they can be very painful; in addition to

obstructing the flow of urine, they may also reduce or eliminate filtration in the affected kidney.

### The Urinary Bladder

The urinary bladder, shown in **Figure 30.19**, is a hollow muscular organ that stores urine. Its dimensions vary with the volume of stored urine, but a full urinary bladder can contain up to 800 milliliters of urine, which is about 27 ounces. The area surrounding the urethral entrance, called the neck of the urinary bladder, contains an internal sphincter muscle that provides involuntary control over the discharge of urine from the bladder.

### The Urethra

The urethra extends from the urinary bladder to the exterior of the body. Its length and functions differ in males and females. In the male (shown in Figure 30.19), the urethra extends to the tip of the penis and is about 18 to 20 centimeters (7 to 8 inches) in length. The initial portion of the male urethra is surrounded by the prostate gland. In contrast, the female urethra is very short, extending 2.5 to 3.0 centimeters (about 1 inch). In both sexes, the urethra contains a circular band of skeletal muscle that forms an external sphincter. Unlike the internal sphincter, the external sphincter is under voluntary control.

### Urination

The urge to urinate usually appears when the bladder contains about 200 milliliters, or about 7 ounces, of urine. We become aware of this need through nerve impulses sent to the brain from stretch receptors in the wall of the urinary bladder. The stimulation of these receptors also results in involuntary contractions of the urinary bladder that increase the fluid pressure inside the bladder (which is why we "have to go" to the bathroom). Urine ejection cannot occur, however, unless both the internal and external sphincters are relaxed. We control the time and place of urination by voluntarily relaxing the external sphincter. When this sphincter relaxes, so does the internal sphincter. If the external sphincter does not relax, the internal sphincter remains closed, and the bladder gradually relaxes. A further increase in bladder volume begins the cycle again, usually within an hour. Once the volume of the urinary bladder gets large enough, sufficient pressure is generated to force open the internal sphincter. This leads to an uncontrollable relaxation in the external sphincter, and urination occurs despite voluntary opposition. Normally, less than 10 milliliters (about a third of an ounce) of urine

**Figure 30.19**
**The Urinary Bladder in a Male**

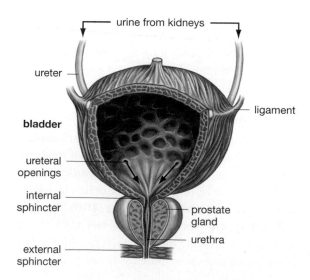

urine from kidneys

ureter

bladder

ligament

ureteral openings

internal sphincter

prostate gland

urethra

external sphincter

remain in the bladder after urination. As noted earlier, an average of about 1.8 liters (half a gallon) of urine is excreted daily.

## On to Development and Reproduction

Over the last five chapters, we've reviewed the ways human beings breathe, think, digest, fight off invaders, remove wastes, and more. While all of these operations are critical to maintaining life, they don't touch on one of the central abilities of living things, which is the ability to pass life on. Each generation of human beings gives rise to another, of course, through the workings of a specialized system, the reproductive system, which comes in two versions: male and female. In essence, these systems must produce reproductive cells—egg and sperm—that come together to produce a single cell that, over the course of 9 months, develops into a 10-trillion-celled baby in full cry. So, how does this process work? How are egg and sperm produced in the first place, how do they come together, and how does the cell that results from their union develop into a new generation of living thing? These are the subjects of the next two chapters.

---

### SO FAR...

1. The "input" into each kidney is _____ that flows through one of the body's _____. The "output" from each kidney is _____ that flows into one of the body's _____ and urine that flows into the _____.

2. The working unit of the kidneys is the nephron, composed of a nephron tubule, its associated _____, and the _____ in which both are immersed.

3. Nephrons work by means of retaining _____ within the nephron tubules while returning _____ and _____ to the circulatory system.

••• ANSWERS

1. blood; renal arteries; blood; renal veins; bladder
2. blood vessels; fluid
3. waste; water; nutrients

# CHAPTER 30 REVIEW

For study help, activities, and more quiz questions, visit **www.aw.com/krogh4**.

## SUMMARY

### 30.1 The Digestive System

- The central functions of the digestive system are (1) to get the foods the body ingests into a form the body can use, (2) to move food in this form out of the digestive tract and into blood circulation, (3) to retain the waste that remains after nutrients from the food have been transferred, and (4) to eliminate this waste from the body. The central structure of the digestive system is a muscular tube called the digestive tract that passes through the body from the mouth to the anus and that receives input along its length from accessory digestive organs such as the gallbladder, liver, and pancreas.

- Digestion is a process of breaking down foods—first into small pieces and then into small molecules. These small molecules leave the digestive system for transport through the rest of the body.

### 30.2 Structure of the Digestive System

- The digestive tract begins with the mouth (or oral cavity) and continues through the pharynx, esophagus, stomach, small intestine, and large intestine before ending at the rectum and anus. Through most of its length, the digestive tract is a tube composed of four layers: the serosa, or outermost layer; the muscularis externa, whose two sets of muscles produce the waves of contraction, called peristalsis, that move food along; the submucosa, which contains blood vessels and nerve tissue that control digestion; and the mucosa, which absorbs digested food. Absorbed food is taken up by an adjacent capillary (in the case of carbohydrates and proteins) or a lymphatic vessel (in the case of fats).

### 30.3 Steps in Digestion

- Digestion begins in the mouth through both mechanical means (chewing) and chemical means (enzymes breaking down carbohydrates). Food is pushed by the tongue into the pharynx (or throat) and moves by muscle contraction from there to the esophagus whose own muscle contractions work with gravity to push food into the stomach.

- The stomach digests food partly by the mechanical means of churning it and partly by chemical means, with proteins a primary target of the stomach's digestive juices. The stomach's contents are very acidic—a quality that is valuable both for breaking food down and for killing microorganisms that come in with the food. The material that leaves the stomach is a mixture of food and digestive juices called chyme.

- Eighty percent of the digestive tract's absorption of nutrients takes place within the small intestine—the portion of the digestive tract that begins at the stomach and ends at the large intestine.

- In digestion, the pancreas is an organ that secretes, into the small intestine, three classes of digestive enzymes that help break down fats, proteins, and carbohydrates. It also secretes "buffers" that raise the pH of the acidic chyme coming from the stomach.

- The gallbladder stores and concentrates bile, a substance produced by the liver that aids in the digestion of fats. The liver is a large organ that plays a central role in digestion. All blood carrying nutrients from the digestive tract is channeled through the hepatic portal vein to the liver, making the liver a first stop for much digested material. The liver controls which nutrients it will store and which nutrients it will send to the rest of the body. The liver is also the first stop in the body for ingested toxins such as alcohol, making it vulnerable to damage.

- The large intestine holds and compacts material left over from digestion, turning it into feces. It also returns water to general circulation and absorbs vitamins produced by resident bacteria. One of the large intestine's regions, the rectum, is usually empty except when peristaltic contractions force fecal materials into it. The resulting stretching triggers the urge to defecate.

### 30.4 Human Nutrition

- Nutrition is the study of the relationship between food and health. A nutrient is a substance, contained in food, that does at least one of three things: provides energy, provides a structural building block for the body, or helps regulate a process in the body. There are six classes of nutrients: water, minerals, and vitamins; and proteins, carbohydrates, and lipids.

### 30.5 Water, Minerals, and Vitamins

- Water makes up 66 percent of the human body and is vital to health. People generally do not need to be concerned about whether they are getting too much or too little water in their diet, however, as recent research has shown that thirst is the best guide as to whether the body is sufficiently hydrated.

- Minerals are chemical elements needed by the body either to help form bodily structures or to facilitate chemical reactions. Dietary minerals are divided into two categories: major minerals, which are needed in amounts of 100 milligrams or more per day; and trace minerals, which are needed in amounts of less than 100 milligrams per day.

- Vitamins are chemical compounds that are needed in small amounts in the diet to facilitate chemical reactions in the body. Vitamins differ from minerals in that vitamins are compounds, rather than elements. In addition, vitamins do not help form bodily structures and thus are needed strictly in small quantities—sometimes milligram quantities per day, but often microgram quantities. There are 13 vitamins that the body must have to function properly. Of these, eight are B vitamins that facilitate chemical reactions by acting as coenzymes or parts of coenzymes.

- Most nutritionists have concluded that, while a well-balanced diet can supply all the vitamins and minerals the average American needs, there is little harm to be had, and possibly some gain to be realized, from taking a one-a day vitamin-and-mineral supplement. Nutritionists do not, however, recommend taking large "megadoses" of any vitamin. The average American diet may be deficient in vitamins A and C and often seems to be deficient in the minerals calcium and iron.

## 30.6  Calories and the Energy-Yielding Nutrients

- All the calories in the human diet are provided by three classes of nutrients: proteins, carbohydrates, and lipids. Calories are units of energy, not units of weight. In nutrition, a calorie (sometimes written as Calorie) is the amount of energy it takes to raise the temperature of 1,000 grams of water by 1 degree Celsius. Proteins, carbohydrates, and lipids sometimes are referred to as the energy-yielding nutrients in recognition of the fact that they alone provide the body with energy.

- Quantities of energy-yielding nutrients that we consume but do not burn up end up being stored within us, mostly as fat tissue. Foods are caloric in accordance with how much energy they contain per unit of weight. Lipids are much more caloric than either proteins or carbohydrates. Each gram of lipids yields 9 calories of energy, while each gram of proteins or carbohydrates yields 4 calories of energy.

## 30.7  Proteins

- Proteins are composed of chemical building blocks called amino acids. Tens of thousands of human proteins are put together from a starting set of only 20 amino acids.

- Proteins have many structural and regulatory functions in the body, but they provide very little energy compared to carbohydrates or lipids. This is so in part because the body has no stored reservoir of proteins. As such, the body can only derive energy from proteins by breaking apart proteins already in use—an inefficient process that the body turns to only when its supplies of carbohydrates and lipids are running low.

- Proteins consumed in food are broken down in the digestive tract into their amino acid building blocks; it is these amino acids, rather than fully formed proteins, that move into circulation. Eleven of the 20 amino acids needed for protein synthesis can be produced by the body and are called nonessential amino acids. There are, however, nine amino acids that cannot be produced by the body and that must be obtained from food; these are called the essential amino acids. Nearly every source of animal protein

supplies a complete, balanced set of essential amino acids. Among plant proteins, however, only soy protein contains all nine essential amino acids in their proper proportions.

## 30.8  Carbohydrates

- The primary function of dietary carbohydrates is to provide energy. The building blocks of carbohydrates are molecules called monosaccharides. A single monosaccharide can be linked to a second monosaccharide to form a molecule called a disaccharide. Monosaccharides and disaccharides form one of the three principal classes of dietary carbohydrates, the simple sugars. The other two classes of dietary carbohydrates, collectively known as complex carbohydrates, are starches, defined as complex carbohydrates that are digestible; and fibers, defined as complex carbohydrates that are indigestible. Both are composed of thousands of linked units of the monosaccharide glucose.

- Nutritionists recommend a diet that is low in simple sugars that have been added to processed foods but high in complex carbohydrates that come from fresh fruits, vegetables, and whole-grained cereals. One of the reasons for this advice is that foods high in added simple sugars often are sources of "empty" calories, meaning calories that are not accompanied by vitamins or minerals.

- The consumption of large quantities of simple sugars can result in a surge of glucose into the bloodstream, whereas the consumption of fresh and whole-grained complex carbohydrates results in a slow, steady entry of glucose into the bloodstream. Glucose surges result in surges in blood levels of insulin, a phenomenon that has been linked to increased hunger, higher levels of lipids in the bloodstream, and a lowering of levels of "good" cholesterol, among other effects. One means of measuring the effect of different carbohydrates on blood levels of glucose is the glycemic load: a measure of how blood glucose levels are affected by defined portions of given carbohydrates. Carbohydrates with low glycemic loads are in general more healthy than carbohydrates with high glycemic loads.

- The indigestible complex carbohydrates known as fibers have several good health effects, among them a lowering of blood levels of cholesterol and a reduction in glycemic load. Whole, unrefined grains tend to be high in fiber, while processed, "white" grains have less fiber, and added simple sugars have no fiber at all.

- Most carbohydrates come from plant sources. Plants in general—and fresh fruits and vegetables in particular—are rich sources of phytochemicals: nonnutritive substances found in plants that promote health. Phytochemicals seem to protect against heart disease, cancer, and age-related macular degeneration. Nutritionists recommend choosing from among a "rainbow" spectrum of brightly colored fruits and vegetables to ensure an adequate intake of phytochemicals.

## 30.9  Lipids

- Lipids are a great source of energy, but they also have significant structural functions in the body as they are major components of all cell membranes. In addition, lipids are used to make hormones, and they provide insulation and serve a shock-absorbing function.

- The building blocks of most dietary lipids are molecules called triglycerides that are composed of a "head" composed of glycerol and three attached fatty acid chains. Triglycerides can be broken down into these component parts to yield energy or can be stored away in fat cells. When the body has sufficient energy, carbohydrates and proteins are transformed into triglycerides that are then stored away.

- In everyday speech, all lipids are referred to as fats. Nutritionists, however, make a distinction between oils, which are dietary lipids that are liquid at room temperature; and fats, which are dietary lipids that are solid at room temperature.

- Fats are saturated or unsaturated in accordance with their fatty acid makeup. A "fat"—meaning a fat or oil—that is saturated is predominantly composed of saturated fatty acids, while a fat that is unsaturated is predominantly composed of unsaturated fatty acids. Fatty acids are defined in a saturated-to-unsaturated spectrum in accordance with the number of carbon double bonds they have in their hydrocarbon chains. A saturated fatty acid is a fatty acid with no double bonds between the carbon atoms of its hydrocarbon chain; a monounsaturated fatty acid is a fatty acid with one double bond between carbon atoms, and a polyunsaturated fatty acid is a fatty acid with two or more double bonds between carbon atoms. Lipids that contain a high proportion of unsaturated fatty acids tend to be oils, while those that contain a high proportion of saturated fatty acids tend to be fats.

- In general, lipids are healthy to the extent that they are unsaturated rather than saturated. Trans fats and saturated fats increase the long-term risk of heart disease; polyunsaturated and monounsaturated fats are at least neutral with respect to it, and polyunsaturated fats containing omega-3 fatty acids seem to protect against it.

- Omega-3 fats are found in greatest abundance in certain kinds of fatty fish; the usual source for other polyunsaturated and monounsaturated fats is plant-based oils; saturated fats normally are found in animal-based fats; and trans fats are found in a number of packaged and fast foods. Trans fats are the product of an industrial process in which naturally occurring oils are turned into fats by partially hydrogenating them—by infusing them with hydrogen atoms.

## 30.10  Elements of a Healthy Diet

- Diets can be judged as healthy to the extent that they contain a high proportion of healthy foods. Food pyramids are one means of getting a sense of proportionality in food intake.

## 30.11  The Urinary System in Overview

- The urinary system filters waste materials from the blood, regulates blood volume, and conserves useful materials, such as water, nutrients, and ions.

## 30.12  Structure of the Urinary System

- The urinary system consists of two kidneys that produce urine; the left and right ureters that the urine travels through on leaving the kidneys; the muscular urinary bladder, which receives the urine from the ureters and temporarily stores it; and the tube called the urethra through which urine passes from the bladder out of the body.

- The working unit of the kidneys is the nephron, composed of a nephron tubule, its associated blood vessels, and the interstitial fluid in which both are immersed. Several nephron tubules empty into a common collecting duct, which will merge with other such ducts to feed into the renal pelvis, which feeds into the ureter.

## 30.13  How the Kidneys Function

- A knotted network of capillaries within a nephron, the glomerulus, receives arterial blood but is porous enough to allow much of the fluid portion of the blood to flow out of it along with smaller molecules such as vitamins, nutrients, and waste products. These materials enter the surrounding Bowman's capsule, thus moving into the nephron's tubule as a fluid called filtrate. At the nephron's next structure, called the proximal tubule, much of the original water and almost all the original nutrients are moved back into blood circulation. Waste products remain in the nephron tubule, however, because of their chemical composition. This general process continues over the length of the nephron tubule: Water and nutrients move back into circulation, while waste products become ever more concentrated within the tubule. By the time the filtrate has reached the collecting duct, it has become urine.

- The body is able to control how much water the kidneys send to the bladder (in urine) or retain in circulation. Retention of water by the kidneys is regulated by a part of the brain, called the hypothalamus, that controls the secretion of antidiuretic hormone (ADH). An increased secretion of ADH means that more water will move out of the kidney's tubules and collecting ducts and back into circulation.

## 30.14  Urine Storage and Excretion

- Waves of muscle contraction squeeze urine out of the renal pelvis, thus moving it through the ureters and into temporary storage in the urinary bladder. The tube called the urethra then carries the urine from the urinary bladder to the exterior of the body. An internal sphincter muscle provides involuntary control over the discharge of urine. The urethra also contains an external sphincter that is under voluntary control. We become aware of the need to urinate when the bladder is stretched beyond a certain threshold. We then relax the voluntary, external sphincter, which relaxes the involuntary internal sphincter, and the urine moves out of the bladder and the body.

**Web Animation 30.1:** The Urinary System

## KEY TERMS

| | |
|---|---|
| **antidiuretic hormone (ADH)** 645 | **nonessential amino acids** 636 |
| **bile** 627 | **nutrient** 629 |
| **calorie (nutritional)** 634 | **nutrition** 629 |
| **chyme** 626 | **oil** 640 |
| **digestive tract** 623 | **pancreas** 627 |
| **essential amino acids** 636 | **peristalsis** 624 |
| **fat** 640 | **pharynx** 625 |
| **fibers** 637 | **phytochemical** 639 |
| **gallbladder** 627 | **polyunsaturated fatty acid** 641 |
| **glomerulus** 645 | |
| **glycemic load** 638 | **saturated fatty acid** 641 |
| **kidneys** 643 | **simple sugars** 636 |
| **large intestine** 628 | **small intestine** 626 |
| **liver** 627 | **starches** 637 |
| **mineral** 629 | **stomach** 625 |
| **monounsaturated fatty acid** 641 | **ureters** 643 |
| | **urethra** 643 |
| **nephron** 645 | **urinary bladder** 643 |
| | **vitamin** 631 |

## BRIEF REVIEW

*Answers to Brief Review questions are in the back of the book. For multiple-choice quiz questions, go to* **www.aw.com/krogh4**.

1. Why are most nutrients ready to enter the bloodstream from the small intestine, but not ready to enter from the stomach?

2. What are some of the ways that the liver is important in digestion?

3. Apart from waste storage and compaction, what does the large intestine do?

4. What are the differences between vitamins and minerals?

5. Name four nutritional reasons to prefer a food rich in unprocessed whole grains to a food rich in added simple sugars.

6. What is a saturated fatty acid saturated with? What kinds of foods are predominantly composed of saturated fatty acids? What kinds of foods are predominantly composed of unsaturated fatty acids?

7. Why don't blood cells and protein molecules get passed into the filtrate that moves through the kidneys' nephrons?

8. What is the small signal of discomfort that tells us when we need to urinate?

## APPLYING YOUR KNOWLEDGE

1. Some weightlifters take protein supplements in the belief that these supplements will provide them with additional energy. Based on what you learned about the dietary role of proteins, are these supplements likely to be doing weightlifters any good?

2. Contrary to popular belief, urine normally has no bacteria in it at all. Indeed, the movement of urine down the urinary tract is one reason the kidneys and the bladder tend to stay free from infection—the flow of urine washes away any encroaching bacteria. Keeping this and the difference between male and female urethras in mind, why is it that women are more prone to bladder infections than men?

3. Suppose that instead of returning nearly all nutrients and 98 percent of fluids back into circulation, the kidneys could return only half these proportions. Could human beings function? How would it change the way they live?

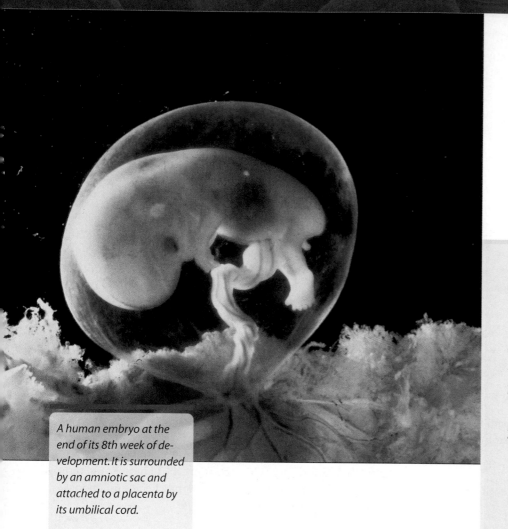

A human embryo at the end of its 8th week of development. It is surrounded by an amniotic sac and attached to a placenta by its umbilical cord.

**31.1**    General Processes in Development    **653**

**31.2**    What Factors Underlie Development?    **658**

**31.3**    Unity in Development: Homeobox Genes    **661**

**31.4**    Developmental Tools: Sculpting the Body    **662**

**31.5**    Development through Life    **663**

How can a living thing start out as a single cell and end up as a fully formed animal? It goes through a process called development.

# AN AMAZINGLY DETAILED SCRIPT:
## Animal Development

**O**f all the marvels in the natural world, the most spectacular may be a complex process that goes under a simple name: development. The set of steps by which, for example, a human embryo goes from being

a speck of microscopic cells to a newborn cradled in its mother's arms is so remarkable that even the old adage about "seeing is believing" is turned on its head. People who view pictures of such a work in progress can scarcely believe what they're seeing. It's not uncommon, months after the birth of a child, for parents to find themselves looking at the baby and then at each other in amazement, still wondering how such a thing is possible.

This process is all the more notable because it takes place through self-assembly. There is no outside hand here, no master artisan watching over each step and intervening if necessary. There is instead a starting set of instructions supplied by maternal and paternal DNA and a starting set of proteins and other molecules that come in the maternal egg. From these elements, an embryo fashions itself, aided only by the nurturing environment of egg or womb.

How can such an event take place? As you'll see, in species as diverse as frogs, insects, and human beings, a common, core set of steps is followed in development. Moreover, nature has provided a number of "tools" that are employed in development, and these likewise tend to be used across the different species.

## 31.1 General Processes in Development

Let's now begin by looking at the developmental progression that is common to much of the animal world, noting first what happens in development and then going over *how* these things happen. We'll be looking at several species, rather than just humans, as this will provide a clearer picture of what takes place.

### Two Cells Become One: Fertilization

When a sperm encounters an egg, these two cells fuse in the moment generally referred to as **fertilization**, but more poetically known as conception. The fertilized egg brought about by the fusion of egg and sperm, a **zygote**, is itself a cell, although a very special one in that an entire living thing will develop from it. Development at its simplest level is a process by which this single, original cell divides to become two cells, after which the two become four, the four eight, and so on. Once the zygote starts dividing, we call it an **embryo**.

## Three Phases of Early Embryonic Development

Early embryonic development in animals can be divided into three phases:

- *Cleavage.* Cellular division results in a ball of cells with an interior cavity.

- *Gastrulation.* These cells rearrange themselves into three layers that will give rise to specific organs and tissues.

- *Organogenesis.* Organs begin to form.

### First Phase: Cleavage

**Cleavage** is a developmental process in which a zygote is repeatedly divided into smaller individual cells through cell division. The reason the cells get progressively smaller is that none of the usual cell growth occurs between these cell divisions.

The product of this process of cleavage is a **morula** (from the Latin for "mulberry"), a tightly packed ball of early embryonic cells. This configuration soon changes, however, as the cells arrange themselves into a **blastula**: an early embryonic structure composed of one or more layers of cells surrounding a liquid-filled cavity (the blastocoel). The progression from zygote to blastula is illustrated in **Figure 31.1**. It's worth noting that the human form of a blastula, called a **blastocyst**, is partly composed of cells you may have heard of, *embryonic stem cells*—the cells that hold out promise for treating conditions such as Parkinson's disease and spinal cord injuries. You can read more about these cells in Chapter 15, beginning on page 277.

An egg that will be fertilized by sperm often contains a quantity of yolk, which helps give the resulting zygotes of many species a *polarity*, meaning a difference between one end and the other. In many zygotes, one end contains a relatively greater proportion of yolk, and this end is called the **vegetal pole**. Meanwhile, the other end has relatively less yolk and lies closer to the cell's nucleus, and this end is called the **animal pole**. As you'll see, this polarity helps define the way the embryo develops.

### Second Phase: Gastrulation

Anyone who has ever watched a play in rehearsal probably has witnessed a phenomenon that bears some comparison with the second phase in embryonic development, *gastrulation*. It's the moment in which the director says to the assembled cast, "Take your places." What happens then is a carefully directed movement of actors; some move only a little, whereas others go from one side of the stage to the other. After all this movement is done, the actors—having assumed their places—are ready to take on their assigned roles.

Gastrulation is likewise a process of directed movement, in this case of groups of cells that move to particular places in the developing embryo. Once this movement is completed, the cells will have formed themselves into three different layers that are ready to take on their assigned roles. These layers—the endoderm, mesoderm, and ectoderm—are "germ" layers

**(a)**

**(b)**

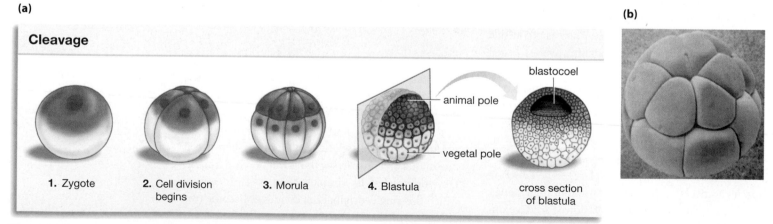

**Figure 31.1**
**Cleavage of a Frog Egg**

**(a)** 1. Like all animals, the frog starts out as a single-celled zygote. 2. Cell division begins and the zygote is now called an embryo. 3. Repeated cell division of the embryo results in a tightly packed ball of smaller cells, called a morula. 4. This configuration gives way to a blastula, in which the cells are arranged around a liquid-filled cavity, the blastocoel. The yolk at the vegetal end of the zygote impedes cleavage, so that fewer but larger cells result at the vegetal pole than at the animal pole of both the morula and blastula. The drawing at far right shows a cross section of the blastula, with its blastocoel visible.

**(b)** An artificially colored micrograph of a frog embryo at the 16-cell stage of development.

of cells, meaning layers that *lead to* other kinds of structures (**Table 31.1**). As it turns out, endoderm, mesoderm, and ectoderm have a kind of "inside/middle/outside" quality to them, and this is reflected in the organs and tissues that develop from them. In gastrulation, endoderm takes its place as the innermost layer, mesoderm as the middle layer, and ectoderm as the outer layer. In line with this, endoderm gives rise to interior tissues (the liver and bladder, for example), mesoderm to tissues exterior to these (muscle and skeletal systems), and ectoderm to tissues that are more exterior yet (the skin and nervous system, which first appear in the embryo's outermost layer.) **Gastrulation** can thus be defined as a process in early development in which an embryo's cells migrate to form three layers of tissue: the endoderm, the mesoderm, and the ectoderm.

### One Example of Gastrulation: The Sea Urchin

Let's now look at the process of gastrulation in one organism, the sea urchin, chosen because its gastrulation is so easy to visualize (**Figure 31.2**). By the time it is ready for gastrulation, the sea urchin blastula amounts to a single layer of about a thousand cells arranged in a hollow ball, the interior of which is the liquid-filled blastocoel. Gastrulation begins at the blastula's vegetal pole. Some cells at the pole, called *mesenchyme cells*, detach from the ring of blastula cells and move into

the blastocoel, and with their loss, the vegetal pole begins to pinch inward. This process, called invagination, eventually will carry the vegetal pole as much as halfway toward the blastula's animal pole. A primitive gut, called the archenteron, starts forming around this indentation. Cells growing at the tip of the archenteron begin to send out slender extensions of themselves that adhere to selected cells that lie "above" them on the rim of the blastocoel. These extensions then contract, effectively pulling the archenteron toward the animal pole and eventually moving the archenteron all the way to the pole itself.

| Table 31.1 Gastrulation's Three Layers of Cells And Some of the Structures They Give Rise to in Mammals | |
|---|---|
| **Cell layer** | **Organs or types of tissue it gives rise to** |
| Endoderm | Lungs, liver, pancreas, lining of digestive tract, lining of bladder, and several glands |
| Mesoderm | Much of the body mass of more complex animals, including most of the muscle and skeletal systems; blood, blood vessels, and bone marrow; the uterus and the sperm- and egg-producing gonads; the kidneys; and the cortex of the adrenal gland |
| Ectoderm | Entire nervous system; lenses of eyes, tooth enamel, and the outer or epidermal layer of skin |

## Gastrulation

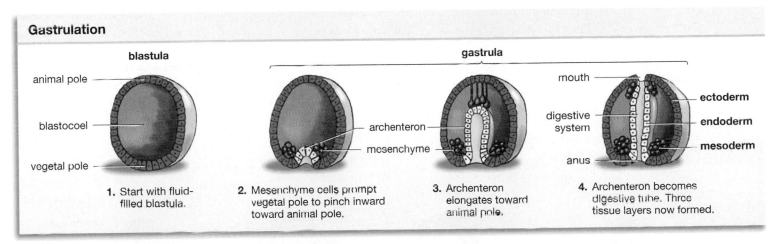

1. Start with fluid-filled blastula.
2. Mesenchyme cells prompt vegetal pole to pinch inward toward animal pole.
3. Archenteron elongates toward animal pole.
4. Archenteron becomes digestive tube. Three tissue layers now formed.

**Figure 31.2**
**Gastrulation in the Sea Urchin**

**1.** The sea urchin blastula amounts to a few thousand cells, arranged in a single layer around the liquid-filled blastocoel.

**2.** Mesenchyme cells at the blastula's vegetal pole move into the blastocoel, prompting the vegetal pole to pinch inward, moving toward the animal pole. The structure that starts forming from this indentation is the archenteron, or primitive gut.

**3.** Mesenchyme cells, growing at the tip of the elongating archenteron, send out slender extensions of themselves that adhere to cells on the rim of the blastocoel. These extensions contract, moving the archenteron to the animal pole.

**4.** The mouth of the sea urchin forms where the archenteron contacts the blastocoel lining. A digestive tube then runs from this mouth to the rim of the vegetal pole, which develops into an anus. These cells constitute the inner layer or endoderm of the sea urchin. Mesenchyme cells move back down into the blastocoel, forming the middle layer or mesoderm of the sea urchin. The outer cells of the blastula form the sea urchin's outer layer or ectoderm.

Now, how does all of this yield the three layers of cells we just talked about? When the archenteron reaches the rim of the blastocoel, a fusion of the two structures takes place that initiates formation of the mouth of the sea urchin. A tube then develops from the mouth that runs the length of the sea urchin, right down to the original vegetal pole. There, the rim of the invagination you saw earlier will develop into an anus. As you might guess, this entire feeding and digestive apparatus is the inner cell layer or endoderm of the sea urchin. The cells whose slender extensions reached the blastocoel are also mesenchyme cells; once they get the archenteron to the top of the blastocoel, they break away from the tip of the archenteron and then move back down into the blastocoel cavity. There they will form the mesoderm, or middle layer, of the sea urchin—eventually its skeleton and muscles. Finally, the outer cells of the blastula form the outer cell layer or ectoderm of the organism, eventually becoming its skin and related organs.

### Third Phase: Organogenesis

In vertebrates, an interplay of two of these three germ layers marks the first steps on the path to the development of organs—in this case, organs of the nervous system.

In talking about development in any organism, an important concept is *body planes*—imagined divisions that pass through the center of an animal, dividing it into two parts of roughly equal size. In **Figure 31.3**, step 1, you can see that a frog embryo has been divided with two such planes. The first yields dorsal and ventral portions of the embryo; the second yields anterior and posterior ends. In the case

of a frog, dorsal eventually will mean its back, ventral its belly; anterior will mean its head and posterior its hindquarters.

Looking at steps 2 and 3 in Figure 31.3, you can see what takes place toward the frog embryo's dorsal surface as gastrulation is nearing completion. First, there is the development of a rod-shaped support organ found in all embryonic vertebrates, the **notochord**. In addition to support, one of the notochord's functions is to *induce*, or bring about, development in the ectodermal tissue that lies above it, toward the dorsal surface. This development begins with an elongation of the ectodermal cells to form a flattened surface called the *neural plate*, which curls up and eventually folds all the way over on itself to form an enclosed, hollow tube that will lie below the dorsal surface. Developing from anterior to posterior, this **neural tube** is a dorsal, ectodermal structure that gives rise in vertebrates to both the brain and spinal cord.

Equally important are **neural crest cells**: cells that break away from the top of the vertebrate neural tube as it folds together and then migrate to varying parts of the embryo, giving rise to various tissues and organs. For example, the medulla or inner portion of the adrenal glands (which in humans sit atop the kidneys) is derived from neural crest cells. You can see the routes of these migrations and some of their outcomes in **Figure 31.4**. One consequence of a failure of neural crest-derived tissue to develop properly is a gap that occasionally occurs between the nose and mouth in humans, which is known as a cleft lip.

As this ectodermal movement is taking place, mesodermal tissue on both sides of the notochord is

**Figure 31.3**
**Body Planes in a Frog and Formation of the Frog Central Nervous System**

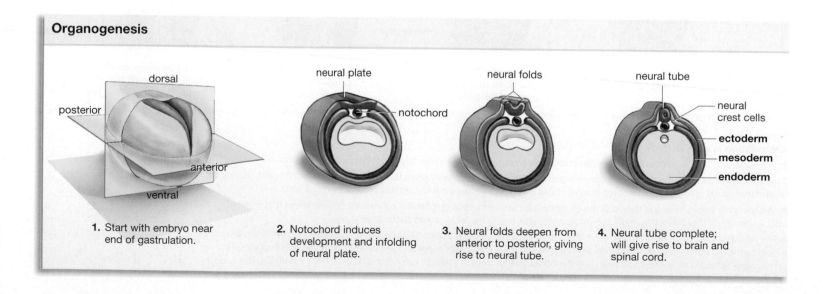

## Organogenesis

1. Start with embryo near end of gastrulation.

2. Notochord induces development and infolding of neural plate.

3. Neural folds deepen from anterior to posterior, giving rise to neural tube.

4. Neural tube complete; will give rise to brain and spinal cord.

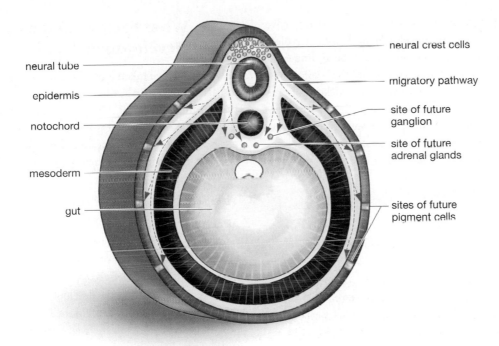

neural crest cells

neural tube

migratory pathway

epidermis

site of future ganglion

notochord

site of future adrenal glands

mesoderm

gut

sites of future pigment cells

**Figure 31.4**
**Neural Crest Migration in a Frog Embryo**
This cross section through a frog embryo shows some of the pathways of neural crest cells as they migrate to various locations, prompting the development of different tissues and organs.

developing into repeating blocks of tissue, called **somites** (**Figure 31.5**). A good part of the human body is built of repeating units—think of each of the vertebra that make up our spinal column, the muscles that attach to each of these vertebrae, and the ribs that extend from the vertebrae. All these structures develop at least in part from the somites.

## Themes in Development: From General to Specific; Retention of Structures and Processes

As complicated as all this may seem, it amounts to a brief look at important, but early, periods of development in animals. Looking at the larger picture, notice that in the early stages of development there is a movement from the general to the specific. The animal goes from a small group of superficially similar cells, to three *layers* of cells, to tissues that proceed from these layers (notochord, somites), to actual organs formed from these tissues. To put a little finer point on this, think of the somites. They arise from undifferentiated mesodermal tissue on each side of the notochord, and they take shape as about 40 blocks of mesodermal tissue. These blocks then go on to give rise to several structures that retain the somites' repeating nature but that are much more specialized (vertebrae, ribs).

It's also worth noting that the three most basic developmental processes we looked at—cleavage, gastrulation, and organogenesis—are processes that take place in all animals, not just a selected few. In a similar vein, if we were to look in detail at nervous system development in vertebrates as different as frogs, chickens, and human beings, we would see the

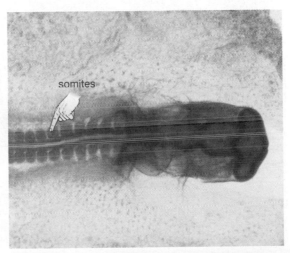

somites

**Figure 31.5**
**Somites**
These regular, repeating blocks of mesodermal tissue, visible in this case on either side of the notochord of a chick embryo, will give rise to muscles and the vertebrae that enclose the spinal cord.

same sequence of events taking place: the supportive notochord inducing development of a neural plate that folds over on itself, eventually becoming both spinal cord and brain. It is often said that evolution is a "tinkerer" that retains what has worked for one species and then modifies it only slightly for later-evolving species. In development, we can see clear examples of just how much is retained from one species to the next (**Figure 31.6**).

---

**SO FAR...**

1. Sperm and egg fuse to produce a cell called a _____. Once this cell starts dividing, it is referred to as an _____.

2. In animal development, the process of cleavage produces a ball of cells with an interior cavity. In the succeeding developmental process, called gastrulation, the cells in the embryo migrate to form three layers of cells, the _____, _____, and _____, which in terms of location are the _____, _____, and _____ layers of cells in the organism.

3. In vertebrates, a support structure called a notochord serves to _____ or bring about the development of a structure called the neural tube, which gives rise to both the _____ and _____.

---

## 31.2 What Factors Underlie Development?

Now that you've taken a brief look at what happens in early development, the next question is: *How* do these things happen? How do vegetal pole cells "know" they are vegetal pole cells? How do neural crest cells take different routes through an embryo and end up serving as different kinds of cells?

### The Process of Induction

As noted, the notochord *induces* or prompts the development of the ectodermal tissue that lies above it. Such **induction** is an important process in development in general. It can be defined as the capacity of some embryonic cells to direct the development of other embryonic cells. If you look at **Figure 31.7**, you can see how induction works in the well-studied example of chick wings. In essence, there is a relatively small group of cells that gives out information for the development of a portion of the wing. Researchers learned this by working with the "limb bud" that is transformed into the chick wing. They took cells from a posterior "zone of polarizing activity" (ZPA) in one limb bud and transplanted them to the *anterior* portion of a second limb bud. The result was activity in this bud by both the natural and transplanted ZPA; this resulted in two sets

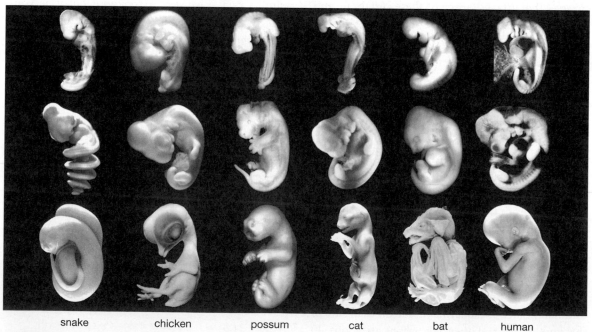

snake   chicken   possum   cat   bat   human

**Figure 31.6**
**Embryonic Development in Six Vertebrates**

Pictured are six vertebrates in early, middle, and late stages of their development. Note the general similarity in vertebrate shape in the early stage of development, which roughly corresponds to the period following somite formation. The embryos are not to scale—in reality some are bigger than others. Photo courtesy of Michael K. Richardson. Human embryo from R. O'Rahilly.

of developing digits in wings, spreading out like two sets of human fingers developing from joined hands. In short, the cells in the ZPA induced the development of other cells, sending out a message that said, in effect, "begin digit development here."

## The Interaction of Genes and Proteins

This induction process raises the question of what lies at the root of development. There are lots of cells in an embryo. How can some of them (such as the ZPA cells) control the development of others? It turns out that, at its most basic level, development is controlled by the interaction of genes and proteins—

the genetic regulation reviewed in Chapter 14. Biologists' understanding of how this interaction works in development has grown dramatically in the period from the early 1980s to the present. Although there have been several important "model" organisms in this research, the most important of them has been the tiny fruit fly, *Drosophila melanogaster*. Let's look at a little of what the *Drosophila* research has uncovered.

## Three Lessons in One Gene

There is a *Drosophila* gene, called *bicoid*, that codes for a protein that is also called bicoid. This single

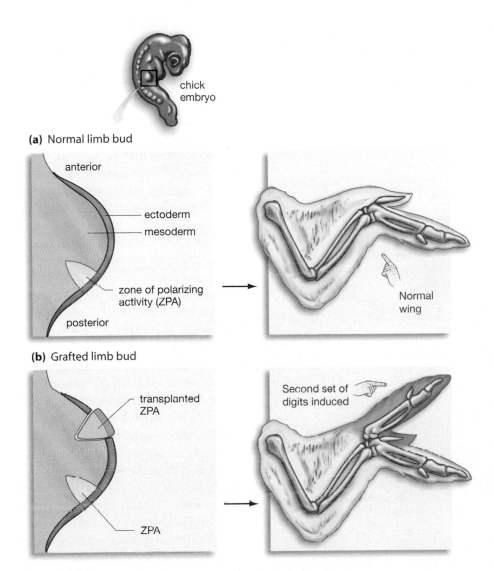

**Figure 31.7**
**Cells Can Induce Development in Other Cells**
Transplantation experiments with chick embryos revealed that a group of cells on the posterior portion of the chick's limb bud give out positional information for the development of digits in the chick wing.
**(a)** When cells from a zone of polarizing activity (ZPA) were left in their normal posterior location, the chick wing developed as usual.
**(b)** When limb bud ZPA cells were transplanted to the anterior portion of a second limb bud, the result was mirror images of two sets of developing posterior wing digits, much like two sets of human fingers extending from joined hands.

gene offers a clear example of three principles that operate in development:

- A critical early step in development is the establishment of positional information in the embryo.

- Substances called morphogens are important in establishing this positional information.

- Development operates from the general to the specific through a genetic cascade.

Look at the unfertilized *Drosophila* egg pictured in **Figure 31.8** to see how these principles play out with respect to the bicoid protein. Recall first the body planes described earlier. In Figure 31.8a, the top of the egg is its dorsal portion and the bottom its ventral portion, while the left side corresponds to its anterior portion and the right side to its posterior. Eventually, the anterior portion will be the head of the fly, the posterior its hindquarters, the dorsal portion its back, and the ventral portion its belly.

While the egg is still in the mother, a group of "nurse cells" surrounds its anterior portion and pumps molecular products into it. These products include bicoid messenger RNAs (mRNAs)—genetic instructions copied from the *bicoid* gene. Eventually, these mRNA instructions will lead to the bicoid protein. Looking at the figure, you can see that the *bicoid* mRNAs are found at the anterior end of the egg. These mRNAs eventually start to diffuse through the egg, but because they are deposited only at the anterior end, they will exist in a *concentration gradient*, with the greatest concentration being at the anterior end, a lesser concentration toward the center, and none at the posterior end. This leads to the concentration gradient of the bicoid protein you can see in Figure 31.8b.

## Positional Information

With this general background, you can now see how *bicoid* exemplifies one of the principles of development noted earlier. What does the bicoid protein do? It imparts positional information to the embryo and is critical in setting in motion the development of structures in the anterior and center portions of the embryo. Recall that the far-anterior portion of the egg eventually becomes the head of the *Drosophila* fly. In experiments, however, scientists have taken bicoid-laden cytoplasm from the anterior portion of one egg, placed it at the *center* of an egg with no bicoid in it, and watched as the head and related structures developed at the center. Meanwhile, an egg with no bicoid protein gives rise to a fly that never develops a head, although its posterior structures start to develop normally.

## Morphogens

The bicoid protein is a **morphogen**: a diffusable substance whose local concentration affects the course of local development in an organism. Several other morphogens are operating in *Drosophila* in early development. Indeed, several positional "systems" are at work in the fly. What's important to note is that, with these systems, a three-dimensional grid is set up in which the destiny of various cells is determined by their position within this grid. How? Their exposure to different morphogens varies according to their position.

But what is it that morphogens are doing to affect cells? You may remember learning in Chapter 14 about proteins that feed back on DNA, binding with it and thus controlling its transcription, which is to say turning its genes on or off. These DNA-regulating proteins are called **transcription factors**. Morphogens are cell-signaling molecules that bring about the production of transcription factors. Diffusing out from one group of cells, morphogens

**(a)** Bicoid mRNAs are deposited in unfertilized egg.

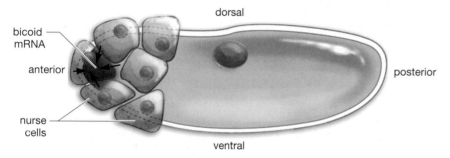

**(b)** Bicoid proteins diffuse through embryo.

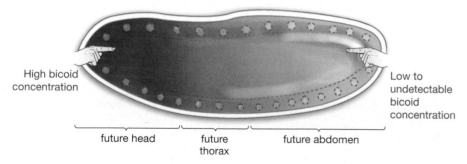

**Figure 31.8**
***Drosophila* Egg and Bicoid Gradient**
A fly begins to develop.

**(a)** Nurse cells surround the anterior portion of an unfertilized *Drosophila* egg. The cells pump messenger RNA molecules for the bicoid protein into the egg.

**(b)** The messenger RNAs are translated into bicoid proteins that diffuse through the now-fertilized egg. These proteins exist, however, in a concentration gradient—in high concentration at the anterior end of the embryo but in low-to-undetectable concentration at the posterior end. The bicoid protein is a morphogen that imparts information regarding development of anterior and center-position structures in the fly. An egg without bicoid in it gives rise to a fly that never develops a head.

provide a signal to other cells to start producing transcription factors.

We can see an example of this effect in the chick wing development noted earlier. Recall that ZPA cells in the limb buds were able to induce the development of digits in the chick wing. ZPA cells can play their controlling role because they produce a morphogen, called *sonic hedgehog*, that acts on surrounding cells: It diffuses to them and prompts them to begin producing transcription factors. Moreover, like all morphogens, sonic hedgehog is having its effects through a concentration gradient. The order of the digits in the chick wing—the equivalent of the order of our digits from thumb to little finger—is specified by how much sonic hedgehog the receiving cells are exposed to. The chick's posterior digit (the equivalent of our little finger) forms from cells closest to the ZPA—the cells exposed to the greatest amount of sonic hedgehog. Farther out from the ZPA the concentration of sonic hedgehog lessens, and the result is cells that become the chick's *anterior* digit (our thumb). In a similar way, the placement of the head and center structures of *Drosophila* is determined by the concentration gradient of the bicoid morphogen. Bicoid exists in greatest concentration at the anterior portion of the embryo, and this is where the fly's head develops; it exists in lesser concentration in the middle of the fly, and this is where its center-portion structures develop.

**The Genetic Cascade**

The developmental effects we've been reviewing work through what is sometimes called a *genetic cascade*. Like water flowing from one level to the next, gene products flow from one developmental level to the next. An initial set of genes is turned on, perhaps by getting a signal from a morphogen, and in time these genes regulate *other* genes, which in turn may regulate still more genes. This explains how development moves from the general to the specific. In *Drosophila*, for example, the first level of genes specifies broad positional patterns across the embryo—anterior and posterior, dorsal and ventral. Then, through the cascade, this positioning gets finer grained; now, other genes are switched on, and development is being specified *within* sections of the embryo. Finally, control is exercised over the development of specific body *structures*, such as wings or antennae. It's worth noting that the genetic cascade turns genes off as well as on. To take one example, once somites have given rise to ribs, vertebrae, and so forth, the genes for building somites will be turned off permanently.

## 31.3 Unity in Development: Homeobox Genes

Recall how you saw earlier that species as diverse as fruit flies, frogs, and human beings have similar processes going on in their development. Now that you have some sense of how genes control development, you won't be surprised to learn that it is similar developmental genes that underlie these similar processes. The remarkable thing, however, is how similar these genes are across species, how widely they are distributed across the living world, and how long they have existed. As scientists would put it, these genes have been highly conserved—portions of them have been passed on intact, like prized bits of knowledge, down long evolutionary lines.

The best example of this conservation comes with a group of genes whose similarity was discovered in *Drosophila* in the 1980s. Scientists were then tracing development in *Drosophila* by mutating the developmental genes of one generation of flies and then watching to see what effect this had on the next generation. Some of the changes they got by doing this were dramatic. One mutation resulted in flies that had legs where their eyes were supposed to be. What the scientists learned from this exercise was that gene X had developmental effect Y.

In carrying out this work, the scientists were sequencing these genes—they were learning the order of their building-block As, Ts, Cs, and Gs. As they did this, they found that every one of these developmental genes had within it an identical sequence of about 180 letters (or base pairs). The real shock came, however, when this so-called homeobox sequence was found to exist in all animals studied—in frogs, humans, and mice, for example, as well as flies. This was surprising because mice and flies went down separate evolutionary paths about 550 million years ago. The implication was that a common ancestor of both animals (probably a variety of worm) possessed homeobox genes at least 550 million years ago. The "at least" comes because versions of homeobox-containing genes have now been found in fungi and in *plants*, which diverged from animals about 1.2 billion years ago.

Not surprisingly, homeobox genes produce transcription factors. One well-studied group of homeobox genes, named *Hox* genes in vertebrates and *HOM* genes in flies, gives out anterior-posterior positioning information—these genes specify where different structures should go along the anterior-posterior axis. (Remember the *Drosophila* flies that got legs where their eyes were supposed to be? This came about because scientists mutated one of their *HOM* genes, thus altering their legs' positioning.) So conserved are the *HOM/Hox* genes that they not only have the same general function in humans and in flies but are arranged in the same order along their respective chromosomes. Even more surprising, some homeobox-containing genes are partly interchangeable across species. Scientists have spliced into the *Drosophila* genome a mouse gene that codes for eye development in the mouse. They then watched as this gene brought about normal eye development in the *fly*. Not mouse eyes in the fly, mind you—fly eyes.

What all this means is that there is a general set of genetic instructions for development that are so important they have changed very little over time and across species. Fly eyes and mouse eyes are structurally quite different, but there is a general set of genetic instructions for "eyes" that seems to trigger the development of *specific* kinds of eyes in specific animals.

Spurred on by these kinds of findings, scientists today are busy studying how homeobox genes and their proteins differ among the species. This is important work because it stands to tell us a great deal not only about development, but about evolution. How does one species evolve from another—how does it branch off? Developmental genes often are involved for the simple reason that one organism will not physically differ much from another without a change in these genes. Insects have six legs, rather than the multitude that many of their fellow arthropods have, because a homeobox gene in one of their ancestors underwent a mutation. In its mutated state, this gene produced a protein that turned *off* another gene—a gene that brings about multiple leg formation. By tracking developmental gene differences in

modern animals, scientists can learn how these various animals evolved in the first place.

## 31.4 Developmental Tools: Sculpting the Body

Genes may lie at the root of development, but the shaping of an organism takes place through genetic instructions that prompt whole cells to take actions. You may wonder how a hand, for example, can be shaped from a seemingly formless mass of cells. The answer lies in capabilities cells have that allow for this sort of development. Three examples of these capabilities are cell movement, cell adhesion, and cell death.

Earlier, you saw examples of cell movement and cell adhesion when reviewing the process of gastrulation in the sea urchin. Recall that cells of the primitive gut were able to move up into the blastocoel, after which cells peripheral to them sent out slender extensions of themselves that adhered to selected cells that lay "above" them in the blastocoel. The first part of this process, cell movement, comes about in various ways. Often, cells use the tent-pole-like structures called microfilaments that grow very rapidly at one end of the cell—in the direction of movement—while decomposing at the other.

Cell adhesion is accomplished through several families of proteins, collectively called *cell adhesion molecules*, that protrude from the surface of cells, allowing them to stick selectively onto other cells. Different groups of cells have different adhesion "affinities" for each other—depending on the proteins they put on their surfaces—and this explains how the *layers* of cells that you've seen come about. Imagine two groups of cells, A and B, that are randomly mixed in with one another. Now imagine that, through genetic control, the A cells start producing a surface protein that makes them strongly adhere to other A cells. By process of random cellular movement, the A and B cells will now start to segregate into two separate groups.

Cell death may not sound like a promising means of developing an organism, but consider that a sculptor works not by adding material, but by chipping away at an initial mass of it. Thus it is in development, where cell death works to create spaces that define shapes. An example of this is the human hand, which but for the death of large numbers of cells between future fingers, would look more like the webbed foot of a duck (**Figure 31.9**). Cells that die in this fashion are not attacked by other cells. Instead, they commit suicide in a process known as *apoptosis* or *programmed cell death*. Apoptosis is a very orderly process, akin to a

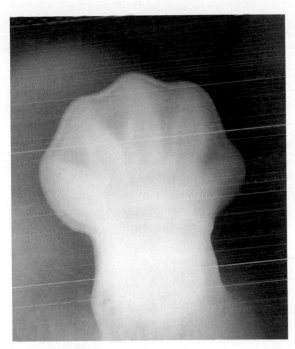

**Figure 31.9**
**Soon to Be Different**
The webbed hand of a human embryo 40 days after conception. The programmed death of cells in the hand will bring about the emergence of five separate fingers.

person dying in a hospital rather than a car crash. The cell undergoing it fades away in a sense; it shuts down its processes, shrinks, and then dies, after which its remains are ingested by neighboring cells.

## 31.5 Development through Life

It is convenient to think of animal development as something that happens before birth. But if this is the case, what are we to call the loss of "baby" teeth in childhood, the onset of puberty in the teenage years, or the cessation of menstruation (menopause) among middle-aged women? Development never stops throughout the life span. In essence, it is governed by the same factors that control the embryonic development you've looked at so far: the interaction of genes and proteins and the influence of environment on them.

## On to Human Reproduction

The steps of development that you have just looked at form a kind of middle phase of the process you'll be looking at next chapter, which is reproduction—specifically human reproduction. Coming before the development you've reviewed are the formation of the egg and sperm that come together to create an embryo. Coming after it are late-stage human embryonic growth and then birth itself. With this chapter, you know something about development; in Chapter 32, you'll see how development fits into the larger scheme of human reproduction.

**SO FAR...**

1. The developmental genes called homeobox genes have been highly _____, meaning they have remained relatively unchanged though the course of evolution. Homeobox genes produce _____.

2. Three tools that work in development across animal species are cell _____, cell _____, and programmed cell _____.

3. Development does not stop at birth but continues through _____.

**••• ANSWERS**

1. conserved; transcription factors

2. movement; adhesion; death

3. the human life span

# CHAPTER 31 REVIEW

For study help, activities, and more quiz questions, visit **www.aw.com/krogh4**.

## SUMMARY

### 31.1 General Processes in Development

- Development in animals begins with the formation of a zygote or fertilized egg, created when the male sperm fuses with the female egg. Once a zygote starts to divide, it is called an embryo.

- There are three phases of early embryonic development. The first phase is cleavage, which produces a ball of cells, called a blastula, with an interior, liquid-filled cavity. The second is gastrulation, in which the blastula's cells rearrange themselves into three layers that give rise to specific organs and tissues. The third is organogenesis, in which organs begin to form.

- The first product of cleavage is a tightly packed ball of cells called a morula. The cells in the morula rearrange themselves into a structure composed of one or more layers of cells surrounding a liquid-filled cavity. This structure is the blastula; its interior cavity is called the blastocoel.

- Gastrulation produces an embryo composed of three layers of cells: the endoderm, the mesoderm, and the ectoderm. The endoderm gives rise to interior tissues (the liver and bladder, for example), the mesoderm to tissues exterior to these (muscle and skeletal systems), and the ectoderm to tissues that are more exterior yet (the skin and nervous system).

- In early organogenesis in vertebrates, a rod-shaped support organ, called a notochord, develops near the dorsal surface and induces the development of ectodermal tissue that lies above it. A neural plate develops from this tissue, folding over on itself to form a hollow neural tube that gives rise to both the brain and the spinal cord. Groups of neural crest cells break away from the top of the neural tube during its development and then migrate to different locations throughout the embryo. These groups go on to develop into different organs or tissues.

- The early stages of development represent a transition from the general to the specific. A small group of superficially similar cells gives rise to three different layers of cells, which give rise to tissues, which in turn give rise to specific organs.

  **Web Animation 31.1:** Embryonic Development

### 31.2 What Factors Underlie Development?

- Induction is a process in which some embryonic cells direct the development of other embryonic cells.

- Development is fundamentally controlled by the interaction of genes and proteins. A key model organism in scientific understanding of these interactions is the fruit fly *Drosophila melanogaster*.

- A critical early step in development is the establishment of positional information—for example, the establishment of anterior and posterior in a growing embryo. Positional information in *Drosophila* is provided in part by a protein called bicoid that exists in a concentration gradient in the fly embryo, the greatest concentration being at the anterior end of the embryo, a lesser concentration toward the center, and none at the posterior end. Bicoid helps bring about anterior structures, such as the head of the fly and center-portion structures as well.

- Bicoid is an example of a morphogen: a diffusable substance whose local concentration affects the course of local development in an organism. In general, morphogens diffuse from one set of cells to a second set of cells. Binding with these latter cells, morphogens prompt them to begin producing transcription factors: proteins that regulate DNA by binding with it, thus turning selected genes on or off. All morphogens work through the kind of concentration gradient seen with bicoid. Exposure to a greater concentration of a morphogen brings about one effect (a head in *Drosophila*), while exposure to a lesser concentration brings about another effect (center-portion structures in the *Drosophila*).

- Development works through a genetic cascade in which one set of genes regulates a subsequent set of genes. This cascade explains how development proceeds from the general to the specific. A first level of genetic instruction specifies broad positional patterns in an embryo (for example, the head and center-section portions of *Drosophila*). Subsequent levels of genetic instruction specify development within these body sections. Finally, control is exercised over the development of specific structures, such as organs.

### 31.3 Unity in Development: Homeobox Genes

- Within many of the developmental genes of animals as different as flies, mice, and human beings, there exists an identical sequence of about 180 DNA bases. Called a homeobox, this sequence has been highly conserved—it has remained largely unchanged over hundreds of millions of years of evolution in all animal species studied. As a result of this conservation, similar basic developmental processes are at work in all animals and possibly in fungi and plants as well. Scientists are now looking at the differences in homeobox genes and their proteins as a means of learning more about both animal development and animal evolution.

### 31.4 Developmental Tools: Sculpting the Body

- Three cell capabilities help shape the animal body in development. The first is cell movement. The second is cell adhesion, in which cells produce proteins, which protrude from their surface, that selectively adhere to other cells. The third is programmed cell death, or apoptosis, in which spaces in tissues are created by means of cells dying.

### 31.5 Development through Life

• Development is sometimes thought of as ending at birth, but it continues through the life span of an organism. It is exemplified in human beings in such processes as puberty and menopause.

## KEY TERMS

| | |
|---|---|
| **animal pole**  654 | **morula**  654 |
| **blastocyst**  654 | **neural crest cells**  656 |
| **blastula**  654 | **neural tube**  656 |
| **cleavage**  654 | **notochord**  656 |
| **embryo**  653 | **somite**  657 |
| **fertilization**  653 | **transcription factor**  660 |
| **gastrulation**  655 | **vegetal pole**  654 |
| **induction**  658 | **zygote**  653 |
| **morphogen**  660 | |

## BRIEF REVIEW

*Answers to Brief Review questions are in the back of the book.*
*For multiple-choice quiz questions, go to* **www.aw.com/krogh4**.

1. Briefly describe the order of events that occur after fertilization to produce an embryo with three layers of cells.

2. What is a morphogen concentration gradient, and how does it affect early development of the fruit fly, *Drosophila*?

3. Outline the steps involved in the production of the chick wing.

4. The same homeobox DNA sequence is shared in the developmental genes of mice and fruit flies, even though these two animals diverged from each other evolutionarily about 550 million years ago. What does this high level of conservation tells us about the homeobox sequence?

5. How do cell adhesion molecules influence early events during development?

6. Describe the role cell death plays during the development of an organism.

## APPLYING YOUR KNOWLEDGE

1. Why would most major evolutionary changes involve a change in developmental genes?

2. It is sometimes said that the most important event in a human life is gastrulation. Why should gastrulation be thought of in this way?

3. Explain how the bicoid protein acts as a morphogen. What would happen to the developing *Drosophila* embryo if bicoid mRNA was introduced into the posterior region of the egg prior to fertilization?

*Baby makes three at the end of a long and complicated process.*

**32.1**    **Overview of Human Reproduction and Development**    **667**

**32.2**    **The Female Reproductive System**    **669**

**32.3**    **The Male Reproductive System**    **675**

**32.4**    **The Union of Sperm and Egg**    **678**

**32.5**    **Human Development Prior to Birth**    **679**

**32.6**    **The Birth of the Baby**    **685**

**Essays**

Hormones and the Female Reproductive Cycle    **672**

Methods of Contraception    **677**

Sexually Transmitted Disease    **682**

To produce a child, human males and females must first produce sperm and eggs, which come together in an act of fusion known as conception. The fertilized egg that results will, over the next 38 weeks, develop into a fully formed baby.

# HOW THE BABY CAME TO BE:
# Human Reproduction

**M**emorable events mark every human life, but for parents, one event that is sure to be indelibly etched into memory is the birth of a child. The mother's labor, the final "pushing" that brings the baby forth,

the first real look at the infant, and the infant's first look at the world: these are moments that, once experienced, are never forgotten.

The birth of a child is a beginning, of course, and yet it is also a culmination. Human beings must grow to a certain level of biological maturity before being capable of reproduction. After this, finding a partner may be difficult and getting pregnant more difficult yet. Eventually, however, two very special "seeds"—*this* egg and *this* sperm, out of all the millions a man and a woman produce in a lifetime, fuse to create this child.

In this chapter, you'll trace the biology of human reproduction, which might be defined as all the biological steps that must fall into place before a child can be born. The process of development you studied in Chapter 31 represents part of what you'll be looking at, but other elements are involved as well.

## 32.1 Overview of Human Reproduction and Development

Taking the story of reproduction from the beginning means picking up a narrative that was begun in Chapter 10, when you went over the formation of the reproductive cells called *gametes*—eggs in females, sperm in males. In this chapter, the focus widens,

taking into account the physical setting in which egg and sperm are formed and the steps that bring these cells together. Finally, you will look at the human development that takes place once egg and sperm have merged. For information on some of the methods humans use to *prevent* conception, see "Methods of Contraception" on page 677.

### Reproduction in Outline

The first part of reproduction, the formation and delivery of egg and sperm, is a straightforward story in outline. The eggs a woman possesses exist in her in a precursor form, called oocytes, that lie within two walnut-shaped structures called ovaries. These exist just right and left of center within the female pelvic area (**Figure 32.1** on the next page). Each oocyte sits in the ovaries within a nurturing complex of cells and fluids called a **follicle**. On average, once every 28 days one follicle-oocyte complex is brought to a state of maturity such that the oocyte and some accessory cells that surround it rupture from their ovary and begin a slow journey to the uterus. This trip takes place inside the structure called the **uterine tube** (or Fallopian tube). The release of the oocyte is called **ovulation**. If, during this trip, the oocyte encounters a male sperm, the two may fuse, in which case conception has occurred.

Sperm in men are produced, in unfinished form, in sets of tubules in the two male testes (**Figure 32.2**). They then are transported to a structure called the epididymis (one for each testis) for further development. Then, with sexual excitation, sperm are transported in a loop: up and over the urinary bladder through two ducts, each called a vas deferens. The contents of these ducts then empty into a single duct, called the urethra, from which they are ejaculated into the female vagina. Along the way, materials secreted by several glands join the sperm in a process that can be likened to tributaries feeding into a common stream. The resulting mixture of sperm and glandular materials is called **semen**.

Whereas the female releases an average of one oocyte per month, the male releases an average of

**Figure 32.1**
**Reproductive Anatomy of the Human Female**

Eggs develop in the ovaries within nurturing complexes called follicles. Once every 28 days, on average, an egg is expelled from one ovary or the other and journeys through the uterine tube to the uterus. If the egg encounters sperm on the way, pregnancy will result when egg and sperm fuse. The fertilized egg then continues traveling through the uterine tube and implants in the lining of the uterus, the endometrium.

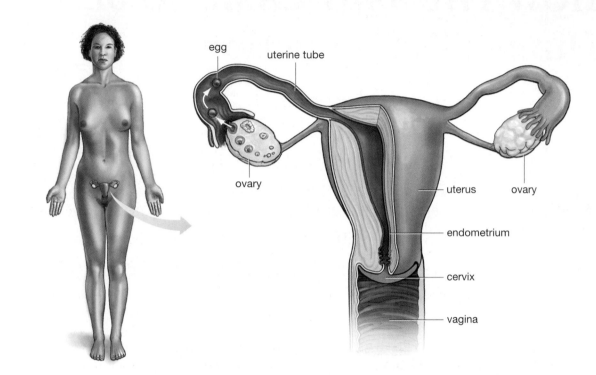

**Figure 32.2**
**Reproductive Anatomy of the Human Male**

Sperm are produced in the testes and transported to the epididymis, where they mature and are stored. With orgasm, the sperm are transported through the vas deferens and then the urethra, from which they are ejaculated. Along the way, they are joined by materials secreted by several accessory glands.

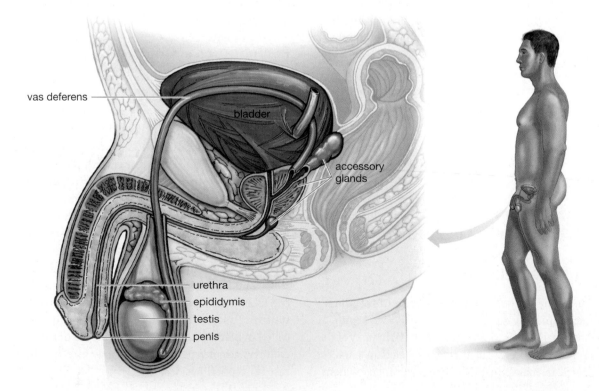

200 million sperm per ejaculation. Of the millions of sperm that begin the journey, a few dozen may encounter the female oocyte as it continues its journey from ovary to uterus (**Figure 32.3**). The first sperm to breach the coating that surrounds the oocyte fuses with it, while the rest are shut out through a kind of gate-slamming mechanism. When this one sperm has joined with an oocyte, conception has occurred. The now-fertilized egg still floats free in the uterine tube, where it begins to develop in the ways described in Chapter 31: One cell becomes two, two become four, and so on. Four days after fertilization, the developing embryo enters the uterus; 3 days more, and it implants itself in the uterine wall and continues to develop.

Both the male and female reproductive systems are greatly influenced by the actions of hormones—chemical signaling molecules that move through the bloodstream. Indeed, the reproductive organs are primary sites of production for some of the most important of these hormones, testosterone in men and estrogen and progesterone in women.

With this outline of human reproduction as background, let's move on now to take a closer look at both the female and male reproductive systems, first as they function separately, and then function together to produce a baby.

## 32.2 The Female Reproductive System

### The Female Reproductive Cycle

Although hormones operate in both sexes, in females they bring about a reproductive *cycle* that repeats itself about once every 28 days (although it can range from 21 to 42 days). In males, conversely, there is no cycle, but instead a steady production of sperm and a destruction or removal of the sperm that are not ejaculated. Why a cycle for females and not for males? The answer lies partly in the two roles the female reproductive system must fulfill. It must form an egg *and* create an environment in the uterus in which a fertilized egg can develop. An important part of this uterine environment is essentially built up and then dismantled once every 28 days—except in those rare instances in which an oocyte is fertilized. The end result of the dismantling of the uterine environment is **menstruation:** a cyclical release of blood and the inner portion of a specialized uterine tissue, the **endometrium,** from the female reproductive tract. The endometrium is tissue that lines the interior of the uterus; it is tissue

**Figure 32.3**
**Oocyte and Sperm**
Numerous sperm surround a single oocyte.

that *would have* housed an embryo had one been implanted there (see Figure 32.1).

### Why Does Menstruation Exist?

From this, it's reasonable to ask: Why do women have this monthly cycle of buildup and destruction rather than just a permanent environment for an embryo? After all, menstruation has some obvious costs attached to it; apart from discomfort, there is a loss of blood and other tissue. (On average, women lose about 40 milliliters—or about 1.4 fluid ounces—of blood in a menstrual cycle and about that much other tissue as well.) So, in the face of these costs, why has menstruation been preserved throughout human evolution? Scientists have no firm answers here, but they do have a couple of favored hypotheses. One is that, costly as menstruation may be, in terms of expended energy it is not *as* costly as the alternative of maintaining a permanent embryonic environment in the uterus. The other has to do with a change in form that the endometrium undergoes each month. As the time approaches in which an oocyte might get fertilized and then implanted, the endometrium goes from being a tissue whose primary activity is growing (in its "proliferative" phase) to one whose primary activity is secreting a nutrient-rich mucus (in its "secretory" phase). The assumption is that a fertilized egg can implant itself only in a secretory-phase endometrium, but that a secretory-phase endometrium cannot perpetuate itself—it must develop from a proliferative-phase endometrium, after which it cannot stay in secretory phase for long. Given this, it makes sense that the endometrium washes a part of itself out every 28 days and starts its growth anew.

## How Does an Egg Develop?

As noted earlier, eggs develop from precursor cells called **oocytes**. If you look at **Figure 32.4**, you can track the process by which oocytes develop in the outer portion of each ovary. Any oocyte that moves very far through the process of development ends up being surrounded by a sphere of accessory cells, called *follicle cells*, that eventually provide the oocyte with nutrients. The complex of an oocyte and its accessory cells and fluids is called an **ovarian follicle**; it is the basic unit of oocyte development.

### Follicle Development and Ovulation

Every oocyte that matures into an egg must go through the developmental steps you see in Figure 32.4. This is a monthly process of selection, running from a large starting set of follicles, through fewer *primary follicles* that develop from them, to a greatly reduced number of *secondary follicles* that continue development, and finally to a single *tertiary follicle* that nurtures an oocyte through its time in the ovary. Responding to hormonal influence midway through the menstrual cycle, the

tertiary follicle ruptures, thus expelling the oocyte and some surrounding accessory cells not only from the follicle but from the ovary as well. With this event, ovulation has occurred. It takes place in this way in one ovary each month, with one ovary or the other selected through an unknown process.

### Movement through the Uterine Tube

Freed from its ovarian housing but still surrounded by an exterior coat and a collection of accessory cells, the oocyte is now swept into the uterine tube through the actions of the finger-like *fimbriae* lying at the tube's funnel-shaped end (see **Figure 32.5**). Propelled by liquid currents and the uterine tube's whip-like cilia, the oocyte—technically not yet an egg—makes a journey of about 10 centimeters (or 4 inches) through the uterine tube over the next 4 days. While within the tube, the oocyte may encounter sperm and become fertilized prior to entering the uterus.

Meanwhile, the ruptured tertiary follicle now takes on a new role: It develops into a body called the **corpus luteum** ("yellow body"), a structure that

**Figure 32.4**
**Steps Leading to an Egg**

An egg develops from precursor cells called oocytes that exist near the surface of the ovary. Oocytes join with nurturing complexes of cells and fluids to form ovarian follicles. On average, only one of these follicles matures—once every 28 days—through the stages of primary, secondary, and tertiary follicles. The oocyte contained in the tertiary follicle is expelled at the end of the process and then begins a journey through the uterine tube. With the oocyte's expulsion, the tertiary follicle is transformed into the corpus luteum, which for a time produces hormones that facilitate pregnancy.

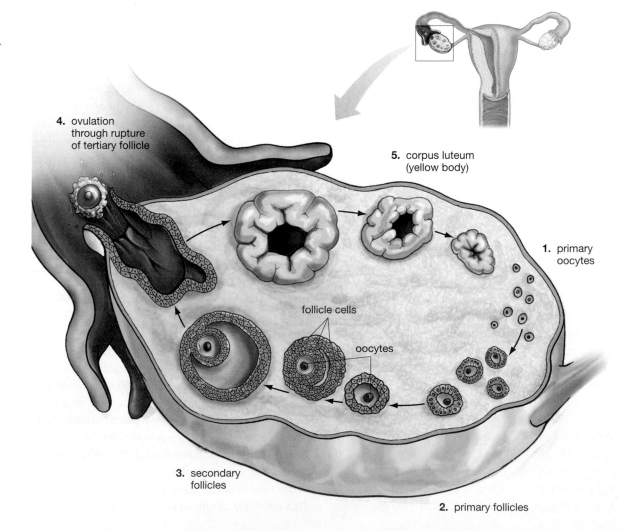

4. ovulation through rupture of tertiary follicle

5. corpus luteum (yellow body)

1. primary oocytes

follicle cells

oocytes

3. secondary follicles

2. primary follicles

**(a)**

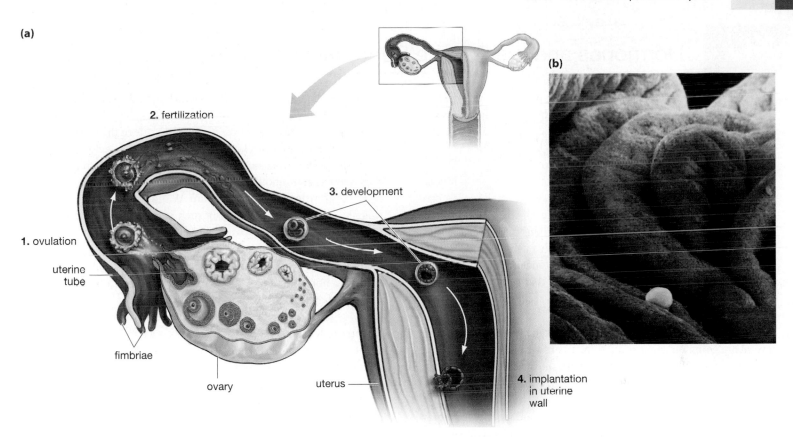

2. fertilization

3. development

1. ovulation

uterine
tube

fimbriae

ovary

uterus

4. implantation
in uterine
wall

**(b)**

secretes hormones that first prepare the female reproductive tract for pregnancy and that then maintain the tract during the early phases of pregnancy. If pregnancy does not occur, the corpus luteum begins degenerating after about 12 days. With the corpus luteum, you can see again the part that hormones play in the monthly female reproductive cycle. For details on this hormonal regulation, see "Hormones and the Female Reproductive Cycle," on page 672.

## Changes through the Female Life Span

In between the creation of oocytes and the journey of one oocyte down the uterine tube lies a multistep process that could be characterized by the phrase: "Many are called, but few are chosen." A 6-month-old female fetus has in its ovaries about 7 million oocytes that have become part of follicle complexes. At birth, however, this number has been reduced to about 800,000, primarily because of the process of *atresia*, or the natural degeneration of follicles. By puberty, the number has declined to about 250,000. A small number of these follicles are "recruited" monthly for further development, but as you've seen, only *one* of these will go on to develop into the oocyte that bursts forth from the ovary each month. The only exception to this rule comes in the 1 percent or so of ovarian cycles that result in *multiple* ovulations, in which case fraternal twins, triplets, and other multiple births are possible. In sum, the oocyte that begins the jour-

ney down the uterine tube each month has been selected from more than 7 million initially created, and it is one of only 500 or so that will make this trip in a woman's lifetime (about 13 a year for 35 or 40 years).

## Follicle Loss and Female Fertility

Given that a woman begins puberty with about 250,000 follicles, you might think that her supply would never run out. This is not the case, however, because the process of atresia continues throughout a woman's lifetime. By the time a woman reaches her early 50s, perhaps a thousand follicles remain; this scarcity is the primary factor that brings about **menopause**, or the cessation of the monthly ovarian cycle.

Taking a step back from this story, you can see that its underlying assumption is that, while a woman will lose follicles over the course of her lifetime, she will not gain any. Indeed, the assumption is that all the follicles a woman will ever have are produced in her prior to her birth. It's clear that men remain fertile throughout their lives because their sperm are produced each day by stem cells, which is to say cells that never lose their ability to produce more sperm. But for decades a bedrock of reproductive science has been that women have no reproductive stem cells. In 2004, however, a research team from Harvard stunned the medical world when it

**Figure 32.5**
**Journey to Pregnancy**

**(a)** Surrounded by accessory cells, an oocyte is expelled from the ovary into the uterine tube, where it encounters sperm and is fertilized by one of them. Now a fertilized egg (or zygote), it begins the process of development and completes its journey with implantation in the endometrium of the uterine wall.

**(b)** A micrograph of an unfertilized oocyte moving through folds of tissue in a uterine tube.

# Hormones and the Female Reproductive Cycle

In females, reproduction is carried out through a *cycle*, which is to say a series of events that repeat themselves regularly—on average, once every 28 days. But, how can the female reproductive system "keep time" like this? What is the clock that governs the cycle?

A real clock is actually a good analogy for what happens. If you look at an old-fashioned mechanical watch, what you'll see is a central spring that exerts a force through a series of gears. Given these parts, it takes 12 hours for the physical mechanism of the "small hand" on the watch to make its way through a complete cycle—to end up where it began. In women, the substances called hormones set in motion a group of physical processes that take about 28 days to complete. The last of these processes leaves the body right where it began, at the start of another cycle.

We actually should speak of female reproductive *cycles*, because several are going on in reproduction, all of them interlinked. There is the ovarian cycle, which has to do with the development and release of oocytes. Then there is the menstrual or "uterine" cycle, which concerns the regular buildup and then breakdown of tissue in the inner lining of the uterus. Finally, there are several hormonal cycles, which help govern both menstruation and egg creation.

If you look at **Figure 1**, you can see a timeline for all these cycles as tracked over a period of 28 days, which you can see marked off at the bottom. To give you a feel for the processes at work, let's follow just the menstrual cycle, whose "first day," over on the far left, is defined as the first day of menstruation. What the graph tracks is the state of the tissue (the endometrium) lining the inside of the uterus. As you can see, beginning on day 1, this tissue goes through a breakdown that lasts about 6 days, after which it begins a steady buildup that continues all the way to about day 21—at which point it starts to break down again. Now look at what is going on in the ovarian cycle (the second one down from the top). This part of the graph is following the single "successful" oocyte (of many that begin development) through its progression into the oocyte that bursts from the ovary on day 14.

The real lesson, however, is to think about the menstrual and ovarian cycles as they relate to *each other*: The endometrium is built up in the uterus as a means of creating an environment in which an egg could implant itself should one be fertilized. If such an egg begins its journey on day 14 and then takes another 7 days or so to implant itself in the uterus, you can see what stage the *endometrium* will be in at this point: near its maximum thickness (and at a highly productive stage of synthesizing a blood vessel

system). Such an environment is just what an implanted embryo would need.

And what of the figures over on the right side of the ovarian cycle? They represent the corpus luteum, the "remains" of the follicle that the successful oocyte broke out of. The corpus luteum is much more than a spent vessel at this point, however. Hormones it secretes will promote the growth of the endometrium. Without a pregnancy occurring, the corpus luteum carries out such secretion only for about a week. After that, it begins to degenerate, and *this* means the degeneration of the endometrium, which washes out, along with some of the blood that supplied it, in the process called menstruation.

Which hormones are secreted by the corpus luteum? Look now at the ovarian hormone cycle, and you'll see a line for "progestins"—a family of hormones whose most important member is progesterone. As you can see, the level of progestins takes a sharp turn upward just as the corpus luteum is formed—no surprise because it is the corpus luteum that is secreting them.

There is a circularity to all this hormonal control, with the hormones involved influencing each other in different ways as the cycle goes on. Let's arbitrarily take as a starting point, however, first the development of a follicle and then the release of an egg from it.

One important hormone in the cycle is produced in neither the ovaries nor the uterus, but in the brain. A brain structure called the hypothalamus releases vanishingly small amounts of a hormone called **gonadotropin-releasing hormone**, whose function is spelled out in its name: It controls the release of *other* hormones active in reproduction, the gonadotropins.

Two gonadotropins are spurred into production in this way, **follicle-stimulating hormone (FSH)** and **luteinizing hormone (LH)**, both of which are produced in a gland located beneath the hypothalamus, the anterior pituitary gland. You can see both FSH and LH levels charted in the gonadotropic hormone cycle at the top of Figure 1.

After being secreted by the anterior pituitary, both FSH and LH have their primary early-cycle effects at the site of the ovarian follicles. FSH promotes the development *of* the follicles, while LH stimulates the synthesis of a hormone *by* the follicles, estrogen (which actually is a family of hormones). The more the follicles develop, the more estrogen they secrete. The result, as you can see in Figure 1, is that estrogen levels slowly rise through the first week or so of the cycle. **Estrogen** is a hormone that is doing two things in the female reproductive cycle: It is promoting the growth of the endometrium in the uterus, and it is controlling the release of both

*continued on page 674*

announced that female mice do possess stem cells that are capable of giving rise to eggs, and that these stem cells remain active throughout life. The hunt is now on, therefore, to see whether human females likewise possess such stem cells. If the answer is yes,

then our ideas about female fertility may change fundamentally. If female reproductive stem cells exist—and if their power can be harnessed—fertility might be prolonged for older females or enhanced for females of any age.

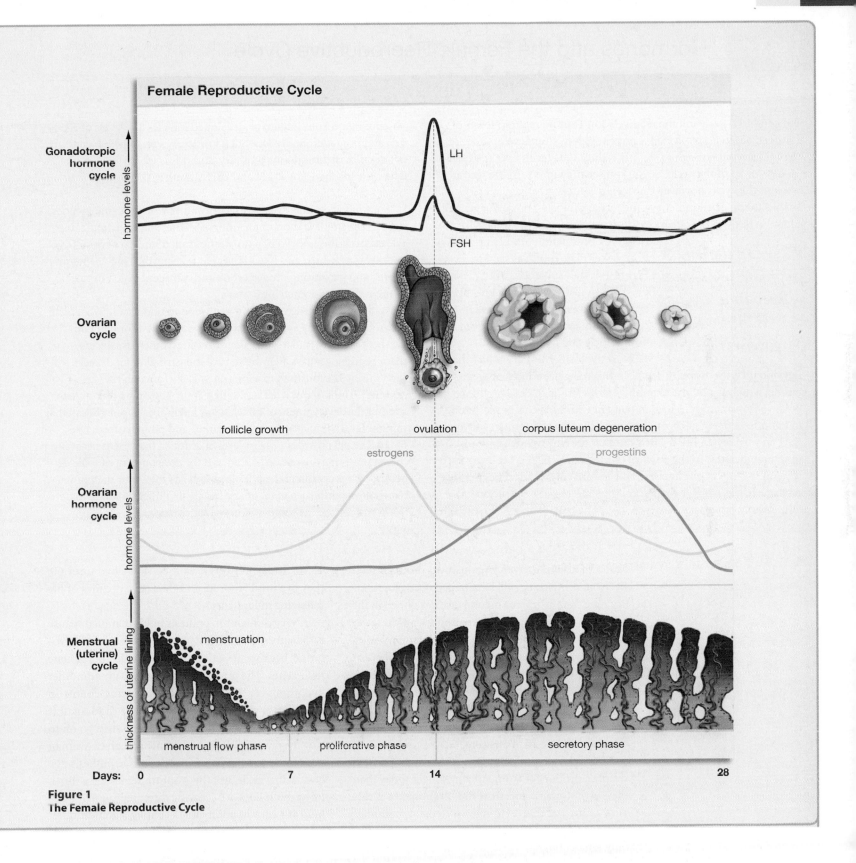

**Female Reproductive Cycle**

Gonadotropic hormone cycle

hormone levels

LH

FSH

Ovarian cycle

follicle growth          ovulation          corpus luteum degeneration

Ovarian hormone cycle

hormone levels

estrogens          progestins

Menstrual (uterine) cycle

thickness of uterine lining

menstruation

menstrual flow phase          proliferative phase          secretory phase

Days:     0                    7                    14                                    28

**Figure 1**
**The Female Reproductive Cycle**

## The Mystery of Menopause

Irrespective of whether female reproductive stem cells exist, the massive loss of follicles over the female life span is real, and it presents science with a mystery: Why should women steadily lose follicles and, as a result, become infertile in midlife? Put another way, why should menopause exist? The living world is shaped by differential success in reproduction.

ESSAY

# Hormones and the Female Reproductive Cycle

*continued from page 672*

FSH and LH. Early on, the relatively low (but increasing) levels of estrogen act to *decrease* FSH—not enough to stop development of the follicles but enough to slightly reduce FSH levels over the first 10 days or so of the cycle, as you can see in Figure 1. In the second week of the cycle, however, estrogen output is greatly increased thanks to the development of the tertiary follicle, whose oocyte will burst forth from the ovary. At this level, estrogen acts to *increase* not only FSH production but LH production as well. The result, as seen in Figure 1, is the "spike" of LH production just before day 14 of the cycle. This hormonal surge is what causes the oocyte to break out from both follicle and ovary.

*How can the female reproductive system "keep time"? What is the clock that governs the cycle?*

Now the former tertiary follicle has a new identity, the corpus luteum. In this role, it produces progesterone, which as noted acts on the uterine lining to develop it for embryo implantation. Then, in tandem with estrogen, progesterone has another effect. It acts on the hypothalamus and pituitary to *lower* FSH and LH levels. But it turns out that the corpus luteum needs LH to remain healthy. In the absence of pregnancy, falling LH levels spell the end for the corpus luteum. This in turn causes a decline in progesterone levels, which brings about the degeneration of the endometrium, essentially by depriving it of a blood supply. Now, in the absence of a corpus luteum or developed follicles, the levels of estrogen and progesterone are *low*. Such low levels *stimulate* FSH and LH production in the anterior pituitary, and this is where we started: The cycle begins again as LH and FSH stimulate the growth of another set of follicles.

So, what keeps this cycle from continuing when pregnancy does occur? The implantation of an embryo in the endometrium causes release from the embryo of a hormone (called human chorionic gonadotropin), which acts like LH in that it keeps the corpus luteum robust and generating progesterone and estrogen, which in turn maintains the endometrium.

With this knowledge, you're in a position to understand what birth control pills do. In general, they are compounds that contain both synthetic estrogen and a synthetic progestin. Women usually take them beginning 5 days after the start of menstruation for 3 weeks (and during the fourth week take a chemically functionless "spacer" or nothing). Think about what is required at the beginning of a cycle to prompt the development of follicles: *low* levels of estrogen. (Recall that this low level stimulates the hypothalamus and pituitary to secrete FSH.) The pill provides a *higher* level of estrogen than exists naturally, thus suppressing the release of FSH and the development of follicles. Second, the progestin in the pill suppresses LH release by the pituitary, which prevents the *ovulation* of any follicle that might manage to develop because the LH surge is required for ovulation.

Traits that help a given organism to produce relatively *more* offspring will be retained over evolutionary time. Yet the trait of menopause developed in the course of evolution, and it seemingly leads to fewer offspring for women. Through it, contemporary women are taken out of reproduction altogether for the last 20 or 30 years of their lives, during which time they suffer from such menopause-related maladies as osteoporosis. Given these factors, why does menopause exist? As was the case with menstruation, we have no firm answers, but we do have several candidate hypotheses. One is that menopause is simply a product of women living longer today than their evolutionary ancestors did. Proponents of this view point out that a number of female mammals undergo what amounts to menopause, but then live only long enough to raise one additional generation of offspring after this loss of reproductive capacity sets in. Human females may once have followed this pattern and lived to, say, 60 years old—just long enough to give birth to a final child in their late 40s and then raise that child to a self-sustaining age of 10 or so. Today, however, women don't live until 60; they live until 75 or 85 and thus have decades of life following menopause.

A very different hypothesis holds that menopause is "adaptive," as biologists say—it arose through evolution because it actively helps females perpetuate their genes. The starting point for this "grandmother hypothesis" is the observation that what counts in evolution is not just whether you have children; it is whether you have children who themselves go on to have healthy children. Now, how is a given woman most likely to achieve this? Under this hypothesis, she best achieves it not by continuing to give birth throughout life but by leaving direct reproduction behind at a certain point and becoming a grandmother who provides both for her children and her grandchildren. We now have evidence, from eighteenth- and nineteenth-century groups in Finland and Canada, that things actually can work this way—that an increased duration of "postmenopausal" life in women can positively affect the reproductive success of their children.

**SO FAR...**

1. While they are in development, the reproductive cells a woman produces are known as _____ and are produced in the organs called _____.

2. Every 28 days, on average, one of these reproductive cells is released in the process called _____ and then begins a journey down one of the woman's _____.

3. Menstruation, the release of blood and a portion of the uterine tissue called the _____, occurs when there has been no _____ during the reproductive cycle.

## 32.3 The Male Reproductive System

Switching now to males, if you look at **Figure 32.6**, you can see another view of the male reproductive system. Here you see the large-scale structures of the system. There is the **testis**, in which sperm begin development; the **epididymis**, the tubule in which sperm mature and are stored; and the **vas deferens**, the tube through which sperm move in the process of ejaculation. (All three of these structures exist in pairs, although the second member of each pair is not visible in the figure.) Then, there is the single **urethra**, the tube each vas deferens empties into and through which sperm

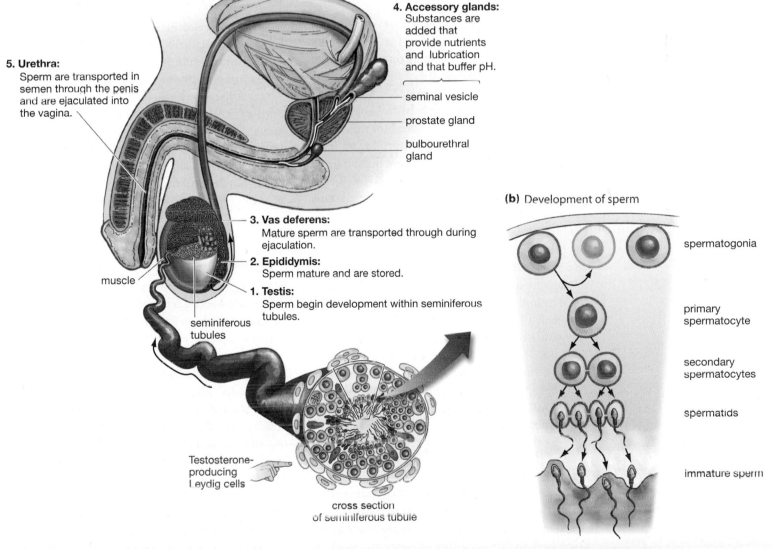

**(a)** Delivery of sperm

**4. Accessory glands:** Substances are added that provide nutrients and lubrication and that buffer pH.

seminal vesicle

prostate gland

bulbourethral gland

**5. Urethra:** Sperm are transported in semen through the penis and are ejaculated into the vagina.

muscle

seminiferous tubules

**3. Vas deferens:** Mature sperm are transported through during ejaculation.

**2. Epididymis:** Sperm mature and are stored.

**1. Testis:** Sperm begin development within seminiferous tubules.

Testosterone-producing Leydig cells

cross section of seminiferous tubule

**(b)** Development of sperm

spermatogonia

primary spermatocyte

secondary spermatocytes

spermatids

immature sperm

**Figure 32.6**
**Development and Delivery of Sperm**

**(a)** After the sperm develop in the testes and mature in the epididymis, orgasm prompts them to be transported through the vas deferens and then the urethra. Secretions from accessory glands (the seminal vesicles, the prostate gland, and the bulbourethral glands) join with the sperm to form semen before ejaculation.

**(b)** How sperm mature. Sperm development begins within seminiferous tubules, located in the testes. Sperm start out as spermatogonia at the periphery of the tubules and end up as immature sperm in the interior cavity of the tubules. From there, they are transported to the epididymis for further development and storage. Outside the seminiferous tubules are Leydig cells, which produce testosterone.

••• ANSWERS

1. oocytes; ovaries

2. ovulation; uterine (or Fallopian) tubes

3. endometrium; fertilization of an egg

leave the male body. There are also several accessory glands whose contributions are added to sperm as it proceeds: the seminal vesicles, the prostate gland, and the bulbourethral glands. (The seminal vesicles and bulbourethral glands also exist in pairs.)

**Web Animation 32.2**
The Male Reproductive System

## Structure of the Testes

The testes are surrounded by a muscle whose contractions help to regulate the temperature of the sperm inside. Sperm development requires temperatures somewhat cooler than those found in the rest of the body, which is why the testes are *external* to the body's torso. Sperm cannot develop in temperatures that are too cold, however. Thus, when a male steps into, say, a cold swimming pool, the muscle surrounding the testes will contract, drawing the testes up closer to the torso—and warming the sperm.

Figure 32.6a shows that the testes are divided into a small series of lobes, and that within each lobe there are a number of highly convoluted tubes, the **seminiferous tubules**: structures that are the sites of initial sperm development in men. If you now look at a single seminiferous tubule in cross section, in Figure 32.6b, you'll see how this development proceeds: from the outside of each tubule—where sperm precursors are in their *least*-developed state—toward the interior cavity of each tubule, where sperm are in a more developed state. Eventually, the immature sperm are released into the interior cavity and are transported up it, moving then to the epididymis for further processing and storage. As you can see, there is an assembly-line quality to sperm production, starting with initial development at the periphery of the tubules, continuing in the interior of the tubules, and ending with storage in the epididymis. Older sperm that are not ejaculated are destroyed (by other cells) or eliminated in the urine. Inside the testes, surrounding the seminiferous tubules, there exist numerous *Leydig cells*, which are the cells that produce testosterone.

## Male and Female Gamete Production Compared

The production of sperm bears an interesting comparison with the development of eggs. The cells from which sperm develop are called **spermatocytes**. In going back to the precursors of *these* cells, however, we see a significant difference between males and females. The cells that give rise to spermatocytes are the stem cells called **spermatogonia**; they exist at the periphery of the seminiferous tubules, and through normal cell division, they actually lead to two kinds of cells. One is the primary spermatocyte, which goes on to develop into a sperm; the other, however, is another spermatogonium. Thus, spermatogonia are "sperm factories," if you will, that give rise not only to sperm, but to more sperm factories as well. This is the reason males *keep* producing sperm throughout their lives while, as we've noted, females may have a fixed quantity of oocytes. Although male sex hormone production lessens with age and sperm decline in quality, it is possible for male reproductive capability to remain intact until death. If you look at Figure 32.6b, you can see how the male sperm progression works in the testes: from spermatogonium, through spermatocyte to spermatid, and finally to immature sperm.

## Further Development of Sperm

As you can see in Figure 32.6a, the seminiferous tubules eventually feed into the epididymis, a single convoluted tubule (one for each testis) that is about 7 meters (about 23 feet) long. This structure serves as the site of storage and final sperm development and as a kind of materials recycling site for damaged sperm. By the time sperm arrive at the bottom portion of the epididymis, they are fairly mature but not yet fully mature. They won't become mobile until fluid from a type of supporting gland noted earlier—the seminal vesicles—mixes with them on ejaculation, thus supplying them with nutrients.

### Time Sequence and Sperm Competition

How long does this whole process of sperm development take? About 2.5 months from the start of development to storage in the epididymis. About 200 million sperm are produced in this assembly line each day, which you may recall is about the number that are ejaculated in an average orgasm.

As noted, the watchword for oocyte selection could be "many are called but few are chosen." How much more extreme is the selection that goes on with sperm! The number expelled with *each ejaculation* vastly exceeds the number of oocytes a woman ever has stored in her ovaries. On average, however, only a single sperm per ejaculation will be able to fertilize an egg—assuming an egg is present in the uterine tube at the right time.

Why is there such an enormous difference between the number of sperm produced and the number that will fertilize eggs? The essential answer here seems pretty clear: Just as males often are in competition with other males to mate with females, so the sperm of males is in competition with the sperm of other

# Methods of Contraception

 **S**elected methods of contraception are presented here. The "effectiveness" for each method refers to how effective the method has been shown to be in preventing unwanted pregnancies during an initial year of typical use. Such typical use stands in contrast to "perfect use," in which a method is used correctly either during every instance of sexual intercourse (in the case of a method such as condoms) or over the course of every month (in the case of birth control pills). Note that only two of the methods offer even limited protection against sexually transmitted disease. See "Hormones and the Female Reproductive Cycle" on page 672 for information on how birth control pills work.

## Table 32.1 Methods of Contraception

| | | |
|---|---|---|
|  | **Birth control pill** | A set of pills containing a combination of hormones that suppress the female ovulatory cycle.<br><br>Effectiveness: 92 percent. |
| | **Condom** | Made of latex, plastic, or animal tissue, condoms are put on the erect penis prior to intercourse and catch sperm after it is ejaculated. Condoms are the only method of contraception that helps protect both partners from some sexually transmitted diseases.<br><br>Effectiveness: 85 percent. |
| | **Diaphragm** | A soft, circular rubber barrier inserted into the woman's vagina before sex. Covering the cervix, it prevents sperm from reaching the egg. Along with the related cervical caps, diaphragms offer females some protection against some sexually transmitted diseases, such as gonorrhea and chlamydia.<br><br>Effectiveness: 84 percent. |
|  | **Intrauterine device (IUD)** | A small device made in some instances of plastic and copper that is inserted into a woman's uterus by a physician. Other IUDs are made of plastic and contain a supply of a progestin hormone that is continually released over a five-year period. Both types of IUDs help prevent fertilization of egg by sperm in a process that may involve altering the movement of each type of gamete. Both types of IUDs also work to prevent implantation in the uterus of any embryo that might be produced.<br><br>Effectiveness: hormone-releasing IUD, 99.9 percent. |
| 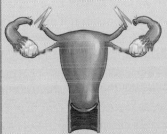 | **Tubal ligation** | A surgical procedure that generally results in the sterilization of a woman, meaning permanent infertility. In it, the uterine tubes are first cut and then tied, thus blocking eggs from moving to a position in which they could be fertilized by sperm. Does not affect feminine characteristics, hormonal production, or sexual performance. Very rarely, tubes will reconnect themselves, which accounts for the (low) failure rate of the procedure. It is possible to reverse the operation in some instances.<br><br>Effectiveness: 99.5 percent. |
| 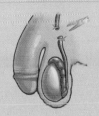 | **Vasectomy** | A surgical procedure that can result in the sterilization of a man, though vasectomies are sometimes reversible. In the procedure, the vas deferens are cut and tied, thus blocking the movement of sperm during ejaculation. Does not affect male hormonal production, physical capacity for erection, or ability to ejaculate. (Sperm make up a small portion of the semen that is ejaculated.) In rare instances, vas deferens will grow back together again, which accounts for the (low) failure rate of the procedure.<br><br>Effectiveness: 99.8 percent. |

males to *fertilize* female eggs. This general phenomenon of *sperm competition* goes on throughout the animal kingdom—in insects, birds, and reptiles as well as in mammals. Its root cause is that there are very few animal species in which females or males are truly monogamous. As a result, mating by itself does not guarantee offspring. If a female who is fertile over a period of several days mates with a number of males during these days, then each male's chance of fatherhood is enhanced to the degree that his sperm can outcompete the sperm of the other males. This is the basis for a sperm "arms race" that has taken place over evolutionary time. Looking at various species, we find that their sperm has gotten longer, or more "motile" (active), or has been produced in greater numbers—all as a means of gaining a reproductive edge. This is why men produce so many sperm. Their sperm production has followed the lottery principle: The more tickets you have, the more likely you are to win.

### From Vas Deferens to Ejaculation

Maturing sperm are stored in both the epididymis and the vas deferens. With sufficient sexual stimulation, ejaculation takes place through the contractions of muscles that surround the penis at its base. This contraction pushes the sperm through the urethra and out the urethral opening at the tip of the penis.

### Supporting Glands

As noted, several other kinds of materials join with sperm before ejaculation to form the substance known as semen. Indeed, given the number of sperm ejaculated, it may be surprising to learn that sperm don't account for much of the volume of semen; some 95 percent of the ejaculated material comes from the accessory glands already mentioned: two seminal vesicles, the prostate gland, and two bulbourethral glands. It's helpful to think of sperm as getting "outfitted" for their travels by these glands just as they are exiting from the body. In addition to the nutrients the seminal vesicles supply, accessory glands provide alkaline and lubricating substances (which neutralize the acidic environment of the vagina and facilitate transportation).

One of these supporting glands may sound familiar. The **prostate gland** is a structure that surrounds the male urethra near the bladder and contributes a substantial amount of material to semen. Later in life, a significant percentage of men develop either enlarged prostate glands or, worse, the growth of cancerous cells in the prostate, which is to say prostate cancer. Given that the prostate encircles the urethra—and that the urethra transmits not only semen but

urine as well—you can see why a warning sign of these conditions is reduced urine flow. But with both urine and semen traveling through the urethra, how does the body make sure that one substance doesn't mix with the other? Traffic control is provided by the prostate gland. During ejaculation, a circular muscle in it contracts, thus sealing off the contents of the bladder.

---

### SO FAR...

1. Sperm begin development in the _____ and continue development and are stored in the convoluted tubules called the _____.

2. The material that is ejaculated from the penis, _____, is composed of both _____ and fluids secreted by three supporting _____.

3. Males are able to remain fertile throughout their lives because the cells called spermatogonia are reproductive _____ that give rise to more of themselves and to the _____ that develop into sperm.

---

## 32.4 The Union of Sperm and Egg

To this point, we've reviewed the process by which male sperm and female egg are first created and then *positioned* so that they can come together. Now let's look at the final steps that lead to the fusion of these two cells.

When the oocyte is expelled from the ovaries, there still is no solitary oocyte in transit, but rather an oocyte surrounded by an extracellular coat that in turn is surrounded by a group of accessory cells (**Figure 32.7b**). When we look at sperm, we find that each one has a "head" that contains not only the nucleus and its vital complement of chromosomes, but an outer compartment, called an **acrosome**, which contains enzymes (**Figure 32.7a**). These enzymes, which are released externally, have the ability to dissolve the bonds between the accessory cells that surround the oocyte. It takes dozens of sperm to release enough enzymes such that *one* sperm can move between the accessory cells and arrive at the external coat that is the last barrier between sperm and oocyte. When this sperm makes contact with the coat, it binds with receptors on it, an event that triggers the release of this sperm's own complement of digestive enzymes. These enzymes clear a path through the coat that the sperm moves through. With this, the sperm reaches the oocyte, at which

**(a)** Structure of a sperm cell

nucleus

tail

acrosome

"head"  mitochondria

**sperm**

**(b)** How sperm fertilizes egg

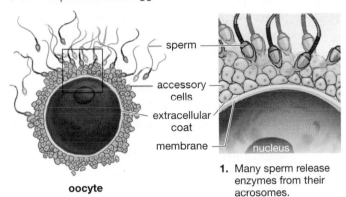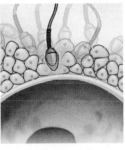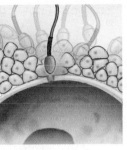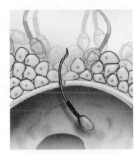

sperm

accessory
cells

extracellular
coat

membrane

nucleus

**oocyte**

1. Many sperm release enzymes from their acrosomes.

2. The enzymes dissolve bonds between accessory cells, enabling one sperm to pass through.

3. The one sperm uses its own enzymes to clear a path through the extracellular coat.

4. The sperm and oocyte membranes fuse, allowing the sperm to move inside the oocyte.

**Figure 32.7**
**Anatomy of a Sperm and Steps in Fertilization**
**(a)** The "head" of each sperm includes not only its complement of chromosomes but also an acrosome, which contains enzymes that, when released externally, break down the bonds between accessory cells that surround the oocyte. Behind the sperm head are the mitochondria that supply the energy for the motion of the tail that propels the sperm.
**(b)** The process by which a sperm fertilizes an oocyte.

point the outer membranes of sperm and oocyte fuse—the two membranes become one, much as two soap bubbles might become one when they come into contact. With this, the contents of the sperm cell move inside the oocyte, which means that the two cells have effectively become one. Once this happens, the nuclei of oocyte and sperm combine, the oocyte at last becomes an egg, and fertilization is complete.

### How Latecomers Are Kept Out

But, what of the other dozens of sperm surrounding the oocyte? None of them can be allowed entry because an embryo could not survive getting two sets of male chromosomes, but only one set of female chromosomes. To guard against this possibility, a kind of gate-slamming process is put into motion. The fusion of the first sperm with the oocyte causes the oocyte to release thousands of tiny granules into the space just outside itself. These granules release substances that harden the membrane immediately outside the oocyte while inactivating the sperm receptors on it. The result is that, with rare exceptions, only one sperm gets in. (The sexual activity that is so important to human reproduction also carries with it the risk of sexually transmitted disease, which you can read about in the essay on page 682.)

## 32.5 Human Development Prior to Birth

Having seen how sperm and egg form and get together, you have reached a point in human reproduction that should be familiar: *development*, meaning the process by which a fertilized egg is transformed, step by step, into a functioning organism. As you'll see, processes you looked at in Chapter 31 in connection with frogs and flies have counterparts in human development. What may be surprising is how *early* in human development many of the critical events of development occur. Here is a list of some of the key developmental processes discussed in Chapter 31.

- Formation of the blastula, or hollow body of cells (called a *blastocyst* in humans and other mammals)

- Gastrulation, in which the three primary layers of cells are formed

- Formation of the neural tube (which gives rise to the brain and nervous system)

- Organogenesis, in which the organs of the body start to take shape

Now, keeping in mind that a human pregnancy lasts an average of 38 weeks, consider the time-line for these major developmental events. Blastocyst

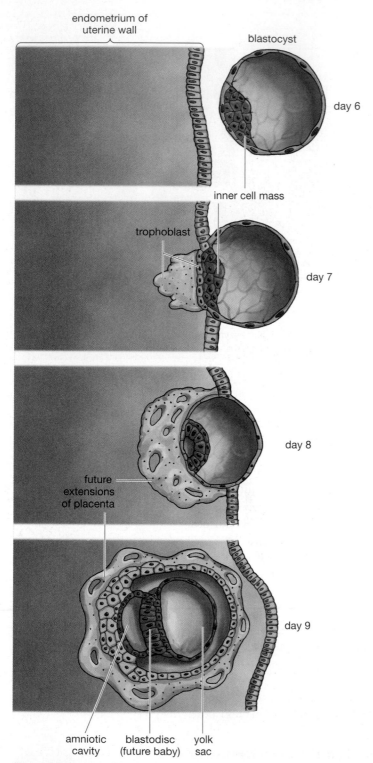

**Figure 32.8**
**Implantation of the Embryo in the Uterus**

Six days after conception, the embryo, now in the blastocyst stage of development, arrives at the uterine wall. It is composed of both an inner cell mass, which develops into the baby, and a group of cells called a trophoblast, which carves out a cavity for the embryo in the uterine wall and then begins establishing physical links with the maternal tissue. The outgrowth of this linkage is the placenta.

formation begins 4 or 5 days after conception; gastrulation begins 16 days after conception; neural tube formation begins in the third week after conception; and organogenesis begins in the fourth week. By the end of 12 weeks, immature versions of all the major organ systems have formed. To place some of these events in the context of what the mother experiences, organogenesis is occurring about the time the first menstrual period is missed, and formation of the immature organ systems is being completed at about the same 12-week point at which morning sickness generally ends.

The embryo's development over time can be contrasted with its *growth* over time. By the end of the third week—after blastocyst formation and gastrulation—a human embryo is perhaps 2 millimeters (or less than a tenth of an inch) in diameter. Skip to the twelfth week after conception, when major portions of organogenesis are concluding, and the fetus may still be only about 13 centimeters, or a little over 5 inches in length.

The message here is that development is concentrated early in pregnancy, while the simple growth of what *has* developed comes later. So pronounced are the changes during development that biologists use different terms for the growing organism over time. The original cell—the fertilized egg—is a **zygote**. In humans, about 30 hours after the zygote is formed, it undergoes its first cell division; at this point, the developing organism is referred to as an **embryo**, a name that will be applied to it through the eighth week of development. From the ninth week of development to birth, the organism is referred to as a **fetus**. The typical 9-month pregnancy is divided into periods of roughly 3 months each, called **trimesters**. The first trimester ends at about the time organogenesis does.

### Early Development

You saw earlier that the zygote/embryo moves from ovary to uterus during the first 4 days after fertilization. It arrives in the uterus on about day 4 as the *morula*, or tightly packed ball of cells noted in Chapter 31 (see Figure 31.1, page 654). Over the next 2 days, the morula becomes the blastocyst, which implants itself in the endometrial lining. Actually, this "implantation" is more like an invasion. The blastocyst carves out a cavity for itself by releasing enzymes that effectively digest endometrial cells (**Figure 32.8**). Implantation normally occurs in the dorsal wall of the uterus (**Figure 32.9**). Occasionally, however, a blastocyst attaches itself inside the uterine tube, or sometimes to the **cervix**, which is the lower part of the uterus that opens onto the vagina. This is an **ectopic pregnancy**:

a pregnancy in which the embryo has implanted in a location other than the uterine wall. In general, embryos do not survive such an implantation.

## Two Parts to the Blastocyst

Figure 32.8 shows you the arrangement of cells in the blastocyst. On one side, you can see a group of cells called the **inner cell mass**: the portion of the blastocyst that develops into the baby. The periphery of the blastocyst, however, is formed by cells called the **trophoblast**. These are the cells that etch their way into the uterine wall; with implantation, they begin extending farther into the endometrium, establishing indirect links with the maternal blood supply. In time, these cells grow into the fetal portion of a well-known structure in pregnancy, the **placenta**: a network of maternal and embryonic blood vessels and membranes that allows nutrients and oxygen to diffuse *to* the embryo (from the mother), while allowing carbon dioxide and other wastes to diffuse *from* the embryo (to the maternal blood system). See **Figure 32.10** for a look at how the placenta functions within a uterine environment as it exists at about 2 months. The tissue that links the fetus with the placenta is the **umbilical cord**, which houses fetal blood vessels that move blood between the fetus and the placenta. So pressing is the embryo's need for nutrients and waste removal that its circulatory

**Figure 32.9**
**Implantation Accomplished**
Twelve days after conception, the embryo, now in its blastocyst stage, has implanted in the uterine wall.

system, which includes the umbilical cord, becomes functional before any other organ system.

## The Origins of Twins

Having learned a little about fetal development, you're now in a position to understand something about how twins come about. Identical twins can develop at any time in the very early stages in pregnancy—from the time the zygote undergoes its first division through the

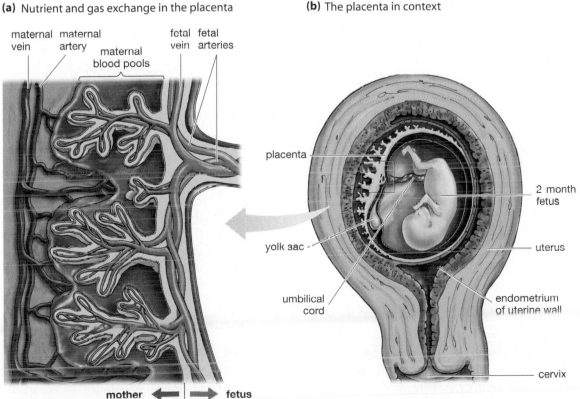

**(a)** Nutrient and gas exchange in the placenta

maternal vein   maternal artery   maternal blood pools   fetal vein   fetal arteries

placenta

yolk sac

umbilical cord

mother ⟵ | ⟶ fetus

**(b)** The placenta in context

2 month fetus

uterus

endometrium of uterine wall

cervix

**Figure 32.10**
**Nutrient and Gas Exchange in the Placenta**

**(a)** Oxygen- and nutrient-rich blood comes from the mother through the maternal arteries and moves, through smaller arterial vessels, into the maternal blood pools in the placenta. Oxygen-poor blood from the embryo, transported through the umbilical cord's fetal arteries, does not flow into the maternal blood pools but instead stays within the fetal blood vessels, which project into the maternal blood pools. Oxygen and nutrients in the maternal blood pools move into the fetal blood vessels, however, through such processes as diffusion. The resulting oxygen- and nutrient-rich blood is carried back to the embryo through the umbilical cord's fetal vein. Meanwhile, carbon dioxide from the embryo moves the other way—into the maternal blood pools—to be carried away through maternal veins.

**(b)** Larger structure of the placenta and uterine environment at 2 months.

# Sexually Transmitted Disease

**G**iven the necessity of sex for the continuation of human life, it may seem surprising that so much risk can be associated with it, but such is the case. The root of the problem is that microorganisms that literally couldn't stand the light of day—that is, that couldn't exist for long in an exterior environment—are quite well adapted to life inside human tissue. More important, these microbes can be transmitted from one human being to another during the act of sex.

The risk of sexually transmitted diseases (STDs) varies in accordance with lots of things: the disease itself, the type of sexual activity a person engages in, and the precautions that are taken to avoid trouble. But one rule that holds true for all STDs and all behaviors is that as the number of partners goes up, so does risk. Here are the most common STDs, their prevalence in the United States, and the methods of treatment for them.

## Table 32.2  Sexually Transmitted Disease

| Disease | Caused by | Possible consequences | New cases in the United States in 2004 | Treatment |
|---|---|---|---|---|
| Chlamydia | *Chlamydia trachomatis* bacterium | Spread to uterine tubes in women, causing pelvic inflammatory disease (PID), which can lead to sterility | 3 million | Curable with timely use of antibiotics |
| Gonorrhea | *Neisseria gonorrhoeae* bacterium | Arthritis, pelvic inflammatory disease | 650,000 | Curable with timely use of antibiotics, though antibiotic-resistant strains now exist |
| Syphilis | *Treponema pallidum* bacterium | Heart, nervous system, bone damage if allowed to progress | 70,000 | Curable with timely use of antibiotics, though antibiotic-resistant strains now exist |
| Genital herpes | Herpes simplex virus | Recurrent skin lesions, eye damage, pregnancy complications | 1 million | Occurence and severity of outbreaks can be lessened with treatment, but no cure exists |
| Genital warts | Human papilloma virus | Cervical infection in women, association with increased risk of cervical and penile cancer | 5.5 million | Gardasil vaccine for several strains approved in 2006 for females aged 11–26 |
| AIDS | Human immunodeficiency virus (HIV) | Death, dementia, injury from a variety of opportunistic infections | 40,000 new infections; 800,000 persons living with HIV in the U.S. | No cure; treatments to reduce symptoms exist |

seventh day after fertilization. In all cases, identical twins result from two *parts* of the embryo developing into separate human beings. This may be a matter of the two cells produced by the zygote's first division each developing as separate embryos, or it may entail the inner cell mass of the blastocyst separating into two masses, with each then giving rise to a different person. In contrast, **fraternal twins** (or triplets, and so on) are twins who are produced first through multiple ovulations in the mother, followed by multiple fertilizations from separate sperm of the father, and then multiple implantations of the resulting embryos in the uterus.

This means that fraternal twins are like any other pair of siblings except that they happen to develop at the *same time*. Meanwhile, **identical twins** are twins who develop from a single zygote. Strictly speaking, they are a single organism at one point in their development and as such have exactly the same genetic makeup.

## Development through the Trimesters

If you look at the series of pictures and diagrams that make up **Figure 32.11**, you can see how an embryo and then fetus develop at selected points in the 38 weeks of pregnancy. Here's an overview of what

## (a) Embryo Development:

Vertical lines show embryo's actual size; measurements are head-to-rump lengths

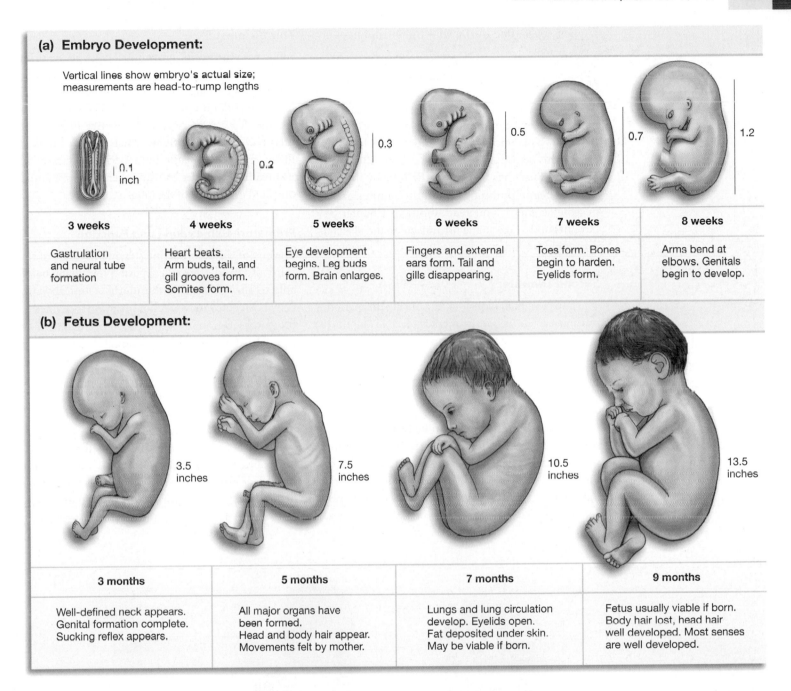

| 3 weeks | 4 weeks | 5 weeks | 6 weeks | 7 weeks | 8 weeks |
|---|---|---|---|---|---|
| Gastrulation and neural tube formation | Heart beats. Arm buds, tail, and gill grooves form. Somites form. | Eye development begins. Leg buds form. Brain enlarges. | Fingers and external ears form. Tail and gills disappearing. | Toes form. Bones begin to harden. Eyelids form. | Arms bend at elbows. Genitals begin to develop. |

## (b) Fetus Development:

| 3 months | 5 months | 7 months | 9 months |
|---|---|---|---|
| Well-defined neck appears. Genital formation complete. Sucking reflex appears. | All major organs have been formed. Head and body hair appear. Movements felt by mother. | Lungs and lung circulation develop. Eyelids open. Fat deposited under skin. May be viable if born. | Fetus usually viable if born. Body hair lost, head hair well developed. Most senses are well developed. |

**Figure 32.11**
**Development through the Trimesters**

(a) Development of the human embryo (3 weeks to 8 weeks).

(b) Development of the human fetus (3 months to 9 months).

(c) Photos of a human embryo at 6 weeks (left), 11 weeks (middle), and 5 months (right).

**(c)**

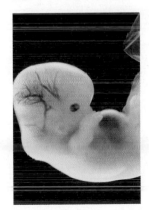

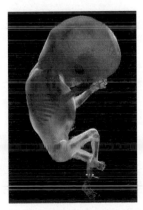

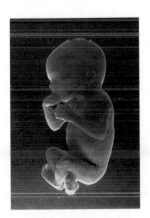

6 weeks

11 weeks

5 months

happens over time during the three trimesters of a pregnancy.

**Web Animation 32.3**
Human Development
Prior to Birth

### First Trimester

As noted, blastocyst formation, gastrulation, and neural tube development all take place within the first month of conception; but much more happens during this period as well. By the twenty-fifth day, a primitive heart has formed in the human embryo and has begun pumping blood; the somites or repeating segments you read about in Chapter 31 have begun to take shape for muscular and skeletal formation.

### Second Trimester

The fetus lengthens considerably during this period, but most of its weight gain takes place in the third trimester. Midway through a pregnancy, a fetus has achieved about half of its birth length (which is about 50 centimeters or 20 inches on average), but only about 15 percent of its birth weight (which averages 3.5 kilograms or 7.7 pounds).

The heartbeat of a fetus can be heard through a stethoscope during the second trimester, and the mother can feel the baby begin kicking by the end of the fourth month. Kicking is only one of several movements the fetus is capable of at this point; others include facial movements and the sucking reflex.

### Third Trimester

Along with the terrific growth that takes place during the third trimester, fetuses are running through a series of activities that can be thought of as practice for life outside the uterus (or womb). The fetus is surrounded by a protective and nutritive fluid in the uterus, the **amniotic fluid**, which fills its lungs. It therefore doesn't breathe, but it periodically exercises its breathing muscles by using them to move amniotic fluid in and out of the lungs.

### Premature Babies and Lung Function

The end of the second trimester and the start of the third also mark the critical period for a baby being able to survive outside the uterus. Clinically, any baby born before 37 weeks of gestation is defined as premature, but many babies born between 35 and 37 weeks after conception require little special care. The more premature a baby is, however, the greater its risk of having difficulties with physical and cognitive development, at least for the first few years of life. In very premature babies, survival itself is called into question (**Figure 32.12**).

A critical factor in determining whether a premature baby can live outside the uterus is lung function. As it turns out, the respiratory system is the last one to develop in a fetus. Working against the survival of very early "preemies" is the fact that, when they

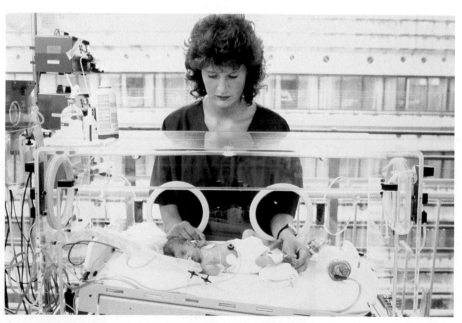

**Figure 32.12**
**Tough Going Early on**
A pediatric nurse attends to a premature baby in an incubator.

exhale, the alveoli or air sacs in their lungs tend to collapse and then stick together, after which they can be reopened only with great effort. Why doesn't this happen with the lungs of a full-term 38-week fetus? Because, beginning late in pregnancy, fetuses begin synthesizing a lung surfactant—a soap-like fatty substance that keeps lung sacs from collapsing. Babies born before the twenty-sixth week of pregnancy must undergo intensive lung therapy to survive because they have not produced much lung surfactant up to this point. Usually, their therapy includes the administration of a manufactured surfactant.

## 32.6 The Birth of the Baby

As the fetus grows, the uterus becomes an increasingly cramped place, and the fetus's movements become more restricted. In late pregnancy, the fetus's head normally will lodge near the base of the mother's spine (**Figure 32.13**). In this "upside-down" position, the fetus is ready to make its entrance into the world. What triggers this journey? The immediate cause is **labor**: the regular contractions of uterine muscles that sweep over the fetus from its legs to its head. The pressure this generates opens or "dilates" the cervix (ultimately to approximately 10 centimeters or about 4 inches). This creates an opening large enough for the baby to pass through. These things must occur in stages, however. The first phase of the uterine contractions is given over to cervical dilation, then comes the expulsion of the baby and finally expulsion of the placenta (also called the *afterbirth*, for obvious reasons).

But what sets this train of events into motion in the first place? Put another way, what is the trigger that initiates labor? Despite a great deal of research on this question, we don't know. One hypothesis is that, as the fetus grows larger, it reaches the point at which even the highly efficient placental system can no longer supply it with enough nutrients. As a result, it secretes hormones that break down nutrients for increased energy, and this secretion triggers a hormonal cascade that brings on labor.

What happens if 38 weeks have passed and labor still doesn't begin? After a period of watching and waiting, a physician may recommend inducing labor through use of a synthetic form of the hormone oxytocin. From Chapter 27, recall that, when labor begins naturally, it is accelerated by oxytocin that is released from the mother's pituitary gland. (This oxytocin brings about muscle contractions that push the fetus down further into the uterus, and

**(a)** Fetus's position prior to birth

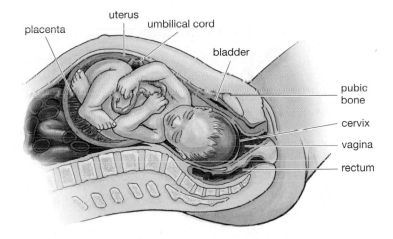

**(b)** Movement through birth canal

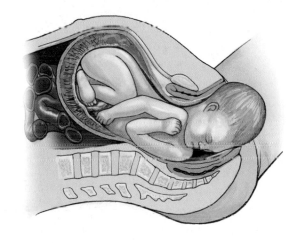

**(c)** Structure of "afterbirth"

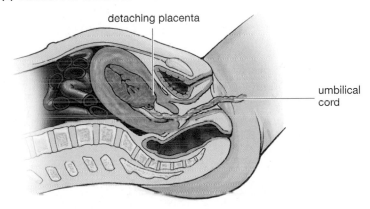

**Figure 32.13**
**Birth of the Baby**
With uterine contractions, the cervix dilates, allowing room for the baby to pass through the birth canal and to the outside world. The placenta and its associated fluids and tissues are expelled shortly afterward as the "afterbirth."

this pushing brings about the production of more oxytocin.) When a baby is overdue, synthetic oxytocin can help get labor started.

# On to Ecology

In this chapter and the six that preceded it, you followed the workings of the cells and tissues that allow animals to function in their marvelously diverse ways. But animals—and for that matter all the other varieties of living things—are not self-contained units. Rather, they are constantly *interacting* with each other and with the nonliving world around them. In the next three chapters, you'll be looking at life at the level of this integrated web. What you will be studying is ecology.

••• ANSWERS

1. one; chromosomes from the mother and father
2. uterus; placenta
3. contractions; muscles

**SO FAR...**

1. The number of sperm that can fuse with an oocyte is limited to _____. This is vital in order to make sure that a fertilized egg has a balanced number of _____.

2. With the implantation of the fertilized egg in the wall of the _____, cells in the egg grow into the fetal portion of a network of blood vessels and membranes that support the fetus. This is the _____.

3. Labor can be defined as regular _____ of uterine _____ that help bring about the birth of the baby.

# CHAPTER 32 REVIEW

For study help, activities, and more quiz questions, go to **www.aw.com/krogh4**.

## SUMMARY

### 32.1 Overview of Human Reproduction and Development

- Both the human male and female reproductive systems must produce their respective gametes (sperm and eggs) and then deliver them to a place where they can fuse, in the moment of fertilization.

- The precursors to eggs, called oocytes, exist in the ovaries, which lie just left and right of center in the female pelvic area. Oocytes mature within nurturing complexes of cells and fluids called follicles. On average, once every 28 days, one follicle-oocyte complex is brought to a state of maturity. The oocyte and some of its accessory cells then rupture from the follicle and ovary and begin a journey through the uterine tube to the uterus. The release of the oocyte is called ovulation. Conception can occur if the oocyte encounters sperm on the way to the uterus.

- Sperm are produced in the male testes and then develop further in a structure called the epididymis. With sexual stimulation, they move through two types of ducts, the vas deferens and the urethra, after which they are ejaculated from the penis into the female vagina. Materials secreted by three separate sets of glands will join with the sperm before ejaculation. The resulting mixture of sperm and glandular materials is called semen. The first of the sperm to reach the oocyte moving through the uterine tube may fertilize it. Development of the fertilized egg, called a zygote, then ensues.

### 32.2 The Female Reproductive System

- A portion of the female reproductive system is built up and then dismantled in a reproductive cycle that takes approximately 28 days to complete. The result of the dismantling is menstruation, which is a release of blood and the specialized uterine tissue it infuses, the endometrium. The endometrium, which lines the uterus, is tissue that an embryo implants itself in when pregnancy occurs. Given the obvious costs of menstruation, it is not clear why it has been preserved in evolution, although several hypotheses have been put forward to explain it.

- The basic unit of oocyte development is the ovarian follicle, a complex that includes the oocyte that develops into an egg and its surrounding accessory cells and fluids.

- On average, a single follicle each month will develop into a tertiary follicle that ruptures, expelling the oocyte and its accessory cells from the ovary. This complex of cells then begins a journey through the uterine tube to the uterus. The ruptured tertiary fol-

licle develops into the corpus luteum, which secretes hormones that prepare the reproductive tract for pregnancy and that maintain it during the early phases of pregnancy.

- The number of follicles a woman possesses steadily decreases throughout her lifetime, primarily through the process of atresia or follicle degeneration. The scarcity of follicles a woman has in middle age is the primary factor that brings about menopause, the cessation of the monthly ovarian cycle. Recent research has, however, called into question the long-standing assumption that all the follicles a woman possesses develop in her prior to birth. It may be that, like men, women possess reproductive stem cells that are capable of generating new follicles. It is not clear why menopause exists since it seems to carry a number of costs to females, but several hypotheses have been put forth to account for it.

Web Animation 32.1: The Female Reproductive System

### 32.3 The Male Reproductive System

- Sperm begin development in the testes and continue development and are stored in the adjacent epididymis. Sperm move from the epididymis through the vas deferens and then the urethra in ejaculation. The accessory glands that contribute material to the sperm before ejaculation are two seminal vesicles, the prostate gland, and two bulbourethral glands.

- Within the testes, sperm development takes place in the seminiferous tubules. Development begins at the periphery of a given tubule, with sperm becoming more developed toward the interior of the tubule. Eventually, developing sperm are transported from the interior of a tubule to the epididymis for further development and storage.

- Sperm development in the testes begins with cells called spermatogonia, which develop into spermatocytes. These in turn give rise to spermatids and then to immature sperm. In cell division, spermatogonia give rise not only to spermatocytes, but to more spermatogonia. This self-generation of sperm precursors is the reason men keep producing sperm throughout their lifetimes.

- About 200 million sperm are produced each day—about the number of sperm ejaculated with each orgasm. Males produce such a huge number of sperm because of sperm competition: In species in which females mate with multiple partners, copulation with a female is not enough to ensure paternity; the sperm of a given male must be able to outcompete the sperm of any other male who mates with the female. Thus, through evolution, sperm have taken on various characteristics that serve to provide an advantage over other sperm in reaching and fertilizing oocytes. One of these characteristics is sheer numbers of sperm produced.

- Sperm make up only 5 percent of the semen that is ejaculated. Materials secreted by the seminal vesicles, the prostate gland, and the bulbourethral glands make up the other 95 percent of semen. These materials provide nutrients, pH neutralizers, and lubricating fluids for the sperm.

Web Animation 32.2: The Male Reproductive System

## 32.4 The Union of Sperm and Egg

- An unfertilized oocyte moving through the uterine tube is surrounded by an external coat that in turn is surrounded by a group of accessory cells. Each sperm has an outer compartment, called an acrosome, that can release enzymes capable of dissolving the bonds between the oocyte's accessory cells. Enzymes released by dozens of sperm allow a single sperm to move between the accessory cells and reach the oocyte's external coat. This sperm binds with receptors on the coat and releases its own complement of enzymes, which clear a path through the coat that the sperm moves through. With this, the outer membranes of sperm and oocyte make contact and fuse, an action that allows the contents of the sperm cell to move inside the oocyte. With this, the chromosomes of oocyte and sperm combine, and fertilization is complete.

- Fusion of the sperm and oocyte cells brings about release of substances from the oocyte that block the entry of any other sperm—an important action since the fusion of two sperm with one egg would be fatal to the resulting embryo.

## 32.5 Human Development Prior to Birth

- Human development proceeds through the stages of animal development noted in Chapter 31: formation of the blastocyst, gastrulation, formation of the neural tube, and organogenesis or the formation of organs. Many of the most important stages in development are completed by the end of 12 weeks of pregnancy. The developing organism is called a zygote at conception, an embryo from the zygote's first division through the eighth week of pregnancy, and a fetus from the ninth week of pregnancy to the birth of the baby.

- The blastocyst implants itself in the endometrial lining of the uterus during the first week of pregnancy.

- The blastocyst initially consists of an inner cell mass, which develops into the baby, and a peripheral group of cells called a trophoblast. The trophoblast is active in uterine implantation and the production of cells that first extend into the endometrium and then establish links with the maternal blood supply. These cells grow into the fetal portion of the placenta, a complex network of maternal and embryonic blood vessels and membranes. The placenta allows for the movement of nutrients, wastes, and gases between the developing embryo and the mother.

Web Animation 32.3: Human Development Prior to Birth

## 32.6 The Birth of the Baby

- The immediate cause of birth is labor, meaning the regular contractions of uterine muscles that sweep over the fetus from its legs to its head. The pressure that results from these contractions opens or dilates the cervix, thus creating an opening large enough for the baby to fit through. It is not clear what the triggering mechanism is for the initiation of labor. When labor does not begin naturally following a full-term 38-week pregnancy, it can be induced by administering a synthetic form of the hormone oxytocin.

## KEY TERMS

| | |
|---|---|
| acrosome 678 | menopause 671 |
| amniotic fluid 684 | menstruation 669 |
| cervix 680 | oocyte 670 |
| corpus luteum 670 | ovarian follicle 670 |
| ectopic pregnancy 680 | ovulation 667 |
| embryo 680 | placenta 681 |
| endometrium 669 | prostate gland 678 |
| epididymis 675 | semen 668 |
| estrogen 672 | seminiferous tubule 676 |
| fetus 680 | |
| follicle 667 | spermatocyte 676 |
| follicle-stimulating hormone (FSH) 672 | spermatogonia 676 |
| | testis 675 |
| fraternal twin 682 | trimester 680 |
| gonadotropin-releasing hormone 672 | trophoblast 681 |
| | umbilical cord 681 |
| identical twin 682 | urethra 675 |
| inner cell mass 681 | uterine tube 667 |
| labor 685 | vas deferens 675 |
| luteinizing hormone (LH) 672 | zygote 680 |

## BRIEF REVIEW

*Answers to Brief Review questions are in the back of the book.*
*For multiple-choice quiz questions, go to* **www.aw.com/krogh4**.

1. Describe the hormonal changes that regulate the female reproductive cycle.

2. How do the testes act to protect developing sperm?

3. Menstruation is costly to females in some ways. Why do scientists believe it occurs?

4. Compare and contrast the tertiary follicle and the corpus luteum.

5. Which mechanism ensures that only one sperm will be able to fuse with a given oocyte?

6. Outline the events in human development that occur up through the time the embryo implants in the uterine wall.

7. Define the term *premature* with respect to human development.

8. How do birth control pills work?

## APPLYING YOUR KNOWLEDGE

1. When a couple is having trouble conceiving a child, the first thing fertility specialists generally will look at is the fertility of the prospective father. What do you think a specialist would be looking for?

2. Most mammals are "placental," which is to say their embryos get nourishment within the mother through the placental system of blood vessels reviewed in the text. By contrast, amphibian and reptilian embryos are nourished by substances contained in eggs the mother lays. Can you think of any advantages mammals derive from the placental means of development? Can you think of any disadvantages?

3. Forty years after birth control pills for women came into widespread use, there is still no birth control pill for men. One reason for this is the basic difference between female production of an oocyte each month and the male production of sperm. Why should it be easier to suppress fertility in women than in men?

*A school of Goldman's sweetlips fish (Plectorhinchus goldmanni) photographed in the coastal waters of Papua New Guinea.*

CHAPTER 33

33.1    The Study of Ecology                        692

33.2    Populations: Size and Dynamics              694

33.3    *r*-Selected and *K*-Selected Species       698

33.4    Thinking about Human Populations            700

Ecology is the study of how living things interact—with each other and with their physical environment. One way of studying these interactions is to focus on populations, meaning all the members of a single species living in one area.

# AN INTERACTIVE LIVING WORLD 1:
# Populations in Ecology

**T**he Audubon Society is generally known for a kindly attitude toward wildlife. So, how did it happen that in November 2004 Kristen Berry of the National Audubon Society ended up writing in an editorial,

"If You Love Birds, Shoot Deer"? Four months later, the New Jersey Audubon Society called for reducing its state's white-tailed deer population by hunting—by bringing in sharpshooters if necessary. And this wasn't some hastily taken position. The group studied the problem for a year and concluded that the threat deer pose to other wildlife is so severe that increased hunting is the best solution to the state's deer-driven "disaster."

What disaster is this? Well, for the Audubon Society, one important issue is that deer eat small-plant species that birds live in or feed on. Indeed, the National Audubon Society reports that bird species reliably get decimated in accordance with the number of deer per square mile. Beyond the problem with birds, when deer mow down native plant species, they leave an opening for invading alien plant species. In New Jersey, the native witch-hazel and spicebush forest shrubs are being replaced by plants such as Japanese stiltgrass. In some places, deer are eating so many saplings that forests themselves are threatened—no new trees are coming along to replace old trees as they die out. And, of course, it's not just wildlife that deer impact. Food crops, timber, and cultivated flowers all get damaged

by them. In 2003, there were an estimated 1.5 million collisions between motor vehicles and deer in the United States.

How did things come to this pass? How did the nation's deer population go from half a million in the early 1900s to at least 25 million today? Well, here's a recipe for how to explode a national deer population. First, get rid of a key deer predator, the wolf, by driving it to extinction in most states. Then, build houses

**In the Danger Zone**
The deer population in the United States has soared in recent decades, creating numerous instances of conflict between deer and human beings.

farther and farther out into the suburbs, on land adjacent to fields and forests. Now the deer have plenty of land to graze on (suburban lawns) that doesn't even have *human* predators on it. Then couple a growing sensibility about animal rights with a decline in hunting nationally. (Hunters are a potent force in this issue; in Wisconsin alone, they killed 484,000 deer in 2003.) The result? Deer right in your back yard, on your streets, ruining your forest preserve. You might think that the silver lining to this is that the *deer* are at least coming out of it fat and content, but consider that in Ohio's Cuyahoga Valley National Park, deer have been starving to death in recent winters because their habitat has been picked clean by . . . deer.

For our purposes, the story of deer in the United States drives home the important message that living things do not live in isolation from one another. What happens with one species affects another. What happens with wolves affects deer, then deer affect plants, and plants affect birds. Predators and prey, native and alien species, plants and animals—all these things are part of an *interrelated* web of life. Moreover, this is a web of life that exists in a larger context yet: the *non*living world and universe, which is to say the Earth's climate and atmospheric gases, its sea currents and volcanoes, and the energy it gets from the sun. Life doesn't exist apart from these things. Instead, it is very much conditioned by them—although, as you'll see, some of these things are also conditioned by life.

## 33.1 The Study of Ecology

In your walk through biology so far, you have made it through the molecular world (DNA, for example) and through the world of whole organisms, such as fungi and plants. Now it is time to take the last steps up; in this chapter and the two that follow, you will consider biology's largest realm of all, **ecology**, defined as the study of the interactions that living things have with each other and with their environment. The final chapter in the book, on animal behavior, is also placed within the ecology unit because it too is concerned with interactions in the living world.

### Ecology Is Not Environmentalism

In common speech, the word *ecology* has become synonymous with the conservation of natural resources or with the "environmental movement" in all its political dimensions. Under this view, you might assume that the job of an ecologist is to help preserve the natural world. In fact, however, the job of an ecologist is to *describe* the natural world in its

largest scale; to say what is, rather than what should be. Are ecologists environmentalists? Most are, but this shouldn't be surprising. We would no more expect ecologists to be indifferent to the environment than we would expect art historians to be indifferent to paintings. Moreover, there is a branch of biology, conservation biology, that is directly concerned with preserving natural resources. But the business of ecology per se is to tell us about the large-scale interactions that go on in the natural world. This work has an important use in that it provides the information base that society can use to make decisions about the environment. Does species diversity make for healthier natural communities? How many gray whales are there in the world? How important are prairie dogs to the prairie? To answer these questions, ecologists end up working in some places you might expect—in tents near the Arctic Circle or at treetop level in tropical rain forests (**Figure 33.1**). They also work in some locations you might not expect, however, such as in offices, hunched over computers, seeing how mathematical models of global warming work out.

**Figure 33.1**
**An Ecologist on the Job**
Ecologists spend a good deal of time going out into nature to study the conditions that exist in various ecosystems—however inaccessible those ecosystems may be. Here, a researcher is studying tree branch growth in the rain-forest canopy in Costa Rica.

## Path of Study

In this chapter and the next two, you'll follow a bottom-to-top approach by moving from smaller ecological units to larger ones. What are the scales of life that concern ecology? The smallest is that of the physical functioning, or physiology, of given organisms. At this level, ecologists are examining individual living things, but at all other scales they are looking at groups of organisms or species. For the ecologist, life is conceptualized as being organized into

- Populations
- Communities
- Ecosystems
- The Biosphere

### From Individuals to a Population

The smallest level of group organization is a **population**, defined in Chapter 17 as all the members of a single species that live together in a specified geographic region. The North American bullfrog (*Rana catesbeiana*) is a species that can be found from southern Canada to central Florida, but all the *R. catesbeiana* in a given pond constitute a single population of this species. Of course, it's possible to define our population to be all the bullfrogs in *two* ponds, or two states, or in two countries—whatever geographic region is most useful for the question an ecologist is asking.

### From Populations to a Community

Going up the scale, if you take the populations of *all* species living in a single region, you have a **community**. Often, the term community is used more restrictively to mean populations in a given area that potentially interact with one another.

### From a Community to an Ecosystem

If you add to the community all the *non*living elements that interact with it—rainfall, chemical nutrients, soil—you have an ecosystem. More formally, an **ecosystem** is a community of organisms and the physical environment with which they interact. Ecosystems can be of various sizes; you'll be looking at a range that goes from fairly small (a single field, for example) up through the enormous ecosystems called biomes that may take up half a continent.

### From Ecosystems to the Biosphere

The largest scale of life is the **biosphere**, which can be thought of as the interactive collection of all the Earth's ecosystems. Given what you've reviewed about ecosystems, this means all life on Earth and all the nonliving elements that interact with life. Sometimes, however, the term *biosphere* is used purely in a territorial sense—to mean that portion of the Earth that supports life. If you look at **Figure 33.2**, you can see a graphic representation of ecology's scales of life. We'll now begin to look at ecology

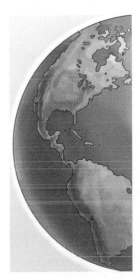

| **organism** (sea lion) | **population** (colony) | **community** (giant kelp forest) | **ecosystem** (Southern California coast) | **biosphere** (Earth) |

**Figure 33.2**
**The Scales of Life That Concern Ecology**
A population is made up of all the organisms of a single species that live together in a defined region—in the example shown, a population of sea lions living together off the California coast. A community is a more inclusive grouping: all the members of all species that live together in a defined region. In the example, this means not only the sea lions, but giant kelp, fish, and a multitude of other species. An ecosystem includes living community members and the nonliving factors that interact with them (such as the tides and wind currents that affect the California coast). The biosphere is then the interactive collection of all the Earth's ecosystems.

through these levels of biological organization, starting small, with the population.

## 33.2 Populations: Size and Dynamics

So, what is it ecologists want to know about a population? Well, they need to know how to count it, how and why it is distributed over its geographical area, and how and why its size changes over time—what its *population dynamics* are, to use the term employed by ecologists.

### Estimating the Size of a Population

The reason ecologists want to count the members of a given population is straightforward: Without such a count, there is no way to answer a question such as how much territory a group of cheetahs must have to flourish or how fast a population of finches recovers from a drought. (How would you know about the latter unless you could compare the population after the drought with the population before it?)

With large, immobile species such as trees, taking a census can sometimes be easy; just mark off an area and count. Things become more difficult when the area under consideration is so large that not all individuals in it can be counted, but rather must be estimated based on a population counted in a smaller representative area. Estimating becomes more difficult yet when the individuals are numerous and

mobile, as with birds. Ecologists employ various means to estimate such populations; they count animal droppings within a defined area or survey bird populations as they migrate, for example.

### Growth and Decline of Populations over Time

How is it that populations *change* size? As you begin to think about this question, it's worth going over a more general concept, which is the way a population of anything might increase in number. If you look at, say, the number of cars coming off the end of a production line, there might be 1,000 on Monday, then 1,000 on Tuesday, and so forth. The important thing to note is that the number of cars produced on Tuesday is not related to the number produced on Monday. The increase in the number of cars is thus an **arithmetical increase**: Over each interval of time (a day in this case), an unvarying number of new units (cars) is added to the population.

Now contrast this with what happens to living things. In most cases, each new unit (each living thing) is capable of playing a part in giving rise to more units, which certainly is not the case with cars. Population increases for living things are thus *proportional to* the number of organisms that already exist. Thus, the increase in a population of organisms comes about through a different sort of increase, an **exponential increase**, which occurs when, over an interval of time, the number of new units added to a population is proportional to the number of units that exist. **Figure 33.3** gives an example of the difference between the two kinds of growth over a period of weeks, using cars for arithmetic growth and the tiny water flea *Daphnia* for exponential growth. Let's assume we start out at the end of day 1 with the 1,000 cars produced that day and with 1,000 water fleas existing in an optimal laboratory environment. As you can see, *Daphnia* is a relatively slow starter, but its population quickly overwhelms that of the car population—not surprising, because the *Daphnia* population doubles every 3 days.

### Population Growth in the Real World

As noted, the *Daphnia* were growing in an "optimal" laboratory environment—plenty of food, habitat kept clean, no predators. Thanks to human intervention, this was a kind of paradise for water fleas, in other words. But could any population ever grow like this in the real world? In all cases, the answer is "not forever," and in most cases the answer is "not for

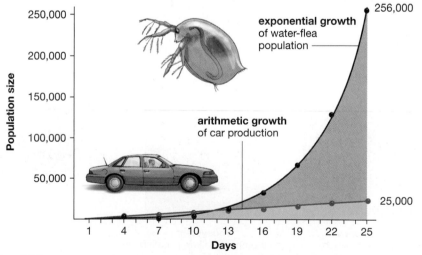

**Figure 33.3**
**Arithmetic and Exponential Growth**
When the same number of objects is produced in a given interval of time—in this case, cars from a factory each day—arithmetic growth is at work. By contrast, the populations of most organisms, such as the water flea *Daphnia*, can exhibit exponential growth, at least for a time, meaning the population grows in proportion to its own size.

long," but some population growths can be dramatic while they last.

In 1859, Thomas Austin released 24 wild European rabbits (*Oryctolagus cuniculus*) onto his estate in southern Australia, near Melbourne, to provide more game for sport hunting. His was not the first attempt to introduce rabbits to Australia, but it was by far the most successful—or perhaps the most fateful. Having no natural predators in Australia, this alien species spread north and west like wildfire; within 16 years, it had expanded its range through the entire latitude of the continent, a distance of almost 1,800 kilometers or 1,100 miles. And this invasion was not a matter of a rabbit here and a rabbit there; the European rabbits became a scourge that only was brought under partial control, beginning in the mid-1990s, through use of a virus (**Figure 33.4**). Here is Eric Rolls' description in *They All Ran Wild* of the rabbit problem east of Peterborough in South Australia in 1887:

> For over a hundred miles most trees and plants had been killed. Outalpa head-station was a thousand square miles of rabbits . . . . Rabbits crawled like possums in branches several feet off the ground. They had burrowed under the beautiful granite homestead; they had taken possession of the cellar; they were in the turkeys' coop; they ran in and out of the men's huts apparently without fear; they took no notice of the dogs nor the dogs of them.

There are lots of instances of such seemingly uncontrolled growth. Charles Darwin devoted some thought to the issue as a theoretical matter. He wrote that "There is no exception to the rule that every organic being naturally increases at so high a rate, that, if not destroyed, the Earth would soon be covered by the progeny of a single pair." To put a little finer point on this, a single pair of carrion flies (which are a little larger than a housefly) could in theory give rise to enough flies in a single year to cover the state of New Jersey to a depth of 1 meter (3 feet).

### The Shape of Growth Curves

If you look at **Figure 33.5a** on the next page, you can see what growth of this type, **exponential growth,** looks like when plotted on a graph. This is called the *J-shaped growth curve* because it can be so extreme that when plotted it looks like the letter J. It may look familiar because it is the growth form you just saw with *Daphnia*. As with *Daphnia,* there is relatively slow growth at first, then faster growth, then faster yet. This is growth, in other words, in which the *rate* of increase keeps accelerating. In the real world, the

**Figure 33.4**
**Out of Control Down Under**
Rabbits have been a huge ecological problem in Australia for 120 years. Although their numbers have been reduced in recent decades, they are still destructive to both Australian agriculture and wildlife. Here, part of a large rabbit population in southern Australia can be seen running on one side of a "dingo fence."

**(a)** Exponential growth

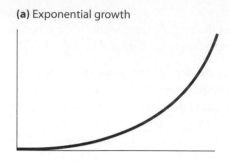

**(b)** Logistic growth

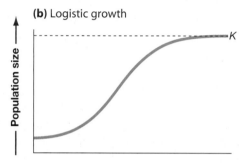

**(c)** More complex growth

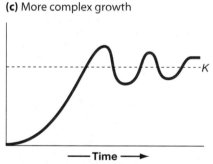

**← Time →**

**Figure 33.5**
**Three Models of Growth for Natural Populations**

curve eventually drops off, often sharply, for reasons you'll go over soon.

Now, look at **Figure 33.5b** to see a second model of growth for natural populations, the *S-shaped growth curve*, which tracks **logistic growth**. This starts like exponential growth, but eventually the rate of growth slows and finally ceases altogether, stabilizing at a certain level, denoted here as *K* (which you'll learn the significance of shortly).

What has intervened to account for the difference between the dizzying increase of the J-shaped curve and the moderate increase of the S-shaped curve? All the forces of the environment that act to limit population growth, which in ecology are known as **environmental resistance**. Organisms will run out of food or have their sunlight blocked; greater numbers of predators will discover the population; the wastes produced by the organisms will begin to be toxic to the population.

There are populations whose dynamics over time look something like either the J- or S-shaped curve, but in the real world, population size is likely to vary in more complex patterns. In **Figure 33.5c**, you can see one example of such a pattern.

## Calculating Exponential Growth in a Population

You've seen that ecologists have ways to count natural populations, but is it possible for them to *predict* the size of populations given certain pieces of starting information? The answer is yes, as you'll now see. Let's look first at a simplified means of calculating exponential growth, meaning the J-shaped curve. For now, assume the population is isolated—that is, that no individuals are moving in or out of it.

What first needs to be done is to pick a time period to evaluate population change. This period may be 40 minutes (for bacterial populations) or 50 years (for elephants), but some unit of time needs to be chosen for the analysis. The next step is to learn the difference between the birth rate and the death rate in the population over the time period you've picked. This, of course, raises the question: What's a birth rate? It is the number of individuals born in a given time period, expressed as a proportion of the whole population. Thus, if a population of 1,000 had 100 births in a year, its birth rate would be 0.10 or 10 percent per year. Figuring the death rate is similar; if a population of 1,000 had 80 deaths per year, its death rate would be 0.08 or 8 percent per year.

The *difference* between these birth and death rates, of course, is 2 percent. This number is the population's growth rate in an optimal setting, denoted as *r*—standing for the *intrinsic rate of increase*—and it is very important. With it in hand, you can predict what the population's size will be in the future. Growing at 2 percent per year, the population of 1,000 will increase to 1,020 in the first year and to about 1,480 in 20 years.

Growth rate (*r*) is critical because a small change in it can mean a big change in population over time. Take a population of another species, once again with 1,000 individuals in total, but this time with an *r* of 6 percent rather than 2 percent. In the first year, this population would grow by 60 individuals, giving it 1,060 members as opposed to the first population's 1,020. Go 20 years out, however, and this population has about 3,200 individuals in it—it's now more than twice the size of the first population even though both were the same at the start. Meanwhile, if a population's *r* is less than zero—that is, if the number of deaths exceeds the number of births—the population is shrinking. If $r$ is zero on the nose, meaning that births exactly match deaths in a given period, the population is at **zero population growth** (**Figure 33.6**).

As noted, the $r$ figure goes by another name, which is a population's **Intrinsic rate of increase**, the rate at which a population would grow if there were no external limits on its growth. You could think of it as a population's potential for growth. Importantly, populations of each species have characteristic rates of increase. It won't surprise you to learn that this rate is higher for *Daphnia* populations than it is for whale populations; you can see some examples of an important factor in rate of increase, the time it takes to produce a new generation, in **Figure 33.7**.

## Logistical Growth of Populations: Reality Makes an Appearance

You've now observed exponential or J-shaped growth, but you've seen that such growth never occurs for long in nature, since environmental resistance will always begin to assert itself. In some instances, there will be a flattening out of the population increase and in time perhaps a complex pattern with the curve moving above and below the line denoted earlier as $K$. But, what is $K$? What is this point around which the population hovers? It is the maximum population density of a given species that a defined geographical area can sustain over time. The term for this measure is an area's **carrying capacity ($K$)**.

### A Real-Life Example of Carrying Capacity

There is an island in the San Francisco Bay called Angel Island, just a short boat ride from the Golden Gate side of the Bay, that had a small population of deer introduced to it by humans early in the twentieth century. The problem was that the island lacked any natural predators for the deer, with the result that the deer population proliferated way beyond what the island's vegetation could support. Had this taken place in a purely natural environment, the deer population would have shrunk in accordance with the amount of available vegetation. But there were human hikers on Angel Island, and seeing emaciated deer, they did what you might expect: They fed the deer with what they brought along in their backpacks. Surviving through this artificial means, however, *more* deer came to live on the island and stripped it of its vegetation, thus reducing its carrying capacity.

The main lesson here is that there are limits to how many members of a given species an expanse of

territory can support. Note also, however, that carrying capacity is not fixed; in the case of Angel Island, it was lowered as the deer, artificially fed by human beings, stripped the island of vegetation. In other

**Web Animation 33.1**
Population Growth

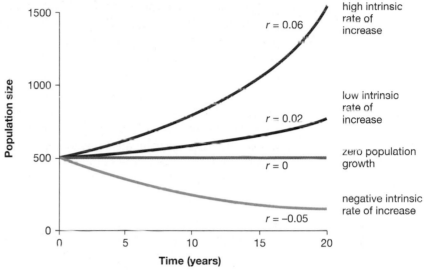

**Figure 33.6**
**A Population's Potential for Growth: The Intrinsic Rate of Increase ($r$)**
If the birth rate exceeds the death rate, then the population size will increase over time ($r = 0.06$ in the example). If the birth rate equals the death rate, then the population size will stay constant ($r = 0$). Finally, if the birth rate is lower than the death rate, the population size will decrease over time ($r = -0.05$). Note what dramatic effects a small change in $r$ can have over time, in this case on a population whose starting size is 500 individuals.

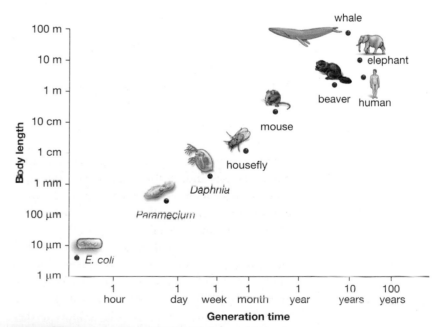

**Figure 33.7**
**How Long between Generations?**
An important factor in the population dynamics of any species is its generation time, meaning the time that elapses between the birth of one generation and the birth of the next. Shown on the graph are the minimum generation times for a few selected species, along with a measurement of the length of each organism at the time it reproduces. The larger a species is, the longer its generation time tends to be. (Redrawn from J. T. Bonner, *Size and Cycle: An Essay on the Structure of Biology*, Princeton University Press, 1965.)

instances, carrying capacity might be raised for given periods (perhaps through abundant rainfall). The environments that many species live in tend to be stable over time, however, meaning that their carrying capacities will tend to be stable as well.

The $K$ line can generally be established for a given population by estimating its numbers over an extended period of time. For many species, their population may exceed or "overshoot" $K$ at times, but this excess will be offset by periods during which the population is under $K$. Once $K$ is known, it is valuable in predicting the change in a population's size because it often serves as a factor that can be used in mathematical equations to calculate the limits to population growth. The higher a population is relative to $K$, the more severely its growth will be limited.

---

### SO FAR...

1. A population is all the members of a species that live together in a _____. A community is all the populations of all _____ living in a single region. An ecosystem is a community of organisms and the _____ with which they interact.

2. Environmental resistance can be defined as all the forces of the environment that act to _____.

3. Carrying capacity can be defined as the maximum _____ of a given species that a defined geographical region can _____.

---

## 33.3  *r*-Selected and *K*-Selected Species

It's well known that for human beings, about 9 months will elapse between conception and birth, but it's less well known that for an elephant this same process takes almost 2 years. For any given human or elephant female, then, births are relatively few and far between. And once the offspring come, the amount of attention lavished on them is great indeed. It's not unusual for an elephant calf to be nurtured by its mother and a larger kinship group for at least the first 10 years of its life. In the first year of this period, a mother might not ever let her calf wander farther than 20 meters away from her. Meanwhile, humans may lavish 18 years of care and concern on a child before the domestic bonds are finally broken. In contrast, if we look at the common housefly, we find that it can produce a new generation once a month, and that the parental generation provides no care whatsoever to the offspring.

Houseflies and elephants lie at opposite ends of a continuum of reproductive strategies, meaning characteristics that have the effect of increasing the number of fertile offspring an organism bears. So, how does the strategy of elephants differ from the strategy of houseflies? This may be apparent from what you've observed already. For elephants, the strategy is to bring forth few offspring but lavish attention on them. For houseflies, it is to bring forth a multitude of offspring but give no attention to any of them. These strategies are in turn related to other characteristics of these species, as you'll now see (**Figure 33.8**).

### *K*-Selected, or Equilibrium, Species

Elephants will not seek out a totally new environment; they experience their environment as a relatively stable entity; and they compete among themselves and with other species for resources within it. Given this stability, species like elephants are known as an *equilibrium species*. In line with this, the elephant population stays relatively stable compared to a fly population; its numbers will fluctuate in a relatively narrow range above and below the environment's carrying capacity. This is another way of saying that the population regularly bumps up against carrying capacity ($K$), which is the density of population that a unit of living space can support. Two things follow from this. First, elephants are said to be a **K-selected species**: one whose population sizes tend to be limited by carrying capacity. The second point, which follows from this definition, is that the pressures on the elephant population are **density dependent**: As the density of the population goes up, the factors that limit the population—food supply, living space—assert themselves ever more strongly.

### *r*-Selected, or Opportunist, Species

In contrast to elephants, houseflies are known as an *opportunist species*—a species whose population size tends to fluctuate greatly in reaction to variations in its environment. Should favorable weather suddenly arrive or a food supply suddenly appear, the fly population in the area will skyrocket. When the food is gone or if the temperature suddenly changes, this same population will plunge in number. In short, environment tends to be highly variable for fly populations, and the flies have a high population growth rate that allows them to take advantage of environmental opportunities. For these animals, the strategy is to produce a multitude of offspring very

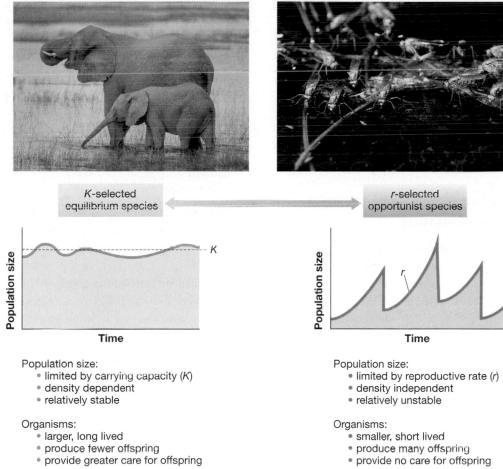

**Figure 33.8**
***K*-Selected and *r*-Selected Species**

The elephants on the left are a *K*-selected species. Here, a mother stays close to her calf, providing the kind of careful attention typical of *K*-selected species. The pond flies on the right are an *r*-selected species, bearing many young but giving no attention to them after birth.

fast. Flies are therefore said to be an **r-selected species**, meaning one whose population sizes tend to be limited by reproductive rate.

This definition implies that fly population numbers have nothing to do with fly population *density*, at least for a time. The pressures on the population are thus said to be **density independent**. The forces at work in these populations tend to be *physical* forces—frost, temperature, rain—as opposed to the *biological* forces that operate in density-dependent control (competition for food, disease). Given this, the *r*-selected strategy of many offspring/little attention is understandable. When you exist at the whim of environmental forces, "life is a lottery and it makes sense simply to buy many tickets," as ecologists R. M. May and D. I. Rubinstein have phrased it.

The *r*- and *K*-selected groupings are categories invented by human beings, and as usual, nature refuses to fit neatly into such boxes. It is true that small, short-lived species tend to lie at the *r*-end of the continuum, and that large, long-lived species lie at the *K*-end, but sea turtles are long lived, and they give no care at all to their newborns. Meanwhile, flies can be limited in a density-dependent way for at least part of their population cycle—by predators and by competition among themselves. All sorts of variations exist on this theme. We've looked at animals as examples of *r*- and *K*-selection, but the concept applies to plants as well. A plant such as ragweed is at the *r*-selected extreme in the plant world, while oak trees and cacti are at the *K*-selected extreme. The ragweed grows rapidly and has a short life span, while the oak grows slowly and lives a very long time.

## Survivorship Curves: At What Point Does Death Come in the Life Span?

*K*- and *r*-strategies are related to the concept of "survivorship curves," which as you can see in

••• ANSWERS

1. defined geographical region; species; physical environment

2. limit the growth of a population

3. density; support

**Figure 33.9**, are thought of in terms of three ideal types: late loss, constant loss, and early loss (sometimes referred to as types I, II, and III). Humans and most of our fellow *K*-selected mammals fall into type I because we tend to survive into old age (our lives are lost late). Insects and many amphibians are type III because their death rates are very high early in life, but level out thereafter. Other types of living things, such as birds, fall into type II because they die off at a constant rate through their life span.

## 33.4 Thinking about Human Populations

We'll look at the final elements in population dynamics in connection with the human population. Some of the concepts involved could be applied to any population of living things. But here, the focus is on human beings, so that you can consider population principles along with the real-world issue of human population growth, which figures so prominently in environmental issues.

### Survivorship Curves Are Constructed from Life Tables

The survivorship curves you just looked at are constructed by developing what are known as **life tables** for the species in question. To create these, scientists divide the species' life span into suitable units of time—days for fruit flies, years for human beings—and see how likely it is that an average species member will be alive after a given number of days or years. The

technique for doing this was not invented by biologists who wanted to know about the survival of flies, but by a nineteenth-century British actuary—a person trained to calculate probabilities—who knew the information would be useful for life insurance companies. How likely is it that a person will be alive at 10, or 20, or 80 years of age? If you look at **Table 33.1**, you can see the answer for an average group of 100,000 persons born in the United States in the year 2003.

### Population Pyramids: What Proportion of a Population Is Young?

You've seen that population growth in the natural world can be exponential (for a time) because a population of living things grows in proportion to its own size. However, all members of a population do not count equally in calculating this growth. If scientists want to peer into the future of a given population, what they want to know is: What proportion of the population is *past* the age of reproduction as opposed to the proportion that is, or will be, of reproductive age? If you look at **Figure 33.10**, you can see the answer to this question, expressed as a population pyramid for two countries in the year 2006. Each bar on the graph represents a 5-year age grouping, or "cohort" of the population (those who are 0–4 years old, 5–9, and so on), with the length of each bar representing the size of the cohort. The reproductive age range for humans generally is considered to be between 15 and 45. If you look at the graph for Kenya, you can see that it has many more individuals of reproductive and pre-reproductive age than of post-reproductive age.

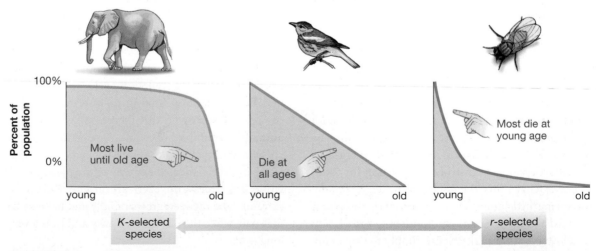

**Figure 33.9**
**When Is Death Likely to Come?**
An organism's chance of living a long life is related to the reproductive strategy of its species, as reflected in these survivorship curves. Elephants and humans (late-loss species) produce few young, but most survive until old age. At the other extreme, flies (early-loss species) produce many young, but most die young. In between are constant-loss species; they produce a moderate number of young, which die at a fairly constant rate during a typical life span.

## Table 33.1  A Life Table for the United States

Taking a hypothetical group of 100,000 persons born in the United States in 2003, the table shows the number likely to still be living at the ages indicated and the average remaining lifetime for persons at each age. The numbers are averaged for men and women, a choice that masks significant differences between the sexes in older age cohorts. Of 100,000 women born in 2003, for example, about 60,000 are likely to be alive at age 80, whereas for men the figure is about 45,000. Note that life expectancy does not drop 10 years for each 10 years lived. The average person surviving to age 70 can expect to live almost to age 85, whereas the average person at age 10 could expect to live only to about age 78.

| At age | Number still living | Average remaining lifetime in years |
|--------|---------------------|-------------------------------------|
| 10 | 99,116 | 68.2 |
| 20 | 98,693 | 58.4 |
| 30 | 97,752 | 48.9 |
| 40 | 96,444 | 39.5 |
| 50 | 93,585 | 30.6 |
| 60 | 87,760 | 22.2 |
| 70 | 75,535 | 14.9 |
| 80 | 52,741 | 9.0 |
| 90 | 21,340 | 5.0 |
| 100 | 2,363 | 2.6 |

Source:  National Center for Health Statistics

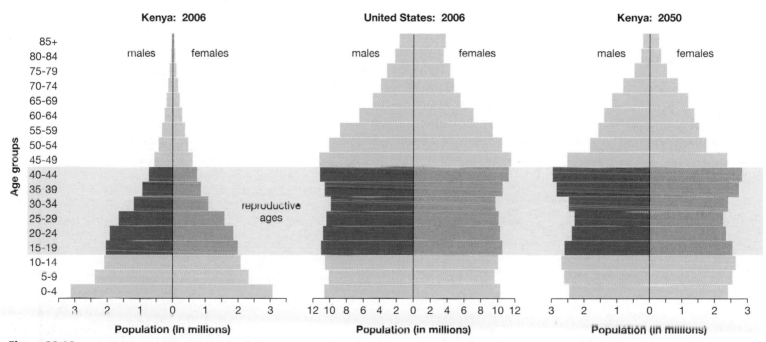

**Figure 33.10**
**Age Structure in Populations: Kenya and the United States**

Each bar on the graphs represents a 5-year age grouping of the population. Note the greater proportion of the 2003 Kenya population that is of "prereproductive" age—an indication of greater population increases in the future than is the case with the United States. The graph on the right shows the change expected to come about in Kenya's age structure by the middle of the century. (Data from the U.S. Census Bureau.)

What does the future hold for Kenya? In all probability, a large growth in the population. In contrast, the age-structure diagram for the United States shows that a much greater proportion of its population is at or beyond reproductive age.

## The World's Human Population: Finally Stabilizing

You probably noticed that there is a third graph in Figure 33.10 that displays Kenya's expected population pyramid for the year 2050. It's easy to see that the projected age distribution for Kenya looks something like the current age distribution for the United States. As a result, Kenya can look forward to slower population growth in the years after 2050. This Kenyan projection actually is just one example of a momentous shift that is going on with human population around the world today. Only in the last few years has it become clear that, following centuries of explosive growth, the human population finally appears to be stabilizing.

To put this change into a larger context, look at **Figure 33.11**, which shows you how human population has grown during "historical time," which is to say the last 10,000 years. For centuries, human numbers grew relatively little but then began an upward climb about 1700. Improved sanitation, better medical care, and increases in the food supply came together to produce the rate of growth you see. The Earth's human population didn't pass the 1-billion mark until 1804; it then took 123 years to double to 2 billion (in 1927), then 48 years to double to 4 billion (in 1974), and now it has reached 6.5 billion. The curve you see that resulted from this growth may look familiar because it is a more extreme version of the J-shaped curve, although greatly elongated on its left side. Were we

looking at any species other than our own and saw this kind of growth curve, we would predict that this kind of increase could not go on for long; that environmental resistance would assert itself, and that the population might even crash in one catastrophic sense or another.

But something has changed. In 1992, demographers in the United Nations were predicting that, by 2050, the Earth would have 12 billion people on it. But in reports issued in 2003 and 2005, the United Nations predicted that the human population will be about 9.1 billion by 2050, and that it will peak out at 9.2 billion people in 2075, after which it will stabilize at something just under this level. To be sure, 9.2 billion represents a big increase from today's 6.5 billion. But this jump is coming about via a *rate* of increase that is steadily declining. From 1990 to 1995, world population was increasing by 82 million people per year; by 2050, it should be down to 34 million people a year.

What has made the difference? The key measurement used by demographers to predict human population change is the **total fertility rate**, which can be informally defined as the average number of children born to each woman in a population. In more developed countries, the "replacement" fertility rate is 2.1. If fertility is above 2.1, the population will grow; if it's below 2.1, the population will shrink. (Why 2.1 instead of 2.0? Some children will not live to have children of their own.) With this in mind, look at **Figure 33.12**. Pictured are fertility levels for a selected group of countries in two periods: 1960–65 and 2000–2005. The dramatic declines you can see are not being repeated in all countries. Indeed, in some countries fertility isn't dropping at all. But when all the world's countries are averaged together, we find that that global fertility has been

**Figure 33.11**
**Human Population Increase through the Centuries**

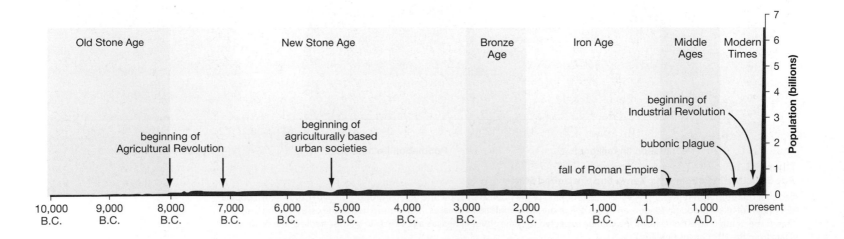

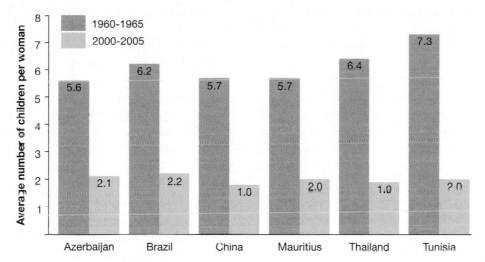

**Figure 33.12**
**Big Changes in Fertility**
Pictured are female fertility rates in six countries during two periods: first during the 1960s and then for the period 2000–2005. Changes such as these have brought about a remarkable decline in the worldwide total fertility rate. (Source: Population Reference Bureau.)

cut nearly in half over the past 50 years. Whereas it stood at 5 in the period 1950–55, it now stands at 2.65.

This global average, however, masks some tremendous differences between more-developed and less-developed countries. Think of it this way. In 2005, the world's population was increasing by 154 people per *minute*. And, of these 154 additional people, 151 were born in less-developed countries. The South Asian country of Bangladesh, which is about the size of Iowa, is expected to have a population increase of more than 100 million people between 2005 and 2050. Meanwhile, in the countries of Europe, the fertility rate is now so low—at 1.4, on average—that the population is expected to decline by about 75 million people during this same period.

Where does the United States stand in this spectrum? In the coming decades, it is expected to have more population growth than any other large, developed country, both in proportional and absolute terms. The U.S. population is expected to go from 300 million people in 2006 to 420 million people in 2050—a 40 percent increase. What makes the United States different from the other developed countries? Partly the American fertility rate, which at 2.03 is not quite at replacement level, but is much higher than those of European countries. The biggest single factor driving the U.S. population increase, however, is immigration. About 1.5 million people immigrate to the United States each year.

## Human Population and the Environment

Almost everyone is aware of a linkage between the quality of the environment and the size of human populations. Some scientists argue that there is no greater single environmental threat than the continued growth of the human population. The basis for this conclusion is that population affects so many environmental issues: the use of natural resources, the amount of waste that is pumped into the environment daily, the reduction of species habitat, the decimation of species through hunting and fishing. Look at almost any environmental problem, and you are likely to find human population growth playing a part in it.

Given that the human population is still growing—but that stabilization of it is on the horizon—it's possible to see the twenty-first century as a crucial moment in time for the environment. The biologist Edward O. Wilson has characterized our century as an environmental *bottleneck*: as a period in which the Earth's ecosystems will be under the greatest assault they are ever likely to face. On the other side of the bottleneck, with human population stabilized in general and declining in some regions, the pressures on the environment should ease. The question is: How many species and habitats can make it *through* the bottleneck? How many primate species can be preserved? How many acres of rain forest can be kept intact? How many acres of American wetlands can remain as they are?

Some experts argue that it is not population per se but rather the use of resources *per person* that is of most pressing concern. Consider that the United States used less water in 2000 than it did in 1980 even though the U.S. population grew by 24 percent during the period. Another perspective on this is provided by the use of resources in more-developed, as opposed to less-developed, countries. If you look at **Figure 33.13**, you can get a sense of the differences that exist. The carbon dioxide ($CO_2$) tracked in the figure is a gas that is found naturally in the atmosphere, but that is also put into the atmosphere when human beings burn fossil fuels, such as coal or gas, to power car engines or run power plants. $CO_2$ produced by these human activities is now regarded as the most significant cause of global warming—the rise in Earth's surface temperature that has occurred over the past century.

The Figure 33.13 graph on the left shows the amount of $CO_2$ produced per person (or "per capita") by human activity in the United States and in China in two different years, 1997 and 2003. A quick glance at the figure shows how little $CO_2$ China produced per capita in either year compared to the United States. In 2003, $CO_2$ emissions in the United States totaled more than 5 metric tons per person, while in China they came to less than a ton per person. Also note, however, that China's per capita emissions grew by 19 percent between 1997 and 2003, while U.S. emissions declined slightly. To see the effects of this change, look at the graph on the right. As it shows, the $CO_2$ emissions for China

as a *country* went from 59 percent of those in the United States in 1997 to 72 percent of those in the United States in 2003. And this is a shift that's only getting started. The International Energy Agency has predicted that, by 2009, China will *surpass* the United States in total $CO_2$ emissions. What's going on in China? Very little population growth, at least in proportional terms. What is going on is development, which drives the per capita use of resources. China is industrializing at a rapid rate, and it is fueling its industrialization largely with coal. The $CO_2$ output that is resulting from this industrialization provides a clear, if depressing, lesson: When a large human population increases its per capita use of resources even by a little, the result can be big trouble for the environment.

## On to Communities

The populations we've looked at this chapter have been abstractions, in a sense, in that we've thought of each of them as existing in a kind of isolation. To be sure, populations do have their own internal dynamics. In an elephant population, for example, only a few young will be born, great care will be lavished on those who are born, and so forth. Yet every elephant population is constantly interacting with populations of other species. Consider that, in the wild, an African elephant can consume up to 200 kilograms or 440 pounds of grasses, leaves, and roots each day. Moreover, it's not uncommon for elephants to pull down small trees as a means of getting at some of this

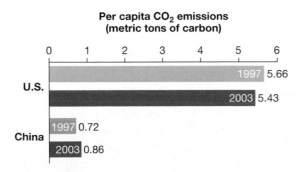

**Per capita CO₂ emissions**
**(metric tons of carbon)**

U.S. 1997 5.66
U.S. 2003 5.43
China 1997 0.72
China 2003 0.86

U.S. per capita $CO_2$ emissions far exceed those in China, but China's per capita emissions grew by 19% between 1997 and 2003 . . .

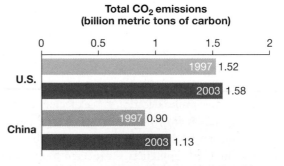

**Total CO₂ emissions**
**(billion metric tons of carbon)**

U.S. 1997 1.52
U.S. 2003 1.58
China 1997 0.90
China 2003 1.13

. . . When coupled with China's large population, this per capita growth meant that China's total $CO_2$ emissions went from 59 percent of those in U.S. in 1997 to 72 percent of those in U.S. in 2003.

**Figure 33.13**
**Per Capita and Total Carbon Emissions**
The use of natural resources per person (or "per capita") varies greatly from one country to the next. The average resident of a developed country, such as the United States, uses far more resources than the average resident of a less-developed country, such as China, and this has environmental consequences. Some developing countries are, however, starting to use more resources. The increase in China's per capita $CO_2$ emissions between 1997 and 2003 brought about a big change in China's total $CO_2$ emissions relative to those of the United States. (Data from G. Marland, T. A. Boden, and R. J. Andres, "Global, Regional, and National $CO_2$ Emissions," in *Trends: A Compendium of Data on Global Change*, Carbon Dioxide Information Analysis Center, Oak Ridge National Laboratory, U.S. Department of Energy, Oak Ridge, Tennessee, 2006.)

material. So African elephant populations can have a significant negative impact on African tree populations. On the other hand, the only way the seeds of some African trees get dispersed is through elephants—*literally* through elephants, in that the seeds pass through the elephant digestive system, after which they're deposited, in elephant dung, at new sites for germination and growth. African elephants and trees thus interact in some complex ways. When we start thinking about such interactions, we are thinking about life at the level of a community, meaning all the populations of all species that live in a defined geographical area. So, in what ways can community members interact? Beyond this, how do communities get established and then become stabilized? And are some species more important than others in preserving a community's structure? These are some of the questions we'll take up in Chapter 34.

## SO FAR...

1. *K*-selected or equilibrium species tend to have _____ offspring to whom they provide _____ care. Conversely, *r*-selected or opportunist species tend to have _____ offspring to whom they provide _____ care.

2. A country whose population pyramid is widest at the bottom can expect to have _____ population growth in the near future than a country whose population pyramid is of uniform width throughout.

3. A country's human population changes can be predicted by looking at its total fertility rate, informally defined as the _____. The "replacement" level for fertility in more-developed countries is _____.

**... ANSWERS**

1. few; abundant; many; little or no

2. greater

3. average number of children born to each woman; 2.1 births per woman

# CHAPTER 33 REVIEW

For study help, activities, and more quiz questions, go to **www.aw.com/krogh4**.

## SUMMARY

### 33.1 The Study of Ecology

- Ecology is the study of the interactions living things have with each other and with their environment.

- Ecology is not the same thing as environmentalism. The function of ecology is to describe interactions that affect the living world, not to work on behalf of the environment.

- There are five scales of life that concern ecology: physiology, populations, communities, ecosystems, and the biosphere. A population is all the members of a single species that live together in a specified geographical area; a community is all the members of all species that live in a single area; an ecosystem is a community and all the nonliving elements that interact with it; the biosphere is the interactive collection of all the Earth's ecosystems.

### 33.2 Populations: Size and Dynamics

- Ecologists employ several means to estimate the size of populations of living things, among them counting animal droppings or surveying bird populations as they migrate.

- An arithmetical increase occurs when, over a given interval of time, an unvarying number of new units is added to a population. An exponential increase occurs when the number of new units added to a population is proportional to the number of units that exists. Populations of living things are capable of increasing exponentially because living things are capable of giving rise to more living things.

- The rapid growth that sometimes characterizes living populations is referred to as exponential growth or as the J-shaped growth curve. Populations that initially grow, but whose growth later levels out, have experienced logistic growth, sometimes referred to as the S-shaped growth curve.

- The size of living populations is kept in check by environmental resistance, defined as all the forces of the environment that act to limit population growth.

- Exponential growth in living populations can be calculated by subtracting a population's death rate from its birth rate, which yields the population's growth rate. Denoted as $r$, this rate is also known as the population's intrinsic rate of increase. It can be thought of as the population's potential for growth.

- Carrying capacity, denoted as $K$, is the maximum population density of a given species that can be sustained within a defined geographical area over an extended period of time.

**Web Animation 33.1:** Population Growth

### 33.3 r-Selected and K-Selected Species

- Different species have different reproductive strategies, meaning characteristics that have the effect of increasing the number of fertile offspring they bear.

- Some species are said to be $K$-selected, or equilibrium, species. These species tend to be physically large, to experience their environment as relatively stable, and to lavish a good deal of attention on relatively few offspring. The pressures on $K$-selected species tend to be density dependent, meaning that as a population's density goes up, factors that limit the population's growth assert themselves ever more strongly.

- Other species are said to be $r$-selected or opportunist species. These species tend to be physically small, to experience their environment as relatively unstable, and to give little or no attention to the numerous offspring they produce. The pressures on $r$-selected species tend to be density independent, meaning pressures that are unrelated to the population's density.

- Survivorship curves describe how soon species members tend to die within the species' life span. There are three idealized types of survivorship curves: late loss, constant loss, and early loss, also known as types I, II, and III, respectively. Members of late-loss species tend to survive into old age; members of constant-loss species tend to die off at a nearly constant rate throughout their lifespan; and members of early-loss species tend to have high death rates early in life, with these rates leveling out thereafter.

### 33.4 Thinking about Human Populations

- Survivorship curves are created from life tables, which set forth the probabilities of a member of a species being alive after given intervals of time.

- An important step in calculating the future growth of human populations is to learn what proportion of the population is at or under reproductive age. A population pyramid displays this proportion. Populations whose pyramids are heavily weighted toward younger age groups are likely to experience relatively large growth.

- Following decades of explosive increase, the world's human population is projected to stabilize in the coming decades, going from about 6.5 billion now to a maximum of about 9.2 billion just past mid-century. This stabilization is being brought about by a decrease in the total fertility rate, informally defined as the number of children born, on average, to each woman in a population.

- The global reduction in fertility masks enormous, ongoing differences between fertility in more-developed and less-developed countries. Fertility in less-developed countries tends to be much higher than that in more-developed countries. The fertility in most European nations is now so low that the continent's population stands to shrink significantly by mid-century. The population of the United States, however, is projected to grow significantly during this same period. The primary factors bringing the U.S. increase about are immigration and a high total fertility rate relative to other developed nations.

- Some scientists believe that there is no greater single threat to the environment than the continued growth of the human population. Others argue that a more important concern is the use of natural resources per person.

## KEY TERMS

| | |
|---|---|
| arithmetical increase   694 | exponential increase   694 |
| biosphere   693 | intrinsic rate of |
| carrying capacity (*K*)   697 |    increase (*r*)   697 |
| community   693 | *K*-selected species   698 |
| density dependent   698 | life table   700 |
| density independent   699 | logistic growth   696 |
| ecology   692 | population   693 |
| ecosystem   693 | *r*-selected species   699 |
| environmental | total fertility rate   702 |
|    resistance   696 | zero population |
| exponential growth   695 |    growth   697 |

## BRIEF REVIEW

*Answers to Brief Review questions are in the back of the book. For multiple-choice quiz questions, go to* **www.aw.com/krogh4**.

1. Name the four levels of group organization of living things in order of increasing complexity.

2. Compare and contrast *r*-selected and *K*-selected species and give an example of each. Are humans *r*-selected or *K*-selected?

3. Explain what is meant by environmental resistance and its relationship to population growth.

4. What kinds of forces tend to limit the growth of *K*-selected species?

5. Which factor has been most important in bringing about the apparent long-term stabilization of Earth's human population? Is this stabilization expected to occur soon?

## APPLYING YOUR KNOWLEDGE

1. Does the concept of carrying capacity apply to human beings? Does the planet as a whole have a human carrying capacity, or do human beings have an infinite capacity to increase resources through such means as obtaining greater crop yields and becoming more energy efficient?

2. Is human disease a form of environmental resistance?

3. Some scientists argue that no single factor poses a greater threat to the environment than the continued growth of the human population, while others believe the use of resources per person is a more critical factor. What arguments can you think of for or against either proposition?

A watering hole is a natural congregation spot for members of a community in Botswana's Okavango Delta.

| 34.1 | Structure in Communities | 709 |
| 34.2 | Types of Interaction among Community Members | 712 |
| 34.3 | Succession in Communities | 719 |

**Essays**

Purring Predators: House Cats and Their Prey — 715
Why Do Rabid Animals Go Crazy? — 722

A community is all the populations of all the species living in one area. Members of different species within a community interact with each other in a variety of ways, some of them competitive, others supportive. Communities establish themselves through a long-term process called succession.

# AN INTERACTIVE LIVING WORLD 2:
# Communities in Ecology

**M**ost people are aware that tropical rain forests are places of great biological diversity, but some years ago Smithsonian biologist Terry Erwin was able to attach a startling number to this diversity. While doing research in Panama, Erwin examined a single tree and found an estimated 1,700 species of beetle living in it. Not 1,700 beetles, mind you—1,700 *species* of beetle. Of course, these beetles were only one part of a larger group of insects that were living in the tree, and the insects in turn were only one variety of animal living in it. Joining these animals were the fungi that did their best to extend their filaments into the tree's bark each day and the bacteria that lived by the billions in or on all the other organisms that inhabited the tree.

We saw last chapter that a **community** can be defined as all the populations of all the species that inhabit a defined area. This means all the plants, animals, fungi, bacteria, protists—every living thing, in other words. If we think of our tree in Panama as being a defined area, we can get a sense of the mind-boggling diversity that can exist within a single community. And yet, by examining communities from the tropics to the poles, ecologists have found that there is a standard set of questions that can be asked about any community. How diverse is it? Which species in it are dependant on which others? Which species in it are the predators and which are the prey? If we look for a common thread among these questions, it's not hard to find. The study of communities is first and foremost the study of *relationships* among species.

When ecologists ask about predators and prey in a community, it's not the species themselves they're chiefly concerned with; it's the relationships between these varieties of species that interests them. Accordingly, relationships are what we'll primarily be looking at this chapter. How do species relate to each other when living in a common area? Put another way, what makes a community tick? Let's find out.

## 34.1 Structure in Communities

### Large Numbers of a Few Species: Ecological Dominants

If we think about the difference between a tropical rain forest, a prairie, and a desert, it's easy to see that there's a tremendous variability in the mix of species found in different communities. A common feature of many types of communities, however, is that they are dominated by only a few species. Forests tend to be populated by certain kinds of trees, stretches of prairie by certain kinds of grasses, deserts by certain kinds of shrubs or cacti. The small number of species that are abundant in a given community are called **ecological dominants** (**Figure 34.1** on the next page). These generally are plants, but they can be other life-forms, as is the case with the tiny animals known as corals that are the ecological dominants in coral reefs.

## Importance beyond Numbers: Keystone Species

Ecologists have long recognized that there are also species who may not be numerous in a given area, but who play a role in a community that cannot be assumed by any other community member. Each of these is a **keystone species**: a species whose absence from a community would bring about significant change in that community. The concept of the keystone species was introduced in the 1960s by marine ecologist Robert Paine, who went with his students to a shallow-water zone of the Pacific Ocean in Washington State and for 6 years regularly removed all sea stars in the genus *Pisaster* from a small area. Such a sea star may sound harmless enough, but *Pisaster* was, in fact, the **top predator** in the area. It preyed on other species, but no species preyed on it (**Figure 34.2**). The impact of the *Pisaster* removal was big: Before the change, there had been 15 species of marine animals in the area; after it, there were 8. One species of mussel, freed from its former predator's control, took over much of the attachment space in the area, crowding out other animals, such as barnacles.

Over time, the keystone species concept has undergone some modification. Where once scientists thought of keystones as always sitting at the top of food chains,

they now recognize that organisms in other positions can take on a keystone role—beavers building dams, for example, or even lichens that are critical in getting communities going in the desert. Beyond this, it turns out that there are communities without keystones; remove any one species from such a community, and its role will be taken over by another species.

## Variety in Communities: What Is Biodiversity?

Apart from ecological dominants and keystones, a third element important to any community is the range of species found within it. This touches on the more general concept of biodiversity, which in the broadest sense simply means variety among living things. In everyday speech, **biodiversity** is thought of as being the same thing as species diversity—to the average person, biodiversity means a diversity *of species* in a given area. This definition of biodiversity is too limited, however. While species diversity is indeed an important type of biodiversity, it is only one of three principal types that exist. To understand the importance of a second type, consider an imaginary experiment that has been put forward by the ecologist Paul Ehrlich.

Suppose that you could get a few members of every species on Earth, but that you restricted each

**Figure 34.1
Ecological Dominants**

**(a)** German forest dominated by single species of tree

**(b)** Kansas prairie dominated by tall-grass

**Figure 34.2**
**Keystone Species**

When the predatory sea star *Pisaster ochraceus* was removed from a small area of rocky shore along the Pacific coast of the United States, the species composition of the community changed drastically—more so than if one of the other species had been removed. Pictured are several *Pisaster* sea stars on the Pacific Northwest coast.

species to a single population housed somewhere (in a single zoo or an aquarium or botanical garden). Species diversity would not drop at all in such a scenario—you'd still have some of each kind of creature, after all—but the Earth quickly would become barren because what's needed is a rich distribution of species in populations across the planet. This geographical distribution of species populations is the second type of biodiversity. The third type exists *within* populations of different species. It is genetic diversity, meaning a diversity of alleles or variants of genes in a population. Without such diversity, populations are vulnerable to disease; their members may die young or suffer from a variety of inherited mental and physical afflictions (see "Lessons from the Cocker Spaniel" in Chapter 17). In summary, there are three principal types of biodiversity or variety among living things: species diversity, geographic diversity, and genetic diversity (**Figure 34.3**).

## What Are the Effects of Species Diversity?

In the last 10 years, two of the hottest questions in ecology have concerned species diversity. The first of these questions is: Do diverse communities tend to be more productive communities? In ecology, productivity is a measure of how much of the sun's energy plants and other photosynthesizers are able to "capture" and then turn into living material or "biomass." A really productive community is one that is rich with biomass—rich in leaves, seeds, and roots and the animals that feed on them. So, does a diverse community tend to be more

productive than one that is not diverse? The answer seems to be yes. The reason for this may be that, in diverse communities, each community member is able to exploit a different portion of the resources available. Some plants will bloom early, while others bloom late; some plants will do well with only a little water, while other plants will flourish with more. In this respect, communities may be something like Jack Sprat and his wife from the old nursery rhyme. Recall that Jack could eat no fat and that his wife could eat no lean, but that between them they licked the platter clean.

A second question about species diversity has been more difficult to answer. Does species diversity lead to greater stability? Is a diverse community better able to recover from, say, a drought than a community that is not diverse? The preponderance of studies seems to indicate the answer is yes, but other studies indicate the answer is no. This question remains a compelling one to ecologists because it has practical ramifications. Human beings are constantly changing the environments around them—fragmenting habitats, clearing forests for farmland, and so forth. When this happens, the number of species in a given area often decreases. The questions are: Will a community whose diversity has been reduced retain its ability to "bounce back" from an environmental disruption? Will it remain as stable as before when confronted with a prolonged freeze or

**Figure 34.3**
**Three Types of Biodiversity**

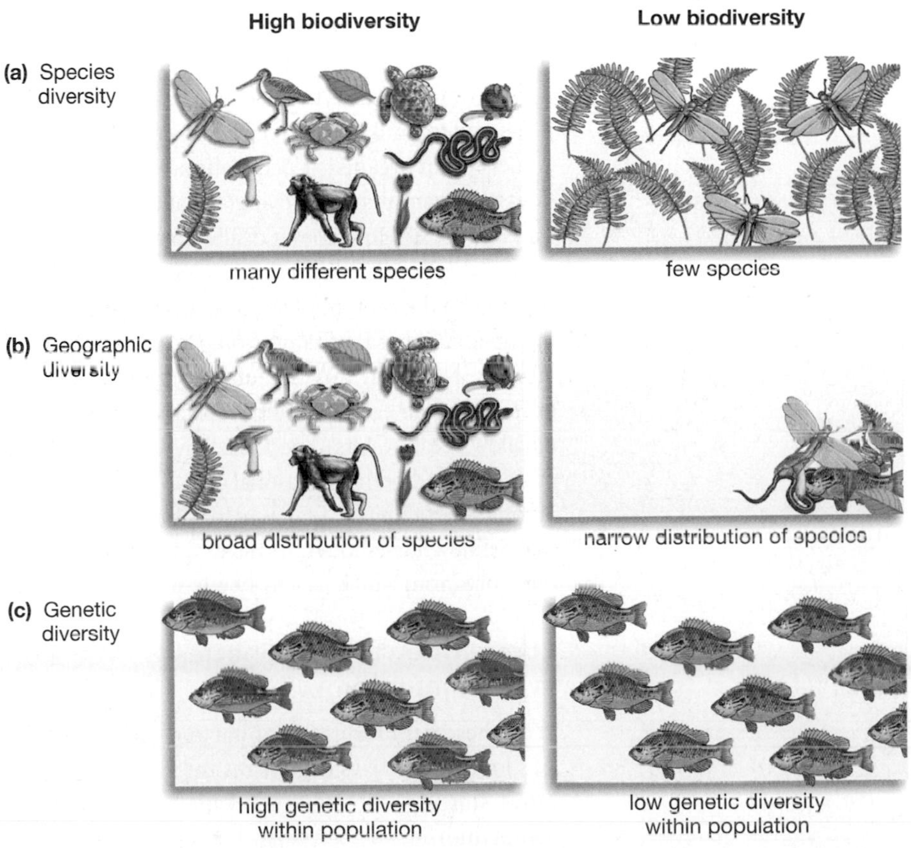

| High biodiversity | Low biodiversity |
|---|---|
| **(a)** Species diversity | |
| many different species | few species |
| **(b)** Geographic diversity | |
| broad distribution of species | narrow distribution of species |
| **(c)** Genetic diversity | |
| high genetic diversity within population | low genetic diversity within population |

an insect invasion or a flood? If the answer is no, then societies might make different decisions about how to use land than if the answer is yes.

## 34.2 Types of Interaction among Community Members

Having considered some structural issues regarding communities, let's now turn to the subject of how members of a community might interact with each other. Here are the four primary modes of interaction:

- Competition
- Predation and parasitism
- Mutualism
- Commensalism

Before continuing with the exploration of these modes, let's go over a couple of concepts that will apply to them all.

### Two Important Community Concepts: Habitat and Niche

A **habitat** is the physical surroundings in which a species can normally be found. Although two populations of a species may be widely separated, they can generally be found dwelling in similar natural surroundings. A habitat is sometimes described as a species' "address," but a more accurate metaphor might be a species' preferred type of neighborhood.

The word **niche** has been defined in several ways, but it is useful to think of it in terms of a simple metaphor: A niche is an organism's occupation. How and where does the organism make a living? What does it do to obtain resources? How does it deal with competition for these resources? The horseshoe crab has found a niche walking on the bottom of shallow coastal waters, feeding on food items that range from algae to small invertebrates. Note that this is about more than what the horseshoe crab eats. It includes specific surroundings (shallow ocean waters), specific behaviors (ocean-floor crawling), and perhaps seasonal or daily feeding times, among other things. If you were to specify all the things that define a horseshoe crab's niche, the odds are that no other organism would exactly fit into it.

### Competition among Species in a Community

With these definitions of habitat and niche in mind, let's look at the ways organisms interact in communities, starting with a familiar type of interaction—competition.

Even though niches tend to be specific to given organisms, some species—particularly closely related species—have niches that *overlap* to some degree in a community. A large proportion of both species' diet may be made up of a given organism, or both species may occupy similar kinds of spaces on rocks, branches, or pond surfaces. What arises in such instances is **interspecific competition**, meaning competition between two or more species.

What may come to mind with interspecific competition is a never-ending series of physical battles between two species, but things seldom work like this. For one thing, competition tends to be indirect. It often is a competition for *resources*, in which the winner generally triumphs not by fighting but by being more efficient at doing something, such as acquiring food. Lots of animals, including birds and many large mammals, are territorial—meaning they will attempt to keep other creatures out of a territory they have laid claim to—but this usually is a matter of them trying to repel members of their own species.

### No Two Species Can Share the Same Vital Resource for Long

When the niches of two species greatly overlap, it's unlikely that you'd see long-standing competition among them, for the simple reason that one species or the other is likely to win such a competition in fairly short order. Because the competition is for nothing less than vital resources, the result is that the losing species will be driven to local extinction.

It was a laboratory experiment, performed by the Russian biologist G. F. Gause in the 1930s, that pointed the way on this latter principle. Gause grew two species of *Paramecium* protists, *P. caudatum* and *P. aurelia*, in culture and found that in time, *P. aurelia* was always the sole survivor. True to what was just noted, *P. aurelia* wasn't eating or wounding *P. caudatum*; instead, it grew faster and thus used more of the surrounding resources. Nevertheless, the outcome was always the death of all the *P. caudatum*. This led Gause to formulate what came to be called the **competitive exclusion principle**: When two species compete for the same limited, vital resource, one will always outcompete the other and thus bring about the latter's local extinction (**Figure 34.4a**).

Ecologists wouldn't expect to witness the competitive exclusion principle operating much in nature, however. This is because, as noted, the *Paramecium* scenario is likely to be played out quickly. Nevertheless, competitive exclusion has been observed in nature, often when humans have a hand in things. For example, humans introduced a Southeast Asian vine,

**Figure 34.4**
**Competition for Resources among Species**

Laboratory experiments by G. F. Gause showed that competition for resources between two species can have two possible outcomes.

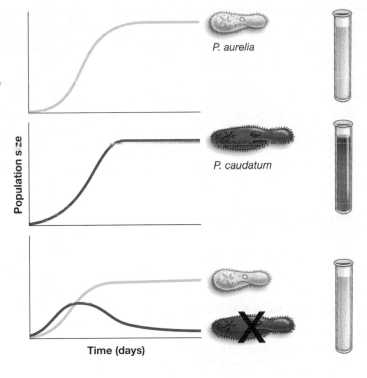

**(a)** Competitive exclusion

When two species compete for the same limited, vital resource, one will always drive the other to local extinction—as the paramecium *P. aurelia* did to the paramecium *P. caudatum*. This is the competitive exclusion principle at work.

*P. aurelia*

*P. caudatum*

Time (days)

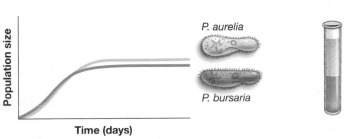

**(b)** Resource partitioning

Conversely, when Gause put *P. aurelia* together with another paramecium, *P. bursaria*, the two species divided up the habitat, and both survived. This is a demonstration of resource partitioning.

*P. aurelia*

*P. bursaria*

Time (days)

the kudzu, to the American South on a large scale in the 1930s. Growing at up to 30 centimeters or 1 foot per day, kudzu has now taken over millions of acres in the South, locally eliminating many plants in its path in a process of competitive exclusion (**Figure 34.5**).

## Resource Partitioning Is Common in Natural Environments

Competitive exclusion notwithstanding, there are instances in nature in which two related species will use the same kinds of resources from the same habitat over a long period of time. So, why isn't one of them eliminated? The answer is contained in another experiment conducted by Gause. He took the successful species from the earlier experiment, *P. aurelia*, and placed it in a test tube with a different paramecium, *P. bursaria*. This time, neither species was eliminated (see **Figure 34.4b**). Instead, the two species divided up the habitat, *P. aurelia* feeding in the upper part of the test tube, and *P. bursaria* flourishing in the lower part. (*P. bursaria* had an advantage in the lower, oxygen-depleted water because it has symbiotic algae that grow with it, producing

**Figure 34.5**
**Kudzu Vines Run Wild**

The kudzu plant was introduced in the American South in the 1930s and has since spread at a rapid rate, locally eliminating many plants in its path. This is competitive exclusion in action. Here, kudzu has overgrown an abandoned house in Mississippi.

oxygen.) In a nutshell, this result describes a situation that often exists in nature: **coexistence** (meaning a sharing of habitat) through a practice called **resource partitioning**, which can be defined as a dividing up of scarce resources among species that have similar requirements. If you look at **Figure 34.6**, you can see how this works among some species that are a little more familiar—several varieties of warbler. The main message here is that species often flourish by specializing; they derive resources by feeding from different locations (or during different times) within the same habitat.

---

### SO FAR...

1. A keystone species can be defined as any species whose _____ a community would bring about significant _____ in it.

2. There are three principal types of biodiversity: _____ , _____ , and _____ .

3. A habitat is the _____ in which a species can _____ , while a niche can be thought of as an organism's _____ .

4. The competitive exclusion principle states that when two populations compete for the same _____ , one will always _____ the other and thus bring about the latter's _____ .

---

**∙∙∙ ANSWERS**

1. absence from; change
2. species diversity; a geographic diversity of species populations; genetic diversity within species' populations
3. physical surroundings; normally be found; occupation
4. limited, vital resource; outcompete; local extinction

## Other Modes of Interaction: Predation and Parasitism

It is one thing for two species to compete for resources; it is another for one species to *be* a resource for another—to be eaten or used by another species. Thus do ecologists distinguish one mode of interaction in communities, competition, from a second mode, predation.

**Predation** can be defined as one organism feeding on parts or all of a second organism. Note that the prey here can be plants, protists, animals—whatever is preyed on. Predation is generally thought of in terms of animals killing other animals, but note that the definition given includes such things as animals consuming whole plants or their seeds. **Parasitism** is a variety of predation in which the predator feeds on prey but does not kill it immediately—and may not ever kill it. Parasites that come to mind tend to be animals, but there are an estimated 3,000 species of parasitic plants, one of which can be seen in **Figure 34.7**.

### The Value of Predation and Parasitism

Predation and parasitism are the features of nature that many nature lovers can't stand, perhaps because these practices seem cruel or unfair from a human point of view. These modes of interaction have

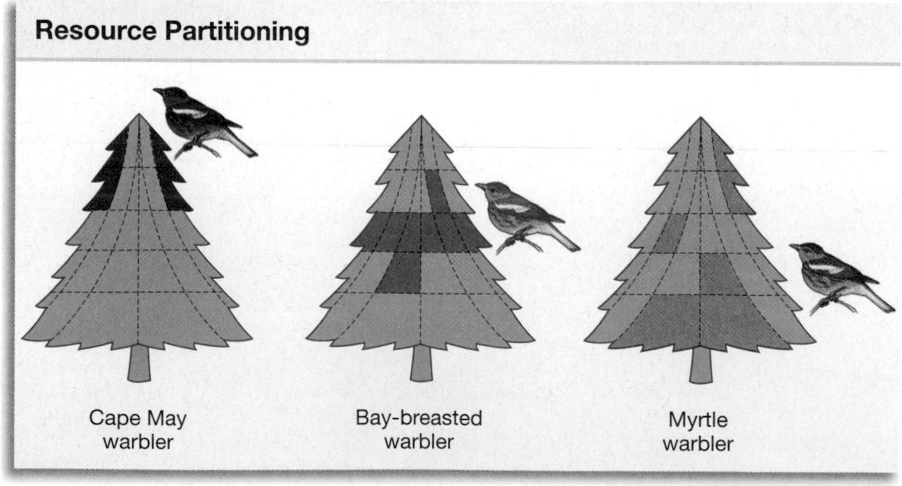

**Figure 34.7**
**Plants Parasitizing Plants**

The orange tendrils in the picture belong to a species of dodder, *Cuscuta pentagona*, a parasitic plant that survives by tapping into the food reserves of other plants, thus feeding off them. The dodder is orange, rather than green, because it lacks the chlorophyll that is used in photosynthesis.

---

**Resource Partitioning**

| Cape May warbler | Bay-breasted warbler | Myrtle warbler |

**Figure 34.6**
**Resource Partitioning**

Ecologist Robert MacArthur spent long stretches of time over several years in the 1950s observing the feeding patterns of several species of warblers. All of them ate caterpillars, but from substantially different, though overlapping, parts of the tree.

# Purring Predators: House Cats and Their Prey

House cats are famous for bringing home animals they've killed or captured, of course; intrigued by the carnage brought into their own homes in the 1980s, Peter Churcher and John Lawton decided to make a scientific study of the predatory behavior of the domestic cat. To do so, they enlisted the help of 172 households in the small English village of Bedfordshire, where Churcher lived. The two researchers asked the locals to "bag the remains of any animal the cat caught" and turn the evidence over to them once a week. This process went on for a year with a high degree of cooperation from the cat owners.

*What was remarkable was the scale of the killing.*

Some of the results were not surprising. Young cats hunt more than older cats; small mammals, such as mice, are the favored prey; and cat hunting is not based purely on hunger—all the cats were fed by their owners and yet hunting was widespread.

What was remarkable, however, was the *scale* of the killing. Concentrating on the village's house sparrow population, Churcher and Lawton found that cats were responsible for between one-third and one-half of all sparrow deaths in the village, a figure they believed no other single predator could match. When they looked at all animals killed and projected the village figures onto the whole of Great Britain, the researchers calculated that cats were responsible for about 70 million deaths a year.

Given that birds account for somewhere between 30 and 50 percent of these kills, this means that cats kill at least 20 million birds a year in Britain. The "at least" here may be an important qualifier. An American biologist, the researchers noted, found that cats bring home only about half of the food they catch.

**Predator on the Loose**
A domestic cat carrying its prey.

arguably been valuable, however, in that they have stimulated the "arms race" that has resulted in evolutionary adaptations such as vision and flight. (One organism's predatory adaptation spurs the development of another organism's defensive adaptation and vice versa.) Whatever we may think of it, predation is simply a fact of nature, and it is impossible to imagine life without it. But, it's also true that to admire the beauty of nature is in significant part to admire the handiwork of predation since it has done so much to shape the living world. You can read about a very familiar predator, the house cat, in "Purring Predators: House Cats and Their Prey," above.

## Predator-Prey Dynamics

It may be obvious that although predators attack prey, predators are also *dependent* on prey as a food source. An ongoing question in ecological research is how tight this linkage is. To what extent do predator and prey population sizes tend to move up and down together? It's clear that the population dynamics of some predator and prey species are very tightly linked. The small lemmings of northeast Greenland (*Dicrostonyx groenlandicus*) are fed on by only four predators: the stoat, the arctic fox, the snowy owl, and another bird called the long-tailed skua. Of these, the stoat is a "specialist" predator in that it feeds almost entirely on lemmings. If you look at **Figure 34.8**, you can see that both the lemming and stoat populations predictably go through 4-year

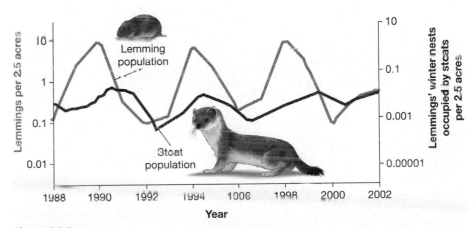

**Figure 34.8**
**Predator and Prey Populations**
The population dynamics of a prey species (the lemming) and one of its predators (the stoat) in northeast Greenland over a period of 14 years. (Reprinted with permission from "Cyclic Dynamics in a Simple Vertebrate Predator-Prey Community," by O. Gliq et al., *Science*, 302:867, 2003.)

up-and-down population cycles. Moreover, note that the movement up or down in stoat population *lags* the change in lemming population, generally by a few months. What happens is that a higher lemming population means more food for the stoats, which then increase in population. This increase, of course, then means more predation of the lemmings, but the fall you can see in lemming population is not brought about solely by the increase in stoat numbers. When the lemmings' density reaches high levels, the arctic fox, snowy owl, and long-tailed skua begin to prey on them. It is this predation, in combination with the increased stoat predation, that then drives the lemming numbers down. Once the lemmings exist in low densities again, the stoat becomes the only predator, lemming numbers go up, and the cycle repeats.

The Finnish and German researchers who reported these findings found that declines in lemming numbers could be explained *entirely* by predation; neither the food nor the space available to the lemmings, for example, played any part in reductions of their population. (And, what about the famous mass suicides of the lemmings? Don't they help keep lemming numbers down? The answer is no, because there are no lemming mass suicides; the whole idea is a myth.) The crystal-clear connection between predation and the size of the lemming population naturally raises the question of how *common* such a linkage is in nature. What we can say for certain is that, in most species, population dynamics involve more than just predator-prey relationships. Environmental resistance can take many forms, of which in-

creased predation is only one. You might say that the connection between lemmings and their predators serves as an example of how strong predator-prey linkage *can* be, but it is not representative of how strong such linkage is likely to be.

## Parasites: Making a Living from the Living

Let's look now at the special variety of predation mentioned earlier, parasitism. Some familiar parasites, such as leeches and ticks, have a straightforward strategy: Get on, hold on, and consume material from the prey, known as the **host**. But the relationship between parasites and hosts can be much more sophisticated than this. Take, for example, the barely visible roundworm *Strongyloides stercoralis*, which lives throughout the tropics and has, as one of its hosts, human beings in the southeastern United States. The worm comes into the human body from the soil in a thread-like larval form, then makes its way into the veins, which it travels through to get to the lungs. From there, it moves up the trachea and then down into the digestive system, ending up in the small intestine. There it lays eggs that develop into larvae that this time move out, with feces, back into the Earth to start the whole life cycle over again.

*Strongyloides* is not just using humans as an incubator, however; it can worm its way, you might say, directly from the small intestines into the bloodstream, bringing about a continuous cycle of infection whose symptoms range from severe diarrhea to lung problems. (Lest all of this seem too disturbing, take heart; there is a medicine that can clear this pest from human bodies.) So, note the complexity here: *Strongyloides* not only feeds on a host but uses the host in its reproduction cycle, something that actually is fairly common among parasitic worms and protozoans. Were you to start examining the range of host-parasite dynamics, you would find lots of interactions as elaborate—and chilling—as this. As you can see in "Why Do Rabid Animals Go Crazy?" on page 722, parasites sometimes alter the behavior of their hosts.

### The Effect of Predator-Prey Interactions on Evolution

Earlier, we noted the "arms race" spurred on by predator-prey interactions: One organism's predatory adaptation spurs the development of another organism's defensive adaptation and vice versa. So, what forms do these evolutionary adaptations take? To see an example on the predator side, look at the warm-water "frogfish" in **Figure 34.9**. These fish have evolved a spine on their dorsal fins that is

**Figure 34.9**
**Fooling Predators about Prey**

The predatory frogfish uses a modified spine resembling a tasty worm to lure its prey. Here, a tasseled frogfish clearly shows its lure while swimming near Edithburg, South Australia.

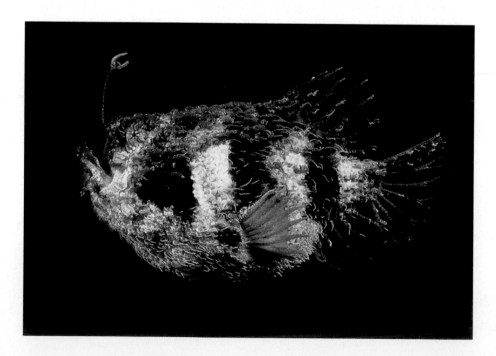

**(a)** A spanworm looking like a twig on a maple tree

**(b)** A casque-head chameleon against a tree in Kenya

**Figure 34.10**
**Avoiding Predation through Camouflage**

tipped with a piece of flesh that looks for all the world like a small worm floating free in the ocean. When a would-be predator tries to snatch up this "worm," however, it finds *itself* the prey. On the prey side, look at **Figure 34.10** to see the amazing kinds of camouflage that animals use to keep themselves hidden from the eyes of predators. Another example of prey adaptation may be as close as your own skin. Why are human beings the only primates whose bodies are not covered with a coat of hair? One hypothesis holds that, while hair warmed our evolutionary ancestors, it also provided a home for many parasites that preyed on them—lice, fleas, and ticks. The adaptation of losing our hair allowed us to lose these parasites to a great extent. Needless to say, predator-prey evolution goes on within all the kingdoms of life. Plants have developed not only spines, thorns, and other protective structures but also a vast array of chemical compounds—more than 15,000 have been characterized so far—that protect them from predators.

### Mimicry Is a Theme in Predator-Prey Evolution

One special form of evolutionary adaptation brought on by predator-prey interaction is **mimicry**, a phenomenon through which one species has evolved to assume the appearance of another. One type of mimicry involves three participants: a model, a mimic,

and a dupe. If you look at **Figure 34.11**, you can see a model on the left, the yellow jacket wasp, which obviously can provide a painful lesson to any animal that tries to eat it. On the right you can see the mimic, the clearwing moth, which is harmless but looks a great deal like a wasp. For the dupe, you could select any predator species that sees a clearwing moth and passes it by, believing it to be a dangerous yellow jacket. This is called **Batesian mimicry**: the evolution of one species to resemble a species that has a superior protective capability.

In a second type of mimicry, **Müllerian mimicry**, several species that *have* protection against predators

**Figure 34.11**
**Harmless, but Looking Dangerous**

In an example of Batesian mimicry, the clearwing moth (on the right) has no sting but has evolved to look like an insect that does, the yellow jacket wasp (on the left). The yellow jacket is the model, and the moth is the mimic.

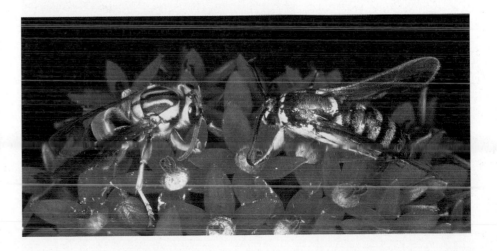

**Figure 34.12**
**Müllerian Mimicry**

The South American butterflies *Heliconius cydno* (top) and *Heliconius sapho* are different in some ways, but they share the characteristic of tasting bad to predators. Over time, they evolved to look like each other, with their appearance serving as a warning to would-be predators. Natural selection favors this Müllerian mimicry because predators can learn about the butterflies' unpalatable taste from both species. The result is that fewer members of either species are likely to be killed or bothered by predators.

come to resemble each other (**Figure 34.12**). This creates a visual warning that becomes known to an array of predators. Given this, a would-be predator learns, by interacting with an individual from one species, to keep away from all individuals in any look-alike species. The result is that fewer individuals in these species will be disturbed or killed.

### Beneficial Interactions: Mutualism and Commensalism

The competitive, predatory, and parasitic modes of interaction we've been reviewing have an each-organism-against-all quality to them, but other kinds of relationships also exist within communities. Consider **mutualism**, meaning an interaction between individuals of two species that is beneficial to both individuals. There's a good example of this in Chapter 22 (page 427), with the linkage between plant roots and the slender, below-ground extensions of fungi called hyphae. What the fungi get from growing into the plant roots is food that

comes from the photosynthesizing plant; what the plants get is minerals and water, absorbed by the network of hyphae. By some estimates, 90 percent of seed plants have this relationship with fungi. Perhaps the most famous example of mutualism, however, is between the rhinoceros and oxpecker birds. Sitting on the back of the rhino, the oxpecker removes parasites and pests (ticks and flies), thus getting food and a safe place to perch, while the rhino gets relief from its tiny adversaries (**Figure 34.13**).

Our final variety of community interaction is **commensalism**, meaning an interaction between individuals in two species in which one benefits while the other is neither harmed nor helped. Birds can make their nests in trees, and benefit from this, but generally don't affect the trees in any way. **Table 34.1** summarizes all the kinds of interactions you've gone over in this chapter.

### Coevolution: Species Driving Each Other's Evolution

Earlier, you read about the effect that predator-prey interactions have had on the evolutionary arms race among species. Yet it's clear that organisms have affected each other's evolution through mutualism as well. Indeed, species that are tightly linked in any way over a long period of time are likely to shape one another's evolution. Perceiving this, scientists Paul Ehrlich and Peter Raven developed the concept of **coevolution**, meaning the interdependent evolution of two or more species. Some of the clearest examples of coevolution involve flowering plants and the animals that pollinate them. Recall from Chapter 22 that many flowers have both fragrances that attract insects and ultraviolet color patterns (invisible to us) that guide

**Figure 34.13**
**Mutually Beneficial**

**(a)** Rhinoceros and oxpecker birds

Several oxpecker birds sit atop a black rhinoceros, ridding the rhino of ticks and other pests, while the rhino provides a safe habitat for the birds. This is a demonstration of mutualism—an interaction between two species that is beneficial to both.

**(b)** Snapping shrimp and shrimp goby

The orange–spotted shrimp goby, on the right, and the snapping shrimp at left also exhibit mutualism. Here the shrimp goby stands guard near the snapping shrimp, which digs out the burrow on the sea floor that the two creatures share.

the insects to the proper spot for pollination. These features are nature's own homing signals and landing lights, respectively communicating the messages "Food lies this way" and "Land here" (**Figure 34.14**). Now, one of the organisms linked with flowering plants in this way is the honeybee. And it turns out that the color vision of honeybees is most sensitive to the colors that exist in the very flowering plants they pollinate. What seems likely is that honeybee eyesight evolved in response to plant coloration, and that plant coloration evolved to be maximally attractive to the pollinating insects. These groups of organisms coevolved, in other words.

Having considered how community members interact with one another, let's now go on to consider how communities change over time.

**SO FAR...**

1. In the interaction known as _____, one species feeds on all or parts of a second organism; in the interaction known as _____, a predator feeds on prey but does not kill it immediately, if ever.

2. In predator-prey interactions, Batesian mimicry has occurred when one species has evolved to _____ another species that has _____. In Müllerian mimicry, several species that _____ come to _____.

3. Mutualism is an interaction between individuals of two species that is _____. Commensalism is an interaction between individuals of two species in which one is _____ while the other is _____.

## 34.3 Succession in Communities

When Washington State's Mount St. Helens volcano erupted in May 1980, it first collapsed inward, thus sending most of the mountain's north face sliding downhill in the largest avalanche in recorded history. Then came the actual eruption, which sent a huge volume of rock and ash hurtling not straight up but out at an angle, toward the north. Some stands of forest in the path of this blast were instantly incinerated down to bare rock; another 35,000 hectares (86,000 acres) of trees were snapped in two like so many twigs. Then came the mudslides caused by the vast expanse of snow and ice melted in the explosion; then came the fall of hundreds of millions of tons of ash, some of it landing as far away as Wyoming. After viewing the devastated area around the blast, Presi-

**Table 34.1 Community Interactions**

| In this type of interaction... | One organism... | While the other... |
| --- | --- | --- |
| Competition | Is harmed | Is harmed |
| Predation and parasitism | Gains | Is harmed |
| Mutualism | Gains | Gains |
| Commensalism | Gains | Is unaffected |

**(a)** What we see

**(b)** What a bee sees

**Figure 34.14**
**Coevolution of Plants and Their Pollinators**

This flower has an ultraviolet color pattern that is not normally visible to humans, but that attracts bees. It is likely that these color patterns evolved in the flowers because they aided in attracting the bees. Meanwhile, the bees developed vision that was sensitive to the colors exhibited by these same plants.

dent Jimmy Carter said that it "makes the surface of the moon look like a golf course."

But today? No one would claim that the Mount St. Helens area has returned to anything like its former state, but look at the pictures in **Figure 34.15**, on the next page, to get an idea of the transition that has occurred at one location around Mount St. Helens.

The exact form of this rejuvenation may surprise us, but don't we intuitively expect something like this? Just from our everyday experience of watching, say, an abandoned urban lot becoming progressively weed filled, don't we expect this would generally be the case? As it turns out, our intuition is right because almost any parcel of land or water that has been either abandoned by humans or devastated by physical forces will be "reclaimed" by nature, at least to some degree.

The question is, how does this reclamation proceed? Areas that are rebounding don't just get one type of vegetation and then retain it. Instead, the vegetation changes through time—one type of growth *succeeds* another—with a general movement toward more and larger greenery as time goes by. Ecologists call this phenomenon **succession**, meaning a series of replacements of community members at a given location until a

••• ANSWERS

1. predation; parasitism
2. resemble; superior protective ability; have protection against predators; resemble each other
3. beneficial to both; helped; neither helped nor harmed

**Figure 34.15**
**Rebound from Disaster**
Mount St. Helens, with its crater in the distance, as it appeared in June 1980, 1 month after the blast, and as it appeared in 1998. This is an example of ecological succession in action.

relatively stable final state is reached. Within this framework, there are two kinds of succession. The first is **primary succession**, meaning succession in which the starting state is one of little or no life and a soil that lacks nutrients. Then there is **secondary succession**, which is succession in which the final state of a habitat has been disturbed by some force, but life remains, and the soil has nutrients. The classic example of this is land that has first been cleared for farming but later abandoned by the farmer.

The relatively stable community that develops at the end of any process of succession is called a **climax community**. There will be some shifts over time in a climax community, just as there are in any other. A prolonged drought will cause some change in the mix of the community's animal and plant life, for example. But, with some exceptions, what does not happen is a shift to a *fundamentally* different mix of life-forms, barring a major change in climate. Grassland stays grassland, and forest stays forest.

Succession can be fairly predictable within small geographical areas. Two abandoned farm plots that lie close together will have very similar kinds of succession—so much so that local farmers or naturalists can sometimes tell to within a few years how long it's been since a field was abandoned. As you may be able to tell, succession generally is thought of in terms of the various *plant* communities that succeed one

another, although modern ecology is attempting to bring animals into the picture as well.

## An Example of Primary Succession: Alaska's Glacier Bay

One of the best examples of primary succession comes from Alaska's 65-mile-long Glacier Bay, whose lower-lying areas were covered by a glacier until about 240 years ago, when the glacier began retreating. Such withdrawal initially leaves behind a rocky terrain that is completely devoid of organic matter—no fallen leaves or branches, no decomposing animals. Indeed, what's left is not soil, but a pulverized rock (called till) that has a high pH and no available nitrogen—not a promising environment for life to establish itself. But establish itself it does. In the case of Glacier Bay, its succession particulars—what species succeeded what—are not important to us. What is important is the general process that occurred. Here's what happened in the eastern arm of Glacier Bay after the glacier retreated (**Figure 34.16**).

In the 20 years immediately following the retreat, the first "pioneer" species to come were lichens and photosynthesizing bacteria, which essentially have the ability to grow on rock. In addition, there were some primitive plants called horsetails and liverworts. Over the next 10 years, a gradual transition took place in which an Arctic dwarf shrub, called *Dryas*, took over from the pioneers, growing in a mat that was broken up only by occasional trees. *Dryas* and other early-stage plants have a capability that is critical not only to their own growth in these environments, but to the growth of plants that come after them. These plants harbor bacteria in their roots that can *fix nitrogen*—that can take nitrogen from the air and convert it into a form that plants can use. This is how the larger plant community could get started on the till in Glacier Bay, which *lacked* nitrogen.

The *Dryas* community was succeeded by a plant called the alder bush that excluded almost everything else—it was a highly successful dominant, in other words, and its success in the environment meant the end of the *Dryas*. Fifty years after the glacier's retreat, alders were growing in stands that were 10 meters, or about 33 feet, tall. With their growth, the terrain got a great deal of nitrogen and, through acidic alder leaves that fell to the ground, a rapid lowering of pH.

The alder sowed the seeds of its own elimination, however. With the lowered soil pH, an extensive population of Sitka spruce trees could develop; when they did, they "shaded out" the smaller alders by blocking their sunlight. About 100 years after the glacier's retreat,

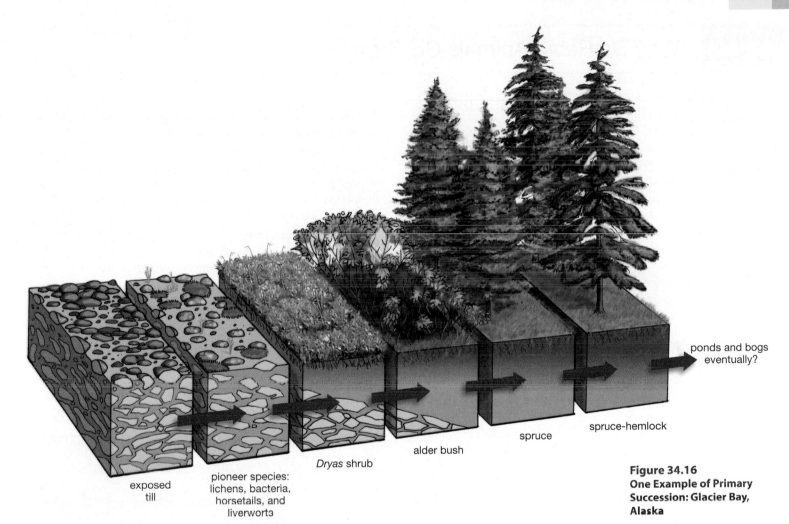

exposed
till

pioneer species:
lichens, bacteria,
horsetails, and
liverworts

*Dryas* shrub

alder bush

spruce

spruce-hemlock

ponds and bogs
eventually?

**Figure 34.16**
**One Example of Primary**
**Succession: Glacier Bay,**
**Alaska**

a spruce forest existed—one that, within 30 more years, started becoming a combined spruce-hemlock forest.

For most of Glacier Bay, the spruce-hemlock forest appears to be the end of the line—the climax community. It's worth noting, however, that on the outer coast of the area there has been one more successional stage. Water-absorbing mosses started to grow and, with them, the forest floor became watery, the tree roots couldn't absorb the oxygen they needed, and the trees died. What eventually was left was land that was filled with ponds and watery bogs. Ecologists have speculated that all of Glacier Bay may be headed in this direction, although this transition could take thousands of years.

## Common Elements in Primary Succession

In the Glacier Bay succession, you can see developments that apply to many primary successions, namely:

- An arrival of photosynthesizing pioneer species—in this case, the bacteria and lichens—that can establish themselves in the most barren environments.

- A steady increase in biomass compared to the starting state. This increase eventually stops in all successions. In parts of Glacier Bay, it actually was reversed with the appearance of the late-stage mosses.

- A general movement toward longer-lived species. *Dryas* can live for 50 years, but alders can live for 100 years and spruce trees for 700.

- A general trend toward an increase in species diversity, although not without some reversals (the alder bushes, for example).

- The facilitation of the growth of some later species (the Sitka spruce here) through the actions of earlier species (the shedding of leaves by the alders). More formally, **facilitation** is a process within ecological succession in which the qualities or actions of some early-arriving species enable the success of later-arriving species.

- The competitive driving out of some species (the alder bush) by the actions of later species (the blocking of sunlight by the spruce trees).

## ESSAY | Why Do Rabid Animals Go Crazy?

There is a tiny worm, called *Plagiorhynchus cylindraceus*, that lives its adult life as a parasite in the intestines of starlings and other songbirds. While in a starling, this worm produces eggs that pass out of the bird in its feces. After this, both eggs and feces may be eaten by pillbugs, which are the familiar little arthropods that "roll up into a ball." The worm eggs, however, are not digested by the pillbugs. As a result, a worm *hatches* inside a pillbug, which makes the bug the second "host" for this parasite. And here is where things get interesting, as researcher Janice Moore has shown.

The pillbug is dark, but once infected by a worm, it starts spending more time on light-colored surfaces. What's more, it stops sheltering itself under overhanging objects such as leaves. And the females among the bugs start moving around more and resting less. The effect of all this is that worm-infected pillbugs are more visible to their predators and thus are more likely to be eaten by them. And who are these predators? Some of them are starlings.

Now, when a starling eats a pillbug, the pillbug dies, but the worm inside the bug remains unharmed. Indeed, this is the way the worm gets back inside a starling host (where it will develop into an adult, lay eggs, and so forth). And that's just the point. It appears that the worm *engineers* this outcome by bringing about the self-destructive behavior of the pillbug. Through some unknown mechanism, this parasite prompts the pillbug to become an accomplice in its own death. When the pillbug is infected, its behaviors are such that it might as well have a sign on it saying, "Please eat me." The starling is happy to oblige, but the real winner is the *P. cylindraceus* worm, which gets to move from one of its hosts to another.

It would be nice to report that such behavioral takeovers by parasites are rare, but in fact they are common. "Horsehair" worms need to mate in the water, but they spend most of their lives inside arthropods such as ants, whose abdomens they feed on. Just before an ant dies from this, however, it aids the worm in one final way—it makes a journey to water. Indeed, a worm-infected ant seemingly has an uncontrollable urge to get wet as it will return repeatedly to water if blocked. Once it wades into a pool or puddle, however, things proceed quickly. In a matter of minutes, the worm inside it will emerge, ready to mate with another worm.

Now, think about all this in connection with one of the things children are taught to fear from an early age: a rabid animal. Rabies is caused by a virus that spreads through animal saliva. In the "furious" form of the disease, the virus is passed on by means of one animal biting another, thus injecting its saliva into the second animal. Not surprisingly, one of the things the virus does is affect salivary tissue in an infected animal's head and neck. This ensures that there is plenty of saliva to be transmitted. The virus also inhibits swallowing in an infected animal, which ensures more saliva. Finally, the virus affects the animal's central nervous system in such a way that it brings about the famous *behavioral* change in infected animals—frenzied, unprovoked attacks on other animals. Thus, rabid raccoons or dogs are not simply "going crazy" because of their disease. Their behavior has a function. To be sure, it's a *perverse* function from a human perspective, but then again, aren't the behaviors of infected pillbugs or ants disturbing in a similar sort of way? The general message here is straightforward but hard to take: Living things can be forced to act in ways that benefit the very organisms that have infected them.

**Accomplice to Its Own Death**
The common pillbug *Armadillidium vulgare*, whose behavior can be manipulated by the parasitical worm *P. cylindraceus*.

## Lessons in Succession from Mount St. Helens

Ecologists have been studying succession for about 100 years now, but it remains a field filled with basic questions. Spirited debate has gone on for years about which factors hold true in all instances of succession and which models of development best describe succession. (Is facilitation more important than competition?)

It turns out, however, that in just the past 30 years, Washington's Mount St. Helens has provided a treasure trove of insight into how primary succession works. If there is a silver lining behind the Mount St. Helens explosion, it is that it has given ecologists an unparalleled opportunity to understand succession. This is so because the Mount St. Helens succession began on a crystal-clear starting date (May 18, 1980) and proceeded from such well-defined starting states—obliteration of some habitats and major or minor damage to others.

So, what have we learned from Mount St. Helens? Perhaps the most important lesson is this: What is left from the past is crucial to the future. When trees were knocked over in the blast, their roots were yanked above ground. In this moment of destruction, however, these roots literally carried the seeds of rejuvenation with them. Soil, small plants, and seeds were pulled up with the roots, and they stayed on them—above the deadening layer of ash below. Dead wood, left strewn about after the blast, provided habitat and food for wood-boring insects. These insects in turn attracted birds, which were able to build nests in "snags," meaning trees that were dead but standing. Gophers survived the blast in large numbers because they live below ground, where they feed on roots. They actually flourished in the years following the explosion, and their activity not only turned over the soil—mixing fertile soil below with ash on top—but also provided a kind of subway system for other animals, which were migrating among patches of viable habitat (**Figure 34.17**). So important were the gophers, snags, and surviving seeds to renewal at Mount St. Helens that a collective term for them has now entered the ecological lexicon. They are examples of **biological legacies**: living things, or products of living things, that survive a major ecological disturbance.

Succession at Mount St. Helens has yielded some other surprises as well. "Classical" primary succession, as outlined in the Glacier Bay example, assumes that life will come to a devastated area from the outside. Seeds and spores will be carried in on the wind, and succession will proceed, in a general way, from the periphery of the barren area to its interior. Although this happened to some degree at Mount St. Helens, in the main recovery was internal: Thousands of small patches of plant habitat rejuvenated without being seeded from the outside. When organisms did come in from the outside, many were the "pioneer" species predicted by classical succession theory. However, ecologists also found solitary tree seedlings—supposedly latecomers in succession—growing right out of some of the most devastated terrain in the blast area, with no pioneer species setting the stage for them. And some of the external help that did arrive took an unexpected form. In the years immediately following the blast, insects and spiders were blown in on the wind in significant quantity. Once these creatures landed, most of them died, but they themselves went on to serve as food for indigenous ants and other species.

**Figure 34.17**
**Important Survivor at Mount St. Helens**
The pocket gopher (*Thomomys talpoides*) played a role in the restoration of the Mount St. Helens environment. The gopher's underground habitat allowed it to survive the blast, after which it continued to feed on roots, turning over soil in the process. In addition, its tunnels provided a passageway for other animals that, in the years following the explosion, were making their way to viable habitats. This gopher is in Wyoming.

Given all this, another lesson from Mount St. Helens seems to be that succession is more varied and habitat-specific than our theories had predicted. Put another way, we still have a lot to learn about succession.

## On to Ecosystems and Biomes

In the last two chapters, you have looked at populations and communities. In populations, you saw the building blocks of communities. In communities, you looked through a lens of the interactions of different species. But it's obvious that community members interact with *nonliving* forces and factors, such as weather and soil composition. When living things and nonliving factors are considered together, the resulting whole is called an *ecosystem*—the subject of the first part of Chapter 35. Very large-scale ecosystems are called *biomes*; a review of them forms the concluding portion of Chapter 35 and will bring to a close this book's coverage of ecology.

**SO FAR...**

1. Succession can be defined as a series of _____ of community members at a given location until a relatively _____ is reached.

2. In primary succession, over time, biomass tends to (pick the correct term from the two) decrease/increase; species tend to be shorter lived/longer lived; and species diversity tends to decrease/increase.

3. Succession at Mount St. Helens revealed the importance of biological legacies, meaning living things or products of living things that _____.

**••• ANSWERS**

1. replacements; stable state

2. increase; longer lived; increase

3. survive a major ecological disturbance

# CHAPTER 34 REVIEW

## SUMMARY

### 34.1 Structure in Communities

- An ecological community is all the populations of all species that inhabit a given area.

- Many communities are dominated by only a few species; the few species that are abundant in a given area are called ecological dominants.

- A keystone species is a species whose absence from a community would bring about significant change in that community. Some keystone species are top predators in a community, meaning species that prey on other species but that are not preyed on themselves. Keystone species need not be top predators, however, and every community does not have a keystone species.

- Biodiversity, defined as variety among living things, takes three primary forms: a diversity of species in a given area; a geographic distribution of species populations; and genetic diversity within species populations.

- Recent research indicates that species diversity tends to enhance a community's productivity, defined as the amount of solar energy that photosynthesizing organisms are able to capture and transform into living material or biomass. There is disagreement about whether species diversity also enhances community stability, meaning the ability of a community to retain its characteristics in the face of environmental disruption.

### 34.2 Types of Interaction among Community Members

- There are four primary types of interaction among community members: competition; predation (and a special variety of it, parasitism); mutualism; and commensalism.

- Habitat can be thought of as the physical surroundings in which a species normally can be found. Niche can be defined metaphorically as an organism's occupation, meaning what the organism does to obtain the resources it needs to live.

- The competitive exclusion principle states that when two species compete for the same limited, vital resource, one always outcompetes the other and thus brings about the latter's local extinction.

- There are numerous instances in nature in which two related species use the same kinds of resources from the same habitat over an extended period of time but will divide the resources up such that neither of the species undergoes local extinction. This phenomenon is called coexistence through resource partitioning.

- Predation is defined as one free-standing organism feeding on parts or all of a second organism. Parasitism is a variety of predation in which the predator feeds on prey but does not kill it immediately and may not kill it ever.

- Predator and prey population sizes can move up and down together in a fairly tight linkage, but predator-prey interaction generally is only one of several factors that control the population level of either predators or prey.

- The prey of a parasite is known as the host. A parasite can use a host not only as a food source but as a vehicle to facilitate its reproduction.

- Over evolutionary time, predator-prey interactions have spurred physical modifications in both predator and prey species. One form that such modifications take is mimicry: a phenomenon in which one species has evolved to assume the appearance of another. One form of mimicry, Batesian mimicry, occurs when one species evolves to resemble a species that has superior protective capability. Batesian mimicry always includes three players: a mimic, a model, and a dupe. The mimic species evolves to match the appearance of the model species, which has superior protective ability. The dupe species is then deceived into believing the mimic species is the model species. In a second form of mimicry, Müllerian mimicry, several species that have protection against predators come to resemble each other.

- Mutualism is an interaction between individuals of two species that is beneficial to both individuals. Commensalism is an interaction in which an individual from one species benefits while an individual from another species is neither harmed nor helped.

- Coevolution is the interdependent evolution of two or more species. Flowers have evolved colors and fragrances that attract bees, for example, while bees have evolved vision that is most sensitive to the colors of the flowers they pollinate.

### 34.3 Succession in Communities

- Parcels of land or water that have been abandoned by humans or devastated by physical forces will almost always be reclaimed by nature to some degree. The process by which this takes place is called succession: a series of replacements of community members at a given location until a relatively stable final state is reached.

- Primary succession proceeds from an original state of little or no life and soil that lacks nutrients. Secondary succession occurs when a final state of habitat is first disturbed by some outside force, but life remains, and the soil has nutrients. The final community in any process of succession is known as the climax community.

- A common set of developments occurs in most instances of primary succession. These developments include the arrival of "pioneer" photosynthesizers, facilitation of the growth of some later species through the actions of earlier species, and the competitive driving out of some earlier species by the actions of later species. As succession proceeds, species diversity tends to increase within communities and smaller, shorter-lived species tend to be replaced by larger, longer-lived species.

- The rejuvenation of the Mount St. Helens area that has occurred since 1980 has provided ecologists with a wealth of information regarding both primary and secondary succession. One of the chief lessons learned concerns the degree to which succession can be facilitated by biological legacies, defined as living things, or products of living things, that survive a major ecological disturbance.

## KEY TERMS

| | |
|---|---|
| **Batesian mimicry**  717 | **interspecific competition**  712 |
| **biodiversity**  711 | **keystone species**  710 |
| **biological legacies**  723 | **mimicry**  717 |
| **climax community**  720 | **Müllerian mimicry**  717 |
| **coevolution**  718 | **mutualism**  718 |
| **coexistence**  714 | **niche**  712 |
| **commensalism**  718 | **parasitism**  714 |
| **community**  709 | **predation**  714 |
| **competitive exclusion principle**  712 | **primary succession**  720 |
| **ecological dominants**  709 | **resource partitioning**  714 |
| **facilitation**  721 | **secondary succession**  720 |
| **habitat**  712 | **succession**  719 |
| **host**  716 | **top predator**  710 |

## BRIEF REVIEW

*Answers to Brief Review questions are in the back of the book. For multiple-choice quiz questions, go to* **www.aw.com/krogh4**.

1. In architecture, a keystone is a stone at the top of an arch that holds the entire arch in place. In what way can certain species be thought of as keystones for their communities?

2. Name the four primary modes of interaction in a community.

3. A wasp from the group known as the ichneumons sometimes will paralyze a caterpillar by injecting a toxin into it, after which the wasp deposits its eggs into or onto the caterpillar's body. When the eggs hatch, the resulting wasp larvae live by consuming the caterpillar, which remains alive while paralyzed. What mode of interaction are the wasp and the caterpillar practicing? What is the role of the caterpillar in this interaction?

4. Lichens are composite organisms, usually made up of upper and lower layers of fungi, with a layer of algae in between. Fungi benefit from their association with the photosynthesizing algae in that they receive food from them. At one time, however, scientists questioned whether the algae were in turn benefiting from their relationship with fungi. It now appears, however, that algae do benefit from their association with fungi in that they receive water, carbon dioxide, and protection from the elements from them. For the sake of argument, assume that the fungi are not providing any benefit to the algae; which mode of interaction would the two types of organisms be practicing? In light of the benefits the fungi are providing to the algae, which mode of interaction are they actually practicing?

5. The red-spotted newt *Notophthalmus viridescens* secretes toxins from its skin that make predators avoid it. The red salamander *Pseudotriton ruber* has no such secretions but resembles the red-spotted newt and gains protection from this resemblance. What is this an example of? Which is the model, the mimic, and the dupe?

6. A volcano erupted in Hawaii, and lava covered land that had been farmed for centuries. Eventually, new lichens, and then plants, grew on the lava. Is this an example of primary or secondary succession?

## APPLYING YOUR KNOWLEDGE

1. In the movie *The Godfather*, Michael Corleone and Enzo the baker stand on the steps outside the hospital where Michael's father is recovering from gunshot wounds and pretend to be holding guns inside their coats as a means of keeping assassins from entering the hospital and killing Michael's father. Allowing for some differences between the natural world and human society, what form of mimicry were Michael Corleone and Enzo the baker practicing?

2. The world's first farmers were not human beings, but the leaf-cutter ants of the genus *Atta*. These ants do not eat the leaves they carry off. Instead, they bring them down into their nests, where they process the leaves into a pulp and then spread the pulp onto a fungus that grows only in their nests. The fungus then feeds on the food that has been brought to it and grows, after which the ants harvest the knob-like stalks of the fungus and consume them. What is the mode of interaction between the ants and the fungus in this community?

3. In the Great Lakes, lake trout in the genus *Salvelinus* feed on other fish as well as on insects and crustaceans. Meanwhile, the eel-like sea lamprey (*Petromyzon marinus*), introduced to the Great Lakes by human beings, parasitizes lake trout by feeding on them over a long period after attaching to them with its sucking disc. Human beings don't seem to be disturbed by the predation that lake trout practice on other animals, but they do seem to be repelled by the lamprey's parasitism of lake trout. What accounts for this difference? Why do human beings seem to be bothered more by parasitism than by predation?

Bromeliads grow along a stream in the tropical rain forest of Brazil's Bocaina National Park.

The living world is affected by the larger physical world around it—and vice versa. The interconnected web of living things is being damaged in many ways by human activity.

| 35.1 | The Ecosystem | 727 |
| 35.2 | Abiotic Factors Are a Major Component of Any Ecosystem | 727 |
| 35.3 | How Energy Flows through Ecosystems | 734 |
| 35.4 | Earth's Physical Environment | 740 |
| 35.5 | Global Warming | 742 |
| 35.6 | Earth's Climate | 746 |
| 35.7 | Earth's Biomes | 750 |
| 35.8 | Life in the Water: Aquatic Ecosystems | 755 |

**Essays**

| A Cut for the Middleman: Livestock and Food | 741 |
| **THE PROCESS OF SCIENCE:** Global Warming and the Harlequin Frogs | 746 |
| Good News about the Environment | 756 |
| Our Overfished Oceans | 758 |

# AN INTERACTIVE LIVING WORLD 3:
## Ecosystems and Biomes

T he word *system* is much in use today. We hear about a given country's political system or its economic system. The term was first used in this context by academics who needed a way to refer to complex entities composed of many interacting parts. This way of looking at things is, however, as applicable to ecology as it is to economics or politics. In Chapter 34, when you went over ecological communities, you actually were studying something in isolation: the interactions of all the *living* things in a given area. It goes without saying, however, that these living things also interact with many nonliving entities. To get a complete picture of how the living world functions, then, it's necessary to understand how living things interact not only with each other, but with these nonliving elements.

living or **biotic** factors and nonliving or **abiotic** factors to form a whole.

Given this, an important unit in ecology is the **ecosystem**, which can be defined as a community of organisms and the physical environment with which they interact. This chapter begins with a look at small-scale ecosystems and concludes with a look at the very large-scale ecosystems called biomes. In today's world, it's impossible to note the ways in which ecosystems function without also noting the ways in which human activity is damaging them.

## 35.1 The Ecosystem

What do nonliving elements add to life? All communities need an original source of energy, generally meaning the sun; they all need a supply of water and a supply of nutrients, such as nitrogen and phosphorus, and all organisms need to exchange gases with their environment (think of your own breathing). The sun's energy is making a one-way trip through the community, ultimately being transformed into heat, but water and nutrients are being *cycled*—from the Earth into organisms and then back again into the Earth. Looking at this, we can begin to perceive the reality of ecological *systems*: of functional units that tightly link both

## 35.2 Abiotic Factors Are a Major Component of Any Ecosystem

In Chapters 33 and 34, you learned a good deal about two of the biotic factors of the ecosystem, namely individual populations of living things and the communities that are made up of these populations. This chapter focuses on some of the abiotic factors in an ecosystem. In overview, these abiotic factors fall into two categories. First are the *resources* that exist in the ecosystem, such as water and nutrients; second are the *conditions* in which an ecosystem exists, such as average temperature. Resources are considered first in this chapter and conditions later.

## The Cycling of Ecosystem Resources

Those of you who went through Chapter 2, on chemistry, know what a chemical **element** is: a substance that is "pure" in that it cannot be broken down into any set of component substances by chemical means. The 92 stable elements include such familiar substances as gold and helium, but only 30 or so of these elements are vital to life and are thus called **nutrients**. Some nutrients, such as iron or iodine, are needed only in small or "trace" quantities by living things, but others are needed in large quantities, among them carbon, oxygen, nitrogen, and phosphorus. Living things also have a great need for water, which is not an element but a molecule composed of two elements, hydrogen and oxygen. Water and nutrients move back and forth between the biotic and abiotic realms, with the term for this movement being a mouthful: **biogeochemical cycling**. Let's look now at the cycling of two elements, carbon and nitrogen, and then at the cycling of water.

### Carbon as One Example of Ecosystem Cycling

A certain amount of cosmic debris makes its way to Earth's surface in the form of meteorites, but this material is pretty sparse. What else comes to Earth from space? The sun's rays, to be sure, as well as light from distant stars, and some other forms of radiation. But when we're talking about chemical elements, the Earth really is a spaceship: It carries a fixed amount of resources with it. Nothing much comes to Earth *from* the outside, and relatively little leaves Earth *for* the outside. For an ecologist, this self-containment simplifies things: What we possess in terms of elements is all we'll ever possess. Thus, the question becomes not so much what we have, but where we have it.

In recent years, there has been considerable concern about the buildup of the heat-trapping gas carbon dioxide ($CO_2$) in the Earth's atmosphere as this increase is now known to be the primary cause of global warming. Well, if Earth has this fixed quantity of elements—in this case, the carbon and oxygen that make up $CO_2$—how could there be a buildup of carbon dioxide in the atmosphere? The answer is that carbon has been *transferred* from one place on Earth to another. It has moved from the "fossil fuels" of coal and oil, where it was stored, into the atmosphere.

Storage and transfer; these two concepts are fundamental to biogeochemical cycling. When we look a little deeper into carbon cycling, we find other players involved in carbon's storage and transfer.

Plants need $CO_2$ to perform photosynthesis, and they take in great quantities of it for that purpose, producing their own food as a result. While the plants are alive, they use some of the carbon they take in to grow—it becomes part of their leaves and stems and roots. And as long as these structures exist, they will store carbon within them, just as coal and oil do. (Thus, one of the proposed solutions to global warming is to grow a huge number of trees, each of which could be thought of as a kind of piggy bank for carbon.) Eventually, the plants die, of course, after which their decomposition by bacteria and fungi releases carbon into the soil and atmosphere, again as $CO_2$. With this, you can begin to see the whole of the carbon cycle in the natural world (**Figure 35.1**).

Some plants will, of course, be eaten by *animals*, who need carbon-based molecules for tissues and energy, just as plants do. The difference is that animals get these molecules *from plants*—in the form of seeds and leaves and roots—rather than from the air. These animals will likewise lock up carbon for a time, only to return it to the soil with their death and decomposition. Note, however, the critical difference between animals and plants: Carbon comes into the living world only through plants (and their fellow photosynthesizers algae and some bacteria). An animal might take in some carbon dioxide in breathing, but nothing life sustaining happens as a result. Conversely, a plant that takes in carbon dioxide can *incorporate* it into a life-sustaining molecule, in this case a sugar that is the initial product of photosynthesis. With this sugar as a food source, plants grow, and this bounty ultimately feeds everything else. Look at this simplified cycle:

$$\text{atmospheric } CO_2 \rightarrow \text{plants} \rightarrow \text{animals} \rightarrow$$
$$\text{decomposers} \rightarrow \text{atmospheric } CO_2$$

What does it track, the path of carbon or the path of food? The answer is both. This is one of the reasons that biologists refer to life on Earth as "carbon based."

### How Much Atmospheric $CO_2$?

In following the carbon trail, it's probably apparent that, just as money is a medium of exchange in our economy, $CO_2$ is the living world's medium of exchange for carbon; photosynthesizers take *in* carbon as $CO_2$ and decomposers yield *up* carbon as $CO_2$. (All organisms also release $CO_2$ as a by-product of carrying out life's basic processes, as with your own exhalations in breathing.) Given how much

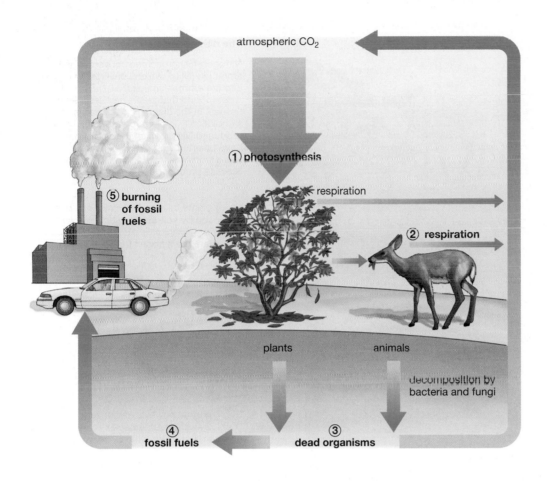

**The carbon cycle**

1. Plants and other photosynthesizing organisms take in atmospheric carbon dioxide ($CO_2$) and convert or "fix" it into molecules that become part of the plant.

2. The physical functioning or respiration of organisms converts the carbon in their tissues back into $CO_2$.

3. Plants and animals die and are decomposed by fungi and bacteria. Some $CO_2$ results, which moves back into the atmosphere.

4. Some of the carbon in the remains of dead organisms becomes locked up in carbon-based compounds such as coal or oil.

5. The burning of these fossil fuels puts this carbon into the atmosphere in the form of $CO_2$.

**Figure 35.1**
**The Carbon Cycle**

biomass or material produced by living things there is in the world, you might think there would have to be a great deal of $CO_2$ in the atmosphere. But as a proportion of atmospheric gases, $CO_2$ is a bit player, making up about 0.035 percent of the atmosphere. But what an important player it is! It's critical in feeding us all, and relatively small changes in its atmospheric concentration are now leading to significant changes in global temperature.

### The Nitrogen Cycle

Like carbon, nitrogen is an element; and like carbon, nitrogen cycles between the biotic and abiotic domains. Nitrogen makes up only a small proportion of the tissues of living things, but it is a critical proportion because it is required for DNA, RNA, and all proteins. All living things need this element, then. The question is, how do they get it?

The source for all the nitrogen that enters the living world is atmospheric nitrogen, which exists in great abundance. You've seen that carbon dioxide makes up less than 1 percent of the atmosphere, and it turns out that oxygen makes up about 21 percent. But nitrogen makes up 78 percent of the atmosphere. So rich is the atmosphere in this element that a typical garden plot has tons hovering above it. The problem?

All this nitrogen is in the form of pairs of nitrogen atoms ($N_2$) that have a great tendency to stay together, rather than combining with anything else. Thus, in a case of so near and yet so far, plants have no direct access to atmospheric nitrogen. So, how do they take it up? In the natural world, the answer essentially is, through the actions of bacteria. It is bacteria that are carrying out the process of **nitrogen fixation**, meaning the conversion of atmospheric nitrogen into a form that can be taken up and used by living things.

### Bacterial Fixation of Nitrogen

You can see nitrogen fixation and the rest of the nitrogen cycle diagrammed in **Figure 35.2** on the next page. The essence of the cycle is that several types of nitrogen-fixing bacteria take in atmospheric nitrogen ($N_2$) and convert it into ammonia ($NH_3$). This ammonia then is converted (either in water or by other microbes) into two types of nitrogen-containing compounds that plants can *assimilate*—can take up and use. And, as you've seen, what comes into the plant world will come into the animal world when animals eat plants. The two usable nitrogen compounds produced in the cycle are the ammonium ion ($NH_4^+$) and nitrate ($NO_3^-$). Some of the nitrate that results from this process is used by yet another kind of bacteria,

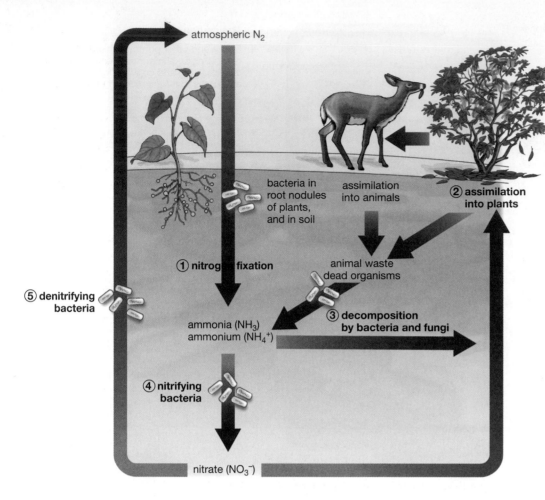

1. Nitrogen-fixing bacteria convert $N_2$ into ammonia ($NH_3$), which converts in water into the ammonium ion ($NH_4^+$). The latter is a compound that plants can assimilate into tissues. In the diagram, bacteria living symbiotically in plant root nodules have produced $NH_4^+$, which their plant partners have taken up and used. Meanwhile, free-standing bacteria living in the soil have likewise produced $NH_4^+$.

2. Other plants take up $NH_4^+$ that has been produced by soil-dwelling bacteria and assimilate it. Animals eat plants and assimilate the nitrogen from the plants.

3. Animal waste and the tissues of dead animals are decomposed by fungi and by other bacteria, which turn organic nitrogen back into $NH_4^+$.

4. Other "nitrifying" bacteria convert $NH_4^+$ into nitrate ($NO_3^-$), which likewise can be assimilated by plants.

5. Some nitrate, however, is converted by "denitrifying" bacteria back into atmospheric nitrogen, completing the cycle.

**Figure 35.2**
**The Nitrogen Cycle**

Some bacteria "fix" nitrogen, meaning they convert atmospheric nitrogen ($N_2$) into an organic form that can be taken up and used by other living things.

denitrifying bacteria, which can convert nitrate back into atmospheric nitrogen, completing the cycle.

### Nitrogen as a Limiting Factor in Food Production

If you were to look at the nitrogen story 100 years ago, there would be little more to it than what you've just gone over. In other words, through most of human history, nitrogen was fixed solely by bacteria (with a small assist from lightning, which can produce a usable form of nitrogen that falls to the Earth in rain). Human beings had a problem with this single route to nitrogen, however, because nitrogen is critical for all plant growth, including *agricultural* plant growth. Indeed, for centuries a lack of nitrogen was a primary limiting factor in getting more food from a given amount of land. Long before anyone knew what nitrogen was, farmers were trying to get more of it to crops in two ways. One was by applying organic fertilizer, such as rotting organic material, to their fields. (Remember how the Pilgrims were taught by Native Americans to bury dead fish around their corn plants?) The second was by planting crops that carried their own nitrogen-fixing bacteria within them. This is the case with certain legumes, such as soybeans, which carry symbiotic bacteria in

their root nodules. So great was the need for nitrogen that nitrogen-fixing plants were sometimes grown as "green manures"—as crops that were plowed right back into the ground to serve as fertilizers.

Early in the twentieth century, however, a momentous change came about in the use of nitrogen when the German chemist Fritz Haber developed an *industrial* process for turning atmospheric nitrogen into ammonia. In essence, human beings became nitrogen fixers, and by the 1960s, they had taken up this activity on a grand scale. In 2005, almost 120 million metric tons of biologically active nitrogen were manufactured for use as fertilizer (**Figure 35.3**). Meanwhile, by the mid-1990s, human beings were planting crops that resulted in the production of another 40 million tons of nitrogen per year. Put together, these forms of human-driven nitrogen production exceed the amount of terrestrial nitrogen that all of nature produces on its own.

What are the consequences of this revolution in nitrogen fixation? One is that much more food can be grown on Earth's arable land. Indeed, the world's human population could not be fed without the nitrogen fixation that human beings carry out. This massive human intervention in the nitrogen cycle has

**Figure 35.3**
**Nutrients Beyond What Nature Provides**
A helicopter applies fertilizer to a sugar-beet crop in California.

a downside as well, however. Remember that human-manufactured nitrogen ends up being used in a very *concentrated* way—it is poured onto farmland. A good deal of this nitrogen does not end up in the crops it was intended for, however, but instead departs as runoff. The amount of runoff generated by each farm adds up, and the end result can be large amounts of nitrogen flowing into creeks, ponds, and rivers. On a regional basis, such nitrogen flows can be enormous. The Mississippi River carries an estimated 1.5 million metric tons of nitrogen into the Gulf of Mexico each year.

You might think that since nitrogen is a nutrient this would be a good thing for the Gulf waters, but the opposite is true. The nitrogen brings on a seasonal "bloom" of algae, the algae feed a huge bacterial population, and the bacteria use up most of the available oxygen at the water's lower depths. The result is a giant "dead zone" that drives away mobile fish and that kills immobile, bottom-dwelling sea creatures (**Figure 35.4**). This problem of *nutrient pollution* is not limited to the Gulf of Mexico, however; it is played out on a smaller scale in estuaries and ponds around the world.

### The Cycling of Water

Now, let's turn from nutrient to water cycling. Those who went through the material on water in Chapter 2 know how important water is to life. The human body is about 66 percent water by weight, and any living thing that does not have a supply of water is doomed in the long run. As with nitrogen or carbon, all water is either being cycled or is locked up. Carbon may be stored in coal or trees, but water can be stored in ice located on land—in glaciers or in polar ice, for example.

**Web Animation 35.2**
The Nitrogen Cycle

**Figure 35.4**
**The Gulf of Mexico "Dead Zone"**

**(a)** In 2002, the seasonal "dead zone" in the Gulf of Mexico stretched from the mouth of the Mississippi River to eastern Texas. The waters in the zone are hypoxic, or oxygen depleted, particularly at their lower depths. The primary cause is the tremendous amount of nitrogen carried into the Gulf from both the Mississippi and the Atchafalaya Rivers. This nitrogen originates as fertilizer that is applied to farmland all the way up the Mississippi River basin.

**(b)** Why is this hypoxic zone so extensive? One factor at work is prevailing winds, which carry the discharge of the Mississippi and Atchafalaya Rivers westward, toward Texas. This satellite image of the region was taken in 2007. (Image courtesy of the Louisiana State University Earth Scan Laboratory.)

**(a)** Runoff from the Atchafalaya and Mississippi Rivers...

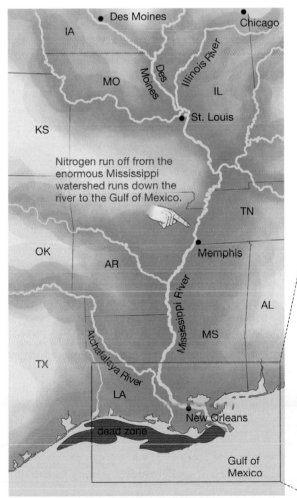

Nitrogen run off from the enormous Mississippi watershed runs down the river to the Gulf of Mexico.

**(b)** ...is given a wide distribution westward in the Gulf of Mexico.

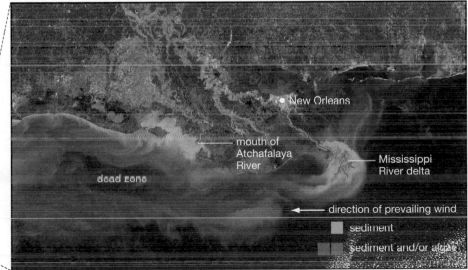

New Orleans

mouth of Atchafalaya River

Mississippi River delta

dead zone

direction of prevailing wind

sediment

sediment and/or algae

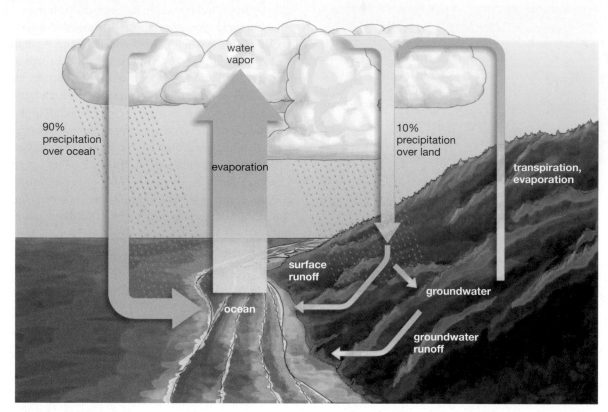

**Figure 35.5**
**The Hydrologic Cycle**

More than 95 percent of Earth's water is stored in the oceans. When water evaporates from the ocean, 90 percent returns directly to the ocean by way of precipitation. The other 10 percent falls on land. There, the water runs back into the ocean, is stored in such structures as glaciers, or is moved by transpiration and evaporation back into the atmosphere.

The oceans hold more than 95 percent of Earth's water, but thankfully for us, when the oceans' salt water evaporates, it falls back to Earth as freshwater. We can also be grateful that not all ocean water falls back to Earth on the oceans; instead, about 10 percent falls on land (**Figure 35.5**). This water represents about 40 percent of the precipitation that land areas get; the other 60 percent comes from a form of cycling reviewed in Chapter 24, *transpiration*, meaning the process by which water is taken up by the roots of land plants, then moved up through their stems and out through their leaves as water vapor. After evaporating into the atmosphere, water returns to Earth as rain, fog, snow—all the forms of precipitation possible. The driving force behind the hydrologic cycle is the sun, whose heat powers both evaporation and transpiration.

With more than 95 percent of Earth's water in the oceans, perhaps as little as 2.5 percent of the Earth's water is freshwater at any given time. Of this freshwater, more than 75 percent is locked up in glaciers and other forms of ice (**Figure 35.6**). Thus, any snapshot we would take of Earth would show that as little as 0.5 percent of its water is available as liquid freshwater, but even here we have to qualify things. About 25 percent of this freshwater is **groundwater**, meaning water that moves down through the soil until it reaches porous rock that is

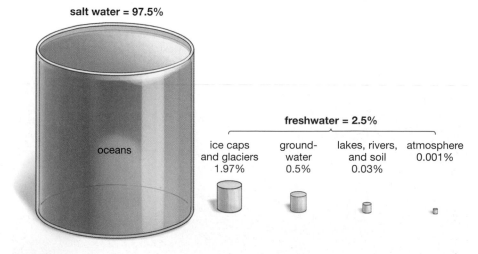

**Figure 35.6**
**Earth's Water**

Although there is a tremendous amount of water on our planet, only a small fraction of it is available to us as freshwater.

saturated with water. In line with this, about 20 percent of the water used in the United States comes from groundwater, and at least a quarter of the world's population depends on groundwater to satisfy basic needs. Groundwater is stored in porous rock called an **aquifer**, an example of which is pictured in **Figure 35.7**. Major aquifers can be enormous, as you can see from the map of the High Plains Aquifer in **Figure 35.8**. Even an aquifer of this size, however, can lose more water through pumping than it gains through replenishment from rain. In Texas, the depth of the High Plains Aquifer declined by more than a foot per year from 1940 to 1980 and then by a little less than half a foot per year between 1980 and 1998.

### Global Human Use of Water

Given the sheer number of people on Earth and the importance of water to each of them, it will not surprise you to learn that human beings are laying claim to a great deal of Earth's freshwater; civilization now uses more than half the world's accessible supply. But from what you've seen so far, think of what this means. There is no less *water* than there ever was. Like nitrogen or carbon, all the water we have is either being recycled or is stored. What's changed is that the human population is using an ever-greater proportion of the water that's available. Human beings fight among themselves for water, of course, because it is a scarce commodity in locations throughout the world. Indeed, sanitary water is so scarce that an estimated 1 billion people worldwide lack access to it (**Figure 35.9** on the next page). A large part of the problem is that there is a mismatch between the location of freshwater and the location of human populations. Sixty percent of the world's population lives in Asia, while only 36 percent of Earth's precipitation falls there. Meanwhile, 6 percent of the world's people live in South America, but 26 percent of the world's precipitation falls there. (The Amazon River alone discharges 200,000 cubic meters of freshwater per *second* into the Atlantic Ocean, an amount so large that it constitutes 20 percent of the entire planet's freshwater runoff.)

Another part of the human water problem, water expert Peter Glieck has noted, is that humans make such poor use of the water they capture. Mexico City's leaky water system *loses* enough water to meet the needs of a city the size of Rome. If the toilet in your house is more than 15 years old, chances are it uses 6 gallons of water with each flush. If it was built

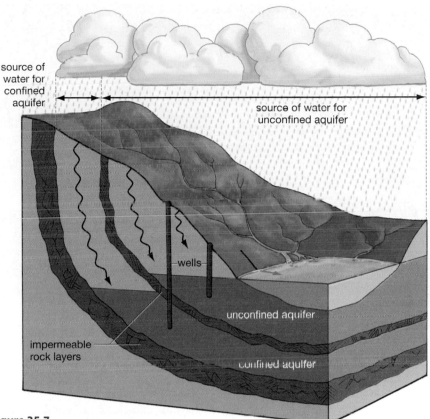

**Figure 35.7**
**Aquifers Store Groundwater**

When water seeps into the ground, it moves freely through layers of sand and porous rock but moves either extremely slowly or not at all through layers of impermeable rock. Water thus becomes trapped in different underground layers, from which humans can draw water. Unconfined aquifers receive water directly from a large surface area and therefore tend to be more vulnerable to pollution by chemicals such as fertilizers and pesticides. Confined aquifers are located between layers of impermeable rock. They are replenished more slowly but tend to contain purer water.

**Figure 35.8**
**Enormous Store of Underground Water**

The High Plains Aquifer, perhaps the largest aquifer in the world, underlies an expanse of land stretching across eight states. Billions of gallons of water are withdrawn from it each day, mostly for agriculture. Since the 1940s, more than 200,000 wells have been drilled into it for this purpose. (Modified from E. D. Gutentag, F. J. Heimes, N. C. Krothe, R. R. Luckey, and J. B. Weeks, "Geohydrology of the High Plains Aquifer in Part of Colorado, Kansas, Nebraska, New Mexico, Oklahoma, South Dakota, Texas, and Wyoming," U.S. Geological Survey Professional Paper 1400-B, 1984.)

**Figure 35.9**
**Freshwater Is a Limited Resource**
Women carry drinking water through a polluted slum in Port-au-Prince, Haiti.

**Web Animation 35.3**
The Hydrologic Cycle

## Human Beings Are Not Separate from the Earth on Which They Live

One final thing about nutrient and water cycling. The average person, who has no knowledge of these cycling processes, may regard human beings as a kind of independent line that stretches back in time. Under this view, water and food may pass through us, but they are as separate from us as gasoline is from a car engine.

From your reading of this section, however, it may now be apparent to you that this is not true. Some of the carbon in the bread you ate this morning could have been contained in a tree that decomposed in China 100 years ago. After moving into the atmosphere as $CO_2$ and staying there for decades, this carbon was taken up by a wheat plant in America whose seeds were ground up, yielding flour that ended in the bread on your shelf. This carbon may be burned by your body to supply the energy it needs, it's true, but it may also become part of your bones or your nerves. Then, when you die, the material you are made of will go *back* to the Earth to be recycled in the ways you've been reading about, perhaps to be part of a tree or a person in a future generation. We are intertwined with the Earth in some basic, physical ways.

since the 1990s, following imposition of new federal standards for low-flow toilets, it probably uses less than a third of that—a mere 1.6 gallons.

When we think of societal use of water, however, household use actually is relatively small. About 70 percent of the water we humans withdraw from rivers, lakes, and groundwater goes to agricultural irrigation. People need food, of course, but the *way* irrigation is carried out in much of the world is extremely wasteful: Water moves from, say, a river to crops by means of an open channel. En route, much of this water either evaporates or is taken up by land that isn't being irrigated. An efficient alternative is "micro-irrigation," in which water gets piped to crops.

Human beings are, of course, capable of diverting water from species that have no say in this division of resources. More than 20 percent of all freshwater fish species are estimated to be threatened or endangered precisely because their water has been dammed or diverted for human use. California's natural populations of coho salmon have been reduced by some 94 percent since the 1940s, in significant part through diversions of the freshwater streams that the salmon swim up in order to spawn. If you look at **Figure 35.10**, you can get a quick idea of humanity's impact on water and on the two elements we just reviewed, carbon and nitrogen.

---

**SO FAR...**

1. An ecosystem includes not only a community of organisms, but the _____ with which they _____.

2. The carbon that helps make up the living world moves into it by means of organisms that perform _____. Carbon then returns to the soil and atmosphere through means of the _____ of organic material by the organisms called _____ and _____.

3. Almost all the nitrogen that enters the living world by natural means does so through the actions of _____. Nitrogen is then returned to the atmosphere by the actions of _____.

---

## 35.3 How Energy Flows through Ecosystems

Having looked at the abiotic elements of nutrients and water, let's now turn to another important component of any ecosystem, energy. Those of you who read through Chapter 6 know that one of energy's inflexible laws—the first law of thermodynamics—is that energy is never gained or lost, but only transformed. The sun's energy is not used up by green plants; rather, some of it is *converted* by plants into chemical form—initially into the chemical bonds of the sugar that plants make in photosynthesis.

The second law of thermodynamics is that energy spontaneously flows in only one direction: from more ordered to less ordered. The carbohydrate molecules in bread are very ordered things—so many atoms of carbon, oxygen, and hydrogen bonded together in a precise spatial relationship to one another. Contrast this with another form of energy, heat, which is the *random* motion of molecules—clearly a disordered form of energy compared to the chemical bonds in a carbohydrate. Indeed, heat is the last stop on the energy line because it is the least-ordered form of energy. And, because all energy spontaneously moves toward less order, heat is the ultimate *fate* of all energy. Energy relentlessly ratchets down toward the form of heat because, like a casino that must get a "cut" of each bet laid down, heat gets a cut of each energy transaction. The chemical bonds in gasoline can be broken through combustion and thus be used to power a car engine, but not all the energy released from the combustion drives the engine. Some dissipates as heat, and this happens *every* time energy is used, whether we're talking about a piston firing or a cell dividing.

All of this provides a framework to conceptualize an important part of ecosystems, which is how energy flows through them. Think of the sun, with the energetic rays that leave it as a starting point; now, think of the ultimate destiny of all this energy, which is heat, randomly dispersed in the universe. Looked at one way, life on Earth *intervenes* in this flow. Life is an enormous energy collection and storage enterprise, gathering some of the sun's energy and locking it up for a time in the form of chemical bonds. These bonds can then be broken—think of digesting a muffin—which *releases* this stored energy so that an organism can grow and reproduce. Life does not stop the march of the sun's energy into heat, but it does intervene in it by transforming it into chemical bonds that have order and stability.

Once we see life in these terms, we can begin to think of the *flow* of energy through it. That's what you'll be looking at in the sections that follow.

**(a)** Atmospheric $CO_2$ concentration

| 76% naturally occurring | 24% human-caused |

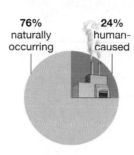

**(b)** Terrestrial nitrogen fixation

| 42% naturally occurring | 58% human-caused |

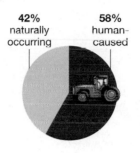

**(c)** Accessible surface water

| 46% available | 54% used |

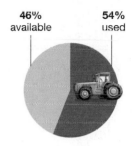

**Figure 35.10**
**Human Impact on Resources**

**(a)** Nearly a quarter of the current atmospheric carbon dioxide concentration is produced by human activity, primarily through the burning of fossil fuels.

**(b)** Over half of terrestrial nitrogen fixation comes about because of human activity, including the manufacture of fertilizer for agriculture.

**(c)** Over half of the Earth's accessible surface water is now diverted for use by humans, mostly for agriculture. Percentages are approximate.

## Producers, Consumers, and Trophic Levels

One way to look at food and energy is in terms of production and consumption. Plants and other photosynthesizers (algae and some bacteria) are an ecosystem's **producers**, while the organisms that eat plants are one kind of **consumer**: an organism that eats other organisms, rather than producing its own food. But as you know, there are consumers *of* consumers—grass is eaten by zebras, but zebras are eaten by lions. Thus do we get to the concept of feeding levels or, as ecologists put it, trophic levels. More formally, a **trophic level** is a position in an ecosystem's food chain or web, with each level defined by a transfer of energy between one kind of organism and another. Producers are one level, and then there are several levels of consumers. Here's a list of trophic levels through four stages:

| First trophic level: | **Producers—** (photosynthesizers) |
| Second trophic level: | **Primary consumers—** plant predators (herbivores) |
| Third trophic level: | **Secondary consumers—** herbivore predators (carnivores) |
| Fourth trophic level: | **Tertiary consumers—** organisms that feed on secondary consumers (carnivores) |

••• ANSWERS

1. physical environment; interact

2. photosynthesis; decomposition; bacteria; fungi

3. bacteria; bacteria

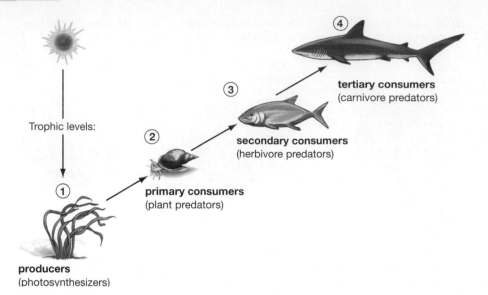

**Trophic levels:**

④ tertiary consumers
(carnivore predators)

③ secondary consumers
(herbivore predators)

② primary consumers
(plant predators)

① producers
(photosynthesizers)

**Figure 35.11**
**Trophic Levels**

Plants, algae, and some bacteria are called producers because they use the sun's energy to produce their own food through photosynthesis. Organisms at all other trophic levels ultimately derive their energy from these photosynthesizers.

It is the sun's energy that is locked up at the first trophic level, in leaves and stems, and it is this energy that is being passed along through all the other trophic levels. Note that the levels are not just categories that could be put in any order. We must have primary consumers to have secondary consumers; sharks cannot exist on algae (**Figure 35.11**). You may remember that an animal that eats only plants or algae is a **herbivore**; then there are **carnivores**, which eat only meat, and **omnivores**, which eat both plants and meat. Of course, many organisms cannot be assigned to just a single trophic level. Most human beings, for example, are primary, secondary, and tertiary consumers.

You have thus far thought in terms of food chains, which is to say trophic *lines* in which a single organism follows another. In reality, of course, nature is much more complex than this; what really exists are feeding patterns called *food webs*, one example of which you can see in **Figure 35.12**.

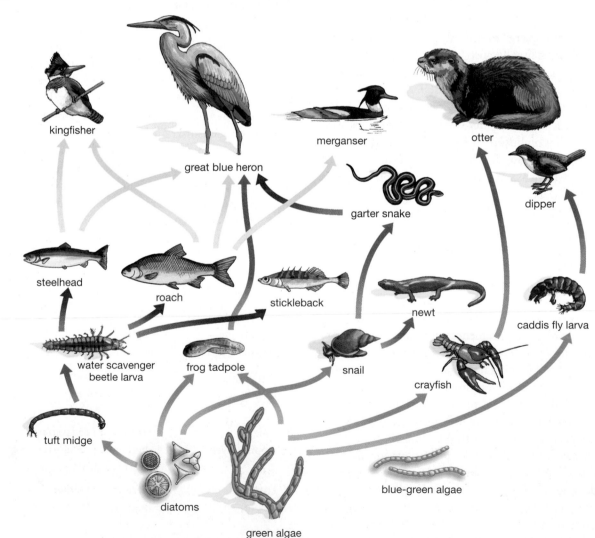

**Figure 35.12**
**A River Food Web**

The diagram indicates who eats whom along a portion of the Eel River in northern California. The arrows have been color coded by trophic level. Note that there are up to five trophic levels in the web. (Adapted with permission from an original drawing by Mary E. Power, University of California, Berkeley.)

kingfisher

great blue heron

merganser

otter

garter snake

dipper

steelhead

roach

stickleback

newt

caddis fly larva

water scavenger beetle larva

frog tadpole

snail

crayfish

tuft midge

diatoms

green algae

blue-green algae

### A Special Class of Consumers: Detritivores

To this scheme of trophic levels, ecologists add a special class of consumers, the **detritivores**. These are consumers that feed on detritus, which in normal usage simply means a collection of debris. In ecology, however, detritus is the remains of dead organisms or cast-off material from living organisms (a fallen branch from a living tree, for example). A worm or a dung beetle feeding on such material would fit this description of a detritivore.

Of particular interest, however, is a special kind of detritivore: a **decomposer**, which is an organism that, in feeding on dead or cast-off organic material, breaks it down into its inorganic components, which then can be recycled through the ecosystem. *Inorganic* here can be thought of as building blocks. (More technically, it means an element or compound that does not contain carbon bound covalently to hydrogen.) The most important decomposers are fungi and bacteria. A number of detritivores usually are responsible for breaking up organic material, first into pieces and then into the inorganic components that are recycled into the Earth and the atmosphere (**Figure 35.13**).

Stepping back from this, it's worth noting what complete use the living world makes of its own materials. Think of it this way: What does nature produce that it does not eventually decompose and recycle? The answer is, nothing at all. Every twig and beetle and bone that comes into existence eventually will be broken down into its component materials, which

then will be used again. This is all to the good, actually, because if there were any natural garbage, spaceship Earth would sooner or later be filled up with this useless material. This very thought ought to give us some pause about the compounds human beings currently are putting into the environment that are not **biodegradable**, meaning capable of being decomposed by living organisms. Of particular concern here are plastics, many of which will remain intact for hundreds of years following their production.

## Accounting for Energy Flow through the Trophic Levels

All natural substances may be recycled in ecosystems, but this is not true of energy, which is just passing through, on its way to becoming heat. Two interesting questions arise from this: How *much* of the sun's energy does the living world collect in the first place, and then how much of the collected energy makes it through to each successive trophic level?

This may seem like a routine set of questions, but the first ecologists who thought in these terms brought about a revolution in ecology. Working in the 1940s at Minnesota's Cedar Bog Lake, ecologist Raymond Lindeman was the first to conceptualize ecosystems as units in which energy is first captured by given organisms and then transferred on to others. This conceptualization is today known as the **energy-flow model** of ecosystems. Part of the impetus for Lindeman's research was to answer a very

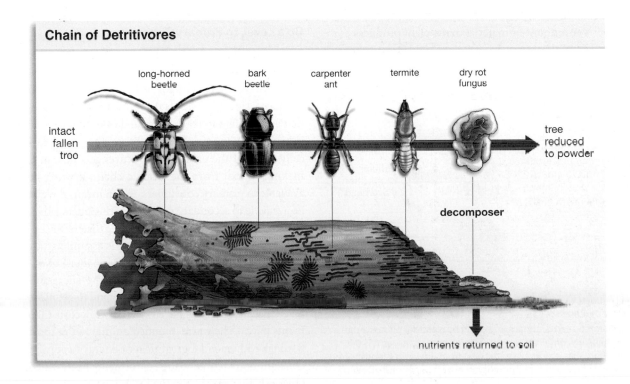

**Chain of Detritivores**

long-horned beetle · bark beetle · carpenter ant · termite · dry rot fungus

intact fallen tree → tree reduced to powder

decomposer

nutrients returned to soil

**Figure 35.13**
**How Organic Material Is Broken Down**

Different kinds of detritivores consume the organic material in a branch that has fallen from a tree. Most of these animals are consuming pieces of the log, but the detritivore called a decomposer (a fungus in the figure) is breaking organic materials into their inorganic components, thus returning nutrients to the soil.

down-to-earth question, succinctly expressed by ecologist Paul Colinvaux: Why are big, fierce animals rare? (If you doubt that they are, think of this. It's utterly routine to see plants, insects, or small birds, but it's often the experience of a lifetime to see a bear or a shark.) The energy-flow model explained the small numbers of big, fierce animals, as you'll see, but it then went on to shed light on all kinds of ecological relationships. Why? Because energy is a kind of currency in the natural world; all organisms use it, and it is transferred among them. Moreover, it can be measured, usually in units called **kilocalories** (the amount of heat it takes to raise 1 kilogram of water 1 degree Celsius). If we know that a field mouse must use 68 percent of the kilocalories it consumes just to stay alive—to move, to keep a constant body temperature—but that a weasel must use 93 percent of its kilocalories for this purpose, that gives us a meaningful basis for comparison between these two animals. Beyond this, weasels *eat* field mice, and with the energy-flow model we have some basis for tracking this important commodity as it moves through an ecosystem.

### The Capture of Energy by Plants

So, what has energy-flow research told us? To take things from the beginning, how much of the sun's energy do plants capture? As you can see in **Figure 35.14**, very little; only about 2 percent of the sun's energy that falls around plants is taken in for photosynthesis. This raises an important point, which is noting the difference between the energy that *comes to* an organism and what the organism *does with* that energy.

The amount of material that a plant produces as a result of photosynthesis is known as its **gross primary**

**production**. Those of you who have been through Chapter 8 know that the initial product of photosynthesis is a sugar that serves as a starting "building block" for all the material in the plant. Every time a molecule of this sugar is produced in photosynthesis, the plant has added to its gross primary production. But plants don't get to retain all the material they produce in this way. Why? Because plants incur energy *costs* in just staying alive. It takes energy to perform photosynthesis, to move sap from one end of the plant to the other, and so forth. Such costs are subsumed under the heading of "cellular respiration," but you could also think of them as the plant's overhead; they are the price of doing business. Apart from this, we know that, in every energy transaction, some of the energy is lost as heat. After subtracting these various costs, the plant is left with its **net primary production**: the amount of material a plant accumulates as a result of photosynthesis. This is the material that serves to build the plant up—its stems, leaves, and roots.

One question that arises from this framework is: What fraction of the energy a plant receives from the sun ends up as net primary production? How efficient are plants, in other words, at turning solar energy into stems, leaves, and roots? Of the solar energy that plants receive, somewhere between 30 and 85 percent is transferred into net primary production. In other words, at the low end of this range, only 30 percent of the solar energy a plant assimilates is transferred into material that is even potentially available to the next trophic level up.

### Up a Level, to Primary Consumers

Now, what happens when you go up to that trophic level—to those primary consumers? Well, in the first place, they can't *get to* a lot of the tissue that's been produced, since it's in roots that are buried or in leaves that are too high to reach. A good deal of the material is inedible in any event, except by detritivores. Critically, most of the energy these scavengers take in is making an exit from the trophic chain; it won't be available to primary consumers, which means it won't be available to secondary or tertiary consumers either. Adding all this up, you can begin to see that, of the energy locked up by plants, very little is assimilated by organisms at the next step up. And this pattern of energy reduction is then repeated in the successive trophic levels. Indeed, the reductions tend to get more severe because animals have higher energy costs than plants have. (The warm-blooded animals we're most familiar with spend a great deal of energy keeping a constant body temperature—producing heat, which obviously is energy lost to the food chain.)

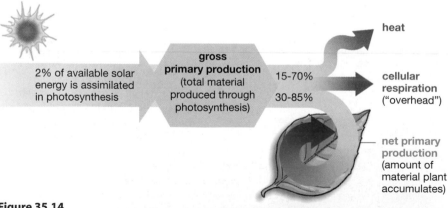

**Figure 35.14**
**Absorption and Use of Solar Energy by Plants**
Only 2 percent of the solar energy reaching the Earth's surface is taken in by plants for photosynthesis. Of this energy, only a fraction ends up being contained in plant material, such as tissues. It is this material (the plant's net primary production) that is available to the next trophic level in a food chain. The remaining energy is either dissipated as heat or used by photosynthesizers to power their own physical processes.

All of this leads to what you see in **Figure 35.15**: a reduction in available energy at each trophic level. The amount that is passed on between levels varies by community, but there is a rule of thumb in ecology that for each jump up in trophic level, the amount of available energy drops by 90 percent. Take all the energy converted into tissue by, say, second-level herbivores (rabbits, mice), and of this, only 10 percent ends up being converted into tissue by third-level carnivores (foxes, weasels). There is a great variability in conversion rates, however, and 10 percent probably is more like a maximum passed on.

### The Effects of Energy Loss through Trophic Levels

So, what are the consequences of this energy reduction at trophic levels? First, *this* is the reason that big, fierce animals are rare. If you had a field that could feed 100 field mice, how many weasels could the field support if those weasels were existing solely on the mice? Remembering that, at most, only 10 percent of a trophic level's energy is likely to be passed along to the next level, you might guess that 10 weasels could make a living in the field, but even this number would be too high. For 10 predators to make a living from 100 mice, these predators would have to be no *bigger* than the mice, which is certainly not true of weasels. Extending this further, not even a single hawk could survive on the available weasels in the field. Why are big, fierce animals rare? Because of the loss of available energy at each step in the trophic chain. Because weasels are not about to start eating the same things as mice, weasels are locked into a trophic level with a more limited energy supply. When there's only so much energy, there can be only so many weasels.

Beyond this, the energy flow model allows us to see the limits to top predators. Why couldn't we have a land carnivore bigger and fiercer than a grizzly bear? Well, imagine that you found some awesome creature that fit this bill somewhere on Earth, and that you plopped it down in Alaska. What would happen? If it could feed only on grizzly bears, it would starve to death. There simply is not enough energy locked up in grizzly bears for there to be a trophic level above them. This is why, in most food webs, there are so few trophic levels above primary producers; the energy simply runs out. This question of energy loss with each trophic level has relevance for a human issue, which is eating meat that comes from grain-fed cattle, as you can see in "A Cut for the Middleman: Livestock and Food," on page 741.

### Primary Productivity Varies across the Earth by Region

Primary productivity concerns *biomass* or material produced by first-level photosynthesizers, as measured by weight. It will come as no surprise that some areas of the Earth are more productive than others in this sense. If you look at **Figure 35.16**, you can get an idea of the differences around the globe. The reasons for these differences are explored in this chapter's section on the large ecosystems called biomes. For now, suffice it to say that much of what you see in Figure 35.16 is intuitive. Productivity is very low in such desert areas as North Africa and very high in such tropical rain forest areas as equatorial Africa. Yet why

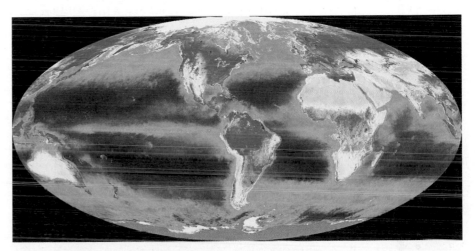

**Figure 35.16**
**Primary Productivity by Geographic Region**

Primary productivity around the globe. This composite satellite image shows how the concentration of plant life varies on both land and sea. On land, the areas of lowest productivity are tan colored, as with the enormous swath of land that begins with the eastern Sahara Desert in North Africa. More productive land areas are yellow, more productive yet are light green, and the most productive of all are dark green. On the oceans, the gradient runs from dark blue (the least productive), through lighter blue, green, yellow, and orange-red (the most productive). The ocean measurements are of concentrations of phytoplankton—the tiny, photosynthesizing organisms that drift on the water and form the base of the marine food web. Note the high productivity of the oceans in both the far northern and southern latitudes. Antarctica itself is gray because the imaging technique did not work well for it.

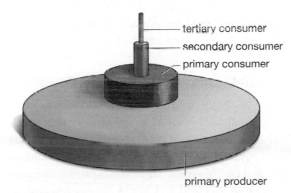

tertiary consumer
secondary consumer
primary consumer

primary producer

**Figure 35.15**
**Energy Pyramid**

Only a small fraction (10 percent or less) of energy at each trophic level is available to the next higher trophic level.

should one part of the Earth be a rain forest and another part a desert? The answer has to do with Earth's large-scale physical environment.

---

### SO FAR...

1. All the organisms at the first trophic level perform _____ and thus are referred to as _____. Any organism that feeds solely on first-trophic-level organisms would have to be a (choose one) carnivore/herbivore/omnivore.

2. All material produced by living things is capable of being broken down into its inorganic components by the class of living things called _____. This material is thus said to be _____.

3. There are fewer predator than prey animals due to drastic reductions of _____ at each _____.

---

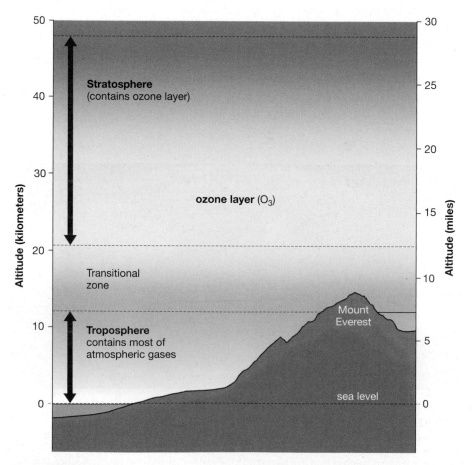

**Figure 35.17**
**Earth's Atmosphere**
The two lowest layers of the atmosphere are shown, with Mount Everest (the tallest mountain) included for vertical scale. The troposphere contains most of the atmosphere's gases—mostly nitrogen and oxygen, with some carbon dioxide, methane, and other gases. The stratosphere contains the ozone layer, which blocks 99 percent of the sun's harmful ultraviolet radiation.

## 35.4 Earth's Physical Environment

### Earth's Atmosphere

Earth itself exists within an environment, called the **atmosphere**, which is the layer of gases surrounding the Earth. One of the main things to realize about Earth's atmosphere is how it differs from outer space. There's something *to* Earth's atmosphere (a mix of gases), while outer space really is *space*; there's almost nothing in it.

The atmosphere is divided into several layers, but here you need be concerned with only two of them, which you can see in **Figure 35.17**. The lowest layer of the atmosphere, called the **troposphere**, starts at sea level and extends upward about 12 kilometers (or 7.4 miles), and it contains the bulk of the gases in the atmosphere. Nitrogen and oxygen make up 99 percent of the troposphere, but as you've seen, carbon dioxide exists there in small amounts, as do some other gases, such as argon and methane. After a transitional zone, the next layer above the troposphere is the **stratosphere**. Of greatest concern in it is a gas that reaches its greatest density at about 20 to 35 kilometers (13 to 21 miles) above sea level. The oxygen we breathe comes in the form of two oxygen molecules bonded together ($O_2$). When ultraviolet sunlight strikes $O_2$ in the stratosphere, however, it can put it in a form of three oxygen molecules bonded together ($O_3$). This is the gas known as **ozone**. Created initially as a *product* of life—since all atmospheric oxygen came from photosynthesis—this ozone layer ultimately came to *protect* life by blocking some 99 percent of the ultraviolet (UV) radiation the sun showers on the Earth. It was this blockage that allowed life to come onto land some 460 million years ago, and it is this blockage that protects life now. The UV radiation that pours from the sun can wreak havoc on living tissue, bringing about cancer and immune-system problems in people, and damaging vegetation as well.

### The Worrisome Issue of Ozone Depletion

Given the importance of the ozone layer, it is sobering to contemplate how fragile it is. Some human-made chemical compounds have the effect of destroying stratospheric ozone, and such destruction went on unchecked for years until atmospheric chemists Sherwood Rowland and Mario Molina revealed, in 1974, that compounds called chlorofluorocarbons posed a direct threat to the ozone layer. Chlorofluorocarbons or **CFCs**—found at one time in spray cans,

| ESSAY | # A Cut for the Middleman: Livestock and Food |

I t may be difficult to think of cows in a field as occupants of one of the trophic levels you've been looking at, but that's just what they are, with some provocative consequences for human beings and food. People can't survive on grass the way cows and sheep can, and it follows that these animals are turning food that is useless to human beings into products that are useful to human beings—meat, wool, and milk. In this sense, then, these animals are a boon to the efficiency of worldwide agriculture in that they allow us to get more human food and other products from the same amount of land.

*It takes 9 pounds of feed to produce 1 pound of beef ready for human consumption.*

There is another side to this story, however, having to do with the *loss* of food that occurs as we move up through the trophic levels. Recall that, as a rough rule of thumb, 90 percent of the energy that is locked up in one trophic level is lost to the organisms in the next trophic level. If cattle and other domesticated animals were doing nothing but *foraging*—if they were being fed solely on the grass available in pastures—this would not be a matter of concern because the grass is not useful to humans as a food source anyway. But in developed countries such as the United States, much of the diet of domesticated animals comes from *grain* that is grown on farms. A large portion of this animal feed could go directly to human beings, but it instead passes through the trophic level of animals, with a predictable result: It takes 9 pounds of feed to produce 1 pound of beef ready for human consumption. What does this mean in practice? A harvest of 2,000 pounds of grain will feed five people for a year. Take roughly this same harvest, feed it to livestock, and then feed the livestock products to people, and you will feed only *one* person for a year.

This calculation takes on added importance when considering the scale of the grain-fed operation. In the feedlot, while being fat-tened for slaughter, an average steer consumes up to 2,700 pounds of grain. It may be apparent that such consumption adds up quickly; some 70 percent of the grain produced in the United States is fed to livestock. There are about 70 million cattle in the United States and Canada alone.

But does this mean that human hunger would be reduced if domestic animals were eliminated as a "middleman"? This is not necessarily the case. In a simple calculation of calories produced versus calories needed, the world *already* produces enough food to feed all its people. The current problem, then, is one of food distribution, not food production. As an often-used phrase has it, "Hunger is real, scarcity is not." The burden that grain-fed animals pose may have to do with the future. As countries get richer, their human populations will eat more meat, which would mean less total food available for humans worldwide.

**Fattening at the Feedlot**
About 40,000 head of cattle are fed for meat production at this commercial feedlot in Imperial Valley, California.

refrigerators, and plastic foams—are undoubtedly the most famous of the ozone-depleting compounds, but they are by no means the only chemicals that have this effect. Other harmful compounds include methyl bromide, which is used as a pesticide and herbicide, notably on tomato and strawberry crops; and bromine, which is found in a group of fire-extinguishing chemicals. The damage that these compounds have done to stratospheric ozone has brought about the spectacle of annual reports on how big the "ozone hole" is over the South Pole, along with a concern about a general thinning of the ozone layer.

Despite the ominous nature of this news, the long-term outlook on ozone depletion actually is good. In an agreement signed in Montreal in 1987, many of the world's nations pledged to phase out production of ozone-depleting chemicals, and this agreement is working. Production of CFCs was banned in the United States even before the Montreal agreement was signed and has dropped dramatically throughout the world in recent years. One variety of ozone-depleting chemical does continue to be manufactured in large quantities: Methyl bromide was supposed to have been phased out in industrialized nations by 2005, but the United States and other

••• ANSWERS
1. photosynthesis; producers; herbivore
2. decomposers; biodegradable
3. energy; trophic level

countries continue to ask for "exemptions" that allow the use of it, on grounds that there is no economically feasible substitute for it. As a result, tens of thousands of tons of this fumigant continue to be applied to farmland worldwide each year, with farmers in the United States being by far the biggest single users of it. This issue notwithstanding, the overall production of ozone-depleting compounds has dropped significantly since 1987, and this reduction has brought about a reduction in the levels of ozone-depleting compounds in Earth's atmosphere. As a consequence, the thickness of Earth's ozone layer stabilized beginning in the mid-1990s, according to a United Nations' assessment of the issue published in 2006. If current trends continue, the Earth's ozone layer outside the polar regions should return to its pre-1980 levels about the middle of this century. The annual Antarctic ozone hole will be a feature of life for decades, but even the Antarctic ozone layer should return to its pre-1980 thickness by 2060 to 2075, according to the United Nations.

## 35.5   Global Warming

Scientists are not nearly so hopeful about a second environmental issue that concerns Earth's atmosphere—the issue of global warming. Identified as a concern as far back as the 1950s, global warming has emerged as perhaps the most pressing environmental problem of our time, a status that's not surprising given that its reach really is global, and that its effects stand to be profound: Residents of low-lying areas from Florida to Bangladesh must worry about its ability to raise sea levels; residents of coastal areas from Texas to the Philippines must worry about its ability to increase the intensity of tropical storms; and farmers from Nebraska to Argentina must worry about its ability to change rainfall patterns.

The seriousness of these issues has prompted an intense worldwide research effort that has managed to answer some questions about global warming, while leaving other questions open. Uncertainty is a given in almost any analysis of global warming, however, because the most crucial questions about it concern not the effects it's having now but rather the effects it stands to have in the future.

### Warming for Certain, Caused by Human Activity

Our most authoritative judgments on global warming come from a United Nations group called the Intergovernmental Panel on Climate Change, or IPCC, which has issued four reports since 1990, each of

them the work of hundreds of scientists who have reviewed the findings of thousands of their fellow researchers. In its latest report, released in 2007, the panel seemed to put to rest, once and for all, the most basic question about global warming: Is it taking place? The evidence is now "unequivocal" that it is, the panel said. In the 100 years that elapsed from 1906 through 2005, Earth's surface temperature increased by about 0.74 degrees Celsius, or about 1.3 degrees Fahrenheit, the IPCC concluded. Moreover, the warming trend over the last 50 years is nearly twice that for the last 100 years, and 11 of the 12 years between 1995 and 2006 were the warmest ever recorded since such measurements started being taken in the middle of the nineteenth century.

So, what is causing this warming? For years, this was a contentious question as a number of well-informed scientists held that natural fluctuations in Earth's climate—rather than human activity—could account for the temperature rise that's occurred. The 2007 IPCC report, however, seemed to put this issue to rest as well. The panel said that there is a 90 to 95 percent likelihood that most of the warming observed since 1950 has been caused by human activity. What activity is this? We human beings are moving heat-trapping gases into the atmosphere through several means. First and foremost, we are burning fossil fuels, such as oil and coal, which results in a massive transfer of carbon dioxide into the atmosphere. Next, the cattle we raise and the rice we grow put large amounts of methane into the atmosphere. Third, the deforestation we are carrying out—particularly in the tropics—either puts $CO_2$ into the atmosphere directly, when trees are burned, or results in less $CO_2$ being taken *out* of the atmosphere since trees that are destroyed no longer take up $CO_2$ to perform photosynthesis.

So, how significantly have these activities altered Earth's atmosphere? To look just at $CO_2$, it turns out that a greater concentration of it exists in the atmosphere now than at any time in the last 650,000 years. In 2005, the Earth's atmospheric $CO_2$ concentration was 379 *parts per million*—for every million molecules of dry air, there were 379 molecules of $CO_2$. In contrast, the highest $CO_2$ concentration ever recorded over the previous 650,000 years was 300 parts per million. (Scientists can look this far back in time by analyzing the $CO_2$ in ancient ice cores.) If you look at **Figure 35.18**, you can see Earth's atmospheric $CO_2$ change graphed over two time frames: first, over the last 10,000 years, and then over the 255 years that have elapsed since the industrial revolution began.

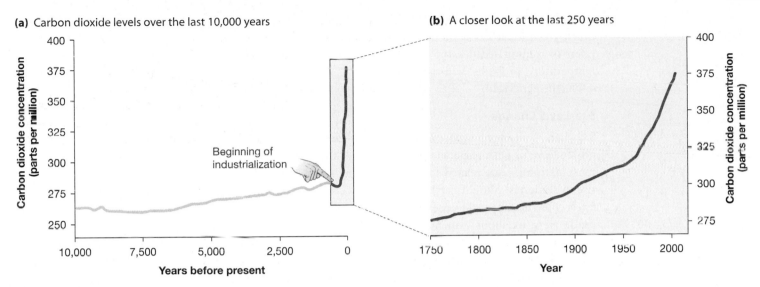

**(a)** Carbon dioxide levels over the last 10,000 years

**(b)** A closer look at the last 250 years

Beginning of industrialization

Years before present

Year

**Figure 35.18**
**A Growing Concentration of Atmospheric CO₂**

Graphs constructed by IPCC scientists show a dramatic upturn in atmospheric $CO_2$ concentrations beginning about the time the Industrial Revolution began, in 1750. $CO_2$ concentrations have continued to increase from that time through the present but not at a steady rate. They took an additional upward turn just after 1950, as can be seen in the graph on the right. (Adapted from IPCC, *Contribution of Working Group I to the Fourth Assessment Report of the Intergovernmental Panel on Climate Change, Climate Change 2007: The Physical Science Basis, Summary for Policymakers*, 2007, Figure SPM-1.)

### What Is the Greenhouse Effect?

But why should the atmospheric concentration of gases such as $CO_2$ have anything to do with a warmer planet? Sunlight that is not filtered out by ozone comes to Earth in the form of very energetic, short waves that can easily pass through the atmosphere (**Figure 35.19**). Once this energy reaches the land and ocean, most of it is quickly transformed into heat that does all the things you've just read about: warms the planet, drives the water cycle, and so forth. This heat ultimately is radiated back toward space. But heat is not short-wave radiation; it is *long*-wave radiation, and it can be trapped by certain compounds, among them carbon dioxide and methane.

### What Are the Likely Consequences of Global Warming?

All of this leads to the most critical question about global warming: What will its consequences be? A prerequisite for answering this question is to come to an understanding of what's called "climate sensitivity," meaning how much Earth's atmosphere stands to be warmed by a given increase in greenhouse gases. The answer the IPCC provided in 2007 managed to allay fears of a scorched Earth but was disturbing nevertheless. A doubling of $CO_2$ levels— which could happen by century's end— would most

likely lead atmospheric temperatures to rise by 3 degrees Celsius, or 5.4 degrees Fahrenheit, the panel said. This is a large increase indeed when compared to the 0.74 degrees Celsius jump Earth has experienced so far, but it's small compared to worst-case scenarios, considered plausible only a few years ago, of a doubling of $CO_2$ leading to temperature increases of up to 11 degrees Celsius, or

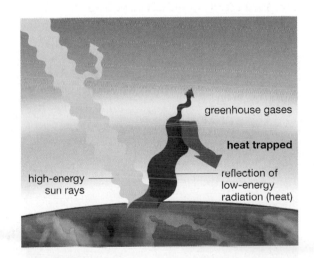

greenhouse gases

heat trapped

high-energy sun rays

reflection of low-energy radiation (heat)

**Figure 35.19**
**The Greenhouse Effect**

The high-energy rays of the sun can easily penetrate the layer of gases in the troposphere. However, the lower-energy radiation (heat) that reflects from Earth's surface cannot penetrate the layer of gases as easily. Carbon dioxide and methane thus take on the role of glass panes in a greenhouse: They let solar energy in, but they retain a good deal of heat.

about 20 degrees Fahrenheit. Depending on how much greenhouse gas is pumped into the atmosphere in coming decades, the actual rise in Earth's temperature probably will range from 1.8°C to 4.0°C, the panel said.

### Sea-Level Change

So, where would temperature increases of this magnitude leave us with respect to what is perhaps the most alarming issue related to global warming—a rise in sea levels? Should Earth warm by 1.8°C, sea levels could be expected to rise by a minimum of 18 centimeters, or about 7 inches, by century's end, according to the IPCC. Should the temperature increases come in at the high end of 4.0°C, we could expect sea levels to rise by up to 59 centimeters, which is about 1.9 feet. To get a feel for what this latter figure would mean, the U.S. Environmental Protection Agency has estimated that a 2-foot rise in sea levels would eliminate about 10,000 square miles of land in the United States, with coastal regions in the southeast hardest hit.

Global warming has the power to raise sea levels to this extent because it affects them through two processes. First, as seawater warms it expands, which raises its level. Second, recall that a great proportion of Earth's freshwater is locked up in glaciers and ice sheets. Global warming is melting glaciers generally, as you can see in **Figure 35.20**, and is bringing about some melting of the world's two greatest ice sheets, those in Greenland and Antarctica. Equally disturbing, the speed with which ice from these sheets is *moving*—into the oceans—has increased greatly in the last 10 years or so, a factor that seems to be of paramount importance in the net loss of ice from Greenland. For the future, the IPCC predicts a continuing loss of ice for Greenland but a net gain of ice for the Antarctic ice sheet due to increased snowfall there.

### Other Effects of Warming

What else can we expect from global warming? According to the IPCC, more intense tropical storms, less rain in the world's already dry "subtropical" regions—those just north and south of the tropics—more rain at very high latitudes, less snow cover globally, more heat waves and instances of heavy rainfall, and a permafrost that thaws to increasing depths. In addition, sea ice will shrink in the Arctic and Antarctic, with the Arctic ice cover possibly disappearing almost entirely during late summer months by the end of the century.

### Warming and Habitat Change

If you look at **Figure 35.21**, you can see the extent to which summer Arctic ice has already disappeared. When such ice melts, it does not raise sea levels—since the ice is already a component of the sea—but this change does spell trouble for creatures who live in the Arctic, such as polar bears. This is but one instance of how global warming stands to profoundly alter natural habitats worldwide. To look at just one other example, in the world's high-mountain or "alpine" regions, native plants and animals are being driven to ever-higher elevations as temperatures warm. The treeless alpine ecosystems are in effect islands of cool habitat sitting above an "ocean" of warmer habitat at lower elevations. But as the ocean rises, the alpine habitats shrink. One study of New Zealand's alpine islands predicts that 80 percent of them will be eliminated in the next 100 years. Not all of global warming's effects on habitat lie in the

**Figure 35.20**
**Warming Planet, Disappearing Glaciers**

Global warming is melting glaciers around the world. Peru's Quelccaya ice cap, in the southern Andes, has shrunk by at least 20 percent since 1963. One of the main glaciers flowing from the cap, Qori Kalis, is shown here as it existed in 1978 and then in 2000. A 10-acre lake now exists where the glacier once extended. Its rate of retreat has reached 155 meters or 509 feet per year. This is three times greater than its rate of shrinkage from 1995 to 1998.

**(a)** The Qori Kalis glacier, 1978

**(b)** The Qori Kalis glacier, 2000

**(a)** Minimum Artic ice, 1979

**(b)** Minimum Artic ice, 2005

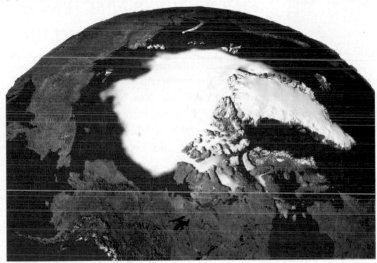

future, however; the warming of our planet has brought about some animal extinctions already—something you can read more about in "Global Warming and the Harlequin Frogs" on page 746

## A Warming We Cannot Stop but Can Lessen

Apart from its specific facts and figures, the 2007 IPCC report delivered a general message that was hard to take: Earth will continue to warm, and its seas will continue to rise, no matter what the world's nations do. As the report puts it, human-caused "warming and sea level rise would continue for centuries due to the timescales associated with climate processes and feedbacks, even if greenhouse gas concentrations were to be stabilized." Why should this be? In large part, it's a reflection of the fact that the greenhouse gases that humanity has already put into the atmosphere will remain there for a long time.

Dispiriting as this news may be, it does not mean that human beings are powerless to alter the course of global warming. This is a trend we cannot stop, but it is a trend we can *lessen*, and the stakes are high. If we continue to increase the amount of greenhouse gases we pump into the atmosphere, then all the trends predicted by the IPCC stand to unfold in their worst forms. If, on the other hand, we can rein in our emissions of these gases while limiting deforestation, then we stand to minimize all the kinds of changes that fall out from a warmer Earth.

There is no shortage of ideas about how to reduce atmospheric greenhouse gas concentrations; economists, scientists, and politicians have put forward a long list of proposals. And there is broad agreement about the kinds of steps we must take to achieve this

reduction: We have to use less energy; reduce the degree to which we use fossil fuels as our energy sources; find a way to safely store or "sequester" some of the carbon dioxide that continues to be produced; and slow the destruction of Earth's tropical rain forests. All these ideas, however, involve economic tradeoffs. For centuries, human societies have been powered mostly by fossil fuels because these have been the cheapest fuels available. To change now means turning to alternatives that are more costly in an economic sense but *less* costly in an environmental sense. The 2007 IPCC report marked the end of any uncertainty about whether global warming is taking place, whether human activity is helping cause it, and whether its effects stand to be significant. Scientists have done their jobs in supplying information on all these issues; now we'll see if human societies can act on what they've been told.

**Figure 35.21**
**Disappearing Arctic Ice**

The Arctic Ocean's so-called perennial sea ice—that portion of the ice cover that never melts, even during the warmest months—has been decreasing at a rate of 9.6 percent per decade. The images above, constructed by NASA from satellite data, show the difference between the Arctic's perennial ice cover in 1979 and this ice cover in 2005. Should this trend continue, there may be no Arctic summer ice by century's end.

### SO FAR...

1. Ozone is a gas composed entirely of the element _____ that reaches its greatest density in the layer of the atmosphere known as the _____ and that has the effect of keeping _____ from reaching Earth's surface.

2. Most of the energetic short-wave radiation from the sun is transformed into _____, which can be trapped by greenhouse gases such as _____ and _____.

3. Global warming may increase sea levels through two means: the _____ of sea water as it warms and the melting of large ice sheets, particularly those of _____ and _____.

••• ANSWERS

1. oxygen; stratosphere; ultraviolet radiation

2. heat; carbon dioxide; methane

3. expansion; Greenland; Antarctica

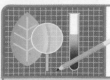

# THE PROCESS *of* SCIENCE
## Global Warming and the Harlequin Frogs

The power that global warming will have to reshape life on Earth came into sharp focus in 2006 when ecologist J. Alan Pounds and his colleagues became the first researchers to definitively tie the warming of the planet to the extinction of some of its individual species. Looking at a group of brightly colored frogs that have flourished for millions of years in Central and South America, the researchers first revealed what has become of these creatures—at least 67 species of them have gone extinct since the 1980s—and then went on to demonstrate that the frogs' loss may have come about because of an enemy's gain: A fungus that attacks the frogs may have become more lethal to them because of global warming.

Pounds is an American ecologist, trained at the University of Florida, who has worked since the early 1980s in Costa Rica's Monteverde Cloud Forest Preserve. Speaking by telephone from the preserve's Golden Toad Laboratory for Conservation, he noted that frog species began disappearing from the Monteverde area as early as the 1980s—something that at the time he thought was related to another climatic change, the El Niño weather disturbances of the period. Yet long after El Niño's importance had been reduced, frog species kept disappearing from numerous locations in Latin America, as researchers in the region learned when they compared notes with each other. So consistent were the reports the researchers were sharing that they decided to create a network called RANA (Spanish for frog), charged with understanding the reasons for the declining amphibian populations in Latin America. Eventually, researchers from this network joined with others in supplying Pounds with information on a particularly hard-hit variety of frog, the harlequin toads of the genus *Atelopus* (see **Figure 1**).

In a 2006 article on these frogs that appeared in the journal *Nature*, Pounds and his colleagues started by laying out the facts about the harlequins and their decline:

> Brightly colored and active during the day near streams, most are readily observed and identified. For the first time, data indicate when each of numerous species was seen for the last time.

This, then, is how we know when a species has slipped from existence: Professionals who are used to observing it in the field simply don't see it anymore. When was the last time anyone saw the harlequin frog *Atelopus chiriquiensis*? The answer is 1996, as a report from a RANA scientist indicated. And *A. ignescens*? Last seen in 1988. In all, some two-thirds of 110 known harlequin species have not been observed since 1998, Pounds reported, and are thus presumed to have become extinct. Amphibians are in decline throughout the world, but the harlequins have the unfortunate distinction of being the amphibian genus in the steepest decline of all. In just collecting information about their fate, Pounds and his colleagues were taking on an important, time-honored scientific role: that of being society's eyes and ears on nature. How would anyone know about the global condition of the harlequins without the systematic observation and reporting of the scientists in Pounds' network? Average citizens in the mountains of Costa Rica may have realized that they were no longer seeing the black-and-red Monteverde harlequin frogs that once flourished in local streams, but no average citizen could have connected this with the near-simultaneous disappearance of other harlequin species from, say, Venezuela or Peru.

For Pounds, the facts about the harlequins' disappearance became the departure point for a deeper investigation. *Why* have the

**Figure 1**
**Missing and Perhaps Extinct**
Roughly two-thirds of the 110 known species of frogs known as harlequins have not been seen in nature since 1998 and are thus presumed to have become extinct. Pictured is a frog from the harlequin species *Atelopus varius,* once found in Panama and Costa Rica but not observed in either country since 2003.

## 35.6 Earth's Climate

### Why Are Some Areas Wet and Some Dry, Some Hot and Some Cold?

Did you ever look at a globe of the Earth and wonder why it is tilted? Earth exists in space, after all, so what could it be tilted against? When we say that a rod stuck in the ground is tilted, we mean that it is not perpendicular to the ground. Here, of course, we're thinking of the ground as a flat surface—a plane. Earth, too, can be viewed as existing on a plane, in this case the plane of its orbit around the sun. (We generally think of this orbit as looking like a

harlequins been dying out, he asked? What factors led to the seeming extinction of so many harlequin species in such a short period of time? At the end of a long trail of investigation and analysis, Pounds saw clearly that global warming had to be playing a role. For starters, it was clear that the tropics have been warming in recent decades. From 1975 through 2000, the average temperature from 30° N to 30° S of the equator rose by 0.18°C per decade, which was triple the average temperature rise for the twentieth century as a whole. Within this context, a key discovery for Pounds was that 80 percent of the *Atelopus* species that disappeared did so right after a relatively warm year. **Figure 2** provides an idea of how close the correlation was between the years when average temperatures spiked in the tropics and the years in which harlequin species were seen for the last time. The odds that this linkage could have occurred by chance turned out to be less than 1 in 1,000.

But what was the mechanism by which temperature change was fostering the disappearance of the harlequins? Just to form a plausible hypothesis, Pounds had to fit together pieces of data that at first seemed contradictory. For one thing, he saw that the proportion of disappeared harlequin species increased at 200 meters above sea level, then increased again at 1,000 meters, and then decreased at 2,400 meters. Elevation is, of course, correlated with *temperature*, and the pattern seen with the harlequin extinctions suggested that whatever was killing the frogs functioned best at an intermediate

temperature range—it flourished neither in the warmer temperatures seen near sea level nor in the cooler temperatures seen at higher elevations. In recent years, a prime suspect in the decimation of any amphibian population has been an aquatic fungus called *Batrachochytrium dendrobatidis* that grows on the skin of amphibians, often killing them. The problem in fingering *B. dendrobatidis* in the harlequin extinctions was that it tends to become more damaging to amphibians under *cool* conditions—at higher elevations or during cooler seasons—whereas the harlequins were being killed off after particularly warm years.

The key to resolving the paradox lay in looking at the *details* of what global warming has done to tropical climates. As Pounds noted in his *Nature* article, global warming brings about increased water evaporation, and this in turn can translate into increased cloud cover over a region. Such cloud cover brings about an overall *moderation* of temperatures—cooler days but warmer nights. During the day, clouds are increasingly blocking out direct sunlight, thus holding temperatures down, but at night, the moist air provided by the cloud cover helps retain daytime heat, keeping temperatures up. In line with this, daytime high temperatures have been falling at Monteverde, while nighttime low temperatures have been rising. And, as Pounds discovered, this same trend was recorded at 11 weather stations in Colombia and Venezuela in the period 1981 to 1990, as compared to 1941 to 1970.

What this temperature moderation may have led to is an expanded set of environments in which the fungal predator *B. dendrobatidis* can mount lethal attacks on harlequin frogs. Many lower elevations may no longer be too hot for it to function optimally, while many upper-middle elevations may no longer be too cold for it to do so. The higher nighttime temperatures in the tropics could explain why harlequin species have tended to die out after particularly warm years.

Consistent as his data on this subject are, Pounds regards his ideas on the connection between *B. dendrobatidis* and the frog extinctions as a working hypothesis in need of more evidence. He is certain that global warming has played a part in the demise of the harlequin species; the question is whether the effects of this warming are indeed being channeled through *B. dendrobatidis*. At the very least, however, Pounds' work on global warming and the harlequins has meant that these frogs have not gone quietly into extinction. Their disappearance has sent out a message that global warming is not a hypothetical threat sitting off in the distance but instead is a force that is affecting Earth's living things right now.

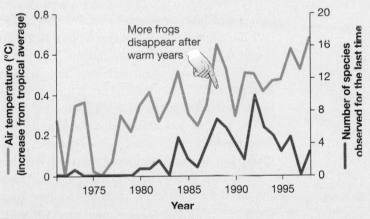

More frogs disappear after warm years

**Figure 2**
**After Warm Years, Frog Species Disappear**
J. Alan Pounds and his colleagues found a close correlation between years in which air temperature increased in the tropics and years in which harlequin frog species were seen for the last time. (Data from J. Alan Pounds et al.,"Widespread Amphibian Extinctions from Epidemic Disease Driven by Global Warming," *Nature* 439:161–167, 2006, Fig. 3b)

large hula hoop with the sun as a yellow ball in the middle.) Now think of the spherical Earth on this plane as having a rod sticking through it in the form of its north-south axis—the imaginary line that runs from the North Pole straight through to the South Pole. The critical thing is that this rod does not stick

straight up and down with respect to the plane of the Earth's orbit; it is tilted, at an angle of 23.5 degrees. This tilt dictates a good deal about Earth's climate, and Earth's climate dictates a great deal about life on Earth. Once again, in other words, the sun determines the basic conditions for life.

You can see the effects of Earth's tilt in **Figure 35.22**: In June, our Northern Hemisphere tilts toward the sun, while in January it tilts away from it. Thus, the sun's rays strike us more directly in June, and the days are warmer (and longer) than days in January. The angle at which the sun's rays strike a given portion of the Earth is very important. Relative to, say, Brazil,

sunlight strikes the Arctic at an indirect or "oblique" angle. Thus, the far north gets less of the solar energy that powers photosynthesis.

## The Circulation of the Atmosphere and Its Relation to Rain

The variation in temperature caused by these sunlight differences is the most important factor in the circulation of Earth's atmosphere. Near the equator, Earth simply gets more warmth, and critically, warm air rises. Warm air also can retain more *moisture* than cold air. You've seen an example of this whenever you've looked at the beads of "sweat" that form on the outside of a cold glass in summer. Moisture-laden warm air comes into contact with the cooler air immediately around the glass; thus cooled, the air cannot hold as much moisture and releases it onto the glass.

This same thing happens with the warm air that rises on both sides of the equator. Because of the heat of the tropics, air is rising in quantity from the tropical oceans; because this air is warm, it is carrying with it a great deal of moisture. The air cools as it rises, however, and then drops much of its moisture on the tropics, which is why they're so wet.

Following this a step further, this volume of air is now cooler, drier, and moving toward the poles in both directions from the equator (**Figure 35.23**). At about 30° N and 30° S of the equator, it descends, warming as it drops and actually absorbing moisture *from* the land. The land will be dry at these latitudes because this is where the dry, hot air descends.

The air that has descended now flows in two directions from the 30° point. In the Northern Hemisphere, this means north toward the North Pole and south toward the equator. Traveling at a fairly low altitude, this air eventually picks up moisture, rises, and deposits its moisture, this time at about 60° N and near the equator, with the same events happening at the same locations in the Southern Hemisphere.

What you get from this rising and falling is what you see in Figure 35.23: a set of interrelated "circulation cells" of moving air, each existing all the way around the globe at its latitude, and each acting like a conveyor belt that is dropping rain on the Earth where it rises but drying the Earth where it descends.

## The Impact of Earth's Circulation Cells

The full meaning of this becomes obvious only when you look at a map such as the one in **Figure 35.24** (or better yet, a globe). Draw your finger across the equatorial latitude on the map and look at how much green there is—Southeast Asia, equatorial

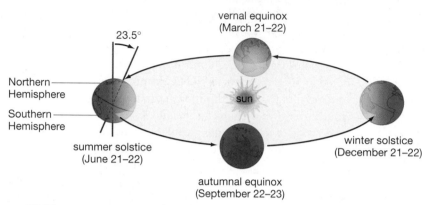

**Figure 35.22**
**Earth's Tilt and the Seasons**

The Northern Hemisphere gets more sunlight in June and less in January, while the reverse is true for the Southern Hemisphere. This is the reason for seasonal climate variations over large portions of the globe.

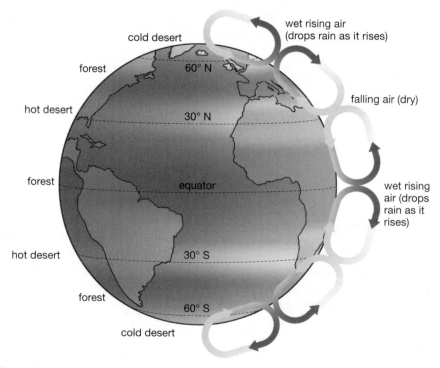

**Figure 35.23**
**Earth's Atmospheric Circulation Cells**

Because the Earth receives its most direct sunlight at the equator, this region is warm—and warm air holds moisture and rises. This air moves north and south from the equator, cooling as it rises, and then loses its moisture as rain, with the result that the land and ocean immediately north and south of the equator get lots of rain. As the now-dry air moves farther north and south, it cools and sinks again at about 30° in both hemispheres. Thus, the land at these latitudes tends to be dry. Two other bands of circulation cells exist at higher latitudes, operating under the same principles.

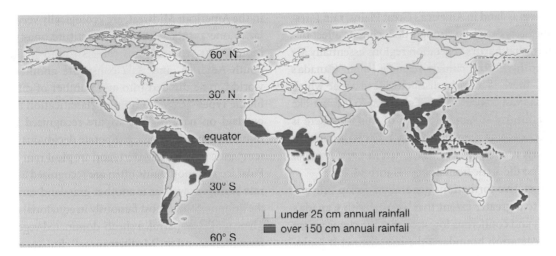

**Figure 35.24**
**Earth's Atmospheric Circulation Cells and Precipitation**
The wettest regions of the Earth lie along the equator, while the driest regions lie along the latitudes 30° N and 30° S. Compare these regions to the circulation cells shown in Figure 35.23. The variations of rainfall patterns within cells are explained by factors such as the presence of mountain ranges.

Africa, the Amazon. This is no surprise because we expect the equator to be wet and green. Now, however, do the same thing at the dry 30° N latitude. Look at how much desert there is at this latitude all around the globe—it's where we find the Sahara, the Arabian, and the Sonoran deserts, among others. And there's more desert territory at 30° S, with the Australian desert and Africa's Namib. From this, you can see that the climatic fate of Earth's various regions is largely written in our globe's large-scale wind patterns.

## Mountain Chains Affect Precipitation Patterns

Latitude doesn't tell the whole story about precipitation on Earth, however. Mountain ranges force air upward, and this air also cools as it rises, dropping its moisture on the windward side of the range. Then the air descends on the opposite (leeward) side of the range, only this time it is dry air and is thus picking up moisture from the ground rather than depositing it (**Figure 35.25**). A dramatic example of this "rain-shadow" effect can be seen on the Pacific Coast of the United States in southern Oregon, where the western side of the Cascade Mountains is lush with greenery all year round, while the basin below the eastern slopes is a desert.

Beyond this, we know that it's possible to be in a warm latitude, such as that of Ecuador in South America, and still be very cold—if you go to the top of the Ecuadorian Andes. As altitude increases, temperature drops, in other words. The changes in climate that occur as you move from flatland to

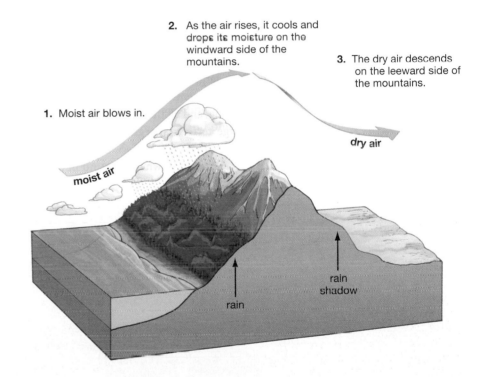

2. As the air rises, it cools and drops its moisture on the windward side of the mountains.

3. The dry air descends on the leeward side of the mountains.

1. Moist air blows in.

dry air

moist air

rain

rain shadow

mountain peak are similar to the changes that occur as you move from the equator to one of the poles.

## The Importance of Climate to Life

The factors you have just reviewed are among the most important in shaping the large-scale **climates** or average weather conditions we find on Earth. It's obvious that climate has a great deal of influence on life since we never see polar bears in Panama or monkeys in Montana, but this influence is even greater than you might think. Looking over the globe, you see different forms of dominant vegetation in different

**Figure 35.25**
**Rain Shadows**
Mountain ranges force moving air to rise. This cools the air, causing it to lose its moisture on the windward side of the range. This creates a rain shadow, meaning a lack of precipitation, on the leeward side of the range.

**Figure 35.29**
**Taiga**

Alaskan caribou in winter in the northern-forest biome known as taiga.

**Figure 35.30**
**Temperate Deciduous Forest**

Plenty of water and seasonally warm temperatures lead to abundant plant and animal life in the biome known as temperate deciduous forest. This forest is in the Great Smoky Mountains National Park, which straddles Tennessee and North Carolina.

and caribou (**Figure 35.29**). When thinking of the fur trade, you might have in mind some freezing, coniferous territory in the far north. The taiga turns out to be it. Sables, minks, beavers, arctic hares, wolverines—this harsh biome has an abundance of these furred creatures.

## Hot in Summer, Cold in Winter: Temperate Deciduous Forest

Temperate deciduous forest is a biome familiar throughout much of the United States since it exists roughly from the Great Lakes nearly to the Gulf of Mexico and from the western Great Lakes to the Atlantic Ocean. The pine forests that cover large parts of the southeastern United States generally are considered to be part of the temperate deciduous biome because the pines are a successional stage *leading to* deciduous forest. (Remember, from Chapter 34, the concept of ecological succession, under which species in a community are replaced by other species over time until a final, stable state is reached.) Meanwhile, the warm Gulf Coast is not temperate deciduous forest per se but a variant on it called temperate evergreen forest. Much of Europe and eastern China fit within the temperate deciduous classification, although large parts of Europe have few forests left, and the same is true of China.

Temperate deciduous forests are the forests of maple, beech, oak, and hickory (**Figure 35.30**). The *deciduous* in the name refers to trees that exhibit a pattern of loss of leaves in the fall and regrowth of them in the spring. Trees do not dominate these forests to the extent they do the taiga, however. Making use of the good soil, a robust "understory" of woody and herbaceous plants will spring up on the floor of the deciduous forest. Some of this activity occurs early in the growing season, before the trees have "leafed out" and blocked the sunlight. Temperate forests grow where there is plenty of water—between 75 and 200 centimeters (or 30 to 78 inches) per year, with the lower amount being three times what the tundra gets. It's worth noting that while the tundra has no reptiles or amphibians and the taiga has only a few, the temperate deciduous forest has an environment that is wet and warm enough to support a variety of both animal forms.

When Europeans first came to North America in significant numbers, in the seventeenth century, there probably were more than a billion acres of forest in what would become the United States. By 1907, about 250 million acres of this forest had been cleared, primarily for agricultural use. The temperate deciduous forest cover in the eastern United States reached its lowest level about 1860, after which it began to rebound because of the westward movement of farming into the Midwest. Between 1950 and 1980, forested land in the United States as a whole actually increased by about 7.5 million acres.

## Dry but Sometimes Very Fertile: Grassland

There is an irregular line in western Indiana that forms the boundary between deciduous forest and prairie. What is "prairie" in the United States is known as "steppes" in Russia, "pampas" in Argentina, and "veldt" in South Africa. These are all names for the biome that ecologists call *temperate grassland* (**Figure 35.31**).

Looked at one way, grasslands are what often literally lie between forests and deserts. Whereas the precipitation range for deciduous forests is from 75 to 200 centimeters per year (30 to 78 inches), for grassland the range is from 25 to 100 centimeters per year (10 to 39 inches). The North American grassland has its own internal division based on the amount of rainfall. It ranges from tallgrass prairie in Illinois to a drier mixed-grass prairie that begins at about the middle of Kansas and Nebraska—this is where the Great Plains begin—to a short-grass prairie that lies in the rain shadow of the Rocky Mountains. (This same tallgrass/short-grass gradient exists in the steppes of Asia, only there the gradient runs from north to south.) Other American grassland areas can be found in California's Central Valley and eastern Washington and Oregon.

A type of tropical grassland, the **tropical savanna**, exists in Australia, Africa, and South America. Tropical savanna is characterized by seasonal drought, small changes in the generally warm temperatures throughout the year, and stands of naturally occurring trees that punctuate the grassland. This is the homeland of many of the great mammals we associate with Africa, such as lions and zebras.

When Europeans first beheld the American grasslands, they could scarcely believe what they saw: a territory that had essentially no trees but instead a sea of grasses that could grow up to 3 meters or 10 feet tall in the wetter tallgrass prairie. Prairie soil turned out to be some of the most fertile on Earth, however, and that fact eventually spelled the end of natural prairie in the United States and most of the rest of the world. Almost all the tallgrass and mixed-grass prairie in the United States eventually was plowed under for cultivation, while the short-grass prairie is now used for crops and cattle grazing. Prior to the coming of Europeans, 22 million acres of Illinois were prairie; today, that figure stands at slightly more than 2,000 acres. From the former prairie, however, the United States feeds itself and some of the world outside its borders. Illinois, Kansas, and Nebraska are still grasslands, but the grasses that grow there now are not the native big bluestem or needlegrass; they are corn and wheat.

## Chaparral: Rainy Winters, Dry Summers

Chaparral is not grassland but instead is a biome dominated by evergreen shrub vegetation that is dotted with pine and scrub oak trees. Chaparral is found in areas that have a so-called Mediterranean climate. There are five of these areas in the world, all very similar in appearance. One of them is the Mediterranean basin itself; then there is southwestern Australia, central Chile, a small part of western South Africa, and parts of coastal California. What these areas have in common is that they are on the west coast of a landmass. Thus, they lie in the path of ocean winds that bring wet, mild winters and very dry summers.

## The Challenge of Water: Deserts

**Deserts** can be defined as biomes in which rainfall is less than 25 centimeters or 10 inches per year, and water evaporation rates are high relative to rainfall. Note that deserts are not defined by temperature. There can be temperate deserts (the Mojave in Southern California), cold deserts (the Gobi in China), and hot-weather

**Figure 35.31**
**Temperate Grassland**
A white-tailed deer fawn stands among the grass and flowers in a tallgrass prairie in Missouri. Tallgrass is one kind of vegetation found in the biome known as temperate grassland.

deserts, like the one in **Figure 35.32**. Around the globe, some 20 major deserts collectively cover about 30 percent of Earth's land surface. What we think of as desert can transition imperceptibly into dry grassland.

Precipitation is not only low in the desert, it is sporadic; that is, most of the moisture a desert gets will come in just a few days during the year. Given this, desert life has relentlessly been shaped by one overriding requirement: Collect water and then conserve it. The flowering plant grouping called bromeliads have a species, *Tillandsia straminea*, that is blown around on the deserts of coastal Peru much like a dead tumbleweed; only this is a living plant, nearly rootless, that gets all its moisture from *fog* that it is able to absorb.

Desert animals are no less remarkable in this respect. The kangaroo rat (genus *Dipodomys*) can be found in many parts of the American Southwest. It comes out of its underground burrow only at night when temperatures fall, it has no sweat glands, and it generally does not drink standing water (there being almost none available). Instead, the kangaroo rat gets by on the water produced by metabolizing the seeds it eats.

## Lush Life, Now Threatened: Tropical Rain Forests

Many people have an idea that the Earth's **tropical rain forests** are important, but they're not quite sure why. There are two main reasons. First, along with coral reefs and algal beds, they are the most productive biome type in the world, meaning they produce more biomass per square meter of territory than any other. Second, no other ecosystem can touch them in terms of species diversity. As noted earlier, for mile after mile in the taiga, trees might be limited to one or two species. In the tropical forests of the Amazon, more than 300 species of trees have been identified in a single hectare of land, which is about 2.5 acres. In Chapter 34, you read about a researcher who found an estimated 1,700 species of beetle in one tree in the Panamanian rain forest. In a sense, this isn't surprising since half of all the identified plant and animal species on Earth reside in tropical rain forests. This despite the fact that, compared to the other biomes you've looked at, they don't occupy a lot of territory: about 7 percent of the Earth's surface.

The problem, as most people know, is that the rain forests are disappearing—significant portions of them are being either burned down or cut down. The period of great temperate forest loss came before the twentieth century, but we are living in the period of great tropical forest loss. How significant is the reduction? Satellite imaging of humid tropical forests indicates that, between 1990 and 1997, about 14 million acres were lost each year—an area that is about twice the size of Maryland. South America has what is by far the world's largest tropical rain forest, the Amazon Basin Forest, which runs from Colombia through Bolivia and at one point nearly from the Atlantic to the Pacific Ocean. Significant rain forests also can be found in Central America, equatorial Africa, and regions of Southeast Asia near the equator. In recent years, the greatest rain forest losses have occurred in Asia, but in 2002, an estimated 6.4 million forest acres were burned or cut in the Brazilian Amazon. (For a different perspective on these kinds of environmental issues, see "Good News about the Environment" on page 756.)

### Poor Soil, Bountiful Covering

The rain forests are lands of rain and stable, warm temperatures (**Figure 35.33**). The norm for rainfall is between 200 and 450 centimeters or about 78 to 175 inches a year, but some areas get up to 1,000 centimeters annually, which is 390 inches.

You've seen that much of the richness of the temperate grasslands can be found in their soil. In the tropical rain forest, almost all the richness is on display above ground (much of it 20 to 40 meters above ground in the extensive forest "canopy"). The rain forest has such a large and efficient group of detritivores that organic material dropping to the forest floor is decomposed by them and absorbed almost

**Figure 35.32 Desert**

When water gets sparse, life does, too. Shown is a desert biome in Monument Valley, Arizona.

immediately by plant roots rather than by the surrounding soil. This means the soil never gets built up; it is nutrient poor and often acidic.

How can this be squared with this biome's fantastic productivity? Note what is going on in the example: The nutrients are simply bypassing the soil, going from plant to detritivore and back to plant. Some rain forest trees have roots that grow *up*, onto tree trunks, rather than down into the soil, giving them the ability to absorb nutrients from the water that washes down the trunks. Farmers on the Illinois prairie could plow under the natural vegetation and grow crops there because perhaps 90 percent of the prairie's nutrients are held in the soil. What happens, though, when a rain forest is cleared for agriculture? The nutrient-poor land will support a year or two of crops and then that's it: No more rain forest, no more crops. (The new crops don't build up the soil because nutrients are leached out by the heavy rains.) With the loss of rain forest comes the loss of habitat for its animals, resulting in a greatly elevated rate of extinction in this biome.

## SO FAR...

1. The warm surface air of the tropics _____ as it rises, causing it to _____ moisture. At about 30° N and 30° S of the tropics, however, air descends and _____ moisture from the land.

2. Earth's large vegetation regions are essentially defined by _____.

3. Starting at the North Pole and going south, list the order in which you would be likely to find the following biomes: temperate deciduous forest, tundra, tropical rain forest, taiga.

mark and on the other by its high-tide mark. This is the **intertidal zone**.

If you look next at the ocean strictly in its vertical dimension, all of the water from the ocean's surface to its floor is called the *pelagic zone*, while the ocean floor itself is called the *benthic zone*. Within the pelagic zone, from sea level down to a depth of at least 100 meters (about 330 feet), the sun's rays can penetrate strongly enough to drive photosynthesis. This zone of photosynthesis is known as the *photic zone*.

**Figure 35.33**
**Tropical Rain Forest**
Abundant rain and warm weather mean abundant growth and a great diversity in life-forms in the biome known as the tropical rain forest. Shown is a lowland rain forest on the Segama River in Borneo.

## 35.8 Life in the Water: Aquatic Ecosystems

### Marine Ecosystems

We turn now from life on land to life in the water. Because we humans are land creatures, it might be good to get our bearings by thinking about how one kind of aquatic environment, the ocean, can be divided up in terms of the life in it. The ocean's **coastal zone** extends from the point on the shore where the ocean's waves reach at high tide to a point out at sea where an ocean-floor formation called the continental shelf drops off (**Figure 35.34**). Beyond this point, there is the *open sea*. Within the coastal zone, there is an area bordered on one side by the ocean's low-tide

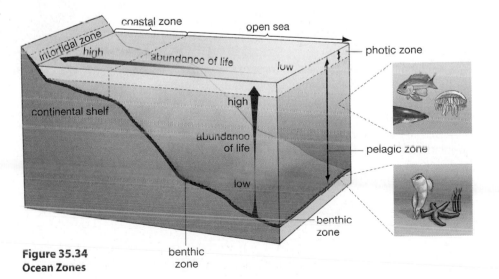

**Figure 35.34**
**Ocean Zones**

Good News about the Environment

A sense of gloom about the environment may arise from between the lines of these pages as the list of current or potential environmental problems piles up: global warming, nitrogen buildup, overuse of water resources, tropical rain forest destruction. These are real problems, but their existence should not blind us to the fact that there has been a great deal of good news about the environment in the last 35 years, much of it showing how effective government can be when prodded into action by its citizens.

If there was a time for unrestrained gloom about the environment in the United States, it probably would have been in the late 1960s, when the country had plenty of endangered species but no Endangered Species Act; when levels of smog in Los Angeles were almost twice what they are now; when there was plenty of household trash to be recycled, but almost no recycling going on; when only a third of the bodies of water in the country were safe for fishing and swimming, whereas today two-thirds are.

*Environmentalism has entered not only our law books but, more important, our consciousness.*

As a symbol of change, consider Cleveland, Ohio, whose Cuyahoga River was so polluted in 1969 that it *caught fire*, while southwestern Lake Erie, into which the Cuyahoga flows, was essentially a dead zone for fish during the late summer. The lake came to this low state in part because of the nearly 80 metric tons of phosphorus—then regularly added to washing detergents—that were dumped into it each day. That figure has now been reduced by perhaps 85 percent, and Lake Erie is biologically very active and a major freshwater fishery. Meanwhile, the Cuyahoga River has been cleaned up enough that an area near the river's mouth has been transformed into an entertainment district complete with restaurant tables near the water's edge—not bad for a small river running through a big city. This does not mean that the Cuyahoga now looks as it did when Europeans first saw it; toxic sediment from the river, mostly the legacy of the past, still flows into Lake Erie. But if we look at the *trend*, there is no doubt that things have gotten better in Cleveland, not worse.

This same thing could be said of most of the United States. Simply put, there has been a profound change with respect to the environment over the past 35 years. From pollution controls to species protection to personal recycling, environmentalism has entered not only our law books but, more important, our consciousness. When populations of Atlantic swordfish were observed to be declining in the late 1990s, a campaign titled "Give Swordfish a Break" was organized that quickly enlisted 27 top New York chefs. What could chefs do about swordfish populations? They pledged to take swordfish off their restaurants' menus. It was government action that brought the Atlantic swordfish population back to healthy levels, but initial public awareness of the swordfish campaign came about because of *personal* sensitivity to the environment—in this case, the sensitivity of a group of American chefs. And sensitivity to the environment has grown not just in America but in most developed countries. The cleanup of Lake Erie required the cooperation of the United States and Canada; the heartening trend in preservation of the Earth's ozone layer is the product of an international agreement.

None of this is to say that mechanisms are in place such that we need not worry about the environment. Far from it: Largely because of the growth in human population and the greater per capita use of natural resources, the environment is threatened on thousands of fronts and will remain so as far out as we can see into the future. But the record of the last 35 years is not just one of environmental loss; it is one of great environmental accomplishment as well. Perhaps the period's most important lesson is that there is nothing inevitable about environmental degradation. Concerned, hardworking people have made a great deal of difference in this issue.

**Back to Healthy Numbers**

Populations of Atlantic swordfish (*Xiaphas gladius*) were severely depleted by the late 1990s. A campaign spearheaded by environmentalists spurred governmental action that brought the swordfish numbers back to healthy levels by 2002—just three years after an international conservation program was put in place.

## More Productive Near the Coasts, Less Productive on the Open Sea

Each of these zones marks off an area that is meaningful in relation to life. Start first at the intertidal zone. This is a world *in between*, we might say, because the creatures in it live part of the time submerged in the ocean and part of the time exposed to the air. It turns out that the intertidal zone is extremely productive for a reason that you might think would make it unproductive: The ocean waves, which bring in nutrients, carry away wastes and expose more of the surface area of photosynthesizers to sunlight.

Ocean life continues to be abundant out past the intertidal zone and in general remains so all the way to the end of the coastal zone. In short, ocean life is most productive in its shallower depths (**Figure 35.35**). Once past the continental shelf and into the open ocean, productivity can drop off to levels that are less than those of a desert. Why should this be so? Remember that sunlight strong enough to power photosynthesis penetrates ocean waters only to a depth of 100 meters or so. Meanwhile, the average depth of the ocean is about 3,000 meters or 1.8 miles. Thus there is a vast depth of open ocean that simply isn't very productive. Indeed, life at greater ocean depths essentially depends on organic particles, referred to as "marine snow," that rain down from above. One measure of the coastal zone's importance is that it is the location of all the world's great fisheries. Fish do exist in the open ocean as well, however, and concern is increasingly being raised about the health of fish populations in both kinds of areas, as you can see in "Our Overfished Oceans" on page 758.

## More Productive Near the Poles, Less Productive Away from Them

Apart from the shallow-to-deep transition, the ocean also has one other gradient with respect to the amount of life it harbors. In the pelagic zone, there is relatively more ocean life toward the poles, with this concentration decreasing with movement toward the equator—the exact opposite of what we find with

**Figure 35.35**
**Diversity in Ocean Life**

**(a)** A man and two children investigate tide pools, a part of the ocean's intertidal zone. The deeper water beyond them can be seen both above and below the surface.

**(b)** King penguins swim underwater near a breeding colony in the fertile waters off the coast of Antarctica.

**(c)** Fish called sweepers and fairy basslets feed in the currents above a coral plate covered with animals called crinoids. The structure is off the coast of Papua New Guinea.

**(a)** A man and two children investigate tide pools

**(b)** King penguins swim underwater

**(c)** Sweepers and fairy basslets feed in currents

## ESSAY  Our Overfished Oceans

At one time, marine wildlife experts thought that the bounty of fish in the world's oceans was so immense it could never be exhausted. But that idea began to be challenged in the late 1980s, and by the mid-1990s it was effectively dead as it became clear that the numbers of certain kinds of fish had not just dwindled but had dropped to alarmingly low levels. The news has not gotten better since then. In 2003, researchers reported that the global ocean has lost 90 percent of the fish in its big predatory species—bluefin tuna, blue marlin, Antarctic cod, and sharks among them (see **Figure 1**). Then, in 2006, in the most gloomy assessment yet, a team of 14 scientists from four countries concluded that, if current practices continue, commercial fishing will come to an end by the middle of this century because there will be almost no commercial fish left to catch.

The team arrived at this conclusion by looking at the trends in commercial catches from 1950 through 2003 in all of the world's "LMEs" or large marine ecosystems—distinct ecological regions in the oceans, such as the Gulf of Mexico and the Bering Sea. The researchers labeled a species of fish as "collapsed" if catches of it within LMEs have fallen to less than 10 percent of their peak levels. Using that criterion, 29 percent of all commercial species underwent collapse between 1950 and 2003. If the rate of decline between those 2 years persists in coming years, then all species of fish that are harvested will undergo collapse by 2048 according to the researchers' calculations. Not surprisingly, the decimation of

**Figure 1**
**One of a Dwindling Number**

A blue marlin (*Makaira nigricans*), one of the large predatory fishes whose numbers have been reduced by 90 percent in recent decades, according to a global assessment.

individual species that has already occurred is being felt now in global fish harvests. Despite large increases in fishing efforts in recent years, actual global catches of fish peaked out in 1994, the researchers said, and have declined by 13 percent since then. As lead researcher Boris Worm of Dalhousie University told News@Nature.com, "It's like a lemon. . . . We have to press harder and harder to get juice out of it. At some point we just can't force more out—we're going to start running out of species."

To be sure, not everyone shares the researchers' pessimism. In the United States, about 20 percent of fish stocks are "overfished," but that proportion is not increasing—indeed, it may be decreasing, according to Steve Murawski, the chief scientist of the Fisheries Service of the National Oceanic and Atmospheric Administration. Nevertheless, if the research team's global findings are anywhere close to the mark, it's sobering to think that we are starting to glimpse "the bottom of the barrel," as researcher Worm puts it.

How did things come to this pass? The answer seems to be that, in recent decades, world fishing fleets have gotten more efficient at scouring the Earth for large populations of fish. By using ever-more-sophisticated technology, the fleets have managed to find fish no matter what kind of terrain they are hidden in. The result, however, is that, in an increasing number of instances, there are no large populations left to find.

Fisheries scientist Daniel Pauly and his colleagues have pointed out that this turn of events actually should not be surprising. We would not expect industrialized *hunting*—of, say, deer or buffalo—to be sustainable over the long run if left unregulated. Why should we expect anything different with industrialized fishing? Many scientists believe that a three-pronged approach has to be taken if global fish populations are to bounce back. First, global fishing must be reduced in general. Second, countries need to restrict the use of "bottom trawling" techniques in which nets and accompanying gear—sometimes weighing thousands of pounds—are dragged across the ocean floor, destroying habitat as well as fish. Third, "no take" zones need to be created so that specific fish populations can have a chance to establish themselves.

The importance of this last point was underscored in an additional analysis performed by the international research team that examined the world's LMEs. Looking at a series of marine reserves and at fisheries that have been closed, they found that conservation works, at least on local and regional scales. In the protected areas they examined, species diversity increased by an average of 23 percent, and this change was accompanied by a fourfold increase in the amount of fish *caught*—in waters surrounding the reserves. In short, conservation can benefit not only fish themselves, but the fishermen who harvest them and, by extension, everyone else.

land life. The oceans that surround Antarctica are filled with life. **Phytoplankton**, meaning floating, microscopic photosynthesizers such as algae and cyanobacteria, are the base producers for it all. These are consumed by the tiny floating animals known as **zooplankton**. Shrimp-like crustaceans called krill feed mostly on the phytoplankton and serve as the principal source of food for the whales and seals we associate with this area.

### Cities of Productivity: Coral Reefs

Despite the richness of the colder oceans, the tropical oceans do boast marine communities that are as productive as any in the ocean—indeed, as any on Earth. These are the coral reefs, which lie in shallow waters at the edge of continents or islands at tropical latitudes. **Coral reefs** are warm-water ocean structures composed of the remains of coral animals. Each animal, known as a polyp is a tiny relative of the more familiar sea anemones. Coral polyps secrete calcium carbonate, better known as limestone, a practice they share with a type of red algae with whom they share their habitat. The limestone forms an external skeleton for the corals, and when each coral dies, its limestone skeleton remains. Thus, coral reefs are composed largely of the stacked-up remains of countless generations of coral polyps along with the remains of their neighboring red algae. But the thin outer veneer of each reef is composed of *living* algae and the latest generation of pinhead-sized corals.

You can get an idea of how many generations of coral are piled up by considering that most reefs are between 5,000 and 8,000 years old—and some are millions of years old. What the reef as a whole creates is *habitat*: nooks and crannies and tunnels and surfaces that are home to an amazing array of living things. So rich is this habitat that it covers only 2 percent of the ocean's floor but is home to perhaps 25 percent of the ocean's species. One coral reef, Australia's Great Barrier Reef, also is one of the largest structures created by any living thing, including humans. It measures some 2,000 kilometers or 1,200 miles in length.

Sadly, coral reefs are now imperiled cities of biological productivity. By one estimate, 27 percent of the world's coral reefs have now been lost as functioning ecosystems, although with coral reefs, loss of activity is not always permanent. The greatest single source of this destruction was the El Niño-caused ocean warming of 1997–1998, which brought about a phenomenon known as *coral bleaching*, meaning a discoloration of the reefs due to the death of symbiotic algae that live within the coral animals. Temporary weather events aside, the longer-term threat to

coral reefs is human activity. Oil is dumped into the ocean, sewage and runoff from cities harms the reefs, divers may trample them. Perhaps most disturbing, thousands of pounds of the poison sodium cyanide are dumped each year on coral reefs in the Philippines and other parts of Asia as a means of capturing colorful aquarium fish. (The cyanide stuns the fish but kills the coral animals.) Global warming has now been tied to the threat to coral reefs as well; both ocean warming and increased ocean $CO_2$ concentrations are harmful to the reefs.

### Freshwater Systems

Whereas the oceans cover almost three-quarters of Earth's surface, freshwater ecosystems—inland lakes, rivers, and other running water—together cover only about 2.1 percent of Earth's surface. Like the ocean, lakes can be divided into biological zones. The most productive zone in a lake is the shallow-water area along its edge, the *littoral zone*, whose outer boundary is defined as the point at which the water is so deep that rooted plants can no longer grow. Then there is the *photic zone*, which starts at the surface of the lake and extends down to the point at which sunlight no longer penetrates strongly enough to drive photosynthesis. Finally, there is the area beneath the photic zone, the *profundal zone*, in which photosynthesis can't be performed (**Figure 35.36**).

**Figure 35.36**
**Zones in a Lake**
The littoral zone starts at water's edge and extends out to the point at which rooted plants can no longer grow. The photic zone starts at the lake's surface and extends down to the point at which sunlight no longer drives photosynthesis. The profundal zone is the area beneath the photic zone in which photosynthesis cannot be performed.

### Nutrients in Lakes

Lakes can be naturally *eutrophic*, meaning nutrient rich, or *oligotrophic*, meaning nutrient poor. A lake that has few nutrients will have relatively few photosynthesizers, and this in turn means few animals. Oligotrophic lakes generally are clear (and deep), while eutrophic lakes often have an abundant algal cover.

Nutrient enrichment can happen naturally over time, or it can take place because of human intervention. The latter variety is known as *artificial eutrophication*, a process that sometimes is undertaken intentionally just to increase the yield of fish from a lake or pond. There is such a thing as having too *many* human-added nutrients in a body of water, however. We touched on this earlier in connection with the Gulf of Mexico's "dead zone," which results from the tons of nitrogen that flow into it. But smaller bodies of water can undergo eutrophication as well. An overabundance of such nutrients as phosphorus or nitrogen brings on an overabundance of algae—a so-called **algal bloom**—and this eventually means an overabundance of dead algae falling to the bottom of the lake (**Figure 35.37**). When this happens, decomposing bacteria flourish, and they are using oxygen—so much of it that the fish in a lake can suffocate. How do these harmful levels of phosphorus or nitrogen get into freshwater aquatic ecosystems? Fertilizers from lawns or agriculture and construction that disturbs bedrock, sewage, and detergents are some of the sources (although in many states phosphate-containing detergents have been banned for decades). This is one form of human environmental impact that can be reversible in a straightforward way: reduce the nutrients that are flowing into the ecosystem.

### Estuaries and Wetlands: Two Very Productive Bodies of Water

There is one important type of aquatic water system that always straddles the line between a freshwater and saltwater habitat. This is the **estuary**, an area where a stream or river flows into the ocean. Estuaries rank with tropical forests and coral reefs in being the most productive ecosystems on Earth. The cause of this productivity is the constant movement of water—the same force at work in the ocean's intertidal zone. Ocean tides and river flow are constantly stirring up estuary silt that is rich in nutrients. Plants and algae thus get an abundance of these substances, grow in abundance, and pass their bounty up the food chain. In **Figure 35.38**, you can see an image of an important estuary in the United States, the Chesapeake Bay.

Another important type of aquatic ecosystem is **wetlands**, which are just what they sound like—lands that are wet for at least part of the year. "Wet" here covers a variety of conditions, from soil that is merely temporarily waterlogged—a prairie "pothole" in Minnesota, for example—to a cover of permanent water that may be several feet in depth (the bayous of Louisiana). Wetlands go by such specific names as

**Figure 35.37**
**Too Many Nutrients**
A pond with an overabundance of nutrients has experienced an overgrowth of algae—an algal bloom—that may be detrimental to other life-forms in the pond.

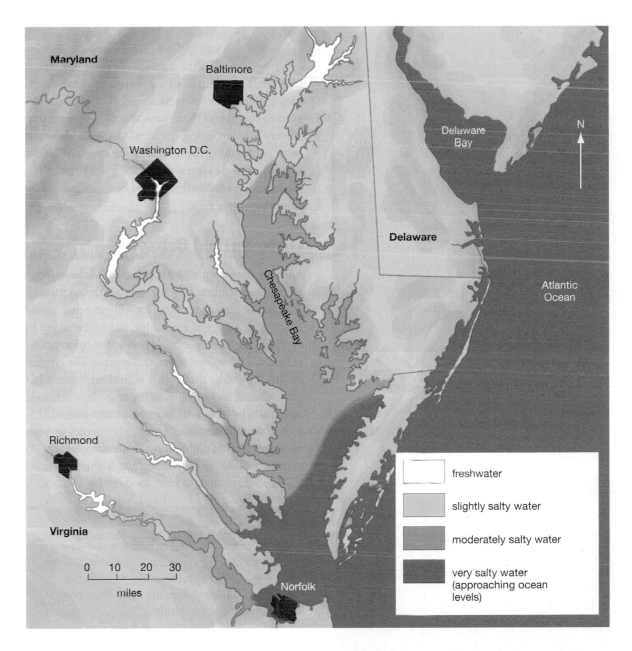

**Figure 35.38**
**Where Salt Water and Freshwater Meet**
The Chesapeake Bay is one of the largest and most productive estuaries in the world. In it, the freshwater of 19 principal rivers meets the salt water of the ocean, forming zones of different degrees of saltiness. Each zone supports specific communities of organisms. (Redrawn from A. J. Lippson and R. L. Lippson, *Life in the Chesapeake Bay,* Johns Hopkins University Press, Baltimore, 1984.)

freshwater

slightly salty water

moderately salty water

very salty water (approaching ocean levels)

swamps, bogs, marshes, and tidal marshes; the overwhelming majority of them are freshwater inland wetlands rather than the ocean-abutting coastal wetlands, which can be freshwater or salt water (**Figure 35.39** on the next page). Wetlands are sites of great biological productivity, and they serve as vital habitat for migratory birds. It's been estimated that the amount of wetlands in the United States has been reduced by 55 percent since the arrival of Europeans—120 million of the 215 million acres that were originally wetlands have been drained or paved over. Currently, the state of Florida and the federal government are working on one of the most far-reaching and expensive habitat restoration efforts in history as they try to bring the largest expanse of wetlands in the nation, the Florida Everglades, to a healthier condition over the next 20 years.

## Life's Largest Scale: The Biosphere

In this chapter, you've been reviewing life on a very large scale, which is to say the aquatic ecosystems and biomes of the Earth. But there is one larger scale of life yet, which is the collection of all the world's ecosystems, the **biosphere**, which is also thought of as that portion of the Earth that supports life. Much could be said about the biosphere, but here's just one observation about life on this global scale. The Earth is about 24,000 miles in circumference, with life existing along any line you could draw all around the globe. If we look at life in its *vertical* dimension, however, it has been found to exist only about 1.8 miles below Earth's surface and about 5.2 miles above sea level (in the upper Himalayas). This means that all of life on Earth exists in a band that is at most about 7 miles thick, with the vast bulk existing in an

## 35.5 Global Warming

- The Earth's atmospheric temperature is increasing through the phenomenon known as global warming. Scientists can now state with at least 90 percent certainty that most of the global warming seen since 1950 has come about because of the human activities of deforestation and the emission of such gases as carbon dioxide and methane into the atmosphere. Both $CO_2$ and methane are referred to as greenhouse gases since both trap additional quantities of the heat that comes to the Earth from the sun. Deforestation—today, primarily of tropical forests—either puts $CO_2$ into the atmosphere directly, when trees are burned, or results in less $CO_2$ being taken out of the atmosphere, since trees that are destroyed no longer take up $CO_2$ to perform photosynthesis.

- The Earth appears to have warmed by 0.74°C or about 1.3°F in the 100 years that ended in 2005, according to our most authoritative source on global warming, the Intergovernmental Panel on Climate Change (IPCC). In 2007, the panel predicted that, over the course of the twenty-first century, the Earth is likely to warm by an additional 1.8°C to 4.0°C, depending on the additional quantity of greenhouse gases put into the atmosphere. The consequences of this warming will include a rise in sea levels and are likely to include more intense tropical storms, less snow cover globally, and a permafrost that thaws to increasing depths. Global warming cannot be stopped, the IPCC said, but it can be lessened in accordance with future human activities.

## 35.6 Earth's Climate

- Earth is tilted at an angle of 23.5° relative to the plane of its orbit around the sun, a fact that dictates much about climate on Earth, which in turn dictates much about life on Earth.

- Sunlight strikes the equatorial region of the Earth more directly than the polar regions. The differential warming that results produces a set of enormous interrelated circulation cells of moving air, each existing all the way around the globe at its latitude. Each of these cells drops rain on the Earth where it rises but dries the Earth where it descends. This is why some regions of the Earth get so much more rainfall than others.

- More rain will be deposited on the windward side of a mountain range than on the opposite leeward side. This rain-shadow effect can cause opposite sides of a mountain range to have dramatically different vegetation patterns.

- A climate is an average weather condition in a given area. Large vegetative formations essentially are defined by climate regions.

## 35.7 Earth's Biomes

- Biomes are large terrestrial regions of the Earth that have similar climates and hence similar vegetative formations. Six types of biomes are recognized at a minimum: tundra, taiga, temperate deciduous forest, temperate grassland, desert, and tropical rain forest. Polar ice and mountains often are recognized as separate biomes, as are the tropical grasslands called tropical savannas and the dry, shrub-dominated formations called chaparral.

- Tundra is the biome of the far north, frozen much of the year but with a seasonal vegetation formation of low shrubs, mosses, lichens, grasses, and the grass-like sedges.

- Taiga is another biome of the north; it includes the enormous expanse of coniferous trees that lies south of the tundra at northern latitudes. The taiga exhibits a great deal of species uniformity, with only a few types of trees—spruce, fir, and pine—serving as ecological dominants. The region supports large populations of fur-bearing animals.

- Temperate deciduous forests grow in regions of greater warmth and rainfall than is the case with tundra or taiga. These forests, existing over much of the eastern United States, are composed of an abundance of trees such as maple and oak, complemented by a robust understory of woody and herbaceous plants.

- Temperate grassland goes by several names around the world, including prairie and steppes. This biome is characterized by less rainfall than that of temperate forest and by grasses as the dominant vegetation formation. Such regions can be very fertile agricultural land.

- Chaparral is a biome dominated by evergreen shrub vegetation. It is found in a few, relatively small regions of the world that have a Mediterranean climate, among them parts of coastal California. These regions have mild, rainy winters and very dry summers.

- Deserts are characterized by both low rainfall and water evaporation rates that are high relative to rainfall. Deserts may be hot, cold, or temperate, but all desert life is shaped by the need to collect and conserve water.

- The tropical rain forest biome is characterized by warm, stable temperatures, abundant moisture, great biological productivity, and great species diversity. Rain forest productivity is concentrated above the forest floor, often in the canopy high above ground. Found in Earth's equatorial region, tropical rain forests are being greatly reduced in size through cutting and burning.

## 35.8 Life in the Water: Aquatic Ecosystems

- Ocean or marine ecosystems are most biologically productive near the coasts, with the deep open oceans having a productivity that can be less than that of deserts. Coastal areas benefit from wave actions that bring in nutrients, carry away wastes, and expose more of the surface area of photosynthesizers to sunlight.

- Modern fishing techniques have seriously depleted many once-abundant commercial fish species.

- There is relatively more ocean life toward the poles, with this concentration decreasing as one moves toward the equator. The food webs in the oceans surrounding Antarctica are based on photosynthesizing phytoplankton, meaning floating microscopic organisms such as algae and cyanobacteria.

- Coral reefs are warm-water marine structures composed of the piled-up remains of generations of coral animals and their associated algae. Coral reefs provide a habitat that results in a rich species diversity.

- Freshwater ecosystems, which include inland lakes, rivers, and other running water, cover only about 2.1 percent of Earth's surface. Freshwater lakes are most productive near their shores and near their surface. Lakes can be naturally eutrophic, meaning nutrient rich, or oligotrophic, meaning nutrient poor. The human activity known as artificial eutrophication can sometimes introduce too many nutrients into a lake, which can cause some species to flourish (algae, bacteria) while harming other species (some fish).

- Estuaries are areas where streams or rivers flow into the ocean. They are characterized by high biological productivity because of the constant movement of water within them.

- Wetlands, also known as swamps or marshes, are lands that are wet for at least part of the year. Wetland soil may merely be waterlogged for part of the year or under a permanent, relatively deep cover of water. Wetlands are very productive and are important habitats for migratory birds. The amount of wetlands in the United States is estimated to have been reduced by 55 percent since the arrival of Europeans.

## KEY TERMS

| | | |
|---|---|---|
| abiotic   727 | groundwater   732 | |
| algal bloom   760 | herbivore   736 | |
| aquifer   733 | intertidal zone   755 | |
| atmosphere   740 | kilocalorie   738 | |
| biodegradable   737 | net primary production   738 | |
| biogeochemical cycling   728 | nitrogen fixation   729 | |
| biomass   729 | nutrients   728 | |
| biome   750 | omnivore   736 | |
| biosphere   761 | ozone   740 | |
| biotic   727 | permafrost   750 | |
| carnivore   736 | phytoplankton   759 | |
| CFCs   740 | primary consumer   735 | |
| climate   749 | producer   735 | |
| coastal zone   755 | secondary consumer   735 | |
| consumer   735 | stratosphere   740 | |
| coral reef   759 | taiga   751 | |
| decomposer   737 | tertiary consumer   735 | |
| desert   753 | trophic level   735 | |
| detritivore   737 | tropical rain forest   754 | |
| ecosystem   727 | tropical savanna   753 | |
| element   728 | troposphere   740 | |
| energy-flow model   737 | tundra   750 | |
| estuary   760 | wetlands   760 | |
| gross primary production   738 | zooplankton   759 | |

## BRIEF REVIEW

*Answers to Brief Review questions are in the back of the book.*
*For multiple-choice quiz questions, go to* **www.aw.com/krogh4**.

1. In a large reserve in Idaho, Forest Service wildlife scientists measured wolf populations as well as elk populations and found that the wolf population was much lower than that of the elk population. If wolves feed on elk, why aren't the population sizes similar?

2. Explain why most deserts across the globe occur at about 30° N and 30° S of the equator.

3. What relationship does climate have to the large-scale vegetation patterns that are seen on Earth?

4. A great many dead fish are found floating on the surface of a pond. What might be the cause of such a fish kill?

5. Name several ways in which the tundra and taiga biomes differ.

## APPLYING YOUR KNOWLEDGE

1. The filmmakers in the movie *King Kong* land on the imaginary Skull Island, which is the home not only of the giant ape Kong, but of a host of meat-eating dinosaurs as well. Setting aside the impossibility of an ape as big as Kong, from an ecological perspective, why could a real Skull Island never have existed, even when dinosaurs roamed the Earth?

2. Bacteria and fungi are to the living world as a foundation is to a house: not often thought of, but vitally important. If all bacteria and fungi were somehow instantly eliminated, almost no life-forms would survive for long. This is so because of the role bacteria and fungi play in ecological processes. Can you name two of these processes?

3. Tundra and desert are in some ways more fragile biomes than are grasslands or forests. Why should this be so?

questions: Why does an animal behave as it does, and how does it manage to exhibit these behaviors?

Answering "what" questions requires observation. Scientists must look at the animals, listen to them, or otherwise measure their behavior in some way. This can be fairly easy to do with animals that can be studied in a laboratory but difficult and time consuming for animals that can be studied only in the wild. There is the difficulty of *getting to* the habitats of some of these latter animals—think of whales or moles—and then there is the problem of observing them without disturbing them (**Figure 36.1**). Scientists who study eagles often must be hidden away in tree-top observation huts, while those who study chimpanzees might have to spend years around a given group before the animals will accept them as a part of their surroundings.

Animal behavior has been answering "what" questions for hundreds of years, but only since the 1930s has it come into its own as a branch of science by systematically pursuing "why" and "how" questions. Two researchers, Konrad Lorenz of Austria and Niko Tinbergen of the Netherlands, led the way in this effort by observing animals in natural conditions and then *manipulating* these conditions as a means of getting at the truth. Tinbergen, for example, observed a species of wasp called a beewolf, so named because it paralyzes honeybees with its sting and then drags the bees down into its nest, where they will serve as food for the wasp offspring. The beewolf's nest lies at the end of a tunnel that it digs in the sand, and the wasps Tinbergen watched made a round trip of about a mile to get the bees and bring them back home.

Watching all this, Tinbergen hit on a question that fascinated him: When returning with prey, how does the beewolf pick out *its* nest from among the multitude

that exist? The nests themselves are underground, after all, and the beewolves cover up the entrances to them on departing to look for bees. He resolved to find out, but this meant he had to have some means of telling one wasp from another. He therefore caught wasps, one by one, and marked them with colored dots. Doing so gave him a feeling that many animal behavior researchers have experienced. As he wrote:

> It was remarkable how this simple trick of marking my wasps changed my whole attitude to them. From members of the species *Philanthus triangulum* they were transformed into personal acquaintances, whose lives from that very moment became affairs of the most personal interest and concern to me.

Tinbergen had observed that beewolves circled their nests in ever-widening and ever-higher loops before leaving on a hunting expedition. With this in mind, he hit on a possible explanation for how they found their nests upon returning. As they circled, perhaps they were putting into memory the position of their nests relative to surrounding landmarks. If so, he reasoned, *disturbing* these landmarks after a wasp left should interfere with its ability to locate its nest on return. Tinbergen did just this, moving the sticks and rocks around a given nest, while leaving the nest entrance unaltered. When the beewolves returned to these changed surroundings, they were disoriented; 4 feet or so above the ground, they would stop, zigzag back and forth, hang in the air, and then circle in a wide loop, starting the whole process over again. This test and others convinced Tinbergen that the wasps were locating their nests through the use of visual landmarks.

Setting aside Tinbergen's specific finding, note how this one piece of research answered all three types of questions fundamental to animal behavior studies. *What* does the beewolf do? (Circles its nest before leaving.) *Why* does the beewolf circle its nest before leaving? (So it can find its nest on returning.) *How* does the beewolf know where its nest is? (It associates landmarks with its nest.) From Tinbergen's time to the present, observation has been joined to ever-more-sophisticated experimental methods. Scientists now make videos of their animal subjects, sequence their DNA, and track them with Earth-orbiting satellites (**Figure 36.2**).

## Proximate and Ultimate Causes

When we look at the "why" questions in animal behavior, a helpful concept is that of proximate and ultimate causes of behavior. *Proximate* can be thought

**Web Animation 36.1**
Behavioral Biology

**Figure 36.1
Observing without Disturbing**

A biologist studies mountain goats, but from a distance so as not to interfere with their behavior.

of as any cause of behavior that involves the physiology or physical functioning of an animal. It is a "triggering" cause of a behavior. Meanwhile, the *ultimate* cause of a behavior generally is linked to survival and reproduction—to evolution in the final analysis, as you'll see.

To take one example of this proximate/ultimate distinction, consider the behavior of the small Belding's ground squirrels (*Spermophilus beldingi*) found in the Sierra Nevada of California (**Figure 36.3**). About 2 months after they are born, the males among these squirrels leave their burrow and never return to it for the rest of their lives. The *sisters* of these travelers, meanwhile, remain close to their "natal" burrow for their whole lives. So, why do the males leave while the females pretty much stay put? To find out, researchers Kay Holekamp and Paul Sherman tested a number of hypotheses, but the evidence they sifted through pointed to only two factors: hormonal influence early in life and growing to a certain minimal body size. To test the first idea, they injected a group of *female* squirrels with the "male" hormone testosterone soon after birth, after which these squirrels and their mother were returned to the wild. The result? The injected females showed the same kind of leaving-home behavior as males. With respect to the body-size idea, the researchers found that males tended to leave about the time their weight reached a minimum of 125 grams, which presumably allowed them to survive on their own.

On one level, then, what seems to cause male Belding's squirrel dispersal is early hormonal influence and the achievement of a certain minimum weight. But these are proximate causes in that they are "triggering" and involve the physiology of the squirrels (hormones, weight). They don't tell us much about the *ultimate* causes of the dispersal. Why should a young male leave home? The researchers tested a number of hypotheses—that the males could avoid food shortages by leaving, or that they would have better mating prospects away from home—but none of these ideas was supported by the evidence.

Another possibility, however, was that the males were dispersing to avoid the possibility of inbreeding, meaning mating with close relatives—a behavior that can lead to offspring that either die in the womb or have abnormalities when born. Here, the evidence was supportive. The movement of males away from the home burrows resulted in a complete absence of mating among kin, the researchers observed. Moreover, after initial dispersal, those males that went on to mate the most then dispersed farthest away from the breeding area—an excellent strategy for avoiding mating

**Figure 36.2**
**High-Tech Tracking**
An Atlantic loggerhead turtle, outfitted with a satellite radio transmitter, is prepared for release into the ocean. Researchers use satellite transmitters to track the migratory movements of animals.

**Figure 36.3**
**Much-Studied Squirrel**

An adult Belding's ground squirrel surrounded by three young on the eastern slopes of California's Sierra Nevada. Young males of the species leave home at an early age, while young females remain near the burrows in which they were born.

with one of your own offspring at a later date. Why do male Belding's squirrels disperse? The *ultimate* cause may be to avoid inbreeding.

## Ultimate Cause and Natural Selection

Now, male Belding's squirrels obviously do not sit around and think: "Well, I'm about 2 months old now; I'd better leave the burrow, or I might breed with one of my relatives." They do not understand what inbreeding is, and yet they behave in such a way that they avoid it. How did this kind of motivation come to be established in them? Here's how.

Squirrels will be motivated to act in a certain way in large part because of genes they inherit. Let us

**Figure 36.14**
**Social Animal**

Zebras are social animals: they live in groups. The members of a zebra family groom each other and warn one another of predators—behaviors that benefit each individual. Living in groups also has costs, such as the spread of disease. These zebras are part of a herd in Kenya.

**Figure 36.15**
**Defense through Cooperation**

Group defense can be one of the benefits of group living. Male musk oxen form a circle like this one when predators threaten. These oxen are in Canada.

stallion, up to six mares, and offspring (**Figure 36.14**). Young males leave these families to join bachelor groups of up to 10 individuals. These two basic zebra units may then become part of herds of zebras that number into the tens of thousands. Zebras do more than just live in close proximity, however. Family members groom one another, play with one another when young, keep lookout for predators to protect sleeping family members, and show what appears to be great loyalty to one another. (Zebras move at the pace of the *slowest* family member, and they may attempt rescue missions for family members who have become separated from a group that is under attack.)

## Why Live Alone—or Together?

From the human perspective, the life of the mason wasp seems lonely and difficult compared to that of a zebra, but of course we have no indication that wasps or zebras would see things this way. It will come as no surprise that the unsentimental logic of natural selection is at work in channeling animals toward solitary or social living. A species will exhibit social behavior to the degree that such behavior aids in the survival and reproduction of individuals in that species. Social behavior always has *costs* as well as benefits; animals will be social to the extent that the costs are outweighed by the benefits. There is a cost to many animals in merely living together in that parasites and diseases can move easily among them because of the physical contact they have. This very factor is decimating North American populations of honeybees, *Apis mellifera*, which pass among themselves a tiny mite that makes its home in their respiratory tubes, suffocating them. Beyond this, think of a hungry zebra moving to a better grazing territory. It incurs a cost by having to move at the speed of the slowest member of its family.

Against the costs, many benefits can come from group living. There is group defense, which you can see a vivid example of in the musk oxen pictured in **Figure 36.15**. One of the oxen's predators, meanwhile, is the wolf, which benefits from hunting in groups. The V-pattern that geese maintain while migrating reduces the amount of energy each animal needs to expend in flight. Pigs and penguins conserve heat by keeping close together in winter.

## Dominance Hierarchies

One form of organization that often arises among social animals is the **dominance hierarchy**: a persistent power ranking in an animal population that gives those of higher rank the ability to control some aspect of the behavior of those of lower rank. This control often results in higher-ranking individuals gaining better access to resources, such as food and mates. If you look at **Figure 36.16**, you can see an exact accounting of a dominance hierarchy in chickens. This is, literally, a pecking order, documented by a researcher who observed a group of 12 hens. It is indeed well-ordered in that the hen at the top, Y, pecked any and all beneath her, while the hen at the bottom, BR, pecked no other hens but *got* pecked by every other hen.

Dominance hierarchies usually are not this rigid. There may be a single dominant individual and then "all others" who have no ranking among them; animal A may be dominant over B, but not over C, and so forth. Wolves (*Canis lupus*) employ a caste system, with each wolf pack composed of a dominant male and female that get privileged access to food and that are the only members of the group who breed. Then, in the middle, come adults that do not breed. Finally, at the bottom are other adults that can literally live on the periphery of the group.

Arrangements such as this may seem unfair from the human point of view, but fairness doesn't count in the struggle for survival and reproduction. At root, dominant hierarchies seem to exist because certain individuals in a group—the dominant ones—benefit from such hierarchies and can enforce them.

### Territoriality

**Territoriality** can be defined as the effort an animal makes to keep other animals out of a given area. In general, territoriality refers to efforts to keep members of one's own species from entering an area. Why would a robin be more concerned about another robin than about a rabbit? Because fellow robins will be competing for the same resources—food, nesting space, and mating partners. Members of the same species are not necessarily territorial, however; any two may inhabit so-called home ranges, meaning relatively large, and sometimes overlapping, areas that neither lays claim to.

The term *territoriality* may bring big, fierce creatures to mind, but most songbirds are territorial, as are animals as small as aphids. Territoriality can be seasonal; birds that are territorial during mating season may congregate in large flocks at other times of the year, completely unconcerned about "turf." Mating partners or larger groups of animals may defend territories, but in general it is individual males that carry out the practice. The songs that male songbirds use to attract females often serve the dual purpose of warning other males to stay away.

A given species can move back and forth between territoriality and dominance hierarchies. Some lizard species defend territories during breeding season, but only until their population reaches a certain density level; then they abandon territorial defense and shift to a dominance hierarchy.

### Eusociality: Life in Animal Societies

Some animals live in such highly organized, cooperative groups that they are referred to as *eusocial* or "truly social" species. Ants and termites, along with some bees and wasps, provide the most familiar examples of eusocial species. However, two species of mammals,

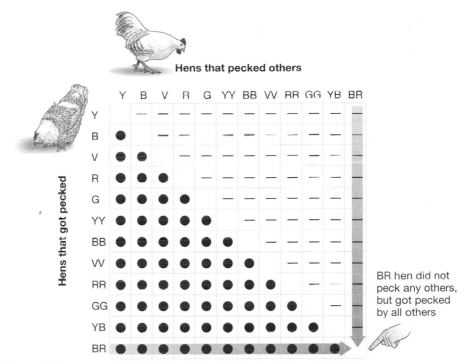

**Figure 36.16**
**Pecking Order**

By recording which hens pecked others in a group of 12 hens, A. M. Guhl documented a pecking order. It is a "linear" pecking order because dominance in it runs in a straight line: Not one hen ever pecked a hen above her in the hierarchy, but all hens pecked hens beneath them. Thus, hen Y pecked hen B and V and so on, down the line, but got pecked by no other hen. Meanwhile hen BR, at the bottom, got pecked by every hen in the group while pecking no other hen. Pecking orders are one form of dominance hierarchy; it is rare in nature, however, for a dominance hierarchy to be this rigid. (Adapted from Aubrey Manning and Marian Stamp Dawkins, *Introduction to Animal Behaviour*, 5th ed., Cambridge University Press, London, 1998.)

one of them an unpleasant-looking East African rodent called the naked mole rat, have been shown to be eusocial (**Figure 36.17**). Lots of species exhibit complex cooperative relationships, but eusociality means something more than this. To be **eusocial**, a species must be organized into a caste system in which there is

**Figure 36.17**
**Extremely Social Society**

Naked mole rats are unusual mammals in that they have a eusocial system similar to that of bees and some other insects. In each colony, a single dominant female mates with a few males. The remaining members of the colony forage for food and maintain the burrow system. Reproduction in subordinate females may be suppressed by pheromones in the dominant female's urine.

# APPENDIX

## Metric-English System Conversions

### Length

1 inch (in.) − 2.54 centimeters (cm)
1 centimeter (cm) = 0.3937 inch (in.)
1 foot (ft) = 0.3048 meter (m)
1 meter (m) = 3.2808 feet (ft) = 1.0936 (yd)
1 mile (mi) = 1.6904 kilometer
1 kilometer (km) = 0.6214 mile (mi)

### Volume

1 cubic inch (in.$^3$) = 16.39 cubic centimeters (cm$^3$ or cc)
1 cubic centimeter (cm$^3$ or cc) = 0.06 cubic inch (in.$^3$)
1 cubic foot (ft$^3$) = 0.028 cubic meter (m$^3$)
1 cubic meter (m$^3$) = 35.30 cubic feet (ft$^3$) = 1.3079 cubic yards (yd$^3$)
1 fluid ounce (oz) = 29.6 milliliters (mL) = 0.03 liter (L)
1 milliliter (mL) = 0.03 fluid ounce (oz) = 1/4 teaspoon (approximate)
1 pint (pt) = 473 milliliters (mL) = 0.47 liter (L)
1 quart (qt) = 946 milliliters (mL) = 0.9463 liter (L)
1 gallon (gal) = 3.79 liters (L)
1 liter (L) = 1.0567 quarts (qt) = 0.26 gallon (gal)

### Area

1 square inch (in.$^2$) = 6.45 square centimeters (cm$^2$)
1 square centimeter (cm$^2$) = 0.155 square inch (in.$^2$)
1 square foot (ft$^2$) = 0.0929 square meter (m$^2$)
1 square meter (m$^2$) = 10.7639 square feet (ft$^2$) = 1.1960 yards (yd$^2$)
1 square mile (mi$^2$) = 2.5900 square kilometers (km$^2$)
1 acre (a) = 0.4047 hectare (ha)
1 hectare (ha) = 2.4710 acres (a) = 10,000 square meters (m$^2$)

### Mass

1 ounce (oz) − 28.3496 grams (g)
1 gram (g) = 0.03527 ounce (oz)
1 pound (lb) = 0.4536 kilogram (kg)
1 kilogram = 2.2046 pounds (lb)
1 ton (tn), U.S. = 0.91 metric ton (t or tonne)
1 metric ton = 1.10 tons (tn), U.S.

## Metric Prefixes

| Prefix | Abbreviation | Meaning |
|--------|--------------|---------|
| giga- | G | $10^9 = 1,000,000,000$ |
| mega- | M | $10^6 = 1,000,000$ |
| kilo- | k | $10^3 = 1,000$ |
| hecto- | h | $10^2 = 100$ |
| deka- | da | $10^1 = 10$ |
| | | $10^0 = 1$ |
| deci- | d | $10^{-1} = 0.1$ |
| centi- | c | $10^{-2} = 0.01$ |
| milli- | m | $10^{-3} = 0.001$ |
| micro- | μ | $10^{-6} = 0.000001$ |

These reaction center compounds receive both the solar energy from the antennae pigments and electrons derived from the splitting of water.

4. In a bright, moderately hot environment, the $C_4$ plant would grow most rapidly. The $C_3$ plant would grow fastest in cloudy or shady conditions, in which a $C_4$ plant would be spending too much energy moving compounds around in the plant leaf. The CAM plant would grow most rapidly in the desert. In other words, each plant grows most rapidly in the area to which it is best adapted.

5. The ATP is used to power the steps of the Calvin cycle. ATP is used in two separate steps during the cycle: once when 3-PGA (3-phosphoglyceric acid) is being energized by electrons produced by the light reactions and again when G3P is being transformed into the RuBP that begins the Calvin cycle.

## CHAPTER 9
### Brief Review

1. DNA contains information for the production of proteins. These proteins are "workers" that carry out a vast array of tasks in organisms such as ourselves. Through the work of proteins, our bodies develop and carry out everyday functions.

2. The cell is in metaphase. There are eight chromatids present in this stage, and each daughter cell will have four chromosomes.

3. Mitosis serves to equally divide a cell's genetic material into two portions of a parent cell that will become daughter cells. Cytokinesis is the physical separation of a parent cell into two daughter cells.

4. Prophase—Duplicated chromosomes condense and become visible; the centrosome duplicates and centrosomes migrate to the poles; the nuclear membrane breaks down; and the mitotic spindle is formed. Metaphase—Chromosome pairs align along the cell midline or metaphase plate and attach to the microtubules extending from the cellular poles.
Anaphase—The sister chromatids paired along the metaphase plate separate and then migrate to the two poles of the cell.
Telophase—The condensed chromosomes unwind at their respective poles, and new nuclear membranes are formed around the chromosomes. At this point, division of genetic material is complete. The cell is ready for cytokinesis.

5. Plant cells have cell walls, while animal cells do not. In plant cells, membrane-lined vesicles, filled with a complex sugar, migrate to an area near the metaphase plate and begin fusing together, eventually forming a cell plate that runs from one side of the parent cell to the other. The membrane portion of the plate fuses with the original parent-cell plasma membrane, thus dividing the parent cell in half via formation of two new plasma membranes. The material inside the cell-plate membrane then forms the foundation of the new cell walls of the daughter cells.

6. The 23 pairs of human chromosomes are matched (or homologous) in the sense that the two members of each pair contain information about similar functions, such as hair color, metabolic processes, and so forth. The one exception to the matched-pair rule comes in human males, who have 22 pairs of matched chromosomes, but then one X chromosome and one Y chromosome, which are not matched. Human females have 23 matched pairs, including two X chromosomes.

7. (a) Four different bases—adenine, thymine, guanine, and cytosine, or A, T, G, and C, respectively; (b) in the nucleus; (c) messenger; (d) mRNA is used as a cellular messenger; DNA's information is copied onto it in the nucleus, after which it carries this information to the cytoplasm, where the information is used within ribosomes to put together sequences of amino acids that form proteins; (e) amino acids; (f) in ribosomes residing in the cytoplasm.

## CHAPTER 10
### Brief Review

1. Chromosome reduction takes place during meiosis I.

2. No, asexual reproduction takes place in all the varieties of organisms in the world with the exception of mammals and birds. It is more common, however, among simple organisms than among more complex organisms.

3. The cells in males that go through meiosis, the primary spermatocytes, are themselves the product of cells called spermatogonia. These spermatogonia are stem cells in that, with each cellular division, they give rise not only to primary spermatocytes but to more spermatogonia as well. It is this dual capability in spermatogonia that allows males to keep producing sperm throughout their lifetimes.

4. Female meiotic cytokinesis differs from mitotic cytokinesis in the selective shunting of cytoplasm. In mitosis, cytokinesis divides the cytoplasmic material of a parent cell equally among daughter cells. In female meiosis, cytokinesis preferentially shunts cytoplasmic material to one daughter cell—the cell destined to become the ovum—and away from the other cells, which become polar bodies.

5. Somatic cells are diploid cells that undergo mitosis. They include every cell in the body except the body's reproductive cells, its gametes, which are sperm in males and eggs in females. Gametes are haploid cells that are derived from diploid cells through the process of meiosis.

6. (a) metaphase I, meiosis I; (b) anaphase II, meiosis II; (c) anaphase, mitosis; (d) metaphase II, meiosis II; (e) metaphase, mitosis.

7. Meiosis ensures genetic diversity in two ways. In the process called crossing over or recombination, homologous chromosomes swap reciprocal portions of themselves with each other. Then, in the process called independent assortment, homologous chromosome pairs line up in a random way, relative to one another, along the metaphase plate.

8. A sperm has 23 chromosomes and an egg has 23 chromosomes; when sperm fertilizes egg, the result is a zygote with 46 chromosomes—22 homologous pairs of chromosomes and either two X chromosomes (if the zygote is female) or an X and a Y chromosome (if the zygote is male).

## CHAPTER 11
### Brief Review

1. (a) Phenotype is a physiological feature, behavior, or physical feature of an organism, while genotype is the genetic composition responsible, at least in part, for the phenotype. In an example from the text, the genotypes *YY* or *Yy* resulted in the phenotype of yellow peas, while the genotype *yy* resulted in the phenotype of green peas. (b) The term dominant is used to refer to an allele that is expressed in the heterozygous genotype. In the heterozygous genotype *Yy*, for example, *Y* is expressed (as yellow color) and thus is said to be dominant over *y*. The term recessive is used to describe an allele that is not expressed in the heterozygous genotype. Looking again at the heterozygous genotype *Yy*, the green y allele is not expressed, since any *Yy* seed will be yellow. The *y* allele is thus is said to be recessive to *Y*. (c) Codominance refers to the condition in which a heterozygote simultaneously exhibits the phenotypes associated with each of the alleles in its genotype. This is seen in blood types, where an AB individual produces both A and B cell surface molecules. Incomplete dominance is displayed when the heterozygote displays a phenotype that is intermediate between the dominant and recessive phenotypes. An example is the generation of pink flowers resulting from a parental cross of one red-flowering and one white-flowering snapdragon.

2. True.

3. The fragile-X syndrome demonstrates the genetic principle of pleiotropy, in which one gene has more than one effect. This differs from what Mendel observed in that, in his pea plants, one gene had one effect. There was a gene for flower color, for example, that came in alleles that brought about either purple or white flowers; but that gene did not then go on to affect other characters in the plant. By contrast, the gene connected to fragile-X syndrome has multiple effects—on intelligence and various physical characteristics.

4. In Mendelian genetics, two different genotypes may bring about a dominant phenotype; for example, *YY* and *Yy* bring about yellow seed color.

5. About 50 percent of the progeny will be yellow peas.

6. Mendel's "element of inheritance" is now referred to as the gene.

7. Mendel's first law (law of segregation): Members of a gene pair will segregate from each other during gamete formation such that each gamete receives only one of the original two genes from the gene pair. Mendel's second law (law of independent assortment): During gamete formation, gene pairs assort independently of one another.

8. (a) Aa × aa; (b) Aa × Aa; (c) Aabb × aaBb; (d) AaBb × AaBb.

### Genetics Problems

1. a
2. b
3. b
4. b
5. c
6. e
7. c
8. c

9. Since the purple allele is dominant, flowers will be purple whether the allele is present in two identical copies (homozygous state) or in one copy (heterozygous state). In contrast, expression of the recessive white allele is masked if a purple allele is present in the genotype. Therefore, it must be in homozygous form for the development of white flowers.

10. **b**

11. (a) 3. (b) The rule of multiplication. (c) 18.75 percent of the offspring should be purple flowered with green seeds. This answer is the product of the joint occurrence of two independent events: the probability of purple flower color (0.75) × the probability of green seed color (0.25).

12. It is possible to unambiguously determine genotype for snapdragons because incomplete dominance governs their color; every genotype manifests as a different phenotype. In the case of flower color in peas, phenotype cannot always unambiguously tell you about genotype. For white flowers, genotype can be assigned from phenotype, but for purple flowers, there are two possible genotypes, *PP* and *Pp*.

13. **d**
14. **c**

## CHAPTER 12

### Brief Review

1. Aneuploidy goes widely unrecognized because, in most instances, it is directly affecting embryos rather than children or adults. The most common result of aneuploidy is a miscarriage of the embryo. In many instances, prospective parents will not realize that a miscarriage has taken place, and even when a miscarriage is recognized as such, prospective parents would rarely be able to identify aneuploidy as its cause.

2. Male. Sex-linked diseases usually are associated with the X chromosome, which females have two copies of. A female can carry one faulty version (or allele) of an X chromosome gene and still be protected by a functional allele on her second X chromosome. Males, however, have only a single copy of the X chromosome. Should an allele on it be faulty, they have no second, protective copy of it.

3. An aneuploidy that results in cancer would take place during mitosis, or the division of a regular, nonsex cell in the body. In consequence, not every cell in the body would be affected by this aneuploidy; only the line of cells stemming from the original aneuploid cells would be affected. Conversely, an aneuploidy that gives rise to a Down syndrome child takes place during meiosis—the formation of eggs or sperm—and hence affects every cell in the body.

4. Deletions—A portion of the chromosome is completely lost.
   Inversions—A portion of the chromosome is broken off and rejoins the original chromosome, but in a reversed order.
   Translocations—Occur when two chromosomes that are not homologous exchange pieces, leaving both with improper gene sequences.
   Duplications—A portion of a chromosome is duplicated, in some instances by gaining a chromosome piece during abnormal crossing over.

5. (a) nondisjunction; (b) polyploidy.

6. Because people who have Klinefelter syndrome have a Y chromosome, which confers the male sex, but they also have two copies of the X chromosome, rather than the one copy that males normally possess. Females normally have two copies of the X chromosome. Thus, it makes sense that those with Klinefelter syndrome would exhibit some female characteristics.

7. A heterozygous genotype for hemoglobin would be advantageous in areas of mosquito infestation, while neither homozygous genotype would be advantageous in such areas. Those who are homozygous for hemoglobin A would not have the increased resistance to malaria; those who are homozygous for hemoglobin S would inherit sickle-cell anemia. Those who are heterozygous would have malaria resistance and yet would not inherit sickle-cell anemia.

### Genetics Problems

1. None will be affected (although half, on average, will be carriers).

2. One-half of her sons can be expected to be affected. The chance that her first child is affected is ¼ since there is a ½ chance the first child is a boy and a ½ chance that any boy is affected (another application of the rule of multiplication).

3. **d**
4. Recessive. Neither of the parents shows the autosomal trait, yet one of their offspring does.
5. His genotype is *aa*.
6. Her genotype must be *Aa*. (Reasoning: if she were *AA*, none of her children would be affected, whereas one is; if she were *aa*, she would manifest the trait herself, which she does not.)
7. One-half can be expected to be affected.
8. **d**

9. Her genotype is $X^+X^p$. (Reasoning: She is not affected, her mate is not affected, yet they have affected sons. This is possible only if she is a carrier.)

10. One-quarter of children and one half of their sons will show this trait (no daughters will have the trait).

11.

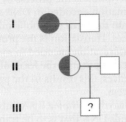

The chance that the son is color blind is ½. This is because the woman in question must be a carrier.

12. One-half of sons and daughters will have hyperphosphatemia. This is in sharp contrast to what would be observed if the trait were X-linked recessive. In this case, half the sons and none of the daughters would be affected.

## CHAPTER 13

### Brief Review

1. "Something old, something new" is a shorthand means of conveying that, in DNA replication, each newly formed DNA double helix is a mixture of the old and new. Each one is made of an existing "template" strand of DNA and a new strand that is complementary to it.

2. The structure of DNA suggested an answer to the question of how genetic information can be passed on from one cell to the next, or from one generation to the next. The answer was: through base-pairing between template-strand DNA nucleotides and nucleotides that are complementary to them.

3. True—DNA polymerases can remove a mismatched nucleotide and replace it with a correct nucleotide.

4. A mutation is a permanent alteration in a DNA base sequence. A change in a DNA base sequence only becomes permanent when it goes uncorrected by a cell's DNA error-correcting mechanisms, meaning the change will be copied over and over during DNA replication. Not all mutations are harmful. Most have no effect at all, and a very small proportion actually are useful to organisms as they result in the production of new, functional proteins.

5. Huntington disease results from an abnormal number of repeats of the bases CAG in the gene that codes for the huntingtin protein. The faulty protein clogs up nerve cells, eventually killing them.

## CHAPTER 14

### Brief Review

1. The replication of DNA and the production of proteins from DNA.

do—and all are composed of cells that lack cell walls. It would be difficult to imagine a feature as specific as a blastula stage of development arising more than once in evolution.

3. The benefit of endothermy is that animals that possess it are fully functional in a wider range of temperatures than are ectotherms, as endotherms do not have to rely on their external environment to control their internal temperature. Endothermic animals also are able to live in colder climates than ectothermic animals. The cost of endothermy is higher energy expenditure; to obtain the energy they need, endothermic animals have to eat more, per unit of body weight, than do ectothermic animals. Only mammals and birds are truly endothermic.

4. Sponges have no brains or organs of any sort, no tissues, and each of their cells functions with a great deal of independence. Thus, the cells in a sponge have a lower level of specialization and integration with one another than is the case in more complex animals.

5. Coral reefs are the skeletal remains of many generations of a variety of cnidarian—the reef-building coral—along with the remains of an associated type of algae.

6. The mollusc's unitary foot evolved over time into the arms and tentacles of the squid.

7. A dust mite; both the spider and mite are part of the arthropod Subphylum Chelicerata, while the grasshopper is part of the arthropod Subphylum Uniramia.

8. The mother must be a monotreme mammal because only mammals provide milk for their young, and only monotreme mammals are hatched from eggs.

## CHAPTER 24

### Brief Review

1. The stomata allow the movement of carbon dioxide into the plant and oxygen out of it. These same openings also allow water to escape from the plant as water vapor. The plant solves the resulting problem of water loss by moving large quantities of water up through the roots and stem and out through the stomata.

2. Roots serve to anchor the plant (usually in soil), and they are the main absorptive organs of the plant. They absorb water and minerals from their surroundings and make them available to the above-ground parts of the plant. In some species, roots also store substantial reserves of food. Taproots and fibrous roots are the two main types of roots produced by flowering plants.

3.

| Flower Part | Function |
|---|---|
| pedicel | Flower stalk; lifts flower up to make it noticeable to pollinators. |
| sepals | Leaf-like structures that protect flower buds before they open; commonly, also augment food supplies because they are usually green and photosynthetic. |
| petals | Generally most conspicuous part of the flower; often brightly colored; sometimes fragrant; attract pollinators. |
| stamens | Male reproductive structures; consist of a stalk called a filament, which holds aloft the pollen-producing sacs called anthers. |
| pollen | The male gametophyte; the vehicle that conducts male sperm to the female reproductive structures; composed at maturity of an outer coat, one tube cell and two sperm cells. |
| carpel | Female reproductive structure; consists of the ovary, style, and stigma—the ovary produces and protects the egg-containing female gametophyte, and after fertilization turns into the fruit; the stigma is the receptive surface that captures pollen and promotes germination of the right kind of pollen; the slender style raises the stigma to make it accessible to pollinators. |

4. While people grow over their whole body, plants confine their vertical growth to distinct zones at the tips of their roots and shoots. Because these zones can be active throughout the lifespan of a plant, plants continue to grow all through their lives. People, in contrast, typically cease growth on reaching maturity.

5. The roots of some plants—those of legumes such as beans and peas, for example—harbor bacteria that receive carbohydrates the plant produces through photosynthesis. In return, the bacteria "fix" atmospheric nitrogen, transforming it into a form the plant can use, and supply the plant with this nitrogen. The roots of most plants are colonized by certain types of fungi (mycorrhizal fungi), which send water and minerals to the plant through their underground hyphae. In return, these fungi receive carbohydrates produced by the plant.

6. The gametophyte is the multicellular gamete-producing generation of a flowering plant, meaning the generation that produces eggs and sperm. The sporophyte is the multicellular spore-producing generation of plant, meaning the generation the produces the microspores and megaspores.

## CHAPTER 25

### Brief Review

1. Differences between monocots and dicots:

| | Monocots | Dicots |
|---|---|---|
| number of cotyledons | one | two |
| Arrangement of flower parts | sets of three | sets of four or five |
| Leaf vein pattern | parallel | netlike veins veins |
| Arrangement of vascular bundles | scattered | in a discrete ring |
| Type of root system | fibrous root | taproot |

2.

| Cell type | Function |
|---|---|
| parenchyma | Living cells; performs a wide variety of metabolic functions such as photosynthesis, food storage, wound repair. |
| collenchyma | Living cells with irregularly thickened walls; performs some metabolic functions such as photosynthesis but also imparts mechanical strength to tissues experiencing active growth. |
| sclerenchyma | Usually dead at maturity; main role is structural support for the plant body and water conduction in some cases. |

3. Meristematic tissue gives rise to all other tissue types in the plant. Meristematic cells remain perpetually embryonic: They are able to keep giving rise to cells that differentiate into various tissue types. Apical meristems are located at the tips of shoots and roots of all angiosperms and contribute to the length of the plant. Woody angiosperms then have two additional types of meristematic tissue that produce lateral or secondary growth. These are the vascular cambium, which develops in between the primary xylem and phloem tissue and which gives rise to secondary xylem and phloem; and the cork cambium, which together with two layers of tissue it produces—cork and phelloderm—forms most of the bark of woody plants.

4. The water-conducting xylem cells come in two types: tracheids and vessel elements. Both are thick walled, and both die at maturity, after which their contents are cleared out. Because they are stacked on top of one another, lines of these cells become water-conducting tubes. Tracheids are tapered at their ends, and any two pairs of tracheid cells have matched pairs of perforations that line up with one another, allowing water to move from one cell to another. Vessel elements may lose part or all of the barriers between one cell and the next, allowing for a greater capacity to transport water than is the case with the tracheids. The food-conducting phloem cells, called sieve elements, are alive at maturity but lack nuclei. Their housekeeping needs are taken care of by one or more adjacent companion cells. A collection of sieve element cells makes up a sieve tube, which transports sucrose, along with hormones and other compounds.

5. The line that we see as a growth ring represents a collection of narrower, darker cells that grew in the tree in late summer, when water was relatively scarce. These cells are juxtaposed with wider, lighter-colored cells that grew in the spring, when water was more plentiful. We can count a tree's age by

its rings because each ring of summer cells is laid down only once each year.

6. When the two sperm cells from a pollen grain enter the embryo sac of a plant, two fertilization events take place. In the first of these, a sperm cell fuses with the embryo sac's egg cell, thus setting in motion development of a new plant embryo. In the second fertilization event, the second sperm cell fuses with the two nuclei of the embryo sac's central cell. This event sets in motion the development of endosperm—food for the growing embryo.

## CHAPTER 26
### Brief Review

1. In negative feedback, an element that brings about a response is affected negatively by that response. Put another way, the activity of the element is reduced by the response. In the example in the text, the hormone-secreting activity of the parathyroid glands caused a release of calcium that fed back negatively on the parathyroids—the activity of the glands was reduced once their activity brought about an increase in blood levels of calcium.
2. Lymphatic organs such as the spleen have high concentrations of immune cells within them that monitor passing lymphatic fluid for signs of invading microorganisms.
3. Our skin is water resistant because the surface epidermal layer of the skin is formed of tightly interlocking cells made almost entirely of the protein keratin, which is water resistant.
4. Subcutaneous fat is found in a layer, varying in thickness throughout the body, that exists just beneath the skin. Visceral fat is found surrounding organs deep within the abdomen.
5. Three functions of the skeletal system are support, storage of minerals and fats, and blood cell production.
6. Hair follicles can shrink later in life, meaning a shorter growing phase for each hair, thus yielding hair that is either short and fine or nonexistent.
7. The sample is from the end (epiphysis) of a long bone.
8. Ligaments hold bones together in stable arrangements. The anterior cruciate ligament (ACL) links the back of the femur (thigh bone) to the front of the tibia (shin bone). If the ACL were damaged, the two bones would not be held in place properly.

## CHAPTER 27
### Brief Review

1. The sensory neuron receives input about conditions inside and outside the body and conveys this received information to the central nervous system (CNS). The motor neuron cell body exists in the CNS, but its axon extends into the peripheral nervous system (PNS); it thus can send instructions from the CNS to such structures as muscles or glands. The interneuron exists solely in the CNS and serves to interconnect neurons and integrate signals from them.

2. Neurons have special extensions of themselves, called axons and dendrites, that respectively carry nervous system signals away from and toward the neuron cell body. In addition, the outer or plasma membrane of each neuron has a charge difference, from one side to the other, that can change rapidly with the proper stimulation, going from negative on the membrane's interior to briefly positive, then back to negative. This change in membrane potential can be propagated down the length of the neuron's axon, after which stimulation of an adjacent neuron can take place through release of neurotransmitters that diffuse from the sending neuron to the receiving neuron across a synaptic cleft.
3. A simple reflex arc works by means of a sensory neuron sending an afferent message about some stimulation to the spinal cord. There, the sensory neuron will synapse with a motor neuron (or perhaps an interneuron), after which the motor neuron sends a signal that brings about some action at an effector such as a muscle. The brain is not involved in such an arc.
4. In the senses of hearing, touch, taste, and vision, various types of sensory receptors responded to their respective forms of stimulation and then acted as transducers, transforming the stimulation they received into the electrical impulses of the nervous system. They did the latter by releasing neurotransmitters to receiving neurons (or by not releasing neurotransmitters, in the case of vision). In each case, the resulting nerve signal went first to the brain's thalamus and then to different areas of its cerebral cortex for processing.
5. The hormones of the endocrine system take at least several seconds to have any effect. This time frame is fine for maintaining fluid levels but of no use in carrying out the rapid, complex movements required to shoot a basketball—the nervous system handles those.
6. A given hormone does not actually target individual cells. Rather, a "target cell" is any cell that has on it, or in it, the receptors necessary to bind with a given hormone.
7. Amino-acid-based hormones are derived from a chemical modification of a single amino acid. Peptide hormones are built from chains of amino acids. The building block of all steroid hormones is the cholesterol molecule. Amino-acid-based and peptide hormones generally bind with receptors that are on the surface of target cells. Steroid hormones can do this, but they more often pass through the outer membrane of a target cell and bind with a receptor inside the cell.

## CHAPTER 28
### Brief Review

1. The immune system increases blood flow near the site of injury and increases the permeability of capillaries near the site (so

that more substances can be transported out); it delivers cells and proteins to the site that kill the invaders; and it releases substances that help seal off the site of injury, thus limiting the area of the infection.
2. A dendritic cell is the most important of three types of antigen-presenting cells (APCs). Dendritic cells destroy cells infected by an invader and then display antigens from the invader on their surface. They then migrate to a lymph node or to the spleen and present the invader's antigens to helper T cells, a small fraction of which will have receptors specific to both the APC and the antigens. The interaction between the dendritic and the helper T cell activates the helper T cell, thus spurring the creation of a clone of helper T cells specific to the invader.
3. HIV has so many effects because infection by it weakens the immune system, which normally fights off so many kinds of invaders. As an example of what's at work, the cancers mentioned in the question are linked to viruses that would be taken care of by a healthy immune system.
4. The immune system regards any transplanted organ as nonself and thus will initiate an attack against it.
5. HIV's rapid mutation has the effect of altering the recognition proteins on the surface of the virus. Immune system antibodies work by binding with such surface proteins, while T cells work by binding with such proteins when they have been displayed on the surface of infected cells. With HIV, there is no fixed set of surface proteins to bind to.

## CHAPTER 29
### Brief Review

1. Red blood cells transport oxygen and carbon dioxide; white blood cells are central to the body's immune system function; and platelets are active in the process of blood clotting.
2. The left and right atrial chambers of the heart pump blood only into their respective ventricles, while the ventricles pump blood to locations outside the heart—the right ventricle to the lungs, the left ventricle to all the tissues of the body.
3. Capillaries are the only blood vessels whose walls are thin enough to permit diffusion of substances in and out; they are composed of but a single layer of cells, whereas arteries and veins have a three-layered structure.
4. The pulmonary arteries carry deoxygenated blood; they are carrying blood to the lungs to be oxygenated.
5. Eating animal fats raises the levels of the low-density lipoproteins (LDLs) that lodge within the coronary artery tissue; smoking may damage LDLs in a way that triggers the immune response against them; exercise lowers LDL levels while raising the levels of high-density lipoproteins or HDLs, which help clear or neutralize LDLs.
6. Anatomical structures of the respiratory system include the lungs, the nose, the nasal cavity, the sinuses, the pharynx (upper

# GLOSSARY

**2n** In a living cell, the condition of being diploid: of having two sets of chromosomes. (Contrast with **n:** In a living cell, the condition of being haploid; of having a single set of chromosomes.)

## A

**abiotic** Pertaining to nonliving things.

**acid** Any substance that yields hydrogen ions in solution. An acid has a number lower than 7 on the pH scale.

**acid rain** rain whose pH level has been skewed toward the acidic side of the pH scale, with air pollution the cause of this shift.

**acrosome** A structure located on the front end of a vertebrate sperm cell; contains enzymes that help the sperm penetrate the accessory cells surrounding the oocyte.

**action potential** A temporary reversal of neuronal cell membrane potential at one location that results in a conducted nerve impulse down an axon.

**activation energy** The energy required to initiate a chemical reaction. Enzymes lower the activation energy of a reaction, thereby greatly speeding up the rate of the reaction.

**active site** The portion of an enzyme that binds with a substrate, thus helping transform it.

**active transport** Transport of materials across the plasma membrane in which energy is expended. Through active transport, solutes can be moved against their concentration and electrical gradients. The sodium-potassium pump is an example of active transport.

**actively acquired immunity** Immunity developed as a result of accidental or deliberate exposure to an antigen. Accidental exposure: coming into contact with someone who has a transmissible disease. Deliberate exposure: being vaccinated.

**adaptation** a modification in the form, physical functioning, or behavior of organisms in a population over generations in response to environmental change.

**adaptive radiation** The rapid emergence of many species from a single species that has been introduced to a new environment. The different species specialize to fill available niches in the new environment.

**adenosine triphosphate (ATP)** A nucleotide that serves as the most important energy-transfer molecule in living things. ATP powers a broad range of chemical reactions by donating one of its three phosphate groups to these reactions. In the process, it becomes adenosine diphosphate (ADP), which reverts to being ATP when a third phosphate is added to it.

**afferent division** The division of the peripheral nervous system that carries sensory information toward the central nervous system, having gathered information about the body or environment.

**algae** Protists that perform photosynthesis.

**algal bloom** An overabundance of algae in a body of water, resulting from an excess of nutrients. The many dead algae that fall to the bottom allow decomposing bacteria to flourish, using up so much oxygen that fish can suffocate.

**alkaline** Basic, as in solutions. Alkaline (basic) solutions have numbers above 7 on the pH scale.

**allele** One of the alternative forms of a single gene. In pea plants, a single gene codes for seed color, and it comes in two alleles—one codes for yellow seeds, the other for green seeds.

**allergen** A foreign substance that triggers an allergic reaction. These substances are usually derived from living things, including pollen, dust mites, foods, and fur.

**allergy** An immune system overreaction to an antigen that results in the release of histamine.

**allopatric speciation** Speciation that involves the geographic separation of populations. Most speciation involves geographic separation, followed by the development of intrinsic isolating mechanisms in the separated populations.

**allosteric regulation** The regulation of an enzyme's activity by means of a molecule binding to a site on the enzyme other than its active site.

**alternation of generations** A life cycle practiced by plants in which successive plant generations alternate between the diploid sporophyte condition and the haploid gametophyte condition.

**alternative splicing** A process in genetics in which a single primary transcript can be edited in different ways to yield multiple messenger RNAs, which in turn yield multiple proteins.

**altruism** A costly or risky behavior carried out by one animal for the benefit of another animal.

**alveoli** Tiny, hollow air-exchange sacs that exist in clusters at the end of each of the air-conducting passageways in the lungs, the bronchioles.

**amino-acid-based hormones** Hormones that are derived from a single amino acid. One of three principal classes of hormones, the other two being peptide and steroid hormones.

**amniotic egg** An egg with a hard outer casing and an inner series of membranes and fluids that provide protection, nutrients, and waste disposal for a growing embryo. The evolution of the amniotic egg, in reptiles, freed them from the constraint of having to reproduce near water.

**amniotic fluid** A protective and nutritive fluid that surrounds the fetus of mammals, including filling the lungs.

**analogy** A structure found in different organisms that is similar in function and appearance but is not the result of shared ancestry. Analogies must be distinguished from homologies to get a true picture of evolutionary relationships.

**ancestral character** A character that existed in the common ancestor of a group of organisms. Cladistics distinguishes ancestral from derived characters and uses these characters to determine evolutionary relationships.

**aneuploidy** A condition in which an individual organism has either more or fewer chromosomes than is normally found in its species' full set. Down syndrome is the result of aneuploidy—generally three copies of chromosome 21, rather than the standard two.

**angiosperm** A flowering seed plant whose seeds are enclosed within the tissue called fruit. Angiosperms are the most dominant and diverse of the four principal types of plants. Examples include roses, cacti, corn, and deciduous trees.

**animal behavior** A subdiscipline of biology concerned with the study of the behavior of animals.

**animal pole** The end of a zygote with relatively less yolk and lying closer to the cell's nucleus. The location of the egg's poles defines the orientation in which the embryo develops.

**annual** A type of plant that goes through its entire life cycle—from germination of the seed through growth, flowering, and death—in one year.

**anterior pituitary** An endocrine gland that releases two hormones that work directly on target cells and four other hormones that regulate the production of hormones by other endocrine glands.

**anther** The part of a flower that produces pollen grains. The anther is on top of a filament, and together they make up the flower's stamen.

**antibiotics** Chemical compounds produced by one microorganism that are toxic to another microorganism.

**antibody** A protein of the immune system that is found on the surface of B cells, or that is exported by them, that is able to bind to a specific antigen, thus playing a role in eliminating it from the body.

**antibody-mediated immunity** An immune system capability that works through the production of proteins called antibodies.

**anticodon** The end of the transfer RNA molecule that can bind with a particular codon on the mRNA transcript.

**antidiuretic hormone (ADH)** Substance that helps control how much water is either sent to the bladder (in urine) by the kidneys or retained in circulation. In a release controlled by the

**G1**

brain's hypothalamus, ADH increases the permeability of both the distal nephron tubule and the nephron's collecting duct to water, thus conserving it.

**antigen** Any foreign substance that elicits a response by the immune system. Certain proteins on the surface of an invading bacterial cell, for instance, act as antigens that trigger an immune response.

**antigen-presenting cell (APC)** Any immune system cell that presents, on its surface, fragments of an antigen that it has ingested. Dendritic cells, macrophages, and B cells are the immune system's three classes of antigen-presenting cells.

**aorta** The enormous artery extending from the heart that receives all the blood pumped by the heart's left ventricle. Branches stemming from the aorta supply oxygenated blood to all the tissues in the body.

**apical dominance** Suppression of the growth of the lateral branches of a plant through the activity of apical meristems.

**apical meristem** The group of plant cells at the tips of the roots and shoots that gives rise to all tissues in the plant.

**appendicular skeleton** The division of the skeletal system consisting of the bones of the paired appendages, including the pelvic and pectoral girdles to which they are attached.

**aquifer** Porous underground rock in which groundwater is stored.

**Archaea** With Bacteria and Eukarya, one of three domains of the living world, composed solely of microscopic, single-celled organisms superficially similar to bacteria but genetically quite different. Many of the Archaea live in extreme environments, such as boiling-hot vents on the ocean floor.

**arithmetical increase** An increase in numbers by an addition of a fixed number in each time period.

**artery** A blood vessel that carries blood away from the heart.

**asexual reproduction** Reproduction that occurs without the union of two reproductive cells (sexual reproduction). Offspring produced through asexual reproduction are genetically identical to their parent organism.

**atmosphere** The layer of gases that surrounds the Earth.

**atomic number** The number of protons in the nucleus of an atom. Gold has 79 protons in its nucleus and thus has the atomic number 79. All elements are ordered on the periodic table according to atomic number.

**ATP synthase** An enzyme functioning in cellular respiration that brings together ADP and inorganic phosphate molecules to produce ATP.

**autoimmune disorder** An attack by the immune system on the body's own tissues.

**autonomic nervous system** That portion of the peripheral nervous system's efferent division that provides involuntary regulation of smooth muscle, cardiac muscle, and glands.

**autosomal dominant disorder** A genetic disorder caused by a single faulty allele located on an autoso-

mal (non-sex chromosome). Huntington disease is one example.

**autosomal recessive disorder** A recessive dysfunction caused by a faulty allele on an autosome (non-sex chromosome). Sickle-cell anemia is one example.

**autotroph** Any organism that manufactures its own food. Almost all plants and algae, and certain bacteria, are autotrophs.

**axial skeleton** The division of the skeletal system that forms the central column, including the skull, vertebral column, and rib cage.

**axon** A single, large extension of the cell body of a neuron that carries signals away from the cell body toward other cells.

## B

**Bacteria** With Archaea and Eukarya, one of three domains of the living world, composed solely of single-celled, microscopic organisms that superficially resemble archaea but are genetically quite different.

**ball-and-stick model** A diagram showing the three dimensional structure of a molecule, with the atoms drawn as balls and the bonds between them drawn as sticks. This type of representation clearly shows the spatial relationship of the atoms to each other.

**bark** In woody plants, all the tissue layers outside the vascular cambium: from interior to exterior, the secondary phloem, phelloderm, cork cambium, and cork.

**base** Any substance that accepts hydrogen ions in solution. A base has a number higher than 7 on the pH scale.

**Batesian mimicry** A type of mimicry in which one species evolves to resemble a species that has superior protection against predators.

**bell curve** A distribution of values that is symmetrically largest around the average.

**biennial** A type of plant that goes through its life cycle in about two years, flowering in the second year.

**bilateral symmetry** A bodily symmetry in which opposite sides of a sagittal plane are mirror images of one another. Animals generally are bilaterally symmetrical.

**bile** Substance produced by the liver that facilitates the digestion of fats. Bile can be released either directly by the liver or by the gallbladder, which stores and concentrates this substance.

**binary fission** The form of reproduction carried out by prokaryotic cells in which the chromosome replicates and the cell pinches between the attachment points of the two resulting chromosomes to form two new cells. In this type of simple cell splitting, each pair of daughter cells is an exact replica of the parental cell.

**binomial nomenclature** The system of naming species that uses two names (genus and species) for each species. This system helps identify groupings among living things.

**biodegradable** Capable of being broken down by living organisms.

**biodiversity** Variety among living things. There are three principal types of biodiversity: species diversity, geographic distribution of species populations, and genetic diversity within species populations.

**biogeochemical cycling** The movement of water and nutrients back and forth between biotic (living) and abiotic (nonliving) realms.

**biogeography** The geographic distribution of living things.

**biological legacies** A living thing, or product of a living thing, that survives a major ecological disturbance. Biological legacies proved to be crucial in the process of succession that occurred at Washington State's Mount St. Helens following its eruption in 1980.

**biological species concept** A definition of species that relies on the breeding behavior of populations in nature. It defines species as groups of actually or potentially interbreeding natural populations that are reproductively isolated from other such groups.

**biology** The study of life.

**biomass** Material produced by living things, generally measured by dry weight.

**biome** Large, terrestrial regions of the Earth that have similar climates and hence similar vegetative formations. Six biome types are recognized at a minimum: tundra, taiga, temperate deciduous forest, temperate grassland, desert, and tropical rain forest.

**biosphere** The interactive collection of all the world's ecosystems. Also thought of as that portion of the Earth that supports life.

**biotechnology** The use of technology to control biological processes as a means of meeting societal needs.

**biotic** Pertaining to living things.

**bipedalism** Among primates, the trait of walking upright, on two legs. Along with tooth structure, the most important trait in defining species who are classified as hominins, meaning members of the taxon of human-like primates.

**bivalves** A class of molluscs that includes mussels, clams, and oysters, among other organisms.

**blade** In plants, the major, broad part of a leaf.

**blastocyst** Hollow, fluid-filled ball of cells that is formed in the early stages of the embryonic development of humans and other mammals. In non-mammalian animals, the blastocyst is known as the blastula.

**blastula** An early-stage animal embryo composed of one or more layers of cells surrounding a liquid-filled cavity known as a blastocoel. A key feature in defining animals as all animals go through a blastula stage in development, but no other organisms do. (In vertebrates, the blastula is referred to as the blastocyst.)

**B-lymphocyte cell (B cell)** The central cells of antibody-mediated immune system function. B cells produce antibodies, called antigen receptors, that bind with specific antigens while remaining embedded in the B cell. Conversely, these antibodies may be exported from B cells as freestanding entities to fight antigens.

# ESSAYS

## ESSAYS

**CHAPTER 1**
Lung Cancer, Smoking, and
   Statistics in Science     10

**CHAPTER 2**
Getting to Know Chemistry's Symbols     25
Free Radicals     27

**CHAPTER 3**
From Trans Fats to Omega-3s: Fats and
   Health     52

**CHAPTER 4**
The Size of Cells     66
The Stranger Within: Endosymbiosis     78

**CHAPTER 7**
When Energy Harvesting Ends at
   Glycolysis, Beer Can Be the Result     130
Energy and Exercise     136

**CHAPTER 9**
When the Cell Cycle Runs Amok:
   Cancer     172

**CHAPTER 11**
Proportions and Their Causes: The
   Rules Of Multiplication and Addition     201
Why So Unrecognized?     204

**CHAPTER 12**
PGD: Screening for a Healthy Child     226

**CHAPTER 14**
Cracking the Genetic Code     256

**CHAPTER 15**
Reading DNA Profiles     278

**CHAPTER 16**
The Evolution of Human Skin Color     304

**CHAPTER 17**
Lessons From the Cocker Spaniel:
   The Price of Inbreeding     318
Detecting Evolution: The
   Hardy-Weinberg Principle     325

**CHAPTER 18**
New Species Through Genetic
   Accidents: Polyploidy     338

**CHAPTER 19**
Physical Forces and Evolution     356

Who Gets a Kingdom? Evolution
   within Life's Categories     362

**CHAPTER 20**
Sequencing the Neanderthal Genome     386

**CHAPTER 21**
Unwanted Guest: The Persistence
   of Herpes     398
Modes of Nutrition: How Organisms
   Get What They Need to Survive     400

**CHAPTER 22**
A Psychedelic Drug from an
   Ancient Source     425

**CHAPTER 24**
What is Plant Food?     484
Keeping Cut Flowers Fresh     488
Ripening Fruit is a Gas     490

**CHAPTER 25**
A Tree's History Can Be Seen in Its Wood     512
The Syrup for Your Pancakes Comes
   From Xylem     516

**CHAPTER 26**
Why Fat Matters and Where It
   Matters Most     540
There Is No Such Thing as a Fabulous
   Tan     542

**CHAPTER 27**
Spinal Cord Injuries     565
Too Loud: Hair Cell Loss and Hearing     571
When Blood Sugar Stays in the Blood:
   Diabetes     584

**CHAPTER 29**
Blood Doping: A Dangerous Way
   to Cheat     610
Listening in on Blood Pressure     614

**CHAPTER 32**
Hormones and the Female
   Reproductive Cycle     672
Methods of Contraception     677
Sexually Transmitted Disease     682

**CHAPTER 34**
Purring Predators: House Cats and
   Their Prey     715
Why Do Rabid Animals Go Crazy?     722

**CHAPTER 35**
A Cut for the Middleman: Livestock
   and Food     741
Good News about the Environment     756

Our Overfished Oceans     758

**CHAPTER 36**
Biological Rhythms and Sports     774
Are Men "Naturally" Promiscuous and
   Women Reserved?     782

## THE PROCESS of SCIENCE ESSAYS

**CHAPTER 4**
The Process Of Science: First Sightings:
   Anton van Leeuwenhoek     86

**CHAPTER 5**
The Process of Science: The Fluid-Mosaic
   Model of the Plasma Membrane     104

**CHAPTER 8**
The Process of Science: Plants Make
   Their Own Food, But How?     154

**CHAPTER 12**
The Process of Science: Thomas Hunt
   Morgan: Using Fruit Flies to Look
   More Deeply into Genetics     228

**CHAPTER 13**
The Process of Science: Getting Clear
   About What Genes Do: Beadle
   and Tatum     242

**CHAPTER 15**
The Process of Science: Making
   Embryonic Stem Cells Work     282

**CHAPTER 19**
The Process of Science: Going after
   the Fossils     370

**CHAPTER 21**
The Process of Science: The Discovery
   of Penicillin     410

**CHAPTER 35**
The Process of Science: Global
   Warming and the Harlequin Frogs     746

**CHAPTER 36**
The Process of Science: How Do Sea
   Turtles Find Their Way?     788

# WEB ANIMATIONS

Each Web Animation is aimed at enhancing basic concepts presented in the text, and consists of an overview, the animation itself, learning exercises, and a computer-graded quiz. Web animations can be accessed on the World Wide Web by going to **www.aw.com/krogh4**.

In the text, Web Animations are denoted in the margins of each chapter with the icon shown above.

What follows is a list of the animations and the chapter sub-sections they are linked to.

| Web Animation Number and Name | Chapter Section Number | | Web Animation Number and Name | Chapter Section Number |
|---|---|---|---|---|
| 1.1 Scientific Method | 1.2 | | 15.2 Recombinant DNA | 15.2 |
| | | | 15.3 Copying DNA through PCR | 15.3 |
| 2.1 Structure of Atoms, Elements, Isotopes | 2.1 | | 16.1 Principles of Evolution | 16.9 |
| 2.2 Chemical Bonding | 2.2 | | | |
| 2.3 Geometry, Chemistry, and Biology | 2.3 | | 17.1 Evolution and Genetics | 17.3 |
| 2.4 Water and pH | 2.5 | | 17.2 Three Modes of Natural Selection | 17.4 |
| 3.1 The Chemistry of Carbon | 3.1 | | 18.1 Speciation and Polyploidy | 18.2 |
| 3.2 Monomers and Polymers | 3.3 | | 19.1 Evolutionary Timescales | 19.1 |
| 3.3 Carbohydrates | 3.3 | | 19.2 RNA-like Self-Replication | 19.1 |
| 3.4 Lipids | 3.4 | | 22.1 Life Cycle of Fungi | 22.3 |
| 3.5 Proteins | 3.5 | | | |
| 3.6 Nucleic Acids | 3.6 | | 24.1 Leaves. Reproduction, and Fluid Transport | 24.2 |
| 4.1 The Structure of Cells | 4.6 | | 24.2 Phototropism | 24.4 |
| 5.1 Plasma Membranes and Diffusion | 5.2 | | 25.1 Plant Tissue and Growth | 25.2 |
| | | | 25.2 The Vascular System for Plants | 25.6 |
| 6.1 Energy and Biology | 6.4 | | 25.3 Plant Reproduction | 25.7 |
| 6.2 Enzyme | 6.6 | | 27.1 The Nervous System | 27.3 |
| 7.1 Oxidation and Reduction | 7.2 | | 27.2 The Endocrine System | 27.14 |
| 8.1 Properties of Light | 8.2 | | 28.1 HIV Replication in Detail | 28.6 |
| 8.2 Different Kinds of Photosynthesis | 8.7 | | 29.1 The Cardiovascular System | 29.4 |
| 8.3 The Calvin Experiments | 8.7 | | 29.2 The Respiratory System | 29.8 |
| 9.1 Cell Division for Bacteria | 9.5 | | 30.1 The Urinary System | 30.14 |
| 10.1 Meiosis and Sex Outcome | 10.3 | | 31.1 Embryonic Development | 31.1 |
| 11.1 Mendel's Experiments and Probability | 11.5 | | 32.1 The Female Reproductive System | 32.2 |
| 11.2 Variations on Mendel | 11.7 | | 32.2 The Male Reproductive System | 32.3 |
| 12.1 X-linked Recessive Traits | 12.1 | | 32.3 Human Development Prior to Birth | 32.5 |
| 12.2 Some Human Genetic Disorders | 12.2 | | 33.1 Population Growth | 33.2 |
| 12.3 Morgan's Flies | 12.6 | | 35.1 The Carbon Cycle | 35.2 |
| 13.1 DNA | 13.1 | | 35.2 The Nitrogen Cycle | 35.2 |
| 13.2 Mutations | 13.2 | | 35.3 The Hydrologic Cycle | 35.2 |
| 13.3 One Gene, One Enzyme Hypothesis | 13.2 | | 36.1 Behavioral Biology | 36.1 |
| 14.1 Structure of Proteins | 14.1 | | 36.2 Proximate Causes of Behavior | 36.5 |
| 15.1 Action of Restriction Enzymes and Plasmids | 15.2 | | | |